REMOTE SENSING HANDBOOK
VOLUME III

REMOTE SENSING OF WATER RESOURCES, DISASTERS, AND URBAN STUDIES

Remote Sensing Handbook

Remotely Sensed Data Characterization, Classification, and Accuracies

Land Resources Monitoring, Modeling, and Mapping with Remote Sensing

Remote Sensing of Water Resources, Disasters, and Urban Studies

REMOTE SENSING HANDBOOK
VOLUME III

REMOTE SENSING OF WATER RESOURCES, DISASTERS, AND URBAN STUDIES

Edited by
Prasad S. Thenkabail, PhD
United States Geological Survey (USGS)

CRC Press
Taylor & Francis Group
Boca Raton London New York

CRC Press is an imprint of the
Taylor & Francis Group, an **informa** business

CRC Press
Taylor & Francis Group
6000 Broken Sound Parkway NW, Suite 300
Boca Raton, FL 33487-2742

First issued in paperback 2019

© 2016 by Taylor & Francis Group, LLC
CRC Press is an imprint of Taylor & Francis Group, an Informa business

No claim to original U.S. Government works

ISBN-13: 978-1-4822-1791-9 (hbk)
ISBN-13: 978-0-367-86896-3 (pbk)

Library of Congress Cataloging-in-Publication Data

Remote sensing of water resources, disasters, and urban studies / editor, Prasad S. Thenkabail.
        pages cm
    Includes bibliographical references and index.
    ISBN 978-1-4822-1791-9 (alk. paper)
    1. Water-supply--Remote sensing. 2. Natural disasters--Remote sensing. 3. Cities and towns--Remote sensing. 4. Artificial satellites in remote sensing. I. Thenkabail, Prasad Srinivasa, 1958- editor of compilation.

GB1001.72.R42R48 2015
621.36'78--dc23   2015002477

Visit the Taylor & Francis Web site at
http://www.taylorandfrancis.com

and the CRC Press Web site at
http://www.crcpress.com

I dedicate this work to my revered parents whose sacrifices gave me an education, as well as to all those teachers from whom I learned remote sensing over the years.

Contents

SECTION I Hydrology and Water Resources

SECTION II Water Use and Water Productivity

SECTION III Floods

SECTION IV Wetlands

SECTION V Snow and Ice

SECTION VI Nightlights

SECTION VII Geomorphology

SECTION VIII Droughts and Drylands

SECTION IX Disasters

Foreword: Satellite Remote Sensing Beyond 2015

Satellite remote sensing has progressed tremendously since Landsat 1 was launched on June 23, 1972. Since the 1970s, satellite remote sensing and associated airborne and in situ measurements have resulted in vital and indispensible observations for understanding our planet through time. These observations have also led to dramatic improvements in numerical simulation models of the coupled atmosphere–land–ocean systems at increasing accuracies and predictive capabilities. The same observations document the Earth's climate and are driving the consensus that *Homo sapiens* is changing our climate through greenhouse gas (GHG) emissions.

These accomplishments are the combined work of many scientists from many countries and a dedicated cadre of engineers who build the instruments and satellites that collect Earth observation (EO) data from satellites, all working toward the goal of improving our understanding of the Earth. This edition of the Remote Sensing Handbook (*Remotely Sensed Data Characterization, Classification, and Accuracies*; *Land Resources Monitoring, Modeling, and Mapping with Remote Sensing*; and *Remote Sensing of Water Resources, Disasters, and Urban Studies*) is a compendium of information for many research areas of our planet that have contributed to our substantial progress since the 1970s. The remote sensing community is now using multiple sources of satellite and in situ data to advance our studies, whatever they may be. In the following paragraphs, I will illustrate how valuable and pivotal satellite remote sensing has been in climate system study over the last five decades. The chapters in the handbook provide many other specific studies on land, water, and other applications using EO data of the last five decades.

The Landsat system of Earth-observing satellites has led the way in pioneering sustained observations of our planet. From 1972 to the present, at least one and sometimes two Landsat satellites have been in operation (Irons et al. 2012). Starting with the launch of the first NOAA–NASA Polar Orbiting Environmental Satellites NOAA-6 in 1978, improved imaging of land, clouds, and oceans and atmospheric soundings of temperature was accomplished. The NOAA system of polar-orbiting meteorological satellites has continued uninterrupted since that time, providing vital observations for numerical weather prediction. These same satellites are also responsible for the remarkable records of sea surface temperature and land vegetation index

from the advanced very-high-resolution radiometers (AVHRRs) that now span more than 33 years, although no one anticipated these valuable climate records from this instrument before the launch of NOAA-7 in 1981 (Cracknell 1997).

The success of data from the AVHRR led to the design of the moderate-resolution imaging spectroradiometer (MODIS) instruments on NASA's Earth-Observing System (EOS) of satellite platforms that improved substantially upon the AVHRR. The first of the EOS platforms, Terra, was launched in 2000; and the second of these platforms, Aqua, was launched in 2002. Both of these platforms are nearing their operational life, and many of the climate data records from MODIS will be continued with the visible infrared imaging radiometer suite (VIIRS) instrument on the polar orbiting meteorological satellites of NOAA. The first of these missions, the NPOES Preparation Project (NPP), was launched in 2012 with the first VIIRS instrument that is operating currently among several other instruments on this satellite. Continuity of observations is crucial for advancing our understanding of the Earth's climate system. Many scientists feel that the crucial climate observations provided by remote sensing satellites are among the most important satellite measurements because they contribute to documenting the current state of our climate and how it is evolving. These key satellite observations of our climate are second in importance only to the polar orbiting and geostationary satellites needed for numerical weather prediction.

The current state of the art for remote sensing is to combine different satellite observations in a complementary fashion for what is being studied. Let us review climate change as an excellent example of using disparate observations from multiple satellite and in situ sources to observe climate change, verify that it is occurring, and understand the various component processes:

1. *Warming of the planet, quantified by radar altimetry from space*: Remotely sensed climate observations provide the data to understand our planet and what forces our climate. The primary climate observation comes from radar altimetry that started in late 1992 with Topex/Poseidon and has been continued by Jason-1 and Jason-2 to provide an uninterrupted record of global sea level. Changes in global sea level provide unequivocal evidence if our planet is warming,

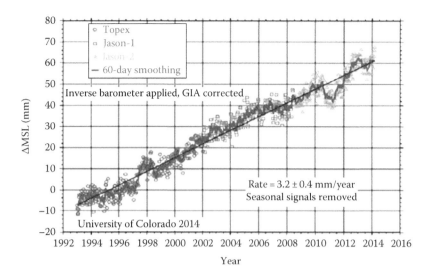

FIGURE F.1   Warming of the planet quantified by radar altimetry from space. Sea level determined from three radar altimeters from late 1992 to the present shows global sea level increases of ~3 mm/year. Sea level is the unequivocal indicator of the Earth's climate—when sea level rises, the planet is warming; when sea level falls, the planet is cooling. (From Gregory, J.M. et al., *J. Climate*, 26(13), 4476, 2013.)

cooling, or staying at the same temperature. Radar altimetry from 1992 to date has shown global sea level increases of ~3 mm/year, and hence, our planet is warming (Figure F.1). Sea level rise has two components, thermal expansion and ice melting in the ice sheets of Greenland and Antarctica, and to a much lesser extent, in glaciers.

2. The Sun is not to blame for global warming, based on solar irradiance data from satellites. Next, we consider two very different satellite observations and one in situ observing system that enable us to understand the causes of sea level variations: total solar irradiance, variations in the Earth's gravity field, and the Argo floats that record ocean temperature and salinity with depth, respectively.

Observations of total solar irradiance have been made from satellites since 1979 and show total solar irradiance has varied only ±1 part in 500 over the past 35 years, establishing that our Sun is not to blame for global warming

(Figure F.2). Thus, we must look to other remotely sensed climate observations to explain and confirm sea level rise.

3. Sea level rise of 60% is explained by a mass balance of melting of ice measured by GRACE satellites. Since 2002, we have measured gravity anomalies from the Gravity Recovery and Climate Experiment Satellite (GRACE) dual satellite system. GRACE data quantify ice mass changes from the Antarctic and Greenland ice sheets (AIS and GIS) and concentrations of glaciers, such as in the Gulf of Alaska (GOA) (Luthcke et al. 2013). GRACE data are truly remarkable—their retrieval of variations in the Earth's gravity field is quantitatively and directly linked to mass variations. With GRACE data, we are able to determine for the first time the mass balance with time of the AIS and GIS and concentrations of glaciers on land. GRACE data show sea level rise of 60% explained by ice loss from

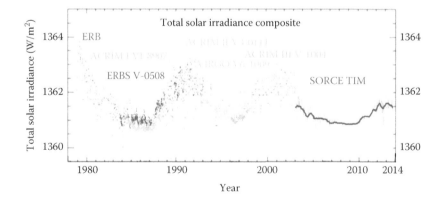

FIGURE F.2   The Sun is not to blame for global warming, based on solar irradiance data from satellites. Total solar irradiance reconstructed from multiple instruments dates back to 1979. The luminosity of our Sun varies only 0.1% over the course of the 11-year solar cycle. (From Froehlich, C., *Space Sci. Rev.*, 176(1–4), 237, 2013.)

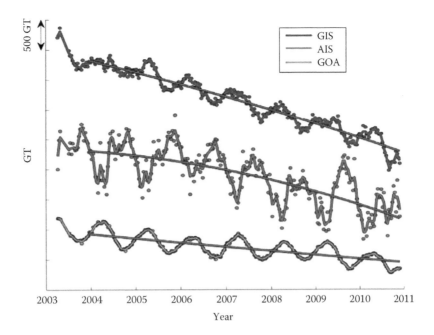

FIGURE F.3 Sea level rise of 60% explained by mass balance of melting of ice measured by GRACE satellites. Ice mass variations from 2003 to 2010 for the Antarctic ice sheets (AIS), Greenland ice sheets (GIS) and the Gulf of Alaska (GOA) glaciers using GRACE gravity data. (From Luthcke, S.B. et al., *J. Glaciol.*, 59(216), 613, 2013.)

land (Figure F.3). GRACE data have many other uses, such as indicating changes in groundwater storage, and readers are directed to the GRACE project's website if interested (http://www.csr.utexas.edu/grace/).

4. Sea level rise of 40% is explained by thermal expansion in the planet's oceans measured by in situ ~3700 drifting floats. The other contributor to sea level rise is thermal expansion in the planet's oceans. This necessitates using diving and drifting floats in the Argo network to record temperature with depth (Roemmich et al. 2009 and Figure F.4). Argo floats are deployed from ships; they then submerge and descend slowly to 1000 m depth, recording temperature, pressure, and salinity as they

descend. At 1000 m depth, they drift for 10 days continuing their measurements of temperature and salinity. After 10 days, they slowly descend to 3000 m and then ascend to the surface, all the time recording their measurements. At the surface, each float transmits all the data collected on the most recent excursion to a geostationary satellite and then descends again to repeat this process.

Argo temperature data show that 40% of sea level rise results from the warming and thermal expansion of our oceans. Combining radar altimeter data, GRACE data, and Argo data provides a confirmation of sea level rise and shows what is

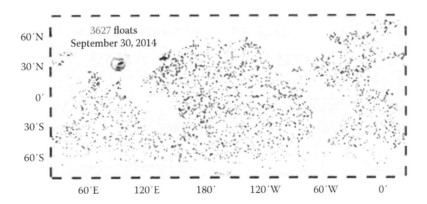

FIGURE F.4 Sea level rise of 40% explained by thermal expansion in the planet's oceans measured by in situ ~3700 drifting floats. This is the latest picture of the 3627 Argo floats that were in operation on September 30, 2014. These floats provide the data needed to document thermal expansion of the oceans. (From http://www.argo.ucsd.edu/.)

responsible for it and in what proportions. With total solar irradiance being near constant, what is driving global warming can be determined. The analysis of surface in situ air temperature coupled with lower tropospheric air temperature and stratospheric temperature data from remote sensing infrared and microwave sounders shows that the surface and near surface are warming while the stratosphere is cooling. This is an unequivocal confirmation that greenhouse gases are warming the planet.

Many scientists are actively working to study the Earth's carbon cycle, and there are several chapters in the handbook that deal with the components of this undertaking. Much like simultaneous observations of sea level, total solar irradiance, the gravity field, ocean temperature, surface temperature, and atmospheric temperatures were required to determine if the Earth is warming and what is responsible; the carbon cycle (Figure F.5) will require several complementary satellite and in situ observations (Cias et al. 2014).

Carbon cycles through reservoirs on the Earth's surface in plants and soils exist in the atmosphere as gases, such as carbon dioxide (CO_2) and methane (CH_4), and in ocean water in phytoplankton and marine sediments. CO_2 and CH_4 are released into the atmosphere by the combustion of fossil fuels, land cover changes on the Earth's surface, respiration of green plants, and decomposition of carbon in dead vegetation and in soils, including carbon in permafrost. The atmospheric concentrations of CO_2 and CH_4 control atmospheric and oceanic temperatures through their absorption of outgoing long-wave radiation and thus also indirectly control sea level via the regulation of planetary ice volumes.

Satellite-borne sensors provide simultaneous global carbon cycle observations needed for quantifying carbon cycle processes, that is, to measure atmospheric CO_2 concentrations and emission sources, to measure land and ocean photosynthesis, to measure the reservoir of carbon in plants on land and its change, to measure the extent of biomass burning of vegetation on land, and to measure soil respiration and decomposition, including decomposing carbon in permafrost. In addition to the required satellite observations, in situ observations are needed to confirm satellite-measured CO_2 concentrations and determine soil and vegetation carbon quantities. Understanding the carbon cycle requires a full court press of satellite and in situ observations because all of these observations must be made at the same time. Many of these measurements have been made over the past 30–40 years, but new measurements are needed to quantify carbon storage in vegetation, atmospheric measurements are needed to quantify CH_4 and CO_2 sources and sinks, better measurements are needed to quantify land respiration, and more explicit numerical carbon models need to be developed.

Similar work needs to be performed for the role of clouds and aerosols in climate because these are fundamental to understanding our radiation budget. We also need to improve our understanding of the global hydrological cycle.

The remote sensing community has made tremendous progress over the last five decades as discussed in this edition of the handbook. Chapters on aerosols in climate, because these are fundamental, provide comprehensive understanding of land and water studies through detailed methods, approaches, algorithms, synthesis, and key references. Every type of remote

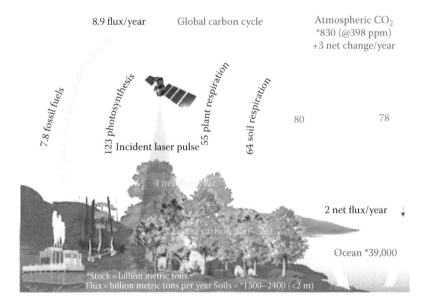

FIGURE F.5 Global carbon cycle measurements from a multitude of satellite sensors. A representation of the global carbon cycle showing our best estimates of carbon fluxes and carbon reservoirs. A series of satellite observations are needed simultaneously to understand the carbon cycle and its role in the Earth's climate system. (From Cias, P. et al., *Biogeosciences*, 11(13), 3547, 2014.)

sensing data obtained from systems such as optical, radar, light detection and ranging (LiDAR), hyperspectral, and hyperspatial is presented and discussed in different chapters. *Remotely Sensed Data Characterization, Classification, and Accuracies* sets the stage with chapters in this book addressing remote sensing data characteristics, within and between sensor calibrations, classification methods, and accuracies taking a wide array of remote sensing data from a wide array of platforms over the last five decades. *Remotely Sensed Data Characterization, Classification, and Accuracies* also brings in technologies closely linked with remote sensing such as global positioning system (GPS), global navigation satellite system (GNSS), crowdsourcing, cloud computing, and remote sensing law. In all, the 82 chapters in the 3 volumes of the handbook are written by leading and well-accomplished remote sensing scientists of the world and competently edited by Dr. Prasad S. Thenkabail, Research Geographer-15, at the United States Geological Survey (USGS).

We can look forward in the next 10–20 years to improving our quantitative understanding of the global carbon cycle, understanding the interaction of clouds and aerosols in our radiation budget, and understanding the global hydrological cycle. There is much work to do. Existing key climate observations must be continued and new satellite observations will be needed (e.g., the recently launched NASA's Orbiting Carbon Observatory-2 for atmospheric CO_2 measurements), and we have many well-trained scientists to undertake this work and continue the legacy of the past five decades.

References

Ciais, P. et al. 2014. Current systematic carbon-cycle observations and the need for implementing a policy-relevant carbon observing system. *Biogeosciences* 11(13): 3547–3602.

Cracknell, A. P. 1997. *The Advanced Very High Resolution Radiometer (AVHRR)*. Taylor & Francis, U.K., 534pp.

Froehlich, C. 2013. Total solar irradiance: What have we learned from the last three cycles and the recent minimum? *Space Science Reviews* 176(1–4): 237–252.

Gregory, J. M. et al. 2013. Twentieth-century global-mean sea level rise: Is the whole greater than the sum of the parts? *Journal of Climate* 26(13): 4476–4499.

Irons, J. R., Dwyer, J. L., and Barsi, J. A. 2012. The next Landsat satellite: The Landsat data continuity mission. *Remote Sensing of Environment* 122: 11–21. doi:10.1016/j.rse.2011.08.026.

Luthcke, S. B., Sabaka, T. J., Loomis, B. D. Arendt, A. A., McCarthy, J. J., and Camp, J. 2013. Antarctica, Greenland, and Gulf of Alaska land-ice evolution from an iterated GRACE global mascon solution. *Journal of Glaciology* 59(216): 613–631.

Roemmich, D. and the Argo Steering Team, 2009. Argo—The challenge of continuing 10 years of progress. *Oceanography* 22(3): 46–55.

Compton J. Tucker
Earth Science Division
Goddard Space Flight Center
National Aeronautics and Space Administration
Greenbelt, Maryland

Preface: Remote Sensing Advances of the Last 50 Years and a Vision for the Future

The overarching goal of the Remote Sensing Handbook (*Remotely Sensed Data Characterization, Classification, and Accuracies*; *Land Resources Monitoring, Modeling,* and *Mapping with Remote Sensing*; and *Remote Sensing of Water Resources, Disasters, and Urban Studies*), with 82 chapters and about 2500 pages, was to capture and provide the most comprehensive state of the art of remote sensing science and technology development and advancement in the last 50 years, by clearly demonstrating the (1) scientific advances, (2) methodological advances, and (3) societal benefits achieved during this period, as well as to provide a vision of what is to come in the years ahead. The three books are, to date and to the best of my knowledge, the most comprehensive documentation of the scientific and methodological advances that have taken place in understanding remote sensing data, methods, and a wide array of land and water applications. Written by 300+ leading global experts in the area, each chapter (1) focuses on a specific topic (e.g., data, methods, and applications), (2) reviews the existing state-of-the-art knowledge, (3) highlights the advances made, and (4) provides guidance for areas requiring future development. Chapters in the books cover a wide array of subject matter of remote sensing applications. The Remote Sensing Handbook is planned as a reference material for remote sensing scientists, land and water resource practitioners, natural and environmental practitioners, professors, students, and decision makers. The special features of the Remote Sensing Handbook include the following:

1. Participation of an outstanding group of remote sensing experts, an unparalleled team of writers for such a book project
2. Exhaustive coverage of a wide array of remote sensing science: data, methods, and applications
3. Each chapter being led by a luminary and most chapters written by teams who further enriched the chapters
4. Broadening the scope of the book to make it ideal for expert practitioners as well as students
5. Global team of writers, global geographic coverage of study areas, and a wide array of satellites and sensors
6. Plenty of color illustrations

Chapters in the books cover the following aspects of remote sensing:

State of the art
Methods and techniques
Wide array of land and water applications
Scientific achievements and advancements over the last 50 years
Societal benefits
Knowledge gaps
Future possibilities in the twenty-first century

Great advances have taken place over the last 50 years using remote sensing in the study of the planet Earth, especially using data gathered from a multitude of Earth observation (EO) satellites launched by various governments as well as private entities. A large part of the initial remote sensing technology was developed and tested during the two world wars. In the 1950s, remote sensing slowly began its foray into civilian applications. During the years of the Cold War, remote sensing applications, both civilian and military, increased swiftly. But it was also an age when remote sensing was the domain of a very few top experts and major national institutes, having multiple skills in engineering, science, and computer technology. From the 1960s onward, there have been many governmental agencies that have initiated civilian remote sensing. The National Aeronautics and Space Administration (NASA) and the United States Geological Survey (USGS) have been in the forefront of many of these efforts. Others who have provided leadership in civilian remote sensing include, but are not limited to, the European Space Agency (ESA) of the European Union, the Indian Space Research Organization (ISRO), the Centre National d'Études Spatiales (CNES) of France, the Canadian Space Agency (CSA), the Japan Aerospace Exploration Agency (JAXA), the German Aerospace Center (DLR), the China National Space Administration (CNSA), the United Kingdom Space Agency (UKSA), and the Instituto Nacional de Pesquisas Espaciais (INPE) of Brazil. Many private entities have launched and operated satellites. These government and private agencies and enterprises launched and operated a wide array of satellites and sensors that captured the data of the planet Earth in various regions of the electromagnetic spectrum and in various spatial, radiometric, and temporal resolutions, routinely and repeatedly. However, the real thrust for remote sensing advancement came during the last decade of the twentieth century and the beginning of the twenty-first century. These initiatives included a launch of a series of new-generation EO satellites to gather data more frequently and routinely, release of pathfinder datasets, web enabling the data for free by many agencies (e.g., USGS release of the entire Landsat archives as well as real-time acquisitions of the world for free dissemination by web-enabling), and providing processed data ready to users (e.g., surface reflectance products of moderate-resolution imaging spectroradiometer [MODIS]). Other efforts like Google Earth made remote sensing more popular and brought in a new platform for easy visualization and navigation of remote sensing data. Advances in computer hardware and software made it possible to handle Big Data. Crowdsourcing, web access, cloud computing, and mobile platforms added a new dimension to how remote sensing data are used. Integration with global positioning systems (GPS) and global navigation satellite systems (GNSS) and inclusion of digital secondary data (e.g., digital elevation, precipitation, temperature) in analysis have made remote sensing much more powerful. Collectively, these initiatives provided a new vision in making remote sensing data more popular, widely understood, and increasingly used for diverse applications, hitherto considered difficult. The free availability of archival data when combined with more recent acquisitions has also enabled quantitative studies of change over space and time. The Remote Sensing Handbook is targeted to capture these vast advances in data, methods, and applications, so a remote sensing student, scientist, or a professional practitioner will have the most comprehensive, all-encompassing reference material in one place.

Modern-day remote sensing technology, science, and applications are growing exponentially. This growth is a result of a combination of factors that include (1) advances and innovations in data capture, access, and delivery (e.g., web enabling, cloud computing, crowdsourcing); (2) an increasing number of satellites and sensors gathering data of the planet, repeatedly and routinely, in various portions of the electromagnetic spectrum as well as in an array of spatial, radiometric, and temporal resolutions; (3) efforts at integrating data from multiple satellites and sensors (e.g., sentinels with Landsat); (4) advances in data normalization, standardization, and harmonization (e.g., delivery of data in surface reflectance, intersensor calibration); (5) methods and techniques for handling very large data volumes (e.g., global mosaics); (6) quantum leap in computer hardware and software capabilities (e.g., ability to process several terabytes of data); (7) innovation in methods, approaches, and techniques leading to sophisticated algorithms (e.g., spectral

matching techniques, and automated cropland classification algorithms); and (8) development of new spectral indices to quantify and study specific land and water parameters (e.g., hyperspectral vegetation indices or HVIs). As a result of these all-around developments, remote sensing science is today very mature and is widely used in virtually every discipline of the earth sciences for quantifying, mapping, modeling, and monitoring our planet Earth. Such rapid advances are captured in a number of remote sensing and earth science journals. However, students, scientists, and practitioners of remote sensing science and applications have significant difficulty gathering a complete understanding of the various developments and advances that have taken place as a result of their vastness spread across the last 50 years. Therefore, the chapters in the Remote Sensing Handbook are designed to give a whole picture of scientific and technological advances of the last 50 years.

Today, the science, art, and technology of remote sensing are truly ubiquitous and increasingly part of everyone's everyday life, often without the user knowing it. Whether looking at your own home or farm (e.g., see the following figure), helping you navigate when you drive, visualizing a phenomenon occurring in a distant part of the world (e.g., see the following figure), monitoring events such as droughts and floods, reporting weather, detecting and monitoring troop movements or nuclear sites, studying deforestation, assessing biomass carbon, addressing disasters such as earthquakes or tsunamis, and a host of other applications (e.g., precision farming, crop productivity, water productivity, deforestation, desertification, water resources management), remote sensing plays a pivotal role. Already, many new innovations are taking place. Companies such as the Planet Labs and Skybox are planning to capture very-high-spatial-resolution imagery (typically, sub-meter to 5 meters), even videos from space using a large number of microsatellite constellations. There are others planning to launch a constellation of hyperspectral or other sensors. Just as the smartphone and social media connected the world, remote sensing is making the world our backyard. No place goes unobserved and no event gets reported without a satellite or other kinds of remote sensing images or their derivatives. This is how true liberation for any technology and science occurs.

Google Earth can be used to seamlessly navigate and precisely locate any place on Earth, often with very-high-spatial-resolution data (VHRI; submeters to 5 m) from satellites such as IKONOS, QuickBird, and GeoEye (Note: the image below is from one of the VHRI). Here, the editor-in-chief (EiC) of this handbook located his village home (Thenkabail) and surroundings that have land covers such as secondary rainforests, lowland paddy farms, areca nut plantations, coconut plantations, minor roads, walking routes, open grazing lands, and minor streams (typically, first and second order) (note: land cover detailed is based on the ground knowledge of the EiC). The first primary school attended by him is located precisely. Precise coordinates (13 degree 45 minutes 39.22 seconds northern latitude, 75 degrees 06 minutes 56.03 seconds eastern longitude) of Thenkabail's village house on the planet and the date

of image acquisition (March 1, 2014) are noted. Google Earth images are used for visualization as well as for numerous science applications such as accuracy assessment, reconnaissance, determining land cover, and establishing land use for various ground surveys. It is widely used by lay people who often have no idea on how it all comes together but understand the information provided intuitively. This is already happening. These developments make it clear that we not only need to understand the state of the art but also have a vision of where the future of remote sensing is headed. Therefore, in a nutshell, the goal of this handbook is to cover the developments and advancement of six distinct eras in terms of data characterization and processing as well as myriad land and water applications:

1. *Pre–civilian remote sensing era of the pre-1950s*: World War I and II when remote sensing was a military tool

2. *Technology demonstration era of the 1950s and 1960s*: Sputnik-I and NOAA AVHRR era of the 1950s and 1960s

3. *Landsat era of the 1970s*: when the first truly operational land remote sensing satellite (Earth Resources Technology Satellite or ERTS, later renamed Landsat) was launched and operated in the 1970s and early 1980s by United States

4. *Earth observation era of the 1980s and 1990s*: when a number of space agencies began launching and operating satellites (e.g., Landsat 4,5 by the United States; SPOT-1,2 by France; IRS-1a, 1b by India) from the middle to late 1980s onward till the middle of 1990s

5. *Earth observation and the first decade of the New Millennium era of the 2000s*: when data dissemination to users became as important as launching, operating, and capturing data (e.g., MODIS Terra\Aqua, Landsat-8, Resourcesat) in the late 1990 and the first decade of the 2000s

6. *Second decade of the New Millennium era starting in the 2010s*: when new-generation micro-\nanosatellites (e.g., PlanetLabs, Skybox) are added to the increasing constellation of multiagency sensors (e.g., Sentinels, and the next generation of satellites such as SMAP, hyperspectral satellites like NASA's HyspIRI and others from private industry)

Motivation for the Remote Sensing Handbook started with a simple conversation with Irma Shagla-Britton, acquisitior editor for remote sensing and GIS books of Taylor & Francis Group/ CRC Press, way back in early 2013. Irma was informally getting my advice about "doing a new and unique book" on remote sensing. Neither the specific subject nor the editor was identified. What was clear to me though was that I certainly did not want to lead the effort. I was nearing the end of my third year of recovery from colon cancer, and the last thing I wanted to do was to take any book project, forget a multivolume remote sensing magnum opus, as it ultimately turned out. However, mostly out of courtesy for Irma, I did some preliminary research. I tried to identify a specific topic within remote sensing where there was a sufficient need for a full-fledged book. My research showed that there was not a single book that would provide a complete and comprehensive coverage of the entire subject of remote sensing starting from data capture, to data preprocessing, to data analysis, to myriad land and water applications. There are, of course, numerous excellent books on remote sensing, each covering a specific subject matter. However, if a student, scientist, or practitioner of remote sensing wanted a standard reference on the subject, he or she would have to look for numerous books or journal articles and often a coherence of these topics would still be left uncovered or difficult to comprehend for students and even for many experts with less experience. Guidance on how to approach the study of remote sensing and capture its state of the art and advances remained hazy and often required referring to a multitude of references that may or may not be immediately available, and if available, how to go about it was still hazy to most. During this process, I asked myself, several times, what remote sensing book will be most interesting, productive, and useful to a broad audience? The answer, each time, was very clear: "A complete and comprehensive coverage of the state-of-the-art remote sensing, capturing the advances that have taken place over the last 50 years, which will set the stage for a vision for the future." When this became clear, I started putting together the needed topics to achieve such a goal. Soon I realized that the only way to achieve this goal was through a multivolume book on remote sensing. Because the number of chapters was more than 80, this appeared to be too daunting, too overwhelming, and too big a project to accomplish. Yet I sent the initial idea to Irma, who I thought would say "forget it" and ask me to focus on a single-volume book. But to my surprise, Irma not only encouraged the idea but also had a number of useful suggestions. So what started as intellectual curiosity turned into this full-fledged multivolume Remote Sensing Handbook.

However, what worried me greatly was the virtual impossibility (my thought at that time) of gathering the best authors. What was also crystal clear to me was that unless the very best were attracted to the book project, it was simply not worth the effort. I had made up my mind to give up the book project, unless I got the full support of a large number of the finest practitioners of remote sensing from around the world. So, I spent a few weeks researching the best authors to lead each chapter and wrote to them to participate in the Remote Sensing Handbook project. What really surprised me was that almost all the authors I contacted agreed to lead and write a chapter. This was truly surreal. These are extremely busy people of great scientific reputation and achievements. For them to spend the time, intellect, and energy to write an in-depth and insightful book chapter spread across a year or more is truly amazing. Most also agreed to put together a writing team, as I had requested, to ensure greater perspective for each chapter. In the end, we had 300+ authors writing 82 chapters.

At this stage, I was somewhat drawn into the project as if by destiny and felt compelled to go ahead. One of the authors who agreed to lead the chapter mentioned "…..whether it was even possible." This is exactly what I felt, too. But I had reached the stage of no return, and I took on the book project with all the seriousness it deserved. It required some real changes to my lifestyle: professional and personal. Travel was reduced to bare minimum during most of the book project. Most weekends were spent editing, writing, and organizing, and other social activities were reduced. Accomplishing such complex work requires the highest levels of discipline, planning, and strategy. But, above all, I felt blessed with good health. By the time the book is published, I will have completed about 5 years from my colon cancer surgery and chemotherapy. So I am as happy to see this book released as I am with the miracle of cancer cure (I feel confident to say so).

But it is the chapter authors who made it all feasible. They amazed me throughout the book project. First, the quality and content of each of the chapters were of the highest standards. Second, with very few exceptions, chapters were delivered on time. Third, edited chapters were revised thoroughly and returned on time. Fourth, all my requests on various formatting and quality enhancements were addressed. This is what made the three-volume Remote Sensing Handbook possible and if I may say so, a true *magnum opus* on the subject. My heartfelt gratitude to these great authors for their dedication. It has been my great honor to work with these dedicated legends. Indeed, I call them my *heroes* in a true sense.

Overall, the preparation of the Remote Sensing Handbook took two and a half years, from the time book chapters and authors were being identified to its final publication. The three books are designed in such a way that a reader can have all three books as a standard reference or have individual books to study specific subject areas. The three books of Remote Sensing Handbook are

Remotely Sensed Data Characterization, Classification, and Accuracies: 31 Chapters
Land Resources Monitoring, Modeling, and Mapping with Remote Sensing: 28 Chapters
Remote Sensing of Water Resources, Disasters, and Urban Studies: 27 Chapters

There are about 2500 pages in the 3 volumes.

The wide array of topics covered is very comprehensive. The topics covered in *Remotely Sensed Data Characterization, Classification, and Accuracies* include (1) satellites and sensors; (2) remote sensing fundamentals; (3) data normalization, harmonization, and standardization; (4) vegetation indices and their within- and across-sensor calibration; (5) image classification methods and approaches; (6) change detection; (7) integrating remote sensing with other spatial data; (8) GNSS; (9) crowdsourcing; (10) cloud computing; (11) Google Earth remote sensing; (12) accuracy assessments; and (13) remote sensing law.

The topics covered in *Land Resources Monitoring, Modeling, and Mapping with Remote Sensing* include (1) vegetation and biomass, (2) agricultural croplands, (3) rangelands, (4) phenology and food security, (5) forests, (6) biodiversity, (7) ecology, (8) land use/land cover, (9) carbon, and (10) soils.

The topics covered in *Remote Sensing of Water Resources, Disasters, and Urban Studies* include (1) hydrology and water resources; (2) water use and water productivity; (3) floods; (4) wetlands; (5) snow and ice; (6) glaciers, permafrost, and ice; (7) geomorphology; (8) droughts and drylands; (9) disasters; (10) volcanoes; (11) fire; (12) urban areas; and (13) nightlights.

There are many ways to use the Remote Sensing Handbook. A lot of thought went into organizing the books and chapters. So you will see a *flow* from chapter to chapter and book to book. As you read through the chapters, you will see how they are interconnected and how reading all of them provides you with greater in-depth understanding. Some of you may be more interested in a particular volume. Often, having all three books as reference material is ideal for most remote sensing experts, practitioners, or students; however, you can also refer to individual books based on your interest. We have also made attempts to ensure the chapters are self-contained. That way you can focus on a chapter and read it through, without having to be overly dependent on other chapters. Taking this perspective, there is a slight (~5%–10%) material that may be repeated in some of the chapters. This is done deliberately. For example, when you are reading a chapter on LiDAR or radar, you don't want to go all the way back to another chapter (e.g., Chapter 1, *Remotely Sensed Data Characterization, Classification, and Accuracies*) to understand the characteristics of these sensors.

Similarly, certain indices (e.g., vegetation condition index [VCI], temperature condition index [TCI]) that are defined in one chapter (e.g., on drought) may be repeated in another chapter (also on drought). Such minor overlaps are helpful to the reader to avoid going back to another chapter to understand a phenomenon or an index or a characteristic of a sensor. However, if you want a lot of details on these sensors or indices or phenomena or if you are someone who has yet to gain sufficient expertise in the field of remote sensing, then you will have to read the appropriate chapter where there is in-depth coverage of the topic.

Each book has a summary chapter (the last chapter of each book). The summary chapter can be read two ways: (1) either as a last chapter to recapture the main points of each of the previous chapters or (2) as an initial overview to get a feeling for what is in the book. I suggest the readers do it both ways: Read it first before going into the details and then read it at the end to recollect what was said in the chapters.

It has been a great honor as well as a humbling experience to edit the Remote Sensing Handbook (*Remotely Sensed Data Characterization, Classification, and Accuracies*; *Land Resources Monitoring, Modeling, and Mapping with Remote Sensing*; and *Remote Sensing of Water Resources, Disasters, and Urban Studies*). I truly enjoyed the effort. What an honor to work with luminaries in this field of expertise. I learned a lot from them and am very grateful for their support, encouragement, and deep insights. Also, it has been a pleasure working with outstanding professionals of Taylor & Francis Group/CRC Press. There is no joy greater than being immersed in pursuit of excellence, knowledge gain, and knowledge capture. At the same time, I am happy it is over. The biggest lesson I learned during this project was that if you set yourself to a task with dedication, sincerity, persistence, and belief, you will have the job accomplished, no matter how daunting.

I expect the books to be standard references of immense value to any student, scientist, professional, and practical practitioner of remote sensing.

Prasad S. Thenkabail, PhD
Editor-in-Chief

Acknowledgments

The Remote Sensing Handbook (*Remotely Sensed Data Characterization, Classification, and Accuracies*; *Land Resources Monitoring, Modeling, and Mapping with Remote Sensing*; and *Remote Sensing of Water Resources, Disasters, and Urban Studies*) brought together a galaxy of remote sensing legends. The lead authors and coauthors of each chapter are internationally recognized experts of the highest merit on the subject about which they have written. The lead authors were chosen carefully by me after much thought and discussions, who then chose their coauthors. The overwhelming numbers of chapters were written over a period of one year. All chapters were edited and revised over the subsequent year and a half.

Gathering such a galaxy of authors was the biggest challenge. These are all extremely busy people, and committing to a book project that requires a substantial work load is never easy. However, almost all those whom I asked agreed to write the chapter, and only had to convince a few. The quality of the chapters should convince readers why these authors are such highly rated professionals and why they are so successful and accomplished in their field of expertise. They not only wrote very high quality chapters but delivered on time, addressed any editorial comments timely without complaints, and were extremely humble and helpful. What was also most impressive was the commitment of these authors for quality science. Three lead authors had serious health issues and yet they delivered very high quality chapters in the end, and there were few others who had unexpected situations (e.g., family health issues) and yet delivered the chapters on time. Even when I offered them the option to drop out, almost all of them wanted to stay. They only asked for a few extra weeks or months but in the end honored their commitment. I am truly honored to have worked with such great professionals.

In the following list are the names of everyone who contributed and made possible the Remote Sensing Handbook. In the end, we had 82 chapters, a little over 2500 pages, and a little over 300 authors.

My gratitude to the following authors of chapters in *Remotely Sensed Data Characterization, Classification, and Accuracies*. The authors are listed in chapter order starting with the lead author.

- Chapter 1, Drs. Sudhanshu S. Panda, Mahesh Rao, Prasad S. Thenkabail, and James P. Fitzerald
- Chapter 2, Natascha Oppelt, Rolf Scheiber, Peter Gege, Martin Wegmann, Hannes Taubenboeck, and Michael Berger
- Chapter 3, Philippe M. Teillet
- Chapter 4, Philippe M. Teillet and Gyanesh Chander
- Chapter 5, Rudiger Gens and Jordi Cristóbal Rosselló
- Chapter 6, Dongdong Wang
- Chapter 7, Tomoaki Miura, Kenta Obata, Javzandulam T. Azuma, Alfredo Huete, and Hiroki Yoshioka
- Chapter 8, Michael D. Steven, Timothy Malthus, and Frédéric Baret
- Chapter 9, Sunil Narumalani and Paul Merani
- Chapter 10, Soe W. Myint, Victor Mesev, Dale Quattrochi, and Elizabeth A. Wentz
- Chapter 11, Mutlu Ozdogan
- Chapter 12, Jun Li and Antonio Plaza
- Chapter 13, Claudia Kuenzer, Jianzhong Zhang, and Stefan Dech
- Chapter 14, Thomas Blaschke, Maggi Kelly, and Helena Merschdorf
- Chapter 15, Stefan Lang and Dirk Tiede
- Chapter 16, James C. Tilton, Selim Aksoy, and Yuliya Tarabalka
- Chapter 17, Shih-Hong Chio, Tzu-Yi Chuang, Pai-Hui Hsu, Jen-Jer Jaw, Shih-Yuan Lin, Yu-Ching Lin, Tee-Ann Teo, Fuan Tsai, Yi-Hsing Tseng, Cheng-Kai Wang, Chi-Kuei Wang, Miao Wang, and Ming-Der Yang
- Chapter 18, Daniela Anjos, Dengsheng Lu, Luciano Dutra, and Sidnei Sant'Anna
- Chapter 19, Jason A. Tullis, Jackson D. Cothren, David P. Lanter, Xuan Shi, W. Fredrick Limp, Rachel F. Linck, Sean G. Young, and Tareefa S. Alsumaiti
- Chapter 20, Gaurav Sinha, Barry J. Kronenfeld, and Jeffrey C. Brunskill
- Chapter 21, May Yuan
- Chapter 22, Stefan Lang, Stefan Kienberger, Michael Hagenlocher, and Lena Pernkopf
- Chapter 23, Mohinder S. Grewal
- Chapter 24, Kegen Yu, Chris Rizos, and Andrew Dempster
- Chapter 25, D. Myszor, O. Antemijczuk, M. Grygierek, M. Wierzchanowski, and K.A. Cyran
- Chapter 26, Fabio Dell'Acqua
- Chapter 27, Ramanathan Sugumaran, James W. Hegeman, Vivek B. Sardeshmukh, and Marc P. Armstrong
- Chapter 28, John Bailey
- Chapter 29, Russell G. Congalton

- Chapter 30, P.J. Blount
- Chapter 31, Prasad S. Thenkabail

My gratitude to the following authors of chapters in *Land Resources Monitoring, Modeling, and Mapping with Remote Sensing*. The authors are listed in chapter order starting with the lead author.

- Chapter 1, Alfredo Huete, Guillermo Ponce-Campos, Yongguang Zhang, Natalia Restrepo-Coupe, Xuanlong Ma, and Mary-Susan Moran
- Chapter 2, Frédéric Baret
- Chapter 3, Wenge Ni-Meister
- Chapter 4, Clement Atzberger, Francesco Vuolo, Anja Klisch, Felix Rembold, Michele Meroni, Marcio Pupin Mello, and Antonio Formaggio
- Chapter 5, Agnès Bégué, Damien Arvor, Camille Lelong, Elodie Vintrou, and Margareth Simoes
- Chapter 6, Pardhasaradhi Teluguntla, Prasad S. Thenkabail, Jun Xiong, Murali Krishna Gumma, Chandra Giri, Cristina Milesi, Mutlu Ozdogan, Russell G. Congalton, James Tilton, Temuulen Tsagaan Sankey, Richard Massey, Aparna Phalke, and Kamini Yadav
- Chapter 7, David J. Mulla, and Yuxin Miao
- Chapter 8, Baojuan Zheng, James B. Campbell, Guy Serbin, Craig S.T. Daughtry, Heather McNairn, and Anna Pacheco
- Chapter 9, Prasad S. Thenkabail, Pardhasaradhi Teluguntla, Murali Krishna Gumma, and Venkateswarlu Dheeravath
- Chapter 10, Matthew Clark Reeves, Robert A. Washington-Allen, Jay Angerer, E. Raymond Hunt, Jr., Ranjani Wasantha Kulawardhana, Lalit Kumar, Tatiana Loboda, Thomas Loveland, Graciela Metternicht, and R. Douglas Ramsey
- Chapter 11, E. Raymond Hunt, Jr., Cuizhen Wang, D. Terrance Booth, Samuel E. Cox, Lalit Kumar, and Matthew C. Reeves
- Chapter 12, Lalit Kumar, Priyakant Sinha, Jesslyn F. Brown, R. Douglas Ramsey, Matthew Rigge, Carson A. Stam, Alexander J. Hernandez, E. Raymond Hunt, Jr., and Matt Reeves
- Chapter 13, Molly E. Brown, Kirsten M. de Beurs, and Kathryn Grace
- Chapter 14, E.H. Helmer, Nicholas R. Goodwin, Valéry Gond, Carlos M. Souza, Jr., and Gregory P. Asner
- Chapter 15, Juha Hyyppä, Mika Karjalainen, Xinlian Liang, Anttoni Jaakkola, Xiaowei Yu, Mike Wulder, Markus Hollaus, Joanne C. White, Mikko Vastaranta, Kirsi Karila, Harri Kaartinen, Matti Vaaja, Ville Kankare, Antero Kukko, Markus Holopainen, Hannu Hyyppä, and Masato Katoh
- Chapter 16, Gregory P. Asner, Susan L. Ustin, Philip A. Townsend, Roberta E. Martin, and K. Dana Chadwick
- Chapter 17, Sylvie Durrieu, Cédric Véga, Marc Bouvier, Frédéric Gosselin, and Jean-Pierre Renaud Laurent Saint-André

- Chapter 18, Thomas W. Gillespie, Andrew Fricker, Chelsea Robinson, and Duccio Rocchini
- Chapter 19, Stefan Lang, Christina Corbane, Palma Blonda, Kyle Pipkins, and Michael Förster
- Chapter 20, Conghe Song, Jing Ming Chen, Taehee Hwang, Alemu Gonsamo, Holly Croft, Quanfa Zhang, Matthew Dannenberg, Yulong Zhang, Christopher Hakkenberg, Juxiang Li
- Chapter 21, John Rogan and Nathan Mietkiewicz
- Chapter 22, Zhixin Qi, Anthony Gar-On Yeh, and Xia Li
- Chapter 23, Richard A. Houghton
- Chapter 24, José A.M. Demattê, Cristine L.S. Morgan, Sabine Chabrillat, Rodnei Rizzo, Marston H.D. Franceschini, Fabrício da S. Terra, Gustavo M. Vasques, and Johanna Wetterlind
- Chapter 25, E. Ben-Dor and José A.M. Demattê
- Chapter 26, Prasad S. Thenkabail

My gratitude to the following authors of chapters in *Remote Sensing of Water Resources, Disasters, and Urban Studies*. The authors are listed in chapter order starting with the lead author.

- Chapter 1, Sadiq I. Khan, Ni-Bin Chang, Yang Hong, Xianwu Xue, and Yu Zhang
- Chapter 2, Santhosh Kumar Seelan
- Chapter 3, Trent W. Biggs, George P. Petropoulos, Naga Manohar Velpuri, Michael Marshall, Edward P. Glenn, Pamela Nagler, and Alex Messina
- Chapter 4, Antônio de C. Teixeira, Fernando B. T. Hernandez, Morris Scherer-Warren, Ricardo G. Andrade, Janice F. Leivas, Daniel C. Victoria, Edson L. Bolfe, Prasad S. Thenkabail, and Renato A. M. Franco
- Chapter 5, Allan S. Arnesen, Frederico T. Genofre, Marcelo P. Curtarelli, and Matheus Z. Francisco
- Chapter 6, Sandro Martinis, Claudia Kuenzer, and André Twele
- Chapter 7, Chandra Giri
- Chapter 8, D. R. Mishra, Shuvankar Ghosh, C. Hladik, Jessica L. O'Connell, and H. J. Cho
- Chapter 9, Murali Krishna Gumma, Prasad S. Thenkabail, Irshad A. Mohammed, Pardhasaradhi Teluguntla, and Venkateswarlu Dheeravath
- Chapter 10, Hongjie Xie, Tiangang Liang, Xianwei Wang, and Guoqing Zhang
- Chapter 11, Qingling Zhang, Noam Levin, Christos Chalkias, and Husi Letu
- Chapter 12, James B. Campbell and Lynn M. Resler
- Chapter 13, Felix Kogan and Wei Guo
- Chapter 14, Felix Rembold, Michele Meroni, Oscar Rojas, Clement Atzberger, Frederic Ham, and Erwann Fillol
- Chapter 15, Brian Wardlow, Martha Anderson, Tsegaye Tadesse, Chris Hain, Wade T. Crow, and Matt Rodell
- Chapter 16, Jinyoung Rhee, Jungho Im, and Seonyoung Park
- Chapter 17, Marion Stellmes, Ruth Sonnenschein, Achim Röder, Thomas Udelhoven, Stefan Sommer, and Joachim Hill

- Chapter 18, Norman Kerle
- Chapter 19, Stefan Lang, Petra Füreder, Olaf Kranz, Brittany Card, Shadrock Roberts, and Andreas Papp
- Chapter 20, Robert Wright
- Chapter 21, Krishna Prasad Vadrevu and Kristofer Lasko
- Chapter 22, Anupma Prakash and Claudia Kuenzer
- Chapter 23, Hasi Bagan and Yoshiki Yamagata
- Chapter 24, Yoshiki Yamagata, Daisuke Murakami, and Hajime Seya
- Chapter 25, Prasad S. Thenkabail

These authors are "who is who" in remote sensing and come from premier institutions of the world. For author affiliations, please see "Contributors" list provided a few pages after this. My deepest apologies if I have missed any name. But, I am sure those names are properly credited and acknowledged in individual chapters.

The authors not only delivered excellent chapters, they provided valuable insights and inputs for me in many ways throughout the book project.

I was delighted when Dr. Compton J. Tucker, senior Earth scientist, Earth Sciences Division, Science and Exploration Directorate, NASA Goddard Space Flight Center (GSFC), agreed to write the foreword for the book. For anyone practicing remote sensing, Dr. Tucker needs no introduction. He has been a *godfather* of remote sensing and has inspired a generation of scientists. I have been a student of his without ever really being one. I mean, I have not been his student in a classroom but have followed his legendary work throughout my career. I remember reading his highly cited paper (now with citations nearing 4000!):

- Tucker, C.J. (1979) Red and photographic infrared linear combinations for monitoring vegetation, *Remote Sensing of Environment*, **8(2)**,127–150.

That was in 1986 when I had just joined the National Remote Sensing Agency (NRSA; now NRSC), Indian Space Research Organization (ISRO). After earning his PhD from the Colorado State University in 1975, Dr. Tucker joined NASA GSFC as a postdoctoral fellow and became a full-time NASA employee in 1977. Since then, he has conducted path-finding research. He has used NOAA AVHRR, MODIS, SPOT Vegetation, and Landsat satellite data for studying deforestation, habitat fragmentation, desert boundary determination, ecologically coupled diseases, terrestrial primary production, glacier extent, and how climate affects global vegetation. He has authored or coauthored more than 170 journal articles that have been cited more than 20,000 times, is an adjunct professor at the University of Maryland, is a consulting scholar at the University of Pennsylvania's Museum of Archaeology and Anthropology, and has appeared on more than twenty radio and TV programs. He is a fellow of the American Geophysical Union and has been awarded several medals and honors, including NASA's Exceptional Scientific Achievement Medal, the Pecora Award from the U.S. Geological Survey (USGS), the National Air and Space Museum Trophy, the Henry Shaw Medal from the Missouri Botanical Garden, the Galathea Medal from the Royal Danish Geographical Society, and the Vega Medal from the Swedish Society of Anthropology and Geography. He was the NASA representative to the U.S. Global Change Research Program from 2006 to 2009. He was instrumental in releasing the AVHRR 32-year (1982–2013) Global Inventory Monitoring and Modeling Studies (GIMMS) data. I strongly recommend that everyone read his excellent foreword before reading the book. In the foreword, Dr. Tucker demonstrates the importance of data from EO sensors from orbiting satellites to maintaining a reliable and consistent climate record. Dr. Tucker further highlights the importance of continued measurements of these variables of our planet in the new millennium through new, improved, and innovative EO sensors from Sun-synchronous and/or geostationary satellites.

I am very thankful to my USGS colleagues for their encouragement and support. In particular, I mention Edwin Pfeifer, Dr. Susan Benjamin, Dr. Dennis Dye, Larry Gaffney, Miguel Velasco, Dr. Chandra Giri, Dr. Terrance Slonecker, Dr. Jonathan Smith, and Dr. Thomas Loveland. There are many other colleagues who made my job at USGS that much easier. My thanks to them all.

I am very thankful to Irma Shagla-Britton, acquisition editor for remote sensing and GIS books at Taylor & Francis Group/CRC Press. Without her initial nudge, this book would never have even been completed. Thank you, Irma. You are doing a great job.

I am very grateful to my wife (Sharmila Prasad) and daughter (Spandana Thenkabail) for their usual unconditional love, understanding, and support. They are always the pillars of my life. I learned the values of hard work and dedication from my revered parents. This work wouldn't have come about without their sacrifices to educate their children and their silent blessings. I am ever grateful to my former professors at The Ohio State University, Columbus, Ohio, United States: Prof. John G. Lyon, Dr. Andrew D. Ward, Prof. (Late) Carolyn Merry, Dr. Duane Marble, and Dr. Michael Demers. They have taught, encouraged, inspired, and given me opportunities at the right time. The opportunity to work for six years at the Center for Earth Observation of Yale University (YCEO) was incredibly important. I am thankful to Prof. Ronald G. Smith, director of YCEO, for his kindness. At YCEO, I learned and advanced myself as a remote sensing scientist. The opportunities I got from working for the International Institute of Tropical Agriculture (IITA), Africa and International Water Management Institute (IWMI) that had a global mandate for water were very important, especially from the point of view of understanding the real issues on the ground. I learned my basics of remote sensing mainly working with Dr. Thiruvengadachari of the National Remote Sensing Agency/Center (NRSA/NRSC), Indian Space Research Organization (ISRO), India, where I started my remote sensing career as a young scientist. I was just 25 years old then and had joined NRSA after earning my masters of engineering (hydraulics and water resources) and bachelors of engineering (civil engineering). During my first day in the office, Dr. Thiruvengadachari asked me how much remote sensing did I know. I said, "zero" and instantly thought that I would be thrown out of the room. But he said "very good" and gave me a manual on remote sensing from the Laboratory for Applications of Remote Sensing (LARS), Purdue. Those were the days where there was no formal training in remote sensing in any Indian universities. So my remote sensing lessons began working practically on projects and one of our first projects was "drought monitoring for India using NOAA AVHRR data." This was an intense period of learning remote sensing by actually practicing it on a daily basis. Data came on 9 mm tapes; data were read on massive computing systems; image processing was done, mostly working on night shifts by booking time on centralized computing; field work was conducted using false color composite outputs and topographic maps (not the days of global positioning systems); geographic information system was in its infancy; and a lot of calculations were done using calculators. So when I decided to resign my NRSA job and go to the United States to do my PhD, Dr. Thiruvengadachari told me, "Prasad, I am losing my right hand, but you can't miss opportunity." Those initial wonderful days of learning from Dr. Thiruvengadachari will remain etched in my memory. Prof. G. Ranganna of the Karnataka Regional Engineering College (KREC; now National Institute of Technology), Karnataka, India, was/is one of my most revered gurus. I have learned a lot observing him, professionally and personally, and he has always been an inspiration. Prof. E.J. James, former director of the Center for Water Resources Development and Management (CWRDM), was another original guru from whom I have learned the values of a true professional. I am also thankful to my good old friend Shri C. J. Jagadeesha, who is still working for ISRO as a senior scientist. He was my colleague at NRSA/NRSC, ISRO, and encouraged me to grow as a scientist. This Remote Sensing Handbook is a blessing from the most special ones dear to me. Of course, there are many, many others to thank especially many of my dedicated students over the years, but they are too many to mention here. I thank the truly outstanding editing work performed by Arunkumar Aranganathan and his team at SPi Global.

It has been my deep honor and great privilege to have edited the Remote Sensing Handbook. I am sure that I won't be taking on any such huge endeavors in the future. I will need time for myself, to look inside, understand, and grow. So thank you all, for making this possible.

Prasad S. Thenkabail, PhD
Editor-in-Chief

Editor

Prasad S. Thenkabail, PhD, is currently working as a research geographer-15 with the U.S. Geological Survey (USGS), United States. Currently, at USGS, Prasad leads a multi-institutional NASA MEaSUREs (Making Earth System Data Records for Use in Research Environments) project, funded through NASA ROSES solicitation. The project is entitled Global Food Security-Support Analysis Data at 30 m (GFSAD30) (http://geography.wr.usgs.gov/science/croplands/index.html also see https://www.croplands.org/). He is also an adjunct professor at three U.S. universities: (1) Department of Soil, Water, and Environmental Science (SWES), University of Arizona (UoA); (2) Department of Space Studies, University of North Dakota (UND); and (3) School of Earth Sciences and Environmental Sustainability (SESES), Northern Arizona University (NAU), Flagstaff, Arizona.

Dr. Thenkabail has conducted pioneering scientific research work in two major areas:

1. Hyperspectral remote sensing of vegetation
2. Global irrigated and rainfed cropland mapping using spaceborne remote sensing

His research papers on these topics are widely quoted. His hyperspectral work also led to his working on the scientific advisory board of Rapideye (2001), a German private industry satellite. Prasad was consulted on the design of spectral wavebands.

In hyperspectral research, Prasad pioneered in the following:

1. The design of optimal hyperspectral narrowbands (HNBs) and hyperspectral vegetation indices (HVIs) for agriculture and vegetation studies.
2. Certain hyperspectral data mining and data reduction techniques such as now widely used concepts of lambda by lambda plots.
3. Certain hyperspectral data classification methods. This included the use of a series of methods (e.g., discriminant model, Wilk's lambda, Pillai trace) that demonstrate significant increases in classification accuracies of land cover and vegetation classes as determined using HNBs as opposed to multispectral broadbands.

In global croplands, Prasad conducted seminal research that led to the first global map of irrigated and rainfed cropland areas using multitemporal, multisensor remote sensing, one book, and a series of more than ten novel peer-reviewed papers.

In 2008, for one of these papers, Prasad (lead author) and coauthors (Pardhasaradhi Teluguntala, Trent Biggs, Murali Krishna Gumma, and Hugh Turral) were the second-place recipients of the 2008 John I. Davidson American Society of Photogrammetry and Remote Sensing (ASPRS) President's Award for practical papers. The paper proposed a novel spectral matching technique (SMT) for cropland classification. Earlier, Prasad (lead author) and coauthors (Andy Ward, John Lyon, and Carolyn Merry), won the 1994 Autometric Award for outstanding paper on remote sensing of agriculture from ASPRS. Recently, Prasad (second author) with Michael Marshall (lead author), won the ASPRS ERDAS award for best scientific paper on remote sensing for their hyperspectral remote sensing work.

Earlier to this **path-breaking Remote Sensing Handbook**, Prasad has published two seminal books (both published by Taylor & Francis Group/CRC Press) related to hyperspectral remote sensing and global croplands:

- Thenkabail, P.S., Lyon, G.J., and Huete, A. 2011. *Hyperspectral Remote Sensing of Vegetation*. CRC Press/Taylor & Francis Group, Boca Raton, FL, 781pp.

Reviews of this book:

- http://www.crcpress.com/product/isbn/9781439845370.
- Thenkabail, P., Lyon, G.J., Turral, H., and Biradar, C.M. 2009. *Remote Sensing of Global Croplands for Food Security*. CRC Press/Taylor & Francis Group, Boca Raton, FL, 556pp (48 pages in color).

Reviews of this book:

- http://www.crcpress.com/product/isbn/9781420090093.
- http://gfmt.blogspot.com/2011/05/review-remote-sensing-of-global.html.

He has guest edited two special issues for the American Society of Photogrammetry and Remote Sensing (PE&RS):

- Thenkabail, P.S. 2014. Guest editor of special issue on "Hyperspectral remote sensing of vegetation and agricultural crops." *Photogrammetric Engineering and Remote Sensing* 80(4).
- Thenkabail, P.S. 2012. Guest editor for Global croplands special issue. *Photogrammetric Engineering and Remote Sensing* 78(8).

He has also guest edited a special issue on global croplands for the *Remote Sensing Open Access Journal* (ISSN 2072-4292):

- Thenkabail, P.S. 2010. Guest editor: Special issue on "Global croplands" for the MDPI remote sensing open access journal. Total: 22 papers. http://www.mdpi.com/journal/remotesensing/special_issues/croplands/.

Prasad is, currently editor-in-chief, *Remote Sensing Open Access Journal,* an on-line journal, published by MDPI; editorial board member, *Remote Sensing of Environment;* editorial advisory board member, *ISPRS Journal of Photogrammetry and Remote Sensing.*

Prior to joining USGS in October 2008, Dr. Thenkabail was a leader of the remote sensing programs of leading institutes International Water Management Institute (IWMI), 2003–2008; International Center for Integrated Mountain Development (ICIMOD), 1995–1997; International Institute of Tropical Agriculture (IITA), 1992–1995.

He also worked as a key remote sensing scientist for Yale Center for Earth Observation (YCEO), 1997–2003; Ohio State University (OSU), 1988–1992; National Remote Sensing Agency (NRSA) (now NRSC), Indian Space Research organization (ISRO), 1986–1988.

Over the years, he has been a principal investigator (PI) of NASA, USGS, IEEE, and other funded projects such as inland valley wetland mapping of African nations, characterization of eco-regions of Africa (CERA), which involved both African savannas and rainforests, global cropland water use for food security in the twenty-first century, automated cropland classification algorithm (ACCA) within WaterSMART (Sustain and Manage America's Resources for Tomorrow) project, water productivity mapping in the irrigated croplands of California and Uzbekistan using multisensor remote sensing, IEEE Water for the World Project, and drought monitoring in India, Pakistan, and Afghanistan.

The USGS and NASA selected Dr. Thenkabail to be on the Landsat Science Team (2007–2011) for a period of five years (http://landsat.gsfc.nasa.gov/news/news-archive/pol_0005.html; http://ldcm.usgs.gov/intro.php). In June 2007, his team was recognized by the Environmental System Research Institute (ESRI) for "special achievement in GIS" (SAG award) for their tsunami-related work (tsdc.iwmi.org) and for their innovative spatial data portals (http://waterdata.iwmi.org/dtView-Common.php; earlier http://www.iwmidsp.org). Currently, he is also a global coordinator for the Agriculture Societal Beneficial Area (SBA) of the Committee for Earth Observation

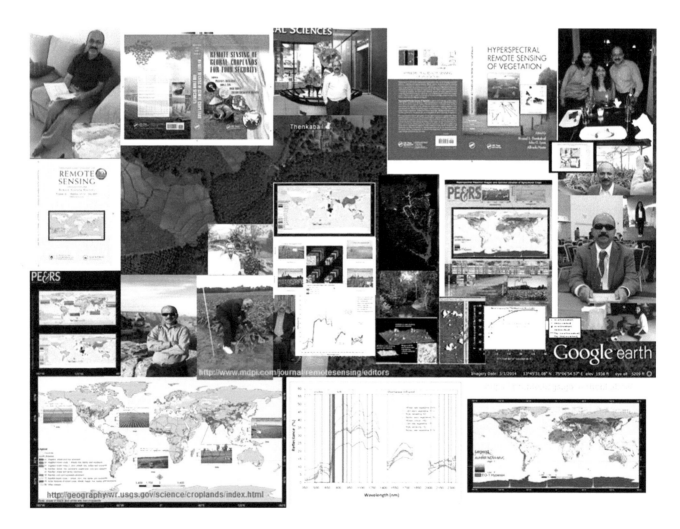

Systems (CEOS). He is active in the Group on Earth Observation (GEO) agriculture and water efforts through Earth observation. He was a co-lead of the Water for the World Project (IEEE effort). He is the current chair of the International Society of Photogrammetry and Remote Sensing (ISPRS) Working Group WG VIII/7: "Land Cover and Its Dynamics, including Agricultural & Urban Land Use" for the period 2013–2016. Thenkabail earned his PhD from The Ohio State University (1992). His master's degree in hydraulics and water resources engineering (1984) and bachelor's degree in civil engineering (1981) were from India. He began his professional career as a lecturer in hydrology, water resources, hydraulics, and open channel in India. He has 100+ publications, mostly peer-reviewed research papers in major international remote sensing journals: http://scholar.google.com/citations?user=9IO5Y7 YAAAAJ&hl=en. Prasad has about 30 years' experience working as a well-recognized international expert in remote sensing and geographic information systems (RS/GIS) and their application to agriculture, wetlands, natural resource management, water resources, forests, sustainable development, and environmental studies. His work experience spans over 25+ countries spread across West and Central Africa, Southern Africa, South Asia, Southeast Asia, the Middle East, East Asia, Central Asia, North America, South America, and the Pacific.

Contributors

Martha Anderson
Hydrology and Remote Sensing
    Laboratory
Agricultural Research Service
United States Department of Agriculture
Beltsville, Maryland

Ricardo G. Andrade
Embrapa Satellite Monitoring
Campinas, Brazil

Allan S. Arnesen
Water and Sanitation Company of the
    State of São Paulo
São Paulo, Brazil

Clement Atzberger
Institute for Surveying, Remote Sensing
    and Land Information
University of Natural Resources and Life
    Sciences (BOKU)
Vienna, Austria

Hasi Bagan
Center for Global Environmental
    Research
National Institute for Environmental
    Studies
Tsukuba, Japan

Trent W. Biggs
Department of Geography
San Diego State University
San Diego, California

Edson L. Bolfe
Embrapa Satellite Monitoring
Campinas, Brazil

James B. Campbell
Department of Geography
Virginia Tech
Blacksburg, Virginia

Brittany Card
Harvard Humanitarian Initiative
Cambridge, Massachusetts

Christos Chalkias
Department of Geography
Harokopio University
Athens, Greece

Ni-Bin Chang
Department of Civil, Environmental and
    Construction Engineering
and
Stormwater Management Academy
University of Central Florida
Orlando, Florida

H.J. Cho
Department of Integrated Environmental
    Science
Bethune-Cookman University
Daytona Beach, Florida

Wade T. Crow
Hydrology and Remote Sensing
    Laboratory
Agricultural Research Service
United States Department of Agriculture
Beltsville, Maryland

Marcelo P. Curtarelli
National Institute for Space Research
São José dos Campos, Brazil

Venkateswarlu Dheeravath
World Food Program
Juba, South Sudan

Erwann Fillol
Action Contre la Faim—International
ACF West Africa Regional Office
Dakar, Senegal

Matheus Z. Francisco
Environmental Science and
    Technology—CTTMar
UNIVALI
Itajaí, Brazil

Renato A.M. Franco
São Paulo State University
Ilha Solteira, Brazil

Petra Füreder
Interfaculty Department of
    Geoinformatics
University of Salzburg
Salzburg, Austria

Frederico T. Genofre
Environmental Engineer
Santa Catarina, Brazil

Shuvankar Ghosh
Department of Geography
University of Georgia
Athens, Georgia

Chandra Giri
Earth Resources Observation and
    Science Center
United States Geological Survey
Sioux Falls, South Dakota

Edward P. Glenn
Department of Soil, Water and
    Environmental Science
University of Arizona
Tucson, Arizona

Murali Krishna Gumma
International Crops Research Institute
    for the Semi Arid Tropics
Hyderabad, India

Wei Guo
Department of Geography
University of Maryland
College Park, Maryland

xxxii

Contributors

Chris Hain
Earth System Science Interdisciplinary
    Center
University of Maryland
College Park, Maryland

Frederic Ham
Action Contre la Faim—International
ACF West Africa Regional Office
Dakar, Senegal

Fernando B.T. Hernandez
São Paulo State University
Ilha Solteira, Brazil

Joachim Hill
Faculty of Regional and Environmental
    Sciences
Department of Environmental Remote
    Sensing and Geoinformatics
Trier University
Trier, Germany

C. Hladik
Department of Geography
Georgia Southern University
Statesboro, Georgia

Yang Hong
Hydrometeorology and Remote Sensing
    Laboratory
School of Civil Engineering and
    Environmental Sciences
University of Oklahoma
Norman, Oklahoma

Jungho Im
Ulsan National Institute of Science and
    Technology
Ulsan, Republic of Korea

Norman Kerle
Faculty of Geo-Information Science and
    Earth Observation (ITC)
University of Twente
Twente, the Netherlands

Sadiq I. Khan
Hydrometeorology and Remote Sensing
    Laboratory
School of Civil Engineering and
    Environmental Sciences
University of Oklahoma
Norman, Oklahoma

Felix Kogan
National Environmental Satellite Data
    and Information Services
Center for Satellite Applications and
    Research
National Oceanic and Atmospheric
    Administration
College Park, Maryland

Olaf Kranz
Research Section
Helmholtz-Association
Berlin, Germany

Claudia Kuenzer
German Aerospace Center (DLR)
and
Earth Observation Center (EOC)
and
German Remote Sensing Data Center
    (DFD)
Wessling, Germany

Stefan Lang
Interfaculty Department of
    Geoinformatics
University of Salzburg
Salzburg, Austria

Kristofer Lasko
Department of Geographical Sciences
University of Maryland
College Park, Maryland

Janice F. Leivas
Embrapa Satellite Monitoring
Campinas, Brazil

Husi Letu
Research and Information Center
Tokai University
Tokyo, Japan

Noam Levin
Department of Geography
The Hebrew University of Jerusalem
Mount Scopus, Jerusalem

Tiangang Liang
College of Pastoral Agriculture Science
    and Technology
Lanzhou University
Lanzhou, Gansu, People's Republic of China

Michael Marshall
Research Program on Climate Change,
    Agriculture and Food Security
World Agroforestry Center
Nairobi, Kenya

Sandro Martinis
German Aerospace Center (DLR)
and
Earth Observation Center (EOC)
and
German Remote Sensing Data Center
    (DFD)
Wessling, Germany

Michele Meroni
Institute for Environment and
    Sustainability
Joint Research Centre
European Commission
Ispra, Italy

Alex Messina
Department of Geography
San Diego State University
San Diego, California

D.R. Mishra
Department of Geography
University of Georgia
Athens, Georgia

Irshad A. Mohammed
International Crops Research Institute
    for the Semi Arid Tropics
Hyderabad, India

Daisuke Murakami
Center for Global Environmental
    Research
National Institute for Environmental
    Studies
Tsukuba, Japan

Pamela Nagler
Sonoran Desert Research Station
Southwest Biological Science Center
U.S. Geological Survey
Tucson, Arizona

Jessica L. O'Connell
Department of Marine Science
University of Georgia
Athens, Georgia

Andreas Papp
Médecins Sans Frontières (MSF) Austria
Vienna, Austria

Seonyoung Park
Ulsan National Institute of Science and
    Technology
Ulsan, Republic of Korea

George P. Petropoulos
Department of Geography and Earth
  Sciences
Aberystwyth University
Wales, United Kingdom

Anupma Prakash
Geophysical Institute
University of Alaska Fairbanks
Fairbanks, Alaska

Felix Rembold
Institute for Environment and Sustainability
Joint Research Centre
European Commission
Ispra, Italy

Lynn M. Resler
Department of Geography
Virginia Tech
Blacksburg, Virginia

Jinyoung Rhee
APEC Climate Center
Busan, Republic of Korea

Shadrock Roberts
Center for Geospatial Research
University of Georgia
Athens, Georgia

Matt Rodell
Hydrologic Science Laboratory
Goddard Space Flight Center
National Aeronautics and Space
  Administration
Greenbelt, Maryland

Achim Röder
Faculty of Regional and Environmental
  Sciences
Department of Environmental Remote
  Sensing and Geoinformatics
Trier University
Trier, Germany

Oscar Rojas
Natural Resources Management and
  Environment Department
Food and Agriculture Organization of
  the United Nations
Rome, Italy

Morris Scherer-Warren
National Water Agency
Brasilia, Brazil

Santhosh Kumar Seelan
Department of Space Studies
University of North Dakota
Grand Forks, North Dakota

Hajime Seya
Graduate School for International
  Development and Cooperation
Hiroshima University
Higashihiroshima, Japan

Stefan Sommer
Institute for Environment and
  Sustainability
Joint Research Centre
European Commission
Ispra, Italy

Ruth Sonnenschein
Institute for Applied Remote Sensing
European Academy of Bozen/Bolzano
Bolzano, Italy

Marion Stellmes
Faculty of Regional and Environmental
  Sciences
Department of Environmental Remote
  Sensing and Geoinformatics
Trier University
Trier, Germany

Tsegaye Tadesse
National Drought Mitigation Center
University of Nebraska-Lincoln
Lincoln, Nebraska

Antônio de C. Teixeira
Embrapa Satellite Monitoring
Campinas, Brazil

Pardhasaradhi Teluguntla
United States Geological Survey
Flagstaff, Arizona

and

Bay Area Environmental Research
  Institute
West Sonoma, California

Prasad S. Thenkabail
Southwest Geographic Science
  Center
United States Geological Survey (USGS)
Flagstaff, Arizona

André Twele
German Aerospace Center (DLR)
and
Earth Observation Center (EOC)
and
German Remote Sensing Data Center
  (DFD)
Wessling, Germany

Thomas Udelhoven
Department of Environmental Remote
  Sensing and Geoinformatics
Faculty of Regional and Environmental
  Sciences
Trier University
Trier, Germany

Krishna Prasad Vadrevu
Department of Geographical Sciences
University of Maryland
College Park, Maryland

Naga Manohar Velpuri
Center for Earth Resources Observation
  and Science
United States Geological Survey
Sioux Falls, South Dakota

Daniel C. Victoria
Embrapa Satellite Monitoring
Campinas, Brazil

Xianwei Wang
Center of Integrated Geographic
  Information Analysis
School of Geography and Planning
Sun Yat-Sen University
Guangzhou, People's Republic of China

Brian Wardlow
Center for Advanced Land Management
  Technologies
and
National Drought Mitigation Center
School of Natural Resources
University of Nebraska-Lincoln
Lincoln, Nebraska

Robert Wright
Hawaii Institute of Geophysics and
  Planetology
University of Hawaii at Mānoa
Honolulu, Hawaii

Hongjie Xie
Laboratory for Remote Sensing and
    Geoinformatics
Department of Geological Sciences
University of Texas at San Antonio
San Antonio, Texas

Xianwu Xue
Hydrometeorology and Remote
    Sensing Laboratory
School of Civil Engineering and
    Environmental Sciences
University of Oklahoma
Norman, Oklahoma

Yoshiki Yamagata
Center for Global Environmental
    Research
National Institute for the Environmental
    Studies
Tsukuba, Japan

Guoqing Zhang
Key Laboratory of Tibetan
    Environmental Changes and Land
    Surface Processes
Institute of Tibetan Plateau Research
Chinese Academy of Sciences
Beijing, People's Republic of China

Qingling Zhang
Department of Forestry and
    Environmental Sciences
Yale University
New Haven, Connecticut

and

Chinese Academy of Sciences
Shenzhen Institutes of Advanced
    Technology
Shenzhen, People's Republic of China

Yu Zhang
Hydrometeorology and Remote Sensing
    Laboratory
School of Civil Engineering and
    Environmental Sciences
University of Oklahoma
Norman, Oklahoma

Hydrology and Water Resources

Remote Sensing Technologies for Multiscale Hydrological Studies: Advances and Perspectives

Sadiq I. Khan
University of Oklahoma

Ni-Bin Chang
University of Central Florida

Yang Hong
University of Oklahoma

Xianwu Xue
University of Oklahoma

Yu Zhang
University of Oklahoma

Acronyms and Definitions

ABI	Advanced Baseline Imager
AIRS	Atmospheric Infrared Sounder
AMSR-E	Advanced Microwave Scanning Radiometer-EOS
ARM	Atmospheric Radiation Measurement
ATMS	Advanced Technology Microwave Sounder
AVHRR	Advanced Very High Resolution Radiometer
CERES	Clouds and the Earth's Radiant Energy System
CMORPH	Climate Prediction Center Morphing algorithm
CrIS	Cross-track Infrared Sounder
DMSP	Defense Meteorological Satellite Program
DPR	Dual-frequency Precipitation Radar
EDRs	Environmental Data Records
EM	Electromagnetic
EOS	Earth Observing System
EROS	Earth Resources Observation and Science
ESA	European Space Agency
ET	Evapotranspiration
ETMA	ET mapping algorithm
EXIS	Extreme Ultraviolet and X–Ray Irradiance Sensors
GLM	Geostationary Lightning Mapper
GMI	GPM Microwave Imager
GOES	Geostationary Operational Environmental Satellite
GOES-R	Geostationary Operational Environmental Satellite-R Series
GPCP	Global Precipitation Climatology Project
GPM	Global Precipitation Measurement
H	Horizontal
iGOS	Integrated Grassland Observing Site
IR	Infrared
LP DAAC	Land Processes Distributed Active Archive Center
METRIC	Mapping EvapoTranspiration at high Resolution with Internalized Calibration
MIRAS	Microwave Imaging Radiometer with Aperture Synthesis
MIS	Microwave Imager/Sounder
MODIS	Moderate Resolution Imaging Spectroradiometer
MOISST	Marena Oklahoma In Situ Sensor Testbed
NEXRAD	Next-Generation Radar
NIR	Near infrared
NPOESS	National Polar-orbiting Operational Environmental Satellite System
NPP	NPOESS Preparatory Project
NTSG	Numerical Terradynamic Simulation Group
OMPS	Ozone Mapping and Profiler Suite

P	Precipitation
PSD	Particle size distribution
PSU	Practical salinity unit
R	Runoff
RADAR	Radio detection and ranging
Rn	Net radiation
SEB	Surface energy balance
SEBAL	Surface Energy Balance Algorithm for Land
SEBS	Surface Energy Balance System
SEISS	Space Environment In-Situ Suite
SEM	Space environment monitor
SGP	Southern Great Plains
SMAP	Soil Moisture Active and Passive
SMOS	Soil and the Moisture and Ocean Salinity
S-SEBI	Simplified Surface Energy Balance Index
SUVI	Solar Ultraviolet Imager
SWIR	Shortwave Infrared
SWOT	Surface Water and Ocean Topography
Tb	Brightness temperatures
TIR	Thermal infrared
TMI	TRMM Microwave Imager
TMPA	TRMM Multisatellite Precipitation Analysis
TRMM	Tropical Rainfall Measuring Mission
TSIS	Total and Spectral Solar Irradiance Sensor
TWS	Terrestrial water storage
V	Vertical
VIIRS	Visible Infrared Imaging Radiometer Suite
VIS	Visible
VSM	Volumetric soil moisture
ΔS	Overall change of water storage

1.1 Introduction

Remote sensing measurements of hydrologic variables and processes represent one of the most challenging research problems in Earth system sciences. Hydrologic processes in the nexus of energy and water cycles include precipitation, runoff, infiltration and soil water contents, evapotranspiration (ET), and groundwater movement. Remote sensing estimates provide water budget information of the temporal and spatial distribution of water fluxes, often difficult to capture by conventional ground-based sparse sensor network with point-measurement instruments. Many hydrological processes can be observed from satellite- to ground-based remote sensing technologies individually. However, integrated approach to measure all relevant hydrological processes at the same time to completely address the changing hydrological states during the cycling of water remains a scientific challenge. Scale linkage of water fluxes integrating hydrological processes observed from small scale to regional and to continental scale has been contemporary challenges for decades.

The terrestrial water budget, commonly defined by the water balance, is the overall change of water storage (ΔS) and the difference between the incoming amount of precipitation (P) and subtracted amount of water in the form of ET and runoff (R) with respect to time:

$$\Delta S = P - ET - R \qquad (1.1)$$

The water budget variables in Equation 1.1 have a different spatiotemporal footprint. For example, precipitation follows a different timescale than groundwater recharge. However, all these processes are interconnected, and several satellite sensors discussed throughout this chapter can be used to estimate these processes at different spatiotemporal scales. To accomplish each independent task, satellite and/or airborne sensors are designed to utilize a wide range of electromagnetic (EM) spectrum that can provide unique information of radiative reflectance regarding the targeted hydrologic processes. Besides the EM sensors, microgravity sensors are used to measure the spatiotemporal variations in the terrestrial water storage (TWS) (Tapley et al., 2004; Syed et al., 2008).

At both regional and global scales, the real-time operational hydrological prediction can be increasingly appreciated and supported by the current and future Earth observation missions. Most of the processes in the terrestrial water budget such as precipitation, ET, soil moisture, and water storage changes can be observed at different spatiotemporal resolutions and accuracy through remote sensing. The current Earth Observing System (EOS) with coarse spatiotemporal resolution can be a limiting factor for small watershed scale studies; however, it should be noted that satellite missions with advanced sensor technologies (some missions discussed in the following subsections) and this limitation are expected to become less important in the near future. The new satellite missions are anticipated to provide better precipitation and soil moisture data in terms of coverage, accuracy, and resolutions. The upcoming soil moisture mission is expected to significantly enhance our understating of wet and dry condition of land surface and therefore may improve numerical modeling efforts. Some of the important satellite missions that will provide better estimates of Earth's water cycle are discussed in this chapter.

This chapter introduces advances and perspectives of multiscale hydrological studies from spaceborne to airborne and to ground-based remote sensing observations. The detailed discussion in remote sensing hydrology provides an overview of various orbital satellite platforms/sensors focusing on the Earth's hydrological cycle within major global satellite missions associated with some local and regional applications in the world. Soil moisture, precipitation, and ET are emphasized in different sections. The principles of physics associated with each of these three important hydrologic variables are entailed individually with an emphasis on the use of remote sensing data products and their availability without involving sensor design details and specific instrumentation. Particular emphasis is also given to science and technology used for spaceborne data estimation, validation, and its applications.

Some of the current remote sensing sensors that are employed to study the water cycle are described throughout the chapter.

Along this line, Section 1.2 details the evolution of different quantitative precipitation estimation methods, Section 1.3 is related to ET estimation using different surface energy balance (SEB) algorithms, and Section 1.4 is linked to satellite remote sensing methods for soil moisture estimation. In the end of this chapter, the focus is placed on how remote sensing estimates can be utilized to concatenate the fundamental hydrologic processes such as precipitation, ET, and soil moisture and how future mission-oriented remote sensing programs are angled to close the hydrological cycle. The last section encompasses perspectives and conclusions.

1.2 Remote Sensing Advances in Hydrology

Advances in hydrology have been made possible with the introduction of research satellite platforms such as the Landsat program, Terra and Aqua Missions, and operational platforms such as Tropical Rainfall Measuring Mission (TRMM) and Geostationary Operational Environmental Satellite (GOES) series, as well as many others. Satellite remote sensing offers a framework to complement and provides better understanding of the regional- as well as global-scale hydrosystem processes. Many hydrological state variables and fluxes can be evaluated through satellite remote sensing. Optical sensors as well as both active and passive sensors are used as source of observations, particularly in regions where ground-based observation is sparsely available or sometimes unavailable. For instance, remote sensors are used to acquire data on topography that can be represented in the form of 3D data, referred as the Digital Elevation Model. Multispectral and hyperspectral remote sensing detect reflected or emitted energy from an object in a number of different spectral bands of the EM spectrum. In remote sensing, this spectral signature is the best diagnostic tool in remotely identifying the composition of an object. Generally, there is a trade-off between the spectral resolution, spectral coverage, radiometric resolution, and temporal resolution.

Precipitation is one of the major drivers of the hydrologic cycle and requires accurate measurement in order to help access the overland runoff fluxes and other hydrological state variables. With the scarcity and poor quality of ground-based observations, satellite remote sensing estimates can be a viable source for precipitation estimation. Microwave techniques are used to directly observe the rain rates as microwave radiation relates strongly with different hydrometeors. Some of the well-researched remote sensing precipitation products include Climate Prediction Center Morphing algorithm (CMORPH) (Joyce et al., 2004), Precipitation Estimation from Remotely Sensed Information Using Artificial Neural Networks (PERSIANN) (Sorooshian et al., 2000), and TRMM Multisatellite Precipitation Analysis (TMPA) (Huffman et al., 2007). The TRMM satellite, with TRMM Microwave Imager (TMI) onboard and Advanced Microwave Scanning Radiometer-EOS (AMSR-E) onboard the EOS Aqua satellite, is providing precipitation estimates.

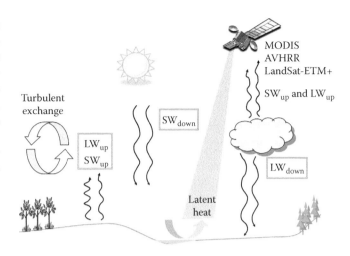

FIGURE 1.1 Remotely sensed surface energy budget components.

ET is one of the most important components of the water balance analysis. ET estimation using ground network-based methods such as pan measurement, Bowen ratio, eddy correlation, and weighing lysimeter and scintillometer are complex techniques that require a lot of data. In comparison to the conventional point-based measurements, remote sensing techniques have the following advantages: (1) reasonable spatial and temporal coverage, (2) economic viability, (3) functional in complex terrains and ungauged areas. Numerous ET models have been used for more than three decades to make use of visible (VIS), near infrared (IR), shortwave IR, and most importantly, thermal data acquired by sensors on airborne and satellite platforms (Figure 1.1).

It is difficult to measure ET directly from remote sensing methods. However, a number of algorithms are available to estimate ET from remotely sensed estimates of surface energy fluxes (Figure 1.1). These models include the Simplified SEB Index (Roerink et al., 2000), the SEB System (Su, 2002), and the SEB Algorithm for Land (Bastiaanssen et al., 1998a,b), ET mapping algorithm (Loheide and Gorelick, 2005), and Mapping EvapoTranspiration at high Resolution with Internalized Calibration (METRIC) (Allen et al., 2007). There are ongoing research efforts to improve the spatial resolution of ET estimation methods to integrate these products into hydrologic models.

Soil moisture is another important hydrologic variable for understanding water cycle. The distributed information of soil moisture is of great importance in the hydrology and climate studies due to its high spatial and temporal variability. Microwave remote sensing provides a unique capability for direct observation of soil moisture. Remote measurements from space afford the possibility of obtaining frequent, global sampling of soil moisture over a large fraction of the Earth's land surface. Integrating estimates from both passive and active microwave sensors promotes potential for improved surface soil moisture estimates. Such methods have been evaluated with in situ observations and validated in several field experiments

TABLE 1.1   Some of the NASA's Future Missions, Hydrologic Observations, and Societal Benefits[a]

Missions	Launch Date	Partner	Observed Quantities	Societal Benefit
LDCM	May 2013	USGS	Land cover/land use, thermal and IR surface properties	Land cover, ecosystem composition, algal blooms, waterborne and zoonotic disease
GPM-Core	July 2013	JAXA	Precipitation	Heat stress and drought, precipitation, all-weather temperature and humidity, surface water and ocean topography
SAGE III/ISS	May 2014	SOMD	Vertical profiles of ozone, water vapor atmosphere	Ozone processes (health), clouds and aerosols (climate)
SMAP	November 2014		Soil moisture, freeze–thaw state	Algal blooms and waterborne infectious disease, soil moisture, surface water, and ocean topography
ICESat 2	October 2015		Ice sheet thickness, vegetation height	Clouds, aerosols, ice, and carbon, glacier surface elevation, glacier retreat
GRACE II	2016	DLR	Time variable gravity, including mass of water/ice	Ocean circulation, heat storage, and climate forcing, groundwater storage, ice sheet mass balance, ocean mass
DESDynI Radar and Lidar	November 2017	DLR	Earth surface deformation; vegetation height, canopy volume	Ice dynamics, ecosystem structure, biomass, and biodiversity, heat stress and drought, glacier velocity
SWOT	2020	CNES/USGS	Lake levels (three million lakes), ocean topography	Surface water and ocean topography, ocean circulation and heat storage, river discharges
HysPIRI	2020	USGS	Aquatic ecosystems, volcanic hazards	Heat stress and drought, vector-borne and zoonotic diseases (health), surface composition and thermal properties

[a] http://science.nasa.gov/media/medialibrary/2010/07/01/Climate_Architecture_Final.pdf.

(O'Neill et al., 1996; Chauhan, 1997; Njoku et al., 2002; Bolton et al., 2003). The AMSR-E sensor on Aqua satellite (Njoku et al., 2003) and the Soil and the Moisture and Ocean Salinity (SMOS) mission launched by the European Space Agency (Kerr, 2001) are providing low-frequency-based soil moisture estimates. The spatial resolution of these sensors is 56 and 37 km, respectively.

Understanding how all these variables interact within the geosphere is critical to link the physical drivers of water variability to the health and sustainability of the ecosystem. Furthermore, by coupling unconventional datasets and robust computer models, new contributions can be made in understanding the relationships between precipitation, surface water movement, groundwater storage, aquifer recharge and drawdown, and ecosystem services. Some of the short-term and long-term future satellite missions are listed in Table 1.1.

1.3 Precipitation (mm) Estimation Based on Multisensor Data

1.3.1 Physical Principles of Radar Precipitation Estimation

Microwave remote sensing, that is, a part of EM sensing, has opened new frontiers in hydrologic science. Radio detection and ranging (radar), with its active sensing capabilities, allows data acquisition during both day and night in all weather conditions. Radar sensors are able to provide high-resolution data on the horizontal (1 m) and vertical (10 cm) movement of the Earth's surface (Tronin, 2006). Radar consists of the basic components such as the transmitter, which generates EM radiation as a pulse or continuous wave. Weather radars transmit EM energy in the microwave spectrum that travel at the speed of light in a vacuum at 3×10^8 meters per second (m/s). The relationship between

radio frequency (f), wavelength (λ), and velocity at the speed of light (c) is shown by Equation 1.2:

$$c = f\lambda \tag{1.2}$$

where

c is 3×10^8 m/s
f is in cycles per second, or Hertz (Hz)
λ is in meter

Table 1.2 shows the most common bands, frequencies, and associated wavelengths that correspond to radars that have hydrologic applications. Note that the typical values for radar microwave frequencies are on the order of 10^7–10^{11} Hz; thus, it is convenient to use mega (10^6) and giga (10^9) prefixes, or MHz and GHz. The corresponding radar wavelengths span a few millimeters (mm) up to m. The radar wavelength and diameter (d) of the parabolic dish dictate the angular width of the radar beam, or beamwidth (θ), as follows in Equation 1.3:

$$\theta = \frac{73\lambda}{d} \tag{1.3}$$

where λ and d are both in the same distance units and θ is in °. In the case of the Weather Surveillance Radar-1988 Doppler (WSR-88D) radar, it operates at an approximate 10.7 cm wavelength and has an 8.5 m diameter dish. This corresponds to a beamwidth of approximately 0.92° (in both azimuth and elevation directions). Targets with horizontal cross sections (for a horizontally polarized wave) less than $\lambda/16$, or approximately 7 mm for the WSR-88D, are Rayleigh scatters and thus have predictable radar signatures for different-sized raindrops. The targets are assumed to produce scattering equal in all directions

TABLE 1.2   List of Most Common Radar Bands, Frequencies, and Associated Wavelengths with Their Hydrologic Applications

Band	Frequency	Wavelength	Hydrologic Applications
W	75–110 GHz	2.7–4.0 mm	Detection of cloud droplets
Ka	24–40 GHz	0.8–1.1 cm	Precipitation estimation from spaceborne radar
Ku	12–18 GHz	1.7–2.5 cm	Precipitation estimation from spaceborne radar
X	8–12 GHz	2.5–3.8 cm	High-resolution precipitation and microphysical studies
C	4–8 GHz	3.8–7.5 cm	Precipitation estimation from operational systems
S	2–4 GHz	7.5–15 cm	Precipitation estimation from operational systems
L	1–2 GHz	15–30 cm	Top-layer soil moisture
UHF	300–1000 MHz	0.3–1 m	Ground-penetrating radar for water table
VHF	30–300 MHz	1–10 m	Ground-penetrating radar for water table

called isotropic scattering. The radar detects the component of scattering that comes back to the radar (backscatter). Shorter wavelength radars at X-band and shorter have a lower upper limit on the diameter of targets that cause Rayleigh scattering. But, these smaller wavelength radars do not require such large dishes to maintain a small beamwidth desirable for high-resolution precipitation measurements and thus are more amenable to spaceborne, transportable, and mobile radar platforms. Table 1.2 lists the most common radar bands, frequencies, and associated wavelengths with their hydrologic applications.

1.3.2 Next-Generation Radar

While weather radar applications are diverse and far reaching, the focus in this section is on the hydrologic use of weather radar. In many cases, examples and typical values for the variables will be provided for the Weather Surveillance Radar-1988 Doppler (WSR-88D), the radar that constitutes the Next-Generation Radar (NEXRAD) network (REF) in operation across the United States.

In conventional single-polarization radar detection, the raindrops scatter the incident wave back to the radar antenna. It uses a pulse that is typically polarized about the horizontal plane (H), and the primary measurement used for quantitative precipitation estimation (QPE) is radar reflectivity, Z. Single-polarization radar can receive one polarization horizontal (H) or vertical (V) wave, while a polarimetric radar can transmit/receive waves in both polarizations, hence obtaining more information by which the target medium can be better characterized. In addition to challenges with data quality (i.e., contamination by nonweather scatterers), many studies have shown that Z alone is insufficient to reveal the natural variability of precipitation (Battan, 1973; Rosenfeld and Ulbrich, 2003). The drop size distribution exhibits variability and thus cannot be adequately described using a single reflectivity-to-rainfall rate relation. Polarimetric radar variables are signatures of EM wave scattering from a targeted medium. For hydrometeors, the radar echo depends on their size, shape, orientation, and density. If the scattering amplitude of each hydrometeor and the particle size distribution are known, polarimetric radar variables can be calculated by the integration of scattering amplitude over all the sizes. Depending on their measurement effects and physical meanings, polarimetric radar variables have different applications. Reflectivity factor and differential reflectivity, related to the power of the radar echo, are widely used in various polarimetric radar algorithms. The correlation coefficient is a primary parameter for radar data quality control. Specific differential phase can be used for rain estimation and attenuation correction. Specific attenuation and specific differential attenuation are usually applied by algorithms of attenuation correction.

With the development of dual-polarization radar (also called polarimetric radar), the accuracy of QPE has been improved through the use of polarimetric variables (Bringi and Chandrasekar, 2001). The U.S. NEXRAD network (Figure 1.2) has been upgraded with dual-polarization technology, and similar upgrades are being conducted or are planned in many other countries.

1.3.3 Global Precipitation Estimation

Precipitation estimates have been successful using microwave remote sensing. Microwave remote sensing may be split into active remote sensing, when a signal is first emitted from aircraft or satellites and then received by onboard sensors, or passive remote sensing when reflective radiation information from sunlight is merely recorded by sensors onboard aircraft or satellites.

Precipitation estimates at varying spatial and temporal scales are vital for climatic and hydrologic studies. The underlying principle in global precipitation estimation from remote sensing is to observe the backscatter from different hydrometeor types (rain, hail, snow, ice crystals) in the atmosphere. Satellite-based precipitation estimation started with the use of VIS and IR instruments, by looking at the cloud-top temperatures (Petty, 1995; Ba and Gruber, 2001; Kuligowski, 2002; Bellerby, 2004; Yan and Yang, 2007; Thies et al., 2008). Since the late 1970s, IR satellite remote sensing techniques were first used for precipitation estimation (Arkin and Meisner, 1987). The majority of algorithms attempt to correlate the surface rain rate with IR cloud-top brightness temperatures (Tb) using the information obtained from IR imagery. The algorithms developed to date may be classified into three groups depending on the level of information extracted from the IR cloud images: cloud pixel based, cloud window based, and cloud patch based (Yang et al., 2011). Several examples of these algorithms may clarify this classification further.

Advancement in global precipitation products through improved accuracy, coverage, and spatiotemporal resolution was materialized by combining data from geostationary VIS/IR and low Earth-orbiting microwave sensors. The first such merging algorithm was performed at a relatively coarse scale to ensure reasonable error characteristics. For instance, the first

FIGURE 1.2   Map showing NEXRAD network stations over the contiguous United States.

such multisensor blending algorithm is the Global Precipitation Climatology Project; a multisensor combination is computed on a monthly 2.5° latitude–longitude grid (Adler et al., 2003) and at 1° daily (Huffman et al., 2001).

In the past decade, a number of quasi-global-scale estimates have been developed, including the TMPA (Huffman et al., 2007), the Naval Research Laboratory Global Blended Statistical Precipitation Analysis (Turk and Miller, 2005), CMORPH (Joyce et al., 2004), PERSIANN cloud classification system (Hong et al., 2004, 2005), and University of California, Irvine, PERSIANN (Hsu et al., 1997; Sorooshian et al., 2000). To date, the most commonly available satellite global rain products are summarized in the Table 1.3. These quasi-global precipitation products are discussed in terms of accuracy and Earth science applications throughout the scientific literature; some of these products are mentioned in Table 1.3.

1.3.4 Future Frontier Missions on Precipitation Estimation

1.3.4.1 National Polar-Orbiting Operational Environmental Satellite System

National Polar-orbiting Operational Environmental Satellite System (NPOESS) is the nation's next-generation environmental satellite system. It will replace National Oceanic and Atmospheric Administration's (NOAA) current Polar-orbiting Operational Environmental Satellite and Defense Meteorological Satellite Program spacecraft that have provided global data for weather forecasting and environmental monitoring for over 45 years. A predecessor to NPOESS, the NPOESS Preparatory Project (NPP) is scheduled to be launched in September 2011. NPOESS spacecraft will carry the following four primary instruments: Visible Infrared Imaging Radiometer

TABLE 1.3   Summary of Satellite-Based Global Rainfall Products for Hydrology, Meteorology, and Climate Studies

Product Name	Agency/Country	Scale	Period	Developer
GPCP	NASA/United States	2.5°/month	1979~	Adler et al. (2003)
CMAP	NOAA/United States	2.5°/5 day	1979~	Xie et al. (2003)
GPCP IDD	NASA/United States	1°/day	1998~	Huffman et al. (2001)
TMPA	NASA-GSFC/United States	25 km/3 h	1998~	Huffman et al. (2007)
CMORPH	NOAA-CPC/United States	25 km/3 h	2002~	Joyce et al. (2004)
PERSIANN	University of Arizona	25 km/6 h	2002~	Sorooshian et al. (2000)
NRL-Blend	Naval Research Lab/United States	10 km/3 h	2003~	Turk and Miller (2005)
GSMAP	JAXA/Japan	10 km/h	2005~	sharaku.eor.jaxa.jp
UBham	University of Birmingham/United Kingdom	10 km/h	2002~	Kidd et al. (2003)
PERSIANN-CCS	University of California, Irvine/United States	4 km/half hour	2006~	Hong et al. (2004)
HE	NOAA/NESDIS	4 km/half hour	2004~	Scofield and Kuligowski (2003)

Source: Yang, H. et al., Global precipitation estimation and applications, in: *Multiscale Hydrologic Remote Sensing*, CRC Press, pp. 371–386, 2011.

Suite (VIIRS), Cross-track Infrared Sounder (CrIS), Advanced Technology Microwave Sounder (ATMS), and Ozone Mapping and Profiler Suite. In addition, the afternoon NPOESS spacecraft will carry the Microwave Imager/Sounder (C3 only), space environment monitor, Clouds and the Earth's Radiant Energy System, and Total and Spectral Solar Irradiance Sensor.

The 22-channel VIIRS will collect calibrated VIS/IR radiances to produce about 20 different Environmental Data Records including imagery, cloud and aerosol properties, albedo, land surface type, vegetation index, ocean color, and land and sea surface temperature to fulfill functions similar to what the Moderate Resolution Imaging Spectroradiometer (MODIS) does for National Aeronautics and Space Administration's (NASA) EOS Terra and Aqua missions. VIIRS will provide complete daily global coverage over the VIS, short/medium IR, and longwave IR spectrum at horizontal spatial resolutions of 370 and 740 m at nadir. VIIRS will image at a near constant horizontal resolution across its ~3000 km swath (i.e., from 370 at nadir to ~800 m at edge of scan), a significant improvement over NOAA's Advanced Very High Resolution Radiometer and NASA's MODIS instruments.

The CrIS and the ATMS will provide vertical profiles of atmospheric temperature, humidity, and pressure from the surface to the top of the atmosphere. CrIS senses upwelled IR radiances from 3 to 16 μm at very high spectral resolution (~1300 spectral channels) to determine the vertical atmospheric distribution of temperature and moisture from the surface to the top of the atmosphere across a swath width of 2200 km. CrIS will succeed the Atmospheric Infrared Sounder, which is on NASA's EOS Aqua spacecraft.

1.3.4.2 Global Precipitation Mission

The Global Precipitation Measurement (GPM) is an international initiative, as a follow up of the TRMM. According to the GPM mission documentation and information published online at http://pmm.nasa.gov/GPM, the two main sensors in the GPM Core are the GPM Microwave Imager (GMI), and the Dual-frequency Precipitation Radar (DPR). The GMI will be a conical-scan, nine-channel passive microwave radiometer. The configuration of this instrument provides a broad measurement swath (850 km) and, like the TMI, maintains a constant Earth incidence angle of 52.8° and a constant footprint size for each measurement channel regardless of scan position. The GMI will have a 1.2 m diameter main reflector, which is twice the diameter of the TMI. The DPR will provide high-resolution (approximately 4 km), high-precision measurements of rainfall, rainfall processes, and cloud dynamics.

The DPR is essentially two radars, the Ku-band precipitation radar (Ku-PR) and the Ka-PR. The Ku-PR operates at 13.6 GHz and is of similar design to the TRMM-PR. The Ka-PR is based upon the PR's design but operates at 35.55 GHz, and the size of its antenna has been selected such that its measurement footprint matches the footprint size of the Ku-PR. Each radar uses a phased array, slotted wave guide antenna. The Ka-PR also has a selectable high-sensitivity mode, which provides an interlacing scan with a swath width of 120 km; this high-sensitivity mode will aid in the measurement of light rain and snow. The two phased array antennas assure better precipitation estimation in terms of higher spatial resolution (Flaming, 2005).

In addition to GPM, the Geostationary Operational Environmental Satellite-R Series (GOES-R) will be an important future program of the NOAA operations. The first launch of the GOES-R series satellite is scheduled for 2015. GOES-R mission will play a key role for weather monitoring, warning, and forecasting. The GOES-R satellites will consist of the following instrument: (1) The 16-channel Advanced Baseline Imager for viewing of Earth's clouds, atmosphere, and surface, (2) the Geostationary Lightning Mapper for monitoring hemispheric lightning flashes, (3) the Extreme Ultraviolet and X–Ray Irradiance Sensors for measuring solar particles, (4) the Solar Ultraviolet Imager for imaging the Sun, and (5) the space environment monitoring suite that includes the Space Environment In-Situ Suite and magnetometer for monitoring Earth's space environment and geomagnetic storms.

These instruments on the GOES-R satellite series will produce more than 50 times the information provided by the current GOES system and will offer a wide variety of unique observations of the environment, with particular emphasis on hazardous weather in the western hemisphere and space weather impacts. Also onboard GOES-R will be improved in communication systems with higher data rates to ensure a continuous and reliable flow of remote sensing products and relay of other environmental and emergency services information critical to a broad range of users and interests.

1.4 Evapotranspiration Estimation Using Multisensor Data

1.4.1 Physical Principles of Surface Energy Balance Method

An overview of remote sensing principles and technologies useful for hydrological studies are presented, with additional reviews of precipitation retrieval models and case studies on a regional and field scale. The remote sensing fundamental of precipitation retrieval is to obtain the latent heat flux (i.e., ET) as a residual of the SEB budget by solving the main components of the SEB equation (Equation 1.4), which includes the sensible heat flux (H) and ground heat flux (G).

The latent heat accompanying ET is λ, where λ is the latent heat of vaporization. The ET process requires a source of heat energy to convert water from the liquid to vapor state. This is ultimately supplied by net radiation (Rn), the amount of incident solar radiation (Rs) that is absorbed at the Earth's surface; a simplified equation for the SEB is

$$\lambda ET = Rn - H - G \qquad (1.4)$$

where the available net radiant energy Rn (W/m^2) is combined between the soil heat flux G and the atmospheric convective

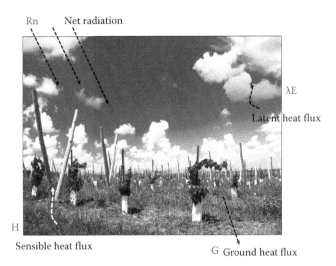

FIGURE 1.3 Energy flux process with radiation budget components.

fluxes (sensible heat flux H) and latent heat flux LE, which is readily converted to ET. The Rn and other variables (H and G) of the Equation 1.3 can be solved using remotely sensed data of surface characteristics such as vegetation cover, surface temperature albedo, and leaf area index. Energy coming from the sun less any radiation gets reflected (or emitted as thermal IR radiation) back to the atmosphere. Some of that Rn is the sensible heat flux (H); some of it is stored in the soil and other objects such as woody material and the rest of the energy is absorbed by water, which can be converted to water vapor for ET (Figure 1.3).

A certain amount of energy per mass of water is required to vaporize moisture, and this is called the latent heat of vaporization. Energy coming from the sun less any radiation that gets reflected (or emitted as thermal IR radiation) back to the atmosphere, or net radiation (Rn), is energy available for actual ET. The remote sensing fundamental of ET retrieval is to obtain the latent heat flux (i.e., ET) as a residual of the SEB budget by solving the main components of the SEB equation that includes the sensible heat flux (H) and ground heat flux (G). Most of these remote sensing ET estimation models use remotely sensed data of surface radiation, temperature, and vegetation properties, which can be retrieved from multispectral sensors from VIS to thermal IR (TIR) bands listed in Table 1.4.

1.4.2 Regional-, Local-, and Field-Scale Studies

A regional-scale modified form of the SEB method is applied to estimate actual ET. Energy balance algorithms are developed by applying satellite-derived surface radiation, meteorological parameters, and vegetation characteristics. The fundamentals of SEB are used to calculate ET as the residual of the energy balance. This method calculates the actual ET by assimilating the MODIS daily dataset and meteorological observations from the Oklahoma Mesonet network. The algorithm is named as the MODIS/Mesonet (M/M-ET). Thereinafter, M/M-ET was developed and evaluated for southern Plains in the United States, particularly for the state of Oklahoma. The M/M-ET measurements are assessed

on daily, 8-day, and regular basis at both field and watershed scales. Two distinctive field sources explained in the succeeding text are used to compare the assessed results: one with meteorological towers for latent heat flux observation and another with the Mesonet sites for crop ET. Latent heat flux observations from the two Atmospheric Radiation Measurements (ARM) (http://www.arm.gov) AmeriFlux eddy covariance tower sites were used for the comparison. These sites are located at the ARM Southern Great Plains (SGP) extended facilities in Lamont and El Reno, Oklahoma. The two Mesonet sites at El Reno and Medford with crop ET data are also selected for the evaluation. The estimated ET is also compared with the crop ET at the selected sites during the wheat-growing season for multiple years. Figure 1.4a and b shows the daily time series and scatter plots for El Reno sites for year 2005 and 2006, respectively.

The estimated ET is in good agreement with the observed crop ET, with an underestimation of –7% and overestimation with 6% for 2005 and 2006, respectively. Correspondingly, the correlation coefficient values indicate that ET values form the model related strongly with values of 0.86 and 0.75 for 2005 and 2006 observations. The correlation coefficient values also indicate that the ET estimates correlate measurements at El Reno site relatively agreed with AmeriFlux observations. The M/M-ET estimation algorithm is implemented for the entire Oklahoma State (Figure 1.5). Khan et al. (2010) provide a detailed evaluation of these results.

Numerous studies have examined the SEB approach at local scale to retrieve remotely sensed ET (Bastiaanssen et al., 1998a, 2005; Su, 2002; Allen et al., 2007). Liu et al. (2010) evaluated the spatiotemporal variations of actual ET for four types of land use and land cover (LULC) in urban settings in Oklahoma. Landsat 5 data and Oklahoma Mesonet (http://www.mesonet.org/) data were used for actual ET estimation. A meteorological gauged river basin in central Oklahoma, near Lovell, Oklahoma, with an area of approximately 1033.59 km² was studied. The catchment includes very diverse LULC from agriculture (Garfield County) to urban (Oklahoma County). The annual precipitation is around 870 (mm), and average high and low temperatures are 21°C and 8°C, respectively. The study watershed is in a semiarid region where agriculture activities were mainly sustained by irrigation. This study evaluates the possible closure of the heat balance equation using unique environmental monitoring network to estimate actual ET and determine the variation with regard to varying types of LULC in urban settings (Figure 1.6). Overall, wetlands have the highest ET, wetlands and forests present a higher rate of ET than grass and agricultural lands, and the highly urbanized areas have the lowest ET (Figure 1.6).

1.4.3 Global Evapotranspiration Estimation

Until recently, long-term changes in ET have been evaluated by studying reference evaporation using measurements of pan evaporation. A research work published earlier showed that, on average, pan evaporation had decreased over the north America, Europe, and Asia from the beginning of the 1950s–1990s (Peterson et al., 1995; Hobbins et al., 2004). Recent studies have reiterated this to be an overall trend throughout the northern latitudes. For instance, over the last half century, decreases in pan evaporation have been reported in India (Chattopadhyay and Hulme, 1997), China (Thomas, 2000), and parts of Europe (Moonen et al., 2002), albeit some mixed trends have also been reported, for example, east Asia (Xu, 2001), and similar anomaly in the Middle East (Cohen et al., 2002; Eslamian et al., 2011). Another comprehensive study on the decrease on pan evaporation over the conterminous United States for the past half century is presented by Hobbins et al. (2004). One of the critical points in most of these studies is that mean observations are used over a wide area with some sites showing decreasing trend while others with increasing anomalies.

The advent of remote sensing estimation of land cover, surface temperature (Ts) and reflectance, vegetation indices (VI), emissivity, and surface albedo lead to the development of ET algorithms based on SEB approach (Su, 2002; Bastiaanssen et al., 2005; Allen et al., 2007). The three main remote sensing based ET estimation methods can be categorized as (1) empirical methods, which integrates remotely sensed VI with measured ET (Nagler et al., 2005; Glenn et al., 2008a,b; Jung et al., 2010),

TABLE 1.4 List of Satellite Precipitation Products and Their Characteristics

Precipitation Product	Satellite Sensors	Spatiotemporal Resolution	Areal Coverage/Start Date	Producer URL
GPI	GEO-IR, LEO-IR	2.5/month	Global—40°N–S/1986–February 2004	NOAA/NWS CPC [1]
	GEO-, LEO-IR	2.5/pentad	Global—40°N–S/1986–November 2004	NOAA/NWS CPC [2]
	GEO-, LEO-IR	1/day	Global—40°N–S/October 1996	NOAA/NWS CPC [3]
GPROF2004	AMSR-E	0.5/orbits	Global—70°N–S/June 2002	NSIDC [4]
GPROF2010	AMSR-E	0.25/day 0.25/month	Global—70°N–S/June 2002	Colorado State University [5]
GPROF2010	SSM/I	0.25/day 0.25/month	Global—70°N–S/July 1987–November 2009	Colorado State University [5]
	Level 2 (swath/pixel/orbit)	Global—70°N–S/ July 1987–November 2009	—	Colorado State University [6]
GPROF2010	SSMIS	0.25/day 0.25/month	Global—70°N–S/October 2003	Colorado State University [5]
GPROF2010	TMI	0.25/day 0.25/month	Global—40°N–S/December 1997	Colorado State University [5]
GPROF2010 (3G68)	TMI	0.5/h	0.1/h land	NASA/GSFC PPS [7]
Hydro-estimator	GEO-IR	4 km/h	Global—60°N–S/March 2007	NOAA/NESDIS/STAR [8]
METH	SSM/I, SSMIS	2.5/month	Global ocean—60°N–S/July 1987–2010	George Mason University [9]
METH (3A11)	TMI	5/month	Global ocean—40°N–S/January 1998	NASA/GSFC PPS [7]
MiRS	AMSU/MHS, SSMIS	Swath	Global—August 2007	NOAA OSDPD [10]
NESDIS/FNMOC scattering index	SSM/I	1.0/month 2.5/pentad, month	Global—July 1987–November 2009	NESDIS/STAR [11]
NESDIS high frequency	AMSU/MHS	0.25/day 1.0/pentad, month 2.5/pentad, month	Global—2000	NESDIS/STAR [12]
OPI	AVHRR	2.5/day	Global—1979	NOAA/NWS CPC [13]
RSS	TMI, AMSRE, SSM/I, SSMIS, QSCAT	0.25/1, 3, 7 days	Monthly	RSS [14]
TRMM PR Precip (3 G68)	PR	0.5/h	Global—37°N–S/December 1997	NASA/GSFC PPS [7]
AIRS	AIRS sounding retrievals	Swath/orbit segments	Global—May 2002	NASA/GSFC 610 [15]
CMORPH	TMI, AMSR-E, SSM/I, AMSU, IR vectors	8 km/30 min	50°N–S/1998	NOAA/CPC [16]
GSMaP NRT	TMI, AMSR-E, SSM/I, SSMIS, AMSU, IR vectors	0.1°/h	60°N–S/October 2007	JAXA [17]
GSMaP MWR	TMI, AMSR-E, AMSR, SSM/I, IR vectors	0.25°/h, day, month	60°N–S/1998–2006	JAXA [18]
GSMaP MVK+	TMI, AMSR-E, AMSR, SSM/I, AMSU, IR vectors	0.1°/h	60°N–S/2003–2006	JAXA [18]
NRL real time	SSM/I-cal PMM (IR)	0.25°/h	Global—40°N–S/July 2000	NRL Monterey [19]
TCI (3G68)	PR, TMI	0.5°/h	Global—37°N–S/December 1997	NASA/GSFC PPS [20]
TOVS	HIRS, MSU	1°/day	Global—1979—April 2005	NASA/GSFC 610 [15]
TRMM real-time HQ (3B40RT)	TMI, TMI-SSM/I, TMI-AMSU	0.25°/3 h	Global—70°N–S/February 2005	NASA/GSFC PPS [21]

(Continued)

TABLE 1.4 (*Continued*)   List of Satellite Precipitation Products and Their Characteristics

Precipitation Product	Satellite Sensors	Spatiotemporal Resolution	Areal Coverage/Start Date	Producer URL
TRMM real-time VAR (3B41RT)	MW-VAR	0.25°/h	Global—50°N–S/February 2005	NASA/GSFC PPS [22]
TRMM real-time HQVAR (3B42RT)	HQ, MW-VAR	0.25°/3 h	Global—50°N–S/February 2005	NASA/GSFC PPS [23]

Source: Modified from Tapiador, F. J. et al., *Atmos. Res.*, 104, 70, 2012.

List of URLs

[1] ftp://ftp.cpc.ncep.noaa.gov/precip/gpi/monthly/.
[2] ftp://ftp.cpc.ncep.noaa.gov/precip/gpi/pentad/.
[3] ftp://ftp.cpc.ncep.noaa.gov/precip/gpi/daily/.
[4] http://nsidc.org/data/ae_rain.html.
[5] http://rain.atmos.colostate.edu/RAINMAP10/.
[6] berg@atmos.colostate.edu; Dr. Wesley Berg.
[7] http://pps.gsfc.nasa.gov.
[8] http://www.star.nesdis.noaa.gov/smcd/emb/ff/digGlobalData.php.
[9] ftp://gpcp-pspdc.gmu.edu/V6/2.5/.
[10] http://mirs.nesdis.noaa.gov; http://www.osdpd.noaa.gov/ml/mirs.
[11] http://www.ncdc.noaa.gov/oa/rsad/ssmi/gridded/index.php?name = data_access.
[12] http://www.star.nesdis.noaa.gov/corp/scsb/mspps/main.html; http://www.osdpd.noaa.gov/ml/mspps/index.html.
[13] pingping.xie@noaa.gov; Dr. Pingping Xie.
[14] http://www.ssmi.com.
[15] joel.susskind-1@nasa.gov; Dr. Joel Susskind.
[16] http://www.cpc.ncep.noaa.gov/products/janowiak/cmorph_description.html.
[17] http://sharaku.eorc.jaxa.jp/GSMaP/.
[18] http://sharaku.eorc.jaxa.jp/GSMaP_crest/.
[19] song.yang@nrlmry.navy.mil; Dr. Song Yang.
[20] ftp://pps.gsfc.nasa.gov/pub/trmmdata/3G/3G68/.
[21] ftp://trmmopen.nascom.nasa.gov/pub/merged/combinedMicro/.
[22] ftp://trmmopen.nascom.nasa.gov/pub/merged/calibratedIR/.
[23] ftp://trmmopen.nascom.nasa.gov/pub/merged/mergeIRMicro/.

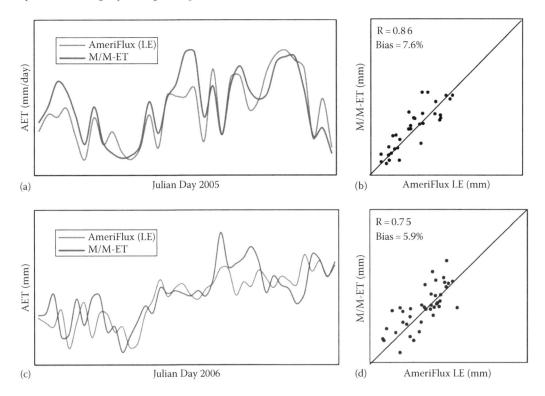

FIGURE 1.4   Comparisons of actual ET from AmeriFlux tower observations and SEB-based M/M-ET estimates at ARM SGP El Reno site. Panels (a and b) show the daily time series and scatter plot comparison for 2005; panels (c and d) are for 2006. (From Khan, S.I. et al., *Int. J. Remote Sens.*, 31(14), 3799, 2010.)

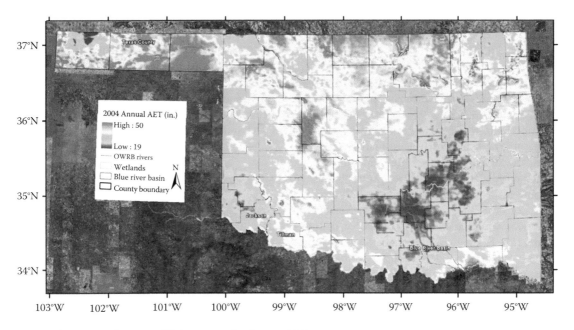

FIGURE 1.5   Annual AET over the entire Oklahoma state for the year 2004.

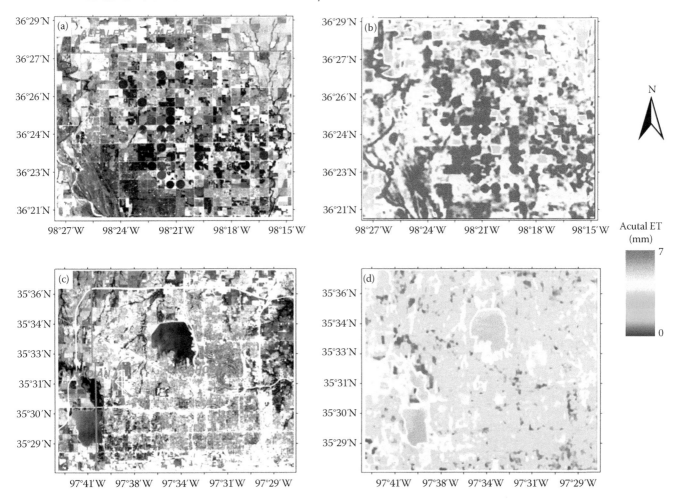

FIGURE 1.6   (a) Landsat false color image of agricultural areas, (b) spatial variations of the AET (mm) over the study area on July 31, 2005, (c) Landsat false color image of urban areas and nearby water body, and (d) distributed AET at urban areas and nearby water body. (Modified from Liu, W. et al., *J. Appl. Remote Sens.*, 4(1), 041873, 2010.)

(2) integrated remote sensing and classical Penman–Monteith approach (Monteith, 1965; Cleugh et al., 2007; Mu et al., 2007), and (3) physical-based SEB models (Bastiaanssen et al., 2005; Overgaard et al., 2006; Allen et al., 2007; Kustas and Anderson, 2009). For physical-based SEB ET models, thermal IR-based land surface temperature is a critical remote sensing variable (Bastiaanssen et al., 1998a,b; Nishida et al., 2003; Su, 2002).

Global terrestrial ET retrieval at a fine spatial scale was never achieved before until the satellite-driven estimation of the terrestrial ET, using the MODIS satellite sensor onboard the Aqua satellite launched on May 4, 2002, with 36 spectral bands, 20 reflective solar, and 16 thermal emissive bands. MODIS provides exceptional data for observing vegetation and surface energy (Justice et al., 2002), which is utilized to develop a remotely sensed ET model (Mu et al., 2007). The moderated resolution global terrestrial ET algorithm is developed and refined by the Numerical Terradynamic Simulation Group is a research laboratory at the University of Montana in Missoula. The data product derived known as the MODIS Global Evapotranspiration Project (MOD16) is defined as an "evaporation fraction," the energy budget equivalent of an index of actual to potential ET, over the global land surface with 1 km resolution every 8 days (Mu, 2007; Mu et al., 2011). MODIS data related to ET retrieval are land surface temperature and emissivity (MOD11), surface reflectance products (MOD09), vegetation index (MOD13), and albedo (MOD43B3) obtained by assimilating bi-hemispherical reflectance data modeled. These datasets were acquired from the Land Processes Distributed Active Archive Center at the U.S. Geological Survey Earth Resources Observation and Science center, with the standard Hierarchical Data Format (http://LPDAAC.usgs.gov). For more information on MODIS, please refer to http://modis.gsfc.nasa.gov.

1.5 Soil Moisture Estimation Based on Multisensor Data

Soil moisture is one of the most crucial hydrological components in the water cycle. While satellite remote sensing is the most cost-effective approach to estimate the soil moister consistently on a large spatial scale (e.g., over the globe), the in situ soil moisture provides more accurate ground validation reference for the development of the satellite soil moisture estimation. Table 1.5 lists the sensors applied in both in situ and remote sensing soil moisture measurement from the past (AMSR-E) to the current (AMSR2 and SMOS) and the future (Soil Moisture Active and Passive [SMAP]). The in situ soil moisture measurement will be elaborated in Section 1.5.1, and the SMOS and SMAP soil moisture mission will be introduced in Section 1.5.2.

1.5.1 Regional-, Local-, and Field-Scale Studies

Observation of soil water content has been accomplished worldwide in the past decades (Service, 2002; PIsofSMEX03, 2003; Jackson and Lettenmaier, 2004; Jacobs et al., 2004; Jackson et al., 2005; Narayan et al., 2006; Chert et al., 2008; Choi et al., 2008; Vivoni et al., 2008; Yilmaz et al., 2008; Kabela et al., 2009). The dimensionless variable, volumetric soil moisture (VSM), that refers to the water volume in unit volume (Ulaby et al., 1986) of soil is usually used to quantitatively represent the level of water content in soil.

The SGP of the United States and adjacent regions include multiple arrays of in situ observations that provide estimates of soil moisture at local and regional scales. At the regional scale, in situ observation network such as Oklahoma Mesonet, which has an average station spacing of approximately 30 km, provides high spatiotemporal soil moisture and other meteorological estimates. Table 1.6 lists some of the specific, past successful campaigns and the ongoing experiments.

The Oklahoma Mesonet is a permanent mesoscale-observing network of 120 meteorological stations across Oklahoma (Brock et al., 1995; McPherson et al., 2007). It is managed by the Oklahoma Climatological Survey in partnership with Oklahoma State University and the University of Oklahoma. Each station measures more than 20 environmental variables, including wind at 2 and 10 m, air temperature at 1.5 and 9 m, relative humidity, rainfall, pressure, solar radiation, and soil temperature and moisture at various depths (Illston et al., 2008). Mesonet data are collected and transmitted to a central point every 5 min where they are quality controlled, distributed, and archived (Shafer et al., 2000; McPherson et al., 2007).

TABLE 1.5　Sensors Applied in Both In Situ and Remote Sensing Soil Moisture Measurement

	Satellite Mission/Sensor	Sensor Type	Measurement	Spatial Resolution	Temporal Resolution
In situ (Oklahoma Mesonet)	N/A	BetaTHERM[a] probe Scientific 229-L[b] probe	Soil temperature	Point based	15 min
Satellite remote sensing	AMSR-E (2002–2011)	Radiometer (six bands: 6.9–89 GHz)	Brightness temperature	25 km	Daily
	AMSR2 (2011–)	Radiometer (six bands: 6.9–89 GHz)	Brightness temperature	10/25 km	Daily
	SMOS (2009–)	Radiometer (L-band 1.4 GHz)	Microwave radiation	35–50 km	Daily
	SMAP (2014–)	Radar (L-band 1.26 GHz)	Backscattering	3 km	Daily
		Radiometer (L-band 1.4 GHz)	Brightness temperature	40 km	

[a] Before December 12, 2012.
[b] After December 12, 2012.

TABLE 1.6 Past Successful Soil Moisture Field Campaigns

Campaign	Period	Location	Primary Data Types	Data Availability
SMEX02	June–July 2002	Iowa	Radiometer, SAR, in situ	Online
SMEX03	June–July 2003	Oklahoma, Georgia, Alabama	Radiometer, SAR, in situ	Online
SMEX04	August 2004	Arizona	Radiometer, SAR, in situ	Online
Washita 92	June 1992	Oklahoma	Radiometer, SAR, in situ	Online
SGP97	June–July 1997	Oklahoma	Radiometer, SAR, in situ	Online

(a)

(b)

FIGURE 1.7 Sensor arrays deployed at (a) the MOISST site near Marena, OK, and (b) the iGOS site near El Reno, OK.

Oklahoma Mesonet data have supported a broad range of scientific research with a bibliography of over 500 peer-reviewed articles. In particular, multiple studies have used Oklahoma Mesonet data for analysis of satellite algorithms focused on soil moisture retrieval (Gu et al., 2008; Swenson et al., 2008; Pathe et al., 2009; Collow et al., 2012).

The Marena Oklahoma In Situ Sensor Testbed (MOISST) (illustrated in Figure 1.7a) was installed in May 2010 as part of the calibration and validation program for the SMAP mission. The site includes more than 200 soil, vegetation, and atmospheric sensors installed over an approximately 64 ha pasture in central Oklahoma with four main stations and multiple in situ sensors installed in profiles. In addition, located at the site include a COsmic-ray Soil Moisture Observing System (Hornbuckle et al., 2012), Global Positioning System reflectometers, a passive distributed temperature system, an eddy correlation flux tower (installed November 2011), and a phenocam (installed May 2012). To enhance the observations from the MOISST site and to calibrate the existing sensor, field samples are routinely collected from soil and vegetation. Other in situ measurements of soil moisture and surface atmosphere interactions can be acquired from similar instrumentations at the Integrated Grassland Observing Site (Figure 1.7b).

Field measurements provide direct but only point samples of soil moisture. In contrast, remote sensing sensors on aerial or satellite platforms are capable of measuring surface VSM at surface scale and relatively high temporal intervals but indirectly through retrieval techniques. Among a variety of bands, microwave is most frequently used for soil moisture because soil moisture affects the soil dielectric constant (permittivity) greatly at microwave frequency and thus sensitive to microwave scattering and emission.

1.5.2 Global Soil Moisture Estimation

Spatially distributed soil moisture measurements and freeze/thaw states are needed to improve our understanding of regional and global water cycles. SMOS is one of the missions that provide global observations of soil moisture over Earth's landmasses and salinity over the oceans. It is commonly known as the water mission and is meant to provide new insights into Earth's water cycle and climate. In addition, it aims to monitor snow and ice accumulation and to provide better weather forecasting. The mission has a low-Earth, polar, Sun-synchronous orbit at an altitude of 758 km. An important aspect of this mission is that it carries out a completely new measuring technique: the first polar-orbiting spaceborne 2D interferometric radiometer instrument called the Microwave Imaging Radiometer with Aperture Synthesis (MIRAS). This novel instrument is capable of observing both soil moisture and ocean salinity by capturing images of emitted microwave radiation around the frequency of 1.4 GHz or wavelength of 21 cm (L-band) (Kerr et al., 2001; Bayle et al., 2002; Font et al., 2004; Moran et al., 2004). The science goal of this mission is to measure soil moisture with an accuracy of 4%, VMS at 35–50 km spatial resolution and 1–3 day revisit time, and ocean surface salinity with an accuracy of 0.5–1.5 practical salinity unit for a single observation at 200 km spatial resolution and 10–30 day temporal resolution (Delwart et al., 2008).

1.5.3 Future Satellite Mission on Soil Moisture Estimation

1.5.3.1 Soil Moisture Active and Passive Mission

SMAP is one of the four first-tier missions recommended by the National Research Council Earth science decadal survey report. SMAP will provide global views of Earth's soil moisture and surface freeze/thaw state, introducing a new era in hydrologic applications and providing unprecedented capabilities to investigate the water, energy, and carbon cycles over global land surfaces. Moreover, these estimates are also helpful in understanding terrestrial ecosystems and the processes that interlink the water, energy, and carbon cycles.

Soil moisture and freeze/thaw information provided by SMAP will lead to improved weather, flood and drought forecasts, and predictions of agricultural productivity and climate change. This mission will contribute to the goals of the carbon cycle and ecosystems, weather, and climate variability and change Earth science focus areas as well as to hydrological science.

The SMAP mission, based on one flight system in a sun-synchronous orbit, at an altitude of 670 km with an 8-day repeat cycle, will include synthetic aperture radar operating at L-band (frequency: 1.26 GHz; polarizations: HH, VV, HV) and an L-band radiometer (frequency: 1.41 GHz; polarizations: H, V, U). At this altitude, the antenna scan design yields a 1000 km swath, with a 40 km radiometer resolution and 1–3 km synthetic aperture radar resolution that provides global coverage within 3 days at the equator and 2 days at boreal latitudes (>45°N). The main goal of SMAP is to provide estimates of soil moisture in the top 5 cm of soil with an accuracy of 0.04 cm^3/cm^3 VMS, at 10 km resolution, with 3-day average intervals over the global land area. These measurements will not be suitable for regions with snow and ice, mountainous topography, open water, and vegetation with total water content greater than 5 kg/m² (Entekhabi et al., 2010).

1.5.3.2 Surface Water and Ocean Topography Mission

Surface Water and Ocean Topography (SWOT) is one of the recommended missions during the decadal survey committee for launch during 2020. It will provide observations on lake and river water levels for inland water dynamics. The mission will measure water surface elevations, water surface slope, and the areal extent of lakes, wetlands, reservoirs, floodplains, and rivers globally (Alsdorf et al., 2003, 2007). The core technology in SWOT mission is the wide swath of Ka-band and C-band radar interferometer, which would achieve spatial resolution to the order of tens of meters (Alsdorf et al., 2003; http://swot.jpl.nasa.gov/).

The planned SWOT satellite mission is intended to indirectly estimate river flow fluctuations from remotely sensed river hydrologic features. SWOT mission aims to offer precise river surface slope estimates; moreover, it will provide estimates on water surface elevations and inundated areas for rivers with widths greater than about 100 m (Durand et al., 2010). There are numerous inadequacies of current altimeters ranging from less frequent visits of the sensor and how the size of inland water bodies are measured; the planned SWOT mission will compensate these limitation (Alsdorf et al., 2003, 2007). Preliminary case studies for this approach using synthetic data showed some promising results (Lee et al., 2010). Earth scientists and other researchers are optimistic about the SWOT missions and its application in hydrologic science and water resource management throughout the globe.

1.6 Groundwater Assessment from GRACE

The Gravity Recovery and Climate Experiment (GRACE) satellites have been widely used to assess drought severity and extent in different regions, including the California's Central Valley, southeastern United States, and more recently to develop drought indicators for soil moisture and groundwater using data assimilation with land surface models (Famiglietti et al., 2011; Houborg et al., 2012; Scanlon et al., 2012). The large-scale nature of GRACE water storage changes can be valuable for providing a regional assessment of overall drought impacts and can also be used to estimate water requirements to overcome cumulative drought impacts. While most studies disaggregate total water storage into surface water, soil moisture, and groundwater, TWS is also very valuable in its own right. For instance, reservoir storage is monitored in most states and can be subtracted from GRACE TWS to estimate subsurface water storage changes. Disaggregating subsurface water storage into soil moisture and groundwater is complicated because of limited information on soil moisture and uncertainties in simulated soil moisture from land surface models, as seen from application of GRACE during the 2011 drought in Texas. Texas has one of the most advanced groundwater level monitoring programs of any state, making it feasible to estimate groundwater storage changes from monitoring data and comparing with GRACE data. Soil moisture estimates from SMAP and the Oklahoma Mesonet can be compared with the simulated soil moisture from various land surface models to assess their reliability.

Correspondence between GRACE TWS and Palmer Drought Severity Index in Texas (Figure 1.8) shows that TWS can be valuable as a drought predictor (Long et al., 2013). Therefore, GRACE can provide a valuable tool in the portfolio of seasonal drought predictors (Sun, 2013) for the Texas–Oklahoma region and will build on previous experience in applying this tool in Texas and throughout the High Plains.

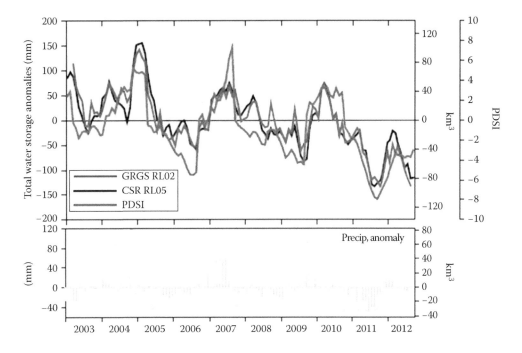

FIGURE 1.8 Relationship between GRACE total water storage from two sources (GRGS and CSR) and PDSI. (From Long, D. et al., *Geophys. Res. Lett.*, 40(13), 3395, 2013.)

1.7 Summary and Conclusions

Global monitoring of geosystem processes with satellite remote sensing improved our understanding of the connections between the water, energy, and carbon cycles. Given the impact of hydrologic extremes in a changing climate, characterization of real-time precipitation, ET, and soil moisture with higher spatiotemporal variability is crucial to monitor hydrologic processes at multiple scales. As discussed in the preceding sections, the fundamental terms in the hydrologic budget equation can now be estimated through satellite sensor. Several remote sensing–based datasets from current and future satellite missions are intended to quantify the water balance components, at different spatial and temporal scales. For instance, global precipitation can be retrieved at a high temporal resolution (3 hourly) by combining microwave and IR-based satellite measurements (Sorooshian et al., 2000; Kummerow et al., 2001; Joyce et al., 2004; Huffman et al., 2007). Energy balance estimation and empirical models have led to moderate resolution and large-scale estimates of ET. Recent research has focused on using radars to determine soil moisture content and the depth of the groundwater table, specifically using L-band and ground-penetrating radars.

The accuracy of water budget estimates at different space and time scales is debatable and therefore an open research question. The retrieval of hydrologic variables based on single sensors might not be as reliable as expected due to the dependence on a number of ground parameters with spatial and temporal variability. The uncertainty in these estimates can be attributed to the satellite instrument type, the retrieval algorithms, and the spatial

spatial–temporal resolution of these different estimates. Another possible reason is that any single satellite instrument does not measure all the water budget components simultaneously (Ferguson et al., 2010; Sheffield et al., 2010). Robust retrieval algorithms as well as data assimilation methods that ingest multisensor measurements to resolve the water budget at local and regional scales are an active area of research. However, hydrologists who have long been counting on the current satellites will be soon able to track down the water budget variations in the specified regions of interest with the new satellites of SWOT and SMAP.

This chapter focused on some of the main remote sensing datasets from satellite sensors such GPM, SMAP NPP, GOES-R, SWOT, and other missions. Specifically, currently used 3 h GOES observations should be increased to at least 1 h from GOES-R observations to better capture the diurnal variation of cloud and radiation fields. Current and future Afternoon Train constellation satellites will provide high-quality seamless and inexpensive Earth observation. Moreover, the NPOESS will ensure the continuity of several key climate measurements. At both regional and global scales, the real-time operational hydrological monitoring is supported by the current and future mission. The new satellite missions are anticipated to provide better precipitation and soil moisture data in terms of coverage, accuracy, and resolutions.

Despite the promise of global high-resolution monitoring of the individual components of terrestrial water budget, there still remain considerable challenges in providing physically consistent and accurate estimates (Sheffield et al., 2009). Hydrologic community currently lacks both sufficient understanding from field studies and quantitative models to make reliable estimates about desired outcomes from management decisions in many hydrologic applications. There are opportunities to study and estimate the contributing variables by employing remotely sensed data and maximizing their use in hydrological studies. The main limitations for estimating local and sometimes regional level hydrologic state variable from remote sensing techniques are the retrieval precision of land surface fluxes and the spatial heterogeneity and scaling issues. Future missions such as the SWOT, GPM, and SMAP will improve our understanding of the terrestrial water cycle with a global coverage of major rivers and watershed. These datasets will supplement in situ observations for better data retrieval combined with robust data assimilation techniques and are expected to significantly benefit water budget estimation at different spatiotemporal scales. We expect an improvement in the sensor technologies, and novel data fusion techniques will lead to better estimation of hydrologic variable and water balance closure.

References

Adler, R. F., G. J. Huffman, A. Chang et al. 2003. The version-2 global precipitation climatology project (GPCP) monthly precipitation analysis (1979–present). *Journal of Hydrometeorology* 4(6):1147–1167.

Allen, R. G., M. Tasumi, A. Morse et al. 2007. Satellite-based energy balance for Mapping Evapotranspiration with Internalized Calibration (METRIC)—Applications. *Journal of Irrigation and Drainage Engineering* 133(4):395–406, doi:10.1061(ASCE)0733–9437.

Alsdorf, D., D. Lettenmaier, and C. Vörösmarty. 2003. The need for global, satellite-based observations of terrestrial surface waters. *Eos, Transactions American Geophysical Union* 84:269–276.

Alsdorf, D. E., E. Rodriguez, and D. P. Lettenmaier. 2007. Measuring surface water from space. *Reviews of Geophysics* 45:RG2002, doi:10.1029/2006RG000197.

Arkin, P. A. and B. N. Meisner. 1987. The relationship between large-scale convective rainfall and cold cloud over the western hemisphere during 1982–84. *Monthly Weather Review* 115:51–74.

Ba, M. B. and A. Gruber. 2001. GOES Multispectral Rainfall Algorithm (GMSRA). *Journal of Applied Meteorology and Climatology* 40:1500–1514.

Bastiaanssen, W. G. M. and N. R. Harshadeep. 2005. Managing scarce water resources in Asia: The nature of the problem and can remote sensing help? Guest Editors Allen and Bastiaanssen. *Irrigation and Drainage Systems* 19:269–284.

Bastiaanssen, W. G. M., M. Menenti, R. A. Feddes, and A. A. M. Holtslag. 1998a. The surface energy balance algorithm for land (SEBAL): Part 2 validation. *Journal of Hydrology* 212–213:213–229.

Bastiaanssen, W. G. M., M. Menenti, R. A. Feddes, and A. A. M. Holtslag. 1998b. A remote sensing surface energy balance algorithm for land (SEBAL): 1. Formulation. *Journal of Hydrology* 212–213:198–212.

Battan, L. J. 1973. *Radar Observation of the Atmosphere*. Chicago, IL: University of Chicago Press.

Bayle, F., J. P. Wigneron, Y. H. Kerr, P. Waldteufel, E. Anterrieu, J. C. Orlhac, A. Chanzy, O. Marloie, M. Bernardini, and S. Sobjaerg. 2002. Two-dimensional synthetic aperture images over a land surface scene. *IEEE Transactions on Geoscience and Remote Sensing* 40(3):710–714.

Bellerby, T. J. 2004. A feature-based approach to satellite precipitation monitoring using geostationary IR imagery. *Journal of Hydrometeorology* 5:910–921.

Bolten, J., V. Lakshmi, and E. Njoku. 2003. A passive-active retrieval of soil moisture for the Southern Great Plains 1999 experiment. *IEEE Transactions on Geoscience and Remote Sensing* 41:2792–2801.

Bringi, V. and V. Chandrasekar. 2001. *Polarimetric Doppler Weather Radar: Principles and Applications*. Cambridge, United Kingdom: Cambridge University Press.

Brock, F. V., K. C. Crawford, R. L. Elliott, G. W. Cuperus, S. J. Stadler, H. L. Johnson, and M. D. Eilts. 1995. The Oklahoma Mesonet: A technical overview. *Journal of Atmospheric and Oceanic Technology* 12(1):5–19.

Chattopadhyay, N. and M. Hulme. 1997. Evaporation and potential evapotranspiration in India under conditions of recent and future climate change. *Agricultural and Forest Meteorology* 87(1):55–73.

Chauhan, N. 1997. Soil moisture estimation under a vegetation cover: Combined active passive microwave remote sensing approach. *International Journal of Remote Sensing* 18:1079–1097.

Chert, Q., Z. Li, L. Wang, Y. Shao, and T. Cheng. 2008. Soil moisture change retrieval using S-band radar data during SGP99 and SMEX02. Paper presented at Geoscience and Remote Sensing Symposium, 2008. IGARSS 2008. IEEE International, July 7–11, 2008.

Choi, M., J. M. Jacobs, and D. D. Bosch. 2008. Remote sensing observatory validation of surface soil moisture using Advanced Microwave Scanning Radiometer E, Common Land Model, and ground based data: Case study in SMEX03 Little River Region, Georgia, US. *Water Resources Research* 44(8):W08421.

Cleugh, H. A., R. Leuning, Q. Mu, and S. W. Running. 2007. Regional evaporation estimates from flux tower and MODIS satellite data. *Remote Sensing of Environment* 106:285–304.

Cohen, S., A. Ianetz, and G. Stanhill. 2002. Evaporative climate changes at Bet Dagan, Israel, 1964–1998. *Agricultural and Forest Meteorology* 111(2):83–91.

Collow, T. W., A. Robock, J. B. Basara, and B. G. Illston. 2012. Evaluation of SMOS retrievals of soil moisture over the central United States with currently available in situ observations. *Journal of Geophysical Research: Atmospheres (1984–2012)* 117(D9):1–15.

Delwart, S., C. Bouzinac, P. Wursteisen, M. Berger, M. Drinkwater, M. Martin-Neira, and Y. H. Kerr. 2008. SMOS validation and the COSMOS campaigns. *IEEE Transactions on Geoscience and Remote Sensing* 46(3):695–704.

Durand, M., L.-L. Fu, D. P. Lettenmaier, D. E. Alsdorf, E. Rodriguez, and D. Esteban-Fernandez. 2010. The surface water and ocean topography mission: Observing terrestrial surface water and oceanic submesoscale eddies. *Proceedings of the IEEE* 98(5):766–779.

Entekhabi, D., E. G. Njoku, P. E. O'Neill et al. May 2010. The soil moisture active passive (SMAP) mission. *Proceedings of the IEEE* 98(5):704, 716, doi: 10.1109/JPROC.2010.2043918.

Eslamian, S., M. J. Khordadi, and J. Abedi-Koupai. 2011. Effects of variations in climatic parameters on evapotranspiration in the arid and semi-arid regions. *Global and Planetary Change* 78(3–4):188–194.

Famiglietti, J., M. Lo, S. Ho et al. 2011. Satellites measure recent rates of groundwater depletion in California's Central Valley. *Geophysical Research Letters* 38:L03403.

Ferguson, C. R., J. Sheffield, E. F. Wood, and H. Gao. 2010. Quantifying uncertainty in a remote sensing-based estimate of evapotranspiration over continental USA. *International Journal of Remote Sensing* 31:3821–3865.

Flaming, G. M. 2005. Global precipitation measurement update. Geoscience and Remote Sensing Symposium, 2005. IGARSS'05. *Proceedings IEEE International.* Vol. 1, 79–82.

Font, J., G. S. E. Lagerloef, D. M. Le Vine, A. Camps, and O. Z. Zanife. 2004. The determination of surface salinity with the European SMOS space mission. *IEEE Transactions on Geoscience and Remote Sensing* 42(10):2196–2205.

Glenn, E. P., A. R. Huete, P. L. Nagler, and S. G. Nelson. 2008a. Relationship between remotely-sensed vegetation indices, canopy attributes and plant physiological processes: What vegetation indices can and cannot tell us about the landscape. *Sensors* 8:2136–2160.

Glenn, E. P., K. Morino, K. Didan et al. 2008b. Scaling sap flux measurements of grazed and ungrazed shrub communities with fine and coarse-resolution remote sensing. *Ecohydrology* 1(4):316–329.

Gu, Y., E. Hunt, B. Wardlow, J. B. Basara, J. F. Brown, and J. P. Verdin. 2008. Evaluation of MODIS NDVI and NDWI for vegetation drought monitoring using Oklahoma Mesonet soil moisture data. *Geophysical Research Letters* 35(22):1–5.

Hobbins, M. T., J. A. Ramírez, and T. C. Brown. 2004. Trends in pan evaporation and actual evapotranspiration across the conterminous US: Paradoxical or complementary. *Geophysical Research Letters* 31(13):1–5.

Hong, Y., K.-L. Hsu, S. Sorooshian, and X. Gao. 2004. Precipitation estimation from remotely sensed imagery using an artificial neural network cloud classification system. *Journal of Applied Meteorology* 43:1834–1853.

Hong, Y., K.-L. Hsu, S. Sorooshian, and X. Gao. 2005. Improved representation of diurnal variability of rainfall retrieved from the tropical rainfall measurement mission microwave imager adjusted precipitation estimation from remotely sensed information using artificial neural networks (PERSIANN) system. *Journal of Geophysical Research* 110:D06102.

Hornbuckle, B., S. Irvin, T. Franz, R. Rosolem, and C. Zweck. 2012. The potential of the COSMOS network to be a source of new soil moisture information for SMOS and SMAP. *Paper read at Geoscience and Remote Sensing Symposium (IGARSS)*, 2012 IEEE International, New York.

Houborg, R., M. Rodell, B. Li et al. 2012. Drought indicators based on model-assimilated Gravity Recovery and Climate Experiment (GRACE) terrestrial water storage observations. *Water Resources Research* 48:W07525.

Hsu, K., X. Gao, S. Sorooshian, and H. V. Gupta. 1997. Precipitation estimation from remotely sensed information using artificial neural networks. *Journal of Applied Meteorology* 36:1176–1190.

Huffman, G. J., R. F. Adler, M. Morrissey, D. T. Bolvin, S. Curtis, R. Joyce, B. McGavock, and J. Susskind. 2001. Global precipitation at one-degree daily resolution from multi-satellite observations. *Journal of Hydrometeorology* 2(1):36–50.

Huffman, G. J., D. T. Bolvin, E. J. Nelkin, D. B. Wolff, R. F. Adler, G. Gu, Y. Hong, K. P. Bowman, and E. F. Stocker. 2007. The TRMM multisatellite precipitation analysis (TMPA): Quasi-global, multiyear, combined-sensor precipitation estimates at fine scales. *Journal of Hydrometeorology* 8(1):38–55.

Illston, B. G., J. B. Basara, D. K. Fischer, R. L. Elliott, C. Fiebrich, K. C. Crawford, K. Humes, and E. Hunt. 2008. Mesoscale monitoring of soil moisture across a statewide network. *Journal of Atmospheric and Oceanic Technology* 25:167–182.

Jackson, T. J., R. Bindlish, A. J. Gasiewski, B. Stankov, M. Klein, E. G. Njoku, D. Bosch, T. L. Coleman, C. A. Laymon, and P. Starks. 2005. Polarimetric scanning radiometer C- and X-band microwave observations during SMEX03. *IEEE Transactions on Geoscience and Remote Sensing* 43(11):2418–2430.

Jackson, T. J. and D. Lettenmaier. 2004. Soil moisture experiments 2004 (SMEX04). Eos Trans. AGU, 85(17), Jt. Assem. Suppl., Abstract H11B-03. Montreal, Quebec, Canada. http://abstractsearch.agu.org/meetings/2004/SM.html

Jacobs, J. M., B. P. Mohanty, E. C. Hsu, and D. Miller. 2004. SMEX02: Field scale variability, time stability and similarity of soil moisture. *Remote Sensing of Environment* 92(4):436–446.

Joyce, R. J., J. E. Janowiak, P. A. Arkin, and P. Xie. 2004. CMORPH: A method that produces global precipitation estimates from passive microwave and infrared data at high spatial and temporal resolution. *Journal of Hydrometeorology* 5(3):487–503.

Jung, M., M. Reichstein, P. Ciais et al. 2010. Recent decline in the global land evapotranspiration trend due to limited moisture supply. *Nature* 467:951–954.

Justice, C., J. Townshend, E. Vermote, E. Masuoka, R. Wolfe, N. Saleous, D. Roy, and J. Morisette. 2002. An overview of MODIS Land data processing and product status. *Remote Sensing of Environment* 83(1–2):3–15.

Kabela, E. D., B. K. Hornbuckle, M. H. Cosh, M. C. Anderson, and M. L. Gleason. 2009. Dew frequency, duration, amount, and distribution in corn and soybean during SMEX05. *Agricultural and Forest Meteorology* 149(1):11–24.

Kerr, Y. H., P. Waldteufel, J. P. Wigneron, J. Martinuzzi, J. Font, and M. Berger. 2001. Soil moisture retrieval from space: The Soil Moisture and Ocean Salinity (SMOS) mission. *IEEE Transactions on Geoscience and Remote Sensing* 39(8):1729–1735.

Khan, S. I., Y. Hong, B. Vieux, and W. Liu. 2010. Development evaluation of an actual evapotranspiration estimation algorithm using satellite remote sensing meteorological observational network in Oklahoma. *International Journal of Remote Sensing* 31(14):3799–3819.

Kidd, C. K., D. R. Kniveton, M. C. Todd, and T. J. Bellerby. 2003. Satellite rainfall estimation using combined passive microwave and infrared algorithms. *Journal of Hydrometeorology* 4:1088–1104.

Kuligowski, R. J. 2002. A self-calibrating real-time GOES rainfall algorithm for short-term rainfall estimates. *Journal of Hydrometeorology* 3:112–130.

Kummerow, C., Y. Hong, W. S. Olson et al. 2001. The evolution of the goddard profiling algorithm (GPROF) for rainfall estimation from passive microwave sensors. *Journal of Applied Meteorology* 40(11):1801–1820.

Kustas, W. and Anderson, M. 2009. Advances in thermal infrared remote sensing for land surface modeling. *Agricultural and Forest Meteorology* 149:2071–2081.

Lee, H., M. Durand, H. Chul Jung et al. 2010. Characterization of surface water storage changes in Arctic lakes using simulated SWOT measurements. *International Journal of Remote Sensing* 31:3931–3953.

Liu, W., Y. Hong, S. I. Khan, M. Huang, B. Vieux, S. Caliskan, and T. Grout. 2010. Actual evapotranspiration estimation for different land use and land cover in urban regions using Landsat 5 data. *Journal of Applied Remote Sensing* 4(1):041873.

Loheide, S. P. and S. M. Gorelick. 2005. A local-scale, high-resolution evapotranspiration mapping algorithm (ETMA) with hydroecological applications at riparian meadow restoration sites. *Remote Sensing of Environment* 98:182–200.

Long, D., B. R. Scanlon, L. Longuevergne et al. 2013. GRACE satellite monitoring of large depletion in water storage in response to the 2011 drought in Texas. *Geophysical Research Letters* 40(13):3395–3401.

McPherson, R. A., C. A. Fiebrich, K. C. Crawford, J. R. Kilby, D. L. Grimsley, J. E. Martinez, J. B. Basara, B. G. Illston, D. A. Morris, and K. A. Kloesel. 2007. Statewide monitoring of the mesoscale environment: A technical update on the Oklahoma Mesonet. *Journal of Atmospheric and Oceanic Technology* 24(3):301–321.

Monteith, J. L. 1965. Evaporation and environment. *Symposium of the Society of Experimental Biology* 19:205–224.

Moonen, A., L. Ercoli, M. Mariotti, and A. Masoni. 2002. Climate change in Italy indicated by agrometeorological indices over 122 years. *Agricultural and Forest Meteorology* 111(1):13–27.

Moran, M. S., C. D. Peters-Lidard, J. M. Watts, and S. McElroy. 2004. Estimating soil moisture at the watershed scale with satellite-based radar and land surface models. *Canadian Journal of Remote Sensing* 30(5):805–826.

Mu, Q., F. A. Heinsch, M. Zhao, and S. W. Running. 2007. Development of a global evapotranspiration algorithm based on MODIS and global meteorology data. *Remote Sensing of Environment* 111:519–536.

Mu, Q., M. Zhao, and S. W. Running. 2011. Improvements to a MODIS global terrestrial evapotranspiration algorithm. *Remote Sensing of Environment* 115:1781–1800.

Nagler, P., J. Cleverly, D. Lampkin, E. Glenn, A. Huete, and Z. Wan. 2005. Predicting riparian evapotranspiration from MODIS vegetation indices and meteorological data. *Remote Sensing of Environment* 94:17–30.

Narayan, U., V. Lakshmi, and T. J. Jackson. 2006. High-resolution change estimation of soil moisture using L-band radiometer and radar observations made during the SMEX02 experiments. *IEEE Transactions on Geoscience and Remote Sensing* 44(6):1545–1554.

Nishida, K., R. Nemani, J. M. Glassy, and S. W. Running. 2003. Development of an evapotranspiration index from Aqua/MODIS for monitoring surface moisture status. *IEEE Transactions on Geoscience and Remote Sensing*, 41(2):493–501.

Njoku, E. G., W. J. Wilson, S. H. Yueh, S. J. Dinardo, F. K. Li, T. J. Jackson, V. Lakshmi, and J. Bolten. 2002. Observations of soil moisture using a passive and active low-frequency microwave airborne sensor during SGP99. *IEEE Transactions on Geoscience and Remote Sensing* 40:2659–2673.

Njoku, E. G., T. J. Jackson, V. Lakshmi, T. K. Chan, and S. V. Nghiem. 2003. Soil moisture retrieval from AMSR-E. *IEEE Transactions on Geoscience and Remote Sensing* 41:215–229.

O'Neill, P., N. Chauhan, and T. Jackson. 1996. Use of active and passive microwave remote sensing for soil moisture estimation through corn. *International Journal of Remote Sensing* 17:1851–1865.

Overgaard, J., D. Rosbjerg, and M. Butts. 2006. Land-surface modelling in hydrological perspective? A review. *Biogeosciences* 3:229–241.

Pathe, C., W. Wagner, D. Sabel, M. Doubkova, and J. B. Basara. 2009. Using ENVISAT ASAR global mode data for surface soil moisture retrieval over Oklahoma, USA. *IEEE Transactions on Geoscience and Remote Sensing* 47(2):468–480.

Peterson, T. C., V. S. Golubev, and P. Y. Groisman. 1995. Evaporation losing its strength. *Nature* 377(6551):687–688.

Petty, G. W. 1995. The status of satellite-based rainfall estimation over land. *Remote Sensing of Environment* 51:125–137.

Roerink, G., Z. Su, and M. Menenti. 2000. S-SEBI: A simple remote sensing algorithm to estimate the surface energy balance. *Physics and Chemistry of the Earth, Part B: Hydrology, Oceans and Atmosphere* 25:147–157.

Rosenfeld, D. and C. W. Ulbrich. 2003. Cloud microphysical properties, processes, and rainfall estimation opportunities. *Meteorological Monographs* 30:237–237.

Scanlon, B. R., C. C. Faunt, L. Longuevergne et al. 2012. Groundwater depletion and sustainability of irrigation in the US High Plains and Central Valley. *Proceedings of the National Academy of Sciences* 109(24):9320–9325.

Scofield, R. A. and R. J. Kuligowski. 2003. Status and outlook of operational satellite precipitation algorithms for extreme-precipitation events. *Monthly Weather Review* 18:1037–1051.

Service, A. R. *SMEX02* (03/24/2002) 2002. Available from http://hydrolab.arsusda.gov/smex02/. (Accessed June 4, 2014.)

Shafer, M. A., C. A. Fiebrich, D. S. Arndt, S. E. Fredrickson, and T. W. Hughes. 2000. Quality assurance procedures in the Oklahoma Mesonetwork. *Journal of Atmospheric and Oceanic Technology* 17(4):474–494.

Sheffield, J., C. R. Ferguson, T. J. Troy, E. F. Wood, and M. F. McCabe. 2009. Closing the terrestrial water budget from satellite remote sensing. *Geophysical Research Letter* 26:L07403, doi:10.1029/2009GL037338.

Sheffield, J., E. F. Wood, and F. Munoz-Arriola. 2010. Long-term regional estimates of evapotranspiration for Mexico based on downscaled ISCCP data. *Journal of Hydrometeorology* 11(2):253–275.

Sorooshian, S., K.-L. Hsu, X. Gao, H. V. Gupta, B. Imam, and D. Braithwaite. 2000. Evaluation of PERSIANN system satellite-based estimates of tropical rainfall. *Bulletin of the American Meteorological Society* 81(9):2035–2046.

Su, Z. 2002. The Surface Energy Balance System (SEBS) for estimation of turbulent heat fluxes. *Hydrology and Earth System Sciences* 6:85–100.

Swenson, S., J. Famiglietti, J. Basara, and J. Wahr. 2008. Estimating profile soil moisture and groundwater variations using GRACE and Oklahoma Mesonet soil moisture data. *Water Resources Research* 44:W01413.

Syed, T. H., J. S. Famiglietti, M. Rodell, J. Chen, and C. R. Wilson. 2008. Analysis of terrestrial water storage changes from GRACE and GLDAS. *Water Resources Research* 44:W02433.

Tapiador, F. J., F. J. Turk, W. Petersen, A. Y. Hou, E. García-Ortega, L. A. T. Machado, C. F. Angelis, P. Salio, C. Kidd, G. J. Huffman, and M. de Castro. 2012. Global precipitation measurement: Methods, datasets and applications. *Atmospheric Research* 104–105:70–97.

Tapley, B. D., S. Bettadpur, J. C. Ries, P. F. Thompson, and M. M. Watkins. 2004. GRACE measurements of mass variability in the Earth system. *Science* 305(5683):503–505.

Thies, B., T. Nauß, and J. Bendix. 2008. Precipitation process and rainfall intensity differentiation using Meteosat Second Generation Spinning Enhanced Visible and Infrared Imager data. *Journal of Geophysical Research* 113:D23206, doi:10.1029/2008JD010464.

Thomas, A. 2000. Spatial and temporal characteristics of potential evapotranspiration trends over China. *International Journal of Climatology* 20(4):381–396.

Tronin, A. 2006. Remote sensing and earthquakes: A review. *Physics and Chemistry of the Earth, Parts A/B/C* 31(4):138–142.

Turk, F. J. and S. D. Miller. 2005. Toward improved characterization of remotely sensed precipitation regimes with MODIS/AMSR-E blended data techniques. *IEEE Transactions on Geoscience and Remote Sensing* 43(5):1059–1069.

Ulaby, F. T., R. K. Moore, and A. K. Fung. 1986. *Microwave Remote Sensing: Active and Passive*. London, U.K.: Artech House, Inc.

Vivoni, E. R., M. Gebremichael, C. J. Watts, R. Bindlish, and T. J. Jackson. 2008. Comparison of ground-based and remotely-sensed surface soil moisture estimates over complex terrain during SMEX04. *Remote Sensing of Environment* 112(2):314–325.

Xie, P., J. E. Janowiak, P. A. Arkin, R. Adler, A. Gruber, R. Ferraro, G. J. Huffman, and S. Curtis. 2003. GPCP pentad precipitation analyses: An experimental data set based on gauge observations and satellite estimates. *Journal of Climate* 16(2):2197–2214.

Xu, J. 2001. An analysis of the climatic changes in eastern Asia using the potential evaporation. *Journal of Japan Society of Hydrology and Water Resources* 14:151–170.

Yan, H. and S. Yang. 2007. A MODIS dual spectral rain algorithm. *Journal of Applied Meteorology and Climatology* 46:1305–1323.

Yang, H., C. Sheng, X. Xianwu, and H. Gina. 2011. Global precipitation estimation and applications. In *Multiscale Hydrologic Remote Sensing*, N.-B. Chang and Y. Hong. (Eds.), pp. 371–386. London: CRC Press.

Yilmaz, M. T., E. R. Hunt, L. D. Goins, S. L. Ustin, V. C. Vanderbilt, and T. J. Jackson. 2008. Vegetation water content during SMEX04 from ground data and Landsat 5 Thematic Mapper imagery. *Remote Sensing of Environment* 112:350–362.

2

Groundwater Targeting Using Remote Sensing

Santhosh Kumar Seelan
University of North Dakota

Acronyms and Definitions

ASTER	Advanced Spaceborne Thermal Emission and Reflection Radiometer
AVHRR	Advanced Very High Resolution Radiometer
DEM	Digital elevation model
ERS	European Remote Sensing satellite
ERTS	Earth Resources Technology Satellite
ETM	Enhanced Thematic Mapper
FCC	False-color composite
GLOBE	Global Land One-km Base Elevation
GRACE	Gravity Recovery and Climate Experiment
IRS	Indian Remote Sensing satellite
LFC	Large-format camera
LISS	Linear Imaging Self-Scanner
MODIS	Moderate Resolution Imaging Spectroradiometer
MSS	Multispectral scanner
NDVI	Normalized difference vegetation index
NOAA	National Oceanic and Atmospheric Administration
PAN	Panchromatic
SAR	Synthetic-aperture radar
SIR	Spaceborne Imaging Radar
SPOT	Satellite Pour l'Observation de la Terre
SRTM	Shuttle Radar Topography Mission
TM	Thematic Mapper

2.1 Introduction

Availability of freshwater has determined the growth of civilizations in the past. As the world's population continues to increase, it is predicted that the availability of freshwater for human needs could be a serious limiting factor in the future. Though the estimated total volume of water on the planet is about 1.4 billion km³, nearly 96.5% of it is held up in the oceans as saltwater. An estimated 10.5 million km³ of freshwater, which is about a third of the total freshwater available, are stored below the surface of the Earth in the form of groundwater. It is widely, but unevenly distributed and is an important source for irrigation and drinking purposes. Groundwater exists whenever water infiltrates beneath the surface, the soils and rocks beneath the surface are porous and permeable enough to hold and transmit this water, and the rate of infiltration is sufficient that these rocks are saturated to an appropriate thickness. Groundwater is a renewable resource and therefore, if located, exploited, and managed carefully, can sustain forever.

Targeting groundwater is a complex, but an essential task, particularly in the arid and semiarid regions, which encompass about one-fifth of the Earth's surface. The increasing number of people continues to expand the use of these regions for food production and living space. The shortage of water for domestic, municipal, and irrigation purposes can be eased in many regions by use of the huge resources of groundwater stored in aquifers underlying these areas—provided we know how to locate it and use it successfully in a sustainable way.

The excavations at Mohenjo Daro, an archeological site in the current Sindh province of Pakistan, have revealed brick-lined, open dug wells existing as early as 3000 BC during the Indus valley civilization. The writings of Vishnu Kautilya, who was a teacher at the ancient Takshashila University during the reign of Chandragupta Maurya, a successful ruler of the Indian subcontinent during 300 BC, indicate that groundwater was being used for irrigation purposes at that time (Regunath, 1987). Use of groundwater, a renewable resource, is an age-old practice in countries like India. The early man located groundwater by using commonsense and by associating certain surface features with groundwater occurrence. The common indicators used were ant hills or termite mounds (the termites need moisture to build the mound), certain types of plants that have a root system extending down to the water table, etc. These ancient methods and thumb rules were carefully recorded in the works of the great scholar Varahamihira—505–587 AD (Prasad, 1980).

The advent of systematic geological mapping in the eighteenth century and its improvements in the subsequent decades opened up the understanding of groundwater regimes in a more scientific way. The introduction of geophysical surveys in the 1930s and their refinement over the subsequent years vastly improved the targeting accuracies, as these techniques provide an insight into the third dimension, namely, depth.

Aerial photographs were first used in groundwater exploration a few decades ago and are now an accepted and fundamental reconnaissance tool, as it greatly reduces the time and cost of groundwater surveys by narrowing down the target areas. Aerial photography too advanced over time with the introduction of multiband, digital cameras and sophisticated analytical tools. During the last four decades or so, remotely sensed data from space platforms have been added to the array of tools available to the groundwater scientist. With refinements in spectral and spatial resolution offered by the newer satellites, there has been an enormous increase in the capacity of the remotely sensed data to provide groundwater information, and the technique has emerged as a major tool. One important factor to be emphasized here is that remote sensing does not replace the existing techniques but is a very useful additional tool. In fact, Varahamihira's thumb rules have withstood the test of time and are still good. But if water is to be provided quickly to the vast parched lands, a faster and cheaper procedure integrating remote sensing, geophysics, and ground-based hydrogeological surveys are a must.

The focus of this chapter is to discuss the relevant targeting keys and parameters for groundwater that can be extracted from remote sensing techniques under various geological settings. A detailed review of literature is also made to trace the developments and current status in this field.

2.2 Role of Remote Sensing in Groundwater Prospecting

While groundwater is a subsurface phenomenon, remote sensing offers information about the features on the surface of the Earth. Therefore, what can essentially be done is to use remote sensing data to identify the surface parameters indicative of groundwater occurrence and movement. These parameters relate to geology, geomorphology, surface hydrology, soils, land use, and natural vegetation (Table 2.1). The greatest advantage of the spaceborne remote sensing is that information on all of the aforementioned

TABLE 2.1   Overview of Groundwater Indicators Discerned from Remote Sensing

No.	Major Parameters	Major Subsets
1	Regional geology	Lithology, trend, faults, fracture lineaments, master joints, dikes/quartz reefs
2	Regional geomorphology	Drainage pattern/drainage density, pediment zones, valley fills, ridges and valleys, cuestas and hogbacks, alluvial fans and bazadas, dunes, lineament-controlled stream channels
3	Regional hydrology	Surface water bodies, springs/seeps, floods/inundated areas/wetlands, runoff and recharge zones, groundwater discharge into streams
4	Soils	Coarse-grained soils, fine-grained soils, soil moisture
5	Hydrologic land use	Surface water–irrigated cropland, urban or built-up land, forest cover
6	Natural vegetation	Vegetation indicative of soil moisture, vegetation indicative of shallow water table
7	Groundwater quality	Saline soils, salt-tolerant vegetation

aspects are obtainable from a single imagery. Features and relationships that would take a long time to put together using the much-larger-scale aerial photography or surface examination can be seen in a single integrated perspective on the space imagery. The conventional surveys usually start with geophysical surveys for selecting a point for drilling. But the question that arises invariably is where to conduct the geophysical survey. As geophysical surveys are manpower intensive and time-consuming, one would always like to start the survey in places where the prospects are higher. This means a reconnaissance-level information has to be generated first so that unfavorable zones can be eliminated straight away and the favorable zones could be examined more closely. This way, the geophysical and other ground surveys can be used more effectively, with better results. Hence, it is ideal to use medium-resolution spaceborne data in the beginning stages of the exploration programs to extract reconnaissance-level information, followed by the more detailed interpretation of high-resolution aerial photographs or satellite imagery and field geological and geophysical studies, before test drilling (Figure 2.1). Ideally, the investigations could be carried out in the following four simplified steps:

1. *Medium-resolution satellite-based regional assessment*: In a regional assessment, the geographical area considered is usually very large, say, a group of districts or state. Satellite data are interpreted for geology, geomorphology, land use, soils, drainage, and vegetation patterns and integrated to arrive at potential groundwater zones on a regional level. This is described in greater detail in subsequent sections.

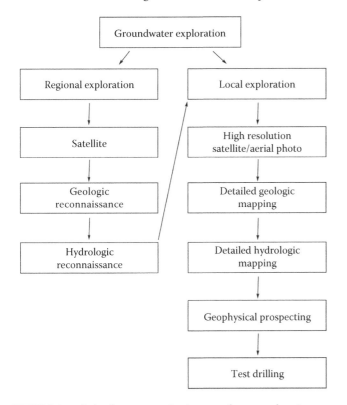

FIGURE 2.1   Role of remote sensing in groundwater exploration.

2. *Aerial photo or high-resolution satellite imagery–based local assessment*: This stage includes (a) flying for aerial photography on a required scale or resolution (using manned or unmanned aircrafts) for areas identified as potential in the reconnaissance study; where photographs are already available, fresh photography is not required; (b) interpretation of aerial photographs for detailed geology, structure, lineaments, geomorphology, soil characteristics, land use patterns, other hydrogeological features, etc.; and (c) collection and corroboration of existing ground-based information such as nature of aquifers, type, and yield characteristics. During this stage, remote sensing–derived information in sufficient detail to identify groundwater-potential zones will be available.

3. *Ground-based hydrogeological and geophysical investigations*: The identified groundwater-potential zones are to be subjected to verification with ground-based investigations. The potential zones demarcated should be identified on the ground for verifying their potential. Detailed hydrogeological surveys to identify the nature and type of aquifers, their yield characteristics, depth of the aquifers, and depth to water table, piezometric gradients, etc., have to be carried out in the identified zones and adjoining areas. With this information, suitable locations for further geophysical surveys can be identified. These selected sites are to be subjected to detailed geophysical surveys to pinpoint the drilling sites. Normally for groundwater investigations, electrical methods such as resistivity surveys are observed to be very successful. To locate certain fracture/fault zones and to identify their dip direction, geophysical resistivity profiles can be carried out. All the hydrogeological and geophysical investigations in the narrow potential zones will help pinpoint the potential drilling sites where substantial yields can be expected.

4. *Exploratory drilling, testing, and well completion*: The various steps involved in the drilling of bore wells include (a) selection of suitable drilling rig such as direct rotary and down-the-hole hammer, depending upon the nature of rock types; (b) drilling of pilot boreholes; (c) borehole logging; (d) well construction; (e) development of the well; (f) pump testing to evaluate the transmissivity, hydraulic conductivity, specific yield, etc.; and (g) installation of a suitable pump such as submersible pump, jet pump, and deep well turbine pump, depending upon the yield.

In this procedure, a favorable zone is picked using remote sensing, investigated in detail using conventional methods, before drilling is taken up. Once the zone is proved good, all the boreholes in the zone become successful. The unsuccessful wells are minimal. This way, there is enormous savings in the cost of ground surveys and drilling. A remote sensing–based survey therefore essentially proves a zone (rather than a point) for its groundwater potential. A remote sensing–based study cannot be independent by itself, but has to be necessarily followed up by conventional surveys.

2.3 Groundwater Information Extraction from Satellite Images

A number of studies have been carried out over the years, on the use of remote sensing in groundwater investigations. Aerial photointerpretation during the 1950s and 1960s laid the foundation for the later decades. Although photography from balloons and aircrafts started during the eighteenth century, it had its potential use only in military reconnaissance. Subsequent to World War II, the photographs found their utility in geological mapping, primarily as an offshoot of oil exploration. Early works of Mollard (1957) and Lattman and Nickelson (1958) assessed the fracture patterns revealed by aerial photographs for oil exploration. Many subsequent researchers have used aerial photographs for groundwater applications and are continuing to do so. Mollard and Patton (1961) outlined what they considered a sound approach to an integrated groundwater investigation using aerial photointerpretation inputs. They indicated the parameters, particularly in unconsolidated sedimentary terrains.

Moore (1967) studied the early satellite pictures from Skylab for mapping lineaments of hydrologic significance. With the launch of Landsat (then Earth Resources Technology Satellite [ERTS]) and the continued availability of data from the satellite, the applications in the area of groundwater grew wider. A detailed study on deciphering groundwater from aerial photographs was made by Nefedov and Popova (1972). Their work included deciphering natural factors such as rocks, relief, and surface water, determining hydrogeological conditions, and also deciphering landscape components dependent on groundwater, such as soil and vegetation. Morphological indicators were also outlined. A very detailed report on the keys to detect aquifers from satellites was first brought out by Moore and Deutsch (1975). The keys to detect various types of aquifers and the procedure to extract groundwater information were outlined. The keys for interpretation also have been summarized in the Manual of Remote Sensing, American Society of Photogrammetry (Solomonson et al., 1983).

2.3.1 Lithology and Lineaments

The preparation of geologic maps showing lithology from reflectance differences, structures on the basis of linear and curvilinear features, and the preparation of geomorphic maps are greatly improved by satellite imagery. Discrimination of rock types based on satellite imagery alone is almost impossible. However, differences in lithology are very well brought out due to their differences in reflectance. The best results are obtained when the available geological maps and data from the literature are used in conjunction with the satellite images. In any case, it must be emphasized that the satellite imagery interpretation must be considered only as a tool that can aid geological mapping. Field checking of selected areas of the interpretation is imperative. Visible, near, and middle infrared (IR) and thermal images and color composites of different band combinations have been

found to be very useful in differentiating rock types. Although no detailed lithological interpretations are possible due to vegetation and soil cover, differences in lithology and the boundaries of various rock outcrops can be marked. This is of direct aid in groundwater studies.

The greatest contribution of satellite images in geologic studies has been in the field of structural analysis. Structural features that are not identifiable on the ground and sometimes even on aerial photographs are easily discernable on satellite images because of their regional coverage. Major faults and fractures that are expressed as linear features on the satellite images are potential zones for groundwater. Mapping of this is very useful, especially in hard rock areas where the occurrence and movement of groundwater are mostly confined to these linear features.

Identification of lineaments is best done on visible and near-IR images and color composites. They appear on the images as an alignment of dark or light soil tones, elongate aligned water bodies, linear stream channels, valleys, etc. An experienced interpreter may also infer the presence of faults and fractures based on the study of topography, geomorphic features, etc.

Tiwari (1993) delineated lineaments using spaceborne and airborne data in the semiarid district of Sirohi in western Rajasthan in India for selection of drilling sites. In all of the six exploratory wells drilled in the lineament zones consisting of crystalline limestone, granite, and schistose rocks, the groundwater yields were consistently high. In order to reduce subjectivity in interpretation, Sander et al. (1997) used multiple interpreters to interpret Landsat Thematic Mapper (TM) and Satellite Pour l'Observation de la Terre (SPOT) data over a project area in Ghana and concluded that 90% of the hydrologically significant lineaments were picked up by all interpreters. Koch and Mather (1997) compared the lineaments extracted from Spaceborne Imaging Radar (SIR)-C L and C band synthetic-aperture radar (SAR) data and stereoscopic large-format camera (LFC) over the Red Sea Hills region of Sudan and concluded that the LFC data were most useful for mapping detailed fracture patterns, while the combination of SIR-C SAR and LFC data was helpful in the location of major deep-seated fracture zones. An imagery integration approach was developed by Bruning et al. (2011) to evaluate satellite imagery for lineament analysis in a volcanic terrain in Nicaragua where human development and vegetation confound imagery interpretation. They used Advanced Spaceborne Thermal Emission and Reflection Radiometer (ASTER), Landsat 7 Enhanced Thematic Mapper Plus (ETM+), Quickbird, and RADARSAT 1 data in conjunction with digital elevation model (DEM) and concluded that nine of the previously mapped lineaments and 26 new ones were identified in this method. They also concluded that RADARSAT -1 products were most suitable for minimizing the anthropogenic features, but they have to be used in conjunction with optical sensor data and DEM for best overall results.

Digitally enhanced images have been found to be very useful in the demarcation of lithological boundaries and in structural mapping (Goetz et al., 1975, Abrams et al., 1977). Seelan (1980) used contrast stretching techniques, where the brightness data distribution was expanded or stretched to fill the dynamic range of the display medium, to demarcate lithological boundaries. Moore and Waltz (1983) used a five-step digital convolution procedure to extract edge and line segments and produce directionally enhanced images, in order to avoid subjectivity in manual interpretation. Rana (1998) used the directional filtering technique suggested by Moore and Waltz (1983) on Indian Remote Sensing satellite (IRS) Linear Imaging Self-Scanner (LISS) I band 4 image over the semiarid Thar Desert region in India to identify subtle lineaments, unnoticed otherwise. The lineaments were recognized as buried channels and zones of coarse sediments representing potential sites for the accumulation of freshwater during rains. Ahmad and Singh (2002) used data-merging techniques to fuse IRS panchromatic (PAN) and LISS III data sets over a study area in the Indo-Gangetic Plain of India to demonstrate the usefulness of such merging in mapping geomorphic features and relating to major tectonic events. Mogaji et al. (2006) used digital enhancement techniques to help extract lineaments on a Landsat 7 ETM + image. This study found that the linear contrast stretch, merging the false-color composite (FCC) and PAN images, and the lineament density map extracted from these were useful for identifying potential groundwater zones in their study area. Pal et al. (2007) fused optical and SAR data to map various lithological units over a part of the Singhbhum Shear Zone and its surroundings in India. In this study, European Remote Sensing (ERS)-SAR data were enhanced using fast Fourier transform–based filtering approach and also using Frost filtering techniques. Both the enhanced SAR imageries were then separately fused with histogram-equalized IRS-1C LISS III image using the principal component analysis technique. Later, they applied the feature-oriented principal components selection technique to generate an FCC from which corresponding geological maps were prepared. A semiautomatic method to infer groundwater flow paths based on the extraction of lineaments from DEMs was developed by Mallast et al. (2011) for the subterranean catchment of the Dead Sea region. They used a combined method of linear filtering and object-based classification to map the lineaments accurately. Subsequently, the lineaments were differentiated into geological and geomorphological lineaments using auxiliary information and finally evaluated in terms of hydrogeological significance. Ali et al. (2012) used contrast stretching, image ratio, image filtering, and intensity–hue–saturation transformation on Landsat ETM+ data over Abidiya area in Sudan to discriminate between mafic and ultramafic granites of ophiolitic origin.

2.3.2 Hydrogeomorphology

Geomorphic studies are indispensable in understanding the occurrence of groundwater, especially areas of late Pleistocene and recent deposits. Certain landscape elements that indicate coarse materials and near-surface water table are identifiable on satellite images. Buried channels, present-day valleys filled with alluvial deposits, underfit valleys, natural levees, alluvial fans,

bazadas, etc., which have their own hydrologic properties, can be mapped. The images are useful in studying the drainage pattern and drainage density, which have a bearing on the recharge conditions and permeability of the rocks.

Where geomorphology exercises a significant control over groundwater movement and occurrence, this relationship has been utilized by several authors to arrive at broad groundwater-potential zones. Moore and Deutsch (1975) identified the limits of an alluvial aquifer in Vancouver, British Columbia, Canada, using FCCs of Landsat (then ERTS). They identified the limits based on the present alluvial valley of the Fraser River and an old valley south of the present river that had cultivated fields. Cultivated fields and noncultivated fields could be separated on the imagery due to the sharp boundaries between them. Small wooded areas between the present and old river valleys had steeper slopes and were inferred to have been underlain by unconsolidated rocks. They recommended geophysical surveys and test drilling to determine the thickness, lithology, and water-bearing potential of the alluvial aquifers in the Fraser area. Seelan (1982) adopted the geomorphological approach over a study area in Central India comprising of hard rocks and alluvium, to arrive at potential groundwater zones. The satellite images were visually analyzed to prepare hydrogeological maps on 1:250,000 scale. This methodology was earlier tried in India through a pioneering, operational remote sensing project conducted jointly by the National Remote Sensing Agency, India, and the Public Works Department of Tamil Nadu, India, during 1977–1979, aimed at delineating potential groundwater zones on a regional level (Thillaigovindarajan et al., 1985). Figure 2.2 illustrates this concept for a part of the study area used in this project. This approach was later adopted in several studies carried out at the National Remote Sensing Agency, India (1986), including the studies under the technology mission on drinking water wherein hydrogeomorphological maps on a 1:250,000 scale were generated for the entire India.

Two buried channels were identified in the area south of Allahabad, India, one west of Tons River and the other east of it, using IRS LISS II FCC imagery by Gautham (1990). Aerial photographs were used for studies on slope, direction, position of natural levees, vegetation density, etc. It was concluded from the configuration of the buried channels that these were initially joined forming one channel that flowed from east to west, although the present master slope of the area is from west to east. The buried channels were confirmed to have good groundwater potential. Some parts of the buried channels were noticed to have been waterlogged due to seepage from the irrigation canals cutting across one of them.

Kumar and Srivastava (1991), in a study carried out in Bihar, India, tried to analyze the spatial distribution of geomorphic classes and depthwise variation in aquifer material within the same class for determining the target horizon for further detailed investigations using remote sensing and electrical sounding. Agarwal and Mishra (1992) delineated different hydrogeomorphological units in and around the immediate environs of Jhansi city in India and attempted correlation between well yields and

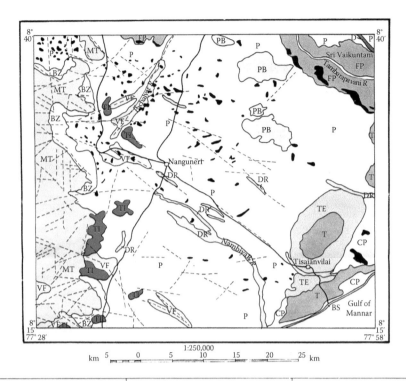

Symbol	Hydrogeomorphic unit	Description	Groundwater potential
MT	Mountainous terrain	Ridge and valley complex. Lineaments prominent. Valleys controlled by lineaments.	Good recharge along valleys controlled by lineaments which can be exploited in the plains.
BZ	Bazada	Coalescence of alluvial fans. Formed along the foot hills of western ghat hills. (mountainous terrain)	Good potential towards the lower part of the fans.
TI	TOR and Inselburg complex	Peripediment zone with tors and inselburgs. Pediment fairly thick in between tors and inselburgs.	Good within the pediment zones.
PB	Pediment with thick black soil cover	Occur as patches over the pediplains. Soil cover is thick and fine grained in nature.	Poor recharge and groundwater conditions.
P	Pediplain	Fairly thick, continuous pediment cover.	Moderate within the pediment zone. Fracture lineaments offer greater potential.
VF	Valley Fill	Thick alluvial fill materials within the fracture controlled valleys.	Good groundwater potential within the alluvium.
DR	Dry river courses	Sandy, dry river beds.	Groundwater occurs as base flow immediately following the monsoon months.
T	Teri soil	Medium to coarse grained, wind blown, thick, deep red soils occuring over the pediplain.	Excellent potential.
TE	Area around teri subjected to erosion	Subjected to severe erosion by wind and water.	Excellent along the eastern and southern parts of the teri where groundwater discharges from the main teri.
D	Dune sands	Narrow strips of sand.	Does not retain water, hence not suitable for groundwater development.
FP	Flood plain	Flood plain deposits of tambaraparani river consisting of sand, silt and clay.	Good groundwater potential.
CP	Coastal plain	Sandy deposits close to the coast.	Excellent aquifers. Cautious development needed.
BS	Beach sands	Narrow sandy stretch along the coast.	Good shallow groundwater. Cautious development needed.
- - -	Lineaments	Fracture lineaments over gneissic terrain.	Excellent potential. Detailed ground investigations needed.
⌐	Water bodies		

FIGURE 2.2 (Top) Hydrogeomorphological map of part of Tamil Nadu, India (based on Landsat multispectral scanner imagery interpretation). (Bottom) Legend for hydrogeomorphological map of part of Tamil Nadu, India.

hydrogeomorphological units prepared based on satellite data interpretation. They observed a good correlation between geomorphological units and well yields. Similarly, Shankarnarayana et al. (1996) delineated broad lithological, morphological, and structural features using Landsat TM imagery and demonstrated that the wells located on the lineament zones produced better yields. They concluded that the wells located based on lineament study yielded 14 times more than the wells located away from the lineaments. The hydrogeomorphological approach to delineate groundwater-potential zones was also used by Panigrahi et al. (1995), Ravindran and Jeyaram (1997), and Rao and Reddy (1999). Dhakate et al. (2008) used ground-based vertical electrical sounding techniques to confirm the potential of geomorphological units derived from satellite imagery interpretations over a granitic terrain in Andhra Pradesh in India. They concluded that the thickness of weathered mantle is higher in areas of higher lineament density and such thicker mantle exhibit higher groundwater potential. Sethupathi et al. (2012) studied the watersheds in northeastern Tamil Nadu region in India where the groundwater discharge rate was greater than the recharge rate. The study involved Landsat 7 ETM+ imagery and delineation of topography, slope, surface cover, soils, vegetation, and lineaments. They identified 194 lineaments and classified them into three categories based on length, minor, medium, and major. Areas with high lineament and high lineament intersection density had high groundwater potential along with valley fills, pediplains, and buried pediments.

Radar data have been successfully used in mapping palaeodrainage. Robinson et al. (2000) determined palaeodrainage directions in the eastern Sahara using high-resolution, multiwavelength, multipolarization SIR-C data and Global Land One-km Base Elevation (GLOBE) DEM data sets. Shuttle Radar Topography Mission (SRTM) data with 90 m resolution were used to delineate mega palaeodrainage in the eastern Sahara by Ghoneim and El-Baz (2007).

2.3.3 Soils, Land Use, and Drainage

Broad textural classes of soils are identifiable based on visual interpretation of satellite data to a great extent. For example, fine-grained soils are generally darker than coarse-grained soils. Computer-aided analysis gives a more detailed classification, but requires adequate ground truth and involves a good amount of field work for accuracy. In a digital analysis, the computer is trained to identify spectral signatures of different soil types and classify them. Interpretation of soils is best done during a season when soils are well exposed and vegetation cover is minimal. Distribution of different textural classes of soils is an important factor in interpreting for groundwater occurrence and recharge. For example, fine-grained clayey soils generally tend to permit less recharge and more runoff, while coarse-grained sandy soils allow more recharge to groundwater.

An understanding of the land use pattern is very important in groundwater studies. For example, a forested area discourages runoff and encourages recharge to the groundwater, while it is the opposite in an urban, built-up area. As land use is a dynamic process, the satellite systems, because of their temporal coverage, are ideally suited for its analysis. The computer analysis of land use almost follows the same process as for soils. The accuracy and details of classification will primarily depend on the type of sensors used and the amount of ground truth available. Broad classes such as forested areas, urban areas, water bodies, and cultivated fields can easily be delineated by visual interpretation itself without any enhancement aids.

Drainage density maps derived from satellite imagery can also provide vital clues to groundwater recharge conditions. Where the drainage density is higher, it is likely that recharge to groundwater is lower and, where it is lower, the recharge and underground flow conditions are higher. Information on land use, soils, and drainage can form important parameters in evaluating groundwater potential of a region as described in the later section on the integrated approach.

Seelan et al. (1983) studied land utilization and landform patterns in parts of southern Uttar Pradesh state in India using Landsat data and established the control exhibited by landforms on groundwater resources situation that in turn controls the land utilization in the region. Ghosh (1993) prepared a groundwater map of a part of Jharia coalfield area in India using soil moisture, vegetation, and morphology as primary indicators based on aerial photointerpretation. The relationship between vegetation growth and groundwater in arid regions is an active area of research, and Xiamei et al. (2007) extracted the vegetation information from the National Oceanic and Atmospheric Administration (NOAA) Advanced Very High Resolution Radiometer (AVHRR) and Moderate Resolution Imaging Spectroradiometer (MODIS) normalized difference vegetation index (NDVI) and established the relationship between phreatophytic vegetation and depth to water table in the Yinchuan Plain in China.

2.3.4 Groundwater Recharge/Discharge

Remotely sensed data have been used in the past to identify groundwater recharge/discharge zones. Thermal data have been found to be particularly useful for this purpose. Surveys made by different workers in different parts of the world have proved the thermal IR sensing to be a very useful tool in locating abnormal water sources such as geysers, hot pools, hot springs, and river seepages. Airborne IR surveys over thermal areas have been carried out in the Yellowstone National Park, USA (Mc Lerran, 1967), Reykjanes and Torfajokull, Iceland (Palmason, 1970), Tampo region of New Zealand (Hochstein and Dickinson, 1970), North Island of New Zealand (Dawson and Dickinson, 1970), and Lake Kinneret and Dead Sea regions of Israel (Seelan, 1975). Sensing from an elevated point on the ground has also been carried out in the Lake Kinneret region of Israel (Otterman, 1971). The objective of most of these surveys was to locate thermal anomalies in regions where thermally active zones were known to occur. Thermal springs could be located successfully in most of these surveys. Palmason's surveys lead to the discovery of dozens of previously unmapped points of thermal activity.

Thermal IR images can be used to locate hot creeks fed by, say, a series of hot springs. If there are two adjacent streams, the one fed by a hot spring appears whiter. Discharges of slightly warmer waters from rivers or drainage canals or springs into lakes could be located. Deutsch (1971) located flow of such warmer waters into Lake Ontario with significant success. Further, the thermal IR images could be used in locating freshwater springs just offshore of saltwater coasts and saline springs discharging into freshwater lakes. Underwater hot springs and saline springs could also be located (Otterman, 1971, Seelan, 1975). When hot saline water flows into a freshwater lake, its boundary is very sharp. This is because the saline water due to its higher density tends to flow under the surface without much mixing on the surface. Images taken at a particular time interval could help in outlining the flow patterns of the underwater springs. The surface thermal patterns determined from IR images should help in understanding the geologic structures that control the upflow of water and, thus, in a better understanding of regional hydrogeology. In the Maligne karst system of the Rocky Mountains, Alberta, the discharge point of the karst polje was unknown, but the thermal images indicated that the water disappearing in the polje was reappearing as a submerged spring under the surface of the Medicine Lake (Ozaray, 1975). Thermal images were also reported to have been used to locate lines of diffuse groundwater discharge along some of the deep ravines of the Lake Pakowki area in Alberta (Ozaray, 1975). Bobba et al. (1992) used Landsat imagery to delineate groundwater discharge, recharge, water table depth, and transition areas in Southern Ontario, Canada. This study used Landsat 1 imagery from March 1974 to July 1974 to represent seasonal differences. Simulation of regional groundwater flow models (recharge, transition, and discharge) was developed.

Tcherepanov et al. (2005) used Landsat TM thermal imagery under different weather conditions from 1989 to 2002, combined with ground-based observations on lake temperatures to identify groundwater recharge into lakes. The study allowed for identification of lake zones with consistently cold temperatures that were inferred to be potential groundwater zones. An integrated approach of remote sensing and geographic information systems (GIS) to identify optimal areas for artificial recharge by flooding techniques was developed by Ghayoumian et al. (2005) for the Meimeh Basin in northern Isfahan Province in Central Iran. They derived information on soil from Landsat TM imagery and integrated this information with other thematic layers such as slope, infiltration rate, transmissivity, water table and aquifer thickness, and water quality. They concluded that 70% of the study area is suitable or very suitable for flood spreading, mostly within the central portions of the Meimeh Basin. Ghayoumian et al. (2007) also applied similar techniques in another site in Iran to identify suitable zones within alluvial fans and pediplains for artificial recharge. Landsat 7 imagery was used to identify groundwater recharge in Indian River and Rehoboth and Indian River bays in Sussex County, Delaware, by Wang et al. (2008). They used PAN, near infra-red, and thermal bands to identify ice patterns and temperature differences in the surface water that are indicative of groundwater discharge in the area. Khalaf and Donoghue (2012) used MODIS Level 3 MOD09Q1 and MOD11A2 products and other information to analyze the relationship between rainfall, evapotranspiration, and soil moisture and recharge rates in the West Bank region.

2.3.5 Groundwater Use for Irrigation

Areas where groundwater is being used for irrigation can be separated using satellite imagery. Deutsch (1974) monitored groundwater use for irrigation in Snake River plain lying between the northern part of Rocky Mountains and the Snake River in United States. From this study, he concluded that further agricultural development was possible, as the aquifer was known to extend far beyond the presently irrigated area. Seelan (1980) mapped the groundwater-irrigated areas in the Bundelkhand granitic terrain in parts of Central India using multi-date satellite images. The groundwater use was conformed to the valley portions of the pediments, which bears the best groundwater potentials in the region. Cropped areas are discernible on the satellite pictures because of their typical high reflectance in the IR region. In India, for example, the monsoon rainfall is restricted between the months of May and September, and the second crop that is grown after the monsoon is mostly irrigated. The areas irrigated by surface water were separable from the groundwater-irrigated areas because of their water impoundments, canal systems, and the distinct command areas.

Groundwater-irrigated areas in regions, such as in the Ogallala Aquifer system in United States, use center pivot irrigation system that are quite easily identified using visual interpretation of satellite imagery. However, this is not so straight forward in some sections of the Ogallala, such as in Nebraska, where the source is a combination of surface water and groundwater. Most of the canal-fed fields are square in shape and thus differ from round shapes exhibited by center pivot groundwater irrigated systems; however, extensive ground truth has shown that there are instances where surface water is used in center pivot irrigation systems as well (Kurz and Seelan, 2009). In addition, there are also groundwater-fed center pivot irrigation systems adjacent to rivers where typically surface water is the source. Thus, a differentiation between surface water–irrigated and groundwater-irrigated fields cannot be done based on the shape of the field alone. A Landsat imagery used in the study is shown in Figure 2.3.

2.3.6 Groundwater Stress

Apart from targeting groundwater source, remote sensing techniques also find some applications related to management of groundwater resource, particularly in detection and monitoring of groundwater systems under stress. However, it must be emphasized that the remotely sensed data provide information only on the objects on the surface of the Earth, while some of the systems under stress or stress-causing factors may not show up on the surface at all. Therefore, like targeting, one has to look for surface indicators of stress, in order to be able to detect them on the satellite imagery or aerial photographs (Seelan, 1986).

Over extraction of groundwater lowers water tables and reduces storage. Rodell and Famiglietti (2002) described the potential of using Gravity Recovery and Climate Experiment (GRACE) satellite's gravity calculation data on groundwater storage levels in the Ogallala Aquifer. Tiwari et al. (2009) analyzed temporal changes in Earth's gravity field in northern India and its surroundings, using data obtained from GRACE, and attributed the large-scale mass loss to excessive extraction of groundwater. Combining GRACE data with hydrological

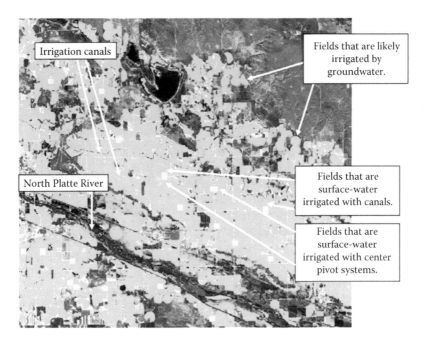

Irrigation canals

Fields that are likely irrigated by groundwater.

North Platte River

Fields that are surface-water irrigated with canals.

Fields that are surface-water irrigated with center pivot systems.

FIGURE 2.3 Surface water–irrigated and groundwater-irrigated fields along the north part of Platte River in Nebraska.

models to remove natural variability, they concluded that the region lost groundwater at the rate of 54 ± 9 km³ per year between April 2002 (the start of GRACE mission) and June 2008. Based on this, they concluded that this is probably the largest rate of groundwater loss in any comparable sized region on Earth and that, if this trend continues, will lead to a major water crisis in the region.

2.3.6.1 Remote Sensing of Stress-Causing Factors

While there are several causative factors, only a few are amenable for monitoring by remote sensing, particularly from satellite altitudes. Satellite data have been used in the past to identify areas irrigated by groundwater as seen in the earlier section. A rapid increase in the area irrigated by groundwater over the years would indicate the tendency toward over exploitation. In the case of aerial photographs on a suitable scale, it is possible to count the number of open wells and correlate the well density with geomorphological units and pinpoint the units that have exceeded the recommended rates of exploitation (Seshubabu and Seelan, 1983).

Pollution sources such as sewage and other solid waste disposal sites can be identified on aerial photographs and high-resolution satellite imagery. The effects of these pollution sources and other industrial waste impoundments on groundwater are well known. In a study reported by Geraghty and Miller Inc. (1972), an aerial survey of portions of Pennsylvania state in the United States was made to determine the extent of impoundments by industry. Impoundments thus located from the air were field checked to identify the owner, content, size, construction, and permit status of the impoundment. One such impoundment was being used by a manufacturer of batteries and cables that was dumping its untreated battery wash into a

limestone quarry. The water quality of the effluent showed pH and lead levels much beyond the permissible levels, and the waste disposal practices of the plant were altered in such a way as to prevent continued contamination of groundwater. The change in the groundwater controls due to unplanned mining activity in Jharia coalfields in India was studied by Ghosh (1993). The groundwater map of the Jharia coalfields area was prepared from aerial photographs using vegetation cover as the criteria.

Agricultural activity influences the quantity of groundwater to a great extent by the use of fertilizers and due to poor management practices such as improper drainage that results in waterlogging conditions. Although the use of fertilizer levels is not possible to gauge from remote sensing, it is possible to delineate areas under irrigation. In countries like India, fertilizer usage is maximum in irrigated areas as compared to nonirrigated agricultural areas. Waterlogged conditions have been demarcated on multispectral data from Landsat satellites (Singh, 1980) introducing the possibility from continuous monitoring for increase or decrease in waterlogged areas. However, the data have limitations in demarcating waterlogged conditions in black soil areas (Venkataratnam, 1984). Visual and digital processing of IRS imagery has been successfully used to delineate the pre- and postmonsoon surface waterlogged areas in the Gangetic Plains of Bihar state in India by Chaterjee et al. (2005) and Chowdary et al. (2008).

Urbanization reduces recharge to groundwater, increases pumpage for domestic and industrial requirements, and thus can affect the quality of groundwater through a deliberate disposal of liquid and solid wastes. Therefore, the growth of urban centers can cause enormous stress on the shallow groundwater systems. A study carried out by Moore and Deutsch (1975) showed the satellite data interpretation of all population centers around the southern tip of Lake Michigan in the United States and suggested monitoring the direction and rate of urbanization, through repetitive imagery to help in determining the location and severity of future problems in aquifer management. The same study also quotes work done to locate changes in both active and newly reclaimed strip mines using satellite imagery, as the acid water drains from some of these mines have influence on the quality of groundwater. It has also been demonstrated that remote sensing can be used to study industrial discharge into lakes and rivers that in turn have an obvious influence on groundwater. Only pollutants that add color and turbidity or alter the temperature of the water can be detected easily by remote sensing. An investigation carried out at the National Remote Sensing Agency in India succeeded in mapping the paper mill effluent disposal pattern in the Godavari River near Rajahmundry town in India using an airborne multispectral scanner (MSS) data (Deekshatulu and Thiruvengadachari, 1981).

2.3.6.2 Remote Sensing of Stress Indicators

Groundwater systems, when subjected to stress, sometimes show up on the surface of the Earth through vegetation, soil, or surface water. When such indications start appearing on the ground, it is possible to diagnose it using remote sensing data. Springs and seeps when characterized by the luxurious growth of vegetation around them can be interpreted on satellite images. Drying of these springs will result in the vanishing of vegetation around and can be used to interpret declining water levels. IR photography has been found suitable for identifying springs (Robinov, 1968). Airborne thermal IR data have been found to be useful in identifying underwater springs (Seelan, 1975, Gandino, 1983). Likewise, drying up of rivers fed by groundwater can also be an indicator of stress. Similarly, reduction in areal spread of water bodies can be successfully monitored. The Landsat satellite picture of March 1983 covering the Madras (now Chennai) city and surrounding areas in India, when compared with the picture of the corresponding period in 1980 (1979–1980 being a normal rainfall year in the region), showed that the major reservoirs in the area, namely, Poondi, Arniar, Cholavaram, Red Hills, and Chembarambakkam, had reduced to 3.7%, 28.6%, 20%, 51.9%, and 21.4%, respectively, in areal spread. This was attributed to the failure of rains in 1982–1982, which resulted in poor recharge to the groundwater and extensive pumping. As these reservoirs supply water to the city, the coastal aquifers south of Madras were subjected to heavy pumping resulting in seawater intrusion in some pockets (Seelan and Narayan, 1984).

Investigations carried out in the Argentine Pampa area by Kruck (1976) revealed that saline groundwater zones can be demarcated on the satellite picture using vegetation as an indicator. Where saline incrustations start appearing on the soils, their high reflectance levels permit easy identification on the imagery. Computer-aided interpretation helps identification of different levels of soil salinity (Venkataratnam, 1984).

2.3.7 Integrated Approach

At times, a single parameter such as geomorphology would play a dominating role and should be sufficient to plan further ground exploration. But many times, it is not possible to arrive at potential groundwater zones based on one or few of these parameters. For example, a single information source such as lithology may not indicate promise, but the situation may change when viewed with corroborative evidence in structure, geomorphology, land use, and soils. Also, it is necessary to incorporate information obtained from sources other than remote sensing. Hence, there is a need for an integrated approach. An integrated approach where the different thematic information on geology, geomorphology, soils, land use, rainfall, drainage, etc., are studies to arrive at potential groundwater zones offers the best results. It also ensures that no relevant information is overlooked. The need for an approach integrating all parameters relating to geomorphology, soils, land use, etc., was advocated by Seelan and Thiruvengadachari (1980, 1981). A part of Tamil Nadu state in India was studied using Landsat MSS data, for geology, geomorphology, soils, and land use that were then superimposed on one another manually using transparent overlays to extract groundwater-potential zones. The geological map was used as a base, and an overlay was prepared indicating potential groundwater zones from the point of view of geology. The overlay was then transferred to the other thematic maps and the procedure was repeated. Finally, in the overlay, the regions where all or most of the themes were found favorable were marked as potential groundwater zones.

Integration of various thematic information to arrive at groundwater potential was extensively used by the National Remote Sensing Agency (1980, 1982a,b, 1983, 1984) during the 1980s. The use of this method was also reported by Sankar (2002) and Gopinath and Seralathan (2004).

A comparison of hydrogeomorphological approach and the integrated approaches was made by Seelan and Thiruvengadachari (1981). It was concluded that in areas where geomorphology exercises a significant control over the groundwater movement and occurrence, the extraction of geomorphic details alone can suffice, and in areas where the control is not significant or where costs and time permit a detailed study, basic resources information with regard to geology, geomorphology, land use, and soils can be obtained and integrated to extract regional-level groundwater-potential zones.

2.3.7.1 Geographic Information Systems

The manual process of integrating various thematic elements to arrive at potential groundwater zones was laborious and subjective to a great extent. With the advent of GIS, it became possible to overlay different thematic information digitally. The subjectivity is reduced as weightages can be assigned to various parameters. It is much faster and a standard procedure could be used under most terrains with suitable modifications. In an early use of GIS in groundwater studies, Rundquist et al. (1991) studied the vulnerability of groundwater systems in Nebraska, USA.

In an attempt to use GIS for targeting groundwater, Seelan (1994) used remote sensing data to derive thematic information on geomorphology, land use, soils, drainage density, dikes, etc., and incorporated information on rainfall, and DEM (Figure 2.4) using a PMAP version 2.22 GIS software to arrive at potential groundwater zones. This study was carried out in a part of Andhra Pradesh state in India, over a hard rock, gneissic terrain where groundwater primarily occurs in weathered pediments and fractured valleys. The land use patterns, soils, and the dikes also control the recharge, occurrence, and movement of groundwater in the region. The six, digitized, thematic maps pertaining to geomorphology, land use, soils, drainage density, drainage and dikes, and rainfall were used in the GIS analysis to extract groundwater-potential zones. Based on the knowledge of the various parameters and their bearing on groundwater,

weightages were assigned for each of the classes in the various maps. The simplified working method for the groundwater prospect model is given in Figure 2.5. The six polygon maps with assigned weightages for each polygon were overlaid digitally to create a final groundwater-potential map (Figure 2.6). The area was classified into five categories ranging from "very high" potential to "very low" potential. The study also concluded that a GIS-based integrated approach is superior to the manual approach as well as the geomorphic approach as this takes into consideration the various parameters and reduces the human bias in integration. However, it is important to reduce bias while assigning weightages.

With improvements in GIS technology over the years, an integrated approach to evaluating groundwater potential has become very popular. A vast number of authors have reported the use of

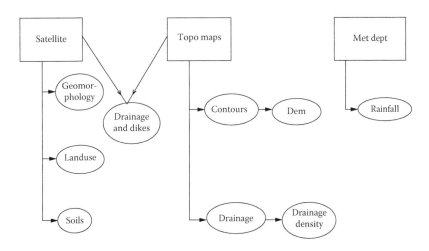

FIGURE 2.4   Sources of data and derivative maps used in geographic information systems study.

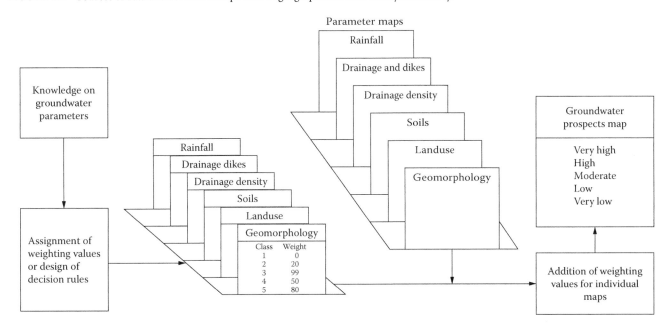

FIGURE 2.5   Simplified remote sensing and geographic information systems–based working model method for the groundwater prospects model.

FIGURE 2.6　Groundwater prospects map derived using remote sensing and geographic information systems.

this technique, or variations of it, in routine groundwater investigations (Sener et al., 2004, Jha et al., 2007, Kumar et al., 2007, Elewa and Qaddah, 2011, Jasmin and Mallikarjuna, 2011, Jagadesha et al., 2012, Lee et al., 2012, Magesh et al., 2012, Bagyaraj et al., 2013, Gumma and Pavelic, 2013, Khodaei and Nassery, 2013).

2.4　Indicators and Interpretation Keys in Unconsolidated Rock Terrain

Unconsolidated sedimentary rocks, also termed as "nonindurated" deposits are composed of particles of gravel, sand, silt, or clay size that are not bound or hardened by mineral cement, by pressure, or by thermal alteration of the grains. Included in this category of rocks are nonindurated sediments of fluvial, aeolian, lacustrine, and marine origin. There is great variation in textural composition, depending on the provenance and conditions of transport and deposition, which is reflected in the wide range in the porosities, specific yields, and permeabilities of the rocks and in the wide differences in the yields of wells. A porosity of 20%–40% is common except for five grained sediments like clays. The specific yield ranges from negligible in clays to about 30% in coarse-textured homogenous sediments. The hydraulic-conductivity values range from less than 1 m/day to as much as 200 m/day (Karanth, 1987).

Fluvial deposits are the materials laid down by physical processes in river channels or on flood plains. The materials are also known as "alluvial deposits." The following sections deal with fluvial materials deposited by nonglacial and glacial environments and materials that are transported by wind, followed by remote sensing parameters in these terrains.

2.4.1　Alluvial Aquifers

The term alluvium is widely used to describe terrestrial sediments of recent geologic age deposited by flowing water. The sediments are composed of clastic material of greatly varying grain size. If the particles are of a fairly uniform size, the material is said to be "well sorted"; if particle sizes are distributed over a wide range, the material is said to be "poorly sorted." A well-sorted material contains better porosity and permeability. Aquifers in alluvial deposits are very common and constitute in many regions the only exploitable source of groundwater. They can be classified according to the environment of deposition into (1) alluvial fans and piedmont deposits, (2) valley fills, (3) alluvial plains, and (4) deltaic terrains.

2.4.1.1　Alluvial Fans and Piedmont Deposits

Alluvial fans form, where a stream leaves its inclined mountain tract and enters the plain, dumping most of the sediment load, because of the sudden decrease flow velocity. The accumulation of a great mass of material forces the river from its point of emergence from the mountains into frequent changes of course, into various directions. Thus, its sediment load is spread over a fan-shaped area. Alluvial fans, distinct near the point of emergence of the valleys from the mountain belt, tend to merge further downstream. The resulting complex of coalescent alluvial fans is often called "piedmont." The continuous belt of fans is also called "bazada."

The sediments of alluvial fans are composed of particles of all sizes, from large boulders and blocks to clay with greatly differing degrees of sorting. The coarsest materials are found near the mountain border, generally mixed with finer fractions, and particle size diminishes toward the lowlands. Stratification is very imperfect in the upper ranges of the piedmont belt. Units of similar lithological composition and sorting are lens shaped in a section across the fan and string like in the direction of the river channels. It is difficult, therefore, to extend stratigraphic correlations over any appreciable distance. In addition, each period of high river flow truncates part of the previous sediments and deposits them further downstream. In the lower part of the alluvial fans, more continuous and better sorted layers are present, and the part of the finer-grained materials increases. Thus, the thickness of the aquiferous beds in a given section is reduced, and simultaneously confined aquifers are formed. A typical section through an alluvial fan is given in Figure 2.7.

2.4.1.1.1　Exploration Parameters in Alluvial Fans and Piedmont Deposits: Humid Regions

Groundwater in alluvial fans is replenished mainly by percolation of river water that may reach a remarkable rate over relatively short stretches, especially near the mountain border. The water may reappear in the form of springs and seepages around the toe of the fan, or it may continue its subsurface flow toward more distant downstream areas. Relatively deep drilling in the downstream part of the fan taps confined aquifers, because of the interstratification if aquiferous and confining beds.

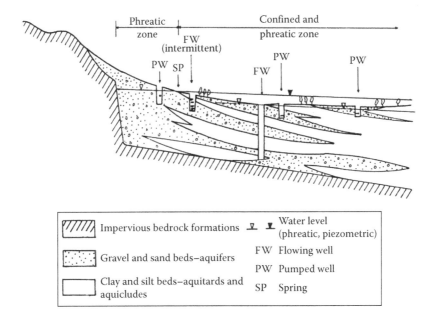

FIGURE 2.7 Typical section through alluvial fan. (After Mandel, S. and Shiftan, Z.L., *Groundwater Resources Investigation and Development*, Academic Press, New York, p. 269, 1981.)

The hydrologic properties of alluvial fans depend on the physical and chemical nature of the constituent rocks. Fine and plastic components may lose their primary porosity by compaction. Calcareous rock debris can be transformed into a compact breccia or conglomerate by alternating solution and precipitation of carbonates. Where the rock material and the climate are conducive to mud flows, fans may be so rich in clay-sized particles as to be practically impervious.

The "Bhabar" belt along the northern fringe of the Gangetic alluvium typifies composite fan deposits that owe their origin to torrential streams emerging out of the Himalayan foothills. The Bhabar belt merges southward with the "Terai" the junction marked by a spring line that separates the flowing well area to the south from the recharge area toward the north.

2.4.1.1.2 Exploration Parameters in Alluvial Fans and Piedmont Deposits: Arid Regions

In arid regions, alluvial fans attain larger dimensions and are more conspicuous as landscape forms, than in humid regions, and also play a more important role as potential aquifers. Most of the infiltration of floodwater takes place along the braided channels of the major streams in the upper and middle reaches of the bazada belt. In the lower parts, percolation is impeded by the presence of greater amounts of clay and silt in the near-surface sediments. Groundwater is phreatic below the higher parts of the piedmont belt, where the water table is often at depths that make exploitation unattractive. In the downslope direction, the water table gradually approaches the surface and finally intersects it as evidenced by the appearance of springs or seepages. In the middle and lower parts of the piedmont belt, groundwater becomes progressively confined because of the increasing number and thickness of semipermeable and impermeable layers and the simultaneous wedging out of the aquiferous and gravel beds.

Some of the fine-grained beds may stem from ancient playas and can be the source of excessive groundwater salinity.

At the lower end of the piedmont belt, or below the adjoining playa or mud flat, groundwater moves upward through semiconfining layers and appears on the surface as dispersed seepages, or occasionally, as more centralized springs. Most of the emerging groundwater is used up by the vegetation or evaporates on the surface.

Best results from the well drillings are to be expected in the middle part of the piedmont belt where the water table is not too deep, a fair amount of aquiferous beds can be expected, and salinity may still be relatively low. The surface features of alluvial fans do not indicate the extent of possible aquifers.

2.4.1.2 Valley Fills

Valley fills are alluvial and colluvial deposits that lie between mountain ranges or between exposures of hard rocks. Coarse materials comprising of sand, gravel, and pebbles form the bulk of the valley fills, which are among the most productive of aquifers. Within the outcrop region of hard rocks, valley fill deposits consist of the coarsest grains, the sediments becoming finer as the distance of transport increases.

Typically, the deposits are characterized by basal gravel and pebbles, succeeded upward by finer materials. Depending on the geologic history, sedimentation may be cyclic with repetition of the gravel–sand–clay sequence. The width of the valley fills ranges from a few tens of meters to tens of kilometers. The thickness of the valley fill varies widely depending on the configuration of the basal rocks and the land surface profile.

In hard rock terrains, the areal spread of the valley fills follows, more or less, a sinuous course of the present course of the stream, lying sometimes to one side or the other, the present ancient courses being the same at places. Usually, chances of the valley fill being thick are remote if the present-day channel deposits comprise coarse-textured materials like pebbles and boulders.

2.4.1.3 Alluvial Plains

Alluvial plains are built up by plastic material deposited by meandering or braided rivers. Of the large amount of sediments carried by streams, the coarsest and most permeable fractions are deposited along the stream channels, while the finer ones are deposited on the flood plains and backswamps. The coarsest-grained gravels and sand make up the traction load of present and ancient buried stream channels. On the inside of meander bends, coarse-grained "point bar" deposits are formed. "Natural levees" often flanking the channels are generally built up by fine sand and silt. During the high-water stages, a slowly moving sheet of water covers the flood plain and deposits silt and clay. Coarser material may reach the flood plain when the levees are pierced during floods.

According to a schematic concept of depositional history, coarser-grained materials should prevail in the upstream part of the plain and also in older, deeper layers that were deposited during early, more vigorously erosive phases. Periodic subsidence of the depositional basin leads to the accumulation of huge thickness of alluvium, as exemplified by the Indo-Gangetic trough that contains over 1000 m of alluvial sediments. The thickness of individual beds of sand, clay, etc., may range from less than a meter to over 100 m. The percentage of coarse granular horizons gradually decreases in the downstream direction. The typical topographic forms and deposits of broad floodplains of large rivers are shown in Figure 2.8.

Though most of the valley deposits have a simple vertical succession from coarse sands and gravel near the bottom of the channels to silts and clays at the top, the relative thickness of the coarse and fine units depends on the type of sediments carried by the river and the geologic history of the river and the point of interest. It is practical to visualize a large alluvial plain as a complex of more or less lens-shaped elongated bodies—or continuous layers—of gravel, sand, silt, and clay including various mixtures of these components. Figure 2.9 shows such a typical section through an alluvial plain and the successful and the unsuccessful wells depending on the sections pierced.

Braiding and meandering of streams give rise to the formation of thick and extensive granular horizons that form productive aquifers. Superimposition of backswamp deposits over meander belt deposits results in the confinement of aquifers by the fine-grained sediments like silt and clay. Cross-bedded sand, which is commonly fine- or medium-grained with a variable content of silt and clay, is deposited on the levees and flood plains that form good aquifers. Point bars, with coarse sands and gravel also hold favorable hydrologic properties.

Because of a great variation in the nature and thickness of aquifers and confining layers, groundwater occurs under confined, semiconfined, and unconfined conditions. Under favorable hydrogeological situations, artesian-flowing conditions may be encountered in shallow wells in low-lying river terraces.

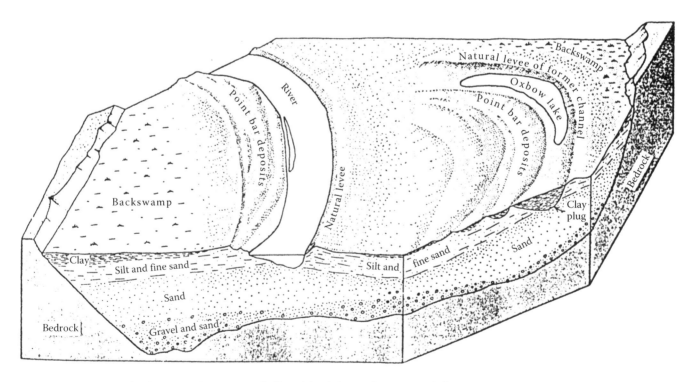

FIGURE 2.8 Topographic forms and deposits typical of broad flood plains of large rivers. (After Davis, S.N. and De Wiest, R.J.M., *Hydrogeology*, John Wiley & Sons, New York, p. 463, 1966.)

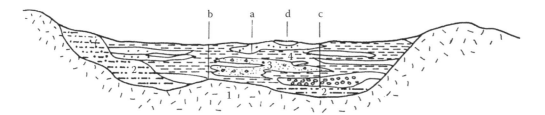

FIGURE 2.9 Section through alluvial plain. (After Mandel, S. and Shiftan, Z.L., *Groundwater Resources Investigation and Development*, Academic Press, New York, p. 269, 1981.) (1) Bedrock; (2) old consolidated terrace deposits; (3) gravel, sands, and silt; and (4) clay. (a) Shallow well, (b) unsuccessful well, (c) successful well, and (d) present river channel.

2.4.1.4 Deltaic Terrains

Deltaic deposits are formed where a river dumps much more material than the sea currents are able to sweep away. Not all rivers are capable of building deltas. Sediments consist predominantly of fine sand and silt. Gravels reach delta area only in rare cases. Coarser sand and sandstones originate from beach sands, dunes, bars and banks, and river channel deposits. Clay and silt are deposited in tidal flats and shore lagoons. As a result of a variety of depositional environment over short distances and time, sedimentary units are lens shaped and discontinuous. Therefore, the appraisal of subsurface conditions presents the same difficulties as on alluvial plains. A typical section through a delta is shown in Figure 2.10.

One of the important features of coastal deposits is the occurrence of groundwater in interbedded alluvial and marine sands, silts, and clays deposited under beach, lagoonal, estuarine, and marine environments. Coarser deposits of gravel and pebbles are commonly found where youthful streams having rocky catchments debouch into the sea. Due to differential compaction, coastal sediments attain a seaward dip. Because of the abrupt reduction in the velocity of streamflow near the coast, the junction between coarse- and fine-grained deposits may be rather abrupt. Palaeodistributaries, particularly from youthful streams, often contain coarse-grained sands.

2.4.1.4.1 Exploration Parameters in Deltaic Terrains

The distribution of freshwater aquifers is controlled by the dynamic equilibrium between hydrostatic heads in freshwater and saline water zones, influx of seawater into streams and a lagoons, and relative movement of the sea in respect to the landmass.

Brackish water may be encountered even in shallow boreholes and in locations far from the sea. On the other hand, the occurrence of freshwater under certain artesian conditions, even below shallower patches of saline water, is not uncommon. It is due to the infiltration into truncated delta deposits that become confined in a downstream direction, the palaeodistributary channels are known to carry fresh groundwater in an otherwise brackish area.

2.4.2 Glacial Terrains

The retreat and advancement of glaciers due to climatic changes have given rise to a complex and often uncertain distribution of beds of sand and gravel. The coarse fractions are generally found close to the ice front or along channels of large streams. Rock debris chiefly of glacial origin is called glacial "drift." Drift includes "till," a highly heterogeneous and mostly nonstratified material consisting of bolder clay deposited directly from ice. The thickness of glacial deposit varies between several meters and several tens of meters, but rarely exceeds 100 m. They often form elongated ridges called moraines.

Glacial till was the most abundant material that was deposited on the land surface during Pleistocene time. In the Precambrian shield region, till is generally sandy, with variable amounts of silt and little clay. Sand till forms local aquifers in some areas. In the regions of sedimentary bedrock, glacial till has considerable silt and clay and therefore low permeability.

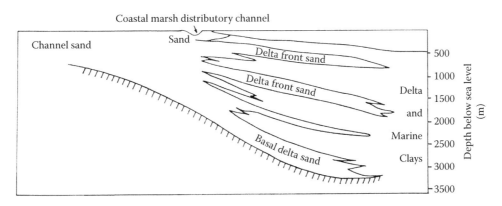

FIGURE 2.10 Hypothetical section through a delta. (From Mandel, S. and Shiftan, Z.L., *Groundwater Resources Investigation and Development*, Academic Press, New York, p. 269, 1981.)

Streams that issued from the ridge of melting ice masses picked up big loads of unconsolidated glacial sediments, dumping coarser constituents at some distance downstream. These outwash gravels occur in the form of outwash fans and outwash terraces and constitute shallow but useful aquifers. Their thickness rarely exceeds several tens of meters. Frequently, the outwash gravels are intimately connected with recent river gravels. In some places, the melt water was dammed up by moraines, thus forming lakes in which delicately layered, fine-grained sediments were deposited. These are aquitards or poor aquifers.

Glacial deposits also express themselves in various typical forms. "Eskers" or winding ridges composed of poorly sorted sand and gravel. They are the most distinctive of the various landforms composed of ice-contact deposits. Eskers are the bed load deposits of former streams that occupied subglacial ice tunnels or, less commonly, streams on the ice surface. Most eskers are formed during the stagnant or near-stagnant phase of glaciation.

"Kame terraces" are formed by the accumulation of glacial debris along the margins of stagnant glacial ice that remains in valley areas. When the ice melts, the debris is left as a terrace along the sides of the valley. The part of the terrace that was once in contact with the ice is strongly affected by the collapse and slumping thus forming irregular borders facing the centers of the valleys. Small patches of ice-contact material may be let down when the glacial ice melts. Small hills formed in this way are known as "kames." Some kames also may be simply erosional remnants of larger masses ice-contact deposits. They consist of poorly sorted sand and gravel.

"Kettle holes" are formed by the collapse of till and ice-contact sediments as isolated masses of residual ice melt. Kettle holes may be found on kame terraces, wide parts of eskers, till plains, and terminal moraines.

2.4.2.1 Exploration Parameters in Glacial Terrains

Location of areas favorable for groundwater development in glacial terrains can be very difficult. In the relatively younger deposits, much of the original topographic expression is retained and therefore is mappable. Groundwater prospection in glacial terrain is concerned with the location of outwash gravels and buried channels. Most of the forms such as moraines, eskers, and kames contain poorly sorted materials and are not favorable.

Aquifers are known to occur as extensive blanket bodies or channel deposits in the surface of buried valleys. The deposits of sand and gravel in buried valleys form aquifers that are generally many tens of kilometers long and several kilometers wide. In many cases, there are no surface indications of the presence of the buried valley aquifers. The overlying till is usually some tens of meters thick or less but occasionally may be of the order of a 100 m thick.

2.4.3 Aeolian Deposits

Materials that are transported and deposited by wind are known as aeolian deposits. The physical process leading to the deposition of aeolian or wind-borne sediments includes deflation whereby soil cover is stripped from surfaces of rock formations and transportation of soil particles through air. Aeolian deposits can be divided into two types: dune sand and loess. The coarser and heavier particles are carried closest to the ground and get deposited as dunes, while the finer particles move further up in the air and get deposited as loess several kilometers away from the source.

Sand dunes form along coasts and in inland areas such as in deserts where the rainfall is scanty and surface sand is available for transportation and deposition. Aeolian sand is characterized by a lack of silt or clay fractions, by uniform texture with particles in the fine- or medium-grain-sized range and by well-sorted rounded grains. The sands are quite homogenous and isotropic. The sorting action of wind tends to produce deposits that are uniform on a local scale and in some cases quite uniform over large areas. Loess, on the other hand, is generally a silt-sized material that like dune sand is well sorted, but unlike dune sand will vary widely in grain size. Because of small amounts of clay and calcium carbonate cement that are almost always present, loess is slightly to moderately cohesive.

2.4.3.1 Exploration Parameters in Aeolian Deposits

Aeolian sands have porosities between 30% and 45% and moderate permeability. Loess deposits have a porosity 40%–50% and low permeability. Loess is not commonly an aquifer because of its low permeability and because where its permeability is the highest it is usually in the high topographic positions where subsurface drainage is good.

Loess deposits in arid regions are cemented to a varying degree to calcium carbonate, causing a reduction in porosity and permeability. Under favorable conditions of topographic and water table slopes, extensive cemented zones may give rise to sufficient confining pressure in the lower zones to cause flowing conditions in wells. In the valleys covered loess and bordered by hills composed of quartzite or other resistant rock types, productive boulder-gravel aquifers occur in basal sections of the valley fill.

Interdunal depressions form receptacles of relatively impermeable material like silt and clay washed down during periods of rainfall. The evaporation of water accumulated in the depressions precipitates calcium carbonate and gypsum in the form of impervious pans. The impervious layers may retain percolating water to give rise to perched groundwater zones occurring above the main water table. Isolated dunes amidst hard rocks may locally give rise to productive sandy aquifer. Also, aeolian action may shift stream courses leaving behind abandoned channels that may form potential aquifers.

2.4.4 Remote Sensing Parameters in Unconsolidated Rock Terrain

From the foregoing paragraphs, it is seen that in an unconsolidated rock terrain, the various geomorphological units and forms have the typical hydrologic properties. Therefore, the first task in the exploration of these terrains is the mapping of these units and forms. As morphology is essentially a surface expression and since remotely sensed data directly provide surface information, the task is accomplished through the use of such data.

The basic interpretation keys are provided by the typical spectral response of these units and forms and also their tone, texture, pattern, shape, size, and association. As spectral signature alone is not conclusive enough delineating these features, a computer-aided classification is not ideal. Best results are obtained through visual interpretation of the imagery where the factors such as tone, texture, pattern, shape, size, and association are taken into consideration. While the satellite imagery of medium to high spatial resolution can provide sufficient information needed in regional exploration, high-resolution satellite imagery or large-scale aerial photographs will be required in identifying objects of smaller size useful in local explorations.

Alluvial fan is reproduced on satellite imagery in the form of a triangle with its apex directed toward the hills and the spread out base touching the surrounding the inclined foothill plain. The tone is lighter and the vegetation is sparse. Where the fans coalesce to form bazadas, the triangular shapes are not obvious but the tone is usually lighter. Spring lines at the base marked by vegetation (red in standard FCC) are often seen.

Stream valleys/valley fills exhibit very clear shapes or form. Where the gradient is low, meandering with large meander wavelength and with broad and only slightly incised valleys is seen. Drainage patterns imply lithology and degree of structural control. Drainage density on humid regions and drainage texture in arid regions imply grain size, compaction, and permeability (Moore, 1978). Valley fills support natural vegetation as well as irrigated agriculture. Agricultural crops in nonrainy seasons are known to show a remarkable correlation with valley fills in certain hard rock terrains in India (Seelan, 1980). Underfit valleys are represented by topographically low, elongate areas with ponded drainage or with a stream meander wavelength smaller than that of the flood plain or terraces.

The natural levees are characteristically associated with stream channels where the gradients are low and seen as sharp lines close to the river channels. The tone is usually lighter, but where the deposits are fine-grained, darker tones are noticed. Typical arcuate shapes characterize oxbow lakes and other associated features such as meander scars and point bars. Elongate lakes, sinuous lakes, and aligned lakes and ponds represent old flood plains. Parallel lines of trees in alluvial plains indicate buried channels. Flood plains are distinguishable by their darker tone. River sands are white in standard FCCs. Backswamps appear dark with little vegetation adjoining the flood plains.

The glacial and aeolian deposits are distinguishable by their unique mesorelief forms and shapes. The sand dunes are elongate and arcuate and have a very light tone. The outwash gravels in the glaciated regions occur in the form of fans. Eskers form winding snakelike ridges. In the snowbound areas, anomalous early melting of snow and greening of vegetation show areas of groundwater recharge.

Soil types are usually indicated by tone. Fine-grained soils commonly are darker than coarse-grained soils. Wet soils are darker than dry soils. Salt-affected soils are very light in tone, and in standard FCCs, they appear as white patches. Distinctive types of native vegetation commonly show upstream extensions of drainage patterns, areas of high soil moisture, and landform outlines. Abrupt changes in land cover type or land use imply landforms that may be hydrologically significant but do not have a characteristic shape.

2.5 Indicators and Interpretation Keys in Semiconsolidated to Consolidated Sedimentary Rock Terrain

Depending on the degree of compaction, cementation, and crystallization, consolidated, semiconsolidated, or unconsolidated sedimentary rocks grade into each other. Sedimentary rocks may occur in any stage of consolidation from incoherent granular materials like silt and sand to firmly held granular rock like siltstone and sandstone. While losing primary porosity, they may develop secondary porosity due to fracturing and weathering. Important stratigraphic and structural features that have a bearing on the occurrence, movement, and availability of groundwater are stratification (in marine environment, the sequence of deposition from bottom upward is limestone, shale, and sandstone), lateral gradation, inclination, folding and faulting of strata, and unconformities.

Sedimentary rocks range from a few meters to thousands of meters in thickness and are sometimes traceable over thousands of square kilometers. When interbedded with clay, sandstones comprise a multiaquifer system. If the beds are inclined, wells located along the dip direction will tap different aquifers, and high-pressure artesian conditions may occur along dip slopes. If there is more than one aquifer, the pressure head may increase with the depth of aquifer.

2.5.1 Sandstone–Shale Aquifers

Sequence of alternating sand, or sandstone and clay, or clay is characteristic of many sedimentary successions. Deposition of such sequences takes place in the marine, deltaic, littoral, and arid continental environment. The main difference between such sequence and alluvium or recent deposits is the greater age and, hence, the more advanced stage of consolidation. The sandstone and shale sequences, however, show more persistent stratigraphy when compared to recent alluvium. The primary porosity of a layer of sandstone is often strongly reduced by compaction and cementation. Zones of secondary porosity are usually found along bedding plains, joints, and fractures.

Alternating sandstone–shale formations occur under variety of geologic conditions. On some of the continental platforms, they fill vast bowl-shaped depressions or basins and constitute large regional, often confined aquifers. In regions dominated by a normal fold pattern, conditions are also favorable for the formation of regional confined aquifers.

Sandstone is the most productive among the semiconsolidated sedimentary rocks. Shale is formed by compaction of clay sediments, and water is found in porous layers, fractures, bedding plains, and weathered zones.

2.5.2 Carbonate Rocks

Carbonate rocks in the form of limestone and dolomite consist mostly of the minerals calcite and dolomite with very minor amounts of clay. Young carbonite rocks have high porosities, but with increasing depth of burial, the soft carbonate minerals are compressed and recrystallized into a denser, less porous rock mass.

Many carbonite rocks have appreciable secondary permeability as a result of fractures or openings along bedding plains. Secondary openings in carbonite rocks caused by changes in the stress conditions may be enlarged as a result of calcite or dolomite dissolution by circulating groundwater. Although some original pore space may be retained in old limestone, other forms of porosity are more important from the stand point of water production. Fractures and secondary solution openings along bedding plains probably transmit the most water.

2.5.3 Exploration Parameters in Sandstone–Shale and Carbonate Rocks

In gently folded regions and in vast bowl-shaped depositional regions, a fairly complete picture of the aquiferous properties of the sandstone–shale formations can be obtained from surface observations. In regions with more or less horizontal strata where outcrops are rare, the task is, however, difficult.

Firmly cemented sandstones with low porosities will yield water to wells along fractures. The same general guiding principles apply to the location of water in these rocks as they apply to the location of groundwater in crystalline rocks of platonic origin. Most favorable areas for development of groundwater are along fault zones and within thoroughly jointed zones. Better wells will be found in broad valleys and on flat upland areas than on hill crests and valley slopes.

Dense shales, devoid of fractures, are practically impervious and form confining layers or barriers. The contact zone of limestone within the underlying shale bed is usually rendered more permeable than the rest of the limestone as the presence of impervious shale limits downward circulation of groundwater, facilitating dissolution of limestone parallel to the bedding plain. However, formations composed of sandwich fashion of thin layers of carbonate rock with alternating shales offer poor prospects for groundwater. The circulation zones for groundwater within the carbonate rocks in such cases are too narrow to facilitate dissolution of limestone.

Though individual limestone beds can be located through structural and stratigraphic studies, the yields of beds are hard to predict. The flow through large caverns resembles that of a surface stream, turbulent and cascading to different levels. Unless the development of secondary porosity is extensive, there is no regular water table. One well may strike good supplies, while another one close by may be a barren one.

Areas of thick limestone or dolomite wells located in valley bottoms are somewhat better than on valley slopes. Water storage in adjacent alluvium, together with a water table that is closer to the surface, accounts for some of this advantage. Wells drilled

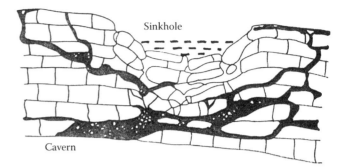

FIGURE 2.11 Sinkhole in limestone. (After Karanth, K.R., *Groundwater Assessment Development and Management*, Tata McGraw Hill publishing company limited, New Delhi, India, p. 720, 1987.)

on broad uplands are also more successful than those drilled on hill slopes. Fractures and solution openings are more abundant along crests of anticlines and within synclinal troughs than they are on the flanks of the folds.

Carbonate rocks are eroded by dissolution in water containing carbon dioxide. The peculiar landscape thus formed is called "karst." Mature karst morphology develops on hard, fissured carbonate rocks under humid to subhumid climatic conditions. It is characterized by the disruption of surficial drainage patterns. River valleys are replaced by arrays of closed depressions (dolines), surface runoff disappears into sinkholes (Figure 2.11), and in extreme cases, the entire rivers flow for some distance in caverns underground. Karst morphology is a strong indication of favorable aquifer zones in limestone.

2.5.4 Remote Sensing Parameters in Semiconsolidated to Consolidated Sedimentary Rock Terrain

Many sedimentary rock formations throughout the world comprise important and extensive aquifers especially those near the surface. Groundwater exploration in sedimentary rock terrains requires accurate geologic mapping as the basic data for exploration for hydrological information such as permeability, storage characteristics, infiltration rates, and specific capacity. Unlike in the unconsolidated rock terrains where geomorphological units and forms play a vital role as remote sensing parameters, here, the lithology and structural information such as folds, faults, dip, and strike play a prominent role. Geomorphology too aids in understanding lithology, structure, land use and land cover patterns, and groundwater recharge conditions. Therefore, remote sensing parameters relating to lithology, structure, geomorphology, and land use are the ones to look for. Standard keys on shapes and patterns can be used while attempting to identify rock types from shapes and patterns on satellite images. However, other factors could be important locally. Previous experience of previous knowledge of the area being interpreted is necessary for good results. Though major rock types could be delineated on satellite images, minor variations within rock types are rarely identifiable.

Landforms and topographic relief often indicate the rock types below. Hard resistant sandstones stand out as serrated and linear ridges, and softer shales and sandstones may form low-lying denudational hills. The outcrop pattern is typically banded type for these sedimentary rocks, often outlined by vegetation in some regions.

The shape of drainage basins is a good indicator with the drainage at times running parallel along the lithological contacts. Drainage pattern and density are an important indicator of the underlying rocks. For example, the sudden disappearance of drainage indicates karst topography. The other indicators could relate to relative abundance, shape, and distribution of lakes and types of native land cover. Tones are difficult to describe as similar rock types may have different tones and vice versa depending on the cover types. However, relative tone variations within a given area can be good indicators of varying lithology.

Color composite images probably are excellent for identification of rock types although drainage patterns are seen best in many areas in IR black and white images (band 4 in Landsat TM and in IRS). Each of the black and white bands has been reported by various authors to be superior to the other bands for lithological delineations; the best single band probably is determined by factors such as atmospheric conditions, time of the year, and type and amount of vegetation cover. The band 7 in Landsat TM is typically meant for lithological discrimination. Rock types are thermally different from each other and the thermal band of TM can be used for this purpose. Little geologic information is lost while making a color composite, and the ideal TM band combinations for lithological discriminations are 2, 4, 7 and 2, 6, 7.

Folded patterns and dip/strike directions are important for understanding the groundwater flow regime. Curved patterns on satellite imagery are indicative of folded beds. Cuestas and hogbacks and asymmetric ridges and valleys are indicative of folded beds. The topography is usually flat on dip slope and irregular on back slope. The vegetation distributed uniformly in dip slope is banded and parallel to ridge crest and back slope. Bazadas are formed on dip slope, and separate alluvial fans are formed on back slope. Trellis, radial, annular, and centripetal drainage patterns are indicative of folded structures. The other indicators are major deflections in drainage channels, change in meander wavelength, or changes from meandering to straight or braided patterns.

Lineaments are indicated by continuous and linear stream channels, valleys, and ridges and elongate or aligned lakes. Identical or opposite deflections in adjacent stream channels, valleys, or ridges and alignment of nearby tributaries and tributary junctions are also indicative of lineaments in sedimentary terrains. Other indicators include alignment of dark or light soil tones, elongate or aligned patterns of native vegetation, and thin strips of or relatively open or dense vegetation.

Folds that may concentrate or block groundwater flows can be detected and delineated with relatively accurate results on satellite images. Vegetation patterns are seen best in color composite images, but IR images may be best for delineating drainage patterns. Vegetation patterns are prominent on green and red bands as well as on color composite images of green, red, and IR. Vegetation patterns can be distracting to the eye when the image is being interpreted for other features. Though vegetation has high reflectance in IR, the vegetation patterns are least obvious in black and white IR images enabling interpretation of other features.

2.6 Indicators and Interpretation Keys in Volcanic Terrain

Volcanic rocks include basalt, rhyolite, and tuff. Commonly, they consist of several successive flows of variable thickness and lateral extent. A typical flow unit consists of a lower dense and massive horizon, passing upward into a vesicular, amygdaloidal, or jointed horizon. The distinctive geohydrogeological feature of volcanic rocks is the significant primary porosity on the form of vesicles, lava tubes, and occasional tunnels formed due to escape of gases. Secondary porosity is developed due to fracturing during cooling of the lavas, tectonic disturbances, and weathering.

2.6.1 Typical Profiles in Volcanic Terrain

From the viewpoint of groundwater occurrence, three principal types of volcanic terrains can be distinguished: basalt plateaus, central volcanic edifices, and mixed pyroclastic–lava terrains.

Basalt plateaus are the result of repeated effusions of low viscosity that have issued from numerous fissures and sometimes contain the conspicuous mineral olivine. Preexisting relief tends to be leveled out and transformed into a flat morphology, or into a steplike one, if successive flows terminate at different distances from the erupting fissures. Clay-rich soils that form by weathering between eruptions, or sands, are sometimes found between successive sheets. The original continuity of basalt plateau is often disrupted by deeply incised valleys cutting into underlying formations or by tectonic disruption, into fault blocks (Figures 2.12 and 2.13). Some volcanic plateaus are built up by ignimbrites–ashes erupted in an incandescent state and welded together by heat.

Central volcanic edifices are cone-, dome-, or shield-shaped volcanoes composed of lava flows and layers of volcanic ash and coarser materials ejected in the solid state. The surrounding lowlands are often covered with thick accumulations of volcanic ash and fine-grained pumice. Only the most copious lava flows reach large distances from the center of eruption (Figure 2.14).

Mixed lava–pyroclastic terrains of regional extent stem from prolonged periods of intensive volcanic activity. If older than Pleistocene, the morphology is often one of the maturely dissected hills that bear little resemblance to the original landscape forms. Where a volcanic activity has persisted into the Pleistocene, better preserved volcanic features are superimposed on the ancient landscape.

Depending upon the period of eruption, volcanic rocks have buried preexisting rocks ranging from the Archaean to the Quaternary period. At places, they are known to rest on highly productive aquifers of sedimentary origin. Flows, nearly conformable to stratification, form effective confining layers (Figure 2.15a and b). Alternating sequences of previous compact

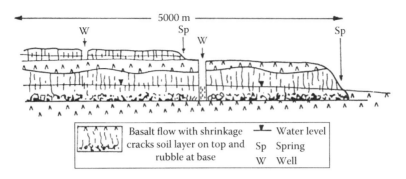

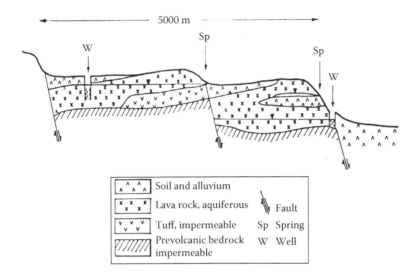

FIGURE 2.12   Groundwater in lava plateau. (After Mandel, S. and Shiftan, Z.L., *Groundwater Resources Investigation and Development*, Academic Press, New York, p. 269, 1981.)

FIGURE 2.13   Groundwater in volcanic, tilted block region. (Mandel, S. and Shiftan, Z.L., *Groundwater Resources Investigation and Development*, Academic Press, New York, p. 269, 1981.)

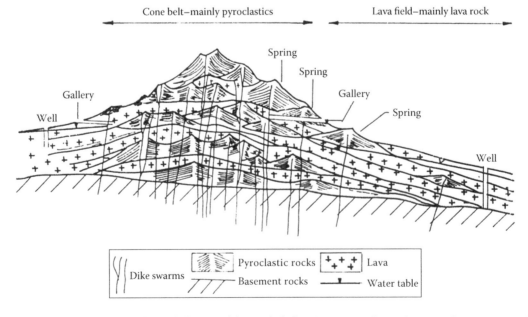

FIGURE 2.14   Groundwater in a central volcano. (After Mandel, S. and Shiftan, Z.L., *Groundwater Resources Investigation and Development*, Academic Press, New York, p. 269, 1981.)

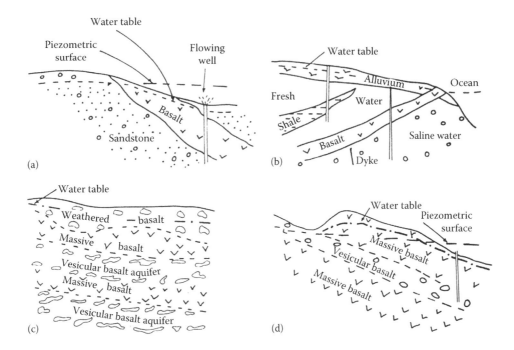

FIGURE 2.15 Basalt as an aquiclude, barrier, and aquifer. (a) Basalt as a confining layer, (b) basalt as a barrier, (c) multi-acquifer system in basalt, (d) confined condition in vascular basalt. (After Karanth, K.R., *Groundwater Assessment Development and Management*, Tata McGraw Hill publishing company limited, New Delhi, India, p. 720, 1987.)

horizons function as a multiaquifer system (Figure 2.15c). If the flow dips at angles gentler than the land surface slope, artesian conditions may result with sufficient pressure to cause free flow in wells (Figure 2.15d).

2.6.2 Groundwater Occurrence and Exploration Parameters in Volcanic Terrain

Volcanic rocks have widely varying hydrologic properties, making predictions about groundwater possibilities uncertain. Some lava flows contain excellent aquifers; others are practically impermeable. The porosity of volcanic rocks varies widely from almost negligible values for dense basalt to over 60% for pumice and other vesicular varieties (Karanth, 1987). Typically, rocks within dikes and sills will have less than 5% porosity, dense massive flow rocks will have values ranging from 1% to 10%, and vesicular volcanic rock will have porosity values ranging from 10% to 50% (Davis and De Wiest, 1966). Although porosity may be high, the permeability is largely a function of other primary and secondary structures within the rock. Many lava flows exhibit vesicular porosity caused by gas bubbles contained in the lava during an eruption. But the pores thus formed are not interconnected. It appears that mainly lavas of fairly recent age (Quaternary, Late Tertiary) are aquiferous, whereas in most of the older lavas, the formation of secondary minerals and the partial disintegration into clay have clogged voids and fissures.

According to Freeze and Cherry (1979), on a large scale, the permeability of the basalt is very anisotropic. The centers of lava flows are generally impervious. Buried soils that produce high permeability develop in the top of cooled lava flows. Stream deposits occur between the flows. The zones of blocky rubble generally run parallel to the flow trend. The direction of the highest permeability is therefore generally parallel to the flows. The permeability is normally greatest in the direction of the steepest original dip of the flows.

Contrary to the belief that the black cotton soils formed on the basaltic plateaus are impervious and help to augment runoff without much percolation to subsurface, there is an exceptionally good percolation down to various depth levels due to varying thickness of soil that covers the weathered zone portion of the lava flows. The percolation is enhanced by the soil profiles that are followed by the underlying horizontally disposed lava flows with step topography on surface and subsurface. An analysis by Dhokarikar (1991) of rainfall runoff indicates watershed having good rainfall nearly 50% of the total precipitation percolates into the ground. In the low rainfall areas, the percolation into subsurface is said to be still higher.

Acid lava rocks containing a high proportion of silica (66% or more) are generally poorer aquifers than the more frequent lavas, such as olivine basalt. Lavas formed by submarine eruptions—so-called pillow lavas—are poor aquifers, or aquifuges, since no voids or fissures are formed because of the rapid chilling and the presence of large amounts of minerals precipitated by steam. Loose pyroclastic rocks (scoria, cinders, pumice, ash) are quite permeable when fresh, but the finer-grained varieties lose much of their permeability through compaction and weathering. Mud flows that owe their origin to torrential rains on steeply inclined soft pyroclastic strata are practically impermeable. They can be discerned by morphologic characteristics and by lack of stratification.

If valleys are near volcanic eruptions, lava will flow down the valleys and bury any alluvium that may be present. Where the valley contains streams from extensive drainage systems, large thickness of gravel may be present, which, on burial, can be important aquifers. Rivers blocked by lava will form lakes that eventually fill with silt, clay, or volcanic ash. These deposits can form important confirming beds for the underlying stream gravel.

In the basaltic plateaus, the hydrologically important porosity of basalt terrains is due to more or less vertical shrinkage cracks and to the essentially horizontal voids and rubble zones left between successive flows. Streams flowing over basalt plateaus frequently lose much water by infiltration. Large springs are formed where the contact between the basalts and an impervious substratum is exposed in incised valleys or along escarpments of the plateau. Smaller springs may issue at various elevations from perched horizons such as ancient soils and tuffs. The presence of major fissures or faults largely determines the occurrence of the larger springs. In some basalt plateaus, permeability of the formation may be as high as in karstic limestone aquifers, and fairly evenly distributed, so that wells stand a fair chance of success where the saturated is thick enough. Under less favorable conditions, well sighting may be guided by the distribution of major fissure systems that are visible on the surface or indicated by hydrologic phenomena.

In the central volcanic edifices, compacted impermeable tuff or soil layers between lava flows and subvertical impermeable dikes divide the subsurface into a number of groundwater compartments, each with its own water level and its own outlet, either into an adjacent lower compartment or into a spring. Thus, the groundwater passes through a number of "steps" down the mountainside into the plain (Figure 2.14). At higher elevations, on the mountain slopes, groundwater can be exploited by galleries, and this has been the traditional method in many volcanic mountains and volcanic islands. Groundwater flow is encountered when a dike is pierced. The initial flow tends to decrease with time; the galleries are therefore extended into additional groundwater compartments to maintain the supply. This practice exploits the groundwater steadily infiltrating into the galleries from above but also entails the exploitation of reserves.

On the lower parts of volcanoes at a greater distance from the center of eruptive activity, dikes are less frequent, and the aquifers pass from "compartment" type to the "layered" type, common in sedimentary layered rocks. The accumulations of volcanic ash, consisting of alternating layers of larger and smaller permeability—together with lava sheets or flows—create confined subartesian and artesian conditions. In these parts, groundwater is exploited by vertical wells. In the coastal volcanic regions, a freshwater–seawater interface problem has to be faced. In some cases, impermeable layers may isolate sections of the aquifer from contact with seawater and thus facilitate exploitation.

The best aquifers in the mixed lava–pyroclastic terrains are the most recent lavas, coarse pyroclastic (scoria, cinders), and some alluvial deposits. Fine-grained pyroclastics tend to become little permeable or impermeable through compaction. The alluvial fill of plains and valleys, consisting of gravel, sand, and clay

of volcanic origin may present important groundwater possibilities, but the alluvial deposits are as diversified and variable as those in any region.

2.6.3 Remote Sensing Parameters in Volcanic Terrain

Volcanic terrains pose many challenges to the hydrogeologist during groundwater investigations. They behave like sedimentaries, where layered, like karstic where the water flows through vesicles and hard rocks where they are massive. Groundwater occurs in confined, semiconfined, and leaky conditions. Satellite data are extremely useful in mapping the flows, geomorphic disposition, and the fracture patterns.

In the Deccan Trap regions of India, hard and soft flows occur alternatively on the standard FCC; the harder flows are generally lighter and devoid of vegetation. The softer flows support vegetation and are red in FCC. The flows are mappable where steep slopes occur but are difficult in flatter terrains. Stereo viewing aids in mapping the flows.

Basalts at land surface commonly can be recognized by dark tones on black and white images and by dark (generally gray to bluish gray) hues on color composite images. Discharge of groundwater occurs at or near the edge of the outcrop area. The areas of discharge are indicated by patches of red color (on color composite images) due to the presence of vegetation. The dark tone of the basalt is mainly due to the black cotton soil developed on the basalts.

The dissection pattern is easily recognizable on the satellite imagery that helps in geomorphic zonation of the different plateau. In the highly dissected plateau, the stream channels are easily recognizable and are controlled by fracture systems. The soil formation is least and therefore, the tone is lighter. In the undissected plateau, the thickness of the weathered material is more and is darker in tone. The stream channels in the basaltic terrain are normally lineament controlled and are easily mappable on satellite imagery. Very good groundwater potential is also observed in the alluvial plains of major rivers in the basaltic region. Buried channels/old river courses are also identifiable.

Valley fills are the most important and useful information that is available in the basaltic terrains of semiarid regions. On the FCC images, it is seen as distinct red-toned patches invariably following the valleys. The presence of these red patches suggests the growth of dense, healthy, luxuriant vegetation. Survival of dense pockets of vegetation during dry periods will suggest the availability of groundwater. The confinement of such pockets mostly along the valleys will indicate the presence of thick alluvial, colluvial, and weathered material.

2.7 Indicators and Interpretation Keys in Hard Rock Terrain

Locating drilling sites in hard rock terrains is considered to be one of the most difficult tasks in groundwater investigations. Extreme variations of lithology and structure coupled with highly localized water-producing zones make geological and geophysical exploration difficult. The percentage of unsuccessful wells is usually the highest in hard rock areas.

Hard rock terrains comprise a great variety of igneous and metamorphic rocks. But from the hydrological point of view, they are rather homogenous in two respects. They have virtually no primary porosity as compared to sandstones and other sedimentary rocks, but they have a secondary porosity due to weathering and fracturing. The storage and flow of groundwater are restricted to these zones. The general term "hard rocks" is used to describe such igneous and metamorphic rocks. The most common hard rocks are gneisses and granites.

Hard rock is a very general term used in hydrogeology for all kinds of igneous and metamorphic rocks, typical of the shield areas of the Earth. For the purpose of this chapter, the term hard rock excludes volcanic and carbonate rocks as these rocks can have primary porosity and as these have been dealt separately in earlier sections.

The weathering processes have considerable influence on the storage capacity of hard rocks. Mechanical disintegration, chemical solution, deposition, and the weathering effects of climate and vegetation bring about local modifications of the primary rock and its fractures. This action can imply either an increase or a decrease of the secondary porosity of the original fracture pattern of the rock. The transition zone between the weathered layer and the underlying fresh rock can function as a reasonably good aquifer, depending upon the porosity of this zone.

The storage capacity of the unweathered hard rocks below the weathered zone is restricted to the interconnected system of fractures, joints, and fissures in the rock. Such openings are mainly the result of the tectonic phenomenon on the Earth's crust. Figure 2.16 shows a typical subsurface profile in crystalline rocks.

2.7.1 Weathered Hard Rocks

Large regions of the continents are directly underlain by extensive batholiths of granite or by metamorphic complexes of gneiss, schist, quartzite, slate, or other metasediment, metavolcanic, or associated igneous rocks. The groundwater contained in the weathered layers of these rocks is commonly tapped by thousands of wells for supply to villages, farms, and livestock. Also, groundwater discharge from the weathered layer sustains the flow of springs and in the more humid regions the dry period baseflow of the streams.

2.7.1.1 Typical Weathered Layer Profile

The weathering process can be commonly grouped into three broad categories: physical or mechanical, chemical, and biological. The thickness, areal extent, and physical character of the weathering vary from place to place and depend on the nature of the process. Though there are variations, a typical weathered profile can be summarized as follows:

Zone (a): Sandy clays or clay sands often concretionary and generally only a few meters thick

Zone (b): Massive accumulation of secondary minerals (clay) (with its thickness that may reach up to 30 m) and high porosity but low permeability

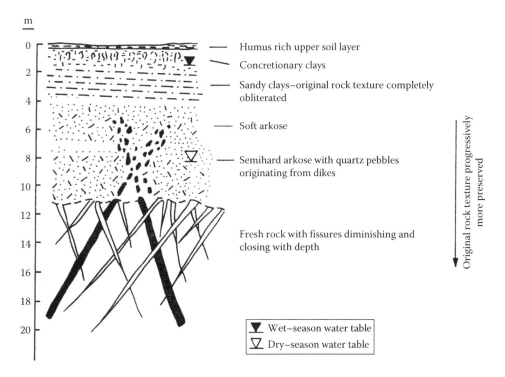

FIGURE 2.16 Profile of subsurface in crystalline terrains of intertropical belt. (After Mandel, S. and Shiftan, Z.L., *Groundwater Resources Investigation and Development*, Academic Press, New York, p. 269, 1981.)

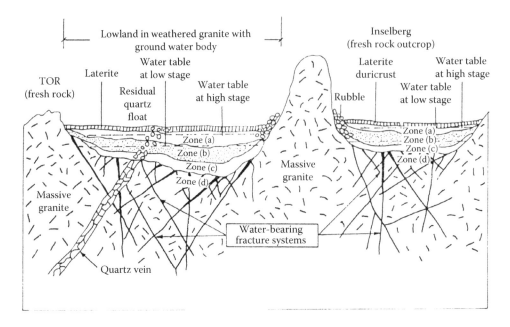

FIGURE 2.17   Idealized cross section through a weathered hard rock terrain. (After Larsson, I., *Groundwater in Hard Rocks*, UNESCO publication, p. 228, 1984.)

Zone (c): Rock that is progressively altered upward to a granular friable layer of disintegrated crystal aggregate and rock fragments and may range from few meters to 30 m in thickness

Zone (d): Fractured and fissured rock, may range from a few tens to several scores of meters in thickness, and low porosity but moderate permeability within the fracture system

A typical cross section of a weathered zone is shown in Figure 2.17.

2.7.1.2 Groundwater Potential in Weathered Zones

In general, the thicker and larger weathered zone, the more productivities as an aquifer. Thin weathered zones are good aquifers only where there is good prevailing recharge, either natural or artificial. But even thin weathered zones act as excellent recharge zones for the deeper fractured systems. Many productive wells tap aquifers in Zone (c) of the weathered profile that averages about 10–20 m in thickness. Most open wells in the Indian shield area, for example, exploit the groundwater in the zone. Where the weathered zones are thin or absent, groundwater in the fracture systems is tapped through bore wells.

2.7.1.3 Exploration Parameters in Weathered Zones

Topography and stage of geomorphic evolution are important in understanding the development of extensive weathered zones. The weathered zones are commonly most extensive and thickest in erosional peneplains of low relief at or near base level where local relief is only a few meters and the slope of the land surface is less than 10%, for example, in the Indian peninsula. Erosional residuals such as tors and inselbergs commonly

take up to about 10%–15% of the gross area of these terrains in the old-age stage of the geomorphic cycle but as much as 50% or more in late mature stage. The residuals are generally devoid of weathered layers and fresh rock outcrops at the surface. The inselbergs may rise as much as 100–500 m above surrounding lowland plains, while the tors may rise only a few meters or few tens of meters above the plains. The weathered layer profile, important from the groundwater point of view, is developed only in the lowland plains between the bare erosional residuals.

The spacing and distribution of fracture systems in the host rocks are also highly important factors in the development of the weathered layer. In granite rock terrains, for example, where the fracture systems are closely spaced, weathering agents may penetrate deeply into the lost rock to form thick weathered layers with permeable zones. On the other hand, the massive and poorly fractured granite resists weathering and forms erosional residuals that may rise from few meters to 100 m or more above intervening lowlands of deeply weathered rock. By this action, discrete groundwater bodies are formed in the weathered layers of the lowland areas separated by unweathered uplands of fresh rock. The lowland plains are usually undulating in nature; the valleys often indicate underlying fracture systems.

2.7.2 Fractured Hard Rocks

The Precambrian shields are among the oldest parts of the Earth's crust. They contain hard rocks of different age, grade of metamorphism, and structure. Many orogenic movements have affected the shields. Faulting processes have had different influences on the rocks of the shields due to differences in strength of the individual rock types. Some rock types are extremely

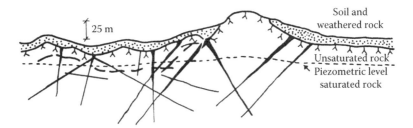

FIGURE 2.18 Typical water-bearing fracture zones in the hard rock. (After Larsson, I., *Groundwater in Hard Rocks*, UNESCO publication, p. 228, 1984.)

fractured, while others are almost undisturbed, even though they belong to the same tectonic environment.

The strength of the rock or its resistance to brittle failure in its crust is rather a complicated matter. Petrographical parameters are involved, that is, grain size, grade of metamorphism, fold structures, and direction of fold axis versus stress orientation. These parameters play a dominant role in rock fracturing, and they are indirectly related to the occurrence of groundwater in hard rocks.

By definition, hard rocks are compact. On the other hand, the fractured pattern of the rocks creates a type of porosity, which is termed fractured porosity. This means that open fractures lying below the water table levels can store water. Figure 2.18 shows typical water-bearing fracture zones in the hard rock.

2.7.2.1 Types of Fractures

From the hydrogeological point of view, three main types of fractures can be identified in hard rock areas, namely, tensile joints, tensile fractures, and shear fractures. Tensile joints have no movement along the sides under undisturbed conditions. The most characteristic feature of these joints in metamorphic rocks is their "en echelon" layout. That means they are not usually interconnected. At each end, they are very narrow, but open up in the middle indicating their tensile origin. Tensile fractures develop parallel to the direction of compression due to parting and dilation perpendicular to compression direction. Shear fractures are a result of differential movement of rock masses along a plane. They can range in length from many kilometers to tiny fractures of few millimeters in length. If two intersection shear fractures develop under the same stress conditions, they are called conjugate shear fractures.

2.7.2.2 Groundwater Potential in Fractured Hard Rocks

Fracturing may create significant porosity and permeability in the hard rocks and is primarily responsible for their groundwater potential. While the unfractured zones have virtually no porosity, in the fractured zones, it may be as high as 30%. The direction of flow, however, is difficult to establish because of the nature of the openings and their possible relationship to fractured pattern in general.

In areas where tensile joints are predominant, the storage capacity is very low due to poor interconnection between the fractures. But in areas where tensile fractures are dominant, the storage capacity is usually high. Here, the fractured systems function as large drain pipes collecting water from minor fractures belonging to the same fracture system. The storage capacity of the shear fractures is a very complex phenomenon. Heavy fracturing of hard rocks is commonly followed by intense weathering. The chemistry of the rocks plays a major role in this process. If two or more thrust faults cut each other, an axis of intersection develops, which can act as an effective drain pipe.

2.7.2.3 Exploration Parameters in Fractured Hard Rocks

The main problem for the hydrogeologist underrating groundwater exploration in hard rock areas is to find fracture pattern with maximum storage capacity. In general, shear zones and tensile fractures would offer good potential. The storage capacity also varies with different rock types. Acid intrusive rocks such as granite, granodiorites, aplites, and pegmatites have a high storage capacity, as they are brittle rocks from a hydrogeological point of view. Fine-grained rocks are generally good aquifers. They have a characteristic type of narrowly spaced fractures.

Pegmatite intrusions are generally very brittle and therefore highly permeable. The vital point is the grain size. The more coarse grained, the more brittle is the pegmatite. The more brittle the rock, the bigger is the potential yield of groundwater. Basic intrusive rocks such as diorites and gabbros have in general low storage capacity. Basic rocks can be considered in field terms as tough rocks and therefore are poor aquifers.

Basic dikes constitute rather poor aquifers because of the weak interconnection between the dike and the country rock. However, the boundary zone between the dike and the country rock often contains open fractures with a high storage capacity. This characteristic usually results from thermal shrinkage at the time of cooling of the dike; consequently, open spaces develop between the dike and the country rock. The fine-grained boundary zone between the two rocks is generally more fractured than the interior part of the dike.

Dikes have another characteristic that may have local importance for groundwater. They commonly act as subterranean dams, dividing the rock into separate hydrologic units. If a mountain slope is cut by a set of dikes trending more or less parallel to the contour lines, the dikes will have a damming effect on the groundwater flow causing the development of springs as shown in Figure 2.19.

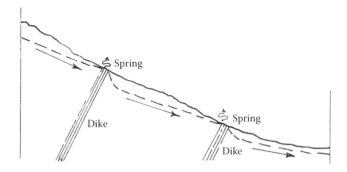

FIGURE 2.19 Damming effect of dikes at the side of a mountain. (After Larsson, I., *Groundwater in Hard Rocks*, UNESCO publication, p. 228, 1984.)

2.7.3 Remote Sensing Parameters in Hard Rocks

As exploration in hard rocks is a very difficult task, any additional clue on the occurrence and movement of groundwater and its recharge conditions is very valuable. Remote sensing provides valuable information on lithology, landforms, lineaments, soils, and land use that control the occurrence movement and recharge of groundwater in hard rocks. Interpretation of lithology and landforms helps in understanding and identifying potential zones in the weathered zones and near-surface conditions. Interpretation and analysis of lineaments help in locating deeper aquifers in the fractured hard rocks. An understanding of soil and land use patterns is crucial in evaluating recharge conditions. Computer-aided classification is possible for soils and land use, whereas lithology, morphology, and lineaments are best done by visual interpretation. As seen in an earlier section, various digital enhancement techniques, however, are useful in enhancing objects of interest before visual interpretation. Remote sensing parameters for interpretation of various geomorphological units such as alluvial fans and valley fills that can also occur on hard rocks are described in an earlier section. The other most important landforms that are indicative of shallow groundwater conditions are inselberg/tor/pediment complexes, pediments, and buried pediments. The rock outcrops in inselberg/tor complexes are readily recognizable on the satellite imagery. The pediment zones within these complexes are characterized by lighter tones. In the case of pediments buried by soil cover, the tone varies with soil cover. The black soil cover is usually represented by a darker tone, while the red soil cover is usually represented by a brighter tone (sometime light yellow) on standard FCCs.

Where the pediment cover is thin or where the declining water levels pose problems, the fracture zones offer the only sustainable source. Satellite remote sensing offers the best possible source for identifying these lineaments. Lineaments are all types of natural straight line features on images. All lineaments are necessarily rock fractures, and they do not necessarily localize groundwater occurrence. On medium-resolution imagery, a few lineaments can be correlated with faults; the physical nature of

most other lineaments must be investigated by indirect means by comparing lineament trends with joint trends. Commonly, residual materials, soils, and vegetation cover many rock outcrops.

Since all lineaments are not necessarily productive, it is imperative to carry out a geophysical survey and test drilling to confirm the usefulness of the lineaments identified on the images. Usually, the productive lineaments follow a certain strike, and therefore, it is necessary to establish the strike direction that offers the best potential.

Many fractures are vertical, and the lineaments may offer favorable locations for drilling in such cases. Most often, the fractures are oblique and have a dip. In such cases, well locations should be offset to intersect the fractures below the water table but at shallow depth (nearly all fracture are progressively smaller at intersecting depths of beyond 100 m). A geophysical survey can help in interpreting the dip amount and direction.

In hard rock areas, many of the stream courses are controlled by lineaments and are being recharged by these streams. This is evidenced by the straight line courses of these streams at these places. Though the streams may deviate after a while, the lineaments can be extended from the straight line courses. Many of these streams are not perennial, and their beds are the potentially favorable sites for test drilling. Lineaments often crisscross each other, and their meeting points are favorable points for exploration. It is not uncommon to see on the satellite images even four or five lineaments converging at a point. Lineaments are picked up more easily on the hilly regions due to their topographic expressions caused by higher erosion along the fractures but are of little significance on the hills where the utilization potential of the groundwater is limited. Such lineaments, when they extend down to the plains, are often covered by pediments and soil cover and are not visible on the images, though the utilization potential is very high on the plains. The lineaments picked up on the hills could be extended down to the plains where further ground surveys could be carried out to confirm their potential.

Though the lineaments are easily identifiable on the images, they are often difficult to identify on the ground. Due to inherent geometric inaccuracies in the satellite images, direct transfer onto the topographic maps does not help in identifying them on the ground. Very careful field investigations across the suspected lineament zones are needed before drilling starts.

Basic dikes and quartz reefs are also linear features that are identifiable easily on satellite images. They do not offer any groundwater potential themselves but act as barriers for the movement of groundwater. A careful analysis of drainage, in conjunction with dikes/quartz reefs, provides a clue as to which side of the feature offers greater potential.

Lineaments express themselves in many ways on the satellite images. It is logical that many fractures that localize the occurrence of groundwater also have an expression at land surface. A fracture that is a plane of weakness for enlargement by groundwater, therefore, may be represented by a topographic depression, a different soil tone, or a vegetation anomaly at land surface.

The lineaments appear on the satellite images as continuous and linear stream channels, valleys, and ridges. Lineaments can also be inferred from elongate or aligned lakes or native vegetation or thin strips of relatively open or dense vegetation. They are also expressed as alignment of dark or light tones soils. Identical or opposite deflections in adjacent stream channels, valleys, or ridges, alignment of nearby tributaries, and tributary junctions are also indicative of lineaments.

The basic dikes appear as dark tones on images. The quartz reefs are usually brighter. The dikes and quartz reefs run for several kilometers at a stretch but disjointed in between. They have a reasonable elevation and, where they are north–south trending, have a distinct shadow zone.

Contrast-enhanced standard FCCs are suited for linear interpretation. Black and white near-IR images are also useful as vegetation patterns are least prominent and distracting on these images. Many lineaments are enhanced by low sun elevation angles.

Distribution of different textural classes of soils is an important factor in interpreting for groundwater recharge in hard rock areas. Fine-grained soils are generally darker than coarse-grained soils and offer less recharge and more runoff relatively. An understanding of land use pattern in hard rock regions is useful in evaluating a particular region for groundwater potential and also in understanding recharge conditions. For example, a forested area or agricultural areas allow more recharge, while a built-up area allows more runoff.

Areas that are already under irrigation by groundwater offer clues on the lineaments and landforms in hard rock areas. Groundwater-irrigated areas often correlate well with old river channels, valley fill zones, and lineaments. Groundwater-irrigated areas can be mapped fairly accurately by visual interpretation of the FCCs where the cropland areas appear pink or reddish pink and are easily separable. To differentiate between areas irrigated by groundwater and surface water sources, good ground information on the existing surface irrigation schemes with their command areas are needed. The surface water bodies are also clearly seen on the images with their command areas immediately downstream. In India, for example, the ideal season for groundwater-irrigated cropland inventory would be in "rabi" (postmonsoon), when the crops generally subsist on irrigation. Images of February/March are ideal as crops by then would develop canopy and are readily identifiable on the images.

2.8 Case Studies and Economic Benefits

The foregoing sections attempted to explain the keys and parameters obtainable from remote sensing for groundwater exploration in various geological terrains and the concept of using the technology in the initial stages of exploration to identify favorable groundwater zones to be followed up by conventional field methods. When remote sensing is used in the beginning stages of exploration to prove water-bearing zones, it not only reduces the cost of exploration but produces a higher rate of success in drilling. This results in cost savings. The following sections describe two case studies where the economic benefits were calculated from the use of remote sensing to target groundwater. These studies have been earlier reported by Seelan et al. (1988) and Seelan (1996).

2.8.1 Case Study I

A remote sensing–based survey was carried out by Seelan et al. (1988) around Sullurpeta town, near the Sriharikota launch station in the coastal area of Andhra Pradesh, India, to augment water supply to the residential quarters of the Department of Space. The town is situated about 80 km north of Chennai (formerly Madras) city along the east coast of India. The Kalangi River flows through the town and meets the Pulicat Lake few kilometers below the town. Pulicat Lake is a saltwater lake with outlets/inlets into the Bay of Bengal. The lake, at sea level, extends to the east and northeast of the Sullurpeta town that has an altitude of 6 m above mean sea level. The terrain is very gently sloping toward the east and almost flat near the lake. The Kalangi River has its distributary system spread over this area. The Kalangi River, draining the western parts of Nagalapuram Hills, passes from the west of Sullurpeta town and continues for about another 10 km before emptying itself into the Pulicat Lake. The Kalangi River is the only major source of water supply to the town, but is dry during the summer months. Geologically, the area consists of a thin layer of deltaic fluvial sediments composed of sand, silt, and clay of various sizes and grades. Below this layer, hard compact marine clay is encountered, but its thickness was unknown. Morphologically, the area forms a delta of the Kalangi River with distributaries draining into the Pulicat Lake. The tidal limit is just a few kilometers below the Sullurpeta town.

A number of earlier systematic investigations have been carried out to identify possible additional sources of water supply to the town. Exploratory boreholes, drilled beyond 300 m by the Central Ground Water Board (Central Ground Water Board, 1978) at a nearby place, showed the formations are predominantly marine clay and the formation waters are brine. The exploration revealed that there is no scope for the development of good quality water at depth excepting tapping the uppermost shallow water table aquifer. Further studies around Sullurpeta by Central Ground Water Board (1983) concluded that the medium to coarse alluvial sands along the Kalangi River, a few kilometers upstream of town, offer potential for exploitation through infiltration galleries. The colony itself suffers from brackish groundwater zones with no perennial surface sources nearby. Accordingly, detailed cost estimates were made for different alternatives for infiltration wells along the Kalangi River, collector wells, pumping mains, and pipelines that worked out to be between $6000 and $9000 (costs during that period) depending upon the alternatives used (Civil Engineering Division, undated). However, before embarking on the civil works, the Civil Engineering Division of the Department of Space commissioned a study using remote sensing to look for nearby sources and cheaper options.

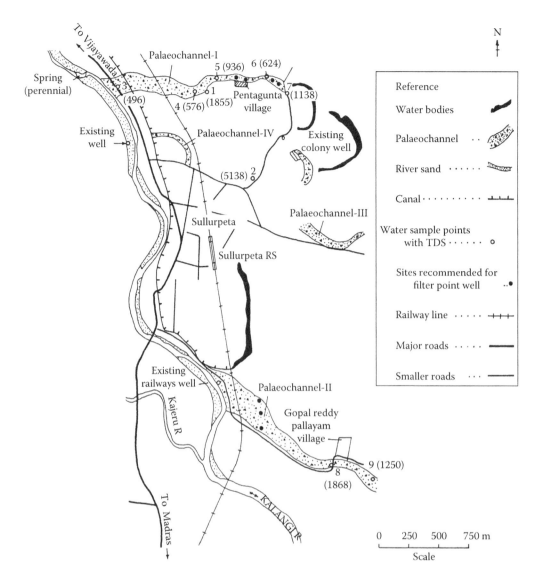

FIGURE 2.20 Map showing palaeochannels around Sullurpeta town, interpreted from aerial photographs on a 1:10,000 scale.

As the area under consideration was too small and the satellite resolutions available at that time were inadequate for local exploration, aerial photographs available for the area on 1:10,000 scale were used in the study. The aerial photographs revealed the presence of four palaeochannels of Kalangi distributaries in the area, one of which within a kilometer from the colony site (Figure 2.20). Having obtained this very useful reconnaissance-level information from aerial photographs, detailed follow-up studies were carried out to prove the availability of fresh potable water within this zone. First, water samples were obtained from the existing shallow open wells from within and outside the palaeochannels. The analysis proved the presence of freshwater within the palaeochannels, while the water was brackish outside the palaeochannels. The increasing salinity along the palaeochannels away from the main Kalangi River showed that the base flow along the palaeochannels is linked to the main Kalangi River, but deteriorated away from the Kalangi. As a next step, geophysical survey was carried out across the palaeochannels.

This revealed a 10 m thick sandy zone along the palaeochannels underlain by nonproductive clay.

Based on these studies, exploratory shallow filter point wells were recommended for channels nearer to the colony site. A successful bore well was drilled to 10 m on the palaeochannel closest to the colony site, and premonsoon and postmonsoon long-duration pump tests were conducted. Based on the test results, a safe discharge of 3000 gallons per hour for 16 h of pumping per day was recommended for the well situated on the palaeochannel to meet the housing colony's requirements. The cost of the bore well, pump house, etc., under this option, worked out to be $2500.

2.8.2 Case Study II

In any natural resources planning and management, it is important that the information input is reliable and available in a timely manner to the planner/decision maker. One of the major advantages of remote sensing is the ability to provide timely information on groundwater-potential zones, which can then be used effectively to identify new water sources during situations like drought.

During a severe drought in 1986 in India, a study was commissioned by the state government of Maharashtra to interpret the satellite images covering the entire state of Maharashtra for identifying favorable indicators of groundwater on a reconnaissance level. Nearly 80% of the state of Maharashtra is covered by Deccan Trap basalts. The 18 Landsat TM scenes covering the state were visually interpreted, with limited ground truth, and groundwater-potential maps were prepared on 1:250,000 scale (Seelan et al., 1986). The fracture lineaments, valley fills, and different plateau regions on the basalts, based on the dissection pattern, were demarcated. The problem villages where severe water shortages were faced were superimposed on the maps. The maps were provided to the state government of Maharashtra during June/July 1986. The cost of satellite images and the preparation of maps, at prevailing prices, was $1750.

During the following 1 year, 19,000 wells were drilled by the government of Maharashtra using the reconnaissance information provided based on Landsat TM and field surveys, out of which 14,000 wells were successful. In was reported that the increase in success rate was 5% more than the previous years when remote sensing–based inputs were not used systematically. This means more than 950 wells, which otherwise might have failed, were converted to successful wells. Assuming the prevailing drilling cost of $250 per well, this resulted in better utilization of funds to the tune of $11,875, in 1 year alone. In addition, there was the intangible benefit of being able to provide water to that many more villages during the severe drought situation, which is hard to quantify.

2.9 Summary

Remote sensing provides very useful reconnaissance-level information in groundwater exploration. It provides qualitative information not only on indicators that help in identifying groundwater-potential zones but also on areas of groundwater overexploitation. The information thus obtained helps in delineating zones where further exploration could be taken up. Various parameters relating to different thematic information relevant to groundwater occurrence and movement are obtainable from satellite imagery. These, when integrated with the help of GIS techniques, can provide information on potential groundwater zones. The occurrence of groundwater, however, varies with geologic setting, and it is important to understand these variations and look for appropriate keys and parameters. It is also emphasized here that remote sensing study, while providing valuable information, has to be followed up with ground-based surveys, as shown in the case studies, for better results.

Satellite remote sensing for groundwater applications have been practiced for the past 40 years or so, and the review showed the availability of a vast amount of literature on the topic. Much of the literature in the area of groundwater targeting has focused on developing countries, especially in the arid and semiarid regions where the technology has been found to be most useful. Not surprisingly, India, which ranks highest in the world in groundwater abstraction rates (i.e., higher than the next two nations on the list, China and the United States put together), has the highest number of case studies reported in the literature. It is also evident from the literature review that in countries like India, where the satellite data products were made available at reasonable costs and as a vast number of hydrogeologists were trained in the use of remote sensing for groundwater targeting, the technology could be adopted early under the nation's technology mission on drinking water, and other operational groundwater projects. With growing world population and increasing demand for water, the remote sensing technology offers great potential, particularly in the developing regions of the world, for careful targeting, extraction, and management of the groundwater resources.

Acknowledgments

The author thanks the Department of Space, India, for the opportunities to work on several national groundwater projects and the flexibility to experiment with new processes and procedures in extracting groundwater information from remote sensing, during his tenure with the department during mid-1970s to mid-1990s. Special thanks are also due to Brett Sergenian, a graduate student at the University of North Dakota and the author's summer intern during 2013, for meticulously collecting and listing many of the literature cited in this chapter.

References

Abrams, M.J., R.P. Ashby, L.C. Rowen, A.F. Goetz, and A.B. Kahale, 1977. Mapping of hydrothermal alteration in cuprite mining district, Nevada, using aircraft scanner images for the spectral region 0.46 to 2.36 μm. *Geology*, 5(12).

Agarwal, A.K. and D. Mishra, 1992. Evaluation of groundwater potential in the environs of Jhansi city Uttar Pradesh, using hydrogeomorphological assessment by satellite remote sensing technique. *Photonirvachak-Journal of the Indian Society of Remote Sensing*, 20(2,3), 121–128.

Ahmad, R. and R.P. Singh, 2002. Comparison of various data fusion for surface features extraction using IRS PAN and LISS III data. *Advanced Space Research*, 29(1), 73–78.

Ali, E.A., S.O. El Khidir, I.A.A. Babikir, and E.M. Abdelrahman, 2012. Landsat ETM+ digital Image processing techniques for lithological and structural lineament enhancement: Case study around Abidiya area, Sudan. *The Open Remote Sensing Journal*, 5, 83–89.

Bagyaraj, M., T. Ramkumar, S. Venkatramanan, and B. Gurugnanam, 2013. Application of remote sensing and GIS analysis for identifying groundwater potential zone in parts of Kodaikanal Taluk, South India. *Frontiers of Earth Science*, 7(1), 65–75.

Bobba, A.G., R.P. Bukata, and J.H. Jerome, 1992. Digitally processed satellite data as tool in detecting potential groundwater flow systems. *Journal of Hydrology*, 131, 25–62.

Bruning, J.N., J.S. Gierke, and A.L. Maclean, 2011. An approach to lineament analysis for groundwater exploration in Nicaragua. *Photogrammetric Engineering and Remote Sensing*, 77(5), 509–519.

Central Ground Water Board, 1978. A report on groundwater exploration at Sriharikota Island, Nellore District, Andhra Pradesh. CGWB report, Southern Region, Hyderabad, India. Unpublished.

Central Ground Water Board, 1983. Hydrogeological investigations to study the feasibility of an infiltration well along Kalangi river for augmentation of water supply to ISRO facilities at Sullurpeta, Nellore District, Andhra Pradesh. CGWB report, Southern Region, Hyderabad, India. Unpublished.

Chaterjee, C., R. Kumar, B. Chakravorty, A.K. Lohani, and S. Kumar, 2005. Integrating remote sensing and GIS techniques with groundwater low modeling for assessment of waterlogged areas. *Water Resources Management*, 19, 539–554.

Chowdary, V.M., R. Vinu Chandran, N. Neeti, R.V. Bothale, Y.K. Srivastava, P. Ingle, D. Ramakrishnan, D. Dutta, A. Jeyaram, J.R. Sharma, and Ravindra Singh, 2008. Assessment of surface and sub-surface waterlogged areas in irrigation command areas of Bihar state using remote sensing and GIS. *Journal of Water Management*, 95, 754–766.

Civil Engineering Divison, undated. Augmentation of water supply to housing at Sullurpeta. Design Review Document, Department of Space, Government of India. Unpublished.

Davis, S.N. and R.J.M. De Wiest, 1966. *Hydrogeology*, John Wiley and Sons, New York, p. 463.

Dawson, G.B. and D.J. Dickinson, 1970. Heat flow studies in thermal areas of North Island of New Zealand. *Geothermics*, 2(part 1), 466–473.

Deekshatulu, B.L. and S. Thiruvengadachari, 1981. Application of remote sensing techniques for water quality monitoring. Final project report, National Remote Sensing Agency, Hyderabad, India, p. 261.

Deutsch, M., 1971. Operational and experimental remote sensing in hydrology. CENTO seminar on the applications of remote sensors in the determination of natural resources, Ankara, Turkey.

Deutsch, M., 1974. Survey of remote sensing applications for groundwater exploration and management. *Proceedings of the Meetings of the American Association for the Advancement of Science.*

Dhakate, R., V.S. Singh, B.C. Negi, S. Chandra, and V.A. Rao, 2008. Geomorphological and geophysical approach for locating favorable groundwater zones in granitic terrain, Andhra Pradesh, India. *Journal of Environmental Management*, 88, 1373–1383.

Dhokarikar, B.G., 1991. *Groundwater Resource Development in Basaltic Rock Terrain of Maharashtra.* Water Industry Publication, Pune, India, p. 275.

Elewa, H. and A.A. Qaddah, 2011. Groundwater potentiality mapping in the Sinai Peninsula, Egypt, using remote sensing and GIS-watershed-based modelling. *Hydrology Journal*, 19, 613–628.

Freeze, A.R. and J.A. Cherry, 1979. *Groundwater.* Prentice-Hall, Inc., Englewood Cliffs, NJ, p. 604.

Gandino, A., 1983. Recent remote sensing technique in freshwater submarine springs monitoring: Qualitative and quantitative approach. *Proceedings of the International Symposium on Methods and Instrumentation for Investigation of Groundwater Systems*, UNESCO, Noordwijkerhout, the Netherlands.

Gautham, A.M., 1990. Application of IRS IA data for delineating buried channels in southern part of Allahabad district of Uttar Pradesh. *Photonirvachak-Journal of Indian Society of Remote Sensing*, 18(3), 52–55.

Geraghty and Miller, Inc., 1972. Groundwater contamination, an explanation of its causes and effects. A special report, Port Washington, NY.

Ghayoumian, J., B. Ghermezcheshme, S. Feiznia, and A.A. Noroozi, 2005. Integrating GIS and DSS for identification of suitable areas of artificial recharge, case study Meimeh Basin, Isfahan, Iran. *Environmental Geology*, 47, 493–500.

Ghayoumian, J., M. Mohseni Saravi, S. Feiznia, B. Nouri, and A. Malekian, 2007. Application of GIS techniques to determine areas most suitable for groundwater recharge in a coastal aquifer in southern Iran. *Journal of Asian Earth Sciences*, 30, 364–374.

Ghoneim, E. and F. El-Baz, 2007. The application of radar topographic data to mapping of a mega palaeo-drainage in the Eastern Sahara. *Journal of Arid Environments*, 69, 658–675.

Ghosh, R., 1993. Remote sensing for analysis of groundwater availability in an area with long unplanned mining history. *Photonirvachak-Journal of Indian Society of Remote Sensing*, 21(3), 119–126.

Goetz, A.F.H., F.C. Billingeley, A.R. Cillespic, M.J. Adams, R.L. Squires, E.M. Shoemaker, I. Lucchitta, and D.P. Elston, 1975. Application of ERTS Images and image processing to regional geologic mapping in Northern Arizona: NASA Jet Propulsion Lab, California, Technical Report 32–1597.

Gopinath, G. and P. Seralathan, 2004. Identification of groundwater prospective zones using IRS ID LISS III and pump test methods. *Photonirvachak-Journal of the Indian Society of Remote Sensing*, 32(4), 329–340.

Gumma, M.K. and P. Pavelic, 2013. Mapping of groundwater potential zones across Ghana using remote sensing, geographic information systems, and spatial modeling. *Environmental Monitoring Assessment*, 185, 3561–3579.

Hochstein, M.P. and D.J. Dickinson, 1970. Infrared sensing of thermal ground in the Tampo region, New Zealand. *Proceedings of the United Nations Symposium on the Development and Utilization of Geothermal Resources*, Pisa, Italy, volume 2, part 1.

Jagadesha, D.S., D. Nagaraju, and B.C. Prabhakar, 2012. Identification of groundwater potential zones through remote sensing and GIS techniques in Muguru Add Halla in Mysore and Chamarajnagar district, Karnataka, India. *International Journal of Earth Sciences and Engineering*, 5(5), 1310–121.

Jasmin, I. and P. Malikarjuna, 2011. Review: Satellite-based remote sensing and geographic information systems and their application in the assessment of groundwater potential, with particular reference to India. *Hydrogeology Journal*, 19, 729–740.

Jha, M.K., A. Chowdhury, V.M. Chowdhury, and S. Peiffer, 2007. Groundwater management and development by integrated remote sensing and geographic information systems: Prospects and constraints. *Water Resources Management*, 21, 427–467.

Karanth, K.R., 1987. *Groundwater Assessment Development and Management*. Tata McGraw Hill publishing company limited, New Delhi, p. 720.

Khalaf, A. and D. Donoghue, 2012. Estimating recharge distribution using remote sensing: A case study from the West Bank. *Journal of Hydrology*, 414–415, 354–363.

Khodaei, K. and H.R. Nassery, 2013. Groundwater exploration using remote sensing and geographic information systems in a semi-arid area (southwest of Urmieh, northwest of Iran). *Arabian Journal of Geosciences*, 6, 1229–1240.

Koch, M. and P.M. Mather, 1997. Lineament mapping for groundwater resource assessment: A comparison of digital Synthetic Aperture Radar (SAR) imagery and stereoscopic Large Format Camera (LFC) photographs in the Red Sea Hills, Sudan. *International Journal of Remote Sensing*, 18(7), 1465–1482.

Kruck, W, 1976. Hydrogeological investigations in the Argentine Pampa using satellite imagery. Natural Resources and Development, vol 3, Institute for Scientific Cooperation, Germany.

Kumar, A. and S.K. Srivastava, 1991. Geomorphological units, their geohydrological characteristic and vertical electrical sounding response near Munger, Bihar. *Photonirvachak-Journal of Indian Society of Remote Sensing*, 19(3), 205–215.

Kumar, P.K.D., G. Gopinath, and P. Serlathan, 2007. Application of remote sensing and GIS for the demarcation of groundwater potential zones in a river basin in Kerala, southwest coast of India. *International Journal of Remote Sensing*, 28(24), 5583–5601.

Kurz, B. and S.K. Seelan, 2009. *Use of Remote Sensing to Map Irrigated Agriculture in Areas Overlying the Ogallala Aquifer, United States*. Chapter 7, Remote Sensing of Global Croplands for Food Security, Taylor & Francis, Boca Raton, FL, p. 476.

Larsson, I., 1984. *Groundwater in Hard Rocks*. UNESCO publication, Paris, p. 228.

Lattman, L.H. and R.P. Nickelson, 1958. Photogeologic fracture-trace mapping in Appalachian Plateau. *Bulletin of American Association of Petroleum Geologists*, 42(9), 2238–2245.

Lee, S., Y.S. Kim, and H.J. Oh, 2012. Applications of a weights-of-evidence method and GIS to regional groundwater productivity potential mapping. *Journal of Environmental Management*, 96, 91–105.

Magesh, N.S., N. Chandrasekar, and J.P. Soundranayagam, 2012. Delineation of groundwater potential zones in Theni district, Tamil Nadu, using remote sensing, GIS, and MIF techniques. *Geosciences Frontiers*, 3(2), 189–196.

Mallast, U., R. Gloaguen, S. Geyer, T. Rodiger, and C. Siebert, 2011. Derivation of groundwater flow paths based on semi-automatic extraction of lineaments from remote sensing data. *Hydrology and Earth System Sciences*, 15, 2665–2676.

Mandel, S. and Z.L. Shiftan, 1981. *Groundwater Resources Investigation and Development*. Academic Press, New York, p. 269.

Mc Lerran, J.K., 1967. Infrared sensing. *Photogrammetric Engineering*, XXXIII(5).

Mogaji, K.A., O.S. Aboyeji, and G.O. Omosuyi, 2006. Mapping of lineaments for groundwater targeting in basement complex area of Ondo State using remotely sensed data. *International Journal of Water Resources and Environmental Engineering*, 3(7), 150–160.

Mollard, J.D., 1957. Aerial mosaics reveal fracture patterns on surface materials in Southern Saskatchewan and Manitoba. *Oil in Canada*, 26, 1840–1864.

Mollard, J.D. and F.D. Patton, 1960. Science and systems in groundwater investigations. *Western Canada Water and Sewage Conference, Proceedings*, pp. 53–72; reprinted in Canadian Municipal Utilities, Canada, June 1961.

Moore, G.K., 1967. Lineaments on Skylab photographs—Detection, mapping and hydrologic significance in Central Tennessee. US Geological Survey—open file report 78–196.

Moore, G.K., 1978. The role of remote sensing in groundwater exploration. *Proceedings of the Indo-US Workshop on Remote Sensing of Water Resources*, Hyderabad, India.

Moore, G.K. and M. Deutsch, 1975. ERTS imagery for groundwater investigations. *Groundwater*, 13(2), 214–226.

Moore, G.K. and F.A. Waltz, 1983. Objective procedures for lineament enhancement and extraction. *Photogrammetric Engineering and Remote Sensing*, 49(5), 641–647.

National Remote Sensing Agency, 1980. Integrated remote sensing survey of natural resources of Tamil Nadu state, India. National Remote Sensing Agency. Project report, National Remote Sensing Agency, Department of Space, Government of India, Hyderabad, India.

National Remote Sensing Agency, 1982a. Integrated remote sensing survey of natural resources of North Karnataka State, India. Project report, National Remote Sensing Agency, Department of Space, Government of India, Hyderabad, India.

National Remote Sensing Agency, 1982b. Integrated remote sensing survey of natural resources of Bundelkhand region, U.P, India. Project report, National Remote Sensing Agency, Department of Space, Government of India, Hyderabad, India.

National Remote Sensing Agency, 1983. Integrated remote sensing survey of natural resources of West Coast region, India. Project report, National Remote Sensing Agency, Department of Space, Government of India, Hyderabad, India.

National Remote Sensing Agency, 1984. Integrated remote sensing survey of natural resources of Upper Barak Watershed, North-East India. Project report, National Remote Sensing Agency, Department of Space, Government of India, Hyderabad, India.

National Remote Sensing Agency, 1986. Report on the groundwater potential maps of drought prone districts of Maharashtra prepared based on visual interpretation of Landsat Thematic Mapper data. Project report, National Remote Sensing Agency, Department of Space, Government of India, Hyderabad, India.

National Remote Sensing Agency, 1986. Report on the groundwater potential maps of drought prone districts of Karnataka prepared based on visual interpretation of Landsat Thematic Mapper data. Project report, National Remote Sensing Agency, Department of Space, Government of India, Hyderabad, India.

Nefedov, K.E. and T.A. Popova, 1972. *Deciphering Groundwater from Aerial Photographs.* Translated from Russian by V.S. Kothekar. Amerind Publication Co, New Delhi, India, p. 191.

Otterman, J., 1971. Thermal mapping of selected sites in the Lake Kinneret region, Israel. *Journal of Earth Sciences,* 20(3).

Ozaray, G., 1975. Remote sensing in hydrogeological mapping (with special respect to India and Canada). *Proceedings of the second World Congress on Water Resources,* New Delhi, India.

Pal, S.K., T.J. Majumdar, and A.K. Bhattacharya, 2007. ERS -2 SAR and IRS-1C LISS III data fusion: A PCA approach to improve remote sensing based geological interpretation. *ISPRS Photogrammetry and Remote Sensing,* 61(2007), 281–297.

Palmason, G., 1970. Aerial infrared surveys of Reykjanes and Torfajokull thermal areas, Iceland, with a section of cost exploration surveys. *Proceedings of the United Nations Symposium on the Development and Utilization of Geothermal Resources,* Pisa, Italy, volume 2, part 1.

Panigrahi, B., A.K. Nayak, and S.D. Sharma, 1995. Application of remote sensing technology for groundwater potential evaluation. *Water Resources Management,* 9, 161–173.

Prasad, E.A.V., 1980. *Groundwater in Varahamihira's Brahat Samhita.* MASSLIT series, Sri Venkateswara University, Tirupati, India, p. 351.

Rana, S.S., 1998. Application of directional filtering in lineament mapping for groundwater prospecting around Bhinmal—A semi-arid part of Thar desert. *Photonirvachak-Journal of the Indian Society of Remote Sensing,* 26(1,2), 35–44.

Rao, N.S. and R.P. Reddy, 1999. Groundwater prospects in a developing satellite township of Andhra Pradesh, India, using remote sensing techniques. *Photonirvachak-Journal of the Indian Society of Remote Sensing,* 27(4), 193–202.

Ravindran, K.V. and A. Jeyaram, 1997. Groundwater prospects of Shahbad Tehsil, Baran District, Eastern Rajasthan: A remote sensing approach. *Photonirvachak-Journal of the Indian Society of Remote Sensing,* 25(4), 239–246.

Regunath, H.M., 1987. *Groundwater,* 2nd edn. Wiley Eastern Limited, New Delhi, India, p. 563.

Robinov, C.J., 1968. The status of remote sensing in hydrology. *Proceedings of the Fifth Symposium on Remote Sensing and Environment,* Michigan University, Ann Arbor, MI.

Robinson, C., F. El-Baz, M. Ozdogan, M. Ledwith, D. Blance, S. Oakley, and J. Inzana, 2000. Use of radar data to delineate palaeodrainage flow directions in the Selima sand sheet, Eastern Sahara. *Photogrammetric Engineering and Remote Sensing,* 66(5), 745–753.

Rodell, M. and J.S. Famiglietti, 2002. The potential for satellite-based monitoring of storage changes using GRACE: The High Plains aquifer, central US. *Journal of Hydrology,* 263(2002) 245–256.

Rundquist, D.C., D.A. Rodekohr, A.J. Peters, R.L. Ehrman, L. Di, and G. Murrey, 1991. Statewide groundwater vulnerability assessment in Nebraska using the DRASTIC/GIS model. *Geocarto International,* 2, 51–58.

Sander, P, T.B. Minor, and M.M. Chesley, 1997. Ground-water exploration based on lineament analysis and reproducibility tests. *Ground Water,* 35(5), 888–895.

Sankar, K., 2002. Evaluation of groundwater potential zones using remote sensing data in Upper Vaigai river basin, Tamil Nadu, India. *Photonirvachak-Journal of the Indian Society of Remote Sensing,* 30(3), 119–128.

Seelan, S.K., 1975. Location of abnormal water resources by thermal infra-red technique. Diploma thesis report. Groundwater Research Center, Hebrew University of Jerusalem. Unpublished.

Seelan, S.K., 1980. Utility of image enhancement in geologic interpretation of remotely sensed data—An Indian example. *Proceedings of the Symposium on Remote Sensing in Subsurface Exploration.* Sixth annual convention of Association of Exploration Geologists, Bangalore, India.

Seelan, S.K., 1982. Landsat image derived geomorphic indicators of groundwater in parts of Central India. *Photonirvachak-Journal of Indian Society of Remote Sensing,* 10(2), 33–37.

Seelan, S.K., 1986. Detection and monitoring of groundwater systems under stress—Can remote sensing help? *Proceedings of the AWRS Conference on Groundwater Systems under Stress,* Brisbane, Australia.

Seelan, S.K., 1994. Remote sensing applications and GIS development for groundwater investigations. Ph.D. Dissertation, Jawaharlal Nehru Technological University, Hyderabad, India. Unpublished.

Seelan, S.K., 1996. Cost benefit aspects of remote sensing for groundwater exploration—Two case studies. Book chapter, *Economics of Remote Sensing,* ISBN 81-86562-04-4, Manek Publications Pvt Ltd., New Delhi, India.

Seelan, S.K., A. Bhattacharya, and R. Venkataraman, 1988. Remote sensing for identification of fresh water zones around Sullurpeta town in Andhra Pradesh. *Proceedings of the National Seminar on Groundwater Development in Coastal Tracts,* Trivandrum, India.

Seelan, S.K., G. Ch. Chennaiah, and N.C. Gautam, 1983. Study of landform control over land utilization pattern in parts of southern U.P.—A remote sensing approach. *Photonirvachak-Journal of Indian Society of Remote Sensing*, 11(1), 49–53.

Seelan, S.K., V.S. Hegde, P.R. Reddy, R.S. Rao, R.K. Sood, A.K. Sharma, A.K. Gupta, G. Raj, A. Perumal, and L.R.A. Narayan, 1986. Groundwater targeting in a drought situation in Maharashtra state, India, using Landsat TM data. *Proceedings of the Seventh Asian Conference on Remote Sensing*, Seoul, Korea.

Seelan, S.K. and L.R.A. Narayan, 1984. Drought study around Madras city. NRSA technical report. National Remote Sensing Agency, Hyderabad, India. Unpublished.

Seelan, S.K. and S. Thiruvengadachari, 1980. An integrated regional approach for delineation of potential groundwater zones using satellite data—An Indian case study. *Abstracts of Proceedings, 26th International Geological Congress*, Paris, France.

Seelan, S.K. and S. Thiruvengadachari, 1981. Satellite sensing for extraction of groundwater resources information. *Proceedings of the 15th International Symposium on Remote Sensing of the Environment*, Ann Arbor, MI.Sener, E., Davraz, A., and Ozcelik, M, 2004. An integration of GIS and remote sensing in groundwater investigations: A case study of Burdur, Turkey. *Hydrogeology Journal*, 13, 826–834.

Seshubabu, K. and S.K. Seelan, 1983. Hydrogeology chapter, Godavari basin test site. Technical report, part B, Indo-FRG technical cooperation programme, National Remote Sensing Agency, Hyderabad, India.

Sethupathi, A.S., C. Lakshmi Narasimhan, and V. Vasanthamohan, 2012. Evaluation of hydrogeomorphological landforms and lineaments using GIS and remote sensing techniques in Bargur-Mathur sub-watersheds, Ponnaiyar River basin, India. *International Journal of Geomatics and Geosciences*, 3(1), 178–190.

Singh, K.P., 1980. Application of Landsat imagery to groundwater studies in parts of Punjab and Haryana states, India. The contribution of space observations to water resources management, COSPAR, Advances in Space Exploration, Bangalore 9, Pergamon Press, pp. 107–111.

Solomonson, V.V., T.J. Jackson, J.R. Lucas, G.K. Moore, A. Rango, T. Schmugge, and D. Scholz, 1983. *Water Resources Assessment. Manual of Remote Sensing*. American Society of Photogrammetry, Falls Church, VA, 2, 1497–1570.

Shankarnarayana, G., N. Lakhmaiah, and P.V. Prakash Goud, 1996. Hydrogeomorphological study based on remote sensing of Mulug Takuk, Warangal district, Andhra Pradesh, India. *Hydrological Sciences*, 42(2), 137–150.

Tcherepanov, E.N., V.A. Zlotnik, and G.M. Henebry, 2005. Using Landsat thermal imagery and GIS for identification of groundwater discharge into shallow groundwater dominated lakes. *International Journal of Remote Sensing*, 26(17), 3649–3661.

Thillaigovindarajan, S., S.K. Seelan, M. Jayaraman, and P. Radhakrishnamoorthy, 1985. The evaluation of hydrogeological conditions in the southern part of Tamil Nadu using remote sensing techniques. *International Journal of Remote Sensing*, 6(3&4), 447–456.

Tiwari, O.N., 1993. Lineament identification for groundwater drilling in a hard rock terrain of Sirohi District, Western Rajasthan. *Photonirvachak-Journal of the Indian Society of Remote Sensing*, 21(1), 14–19.

Tiwari, V.M., J. Wahr, and S. Swenson, 2009. Dwindling groundwater resources in Northern India, from satellite gravity observations. *Geophysical Research Letters*, 36, LI8401, doi:10.1029/2009GL039301.

Venkataratnam, L.V., 1984. Mapping of land/soil degradation using multispectral data. *Proceedings of the Eighth Canadian symposium on remote sensing*, Montreal, Quebec, Canada.

Xiamei, J., W. Li, Z. Youkuan, X. Zhongqi, and Y. Ying, 2007. A study of relationship between vegetation growth and groundwater in the Yinchuan Plain. *Earth Science Frontiers*, 14(3), 197–203.

Wang, L.T., T.E. McKenna, and T. DeLiberty, 2008. Locating groundwater discharge areas in Rehoboth and Indian River Bays and Indian River, Delaware, using Landsat 7 imagery. Report of Investigations no. 74, Delaware Geological Society, Delaware, pp. 1–10.

II

Water Use and Water Productivity

3

Remote Sensing of Actual Evapotranspiration from Croplands

Trent W. Biggs
San Diego State University

George P. Petropoulos
Aberystwyth University

Naga Manohar Velpuri
United States Geological Survey

Michael Marshall
World Agroforestry Center

Edward P. Glenn
University of Arizona

Pamela Nagler
U.S. Geological Survey

Alex Messina
San Diego State University

Acronyms and Definitions

a	Intercept in linear model relating T_R to T_1–T_2 (SEBAL, METRIC)	
a_{ds}	Empirical coefficient in downscaling model (Kustas et al., 2003)	
a_k	Coefficient in empirical coefficient method (Equation 3.8)	
$a_{p,q}$	Coefficients in the T_R-VI-SVAT model (Carlson, 2007)	
α	Broadband, blue-sky albedo	
α_{PT}	Priestley–Taylor coefficient	
b	Slope of linear model relating T_R to T_1–T_2 (SEBAL, METRIC)	
b_{ds}	Empirical coefficient in downscaling model (Kustas et al., 2003)	
b_k	Coefficient in empirical coefficient method (Equation 3.8)	
c	Temperature correction factor (SSEBop)	

c_{ds}	Empirical coefficient in downscaling model (Kustas et al., 2003)
c_k	Coefficient in empirical coefficient method (Equation 3.8)
c_L	Mean potential stomatal conductance per unit leaf area (Mu et al., 2011)
C_p	Specific heat capacity of air
C_{rad}	Adjustment factor for sloped surfaces (Allen et al., 2007)
CWSI	Crop water stress index = 1–ET_f
D	Vapor pressure deficit
d	Zero-plane displacement height ~ 2/3 h
dT	Temperature difference between hot and cold pixels (SSEBop)
e	Vapor pressure
e_{sat}	Saturated vapor pressure
ET	Actual evapotranspiration
ET_f	Reference ET fraction
ET_o	Potential evapotranspiration of a grass reference crop

ET_{o24}	ET_o for a 24 h period
EVI	Enhanced vegetation index
ε_o	Broadband surface emissivity
f_c	Vegetation cover fraction
f_g	Green canopy fraction (Fisher et al., 2008)
f_M	Plant moisture constraint (Fisher et al., 2008)
FPAR	Photosynthetically active radiation
f_{SM}	Soil moisture constraint (Fisher et al., 2008)
f_T	Temperature constraint to ET (Fisher et al., 2008)
f_{wet}	Relative surface wetness (Fisher et al., 2008)
F_{wet}	Water cover fraction (Mu et al., 2011)
G	Ground heat flux
γ	Psychometric constant
H	Sensible heat flux
h	Vegetation height
η	Coefficient in empirical crop coefficient method
K_c	Crop coefficient in FAO-56 method
K_s	Soil moisture stress coefficient in FAO-56 method
LAI	Leaf area index
LST	Land surface temperature, equivalent to T_R
LW↑	Upwelling longwave radiation
LW↓	Downwelling longwave radiation
Λ	Evaporative fraction
Λ_{24}	Λ for 24 h period
Λ_d	Λ for daylight hours
Λ_{op}	Λ at time of overpass
λ	Latent heat of vaporization
λE_I	Latent heat flux from evaporation from wet canopy leaf surfaces
λE_s	Latent heat flux from evaporation from the soil surface
λE_{SP}	Potential latent heat flux from soil evaporation (Mu et al., 2011)
λET	Latent heat flux
λET_c	Latent heat flux from transpiration
m(D)	Multiplier limiting stomatal conductance by D (Mu et al., 2007)
m(Tmin)	Multiplier limiting stomatal conductance by minimum air temperature (Mu et al., 2007)
Mo	Soil moisture
N*	NDVI
NDVI	Normalized difference vegetation index
$NDVI_o$	NDVI for bare soil
$NDVI_s$	NDVI for dense vegetation
Ω	Index of degree of clumping (ALEXI)
σ	Stefan–Boltzmann constant
r_a	Aerodynamic surface resistance
r_{a_s}	Aerodynamic resistance at the soil surface (Mu et al., 2011)
R_{ah}	Aerodynamic resistance to turbulent heat transport between z_1 and z_2
RH	Relative humidity
RMSE	Root-mean-square error
R_n	Net radiation

R_{n24}	Net radiation over 24 h period
R_{ns}	Net radiation at the soil surface
Rs	sensible heat exchange resistance of the soil surface
r_s	Resistance of the land surface or plant canopy to ET
r_{s_c}	Dry canopy resistance to transpiration (Mu et al., 2011)
r_{s_wetC}	Wet canopy resistance to evaporation
Rx	total boundary layer resistance of the canopy
ρ	Air density
s	Slope of the saturation vapor pressure versus temperature curve
SAVI	Soil-adjusted vegetation index
SW↓	Incoming shortwave radiation
T_1	Aerodynamic temperature of the evaporating surface at height z_1
T_{1c}	Vegetation canopy temperature (ALEXI)
T_{1s}	Soil temperature (ALEXI)
T_2	Air temperature at height z_2
T_c	Theoretical surface temperature under cool/moist conditions (SSEBop)
T_h	Theoretical surface temperature under hot/dry conditions (SSEBop)
T_R	Radiometric surface temperature, equivalent to LST
$\hat{T}_{Rhi}$	Predicted T_R at high spatial resolution (Kustas et al., 2003)
$\hat{T}_{Rlow}(NDVI_{low})$	Predicted radiometric temperature using low resolution NDVI
$\hat{T}_{Rlow}(NDVI_{hi})$	Predicted radiometric temperature using high-resolution NDVI (Kustas et al., 2003)
T_{Rmax}	Minimum T_R over vegetation
T_{Rmin}	Maximum T_R over bare soil
T_{scaled}	Scaled T_R
θ	View angle
VI	Vegetation index
VI_{max}	VI value when ET is maximum
VI_{min}	VI value for bare soil
z_1	Height above the ground surface of the evaporating surface, $= d + z_{0m}$
z_2	Height at which air temperature is measured (often 2 or 3 m)
z_{om}	Surface roughness for momentum transport, ~0.03–0.123 h

Organizations, Satellite, and Model Acronyms

ABL	Atmospheric boundary layer
AGRIMET	Agricultural meteorological modeling system
ALEXI	Atmosphere–land exchange inverse model (Anderson et al., 1997)
ASTER	Advanced spaceborne thermal emission and reflection radiometer

AVHRR	Advanced very high resolution radiometer
CERES	Clouds and Earth's Radiant Energy System
CONUS	Conterminous United States
DAIS	Digital airborne imaging spectrometer
DisALEXI	Disaggregation ALEXI model (Norman et al., 2003)
DSTV	diurnal surface temperature variation
DTD	Dual-temperature-difference
ECMWF	European Centre for Medium-Range Weather Forecasts
EO	Earth observation
FIFE	First ISLSCP (International Satellite Land Surface Climatology Project) Field Experiment
FLUXNET	Global network of micrometeorological flux tower sites
GDAS	Global Data Assimilation System
GG model	Granger and Gray (GG) model (Granger and Gray, 1989)
GLDAS	Global Land Data Assimilation System
LSA-SAF	Land Surface Analysis Satellite Applications Facility
MERRA	Modern-Era Retrospective Analysis for Research and Applications
METRIC	Mapping Evapotranspiration at high Resolution with Internalized Calibration
MMR	Modular multispectral radiometer
MOD16	MODIS ET product, also called PM-Mu (Mu et al., 2011)
MOD43B3	MODIS albedo product
MODIS	Moderate Resolution Imaging Spectroradiometer
MSG	Meteosat second generation satellite
NCEP-NCAR	National Centers for Environmental Prediction–National Center for Atmospheric Research
NWS-NOAH	National Weather Service
PoLDER	Polarization and directionality of Earth reflectance instrument
PT-JPL	Priestley–Taylor jet propulsion laboratory model (Fisher et al., 2008)
RMSD	Root-mean-square difference
SEBAL	Surface energy balance algorithm (Bastiaanssen et al., 1998)
SEBS	Surface Energy Balance System (Su, 2002)
SEVIRI	Spinning enhanced visible and infrared imager
SGP	Southern Great Plains
Sim-ReSET	Simple remote sensing evapotranspiration model
SMACEX	Soil Moisture–Atmosphere Coupling Experiment
SRB	Surface radiation budget
SSEB	Simplified surface energy balance
S-SEBI	Simplified surface energy balance index
SSEBop	Operational simplified surface energy balance
STARFM	Spatial and temporal adaptive reflectance fusion model
SVAT	Soil vegetation atmosphere transfer model

TRMM	Tropical Rainfall Measurement Mission
TSM	Two source model
VMC	Vegetation and moisture coefficient, equivalent to ET_f

3.1 Introduction

Agriculture accounted for the majority of human water use and for more than 90% of global freshwater consumption during the twentieth century (Hoekstra and Mekonnen, 2012; Shiklomanov, 2000). Streamflow depletion due to enhanced evapotranspiration (ET) from irrigated crops impacts freshwater ecosystems globally (Foley et al., 2005). Water scarcity limits crop production in many arid and semiarid regions, and water is likely to be a key resource limiting food production and food security in the twenty-first century (Foley et al., 2011; Vorosmarty et al., 2000). Despite this, estimates of the location and temporal dynamics of ET from croplands are often uncertain at a variety of spatial and temporal scales. Better information on ET can be useful in several applications at a range of spatial scales, including water resources, agronomy, and meteorology (e.g., Rivas and Caselles, 2004). At the scale of irrigation projects, maps of ET can assist with irrigation scheduling and demand assessment. Measurements of ET are required for monitoring plant water requirements, plant growth, and productivity, as well as for irrigation management and deciding when to carry out cultivation procedures (e.g., Consolli et al., 2006; Glenn et al., 2007; Yang et al., 2010).

At regional and river basin scales, ET estimates can assist with water allocation decisions to support agriculture and ecosystems, including strategies for drought management. Global ET assessments can help understand, for example, how the global food production system may respond to global climate change. Data on the energy equivalent of ET (latent heat flux, λET) are also of key significance in the numerical modeling and prediction of atmospheric and hydrological cycles and in improving the accuracy of weather forecasting models (Jacob et al., 2002). ET is the single most important mechanism of mass and energy exchange between the hydrosphere, biosphere, and atmosphere, playing a critical role in both water cycle and energy balance (Sellers et al., 1996) and regional circulation patterns (Lee et al., 2009, 2011). Quantitative information on ET is also important for understanding the processes that control ecosystem CO_2 exchange (Scott et al., 2006) as well as the interactions between parameters in different ecosystem processes (Wever et al., 2002).

ET can be measured in the field with different types of instruments including lysimeters and eddy covariance, surface renewal, or flux variance systems (French et al., 2012; Petropoulos et al., 2013; Swinbank, 1951). Field methods estimate ET over a range of spatial scales, from ~1 m for lysimeters and up to ~100–1000 m for eddy covariance towers. While field data are the most direct way to measure ET, their use for the measurement of ET over large areas is limited due to the expense of maintaining the field equipment and to the large spatial variability in ET, particularly in agricultural settings (Ershadi et al., 2013; McCabe and Wood, 2006).

For example, the global inventory of eddy flux towers (FLUXNET) (Baldocchi et al., 2001) reports data for fewer than 500 active towers with most towers concentrated in the United States and Europe and no towers in some countries where knowledge of ET is critical to water management (Jung et al., 2009).

In the absence of in situ data on ET, crop models are often used to estimate ET under different crop and soil moisture conditions (Allen, 2000). Application of crop models to a new location and at regional scales suffers from several difficulties. Required input data, including crop growth stage and cropping calendars, may be difficult to determine in large areas with heterogeneous cover and intraseasonal or interannual variability in cropping patterns. The model parameters are often functions of relative humidity (RH) and wind speed, resulting in corrections of up to 30% for tall crops (2–3 m) growing in conditions of low humidity and high wind speeds compared to the same tall crop in a humid climate with low wind speeds due to its large aerodynamic roughness (Allen et al., 1998, p. 93). Crop models of ET may be difficult to parameterize during the initial growth stages, which are particularly sensitive to environmental conditions like wetting frequency and soil texture, which introduces uncertainty where soil evaporation is an important component of total ET.

Given the limitations of field measurements and the difficulty of estimating the parameters in crop models without additional data on crop calendars and condition, Earth observation (EO) technology provides an opportunity to estimate ET at a variety of spatial and temporal scales. EO is defined here as the collection of imagery from aerial and satellite platforms. EO technology is recognized as the only viable solution for obtaining estimates of ET at the spatiotemporal scales and accuracy levels required by many applications (Glenn et al., 2007; Melesse et al., 2008). Previous reviews of the use of EO data for ET estimation are available (Courault et al., 2005; Gonzalez-Dugo et al., 2009; Kalma et al., 2008; Kustas and Norman, 1996; Petropoulos, 2013; Verstraeten et al., 2008). Gowda et al. (2008) provided an overview of ET estimation in agriculture but focused on energy-based methods, which is one of several methods available (Table 3.1).

The present chapter provides a critical and systematic overview of different modeling approaches to estimate ET from EO data, with a focus on applications in agriculture. The chapter is structured as follows: First, we review methods for estimating net radiation (R_n), which is required by all ET modeling techniques described in the chapter. Descriptions of R_n calculations are also included because, in agricultural applications, special adaptations are often required to calculate outgoing radiation at sufficiently high spatial resolution to capture spatial variability in radiation caused by the heterogeneous land surface. We then discuss three major families of methods used to calculate ET using EO data, with a focus on methods that use vegetation indices and/or radiometric land surface temperature (T_R). T_R is often called the land surface temperature or the surface temperature measured by the satellite (T_s); here, we use the term radiometric land surface temperature and the symbol T_R to emphasize that it is the temperature estimated from thermal radiation detected at the satellite sensor. While microwave (MW)-based methods have been developed and could be particularly useful for areas with cloud cover, the spatial resolution of MW imagery, at 35–65 km², is deemed too coarse for application to agricultural areas (Petropoulos, 2013).

We use common mathematical symbols for each method to facilitate intercomparison, and highlight similarities in the conceptual foundations of the various methods. Sufficient detail is provided to implement and compare some of the most commonly used algorithms and recommendations for the application of each method are also given, discussing any special issues related to estimating ET in agricultural landscapes. The accuracy of the methods is then compared, special problems in application of the methods to crops are discussed, and future research directions are highlighted.

3.2 Overview of Methods for ET Calculation Using Remote Sensing

EO sensors do not directly measure ET or λET. The spectral radiance measures they provide have to be combined in some form of retrieval algorithm or model in order to estimate ET. Several algorithms have been developed in the last four decades for estimating ET using either space- or airborne systems. The available EO-based methods can be broadly grouped into three basic families: (1) vegetation-based methods, (2) radiometric land surface temperature–based methods, and (3) triangle/trapezoid or scatterplot inversion methods.

Several different terms are used to describe the water demand of the atmosphere and the actual use of water by crops (Allen et al., 1998). Potential evapotranspiration is the amount of water lost to both evaporation from the soil surface, wet canopy, and open water surfaces and to transpiration through the leaves of vegetation not experiencing soil moisture stress. Potential ET is a function of climatic variables like net radiation, temperature, wind, and humidity and of vegetation characteristics. For a well-watered crop grown under optimal conditions, potential ET can vary with leaf area, stage of development, photosynthetic pathway, and rooting depth, so a second term, reference ET (ET_o) is defined as the amount of water used by a specific reference crop, usually a well-watered grass with specific characteristics (see Section 3.2.2). Crop ET for a given vegetation or crop type without soil moisture limitation is indicated by the variable ET_c (Allen et al., 1998). Finally, actual ET is the amount of water that is lost via both evaporation from the soil surface and transpiration from a specific vegetation cover under actual field conditions, including limitations to ET caused by soil moisture stress, nutrient limitation, and pathogens. In this chapter, we use the symbol ET and the term evapotranspiration to represent actual evapotranspiration of a given land surface (Glenn et al., 2011). λET is simply the product of ET and the latent heat of vaporization of water at a given

TABLE 3.1 Three Families of Methods to Estimate Evapotranspiration (ET) from Earth Observation Data, and Their Advantages, Disadvantages, Error, and Recommended Uses

Method of ET Estimation	Advantages	Disadvantages	Recommended for	Not Recommended or Untested for	Error	Reference
1. Vegetation-based models						
Empirical	Ease of implementation	Requires ground data for calibration; does not directly estimate soil evaporation	Small geographic regions with ground reference data, riparian and agricultural vegetation, desert, semi-arid rangelands, vegetated wetlands	Regional or Global application without reference data for calibration	Mean error 0.12 for ETf and 5% error for annual ET from flux towers and soil moisture balance (Nagler et al., 2013) RSMD 10–30% compared to mean ET (Glenn et al., 2010) Mean error 4% compared with daily ET from eddy covariance towers (Samani et al., 2009); Mean error 15% compared with daily ET (Duchemin et al., 2006); RMSE 15% compared with daily ET (Hunsaker et al., 2007a, 2007b).	Power function: Nagler et al. (2013); Extreme VI values: Glenn et al. (2010); Beer-Lambert Law: Nagler et al., 2013; CropCoeff-VI: Bausch and Neale, 1987, Neale, 1989, Choudhury et al., 1994; Linear regression on NDVI, SAVI: Samani et al., 2009; Campos et al., 2010; Desert: Glenn et al., 2008b; Wetlands: Glenn, et al., 2013a
PT-JPL	Relative ease of implementation; Minimal ground data requirements; Operational globally; Does not require calibration	May underestimate soil evaporation from irrigated areas in dry climates	Global and regional rainfed systems	Needs further testing in irrigated agriculture with high soil evaporation	Mean error 13% compared with of mean annual ET from eddy covariance towers (Fisher et al., 2008)	Fisher et al., 2008; Alternate formulation: Marshall et al., 2013
MOD16	Minimal ground data requirements; Operational globally; Does not require calibration	May underestimate soil evaporation from irrigated areas in dry climates	Global and regional rainfed systems	Needs further testing in irrigated agriculture with high soil evaporation	RMSE ~20%, MAB 24–25%, compared with daily ET from eddy covariance towers (Mu et al., 2011); ~50% compared with data from 60 eddy covariance towers pooled across CONUS (Velpuri et al., 2013)	Leuning et al., 2008; Mu et al., 2011, 2007, 2013; Nishida and Nemani, 2003
2. Radiometric land surface temperature based models						
Overall					RMSD <50 W/m² and <33 W/m² compared with instantaneous λET and H fluxes respectively (Gonzalez-Dugo et al., 2009)	Gonzalez-Dugo et al. (2009)
SEBAL, METRIC	Minimal ground data; Accurate in semi-arid environments with irrigation; Accurate for wet soil and inundated surfaces	Moderate complexity; Requires extremes of wet and dry to be present in a scene, and calibration for each image; Issues in merging across scenes	Routine application in irrigated agriculture in semi-arid and arid climates; Single scenes	Humid environments; Global applications; Data scarce regions	RMSE 15–20% compared to daily ET from eddy covariance towers (Allen et al., 2011); RMSE ~5% compared to seasonal ET estimates (Bastiaanssen et al., 2005)	Bastiaanssen et al., 2002, 1998; Allen et al., 2007

(Continued)

TABLE 3.1 (*Continued*)   Three Families of Methods to Estimate Evapotranspiration (ET) from Earth Observation Data, and Their Advantages, Disadvantages, Error, and Recommended Uses

Method of ET Estimation	Advantages	Disadvantages	Recommended for	Not Recommended or Untested for	Error	Reference
ALEXI, DisALEXI	Minimal ground data; Accurate in a variety of vegetation types and climates; Accurate for wet soil and inundated surfaces	High complexity of implementation; Requires two images for each daily ET estimate; Coarse resolution before downscaling	Regional and global applications, operational data product	Routine application for applied irrigation systems or management; Data scarce regions	RMSD 40 W/m², MAD 30 W/m², R² 0.77 (Anderson et al., 1997) RMSD 40–50 W/m² (Norman et al., 2003) MAD 15–20% for 30 min avg, 10% daily, and ~5% seasonal (Anderson et al., 2013)	ALEXI: Anderson et al., 1997; DisALEXI: Norman et al., 2003;
SSEB, SSEBop	Minimal ground data; Ease of implementation; Operational application over large areas	Potentially inaccurate in regions with high spatial variability in albedo or elevation; Requires in-situ meteorological data; Does not solve energy balance completely; Physical processes of ET are not fully represented	Regional and global applications, irrigated agriculture	Heterogeneous vegetation, mountainous regions; Regions with high albedo and high emissivity; Data scarce regions	Mean error <30% when compared to individual eddy covariance towers (Senay et al., 2007); Mean error ~60% when compared with data from 60 eddy covariance towers pooled across CONUS (Velpuri et al., 2013).	SSEB: Senay et al., 2007; SSEBop: Senay et al., 2013;

3. Scatterplot or triangle methods

Overall	Ease of implementation, few parameters (except SVAT method)	Require extremes of wet and dry to be present in a scene; subjectivity in selecting wet/dry pixels; issues in merging across scenes for large areas	Routine application in irrigated agriculture in semi-arid and arid climates	Humid environments; Global applications		Petropoulos et al., 2009
T$_R$-VI methods	Computationally straightforward; Relative independence from site specific tuning of model-parameters	Assumes linear relationships between location in T$_R$-VI space and ET; Clouds, standing water, and sloping terrain need to be masked	Local to regional scale sites	Homogenous land cover	RSMD ~30% compared to daily ET (Jiang and Islam, 2001); RMSD 45 Wm⁻², bias of 5.6 Wm⁻², R² = 0.86, compared to daily λET from eddy covariance towers Nishida (2003) and Nishida and Nemani (2003)	Jiang and Islam, 2001; Nishida (2003) and Nishida and Nemani (2003); Tang et al., 2010; Zhang et al., 2006

(*Continued*)

TABLE 3.1 (*Continued*)   Three Families of Methods to Estimate Evapotranspiration (ET) from Earth Observation Data, and Their Advantages, Disadvantages, Error, and Recommended Uses

Method of ET Estimation	Advantages	Disadvantages	Recommended for	Not Recommended or Untested for	Error	Reference
T_R-T_2 difference and VI scatterplot	Insensitive to absolute accuracy of T_R; Requires small number of in-situ observations	Often applicable for homogenous areas (Moran et al. 1994)	Areas of partial vegetation cover		RMSD 29 Wm^{-2} compared to instantaneous λET (Moran 1996); RMSD 59 Wm^{-2}, bias −42 Wm^{-2} compared to λET from eddy covariance towers (Jiang and Islam, 2003) RMSD 0.08 to 0.19 and R^2 0.4–0.7 in Λ from MODIS and AVHRR (Venturini et al., 2004)	Moran et al., 1994
T_R-albedo scatterplot (moved up in table because it is above Day-Night Tr in text)	Requires small number of *in-situ* observations; Realistic assumption (extreme T for the wet and dry conditions vary with changing surface reflectance)	Requirement to identify extreme points in the scatterplot domain	Operational products in small geographic areas		Error 90 Wm^2 for instantaneous ET and 1mm/d daily ET compared to lysimeters (Gomez et al., 2005, Sobrino et al., 2005) RMSD 64 Wm^{-2}, R^2 0.85 compared to daily ET (Zahira et al. (2009); RMSE 25–32% compared to mean daily ET from Bowen ratio tower (Bhattacharya et al., 2010)	S-SEBI: Roerink et al., 2000
Day-night T_R difference and VI scatterplot	Requires small number of *in-situ* observations	Assumes three dominant land cover types; Requires both day and night time observations	Areas with three dominant land cover types	Areas with mixed land cover types	Errors 2.8%–3.9% compared to daily ET from lysimeters (Chen et al 2002); RMSD 0.106, bias −0.002, R^2 0.61 in Λ (Wang et al., 2006)	Chen et al., 2002; Wang et al., 2006
Triangle-SVAT model	Non-linear interpretation of T_R/VI feature space; Potential for deriving additional parameters (soil surface moisture, daytime mean λET and H fluxes)	Large number of input parameters; Requires user expertise	Operational products; Global and regional scale sites	Homogenous land cover	RMSD ±10% for daily ET from FIFE/MONSOON (Gillies et al., 1997); Error ~15–50% for λET from in situ and airborne measurements (Brunsell and Gillies, 2003); RMSD 40 Wm^{-2} compared to λET from CarboEurope	Carlson, 2007

temperature (λ). It is sometimes used instead of ET because it is a rate that can be expressed for a given instant in time and is also the variable measured by several field techniques. We use ET and λET interchangeably depending on the method being described.

3.2.1 Net Radiation

EO-based methods for estimating ET, including all methods reviewed in this chapter, depend on accurate determination of R_n, (in units of W/m^2) which is calculated as

$$R_n = (1-\alpha)SW\downarrow + (LW\downarrow - LW\uparrow) \qquad (3.1)$$

where

α is broadband blue-sky albedo (dimensionless)
SW$\downarrow$ is incoming shortwave radiation (W/m^2)
LW$\downarrow$ is downwelling longwave radiation
LW$\uparrow$ is upwelling longwave radiation (W/m^2)

For field-scale applications, R_n is often estimated with meteorological data alone using a variety of methods reviewed in several publications (e.g., Allen et al., 1998).

For large regions or areas without adequate meteorological data, R_n can be calculated from well-established approaches based primarily on EO data. An excellent review of the different methods for the estimation of the different components of the radiation budget from EO sensors, including the operationally distributed products available, was provided recently by Liang et al. (2013). SW$\downarrow$ and LW$\downarrow$ depend on atmospheric properties that can be estimated accurately with coarse-resolution datasets (1 degree). α and LW$\uparrow$ depend on surface conditions, including reflectivity and temperature, which are more spatially variable. In the following text, we review global datasets for SW$\downarrow$ and LW$\downarrow$ and other methods to estimate α and LW$\uparrow$.

3.2.1.1 Regional and Global Datasets for Net Radiation

At regional and global scales, R_n can be estimated with gridded data from surface climate reanalysis that assimilate remote sensing data or from EO data alone, with errors of $\pm10\%$–20% compared to ground measurements (Bisht et al., 2005). Gridded datasets used for regional to global scale R_n estimation can be separated into two spatiotemporal categories: 1979–present (1° spatial resolution) and 2000–present (0.25° spatial resolution). The higher spatial resolution post-2000 is due to the launch of the moderate resolution imaging spectroradiometer (MODIS) satellites and other EO systems that facilitate the downscaling of the surface energy budget (Gottschalck et al., 2005). Liang et al. (2010, 2013) provide good introductions to commonly used R_n datasets and associated uncertainties. For 1979–present, several coarse resolution and downscaled sources exist. The most commonly used is the global energy and water cycle experiment surface radiation budget (Gupta et al., 1999), which provides three-hourly shortwave and longwave radiation fluxes at one-degree resolution. These data are generated

primarily from the International Satellite Cloud Climatology Project (Rossow and Schiffer, 1991, 1999; Schiffer and Rossow, 1983) and Global Modeling and Assimilation Office (National Aeronautics and Space Administration, 2015) meteorology. The original dataset covering 1983–2007 has been expanded to cover from 1979 to present, as part of the Modern ERA-Retrospective Analysis for Research and Applications (MERRA) (Rienecker et al., 2011) dataset. MERRA is updated regularly with remote sensing and observed data and fed through a land surface catchment hydrology model, which provides additional outputs and further reduces inconsistencies. Another one-degree resolution surface reanalysis dataset is the Global Land Data Assimilation System (GLDAS) (Rodell et al., 2004) product. GLDAS assimilates NOAA/Global Data Assimilation System atmospheric fields, Climate Prediction Center merged analysis of precipitation fields, and observation-driven shortwave and longwave radiation using the Air Force Weather Agency's agricultural meteorological modeling system (Idso, 1981; Shapiro, 1987) to parameterize four land surface realizations: (1) Noah land surface model (Chen et al., 1996, 1997), (2) community land model (Bonan, 1998), (3) mosaic (Koster and Suarez, 1996), and (4) variable infiltration capacity model (Liang et al., 1994). The forcing data for GLDAS, like MERRA, is produced at three hourly intervals at one-degree resolution from 1948 to present (Rienecker et al, 2011). Although at much coarser spatial resolution, two other datasets are commonly used: National Centers for Environmental Prediction–National Center for Atmospheric Research (NCEP-NCAR) (Kalnay et al., 1996) and European Centre for Medium-Range Weather Forecasts (ECMWF) interim reanalysis (Morcrette, 1991, 2002). NCEP-NCAR shortwave and longwave fluxes are available at six hourly intervals from 1948–present at 2.5 degree resolution, while the ECMWF interim shortwave and longwave flux is available at six hourly intervals spanning1979–present at 1.5 degree resolution. Sheffield et al. (2006) use surface elevation to downscale the NCEP-NCAR to one-degree resolution. Other downscaled R_n products can be found on the corresponding Princeton University Terrestrial Hydrology Research Group webpage (http://hydrology.princeton.edu/data.php).

For regional estimation of R_n after 2000, EO data are incorporated directly or indirectly into one of the aforementioned reanalysis datasets. These methods involve several assumptions or ground-based estimates of α, T_R, and emissivity for model calibration (Bisht et al., 2005). Since 2000, MODIS products, including aerosol depth, temperature, emissivity, air temperature, dew point temperature, and α, have been combined and extrapolated from once-a-day measurements to daily flux at a quasi-1 km resolution. The Clouds and Earth's Radiant Energy System initially aboard NASA's Tropical Rainfall Measurement Mission platform and later placed on NASA's Terra and Aqua platforms is a radiometer that collects solar-reflected, Earth-emitted, and total radiation to determine Earth's radiation budget. Data is available from 2000–present at three hourly, monthly average, or monthly average by an hour at one-degree resolution. The Land Surface Analysis Satellite Applications Facility (http://landsaf.meteo.pt/) has also developed a radiation budget for the Africa and Europe using the spinning enhanced visible and infrared imager (SEVIRI) radiometer onboard the Meteosat Second Generation (MSG) satellite. The MSG/SEVIRI platform provides 30 min 3-km resolution α, land surface temperature, emissivity, SW↓, and LW↓ information from 2004 to present.

Satellite methods to estimate components of R_n are continually evolving and being evaluated against each other, including for shortwave (Ma and Pinker, 2012), longwave (Gui et al., 2010), and R_n (Bisht and Bras, 2010), and it can be anticipated that reanalysis datasets that incorporate satellite imagery will continue to be improved.

ET mapping often uses vegetation indices or T_R data that has a higher spatial resolution (e.g., 30 m, 250 m, 1 km) than is available from the global grids of radiation, but not all parts of the radiation budget need to be downscaled in detail. Incoming shortwave (SW↓) and incoming longwave (LW↓) are determined primarily by atmospheric properties and are therefore often assumed homogeneous over a given cell in the global gridded products, so they can be taken directly from the gridded data, though some applications (e.g., MOD16) interpolate to the resolution of the other satellite imagery used to map vegetation indices or radiometric surface temperature in order to avoid abrupt changes at cell boundaries (Mu et al., 2011). MOD16 calculates net longwave as a function of grid-cell average air temperature, which is in turn taken from the global gridded dataset (MERRA) and interpolated to 1 km using nonlinear interpolation on the four nearest neighbors (Mu et al., 2007).

3.2.1.2 Outgoing Shortwave and Longwave at High Spatial Resolution

In contrast to incoming radiation, SW↑ and LW↑ depend strongly on land surface properties and so may exhibit significant spatial variation over short distances, particularly in agricultural areas with sharp boundaries in vegetation with different levels of soil moisture stress. Algorithms for calculating albedo from Landsat imagery are included in several ET estimation models, including mapping evapotranspiration at high resolution with internalized calibration (METRIC) (Allen et al., 2007), and an albedo product is directly available for MODIS (MOD43B3). A review of the methods is also presented in Liang et al. (2013). Outgoing longwave radiation can be calculated at high spatial resolution using surface radiometric temperature from satellite imagery and an estimate of the surface emissivity. This approach is used in the METRIC and surface energy balance algorithm (SEBAL) models (Allen et al., 2007) and in many other applications (Bisht et al., 2005; Tang and Li, 2008):

$$LW \uparrow = \varepsilon_o \sigma T_R^4 \qquad (3.2)$$

where

ε_o is the broadband surface emissivity

σ is the Stefan–Boltzmann constant (5.67×10^{-8} W/m² K)

T_R is the radiometric surface temperature, which is often assumed equal to the surface temperature

Some remote sensing products include estimates of ε_o, including MODIS and advanced spaceborne thermal emission and reflection radiometer (ASTER), though there may be significant errors over heterogeneous landscapes (Liang et al., 2013).

3.2.1.3 Available Energy and the Ground Heat Flux

Available energy is the amount of R_n left over after accounting for the ground heat flux (G; W/m²) and is calculated as R_n–G. G is usually close to zero over 24 h, weekly, or 10 day periods and is assumed to be zero in several methods (e.g., Table 3.2), but G can be significant at the instant of satellite overpass and is particularly important for energy-based methods. Instantaneous G can account for up to 50% of R_n for sparse vegetation and can average 20%–30% for normalized difference vegetation index (NDVI) values up to around 0.6 (Bastiaanssen et al., 1998), so it cannot be neglected unless there is a high canopy cover fraction. Instantaneous G can be calculated with a variety of algorithms (Murray and Verhoef, 2007), the simplest of which is to assume that G is a constant fraction of R_n, usually between 0.2 and 0.5 at midday (Choudhury, 1989), with specific model applications using constant values of, for example, 0.35 (Norman et al., 1995) or 0.31 (Anderson et al., 1997). Other models include those where G/R_n is a function of NDVI, which is applicable over vegetated areas but not water (Morse et al., 2000):

$$G = 0.30\left(1 - 0.98\text{NDVI}^4\right)R_n \qquad (3.3)$$

G can also be estimated by a more complicated empirical equation (Table 3.3) that describes heat transfer using T_R, α and an extinction factor that describes attenuation of radiation through canopies using NDVI (Clothier et al., 1986; Choudhury, 1989; Kustas and Daughtry, 1990; Van Oevelen, 1991). More detailed approaches incorporate soil properties (Murray and Verhoef, 2007). Using empirical equations, the error in G is often around 20%–30% (Petropoulos, 2013). Over water bodies, G is usually larger and requires a different equation, often calibrated to local measurements (Morse et al., 2000). The user is recommended to try a few different equations for G to test its sensitivity to the equation used.

3.2.2 Vegetation-Based Methods for ET Estimation

Vegetation-based methods to estimate ET use an index of vegetation biomass or leaf area index (LAI) to calculate crop ET. One of the most widely available global ET datasets, MOD16, uses a variant of the Penman–Monteith (PM) equation to estimate ET (Leuning et al., 2008; Mu et al., 2007, 2011; Nishida and Nemani, 2003). MOD16 is also referred to as PM-Mu in the literature. MOD16 and several other vegetation-based methods use the PM

TABLE 3.2   Inputs and Calculation Steps for the PT-JPL Model

Inputs	Units	Description	Source or Equation
1. R_n	W/m²	24 h mean net radiation	Equation 3.1
2. r_{NIR}		Near-infrared spectrum reflectance	Imagery
3. r_{VIS}		Visible spectrum reflectance	Imagery
4. T_{max}	°C	Maximum temperature	Meteorological data
5. e_a or RH	kPa	Water vapor pressure	Meteorological data
	—	Relative humidity	Meteorological data
Derived Variables			
1. G	W/m²	Ground heat flux	Assumed zero
2. SAVI	—	Soil-adjusted vegetation index	$1.5(r_{NIR} - r_{VIS})/(r_{NIR} + r_{VIS} + 0.5)$
3. NDVI	—	Normalized difference vegetation index	$(r_{NIR} - r_{VIS})/(r_{NIR} + r_{VIS})$
4. f_{APAR}	—	Fraction of PAR absorbed by green vegetation	m_1SAVI + b_1 $m_1 = 1.2 \times 1.136$ $b_1 = 1.2 \times -0.04$
5. f_{IPAR}	—	Fraction of PAR intercepted by all vegetation cover	m_2NDVI + b_2 $m_2 = 1$, $b_2 = -0.05$
6. LAI	—	Leaf area index	$-\ln(1 - f_{IPAR})/k_{PAR}$ $k_{PAR} = 0.5$
7. R_{ns}	W/m²	Net radiation to the soil	$R_{ns} = R_n(\exp(-k_{Rn}LAI))$ $k_{Rn} = 0.6$
8. f_{wet}	—	Relative surface wetness	RH^4
9. f_g	—	Green canopy fraction	f_{APAR}/f_{IPAR}
10. T_{opt}	°C	Optimum plant growth temperature	T_{max} at max($PARf_{APAR}T_{max}/VPD$)
11. f_T	—	Plant temperature constraint	$\exp\left(-\left(\dfrac{T_{max} - T_{opt}}{T_{opt}}\right)^2\right)$
12. f_{SM}	—	Soil moisture constraint	$RH^{VPD/\beta}$ $\beta = 1.0$ kPa
13. λET_c	W/m²	Transpiration from dry canopy	Equation 3.9
14. λET_I	W/m²	ET from wet canopy	Equation 3.10
15. λET_s	W/m²	Soil evaporation	Equation 3.11

Source: Fisher, J.B. et al., *Remote Sens. Environ.*, 112, 901, 2008. With permission.

equation to calculate ET from crop canopies and soil surfaces (Mu et al., 2011):

$$\lambda ET = \frac{s(R_n - G) + \rho C_p D / r_a}{s + \gamma(1 + r_s / r_a)} \qquad (3.4)$$

where

λET is latent heat flux (W/m²)

ET is mean daily evapotranspiration (mm/s, converted to mm/day by multiplying by 86,400 s/day)

λ is the latent heat of vaporization ($\sim 2.26 \times 10^6$ J/kg)

s (Pa/K) is the slope of the curve relating saturated vapor pressure (e_{sat} in Pa) to air temperature

ρ is air density (kg/m³)

C_p is the specific heat capacity of air (J/kg K)

D is the vapor pressure deficit ($e_{sat}-e$, where e is actual vapor pressure in Pa)

r_a is the aerodynamic surface resistance (s/m)

γ is the psychometric constant (~ 0.066 kPa/K or as calculated in Mu et al., 2007)

r_s is the resistance of the land surface or plant canopy to ET (s/m)

R_n-G is for the 24 h period, so G is usually close to zero

Several of these parameters (s, γ, ρ, C_p, e, e_{sat}) are determined from meteorological data or elevation and do not depend on satellite-derived vegetation characteristics. The meteorological inputs are taken from either local meteorological stations or gridded global meteorological datasets and include air temperature, which is used to calculate e_{sat}, wind speed, and RH, which is used to calculate e. The two main parameters that control ET for different vegetation types and different levels of soil moisture stress are r_a and, often more importantly, r_s. Reference ET (ET$_o$) is calculated with r_a and r_s parameters for a reference surface, for example, a grass 12 cm tall with an r_s of 70 s/m and albedo 0.23 (Allen et al., 1998).

Alternatively, the Priestley–Taylor equation has been used in global models, particularly the Priestly–Taylor jet propulsion laboratory (PT-JPL) model (Fisher et al., 2008) and in several scatterplot methods (Jiang and Islam, 2001):

$$\lambda ET = \alpha_{PT} \frac{s}{s + \gamma} (R_n - G) \qquad (3.5)$$

where α_{PT} (unitless) is an empirical coefficient

TABLE 3.3 Inputs and Calculation Steps for the SEBAL Model

Inputs	Description	Source
1. R_n	Instantaneous net radiation	Eq 1
2. R_{n24}	24-hour mean net radiation	Eq 1
3. NDVI	Normalized difference vegetation index	Imagery
4. T_R	Radiometric land surface temperature (K)	Imagery
5. U	Wind speed (m/s)	Wind Speed from meteorological station or gridded data
6. z_{0m}	Surface roughness (m)	$z_{0m} = 0.123 * h$, where h is vegetation height in meters from land cover map or NDVI
7. Elevation	Surface elevation (m)	Digital elevation model (DEM)

Derived Variables	Description	Equation[a]
1. G	Ground heat flux (W/m²)	$G = R_n * \left(\dfrac{T_R - 273.15}{\alpha} \right)(0.0032\alpha + 0.0062\alpha^2)(1 - 0.978\text{NDVI}^4)$ where α is daytime-average albedo[b]
2. U_x^*	Friction velocity at meteorological station (m/s)	$U_x^* = \dfrac{0.41U}{\ln\left(\dfrac{z_2}{z_{0m}\ \text{at station}}\right)}$
3. U_{200}	Wind speed at blending height (200m) above the meteorological Station (m/s)	$U_{200} = U_x^* \dfrac{\ln\left(\dfrac{200}{z_{0m}\ \text{at station}}\right)}{0.41}$
4. U^*	Initial value of Friction velocity	$U^* = \dfrac{0.41U_{200}}{\ln\left(\frac{200}{z_{0m}}\right)}$
5. R_{ah}	Initial value of Aerodynamic resistance to heat transport	$R_{ah} = \dfrac{\ln\left(\dfrac{z_2}{z_1}\right)}{(0.41U_*)}$
6. Dry pixel	The dry pixel selected for calibration	Selected from the image manually[a], or automatically as pixel with the lowest NDVI from subset of pixels with highest T_R[c]
7. Wet pixel	The wet pixel is selected for calibration	Selected from the image manually[a], or automatically as pixel with the highest NDVI from subset of pixels with lowest T_R[c]
8. $T_{R\,Dry}$, $T_{R\,Wet}$	T_R at dry and wet pixels	Imagery

Iteration loop starts here:

9. T_2	Air temperature at height z_2	$T_2 = T_R - dT$ Assume $T_2 = T_R$ for the first iteration
10. P_a	Air pressure (hPa)	$P_a = 101.3\left(\dfrac{T_2 - 0.0065z}{T_2}\right)^{5.26}$ where z is elevation of pixel or region (m)
11. ρ_{air}	Air density (kg/m³)	$\rho_{air} = \dfrac{1000P_a}{1.01T_2\,287}$
12. a	Calibration coefficient a	$a = \dfrac{dT_{Dry}}{T_{R\,Dry} - T_{R\,Wet}}$ where $dT_{Dry} = \dfrac{H_{Dry}R_{ah\,Dry}}{\rho_{air}C_p}$ and $H_{Dry} = R_{nDry} - G_{Dry}$
13. b	Calibration coefficient b	$b = aT_{R\,Wet}$
14. dT	Temperature difference between z_1 and z_2 (K)	$dT = aT_R - b$
15. H	Sensible heat flux (W/m²)	$H = \dfrac{\rho_{air}C_p dT}{R_{ah}}$ where $C_p = 1004$ J/kg/K
16. L	Monin-Obhukov length (m), atmospheric stability condition	$L = -\dfrac{\rho_{air}C_p U_*^3 T_R}{0.41gH}$ where g = 9.81 m/s²

(Continued)

TABLE 3.3 *(Continued)*　　Inputs and Calculation Steps for the SEBAL Model

Inputs	Description	Source
17. ψH	Stability correction for heat transport under unstable (L < 0), stable (L > 0), and neutral (L = 0) atmospheric conditions[d]	$L < 0: \psi H = 2\ln\left(\dfrac{1+x_{z_2}^2}{2}\right)$ where $x_{z_2} = \left(1-16\left(\dfrac{z_2}{L}\right)\right)^{\frac{1}{4}}$ $L > 0: \psi H = -5\left(\dfrac{z_2}{L}\right)$ $L = 0: \psi H = 0$
18. ψM	Stability correction for momentum transport under Unstable (L < 0), Stable (L > 0), and Neutral (L = 0) atmospheric conditions[d]	$L < 0: \psi M = 2\ln\left(\dfrac{1+x_{200}}{2}\right) + \ln\left(\dfrac{1+x_{200}^2}{2}\right) - 2\arctan(x_{200}) + \dfrac{\pi}{2}$ where $x_{200} = \left(1-16\left(\dfrac{200}{L}\right)\right)^{\frac{1}{4}}$ $L > 0: \psi M = -5\left(\dfrac{z_2}{L}\right)$ $L = 0: \psi M = 0$
19. U_*	Friction velocity with stability correction for momentum transport	$U_* = \dfrac{0.41 \times U_{200}}{\ln\left(\dfrac{200}{z_{0m}}\right) - \psi M}$
20. R_{ah}	Aerodynamic resistance to heat transport with stability correction for heat transport	$R_{ah} = \dfrac{\ln\left(\dfrac{z_2}{z_1}\right) - \psi H}{0.41 U_*}$

Iteration: Repeat steps 9-20 until changes in H are <5%, usually ~ 5-10 iterations

21. Λ_{op}	Evaporative fraction at overpass (dimensionless)	$\Lambda_{op} = \dfrac{R_n - G - H}{R_n - G}$
22. ET_{24}	24 hour evapotranspiration (mm/day)	$ET_{24} = \dfrac{86,400 \Lambda_{op} R_{n24}}{\rho_w \lambda}$ where 86,400 is s/day, ρ_w is density of water (kg/m³), λ is latent heat of vaporization (J/kg).

　[a] Morse et al., 2000.
　[b] Bastiaanssen et al, 1998. Here we assume daytime-average α equals instantaneous α.
　[c] Messina (2012).
　[d] Paulson (1970); Webb (1970).

For open water and vegetation without soil moisture limitation, α_{PT} is 1.26, though adjustments may be applied in different environments. λET can be converted to ET in mm by dividing by the latent heat of vaporization (λ) (Allen et al., 1998) and an appropriate time constant (e.g., to convert from mm/s to mm/day).

Three basic approaches are used in the application of (3.4) or (3.5) to estimate actual ET: (1) empirical crop coefficient methods, (2) physically based coefficient methods, and (3) canopy resistance methods. Each of these three methods is reviewed in the following text.

3.2.2.1　Empirical Vegetation Methods: Crop Coefficients

3.2.2.1.1　Calibration Methods

Crop coefficient methods calculate ET as the product of ET_o and a crop coefficient. In the original FAO-56 model of crop ET (Allen et al., 1998), under conditions of soil moisture limitation, the coefficient describing the effect of crop type and growth stage (K_c) is multiplied by a coefficient quantifying the effect of soil moisture stress (K_s) (Allen et al., 1998, Equation 3.81). Most satellite methods estimate the product ($K_c K_s$), which is also sometimes called the reference evapotranspiration fraction (ET_f). In order to simplify notation, and because the method is often used for other vegetation types besides crops, here, we use the term "reference ET fraction" and the symbol ET_f to represent $K_c K_s$ throughout the chapter. Some references refer to the crop coefficient derived from EO data as K_c or K_{c-VI} (Glenn et al., 2011); here, we call it ET_f to be consistent with recent publications (Nagler et al., 2013), to highlight similarities with other methods that use ET_f (Allen et al., 2007) and reduce confusion between K_c in the original FAO-56 method, which is for conditions of no water stress and estimates potential ET for a given crop, and $ET_f = K_c K_s$, which estimates actual ET and is what EO-based crop coefficient methods determine.

In EO-based methods, ET_f is modeled as a function of a vegetation index (VI) using several possible empirical equations

(Equations 3.6 through 3.8), including as a power function (Nagler et al., 2013):

$$ET_f = a_K VI^{\eta} \qquad (3.6)$$

where

a_K is an empirical coefficient determined by regression

η is a coefficient that varies by the VI

Alternatively, ET_f can be modeled by first normalizing by extreme VI values in the image (Glenn et al., 2010):

$$ET_f = \left[1 - (VI_{max} - VI)/(VI_{max} - VI_{min})\right]^{\eta} \qquad (3.7)$$

where

VI_{max} is the VI value when ET is at a maximum

VI_{min} is the VI of bare soil (VI=0)

η is often close to 1 for some vegetation indices (enhanced vegetation index [EVI], soil-adjusted vegetation index [SAVI]) but may be less than 1 for NDVI due to NDVI's lack of sensitivity for leaf area indices greater than ~3 (Glenn et al., 2011).

Note that this equation assumes evaporation is zero when VI equals VI_{min}, which may not be the case for wet soil, including irrigated fields at initial growth stages. Soil evaporation can be included by introducing a second coefficient K_e that is determined by modeling the soil drying curve after precipitation or irrigation events, which is independent of remote sensing data (Glenn et al., 2010).

Other formulations of the ET_f–VI relationship include those derived from Beer–Lambert law of absorption of light by a canopy, assuming a linear relationship between EVI and LAI (Nagler et al., 2013) (Figure 3.1):

$$ET_f = a_K\left(1 - \exp(-b_K EVI)\right) - c_K \qquad (3.8)$$

where a_K, b_K and c_K the coefficients determined by regression of EVI against observed ET_f in pixels that have ground-level measurements of ET.

The coefficients in Equations 3.6 through 3.8 may vary by crop type, climate, or soil type, so implementation depends on availability of ground data of ET from lysimeters, eddy flux correlation towers, sap flow measurements, or other methods to calibrate the ET_f–VI relationship. If data from lysimeters or towers are used for calibration, the equation estimates ET, and if data from sap flow measurements are used, the equation estimates transpiration only. The method assumes that all crops with identical VI have the same ET_f values. While the coefficients in the ET_f–VI relationship may be constant over several vegetation types (Nagler et al., 2009, 2013), they may vary under different climates, soil types, or soil moisture stress conditions, so the method is typically used to estimate ET over relatively

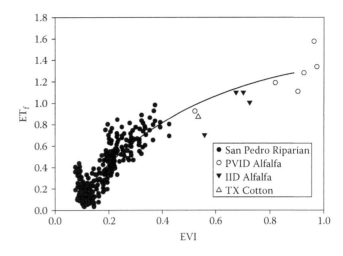

FIGURE 3.1 Reference evapotranspiration fraction (ET_f) from field measurements versus the enhanced vegetation index (EVI). The sites span a range of geographic locations (San Pedro River, AZ; Imperial Irrigation district, CA; Palo Verde irrigation district, CA; Texas). The line is the best fit regression line. (Reprinted from Nagler, P.L. et al., *Remote Sens.*, 5, 3849, 2013. With permission from MDPI.)

small areas with available data. Regional ET_f–VI curves have been constructed for various locations in the western United States, Spain, and Australia with an accuracy of within 5% of measured values on an annual basis (Nagler et al., 2013), suggesting the possibility of monitoring ET without field-by-field knowledge of cropping patterns. The spatial and temporal variability in the ET_f–VI relationships and the size region that can be applied with a given accuracy need further documentation.

The accuracy of empirical crop coefficient methods has been assessed using flux towers, soil water balances, and annual water balances. Nagler et al. (2013) reported a standard mean error in ET_f of 0.12 in the application of Equation 3.8 to MODIS EVI data compared to flux towers (riparian areas and irrigated alfalfa) and soil moisture balance (cotton). Glenn et al. (2010), in a review of numerous applications of empirical ET_f methods, report that the empirical methods have root-mean-square deviation (RMSD) in the range of 10%–30% of mean ET across several different biomes. The main disadvantage of the method is the requirement of meteorological data required to estimate ET_o and some ground-level measurements of ET to calibrate the ET_f–VI equation.

3.2.2.1.2 Application of the Crop Coefficient Approach to Agricultural Crops

Bausch and Neale (1987) and Neale et al. (1989) first established the validity of the empirical crop coefficient (ET_f) method at two experimental farm sites in Colorado. They grew irrigated corn in fields equipped with weighing lysimeters, and ET_f was calculated with alfalfa grown at the same facility as the reference crop (i.e., $ET_f = ET_{corn}/ET_{alfalfa}$). NDVI was measured with radiometers suspended over corn canopies. NDVI was strongly correlated with LAI and fractional vegetation cover, and ET_f derived from radiometric measurements closely tracked measured ET_f over the crop cycle. Choudhury et al. (1994) used a modeling approach to show the theoretical justification for replacing ET_f from ground-based measurements with ET_f estimated from a VI, with ET_f from the VI replacing canopy resistance terms in the PM equation. The ET_f method has since been successfully applied to a wide variety of crops as outlined in the following examples.

Vineyards and orchards are difficult to model with normal crop coefficients due to differences in plant spacing and other crop variables among plantings. By using NDVI and SAVI from satellite images, Campos et al. (2010) accurately predicted actual ET by simple linear regression equations in grape orchards in Spain. Samani et al. (2009) used NDVI from Landsat imagery to develop field-scale ET_f values for pecan orchards in the lower Rio Grande Valley, New Mexico, Unites States. Their estimates of ET were within 4% of values measured at an eddy covariance flux tower. Only 5% of fields were within the range of ET and ET_f set by expert opinion, indicating the potential for significant water savings.

Wheat has been extensively studied due to its importance as an irrigated or dryland crop around the world. Duchemin et al. (2006) used NDVI from Landsat images to map LAI and ET in irrigated wheat fields in Morocco and reproduced ground measurements of ET within 15%. Wheat fields varied widely in ET, opening up the possibility of improving irrigation efficiency by tailoring water applications to actual crop needs determined by satellite imagery. Gontia and Tiwari (2010) developed field-specific crop coefficients for a wheat field in West Bengal, India, using ET_f values determined from NDVI and SAVI from satellite sensors.

Extensive work on wheat has been conducted by the USDA Agricultural Research Service in a desert irrigation district in Maricopa, Arizona, United States (Hunsaker et al., 2005, 2007a,b). They developed NDVI-derived ET_f values that tracked within 5% of measured ET_f for wheat grown in weighing lysimeters. Wheat was then grown in field plots for two seasons under stress and nonstress conditions. The NDVI method gave more accurate predictions of actual irrigation demands than the FAO-56 method under all treatment conditions with a root-mean-square error (RMSE) of about 15% of measured water use with no bias toward under or overestimation across treatments.

Similar studies have been conducted with cotton in the southwestern Unites States (Hunsaker et al., 2003, 2005, 2009) and Spain (González-Dugo and Mateos, 2008). The latter study was conducted in a large irrigation district and the results showed that considerable water savings could be achieved by scheduling irrigations based on NDVI-derived crop models rather than ET_f values determined for crops grown under optimal conditions. ET_f methods have been developed for other crops, including potato (Jayanthi et al., 2007), broccoli (El-Shikha et al., 2007), sugar beet (González-Dugo and Mateos, 2008), soybean (Gonzalez-Dugo et al., 2009), the oilseed crop camelina (Hunsaker et al., 2011), sorghum (Singh and Irmak, 2009), and alfalfa (Singh and Irmak, 2009). All these studies reported positive results and pointed to the possibility of considerable water savings by replacing static FAO-56 crop coefficients with locally derived ET_f values.

Often, the main interest is determining district-wide water demand or consumptive use, which requires estimating ET over mixed crop areas. Choudhury et al. (1994) pointed out that the ET_f–VI relationship was not necessarily crop specific and that the ET_f approach might be used over mixed crops without a serious loss of accuracy. Allen and Pereira (2009) found a reasonable agreement between measured ET_f and vegetation cover fraction (f_c) over a wide range of tree crops, and the relationship was improved by including plant height in the regression. Similar findings were reported by Trout et al. (2008) for tree and vegetable crops, and a close correlation was noted between Landsat-derived f_c based on NDVI and f_c measured on the ground over 30 fields with crops ranging from trees (almonds, pistachios), to vines (grapes), and row crops (onions, tomatoes, cantaloupes, watermelon, beans, pepper, garlic, and lettuce). The only aberrant crop was red lettuce, which has a low NDVI due to reflection of red light. They developed an operational ET-monitoring program for California's irrigation districts based on Landsat-derived NDVI and ET_o from the California Irrigation Management Information System's network of micrometeorological stations (Johnson and Trout, 2012).

3.2.2.2 Physically Based Coefficient Methods: PT-JPL Model

For regional or global application, ground reference data are often not available at sufficient spatial density to support local calibration of the ET_f–VI relationship, so more physically based models that require minimum calibration have been developed for large geographic scales. The PT-JPL model of Fisher et al. (2008) estimates ET directly from satellite imagery with minimal ground data requirements. It is a coefficient method in that it estimates coefficients that are multiplied by ET_o to estimate ET, similar to the empirical ET_f method, but using physically based models to estimate the coefficients from satellite imagery and meteorological data. In the PT-JPL algorithm, λET is calculated as the sum of evaporation of water intercepted by the canopy (λE_I), evaporation from the soil surface (λE_s), and transpiration from the dry canopy (λE_c) (Table 3.2), where a "dry canopy" has no liquid water on the surface of the leaves. ET_f for transpiring vegetation is assumed to be the product of four coefficients that account for variations in surface wetness (f_{wet}), green canopy fraction (f_g), a plant temperature constraint (f_T), and a plant moisture constraint (f_M):

$$\lambda ET_c = \left(1 - f_{wet}\right) f_g f_T f_M \alpha_{PT} \frac{s}{s + \gamma} (R_n - R_{ns}) \tag{3.9}$$

where R_{ns} is R_n at the soil surface and is a function of vegetation cover (Table 3.2).

Alternate formulations are available that use fractional vegetation cover without calculating R_{ns} separately (Marshall et al., 2013). The variable descriptions and their equations are listed in Table 3.2. λE_I is calculated as

$$\lambda E_I = f_{wet} \alpha_{PT} \frac{s}{s + \gamma} (R_n - R_{ns}) \tag{3.10}$$

Soil evaporation (λE_s) is calculated separately but also has coefficients related to surface wetness and soil moisture:

$$\lambda E_s = \left(f_{wet} + f_{SM} \left(1 - f_{wet}\right)\right)\alpha_{PT} \frac{s}{s+\gamma}(R_{ns} - G) \qquad (3.11)$$

Several of the parameters in Table 3.2 may be adjusted according to local conditions. See Fisher et al. (2008) for more detail.

Fisher et al. (2008) compared the PT-JPL model predictions to measurements at FLUXNET eddy covariance towers, and report an RMSE of 16 mm/month and an error in annual ET of 12 mm/yr or 13% of the observed mean. The flux sites covered a range of biomes, including temperate C3/C4 crops but did not include irrigated cropland, which might be expected to have higher error due to high evaporation from inundated and wet soil at the beginning of the growing season (Yilmaz et al., 2014).

3.2.2.3 Physically Based Vegetation Methods: Canopy Resistance and MOD16

Canopy resistance methods predict the resistance parameters in the PM equation (r_a, r_s in Equation 3.4) directly from satellite imagery. Several resistance methods, including MOD16, are based on the model of Cleugh et al. (2007). MOD16 is vegetation based in that the primary inputs driving ET for a given amount of R_n are derived from VI. The fraction of photosynthetically active radiation is used to determine the fraction of the surface covered by crop canopy (f_c) and soil ($1-f_c$). The LAI is used to determine the dry canopy resistance to transpiration (r_{s_c}), the aerodynamic resistance (r_a), and wet canopy resistance to evaporation (r_{s_wetC}).

Dry canopy resistance to transpiration (r_{s_c}) is calculated using LAI, minimum air temperature, and vapor pressure deficit:

$$r_{s\_c} = \frac{1}{LAIc_L m\left(Tmin\right)m(D)} \qquad (3.12)$$

where
  c_L is the mean potential stomatal conductance per unit leaf area
  m(Tmin) and m(D) are multipliers (range 0.1–1) that limit stomatal conductance by minimum air temperature (Tmin) and D

The specific equations for m(Tmin) and m(D) are given in Mu et al. (2007).

3.2.2.4 Vegetation-Based Methods and Soil Evaporation

In both the PT-JPL and MOD16 methods, ET increases with VI. Soil evaporation, including from both saturated and unsaturated surfaces, is assumed to increase with the fourth power of RH (see Table 3.2 and Figure 3.2). In MOD16, evaporation from saturated and unsaturated soils is calculated separately. The

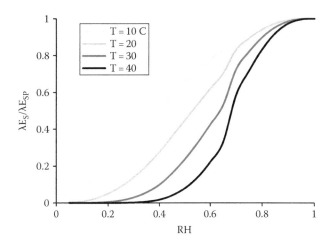

FIGURE 3.2 Soil moisture function.

water cover fraction or fraction of the soil that is saturated at the surface (F_{wet}) is assumed zero in grid cells where RH is less than 70% (0.7) (Mu et al., 2011 Equation 3.15):

$$F_{wet} = \begin{cases} 0.0 & | & RH < 0.7 \\ RH^4 & | & RH \geq 0.7 \end{cases} \qquad (3.13)$$

In MOD16, potential evaporation from soils, both saturated and unsaturated, is

$$\lambda E_{SP} = \frac{\left(s\left(R_n - G\right)+\left(\rho C_p D/r_{a_s}\right)\right)}{s+\gamma\left(1+r_{tot}/r_{a_s}\right)} \qquad (3.14)$$

where
  r_{tot} is the total aerodynamic resistance to vapor transport
  r_{a_s} is the aerodynamic resistance at the soil surface (Mu et al., 2011)

Total evaporation from both saturated and unsaturated soils, based on a rearrangement of Equation 3.27 in Mu et al. (2011), is

$$\lambda E_s = \lambda E_{SP}\left(F_{wet} + \left(1 - F_{wet}\right)f_{SM}\right) \qquad (3.15)$$

where f_{SM} is the same as f_{SM} in the PT-JPL method (Table 3.2).

A sample plot for a range of air temperatures shows that soil evaporation as a fraction of potential soil evaporation is very small (<0.2) for RH less than 0.4, particularly at high temperatures (Figure 3.2). At 40°C and RH = 0.5, for example, soil evaporation is less than 0.1 of the potential. The plot highlights the assumption that soil evaporation is predicted to be very low in arid and semiarid environments with low RH and high daytime temperature, since the model assumes that soil evaporation is a function of the regional climate, with more soil evaporation in humid regions. While this assumption may be accurate in rainfed systems, it may be problematic in irrigated areas in arid and semiarid climates, because soil moisture may be high locally due to irrigation, even if the grid-cell average RH is low. This could be a problem in particular for ET estimation in areas cultivated in rice or other crops that have a significant period of inundation or wet, bare soil.

The authors of the MOD16 algorithm have recognized problems with its performance in irrigated areas and wetland and have updated the algorithm in an application in the Nile River Delta. The revised MOD16 uses radiometric land surface temperature (T_R) to calculate a revised surface resistance to ET (Mu, 2013). In early 2014, NASA has officially approved this as the revised MOD16 product (Mu, 2013), and future evaluations of the algorithm should include irrigated areas and wetlands.

Mu et al. (2011) validated MOD16 against eddy flux covariance towers and report mean absolute bias in daily ET of 0.31–0.40 mm/day, or 24%–25% of observed daily ET when using GMAO MERRA for meteorological input. Their flux towers were located in a range of biomes in the United States and at several sites in the Brazilian Amazon. The validation included two irrigated sites, both of which had higher error (1.2 mm/day, 72%–76% of the observed mean) compared to the mean RMSE for all sites with flux tower data.

3.2.2.5 Comparison of Vegetation-Based Methods

Vegetation-based methods have gained great appeal, because they can be run globally and continuously with remote sensing and surface climate reanalysis at low computational cost. The PT-JPL model was evaluated against several other vegetation-based approaches in the humid tropics, given its simplicity and strong dependence on R_n. In regions where light limits carbon assimilation, such as the humid tropics, R_n is the dominant control on λET (Fisher et al., 2009). The PT-JPL model performed best overall, as PM (resistance based) methods include more parameters and therefore may be inherently more uncertain and more strongly coupled to the atmosphere. A comparison of PT-JPL and MOD16 show that PT-JPL overpredicted ET, while MOD16 had lower bias but higher overall error (Chen et al., 2014). Vinukollu et al. (2011a) and Marshall et al. (2013) point out that the performance of these models has been evaluated primarily with eddy covariance flux tower data and their performance can significantly degrade at larger spatial scales, due to the large uncertainties in surface climate reanalysis products, in particular RH.

3.2.3 Radiometric Land Surface Temperature Methods for ET Estimation

Radiometric land surface temperature (T_R) approaches are based on the fact that ET is a change of state of water that uses energy in the environment for vaporization and reduces surface temperature (Su et al., 2005). A subset of these methods is often called energy balance methods, since they solve the energy balance equation using T_R to partition R_n between the sensible and latent heat fluxes; here, we use the more generic term radiometric land surface temperature (T_R) method.

T_R methods have been used as early as the 1970s, when Stone and Horton (1974) used a thermal scanner to estimate ET, and

Verma et al. (1976) developed a resistance model with thermal imagery inputs. Since then, a variety of methods has been developed, including SEBAL (Bastiaanssen et al., 1998, 2002), METRIC™, the Surface Energy Balance System (SEBS) (Su, 2002), Atmosphere–Land Exchange Inverse (ALEXI) (Anderson et al., 1997), Disaggregation ALEXI (DisALEXI) (Norman et al., 2003), and operational simplified surface energy balance (SSEBop) (Senay et al., 2013). Most methods in this category use T_R to estimate components of the energy balance, though some simplified methods (e.g., simplified surface energy balance [SSEB]) use temperature directly without solving the energy balance. In the following text, we summarize the theoretical foundations of the energy balance methods, describe simplified approaches based on T_R, and highlight key differences in the most used algorithms.

In energy balance methods that use T_R, λET is computed as a residual of the energy balance equation:

$$\lambda ET = R_n - G - H \tag{3.16}$$

where

　G is soil heat flux
　H is the sensible heat flux (all in W/m²)

While there may be some energy exchange from photosynthesis, it is usually a small fraction of R_n, is not easily measured even by ground instrumentation (Wilson et al., 2001), and is assumed to be zero (Meyers and Hollinger, 2004). In vegetation having a significant amount of canopy, such as forests, energy exchange from photosynthesis can become high (7%–15%), particularly over short time intervals (Meyers and Hollinger, 2004).

The most important term in the energy balance equation after R_n–G is H, which can be estimated using either one- or two-source models (TSM) (Figure 3.3). One-source models, including SEBAL (Bastiaanssen et al., 1998), METRIC (Allen et al., 2007), and SEBS (Su, 2002), estimate evapotranspiration from the surface as a whole. TSM (ALEXI/DisALEXI) (Anderson et al., 1997) separate ET into E from soil and ET from the vegetation canopy, which is sometimes further separated into evaporation of intercepted water from a wet canopy and transpiration from a dry canopy as in the PT-JPL and MOD16 models. The separation into two sources results in two additional resistance variables that need to be estimated: Rx, the total boundary layer resistance of the canopy, and Rs, the sensible heat exchange resistance of the soil surface (Figure 3.3).

3.2.3.1 One-Source Models: SEBAL, METRIC, and SEBS

In one-source models (Figure 3.3) H is calculated as

$$H = \frac{\rho_{air} C_p \left(T_1 - T_2 \right)}{R_{ah}} \tag{3.17}$$

where

　ρ_{air} is the density of air (kg/m³)
　C_p is the specific heat of air (J/kg/K)

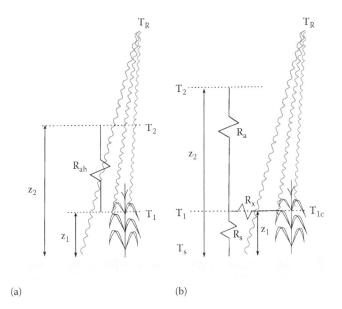

(a)   (b)

FIGURE 3.3   Schematic of one (a) and two-source (b) models for temperature-based ET calculations. T_R is the radiometric surface temperature detected at the satellite, and the grey lines indicate thermal radiation emission from the soil and canopy.

T_1 is the aerodynamic temperature (K) of the evaporating surface at height z_1 which is the height of the zero-plane displacement (d) plus the surface roughness for momentum transport (z_{0m})

T_2 is the air temperature at height z_2, which is usually the height where air temperature is measured (2 or 3 m above the evaporating surface)

R_{ah} is the aerodynamic resistance to turbulent heat transport from z_1 to z_2 (Figure 3.3)

The model assumes that evaporating surfaces have a temperature equal to or hotter than the air above ($T_1 \geq T_2$), resulting in a nonnegative sensible heat flux. The zero-plane displacement height (d) is the mean height where momentum is absorbed by the canopy, typically around 2/3 of the vegetation height (h), and z_{0m} is a relatively small fraction of the height of vegetation (0.03– 0.1 h, or 0.123 h in Morse et al., 2000), and so is around 0.03 m over grassland, 0.10–0.25 m over cropland, and 0.5–1.0 m over forest or shrubland. In practice, z_{0m} is estimated as a function of NDVI or with a land cover map (see Table 3.3).

There are two main uncertain variables in the calculation of H in Equation 3.17. First, the radiometric surface temperature (T_R) may differ from the actual aerodynamic temperature (T_1). Note that neither of the air temperatures in Equation 3.17 are directly sensed by the satellite, which estimates the radiometric surface temperature (T_R) based on thermal radiation reaching the sensor from the combined soil and canopy surfaces. The correspondence between T_R and air temperatures at either z_1 or z_2 varies by surface type, roughness, and crop canopy structure. Different one-source models have different strategies for estimating the temperature difference between z_1 and z_2. Some models include

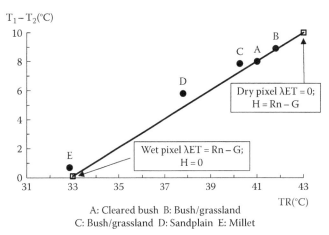

A: Cleared bush   B: Bush/grassland
C: Bush/grassland   D: Sandplain   E: Millet

FIGURE 3.4   Example of the assumption of the linear relationship between radiometric temperature (T_R) and the temperature difference between heights z_1 and z_2. Data are from Bastiaanssen (1998), and dry and wet pixels are added for illustration. (Reprinted from *J. Hydrol.*, 212–213, Bastiaanssen, W.G.M., Menenti, M., Feddes, R.A., and Holtslag, A.A.M., A remote sensing surface energy balance algorithm for land (SEBAL) 1. Formulation, 198–212, Copyright (1998) with permission from Elsevier.)

an extra term in R_{ah}, while others (e.g., SEBAL, METRIC) calibrate an empirical linear model relating T_1–T_2 to T_R. Second, R_{ah} has high spatial variability and may be difficult to predict.

In one-source models (SEBAL, METRIC, SEBS), a linear equation predicts the difference between T_1 and T_2 as a function of the radiometric surface temperature (Figure 3.4 and Table 3.3):

$$T_1 - T_2 = a + bT_R \tag{3.18}$$

where a and b are empirical parameters determined from the imagery in a process called "internalized calibration" (Allen et al., 2007).

Field investigations suggest that Equation 3.18 holds under a variety of conditions (Figure 3.4). Note that in applications of the method, a and b are determined from the wet and dry pixels only, with no field data of air temperatures for calibration and therefore no estimate of error of Equation 3.18.

Combining Equation 3.17 with Equation 3.18 gives

$$H = \frac{\rho_{air} C_p \left(a + bT_R\right)}{R_{ah}} \tag{3.19}$$

Identifying the parameters a, b, and R_{ah} requires calculating ρ_{air} and C_p (Table 3.3) and identifying some pixels where H and T_R are known. First, one pixel is selected that is "wet" or "cold", where H is assumed to be 0 and λE is assumed R_n–G, and another that is "hot" or "dry", where H is assumed to be R_n–G and λE is assumed to be 0 (Figure 3.4). An initial guess of R_{ah} is made for the image without correction for unstable atmospheric conditions. An initial guess of T_1–T_2 is made at the dry pixel by solving for it in Equation 3.17, using the assumption H = R_n–G. Coefficients a and b in (3.18) are then determined from the observed T_R and estimated T_1–T_2 at the wet and dry pixels (Table 3.3). H is then calculated again using (3.19), this time accounting for unstable atmospheric conditions using the Monin–Obukhov equations (Table 3.3) (Allen et al., 2011; Bastiaanssen et al., 1998, 2002). The values of a, b, and R_{ah} are then solved iteratively by updating the values of each until the result converges on H = 0 for the wet pixel and H = R_n–G for the dry pixel. The internal calibration of SEBAL and METRIC allows estimation of ET without knowing either T_1 or T_2, which is an advantage in data-scarce regions.

Sensitivity analysis suggests that T_R at the hot and cold pixels are the most important controls on H and λET estimates for a given image, followed by R_n at the hot pixel (Long et al., 2011). Since H is assumed to be zero at the cold pixel, R_n at the cold pixel does not influence the resulting model parameters and calculated H. The criteria for pixel selection are important, but there is no generally agreed method for selecting them (Long et al., 2011). Past applications of SEBAL and METRIC have used manual pixel selection, since user experience in the study area helps select the appropriate pixels that represent typical field conditions in the image, but manual pixel selection can add significantly to the processing time. The selection procedure can be automated, which reduces variability among users and allows for more rapid implementation (Kjaersgaard et al., 2009; Messina, 2012). In the automation algorithm of Long et al. (2011), the dry pixel is the pixel with the highest T_R in the subset of pixels with specified land use (bare, urban, or dry cropland), and the wet pixel is the pixel with the lowest T_R, after screening for cloud contamination. Automation of pixel selection in METRIC (Allen et al., 2013) uses a combination of NDVI, T_R, and α. Other semiautomated approaches simply select the highest and lowest T_R in a given image, using masks to exclude either clouds or nonrepresentative land covers. The reasons for excluding certain land covers for the wet and dry pixel selection are often not explicit and vary by application. For the wet pixel, some studies advocate excluding water bodies since they have different aerodynamic properties than agricultural fields where ET is being estimated (Conrad et al., 2007; Morse et al., 2000), while others include water bodies, particularly if vegetated pixels have much higher temperatures than open water bodies. For dry pixels, some studies exclude urban environments (Conrad et al., 2007), while others include them (Long et al., 2011). The impact of different selection rules, including which surfaces should or should not be included, has not been determined for a range of surface types and geographic regions.

One-source models have the convenience of being relatively simple to use and are calibrated to wet and dry pixels, reducing the need for meteorological data. However, the calibration is performed on a single image, and the a and b parameters from Equation 3.19 may only be valid for that image. While this may not be a problem for study areas the size of a single scene, areas that cover multiple scenes may suffer from problems of merging along scene boundaries. To the authors' knowledge, at the date of publication, there have not been any efforts to determine the spatial and temporal variability in the a and b parameters or evaluations of the extent and magnitude of scene boundary problems. The SSEBop model (Section 3.2.3.3) was designed to address scene boundary problems by estimating T_1–T_2 for each pixel under dry and wet conditions.

Summaries of the accuracies of SEBAL are available in Bastiaanssen et al. (2005) and Kalma et al. (2008), with numerous case studies (Teixeira et al., 2009). In general, the reported errors are higher for smaller spatial scales and small time intervals and are within the errors of measurements of the device used for validation, which is typically 10%–15%. Reported accuracies from numerous validation exercises using point- and field-scale instruments suggest that one-source models have errors around 50 W/m², or a maximum error of around 15%–30% for daily estimates (Kalma et al., 2008), though the errors may vary with the spatial resolution of the input data. Errors over long time scales, including the seasonal estimates of importance to water managers and assessments of water productivity, are typically lower (RMSE ~5%) due to cancelling out of daily errors (Bastiaanssen et al., 2005).

Most validation sites, both for SEBAL/METRIC and for EO-based ET methods in general, are located in relatively large plots of homogeneous vegetation, which facilitates comparison with satellite imagery but may not assess accuracy well over heterogeneous landscapes. SEBAL, for example, assumes minimal advection of energy among pixels, which is likely valid over large homogenous vegetation but may not be valid in heterogeneous irrigated landscapes in semiarid and arid climates. Advection may double the amount of ET in situations of extreme humidity gradients and high winds (Allen et al., 2011), which motivated the use of ET_o and ET_f in METRIC in place of R_n–G as used in SEBAL. For small irrigated plots in semiarid or arid climates, the assumption of no advection may be especially problematic, though this has not been systematically quantified using one-source models.

Water balance measurements have also been used to validate SEBAL ET at the scales of individual fields, watersheds, or irrigation projects (Bastiaanssen et al., 2002, 2005). Validation using water balances at the watershed scale is difficult in rainfed systems in arid and semiarid environments, since streamflow as a percentage of precipitation is often within the error of ET estimated by any method. Water balance validation is more feasible in surface irrigated systems, where inflows and outflows are large relative to ET (Bastiaanssen et al., 2002).

3.2.3.2 Two-Source Models: ALEXI and DisALEXI

TSMs account for differences in aerodynamic resistance between soil and vegetation, which are lumped into a single resistance parameter in single-source models. TSMs require estimation of the energy balance and therefore of T_1 and T_2 over vegetation and soil separately and so cannot use internal calibration to wet and dry pixels. One popular TSM, the ALEXI model uses the two-source energy balance model (TSEB) of Norman et al. (1995). In ALEXI, temperature of the soil (T_s) and canopy (T_{1c}) are estimated by separating radiometric temperature (T_R) by the vegetation cover fraction:

$$T_R = \left[f_c T_{1c}{}^4 + (1 - f_c T_s{}^4) \right]^{1/4} \tag{3.20}$$

where f_c is the fractional vegetation cover at a given view angle, calculated as

$$f_c = 1 - \exp\left(\frac{-0.5\Omega \text{LAI}}{\cos(\theta)}\right) \qquad (3.21)$$

where

Ω is an index of the degree of clumping from the given view angle

θ is the view angle

T_2 is estimated using an atmospheric boundary layer (ABL) model (Anderson et al., 2013) calibrated to the observed increase in temperature during the morning hours (from 1 to 1.5 h after sunrise to before local noon), which is obtained from geostationary satellites such as the geostationary operational environmental satellite (GOES) (Anderson et al., 2013). Like SEBAL, the use of temperature difference instead of absolute temperature avoids the need for in situ measurements of air temperatures or estimates of atmospheric corrections and is a significant advantage in data-scarce regions. The ABL model used in ALEXI is a relatively simple one that can be programmed as a system of equations.

The spatial resolution of the ALEXI model is constrained by the resolution of imagery from geostationary satellites (5–10 km), so a different algorithm, DisALEXI, uses higher-resolution imagery from MODIS (1 km) or Landsat (30 m) to generate high-resolution ET estimates using ALEXI results (Norman et al., 2003). DisALEXI utilizes the temperature and wind speed at the blending height (~50 m above the land surface) and downwelling short- and longwave radiation from ALEXI as input, assuming those four variables are spatially uniform over the resolution of the ALEXI model (5–10 km). The high-resolution thermal imagery is then adjusted to the view angle of the GOES satellite to ensure consistency in the radiometric temperature. The angle-adjusted radiometric temperature, vegetation cover, and land use maps from the high-resolution imagery are then used to calculate R_n at high resolution, and the TSM run on each high-resolution pixel with the ALEXI-derived temperature at 50 m as the upper boundary condition (Norman et al., 2003). The DisALEXI values are adjusted to match the mean ALEXI values by iteratively altering the air temperature map (T_2) until the aggregated DisALEXI values match the ALEXI ET values, ensuring consistency across scales.

Methods for fusing DisALEXI results from MODIS for daily resolution with Landsat for high spatial resolution have been developed and tested over rainfed (Cammalleri et al., 2013) and irrigated areas (Cammalleri et al., 2014), based on a data fusion strategy for MODIS and Landsat (Gao et al., 2006). The spatial and temporal adaptive reflectance fusion model finds the date with the highest correlation between Landsat- and MODIS-derived ET and uses that correlation structure to predict Landsat-scale ET on dates with only MODIS data available (Figure 3.5). The use of MODIS improves the estimation of ET over the use of a simple spline function on ET using available Landsat dates. The results highlight the importance of having daily MODIS estimates for some locations where vegetation responds to changes in moisture

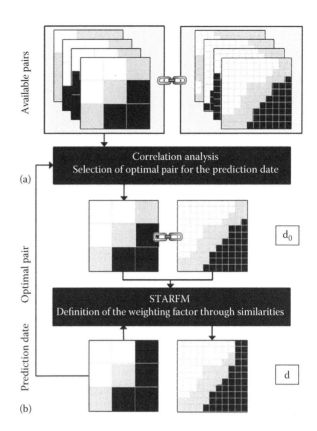

FIGURE 3.5 Schematic of the STARFM method for fusing MODIS and Landsat imagery for high spatial and temporal resolution of ET. In step (a), MODIS ET on the prediction date (d) is correlated with MODIS ET for each date where Landsat ET is also available. The date (d_0) with the highest correlation between the two MODIS ET images is selected for step (b), where a weighting function is calculated using the Landsat–MODIS pair on day d_0, and a map of ET with Landsat resolution is predicted for date d. (Reprinted from *Agric. Forest Meteorol.*, 186, Cammalleri, C., Anderson, M.C., Gao, F., Hain, C.R., and Kustas, W.P., Mapping daily evapotranspiration at field scales over rainfed and irrigated agricultural areas using remote sensing data fusion, 1–11, Copyright (2014), with permission from Elsevier.)

(Figure 3.6). At some sites (162), little difference was observed between the interpolated Landsat ET and the MODIS–Landsat fusion product, but at other sites (161), ET increased rapidly and was higher than the interpolated Landsat values for a 15–20 day period following a rainfall event, showing the importance of high-temporal-resolution data in certain parts of the time series.

3.2.3.2.1 Other TSM Models

Another TSM model, the dual-temperature-difference (DTD) model (Norman et al., 2000) was based on the time rate of change in T_R and T_2, where the equations in ALEXI model (Anderson et al., 1997) were used to form a dual-difference ratio of radiometric and air temperatures. The H flux is then calculated from temporal measurements of T_2, T_R, and wind speed. An ABL model was not required in implementing the model, so the calculations can be made efficiently with minimal ground-based data. Various studies have evaluated the ability

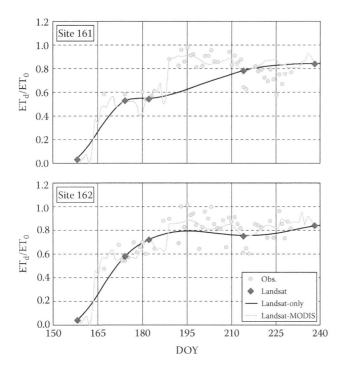

FIGURE 3.6 Time series of ET_d/ET_o estimated using Landsat alone and using MODIS downscaled to Landsat resolution using the STARFM algorithm over rainfed soybeans in Iowa ET_d is day-time mean ET. (Reprinted from Cammalleri, C., Anderson, M.C., Gao, F., Hain, C.R., and Kustas, W.P., A data fusion approach for mapping daily evapotranspiration at field scale, *Water Resour. Res.*, 49, 4672–4686, 2013. With permission from American Geophysical Union.)

of DTD TSM for deriving surface heat fluxes over dissimilar testing conditions. Generally, those studies have indicated an agreement between the model-predicted λET and corresponding ground observations in the order of ~50 W/m² (Gowda et al., 2008).

Sun et al. (2009) proposed another TSM energy balance model, the simple remote sensing evapotranspiration model (Sim-ReSET). The major difference in their model compared to other TSMs was that the aerodynamic resistance (R_{ah}) was calculated using a reference dry bare soil and canopy height, assuming homogeneous wind speed in the upper boundary layer. Unlike ALEXI, Sim-ReSET is based on a single image, and, like SEBAL and METRIC, is based on internal calibration to dry soil and wet vegetated pixels. Evaluation of Sim-ReSET was performed for a region in north China using MODIS data. Comparisons of the predicted λET fluxes by Sim-ReSET versus concurrent ground measurements from 12 experimental days showed an RMSD for instantaneous values of ~42 W/m² and an MAE of 34 W/m². Error in the mean daily ET over a 6 day period was lower (MAD = 0.26 mm/day, RMSD = 0.30 mm/day) as observed in other applications. Two of the key advantages of Sim-ReSET were that it avoids the direct computation of the aerodynamic resistance and that all the model inputs required can be estimated from remote sensing data alone.

3.2.3.3 Simplified T_R-Based Approaches: SSEB and SSEBop

Several other T_R-based methods have been developed, including the SSEB model (Senay et al., 2007). Similar to the crop coefficient methods in Section 3.2.2.1., SSEB calculates ET as the product of ET_o (Equations 3.4 or 3.5) and ET_f. The SSEB model assumes that ET_f for a given pixel can be estimated from the radiometric temperatures at the hot, cold, and observed pixels alone, without explicitly solving the energy balance equation:

$$ET_f = \frac{(T_h - T_R)}{(T_h - T_c)} \qquad (3.22)$$

where

T_h is the radiometric temperature of the hot pixel
T_c is the radiometric temperature of the cold pixel

Since the maximum ET of a crop with a high LAI may be higher than that of the reference grass, the ET_o parameterized for the reference grass is multiplied by a correction factor, usually 1.2 (Senay et al., 2007). Similar to other energy balance methods, the temperatures at the hot and cold pixel are derived from the image, assuming that pixels where ET = 0 and ET = 1.2ET_o exist in the image.

While the original formulation of SSEB is easy to implement and produces estimates for regions with a uniform hydroclimate such as irrigated districts, an improvement was necessary to account for land surface temperature differences caused by spatial variation in elevation and albedo. An enhanced version of SSEB was introduced by Senay et al. (2011) to adjust the radiometric temperature using a lapse-rate correction before using it in Equation 3.22. A similar lapse-rate adjustment is also performed in other temperature-based methods, including SEBAL and METRIC. A comparison between the enhanced SSEB and METRIC showed a strong relationship at elevations below 2000 m (R^2 = 0.91) compared to elevations above 2000 m (R^2 = 0.52) in Central Idaho (Senay et al., 2011). The original SSEB model also calculates ET_o using a fixed value for α (0.23). While SW↓, LW↓, and LW↑ can vary by pixel under the original SSEB method, using the fixed α ignores the impact of spatial variability in α and G on ET, which could result in overestimation of ET at pixels with high albedo and overestimation of ET at pixels with high and positive G. An improved version of SSEB (Senay et al., 2011) adjusts ET_f by a factor that varies with NDVI, and ranges from 0.65 to 1.0 (Equation 3.6 in Senay et al., 2011). The adjustment is designed to account for generally higher albedo and greater ground heat flux for in pixels with low NDVI, though the relationship between the adjustment factor and NDVI may vary geographically and is not derivable from other measurements.

The SSEB model requires the selection of hot and cold pixels in the image, as in the SEBAL and METRIC models. This selection process generally inhibits operational application of such models over large areas and introduces problems along scene boundaries. The SSEBop ET algorithm is an operational parameterization of

TABLE 3.4   Inputs, Parameters, Steps, and Equations Used in the SSEBop Model

No	Inputs	Units	Description	Data Source
1.	T_R	K	Radiometric surface temperature	Satellite imagery (Landsat, MODIS)
2.	T_2	K	Air temperature	PRISM
3.	R_n	W/m²	Clear-sky net radiation	Equations in Allen et al. (1998)
4.	R_{ah}	s/m	Aerodynamic resistance	Constant value (110)
5.	R	J/kg K	Specific gas constant	Constant value (287)
6.	C_p	J/kg K	Specific heat of air at constant pressure	Constant value (1013)
7.	ET_o	mm/day	Reference evapotranspiration	Penman–Monteith Equation
8.	NDVI	—	Normalized difference vegetation index	Satellite imagery (Landsat, MODIS)
9.	Z	m	Surface elevation	Digital elevation model (DEM)
10.	T_{R_cold}	K	T_R at a cold pixel	Satellite imagery (Landsat, MODIS)

Steps	Parameters	Units	Description	Equation
1.	c	—	Temperature correction coefficient	$c = \dfrac{T_{R\_cold}}{T_2}$
2.	P_a	kPa	Atmospheric pressure	$P_a = 101.3 \left(\dfrac{293 - 0.0065 \times Z}{293} \right)^{5.2}$
3.	T_{kv}	K	Virtual temperature	$T_{kv} = 1.01(T_2 + 273)$
4.	ρ_{air}	kg/m³	Density of air	$\rho_{air} = \dfrac{1000P}{T_{kv}R}$
5.	dT	K	Temperature difference	$dT = \dfrac{R_n R_{ah}}{\rho_a C_p}$
6.	T_c	K	Cold reference temperature	$T_c = cT_2$
7.	T_h	K	Hot reference temperature	$T_h = T_c + dT$
8.	ET_f	—	Reference ET fraction	$ET_f = \dfrac{(T_h - T_R)}{dT}$
9.	ET	mm	Actual evapotranspiration	$ET = ET_f \times ET_o$

Source: Senay, G.B. et al., *JAWRA J. Am. Water Res. Assoc.*, 49, 577, 2013. With permission.

the SSEB model (Senay et al., 2007), renamed SSEBop because of its operational capability (Senay et al., 2013) (Table 3.4). SSEBop proposes that the hot and cold temperatures can be determined separately for each pixel, rather than for each image. The dry-minus-wet temperature difference (T_h–T_c = dT) is calculated for each day and pixel by assuming G is zero for a dry surface (R_n = H, ET = 0) and solving Equation 3.17 for dT for dry conditions:

$$T_h - T_c = dT = \frac{R_n \times R_{ah}}{\rho_{air} \times C_p} \qquad (3.23)$$

where the variable definitions are the same as those in Equation 3.19. R_{ah} is assumed to have a constant value of 110 s/m, which corresponds to a bare, dry soil based on model inversion to field data in the western United States (Senay et al., 2013). The constant value of R_{ah} is used for all pixels regardless of actual cover, because dT represents the temperature difference between a hypothetical bare, dry soil surface, and a well-vegetated surface at each pixel. dT generally ranges between 5 and 25 K depending on location and season.

SSEBop assumes that the theoretical cold–wet surface temperature of a given pixel can be estimated using the air temperature, T_2 as

$$T_c = c \times T_2 \qquad (3.24)$$

where

T_2 is the air temperature obtained from a meteorological station or gridded meteorological data

c is a temperature correction factor estimated as the ratio of T_R to T_2 at the time of overpass for a well-vegetated wet surface (high *NDVI*; the specific threshold will depend on the NDVI used)

In previous applications of SSEBop, c is close to 1 (Senay et al., 2013). Equation 3.24 is based on the same premise as the SEBAL and METRIC algorithms, where H = 0 at a cold pixel. When the air temperature corresponding to the satellite overpass is not available, daily maximum air temperature can be used, and c calculated as the ratio of daily maximum air temperature and daily temperature. The use of daily maximum air temperature offers an advantage when modeling over large areas (such as regional to global modeling) as exact time of satellite overpass could be different for different locations.

Once dT and T_c are calculated, the theoretical surface temperature under hot and dry conditions can be estimated for each pixel as T_h = T_c + dT. This simplification permits the estimation of ET_f as

$$ET_f = \frac{(T_h - T_R)}{dT} \qquad (3.25)$$

Evapotranspiration at a given pixel λET is calculated by multiplying ET_f by ET_o. The advantage of this approach is the simplification of the model that enables operational application over large areas and limited data requirements. SSEBop relies on the accuracy of the T_R and dT estimation, so it can produce inaccurate ET estimates on surfaces with high albedo and emissivity values that are different from the soil-vegetation complexes found in agricultural settings. To improve the accuracy of ET estimates, please refer to Senay et al. (2013) for suggested methods to condition T_R over surfaces with high albedo and emissivity.

3.2.3.4 From Instantaneous to Daily ET

Temperature- and energy-based methods provide instantaneous estimates of ET, but instantaneous estimates may be of little use for mapping seasonal crop water use (Petropoulos, 2013). Several approaches may be used to calculate daily total ET from the instantaneous imagery (Crago, 1996; Chávez et al., 2008; Kalma et al., 2008; Petropoulos, 2013), though two types of methods are most commonly applied: the evaporative fraction method and the crop coefficient method. The evaporative fraction (Λ) approach uses the satellite-derived ET to calculate ET as a fraction of available energy (R_n–G):

$$\Lambda_{op} = \frac{\lambda E}{Rn - G} \qquad (3.26)$$

The evaporative fraction at the time of overpass (Λ_{op}) is assumed equal to Λ for the daytime (Λ_d) or during a 24 h period (Λ_{24}). Either daytime or 24 h total ET is calculated as the product of Λ_{op} and the daytime net available energy (R_n–G) or 24 h net radiation (Rn_{24}), since G is assumed to be zero over 24 h. The assumption that Λ_{op} equals Λ_d or Λ_{24} is justified by some field measurements (Hall et al., 1992; Jackson et al., 1983), though clouds can change the temporal stability of Λ (Crago, 1996), and modeling studies suggest there may be diurnal variation in Λ, with minimum values during midday that can result in underestimation of the daily mean Λ of up to 20%–40% when using overpass times between 11 am and 3 pm (Gentine et al., 2007; Lhomme and Elguero, 1999). Clouds reduce R_n–G, but typically, H decreases more than λE, resulting in an increase in Λ during cloudy periods, though this difference was not statistically significant in field studies, and the assumption of constant Λ over daytime hours is "surprisingly robust" (Crago, 1996). Other studies suggest that a correction factor should be applied that varies with time of overpass and soil moisture (Gentine et al., 2007), though in practice, a constant Λ is often assumed.

Either daytime available energy (R_n–G) or 24 h total net radiation (R_{n24}) is used as the multiplier to calculate total ET from Λ_{op}. R_{n24} is most commonly used to estimate ET_{24} given the (near) zero G term, though Van Niel et al. (2011) caution that, in addition to the assumption that the evaporative fraction is constant over a 24 h period, the use of R_{n24} also assumes that net available energy (R_n–G) and latent heat flux (λET) are zero or near zero at night (Van Niel et al., 2011). R_n–G is

commonly negative at night due to longwave emission from the surface. While the latent heat flux can also be negative at night, corresponding to condensation or dew formation, much of the negative available energy changes the sensible heat flux rather than the latent heat flux. The latent heat flux can also be positive at night if sensible heat is advected onto a given location, which can occur where irrigated vegetation may have heat advected to it from surrounding hotter rainfed vegetation. In an irrigated alfalfa plot, nighttime ET was >7% of total daily ET (Tolk et al., 2006). This positive nighttime latent heat flux can result in significant underestimation of daily ET when using R_{n24}, of up to –24% to –38% when using the midmorning value of Λ_{op} and lower errors when using midafternoon values (–5% to –21%). The main contributor to the error of using R_{n24} was the nonzero nighttime available energy flux, which was sometimes nearly equal to the daytime available energy in a wet forest site (Van Niel et al., 2011). The error was smaller at a drier savanna site. More documentation is needed about how the magnitude of errors incurred by using R_{n24} instead of (R_n–G)$_d$ to calculate daily ET depends on season, climate, and vegetation.

Other methods, including the original METRIC model (Allen et al., 2007), use the crop coefficient approach, which calculates the ratio of actual to reference ET at the time of satellite overpass, then multiplies that fraction by reference ET for the day:

$$ET_{24} = C_{rad} \times ET_f \times ET_{o24} \qquad (3.27)$$

where ET_{24} is ET over the 24 h period, C_{rad} is an adjustment applied to sloped surfaces, ET_f is the ratio of actual to reference ET at the time of satellite overpass, and ET_{o24} is reference ET for the 24 h period. C_{rad} is likely to be close to 1 for most crops, which are mostly grown on flat surfaces, but there may be local exceptions for agroforestry crops in mountainous terrain.

The crop coefficient method was advocated over the Λ method by Allen et al. (2007) who suggested that advection, which is not included in the Λ method, is important for heterogeneous irrigated landscapes and is accounted for by the PM equation. A review of field studies suggested that ET_f is relatively constant over a 24 h period in irrigated plots (Romero, 2004, cited in Allen et al., 2007). In one comparison study, the evaporative fraction method had a lower RMSE (7.0%) than the crop coefficient method (16.6%) (Chávez et al., 2008), but the accuracies of each method likely change with meteorological conditions, vegetation, and soil moisture.

3.2.3.5 Comparison of Temperature-Based Methods

As temperature-based methods gained popularity for their simplicity and accuracy in measuring energy fluxes across landscapes, the merits of one-source and two-source approaches were scrutinized. Timmermans et al. (2007) compared two common energy-based methods: one-source (SEBAL) and TSM (ALEXI). SEBAL accuracy declined over hot, dry, heterogeneous terrain, because of the difficulty in selecting a dry end member pixel within the boundaries of the remote sensing image, which is then used to calibrate Equation 3.18. TSM, on the other hand, which rely heavily on vegetation fraction, tended to be less accurate in densely vegetated areas, where small changes in vegetation cover can have a significant impact on canopy temperature estimation. Other studies suggest that TSM may perform better in conditions of either dense or sparse vegetation, or extremes of soil moisture (Anderson et al., 2013, p. 212), and field-scale comparisons suggest that TSM outperforms single-source models (Gonzalez-Dugo et al., 2009), though both were found to produce acceptable results.

Gonzalez-Dugo et al. (2009) used data collected during the SMACEX/SMEX02 field experiments (Kustas et al., 2005) to evaluate instantaneous λET fluxes derived from an empirical one-layer energy balance model (Chávez et al., 2005), METRIC (Allen et al., 2007), and the TSM of Kustas and Norman (1999), the latter being an updated version of the Norman et al. (1995) TSM model that forms the basis of ALEXI. The authors reported an RMSD of less than 50 W/m² and less than 33 W/m² in the estimation of the instantaneous λET and H fluxes, respectively, by all methods. The fluxes predicted by the TSM of Kustas and Norman (1999) had the closest agreement to the ground observations (RMSD of 30 W/m², R^2 = 0.83), followed by METRIC (RMSD of 42 W/m², R^2 = 0.70), and last by the empirical one-layer model (RMSD of 50 W/m², R^2 = 0.70). Gonzalez-Dugo et al. (2009) underlined as a major disadvantage of both the TSM and the empirical one layer model the requirement of both models for accurate emissivity and atmospheric correction of the thermal-infrared (TIR) imagery used subsequently for computing the land surface temperature. METRIC had a very important disadvantage in the requirement for scene internal calibration each time, which, although it reduces the need for accurate temperature retrieval, significantly diminishes the use of this model for operational application and introduces subjectivity in the pixel selection, though recent advances at automated pixel selection may reduce problems with application and subjectivity. Gonzalez-Dugo et al. (2009) also evaluated the performances of three modeling schemes for interpolating instantaneous to daily fluxes. The schemes evaluated included the Λ method (Crago, 1996), the adjusted Λ method (Anderson et al., 1997), and the reference evapotranspiration fraction (Doorembos and Pruitt, 1977—in Gonzalez-Dugo et al., 2009). Authors reported similar accuracy among the three models. The daily λET fluxes by the adjusted Λ method returned the closest agreement to the reference measurements (RMSD = 0.74 mm/day, R^2 = 0.76). The daily λET fluxes predicted by the reference evapotranspiration method were found to be overestimated during conditions of the prolonged dry down period.

Senay et al. (2011) compared λET fractions derived from SSEB and METRIC models. The comparison study was carried out using Landsat images acquired for South Central Idaho during the 2003 growing season. Results indicated that SSEB model compared well with METRIC model output. However, SSEB model was more reliable over wide range of elevation (especially >2000 m) to detect anomalies in space and time for water resources management and monitoring such as for drought early warning systems in data scarce regions.

3.2.4 Scatterplot-Based Methods for ET Estimation

Scatterplot methods (Figure 3.7), also called triangle methods (Gillies et al., 1997) or trapezoidal methods (Moran and Jackson, 1991), combine features of the vegetation-based and energy-based methods. Like the vegetation-based methods, they use a VI but also incorporate T_R to account for spatial variability in soil evaporation and in transpiration from vegetation experiencing different levels of soil moisture stress. They are based on the relationships between T_R and some other satellite-derived variable, often a VI or albedo, when these are plotted in a scatterplot (Figure 3.7). The method places theoretical boundary lines on the observed inverse relationship between T_R and VI or albedo and uses the position of a pixel in the T_R–VI or T_R–α space relative to those boundary lines to calculate either the evaporative fraction (Λ) as in the temperature-based methods or the reference evapotranspiration fraction (ET$_f$) as in the empirical VI methods. A review of these methods including the theoretical basis of the principles underlying the scatterplot methods can be found in Petropoulos et al. (2009).

Briefly, assuming that cloud-contaminated pixels and pixels containing standing water have been masked, per pixel-level values of T_R and VI usually fall within a triangular (or trapezoidal) shape in the T_R–VI feature space (Figure 3.7). In Figure 3.7, each yellow circle represents the measurements from a single image pixel and includes the main properties believed to be represented by the T_R–VI pixel envelope. The triangular or trapezoidal shape in T_R–VI feature space is the result of the low variability of T_R and its relative insensitivity to soil water content variations over areas covered by dense vegetation and its increased sensitivity to soil moisture and larger spatial variation over areas of bare soil. The right-hand side border, the "dry edge" or "warm edge" is defined by the image pixels of highest temperature for the differing amounts of bare soil and vegetation and is assumed to represent conditions of limited surface soil water content and near-zero evaporative flux from the soil. The left-hand border, the so-called "wet edge" or "cold edge," corresponds to cooler pixels with varying amounts of vegetation cover and represents the limit of maximum surface soil water content. Variation along the triangle's base represents pixels of bare soil and is assumed to reflect the combined effects of soil water content variations and topography. The triangle apex equates to full vegetation cover. Similar to single-source temperature-based methods, pixels with minimum T_R represent the strongest evaporative cooling (point A in Figure 3.7), while those with maximum T_R represent the weakest evaporative cooling and low ET (point B in Figure 3.7). The triangle method defines vegetation cover directly from NDVI, whereas the trapezoid method uses fractional cover. Several methods then calculate ET$_f$ as the ratio of distances CB and AB (Jiang and Islam, 2001; Moran et al., 1996), though a variety of methods is used to relate the position of a pixel in the scatterplot to ET.

Scatterplot methods for the estimation of ET can be divided into four groups based on the variables used in the scatterplot,

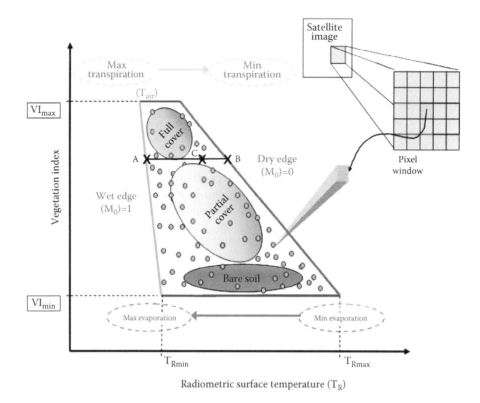

FIGURE 3.7 Example of a scatterplot of T_R versus VI, indicating wet and dry edges. (Reprinted from Petropoulos et al. (2009) with permission from SAGE Publications Ltd.)

namely, (1) T_R–VI scatterplots, (2) surface-to-air temperature difference and VI scatterplots, (3) T_R-albedo scatterplots, and (4) day–night temperature difference and VI scatterplots. A fifth method couples the T_R–VI scatterplot with a soil–vegetation–atmosphere transfer (SVAT) model. In the remainder of this section, each of the aforementioned groups of methods is reviewed, providing some information on the methods' principles and operation, as well as examples from its implementation in different ecosystems. A summary of the strengths and limitations of the different groups of approaches is also provided in Table 3.1.

3.2.4.1 T_R–VI Scatterplot Methods

A number of methods have been proposed to retrieve regional maps of ET from the T_R–VI triangular space over a number of different land cover types. Jiang and Islam (1999, 2001) suggested a technique based on an extension of the Priestley–Taylor equation and a relationship between remotely sensed T_R and VI. Jiang and Islam (2001) and Tang et al. (2010) use the T_R–VI graph to determine the α_{PT} value from Equation 3.5 by assuming it is maximum on the wet edge and minimum on the dry edge, with global maxima at the minimum T_R and maximum T_R. α_{PT} is calculated for a given pixel as 1.26 times the ratio of distance CB to AB in Figure 3.7. A key assumption in the method is that pixels with ET = 0 and ET = ET_o could be identified from the remotely sensed data. Jiang and Islam (2001) showed an RMSD in ET around 30% of the observed mean.

Jiang and Islam (2001) assume that dense vegetation transpires at the potential rate, which may not be the case for soil moisture stress. Nishida (2003) and Nishida and Nemani (2003) addressed this problem by estimating the evaporative fraction (Λ) with MODIS data for vegetation and soil separately, where ET from vegetation is calculated from a combination of the PM equation and the complementary relationship between potential ET and actual ET, and soil evaporation is calculated using the triangle method. Λ was computed every eight days for a range of climate and biome types, and validated at selected AmeriFlux sites, with good agreement (RMSD = 45.1 W/m^2, bias = 5.6 W/m^2, R^2 = 0.86).

Critical to successful implementation of the T_R–VI methods is an identification of the wet and dry edges. Zhang et al. (2006) proposed using the VI for estimating the dry and wet edges of the scatterplot. Tang et al. (2010) emphasized the importance of the accurate determination of the wet and dry edges in the accurate retrievals of the λET fluxes by the T_R–VI method of Jiang and Islam et al. (2001) and proposed a novel technique for determining quantitatively the dry and wet edges over a homogenous agricultural area.

3.2.4.2 Surface-to-Air Temperature Difference and VI Scatterplot Methods

Surface-to-air temperature difference methods are similar to T_R–VI methods but replace T_R with the difference between T_R and air temperature above the evaporating surface (z_2). Moran et al. (1994) introduced the "vegetation index–temperature trapezoid" for estimation of λET from the dT–VI domain in areas of partial vegetation cover, based on the PM Equation 3.4 and the crop water stress index (CWSI = 1–ET_f) (Jackson et al., 1981). The PM equation is inverted following Jackson et al. (1981) to estimate T_R at the four vertices of the dT–VI trapezoid, and 1–ET_f is calculated as ratio of the difference in temperature between a given pixel and the dry temperature (CB in Figure 3.7) to the difference in temperature between the dry edge and wet edge at the pixel's NDVI value (AB in Figure 3.7). Inversion of the PM equation to determine T_R at the four corners of the trapezoid avoids the requirement that there be a pixel in the image where ET = ET_o and another where ET = 0. Validation of the method was carried out using a number of techniques over various land cover types. Moran et al. (1994) performed validation studies over both agricultural and semiarid grasslands in Arizona, United States, using model simulations and airborne data from the modular multispectral radiometer and ground-based measurements, respectively. Moran et al. (1996) used Landsat TM data to estimate ET for all grassland sites within their study region (Arizona, United States) and reported an RMSD of 29 W/m^2 in the estimation of the instantaneous λET and a consistent overestimation of λET in most sites.

Jiang and Islam (2003) modified the Jiang and Islam (1999) method, using T_R–T_2 in place of T_R and also by using the fractional vegetation cover (f_c) parameter, as a proxy for vegetation amount in place of the NDVI. In validation studies in the United States, λET flux estimates were predicted with an RMSD and bias of 58.6 and −42.4 W/m^2, respectively. Venturini et al. (2004) validated the method utilizing the same sensors for a region over South Florida, with RMSD in Λ prediction varying from 0.08 to 0.19, and a R^2 ranging from 0.4 to 0.7. Stisen et al. (2008) modified the method proposed by Jiang and Islam (2003) by combining it with thermal inertia information obtained from the geostationary SEVIRI sensor to estimate regional evaporative fraction (Λ) for a semiarid region in South Senegal in Africa. Comparisons performed by the authors showed a close agreement for both Λ (RMSD of 0.13 and R^2 of 0.63) and the instantaneous λET (RMSD of 41.45 W/m^2 and R^2 of 0.66). Shu et al. (2011) performed further validation of the Stisen et al. (2008) method using observations from the Fengyun-2C satellite in combination with the MODIS satellite products over a subtropical region in the North China Plain. Authors reported an R^2 equal to 0.73 and an RMSD of 0.92 mm/day for daily ET and R^2 = 0.55 and RMSE = 0.14 for Λ. A very important advantage of the Jiang and Islam (2003) method included its independence from absolute accuracy of the T_R measures, since T_R–T_2 equal to zero in their technique always represented the true cold edge of the triangle space where Λ equals zero. Nonetheless, this method also assumed a linear variation in Λ across the triangular/trapezoid domain of f_c–(T_R–T_2) feature space for each class of f_c, which might not be a realistic approximation, an issue that Stisen et al. (2008) tried to address by assuming nonlinear relationships between the biophysical properties encapsulated in the scatterplot.

3.2.4.3 T_R-Albedo Scatterplot Methods

A different group of triangle methods is based on the correlation between T_R and α. Roerink et al. (2000) proposed the simplified surface energy balance index (S-SEBI) method for mapping ET based on the estimation of the evaporative fraction (Λ). S-SEBI calculates Λ using the same equation that SSEB uses to calculate ET_f (3.22), but in S-SEBI, T_h and T_c are linear functions of albedo, where the linear function coefficients are determined from the boundaries of the T_R-albedo plot. Gómez et al. (2005) extended S-SEBI to the retrieval of daily evapotranspiration (λET) from high-spatial-resolution data (20 m) from polarization and directionality of Earth reflectance airborne instrument and a TIR camera (20 m), with an error of 90 W/m^2 and 1 mm/day in the estimates of instantaneous and daily total λET, respectively. The method was evaluated further by Sobrino et al. (2005) using high-spatial-resolution airborne images from the digital airborne imaging spectrometer over agricultural areas in Spain. Authors reported accuracy in daily ET prediction better than 1 mm/day. Sobrino et al. (2007, 2008) subsequently adapted this methodology to the low-spatial-resolution data provided by the advanced very high resolution radiometer (AVHRR) and in the framework of the SEN2FLEX (field measurements, airborne data) project for similar sites in Spain. Authors reported a mean RMSD of ~1 mm/day in the estimation of daily λET by S-SEBI in comparison to ground λET measurements from lysimeters. Garcia et al. (2007) evaluated three operative models for estimating the non-evaporative fraction as an indicator of the surface water deficit in a semiarid area of southeast Spain. Other studies, such as by Zahira et al. (2009) monitored the drought status in Algerian forest covered areas with the combined use of S-SEBI with the visible, near-infrared (NIR) and TIR bands of Landsat enhanced thematic mapper plus (ETM+) imagery. Comparisons performed between the instantaneous λET fluxes predicted by S-SEBI and corresponding in situ data showed a R^2 of 0.85 and an RMSD of 64 W/m^2. Bhattacharya et al. (2010) calculated Λ from the T_R–α plot, validated over an agricultural region in India from the Indian geostationary meteorological satellite Kalapana-1. They reported an overall RMSE in the predicted daily ET estimates in the range of 25%–32% of measured mean. Comparisons of the 8 day ET product over agricultural land uses yielded RMSD of 26% (0.45 mm/day) with r = 0.8 (n = 52,853) as compared to daily ET. An important advantage of the technique of Bhattacharya et al. (2010) was that it could be implemented without the need of any ground observations, making it potentially a very good choice for operational use. Also, in contrast to some other methods discussed so far, their method avoided the H flux computation prior to ET flux estimation. A limitation is the assumption of uniform atmospheric conditions, and, like SEBAL and METRIC, the method cannot be implemented if wet and dry pixels are not present within the image.

3.2.4.4 Day–Night Temperature Difference and VI Scatterplot Methods

Another variant of the triangle method uses the difference between the day and night T_R (diurnal surface temperature variation or DSTV) versus the VI (Chen et al., 2002). The approach is based on the observed relationship between the difference between DSTV and soil moisture and thermal inertia (Engman and Gurney, 1991; Van de Griend et al., 1985). The technique first implements a simple linear mixture model to determine the fractional contribution of vegetation, dry soil surfaces, and wet soil surfaces to the observed values of NDVI and T_R. The vegetation and moisture coefficient (VMC), which is the same as ET_f in the crop coefficient methods, was then expressed at each pixel as the sum of the VMC for each of the three surface types weighted by the fraction contribution at the given pixel. They implemented the algorithm with AVHRR data for a site in South Florida, United States, and showed percentage errors in the prediction of ET compared to lysimeter measurements between 2.8% and 23.9% with RMSD's varying from 3.08 to 5.74 mm/day. An important advantage of the Chen et al. (2002) method was its requirement for only a small number of ground parameters. Limitations included the constraint to assume only three land cover types in the mixture modeling and the need for two T_R images to calculate the DSTV. Similar to Jiang and Islam (2001), this required assuming that ET was the same for all dense vegetation. Also, the fact that the method required identification of areas of the three distinct land use/cover types that are homogeneous and of sufficiently large spatial extent for the VMC to be estimated is a limitation for its implementation.

Wang et al. (2006) proposed a modification to the method of Jiang and Islam (2003) using DSTV and NDVI in place of the daytime T_R. Wang et al. (2006) determined spatial variations of the DSTV–NDVI space using data from MODIS global 1 km daily products collected by the Aqua and Terra satellites, which were used to estimate Λ (parameterized as a function of air temperature and the Priestley–Taylor parameter α_{PT}), that was then compared to observations taken during 16 days in 2004, again at the Southern Great Plains (SGP) site, United States. The Λ retrievals from the DSTV–NDVI plot showed marked improvement compared to those retrieved from the daytime T_R alone. The DSTV method with Aqua day and nighttime images had an RMSD of 0.106, a bias of −0.002, and an R^2 of 0.61 for Λ, which was deemed satisfactory, especially after taking into account the simplicity of the approach and the requirement for only a small number of input parameters for its implementation.

3.2.4.5 Coupling T_R–VI Scatterplots with SVAT Models

The outputs from a SVAT model can be coupled with T_R and VI, where VI is replaced by the fractional vegetation cover (f_c). Overviews of this technique can be found in Carlson (2007) and Petropoulos and Carlson (2011). First, both NDVI and T_R are scaled to the maximum and minimum values in the image:

$$N^* = \frac{NDVI - NDVI_o}{NDVI_s - NDVI_o} \qquad (3.28)$$

where the subscripts s and o indicate dense vegetation and bare soil. Fractional vegetation cover (f_c) is calculated as the square of NDVI* following the methods of Gillies and Carlson (1995) and Choudhury et al. (1994). In an image with a full range of vegetation cover, N^* and f_c will range from zero to one. Use of f_c allows us to plot both the SVAT-simulated and the measured surface radiant temperatures from the satellite sensor on the same scale. Scaled temperature is calculated as

$$T_{scaled} = \frac{T_R - T_{Rmin}}{T_{Rmax} - T_{Rmin}} \qquad (3.29)$$

where T_{Rmax} and T_{Rmin} are the maximum and minimum T_R for the dry, bare soil, and the wet vegetation, respectively, interpolated from the scatterplot bounds.

In the next step, the SVAT model is combined with T_{scaled} and f_c in order to derive the inversion equations that will provide the spatially explicit maps of ET. Initially, the SVAT model is parameterized using the time and geographic location and the site-specific atmospheric, biophysical and geophysical characteristics. Subsequently, the SVAT model is iterated until the simulated and observed extreme values of f_c and T_{scaled} in the T_{scaled}–f_c scatterplot are matched. Initial model simulations aim to align observed T_{scaled} with two end points (NDVIo, NDVIs) where they intersect the "dry" edge. This extrapolation to NDVIo and NDVIs guarantees that the implied temperatures along the "dry" edge for bare soil and full vegetation cover are consistent with simulations for soil moisture (Mo) of zero. Once the model tuning is completed, the simulation time corresponding to the satellite overpass is kept the same as the SVAT model is ran repeatedly, varying f_c and Mo over all possible values (0%–100% and 0–1, respectively), for all possible theoretical combinations of Mo and f_c. The result is a matrix of model outputs for a number of simulated parameters: Mo, f_c, T_{scaled}, λET, and H. Finally, this output matrix is used to derive a series of linear or quadratic equations, relating f_c and T_{scaled} to each of the other variables of interest: H, λET, and Λ. The SVAT model outputs are then used to derive a series of simple, empirical relations relating each of these parameters to f_c and T_{scaled} recorded at that location as quadratic polynomial equations:

$$ET = \sum_{p=0}^{3}\sum_{q=0}^{3} a_{p,q} \left(T_{scaled}\right)^p f_c^q \qquad (3.30)$$

where the coefficients $a_{p,q}$ are derived from nonlinear regression between the matrix values of f_c, T_{scaled}, and ET, and p and q vary from 0 to 3.

Gillies and Carlson (1995) first applied the technique using AVHRR images for a region in the United Kingdom. Carlson et al. (1995) then utilized the technique to derive daily estimates of ET for a site in Pennsylvania, United States, and validated the results using ground-based measurements from the push broom microwave radiometer and the NS001 instrument (30 m spatial resolution). Gillies et al. (1997) validated their method using high-resolution airborne data from the NS001 instrument and observations collected from the first International Satellite Land Surface Climatology Project (ISLSCP) field experiment (Vernekar et al., 2003) and MONSOON 90 (Kustas et al., 1991) field campaigns. The RMSD was ±10% for ET. Brunsell and Gillies (2003) implemented the method using data from the TMS/TIMS airborne (12 m) and coarse AVHRR (1 km) radiometers. Predicted fluxes from the implementation of the triangle method using the different remote sensing data were compared versus in situ observations from the SGP test site (United States). A close agreement for the high-resolution airborne data, within ~15% for λET, was reported, but results from the satellite data were in poor agreement with both the observations and the airborne data (50% difference for λET). Recently, Petropoulos et al. (2010) and Petropoulos and Carlson (2011) evaluated the triangle-SVAT method at several CarboEurope sites using ASTER data. Closer agreements with the ground observations were generally found when comparisons were limited to cloud-free days at flat terrain sites. Under such conditions, the triangle-SVAT method estimated instantaneous λET with a mean RMSD of 27 W/m².

The triangle-SVAT method has several advantages over the other scatterplot methods. First, it provides a nonlinear interpretation of the T_R–VI space, which can be a more realistic assumption than the linear assumption in the empirical triangle methods. The triangle-SVAT method offers the potential for relatively easy transformation of the instantaneous-derived ET fluxes to daytime averages and allows regional estimates of the H flux and the surface soil moisture content together with the ET fluxes, potentially very useful parameters to have information for many practical applications. In addition, it offers the possibility to extrapolate the instantaneous measurements of the computed energy fluxes from one time of day to another (see Brunsell and Gillies, 2003) and to times with clouds. A disadvantage is that the SVAT requires a large number of input parameters and its parameterization also requires user expertise, complicating its implementation over large geographic areas.

3.2.5 Seasonal ET Estimates and Cloud Cover Issues

All three families of methods for estimating ET reviewed in this chapter (vegetation based, temperature or energy based, and scatterplot based) produce daily maps of ET that can be temporally interpolated to estimate seasonal ET, which is often the main output of concern to water managers and agriculturalists. Interpolation is necessary because the satellite platforms that generate high-resolution imagery often have long overpass return periods (e.g., Landsat at 2 weeks) and because of clouds, which compromise the generation of daily ET maps even when daily imagery are potentially available (MODIS). Cloud cover may be less of a problem in arid and some semiarid climates but is a significant obstacle for determining season-total ET in semihumid and humid climates. In regions with Mediterranean climates, the main growing season in summer corresponds to cloud-free conditions, and satellite methods work well for determining seasonal ET from daily values. In other locations, where the growing season coincides with the wet season, such as monsoon-dominated areas, cloud-free imagery is often not available during the main crop growing season, and EO-based methods for ET estimation may need to be supplemented with other modeling approaches like the FAO-56 method (Allen et al., 1998) or more complex SVAT models.

Three methods to interpolate ET maps for days without cloud-free imagery are (1) the evaporative fraction (Λ) method, (2) the crop coefficient (ET_f) method, and (3) the simulation model method (Long and Singh, 2010). The Λ and crop coefficient methods for estimating seasonal ET are very similar to the methods for generating daily estimates from instantaneous estimates, but there are special problems with cloud cover when interpolating over longer time scales that are discussed further in this section.

The Λ method, which is also used to calculate daily ET from instantaneous values of the evaporative fraction (Λ_{op}) (Section 3.2.3.1), calculates ET for a date without cloud-free imagery as the product of R_n for the day without cloud-free imagery and Λ_{op} from the date with cloud-free imagery. The Λ method assumes that Λ_{op} is constant between the dates of cloud-free imagery, which is more likely to be violated over several days than for a single day, especially if cloud cover changes significantly. Farah et al. (2004) found that Λ does not vary with cloud cover over short (weekly) time intervals over woodland and grassland in central Kenya, which encourages the use of the evaporative fraction method (Farah et al., 2004). The method may produce accurate ET values over periods of 5–10 days (Farah et al., 2004), but over many areas, cloudy conditions persist much longer. Others have found that Λ increases during cloudy periods due to a larger proportionate reduction in H than in λET (Van Niel et al., 2011).

The crop coefficient approach calculates the ratio of actual to reference ET on the day of satellite imagery, then multiplies that fraction by reference ET for each day without imagery (Long and Singh, 2010):

$$ET_{period} = \sum_{i=d1}^{n} ET_f ET_{oi} \qquad (3.31)$$

where

ET_f is for the 24 h period on the date with satellite-derived ET
ET_{oi} is reference ET on day i, which does not have imagery
d1 is the beginning day without ET data
n is the number of consecutive days without image-derived ET estimates

The ET_f can be assumed constant between images, or could be linearly varied between available ET images. Allen et al. (2007) suggest that one ET image per month is sufficient to estimate seasonal total ET, though this may not be the case under conditions of rapidly varying soil moisture conditions or surface saturation, as might be expected in irrigated areas. The ET_f method has been applied using the METRIC model (Allen et al., 2007) and in northern China (Li et al., 2008).

The third approach to estimating ET on days without cloud-free imagery, the simulation model method, uses satellite-based ET on clear days to calibrate an SVAT model or some simplified version of an SVAT model, which is then run for all days including cloudy days. Simplified models of the relationship between meteorological conditions and ET, such as the Granger and Gray (GG) model (Granger and Gray, 1989), have been used in combination with SEBAL to estimate ET during cloudy periods (Long and Singh, 2010). The GG model uses the complementary relationship between actual and potential ET to estimate actual ET when imagery is not available. One major limitation of the GG model as currently applied is that the spatial resolution and accuracy of the ET estimates depend on the availability of meteorological data at a comparable resolution to the observed heterogeneity in ET. Other simple models of soil moisture stress have been developed that use ET derived from remote sensing to estimate model states on clear days and extrapolate those state variable values to dates with clouds (Anderson et al., 2007).

3.3 ET Methods Intercomparison Studies

Each of the three families of methods used to estimate ET has different strengths and weaknesses (Table 3.1). Temperature-based methods were developed to estimate ET from irrigated agriculture and therefore are more likely to perform best there but are often sensitive to how they are calibrated and sometimes depend on the existence of extreme values of ET in the image. Vegetation-based methods were developed for global application, with a focus on rainfed systems, and may have lower accuracy in irrigated systems where ET may be decoupled from a VI, particularly on the shoulders of the growing season. Scatterplot methods incorporate both temperature and vegetation but usually require internal calibration and, like some one-source energy-based methods, often depend on extremes of ET to be present in the image. Studies that compare vegetation-, temperature-, and scatterplot-based methods together are not common, and here, we review some recent examples. Such intercomparisons are important, because vegetation- and temperature-based methods have strengths in different environments, and quantitative information on which does better under what climate and land use conditions can guide the user in method selection.

Rafn et al. (2008) compared an NDVI method for deriving ET_f with the energy balance METRIC method over mixed crops in an Idaho irrigation district. They concluded that the empirical coefficient method was a fully objective and repeatable process that was fast, easy, and less costly to employ than the METRIC method.

Similarly, Gonzalez-Dugo et al. (2008) found that NDVI-derived ET_f combined with ET_o measurements predicted ET of corn and soybean crops as well as thermal-band methods in Central Iowa, although it over-estimated ET of corn during a dry-down period.

Vinukollu et al. (2011b) compared three different models: SEBS (Su, 2002), MOD16 (Mu et al., 2007), and PT-JPL (Fisher et al., 2008). The focus of their study was to compare the instantaneous (W/m²) and daily (mm) ET fluxes predicted by the three models implemented with data from sensors on the MODIS–Aqua satellite augmented by AVHRR data for vegetation characterization. Vinukollu et al. (2011b) compared the three models at three spatial scales. At the first spatial scale, λET from the three models was compared to and parameterized by eddy covariance flux tower data. At the second scale, a basin-scale water balance validation was performed using the models parameterized by remote sensing data. Finally, the models were compared against a hydrologic model driven by surface climate reanalysis at a global scale and on a latitudinal basis. For towers where soil evaporation plays an important role following precipitation events, SEBS and PT-JPL showed the highest and similar correlations, though large differences occurred during the primary growing season. Correlations between λET measured at towers in densely vegetated areas, such as evergreen and deciduous broadleaf forests, were highest for PT-JPL, reflecting again the importance of R_n in modulating λET in these areas. At the basin scale, however, the performance of each model was comparable. At the global scale, the vegetation-and energy-based ET methods tended to underestimate simulated soil moisture storage in water-limited (arid) regions of the world.

Ershadi et al. (2014) evaluated the PT-JPL model against SEBS, the 2011 updated MOD16 (Mu et al., 2011), and a complementary approach (advection–aridity model) against ET observed at FLUXNET towers. The PT-JPL model had the closest correlation with the FLUXNET-estimated ET but compared closely with SEBS. The PT-JPL model did particularly well in densely vegetated areas and was comparable to SEBS over croplands and grasslands. On a seasonal basis, all of the models, except the PT-JPL model, exhibited strong seasonality. The poor performance of MOD16 and SEBS in densely vegetated areas was attributed to the uncertainties that arise from a large number of model input requirements and complexity, including the sensitivity of their ET estimates to resistance parameters. All of the models did poorly over shrublands and evergreen needleleaf forests, reflecting the difficulty of NDVI in capturing vegetation dynamics for these land cover types. In conclusion, the authors suggested that, for regional to global studies, an ensemble of models, weighted by the success of contributing models for each land cover type be employed, given that no one single model performs consistently well across all land cover types. Such an approach could prove particularly useful in regional or global applications, where different model output could be used depending on land cover.

Velpuri et al. (2013) compared SSEBop ET (Senay et al., 2013) with point and gridded flux tower observations and water balance ET, gridded FLUXNET ET (Jung et al., 2011) and MOD16 ET (Mu et al., 2011) over the conterminous United States (CONUS). Point-scale validation using data from 60 FLUXNET tower locations against monthly SSEBop and MOD16 ET data aggregated by years revealed that both ET products showed overall comparable annual accuracies with mean errors in the order of 30%–60%. Although both ET products showed comparable results for most land cover types, SSEBop showed lower RMSE than MOD16 for grassland, irrigated cropland, and forest classes, while MOD16 performed better than SSEBop in rainfed croplands, shrublands, and woody savanna classes. Basin-scale validation of MOD16, SSEBop and gridded FLUXNET ET data against water balance ET indicated that both MOD16 and SSEBop ET matched the accuracies of the global gridded FLUXNET ET dataset at different scales. Both MODIS ET products effectively reproduced basin-scale ET response (up to 25% uncertainty) compared to CONUS-wide point-based ET response (up to 50%–60% uncertainty) illustrating the potential for MODIS ET products for basin-scale ET estimation. The apparent CONUS-wide uncertainties (up to 50%–60%) for monthly MODIS ET represented an overall error using data from several FLUXNET stations. The uncertainty for individual stations over time is much lower as shown from previous studies, including Mu et al. (2011) and Senay et al. (2013) who reported uncertainties up to 20% (MOD16) and 30% (SSEBop), on individual station-based FLUXNET validation, and Singh et al. (2013) reported uncertainties as low as 10% for individual stations in the Colorado River basin. Thus, despite an apparent high level of uncertainty at the CONUS-scale, the spatially explicit monthly SSEBop ET products can be useful for localized applications.

Choi et al. (2009) compared three models for estimating spatially distributed λET fluxes over a region in Iowa, United States, using Landsat TM/ETM+ imagery and ancillary observations from the SMACEX 2002 field experiment: TSEB (based on ALEXI), METRIC, and the T_R/VI method of Jiang and Islam (2001). TSEB and METRIC yielded similar and reasonable agreement with measured λET and H fluxes, with RMSD of 50–75 W/m² whereas for the T_R–VI method RMSE was over 100 W/m² (3.8 mm/day). Although TSEB and METRIC were in good agreement at the point comparisons performed, a spatial intercomparison of their results of gridded model output (i.e., comparing output on a pixel-by-pixel basis) revealed significant discrepancies in modeled turbulent heat flux patterns that correlated with vegetation density, particularly for H fluxes.

3.4 Special Problems in Cropped Areas

3.4.1 Landscape Heterogeneity and Spatial Disaggregation

Estimation of ET from croplands using remote sensing is particularly challenging in heterogeneous landscapes where agricultural plots are small (Figure 3.8). In India, for example, there are large areas of homogeneous irrigated cropping in canal-irrigated systems, but more than 50% of the irrigated area is supplied from groundwater wells, which are typically individually owned bore wells supplying small plots (<1 ha). The small groundwater-irrigated plots are often topographically organized, occurring mostly near stream channels where the

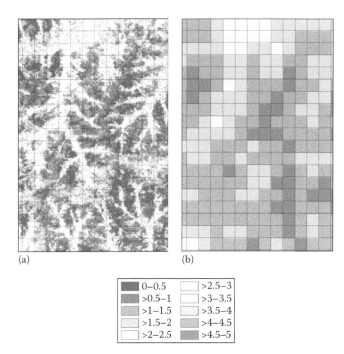

(a)   (b)

■ 0–0.5	□ >2.5–3
■ >0.5–1	□ >3–3.5
▨ >1–1.5	□ >3.5–4
□ >1.5–2	▨ >4–4.5
□ >2–2.5	▨ >4.5–5

FIGURE 3.8 Maps of SEBAL ET (mm/day) in a groundwater irrigated area in southern India at 30 m (a) and aggregated to 1 km (b). The grid in both panels represents 1 km pixels. (Based on data from Ahmad, M. et al., *Water Sci. Technol.*, 53, 83, 2006.)

water table is shallow, resulting in narrow bands of irrigation (Figure 3.8), which requires mapping irrigated areas as fractional cover of 1 km MODIS pixels (Biggs et al., 2006). Most globally available datasets are at a resolution of 1 km (MOD16) or coarser, which is significantly larger than irrigated patches in many areas. Even in the United States, where agricultural fields are large, 1 km resolution can be too coarse to resolve individual fields and to map ET differences by crop (Kustas et al., 2004). Kustas et al. (2004) documented that 250 m resolution was necessary to resolve differences in ET among agricultural fields in Iowa and that the variance in estimated ET decreases with increasing pixel size. Townshend and Justice (1988) estimated that land cover change mapping requires a spatial resolution of 250–500 m. Irrigated landscapes might be expected to require finer resolution than 250 m, especially in developing countries where plots are small.

Global methods (PT-JPL, MOD16) were designed to estimate ET over large spatial scales, often as input to community land surface models, rather than to assess crop-specific ET. In heterogeneous irrigated landscapes in semiarid climates, extreme spatial variability in soil moisture and ET means that extremes of low and high ET may occur in a single 1 km pixel, which significantly reduces ET estimates in the 1 km aggregated average. This could result in an underestimation of ET from irrigated cropland if no further disaggregation technique were applied. High-resolution imagery (<100 m, e.g., Landsat 30 m) is only available at 2 weeks or greater temporal resolution, which makes its application problematic in areas with high cloud cover and dynamic land cover. Some efforts have focused on combining imagery from different platforms to generate high-resolution maps of seasonal ET. Here, we review a few select studies to illustrate the potential for cross-platform downscaling.

Thermal imagery typically has a coarser resolution than visible and near-infrared (VNIR) bands for a given spectroradiometer. For the MODIS sensor, VNIR bands are available at 250 m resolution, while the thermal band is at 1 km. For Landsat, VNIR are at 30 m, while the TIR band is 120 m but has been resampled to either 60 m (before February 25, 2010) or 30 m (after February 25, 2010). The higher-resolution bands can be used to sharpen the coarse resolution thermal band using the inverse relationship between T_R and NDVI (Agam et al., 2007; Kustas et al., 2003). The first step is to coarsen the NDVI image to the resolution of the thermal image. Both the mean NDVI and the coefficient of variation (CV) of the high-resolution NDVI are calculated at the low resolution. Then, a percentage of pixels with the lowest CV with each of several bins of NDVI (e.g., 0–0.25, 0.2–0.5, >0.5) are selected to parameterize a function relating low-resolution T_R to low-resolution NDVI, which is used to predict T_R at low resolution:

$$\hat{T}_{Rlow}\left(NDVI_{low}\right) = a_{ds} + b_{ds}NDVI_{low} + c_{ds}NDVI_{low}^2 \quad (3.32)$$

where

a_{ds}, b_{ds}, and c_{ds} are empirical coefficients determined through least squares regression

ds subscript designates downscaling to differentiate the coefficients from the a and b parameters of the SEBAL algorithm

A linear equation could also be used, and the form of the equation depends on the observed T_R-NDVI relationship. While (3.32) can be used directly to predict T_R at high resolution using high-resolution NDVI, the correlation between T_R and NDVI may break down for pixels with low NDVI, including irrigated areas in either the beginning or end of the growing season, where high soil moisture or standing water and therefore low T_R can occur with low vegetation cover (Figure 3.7). Kustas et al. (2003) proposed an additional correction using the residual of (3.32):

$$\hat{T}_{Rhi} = \hat{T}_{Rlow}\left(NDVI_{hi}\right) + \left(T_{Rlow} - \hat{T}_{Rlow}\left(NDVI_{low}\right)\right) \quad (3.33)$$

where

$\hat{T}_{Rhi}$ is the predicted radiometric temperature at high spatial resolution

$\hat{T}_{Rlow}\left(NDVI_{hi}\right)$ is the predicted radiometric temperature using high-resolution NDVI and (3.32)

T_{Rlow} is the observed radiometric temperature at low resolution

$\hat{T}_{Rlow}\left(NDVI_{low}\right)$ is the predicted radiometric temperature using low-resolution NDVI (3.32)

Based on observed scatterplots of NDVI and T_R (Figure 3.7), this correction will be minimal for pixels with high NDVI and largest for pixels with low NDVI.

More complex algorithms that use all visible and NIR bands to model T_R in a moving window may be more successful in irrigated landscapes (Gao et al., 2012). Other algorithms have been developed to downscale T_R by fusing MODIS and Landsat, including those that use artificial neural network models (Bindhu et al., 2013).

One downscaling method specific to the ALEXI model, DisALEXI, was discussed in Section 3.2.3.2. DisALEXI ensures consistency in aggregated ET across scales by adjusting the high-resolution estimates to match the low-resolution average. Comparisons of ET estimated using algorithms forced by imagery with different resolutions have found disagreement. For example, SEBS-ET was calculated using Landsat, ASTER, and MODIS data (McCabe and Wood, 2006). SEBS-ET was similar for the high-resolution imagery (Landsat and ASTER) but SEBS-ET from Landsat or ASTER differed markedly from SEBS-ET derived from MODIS, even when the Landsat and ASTER ET were aggregated to MODIS resolution (McCabe and Wood, 2006) (Figure 3.9). This is likely due to nonlinear averaging of important inputs to the energy balance equations and needs to be acknowledged when using moderate resolution (1 km) data to estimate ET. The spatial average ET from all three image sources matched to within 10%–15%, suggesting that the low-resolution MODIS imagery were useful for watershed-scale estimates of ET (50 km²), but the MODIS data underestimated the variability present in the landscape. McCabe and Wood (2006) conclude that MODIS data are sufficient for estimating ET at watershed scales but are likely not accurate for estimating crop ET at resolutions that resolve ET from individual fields.

3.4.2 High-Resolution ET Mapping: New and Upcoming Platforms

There is a pressing need for datasets that allow mapping of ET at the spatial resolution of individual fields. In developing countries, field sizes may be very small (<1 ha), requiring high-resolution data for field-scale mapping (~10^1 m). While high-resolution data exist in historical archives (Landsat, 30 m) and contemporary datasets (Landsat 8, ASTER), the overpass frequency (~2 weeks for Landsat) may not be sufficient to capture high-quality data in areas with either dynamic land cover or cloud cover during the cropping season. Data from new remote sensing platforms could prove very useful for mapping ET at spatial and temporal resolutions that more closely approximate actual variability in agricultural landscapes. Satellites such as Sentinels 2 and 3, planned for launch in 2015 by the European Space Agency, and NASA's hyperspectral infrared imager (HyspIRI, launch date unknown at time of publication) (Hook and Green, 2013) hold promise for providing high-resolution data at high temporal frequency. Sentinel-2 will provide data on visible, NIR, and shortwave infrared wavelengths at 10 or 20 m with a revisit time of 5 days at the

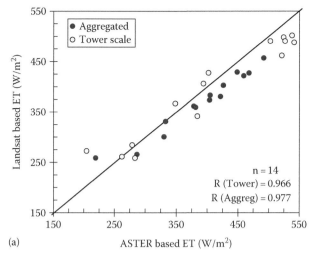

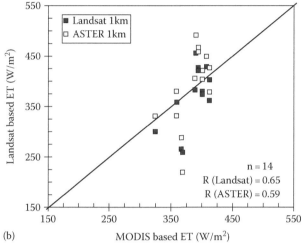

FIGURE 3.9 Comparison of SEBS-ET calculated using ASTER and Landsat data (a) and MODIS, ASTER, and Landsat (b). In the top panel, tower scale refers to individual pixels of Landsat or ASTER, while aggregated are the Landsat and ASTER SEBS-ET aggregated to MODIS resolution (1 km). (Reprinted from *Remote Sens. Environ.*, 105, McCabe, M.F. and Wood, E.F., Scale influences on the remote estimation of evapotranspiration using multiple satellite sensors, 271–285, Copyright (2006), with permission from Elsevier; Glenn, E.P. et al., *Ecohydrology*, 1, 316, 2008b; Glenn, E.P. et al., *Ecol. Eng.*, 59, 176, 2013.)

equator and 2–3 days at midlatitudes. This will provide unprecedented data for vegetation-based ET models, including NDVI and the LAI. Sentinel-3 will provide thermal imagery at 1 km resolution with revisit times of approximately 1 day at the equator. This temporal and spatial resolution is similar to existing MODIS data, so the additional gains for temperature-based methods may come in downscaling the 1 km data using Sentinel-2 data.

Other high-resolution datasets include the Landsat series (30 m), with the latest Landsat 8 data beginning in June 2013, with 14-day overpass. The Landsat archive can provide historical imagery to the late 1970s, and it has been used in many ET applications (Ahmad et al., 2006; Allen et al., 2013; Anderson et al.,

2012; Glenn et al., 2011; Kjaersgaard et al., 2011; Norman et al., 2003) though problems with cloud cover may be encountered, particularly where the rainy season coincides with the cropping season. ASTER generates data with 15 m resolution in the visible and 90 m in the TIR bands, but the image footprints and times of acquisition are often irregular, complicating its use for seasonal ET estimation (Er-Raki et al., 2008; Galleguillos et al., 2011). NASA's HyspIRI mission will provide visible to shortwave infrared (VSWIR: 380–2500 nm) at 60 m resolution with a revisit time of 19 days, and mid-infrared and TIR (3–12 µm) at 60 m resolution and revisit time of 5 days (Hook and Green, 2013). The high-resolution TIR data, in particular, generated by HyspIRI will be valuable for mapping ET from irrigated croplands in heterogeneous landscapes.

3.4.3 Model Complexity, Equifinality, and Sources of Error in ET Models

The EO-based estimates of ET presented in this chapter all rely on models to predict ET from EO data, and so have problems similar to other applications of models to estimate hydrological processes, including equifinality and parameter uncertainty. Equifinality arises in models that have many variables in the determining equations, but few actual data to populate the equations (Beven, 2006a; Franks et al., 1997; Medlyn et al., 2005). These models are frequently calibrated using approximated or assumed values for unmeasured variables. As a result, different models with different assumptions and levels of complexity can converge on the same output values, despite different process representations. Equifinality in hydrological models complicates the choice of a model for a given application, makes the use of models for hypothesis testing difficult, and results in uncertain prediction for times or places that have not been used for calibration (Beven, 2006a).

In the case of remote sensing of ET, problems of equifinality can occur in process-based models for use over mixed landscapes of natural and agricultural areas, for which few spatially distributed ground data are available. Remote sensing data usually consists only of radiance values from two or three bands (VNIR), and an imperfect estimate of T_R from thermal bands. However, the determining equations generally have numerous variables, including fractional cover, LAI, roughness lengths for mass and momentum transfer, albedo, emissivity, net radiation and ground heat flux (e.g., Bastiaanssen et al., 1998; Kustas and Norman, 1999). All of these are related in some way to vegetation cover, so they are often estimated by the use of vegetation indices. However, this introduces the problem of collinearity among what are supposed to be independent variables. For example, Glenn et al. (2008a) gave examples where five different plant stands in a mixed agricultural and riparian environment had the same NDVI values but differed markedly in plant heights (and presumably therefore roughness lengths), fractional cover, and LAI. VIs cannot uniquely determine separate biophysical variables; rather, they give an integrated measure of canopy "greenness" (Baret and Buis, 2008; Glenn et al., 2008a). Equifinality can be suspected to apply to surface energy balance methods as well as to VI methods that attempt to parameterize numerous variables with limited remote sensing data. For example, the SEBAL algorithm depends on the calibration of resistance terms and of the coefficients in the T_R and air temperature (T_2) relationship at the wet and dry pixels, but other representations of the T_R–T_2 relationship may result in different predictions at other locations in the image. McCabe et al. (2005) explicitly include the impact of different model parameterizations on ET estimated by calibrating a hydrological model with remotely sensed data. Explicit treatments of uncertainty in ET estimation should be produced in future efforts.

A problem related to equifinality is parameter uncertainty. In the PT-JPL model, for example, several parameters are based on field studies in specific locations, with unknown applicability in other regions. On the other hand, oversimplification of the process representation may require additional ground-level data or calibration of the simplified model to a specific location, as in the empirical crop coefficient methods. This complicates the use of simplified methods in global applications or in other regions with limited ground-level data for input or calibration. Overall, a balance needs to be struck between model complexity and applicability to unmeasured locations; the ideal model is the simplest one that provides adequate fit to observed data given the limitations in input data availability. There is significant scope for future research in the level of complexity needed for different scales of application, particularly at regional and global scales.

Another problem with application of EO data to estimate ET is error and uncertainty in the ground data used for calibration and validation. Ground measurements usually come from eddy covariance flux tower measurements, and these have typical errors on the order of 15%–30% when compared to lysimeters or other highly accurate measurements of ET (Allen et al., 2011). A particular problem with eddy covariance data is the "energy closure" error, where the sum of measured λE + H does not equal measured Rn−G. λE and H are usually increased to force closure, preserving the λE/H ratio of the measured data (Twine et al., 2000), but usually, there is no way to check if this correction improves the ET estimates. Scott (2010) compared eddy covariance results at three flux tower sites in a semiarid rangeland, at which precipitation, infiltration, and runoff were also measured. Uncorrected eddy covariance data gave ET values close to precipitation minus runoff and infiltration, where data corrected to force closure overestimated ET at each site by 10%–20%.

A further problem is a mismatch between scales of measurement. Evett et al. (2012) compared surface energy balance components used to calculate ET at scales ranging from weighing lysimeters (4.7 m diameter), to small plots, to whole fields (several hundred ha) captured by aerial and satellite imagery. They concluded that even with the best equipment and expertise, it was difficult to measure ET accurately using flux towers. Advection led to underestimates of ET by tower sensor systems

compared to lysimeters even after correcting for energy closure. This inaccuracy affected the interpretation of remote sensing results, which depended on flux tower data for validation. They urged caution in interpreting ET data from semiarid environments with advective conditions, especially those with mixes of irrigated and dryland crops and native vegetation. In dryland areas in Spain, Morillas et al. (2013) found that a TSM had errors of up to 90% in estimating λET compared to eddy covariance because latent heat flux is estimated as a residual and was a small component of the overall energy budget. Glenn et al. (2013b) reported that both energy balance and vegetation-based remote sensing methods overestimated ET of salt-stressed shrubs by 50% or more in a riparian corridor.

As this review shows, there has been a proliferation of remote sensing ET methods, most of which have uncertainty or errors of 10%–30% compared to ground measurements on a monthly basis with lower error (e.g., 5%–10%) on a seasonal or annual basis (Table 3.1) (Allen et al., 2011). As also seen in this review chapter, comparison studies often do not point to a clear choice of methods due to the problem of equifinality and errors and uncertainties in ground data (e.g., Gonzalez-Dugo et al., 2009). Simple methods tend to perform as well as more complex methods (e.g., Jiang et al., 2004), but simple methods often depend on calibration to ground level data and may not be applicable outside of the area of calibration. Reducing the error and uncertainty in remote sensing estimates of ET must depend in part on improving ground methods for measuring ET. Medlyn et al. (2005) and Beven (2006b) recommended more rigorous sensitivity analyses of ET models and explicit representation of uncertainty in model estimates. Medlyn et al. (2005) concluded that simplistic comparisons of ET models with eddy covariance data could lead to errors due to problems of equifinality, insensitivity, and uncertainty in both the models and the ground data. Their main concern was SVAT models, but their conclusions can also be applied to remote sensing methods for estimating ET (Glenn et al., 2008a). A central goal of future research should be to calculate pixel-wise estimates of the uncertainty in ET estimated from satellite imagery that accompany any map of ET, calculated using either a range of plausible parameter values or an intercomparison of models with different assumptions.

3.5 Conclusions

This chapter provided an overview of methods for estimating ET from EO platforms, with a focus on croplands. The chapter used consistent mathematical symbols across all methods, facilitating intercomparison of multiple techniques. The hope in providing a single comparison of multiple methods in one text is that practitioners and researchers can see the similarities among different methods and potentially see how their particular model choice could be extended to include other methods, as well as to more clearly identify the assumptions, strengths, and weaknesses of each family of methods. There are many different methods that are available to use with different degrees of complexity, utilizing EO data acquired from different platforms at different regions of the electromagnetic spectrum. Techniques are also characterized by different strengths and limitations related to their practical implementation and have varying accuracy. Many validation studies have, however, confirmed at least the potential for regional- and global-scale mapping at 1 km spatial resolution, and in some cases, operational implementation. The global models are typically vegetation-based methods, and their ability to perform in irrigated areas, where soil evaporation may occur in dry environments, has not been extensively tested. By contrast, temperature-based methods, especially those requiring internal calibration, have been widely tested in irrigated environments but face challenges in application for geographic scales larger than a single satellite image. On this basis, we encourage further intercomparison of the different EO-based modeling schemes for deriving ET in croplands and, for operational purposes, EO-based model ensembles that integrate the spatiotemporal benefits of each method.

Most studies that evaluate the ability of different techniques to predict ET have been based on direct comparisons between predicted fluxes and corresponding in situ measurements. Other modeling approaches, such as uncertainty or sensitivity analysis, have so far been little incorporated in such studies, despite their importance for any all-inclusive model validation/verification (Petropoulos et al., 2013). Explicit estimation of uncertainty in ET, including pixel-wise estimates of model error, would represent a significant advance.

More validation studies for operationally distributed products need to be conducted in different ecosystems globally. Such studies, if conducted in a systematic way following an acceptable protocol, will identify issues in the algorithmic design of these products, which will improve our capability to operationally estimate ET from EO sensors. More work should be directed toward the development of schemes for the temporal interpolation of the instantaneous ET estimates, as well as of downscaling approaches of ET to the resolution of individual fields where possible. Since no model performs optimally in all land covers, regional or global applications could consider using different models for different land cover types. We encourage further intercomparison of vegetation- and temperature-based methods, and further research on downscaling to the resolution of individual fields. We anticipate that EO data from new satellite platforms planned already to be launched in the next few years alone or in synergistic use to each other will help to meet some but not all of these needs.

Acknowledgments

Thanks to Dr. Gabriel Senay, Dr. Prasad Thenkabail, and one anonymous reviewer for their helpful comments. Dr. Petropoulos' participation in this work was supported by the European Commission under the Marie Curie Career Re-Integration Grant "TRANSFORM-EO" project 531. Any use of trade, product, or firm names is for descriptive purposes only and does not imply endorsement by the U.S. Government.

References

Agam, N., Kustas, W.P., Anderson, M.C., Li, F., and Colaizzi, P.D. 2007. Utility of thermal sharpening over Texas high plains irrigated agricultural fields. *Journal of Geophysical Research* 112, D19110. doi:10.1029/2007JD008407.

Ahmad, M., Biggs, T.W., Turral, H., and Scott, C.A. 2006. Application of SEBAL approach to map the agricultural water use patterns in the data scarce Krishna River Basin of India. *Water Science and Technology* 53, 83–90.

Allen, R., Irmak, A., Trezza, R., Hendrickx, J.M.H., Bastiaanssen, W., and Kjaersgaard, J. 2011. Satellite-based ET estimation in agriculture using SEBAL and METRIC. *Hydrological Processes* 25, 4011–4027. doi:10.1002/hyp.8408.

Allen, R.G. 2000. Using the FAO-56 dual crop coefficient method over an irrigated region as part of an evapotranspiration intercomparison study. *Journal of Hydrology* 229, 27–41.

Allen, R.G., Burnett, B., Kramber, W., Huntington, J., Kjaersgaard, J., Kilic, A., Kelly, C., and Trezza, R. 2013. Automated calibration of the METRIC-landsat evapotranspiration process. *Journal of the American Water Resources Association* 49, 563–576. doi:10.1111/jawr.12056.

Allen, R.G. and Pereira, L.S. 2009. Estimating crop coefficients from fraction of ground cover and height. *Irrigation Science* 28, 17–34.

Allen, R.G., Pereira, L.S., Howell, T.A., and Jensen, M.E. 2011. Evapotranspiration information reporting: I. Factors governing measurement accuracy. *Agricultural Water Management* 98, 899–920.

Allen, R.G., Pereira, L.S., Raes, D., and Smith, M. 1998. Crop evapotranspiration-guidelines for computing crop water requirements-FAO, Irrigation and drainage paper 56. Rome.

Allen, R.G., Tasumi, M., and Trezza, R. 2007. Satellite-based energy balance for mapping evapotranspiration with internalized calibration (METRIC)—Model. *Journal of Irrigation and Drainage Engineering* 133, 380–394. doi:10.1061/(ASCE)0733-9437(2007)133:4(380).

Anderson, M.C., Allen, R.G., Morse, A., and Kustas, W.P. 2012. Use of Landsat thermal imagery in monitoring evapotranspiration and managing water resources. *Remote Sensing of Environment* 122, 50–65. doi:10.1016/j.rse.2011.08.025.

Anderson, M.C., Kustas, W.P., and Hain, C.R. 2013. Mapping surface fluxes and moisture conditions from field to global scales using ALEXI/DisALEXI. In: *Remote Sensing of Energy Fluxes and Soil Moisture Content*, Petropoulos, G.P. (ed.), pp. 207–232. New York: Taylor and Francis.

Anderson, M.C., Norman, J.M., Diak, G.R., Kustas, W.P., and Mecikalski, J.R. 1997. A two-source time-integrated model for estimating surface fluxes using thermal infrared remote sensing. *Remote Sensing of Environment* 60, 195–216. doi:http://dx.doi.org/10.1016/S0034-4257(96)00215-5.

Anderson, M.C., Norman, J.M., Mecikalski, J.R., Otkin, J.A., and Kustas, W.P. 2007. A climatological study of evapotranspiration and moisture stress across the continental United States based on thermal remote sensing: 1. Model formulation. *Journal of Geophysical Research: Atmospheres* 112, D10117. doi:10.1029/2006JD007506.

Baldocchi, D., Falge, E., Gu, L., Olson, R., Hollinger, D., Running, S., Anthoni, P., Bernhofer, C., Davis, K., and Evans, R. 2001. FLUXNET: A new tool to study the temporal and spatial variability of ecosystem-scale carbon dioxide, water vapor, and energy flux densities. *Bulletin of the American Meteorological Society* 82, 2415–2434.

Baret, F. and Buis, S. 2008. Estimating canopy characteristics from remote sensing observations: Review of methods and associated problems. In: *Advances in Land Remote Sensing*, New York, USA, pp. 173–201.

Bastiaanssen, W.G.M., Ahmad, M.D., and Chemin, Y. 2002. Satellite surveillance of evaporative depletion across the Indus Basin. *Water Resources Research* 38, 1273. doi:10.1029/2001WR000386.

Bastiaanssen, W.G.M., Menenti, M., Feddes, R.A., and Holtslag, A.A.M. 1998. A remote sensing surface energy balance algorithm for land (SEBAL) 1. Formulation. *Journal of Hydrology* 212–213, 198–212.

Bastiaanssen, W.G.M., Noordman, E.J.M., Pelgrum, H., Davids, G., Thoreson, B.P., and Allen, R.G. 2005. SEBAL model with remotely sensed data to improve water-resources management under actual field conditions. *Journal of Irrigation and Drainage Engineering* 131, 85–93.

Bausch, W.C. and Neale, C.M.U. 1987. Crop coefficients derived from reflected canopy radiation: A Concept. *American Society of Agricultural Engineers* 30, 703–709.

Beven, K. 2006a. A manifesto for the equifinality thesis. *Journal of Hydrology* 320, 18–36.

Beven, K. 2006b. On undermining the science? *Hydrological Processes* 20, 3141–3146.

Bhattacharya, B.K., Mallick, K., Patel, N.K., and Parihar, J.S. 2010. Regional clear sky evapotranspiration over agricultural land using remote sensing data from Indian geostationary meteorological satellite. *Journal of Hydrology* 387, 65–80.

Biggs, T.W., Thenkabail, P.S., Gumma, M.K., Scott, C. a., Parthasaradhi, G.R., and Turral, H.N. 2006. Irrigated area mapping in heterogeneous landscapes with MODIS time series, ground truth and census data, Krishna Basin, India. *International Journal of Remote Sensing* 27, 4245–4266. doi:10.1080/01431160600851801.

Bindhu, V.M., Narasimhan, B., and Sudheer, K.P. 2013. Development and verification of a non-linear disaggregation method (NL-DisTrad) to downscale MODIS land surface temperature to the spatial scale of Landsat thermal data to estimate evapotranspiration. *Remote Sensing of Environment* 135, 118–129. doi:10.1016/j.rse.2013.03.023.

Bisht, G. and Bras, R.L. 2010. Estimation of net radiation from the MODIS data under all sky conditions: Southern Great Plains case study. *Remote Sensing of Environment* 114, 1522–1534. doi:10.1016/j.rse.2010.02.007.

Bisht, G., Venturini, V., Islam, S., and Jiang, L. 2005. Estimation of the net radiation using MODIS (Moderate Resolution Imaging Spectroradiometer) data for clear sky days. *Remote Sensing of Environment* 97, 52–67.

Bonan, G.B. 1998. The land surface climatology of the NCAR land surface model coupled to the NCAR community climate model. *Journal of Climate* 11, 1307–1326.

Brunsell, N.A. and Gillies, R.R. 2003. Scale issues in land–atmosphere interactions: implications for remote sensing of the surface energy balance. *Agricultural and Forest Meteorology* 117, 203–221.

Cammalleri, C., Anderson, M.C., Gao, F., Hain, C.R., and Kustas, W.P. 2013. A data fusion approach for mapping daily evapotranspiration at field scale. *Water Resources Research* 49, 4672–4686. doi:10.1002/wrcr.20349.

Cammalleri, C., Anderson, M.C., Gao, F., Hain, C.R., and Kustas, W.P. 2014. Mapping daily evapotranspiration at field scales over rainfed and irrigated agricultural areas using remote sensing data fusion. *Agricultural and Forest Meteorology* 186, 1–11. doi:10.1016/j.agrformet.2013.11.001.

Campos, I., Neale, C.M.U., Calera, A., Balbontín, C., and González-Piqueras, J. 2010. Assessing satellite-based basal crop coefficients for irrigated grapes (*Vitis vinifera* L.). *Agricultural Water Management* 98, 45–54. doi:10.1016/j.agwat.2010.07.011.

Carlson, T. 2007. An overview of the"triangle method" for estimating surface evapotranspiration and soil moisture from satellite imagery. *Sensors* 7, 1612–1629.

Carlson, T.N., Gillies, R.R., and Schmugge, T.J. 1995. An interpretation of methodologies for indirect measurement of soil water content. *Agricultural and Forest Meteorology* 77, 191–205. doi:10.1016/0168-1923(95)02261-U.

Chávez, J., Neale, C.U., Prueger, J., and Kustas, W. 2008. Daily evapotranspiration estimates from extrapolating instantaneous airborne remote sensing ET values. *Irrigation Science* 27, 67–81. doi:10.1007/s00271-008-0122-3.

Chávez, J.L., Neale, C.M.U., Hipps, L.E., Prueger, J.H., and Kustas, W.P. 2005. Comparing aircraft-based remotely sensed energy balance fluxes with eddy covariance tower data using heat flux source area functions. *Journal of Hydrometeorology* 6, 923–940.

Chen, F., Janjić, Z., and Mitchell, K. 1997. Impact of atmospheric surface-layer parameterizations in the new land-surface scheme of the NCEP mesoscale Eta model. *Boundary-Layer Meteorology* 85, 391–421.

Chen, F., Mitchell, K., Schaake, J., Xue, Y., Pan, H., Koren, V., Duan, Q.Y., Ek, M., and Betts, A. 1996. Modeling of land surface evaporation by four schemes and comparison with FIFE observations. *Journal of Geophysical Research: Atmospheres* (1984–2012) 101, 7251–7268.

Chen, J.-H., Kan, C.-E., Tan, C.-H., and Shih, S.-F. 2002. Use of spectral information for wetland evapotranspiration assessment. *Agricultural Water Management* 55, 239–248.

Chen, Y., Xia, J., Liang, S., Feng, J., Fisher, J.B., Li, X., Li, X. et al. 2014. Comparison of satellite-based evapotranspiration models over terrestrial ecosystems in China. *Remote Sensing of Environment* 140, 279–293. doi:10.1016/j.rse.2013.08.045.

Choi, M., Kustas, W.P., Anderson, M.C., Allen, R.G., Li, F., and Kjaersgaard, J.H. 2009. An intercomparison of three remote sensing-based surface energy balance algorithms over a corn and soybean production region (Iowa, US) during SMACEX. *Agricultural and Forest Meteorology* 149, 2082–2097.

Choudhury, B.J. 1989. Estimating evaporation and carbon assimilation using infrared temperature data. In: *Theory and Applications of Optical Remote Sensing*, Asrar, G. (ed.), pp. 628–690. New York: Wiley-Interscience.

Choudhury, B.J., Ahmed, N.U., Idso, S.B., Reginato, R.J., and Daughtry, C.S.T. 1994. Relations between evaporation coefficients and vegetation indices studied by model simulations. *Remote Sensing of Environment* 50, 1–17.

Cleugh, H.A., Leuning, R., Mu, Q., and Running, S.W. 2007. Regional evaporation estimates from flux tower and MODIS satellite data. *Remote Sensing of Environment* 106, 285–304.

Clothier, B.E., Clawson, K.L., Pinter, P.J., Moran, M.S., Reginato, R.J., and Jackson, R.D. 1986. Estimation of soil heat flux from net radiation during the growth of alfalfa. *Agricultural and Forest Meteorology* 37, 319–329. doi:10.1016/0168-1923(86)90069-9.

Conrad, C., Dech, S., Hafeez, M., Lamers, J., Martius, C., and Strunz, G. 2007. Mapping and assessing water use in a Central Asian irrigation system by utilizing MODIS remote sensing products. *Irrigation and Drainage Systems* 21, 197–218. doi:10.1007/s10795-007-9029-z.

Courault, D., Seguin, B., and Olioso, A. 2005. Review on estimation of evapotranspiration from remote sensing data: From empirical to numerical modeling approaches. *Irrigation and Drainage Systems* 19, 223–249.

Crago, R.D. 1996. Conservation and variability of the evaporative fraction during the daytime. *Journal of Hydrology* 180, 173–194. doi:10.1016/0022-1694(95)02903-6.

Duchemin, B., Hadria, R., Erraki, S., Boulet, G., Maisongrande, P., Chehbouni, A., Escadafal, R., Ezzahar, J., Hoedjes, J.C.B., and Kharrou, M.H. 2006. Monitoring wheat phenology and irrigation in Central Morocco: On the use of relationships between evapotranspiration, crops coefficients, leaf area index and remotely-sensed vegetation indices. *Agricultural Water Management* 79, 1–27.

El-Shikha, D.M., Waller, P., Hunsaker, D., Clarke, T., and Barnes, E. 2007. Ground-based remote sensing for assessing water and nitrogen status of broccoli. *Agricultural Water Management* 92, 183–193.

Engman, E.T. and Gurney, R.J. 1991. Recent advances and future implications of remote sensing for hydrologic modeling. In: *Recent Advances in the Modeling of Hydrologic Systems*, Bowles, D.S., (ed.), pp. 471–495. Dordrecht, the Netherlands: Kluwer Academic Publishers.

Er-Raki, S., Chehbouni, A., Hoedjes, J., Ezzahar, J., Duchemin, B., and Jacob, F. 2008. Improvement of FAO-56 method for olive orchards through sequential assimilation of thermal infrared-based estimates of ET. *Agricultural Water Management* 95, 309–321.

Ershadi, A., McCabe, M.F., Evans, J.P., Chaney, N.W., and Wood, E.F. 2014. Multi-site evaluation of terrestrial evaporation models using FLUXNET data. *Agricultural and Forest Meteorology* 187, 46–61. doi:10.1016/j.agrformet.2013.11.008.

Ershadi, A., McCabe, M.F., Evans, J.P., and Walker, J.P. 2013. Effects of spatial aggregation on the multi-scale estimation of evapotranspiration. *Remote Sensing of Environment* 131, 51–62. doi:10.1016/j.rse.2012.12.007.

Evett, S.R., Kustas, W.P., Gowda, P.H., Anderson, M.C., Prueger, J.H., and Howell, T.A. 2012. Overview of the bushland evapotranspiration and agricultural remote sensing experiment 2008 (BEAREX08): A field experiment evaluating methods for quantifying ET at multiple scales. *Advances in Water Resources* 50, 4–19.

Farah, H.O., Bastiaanssen, W.G.M., and Feddes, R.A. 2004. Evaluation of the temporal variability of the evaporative fraction in a tropical watershed. *International Journal of Applied Earth Observation and Geoinformation* 5, 129–140.

Fisher, J.B., Tu, K.P., and Baldocchi, D.D. 2008. Global estimates of the land–atmosphere water flux based on monthly AVHRR and ISLSCP-II data, validated at 16 FLUXNET sites. *Remote Sensing of Environment* 112, 901–919.

Foley, J.A., DeFries, R., Asner, G.P., Barford, C., Bonan, G., Carpenter, S.R., Chapin, F.S. et al. 2005. Global consequences of land use. *Science* 309, 570–574. doi:10.1126/science.1111772.

Foley, J.A., Ramankutty, N., Brauman, K.A., Cassidy, E.S., Gerber, J.S., Johnston, M., Mueller, N.D. et al. 2011. Solutions for a cultivated planet. *Nature* 478, 337–342. doi:10.1038/nature10452.

Franks, S.W., Beven, K.J., Quinn, P.F., and Wright, I.R. 1997. On the sensitivity of the soil-vegetation-atmosphere transfer (SVAT) schemes: Equifinality and the problem of robust calibration. *Agricultural and Forest Meteorology* 86, 63–75.

French, A.N., Alfieri, J.G., Kustas, W.P., Prueger, J.H., Hipps, L.E., Chávez, J.L., Evett, S.R. et al. 2012. Estimation of surface energy fluxes using surface renewal and flux variance techniques over an advective irrigated agricultural site. *Advances in Water Resources* 50, 91–105. doi:10.1016/j.advwatres.2012.07.007.

Galleguillos, M., Jacob, F., Prévot, L., French, A., and Lagacherie, P. 2011. Comparison of two temperature differencing methods to estimate daily evapotranspiration over a Mediterranean vineyard watershed from ASTER data. *Remote Sensing of Environment* 115, 1326–1340.

Gao, F., Kustas, W.P., and Anderson, M.C. 2012. A data mining approach for sharpening thermal satellite imagery over land. *Remote Sensing* 4, 3287–3319.

Gao, F., Masek, J., Schwaller, M., and Hall, F. 2006. On the blending of the Landsat and MODIS surface reflectance: Predicting daily Landsat surface reflectance. *Geoscience and Remote Sensing, IEEE Transactions on* 44, 2207–2218.

García, M., Villagarcía, L., Contreras, S., Domingo, F., and Puigdefábregas, J. 2007. Comparison of three operative models for estimating the surface water deficit using ASTER reflective and thermal data. *Sensors* 7, 860–883.

Gentine, P., Entekhabi, D., Chehbouni, A., Boulet, G., and Duchemin, B. 2007. Analysis of evaporative fraction diurnal behaviour. *Agricultural and Forest Meteorology* 143, 13–29. doi:10.1016/j.agrformet.2006.11.002.

Gillies, R.R. and Carlson, T.N. 1995. Thermal remote sensing of surface soil water content with partial vegetation cover for incorporation into climate models. *Journal of Applied Meteorology* 34, 745–756.

Gillies, R.R., Kustas, W.P., and Humes, K.S. 1997. A verification of the 'triangle' method for obtaining surface soil water content and energy fluxes from remote measurements of the Normalized Difference Vegetation Index (NDVI) and surface radiant temperature. *International Journal of Remote Sensing* 18, 3145–3166.

Glenn, E.P., Huete, A.R., Nagler, P.L., Hirschboeck, K.K., and Brown, P. 2007. Integrating remote sensing and ground methods to estimate evapotranspiration. *Critical Reviews in Plant Sciences* 26, 139–168.

Glenn, E.P., Huete, A.R., Nagler, P.L., and Nelson, S.G. 2008a. Relationship between remotely-sensed vegetation indices, canopy attributes and plant physiological processes: What vegetation indices can and cannot tell us about the landscape. *Sensors* 8, 2136–2160.

Glenn, E.P., Mexicano, L., Garcia-Hernandez, J., Nagler, P.L., Gomez-Sapiens, M.M., Tang, D., Lomeli, M.A., Ramirez-Hernandez, J., and Zamora-Arroyo, F. 2013a. Evapotranspiration and water balance of an anthropogenic coastal desert wetland: Responses to fire, inflows and salinities. *Ecological Engineering* 59, 176–184.

Glenn, E.P., Morino, K., Didan, K., Jordan, F., Carroll, K., Nagler, P.L., Hultine, K., Sheader, L., and Waugh, J. 2008b. Scaling sap flux measurements of grazed and ungrazed shrub communities with fine and coarse-resolution remote sensing. *Ecohydrology* 1, 316–329.

Glenn, E.P., Nagler, P.L., and Huete, A.R. 2010. Vegetation index methods for estimating evapotranspiration by remote sensing. *Surveys in Geophysics* 31, 531–555.

Glenn, E.P., Nagler, P.L., Morino, K., and Hultine, K.R. 2013b. Phreatophytes under stress: transpiration and stomatal conductance of saltcedar (*Tamarix* spp.) in a high-salinity environment. *Plant and Soil* 371, 655–672.

Glenn, E.P., Neale, C.M.U., Hunsaker, D.J., and Nagler, P.L. 2011. Vegetation index-based crop coefficients to estimate evapotranspiration by remote sensing in agricultural and natural ecosystems. *Hydrological Processes* 25, 4050–4062. doi:10.1002/hyp.8392.

Gómez, M., Olioso, A., Sobrino, J.A., and Jacob, F. 2005. Retrieval of evapotranspiration over the Alpilles/ReSeDA experimental site using airborne POLDER sensor and a thermal camera. *Remote Sensing of Environment* 96, 399–408.

Gontia, N.K. and Tiwari, K.N. 2010. Estimation of crop coefficient and evapotranspiration of wheat (*Triticum aestivum*) in an irrigation command using remote sensing and GIS. *Water Resources Management* 24, 1399–1414.

González-Dugo, M.P., and Mateos, L. 2008. Spectral vegetation indices for benchmarking water productivity of irrigated cotton and sugarbeet crops. *Agricultural Water Management* 95, 48–58.

Gonzalez-Dugo, M.P., Neale, C.M.U., Mateos, L., Kustas, W.P., Prueger, J.H., Anderson, M.C., and Li, F. 2009. A comparison of operational remote sensing-based models for estimating crop evapotranspiration. *Agricultural and Forest Meteorology* 149, 1843–1853.

Gottschalck, J., Meng, J., Rodell, M., and Houser, P. 2005. Analysis of multiple precipitation products and preliminary assessment of their impact on global land data assimilation system land surface states. *Journal of Hydrometeorology* 6, 573–598. doi:10.1175/JHM437.1.

Gowda, P., Chavez, J., Colaizzi, P., Evett, S., Howell, T., and Tolk, J. 2008. ET mapping for agricultural water management: Present status and challenges. *Irrigation Science* 26, 223–237. doi:10.1007/s00271-007-0088-6.

Granger, R.J. and Gray, D.M. 1989. Evaporation from natural nonsaturated surfaces. *Journal of Hydrology* 111, 21–29.

Gui, S., Liang, S., and Li, L. 2010. Evaluation of satellite-estimated surface longwave radiation using ground-based observations. *Journal of Geophysical Research* 115, D18214. doi:10.1029/2009JD013635.

Gupta, S.K., Ritchey, N.A., Wilber, A.C., Whitlock, C.H., Gibson, G.G., and Stackhouse, P.W.J. 1999. A climatology of surface radiation budget derived from satellite data. *Journal of Climate* 12, 2691–2710.

Hall, F.G., Huemmrich, K.F., Goetz, S.J., Sellers, P.J., and Nickeson, J.E. 1992. Satellite remote sensing of surface energy balance: Success, failures, and unresolved issues in FIFE. *Journal of Geophysical Research: Atmospheres* 97, 19061–19089. doi:10.1029/92JD02189.

Hoekstra, A.Y. and Mekonnen, M.M. 2012. The water footprint of humanity. *Proceedings of the National Academy of Sciences of the United States of America* 109, 3232–3237. doi:10.1073/pnas.1109936109.

Hook, S.J. and Green, R.O. 2013. *2013 HyspIRI Science Workshop: Objectives, Overview and Update*. Pasadena, CA: Jet Propulsion Laboratory, California Institute of Technology.

Hunsaker, D.J., Fitzgerald, G.J., French, A.N., Clarke, T.R., Ottman, M.J., and Pinter Jr., P.J. 2007a. Wheat irrigation management using multispectral crop coefficients. I. Crop evapotranspiration prediction. *Transactions of the American Society of Agricultural and Biological Engineers* 50, 2017–2033.

Hunsaker, D.J., Fitzgerald, G.J., French, A.N., Clarke, T.R., Ottman, M.J., and Pinter Jr., P.J. 2007b. Wheat irrigation management using multispectral crop coefficients: II. Irrigation scheduling performance, grain yield, and water use efficiency. *Transactions of the American Society of Agricultural and Biological Engineers* 50, 2035–2050.

Hunsaker, D.J., French, A.N., Clarke, T.R., and El-Shikha, D.M. 2011. Water use, crop coefficients, and irrigation management criteria for camelina production in arid regions. *Irrigation Science* 29, 27–43. doi:10.1007/s00271-010-0213-9.

Hunsaker, D.J., Pinter Jr, P.J., and Kimball, B.A. 2005. Wheat basal crop coefficients determined by normalized difference vegetation index. *Irrigation Science* 24, 1–14.

Idso, S.B. 1981. A set of equations for full spectrum and 8-to 14-μm and 10.5-to 12.5-μm thermal radiation from cloudless skies. *Water Resources Research* 17, 295–304.

Jackson, R.D., Hatfield, J.L., Reginato, R.J., Idso, S.B., and Pinter Jr, P.J. 1983. Estimation of daily evapotranspiration from one time-of-day measurements. *Agricultural Water Management* 7, 351–362.

Jackson, R.D., Idso, S.B., Reginato, R.J., and Pinter, P.J. 1981. Canopy temperature as a crop water stress indicator. *Water Resources Research* 17, 1133–1138.

Jayanthi, H., Neale, C.M.U., and Wright, J.L. 2007. Development and validation of canopy reflectance-based crop coefficient for potato. *Agricultural Water Management* 88, 235–246. doi:10.1016/j.agwat.2006.10.020.

Jiang, L. and Islam, S. 1999. A methodology for estimation of surface evapotranspiration over large areas using remote sensing observations. *Geophysical Research Letters* 26, 2773–2776.

Jiang, L. and Islam, S. 2003. An intercomparison of regional latent heat flux estimation using remote sensing data. *International Journal of Remote Sensing* 24, 2221–2236. doi:10.1080/01431160210154821.

Jiang, L. and Islam, S. 2001. Estimation of surface evaporation map over southern great plains using remote sensing data. *Water Resources Research* 37, 329–340.

Jiang, L., Islam, S., and Carlson, T.N. 2004. Uncertainties in latent heat flux measurement and estimation: implications for using a simplified approach with remote sensing data. *Canadian Journal of Remote Sensing* 30, 769–787.

Johnson, L.F. and Trout, T.J. 2012. Satellite NDVI assisted monitoring of vegetable crop evapotranspiration in California's San Joaquin Valley. *Remote Sensing* 4, 439–455.

Jung, M., Reichstein, M., and Bondeau, A. 2009. Towards global empirical upscaling of FLUXNET eddy covariance observations: Validation of a model tree ensemble approach using a biosphere model. *Biogeosciences* 6, 2001–2013. doi:10.5194/bg-6-2001-2009.

Jung, M., Reichstein, M., Margolis, H.A., Cescatti, A., Richardson, A.D., Arain, M.A., Arneth, A. et al. 2011. Global patterns of land-atmosphere fluxes of carbon dioxide, latent heat, and sensible heat derived from eddy covariance, satellite, and meteorological observations. *Journal of Geophysical Research* 116, G00J07. doi:10.1029/2010JG001566.

Kalma, J.D., McVicar, T.R., and McCabe, M.F. 2008. Estimating land surface evaporation: A review of methods using remotely sensed surface temperature data. *Surveys in Geophysics* 29, 421–469.

Kalnay, E., Kanamitsu, M., Kistler, R., Collins, W., Deaven, D., Gandin, L., Iredell, M. et al. 1996. The NCEP/NCAR 40-year reanalysis project. *Bulletin of the American Meteorological Society* 77, 437–472.

Kjaersgaard, J., Allen, R., Garcia, M., Kramber, W., and Trezza, R. 2009. Automated selection of anchor pixels for landsat based evapotranspiration estimation. In: *World Environmental and Water Resources Congress 2009*, pp. 1–11. Reston, VA: American Society of Civil Engineers. doi:10.1061/41036(342)442.

Kjaersgaard, J., Allen, R.G., and Irmak, A. 2011. Improved methods for estimating monthly and growing season ET using METRIC applied to moderate resolution satellite imagery. *Hydrological Processes* 25, 4028–4036. doi:10.1002/hyp.8394.

Koster, R.D. and Suarez, M.J. 1996. Energy and water balance calculations in the Mosaic LSM. *NASA Tech. Memo* 104606, 59.

Kustas, W.P. and Daughtry, C. 1990. Estimation of the soil heat flux/net radiation ratio from spectral data. *Agricultural and Forest Meteorology* 49, 205–223. doi:10.1016/0168-1923(90)90033-3.

Kustas, W.P., Hatfield, J.L., and Prueger, J.H. 2005. The soil moisture–atmosphere coupling experiment (SMACEX): Background, hydrometeorological conditions, and preliminary findings. *Journal of Hydrometeorology* 6, 791–804.

Kustas, W.P., Jackson, T.J., Schmugge, T.J., Parry, R., Goodrich, D.C., Amer, S.A., Bach, L.B., Keefer, T.O., Weltz, M.A., and Moran, M.S. 1991. An interdisciplinary field study of the energy and water fluxes in the atmospheric-biosphere system over semiarid rangelands: Description and some preliminary results. *Bulletin of the American Meteorological Society* 72, 1683–1705.

Kustas, W.P., Li, F., Jackson, T.J., Prueger, J.H., MacPherson, J.I., and Wolde, M. 2004. Effects of remote sensing pixel resolution on modeled energy flux variability of croplands in Iowa. *Remote Sensing of Environment* 92, 535–547.

Kustas, W.P. and Norman, J.M. 1999. Evaluation of soil and vegetation heat flux predictions using a simple two-source model with radiometric temperatures for partial canopy cover. *Agricultural and Forest Meteorology* 94, 13–29.

Kustas, W.P. and Norman, J.M. 1996. Use of remote sensing for evapotranspiration monitoring over land surfaces. *Hydrological Sciences Journal* 41, 495–516. doi:10.1080/02626669609491522.

Kustas, W.P., Norman, J.M., Anderson, M.C., and French, A.N. 2003. Estimating subpixel surface temperatures and energy fluxes from the vegetation index–radiometric temperature relationship. *Remote Sensing of Environment* 85, 429–440. doi:10.1016/S0034-4257(03)00036-1.

Lee, E., Chase, T.N., Rajagopalan, B., Barry, R.G., Biggs, T.W., and Lawrence, P.J. 2009. Effects of irrigation and vegetation activity on early Indian summer monsoon variability. *International Journal of Climatology* 29, 573–581.

Lee, E., Sacks, W.J., Chase, T.N., and Foley, J.A. 2011. Simulated impacts of irrigation on the atmospheric circulation over Asia. *Journal of Geophysical Research: Atmospheres* (1984–2012) 116, D08114. doi:10.1029/2010JD014740.

Leuning, R., Zhang, Y.Q., Rajaud, A., Cleugh, H., and Tu, K. 2008. A simple surface conductance model to estimate regional evaporation using MODIS leaf area index and the Penman–Monteith equation. *Water Resources Research* 44, W10419. doi:10.1029/2007WR006562.

Lhomme, J.-P. and Elguero, E. 1999. Examination of evaporative fraction diurnal behaviour using a soil-vegetation model coupled with a mixed-layer model. *Hydrology and Earth System Sciences* 3, 259–270.

Li, H., Zheng, L., Lei, Y., Li, C., Liu, Z., and Zhang, S. 2008. Estimation of water consumption and crop water productivity of winter wheat in North China Plain using remote sensing technology. *Agricultural Water Management* 95, 1271–1278.

Liang, S., Wang, K., Zhang, X., and Wild, M. 2010. Review on estimation of land surface radiation and energy budgets from ground measurement, remote sensing and model simulations. *IEEE Journal of Selected Topics in Applied Earth Observations and Remote Sensing* 3, 225–240.

Liang, S., Zhang, X., He, T., Cheng, J., and Wang, D. 2013. Remote sensing of the land surface radiation budget. In: *Remote Sensing of Energy Fluxes and Soil Moisture Content*, Petropoulos, G.P. (ed.), pp. 121–162. New York: Taylor and Francis.

Liang, X., Lettenmaier, D.P., Wood, E.F., and Burges, S.J. 1994. A simple hydrologically based model of land surface water and energy fluxes for general circulation models. *Journal of Geophysical Research* 99, 14, 414–415, 428.

Long, D. and Singh, V.P. 2010. Integration of the GG model with SEBAL to produce time series of evapotranspiration of high spatial resolution at watershed scales. *Journal of Geophysical Research* 115, D21128. doi:10.1029/2010jd014092.

Long, D., Singh, V.P., and Li, Z.-L. 2011. How sensitive is SEBAL to changes in input variables, domain size and satellite sensor? *Journal of Geophysical Research* 116, D21107. doi:10.1029/2011jd016542.

Ma, Y. and Pinker, R.T. 2012. Modeling shortwave radiative fluxes from satellites. *Journal of Geophysical Research* 117, D23202. doi:10.1029/2012JD018332.

Marshall, M., Tu, K., Funk, C., Michaelsen, J., Williams, P., Williams, C., Ardö, J. et al. 2013. Improving operational land surface model canopy evapotranspiration in Africa using a direct remote sensing approach. *Hydrology and Earth System Science* 17, 1079–1091. doi:10.5194/hess-17-1079-2013.

McCabe, M.F., Kalma, J.D., and Franks, S.W. 2005. Spatial and temporal patterns of land surface fluxes from remotely sensed surface temperatures within an uncertainty modelling framework. *Hydrology and Earth System Sciences* 9, 467–480.

McCabe, M.F. and Wood, E.F. 2006. Scale influences on the remote estimation of evapotranspiration using multiple satellite sensors. *Remote Sensing of Environment* 105, 271–285.

Medlyn, B.E., Robinson, A.P., Clement, R., and McMurtrie, R.E. 2005. On the validation of models of forest CO_2 exchange using eddy covariance data: some perils and pitfalls. *Tree Physiology* 25, 839–857.

Melesse, A.M., Frank, A., Nangia, V., and Hanson, J. 2008. Analysis of energy fluxes and land surface parameters in a grassland ecosystem: A remote sensing perspective. *International Journal of Remote Sensing* 29, 3325–3341.

Messina, A. 2012. Mapping drought in the Krishna Basin with remote sensing. Mapping drought in the Krishna Basin with remote sensing, San Diego State University.

Meyers, T.P. and Hollinger, S.E. 2004. An assessment of storage terms in the surface energy balance of maize and soybean. *Agricultural and Forest Meteorology* 125, 105–115.

Moran, M.S., Clarke, T.R., Inoue, Y., and Vidal, A. 1994. Estimating crop water deficit using the relation between surface-air temperature and spectral vegetation index. *Remote Sensing of Environment* 49, 246–263.

Moran, M.S. and Jackson, R.D. 1991. Assessing the spatial distribution of evapotranspiration using remotely sensed inputs. *Journal of Environment Quality* 20, 725–737. doi:10.2134/jeq1991.00472425002000040003x.

Moran, M.S., Rahman, A.F., Washburne, J.C., Goodrich, D.C., Weltz, M.A., and Kustas, W.P. 1996. Combining the Penman–Monteith equation with measurements of surface temperature and reflectance to estimate evaporation rates of semiarid grassland. *Agricultural and Forest Meteorology* 80, 87–109.

Morcrette, J. 1991. Radiation and cloud radiative properties in the European Centre for Medium Range Weather Forecasts forecasting system. *Journal of Geophysical Research: Atmospheres* (1984–2012) 96, 9121–9132.

Morcrette, J.-J. 2002. The surface downward longwave radiation in the ECMWF forecast system. *Journal of Climate* 15, 1875–1892.

Morillas, L., Leuning, R., Villagarcía, L., García, M., Serrano-Ortiz, P., and Domingo, F. 2013. Improving evapotranspiration estimates in Mediterranean drylands: The role of soil evaporation. *Water Resources Research* 49, 6572–6586. doi:10.1002/wrcr.20468.

Morse, A., Tasumi, M., Allen, R.G., and Kramber, W.J. 2000. Application of the SEBAL methodology for estimating consumptive use of water and streamflow depletion in the Bear River Basin of Idaho through remote sensing. Final Report, Idaho Department of Water Resources, Boise, ID, December 15, 2000.

Mu, Q.M., Zhao, M.S., and Running, S.W. 2013. MOD16 1-km² terrestrial evapotranspiration (ET) product for the Nile Basin; algorithm theoretical basis document. Numerical Terradynamic Simulation Group, College of Forestry and Conservation, University of Montana: Missoula, MT.

Mu, Q., Heinsch, F.A., Zhao, M., and Running, S.W. 2007. Development of a global evapotranspiration algorithm based on MODIS and global meteorology data. *Remote Sensing of Environment* 111, 519–536.

Mu, Q., Zhao, M., and Running, S.W. 2011. Improvements to a MODIS global terrestrial evapotranspiration algorithm. *Remote Sensing of Environment* 115, 1781–1800.

Murray, T. and Verhoef, A. 2007. Moving towards a more mechanistic approach in the determination of soil heat flux from remote measurements: I. A universal approach to calculate thermal inertia. *Agricultural and Forest Meteorology* 147, 80–87. doi:10.1016/j.agrformet.2007.07.004.

Nagler, P.L., Glenn, E.P., Nguyen, U., Scott, R.L., and Doody, T. 2013. Estimating riparian and agricultural actual evapotranspiration by reference evapotranspiration and MODIS enhanced vegetation index. *Remote Sensing* 5, 3849–3871. doi:10.3390/rs5083849.

Nagler, P.L., Morino, K., Murray, R.S., Osterberg, J., and Glenn, E.P. 2009. An empirical algorithm for estimating agricultural and riparian evapotranspiration using MODIS enhanced vegetation index and ground measurements of ET. I. Description of method. *Remote Sensing* 1, 1273–1297.

National Aeronautics and Space Administration. Global Modeling and Assimilation Office, Overview. http://gmao.gsfc.nasa.gov. (Accessed April 30, 2015.)

Neale, C.M.U., Bausch, W.C., and Heermann, D.F. 1989. Development of reflectance-based crop coefficients for corn. *Transactions of the American Society of Agricultural Engineers* 32, 1891–1899.

Nishida, K. 2003. An operational remote sensing algorithm of land surface evaporation. *Journal of Geophysical Research: Atmospheres (1984–2012)* 108, 4270. doi:10.1029/2002JD002062.

Nishida, K. and Nemani, R. 2003. Development of an evapotranspiration index from Aqua/MODIS for monitoring surface moisture status. *IEEE Transactions on Geoscience and Remote Sensing* 41, 493–501.

Norman, J.M., Anderson, M.C., Kustas, W.P., French, A.N., Mecikalski, J., Torn, R., Diak, G.R., Schmugge, T.J., and Tanner, B.C.W. 2003. Remote sensing of surface energy fluxes at 10 1-m pixel resolutions. *Water Resources Research* 39, 1221. doi:10.1029/2002WR001775.

Norman, J.M., Kustas, W.P., and Humes, K.S. 1995. Source approach for estimating soil and vegetation energy fluxes in observations of directional radiometric surface temperature. *Agricultural and Forest Meteorology* 77, 263–293.

Norman, J.M., Kustas, W.P., Prueger, J.H., and Diak, G.R. 2000. Surface flux estimation using radiometric temperature: A dual-temperature-difference method to minimize measurement errors, *Water Resources Research* 36, 2263–2274.

Paulson, C.A. 1970. The Mathematical representation of wind speed and temperature profiles in the unstable atmospheric surface layer. *Applied Meteorology* 9, 857–861.

Petropoulos, G., Carlson, T.N., Wooster, M.J., and Islam, S. 2009. A review of Ts/VI remote sensing based methods for the retrieval of land surface energy fluxes and soil surface moisture. *Progress in Physical Geography* 33, 224–250.

Petropoulos, G.P. 2013. Remote sensing of surface turbulent energy fluxes. In: *Remote Sensing of Energy Fluxes and Soil Moisture Content*, Petropoulos, G.P. (ed.), pp. 49–84. New York: Taylor and Francis.

Petropoulos, G.P. and Carlson, T.N. 2011. Retrievals of turbulent heat fluxes and surface soil water content by remote sensing. In: *Advances in Environmental Remote Sensing: Sensors, Algorithms, and Applications*, Weng, Q. (ed), p. 469. New York: CRC Press.

Petropoulos, G.P., Carlson, T.N., and Griffiths, H. 2013. Turbulent fluxes of heat and moisture at the earth's land surface: Importance, controlling parameters and conventional measurement techniques, In: *Remote Sensing of Energy Fluxes and Soil Moisture Content*, Petropoulos, G.P. (ed.), pp. 3–28. New York: Taylor and Francis.

Rafn, E.B., Contor, B., and Ames, D.P. 2008. Evaluation of a method for estimating irrigated crop-evapotranspiration coefficients from remotely sensed data in Idaho. *Journal of Irrigation and Drainage Engineering* 134, 722–729.

Rienecker, M.M., Suarez, M.J., Gelaro, R., Todling, R., Bacmeister, J., Liu, E., Bosilovich, M.G. et al. 2011. MERRA: NASA's modern-era retrospective analysis for research and applications. *Journal of Climate* 24, 3624–3648.

Rivas, R. and Caselles, V. 2004. A simplified equation to estimate spatial reference evaporation from remote sensing-based surface temperature and local meteorological data. *Remote Sensing of Environment* 93, 68–76.

Rodell, M., Houser, P.R., Jambor, U., Gottschalck, J., Mitchell, K., Meng, C.-J., Arsenault, K. et al. 2004. The global land data assimilation system. *Bulletin of the American Meteorological Society* 85, 381–394. doi:10.1175/BAMS-85-3-381.

Roerink, G., Su, Z., and Menenti, M. 2000. S-SEBI: A simple remote sensing algorithm to estimate the surface energy balance. *Physics and Chemistry of the Earth, Part B: Hydrology, Oceans and Atmosphere* 25, 147–157. doi:10.1016/S1464-1909(99)00128-8.

Romero, M.G. 2004. Daily evapotranspiration estimation by means of evaporative fraction and reference evapotranspiration fraction. Daily evapotranspiration estimation by means of evaporative fraction and reference evapotranspiration fraction, Utah State University, Department of Biological and Irrigation Engineering.

Rossow, W.B. and Schiffer, R.A. 1999. Advances in understanding clouds from ISCCP. *Bulletin of the American Meteorological Society* 80, 2261–2287.

Rossow, W.B. and Schiffer, R.A. 1991. ISCCP cloud data products. *Bulletin of the American Meteorological Society* 72, 2–20.

Samani, Z., Bawazir, a. S., Bleiweiss, M., Skaggs, R., Longworth, J., Tran, V.D., and Pinon, A. 2009. Using remote sensing to evaluate the spatial variability of evapotranspiration and crop coefficient in the lower Rio Grande Valley, New Mexico. *Irrigation Science* 28, 93–100. doi:10.1007/s00271-009-0178-8.

Schiffer, R.A. and Rossow, W.B. 1983. The international satellite cloud climatology project (ISCCP)—The first project of the World Climate Research Programme. *American Meteorological Society, Bulletin* 64, 779–784.

Scott, R.L. 2010. Using watershed water balance to evaluate the accuracy of eddy covariance evaporation measurements for three semiarid ecosystems. *Agricultural and Forest Meteorology* 150, 219–225.

Scott, R.L., Huxman, T.E., Cable, W.L., and Emmerich, W.E. 2006. Partitioning of evapotranspiration and its relation to carbon dioxide exchange in a Chihuahuan Desert shrubland. *Hydrological Processes* 20, 3227–3243.

Sellers, P.J., Randall, D.A., Collatz, G.J., Berry, J.A., Field, C.B., Dazlich, D.A., Zhang, C., Collelo, G.D., and Bounoua, L. 1996. A revised land surface parameterization (SiB2) for atmospheric GCMs. Part I: Model formulation. *Journal of Climate* 9, 676–705.

Senay, G., Budde, M., Verdin, J., and Melesse, A. 2007. A coupled remote sensing and simplified surface energy balance approach to estimate actual evapotranspiration from irrigated fields. *Sensors* 7, 979–1000.

Senay, G.B., Bohms, S., Singh, R.K., Gowda, P.H., Velpuri, N.M., Alemu, H., and Verdin, J.P. 2013. Operational evapotranspiration mapping using remote sensing and weather datasets: A new parameterization for the SSEB approach. *Journal of the American Water Resources Association* 49, 577–591. doi:10.1111/jawr.12057.

Senay, G.B., Budde, M.E., and Verdin, J.P. 2011. Enhancing the simplified surface energy balance (SSEB) approach for estimating landscape ET: Validation with the METRIC model. *Agricultural Water Management* 98, 606–618.

Shapiro, R. 1987. *A Simple Model for the Calculation of the Flux of Direct and Diffuse Solar Radiation through the Atmosphere.* MA: Hanscom Air Force Base.

Shiklomanov, I.A. 2000. Appraisal and assessment of world water resources. *Water International* 25, 11–32.

Shu, Y., Stisen, S., Jensen, K.H., and Sandholt, I. 2011. Estimation of regional evapotranspiration over the North China Plain using geostationary satellite data. *International Journal of Applied Earth Observation and Geoinformation* 13, 192–206.

Singh, R.K. and Irmak, A. 2009. Estimation of crop coefficients using satellite remote sensing. *Journal of Irrigation and Drainage Engineering* 135, 597–608.

Singh, R.K., Senay, G.B., Velpuri, N.M., Bohms, S., Scott, R.L., and Verdin, J.P. 2013. Actual evapotranspiration (water use) assessment of the Colorado River Basin at the Landsat resolution using the operational simplified surface energy balance model. *Remote Sensing* 6, 233–256.

Sobrino, J.A., Gómez, M., Jiménez-Muñoz, J.C., and Olioso, A. 2007. Application of a simple algorithm to estimate daily evapotranspiration from NOAA–AVHRR images for the Iberian Peninsula. *Remote Sensing of Environment* 110, 139–148.

Sobrino, J.A., Gómez, M., Jiménez-Muñoz, J.C., Olioso, A., and Chehbouni, G. 2005. A simple algorithm to estimate evapotranspiration from DAIS data: Application to the DAISEX campaigns. *Journal of Hydrology* 315, 117–125.

Sobrino, J.A., Jiménez-Muñoz, J.C., Sòria, G., Gómez, M., Ortiz, A.B., Romaguera, M., Zaragoza, M., Julien, Y., Cuenca, J., and Atitar, M. 2008. Thermal remote sensing in the framework of the SEN2FLEX project: Field measurements, airborne data and applications. *International Journal of Remote Sensing* 29, 4961–4991.

Stisen, S., Sandholt, I., Nørgaard, A., Fensholt, R., and Jensen, K.H. 2008. Combining the triangle method with thermal inertia to estimate regional evapotranspiration—Applied to MSG-SEVIRI data in the Senegal River basin. *Remote Sensing of Environment* 112, 1242–1255.

Stone, L.R. and Horton, M.L. 1974. Estimating evapotranspiration using canopy temperatures: Field evaluation. *Agronomy Journal* 66, 450–454.

Su, H., McCabe, M.F., Wood, E.F., Su, Z., and Prueger, J.H. 2005. Modeling evapotranspiration during SMACEX: Comparing two approaches for local- and regional-scale prediction. *Journal of Hydrometeorology* 6, 910–922.

Su, Z. 2002. The surface energy balance system (SEBS) for estimation of turbulent heat fluxes. *Hydrology and Earth System Sciences Discussions* 6, 85–100.

Sun, Z., Wang, Q., Matsushita, B., Fukushima, T., Ouyang, Z., and Watanabe, M. 2009. Development of a simple remote sensing evapotranspiration model (Sim-ReSET): Algorithm and model test. *Journal of Hydrology* 376, 476–485.

Swinbank, W.C. 1951. The measurement of vertical transfer of heat and water vapor by eddies in the lower atmosphere. *Journal of Meteorology* 8, 135–145.

Tang, B. and Li, Z.-L. 2008. Estimation of instantaneous net surface longwave radiation from MODIS cloud-free data. *Remote Sensing of Environment* 112, 3482–3492.

Tang, R., Li, Z.-L., and Tang, B. 2010. An application of the Ts-VI triangle method with enhanced edges determination for evapotranspiration estimation from MODIS data in arid and semi-arid regions: Implementation and validation. *Remote Sensing of Environment* 114, 540–551.

Teixeira, A.H. de C., Bastiaanssen, W.G.M., Ahmad, M.D., and Bos, M.G. 2009. Reviewing SEBAL input parameters for assessing evapotranspiration and water productivity for the Low-Middle São Francisco River basin, Brazil: Part A: Calibration and validation. *Agricultural and Forest Meteorology* 149, 462–476.

Timmermans, W.J., Kustas, W.P., Anderson, M.C., and French, A.N. 2007. An intercomparison of the surface energy balance algorithm for land (SEBAL) and the two-source energy balance (TSEB) modeling schemes. *Remote Sensing of Environment* 108, 369–384.

Tolk, J.A., Howell, T.A., and Evett, S.R. 2006. Nighttime evapotranspiration from alfalfa and cotton in a semiarid climate. *Agronomy Journal* 98, 730–736.

Townshend, J.R. and Justice, C.O. 1988. Selecting the spatial resolution of satellite sensors required for global monitoring of land transformations. *International Journal of Remote Sensing* 9, 187–236. doi:10.1080/01431168808954847.

Trout, T.J., Johnson, L.F., and Gartung, J. 2008. Remote sensing of canopy cover in horticultural crops. *Hortscience* 43, 333–337.

Twine, T.E., Kustas, W.P., Norman, J.M., Cook, D.R., Houser, Pr., Meyers, T.P., Prueger, J.H., Starks, P.J., and Wesely, M.L. 2000. Correcting eddy-covariance flux underestimates over a grassland. *Agricultural and Forest Meteorology* 103, 279–300.

Van de Griend, A.A., Camillo, P.J., and Gurney, R.J. 1985. Discrimination of soil physical parameters, thermal inertia, and soil moisture from diurnal surface temperature fluctuations. *Water Resources Research* 21, 997–1009.

Van Niel, T.G., McVicar, T.R., Roderick, M.L., van Dijk, A.I.J.M., Renzullo, L.J., and van Gorsel, E. 2011. Correcting for systematic error in satellite-derived latent heat flux due to assumptions in temporal scaling: Assessment from flux tower observations. *Journal of Hydrology* 409, 140–148. doi:10.1016/j.jhydrol.2011.08.011.

Van Oevelen, P.J. 1991. Determination of the available energy for evapotranspiration with remote sensing. *Determination of the Available Energy for Evapotranspiration with Remote Sensing.* the Netherlands: Wageningen University.

Velpuri, N.M., Senay, G.B., Singh, R.K., Bohms, S., and Verdin, J.P. 2013. A comprehensive evaluation of two MODIS evapotranspiration products over the conterminous United States: Using point and gridded FLUXNET and water balance ET. *Remote Sensing of Environment* 139, 35–49. doi:10.1016/j.rse.2013.07.013.

Venturini, V., Bisht, G., Islam, S., and Jiang, L. 2004. Comparison of evaporative fractions estimated from AVHRR and MODIS sensors over South Florida. *Remote Sensing of Environment* 93, 77–86. doi:10.1016/j.rse.2004.06.020.

Verma, S.B., Rosenberg, N.J., Blad, B.L., and Baradas, M.W. 1976. Resistance-energy balance method for predicting evapotranspiration: Determination of boundary layer resistance and evaluation of error effects. *Agronomy Journal* 68, 776–782. doi:10.2134/agronj1976.00021962006800050023x.

Vernekar, K.G., Sinha, S., Sadani, L.K., Sivaramakrishnan, S., Parasnis, S.S., Mohan, B., Dharmaraj, S. et al. 2003. An overview of the land surface processes experiment (*Laspex*) over a Semi-Arid Region of India. *Boundary-Layer Meteorology* 106, 561–572.

Verstraeten, W.W., Veroustraete, F., and Feyen, J. 2008. Assessment of evapotranspiration and soil moisture content across different scales of observation. *Sensors* 8, 70–117.

Vinukollu, R.K., Meynadier, R., Sheffield, J., and Wood, E.F. 2011. Multi-model, multi-sensor estimates of global evapotranspiration: climatology, uncertainties and trends. *Hydrological Processes* 25, 3993–4010. doi:10.1002/hyp.8393.

Vinukollu, R.K., Wood, E.F., Ferguson, C.R., and Fisher, J.B. 2011. Global estimates of evapotranspiration for climate studies using multi-sensor remote sensing data: Evaluation of three process-based approaches. *Remote Sensing of Environment* 115, 801–823.

Vorosmarty, C.J., Green, P., Salisbury, J., and Lammers, R.B. 2000. Global water resources: vulnerability from climate change and population growth. *Science* 289, 284–288. doi:10.1126/science.289.5477.284.

Wang, K., Li, Z., and Cribb, M. 2006. Estimation of evaporative fraction from a combination of day and night land surface temperatures and NDVI: A new method to determine the Priestley–Taylor parameter. *Remote Sensing of Environment* 102, 293–305.

Webb, E.K. 1970. Profile relationships: The log-linear range, and extension to strong stability. *Quarterly Journal of the Royal Meteorological Society*, 96, 67–90.

Wever, L.A., Flanagan, L.B., and Carlson, P.J. 2002. Seasonal and interannual variation in evapotranspiration, energy balance and surface conductance in a northern temperate grassland. *Agricultural and Forest Meteorology* 112, 31–49.

Wilson, K.B., Hanson, P.J., Mulholland, P.J., Baldocchi, D.D., and Wullschleger, S.D. 2001. A comparison of methods for determining forest evapotranspiration and its components: sap-flow, soil water budget, eddy covariance and catchment water balance. *Agricultural and Forest Meteorology* 106, 153–168.

Yilmaz, M.T., Anderson, M.C., Zaitchik, B., Hain, C.R., Crow, W.T., Ozdogan, M., Chun, J.A., and Evans, J. 2014. Comparison of prognostic and diagnostic surface flux modeling approaches over the Nile River basin. *Water Resources Research* 50, 386–408. doi:10.1002/2013WR014194.

Zahira, S., Abderrahmane, H., Mederbal, K., and Frederic, D. 2009. Mapping latent heat flux in the western forest covered regions of Algeria using remote sensing data and a spatialized model. *Remote Sensing* 1, 795–817.

Zhang, Y., Liu, C., Lei, Y., Tang, Y., Yu, Q., Shen, Y., and Sun, H. 2006. An integrated algorithm for estimating regional latent heat flux and daily evapotranspiration. *International Journal of Remote Sensing* 27, 129–152.

4

Water Productivity Studies from Earth Observation Data: Characterization, Modeling, and Mapping Water Use and Water Productivity

Antônio de C. Teixeira
Embrapa Satellite Monitoring

Fernando B.T. Hernandez
São Paulo State University

Morris Scherer-Warren
National Water Agency

Ricardo G. Andrade
Embrapa Satellite Monitoring

Janice F. Leivas
Embrapa Satellite Monitoring

Daniel C. Victoria
Embrapa Satellite Monitoring

Edson L. Bolfe
Embrapa Satellite Monitoring

Prasad S. Thenkabail
United States Geological Survey (USGS)

Renato A.M. Franco
São Paulo State University

Acronyms and Definitions

APAR	Absorbed photosynthetically active radiation
BA	Bahia
BIO	Biomass production
CWP	Crop water productivity
DD	Degree days
DOY	Day of the year
ET	Evapotranspiration
EVI	Enhanced vegetation index
FAO	Food Agricultural Organization
G	Grain
GPP	Gross primary production
GS	Growing season
HA	Harvested area
HI	Harvest index
I	Ìrrigation
IC	Irrigated crops
Kc	Crop coefficient
L	Land
LANDSAT	Land remote sensing satellite
LUE	Light use efficiency
METRIC	Mapping evapotranspiration with high resolution and internalized calibration

MODIS	Moderate resolution imaging spectroradiometer
NC	Nilo Coelho
NIR	Near infrared
NV	Natural vegetation
P	Precipitation
PAR	Photosynthetically active radiation
PE	Pernambuco
Perc	Percolation
PM	Penman–Monteith
NDVI	Normalized Difference Vegetation Index
S	Silage
SAFER	Simple Algorithm For Evapotranspiration Retrieving
SEBAL	Surface Energy Balance Algorithm for Land
SD	Standard deviation
SUREAL	Surface resistance algorithm
Tr	Transpiration
WD	Water deficit
WP	Water productivity
WS	Water supply

4.1 Introduction

Water demand already exceeds supply in many parts of the world, and as the human population continues to increase, many more areas are expected to experience water scarcity (Smakhtin et al., 2004; Bos et al., 2005; Gourbesville, 2008). Compounding this scarcity is the deterioration of water quality in several developing countries, mainly in river basins with rapid change of land usage. Local solutions for these problems are hampered by a lack of commitment to solving problems related to water and poverty, inadequate and inadequately targeted investment, insufficient human capacity, ineffective institutions, and poor governance (Molden et al., 2007). To provide sustainable water resource development and secure water availability for competing user groups, future water management may observe the water accounting approach (Cai et al., 2002), which recognizes the various water users of a basin and the water flows in terms of net water production or net water consumption.

Better knowledge on water productivity (WP) provides valuable information to achieve local water conservation practices without losing agricultural production. To attain this knowledge, it is extremely important to upscale actual evapotranspiration (ET) and actual crop yield (Y_a) from field to large scale (irrigated command area or river basin). Water resources can boost the rural economy; however, one of the biggest consequences of this development is that other water users will call for an appropriate share of the freshwater resources (Teixeira, 2009). On the other hand, photosynthesis and biomass production (BIO) can act as a carbon sink, affecting the energy exchanges between the vegetated surfaces and the lower atmosphere (Ceschia et al., 2010). In this dynamic situation, the use of tools to quantify WP on a large scale is strongly relevant in the support of policy planning and decision making about water resources.

ET is critically important because it is essential for agricultural production, and its increase means less water available

for ecological and human uses in river basins. Distinctions are made between reference (ET_0), potential (ET_p), and actual (ET) evapotranspiration (Allen et al., 1998). ET_0 is the water flux from a reference surface, not a shortage of water, which in this chapter and in standard practice is considered to be a hypothetical grass surface with specific characteristics. ET_p refers to the water flux from crops that are grown in large fields under optimum soil moisture, excellent management and environmental conditions, achieving full production. ET is the real water flux occurring from vegetation in a specific situation involving all environmental conditions. Due to suboptimal crop management and environmental constraints that affect plant growth and limit the water fluxes, ET is frequently smaller than ET_p in agricultural fields (Figure 4.1).

Field methods for ET measurements provide values for specific sites and are not suitable for large-scale WP analyses. The spatial variability of soil moisture in hydrological basins may be significant and these variations are caused by different amounts of precipitation, seepage, flooding, irrigation, hydraulic characteristics of soils, vegetation types and densities, while the temporal variations in ET can be ascribed to weather conditions and vegetation development (Teixeira, 2012). Direct extrapolation of point measurements to a surrounding

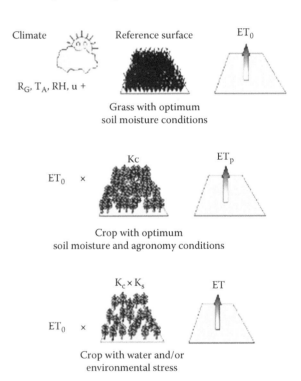

FIGURE 4.1 Different evapotranspiration concepts: reference (ET_0), potential (ET_p), and actual (ET) evapotranspiration, together with crop (Kc) and stress (Ks) coefficients. (Adapted from Allen, R.G. et al., *Crop Evapotranspiration: Guidelines for Computing Crop Water Requirements*, Food and Agriculture Organization of the United Nations, Rome, Italy, 1998.) The climate parameters for ET_0 calculations by Penman–Monteith method are global solar radiation (R_G), air temperature (T_a), relative humidity (RH), and wind speed (u).

landscape environment can lead to inaccurate large-scale estimates because a few sites cannot provide a fair sample of a whole biome (Wylie et al., 2003).

Such difficulties stimulated remote sensing by satellite images as a valuable tool in the determination of WP components, and it is used in distinct climatic regions (Tang et al., 2009; Teixeira, 2010; Miralles et al., 2011; Pôças et al., 2013; Teixeira et al., 2013). Aiming to increase the spatial scales from field measurements in the Brazilian semiarid region, ET has been acquired by means of land remote sensing satellite (Landsat) images with application of the SEBAL (Surface Energy Balance Algorithm for Land) algorithm (Bastiaanssen et al., 1998) following processes of calibration and validation (Teixeira et al., 2009a,b). Although the model performed well, it is not reasonable to ignore ET from all pixels during a whole year because in the rainy season the large-scale soil moisture status in mixed ecosystems is homogeneous. Another problem in relation to the applicability of large-scale remote sensing energy balance models, from the point of view of the end users, is the need for user expertise.

The suitability of applying the Penman–Monteith (PM) equation has been shown by using remotely sensed vegetation indices, such as the Normalized Difference Vegetation Index (NDVI) and the Enhanced Vegetation Index (EVI), together with weather data (Cleugh et al., 2007; Nagler et al., 2013). The PM equation is also highlighted by the use of the Kc approach (Kamble et al., 2013) and by the model Mapping Evapotranspiration with High Resolution and Internalized Calibration (METRIC) (Allen et al., 2007). Considering the simplicity of application and its need of neither crop classification nor extreme conditions, the SAFER (Simple Algorithm For Evapotranspiration Retrieving) algorithm, based on the modeled ratio of ET to ET_0, has been developed and validated in Brazil with field data from four flux stations and Landsat images (Teixeira, 2010; Teixeira et al., 2013, 2014a). Retrieving ET from SAFER together with estimations of the available energy allows BIO and WP acquirements on a large scale.

The energy captured in photosynthesis is represented in part by BIO, which is a key indicator for any agroecosystem (Wu et al., 2010), and its values are also highly variable in both space and time (Teixeira et al., 2009b; Yan et al., 2009; Adak et al., 2013). In water-limited environments, the challenge is to make improvements in BIO through optimized management practices (Adak et al., 2013).

The slope of the linear regression between BIO and cumulative Photosynthetically Active radiation (PAR) intercepted by a crop has been used to determine the light use efficiency (LUE) (ε) (Ceotto and Castelli, 2002; Tesfaye et al., 2006). This relationship is also employed to develop simple crop models. Russell et al. (1989) expressed yield as a function of the photosynthetically active radiation absorbed by the crop (APAR), ε, and the harvest index (HI). HI includes the water content of the freshly harvested product and in most studies does not include roots (Lobell et al., 2003). Comparisons of species with respect to photosynthetic processes indicate that C4 species have higher ε than C3 species (Gosse et al., 1986). Reductions in ε due to water deficits have

been reported (Singh and Sri Rama, 1989); however; in general, ε is stable across environments under optimal growing conditions (Sinclair and Muchow, 1999) because it is a relatively constant property of plants. In addition, light harvesting can be adjusted to the availability of resources needed to use the absorbed light (Russell et al., 1989; Field et al., 1995).

Several more sophisticated crop models have been developed in order to optimize agricultural management, but also to investigate the effect of climatic variability and soil hydrology on Y_a. They in general employ data of plant phenology and physiology, and have been applied over different scales, from point (Eitzinger et al., 2004) to regional scales (Boogaard et al., 2002), resulting in laborious parameterizations and calibrations. To avoid this problem, empirical models have been developed for global-scale applications (Gervois et al., 2004; Bondeau et al., 2007; Osborne et al., 2007). Monteith and Scott (1982) analyzed Y_a accounting temperature effects on leaf area development and crop ontogeny, and solar radiation effects on BIO. This approach was also used in soybean (Spaeth et al., 1987), corn (Muchow, 1990), wheat (Amir and Sinclair, 1991), and rice (Sheehy et al., 2004; Pirmoradian and Sepaskhah, 2005).

To estimate crop yield by satellite data, a commonly applied method is the development of empirical relationships between NDVI and Y_a (e.g., Groten, 1993; Sharma et al., 2000). To obtain the coefficients of these relationships, excessive field measurements need to be done, which at the large scale are difficult and expensive. Hamar et al. (1996) established a linear regression model to estimate corn and wheat yield at a large scale based on vegetation indices computed with Landsat MSS data. Maselli and Rembold (2001) demonstrated the potential of using multiyear NOAA-AVHRR NDVI data to estimate wheat yield in North African countries. Similar relationships were obtained for millet (Rasmussen, 1992) and for wheat (Manjunath and Potdar, 2002).

The LUE model proposed by Monteith (1972) based on incident global solar radiation (R_G) and canopy development can be used together with satellite data (e.g., Kumar and Monteith, 1982; Daughtry et al., 1992; Gower et al., 1999; Bastiaanssen and Ali, 2003; Teixeira et al., 2009b, 2013; Claverie et al., 2012). The model proposes a direct proportional relationship between BIO and APAR. APAR is variable throughout the year and during the crop growing periods (Tesfaye et al., 2006; Teixeira et al., 2007). Although uncertainties arise in connection with ε values, due to their spatiotemporal variability (Zhao et al., 2005), the Monteith LUE model accuracy has been considered acceptable for large-scale applications with different satellite data.

BIO field measurements have been made on oilseed crops under the semiarid conditions of India (Adak et al., 2013), showing spatial variations due to microclimate conditions. In France, field measurements were coupled with high-resolution FORMOSAT satellite images in irrigated maize and rainfed sunflower, where the authors attributed the main spatial differences in BIO to precipitation conditions during the second crop (Claverie et al., 2012).

Satellite remote sensing is an efficient tool for crop area and BIO estimates because it provides spatial and temporal information on the location and state of vegetation (Teixeira et al., 2013), overcoming the lack of extensive observations and/or measurements over large areas (Wu et al., 2010; Ahamed et al., 2011). Moderate resolution imaging spectroradiometer (MODIS) data has been processed in combination with precipitation, temperature, and elevation for BIO mapping in the California forest (Baccine et al., 2004). BIO has also been estimated from MODIS images in Guandong, China, to evaluate the feasibility of setting up new biomass power plants and to optimize the locations of plants (Shi et al., 2008). In Brazil, BIO estimations have been made in São Francisco (Teixeira, 2009) and Amazon (Lu, 2005) hydrological basins using Landsat images.

WP may be defined as the ratio of the net benefits from crop, forestry, fishery, livestock, and mixed agricultural systems to the amount of water required to produce those benefits. Considering vegetation, WP can be BIO per land (L) or per water consumed, including that which originates from rainfall, irrigation, seepage, and changes in soil storage (Molden et al., 2007). Y_a and water consumption are two closely linked processes. The crop water productivity (CWP) may be considered as the ratio of Y_a to the amount of water consumed or applied. Many ways of raising CWP in agriculture are possible for both irrigated and rainfed crops (Teixeira, 2009).

Benchmark CWP values have been summarized for irrigated crops (wheat, rice, cotton, and maize) by Zwart and Bastiaanssen (2004), for dry land crops by Oweis and Hachum (2006), and for rainfed crops by Rockstrom and Barron (2007). Examples of field measurements to quantify CWP in oats, sunflower, legumes, and potato have been given for the semiarid conditions of Mongolia (Yuan et al., 2013). Also under semiarid conditions, but in northern India, CWP has been quantified in oilseed crops (Adak et al., 2013). In the Brazilian semiarid region, this task has been realized in vineyards (Teixeira et al., 2007) and mango orchards (Teixeira et al., 2008a). Considering the estimations of large-scale water variables acquirements by satellite images, CWP mapping, based on remote sensing parameters, can be found for several agroecosystems, combining these parameters with weather variables (Mo et al., 2009; Teixeira et al., 2009b; Zwart et al., 2010).

Despite these site-specific and large-scale studies, research is still needed to further evaluate the combined ET and BIO models, especially for operational applications in different agroecosystems with high temporal and spatial thermo-hydrological inhomogeneity. This chapter highlights the combination of the SAFER algorithm (Teixeira, 2010; Teixeira et al., 2013) and Monteith LUE model (Monteith, 1972) to demonstrate that satellite measurements, together with agrometeorological data, can be used for WP assessments on a large scale. A third model for the surface resistance to water fluxes (r_s), SUREAL (Surface Resistance Algorithm), is used to classify the vegetation into irrigated crops and natural ecosystems (Teixeira, 2010; Teixeira et al., 2013) to retrieve the incremental values (resulting from the replacement of natural vegetation by irrigated crops) of ET and BIO in order to support the rational water resources management under semiarid conditions.

Following the introduction, the study regions and the steps for modeling are described. WP assessments are made using remote sensing methods with satellite images at different spatial and temporal resolutions. These are shown for natural vegetation and agricultural crops, considering both irrigation and rainfed aspects, for some Brazilian agroecosystems.

4.2 Study Regions

Figure 4.2 shows the Brazilian regions and the locations of the Petrolina and Juazeiro municipalities, respectively, in Pernambuco (PE) and Bahia (BA) states in the Brazilian northeast region, together with the Nilo Coelho (NC) irrigation scheme. The agrometeorological stations inside irrigated areas and natural vegetation were used for WP modeling and for the majority of applications with Landsat and MODIS images in this chapter.

In Bebedouro (Petrolina, PE), there are two agrometeorological stations, one automatic and the other conventional. For historical analyses, regression equations between conventional and automatic stations were used to retrieve ET and BIO before 2003 (as automatic stations did not exist prior to 2003 in the modeling region), insuring the spatial variation of the weather variables.

Because of the irrigation, the Petrolina and Juazeiro municipalities, developed considerably. The main reason for this growth is the availability of water for irrigation. These municipalities are separated by the São Francisco River, which flows through six Brazilian states and has a basin size of 636,920 km², divided into four sub-basins: The Upper, the Middle, the Low-Middle, and the Lower São Francisco (Teixeira, 2009).

Petrolina (Pernambuco, PE) and Juazeiro (Bahia, BA) are located in the low-middle sub-basin, which has an area of 115,987 km², elevations from 800 to 200 m, prevailing climate semiarid and arid, with "Caatinga" as a predominant vegetation cover. This natural ecosystem has been rapidly replaced mainly by fruit crops, the most important are vineyards, mango, banana, guava, and coconut.

More than 50% of the territory of the São Francisco River basin is located in the Brazilian semiarid region, which is situated in the northeastern part of Brazil. Disturbed currents of south, north, east, and west influence the climatology of the Brazilian northeast. Excluding the areas of high altitude, all semiarid regions in this sub-basin present annual averaged air temperatures (T_a) higher than 24°C, even larger than 26°C in the depressions at 200–250 m altitude. The average maximum air temperature is 33°C, and the average minimum is 19°C. The average monthly values are in the range from 17°C to 29°C (Teixeira, 2009).

Despite the relatively small thermal annual amplitude due to the proximity of the equator, the increase of T_a together with higher solar radiation outside the rainy season is significant because of the intensification of ET. The warmest months are October and November when the sun is near the zenith position in the region and the coldest months are June and July at the winter solstice in the southern hemisphere.

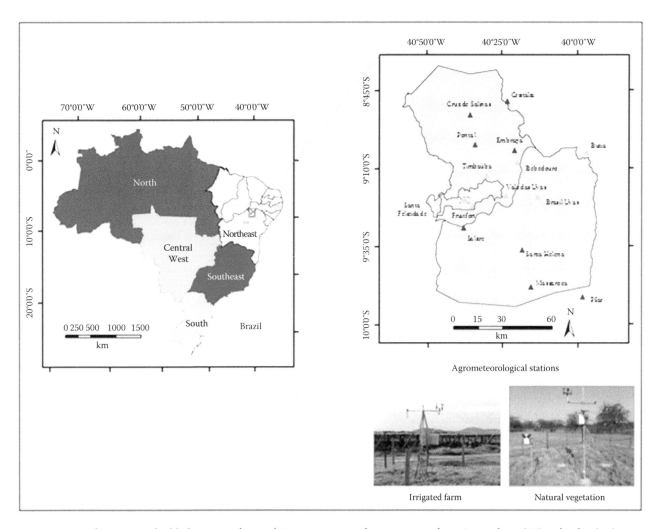

FIGURE 4.2 Brazilian regions, highlighting Petrolina and Juazeiro municipalities, respectively, in Pernambuco (PE) and Bahia (BA) states, in the Brazilian northeast, together with the Nilo Coelho (NC) irrigation scheme. The agrometeorological stations were used for water productivity (WP) modeling and the majority of applications. The stations represented by green triangle are inside irrigated farms while those with brown triangle are in natural vegetation.

The thermal homogeneity strongly contrasts with the spatial and temporal heterogeneity of the rainfall regime. The long-term annual precipitation (P) for the low-middle sub-basin is in average 693 mm with 70% of the rainy period concentrated during the months January–April. March (50–230 mm) and August (0–50 mm) are the wettest and the driest months, respectively. There are water deficits in the climatic water balance along the year, with the exception of March, when the weather conditions are rarely dry (Teixeira, 2009).

Considering the São Francisco River flow throughout the four sub-basins, the average is 2850 m³ s⁻¹. The river accounts for roughly two-thirds of the freshwater available in the entire northeastern region of Brazil. The availability and demand for water by sub-basins is shown in Table 4.1.

The highest water demand (68%) comes from the irrigation areas, followed by urban demand. As a consequence of water use, the principal impacts of interactions between water resources and the environment in the São Francisco River basin are widespread pollution caused by agriculture, uncontrolled discharges, inadequate disposal of solid wastes, and water shortages owing to the intermittent nature of tributaries.

After validating the equations under the semiarid conditions of the low-middle São Francisco sub-basin and becoming more confident with the models, tests and applications are demonstrated in corn crop under irrigation and rainfed conditions in the Brazilian southeast and central–west regions (see Figure 4.2) with Landsat and MODIS images, respectively.

4.3 Modeling WP Components

4.3.1 Actual Evapotranspiration

In this chapter, large-scale ET estimations with SAFER algorithm are shown for both MODIS and Landsat 5 images. The basic remote sensing parameters for modeling are surface albedo (α_0), surface temperature (T_0), and NDVI, which are used together with weather data.

TABLE 4.1   Availability and Demand for Water in the São Francisco River Basin: Urban, Rural, Livestock, Industry, Irrigation and Total

Sub Basin	Area (km²)	Flow (m³ s⁻¹)	Urban (m³ s⁻¹)	Rural (m³ s⁻¹)	Livestock (m³ s⁻¹)	Industry (m³ s⁻¹)	Irrigation (m³ s⁻¹)	Total (m³ s⁻¹)
U	99,387	1189	26.8	2.2	2.5	11.4	14.4	57.3
M	401,559	1522	4.6	2.8	3.2	0.8	58.8	70.2
LM	115,987	111	2.8	2.3	1.4	0.4	50.5	57.4
L	19,987	28	1.1	1.4	0.7	0.3	14.4	17.9
Total	636,920	2850	35.3	8.7	7.8	12.9	138.1	202.8

Note: U, M, LM, and L mean Upper, Middle, Low-Middle and Low São Francisco sub-basins.

From Landsat 5 images, to retrieve α_0, first the planetary albedo for each satellite band ($\alpha_{p_{band}}$) is calculated as

$$\alpha_{p_{band}} = \frac{L_{band}\pi d^2}{R_{a_{band}}\cos\phi} \tag{4.1}$$

where

L_{band} is the spectral radiance for the wavelengths of the band
d is the relative earth–sun distance
$R_{a_{band}}$ is the mean solar irradiance at the top of the atmosphere for each band (W m⁻² μm⁻¹)
ϕ is the solar zenith angle

The planetary albedo for the visible region of the solar spectrum (α_p) is calculated as the total sum of the different narrow-band $\alpha_{p_{band}}$ values according to weights for each band (w_{band}):

$$\alpha_p = \sum w_{band}\alpha_{p_{band}} \tag{4.2}$$

The weights for the different bands are computed as the ratio of the amount of the incoming shortwave radiation from the sun in a particular band and the sum of incoming shortwave radiation for all the bands at the top of the atmosphere.

The brightness temperature (T_{bright}) is calculated as follows:

$$T_{bright} = \frac{K_2}{\ln\left(\frac{K_1}{L_6 + 1}\right)} \tag{4.3}$$

where

L_6 is the uncorrected thermal radiance from the land surface (band 6)
$K_1 = 607.76$ and $K_2 = 1260.56$ are the conversion coefficients for Landsat 5

The results for both α_p and T_{bright} need to be corrected atmospherically to acquire the surface values of albedo (α_0) and temperature (T_0), which has been carried out in the Brazilian northeast by combining satellite and field measurements (Teixeira et al., 2008b, 2009a; Teixeira, 2010).

NDVI is an indicator related to the land cover obtained from satellite images as follows:

$$NDVI = \frac{\alpha_{p(NIR)} - \alpha_{p(RED)}}{\alpha_{p(NIR)} + \alpha_{p(RED)}} \tag{4.4}$$

where $\alpha_{p(NIR)}$ and $\alpha_{p(RED)}$ represent the planetary albedo over the ranges of wavelengths in the near infrared (NIR) and red (RED) regions of the solar spectrum, respectively.

There are several levels of MODIS products available; however, in this chapter, only four of the MODIS's spectral bands are used for the calculations of the input remote sensing parameters. They are the reflective solar bands (bands 1 and 2, red and near infrared) with a spatial resolution of 250 m and the thermal emissive bands (bands 31 and 32) with a spatial resolution of 1000 m.

For MODIS α_0 calculations, the reflectance values for bands 1 and 2 were used, according to the following regression equation (Valiente et al., 1995):

$$\alpha_0 = a + b\alpha_{p(1)} + c\alpha_{p(2)} \tag{4.5}$$

where $\alpha_{p(1)}$ and $\alpha_{p(2)}$ are the planetary albedo for bands 1 and 2 of the MODIS satellite measurements.

a, b, and c are regression coefficients obtained by comparing these measurements with field data (Teixeira et al., 2008b), thus including the atmospheric effects through the radiation path

The values found for the Brazilian northeast were, respectively, 0.08, 0.41, and 0.14 (Teixeira et al., 2014a).

For the MODIS T_0, the thermal bands 31 and 32 were used. Having the aerodynamic T_0 data from the same energy balance experiments as those carried out for α_0 (Teixeira et al., 2008b), a simple regression equation was retrieved with reasonable accuracy in relation to field experimental data:

$$T_0 = dT_{31} + eT_{32} \tag{4.6}$$

where T_{31} and T_{32} are the brightness temperatures from bands 31 and 32; the regression coefficients d and e were equally 0.50 for the Brazilian northeast conditions, also including already the atmospheric effects through the radiation path (Teixeira et al., 2013, 2014a).

Instead of using the thermal band from Landsat 5 (120 m) and MODIS (1000 m), T_0 in the current study was also estimated with only the visible and infrared bands (spatial resolution of 30 and 250 m for Landsat and MODIS, respectively) as a residue in the radiation balance equation (Teixeira et al., 2014b):

$$R_n = R_G - \alpha_0 R_G - \varepsilon_0 \sigma T_0^4 + \varepsilon_a \sigma T_a^4 \qquad (4.7)$$

where

R_G and T_a are respectively the daily values of the incident solar radiation and mean air temperature at the agrometeorological stations

R_n is the daily net radiation

ε_0 and ε_a are respectively the surface and atmospheric emissivities

σ is the Stefan–Boltzmann constant (5.67×10^{-8} W m^{-2} K^{-4})

ε_0 and ε_a were calculated as follows (Teixeira, 2010; Teixeira et al., 2014a):

$$\varepsilon_0 = a_0 \ln \text{NDVI} + b_0 \qquad (4.8)$$

$$\varepsilon_a = a_a \left(-\ln \tau_{sw} \right)^{b_a} \qquad (4.9)$$

where

τ_{sw} is the short-wave transmissivity calculated as the ratio of R_G to the incident solar radiation at the top of the atmosphere (R_a)

a_0, b_0, a_a, and b_a are regression coefficients taken as 0.06, 1.00, 0.94, and 0.10 according to Teixeira et al. (2014a)

Daily R_n can be described by the 24 h values of net shortwave radiation, with a correction term for net long-wave radiation for the same time scale (Teixeira et al., 2008b):

$$R_n = \left(1 - \alpha_0 \right) R_G - a_L \tau_{sw} \qquad (4.10)$$

where a_L is the regression coefficient of the relationship between net long-wave radiation and τ_{sw} on the daily scale.

Because of the thermal influence on long-wave radiation via the Stefan–Boltzmann equation, a previous study investigated whether the variations of the a_L coefficient from Equation 4.10 could be explained by variations in 24-h mean T_a (Teixeira et al., 2008b):

$$a_L = f T_a - g \qquad (4.11)$$

where f and g are regression coefficients found to be 6.99 and 39.93, respectively, for the Brazilian northeast conditions.

The SAFER algorithm is used to model the instantaneous values of the ratio ET/ET_0, which is then multiplied by ET_0 from the agrometeorological stations to estimate the daily ET large-scale values:

$$\frac{ET}{ET_0} = \exp \left[h + i \left(\frac{T_0}{\alpha_0 \text{NDVI}} \right) \right] \qquad (4.12)$$

where h and i are the regressions coefficients, 1.9 and −0.008, respectively, for the Brazilian northeast conditions (Teixeira et al., 2013, 2014a).

4.3.2 Biomass and Crop Production

According to Field et al. (1995), in terms of gas exchange, BIO can be defined as

$$\text{BIO} = \text{GPP} + R_{aut} \qquad (4.13)$$

where

GPP (gross primary production) is the carbon fixed during the photosynthesis

R_{aut} is autotrophic respiration

For obtaining estimates of carbon balance, the light use efficiency (LUE) concept devised by Monteith (1972) and Kumar and Monteith (1981), and modified by Prince (1990), can be applied:

$$\text{BIO} = \varepsilon f_{PAR} \text{PAR} \qquad (4.14)$$

where

ε is the LUE

f_{PAR} is the ratio of APAR to PAR

According to Teixeira (2009), to acquire crop yield (Y) for a growing season (GS), BIO_{GS} is multiplied by the harvest index (HI) and the harvested area (HA):

$$Y_{GS} = \text{BIO}_{GS} \text{HI HA} \qquad (4.15)$$

Although HI is a crop-and variety-specific parameter and can be reduced by water stress, a constant value fine-tuned to the average condition on the estate will provide some first yield estimation at farm level.

For BIO calculations on large scales, the Monteith LUE model is applied in this chapter, including the evaporative fraction (E_f) to take into account the soil moisture effects (Teixeira et al., 2013). E_f is defined as the latent heat flux (λE) divided by the available energy, which in turn is the difference between R_n and soil heat flux (G):

$$E_f = \frac{\lambda E}{R_n - G} \qquad (4.16)$$

where λE is obtained by transforming ET into energy units, with all terms considered in MJ m^{-2} day^{-1} in the SAFER algorithm.

For the daily G values, the equation derived by Teixeira (2010) is used:

$$\frac{G}{R_n} = j\exp(k\alpha_0) \qquad (4.17)$$

where j and k are regression coefficients found to be 3.98 and −25.47, respectively, for the Brazilian northeast conditions.

R_G daily values are used to estimate the large-scale PAR for the daily time scale:

$$PAR = f_S R_G \qquad (4.18)$$

where $f_S = 0.44$ is the constant of the regression equation found under the Brazilian northeast conditions that reflects the portion of R_G that can be used by leaf chlorophyll for photosynthesis (Teixeira et al., 2009a).

APAR can be approximated directly from PAR:

$$APAR = f_{PAR}PAR \qquad (4.19)$$

The factor f_{PAR} is estimated from the NDVI values (Bastiaanssen and Ali, 2003; Teixeira et al., 2009b):

$$f_{PAR} = mNDVI + n \qquad (4.20)$$

The coefficients m and n of 1.257 and −0.161, respectively, reported for a mixture of arable crop types (Bastiaanssen and Ali, 2003) are considered in this chapter and the BIO is quantified as

$$BIO = \varepsilon_{max}E_f APAR\,0.864 \qquad (4.21)$$

where ε_{max} is the maximum LUE, which depends on whether the vegetation is C3 or C4 species and 0.864 is a unit conversion factor (Teixeira, 2009; Teixeira et al., 2009b).

4.3.3 Water Productivity

In this chapter, the WP assessments are done in terms of ET, including capillary rise and soil moisture changes and irrigation (I) (Teixeira, 2009).

Considering all kinds of vegetation, WP based on ET is calculated by

$$WP = \frac{BIO}{ET} \qquad (4.22)$$

Figure 4.3 presents the flowchart of all the steps for WP acquirements based on ET by combining the remote sensing measurements by satellites and weather data.

When considering agriculture, CWP, considered as the ratio of Y_a to the amount of water consumed or applied, can be acquired multiplying BIO by HI. Table 4.2 shows the different WP indicators according to Teixeira and Bassoi (2009).

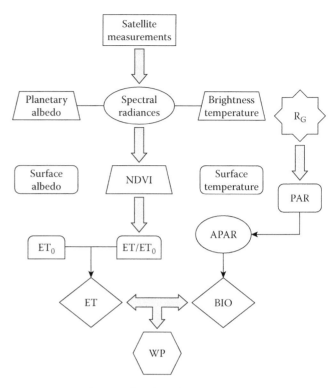

FIGURE 4.3 Schematic flowchart for calculation of the biophysical water productivity (WP) based on actual evapotranspiration (ET).

Organizations responsible for irrigation management are interested in Y_a per unit applied of irrigation water, but the drawback is that not all irrigation water is used for generating crop production. Renault et al. (2001) showed that perennial vegetation at Kirindi Oya system in Sri Lanka has water consumption around the same amount as rice and generates valuable ecosystem services. Moreover, Y_a is also a consequence of rainfall. Early studies commonly expressed CWP in terms of Y_a per unit of applied water, including rainfall and irrigation (Peacock et al., 1977; Araujo et al., 1995; Srinivas et al., 1999); however, it is also important to analyze the CWP in terms ET, including capillary rise and soil moisture changes (e.g., Droogers et al., 2000).

The economic WP is the value derived per unit of water consumed by the crop or applied through irrigation. Its increases may indicate a shift toward higher-valued crops, increase in yields, or a saving in water input. According to Bos et al. (2005), the economic indicators (see Table 4.2) are the standard gross value of production over the irrigation supply, I (CWP$_I$), over actual evapotranspiration, ET (CWP$_{ET}$), or over actual transpiration, Tr (CWP$_{Tr}$).

While there is a fixed relation between BIO and Tr, this is not true for Y_a relative to ET because of differences in soil evaporation, HI, climatic conditions, water stress, pests and diseases, nutritional and soil status, and the nature of agronomic practices. Thus, there seems to be considerable scope for raising CWP before reaching the upper limit. The fact that the variability in CWP is due to crop and water management practices is important because it offers hope of possible improvements (Molden et al., 2007).

AQ

TABLE 4.2   Different Water Productivity Indicators: Biomass Production (BIO), Actual Yield (Y_a), Water Productivity (WP), and Crop Water Productivity (CWP) Based on Land (L), Irrigation (I), Actual Evapotranspiration (ET), and Actual Transpiration (Tr), Together with the Economic Values of These Indices ($)

Output	Land (ha)	I (m³)	ET (m³)	Tr (m³)
BIO (kg)	WP_L (kg ha^{-1})	WP_I (kg m^{-3})	WP_{ET} (kg m^{-3})	WP_{Tr} (kg m^{-3})
Net benefit ($)	$WP\$_L$ ($ ha^{-1})	$WP\$_I$ ($ m^{-3})	$WP\$_{ET}$ ($ m^{-3})	$WP\$_{Tr}$ ($ m^{-3})
Y_a (kg)	CWP_L (kg ha^{-1})	CWP_I (kg m^{-3})	CWP_{ET} (kg m^{-3})	CWP_{Tr} (kg m^{-3})
Gross return ($)	$CWP\$_L$ ($ ha^{-1})	$CWP\$_I$ ($ m^{-3})	$CWP\$_{ET}$ ($ m^{-3})	$CWP\$_{Tr}$ ($ m^{-3})

To retrieve the ET and BIO incremental values, a simplified classification is presented in this chapter. For the separation of irrigated crops and natural vegetation, the SUREAL algorithm is applied in an image obtained from a semiarid region, during the naturally driest period of a year, considering threshold limits for the surface resistance (r_s) to water fluxes (Teixeira et al., 2013):

$$r_s = \exp\left[o\left(\frac{T_0}{\alpha_0}\right)(1 - NDVI) + p \right] \quad (4.23)$$

where o and p are regression coefficients, found to be 0.04 and 2.72, respectively, for the Brazilian northeast conditions.

The average values of the ET/ET_0 ratio without water deficits allow the crop coefficient (Kc) modeling as a function of the accumulated degree-days (DD_{ac}):

$$Kc = qDD_{ac}^2 + rDD_{ac} + s \quad (4.24)$$

where q, r, and s are the crop-specific regression coefficients.

The ET_p values are estimated as

$$ET_p = Kc\, ET_0 \quad (4.25)$$

Following Bastiaanssen et al. (2001) and Teixeira et al. (2008a), the irrigation performance indicators discussed and analyzed in this chapter are the relative evapotranspiration (R_{ET}), relative water supply (R_{WS}), water deficit (WD), CWP_{ET}, and CWP_I:

$$R_{ET} = \frac{ET}{ET_p} \quad (4.26)$$

$$R_{WS} = \frac{V_I + P}{ET_p} \quad (4.27)$$

$$WD = ET_p - ET \quad (4.28)$$

$$CWP_{ET,I} = \frac{Y_a}{ET, V_I} \quad (4.29)$$

where
  V_I is the volume of water applied through irrigation
  P is the precipitation

Discarding corrections for soil storage changes and runoff, the percolation rates (Perc) are also estimated:

$$Perc = P + V_I - ET \quad (4.30)$$

4.4 Retrieving Water Productivity at Different Spatial and Temporal Scales

4.4.1 Municipality Scale

In the Brazilian northeast region, Petrolina and Juazeiro municipalities are two of the main agricultural municipal districts, located respectively in Pernambuco (PE) and Bahia (BA) states. The visible, infrared, and thermal bands from MODIS images, 6 of which were for 2010 and 9 for 2011, together with 14 automatic agrometeorological stations (see Figure 4.2), were used for WP analyses, following field calibrations and spatial and temporal interpolations. Figure 4.4 presents the spatial distribution of the monthly ET values in the mixed agroecosystems inside these municipalities, along the year 2011.

Spatial and temporal ET variations throughout the year are evident, mainly when observing the wettest period from February to April, with the driest period between August and October. The ET maxima were observed in April, with averages of 60 and 45 mm month^{-1} for Petrolina and Juazeiro, respectively. However, the highest pixel values, reaching 200 mm month^{-1}, were from November to December in both municipalities, representing well-irrigated crops. Intermediate ET values in natural vegetation occurred just after the rainy period, from May to June, because antecedent precipitation still keeps the natural vegetation ("Caatinga") brushes wet and green. During this time of the year, the hydrological large-scale uniformity causes "Caatinga" species to have ET rates similar to those from irrigated crops or even higher on some occasions.

Because the largest fractions of the available energy are used as sensitive heat fluxes (H) during the driest period of the year, between August and October, the natural vegetation presented the lowest ET pixel values (bluish pixels), while the irrigated fields showed the highest ones (reddish pixels). Stomata of "Caatinga" species close under these conditions, limiting transpiration and photosynthesis, while, in general, irrigation intervals in crops are short (daily irrigation), with a uniform water supply, reducing the heat losses to the atmosphere.

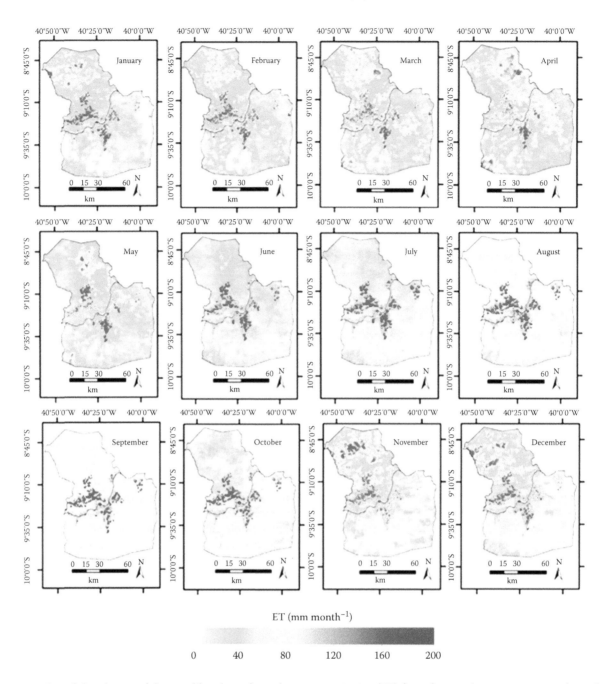

FIGURE 4.4 Spatial distribution of the monthly values of actual evapotranspiration (ET) from the mixed agro-ecosystems of Petrolina and Juazeiro municipalities, respectively, Pernambuco (PE) and Bahia (BA) states, northeast Brazil, during the year 2011.

Table 4.3 shows the ET monthly average values and standard deviations (SD) for irrigated crops (IC) and natural vegetation (NV) in Petrolina and Juazeiro for the year 2011.

Considering only irrigation conditions, differences between the municipalities arose, with November presenting the highest crop ET rates in Petrolina, while in Juazeiro they happened in April. In natural vegetation, the largest ET values were in April for both municipalities, while the lowest rates occurred in September for both ecosystems and municipalities. In relation to the annual incremental ET, irrigated crops consumed 555 mm year[-1] more water than "Caatinga" species in Petrolina, and in the second municipality, this extra consumption was 545 mm year[-1].

Considering both municipalities and ecosystems, the lowest SD values were during the driest period of the year. Under rain-fed or irrigation conditions, the plants are strongly sensitive to the spatial distribution of precipitation and soil water content (Claverie et al., 2012). During the naturally driest periods, the higher ET and SD values for irrigated crops are mainly caused by different levels of fertilization, crop stages, and irrigation (Hatfield et al., 2001; Wu et al., 2010).

TABLE 4.3   Monthly Average Values and Standard Deviations (SD) of Actual Evapotranspiration (ET) from Irrigated Crops (IC) and Natural Vegetation (NV) in the Petrolina and Juazeiro Municipalities, Respectively, Pernambuco (PE) and Bahia (BA) States, Northeast Brazil, during the Year 2011

	PETROLINA, PE ET (mm)		JUAZEIRO, BA ET (mm)	
Months/Year	IC	NV	IC	NV
January	95.1 ± 41.0	42.2 ± 19.9	68.7 ± 38.1	16.6 ± 16.7
February	86.8 ± 31.7	43.6 ± 19.3	69.2 ± 33.2	24.4 ± 20.0
March	94.3 ± 34.0	52.4 ± 24.4	82.1 ± 38.9	34.0 ± 25.6
April	87.3 ± 27.5	57.9 ± 25.0	86.6 ± 35.5	41.7 ± 26.2
May	74.3 ± 37.2	36.6 ± 20.7	67.1 ± 33.8	25.6 ± 19.0
June	66.9 ± 29.5	22.5 ± 12.1	60.9 ± 31.0	14.5 ± 11.3
July	66.1 ± 26.7	14.6 ± 8.4	63.6 ± 35.9	8.9 ± 8.9
August	68.8 ± 34.0	4.5 ± 5.6	58.5 ± 41.4	5.7 ± 8.4
September	49.5 ± 29.7	1.6 ± 3.5	36.8 ± 32.8	1.7 ± 4.3
October	58.8 ± 31.8	11.9 ± 7.9	38.6 ± 33.3	4.1 ± 7.0
November	96.7 ± 42.1	53.4 ± 30.7	52.0 ± 35.7	9.1 ± 12.9
December	90.5 ± 42.0	43.4 ± 27.0	54.4 ± 38.1	7.7 ± 13.4
Year	937.9 ± 351.2	385.0 ± 140.6	738.5 ± 328.0	194.0 ± 140.7

Note: IC, irrigated crops; NV, natural vegetation.

Figure 4.5 presents the spatial variation of the BIO monthly values in the mixed agroecosystems of Petrolina and Juazeiro municipalities throughout the year 2011.

As there is a relation between water consumption and BIO (Yuan et al., 2013), the spatial and temporal variations of BIO are also strong throughout the year, for both irrigated crops and natural vegetation, with the highest values during the rainy season from February to April, and the lowest ones in the driest period, between July and September. The maxima occurred in April, with averages of 1817 and 1306 kg ha^{-1} month^{-1} for Petrolina and Juazeiro, respectively. During April, several areas presented rates higher than 4000 kg ha^{-1} month^{-1}, including both natural vegetation and irrigated crops. The lowest BIO happened in September in both municipalities, with corresponding mean pixel values of 142 and 92 kg ha^{-1} month^{-1}, respectively.

Precipitation from January to April provides enough water storage in the root zones of the "Caatinga" species to maintain their developments, while from June to September, irrigated crops present larger BIO than natural vegetation: as with absence of rains, the soil moisture is close to the field capacity due to the general daily irrigation. Under these conditions, crops are well visible on BIO maps, confirming the effectiveness of coupling the SAFER and the Monteith LUE models. Similarly, irrigated corn crop showed twice the BIO values when comparing with natural alpine meadow in the Heihe River basin (Wang et al., 2012), irrigation being considered the main reason for these marked differences.

Table 4.4 presents the BIO monthly average values and SD for IC and NV in Petrolina and Juazeiro municipalities for the year 2011.

In general, Petrolina presented higher BIO than Juazeiro for both natural vegetation and irrigated crops. However, under irrigation conditions, this difference was only 33%, while for "Caatinga" species it was double. For all agroecosystems, the period with the highest BIO was from March to April. In irrigated crops, the lowest average values were from September to October, while in natural vegetation, they occurred between August and September. The annual incremental BIO rates (differences between irrigated crops and natural vegetation) were 18.6 and 16.7 t ha^{-1} year^{-1} on average for Petrolina and Juazeiro, respectively.

As in the case of ET, the lowest SD values occurred during the driest period of the year, independently of the agroecosystem. However, Petrolina presented slightly higher BIO spatial variation in comparison with Juazeiro.

Table 4.5 shows the WP monthly average values and SD for IC and NV in Petrolina and Juazeiro municipalities for the year 2011.

Analyzing only irrigated crops, WP differences between the municipalities can also be identified. The period from March to April presented the highest values in Petrolina, while in Juazeiro they happened in May. For natural vegetation, the largest WP was in April and the lowest occurred in October for both municipalities. WP from irrigated crops in Juazeiro was 86% of that for Petrolina, while this percentage for natural vegetation was 74%, evidencing the effect of the differences in rainfall and soil water conditions between the municipalities.

The smaller BIO values compensated the lower ET rates in Juazeiro in WP calculation (see Equation 4.22). Considering the annual incremental WP, it was an average 1.0 kg m^{-3} for both municipalities, with the spatial variation in Juazeiro being slightly higher. Although the lowest SD values were during the driest period, as they were for ET and BIO, the monthly WP spatial differences were smaller, showing compensation when taking into account the spatial variations of BIO and ET together in Equation 4.22.

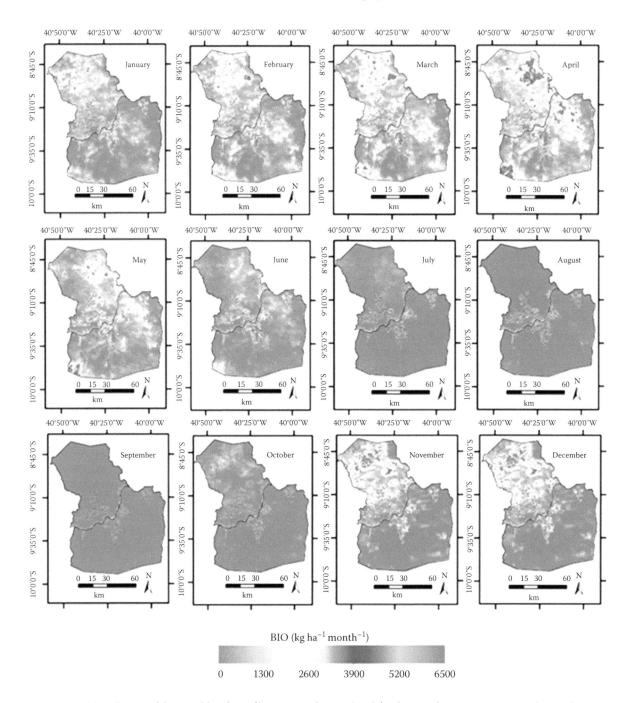

FIGURE 4.5 Spatial distribution of the monthly values of biomass production (BIO) for the mixed agroecosystems inside Petrolina and Juazeiro municipalities, respectively, Pernambuco (PE) and Bahia (BA) states, northeast Brazil, during the year 2011.

Multiplying BIO by HI makes it possible to estimate CWP (Teixeira, 2009, 2012; Teixeira and Bassoi, 2009), which corresponds to the highest WP in the vicinities of the São Francisco River. HI values were found to be around 0.60 for vineyards and 0.80 for mango orchards under these conditions (Teixeira et al., 2009b). In southwest France, HI values were 0.25 for rainfed sunflower and 0.48 for irrigated corn (Claverie et al., 2012), while for wheat, an average of 0.35 was reported as a result of several experiments around the word (Zwart et al., 2010).

4.4.2 Scale of the Irrigation Scheme

On a large scale, agricultural crops in irrigation schemes have rapidly replaced the natural vegetation of the Brazilian semi-arid region, the "Caatinga." Among the most important of these schemes is the Nilo Coelho (NC), located on the left bank of the São Francisco River, with almost its whole area within the Petrolina (PE) municipality. Under these circumstances, it is becoming important to develop and apply tools in order to

TABLE 4.4 Monthly Average Values and Standard Deviations (SD) of Biomass Production (BIO) from Irrigated Crops (IC) and Natural Vegetation (NV) in the Petrolina and Juazeiro Municipalities, Respectively, Pernambuco (PE) and Bahia (BA) States, Northeast Brazil, during the Year 2011

Months/Year	PETROLINA, PE BIO (t ha⁻¹)		JUAZEIRO, BA BIO (t ha⁻¹)	
	IC	NV	IC	NV
January	2.8 ± 1.7	1.1 ± 0.7	1.9 ± 1.5	0.4 ± 0.6
February	2.8 ± 1.4	1.2 ± 0.8	2.2 ± 1.4	0.6 ± 0.7
March	3.2 ± 1.6	1.5 ± 1.0	2.4 ± 1.6	0.8 ± 0.8
April	2.9 ± 1.3	1.7 ± 1.1	2.9 ± 1.7	1.2 ± 1.0
May	2.4 ± 1.4	0.9 ± 0.8	2.4 ± 1.7	0.7 ± 0.8
June	2.0 ± 1.1	0.4 ± 0.5	2.0 ± 1.5	0.4 ± 0.4
July	1.9 ± 1.1	0.1 ± 0.3	0.9 ± 0.8	0.5 ± 0.1
August	1.9 ± 1.4	0.0 ± 0.1	1.7 ± 1.7	0.1 ± 0.2
September	1.3 ± 1.2	0.0 ± 0.1	1.0 ± 1.2	0.0 ± 0.1
October	1.6 ± 1.2	0.2 ± 0.4	0.9 ± 1.2	0.1 ± 0.2
November	2.9 ± 1.7	1.4 ± 1.1	1.3 ± 1.3	0.2 ± 0.4
December	2.6 ± 1.7	1.0 ± 1.0	1.4 ± 1.4	0.2 ± 0.4
Year	28.5 ± 14.4	9.9 ± 4.9	21.3 ± 13.4	4.6 ± 4.7

Note: IC, irrigated crops; NV, natural vegetation.

TABLE 4.5 Monthly Average Values and Standard Deviations (SD) of Water Productivity (WP) from Irrigated Crops (IC) and Natural Vegetation (NV) in the Petrolina and Juazeiro Municipalities, Respectively, Pernambuco (PE) and Bahia (BA) States, Northeast Brazil, during the Year 2011

Months/Year	PETROLINA, PE (kg m⁻³)		JUAZEIRO, BA (kg m⁻³)	
	IC	NV	IC	NV
January	2.8 ± 0.6	2.1 ± 0.5	2.4 ± 0.6	1.3 ± 0.5
February	3.1 ± 0.6	2.4 ± 0.5	1.9 ± 0.6	1.9 ± 0.6
March	3.2 ± 0.6	2.6 ± 0.6	2.9 ± 0.8	2.1 ± 0.7
April	3.2 ± 0.6	2.7 ± 0.6	3.1 ± 0.7	2.4 ± 0.7
May	3.0 ± 0.6	2.3 ± 0.5	3.1 ± 0.8	2.2 ± 0.6
June	2.9 ± 0.5	1.8 ± 0.4	2.9 ± 0.7	1.6 ± 0.4
July	2.7 ± 0.6	1.4 ± 0.3	2.3 ± 0.9	0.7 ± 0.3
August	2.5 ± 0.7	0.9 ± 0.3	2.3 ± 0.8	0.9 ± 0.4
September	2.4 ± 0.7	0.8 ± 0.2	2.0 ± 0.7	0.7 ± 0.4
October	2.3 ± 0.6	0.8 ± 0.3	1.8 ± 0.8	0.5 ± 0.3
November	2.7 ± 0.6	2.1 ± 0.6	2.0 ± 0.7	1.0 ± 0.5
December	2.6 ± 0.6	2.0 ± 0.6	2.1 ± 0.7	0.9 ± 0.5
Year	2.8 ± 0.5	1.8 ± 0.3	2.4 ± 0.6	1.3 ± 0.4

Note: IC, irrigated crops; NV, natural vegetation.

quantify the dynamics of the WP parameters, with analyses of the agroecosystems that characterize the changes in land use. To study these dynamics, data from one conventional and 5 automatic agrometeorological stations were used together with the visible, infrared, and thermal bands from 10 Landsat 5 images during crop growing periods outside the rainy season between 1992 and 2011. The stations used were those closer to the São Francisco River, represented by the edge line between Petrolina and Juazeiro municipalities (see Figure 4.2).

The difficulty of identifying land-use change effects on water and vegetation variables by using only NDVI is the variability of this indicator with the thermohydrological conditions. NDVI will reflect the land use change, but it will be also dependent of the weather conditions, which may be variable along the years. Besides NDVI and accumulated precipitation (P_{ac}), ET and BIO depend also on APAR, which in turn is conditioned by R_G levels. Figure 4.6 shows the variation of the mean values of these parameters along the analyzed days and years in the NC irrigation scheme from 1992 to 2011.

P_{ac} data were from the Bebedouro agrometeorological station (Petrolina, PE), while the other parameters are mean pixel values. Although the years 1992, 1995, and 1997 had lower irrigated areas in relation to the more recent years, high amounts of precipitation during the rainy period left the root zones of the "Caatinga" species and crops wet during these years (Figure 4.6a). On the other hand, lower P_{ac} between the analyzed

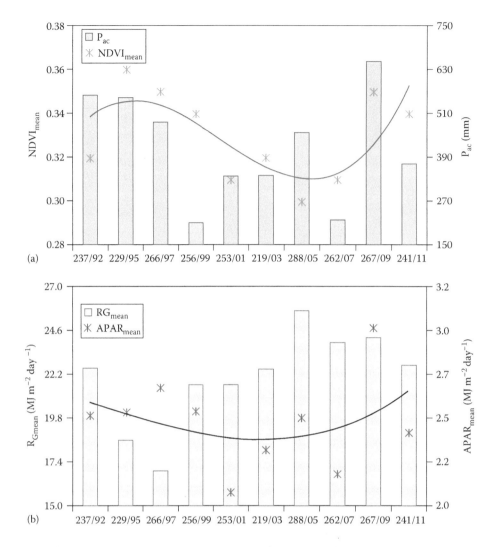

FIGURE 4.6 Mean values of vegetation indicators and weather parameters for the period from 1992 to 2011 for each day/year in the Nilo Coelho (NC) irrigation scheme, northeast Brazil: (a) Normalized Difference Vegetation Index (NDVI) and accumulated precipitation (P_{ac}); (b) Absorbed Photosynthetically Active Radiation (APAR) and incident Global Solar Radiation (R_G).

years 1999 and 2007 was not favorable for vegetation growth. Figure 4.6b reinforces the dependence of plant development on soil moisture conditions, as high APAR values are observed with low R_G levels in 1997, for example, while the highest mean R_G did not correspond to the largest APAR, which occurred in 2009, with both high P_{ac} and high R_G.

Figure 4.7 shows the spatial distribution of the ET daily values, within the driest period of the year from 1992 to 2011 for each Landsat satellite overpass date, for the NC irrigation scheme in northeast Brazil.

Again, one can distinguish irrigated areas from the natural vegetation clearly by the higher ET values under irrigation conditions. Because the largest portion of the available energy used as sensible heat flux (H) by the "Caatinga" species are during the driest and hottest period of the year, their ET values are lower than 1.0 mm day^{-1}, while the corresponding values for irrigated crops are above 3.5 mm day^{-1}. Considering the whole irrigation perimeter, the increments throughout the years are around

350% when comparing 1992 with 2011. However, some variations occurred during the years between due to the combined effects of land-use change and different thermohydrological conditions.

Figure 4.8 presents the spatial distribution of the BIO daily values for the same days as for ET, within the driest period of the year from 1992 to 2011, in the NC irrigation scheme in northeast Brazil.

With low APAR and the absence of rains during the driest periods of the year, natural vegetation species presented low BIO daily values, while irrigated crops showed the highest BIO values, promoting a strong vegetation contrast between these ecosystems. In some irrigated areas, BIO reached values above 200 kg ha^{-1} day^{-1}. The averages for the whole irrigation perimeter ranged from 8 to 38 kg ha^{-1} day^{-1} from 1992 to 2011, accounting for an increment of 475% as a result of increasing irrigated areas. The spatial variation is great during the analyzed period, with SD from 21 to 70 kg ha^{-1} day^{-1}.

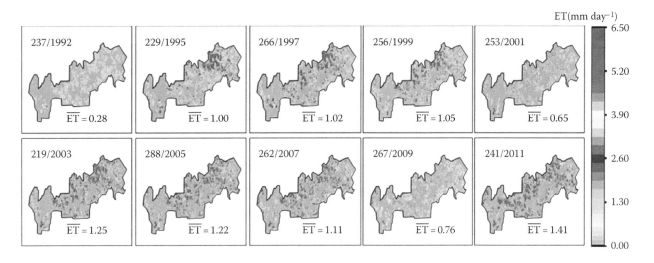

FIGURE 4.7 Spatial distribution of daily actual evapotranspiration (ET) during the driest period of the year for each Landsat satellite overpass date from 1992 to 2011 in the Nilo Coelho (NC) irrigation scheme, northeast Brazil. The bars indicate average values of all pixels.

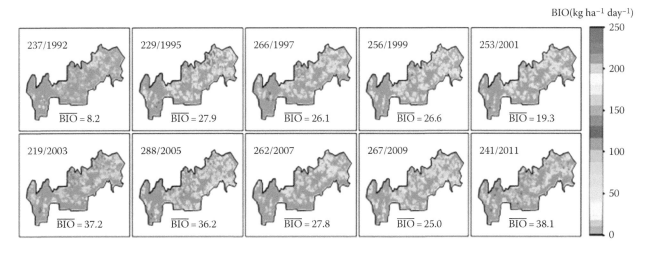

FIGURE 4.8 Spatial distribution of biomass production (BIO) during the driest period of the year for each Landsat satellite overpass date from 1992 to 2011 in the Nilo Coelho (NC) irrigation scheme, northeast Brazil. The bars indicate average values of all pixels.

For irrigated crops and natural vegetation, the variations in the mean pixel values of the WP parameters, together with their SDs, were analyzed throughout the years in the NC irrigation scheme, northeast Brazil (Table 4.6).

Taking into account irrigated areas, the ET rates almost doubled, when comparing 1992 with 2011. The rates for "Caatinga" species were more stable and lower, converting the largest part of the available energy for warming the air near the surface during the driest and hottest periods of the years. Under irrigation conditions, daily mean BIO values ranged from 10 to 100 kg ha^{-1} day^{-1}, with the lower values occurring in 1992. The highest daily rates were in 2009; however, the greatest spatial variations for both WP parameters, as determined by the SD values, occurred in 2011. For "Caatinga," BIO was found to be highly variable, with the values ranging from 1 to 40 kg ha^{-1} day^{-1}, and the lowest values occurred in 1999 and 2005. The year 2011 is highlighted by the largest mean BIO values in natural vegetation.

Incremental ET, represented by the different water fluxes between irrigated crops and natural vegetation, ranged from 1.6 in 1992 to 3.1 mm day^{-1} in 2011, doubling the water consumption because of the introduction of commercial irrigated agriculture. According to the SD values for irrigated crops, one can see a greater heterogeneity of water consumption than for those of natural vegetation due to different crop management and stages.

The main irrigated crops in NC scheme are fruits. The average Kc values from field experiments were applied to ET$_0$ and yield data by Teixeira and Bassoi (2009) to estimate CWP$_{ET}$ in Petrolina, PE, resulting in 2.4, 1.5, 1.1 and 1.3 kg m^{-3} for table grapes, mango, guava, and banana, respectively. Despite the smaller area with guava crop, CWP$_{ET}$ is reasonably high when comparing with that for mango, which is one of the most important commercial fruit crops in this irrigation scheme. Although the production cost for table grape is very high, its CWP$_{ET}$ ranks the best.

TABLE 4.6 Daily Average Values and Standard Deviations (SD) for the Water Productivity
Parameters, during the Period from 1992 to 2011, in the Nilo Coelho (NC) Irrigation Scheme,
Northeast Brazil: Actual Evapotranspiration (ET) and Biomass Production (BIO)

Day/Year	ET (mm day^{-1})		BIO (kg ha^{-1} day^{-1})	
	IC	NV	IC	NV
237/92	1.8 ± 0.8	0.2 ± 0.3	12.6 ± 24.7	8.5 ± 21.9
229/95	2.8 ± 1.8	0.4 ± 0.6	42.5 ± 70.3	23.0 ± 51.8
266/97	3.5 ± 1.6	0.4 ± 0.5	33.3 ± 61.9	25.7 ± 53.2
256/99	2.7 ± 1.5	0.2 ± 0.2	73.2 ± 63.7	2.1 ± 3.2
253/01	2.6 ± 1.0	0.5 ± 0.5	28.8 ± 47.1	21.8 ± 41.3
219/03	3.0 ± 1.8	0.2 ± 0.4	41.0 ± 71.5	34.8 ± 69.0
288/05	2.9 ± 1.6	0.1 ± 0.2	88.5 ± 73.7	1.1 ± 2.8
262/07	3.3 ± 1.8	0.4 ± 0.6	30.2 ± 57.2	28.8 ± 58.7
267/09	2.6 ± 1.1	0.5 ± 0.4	98.0 ± 56.7	11.0 ± 11.0
241/11	3.4 ± 1.9	0.3 ± 0.6	42.1 ± 72.7	37.4 ± 70.2

Note: IC, irrigated crops; NV, natural vegetation.

On commercial farms within the Brazilian semiarid region, Teixeira et al. (2009b), considering the year 2008, found an average value CWP$_{ET}$ of 1.15 L m^{-3} (i.e., 1.44 kg m^{-3} per water consumed) for wine grape with corresponding monetary values (CWP\$$_{ET}$) reaching a maximum of 1.55 US\$ m^{-3}. For table grapes, the average CWP$_{ET}$ value of 2.80 kg m^{-3} returned a CWP\$$_{ET}$ of 8.80 US\$ m^{-3} for the monetary counterpart. For mango orchard, the average physical value was 3.40 kg m^{-3}, corresponding to a CWP\$$_{ET}$ of 5.10 US\$ m^{-3}.

Jairmain et al. (2007) found higher CWP$_{ET}$ values for wine and table grapes in South Africa (4.70 and 3.70 kg m^{-3}, respectively). The CWP$_{ET}$ values for vineyards in Brazil evidenced scope for improvements in water management. The low CWP$_{ET}$ should be related to the lower yields associated with higher daily water consumptions in comparison with South Africa.

4.4.3 Spatial Resolution Effects

Much of the uncertainty in MODIS results involving irrigated areas and natural vegetation is largely due to the use of the 1 km thermal band, in contrast to the spatial resolution of 120 m from the thermal band of Landsat 5 images. MODIS pixels cover a greater mixture of land use, reducing the spatial accuracy for the ET and BIO calculations, which in turn will affect WP results. For comparisons, Figure 4.9 presents the WP spatial distribution and histograms from these two different satellites in the NC irrigation scheme during the driest period of the year 2011.

Because of the absence of Landsat and MODIS images for the same day of the year (DOY), for Landsat, DOY 241 is used, while DOY 231 is taken for MODIS, both in August. Although the image acquisitions were on different days, according to data from Bebedouro agrometeorological station in Petrolina, PE (see Figure 4.2), the weather conditions were similar as shown in Table 4.7.

The SAFER algorithm is strongly based on ET$_0$, and for DOY 241 it has an average daily value of only 0.2 mm day^{-1} more

than that for DOY 231. Thus, with no significant differences in atmospheric demands between the satellite overpass days, WP acquired with different spatial resolutions could be analyzed under this driest condition of the year, when irrigated crops are well distinct from natural vegetation.

It is clear from the pixel values that there are WP overestimations from MODIS, when compared with the Landsat results. For Landsat and MODIS pixel values, in the mixed agroecosystems of the NC irrigation scheme, WP was, on average, 1.2 ± 1.2 kg m^{-3} and 2.0 ± 1.8 kg m^{-3}, respectively. This happens mainly because processed MODIS images wrongly identify some areas under irrigation conditions as more pixels cover a greater mixture of land use than Landsat images.

For the frequency distribution analyses, WP values lower than 0.5 kg m^{-3}, which should not be from vegetated surfaces, were excluded from the histograms depicted in Figure 4.9. While for Landsat satellite measurements there were 31% of the pixels with zero values, for MODIS there was an absence of these null values. The most frequent pixel values were respectively 0.5 and 0.8 kg m^{-3} for Landsat and MODIS images, representing a transition from irrigation conditions to natural vegetation. There was an overestimation around 67% in WP MODIS values when comparing with the Landsat results. Although both satellites detected irrigation conditions by the WP pixel values between 1.5 and 4.5 kg m^{-3}, inside this range, MODIS images showed frequencies that were 75% larger than those for Landsat, thus overestimating irrigated areas.

Aiming to improve the spatial accuracy for WP determined from the MODIS satellite, and considering the good availability of automatic agrometeorological stations, creating the opportunity of using gridded weather parameters, a new methodology was developed to calculate T$_0$ as a residue in the radiation balance equation. This method downscales all WP components to a lower spatial resolution of 250 m by using the visible and infrared bands from the MODIS sensor. In addition, the problems of cloud contamination are reduced, increasing the possibility of using the available MODIS products.

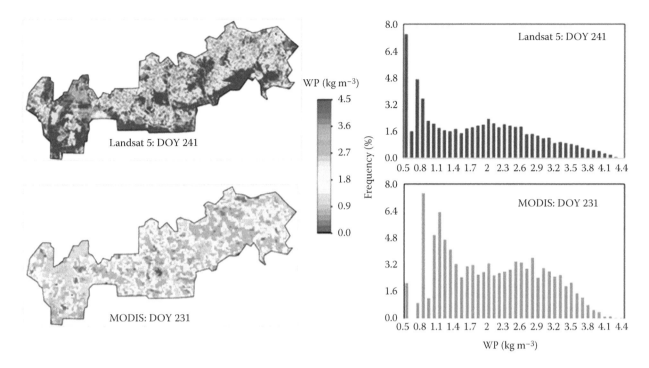

Landsat 5: DOY 241

MODIS: DOY 231

FIGURE 4.9 Spatial distributions and histograms of water productivity (WP) (kg m⁻³) for the days of the year (DOY) 241 and 231 from Landsat 5 and MODIS, respectively, in the Nilo Coelho (NC) irrigation scheme, northeast Brazil.

TABLE 4.7 Weather Variables in the Landsat and MODIS Satellite Overpass Days for, Respectively, the Days of the Year (DOY) 241 and 231 of 2011, in the Nilo Coelho (NC) Irrigation Scheme, Northeast Brazil: Global Solar Radiation (R_G), Mean Air Temperature (T_a) Relative Humidity (RH), Wind Speed (u) and Reference Evapotranspiration (ET_0)

Date/DOY	R_G (MJ m⁻² day⁻¹)	T_a (°C)	RH (%)	u (m s⁻¹)	ET_0 (mm day⁻¹)
08/19/2011 (DOY 231)	21.0	24.8	48	2.6	5.9
08/29/2011 (DOY 241)	20.8	25.1	52	2.9	6.1

In the following sections, this methodology is first tested with corn crop under irrigation conditions using the visible and infrared bands from Landsat images at farm level in the Brazilian Southeast region (see Figure 4.2). Secondly, it is applied to the same crop, but under rainfed conditions with a MODIS product based on the reflectance values of bands 1 and 2 in the Brazilian central–west region (see Figure 4.2).

4.4.4 Irrigated Crop

From the previous sections, it can be concluded that with the availability of weather data, the SAFER algorithm has been powerful for WP analyses under the Brazilian northeast conditions. However, when using Equation 4.12 in other ecosystems, some calibrations should be probably necessary. Hernandez et al. (2004) recently compared the algorithm by using its original regression coefficients with the water management based on the traditional FAO crop coefficient (Kc) method in the southeast region of Brazil (see Figure 4.2). They observed the need to adjust the "h" coefficient from 1.9 to 1.0 in Equation 4.12. With this calibration, irrigation pivots with corn crop in this region were

analyzed in terms of WP, considering two commercial situations, for grains and for silage.

The SAFER algorithm was applied to Landsat satellite images from March 22, April 7, April 23, June 10, June 26, July 12, and August 29 of 2010, together with weather data from one agrometeorological station close to the corn crop, and successive temporal interpolations were performed to cover the complete growing seasons at each pivot. Instead of using the thermal band, T_0 was estimated by Equation 4.7 as an input parameter in Equation 4.12. To see the accuracy of the algorithm without the Landsat thermal band, the ratio ET/ET_0 under optimum soil moisture conditions was compared with standard corn Kc values.

Figure 4.10 shows the spatial distribution of ET/ET_0 involving the commercial Bonança farm, located in the northwestern part of São Paulo state, Brazilian southeast region, including irrigated areas, natural vegetation, and water.

Clearly, one can distinguish irrigation pivots from the natural vegetation by their higher ET/ET_0 values, with some pixels reaching close to 1.40. The Bonança farm has pivots with corn, soybeans, beans, and sugarcane. However, to relate Kc as a function of the accumulated degree-days (DD_{ac}) for corn crop

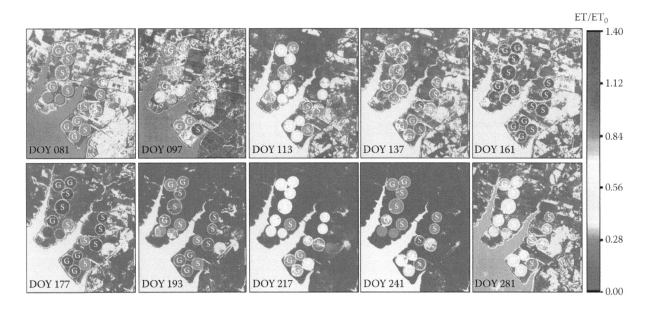

FIGURE 4.10   Spatial distribution of ET/ET_0 in the Bonança farm, located at the northwestern side of São Paulo State, Brazilian southeast region. DOY means days of the year and the letters G and S are irrigation pivots for corn, grain, and silage, respectively.

(Equation 4.24), the values of ET/ET_0 from SAFER results were used, considering the average pixel values inside the buffered area of six pivots of corn for grain (G) and eight for silage (S), under optimum soil moisture conditions and a basal temperature (T_b) of 10°C.

Kc, at different corn crop stages, was between 0.3 and 1.2. This range is in agreement with the values reported by DeJonge et al. (2012) during their ET modeling improvements in Colorado (USA) and with the values of the standard work from Allen et al. (1998), which justified confidence in using the SAFER algorithm with acceptable accuracy and without the thermal band. The

advantage of Equation 4.24 is the possibility of upscaling Kc values to different thermal conditions (Teixeira, 2009). The regression equations were used to estimate ET_p (see Equation 4.25), which together with ET are the key parameters for irrigation performance assessments.

For these assessments, including CWP, ten pivots with corn crop were selected in the Bonança farm (five each for grains (G) and silage (S)). Figure 4.11 shows the spatial variations of the ET totals for a growing season (ET_{GS}) (Figure 4.11a) and the seasonal mean daily pixel and SD values for each of them (Figure 4.11b and c).

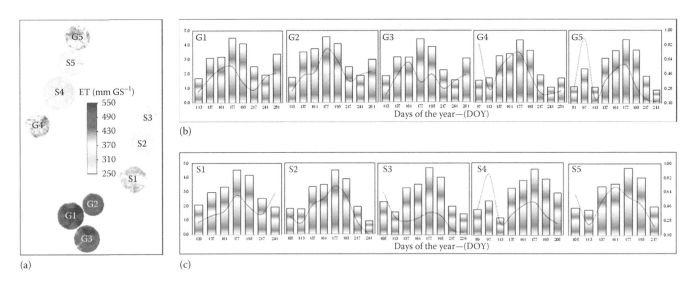

FIGURE 4.11   Actual evapotranspiration (ET) for corn crop irrigation pivots in the Bonança farm, northwestern São Paulo, southeast Brazil in 2010. Spatial variation of the totals for a growing season (GS) including both grains and silage (a); seasonal variation of the daily pixel values for grains (b) and silage (c). G and S indicate grain and silage, respectively, and DOY is day of the year.

For both grains and silage, the highest ET_{GS} values were for the pivots G1, G2, and G3, with several pixels above 450 mm GS^{-1}, while for S1, S2, and S3 most pixels present ET_{GS} below 400 mm GS^{-1}. The larger water consumption for these pivots was due to the higher atmospheric demands involving periods ranging from DOY values of 108–285 (mid-April to mid-October) for grains and DOY values of 105–241 (mid-April to the end of August) for silage. The higher values for grain than for silage are due to the different GS lengths, which were, on average, 160 and 120 days, respectively, and this is bound to affect the WP depending mainly on the rainfall conditions along the crop stages.

According to Figure 4.11b and c, the periods with the maximum daily ET rates were around DOY 177 (end of June), corresponding to the grain-filling stage, while the lowest ones were verified on DOY 241 (end of August), the harvest time at the end of the growing seasons. The daily ET rates were in the range 0.9–4.7 mm day^{-1} for both grains and silage. In the northwest of China, Ding et al. (2013), through field measurements and modeling, found similar daily rates, averaging 3.5 mm day^{-1}.

Knowledge on the water input and output in each pivot allowed the irrigation performance assessments for corn crop, including CWP_{ET} and CWP_I from Equation 4.29. The indicators are summarized in Table 4.8 for grain (a) and for silage (b).

R_{ET} showed a gap between the water demand and water requirements only on the G4 and G5 pivots, when ET was lower than 80% of ET_p. On all other occasions, R_{ET} was close to 1.00, ranging from 0.78 to 1.00 with WD at a maximum of 110 mm GS^{-1}. R_{WS} with values from 1.1 to 1.4 indicated high drainage rates due to sandy soil together with rainfall events. This problem occurred with greater intensity at the pivots for silage than at the pivots for grains. Such R_{WS} values imply that in general 10%–40% more irrigation water was supplied than necessary to meet the corn crop water requirements.

Taking percolation rates as the differences between P, I, and ET without corrections for soil storage changes and discarding runoff (see Equation 4.30), they were on average 159 mm (35% of ET) and 132 mm (37% of ET), for grains and silage, respectively. To reduce these drainage rates, practices of mulching should be used (Ding et al., 2013) to improve CWP. It is important to note that even with R_{WS} higher than 1.0 for G4 and G5 pivots, low R_{ET} and high WD values indicated that, on some occasions, more water should have been supplied by irrigation as it quickly percolated away in the sand soil. Teixeira et al. (2008a) reported a similar gap in crop water requirements in a commercial mango orchard in the Brazilian northeast region.

Considering the pivot areas and the yield for each of them, CWP_L in terms of grains ranged from 7.2 to 10.7 t ha^{-1}, while for silage it was between 31.2 and 48.2 t ha^{-1}. CWP_{ET} showed good return, with values ranging from 1.4 to 2.8 kg m^{-3} for grains and between 8.8 and 14.1 kg m^{-3} for silage. The best values for grains were found with some WD values, which is an indication that CWP_{ET} increases with some degree of water stress. On average, there were no significant differences when the CWP was based on ET or I in the case of grain; however, for silage CWP_I was 86% of CWP_{ET}, indicating much room for water management improvements in this latter case. The main reason for this might be the use of inappropriate Kc for silage.

In the southeast Brazilian region, CWP_{ET} corn values for grains are higher than those for wheat and rice, reported by Zwart and Bastiaanssen (2004), which were inside the range from 0.5 to 1.5 kg m^{-3}. With the availability of prices for grains, the corresponding $CWP\$_{ET}$ and $CWP\$_I$ were respectively from 0.34–0.68 US\$ m^{-3} to 0.41–0.63 US\$ m^{-3}. Sakthivadivel et al. (1999) reported typical $CWP\$_{ET}$ values for arable crops between 0.10 and 0.20 US\$ m^{-3}, lower than those found for corn crop in southeast Brazil.

TABLE 4.8 Irrigation Performance Indicators of Corn Crop for Grain (a) and for Silage (b). Area; Growing Season (GS); Volume of Water Applied through Irrigation (V_I); Precipitation (P), Relative Evapotranspiration (R_{ET}); Water Deficit (WD); Relative Water Supply (R_{WS}); 3 Crop Water Productivity Based on Evapotranspiration (CWP_{ET}) and on Irrigation (CWP_I); and Productivity Crop Water Productivity Based on Land CWP_L

Pivots	Area (ha)	GS (days)	V_I (mm)	P (mm)	R_{ET} (−)	WD (mm)	R_{WS} (−)	CWP_L (t ha^{-1})	CWP_{ET} (kg m^{-3})	CWP_I (kg m^{-3})
(a) Irrigation performance indicators for grain										
G1	108.0	169	436.9	240.0	0.98	11.8	1.3	7.2	1.4	1.7
G2	74.0	155	498.2	48.0	0.96	20.0	1.1	10.3	2.1	2.1
G3	108.0	168	463.7	242.0	0.93	36.5	1.4	8.0	1.6	1.7
G4	91.0	155	495.6	65.0	0.78	110.2	1.1	8.9	2.3	1.8
G5	100.0	158	405.9	160.0	0.79	100.4	1.2	10.7	2.8	2.6
Mean	96.2	161	460.1	151.0	0.89	55.8	1.2	9.0	2.0	2.0
(b) Irrigation performance indicators for silage										
S1	118.0	123	454.9	57.0	0.99	2.6	1.3	33.3	8.8	7.3
S2	77.1	129	443.2	77.0	0.90	40.7	1.3	31.2	8.9	7.0
S3	75.0	124	442.7	77.0	0.95	20.5	1.4	36.5	10.3	8.3
S4	157.2	111	358.6	95.0	0.99	2.6	1.4	46.5	14.1	13.0
S5	105.5	114	361.89	52.0	1.00	0.0	1.2	48.2	13.8	13.3
Mean	96.2	120	412.1	71.6	0.97	13.3	1.3	39.1	11.1	9.5

Note: G, grains; S, silage.

Considering the importance for human and animal feed, mainly in rural environments, the water usage for corn crop should be stimulated with sustainable water management in areas with climatic aptitude in the southeast Brazil.

4.4.5 Rainfed Crop

For WP calculations on rainfed corn crop in the Brazilian central–west region (see Figure 4.2), the reflectances for bands 1 and 2 were extracted from MODIS product MOD13Q1. These images covered an interval of 16 days and used together with a net of 32 agrometeorological stations. The weather data were scaled up to the same MODIS time scale in the state of Mato Grosso.

The models were applied considering a cropland mask, which separates corn crop from other surface types during the growing season from mid-February to the end of July in 2012. The main corn-growing regions, north, southeast, and

northeast, were extracted and analyzed in terms of WP. The spatial variations and the average daily pixel values during the crop stages of ET, BIO, and WP within these growing regions are shown in Figure 4.12.

Considering ET values, the GS totals were, in average, 218, 297, and 210 mm GS^{-1} for the north, southeast, and northeast regions, respectively. With corresponding BIOs of 10.6, 14.0, and 9.5 t ha^{-1} GS^{-1}, these resulted in WP values for rainfed corn crop of 3.9, 4.0, and 3.7 kg m^{-3}.

According to the average daily pixel values during the growing seasons, peaks were observed, around DOY 112 (second half of April) with ET and BIO values above 2.0 mm day^{-1} and 100 kg ha day^{-1}, respectively, resulting in WP values close to 5.0 kg m^{-3}. In all cases, the maximum WP occurred during the crop stages, corresponding to the transition from blooming to grain filling. The southeast growing region presented the highest values for ET, BIO, and WP, while the northeast presented the lowest ones. Throughout the water indicator represented by the ratio of

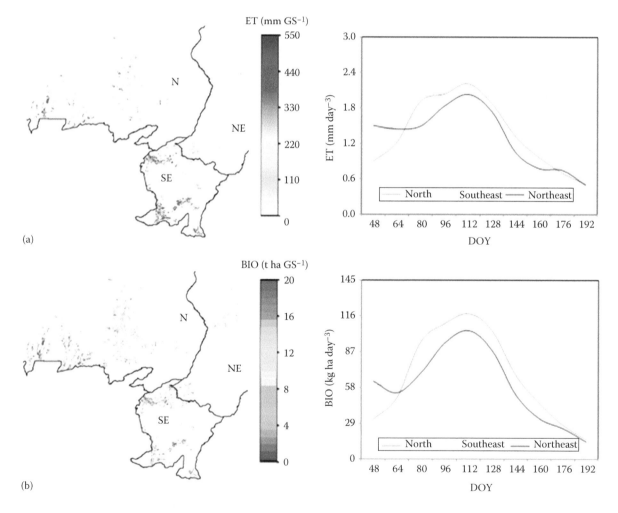

FIGURE 4.12 Spatial variations and the average daily pixel values of water productivity parameters during the corn crop stages and for a growing season in the growing regions north (N), northeast (NE), and southeast (SE) of the Mato Grosso state, Brazilian central–west region: (a) actual evapotranspiration (ET); (b) biomass production (BIO).   (*Continued*)

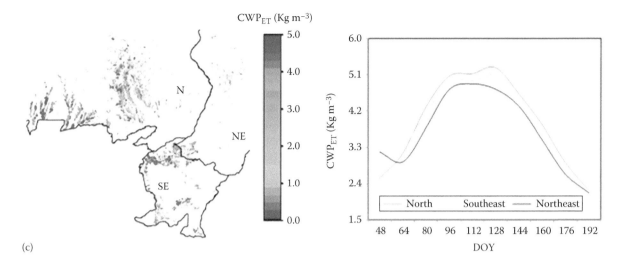

CWP$_{ET}$ (Kg m^{-3})

North   Southeast   —— Northeast

(c)

FIGURE 4.12 (*Continued*) Spatial variations and the average daily pixel values of water productivity parameters during the corn crop stages and for a growing season in the growing regions north (N), northeast (NE), and southeast (SE) of the Mato Grosso state, Brazilian central–west region: (c) Crop water productivity (CWP).

precipitation (P) to ET, it was noted that indeed irrigation was not necessary to meet the crop water requirements, with low P/ET values (0.08–0.22), indicating that P attended only 8%–22% of the water consumption, taking place only at the harvest time, when high humidity is not desirable.

Application of an HI value of 0.48 (Claverie et al., 2012) results in a CWP$_{ET}$ value around 2.0 kg m^{-3} for the rainfed corn crop in the central–west region of Brazil. In the semiarid region of Inner Mongolia, CWP$_{ET}$ values of 1.1–1.3 kg m^{-3} for oats, 1.5–2.6 kg m^{-3} for sunflower, 0.5–1.1 kg m^{-3} for legumes, and 3.1–4.4 kg m^{-3} for potato were reported (Yuan et al., 2013). For oil seed crop, under the semiarid conditions of India, CWP$_{ET}$ ranged from 1.9 to 2.3 kg m^{-3} (Adak et al., 2013). For wheat, analyses of CWP$_{ET}$ around the world resulted in a range of 0.5–1.4 kg m^{-3} (Zwart et al., 2010). Considering the importance for rural environments and the high CWP$_{ET}$ values in comparison with other crops from other ecosystems, the rainfed corn crop is important in Mato Grosso, mainly in the southeast growing region of the state, which presented the best rainfall use efficiency.

4.5 Conclusions and Policy Implications

The models SAFER, Monteith, and SUREAL have been applied to demonstrate the feasibility of computing water variables on large scales. Surface albedo, surface temperature, and NDVI from satellite images, together with agrometeorological data, are the input parameters for the modeling. This combination allowed the large-scale WP assessments from different agroecosystems in Brazil. The analyses may contribute to better understandings of the biophysical parameter dynamics, important for appraisal of the land-use change impacts, key information when Federal and municipal governments plan expansions of the irrigated areas with rational criteria.

It was demonstrated that the large-scale WP components can be analyzed from instantaneous satellite measurements in the visible and infrared bands. This was possible by modeling the ratio of the actual to the reference ET at a satellite overpass time in conjunction with the availability of daily weather variables. These approaches provide large-scale temporal information on growth rates as well as plant responses to dynamic weather under irrigation and rainfed conditions, being useful for monitoring vegetation and water parameters. In the specific case of Brazil, this is particularly relevant because there have been rainfall shortages in several regions and the analyses in this study, therefore, become highly relevant to the application of concepts of evidence-based decision making on water resources use.

The available tools tested here can be operationally implemented to monitor the intensification of agriculture and the adverse impact on downstream water users in changing environments. They can be used with more confidence in other parts of the world, with probably only the need of adjustments in the original equation regression coefficients.

References

Adak, T., Kumar, G., Chakravarty, N.V.K., Katiyar, R.K., and Deshmukh, P.S. 2013. Biomass and biomass water use efficiency in oilseed crop (*Brassica Jnceae* L.) under semi-arid microenvironments. *Biomass and Bioenergy*, 51, 154–162.

Ahamed, T., Tian, L., Zhang, Y., and Ting, K.C. 2011. A review of remote sensing methods for biomass feedstock production. *Biomass and Bioenergy*, 35, 2455–2469.

Allen, R.G., Pereira, L.S., Raes, D., and Smith, M. *Crop Evapotranspiration: Guidelines for Computing Crop Water Requirements*. Food and Agriculture Organization of the United Nations: Rome, Italy, 1998.

Allen, R.G., Tasumi, M., Morse, A., Trezza, R., Wright, J.L., Bastiaanssen, W.G.M., Kramber, W., Lorite, I., and Robison, C.W. 2007. Satellite-based energy balance for mapping evapotranspiration with internalized calibration (METRIC)—Applications. *Journal of Irrigation and Drainage Engineering*, 133, 395–406.

Amir, J. and Sinclair, T.R. 1991. A model of temperature and solar radiation effects on spring wheat growth and yield. *Field Crops Research*, 28, 47–58.

Araujo, F., Williams, L.E., and Mattews, M.A. 1995. A comparative study of young "Thompson Seedless" grapevines (Vitis vinifera L.) under drip and furrow irrigation, II. Growth, water use efficiency, and nitrogen partioning. *Scientia Horticulturae*, 60, 251–265.

Baccine, A., Friedl, M.A., Woodcrock, C.E., and Warbington, R. 2004. Forest biomass estimation over regional scales using multisource data. *Geophysical Research Letters*, 31, 1–4.

Bastiaanssen, W.G.M. and Ali, S. 2003. A new crop yield forecasting model based on satellite measurements applied across the Indus Basin, Pakistan. *Agriculture, Ecosystems & Environment*, 94, 32–340.

Bastiaansssen, W.G.M., Brito, R.A.L., Bos, M.G., Souza, R.A., Cavalcanti, E.B., and Bakker, M.M. 2001. Low cost satellite data for monthly irrigation performance monitoring: Benchmarks from Nilo Coelho, Brazil. *Irrigation and Drainage Systems*, 15, 53–79.

Bastiaanssen, W.G.M., Menenti, M., Feddes, R.A., Roerink, G.J., and Holtslag, A.A.M. 1998. A remote sensing surface energy balance algorithm for land (SEBAL) 1. Formulation. *Journal of Hydrology*, 212–213, 198–212.

Bondeau, A., Smith, P.C., Zaehle, S., Schaphoff, S., Lucht, W., Cramer, W., Gerten, D. et al. 2007. Modelling the role of agriculture for the 20th century global terrestrial carbon balance. *Global Change Biology*, 13, 679–706.

Boogaard, H.L., Eerens, H., Supit, I., Diepen, C.A.v., Piccard, I., and Kempeneers, P., 2002. METAMP, Methodology Assessment of MARS Predictions. Description of the MARS crop yield forecasting system (MCYFS), Joint Research Centre, Study contract number 19226-2002-02-F1FED ISP NL, Report 1/3.

Bos, M.G., Burton, D.J., and Molden, D.J. 2005. *Irrigation and Drainage Performance Assessment: Practical Guidelines*. CABI Publishing: Cambridge, MA, 158pp.

Cai, X., Mckinney, C., and Lasdon, S. 2002. A framework for sustainable analysis in water resources management and application to the Syr Darya Basin. *Water Resources Research*, 38, 21–1/14.

Ceotto, E. and Castelli, F. 2002. Radiation use efficiency in flue-cured tobacco (*Nicotiana tabacum* L.): Response to nitrogen supply, climate variability and sink limitations. *Field Crops Research*, 74, 117–130.

Ceschia, E., Beziat, P., Dejoux, J.F., Aubinet, M., Bernhofer, C., Bodson, B., Carrara, A., Cellier, P. et al. 2010. Management effects on net ecosystem carbon and GHG budgets at European crop sites. *Agriculture, Ecosystems & Environment*, 139, 363–383.

Claverie, M., Demarez, V., Duchemin, B., Hagolle, O., Ducrot, D., Marais-Sicre, C., Dejoux, J.-F. et al. 2012. Maize and sunflower biomass estimation in southwest France using spatial and temporal resolution remote sensing data. *Remote Sensing of Environment*, 124, 884–857.

Cleugh, H.A., Leuning, R., Mu, Q., and Running, S.W. 2007. Regional evaporation estimates from flux tower and MODIS satellite data. *Remote Sensing of Environment*, 106, 285–304.

Daughtry, C.S.T., Gallo, K.P., Goward, S.N., Prince, S.D., and Kustas, W.P. 1992. Spectral estimates of absorbed radiation and phytomass production in corn and soybean canopies. *Remote Sensing of Environment*, 39, 141–152.

DeJonge, K.C., Ascoughi, J.C., Andales, A.A., Hansen, N.C., Garcia, L.A., and Arabi, M. 2012. Improving evapotranspiration simulations in the CERES-Maize model under limited irrigation. *Agricultural Water Management*, 115, 92–103.

Ding, R., Kang, S., Li, F., Zhang, Y., and Tong, L. 2013. Evapotranspiration measurement and estimation using modified Priestly-Taylor model in an irrigated maize field with mulching. *Agricultural and Forest Meteorology*, 168, 140–148.

Droogers, P., Kite, G.W., and Murray-Rust, H. 2000. Use of simulation models to evaluate irrigation performance including water productivity, risk and system analysis. *Irrigation Science*, 19, 139–145.

Eitzinger, J., Trnka, M., Hosch, J., Zalud, Z., and Dubrovsky, M. 2004. Comparison of CERES, WOFOST and SWAP models in simulating soil water content during growing season under different soil conditions. *Ecological Modelling*, 171, 223–246.

Field, C.B., Randerson, J.T., and Malmström, C.M. 1995. Global net primary production: Combining ecology and remote sensing. *Remote Sensing of Environment*, 51(1), 74–88.

Gervois, S., de Noblet-Ducoudre, N., and Viovy, N. 2004. Including croplands in a global biosphere model: Methodology and evaluation at specific sites. *Earth Interactions*, 8, 1–25.

Gosse, G., Varlet-Grancher, C., Bonhomme, R., Chartier, M., Allirand, J.M., and Lemaire, G. 1986. Maximum dry matter production and solar radiation intercepted by a canopy. *Agronomie*, 6, 47–56.

Gourbesville, P. 2008. Challenges for integrated water resources management. *Physics and Chemistry of the Earth*, 33, 284–289.

Gower, S.T., Kucharik, C.J., and Norman, J.M. 1999. Direct and indirect estimation of leaf area index, fAPAR, and net primary production of terrestrial ecosystems. *Remote Sensing of Environment*, 70, 29–51.

Groten, S.M.E. 1993. NDVI-crop monitoring and early yield assessment of Burkina Faso. *International Journal of Remote Sensing*, 14, 1495–1515.

Hamar, D., Ferencz, C., Lichtenberg, J., Tarcsai, G., and Frencz-Arkos, I. 1996. Yield estimation for corn and wheat in the Hungarian Great Plain using Landsat MSS data. *International Journal of Remote Sensing*, 17, 1689–1699.

Hatfield, H.L., Thomas, J.S., and John, H.P. 2001. Managing soil to achieve greater water use efficiency: A review. *Agronomy Journal*, 93, 271–280.

Hernandez, F.B.T, Teixeira, A.H. de C., Neale, C.M.U., and Taghvaeian, S. 2004. Determining large scale actual evapotranspiration using agro-meteorological and remote sensing data in the Northwest of Sao Paulo State, Brazil. *Acta Horticulturae*, 1038, 263–270.

Jairmain, C., Klaasse, A., Bastiaanssen, W.G.M., and Roux, A.S. 2007. Remote sensing tools for water use efficiency of grapes in the Winelands region, Western Cape, *Proceedings of the 13th Sanciahs Symposium*, Capetown, South Africa, September 6–7.

Kamble, B., Kilic A., and Hubard, K. 2013. Estimating crop coefficients using remote sensing-based vegetation index. *Remote Sensing*, 5, 1588–1602.

Kumar, M. and Monteith, J.L. 1981. Remote sensing of crop growth. In: Smith, H. (Ed.), *Plants and the Daylight Spectrum*. Academic Press: London, U.K., pp. 133–144.

Lobell, D.B., Asner, G.P., Ortiz-Monasterio, J.I., and Benning, T.L. 2003. Remote sensing of regional crop production in the Yaqui Valley, Mexico: Estimates and uncertainties. *Agriculture, Ecosystems & Environment*, 94, 205–220.

Lu, D. 2005. Aboveground biomass estimation using Landsat TM data in the Brazilian Amazon basin. *International Journal of Remote Sensing*, 26, 2509–2525.

Manjunath, K.R. and Potdar, M.B. 2002. Large area operational wheat yield model development and validation based on spectral and meteorological data. *International Journal of Remote Sensing*, 23, 3023–3038.

Maselli, F. and Rembold, F. 2001. Analysis of GAC NDVI data for cropland identification and yield forecasting in Mediterranean African countries. *Photogrammetric Engineering & Remote Sensing*, 67, 593–602.

Miralles, D.G., Holmes, T.R.H., De Jeu, R.A.M., Gash, J.H., Meesters, A.G.C.A., and Dolman, A.J. 2011. Global land-surface evaporation estimated from satellite-based observations. *Hydrological and Earth System Science*, 15, 453–469.

Mo, X., Liu, S., Lin, Z., and Guo, R. 2009. Regional crop yield, water consumption and water use efficiency and their responses to climate change in the North China. *Agriculture, Ecosystems & Environment*, 134, 67–78.

Molden, D., Oweis, T., Steduto, P., Kijne, J.W., Hanjra, M.A., and Bindraban, P.S. 2007. Pathways for increasing agricultural water productivity. In: Chapter 7 in *Water for Food, Water for Life: A Comprehensive Assessment of Water Management in Agriculture*, International Water Management Institute: London, Earthscan, Colombo.

Monteith, J.L. 1972. Solar radiation and productivity in tropical ecosystems. *Journal of Applied Ecology*, 9, 747–766.

Monteith, J.L. and Scott, R.K. 1982. Weather and yield variation of crops. In: Blaxter, K. and Fowden, L. (Eds.), *Food, Nutrition and Climate*. Applied Science: Englewood Cliffs, NJ, pp. 127–149.

Muchow, R.C. 1990. Effect of high temperature on grain growth in maize. *Field Crops Research*, 23, 145–158.

Nagler, P.L., Glenn, E.P., Nguyen, U., Scott, R.L., and Doody, T. 2013. Estimating riparian and agricultural actual evapotranspiration by reference evapotranspiration and MODIS enhanced vegetation index. *Remote Sensing*, 5, 3849–3871.

Osborne, T.M., Lawrence, D.M., Challinor, A.J., Slingo, J.M., and Wheeler, T.R. 2007. Development and assessment of a coupled crop-climate model. *Global Change Biology*, 13, 169–183.

Oweis, T. and Hachum, A. 2006. Water harvesting and supplemental irrigation for improved water productivity of dry farming systems in West Asia and North Africa. *Agricultural Water Management*, 80, 57–73.

Peacock, W.L., Rolston, D.E., Aljibury, F.K., and Rauschkolb, R.S. 1977. Evaluating drip, flood, and sprinkler irrigation of wine grapes. *American Journal of Enology and Viticulture*, 28, 193–195.

Pôças, I., Cunha, M., Pereira, L.S., and Allen, R.G. 2013. Using remote sensing energy balance and evapotranspiration to characterize montane landscape vegetation with focus on grass and pasture lands. *International Journal of Applied Earth Observation and Geoinformation*, 21, 159–172.

Pirmoradian, N. and Sepaskhah, A.R. 2005. A very simple model for yield prediction of rice under different water and nitrogen applications. *Biosystems Engineering*, 93(1), 25–34.

Prince, S.D. 1990. High temporal frequency remote sensing of primary production using NOAA/AVHRR. In: Steven, M.D. and Clark, J.A. (Eds.), *Applications of Remote Sensing in Agriculture*. Butterworths: London, U.K., pp. 169–183.

Rasmussen, M.S. 1992. Assessment of millet yields and production in northern Burkina Faso using integrated NDVI from the AVHRR. *International Journal of Remote Sensing*, 13, 3431–3442.

Renaut, D., Hemakumara, M., and Molden, D. 2001. Importance of water consumption by perennial vegetation in irrigated areas of the humid tropics: Evidence from Sri Lanka. *Agricultural Water Management*, 46(3), 215–230.

Rockstrom, J. and Barron, J. 2007. Water productivity in rainfed systems: Overview of challenges and analysis of opportunities in water scarcity prone savannahs. *Irrigation Science*, 25, 299–311.

Russell, G., Jarvis, P.G., and Monteith, J.L. 1989. Absorption of radiation by canopies and stand growth. In: Russell, G., Marshall, B., and Jarvis, P.G. (Eds.), *Plant Canopies: Their Growth, Form and Function*. Cambridge University Press: Cambridge, U.K., pp. 21–40.

Sakthivadivel, R., de Fraiture, C., Molden, D.J., Perry, C., and Kloezen, W. 1999. Indicators of land and water productivity in irrigated agriculture. *International Journal of Water Resource Development*, 15, 161–180.

Sharma, P.K., Chaurasia, R., and Mahey, R.K. 2000. Wheat production forecasts using remote sensing and other techniques-experience of Punjab State. *Indian Journal of Agricultural Econominics*, 55(2), 68–80.

Sheehy, J.E., Peng, S., Doberman, A., and Mitchell, P.L. 2004. Fantastic yields in the system of rice intensification: Fact or fallacy? *Field Crops Research*, 88, 1–8.

Shi, X., Elmore, A., Li, X., Gorence, N.J., Jin, H., and Zhang, X. 2008. Using spatial information technologies to select sites for biomass power plants: A case study in Guangdong, China. *Biomass and Bioenergy*, 32, 35–43.

Sinclair, T.R. and Muchow, R.C. 1999. Radiation use efficiency. *Advances in Agronomy*, 65, 215–265.

Singh, P. and Sri Rama, Y.V. 1989. Influence of water deficit on transpiration and radiation use efficiency of chickpea (*Cicer arietinum* L.). *Agricultural and Forest Meteorology*, 48, 317–330.

Smakthin, V., Revenga, C., and Doll, P. 2004. Taking into account environmental water requirements in global-scale water resources assessments. IWMI: Colombo, Sri Lanka, 24pp. (Comprehensive Assessment of Water Management in Agriculture Research Report 2).

Spaeth, S.C., Sinclair, T.R., Ohunuma, T.K., and Onno, S. 1987. Temperature, radiation and duration dependence of light soybean yields: Measurements and simulation. *Field Crops Research*, 16, 297–307.

Srinivas, K., Shikhamany, S.D., and Reddy, N.N. 1999. Yield and water-use of 'Änab-e Shahi' grape (*Vitis vinifera*) vines under drip and basin irrigation. *Indian Journal of Agricultural Science*, 69, 21–23.

Tang, Q., Rosemberg, E.A., and Letenmaier, D.P. 2009. Use of satellite data to assess the impacts of irrigation withdrawals on Upper Klamath Lake, Oregon. *Hydrological and Earth System Science*, 13, 617–627.

Teixeira, A.H. de C. 2009. *Water Productivity Assessments from Field to Large Scale: A Case Study in the Brazilian Semi-Arid Region*. LAP Lambert Academic Publishing: Saarbrücken, Germany, 226pp.

Teixeira, A.H. de C. 2010. Determining regional actual evapotranspiration of irrigated and natural vegetation in the São Francisco river basin (Brazil) using remote sensing and Penman-Monteith equation. *Remote Sensing*, 2, 1287–1319.

Teixeira, A.H. de C. 2012. Modelling water productivity components in the Low-Middle São Francisco River basin, Brazil. In: *Sustainable Water Management in the Tropics and Subtropics and Case Studies in Brazil*, 1st edn. University of Kassel: Kassel, Germany, Vol. 3, pp. 1077–1100.

Teixeira, A.H. de C. and Bassoi, L.H. 2009. Crop water productivity in semi-arid regions: From field to large scales. *Annals of Arid Zone*, 48, 1–13.

Teixeira, A.H. de C., Bastiaanssen, W.G.M., Ahmad, M.D., and Bos, M.G. 2008b. Analysis of energy fluxes and vegetation-atmosphere parameters in irrigated and natural ecosystems of semi-arid Brazil. *Journal of Hydrology*, 362, 110–127.

Teixeira, A.H. de C., Bastiaanssen, W.G.M, Ahmad, M.D., and Bos, M.G. 2009a. Reviewing SEBAL input parameters for assessing evapotranspiration and water productivity for the Low-Middle São Francisco River basin, Brazil Part A: Calibration and validation. *Agricultural and Forest Meteorology*, 149, 462–476.

Teixeira, A.H. de C., Bastiaanssen, W.G.M., Ahmad, M.D., and Bos, M.G. 2009b. Reviewing SEBAL input parameters for assessing evapotranspiration and water productivity for the Low-Middle São Francisco River basin, Brazil Part B: Application to the large scale. *Agricultural and Forest Meteorology*, 149, 477–490.

Teixeira, A.H. de C., Bastiaanssen, W.G.M., and Bassoi, L.H. 2007. Crop water parameters of irrigated wine and table grapes to support water productivity analysis in Sao Francisco River basin, Brazil. *Agricultural Water Management*, 94, 31–42.

Teixeira, A.H. de C., Bastiaanssen, W.G.M., Moura, M.S.B., Soares, J.M., Ahmad, M.D., and Bos, M.G. 2008a. Energy and water balance measurements for water productivity analysis in irrigated mango trees, Northeast Brazil. *Agricultural and Forest Meteorology*, 148, 1524–1537.

Teixeira, A.H. de C., Hernandez, F.B.T., Andrade, R.G., Leivas, J.F., Victoria, D. de C., and Bolfe, E.L. 2014b. Irrigation performance assessments for corn crop with Landsat images in the São Paulo state, Brazil. In: II International INOVAGRI meeting, 2014, Fortaleza. Proceeding II International INOVAGRI meeting. INOVAGRI: Fortaleza, Brazil, pp. 739–748.

Teixeira, A.H. de C., Hernandez, F.B.T., Lopes, H.L., Scherer-Warren, M., and Bassoi, L.H. 2014a. A comparative study of techniques for modeling the spatiotemporal distribution of heat and moisture fluxes in different agroecosystems in Brazil. In: George G.P. (Ed.) (Org.), *Remote Sensing of Energy Fluxes and Soil Moisture Content*, 1st edn. CRC Group, Taylor & Francis: Boca Raton, FL, pp. 169–191.

Teixeira, A.H. de C., Scherer-Warren, M., Hernandez, F.B.T., Andrade, R.G., and Leivas, J.F. 2013. Large-scale water productivity assessments with MODIS images in a changing semi-arid environment: A Brazilian case study. *Remote Sensing*, 5, 55783–5804.

Tesfaye, K., Walker, S., and Tsubo, M. 2006. Radiation interception and radiation use efficiency of three grain legumes under water deficit conditions in a semi-arid environment. *European Journal of Agronomy*, 25, 60–70.

Valiente, J.A., Nunez, M., Lopez-Baeza, E., and Moreno, J.F. 1995. Narrow-band to broad-band conversion for Meteosat visible channel and broad-band albedo using both AVHRR-1 and -2 channels. *International Journal of Remote Sensing*, 16, 1147–1166.

Wang, X., Ma, M., Huang, G., Veroustraete, F., Zhang, Z., Song, Y., and Tan, J. 2012. Vegetation primary production estimation at maize and alpine meadow over the Heihe River Basin, China. *International Journal of Applied Earth Observation Geoinformation*, 17, 94–101.

Wu, C., Munger, J.W., Niu, Z., and Kuanga, D. 2010. Comparison of multiple models for estimating gross primary production using MODIS and eddy covariance data in Havard Forest. *Remote Sensing of Environment*, 114, 2925–2939.

Wylie, B., Johnson, D., Laca, E., Saliendra, N., Gilmanov, T., Reed, B., Tieszen, L., Bruce B., and Worstell, B. 2003. Calibration of remotely sensed, coarse resolution NDVI to CO_2 fluxes in a sage-brush-steppe ecosystem. *Remote Sensing of Environment*, 85, 243–255.

Yan, H.M., Fu, Y.L., Xiao, X.M., Huang, H.Q., He, H.I., and Ediger, L. 2009. Modeling gross primary productivity for winter wheat-maize double cropping system using MODIS time series and CO_2 eddy flux tower data. *Agriculture, Ecosystems & Environment*, 129, 391–400.

Yuan, M., Zhang, L., Gou, F., Su, Z., Spiertz, J.H.J., and van der Werf, W. 2013. Assessment of crop growth and water productivity for five C3 species in the semi-arid Inner Mongolia. *Agricultural Water Management*, 122, 28–38.

Zhao, M., Heinsch, F.A., Nemani, R.R., and Running, S.W. 2005. Improving of the MODIS terrestrial gross and net primary production global dataset. *Remote Sensing of Environment*, 95, 164–176.

Zwart, S.J. and Bastiaanssen, W.G.M. 2004. Review of measured crop water productivity values for irrigated wheat, rice, cotton and maize. *Agricultural Water Management*, 69, 115–153.

Zwart, S.J., Bastiaanssen, W.G.M., de Fraiture, F., and Molden, D.J. 2010. WATPRO: A remote sensing based model for mapping water productivity of wheat. *Agricultural Water Management*, 97, 1628–1636.

III

Floods

Flood Monitoring Using the Integration of Remote Sensing and Complementary Techniques

Allan S. Arnesen
SABESP

Frederico T. Genofre

Marcelo P. Curtarelli
*National Institute for
Space Research*

Matheus Z. Francisco
UNIVALI

Acronyms and Definitions

1D	One-dimensional
2D	Two-dimensional
ALOS	Advanced land observing satellite
AVHRR	Advanced very high resolution radiometer
COSMO-SkyMed	Constellation of small satellites for Mediterranean basin observation
dB	Decibel
DEM	Digital elevation model
DJF	December–January–February
DMSP	Defense Meteorological Satellite Program
ETM+	Enhanced Thematic Mapper Plus
GIS	Geographic information systems
GOES	Geostationary operational environmental satellite
GPM	Global Precipitation Measurement
HMS	Hydrologic modeling system
JERS-1	Japan Earth resources satellite
MGB-IPH	*Modelo de Grandes Bacias–Instituto de Pesquisas Hidráulicas*
MODIS	Moderate-resolution imaging spectroradiometer
NCEP	National Centers for Environmental Prediction
NOAA	National Oceanic and Atmospheric Administration
NRT	Near real time
PALSAR	Phased array L-band SAR
POES	Polar operational environmental satellites
QPE	Quantitative precipitation estimates
QPF	Quantitative precipitation forecasts
RMSE	Root-mean-square error
SAR	Synthetic aperture radar
SCS	Soil conservation service
SRTM	Shuttle Radar Topography Mission
SSM/I	Special sensor microwave imager
TM	Thematic Mapper
TMPA	TRMM multisatellite precipitation analysis
TRMM	Tropical Rainfall Measuring Mission
ERS-1	European Remote Sensing Satellite
MAM	March–April–May
JJA	June–July–August
FF	Flooded forest
NFF	non Flooded forest
WS	wet/rough soil
EM	emergent macrophyte
ROW	rough open water
DS	dry smooth soil
FM	floating macrophyte
SOW	smooth open water
GFDS	global flood detection system
GFMS	global flood monitoring system

5.1 Introduction

Flood is defined as an overflowing by water of the normal confines of a stream or other water body or accumulation of water by drainage over areas that are not normally submerged (WMO, 2011). It occurs when the flow capacity of the riverbed is insufficient and the water overflows occupying adjacent areas. The frequency and magnitude of floods vary in space and time, depending on the river basin characteristics (e.g., size, land coverage, and topography) and climate conditions.

Floods are one of the greatest challenges to weather prediction because of its devastating effects in the world. Moreover, climate changes are making the flood occurrence more frequent in several countries of the world. Using stream flow measurements and numerical simulations, the study of Milly et al. (2002) verified, in a global approach, that the great river floods had its frequency increased in the twentieth century. Anyway, these disasters may occur due to natural behaviors or consequence of urbanization process.

When long terms precipitation occurs in basins characterized by a high degree of urbanization, the flooding can be extensive, resulting in a great amount of damage and loss of life (Jeyaseelan, 2005). It has been shown that progressive urbanization increases the risk of inundation (Nirupama and Simonovic, 2007). Zhang et al. (2009) analyzed the impact of rapid urban expansion of an important economic zone in China, the Pearl River Delta. They pinpointed that the urbanization in this region was one of the major facts responsible for reducing the rivers capability of buffering floods.

In urban areas, the growth of flood frequency at inner rivers may occur as a consequence of suppression of minor streams and ponds, removal of vegetation and spreading of soil sealing, increment in runoff together with restriction of infiltration and evaporation volumes, increase stream sedimentation, and peak discharges (Zhang et al., 2009; Du et al., 2012).

Although flooding is a major hazard both in urban and rural areas, high-populated areas have most severe economic impacts by this natural disaster where the value of assets at risk is greater (Neal et al., 2009). In the European Union, more than 40 billion dollars per year is spent on flood mitigation, recovery, and compensation, and another 3 billion per year is spent on flood defense structures (van Ree et al., 2011).

The socioeconomic impacts caused by river floods are increasing once there is a lack of flood risk management strategies to reduce flood damages. To minimize flood impacts over socioeconomic activities, stakeholders need reliable information to support them on evaluating flood risk, organizing aid distribution to most severely affected areas, planning mitigation alternatives, and assessing damages. Surface and hydrological observational data are required to support these risk assessments (Tralli et al., 2005).

Remote sensing and geographic information systems (GIS) are powerful tools for delineation of flood zones and preparation of flood hazard maps for the vulnerable areas (Sanyal and Lu, 2004). However, one of the major limitations of remote sensing flood-analyzing approaches is the scarce availability of high-temporal-resolution data, either from passive or active sensors.

The remote sensing instruments can be classified according to the range of the electromagnetic spectrum that they operate. Optical or passive sensors act at the visible wavelengths of the spectrum, depending on the external illumination (i.e., sunlight). The major advantage of this type of data is the easiness of image visual interpretation. However, they are subject to atmospheric conditions, once clouds and shadows cause losses of information.

On the other hand, active or synthetic aperture radar (SAR) instruments overcome the weather condition restriction, because they operate at longer wavelengths (microwave radiation) than optical data allowing the acquisition of useful data even on cloudy days. Besides that, SAR data are independent of solar illumination conditions. Considering that flood events mostly occur on rainy days, these are fundamental characteristics of the remote sensing data used for flood extent delimitation.

Furthermore, new high temporal SAR sensors can allow flood monitoring in all phases of an extreme event. The constellation of small satellites for Mediterranean basin observation (COSMO-SkyMed) mission is the best example of high-temporal-resolution SAR data. It is an Italian mission that offers meter-level spatial resolution, daily acquired X-band radar images for disaster management.

Even though the SAR sensors provide important information for flood monitoring, the integration of other data and tools is fundamental for obtaining accurate flood extent mapping. Optical and topographic data, hydrological models, and forecasting systems can also be explored to better analyze flood events.

In this chapter, we review recent advances on methodological alternatives that combine remote sensing (SAR and ancillary data), GIS, and hydrological model tools for providing flood information that help stakeholders to develop contingency plans and flood forecasting. After an overview of the remote sensing concepts and importance, Section 5.2.2 presents the new SAR data and algorithms used to map flood extent. Section 5.2.3 refers to the recent applications that used the integration of remote sensing–derived information and flood models. Section 5.2.3 presents the main steps of flood forecasting and some applications using land cover and meteorological remote sensing data to minimize flood socioeconomic impacts. Finally, the conclusions gather relevant aspects highlighted during the chapter.

5.2 Remote Sensing in Flood Studies

Remote sensing provides potential information to analyze flood events to support flood management and controlling. One challenge on using remote sensing images for flood studies is the atmospheric conditions. Optical sensors are negatively influenced by weather conditions, and when flood events occur, as a consequence of rainy periods, severe cloudiness leads to completely uselessness of the data.

Active remote sensors provide data that are nearly independent of weather conditions, since the microwave radiation are about 200,000 times longer than the visible wavelengths of optical sensors. For this reason, the interaction between the atmospheric micrometric particles (and water droplets) and the microwave radiation is negligible, allowing the acquisition of useful images even with complete cloud coverage over the area of interest.

This aspect, by itself, justifies the use of radar imagery for flood studies due to the possibility of improved temporal resolution during a specific event. However, the following section presents other considerable advantages of SAR data for flood mapping and monitoring.

5.2.1 Flood Mapping with SAR and Ancillary Data

The following subsections present the SAR data qualities for flood mapping, some relevant studies of different methodologies for flood monitoring, and discuss about the accuracies, errors, and uncertainties of these approaches.

5.2.1.1 SAR Favorable Characteristics for Flood Delimitation

Besides the advantage of the nearly all-weather capability of acquiring useful data, a considerable benefit of SAR data for mapping flood extent is the interaction of the microwave radiation with flooded vegetation. Although interpreting SAR backscatter from flooded vegetation is not a straightforward task and statistical knowledge is required to obtain accurate results, SAR data allow the identification of water below canopies of forest and aquatic vegetation.

The interaction between the microwave radiation and the targets is a special benefit of SAR data to map flood extent. The radiation penetration into flooded forest (FF) and macrophytes favors the identification of the total flood extent (Arnesen et al., 2013). Three scattering mechanisms are dominant in floodplain areas: double bounce, volumetric (or canopy), and surface (or specular) (Figure 5.1).

The double bounce scattering occurs when a wave is incident on the inside corner of two flat surfaces that are joined at 90° and returns back to the radar resulting in high backscattering values (Woodhouse, 2006). The volumetric backscattering occurs when

the radiation interacts with several elements of the target being scattered in multiple directions (Henderson and Lewis, 1998). The surface scattering occurs when the radiation reaches a simple surface (open water, for instance) and, in this case, the return signal depends on its roughness and SAR incidence angle.

The predominance of backscattering mechanisms depends on the SAR wavelength. For FF areas, for example, the longer the wavelength, the higher will be the radiation penetration in the canopy, and consequently, high backscattering values are expected due to double bounce occurrence (Freeman and Durden, 1998). On the other hand, shorter wavelengths (X and C bands) have reduced penetration at the canopy, and volumetric and surface backscattering mechanisms are predominant.

Another SAR sensors parameter capable of influencing the interaction between the electromagnetic radiation and the floodable areas targets is the polarization. To identify FF, for instance, copolarization configuration (HH and VV) data are preferable than cross polarization (HV or VH) (Townsend, 2002).

Among the copolarization data, HH data are most used than VV for flood mapping, especially at rural and wetland areas. Wang et al. (1995) compared the backscattering signal of a FF on HH and VV for the same wavelength and incidence angle and verified that the backscattering ratio between flooded and non-FF (NFF) targets is higher at HH polarization than at VV polarization.

Microwave incidence angle must also be considered for mapping flood extent. Targets of interest may present different responses according to the range of SAR image incidence angle. Several studies consider that steep incidence angles can provide better results when mapping FF than the shallow ones (Wang et al., 1995; Lang and Kasischke, 2008).

However, the response behavior as a function of the incidence angle varies according to the interest target. For instance, at open water areas, steeper incidence angle implies in lower backscattering values, while the gain of roughness increases significantly the signal. A theory that justifies the backscattering variation as a function of the incidence angle is the Bragg resonance model, usually applied to ocean surface water (Robinson, 2004).

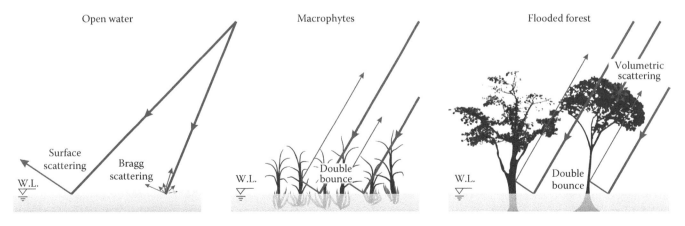

FIGURE 5.1 Resultant scattering mechanisms of the interaction between microwave radiation and usual wetland targets. (Adapted from Arnesen, A.S., Monitoramento da área inundada na Planície de Inundação do Lago Grande de Guruai (PA) por meio de imagens ScanSAR/ALOS e dados auxiliares. Master Science Thesis, Instituto Nacional de Pesquisas Espaciais, São José dos Campos, Brazil, 2012.)

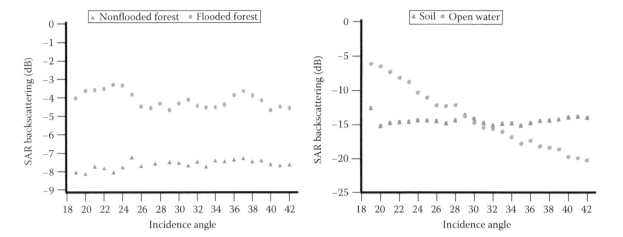

FIGURE 5.2 Samples backscattering variation as a function of the incidence angle for the classes nonflooded forest, flooded forest, open water, and soil. (Adapted from Arnesen, A.S., Monitoramento da área inundada na Planície de Inundação do Lago Grande de Guruai (PA) por meio de imagens ScanSAR/ALOS e dados auxiliares. Master Science Thesis, Instituto Nacional de Pesquisas Espaciais, São José dos Campos, Brazil, 2012.)

According to this model, an almost linear decrease of open water SAR backscatter occurs at the incidence angle range of 20°–70°.

Using L-band ScanSAR wide swath width (360 km) data from phased array L-band SAR (PALSAR) onboard of the advanced land observing satellite (ALOS) sensor, Arnesen (2012) inspected the backscatter variation along the scene range for four different targets, being two of them flooded (FF and open water) and two non-flooded (soil and nonflooded). A convergence was verified among forest classes, once FF presented a small but significant decrease of backscatter from near (19°) to far (42°) range (Figure 5.2). The other two classes investigated presented a higher similarity at the PALSAR ALOS data because both open water and soil are smooth targets that favor surface scattering and signal return at L-band. This similarity is essentially high at some ranges and needs to be overcome in order to provide accurate flood extent maps.

5.2.1.2 Flood Mapping Methodologies Using SAR Data

The scientific community has developed and applied several remote sensing approaches to map flood extent and/or risk making use of SAR data (Table 5.1). Classification accuracies can vary considerably among the SAR flood mapping approaches. According to Schumann et al. (2009), the main classification errors arise from improper image processing algorithms, variation of backscatter characteristics, unsuitable wavelengths or polarizations, unsuccessful speckle filtering, remaining geometric distortions, and inaccurate image geocoding.

The digital image classification approaches can be divided into supervised and unsupervised classifications. While supervised classification includes the classes sample-collecting step (training step), the unsupervised classification doesn't have this step, and a certain algorithm segregates the image into classes according to their spectral similarities (Jensen, 2005).

SAR images studies demonstrate that both supervised and unsupervised classifications approaches can have robust results when multipolarization is available (Cloude and Pottier, 1997; Frery et al., 2007). However, if only a single polarization data is available, radiometric variation within a scene, resulting from acquisition effects, can implicate in lower-accuracy classification results (Richards, 2009).

Thresholding radar backscatter value is frequently used and simple approaches. A binary algorithm is used to distinguish whether the area is flooded or not based on the backscatter value in decibel (dB) (Sanyal and Lu, 2004). However, as backscattering overlaps may occur among flooded and nonflooded targets, this approach can present poor results.

Another way to divide the classification approaches is by the basic processing unit of images: pixels or objects (also called as regions, an aggregation of homogeneous neighbor pixels) (Jensen, 2005).

Several pixel-based studies used not only the backscattering values to discriminate the flooded or floodable areas but also, and specially, the multitemporal criterion (Hess et al., 2003; Frappart et al., 2005; Martinez and Le Toan, 2007). The study of Martinez and Le Toan (2007) utilized a temporal series of 21 Japan Earth resources satellite (JERS-1) SAR images acquired between the period of 1993 and 1997 to map the land cover types and the average duration in which each pixel is exposed to the Amazon River flood. Only the temporal changing estimator thresholds (backscattering average coefficient and temporal change) were used to segregate the classes.

However, the global accuracy of this flood mapping was limited. One of the reasons to the low mapping accuracy was the pixel-based classification approach adopted. The major limitation of pixel-based classifications of SAR data is the high radiometric variation provoked by speckle, which is the resulting effect from the interference among the coherent echoes of the individual scatters within a resolution cell (Woodhouse, 2006). Speckle can be easily explained using the complex vector representation of scattered microwaves: each individual scatter has a different location (phase) from the instrument and a different radar cross section (signal amplitude) in such a way that the coherent sum of all the individual scattered waves is the total return signal.

TABLE 5.1 Summary of Recent Studies Exploring Methodologies on SAR and Ancillary Data Utilization for Flood Extent Monitoring: SAR and Ancillary Data Information, Classification Aspects, and Strength, Limitation, and Accuracy

SAR Data	SAR Band	Polarization	Pixel Dimensions	Flood Classification Aspects	Ancillary Data	Strength	Limitations	Accuracy	Reference
JERS-1 images	L	HH	Approx. 100 m	Temporal changing thresholds: mean backscatter coefficient computed over dry and wet seasons and ratio of both seasons (change value)	Topex/Poseidon (T/P) altimetry satellite and in situ hydrograph stations	Combined use of altimetric water level observations and inundation patterns to determine water volume variations in a wetland area.	Lack of reference data for validating the classification and data acquisition of images did not coincide with the real maximum water level.	90% for FF and 70% for low vegetated themes	Frappart et al. (2005)
JERS-1 images	L	HH	Approx. 100 m	Temporal changing thresholds: backscattering average coefficient and temporal change	Bathymetric transects of the flood plain	Temporal change to monitor the floodplain.	Lack of reference data for validating the classification and data acquisition of images did not coincide with the real maximum water level.	90% for all themes except from lox vegetation and FF of 80%	Martinez and Le Toan (2007)
JERS-1 images	L	HH	Approx. 100 m	Decision tree classification algorithm based on training sites of the land use types	Landsat images, DEM, and ground observation	High overall accuracy.	Limited accuracy for marsh (40%–67%) and uncertainty to locate transition zones due to high annual variations.	>82%	Wang (2004)
ALOS PALSAR images (ScanSAR mode)	L	HH	Approx. 100 m	Data mining tool for extracting backscatter thresholds, temporal changing thresholds, decision tree algorithm, and object-oriented classification	Landsat and MODIS images, DEM and bathymetric topographic data, and ground observation	Integrated use of SAR (single images and temporal aspects) and ancillary data for segregation of dry and wet classes.	Validation of only two dates (high- and low-water stages) because of lack of optical useful data.	78% for low-water stage and 80% for high-water stage	Arnesen et al. (2013)
ERS-1 images	C	VV	12.5 m × 12.5 m	Visual interpretations and adoption of two thresholding techniques	Landsat images, digital topographic map (1:10,000), and color aerial photographs	Temporal change was critical to monitor the floodplain.	Accurate results overcoming the SAR temporal resolution constraint.	96%	Brivio et al. (2002)

(Continued)

TABLE 5.1 (*Continued*) Summary of Recent Studies Exploring Methodologies on SAR and Ancillary Data Utilization for Flood Extent Monitoring: SAR and Ancillary Data Information, Classification Aspects, and Strength, Limitation, and Accuracy

SAR Data	SAR Band	Polarization	Pixel Dimensions	Flood Classification Aspects	Ancillary Data	Strength	Limitations	Accuracy	Reference
RADARSAT-1 images	C	HH	12.5 m × 12.5 m	Image processing with PCI and supervised classification using parallelepiped	Hydrological properties of the drainage basins and Landsat images	Information extracted by remote sensing translated into hydrological models parameters for flood forecasting.	Need of temporal resolution improvement.	80%	Bonn and Dixon (2005)
TerraSAR-X images	C	HH	1.5 m × 1.5 m	Hybrid methodology containing backscatter thresholding, region growing, and change detection information	Very-high-resolution aerial photographics acquired during the flooding event	Completely unsupervised technique for flood extend mapping with satisfactory results.	Requires a reference image with the same imaging properties as the SAR image.	82%	Giustarini et al. (2013)
TerraSAR-X image	C	HH	3 m × 3 m	Snake algorithm applied to SAR and LIDAR data, supervised classification, water height thresholding, and seed region growing	Airborne scanning laser altimetry (LIDAR) data, ASAR image data, and aerial photographs (0.2 m spatial resolution)	Provide flooded urban areas to enable a 2D inundation model to predict flood extent.	SAR invisible areas of the image (shadows) have lower flood mapping accuracy.	76% for pixels visible to TerraSAR-X and 58% to all pixels	Mason et al. (2010)
TerraSAR-X image	C	HH	3 m × 3 m	Automatic algorithm including preprocessing, water height threshold, image segmentation, and object classification	Airborne scanning laser altimetry (LIDAR) data, DEM, and aerial photographs	Presents operational consideration for a near real time using high-resolution SAR data.	Urban flood detection accuracy limited due to radar shadow and layover.	75% for pixels visible to TerraSAR-X and 57% to all pixels	Mason et al. (2012)
COSMO-SkyMed images (spotlight mode)	X	HH	1 m × 1 m	Morphological filtering, unsupervised K-means clustering, segmentation, backscattering analysis, and object classification	DEM and advanced visible and near-infrared radiometer type-2 (AVNIR-2) data	Multitemporal backscattering trends were used to associate image segments to different flood stages.	Absence of a validation procedure for the specific flood event mapping.	—	Pulvirenti et al. (2011)
COSMO-SkyMed images (strip map and spotlight mode)	X	HH	1 m × 1 m	Image segmentation and fuzzy logic classifier	Landsat and MODIS images	Thresholding analysis of land cover types was important for flood mapping.	Absence of a validation procedure for the specific flood event mapping.	—	Pulvirenti et al. (2013)

This effect can be minimized on the object-based classification approaches, in which the analysis elements are image segments (or objects) composed by a group of pixels aggregated by similarity at the segmentation process.

Object-based classification methodologies have been applied to detect flooded areas in several urban studies, once high-spatial-resolution data are becoming available recently, such as TerraSAR-X and COSMO-SkyMed SARs (Mason et al., 2010, 2012; Pulvirenti et al., 2011, 2013; Giustarini et al., 2013; Pierdiccca et al., 2013). For this kind of data, segmentation procedure is essential because the spatial details of the images are considerably smaller than the objects dimensions, which results in a large radiometric variance.

Taking advantage of the high temporal resolution of the COSMO-SkyMed mission, Pulvirenti et al. (2013) developed an algorithm based on an image segmentation method and a fuzzy logic classifier to map a flood occurred on December 2009, in Tuscany, Italy. The segmentation algorithm of this study was also applied by Pulvirenti et al. (2011) and consists in an unsupervised clustering algorithm that considers multitemporal and multiscale features. The array of this segmentation object-based approach, fuzzy logic algorithm, and land cover information derived from ancillary data allowed the authors to generate accurate inundation maps.

A notable advantage of object-based approaches is the inclusion of objects contextual characteristics in the structuration of a hierarchical decision tree. At a decision tree, the classes are organized in a hierarchy and discriminated according to a logic sequence of decision rules that consider not only the SAR radiometric information but also other targets properties such as temporal patterns, geometric information, complementary optical and topographic data, and spatial context (Benz, 2004; Silva et al., 2010).

To assess the flood extent seasonal variation in a wetland area of the Lower Amazon River floodplain, Curuai Lake floodplain, Brazil, Arnesen et al. (2013) developed a hierarchical object-based classification methodology based on L-band PALSAR

ALOS images and auxiliary data (optical images, water level records, field photographs, and topographic information). The classification scheme consisted of four hierarchical levels. The first level (Level 1) is based on the annual flooding pattern of the region, segregating images into three classes: upland (non-floodable areas), floodplain (variable flooding), and permanent open water (open water surface during the lowest water level at the analyzed period). The second level divided floodplain into two classes assumed constant along the hydrological year: forest and nonforest. These first two levels have constant areas for all images acquired during the study period (2006–2010), while the next levels were applied for each image separately.

Therefore, at Level 3 the forest class was divided into "FF" and "NFF," while nonforest was split according to backscattering similarity into two intermediate classes: "bright" (including wet/rough soil ["WS"], emergent macrophyte ["EM"], and rough open water ["ROW"]) and "dark" (with dry smooth soil ["DS"], floating macrophyte ["FM"], and smooth open water ["SOW"]). It is important to note that this segregation required a statistical backscattering analysis, including a data mining tool, in order to identify the main radiometric response overlapping and identify the need of complementary data (such as moderate-resolution imaging spectroradiometer [MODIS] optical images and topographic model).

The last level of this approach (Level 4, flooding status) merged Level 3 classes according to their flood condition, being the flooded class the sum of open water ("ROW" and "SOW"), macrophytes ("FM" and "EM"), and "FF" classes, while non-flooded was represented by soil ("WS" and "DS"), "NFF," and upland classes. The entire hierarchy is presented in Figure 5.3.

The classification results of Arnesen et al. (2013) had high accuracy for the period and data analyzed: 84% and 94% overall accuracies of the flood maps for low and high-water stages (Figure 5.4), respectively. The authors justified the higher accuracy of the methodology for high-water stages to the larger extent of open water, which was easily identified by the algorithm.

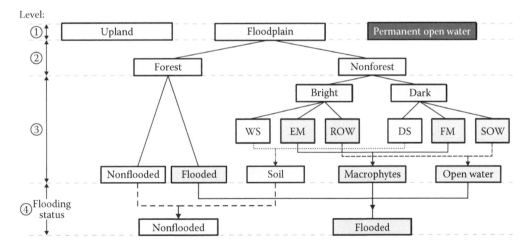

FIGURE 5.3 Classification hierarchy applied by Arnesen et al. (2013) to map flood extent at the Curuai Lake floodplain, Lower Amazon River, Brazil. Dark blue box represents the flooded class for the entire flood pulse; grey boxes are intermediate classes; beige boxes represent the non-flooded classes; and bright blue boxes represent flooded classes at each analyzed image.

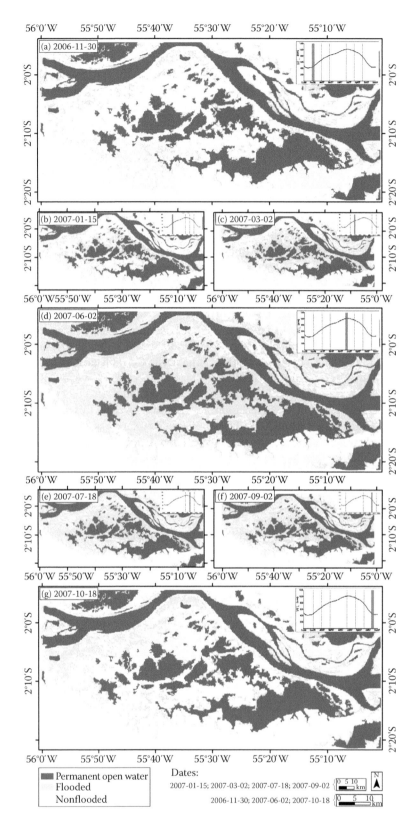

FIGURE 5.4 Flood extent variation during the 2007 Flood Pulse, Amazon River floodplain, Curuai Lake floodplain, Brazil. Graphics of water level at each date representing the water level at Curuai gauge at the image acquisition. Image dates: (a) 2006-11-30; (b) 2007-01-15; (c) 2007-03-02; (d) 2007-06-02; (e) 2007-07-18; (f) 2007-09-02; and (g) 2007-10-18. (Adapted from Arnesen, A.S., Monitoramento da área inundada na Planície de Inundação do Lago Grande de Guruai (PA) por meio de imagens ScanSAR/ALOS e dados auxiliares. Master Science Thesis, Instituto Nacional de Pesquisas Espaciais, São José dos Campos, Brazil, 2012.)

5.2.1.3 Accuracies, Errors, and Uncertainties of Flood Mapping with SAR Data

The methodological approaches for SAR mapping have inherent uncertainties due to these data characteristics. As Table 5.1 shows, the classification errors usually varies between the land use types, because the polarization, incidence angle, spatial resolution, and SAR band have direct implications on the feasibility of distinguishing flooded and nonflooded classes.

While some methodologies made use of temporal variation characteristics, specially the coarse spatial resolution data (such as JERS-1 and ALOS ScanSAR mode data), with satisfactory results for flood delimitation, others have explored their fine spatial resolution to supply this task (e.g., case of COSMO-SkyMed and TerraSAR-X applications).

Nevertheless, all of them are in one way or another subject to SAR data parameters, as the interference of speckle noise, occurrence of shadows and layovers, and great similarity of dry and wet classes according to the interaction between SAR radiation and targets with similar roughness.

Therefore, the best way to overcome these inherent SAR limitations for flood mapping is integrating different data (optical, topographic, and hydrologic) with others tools available for flood monitoring, such as GIS and hydrological models.

5.2.2 Integration of Remote Sensing, GIS, and Hydrological Models

A complementary approach, which has been used to map and understand different aspects of floods in large river basins, as well as the monitoring of hydrological cycle components, is the integration of remote sensing–derived information and hydrological models (Bates, 2004; Bates et al., 2006; Matgen et al., 2007; Schumann et al., 2009; Paiva et al., 2013; Rudorff et al., 2014a, b). One of the main advantages is that this approach allows not only predicting the extent and stage of the flood but also monitoring and simulating different scenarios such as storms, climate changes, different rainfall intensities, and changes in the land use and coverage. This integrated approach also allows the understanding of different environmental drivers of floods and the main mechanisms governing its dynamics.

Schumann et al. (2009) presented an extensive review of the recent progress in the integration of remote sensing–derived flood extent, water stage information, and hydrological models. According to these authors, this integration has only emerged over the last decade as a result of significant advances in SAR remote sensing techniques and high-performance computing encouraging a boost in distributed flood modeling.

According to the literature, there are four main topics addressed in studies on integrating remote sensing–derived information: (1) the retrieval and modeling of flood hydrology information from remote sensing observations, (2) the use of remote sensing data to calibrate and validate hydrological models, (3) the potential of remote sensing to understand and improve model structures, and (4) the utility of remote sensing data assimilation with models.

Despite of the several advantages and the high potential to study the flood dynamics, the integrated use of remote sensing–derived information and flood models requires a deep understanding of the many factors underlying both the remote sensing and modeling parts (Schumann et al., 2009). For this reason, relatively few studies have looked at this complex interplay of these two fields. Table 5.2 shows a summary of some studies that used this approach.

TABLE 5.2 Summary of Recent Studies Which Used Integration between Remote Sensing Data and Numerical Models: Models Used, Remote Sensing Data, and Parameters and Variables Extracted from Remote Sensing Data

Model	Remote Sensing Data	Parameters and Variables Derived from Remote Sensing Data	Reference
LISFLOOD-FP	SRTM/DEM, Radarsat-1/SAR, Landsat/TM, JERS/SAR, ALOS/PALSAR	Floodplain bathymetry, roughness coefficient, flood extent	Rudorff et al. (2014a)
LISFLOOD-FP	SRTM/DEM, Radarsat-1/SAR, Landsat/TM, JERS/SAR, ALOS/PALSAR	Floodplain bathymetry, roughness coefficient, flood extent	Rudorff et al. (2014b)
MGB-IPH	TRMM/3B42 product, SRTM/DEM, ENVISAT/altimeter, SSM/I, ERS/scatterometer, AVHRR, GRACE/RL04 product	Daily precipitation, morphological parameters, water level, flood extent, terrestrial water storage	Paiva et al. (2013)
HEC-RAS	LiDAR, ENVISAT/ASAR, ERS/SAR, aerial photographs	DEM, flood extent, roughness coefficient	Schumann et al. (2007)
HEC-RAS	LiDAR, ENVISAT/ASAR	DEM, flood extent	Matgen et al. (2007)
TELEMAC-2D and Simple Finite Volume Model (SFV)	LiDAR, ENVISAT/ASAR, ESR/SAR	DEM, flood extent	Horritt et al. (2007)
LISFLOOD-FP	LiDAR/ALTM2033, ENVISAT/ASAR, ESR/SAR, Radarsat-1/SAR, ERS-2/SAR	DEM, flood extent	Bates et al. (2006)
LISFLOOD-FP	ERS-1/SAR, aerial photographs	Flood extent	Hunter et al. (2005)
LISFLOOD-FP	ERS-1/SAR	Flood extent	Hall et al. (2005)
HEC-RAS; LISFLOOD-FP and TELEMAC-2D	LiDAR, Radarsat-1/SAR, ERS-1/SAR	DEM, flood extent	Horritt and Bates (2002)
LISFLOOD-FP and TELEMAC-2D	ERS-1/SAR	Flood extent	Horritt and Bates (2001)

Recently, Paiva et al. (2013) combined remote sensing data (SAR and optical data) and a distributed hydrological model to study the hydrodynamic and the flood dynamics of Amazon River. The authors used the *Modelo de Grandes Bacias–Instituto de Pesquisas Hidráulicas* (in Portuguese) (MGB-IPH) model, which is a large-scale distributed hydrological model that uses physical and conceptual equations, to simulate land surface hydrological processes (Collischonn et al., 2007).

The model discretization into river reaches, catchments, hydrodynamic computational cross sections, and parameter estimation was carried out using the digital elevation model (DEM) derived from the Shuttle Radar Topography Mission (SRTM) (Farr et al., 2007), with 15 arc-seconds resolution (approximately 500 m) and GIS-based algorithms described in Paiva et al. (2011). The simulation was validated using remote sensing–derived information (water level and flood extent).

The modeled water level obtained by Paiva et al. (2013) showed good agreement with the water level obtained using ENVISAT satellite altimetry data (Santos da Silva et al., 1996), with a Nash–Sutcliffe coefficient higher than 0.60 in 55% of the stream gauges used and correlation coefficient higher than 0.8 in 80% of the cases.

The overall inundation extent results from the MGB-IPH model obtained by Paiva et al. (2013) was similar to remote sensing estimates from Papa et al. (2010), showing that the model was able to reproduce the seasonal variation of flood extent and the north–south contrast, with flood peaks occurring in December–January–February (DJF) and March–April–May (MAM) at the Bolivian Amazon, in MAM and June–July–August (JJA) at central Amazon and JJA in the north.

Rudorff et al. (2014a,b) used a synergistic approach, which combined in situ data, remote sensing data, and a numerical model to study the hydraulic controls and the interannual and seasonal variability of the floodplain water balance components at Curuai Lake floodplain on the lower Amazon River. These authors used the LISFLOOD-FP model (Bates and De Roo, 2000), a raster-based hydraulic model, which combines one-dimensional (1D) river routing with two-dimensional (2D) overland flow.

The LISFLOOD-FP model uses different schemes to solve the river routing and overland flow equations, while the 1D river routing model used the diffusive wave scheme developed by Trigg et al. (2009); the 2D floodplain flows were solved by the inertia approximation of the St. Venant momentum equation (Bates et al., 2010). The SRTM DEM version 4.1 (Jarvis et al., 2008) was used as input data in the LISFLOOD-FP model. Before the simulations, Rudorff et al. (2014a) performed a series of correction on the SRTM DEM and tested different approaches to combine in situ bathymetric data and the SRTM DEM to generate the final DEM. The authors also used floodplain roughness coefficients derived from RADARSAT and Landsat Thematic Mapper (TM) images as input in the model simulations. The simulations were validated using the water level of Curuai gauging station and flood maps derived from SAR images produced by Hess et al. (2003) and Arnesen et al. (2013).

The results obtained by Rudorff et al. (2014a) showed that the DEM with channel adjustment yielded the best results in the simulation of the water level showing the lowest root-mean-square error (RMSE = 0.17 m). The DEM corrected with channel adjustments and local hydrological flows (RMSE = 0.27 m), the original SRTM DEM (RMSE = 0.38 m) and the DEM without channel adjustment (RMSE = 0.81 m) produced the worse results.

The simulated inundation extents also showed good agreement with SAR-derived maps for high-water conditions and lower agreement for low-water conditions. The inundation maps produced by Hess et al. (2003) based on JERS SAR images are commonly used for assessment of accuracy in studies of inundation modeling in the central Amazon basin. However, the results obtained by Rudorff et al. (2013) showed that these maps overestimate the actual inundation extents on the Curuai Lake floodplain at the time of the image acquisitions. Predictions of total flooded area were 36% and 20% lower in LISFLOOD-FP simulations using the rectified SRTM DEM compared to JERS mapping for high-water and low-water dates, respectively.

According to Rudorff et al. (2014a), the error in the SRTM interferometric baseline is of long wavelength nature and is likely to span a negative elevation bias over wide regions in the Amazon basin. Forest canopies are a source of positive elevation bias. Though upstream reaches of the Amazon floodplain tend to be more predominantly covered by alluvial forest, which then becomes the main source of elevation error, it should be important to account for the interferometric baseline error along with correction for vegetation effect.

Another combined methodology was used by Weng (2001). The author evaluated the changes in runoff in Zhujiang Delta region, the third biggest river delta in China, through the integration of GIS, remote sensing, and hydrological models and has automated the soil conservation service (SCS) model.

With the support of the Landsat TM data, changes in coverage of urban land between 1989 and 1997 period was identified through classifications of land cover techniques. Then, based on SCS model, urban growth patterns were identified to assess the effects on surface runoff parameters. The results of this research indicated that the annual runoff depth had increased by 8.10 mm/year.

Similarly, Du et al. (2012) developed an integrated modeling system to analyze the effect of urbanization on annual runoff and flood events in Qinhuai River basin in China. The authors used Landsat TM and Enhanced Thematic Mapper Plus (ETM+) images to represent land use changes over the time. The urbanization scenarios identified was modeled in hydrologic modeling system (HMS) model and indicated that the values of annual runoff, daily peak flow, and flood level have increased.

As shown briefly in the previous text, the integrated use of remote sensing–derived information and flood models has become established as a powerful tool, the robustness of which needs, however, to be examined further, in particular for flood forecasting. What is certainly to be gained from this development is that fundamental research questions in terms of both model evaluation and digital image processing techniques will be addressed in one way or another. Moreover, it is expected that remote sensing data and flood models that allow direct integration of such data would provide the basis for assimilation of remotely sensed data, for instance, in operational flood-forecasting systems.

5.2.3 Flood Forecasting Using Remote Sensing–Derived Information

Floods usually affect a lot of people, and an increasing of frequency is expected for these extreme events around the world. Therefore, flood-forecasting systems and assessments of potential natural hazards areas are tools continuously desired by decision makers. Flood prediction unfolds distinct phases that are better described in the following items.

5.2.3.1 Flood Preparedness Phase: Flood-Prone/ Risk Zone Identification

The first step for enhancing our knowledge of a particular flood event is through flood-affected population, where results such as damage assessment, recommendations, and other important information must be documented.

Flood risk zone map could be presented in two different ways: a detailed mapping approach, which is required for the production of hazard assessment and updating/creating flood risk maps, or at larger-scale mapping approach that explores the general flood situation within a river basin, identifying areas under greatest risk. In both cases, the integration of remote sensing and ancillary data contributes for mapping the inundated areas, as presented in previous sections (Jeyaseelan, 2005).

5.2.3.2 Flood Prevention Phase

Although meteorological remote sensing flood monitoring can be applied from global scale to local storm scale, the most used is at storm scale associated with hydrodynamic models by monitoring the intensity, movement, and propagation of the precipitation system to determine how much, when, and where the heavy precipitation is going to move during the next 0–3 h (called *nowcasting*). Meteorological satellites can detect various aspects of the hydrological cycle—precipitation (rate and accumulations), wetness transport, and surface/soil moisture (Scofield and Achutuni, 1996).

Satellite optical observations of floods have been hampered by the presence of clouds that resulted in the lack of near-real-time (NRT) data acquisitions. SAR sensors can achieve regular observation of the Earth's surface, even in the presence of thick cloud cover. The advanced very high-resolution radiometer (AVHRR) sensor onboard of the National Oceanic and Atmospheric Administration (NOAA) satellites allows flood monitoring in near real time. High-resolution infrared (10.7 μm) and visible are the principal data sets used in this diagnosis. Soil moisture due to a heavy rainfall event or snowmelt is extremely useful information for flash flood guidance. Data from the special sensor microwave imager sensor onboard

of the Defense Meteorological Satellite Program (DMSP) can also be used in this analysis.

5.2.3.3 Flood Forecasting

An effective real-time flood-forecasting system essentially requires basic structures that need to be linked in an organized manner (WMO, 2011). Parts of this structure are listed as follows:

1. Provision of specific forecasts relating to rainfall (both quantity and timing), for which numerical weather-precipitation models are necessary
2. Establishment of a network of manual or automatic hydrometric stations, linked to a central control by some form of telemetry
3. Flood-forecasting model software, linked to the observing network and operating in real time

Flood forecast can be issued over the areas, in which remote sensing is complementary to direct precipitation and stream flow measurements, and those areas that are not instrumentally monitored (or the instruments are not working or are inaccurate). In this second category, remote sensing represents an essential tool.

Quantitative precipitation estimates (QPE) and quantitative precipitation forecasts (QPF) use satellite data as one source of information to facilitate flood forecasts. New algorithms are being developed that integrate geostationary operational environmental satellite (GOES) precipitation estimates with the more physically based polar operational environmental satellites (POES) microwave estimates. An improvement in rainfall spatial distribution measurements is being achieved by integrating radar, rain gauges, and remote sensing techniques to improve real time flood forecasting (Vicente et al., 1998).

GOES and POES weather satellites can provide climatological information on precipitation especially for those areas not instrumentally monitored. The technique developed by Vicente et al. (1998) uses data from the infrared brightness temperature of channel 4 (10.7 μm) of GOES-8 and GOES-9 and converts from a power ratio with the rate of precipitation estimated by radar. Subsequently, the precipitation rates are adjusted to different moisture regimes, growth of convective cells, and spatial gradients of clouds. Finally, corrections for wet and dry environments are made using the average relative humidity between surface and 500 mbar level and precipitable water at that level, obtained by National Centers for Environmental Prediction (NCEP) Eta model.

Vicente et al. (2002) made other adjustments and the method was called "Hidroestimator." The algorithm fixes the level of convective equilibrium for events hot tops and errors due to terrain and parallax were used to adjust automatically precipitation rate. The components of direction and wind speed are extracted from the NCEP Eta model at the level of 850 mbar and the elevation and topography of the terrain are taken from geographic maps. This allows for improved accuracy of estimated rainfall

TABLE 5.3　Global Flood Monitoring Tools Using Input Data from Tropical Rainfall Measurement Mission: NASA/TRMM

Forecast System	Country	Hydrometeorological Input Data	Methods/Models	Operational Resolution (Temporal and Spatial)	Forecast Lead-Time
ALERT	Australia	Real-time rainfall and water level data	Input data fed hydrological model-magnitude and timing of a flood event	Event-based and catchment scale	Now-cast
Gridded Flash Flood Guidance (GFFG)	Northern America	Multisensor precipitation estimates and forecasts, based on NEXRAD radar, rain gauges, and NWP model outputs	(HL-RMS model) Natural Resource Conservation Service's (NRCS) Curve Number Model NCRS Unit Hydrograph Model combined with a dosing storm to estimate bank-full flow The HL-RMS model is applied to 4 × 4 km² grid in a catchment	Hourly and 100–300 km² catchment scale	3–24 h
Central American Flash Flood Guidance (CAFFG)	Belize, Costa Rica, El Salvador, Guatemala, Honduras, Nicaragua and Panama	Gauged rainfall, satellite-based rainfall derived by Hidro-Estimator, and NWP model (WSETA) output	CAFFG uses saturation-excess physically-based soil moisture deficit in real-time	Hourly and 100–300 km² catchment scale	3–6 h
Flash Flood System in Northern Austria	Northern Austria	Observed discharge, gauged rainfall, radar rainfall, and NWP model outputs	A grid-based distributed model for river flow forecasting Ensemble Kalman Filtering is used for real-time updating of the model states (grid soil moisture based on runoff.	14 min and 1 km² grid	48 h
European Flood Forecasting System (EFFS)	Europe	Gauged rainfall and discharge, GCM outputs downscaled by two NWP models	Input data feed raster-based distributed rainfall-runoff modeling suite LISFLOOD-FF for generating river flow LISFLOOD-FP to model inundation areas	LISFLOOD-FF; hourly and 1 km² grid LISFLOOF-FP: 10–100 m grid	72–120 h
Decision Support System for Flash Flood Warning in Thailand	Thailand (Uttaradit, Sukhothai, and Phrae provinces)	Gauged rainfall and discharge, NWP model outputs and satellite images	Artificial neural network model is used to forecast riverflow Warning are provided by comparing the river flow forecasts with the magnitude of historical flash flood events in a decision support system	Event-based and catchment scale	24 h
GEOREX FLOOD	Malaysia	Gauged rainfall and radar rainfall	The Stream-flow Synthesis and Reservoir Regulation (SSARR) model for river flow forecasting	6 h and catchment scale	Now-cast
Flash Flood Forecasting System od Ayalon Stream	Israel	Gauged rainfall and discharge	A module that autoregresses discharges, A flood routing module, A module that convolutes contributions from rain falling over the area between the stations, And a module that estimates flow recession at the target station. Also the system uses an adaptive mechanism (Box and Jenkins, 1970) to continuously update parameter values	Event-based and catchment scale	30 min–3.5 h

TABLE 5.4 Review of Methods Available for Flash Flood Forecasting

Flood Monitoring Tool	Input Data	Spatial Coverage and Resolution	Reference
NASA Tropical Rainfall Measuring Mission, TRMM NRT	TRMM/TMPA-RT	50°S–50°N 12 km	Huffman et al. (2006)
MODIS NRT	Terra and Aqua/MODIS	50°S–50°N 12 km	Turk and Miller (2005)
Dartmouth Flood Observatory (DFO)	TRMM/TMI and Aqua/AMSR-RT	Global 250 m and 10 km	Brakenridge and Anderson (2005)
Global Flood Detection System (GFDS)	TRMM/TMI and Aqua/AMRS-E	Global 10 km	Kugler and De Grove (2007)
Global Flood Monitoring System (GFMS)	TRMM/TMPA-RT MERRA	Global 10 km	Wu et al. (2014)

Source: Hapuarachchi, H.A.P. and Wang, Q.J., A review of methods and systems available for flash flood forecasting, CSIRO: Water for a Healthy Country National Research Flagship, Water for a Healthy Country Flagship Report series, Melbourne, Victoria, Australia, 2008.

in difficult areas to obtain surface data, both by conventional weather stations and weather radar.

As mentioned before (Section 5.2.2), hydrological models play a major role in assessing and forecasting flood risk. These models requires several types of data as input, such as DEM data, land use, soil type, soil moisture, stream/river base flow, drainage basin size, snowpack characterization, and rainfall amount/intensity.

In 1997, the Tropical Rainfall Measuring Mission (TRMM) satellite was launched. The objective of the TRMM mission was to provide accurate global tropical precipitation estimates by using a unique combination of instruments designed purely for rainfall observation (Simpson et al., 1996; Kummerow et al., 1998).

Just 10 years after TRMM launch, a series of high-resolution, quasi-global, near-real-time, TRMM-based precipitation estimates started to be available for the research community. The proposed Global Precipitation Measurement (GPM) mission, which would be the successor to the TRMM, envisions providing global precipitation products with temporal sampling rates ranging from 3 to 6 h and a spatial resolution of 25–100 km^2 (Smith, 2007). A major GPM science objective is to improve prediction capabilities for floods, landslides, freshwater resources, and other hydrological applications.

The TRMM-based algorithms for flood monitoring use as input data a product named 3B42 that combines precipitation radar and TRMM microwave imager adjusted with surface rain gauge measurements on monthly basis and near-real-time (3B42RT) product (Huffman et al., 2006; Table 5.3). TRMM multisatellite precipitation analysis (TMPA) has improved algorithms during the last years (Sorooshian et al., 2000; Kidd et al., 2003; Joyce et al., 2004; Turk and Miller, 2005).

A first approach to global runoff simulation using satellite rainfall estimation was reported by Hong et al. (2007) as an approximate assessment of quasi-global runoff. The Natural Resources Conservation Service runoff curve number method incorporated rainfall data and other remote sensing products in a rainfall-runoff approach.

Kugler and De Grove (2007) demonstrated in the report the verification and validation of the model that support the global flood detection system (GFDS). GFDS provides a systematic monitoring of ongoing flood events around the world updated every day. The model can also be used to reconstruct historical flood events that were not recorded in a systematic way.

Another satellite-based tool that is running and receiving improvements during the last few years is called global flood monitoring system (GFMS, http://flood.umd.edu). The validation and analysis based on the recent flood events over the upper Mississippi valley from the GFMS real-time system demonstrated that the real-time GFMS had a satisfactory response in flood detection, evolution, and magnitude calculation according to observed daily stream flow data (Wu et al., 2014). Some of the monitoring tools of flooding are shown at Table 5.4.

5.3 Conclusions

The major limitation of remote sensing data for flood monitoring is the low frequency of data acquisition during a specific natural disaster. Cloud coverage at rainy events makes optical remote sensing data nearly impossible to be used in order to extract flood extent. SAR data are almost independent of weather conditions, which may consequently provide more frequent cloud free images, and have other potential advantages for flood monitoring as the interaction of microwave radiation with flooded targets.

Moreover, recently, new remote sensing satellites are providing daily SAR data that allow a certain flood event to be mapped during all stages, such as the COSMO-SkyMed mission. All the SAR sensors characteristics are reflected in the backscatter and studies that explore these properties can achieve high flood mapping accuracies. Nevertheless, overlaps between classes may occur at SAR images, and ancillary data (such as optical images) represent an effective alternative to minimize classification confusion.

There are several digital image processing methodologies available for extracting flood areas from remote sensing data. However, this chapter showed that the most successful of them are techniques that integrate different sources of data and tools. The integration of GIS, hydrological models, and remote sensing is advantageous for flood mapping approaches in urban, rural, and wetland areas since the input data are reliable and accurate (gauging and topographic data, for instance).

As this chapter has presented, the integration of remote sensing, GIS, and hydrological modeling tools can be part of a major flood prevention system. Predictions models of potential flood extent can help stakeholders to develop contingency plans, facilitate a more effective response, and minimize the socioeconomic impacts of these natural disasters.

References

Arnesen, A.S. 2012. Monitoramento da área inundada na Planície de Inundação do Lago Grande de Guruai (PA) por meio de imagens ScanSAR/ALOS e dados auxiliares. Master Science Thesis, Instituto Nacional de Pesquisas Espaciais, São José dos Campos, Brazil.

Arnesen, A.S., Silva, T.S.F., Hess, L.L., Novo, E.M.L.M., Rudorff, C.M., Chapman, B.D., and McDonald, K.C. 2013. Monitoring flood extent in the lower Amazon River floodplain using ALOS/PALSAR ScanSAR images. *Remote Sensing of Environment*, 130, 51–61, doi: 10.1016/j.rse.2012.10.035.

Bates, P.D. 2004. Remote sensing and flood inundation modelling. *Hydrological Processes*, 18, 2593–2597, doi: 10.1002/hyp.5649.

Bates, P.D. and De Roo, A.P.J. 2000. A simple raster-based model for flood inundation simulation. *Journal of Hydrology*, 236, 54–77, doi: 10.1016/S0022-1694(00)00278-X.

Bates, P.D., Horritt, M.S., and Fewtrell, T.J. 2010. A Simple inertial formulation of the shallow water equations for efficient two-dimensional flood inundation modelling. *Journal of Hydrology*, 387, 33–45, doi: 10.1016/j.jhydrol.2010.03.027.

Bates, P.D., Wilson, M.D., Horritt, M.S., Mason, D.C., Holden, N., and Currie, A. 2006. Reach scale floodplain inundation dynamics observed using airborne synthetic aperture radar imagery: Data analysis and modelling. *Journal of Hydrology*, 328, 306–318, doi: 10.1016/j.jhydrol.2005.12.028.

Benz, U. 2004. Multi-resolution, object-oriented fuzzy analysis of remote sensing data for GIS-ready information. *ISPRS Journal of Photogrammetry and Remote Sensing*, 58(3–4), 239–258, doi: 10.1016/j.isprsjprs.2003.10.002.

Bonn, F., and Dixon, R. 2005. Monitoring flood extent and forecasting excess runoff risk with RADARSAT-1 data. *Natural Hazards*, 35(3), 377–393.

Brakenridge, G.B. and Anderson, E. 2005. MODIS-based flood detection, mapping, and measurement: The potential for operational hydrological applications. In: *Transboundary Floods, Proceedings of the NATO Advanced Research Workshop*, Baile Felix—Oradea, Romania, May 4–8, 2005.

Brivio, P.A., Colombo, R., Maggi, M., and Tomasoni, R. 2002. Integration of remote sensing data and GIS for accurate mapping of flooded areas. *International Journal of Remote Sensing*, 23, 429–441, doi: 10.1080/01431160010014729.

Cloude, S.R. and Pottier, E. 1997. An entropy based classification scheme for land applications of polarimetric SAR. *IEEE Transactions on Geoscience and Remote Sensing*, 35(1), 68–78, doi: 10.1109/36.551935.

Collischonn, W., Allasia, D.G., Silva, B.C., and Tucci, C.E.M. 2007. The MGB-IPH model for large-scale rainfall-runoff modeling. *Hydrology Science Journal*, 52, 878–895, doi: 10.1623/hysj.52.5.878.

Du, J., Qian, L., Rui, H., Zuo, T., Zheng, D., Xu, Y., and Xu, C.-Y. 2012. Assessing the effects of urbanization on annual runoff and flood events using an integrated hydrological modeling system for Qinhuai River basin, China. *Journal of Hydrology*, 464–465, 127–139, doi: 10.1016/j.jhydrol.2012.06.057.

Farr, T.G., Rosen, P.A., Caro, E., Crippen, R., Duren, R., Hensley, S., Kobrick, M. et al. 2007. The shuttle radar topography mission. *Reviews of Geophysics*, 45, RG2004, doi: 10.1029/2005RG000183.

Frappart, F., Seyler, F., Martinez, J.-M., León, J.G., and Cazenave, A. 2005. Floodplain water storage in the Negro River basin estimated from microwave remote sensing of inundation area and water levels. *Remote Sensing of Environment*, 99, 387–399, doi: 10.1016/j.rse.2005.08.016.

Freeman, A. and Durden, S.L. 1998. A three-component scattering model for polarimetric SAR data. *Jet Propulsion*, 36, 963–973, doi: 10.1109/36.673687.

Frery, A.C., Correia, A.H., and Freitas, C.C. 2007. Classifying multifrequency fully polarimetric imagery with multiple sources of statistical evidence and contextual information. *IEEE Transactions on Geoscience and Remote Sensing*, 45(10), 3098–3109, doi: 10.1109/TGRS.2007.903828.

Giustarini, L., Hostache, R., Matgen, P., Schumann, G.J.-P., Bates, P.D., and Mason, D.C. 2013. A change detection approach to flood mapping in urban areas using TerraSAR-X. *IEEE Transactions on Geoscience and Remote Sensing*, 51, 2417–2430, doi: 10.1109/TGRS.2012.2210901.

Hall, J.W., Tarantola, S., Bates, P.D., and Horritt, M.S. 2005. Distributed sensitivity analysis of flood inundation model calibration. *Journal of Hydraulic Engineering*, 131, 117–126, doi: 10.1061/(ASCE)0733-9429(2005)131:2(117).

Hapuarachchi, H.A.P. and Wang, Q.J. 2008. A review of methods and systems available for flash flood forecasting. CSIRO: Water for a Healthy Country National Research Flagship. Water for a Healthy Country Flagship Report series, Melbourne, Victoria, Australia.

Henderson, F. and Lewis, A. 1998. *Manual of Remote Sensing: Principles and Applications of Imaging Radar*, 3rd ed. New York: Wiley, 896pp.

Hess, L.L., Melack, J.M., Novo, E.M.L.M., Barbosa, C.C.F., and Gastil, M. 2003. Dual-season mapping of wetland inundation and vegetation for the central Amazon basin. *Remote Sensing of Environment*, 87, 404–428, doi: 10.1016/j.rse.2003.04.001.

Hong, Y., Adler, R.F., Hossain, F., Curtis, S., and Huffman, G.J. 2007. A first approach to global runoff simulation using satellite rainfall estimation. *Water Resources Research*, 43, W08502, doi: 10.1029/2006WR005739.

Horritt, M.S. and Bates, P.D. 2001. Predicting floodplain inundation: Raster-based modelling versus the finite-element approach. *Hydrological Process*, 15, 825–842, doi: 10.1002/hyp.188.

Horritt, M.S. and Bates, P.D. 2002. Evaluation of 1D and 2D numerical models for predicting river flood inundation. *Journal of Hydrology*, 268, 87–99, doi: 10.1016/S0022-1694(02)00121-X.

Horritt, M.S., Di Baldassarre, G., Bates, P.D., and Brath, A. 2007. Comparing the performance of a 2-D finite element and a 2-D finite volume model of floodplain inundation using airborne SAR imagery. *Hydrological Process*, 21, 2745–2759, doi: 10.1002/hyp.6486.

Huffman, G.J., Adler, R.F., Bolvin, D.T., Gu, G., Nelkin, E.J., Bowman, K.P., Stocker, E.F., and Wolff, D.B. 2006. The TRMM multi-satellite precipitation analysis: Quasi-global, multi-year, combined-sensor precipitation estimates at fine scale. *Journal of Hydrometeorology*, 8, 38–55, doi: 10.1175/JHM560.1.

Hunter, N.M., Bates, P.D., Horritt, M.S., de Roo, P.J., and Werner, M.G.F. 2005. Utility of different data types for calibrating flood inundation models within a GLUE framework. *Hydrology and Earth System Sciences*, 9, 412–430, doi: 10.5194/hess-9-412-2005.

Jarvis, A., Reuter, H.I., Nelson, A., and Guevara, E. 2008. *Hole-Filled Seamless SRTM Data V4*. Cali, Colombia: International Center for Tropical Agriculture. Available at: http://srtm.csi.cgiar.org. (Accessed June 15, 2014.)

Jensen, J.R. 2005. *Introductory Digital Image Processing: A Remote Sensing Perspective*, 3rd ed. Englewood Cliffs, NJ: Prentice Hall, 526pp.

Jeyaseelan, A.T. 2005. Droughts and floods assessment and monitoring using remote sensing and GIS. In: *Proceedings of the Satellite Remote Sensing and GIS Applications in Agriculture Meteorology*, Dehra Dun, India, July 7–11, 2003, pp. 291–313.

Joyce, R.J., Janowiak, J.E., Arkin, P.A., and Xie, P.. 2004. CMORPH: A method that produces global precipitation estimates from passive microwave and infrared data at high spatial and temporal resolution. *Journal of Hydrometeor*ology, 5, 487–503, doi: 10.1175/1525-7541(2004)005.

Kidd, C.K., Kniveton, D.R., Todd, M.C., and Bellerby, T.J. 2003. Satellite rainfall estimation using combined passive microwave and infrared algorithms. *Journal of Hydrometeor*ology, 4, 1088–1104, doi: 10.1175/1525-7541(2003)004.

Kugler, Z. and De Groeve, T. 2007. The global flood detection system. Joint Research Centre, Institute for the Protection and Security of the Citizen, European Commission, Ispra, Italy. EUR 23303 EN-2007, 44pp.

Kummerow, C., Barnes, W., Kozu, T., Shiue, J., and Simpson, J. 1998. The tropical rainfall measuring mission TRMM sensor package. *Journal of Atmospheric and Oceanic Technol*ogy, 15, 809–817, doi: 10.1175/1520-0426(1998)015.

Lang, M.W. and Kasischke, E.S. 2008. Using C-band synthetic aperture radar data to monitor forested wetland hydrology in Maryland's Coastal Plain, USA. *Wetlands*, 46, 535–546, doi: 10.1109/TGRS.2007.909950.

Martinez, J.M. and Le Toan, T. 2007. Mapping of flood dynamics and spatial distribution of vegetation in the Amazon floodplain using multitemporal SAR data. *Remote Sensing of Environment*, 108, 209–223, doi: 10.1016/j.rse.2006.11.012.

Mason, D.C., Davenport, I.J., Neal, J.C., Schumann, G.J.-P., and Bates, P.D. 2012. Near real-time flood detection in urban and rural areas using high-resolution synthetic aperture radar images. *IEEE Transactions on Geoscience and Remote Sensing*, 50, 3041–3052, doi: 10.1109/TGRS.2011.2178030.

Mason, D.C., Speck, R., Devereux, B., Schumann, G.J., Member, A., Neal, J.C., and Bates, P.D. 2010. Flood detection in urban areas using TerraSAR-X. *IEEE Transactions on Geoscience and Remote Sensing*, 48, 882–894, doi: 10.1109/TGRS.2009.2029236.

Matgen, P., Schumann, G., Henry, G.-B., Hoffmann, L., and Pfister, L. 2007. Integration of SAR-derived river inundation areas, high-precision topographic data and a river flow model toward near real-time flood management. *International Journal of Applied Earth Observation and Geoinformation*, 9, 247–263, doi: 10.1016/j.jag.2006.03.003.

Milly, P.C.D., Wetherald, R.T., Dunne, K.A., and Delworth, T.L. 2002. Increasing risk of great floods in a changing climate. *Nature*, 415, 514–517, doi: 10.1038/415514a.

Neal, J.C., Bates, P.D., Fewtrell, T.J., Hunter, N.M., Wilson, M.D., and Horritt, M.S. 2009. Distributed whole city water level measurements from the Carlisle 2005 urban flood event and comparison with hydraulic model simulations. *Journal of Hydrology*, 368, 42–55, doi: 10.1016/j.jhydrol.2009.01.026.

Nirupama, N. and Simonovic, S.P. 2007. Increase of flood risk due to urbanisation: A Canadian example. *Natural Hazards*, 40, 25–41, doi: 10.1007/s11069-006-0003-0.

Paiva, R.C.D., Buarque, D.C., Collischonn, W., Bonnet, M.-P., Frappart, F., Calmant, S., and Mendes, C.A.B. 2013. Large-scale hydrologic and hydrodynamic modeling of the Amazon River basin. *Water Resource Research*, 49, 1226–1243, doi: 10.1002/wrcr.20067.

Paiva, R.C.D., Collischonn, W., and Tucci, C.E.M. 2011. Large scale hydrologic and hydrodynamic modeling using limited data and a GIS based approach. *Journal of Hydrology*, 406, 170–181, doi: 10.1016/j.jhydrol.2011.06.007.

Papa, F., Prigent, C., Aires, F., Jimenez, C., Rossow, W.B., and Matthews, E. 2010.Interannual variability of surface water extent at the global scale, 1993–2004. *Journal of Geophysical Research*, 115, D12111, doi: 10.1029/2009JD012674.

Pierdicca, N., Pulvirenti, L., Chini, M., Guerriero, L., and Candela, L. 2013.Observing floods from space: Experience gained from COSMO-SkyMed observations. *Acta Astronautica*, 84, 122–133, doi: 10.1016/j.actaastro.2012.10.034.

Pulvirenti, L., Chini, M., Pierdicca, N., Guerriero, L., and Ferrazzoli, P. 2011. Flood monitoring using multi-temporal COSMO-SkyMed data: Image segmentation and signature interpretation. *Remote Sensing of Environment*, 115, 990–1002, doi: 10.1016/j.rse.2010.12.002.

Pulvirenti, L., Pierdicca, N., Chini, M., Member, S., and Guerriero, L. 2013. Monitoring flood evolution in vegetated areas using COSMO-SkyMed Data: The Tuscany 2009 case study. *IEEE Journal of Selected Topics in Applied Earth Observations and Remote Sensing*, 6, 1807–1816, doi: 10.1109/JSTARS.2012.2219509.

Richards, J.A. 2009. *Remote Sensing with Imaging Radar*, 3rd ed. Berlin, Germany: Springer-Verlag, 361pp.

Robinson, I.S. 2004. *Measuring the Oceans from Space: The Principles and Methods of Satellite Oceanography*. Chichester, U.K.: Springer-Praxis.

Rudorff, C.M., Melack, J.M., and Bates, P.D. 2014a. Flooding dynamics on the lower Amazon floodplain: 1. Hydraulic controls on water elevation, inundation extent, and river-floodplain discharge. *Water Resource Research*, 50, 1–16, doi: 10.1002/2013WR014091.

Rudorff, C.M., Melack, J.M., and Bates, P.D. 2014b. Flooding dynamics on the lower Amazon floodplain: 2. Seasonal and interannual hydrological variability. *Water Resource Research*, 50, 1–15, doi: 10.1002/2013WR014714.

Santos da Silva, J., Calmant, S., Seyler, F., Rotunno Filho, O.C., Cochonneau, G., Scofield, R.A., Zaras, D., Kusselson, S., and Rabin, R.1996. A remote sensing precipitable water product for use in heavy precipitation forecasting. In: *Proceedings of the Eighth Conference on Satellite Meteorology and Oceanography*, Atlanta, GA, January 29–February 2, 1996. Boston, MA: AMS, pp. 74–78.

Sanyal, J. and Lu, X.X. 2004. Application of remote sensing in flood management with special reference to monsoon Asia: A review. *Natural Hazards*, 33, 283–301, doi: 10.1023/B:NHAZ.0000037035.65105.95.

Schumann, G., Bates, P.D., Horritt, M.S., Matgen, P., and Pappenberger, F. 2009. Progress in integration of remote sensing–derived flood extent and stage data and hydraulic models. *Reviews of Geophysics*, 47, RG4001, doi: 10.1029/2008RG000274.

Schumann, G., Matgen, P., Hoffmann, L., Hostache, R., Pappenberger, F., and Pfister, L. 2007. Deriving distributed roughness values from satellite radar data for flood inundation modelling. *Journal of Hydrology*, 344, 96–111, doi: 10.1016/j.jhydrol.2007.06.024.

Scofield, R.A. and Achutuni, R. 1996. The satellite forecasting funnel approach for predicting flash floods. *Remote Sensing Reviews*, 14, 251–282, doi: 10.1080/02757259609532320.

Silva, T.S.F., Costa, M.P.F., and Melack, J.M. 2010. Assessment of two biomass estimation methods for aquatic vegetation growing on the Amazon Floodplain. *Aquatic Botany*, 92, 161–167, doi: 10.1016/j.aquabot.2009.10.015.

Simpson, J., Kummerow, C., Tao, W.-K., and Adler, R.F. 1996. On the tropical rainfall measuring mission (TRMM). *Meteorology Atmospheric Phys*ics, 60, 19–36, doi: 10.1007/BF01029783.

Smith, E. 2007. The international global precipitation measurement (GPM) program and mission: An overview. In: Levizzani, V. and Turk, F.J. (eds.) *Measuring Precipitation from Space: URAINSAT and the Future*. Berlin, Germany: Springer-Verlag, pp. 611–653.

Sorooshian, S., Hsu, K.-L., Gao, X., Gupta, H., Imam, B., and Braithwaite, D. 2000. Evaluation of PERSIANN system satellite-based estimates of tropical rainfall. *Bulletin of the American Meteorological Society*, 81, 2035–2046, doi: 10.1175/1520-0477(2000)081.

Townsend, P.A. 2002. Estimating forest structure in wetlands using multitemporal SAR. *Remote Sensing of Environment*, 79, 288–304, doi: 10.1016/S0034-4257(01)00280-2.

Tralli, D.M., Blom, R.G., Zlotnicki, V., Donnellan, A., and Evans, D.L. 2005. Satellite remote sensing of earthquake, volcano, flood, landslide and coastal inundation hazards. *ISPRS Journal of Photogrammetry and Remote Sensing*, 59, 185–198, doi: 10.1016/j.isprsjprs.2005.02.002.

Trigg, M.A., Wilson, M.D., Bates, P.D., Horritt, M.S., Alsdorf, D.E., Forsberg, B.R., and Vega, M.C. 2009. Amazon flood wave hydraulics. *Journal of Hydrology*, 374, 92–105, doi: 10.1016/j.jhydrol.2009.06.004.

Turk, F.J. and Miller, S.D., 2005. Toward improved characterization of remotely sensed precipitation regimes with MODIS/AMSR-E blended data techniques. *IEEE Transactions on Geoscience and Remote Sensing*, 43, 1059–1069, doi: 10.1109/TGRS.2004.841627.

Van Ree, C.C.D.F., Van, M.A., Heilemann, K., Morris, M.W., Royet, P., and Zevenbergen, C. 2011. FloodProBE: Technologies for improved safety of the built environment in relation to flood events. *Environmental Science & Policy*, 14, 874–883, doi: 10.1016/j.envsci.2011.03.010.

Vicente, G.A., Davenport, J.C., and Scofield, R.A. 2002. The role of orographic and parallax corrections on real time high resolution satellite estimation. *International Journal of Remote Sensing*, 23, 221–230, doi: 10.1080/01431160010006935.

Vicente, G.A., Scofield, R.A., and Menzel, W.P. 1998. The operational GOES infrared rainfall estimation technique. *Bulletin of American Meteorology Society*, 79, 1883–1898, doi: 10.1175/1520-0477(1998)079.

Wang, Y. 2004. Seasonal change in the extent of inundation on floodplains detected by JERS-1 Synthetic Aperture Radar data. *International Journal of Remote Sensing*, 25(13), 2497–2508.

Wang, Y., Hess, L.L., Filoso, S., and Melack, J.M. 1995. Understanding the radar backscattering from flooded and nonflooded Amazonian forests: Results from canopy backscatter modeling. *Remote Sensing of Environment*, 54, 324–332, doi: 10.1016/0034-4257(95)00140-9.

Weng, Q. 2001. Modeling urban growth effects on surface runoff with the integration of remote sensing and GIS. *Environmental Management*, 28, 737–748, doi: 10.1007/s002670010258.

Woodhouse, I.H. 2006. *Introduction to Microwave Remote Sensing*. Boca Raton, FL: CRC Press, 370pp.

World Meteorological Organization (WMO). 2011. *Manual on Flood Forecasting and Warning*, WMO No. 1072. Geneva, Switzerland: WMO.

Wu, H., Adler, R.F., Tian, Y., Huffman, G.J., Li, H., and Wang, J. 2014. Real-time global flood estimation using satellite-based precipitation and a coupled land surface and routing model. *Water Resources Research*, 50, 2693–2717, doi: 10.1002/2013WR014710.

Zhang, S., Na, X., Kong, B., Wang, Z., Jiang, H., Yu, H., and Dale, P. 2009. Identifying wetland change in China's Sanjiang Plain using remote sensing. *Wetlands*, 29, 302–313, doi: 10.1672/08-04.1.

Flood Studies Using Synthetic Aperture Radar Data

Sandro Martinis
German Aerospace Center (DLR)

Claudia Kuenzer
German Aerospace Center (DLR)

André Twele
German Aerospace Center (DLR)

Acronyms and Definitions

ACM	Active contour models
ASTER	Advanced Spaceborne Thermal Emission and Reflection Radiometer
DEM	Digital elevation model
DLR	German Aerospace Center
EEC	Enhanced ellipsoid corrected
EO	Earth observation
EOC	Earth Observation Center
ESA	European Space Agency
FNEA	Fractal net evolution approach
GDEM	Global digital elevation model
GIM	Geocoded incidence angle mask
RaMaFlood	Rapid Mapping of Flooding
SAR	Synthetic aperture radar
SC	ScanSAR
SM	Stripmap
SRTM	Shuttle Radar Topography Mission
SWBD	SRTM water body mask
TFS	TerraSAR-X flood service
WaMaPro	Water mask processor
WSM	Wide swath mode
ZKI	Center for Satellite-Based Crisis Information

6.1 Introduction

The demand for crisis information on natural disasters, humanitarian emergency situations, and civil security issues has substantially increased during recent years worldwide. The use of Earth observation (EO) data in disaster management is essential to provide large-scale crisis information. Especially in the preparedness, emergency and reconstruction stage of the disaster cycle remote sensing data have been proven to be indispensable for various national and international initiatives. For this reason, the European Space Agency (ESA) and the National Centre of Space Research of France initiated the International Charter "Space and Major Disasters" in 1999 (http://www. disasterscharter.org). This consortium of space agencies and satellite data providers aims at providing a unified system of rapid satellite data acquisition and delivery in case of major natural or man-made disasters. A raising awareness of satellite-based crisis information has led to an increase in requests to corresponding value adders to support civil-protection and relief organizations with disaster-related mapping and analysis. Examples for value adders are as follows: Service Régional de Traitement d'Image et de Télédétection (SERTIT), ZKI (Center for Satellite-Based Crisis Information), Information Technology for Humanitarian Assistance, Cooperation and Action (ITHACA), e-Geos, and

United Nations Institute for Training and Research Operational Satellite Applications Programme (UNOSAT).

Flood is not only one of the most widespread natural disasters, which regularly causes large numbers of casualties with rising economic loss, extensive homelessness, and disaster induced disease, but is also the most frequent disaster type. According to statistics of the International Charter "Space and Major Disaster," ~52.3% of the total number of activations between the years 1999–2013 are related to floods.

But not only in the context of disaster situations does knowledge on flood extent play a crucial role. Many places on Earth are subject to water level fluctuations. Especially in tropical and subtropical regimes, large areas of natural and inhabited sphere are frequently inundated due to monsoon-driven rainy seasons. Extreme changes in inundation extent, inundation depth, as well as intra-annual inundation frequency define phenological patterns, animal migration routes, and last but not least human living space including future development plans for the expansion of infrastructure. Some selected examples for regions with extreme variations in inundation are the large inland lakes of China (Poyang lake, Dongting Lake), which frequently double in their extent during flood seasons; the Tonlé Sap Lake, the largest freshwater lake of Southeast Asia, in Cambodia, with annual variations strongly related to the flood pulse of the Mekong; the Okavango inland delta in Botswana, Africa, which, when inundated via annual flood pulse and rainfall, is a major biodiversity hot spot on the African continent; or the Mekong Delta in Vietnam, where the annual rainy seasons bring with it "the beautiful flood" that people are used to and which enables them to fish and to irrigate their paddy rice fields and orchards. Thus, not every situation of flood or inundation is automatically disastrous or a natural hazard. As EO delivers a neutral representation of what is happening on the ground, the categorization into an unwanted or even catastrophic flood, an annual river-pulse-related flood that people are used to, or simply a short term inundation event has to be undertaken by an image analyst.

What all floods and inundation situations have in common is that they often cover large regions, which are difficult to access from the ground. Spaceborne remote sensing data are a well-suited information source to obtain a synoptic view about large-scale flood situations and their spatiotemporal evolution in a time- and cost-efficient manner. This is especially valid in regions where hydrological information is difficult to obtain due to inaccessibility or sparse distribution of gauging stations.

Optical satellite sensors have successfully been used in the past to detect flood areas (e.g., Blasco et al. 1992, Smith 1997, Wang et al. 2002, Peinado et al. 2003, Van der Sande et al. 2003, Ahtonen et al. 2004, Brakenridge and Anderson 2005, Ottinger et al. 2013). A detailed review can be found in Marcus and Fonstad (2008). If available, optical data are the preferred information source for flood and inundation mapping due to their straightforward interpretability and rich information content. However, as flood events often occur during long-lasting periods of precipitation and persistent cloud cover, a systematic monitoring by optical imaging instruments is usually impossible.

This fact drastically decreases the regular utilization of spaceborne optical sensors in an operational rapid mapping context. It is further a particular obstacle in small- to medium-sized watersheds where inundations often recede before meteorological conditions improve (Schumann et al. 2007).

The use of the microwave region (1 mm–1 m) of the electromagnetic spectrum offers some clear advantages compared to sensors operating in the visible, infrared, or thermal range. Being an active monostatic instrument and therefore providing its own source of illumination in the microwave range, synthetic aperture radar (SAR) is characterized by nearly all-weather/day–night imaging capabilities as the emitted radar signal are able to penetrate clouds and the imaging process is independent from solar radiation. Thus, principally all acquired data sets can be used for flood detection. SAR sensors in recent years underwent a striking improvement in spatial and temporal resolution and are therefore an ideal choice for near real-time assessments in emergency situations. An overview for spatial resolution categories of remote sensing sensors is given in Table 6.1.

SAR-derived flood extent maps can be an important information source for an effective flood disaster management by supporting humanitarian relief organizations and decision makers (Voigt et al. 2007). Furthermore, such maps provide valuable distributed calibration and validation data for hydraulic models of river flow processes (e.g., Bates et al. 1997, Horritt 2000, 2006, Aronica et al. 2002, Hunter et al. 2005, Pappenberger et al. 2007, Hostache et al. 2009, Schumann et al. 2009, Matgen et al. 2010) and support the derivation of spatially accurate hazard maps in terms of flood prevention activities, insurance risk management, and spatial planning (e.g., De Moel et al. 2009).

Over the last decades, spaceborne SAR systems (Figure 6.1) have increasingly been used for flood extent mapping. While past and current SAR satellite and space shuttle radar missions with spatial resolutions of the categories HR2 to MR1 (Table 6.1) have a proven track record for large-scale flood and inundation mapping in the X- (SIR-C/X-SAR, Shuttle Radar Topography Mission [SRTM]), C- (ERS-1/2 AMI, Envisat ASAR, RADARSAT-1, RISAT-1, SIR-C/X-SAR), and L-band domain (SEASAT-1, JERS-1, ALOS PALSAR, SIR-A/B/C/X-SAR), their capability for deriving flood parameters in complex and small-scale scenarios is limited.

TABLE 6.1   Categories for Spatial Resolution of Remote Sensing Sensors

Category	Acronym	Resolution (m)
Very high resolution 1	VHR1	≤1
Very high resolution 2	VHR2	>1 and ≤4
High resolution 1	HR1	>4 and ≤10
High resolution 2	HR2	>10 and ≤30
Medium resolution 1	MR1	>30 and ≤100
Medium resolution 2	MR2	>100 and ≤300
Low resolution	LR	>300

Source: European Commission, GMES data access specifications of the Earth observation needs over the period 2011–2013, Brussels, Belgium, 2011.

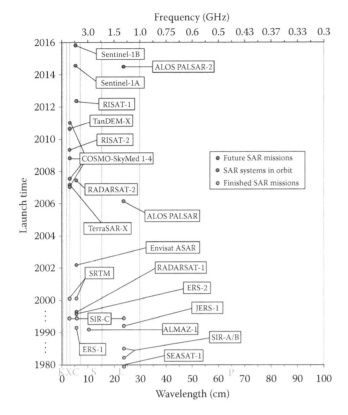

FIGURE 6.1 Launch of past and future civil spaceborne SAR missions between the years 1978 and 2016 in dependence of system's wavelength. Corresponding frequencies and commonly used spectral bands are also illustrated. (Modified based on Martinis, S., Automatic near real-time flood detection in high resolution X-band synthetic aperture radar satellite data using context-based classification on irregular graphs, PhD thesis, Ludwig-Maximilians-University Munich, Munich, Germany, 2010.)

Since 2007, the successful launch of the European platforms TerraSAR-X/TanDEM-X and the COSMO-SkyMed constellation consisting of four satellites marks a new generation of civil X-band SAR systems suitable for flood-mapping purposes. These satellites provide data up to the 0.24 m spatial resolution (TerraSAR-X Staring Spotlight mode), permitting an operational derivation of detailed hydrological parameters from space during rapid mapping activities. The potential of these data has been demonstrated by several studies to support flood emergency situations (e.g., Martinis et al. 2009, 2011, 2013, 2015, Martinis and Twele 2010, Schumann et al. 2010, Matgen et al. 2011, Pulvirenti et al. 2011, 2012, Giustarini et al. 2012, Mason et al. 2012, Kuenzer et al. 2013a,b, Pierdicca et al. 2013).

The upcoming ESA's satellite mission Sentinel-1, a constellation of two polar-orbiting C-Band SAR sensors, will enable a systematic large-scale flood monitoring with a spatial resolution of 5 × 20 m in the standard interferometric wide swath mode and a high temporal resolution of 6 days. Sentinel-1 is designed to operate in a preprogrammed conflict-free mode that ensures a consistent long-term data archive for flood-mapping purposes (Torres et al. 2012). An overview of the key characteristics of

TABLE 6.2 Key Characteristics of Current, Past, and Planned Future Civil Spaceborne SAR Missions

SAR System	Band	Polarization	Look Angle (°)	Swath (km)	Resolution (m)
ALMAZ-1	S	HH	20–70	350	10–30
ALOS PALSAR-1	L	HH, VV, HV, VH	10–51	40–350	6.25–100
ALOS PALSAR-2	L	HH, VV, HV, VH	8–70	25–350	1–100
COSMO-SkyMED 1–4	X	HH, VV, HV, VH	20–59.5	10–200	1–100
Envisat ASAR	C	HH, VV, HV, VH	14–45	58–405	30–1000
ERS-1/2 AMI	C	VV	23	100	30
JERS-1	C	HH	35	75	18
RADARSAT-1	C	HH	10–60	45–500	8–100
RADARSAT-2	C	HH, VV, HV, VH	10–60	10–500	3–100
RISAT-1	C	HH, VV, HV, VH	12–55	10–225	1–50
RISAT-2	X	HH, VV, HV, VH	20–45	10–50	1–8
SEASAT-1	L	HH	20–26	100	25
Sentinel-1A/B	C	HH, VV, HV, VH	20–45	20–400	5–100
SIR-A	L	HH	47–53	40	40
SIR-B	L	HH	15–60	10–60	15–45
SIR-C	X, C, L	HH, VV, HV, VH	15–60	15–90	15–45
SRTM	X, C	HH, VV	45–52	50–250	30, 90
TanDEM-X	X	HH, VV, HV, VH	15–60	5–200	0.24–40
TerraSAR-X	X	HH, VV, HV, VH	15–60	5–200	0.24–40

Source: Based on Lillesand, T.M. et al., *Remote Sensing and Image Interpretation*, 5th edn., John Wiley & Sons, New York, 2004; eoPortal, https://directory.eoportal.org/, 2014.

current, past, and planned future spaceborne SAR systems is given in Table 6.2.

This chapter aims to give an overview of the strengths and limitations of spaceborne SAR to monitor flood and inundation events. First, the physical basics of the interaction of the radar signal with water surfaces under different conditions, as well as the difficulties that may arise in detecting water using SAR data, are described. An overview about the state of the art concerning SAR-based water detection is given. Finally, three operational algorithms from recent literature—all developed by the Earth Observation Center of the German Aerospace Center (DLR)—are presented, which are targeted at different application domains: an object-based algorithm optimized for semiautomatic water detection during flood rapid mapping activities by an active image interpreter (rapid mapping of flooding [RaMaFlood]), an automatic and open-source distributable software tool enabling continuous water monitoring from TerraSAR-X and Envisat ASAR on a local to global level (water mask processor [WaMaPro]), and a fully automatic processing chain for flood mapping based on TerraSAR-X data on a global level (TerraSAR-X flood processor). These different tools all have their own advantages and limitations and will be presented in detail. Furthermore, current trends and open research gaps in SAR-based flood mapping will be elucidated and discussed.

6.2 Interaction of SAR Signal and Water Bodies

This section describes the physical basics of the interaction of the SAR signal with water bodies under various environmental conditions (smooth and rough open water surfaces, partially submerged vegetation areas, urban areas) and addresses the difficulties and limitations of using SAR in water mapping and flood detection.

The interaction between actively emitted microwaves and water bodies depends both on environmental factors and acquisition parameters of the satellite system. In principle, the detectability of water in SAR imagery is controlled by the contrast between water areas and the surrounding land, which is highly influenced by surface roughness characteristics, and the system-specific parameters wavelength λ, incidence angle θ_i, and polarization. The following reflection and scattering types can be observed, which are schematically illustrated in Figure 6.2: specular reflection, corner reflection, diffuse surface scattering, diffuse volume scattering, and Bragg scattering. These effects occur when the radar signal interacts with smooth and rough open water surfaces, partially submerged flooded vegetation, or flooded urban settlements. Factors that lead to misclassifications of the flood extent are summarized in Table 6.3 and partially

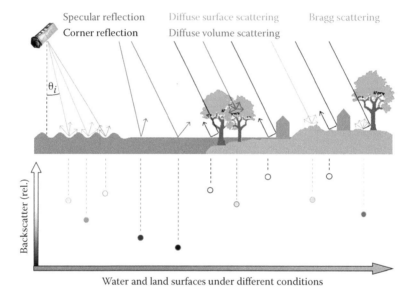

FIGURE 6.2 Scattering mechanisms of water and land surfaces under different environmental conditions as well as specular and diffuse components of surface-scattered radiation as a function of SAR incidence angle and surface roughness. (From Martinis, S., Automatic near real-time flood detection in high resolution X-band synthetic aperture radar satellite data using context-based classification on irregular graphs, PhD thesis, Ludwig-Maximilians-University Munich, Munich, Germany, 2010.)

TABLE 6.3 Factors Leading to Misclassification of Flooding in SAR Data as well as Their Occurrence and Impact on the Flood Classification Result

Overestimation of Flooding		Underestimation of Flooding	
Factor	Occurrence/ Impact	Factor	Occurrence/ Impact
Shadowing effects behind vertical objects (e.g., vegetation, topography, anthropogenic structures)	+++	Volume scattering of partially submerged vegetation and water surfaces completely covered by vegetation	+++
Smooth natural surface features (e.g., sand dunes, salt and clay pans, bare ground)	+++	Double-bounce scattering of partially submerged vegetation	++
Smooth anthropogenic features (e.g., streets, airstrips)	++	Anthropogenic features on the water surface (e.g., ships, debris)	+
Heavy rain cells	+	Roughening of the water surface by wind, heavy rain, or high flow velocity	+
		Layover effects on vertical objects (e.g., topography urban structures, vegetation)	+

Note: Feature range: high +++; medium ++; low +.

Wind/waves—Sveggesund (Norway)
TerraSAR-X SM HH, © DLR 2008
CC: 60.18°N, 5.08°E

Flooded vegetation—Arkansas (USA)
TerraSAR-X SM HH, © DLR 2008
CC: 35.81°N, 91.24°W

Debris/vegetation—Sendai (Japan)
TerraSAR-X SM HH, © DLR 2011
CC: 38.22°N, 140.93°E

Sand dunes—Walvis Bay (Namibia)
TerraSAR-X SM HH, © DLR 2008
CC: 22.94°S, 14.54°E

Radar shadow—Sinabung (Indonesia)
TerraSAR-X HS HH, © DLR 2014
CC: 3.17°N, 98.39°E

Streets/airport—Leipzig (Germany)
TerraSAR-X SM HH, © DLR 2011
CC: 51.41°N, 12.23°E

FIGURE 6.3 Examples of SAR scenes [dB] with potential features leading to under- (upper row) and overestimations (lower row) in water and flood mapping.

TABLE 6.4 Preferred Radar System Parameters for Observing Inundations in Dependence of Environmental Conditions

Flood Type	Band	Inc. Angle	Polarization	Reason
Smooth water	X	Shallow	HH	Higher contrast between nonwater/water surfaces with decreasing wavelength
Rough water	L	Shallow	HH	Lower sensitivity of the SAR signal to roughened water surfaces with increasing wavelength
Flooded dense vegetation	L	Moderate/ steep	HH/VV	Increased probability of double-bounce effects; enhanced contrast between nonflooded/flooded vegetation
Flooded short/ sparse vegetation	X/C	Moderate/ steep	HH/VV	Increased probability of double-bounce effects; enhanced contrast between nonflooded/flooded vegetation
Urban areas	X/C	Steep	HH	Reduced shadowing effects; enhanced contrast between nonwater/water areas

visualized in Figure 6.3. The preferred radar system parameters for monitoring floods in dependence of different environmental parameters are listed in Table 6.4.

6.2.1 Smooth Open Water

Surface roughness, which determines the angular distribution of surface scattering, is the main environmental factor affecting backscattering of the SAR signal, whereas the dielectric properties of a target control the penetration depth and therefore the intensity of the returned energy. Generally, an increasing

moisture content of objects results in enhanced reflectivity and stronger surface scattering (Ulaby et al. 1982).

The ideal case in detecting the extent of a water body is that the water surface is smoother than the surrounding land with respect to wavelength and incident angle of the transmitted pulse. A water surface that is smooth in relation to the surrounding dry land surface with respect to wavelength and incident angle of the SAR radiation facilitates the extraction of the water extent. An open water body may be simplistically modeled as a perfectly smooth boundary separating two semi-infinite media of high dielectric constant (Ulaby et al. 1982), which acts as a specular reflector directing the incident microwave signal away from a side-looking SAR sensor (e.g., Smith 1997, Horritt et al. 2003, Mason et al. 2007). Due to a very low signal return, open smooth water surfaces appear relatively dark in the data. These regions contrast to surrounding nonwater areas of enhanced roughness, which are characterized by increased diffuse surface scattering. According to the Rayleigh criterion, the contrast between water and other land-cover classes rises with increasing local incidence angle θ_{loc} (e.g., Ulaby et al. 1982):

$$\sigma_{rms} = \frac{\lambda}{8\cos\theta_{loc}} \qquad (6.1)$$

where σ_{rms} is the root mean square height of the surface variations.

With decreasing system wavelength, the sensitivity of a smooth water surface to diffuse scattering increases. However, as the number of possible objects on the land that might appear smooth and have a similar backscatter as water is reduced at longer wavelengths, a higher contrast ratio between water and the land areas occurs at higher system frequencies (Drake and Shuchman 1974). Consequently, water monitoring using X-band SAR appears to be more suitable than using longer wavelengths, for example, in the C- and L-band domain. Imaging at low incidence angles also means an increase in the occurrence of radar shadowing effects (Lewis 1998). Radar shadow mainly occurs behind steep vertical features or slopes when the radar beam is no longer able to illuminate the ground surface (Figure 6.3). The increasing incidence angle from near to far range is accompanied by a higher frequency of shadows effects. In high-resolution SAR data, shadowing occurs also behind single vertical objects such as anthropogenic structures (e.g., buildings) and natural features (e.g., vegetation). These areas of low radar backscatter are easily mixed up with smooth open water areas and lead to an overestimation of the water extent. These errors also occur at look-alike areas of low surface roughness such as sand dunes (Figure 6.3), bare ground, agricultural crop land, airport runways (Figure 6.3), and streets (Figure 6.3). A further phenomena primarily occurring in X-band data are cloud shadows, which are generated by an attenuation of the traversing signal due to hydrometeors in rain cells (Danklmayer et al. 2009).

In contrast, imaging with steep incidence angles increases the probability of radar layover, which can be observed if the incidence angle is smaller than the slope of the object facing the sensor. This phenomenon is related to information loss, which is particularly obstructive in identifying narrow water bodies bordered by high trees, banks (Henderson 1987), as well as between buildings. Anthropogenic features on the water surface such as ships and debris (Figure 6.3) generally lead to underestimations of the water extent.

Also, the type of polarization plays an important role in detecting open water bodies, which describes the restriction of electromagnetic waves to a single plane perpendicular to the direction of propagation of the SAR signal. Polarimetric SAR systems transmit either in a horizontal (H) or vertical (V) plane, which also can also be received horizontally or vertically. Thus, there can be two possibilities of like polarization (HH, VV) and cross-polarization (HV, VH). Generally, HH polarization provides the best discrimination between water and nonwater terrain (e.g., Ahtonen et al. 2004, Henry et al. 2006, Schumann et al. 2007). This is caused by a low scattering of the horizontal component of the signal from the smooth open water surface. An increase in surface roughness reduces the ability to discriminate between water and land comparably more using VV than using HH polarization. Over smooth water surfaces, like polarization offers enhanced class separability in comparison to cross-polarization (e.g., Martinis 2010). Several studies showed the superiority of cross-polarization HV (Horritt et al. 2003, Henry et al. 2006) and VH (Schumann et al. 2007) over like polarization VV in terms of a roughened water surface, given the fact that VV-polarized electromagnetic waves are more sensitive to ripples and waves.

6.2.2 Rough Open Water

The influence of wind and heavy rain leads to the appearance of small perturbations (ripples) (Lewis 1998) in the scale of millimeters to centimeters and longer waves with wavelength in the order of meters and kilometers on water surfaces. An increasing roughness causes a higher backscatter of the SAR signal with similar intensities than the surrounding nonwater regions (Figure 6.2). Therefore, the existence of roughened water surfaces potentially causes an underestimation of the water extent in SAR-based water mapping, which depends on the size of the roughened water surface within a SAR scene.

Bragg scattering occurs for slightly rough water surfaces with tiny capillary waves and short gravity waves at incidence angles >30° (Ulaby et al. 1982). It is a special case of scattering and can be expressed by the Bragg equation, which describes the relationship between the wavelengths of periodically spaced surface patterns. If the scatterer positions are oriented in such a way that they have geometric structures aligned with the phase fronts of the illumination and if they are spaced periodically in range, the backscattering strongly increases through constructive inference at certain incidence angles (Raney 1998).

The larger a water body, the more sensible it is for the formation of waves. Narrow rivers enclosed by riparian vegetation are rarely affected by wind, whereas on oceans, roughness structures appear more frequently. Generally, the visible roughness structure of roughened inland water bodies does not show regular wave patterns. Regular patterns are mainly observable on ocean surfaces (Figure 6.3), whereas irregular patterns occur more frequently on the surface of large inland water bodies such as lakes or large inundated areas—for example, rice paddy land in Southeast Asia or other monsoon-influenced regions.

6.2.3 Partially Submerged Vegetation

In addition to the roughening of a water surface also partially submerged vegetation may cause a backscatter increase over water bodies (Figure 6.3). Microwaves have the capability to penetrate into media. Therefore, in comparison to optical sensors, SAR offers the unique opportunity to detect—to a certain extent—inundation beneath vegetation. This is enabled by multiple-bounce effects: The penetrated radar pulse is backscattered from the horizontal water surface and lower sections of the vegetation (trunks and branches). This results in an increased signal return (e.g., Richards et al. 1987, Townsend 2001, Hong et al. 2010) in comparison to nonflood conditions (Figure 6.2) as diffuse scattering on the dry ground reduces the corner reflection effect.

However, the signal return from partially submerged vegetation is very complex and strongly depends on system parameters (e.g., wavelength, incidence angle, and polarization) and environmental parameters (canopy type, structure, and density). Theoretical scattering models can be used to describe the interaction of these parameters (e.g., Ormsby et al. 1985, Richards et al. 1987, Wang et al. 1995, Kasischke and Bourgeau-Chavez 1997).

According to Kasischke and Bourgeau-Chavez (1997) and Townsend (2002), the backscatter coefficient $\sigma_{0,w}$ of wetlands dominated by woody vegetation such as shrubs and trees can be described by

$$\sigma_{0,w} = \sigma_{0,c} + \tau_c^2 \tau_t^2 \left(\sigma_{0,s} + \sigma_{0,t} + \sigma_{0,d} + \sigma_{0,m} \right) \qquad (6.2)$$

where

$\sigma_{0,c}$ is the backscatter coefficient of the vegetation canopy
τ_c is the transmission coefficient of the vegetation canopy
τ_t is the attenuation of the SAR signal by the tree trunks
$\sigma_{0,s}$ is the backscatter from the ground surface
$\sigma_{0,t}$ is the direct backscatter from the tree trunks
$\sigma_{0,d}$ is the double-bounce scattering between the trunks and the water surface
$\sigma_{0,m}$ is the backscatter from multipath scattering between the ground surface and the canopy (Figure 6.4)

With very dense covering vegetation canopies (e.g., a dense mangrove forest canopy above inundated areas), water surface might also remain undetected, if the SAR pulse does not reach the water surface and is caught in volume scattering within the canopy.

Next to water surfaces covered by forest canopies, a very common and special case are the backscatter characteristics of paddy rice—a land use class extensively distributed on Earth (Kuenzer and Knauer 2013). Here, vegetation emerges subaquatically (very low backscatter of a typical water surface) and then reaches the water surface so that rice plant components influence the return pulse (increase in backscatter). During the ripening phase, backscatter then decreases again due to plant geometry (Kuenzer and Knauer 2013).

Floating aquatic plants such as water hyacinths are a major problem in water mapping as the plant structure increases the radar signal in comparison to open water areas. In case that the radar backscatter over water hyacinths is higher than the surrounding nonwater surfaces, classification accuracies can be significantly increased (e.g., Martinis and Twele 2010).

Generally, the capability of microwaves to penetrate into vegetation canopy increases with the system's wavelength. L-band SAR sensors have proven to be very effective to detect flooding in forests (e.g., Ormsby et al. 1985, Richards et al. 1987, Hess et al. 1990, 1995, 2003, Hess and Melack 1994, Townsend and Walsh 1998). In these wavelengths, the double-bounce trunk–ground signal interactions generate bright signatures in the data (Richards et al. 1987). In C-band and especially X-band, canopy attenuation, volume, and surface scattering from the top layer of the forest canopy is usually higher (Richards et al. 1987). This is related to a decreased backscatter ratio between forests with dry and flooded conditions.

Some studies state that also C-band SAR data can be used to map inundation beneath selected floodplain forest canopies (Townsend and Walsh 1998, Townsend 2001, 2002, Lang et al. 2008). A decrease in leaf-area index increases the transmissivity of the crown layer (Townsend 2001, 2002) and therefore increases the amount of microwave energy reaching the forest floor (Lang et al. 2008). Therefore, higher classification accuracies can generally be derived during leaf-off conditions

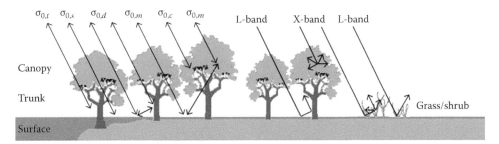

FIGURE 6.4 Conceptual illustration of the major sources of backscatter from partially inundated vegetation and effect of flooded vegetation on X- and L-band SAR. (From Martinis, S., Automatic near real-time flood detection in high resolution X-band synthetic aperture radar satellite data using context-based classification on irregular graphs, PhD thesis, Ludwig-Maximilians-University Munich, Munich, Germany, 2010; based on Kasischke, E.S. et al., *Remote Sens. Environ.*, 59, 141, 1997; Lang, M.W. et al., *Remote Sens. Environ.*, 112, 3898, 2008; Ormsby, J.P. et al., *Photogramm. Eng. Remote Sens.*, 51, 317, 1985.)

(Townsend 2001, 2002). For example, Townsend (2001) achieved differences in classification accuracy of the class flood, which is increased more than 17% between leaf-on and leaf-off RADARSAT-1 data over forests in the lower Roanoke River floodplain in Eastern North Carolina.

Increasing backscatter using C-band SAR over floating aquatic macrophytes and emergent shrubs in floodplain lakes is reported by Alsdorf et al. (2000). Usually, high X-band double-bounce returns from flooded foliated forests only occur at the edges as the SAR signal is not able to the penetrate the vegetation canopy (Henderson 1995).

Double bouncing may also occur with decreasing wavelength at shorter or sparser vegetation with thin branches and small diameter trunks. Horritt et al. (2003) observed backscatter increases over marshlands in C-band rather than in L-band. The reason for this lies in the capability of the C-band signal to penetrate the sparse canopy and to interact with the water surface and lower parts of the vegetation. This causes an enhanced signal return. Sparse vegetation may be very transparent for the SAR signal at L-band (Figure 6.4). In that case, no interaction of the water body and the vegetation occurs. Results reported by Ormsby et al. (1985) and Ramsey (1995) indicate enhanced backscattering in marshland environments in C-Band and even in X-band, respectively. Voormansik et al. (2013) used TerraSAR-X data to successfully map flooding in deciduous and coniferous forests of the temperate zone in Estonia during the no-foliage season. In this study, an average backscatter increase of 3.2 dB could be stated over mixed forests. The difference in average backscatter during nonflood and flood situations offered values of 6.2 and 4.0 dB over deciduous and coniferous forests, respectively. Increased double-bounce mechanisms could also be stated over flood surfaces covered by olive groves and deciduous forests in Italy using COSMO-SkyMed data (Pulvirenti et al. 2012) and covered by grassland and foliated shrubs in Caprivi/Namibia using TerraSAR-X data (Martinis and Twele 2010). Therefore, even in X-band, there lies a certain potential to map flooding beneath vegetation.

Other properties that need to be considered for flood mapping beneath vegetation are incidence angle and polarization. Several studies have shown that steep incidence angles are better suited for flood mapping in forests than shallow ones (e.g., Richards et al. 1987, Hess et al. 1990, Wang and Imhoff 1993, Wang et al. 1995, Bourgeau-Chavez et al. 2001, Lang et al. 2008). This generalization can be attributed to a shorter path length of the SAR signal through the canopy, which increases the transmissivity in the crown layer. Thus, more microwave energy is available for ground–trunk interactions. In contrast, shallower incidence angle signals interact more strongly with the canopy, resulting in increased volume scattering (Hess et al. 1990, Lang et al. 2008). For example, Lang et al. (2008) stated a decrease of 2.45 dB between incidence angles of 23.5° and 47° and of 0.62 dB between incidence angles of 23.5° and 43.5° in RADARSAT-1 data over flooded forests at the Roanoke River, North Carolina.

Radar systems with multiple polarizations provide more information on inundated vegetation areas than single-polarized SAR (Hess and Melack 2003, Horritt et al. 2003, Gebhardt et al. 2012, Kuenzer and Knauer 2013). Many studies employing multipolarized data indicate advantages of like polarization (HH or VV) for separating flooded and nonflooded forests (e.g., Evans et al. 1986, Wu and Sader 1987, Martinis 2010). Backscattering is generally very low for cross-polarization (HV or VH) as depolarization does not occur for ideal corner reflectors (Leckie 1998). According to Wang et al. (1995) and Townsend (2002), the backscatter ratio between flooded and nonflooded forest is higher at HH polarization than at VV polarization.

6.2.4 Flooding in Urban Areas

In urban settlements, the detectability of flooding is strongly reduced in comparison to rural areas. This is a common problem in SAR-based water detection, which is widely discussed in the literature (e.g., Giacomelli et al. 1995, Martinis 2010, Mason et al. 2010, Giustarini et al. 2012, Kuenzer et al. 2013a). Dihedral and trihedral reflection from anthropogenic structures such as buildings as well as the presence of metal surfaces leads to enhanced backscatter and strong contributions from side lobes, which are nearly identical for nonflooded and flooded situations. Also, illumination phenomena that are related to the side-looking geometry of imaging radar systems restrict the ability to detect urban flooding. Frequently, water surfaces might not be visible due to shadowing and layover effects of urban structures. The likelihood to detect flooding in urban areas generally increases with decreasing incidence angle, increasing spatial resolution of the SAR sensor, and increasing distance between anthropogenic structures in range direction. However, as nonflooded roads and other smooth man-made structures commonly also appear dark due to specular reflection, these targets can hardly be separated from smooth water surfaces.

6.3 State of the Art in SAR-Based Water Detection

Various approaches have been proposed in the literature for extracting the inundation extent from SAR data. This section provides an overview of the state of the art regarding SAR-based water mapping algorithms. Using unitemporal data, these techniques are only able to detect water surfaces. The comparison of the derived water extent with auxiliary reference water masks of normal water level conditions or with pre-event remote sensing data sets allows the separation between flooded and permanently standing water areas. The same holds true for the analyses of long intra-annual or multiyear time series, which then allows for the extraction of permanently inundated water bodies.

Classification accuracies of water/flood surfaces vary considerably within the literature and rarely exceed 90%. Beside the method and type of SAR data used, the quality of the classification result depends on the complexity of the flood situation covered by the SAR data. In Table 6.5, the strengths and limitations of the image-processing approaches commonly used in water/flood detection are summarized. Some of these approaches can

TABLE 6.5   Strengths and Limitations of Methods Commonly Applied for SAR-Based Water/Flood Extent Mapping

Flood/Water Detection Method	Subtype	Strengths	Limitations
Visual interpretation	—	Straightforward applicability, high quality	Very subjective, time demanding, requires an experienced image interpreter
Thresholding	Manual trial-and-error thresholding; automatic thresholding	Very fast, high potential of automation, moderate complexity, basis for other methods (e.g., change detection, integration of contextual and auxiliary information)	May fail in case of low contrast between water/nonwater surfaces
Change detection	Postclassification comparison; analysis of feature maps; repeat-pass interferometry	Reduction of water look-alikes, improved detection of flooded vegetation areas, separation between flooded and permanent water areas	Availability of reference data of nonflood conditions, possibly high complexity
Contextual classification	Texture based; object based; ACMs; Markov models (spatial/hierarchical)	Improvement of classification accuracy by considering relationships between pixels	Possibly high complexity and high computational demands

be combined (e.g., automatic thresholding of SAR difference images). Further improved can be achieved by integrating auxiliary data into the classification process (Section 6.3.5).

Flood mapping using SAR data is generally carried out either by visual interpretation or by digital image analysis.

6.3.1 Visual Interpretation

Visual interpretation and manual digitalization of the land/water boundary gives a reasonably accurate assessment of the water extent (Sanyal and Lu 2003). As a major disadvantage, this approach is very time consuming, especially in case of large-scale flood situations. The results are further subjective and hard to reproduce since they can vary according to the experience of the image interpreter.

6.3.2 Thresholding

Digital image-processing techniques classify each pixel of the SAR raster data into water and nonwater areas using spectral properties as well as contextual and auxiliary information.

Thresholding is the most popular image-processing techniques in water detection. Commonly, a threshold value is used to label all image elements of the SAR data as water or nonwater. Due to its simplicity, this method is computationally very fast and therefore suitable for rapid mapping purposes. The quality of thresholding procedures for detecting water using SAR imagery depends on the contrast between water and nonwater areas. In low to moderate roughness conditions of open water surfaces, usually, most of the water extent can be extracted when employing this technique. All image elements of the SAR amplitude or intensity data with gray values lower than a defined threshold are assigned to the water class. The thresholding process may fail in case increased surface roughness conditions reduce the contrast between water and nonwater areas. The occurrence of double-bouncing with partially flooded vegetation areas demands for the definition of an additional threshold value, which labels all pixels above a certain backscatter value to an additional class "flooded vegetation."

An adequate threshold can be determined in a supervised manner using visual inspection of the global scene histogram or manual trial-and-error approaches (e.g., Townsend and Walsh 1998, Townsend 2001, Brivio et al. 2002, Henry et al. 2006, Matgen et al. 2007, Lang et al. 2008, Gstaiger et al. 2012). Several studies also state that empirically defined default threshold values can be successfully used for detecting smooth water surfaces in case areas of stable environmental conditions are repeatedly being monitored or data are acquired with the same SAR system parameters. In this context, Kuenzer et al. (2013a,b) and Gstaiger et al. (2012) performed water extent mapping based on Envisat ASAR wide swath mode (WSM) and TerraSAR-X data in the Mekong Delta. Wendleder et al. (2012) conducted global water body detection based on amplitude and bistatic coherence information of TanDEM-X VV-polarized Stripmap (SM) data. The combined approach based on amplitude and bistatic coherence information results in a significant higher accuracy (98.7%) than the water mask solely based on amplitude information (92.5%) for a test site at river Elbe, Germany. As the bistatic coherence is only generated in the course of the TanDEM-X mission for the generation of a global DEM in a predefined acquisition plan, this information is commonly not available for flood disaster mapping.

Only few approaches for the automatic extraction of water-related threshold values based on image statistics can be found in the literature (Martinis et al. 2009, Schumann et al. 2010, Matgen et al. 2011, Pulvirenti et al. 2012). However, automatic threshold extraction approaches should be favored for near real-time applications and systematic satellite-based flood monitoring systems. Schumann et al. (2010) compute a threshold value of −8.5 dB from the global gray level histogram of ASAR WSM data using Otsu's method (Otsu 1979), which derives a criterion measure to evaluate the between-class variance of water and nonwater areas. Matgen et al. (2011) perform thresholding by modeling the flood class using a nonlinear fitting algorithm under the gamma-distribution assumption. Martinis et al. (2009) present an automatic tile-based thresholding approach that solves the flood-detection problem in large-size TerraSAR-X amplitude data even with small *a priori* class probabilities by applying the KI thresholding approach (Kittler and Illingworth 1986) on selected image tiles, which are likely to represent a bimodal distribution of the classes to be separated. This method is enhanced in robustness and adapted to SAR data radiometrically calibrated to sigma naught in Martinis et al. (2015) and adapted to calibrated COSMO-SkyMed data by Pulvirenti et al. (2012).

6.3.3 Change Detection

Change detection based on multitemporal image analysis is an effective tool for delineating varying inundation extent and inundation frequency. In the context of flood mapping, change detection is usually performed by comparing preflood reference data with in-flood imagery using postclassification comparison (e.g., Herrera-Cruz and Koudogbo 2009) as well as by analyzing feature maps such as difference (e.g., McMillan et al. 2006, Matgen et al. 2011, Giustarini et al. 2012), normalized difference (Nico et al. 2000, Martinis et al. 2011), ratio (e.g., Rémi and Hervé 2007), and log ratio data (e.g., Bazi et al. 2005). Change detection is ideally performed using data acquired with the same sensor and with similar system parameters. Change detection may help to enhance the flood-mapping result obtained from the analysis of a single flood image by reducing overestimations of inundated areas related to water look-alike areas. Also, it allows separating areas permanently covered by water from temporally flooded terrain. Amplitude and coherence change detection are applied in the SAR domain. In the amplitude approach, regions are labeled as flooded where the backscatter has considerably decreased in case of smooth water surfaces or increased in case of double-bouncing vegetation areas from pre- to postdisaster data. Phase information derived from SAR interferometry also has the potential to be used for flood mapping. Various studies (e.g., Marinelli et al. 1997, Dellepiane et al. 2000) state that repeat-pass SAR interferometry can be employed to identify water as regions of low interferometric phase correlation, which can be separated from dry land of higher coherence. Single-pass interferometry can be used to eliminate temporal decorrelation effects and therefore to enhance the quality of water masks (Wendleder et al. 2012, Warth and Martinis 2013).

6.3.4 Contextual Classification

Some approaches integrate spatial-contextual information from a local neighborhood within the flood-detection workflow: Ahtonen et al. (2004) present an automatic surface water procedure, which integrates local textural features into the labeling scheme. This method uses an ML classifier trained by unsupervised thresholding of log-mean data. The classification process is performed on a 3-D feature space composed of logarithmically transformed occurrence measures within a kernel of size 5×5.

A supervised flood-mapping algorithm SAR using Kohonen's self-organizing maps (Kohonen 1995) based on artificial neural networks is proposed by Kussul et al. (2008). For considering spatial connections between neighboring pixels, the network is trained in an unsupervised manner using backscatter values from sliding windows in Envisat ASAR, ERS-2, and RADARSAT-1 data.

In the past, several methodologies based on region growing have been used in waterline detection. Commonly, seeded regions using semiautomatic or automatic algorithms are dilated according to their statistical properties until stopping criteria are reached (Malnes et al. 2002, Mason et al. 2010, 2012, Matgen et al. 2011, Martinis et al. 2015).

In recent years, statistical active contour models (ACMs), so-called snake algorithms, gained in popularity for delineating land/water boundaries in single-polarized SAR data. These sophisticated region-growing procedures make use of a dynamic curvilinear contour to iteratively search through the 2-D image space until they settle upon object boundaries, driven by an energy function that is attracted to edge points. ACMs have proven useful for converting unconnected or noisy image edges into smooth continuous vector boundaries. Therefore, these algorithms are suitable for segmenting speckle-affected SAR imagery. Based on the study of Ivins and Porill (1995), a semiautomatic ACM (Psnake NT) is developed by Horritt (1999). This tool identifies inundated areas as pixels of homogeneous speckle statistics accounting for the gamma-distribution intensity of SAR data. This method was widely applied for river flood delineation in rural areas using SAR data in the HR2 and MR1 resolution domain (e.g., De Roo et al. 1999, Horritt et al. 2001, Ahtonen et al. 2004, Matgen et al. 2007, Schumann et al. 2009). Further, it is successfully applied for computing polygonal approximations of rough seawater surfaces (Horritt et al. 2001). Mason et al. (2007) modify the algorithm in such a way that the snake is conditioned both on SAR backscatter values and on LiDAR digital elevation models (DEMs). Using 3-D rather than 2-D curvatures, the waterline becomes smoothly varying in elevation. One disadvantage of Psnake NT is that this algorithm is dependent on significant user input. Several initializations of the contour line by manually set seed vectors are necessary to obtain satisfying results. Further, as Psnake NT belongs to the groups of parametric ACMs that have a rigid topography, additional seeds are necessary to delineate isolated flood regions (Mason et al. 2010). This, however, is critical in high-resolution SAR data of the categories VHR1-HR1, where, in contrast to data of coarser resolution (HR2-LR), the inundation area is commonly separated in multiple isolated flood regions by, for example, vegetation areas or man-made objects, which prohibit the expansion of the snake. In this case, geometric snake models (e.g., Malladi et al. 1995), which permit topology changes due to flexible level sets to simultaneously detect several water objects seem to be more suitable. Within this context, a semiautomatic flood-detection algorithm based on region-based level sets is proposed by Silveira and Heleno (2009).

A Bayesian segmentation technique to separate land and sea regions in X-band SAR data is proposed by Ferreira and Bioucas-Dias (2008). The class-conditional densities are estimated by a finite mixture of gamma distributions whose parameters are estimated from manually selected training samples. The *a priori* probability of the labels is modeled by a Markov random field, which promotes local continuity of the classification result given a spatial neighborhood system. The maximum *a posteriori* estimation is performed by using graph cuts (Kolmogorov and Zabih 2004).

Several studies present object-based classifications for flood-mapping purposes (e.g., Hess et al. 2003, Herrera-Cruz and Koudogbo 2009, Martinis et al. 2009, 2011, Martinis and Twele 2010, Mason et al. 2012, Pulvirenti et al. 2012). They are based on the concept that important information necessary for image analysis is not always represented in single pixels but in homogeneous image segments and their mutual relations (Baatz and Schäpe 1999; Benz et al. 2004). Within this context, a hybrid multicontextual Markov model for unsupervised near real-time flood detection in X-band SAR data has been developed. The Markov model is initialized by an automatic tile-based thresholding procedure (Martinis et al. 2009). Scale-dependent (Martinis et al. 2011) and optional spatiotemporal contextual information (Martinis and Twele 2010) is integrated into the segment-based classification process by combining causal with noncausal Markov image modeling related to hierarchical directed and planar undirected irregular graphs, respectively.

6.3.5 Integration of Auxiliary Data

The integration of auxiliary data sets can significantly support the flood-mapping process. Some studies make use of digital topographic information to improve classification results by detecting flooding beneath vegetation or by removing look-alike areas according to simplified hydrological assumptions (e.g., Wang et al. 2002, Horritt et al. 2003, Mason et al. 2007, 2010, Martinis et al. 2009). Other hydrologically relevant layers, such as the height above nearest drainage (HAND) index (Rennó et al. 2008), can be used to filter out regions where the probability of flood occurrence is low (Westerhoff et al. 2013).

Within the last years, fuzzy-logic techniques (Zadeh 1965) have increasingly been used in flood monitoring to combine ambiguous information sources by accounting for their uncertainties as opposed to only relying on crisp data sets. Martinis and Twele (2010) apply fuzzy theory for quantifying the uncertainty in the labeling of each image element in flood possibility masks. The proposed method combines marginal posterior entropy-based confidence maps with spatiotemporal relationships of potentially submerged vegetation to smooth open water areas. A pixel- and object-based fuzzy-logic approach for inundation mapping based on Pierdicca et al. (2008) is described in Pulvirenti et al. (2011, 2013). It integrates theoretical SAR scattering models, simplified hydrologic assumptions, and local context in form of intensity, topographical, and land-cover information. Based on the availability of preflood scenes, the semiautomatic algorithm is able to detect both open water areas and submerged vegetation areas. A fully automatic TerraSAR-X-based flood-mapping service is proposed by Martinis et al. (2015), which uses a fuzzy-logic-based algorithm combining SAR backscatter information with digital elevation and slope information as well as the size of water bodies for the refinement of the initial thresholding result. The fuzzy-logic-based postclassification process results in an improvement of the flood mask by mainly reducing water look-alike areas in mountainous regions related to radar shadowing. When applying this approach to a TerraSAR-X ScanSAR (SC) test scene, the users' and producers' accuracies improved from 20.9% to 82.4% and 51.7% to 83.7%, respectively.

6.4 Case Studies

In this chapter, the use of medium- and high-resolution SAR (HR1-MR1) data for flood monitoring is demonstrated. In three test scenarios, three different and recently published operational flood/water detection algorithms are presented in relation to the desired application domain. The first test case describes the application of a semiautomatic object-based water detection algorithm (RaMaFlood) on multitemporal TerraSAR-X data in the Caprivi Region of Namibia to map both open flood water surface and flooded vegetation areas (Section 6.4.1). This method commonly is used in the context of flood-related rapid mapping activities by an active image interpreter. In the second test case, a time-efficient automatic algorithm for continuous water monitoring (WaMaPro) is presented (Section 6.4.2). The algorithm is applied on a time series of Envisat ASAR WSM and TerraSAR-X data sets acquired with similar acquisition parameters, respectively, in the Mekong Delta in Vietnam. The third case study describes the use of a fully automatic and globally applicable TerraSAR-X flood service (TFS), which is designed to deliver robust results for a broad range of biomes and acquisition conditions (Section 6.4.3). The processing chain is exemplarily applied on TerraSAR-X SM and SC scenes acquired during flood situations in Nepal (2008), Germany (2011), and Albania/Montenegro (2013). The main characteristics of these methods are qualitatively compared in Table 6.6.

6.4.1 Semiautomatic Object-Based Flood Detection (RaMaFlood)

In this chapter, an object-based water detection algorithm RaMaFlood is described (Martinis et al. 2009, 2011), which can be

TABLE 6.6 Comparison of the State-of-the-Art Flood and Water Detection Algorithms RaMaFlood, WaMaPro, and TFS

	Semiautomatic Object-Based Water Detection (RaMaFlood)	Automatic Pixel-Based Water Detection (WaMaPro)	Fully Automatic Pixel-Based Flood Detection (TFS)
Accuracy	++	+	+
Look-alike elimination	++	−	+
Automation	0	+	++
Processing time	+	++	++
Multisensor capability	++	++	+
Flood possibility mask	−−	−−	++
Open source	−−	++	−−

Note: Ranking: very high ++, high +, medium 0, low −, and very low −−.

run both automatically and semiautomatically on preprocessed SAR data within the eCognition Developer software. However, as a graphical user interface has been implemented for this method, which allows an image interpreter to control and modify the parameters of each single processing step in a WYSIWYG manner, it is especially useful when applied in a semiautomatic way. Using this method, detailed information about the extent of open water bodies can be derived. Also, partially submerged vegetation areas can be extracted semiautomatically by an active image interpreter. This is hardly feasible in a completely automatic way due to the complexity of double-bounce scattering mechanisms. By intersecting the derived water masks with auxiliary reference water masks of normal water level conditions, the separation between flooded and permanently standing water areas is accomplished. The derived crisis information is integrated into map products generated by GIS experts and disseminated to end users. During numerous flood-related rapid mapping activities of DLR's Center for Satellite-Based Crisis Information (ZKI), this method has proven its effectiveness for SAR data acquired by sensors in the X-band (TerraSAR-X/TanDEM-X, COSMO-SkyMed), C-band (Envisat ASAR, RADARSAT-1/2), and L-Band (ALOS PALSASR) domain.

6.4.1.1 Methodology

RaMaFlood is experimentally applied to a multitemporal data set of four TerraSAR-X SM scenes covering the evolution of a flood situation over a period of 3½ months in the Zambezi floodplain. The study area is mainly situated in the Caprivi Strip in northeastern Namibia, which is surrounded by Zambia in the north and Botswana in the south (Figure 6.5). The Caprivi Strip is regularly affected by flooding related to heavy seasonal rainfalls.

The SAR data are all acquired in HH polarization with the same beam mode (minimum incidence angle, ~29.4°; maximum incidence angle, ~32.4°) in ascending orbit. The pixel spacing is 2.75 m (see Table 6.7).

Before applying the segmentation and classification step, the data are radiometrically calibrated to sigma naught. Even if this step is optional, it may have the advantage of removing topographic effects within the SAR data. For a reduction of the SAR data inherent speckle effect, a Gamma-MAP filter (Lopes et al. 1990) of window size 3 × 3 is used. The filtering step also minimizes the statistical overlap between class distributions and, therefore, causes improved class separability.

Within the sequence of the four acquisitions, the evolution of a large-scale flood situation is visible. In comparison to the surrounding dry land, open water surfaces appear dark due to specular reflection of the incident radar signal. In contrast, flooded vegetation causes very distinct and bright signatures. Indeed, X-band SAR has a strongly reduced ability to detect inundation beneath dense vegetation such as forest due to increased canopy attenuation and volume scattering in comparison to the longer C-, and L-band signals (e.g., Richards et al. 1987). In this study area, however, the emergent vegetation is mainly composed of grassland, foliated shrubs, and aquatic plants, whose structure

causes a multiple-bounce effect, which leads to an interaction of the penetrated radar pulse with the water surface and lower sections of the vegetation. This phenomenon causes a high signal return.

The workflow of the semiautomatic object-based flood-detection approach consists of the following main processing steps:

- Image segmentation
- Thresholding
- Postclassification refinement

The first processing step of the proposed method is the segmentation of the SAR data. Segmentation is the basic step in low-level image processing (Lucchese and Mitra 1998), in which an image is subdivided into disjoint regions, which are uniform with respect to homogeneity criteria such as spectral or textural parameters (Haralick and Sharipo 1985). In comparison to pixel-based applications, this offers the advantage that besides spectral-related information, also object parameters such as contextual information, texture, and object geometry can be used for improving classification accuracy. Object-based image analysis has constantly gained importance in EO applications during the last decade (Baatz and Schäpe 1999, Blaschke and Strobl 2001). This is particularly related to the strongly increased spatial resolution of remote sensing data, which demands image analysis techniques that are specifically adapted to the increased intraclass and decreased interclass variability of images (Bruzzone and Carlin 2006). In particular, for data of the recently launched high-resolution SAR sensors (TerraSAR-X, COSMO-SkyMed, RADARSAT-2), the use of the per-parcel methods appears promising. These data are, in comparison to SAR imagery of coarser resolution, characterized by higher variances in backscattering properties of different land-cover classes due to the reduced mixed pixel phenomenon and the SAR intrinsic speckle effect. Therefore, semantic image information is less represented in a single pixel but in homogeneous image objects and their mutual relations (Baatz and Schäpe 1999). The decomposition of the images can be accomplished by several segmentation techniques described in the literature (Haralick and Sharipo 1985, Zhang 1996, Carleer et al. 2005). This study is based on the fractal net evolution approach (FNEA) (Baatz and Schäpe 1999), which enables a multiscale representation of the data using a bottom-up region merging approach. The multilevel representation of an image Y can be represented as a connected graph Ψ_L with L levels composed of a set of nodes S. An irregular graph with three levels is generated according to the method proposed in Martinis et al. (2011) to integrate the advantages of small-, medium-, and large-scale objects into the classification process (Martinis et al. 2009). In order to prevent over- and undersegmentation of the data, the graph is automatically generated by modeling the segmentation parameters to decompose the image at each level by a mean number of objects intended by the user. This is accomplished by the procedure described in Martinis et al. (2011): First, several subsets of the SAR image are automatically selected to describe the heterogeneity of the backscatter of the SAR data. Then, a presegmentation of the subsets is

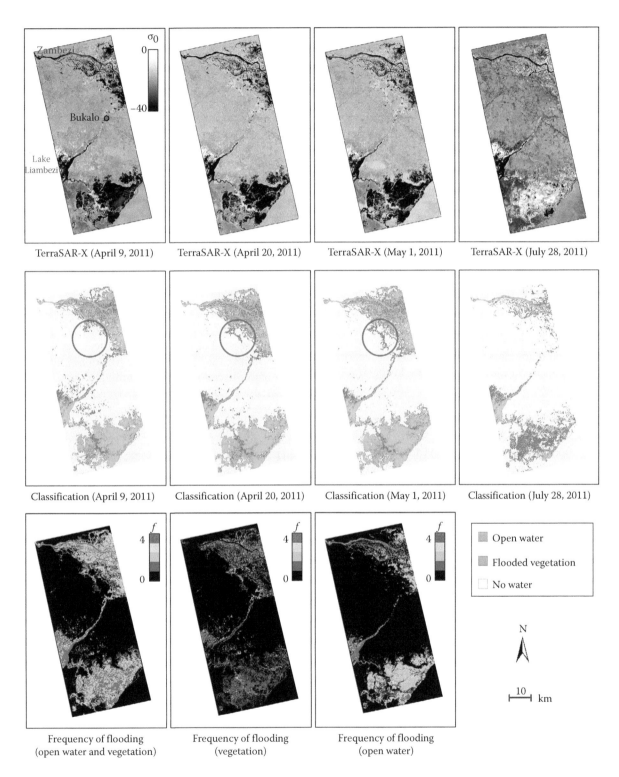

FIGURE 6.5 Time series of TerraSAR-X data (upper row) and respective inundation maps (mid row) in Caprivi/Namibia. Flood frequency (f) maps (lower row): total flooding, flooded vegetation areas, and open water surfaces. Coordinates: UL, 17.450°S, 24.245°E; UR, 17.390°S, 24.539°E; LL, 18.173°S, 24.402°E; LR, 18.111°S, 24.702°E.

TABLE 6.7   Acquisition Parameters of Multitemporal TerraSAR-X SM Scenes Covering the Caprivi Strip in Northeastern Namibia

Acquisition Time	Beam Mode	Polarization	Pixel Spacing (m)
2011-04-09, 16:42 UTC	SM_007R	HH	2.75
2011-04-20, 16:42 UTC	SM_007R	HH	2.75
2011-05-01, 16:42 UTC	SM_007R	HH	2.75
2011-07-28, 16:42 UTC	SM_007R	HH	2.75

performed by the FNEA approach. The homogeneity parameter is estimated, which leads to a decomposition of the entire image with average object sizes of the segments at each level, which come close to those intended by the user. This is accomplished by generating a database, which contains models describing the relationship between homogeneity parameter and object size according to data of different SAR sensor types and image contents. Finally, this model is selected for creating the whole graph, which best fits to the presegmentation result at each segmentation level. The irregular graph is built with a relative object size of ~50% between adjacent graph levels.

The classification of smooth open water areas is initialized by labeling all image elements with a backscatter value lower than a defined threshold to the class "water." Thresholding algorithms only extract adequate threshold values if the histogram is not unimodal. Therefore, the capability of approaches to detect an adequate threshold in the histogram depends on the *a priori* probability of the classes to be separated. If, for example, the spatial extent of the water bodies in large SAR scenes is low, the class distributions cannot be modeled sufficiently. In this study, the threshold value is automatically derived using a tile-based thresholding procedure proposed by Martinis et al. (2009, 2015), which solves the flood-detection problem in even large-size SAR data with small *a priori* class probabilities. The thresholding approach consists of the following processing steps: image tiling, tile selection, and subhistogram-based thresholding of a small number of tiles of the entire SAR image.

First, based on the SAR scene, a bi-level quadtree structure is generated. The SAR data are divided into N quadratic nonoverlapping subimages of user-defined size c^2 on level S^+. Each parent object is represented by four quadratic child objects of size $(c/2)^2$ on level S^-. The variable c is empirically defined to 400 pixels. A limited number of tiles are selected out of N according to the probability of the tiles to contain a bimodal mixture distribution of the classes "open water" and "nonwater." This selection step is based on statistical hierarchical relations between parent and child objects in a bi-level quadtree structure. The parametric Kittler and Illingworth minimum error thresholding approach (Kittler and Illingworth 1986) is used to derive local threshold values using a cost function, which is based on statistical parameterization of the subhistograms of all selected tiles as bimodal Gaussian mixture distributions. One global threshold position is derived by computing the arithmetic mean of the local thresholds. This is used for initially separating open water surfaces and nonwater areas in the SAR data.

For postclassification refinement, multiscale image information is combined with thresholding (Martinis et al. 2009). The initial threshold is first applied to the coarsest segmentation level S_1. Objects on this level contain some variations in the spectral properties of the pixels. Most of the inundation is identified by this step; however, fine tuning is subsequently reached, progressively enforcing the spectral homogeneity constraints of nonflood objects in the neighborhood around open water and flooded vegetation objects.

The preliminary extracted water bodies on S_1 are used as seeds for dilating the water regions based on medium-scale objects on S_2. This is repeated accordingly by performing region growing of small-scale objects on S_3 adjacent to seeds defined by the thresholding process on the S_1 and S_2. Only image elements located in the neighborhood of flood areas identified at level S_{l+1} are scanned. Thus, the risk of detecting flood look-alikes distant from initially labeled flood objects is reduced.

Finally, DEMs can be integrated into the classification process to improve the quality of the flood masks in a hydrological plausible way (Martinis et al. 2009).

6.4.1.2 Results

By applying the RaMaFlood tool to the TerraSAR-X test data set of Namibia, the finest level S_1 is partitioned into an intended mean object size of ~250 m². Accordingly, the mean object size increases to ~500 and ~1000 m² at S_2 and S_3, respectively. Open water areas are derived by automatically extracting the threshold value using a tile-based thresholding procedure. The threshold values are defined by $\tau_1 = -16.2$ dB (2011-04-09), $\tau_1 = -19.4$ dB (2011-04-20), $\tau_1 = -19.0$ dB (2011-05-01), and $\tau_1 = -18.9$ dB (2011-07-28) (Table 6.8). Flooded vegetation areas appear more heterogeneous than open water areas due to the occurrence of different vegetation types and various vegetation densities, which alter the intensity of the double-bounce effect. Therefore, better results are derived by setting the threshold between the class-conditional densities of nonwater areas and flooded vegetation using empirically derived values: $\tau_2 = -8.5$ dB (2011-04-09), $\tau_2 = -8.5$ dB (2011-04-20), $\tau_2 = -9.0$ dB (2011-05-01), and $\tau_2 = -9.0$ dB (2011-07-28) (Table 6.8). The threshold values are used for an initial separation of open water and flooded vegetation areas from nonwater areas using coarse-scale image information.

These regions are grown incorporating information of finer object scales into the classification process. The final postclassification step contains the removing of flood objects with an

TABLE 6.8   Threshold Values Used for Separating Open Water and Nonwater Areas (τ_1) as well as Nonwater and Partially Flooded Vegetation Areas (τ_2)

Acquisition Time	τ_1 (dB)	τ_2 (dB)
2011-04-09, 16:42 UTC	−16.2	−8.5
2011-04-20, 16:42 UTC	−19.4	−8.5
2011-05-01, 16:42 UTC	−19.0	−9.0
2011-07-28, 16:42 UTC	−18.9	−9.0

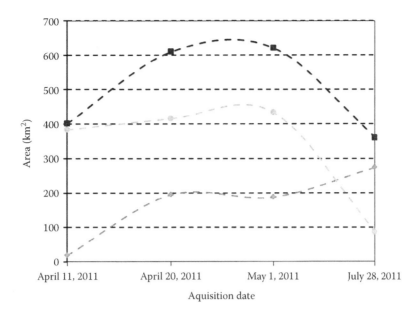

FIGURE 6.6   The evolution of the total flood area, the open water areas, and flooded vegetation areas in a time series of four TerraSAR-X Stripmap data acquired over Namibia/Caprivi.

area lower than 300 m² and the closing of nonwater areas with an area lower than 1000 and 2500 m² completely surrounded by open water areas and flooded vegetation areas, respectively.

The flood extent increases from 2011-04-09 until 2011-05-01 and reaches its lowest extent on 2011-07-28 (see Figures 6.5 and 6.6). The frequency of the flooding for open water surfaces and flooded vegetation areas of the time series is shown in the lower row of Figure 6.5.

The main flood extent consists of open water areas on the data of 2011-04-09 until 2011-05-01. Flooded vegetation areas mainly exist at the borders of lakes. Some smaller rivers and tributaries are completely covered by vegetation. The filling of dry river beds covered by vegetation is clearly apparent in some regions where the backscatter increases due to the transfer from vegetation-related volume scattering to double-bounce scattering (see, e.g., area surrounded by red circle in Figure 6.5). In this area flooded vegetation has a backscatter of ca. −4 to −5 dB. This corresponds to backscatter increase of up to 7.5 dB in comparison to preflood conditions.

The reverse behavior can be observed with decreasing water level between the 2011-05-01 and 2011-07-28. The occurrence of vegetation-induced double-bounce effects is considerable within the scenes. Its proportion of the total flood extent increases from ~5.0% on 2011-04-09 to ~30% on 2011-04-20/2011-05-01. If double-bounce effects would not be considered in the classification process, the flood extent would be considerably underestimated. This would especially be the case for the data acquired on 2011-07-28, where the percentage of the flooded vegetation increases to ~75% of the total flood extent. At this time, the water level decreases. Vegetation completely flooded on 2011-04-09 and on 04-20/05-01 gradually emerges through the water surface on 2011-07-28. The backscatter behavior is therefore reversed changing from specular reflecting surfaces to double-bouncing vegetation.

6.4.2 Automatic Pixel-Based Water Detection (WaMaPro)

In the following section, the tool WaMaPro (Huth et al. 2009, Gstaiger et al. 2012, Kuenzer et al. 2013a,b) is introduced, which is a pixel- and threshold-based, open-source software tool, able to handle TerraSAR-X and Envisat ASAR data of different spatial resolution to derive water masks in a fast and efficient manner. Most of the case study material presented here has been elaborated on in detail in Kuenzer et al. (2013a,b).

6.4.2.1 Methodology

Study area for the WaMaPro tool is the Mekong Delta in the South of Vietnam—one of the world's largest river deltas, covering 39,000 km², located between 8°30′–11°30′N and 104°30′–106°50′E. The Mekong is a single-peak pulsing river, with an annual discharge of 475 km³ and it defines the flood pulse pattern and sediment delivery to the delta. Flood pulse of the Mekong as well as precipitation in the Mekong Delta itself are characterized by accentuated dry and rainy seasons, defined by the Southwest Indian and Northwest Pacific monsoon. Whereas the rainy season lasts from early June to December, the dry season lasts from December to May. During the rainy season, large parts of the delta are frequently flooded. Despite these floods and the fact that most of the delta area is located well below 3 m above sea level, the local inhabitants term the annual flood waters "the beautiful flood." Flooding sets the base for their livelihood: the Mekong Delta is the agricultural base for Vietnam, often termed the country's "rice bowl." The frequent natural floods—paired with fertile soils and warm climate—enable up to three rice harvest per year. Furthermore, fruit tree orchards and aquaculture activities are prominent in the delta. Annual flood waters bring with them nutrient-rich sediments, enable local inhabitants to fish and irrigate, and improve the navigability of 55,000 km of man-made canals. Of course, the floods also have some negative impacts in the region, such as inundation on the ground floors of houses, which are not elevated, and also fatalities, which have occurred in the past. However, the region is—in the first place— not a region facing "disastrous," "hazardous" floods, but rather experiencing an annually returning, natural phenomenon, which is also welcomed. Nevertheless, a good understanding of the flood dynamics in the region are crucial—especially in an environment of rapid socioeconomic development, including a strong increase in urban and settled space, expanding infrastructure networks and industry, and increasing mobility. Planners of the region need to know which areas are frequently flooded and for how long, how floods proceed, and which areas are rarely or never flooded and thus pose a safe ground for, for example, construction and development.

The flood dynamics of the delta were investigated using a 5-year time series of 60 Envisat ASAR WSM data sets at 150 m pixel spacing, as well as multitemporal TerraSAR-X data in SC (five scenes) and SM mode (four scenes) for selected regions within the delta, at 8.25 and 2.5 m, respectively. An overview of this data can be found in Table 6.9.

Before applying WaMaPro, when using Envisat ASAR WSM data, a fully automatic preprocessing code triggers the geocorrection of the SAR data as well as an incidence angle correction in the ESA's NEST software. Data are corrected to

the normalized radar backscatter coefficient (sigma naught) to an incidence angle of 30°. TerraSAR-X data—due to its excellent geolocation accuracy and limited swath width— does not need any pretreatment. WaMaPro is automatically triggered if data availability is indicated via an e-mail of the data provider.

The automatic processing of water masks proceeds as already depicted in Huth et al. (2009), Gstaiger et al. (2012), and Kuenzer et al. (2013a,b). As WaMaPro has been developed in the context of several EO-based research projects with partners in developing and emerging countries, the main requirement for the algorithm was an intuitive simplicity, speedy performance, the ability to process SAR data of different sensors, and especially the independence of licensed software or complex processing infrastructures. Furthermore, it was found to be convenient to embed WaMaPro in a Web Processing Service (WPS). WaMaPro was first coded in MATLAB®, then recoded in C++, and is currently developed toward an open-source plug-in for Q-GIS. It is based on a simple threshold method, which allows for the separation of land and water pixels. As laid out in Gstaiger et al. (2012), "to firstly reduce the typical SAR-inherent speckle noise, the first step of the algorithm is to apply a standard convolution median filter with a kernel size of 5 × 5 pixels, resulting in a filtered and speckle reduced image P1. After this preprocessing two empirically chosen thresholds divide water from nonwater pixels (processed image P5). Here, the first threshold, T1, which has a lower value than the final water threshold, defines confident water areas, leading to P2. The second threshold T2, which has a higher value than the land threshold, classifies confident land areas, leading to image product P3. Then buffer zones of two pixels, which are only generated via dilatation, are applied to P," which results in product P4. The buffers define the transition zone from water to land, also represented by mixed pixels. The second threshold now enables the inclusion of the water pixels within this zone in the initial binary water mask. The temporary results P3 and P4 are now compared, and if coincidence occurs, the value (water or land, 0 or 1) is written to P5. Otherwise, the value from P2 is written to P5 ((P4 and P3) || P2). In this way, overestimated water pixels are excluded (Kuenzer et al. 2013b, p. 694). Isolated pixels are removed via morphological image closing (P6) (Figure 6.7). The elimination of the so-called islands and lakes according to a predefined maximum size (T3, T4) is mainly of relevance for high-resolution SAR data (e.g., TerraSAR-X), but does not affect Envisat ASAR WSM–derived results. The results reached with WaMaPro have been compared with other water mask or flood mask derivation algorithms and could reach or partially also exceed the accuracies of other methods tested, as elaborated in Gstaiger et al. (2012). WaMaPro enables automatic processing of the water masks right at data delivery announced via e-mail by the data provider, and the chain automatically performs all processing steps from the retrieval of the data to product generation—a process, which also has been encapsulated in a WPS. Two exemplary outputs of the water mask processing are presented in Figure 6.8.

TABLE 6.9 Data Analyzed in This Test Case Study

SAR Data Type		Acquired Date
Envisat ASAR WSM data	2007	2007-06-14, 2007-07-03, 2007-07-10, 2007-07-19, 2007-08-07, 2007-08-14, 2007-08-23, 2007-09-11, 2007-09-18, 2007-10-16, 2007-10-23, 2007-11-01, 2007-11-20, 2007-11-27, 2007-12-06
	2008	2008-06-01, 2008-06-17, 2008-06-24, 2008-07-03, 2008-07-22, 2008-08-14, 2008-08-23, 2008-08-26, 2008-09-11, 2008-09-30, 2008-10-07, 2008-10-16, 2008-11-04, 2008-11-11, 2008-11-20, 2008-11-23, 2008-12-16, 2008-12-25
	2009	2009-06-02, 2009-06-18, 2009-07-04, 2009-08-27, 2009-10-01, 2009-12-10, 2009-12-13
	2010	2010-01-14, 2010-01-17, 2010-02-18, 2010-03-25, 2010-04-29, 2010-05-02, 2010-06-03, 2010-07-08, 2010-08-12, 2010-08-15, 2010-09-16, 2010-10-08, 2010-10-21
	2011	2011-01-03, 2011-01-14, 2011-02-02, 2011-03-04, 2011-03-15, 2011-04-03, 2011-06-21
TerraSAR-X ScanSAR data	2008	2008-06-18, 2008-08-23, 2008-09-25, 2008-10-28, 2008-11-30
TerraSAR-X Stripmap data	2008	2008-08-01, 2008-09-03, 2008-10-06, 2008-11-08

Source: Kuenzer, C. et al., *Remote Sens.*, 5, 5122, 2013.

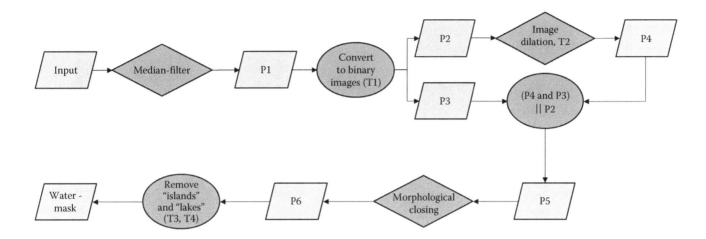

FIGURE 6.7   Processing chain of the histogram-based approach for water mask derivation. (Modified from Gstaiger et al. 2012; Kuenzer, C. et al., *Remote Sens.*, 5, 687, 2013.)

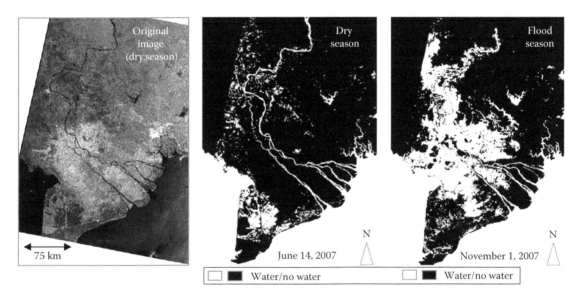

FIGURE 6.8   Envisat ASAR WSM amplitude image of 2007-06-14 (left) and water mask derived from this data set (center). The differing extent of water between the start of the rainy season (center) and the flood peak around the end of the rainy season (water masks derived from a data set in December, right) is obvious. Extent: UL, 12°N, 104°15′E; LR, 8°30′N, 106°50′E. (From Kuenzer, C. et al., *Remote Sens.*, 5, 687, 2013.)

6.4.2.2 Results

The joint visualization of individual water masks from, for example, one time series of rainy season data acquired by Envisat ASAR WSM data, yields results like the one shown in Figure 6.9, presenting the progression of water extent in the Mekong Delta for the year 2007.

Furthermore, the complete Envisat ASAR WSM–derived water mask time series of 60 data sets, covering the years 2007–2011 has been analyzed. Such a unique, long-term data set at a resolution of 150 m enables the visualization of flood frequencies during the time span and is a good source of information for the extraction of areas, which are unusually often flooded or which are hardly ever flooded. In Figure 6.9, we can see that regions, which are always flooded, are of course the arms of the Mekong River, entering the South China Sea. However, also some regions at the southwestern tip of the Mekong Delta (province of Ca Mao) are very frequently flooded.

At the same time, the northern part of the delta is also very often flooded, which reflects the flood pulse rhythm typical for the Mekong Delta. Floods protrude into the delta via overland flow and overbank flow mainly coming from the North. So the flood in the delta moves from the North to the South and radially away from the Mekong's main stem (see also Figure 6.10). The coastal areas as well as regions in the southern center of the delta are least often flooded—these are regions of fruit orchards, slightly elevated land, or coastal forest. The high-resolution snaps from Google Earth depict regions, which are never flooded: such as elevated hills and small mountains (Figure 6.10a), strongly

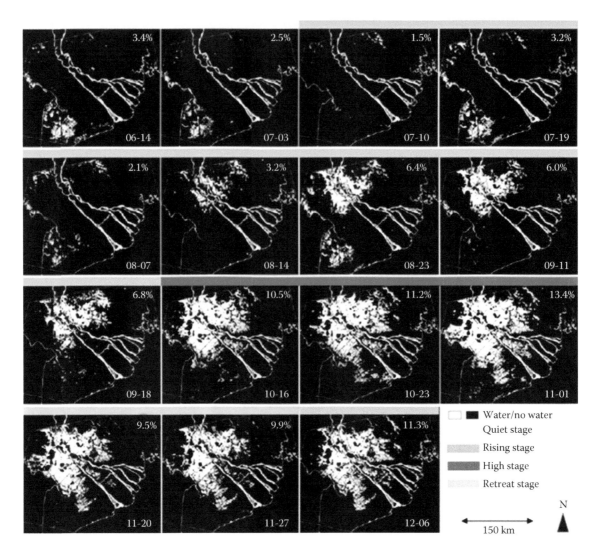

FIGURE 6.9 Flood progression in the Mekong Delta over the course of the year 2007 as derived from 15 amplitude Envisat ASAR WSM data sets using WaMaPro. Black areas indicate regions, which are not flooded. White areas indicate inundated/flooded areas. In the lower right corner, the month and day of the acquisition are depicted. In the upper right corner of each water mask is visualized how much percent of the Mekong Delta, covering 39,000 km², is flooded. Flooded areas usually extend during the rainy season (rising stage) between June and September, flood extent reaches its peak between October and November, and then the annual flood waters recede again. (From Kuenzer, C. et al., *Remote Sens.*, 5, 687, 2013.)

diked land, such as presented in Figure 6.10b (a national park), dike-protected research farms (Figure 6.10c and d), and orchard areas on slightly elevated ground (Figure 6.10f). Only for the very dense mangrove areas in the Mekong Delta, it is not possible to say if the ground below the very dense canopies is flooded or not as the penetrability of Envisat ASAR is limited.

Figure 6.11 depicts the results of a scale-related comparison between water mask derivation from four observations undertaken more or less during the same time in the rainy season of 2008 with multitemporal Envisat ASAR WSM data (Figure 6.11a), TerraSAR-X SC data (Figure 6.11b), and TerraSAR-X SM data (Figure 6.11c). The three subsets depict a region in Can Tho City, located in the center of the Mekong Delta. Their extent is about 2 × 3 km. White little boxes within the black background

are digitized houses, which are usually located alongside canals. Different shades of blue indicate how often a pixel has been detected as flooded during the four observations. It can be observed that the better the spatial resolution, the better and more precise the mapping result of the flood map. While in Envisat ASAR–derived water masks the narrow rivers and canals in the region cannot be extracted, TerraSAR-X SM data allow for the extraction of smallest canals. In the Envisat-derived water masks, it appears like several houses are flooded, whereas it can be clearly noted in the two TerraSAR-X-based results that the areas are not really flooded (problem of mixed pixels in coarser resolution SAR data). For any kind of flood impact assessment, it is thus of utmost importance to evaluate the own results with respect to the observation scale chosen.

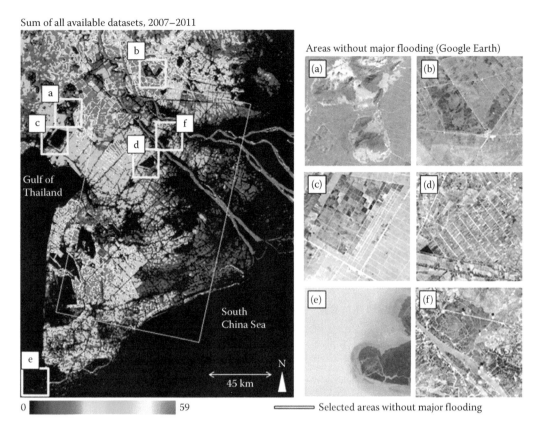

Sum of all available datasets, 2007–2011

Areas without major flooding (Google Earth)

Gulf of Thailand

South China Sea

45 km

N

0   59

Selected areas without major flooding

FIGURE 6.10 Inundation in the Mekong Delta from 2007 to 2011 derived from all available Envisat ASAR WSM data sets, enabling the visualization of spatial patterns of flood frequency. Sixty observations were available and 60 water masks could be derived. Dark blue areas are rarely flooded, while reddish tones depict areas that are always water covered (e.g., the Mekong River branches). It is obvious that the northern and southwestern parts of the Mekong Delta are most frequently flooded. In the northern part of the delta, triple season rice crop is grown. Rarely flooded are the fruit orchard regions in the center and east of the Delta, as well as well-diked areas. The little subsets on the right side of the figure depict regions, which are rarely flooded, such as elevated hills (a), well-diked regions (b–d), fruit orchards at higher elevation (f), or regions where flood water cannot be detected, such as under dense mangrove canopies (e). (From Kuenzer, C. et al., *Remote Sens.*, 5, 687, 2013.)

Thresholds in WaMaPro are defined empirically. This is not as elegant as a threshold automatically derived from the image histogram itself. However, WaMaPro has been applied to study sites not only in the Mekong Delta but also to large areas in West Africa, China's Dongting Lake region, the Chinese Yellow River Delta, as well as to study sites in Russia (Huth et al. 2014). We can see here that an empirical threshold selected for TerraSAR-X processing can work in a "one-fits-all" manner for these regions (T1: 60 DN, T2: 90 DN). Thus, WaMaPro could theoretically be used for a global processing based on TerraSAR-X data. For Envisat ASAR data, a "one-fits-all threshold" can only be transferred from one region to another, if the data is properly corrected geometrically, as well as for incidence angle effects. This challenge also applies for upcoming Sentinel-1 data. Depending on incidence angle and mode (spatial resolution will vary between 20 × 40 m up to 5 × 2 m), swath widths between 400 and 80 km can occur.

Generally, WaMaPro can meet difficulties in regions with heavy terrain. Here, the integration of a DEM into the processing chain (such as presented for the TFS in Section 6.4.3) seems recommendable. However, for the study of typical wetland dynamics or large overland flooding (usually not located in very rugged terrain, but in flat tundras, savannahs, river deltas, coastal or lakeshore, and river floodplain regions), WaMaPro so far worked sufficiently also without a DEM. Problems, which arise in these regions, are mainly related to volume and double bouncing partially submerged vegetation areas, which reduce the detectability of the water extent.

One of the advantages of WaMaPro is its very easy handling. The software is encapsulated in a virtual machine and can be handed on to interested users, which can apply WaMaPro as a ready to use Q-GIS plug-in (Huth and Kuenzer 2014). The software is, therefore, independent of any software package inflicting license costs. WaMaPro runs exceptionally fast: the processing of an Envisat ASAR WSM scene at 150 m resolution and a frame size of about 200 × 200 km to a binary water mask is performed in well below 1 min per scene (30–40 s) on an Intel 8-core CPU with 2.4 GHz and 32 GB of RAM. The processing of a water mask from a TerraSAR-X SC scene at 8.25 m pixel spacing with a frame size of 100 × 150 km or a TerraSAR-X SM scene with 2.5 m pixel spacing and a frame extent of about 30 × 60 km takes about 4–5 min per scene.

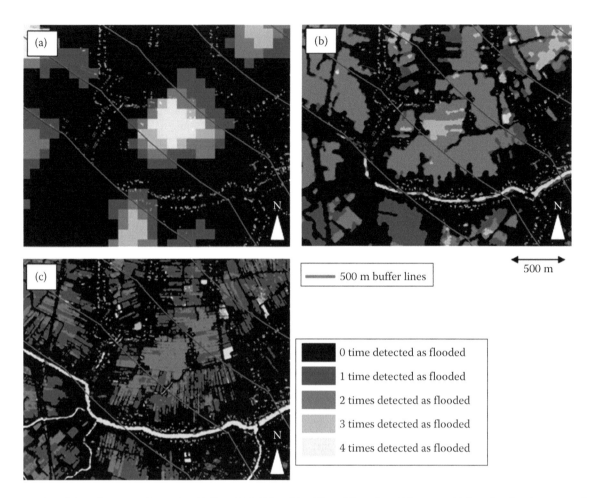

FIGURE 6.11 Capability of Envisat ASAR WSM (a), TerraSAR-X ScanSAR (b), and TerraSAR-X Stripmap (c) for urban flood mapping. (Modified based on Kuenzer, C. et al., *Remote Sens.*, 5, 5122, 2013.)

6.4.3 Fully Automatic Pixel-Based Flood Detection (TerraSAR-X Flood Service)

This section presents a fully automated processing chain for near real-time pixel-based flood detection at a global level. Compared to semiautomatic flood-mapping approaches (Section 6.4.1) commonly applied in the rapid mapping community, automatic processing chains enable to reduce the critical time span from the delivery of satellite data after flood events to the provision of satellite-derived crisis information (i.e., flood extent) to emergency management and decision makers. The thematic results of automatic processing chains can be directly ingested in web mapping applications to visualize and intersect the derived flood information with a number of other relevant geodata (e.g., DEMs, reference water levels, gauge data, hydrological features, and critical infrastructure).

When considering the broad range of different biomes and acquisition scenarios related to a global flood-mapping approach, a major concern in the design and implementation of the algorithm is to reach a high level of robustness. This is a particular challenge when the algorithm is part of an automatic processing chain since a user-based classification refinement using different postprocessing options (see Section 6.4.1) is no longer possible. Hence, the methodology needs to be as universally applicable as possible, delivering satisfying results independent of varying environmental conditions and acquisition parameters (e.g., beam mode, incidence angle). For this purpose, a classification methodology based on previous work of Martinis et al. (2009) was substantially refined and extended for purposes of robustness and transferability (Martinis et al. 2015). The methodology is described in Section 6.4.3.1. The robustness and accuracy of the approach are then tested for several study sites located in different biomes (Section 6.4.3.2).

6.4.3.1 Methodology

The workflow described in this chapter has been tested for a comprehensive set of TerraSAR-X scenes acquired during flood situations all over the world with different sensor configurations (Figure 6.12). Three TerraSAR-X scenes out of this data set are analyzed in detail and the classification result is discussed in Section 6.4.3.2. The scenes correspond to flood events in Nepal (2008), Germany (2011), and Albania/Montenegro (2013). Table 6.10 lists the main acquisition parameters.

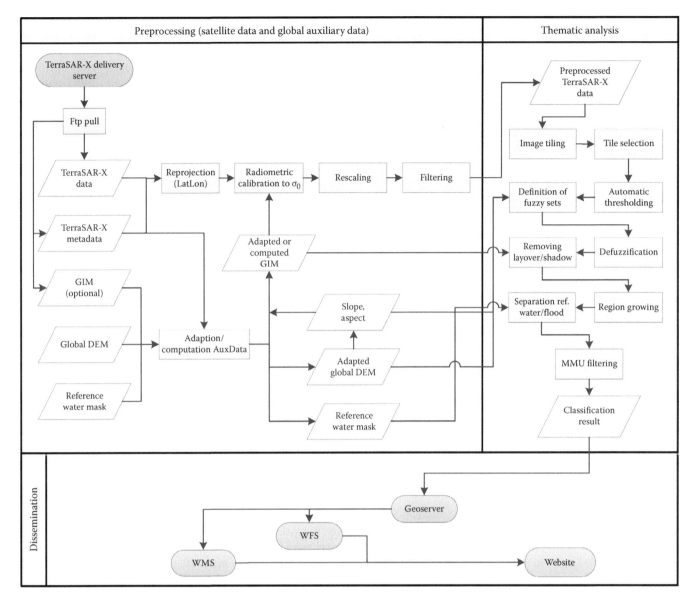

FIGURE 6.12 Workflow of the fully automatic TerraSAR-X Flood Service. (Modified based on Martins et al. 2015.)

While the automatic processing chain has been designed for enhanced ellipsoid corrected (EEC) and ground ellipsoid corrected TerraSAR-X amplitude imagery of different acquisition modes (SpotLight, SM, SC, Wide SC), it can principally be extended to other SAR satellite systems. The TerraSAR-X amplitude data are commonly delivered via an FTP server. In order to ensure immediate data processing, the data download is triggered automatically through a Python script after the reception of a delivery e-mail. When the download to the local file system has been completed, the data are extracted and the corresponding file structure is searched for all relevant files, that is, the SAR data, the metadata file, and, optionally, the so-called geocoded incidence angle mask (GIM). The GIM can be ordered as an optional add-on layer together with the EEC product and provides information on the local incidence angle for each image element of the geocoded SAR scene and on the presence of shadow and layover areas (Infoterra 2014). In case no GIM was ordered jointly with the TerraSAR-X scene, this auxiliary layer is generated automatically during the following preprocessing steps.

In order to ensure the same coordinate systems for all data products on a global level, the delivered TerraSAR-X images are

TABLE 6.10 Acquisition Parameters of TerraSAR-X Scenes Selected from the Test Data Set

Location	Acquisition Time	Beam Mode	Polarization	Pixel Spacing (m)
Nepal	2008-09-05, 12:09 UTC	SC_006R	HH	8.25
Germany	2011-01-17, 16:52 UTC	SM_008R	HH	2.75
Albania/ Montenegro	2013-03-20, 16:32 UTC	SC_008R	HH	8.25

first reprojected to geographical coordinates (lat/lon, WGS84). This target system is also used for all global auxiliary data products. Besides TerraSAR-X satellite imagery, two types of auxiliary data are included into the process: reference water masks and DEMs. Since the SAR-based methodology detects all open water areas irrespective of their context (permanent water bodies, seasonally inundated areas, extreme and disastrous flood events), reference water masks are needed for separating between permanent water bodies and flooded areas. For this purpose, the SRTM water body mask (SWBD) (SWBD 2005) with a horizontal resolution of 30 m is extracted and resampled to fit each respective SAR scene. While not at the same spatial resolution as high-resolution TerraSAR-X data, there is currently no alternative to SWBD with respect to spatial resolution and near global coverage. Since the SWBD is primarily based on data recorded during the Space Shuttle flight in February 2000, it needs to be considered that due to seasonal effects, the derived water body extents can differ considerably from an assumed normal water level. In the future, the inclusion of new global reference water data sets such as the TanDEM-X Water Indication Mask (WAM) (Wendleder et al. 2012) with a horizontal resolution of 12 m might be considered instead of the SWBD. For a refinement of the flood mask, the ASTER Global DEM Version 2 (GDEM V2) (METI and NASA 2011) with a pixel size of one arc second is employed. The same DEM is also used for the optional computation of a GIM in the preprocessing step. For the thematic analysis, the slope information in degrees for each pixel (x, y), that is, the local steepness of the terrain, is computed. Further, the local incidence angle θ_{loc} for each pixel (x, y), that is, the GIM, is used for the radiometric calibration of a SAR scene to sigma naught σ_0 [dB].

The preprocessing of the TerraSAR-X amplitude data includes a radiometric calibration of the data to normalized radar cross section σ_0. This is performed to account for incidence-angle-linked SAR backscatter variations in range direction and for the reduction of topographic effects, which can both negatively influence the automatic threshold derivation. σ_0 is rescaled to a value range of [0, 400] in order to derive positive values during all following processing steps. For the reduction of the SAR-inherent speckle effect, a median filter of kernel size 3 × 3 is finally applied on the rescaled pixel values.

For the unsupervised initialization of the flood processor, a parametric tile-based thresholding procedure is applied (Martinis et al. 2009). This approach was originally developed to automatically detect the inundation extent in SAR amplitude data with even small class *a priori* probabilities in a time-efficient manner. For a detailed description of the algorithm, the reader is directed to Section 6.4.1.1. In the following section, the main emphasis is laid on the description of the enhancements required to meet the robustness and transferability demands of a fully automated processing chain operational for the global scale.

From the automatic thresholding procedure, one global threshold τ_g is obtained by computing the arithmetic mean of the locally derived thresholds. The standard deviation σ_τ of the local

thresholds can be used as an indicator for a successful thresholding. If σ_τ exceeds an empirically derived critical threshold τ_σ (e.g., 5.0 dB) a (sub)histogram merging strategy is applied by computing τ_g directly from a merged histogram, which is a combination of the distributions of the individual tiles. If the tile selection or the derivation of a reasonable threshold value fails (in this study, the maximum possible threshold is set to −10 dB), it can be assumed that (1) either no water areas exist in the covered region, (2) the water extent is very small, or (3) water bodies do not appear as dark backscatter regions due to, for example, wind-induced roughening of the water surface or protruding vegetation leading to volume or double-bounce scattering of the radar signal. In this case, the threshold is approximated by the following equation, which expresses the linear relationship between the global threshold value τ_g [dB] separating water and nonwater areas and the scene center incidence angle θ_c:

$$\tau_g = -0.1002 \times \theta_c - 12.08 \qquad (6.3)$$

This regression describes the backscatter decrease over calm water areas with increasing incidence angle and was derived empirically by analyzing a test data set of 190 TerraSAR-X HH scenes of flood events acquired in different acquisition modes and incidence angles.

After the initial classification result is derived by the application of the global threshold, a fuzzy-logic-based algorithm is employed for postclassification purposes. Fuzzy logic (Zadeh 1965) is a valuable tool for combining ambiguous information sources by accounting for their uncertainties as opposed to only relying on crisp data sets. Over the last years, fuzzy-logic techniques have increasingly been used for the improvement of flood monitoring algorithms. Martinis and Twele (2010) apply the fuzzy theory for the quantification of the uncertainty in the labeling of each image element in flood possibility masks. The algorithm combines marginal posterior entropy-based confidence maps with spatiotemporal relationships of potentially flooded double bouncing vegetation to open water areas. A pixel- and object-based fuzzy-logic approach for flood detection based on Pierdicca et al. (2008) is described in Pulverenti et al. (2011) and Pulverenti et al. (2013), respectively, integrating theoretical electromagnetic scattering models, simplified hydrologic assumptions and contextual information. For the purposes of the underlying study, a fuzzy set of four elements is built consisting of SAR backscatter (σ_0), digital elevation (h), and slope (s) information as well as the extent (a) of water bodies. The elements of the fuzzy set are defined by standard S and Z membership functions (Pal and Rosenfeld 1988), which express the degree of an element's membership m_f to the class water. The degree of membership is defined by real numbers within the interval [0, 1], where 0 denotes minimum and 1 maximum class membership. The membership degree depends on the position of the crossover point x_c (i.e., the half-width of the fuzzy curve), which is defined by the fuzzy thresholds x_1 and x_2.

The fuzzy threshold values for each element are either determined according to statistical computations or are set empirically. Incorrectly labeled water regions are commonly caused by classifying objects with a low surface roughness and therefore low backscatter characteristics similar to calm water surfaces, such as roads, "smooth" agricultural crop land, or radar shadow. DEMs (GDEM V2) are integrated into the post-classification step to improve the classification accuracy through simple hydrological assumptions, that is, by reducing the membership degree of an image element in dependence of the height above the main water area by applying the standard Z membership function. The open water surface is derived by applying the global threshold τ_g to image Y. Subsequently, the separation between flooded regions and standing water areas is performed using a reference water mask.

The fuzzy thresholds of the elevation information are defined as

$$x_{1[h]} = \mu_{h(water)} \tag{6.4}$$

and

$$x_{2[h]} = \mu_{h(water)} + f_\sigma \times \sigma_{h(water)} \tag{6.5}$$

where $\mu_{h(water)}$ and $\sigma_{h(water)}$ are the mean and standard deviation of the elevation of all initially derived water objects. Using this fuzzy set, the number of look-alike areas in regions significantly higher in elevation than the mean water areas is reduced, for example, in mountainous terrain. The factor f_σ is defined by

$$f_\sigma = \sigma_{h(water)} + 3.5 \tag{6.6}$$

This function was integrated to reduce the influence of the DEM in areas of low topography. The minimum value of f_σ is defined by 0.5.

The standard Z function is used for describing the membership degree to open water areas according to the radar backscatter. Full membership is assigned to image elements with a backscatter lower than the fuzzy threshold

$$x_{1[\sigma_0]} = \mu_{\sigma_0(\tau_g)} \tag{6.7}$$

where $\mu_{\sigma_0(\tau_g)}$ is the mean backscatter of the initial flood classification result by applying τ_g to Y. No membership degree [0] is assigned to pixels greater than

$$x_{2[\sigma_0]} = \tau_g \tag{6.8}$$

Topographic slope information derived from globally available digital elevation data is integrated as a third element in the fuzzy system by using the Z membership function with parameters $x_{1[sl]} = 0°$, $x_{2[sl]} = 15°$. Using this auxiliary information layer, water look-alikes in steep terrain are removed.

The S membership function is applied to the size a of the water bodies to reduce the number of dispersed small areas of low backscatter, which are commonly related to water look-alike areas. No membership degree is further assigned to elements with a size lower than $x_{1[a]} = 250$ m^2 and maximum grade to elements with a size greater than $x_{2[a]} = 1000$ m^2.

In order to combine all fuzzy elements into one composite fuzzy set, the average of the membership degrees is computed for each pixel. Subsequently, the flood mask is created through a threshold defuzzification step, which transforms each image element with a membership degree >0.6 into a crisp value, that is, a discrete thematic class.

In order to integrate image elements at the boundary of flood water surfaces and nonflooded regions and to increase the spatial homogeneity of the detected flood plain, a region-growing step is performed. The preliminary extracted water bodies of the defuzzified classification result are used as seeds for dilating the water regions. The water areas are iteratively enlarged until a tolerance criterion is reached. Only image elements located in the neighborhood of the flood areas are scanned to avoid the detection of water look-alikes distant from initially labeled water surfaces. The region-growing tolerance criterion is defined by a relaxed fuzzy threshold of >0.45. Therefore, the region-growing step is controlled by both the SAR backscatter information and auxiliary data (topographic slope, elevation, and size of water bodies).

To eliminate open water look-alikes in areas affected by radar layover and radar shadow, the GIM is integrated into the classification process. Using a minimum mapping unit (MMU) of 30 pixels, small isolated flood objects with a pixel count less than this threshold are removed from the water mask. Small land objects (i.e., islands) that are fully enclosed by water are reclassified to water based on the same MMU value. The classification result is subsequently matched to a global reference water mask (SWBD) to differentiate between flooded areas and standing water bodies.

For dissemination of the results, the final flood mask, the fuzzy mask, which can be used for the quantification of the uncertainty in the labeling of each image element, and satellite footprints are stored in a database and visualized through a web-based user interface. The process chain is based on a framework of WPS standard-compliant to the Open Geospatial Consortium.

6.4.3.2 Results

Due to specular reflection of the incident radar signal, open water areas visually appear dark in the test scenes (see Table 6.10 upper row and Figure 6.13) and can be discerned from land surfaces, where diffuse scattering is predominant. Water look-alike areas comprise features with a low surface roughness (e.g., roads or airport runways) and areas of radar shadow. The scene histograms are hence marked by a bimodal distribution of the classes water and nonwater. The histogram of each tile selected for threshold computation is modeled by statistical parameterization of local bimodal class-conditional density functions and reliable thresholds are derived using minimum error thresholding. Since the standard deviations τ_σ of locally derived thresholds are significantly lower than the critical value of 5.0 dB for each scene, they are combined to global thresholds by computing their arithmetic mean. Accordingly, global threshold values of −16.7 dB (Nepal), −15.9 dB (Germany), and −19.2 dB (Albania/Montenegro) are subsequently employed for initializing the postclassification process.

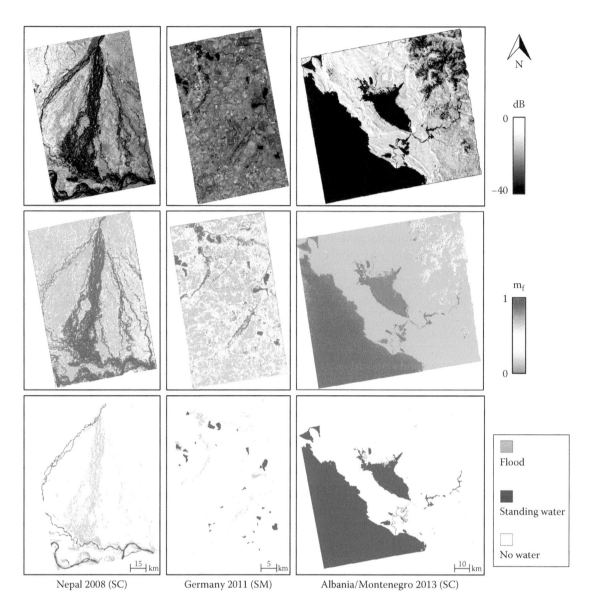

FIGURE 6.13 Radiometrically calibrated TerraSAR-X data (upper row), fuzzy maps (middle row), and final classification results (lower row) for three test areas. (Modified based on Martinis, S. et al., *ISPRS J. Photogramm. Remote Sens.*, 104, 203, 2015.)

The fuzzy-logic-based postclassification refinement is used for combining SAR backscatter information with the fuzzy elements terrain height, slope, and water body size (see Section 6.4.3.1). The fuzzy maps shown in Figure 6.13 indicate the membership degree [0, 1] of the composite fuzzy set for each pixel. The test data acquired for the Albania/Montenegro 2013 floods show mountainous terrain with steep slopes in the northeastern parts of the scene. The areas facing away from the sensor are thus marked by a low signal return. Using a classification approach purely based on SAR backscatter, such areas would potentially be labeled as flooded. However, as depicted on the fuzzy map, the steep slope and high altitude of these areas led to a reduction in the total membership degree of these pixels. In the threshold defuzzification step, which transforms each image element with a membership degree >0.6 into a crisp value, that is, a discrete label, most of these water look-alikes are successfully excluded from the

final flood mask. In contrast to the Albania/Montenegro (2013) data set, the scene recorded during the Germany (2011) floods (mid row) is characterized by much lower differences in altitude. Here, SAR backscatter information has a major influence on the final membership degree of the composite fuzzy set. Typical water look-alikes such as airport runways and roads in flat areas without much height difference in comparison to initially found flood pixels can thus erroneously get labeled as flooded, even after a fuzzy-logic-based postclassification refinement.

Classification problems can also occur due to artifacts in DEM data. In our examples, ASTER GDEM V2 data occasionally contained spikes in regions of normal water bodies. Accordingly, the high slope values surrounding these spikes can lower the membership degree of the composite fuzzy set below 0.6, which would lead to a misclassification of these pixels to the class "nonwater."

The algorithm generally performs very well in rural areas. In urban areas, the detectability of water regions is reduced due to enhanced signal returns with backscatter values frequently higher the extracted threshold. Further, floods in urban areas might simply not be visible to the SAR instrument when they are situated within areas of radar shadow cast by buildings (Mason et al. 2010). The low signal return from these areas can then lead to a potential confusion with flooded areas that exhibit comparable backscatter levels. Although the presence of shadow and layover in urban areas can be simulated using high-resolution digital surface models such as LiDAR data (Soergel et al. 2003), this type of data is expensive and not available on a global scale.

6.5 Conclusion

SAR-derived water and flood extent maps are an important information source for an effective flood disaster management. Various approaches have been proposed in the literature for extracting the inundation extent from SAR data. In this study, three SAR-based operational water and flood-detection algorithms of the recent literature are described, which are developed for different application domains.

Semiautomatic object-based algorithms can be well used to derive very detailed flood maps with high classification accuracy during rapid mapping activities. For example, the overall accuracy of the RaMaFlood algorithm is stated to be 95.44% (producers' accuracy, 82.01%; users' accuracy, 98.65%) for a TerraSAR-X SM scene covering a flood situation in Tewkesbury, United Kingdom, in 2007 (Martinis et al. 2009). Commonly, commercial and noncommercial value-adding entities provide flood maps extracted from remote sensing data in map products upon request by, for example, relief organizations or political decision makers several hours after satellite data delivery. The disadvantage is that in most cases, a certain amount of user interaction is needed for SAR data preprocessing, the collection and adaptation of auxiliary data useful for classification refinement, the thematic analysis, and the preparation and dissemination of the crisis information to end users.

Compared to semiautomatic water/flood-mapping approaches, automatic processing chains significantly reduce the time delay between satellite data delivery and product dissemination.

In this context, an automatic and open-source distributable processing chain (WaMaPro) enabling continuous water monitoring from TerraSAR-X and Envisat ASAR on a local to global level using empirically predefined thresholds is presented. This software has been successfully used for detecting smooth water surfaces in areas of stable environmental conditions, which are repeatedly being monitored with data acquired with similar SAR system parameters.

Further, a global fully automatic TerraSAR-X-based flood-mapping approach is proposed. The processing chain includes the download and preprocessing of TerraSAR-X data, computation and adaptation of global auxiliary data, unsupervised class initialization, postclassification refinement, and dissemination of the flood masks via a web-based user interface.

The computational effort of the whole processing chain is lower than 1 h. The methodology is universally applicable, delivering satisfying results independent from acquisition parameters, and requires no user input. These promising qualitative results have been further confirmed by means of a quantitative accuracy assessment performed for two study sites in Germany and Thailand, achieving encouraging overall accuracies of ~91.6% and ~87.5%, respectively.

Within the last decade, fully automatic algorithms and processing chains have been developed for unsupervised mapping of open water surfaces using various SAR sensors. Further improvements in classification accuracy could be reached by integrating upcoming high-resolution auxiliary data such as DEMs, reference water masks, and land-cover information. Future work should focus on improving and automating the algorithms in complex situations such as flooded urban areas and partially submerged vegetation areas. By adapting the proposed methods to data of the upcoming ESA's Sentinel-1 C-Band SAR mission, which is characterized by a systematic acquisition strategy and therefore enables a continuous monitoring of water surfaces, some of the presented limitations of SAR-based water mapping could be reduced by automatically exploiting the growing archive during the lifetime of the mission.

References

Ahtonen, P., M. Euro, M. Hallikainen, S. Solbø, B. Johansen, and I. Solheim. 2004. SAR and optical based algorithms for estimation of water bodies. Technical Report, FloodMan Project, Helsinki University of Technology, Helsinki, Finland.

Alsdorf, D. E., J. M. Melack, T. Dunne, L. A. K. Mertes, L. L. Hess, and L. C. Smith. 2000. Interferometric radar measurements of water level changes on the Amazon flood plain. *Nature* 404:174–177.

Aronica, G., P. D. Bates, and M. S. Horritt. 2002. Assessing the uncertainty in distributed model predictions using observed binary pattern information within GLUE. *Hydrol Process* 16:2001–2016.

Baatz, M., and A. Schäpe. 1999. Object-oriented and multi-scale image analysis in semantic networks. *Proceedings of the Second International Symposium on Operationalization of Remote Sensing*, August 16–20, Enschede, the Netherlands, pp. 1–7.

Bates, P. D., M. S. Horritt, C. N. Smith, and D. C. Mason. 1997. Integrating remote sensing observations of flood hydrology and hydraulic modelling. *Hydrol Process* 11:1777–1795.

Bazi, Y., L. Bruzzone, and F. Melgani. 2005. An unsupervised approach based on the generalized Gaussian model to automatic change detection in multitemporal SAR images. *IEEE Trans Geosci Remote Sens* 43:874–887.

Benz, U., P. Hofmann, G. Wilhauck, I. Lingenfelder, and M. Heynen. 2004. Multi-resolution, object-oriented fuzzy analysis of remote sensing data for GIS-ready information. *ISPRS J Photogramm Remote Sens* 58:239–258.

Blaschke, T., and J. Strobl. 2001. What's wrong with pixels? Some recent developments interfacing remote sensing and GIS. *GIS Zeitschrift für Geoinformationssysteme* 6:12–17.

Blasco, F., M. F. Bellan, and M. U. Chaudury. 1992. Estimating the extent of floods in Bangladesh using SPOT data. *Remote Sens Environ* 39:167–178.

Bourgeau-Chavez, L. L., E. S. Kasischke, S. M. Brunzell, J. P. Mudd, K. B. Smith, and A. L. Frick. 2001. Analysis of space-borne SAR data for wetland mapping in Virginia riparian ecosystems. *Int J Remote Sens* 22:3665–3687.

Brakenridge, G. R., and E. Anderson. 2005. MODIS-based flood detection, mapping, and measurement: The potential for operational hydrological applications. *Proceedings of the NATO Advanced Research Workshop*, Baile Felix–Oradea, Romania.

Brivio, P. A., R. Colombo, M. Maggi, and R. Tomasoni. 2002. Integration of remote sensing data and GIS for accurate mapping of flooded areas. *Int J Remote Sens* 23:429–441.

Bruzzone, L., and L. Carlin. 2006. A multilevel context-based system for classification of very high spatial resolution images. *IEEE Trans Geosci Remote Sens* 44:2587–2600.

Carleer, A. P., O. Debeir, and E. Wolff. 2005. Assessment of very high spatial resolution satellite image segmentations. *Photogramm Eng Remote Sens* 71:1285–1294.

Danklmayer, A., B. J. Döring, M. Schwerdt, and M. Chandra. 2009. Assessment of atmospheric propagation effects in SAR images. *IEEE Trans Geosci Remote Sens* 47:3507–3518.

De Moel, H., J. Van Alphen, and J. Aerts. 2009. Flood maps in Europe, methods, availability and use. *Nat Hazards Earth Syst Sci* 9:289–301.

De Roo, A., J. Van der Knijff, M. S. Horritt, G. Schmuck, and S. De Jong. 1999. Assessing flood damages of the 1997 Oder flood and the 1995 Meuse flood. *Proceedings of the Second International Symposium on Operationalisation of Remote Sensing*, August 16–20, Enschede, the Netherlands, pp. 1–9.

Dellepiane, S., S. Monni, G. Bo, and C. Buck. 2000. SAR images and interferometric coherence for flood monitoring. *Proceedings of the IEEE Geoscience and Remote Sensing Symposium*, July 24–28, Honolulu, HI, pp. 2608–2610.

Drake, B., and R. A. Shuchman. 1974. Feasibility of using multiplexed SLAR imagery for water resources management and mapping vegetation communities. *Proceedings of the Ninth International Symposium on Remote Sensing of Environment*, April 15–19, Ann Arbor, MI, pp. 714–724.

eoPortal. 2014. https://directory.eoportal.org/. (Accessed April 24, 2015.)

European Commission. 2011. GMES data access specifications of the Earth observation needs over the period 2011–2013. Brussels, Belgium.

Evans, D. C., T. G. Farr, J. P. Forf, T. W. Thompson, and C. L. Werner. 1986. Multipolarization radar images for geologic mapping and vegetation discrimination. *IEEE Trans Geosci Remote Sens* 24:246–257.

Ferreira, J. P. G., and J. M. Bioucas-Dias. 2008. Bayesian land and sea segmentation of SAR imagery. *Proceedings of the Third TerraSAR-X Science Team Meeting*, November, Oberpfaffenhofen, Germany, pp. 1–7.

Gebhardt, S., J. Huth, N. Lam Dao, A. Roth, and C. Kuenzer. 2012. A comparison of TerraSAR-X quadpol backscattering with RapidEye multispectral vegetation indices over rice fields in the Mekong Delta, Vietnam. *Int J Remote Sens* 33:7644–7661.

Giacomelli, A., M. Mancini, and R. Rosso. 1995. Assessment of flooded areas from ERS-1 PRI data: An application to the 1994 flood in Northern Italy. *Phys Chem Earth* 20:469–474.

Giustarini, L., R. Hostache, P. Matgen, G. Schumann, P. D. Bates, and D. C. Mason. 2012. A change detection approach to flood mapping in urban areas using TerraSAR-X. *IEEE Trans Geosci Remote Sens* 51:2417–2430.

Gstaiger, V., J. Huth, S. Gebhardt, T. Wehrmann, and C. Kunezer. 2012. Multi-sensorial and automated derivation of inundated areas using TerraSAR-X and ENVISAT ASAR data. *Int J Remote Sens* 33:7291–7304.

Haralick, R. M., and L. G. Sharipo. 1985. Image segmentation techniques. *Comput Vis Graph Image Process* 29:100–132.

Henderson, F. M. 1987. Consistency of open surface detection with L-band SEASAT SAR imagery and confusion with other hydrologic features. *Proceedings of the 13th Annual Conference of Remote Sensing Society* (*Advances in Digital Image Processing*), September 7–11, University of Nottingham, Nottingham, U.K., pp. 69–78.

Henderson, F. M. 1995. Environmental factors and the detection of open surface water using X-band radar imagery. *Int J Remote Sens* 16:2423–2437.

Henry, J. B., P. Chastanet, K. Fellah, and Y. L. Desnos. 2006. ENVISAT multi-polarised ASAR data for flood mapping. *Int J Remote Sens* 27:1921–1929.

Herrera-Cruz, V., and F. Koudogbo. 2009. TerraSAR-X Rapid mapping for flood events. *Proceedings of the International Society for Photogrammetry and Remote Sensing (Earth Imaging for Geospatial Information)*, June 14–17, Hannover, Germany, pp. 170–175.

Hess, L. L., and J. M. Melack. 1994. Mapping wetland hydrology and vegetation with synthetic aperture radar. *Int J Ecol Environ Sci* 20:197–205.

Hess, L. L., and J. M. Melack. 2003. Remote sensing of vegetation and flooding on Magela Creek floodplain (Northern Territory, Australia) with the SIR-C synthetic aperture radar. *Hydrobiologia* 500:65–82.

Hess, L. L., J. M. Melack, S. Filoso, and Y. Wang. 1995. Delineation of inundated area and vegetation along the Amazon floodplain with the SIR-C synthetic aperture radar. *IEEE Trans Geosci Remote Sens* 33:896–904.

Hess, L. L., J. M. Melack, E. M. Novo, C. C. Barbosa, and M. Gastil. 2003. Dual season mapping of wetland inundation and vegetation for the central Amazon basin. *Remote Sens Environ* 87:404–428.

Hess, L. L., J. M. Melack, and D. S. Simonett. 1990. Radar detection of flooding beneath the forest canopy: A review. *Int J Remote Sens* 11:1313–1325.

Hong, S.-H., S. Wdowinski, and S.-W. Kim. 2010. Evaluation of TerraSAR-X observations for wetland InSAR application. *IEEE Trans Geosci Remote Sens* 48:864–873.

Horritt, M. 1999. A statistical active contour model for SAR image segmentation. *Image Vis Comput* 17:213–224.

Horritt, M. S. 2000. Calibration of a two-dimensional finite element flood flow model using satellite radar imagery. *Water Resour Res* 36:3279–3291.

Horritt, M. S. 2006. A methodology for the validation of uncertain flood inundation models. *J Hydrol* 326:153–165.

Horritt, M. S., D. C. Mason, D. M. Cobby, I. J. Davenport, and P. Bates. 2003. Waterline mapping in flooded vegetation from airborne SAR imagery. *Remote Sens Environ* 85:271–281.

Horritt, M. S., D. C. Mason, and A. J. Luckman. 2001. Flood boundary delineation from synthetic aperture radar imagery using a statistical active contour model. *Int J Remote Sens* 22:2489–2507.

Hostache, R., P. Matgen, G. Schumann, C. Puech, L. Hoffmann, and L. Pfister. 2009. Water level estimation and reduction of hydraulic model calibration uncertainties using satellite SAR images of floods. *IEEE Trans Geosci Remote Sens* 47:431–441.

Hunter, N. M., P. D. Bates, M. S. Horritt, P. J. De Roo, and M. Werner. 2005. Utility of different data types for flood inundation models within a GLUE framework. *Hydrol Earth Syst Sci* 9:412–430.

Huth, J., M. Ahrens, and C. Kuenzer, C. 2014. WaMaPro—The Water Mask Processing Hand Book for WaMaPro Version 2.2.0 (Status April 2014). 38 pp.

Huth, J., S. Gebhardt, T. Wehrmann, I. Schettler, C. Kuenzer, M. Schmidt, and S. Dech. 2009. Automated inundation monitoring using TerraSAR-X multi-temporal imagery. *Proceedings of the European Geosciences Union General Assembly*, April 19–24, Vienna, Austria.

Huth, J., and C. Kuenzer. 2014. WaMaPro Handbook. Version 1.0, Unpublished Software Manual, 24pp.

Infoterra, 2008. Radiometric Calibration of TerraSAR-X Data. http://www.astriumgeo.com/files/pmedia/public/r465_9_tsxx-itd-tn-0049-radiometric_calculations_i1.00.pdf (Accessed April 8, 2013.)

Infoterra, 2014. Radiometric calibration of TerraSAR-X data. URL: http://www2.geo-airbusds.com/files/pmedia/public/r465_9_tsx-x-itd-tn-0049-radiometric_calculations_i3.00.pdf (Accessed April 24, 2015).

Ivins, J., and J. Porill. 1995. Active region models for segmenting textures and colours. *Image Vision Comput* 13:431–438.

Kasischke, E. S., and L. L. Bourgeau-Chavez. 1997. Monitoring south Florida wetlands using ERS-1 SAR imagery. *Photogramm Eng Remote Sens* 33:281–291.

Kasischke, E. S., J. M. Melack, and M. C. Dobson. 1997. The use of imaging radars for ecological applications—A review. *Remote Sens Environ* 59:141–156.

Kittler, J., and J. Illingworth. 1986. Minimum error thresholding. *Pattern Recogn* 19:41–47.

Kohonen, T. 1995. *Self-Organizing Maps*, 3rd edn. Springer-Verlag, Heidelberg, Germany.

Kolmogorov, V., and R. Zabih. 2004. What energy functions can be minimized via graph cuts? *IEEE Trans Pattern Anal Mach Intell* 26:147–159.

Kuenzer, C., H. Guo, I. Schlegel, V. Q. Tuan, X. Li, and S. Dech. 2013a. Varying scale and capability of Envisat ASAR-WSM, TerraSAR-X Scansar and TerraSAR-X Stripmap data to assess urban flood situations: A case study of the Mekong delta in Can Tho province. *Remote Sens* 5:5122–5142.

Kuenzer, C., G. Huadong, J. Huth, P. Leinenkugel, L. Xinwu, and S. Dech. 2013b. Flood mapping and flood dynamics of the Mekong Delta: ENVISAT-ASAR-WSM based time series analyses. *Remote Sens* 5:687–715.

Kuenzer, C., and K. Knauer. 2013. Remote sensing of rice crop areas—A review. *Int J Remote Sens* 34:2101–2139.

Kussul, N., Shelestov, A., and Skakun, S. 2008. Intelligent computations for flood monitoring. In Markov, K., K. Ivanova, I., Mitov (eds.), *Advanced Research in Artificial Intelligence*, book 2, pp. 48–54.

Lang, M. W., P. A. Townsend, and E. S. Kasischke. 2008. Influence of incidence angle on detecting flooded forests using C-HH synthetic aperture radar data. *Remote Sens Environ* 112:3898–3907.

Leckie, D. G. 1998. Forestry applications using imaging radar. In Henderson, F. M., and A. J. Lewis (eds.), *Manual of Remote Sensing: Principles and Applications of Imaging Radar*, 3rd edn. John Wiley & Sons, New York.

Lewis, A. J. 1998. Geomorphic and hydrologic applications of active microwave remote sensing. In Henderson, F. M., and A. J. Lewis (eds.), *Manual of Remote Sensing: Principles and Applications of Imaging Radar*, 3rd edn. John Wiley & Sons, New York.

Lillesand, T. M., R. W. Kiefer, and J. W. Chipman. 2004. *Remote Sensing and Image Interpretation*, 5th edn. John Wiley & Sons, New York.

Lopes, A., E. Nezry, R. Touzi, and H. Laur. 1990. Maximum a posteriori speckle filtering and first order texture models in SAR images. *Proceedings of IEEE International Geoscience and Remote Sensing Symposium (IGARSS 1990)*, College Park, MD, 1990, Vol. 3, pp. 2409–2412.

Lucchese, L., and S. K. Mitra. 1998. An algorithm for unsupervised color image segmentation. *Proceedings of the IEEE Second Workshop Multimedia Signal Process*, December 7–9, Redondo Beach, CA, pp. 33–38.

Malladi, R., J. Sethian, and B. Vemuri. 1995. Shape modeling with front propagation: A level set approach. *IEEE Trans Pattern Anal* 17:158–175.

Malnes, E., T. Guneriussen, and K. A. Høgda. 2002. Mapping of flood-area by RADARSAT in Vannsjø, Norway. *Proceedings of the 29th International Symposium on Remote Sensing of the Environment*, April, Buenos Aires, Argentina, pp. 1–4.

Marcus, W. A., and M. A. Fonstad. 2008. Optical remote mapping of rivers at sub-meter resolution and watershed extents. *Earth Surf Process Landforms* 33:4–24.

Marinelli, L., R. Michel, A. Beaudoin, and J. Astier. 1997. Flood mapping using ERS tandem coherence image: A case study in south France. *Proceedings of the Third ERS Symposium*, March 17–20, Florence, Italy, pp. 531–536.

Martinis, S. 2010. Automatic near real-time flood detection in high resolution X-band synthetic aperture radar satellite data using context-based classification on irregular graphs. PhD thesis, Ludwig-Maximilians-University Munich, Munich, Germany.

Martinis, S., J. Kersten, and A. Twele. 2015. A fully automated TerraSAR-X based flood service. *ISPRS J Photogramm Remote Sens* 104:203–212.

Martinis, S., and A. Twele. 2010. A hierarchical spatio-temporal Markov model for improved flood mapping using multi-temporal X-band SAR data. *Remote Sens* 2:2240–2258.

Martinis, S., A. Twele, C. Strobl, J. Kersten, and E. Stein. 2013. A multi-scale flood monitoring system based on fully automatic MODIS and TerraSAR-X processing chains. *Remote Sens* 5:5598–5619.

Martinis, S., A. Twele, and S. Voigt. 2009. Towards operational near real-time flood detection using a split-based automatic thresholding procedure on high resolution TerraSAR-X data. *Nat Hazards Earth Syst Sci* 9:303–314.

Martinis, S., A. Twele, and S. Voigt. 2011. Unsupervised extraction of flood-induced backscatter changes in SAR data using Markov image modeling on irregular graphs. *IEEE Trans Geosci Remote Sens* 49:251–263.

Mason, D. C., I. J. Davenport, J. C. Neal, G. J.-P. Schumann, and P. D. Bates. 2012. Near real-time flood detection in urban and rural areas using high-resolution synthetic aperture radar images. *IEEE Trans Geosci Remote Sens* 50:3041–3052.

Mason, D. C., M. S. Horritt, J. T. Dall'Amico, T. R. Scott, and P. D. Bates. 2007. Improving river flood extent delineation from synthetic aperture radar using airborne laser altimetry. *IEEE Trans Geosci Remote Sens* 45:3932–3943.

Mason, D. C., R. Speck, B. Devereux, G. J.-P. Schumann, J. C. Neal, and P. D. Bates. 2010. Flood detection in urban areas using TerraSAR-X. *IEEE Trans Geosci Remote Sens* 48:882–894.

Matgen, P., R. Hostache, G. Schumann, L. Pfister, L. Hoffman, and H. H. G. Svanije. 2011. Towards an automated SAR based flood monitoring system: Lessons learned from two case studies. *Phys Chem Earth* 36:241–252.

Matgen, P., M. Montanari, R. Hostache, L. Pfister, L. Hoffmann, D. Plaza, V. R. N. Pauwels, G. J. M. De Lannoy, R. De Keyser, and H. H. G. Savenije. 2010. Towards the sequential assimilation of SAR-derived water stages into hydraulic models using the particle filter: Proof of concept. *Hydrol Earth Syst Sci* 14:1773–1785.

Matgen, P., G. Schumann, J. B. Henry, L. Hoffmann, and L. Pfister. 2007. Integration of SAR derived river inundation areas, high precision topographic data and a river flow model toward near real-time flood management. *Int J Appl Earth Observ Geoinform* 9:247–263.

McMillan, A. J. G. Morley, B. J. Adams, and S. Chesworth. 2006. Identifying optimal SAR imagery specifications for urban flood monitoring: A hurricane Katrina case study. *4th International Workshop on Remote Sensing for Disaster Response*, September 25–26, Cambridge University, Cambridge, England.

Nico, G., M. Pappalepore, G. Pasquariello, S. Refice, and S. Samarelli. 2000. Comparison of SAR amplitude vs. coherence flood detection methods—A GIS application. *Int J Remote Sens* 21:1619–1631.

Ormsby, J. P., B. J. Blanchard, and A. J. Blanchard. 1985. Detection of lowland flooding using active microwave systems. *Photogramm Eng Remote Sens* 51:317–328.

Otsu, N. 1979. A threshold selection method from gray-level histograms. *IEEE Trans Syst Man Cybern* 9:62–66.

Ottinger, M., C. Kuenzer, G. Liu, S. Wang, and S. Dech. 2013. Monitoring land cover dynamics in the Yellow River Delta from 1995 to 2010 based on Landsat 5 TM. *Appl Geogr* 44:53–68.

Pal, S. K., and A. Rosenfeld. 1988. Image enhancement and thresholding by optimization of fuzzy compactness. *Pattern Recogn Lett* 7:77–86.

Pappenberger, F., K. Frodsham, K. Beven, R. Romanowicz, and P. Matgen. 2007. Fuzzy set approach to calibrating distributed flood inundation models using remote sensing observations. *Hydrol Earth Syst Sci* 11:739–752.

Peinado, O., C. Kuenzer, S. Voigt, P. Reinartz, and H. Mehl. 2003. Fernerkundung und GIS im Katastrophenmangement—Die Elbe Flut 2003. In Strobl, J., T. Blaschke, and G. Griesebner (eds.), *Angewandte Geographische Informationsverarbeitung XV. Beitraege zum AGIT-Symposium Salzburg 2003*. Wichmann, Heidelberg, Germany, pp. 342–348.

Pierdicca, N., M. Chini, L. Pulvirenti, and F. Macina. 2008. Integrating physical and topographic information into a fuzzy scheme to map flooded area by SAR. *Sensors* 8:4151–4164.

Pierdicca, N., L. Pulvirenti, M. Chini, L. Guerriero, and L. Candela. 2013. Observing floods from space: Experience gained from COSMO-SkyMed observations. *Acta Astronaut* 84:122–133.

Pulvirenti, L., M. Chini, F. S. Marzano, N. Pierdicca, S. Mori, L. Guerriero, G. Boni, and L. Candela. 2012. Detection of floods and heavy rain using Cosmo-SkyMed data: The event in Northwestern Italy of November 2011. *Proceedings of IEEE International Geoscience and Remote Sensing Symposium (IGARSS 2012)*, July 22–27, Munich, Germany, pp. 3026–3029.

Pulvirenti, L., N. Pierdicca, M. Chini, and L. Guerriero. 2011. An algorithm for operational flood mapping from Synthetic Aperture Radar (SAR) data using fuzzy logic. *Nat Hazards Earth Syst Sci* 11:529–540.

Pulvirenti, L., N. Pierdicca, M. Chini, and L. Guerriero. 2013. Monitoring flood evolution in vegetated areas using COSMO-SkyMed data: The Tuscany 2009 case study. *IEEE J Sel Top Appl Earth Observ Remote Sens* 99:1–10.

Ramsey, E. W. 1995. Monitoring flooding in coastal wetlands by using radar imagery and ground-based measurements. *Int J Remote Sens* 16:2495–2502.

Raney, R. K. 1998. Radar fundamentals: Technical perspective. In Henderson, F. M., and A. J. Lewis (eds.), *Manual of Remote Sensing: Principles and Applications of Imaging Radar*, 3rd edn. John Wiley & Sons, New York, pp. 9–130.

Rémi, A., and Y. Hervé. 2007. Change detection analysis dedicated to flood monitoring using ENVISAT Wide Swath mode data. *Proceedings of the ENVISAT Symposium*, April 23–27, Montreux, Switzerland, SP-636.

Rennó, C. D., A. D. Nobre, L. A. Cuartas, J. V. Soares, M. G. Hodnett, J. Tomasella, and M. J. Waterloo. 2008. HAND, a new terrain descriptor using SRTM-DEM: Mapping terra-firme rainforest environments in Amazonia. *Remote Sens Environ* 112:3469–3481.

Richards, J. A., P. W. Woodgate, and A. K. Skidmore. 1987. An explanation of enhanced radar backscattering from flooded forests. *Int J Remote Sens* 8:1093–1100.

Sanyal, J., and X. X. Lu. 2003. Application of remote sensing in flood management with special reference to Monsoon Asia: A review. *Nat Hazards* 33:283–301.

Schumann, G., G. D. di Baldassarre, and P. D. Paul. 2009. The utility of spaceborne radar to render flood inundation maps based on multialgorithm ensembles. *IEEE Trans Geosci Remote Sens* 47:2801–2807.

Schumann, G., G. D. di Baldassarre, D. Alsdorf, and P. D. Bates. 2010. Near real-time flood wave approximation on large rivers from space: Application to the River Po, Italy. *Water Resour Res* 46:1–8.

Schumann, G., R. Hostache, C. Puech, L. Hoffmann, P. Matgen, F. Pappenberger, and L. Pfister. 2007. High-resolution 3-D flood information from radar imagery for flood hazard management. *IEEE Trans Geosci Remote Sens* 45:1715–1725.

Silveira, M., and S. Heleno. 2009. Separation between water and land in SAR images using region-based level sets. *IEEE Trans Geosci Remote Sens* 6:471–475.

Smith, L. C. 1997. Satellite remote sensing of river inundation area, stage, and discharge: A review. *Hydrol Process* 11:1427–1439.

Soergel, U., U. Thoennessen, and U. Stilla. 2003. Visibility analysis of man-made objects in SAR images. *Proceedings of the Second GRSS/ISPRS Joint Workshop on Data Fusion and Remote Sensing Over Urban Areas*, Berlin, Germany, pp. 120–124.

SWBD, 2005. Shuttle radar topography mission water body dataset. https://lta.cr.usgs.gov/srtm_water_body_dataset. (Accessed April 24, 2015.)

Torres, R., P. Snoeij, D. Geudtner, D. Bibby, M. Davidson, E. Attema, P. Potin et al. 2012. GMES Sentinel-1 mission. *Remote Sens Environ* 120:9–24.

Townsend, P. A. 2001. Mapping seasonal flooding in forested wetlands using multi-temporal RADARSAT SAR. *Photogramm Eng Remote Sens* 67:857–864.

Townsend, P. A. 2002. Relationships between forest structure and the detection of flood inundation in forest wetlands using C-band SAR. *Int J Remote Sens* 23:332–460.

Townsend, P. A., and S. J. Walsh. 1998. Modeling floodplain inundation using an integrated GIS with radar and optical remote sensing. *Geomorphology* 21:295–312.

Ulaby, F. T., R. K. Moore, and A. K. Fung. 1982. *Microwave Remote Sensing: Active and Passive. Vol. II—Radar Remote Sensing and Surface Scattering and Emission Theory*. Addison-Wesley Publishing Company, Advanced Book Program, Reading, MA.

Van der Sande, C. J., S. M. de Jong, and A. P. J. De Roo. 2003. A segmentation and classification approach of IKONOS-2 imagery for land cover mapping to assist flood risk and flood damage assessment. *Int J Appl Earth Observ Geoinform* 4:217–229.

Voigt, S., T. Kemper, T. Riedlinger, R. Kiefl, K. Scholte, and H. Mehl. 2007. Satellite image analysis for disaster and crisis-management support. *IEEE Trans Geosci Remote Sens* 45:1520–1528.

Voormansik, K., J. Praks, O. Antropov, J. Jagomägi, and K. Zalite. 2013. Flood mapping with TerraSAR-X in forested regions in Estonia. *IEEE J Sel Top Appl Earth Observ Remote Sens* 7:562–577.

Wang, Y., J. D. Colby, and K. A. Mulcahy. 2002. An efficient method for mapping flood extent in a coastal flood plain using Landsat TM and DEM data. *Int J Remote Sens* 23:3681–3696.

Wang, Y., L. L. Hess, S. Filoso, and J. M. Melack. 1995. Understanding the radar backscattering from flooded and non-flooded Amazonian forests: Results from canopy backscatter modeling. *Remote Sens Environ* 54:324–332.

Wang, Y., and M. L. Imhoff. 1993. Simulated and observed L-HH radar backscatter from tropical mangrove forests. *Int J Remote Sens* 14:2819–2828.

Warth, G., and S. Martinis. 2013. Improved flood detection by using bistatical coherence data of the TanDEM-X mission. *Proceedings of 4. TanDEM-X Science Team Meeting. 4*, June 10–14, Oberpfaffenhofen, Germany, pp. 1–2.

Wendleder, A., B. Wessel, A. Roth, M. Breunig, K. Martin, and S. Wagenbrenner. 2012. TanDEM-X water indication mask: Generation and first evaluation results. *IEEE J Sel Top Appl Earth Observ Remote Sens* 6:171–179.

Westerhoff, R. S., M. P. H. Kleuskens, H. C. Winsemius, H. J. Huizinga, G. R. Brakenridge, and C. Bishop. 2013. Automated global water mapping based on wide-swath orbital synthetic-aperture radar. *Hydrol Earth Syst Sci* 17:651–663.

Wu, S. T., and S. A. Sader. 1987. Multipolarization SAR data for surface feature delineation and forest vegetation characterization. *IEEE Trans Geosci Remote Sens* 25:67–76.

Zadeh, L. A. 1965. Fuzzy sets. *Inform Control* 8:338–353.

Zhang, Y. J. 1996. A survey on evaluation methods for image segmentation. *Pattern Recogn* 29:1335–1346.

IV

Wetlands

Remote Sensing of Mangrove Wetlands

7

Chandra Giri
United States Geological Survey

Acronyms and Definitions

DEM	Digital elevation model
DOC	Dissolved organic carbon
FAO	Food and Agriculture Organization of the United Nations
GLS	Global Land Survey
Ha	Hectare
KM	Kilometer
MODIS	Moderate resolution imaging spectroradiometer
NDVI	Normalized Difference Vegetation Index
NGA	National Geospatial Agency of the United States
REDD+	Reducing emissions from deforestation and forest degradation plus
RSLR	Relative sea level rise
SRTM	Shuttle Radar Topography Mission
TOA	Top of atmosphere
USGS/EROS	U.S. Geological Survey Earth Resources Observation and Science Center

7.1 Introduction

Mangrove wetlands are distributed in the intertidal region between the sea and the land in the tropical and subtropical regions of the world spanning mostly between 35°N and 35°S latitude (Giri et al., 2011b). They grow in harsh environmental settings such as high salinity, high temperature, extreme tides, high sedimentation, and muddy anaerobic soils (Kathiresan and Bingham, 2001). The total area of mangroves in the year 2000 was 137,760 km² spread across 118 countries and territories (Giri et al., 2011b). At least 35% of the area of mangrove forests—wooded, tropical wetlands—has been deforested from 1980 to 2000 (Valiela et al., 2001). Much of what remains in these deforested areas is in degraded condition (Duke et al., 2007). Coastal habitats across the world are under heavy population and development pressures and are subjected to frequent storms and other natural disturbances.

The continued decline of mangrove forests is caused by conversion to agriculture, aquaculture, tourism, urban development, and overexploitation (Alongi, 2014; Giri et al., 2008). The remaining mangrove forests have been declining at a faster rate than inland tropical forests and coral reefs (Caldeira, 2012). Relative sea-level rise (RSLR) could be the greatest threat to mangroves in the future (Gilman et al., 2008). Predictions suggest that 30%–40% of coastal wetlands (IPCC, 2007) and functionality of mangrove forests could be lost in the next 100 years if the present rate of loss continues. As a consequence, important ecosystem goods and services (e.g., natural barrier, carbon sequestration, biodiversity) provided by mangrove forests will be diminished or lost (Duke et al., 2007).

Mangrove forests are among the most productive and biologically important ecosystems in the world because they provide

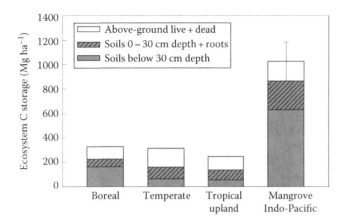

FIGURE 7.1 Carbon stock potential (per hectare) of mangroves compared to other ecosystems (Donato et al. 2011). Although mangroves occupy only ~0.1% of the Earth's surface, they sequester a disproportionately high percentage of above- and belowground carbon.

important and unique ecosystem goods and services to human society and coastal and marine systems. The forests help stabilize shorelines and reduce the devastating impacts of natural disasters such as tsunamis and hurricanes. They also provide breeding and nursing grounds for marine and pelagic species and food, medicine, fuel, and building materials for local communities. Mangroves, including associated soils, can sequester more carbon than other tropical ecosystems (Figure 7.1). Covering only 0.1% of the Earth's continental surface, mangrove forests account for 11% of the total input of terrestrial carbon into the ocean (Jennerjahn and Ittekkot, 2002) and 10% of the terrestrial dissolved organic carbon (DOC) exported to the ocean (Dittmar et al., 2006). Rapid disappearance and degradation of mangroves could have negative consequences on the transfer of materials into the marine systems and influence the atmospheric composition and climate.

Despite the importance of mangrove forests, our understanding of their present status and distributions is inadequate. Previous estimates of the total area of global mangroves range

from 110,000 to 240,000 km² (Wilkie and Fortune, 2003). The Food and Agriculture Organization (FAO) of the United Nations estimate was based on a compilation of disparate and incompatible geospatial and statistical data sources and did not provide spatial information with sufficient detail (Wilkie and Fortune, 2003). Global estimates also have been computed using published literature (Alongi, 2002); however, these estimates were inconsistent across space and time. Local studies to map mangroves are abundant; however, they do not cover large areal extents, often limited to very small study areas that use different remote sensing or non–remote sensing data sources, classification approaches, and classification systems.

Thus, regular mapping of mangroves using consistent data sources and methodologies was needed, especially over large areas and worldwide. To fulfill this need, the U.S. Geological Survey Earth Resources Observation and Science (USGS EROS) Center mangrove research team has been mapping and monitoring global mangroves since 2007. This chapter presents a summary of major activities conducted by the team in the context of how remote sensing can be used for mangrove mapping and monitoring from local to global scales.

7.2 Defining Mangroves for Mapping and Monitoring

A mangrove is a tree, shrub, palm, or ground fern that grows in the coastal intertidal zone in the tropical and subtropical regions of the world. Mangrove species are classified into two categories: true mangroves and mangrove associates. Tomlinson (1986) defined true mangroves with the following features: (1) occurring only in a mangrove environment and not extending into terrestrial communities, (2) presence of morphological specialization (aerial roots, vivipary), (3) possessing a physiological mechanism for salt exclusion and/or salt excretion, and (4) taxonomic isolation from terrestrial relatives. This definition has been accepted widely. True mangroves generally possess a distinct signature in optical remote sensing data (Figure 7.2).

FIGURE 7.2 Mangroves (dark red) in the Sundarbans (Bangladesh and India) in a Landsat false color composite (band combinations 4R, 3G, 2B).

7.3 Remote Sensing Data

A large variety of remote sensing data is being used to obtain information needed for effective management of mangrove forests. Initially, aerial photographs were used to obtain information about mangrove forests, but now a wide variety of remotely sensed data such as Landsat/SPOT/IRS/CBERES, IKONOS, QuickBird, GeoEye, Radar Phased Array type L-band Synthetic Aperture Radar [PALSAR], Radarsat, Shuttle Radar Topography Mission [SRTM]*, hyperspectral data, Moderate Resolution Imaging Spectroradiometer (MODIS), and aerial videography are available for use. In recent years, unmanned aerial vehicle imagery has been used to collect data (Wan et al., 2014). A brief summary of recent applications of remote sensing of mangrove wetlands is presented in Table 7.1.

7.4 Scale Issues

Moderate resolution satellite data such as Landsat contain enough detail to capture mangrove forest distribution and dynamics. However, very small patches (<900 m²) of mangrove forests including newly colonized individual mangrove stands, small island mangroves, or relatively small, fragmented, and linear patches of mangrove stands cannot be mapped using Landsat. High-resolution satellite data (e.g., IKONOS, QuickBird) (Figure 7.3), aerial photographs, or subpixel classification are needed to assess and monitor these small areas. It should be noted, however, that these small areas, while important for many applications, will not make a substantial difference in the global total (Wilkie and Fortune, 2003).

Mangrove mapping and monitoring at 1–2 m resolution will be needed in the future to create precise information on individual land cover features such as species composition and alliances, ecosystem shifts, water courses, infrastructure, and developed areas. High-resolution monitoring, deemed impossible until recently, is becoming a reality due to the increasing availability of high-resolution (<5 m) multispectral satellite data, improvements in image processing algorithms, and advancements in computing resources.

7.5 Methods of Mapping Mangroves Using Remote Sensing

Both visual and digital image classifications using both supervised and unsupervised classification approaches are being used for mapping and monitoring mangrove wetlands from local to global scales. Visual interpretation/on-screen digitizing, pixel-based classification, and object-based classification are some of the common approaches. All of these techniques are being used for interpreting aerial photography, aerial videography, aerial digital imagery, high-resolution imageries (QuickBird, IKONOS), medium-resolution imageries (ASTER, SPOT, Landsat, IRS, CBERS), hyperspectral data, and radar data. The pixel-based classification approach is most commonly used for interpreting (Kuenzer et al., 2011). Collection and use of *in situ* data play a critical role in the classification of remote sensing imageries.

There are numerous image classification and change detection techniques documented in the literature (Civco et al., 2004; Singh, 1989). For change analysis, Civco et al. (2004) compared four techniques: traditional postclassification, cross-correlation analysis, neural networks, and object-oriented classification. They concluded that there are advantages and disadvantages of each method, suggesting there is no single best method. Myint et al. (2014) evaluated and compared the effectiveness of different band combinations and classifiers (unsupervised, supervised, object-oriented nearest neighbor, and object-oriented decision rule) for quantifying mangrove forest changes using multitemporal Landsat data. The study was conducted in three tropical areas in Asia: the Sundarbans (Bangladesh and India), Irrawaddy Delta (Myanmar), and Trang (Thailand). A ranking system of 36 change maps produced by evaluating and comparing the effectiveness and efficiency of 13 change detection approaches was used.

Myint et al. (2014) found that the performance of the change detection approaches varied within and among study areas. The overall accuracy increased with decreasing environmental complexity. Which change detection approach performed best or worst depended on the criterion considered (e.g., overall, producer's or user's accuracy, or the mean of all accuracies).

TABLE 7.1 Non-Exhaustive List of Examples Illustrating How Remote Sensing Can Be Used for Remote Sensing of Mangrove Wetlands

Data	Applications	References
Aerial photographs	Mapping (extent and distribution), monitoring (deforestation and regrowth), forest degradation, species discrimination, forest density	Heenkenda et al. (2014), Jeanson et al. (2014)
Landsat/SPOT/IRS/ CBERES	Mapping (extent and distribution), monitoring (deforestation and regrowth), forest degradation, species discrimination, forest density, disturbance (storms), habitat fragmentation, ecosystem health, mangrove coastal retreat monitoring	Giri et al. (2011b), Gebhardt et al. (2012), Kalubarme (2014), Santos et al. (2014)
IKONOS, QuickBird, GeoEye	Mapping (extent and distribution), monitoring (deforestation and regrowth), forest degradation, species discrimination, forest density, disturbance (storms)	Leempoel et al. (2013), Giri et al. (2011a), Heenkenda et al. (2014)
Radar (PALSAR, Radarsat, SRTM),	Tree height/biomass, disturbance (storms), mapping (extent and distribution)	Hamdan et al. (2014), Lucas et al. (2007)
Hyperspectral data	Mapping (extent and distribution), monitoring (deforestation and regrowth), forest degradation, species discrimination, forest density	Mitchell and Lucas (2001)
MODIS	Mapping (extent and distribution), monitoring (deforestation and regrowth), forest degradation	Rahman et al. (2013)
Aerial video imagery	Species discrimination	Everitt et al. (1991)

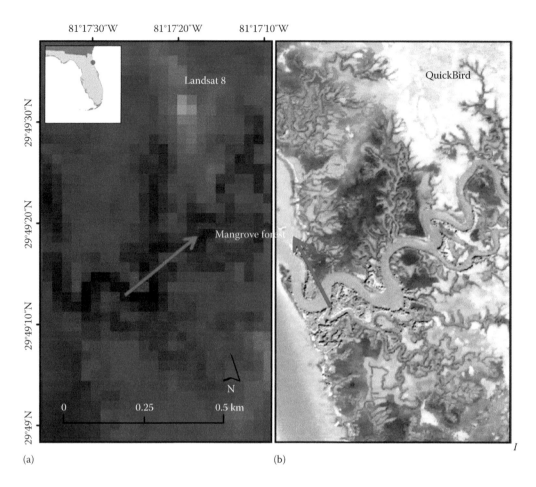

81°17'30"W   81°17'20"W   81°17'10"W

29°49'30"N

29°49'20"N

29°49'10"N

29°49'N

Landsat 8

Mangrove forest

QuickBird

0   0.25   0.5 km

N

(a)   (b)

FIGURE 7.3   Landsat imagery (a) can capture large mangroves (dark red) but its 30 m resolution is too coarse to accurately detect and classify the small, fragmented, and linear patterns of Florida's mangroves. QuickBird imagery (b), with a spatial resolution of 2.44 m, is better suited for mapping and monitoring smaller, fragmented, and linear mangrove stands.

However, several clear patterns emerged when aggregating the results for all study areas and considering overall producer's and user's accuracies simultaneously.

First, the choice of band combination had a greater impact on change detection accuracies than the choice of classifier (e.g., supervised, unsupervised, or object-oriented nearest neighbor). Second, a combination of bands 2, 5, and 7 was the most effective at detecting mangrove loss and persistence, followed by a combination of principal component bands 1, 2, and 3. It was determined that more bands do not necessarily mean better results; in fact, the inclusion of additional bands led to more signature confusion and thus lowered change detection accuracy. Third, discriminant analysis can be effective at identifying the optimal bands for differentiating between mangrove loss and other categories. Finally, alternative approaches such as a decision-rule object-oriented approach based on principal component band 3 are useful for mapping mangrove loss (Myint et al., 2014).

The automated decision-rule approach as well as a composite of bands 2, 5, and 7 with the unsupervised method and the same composite with the object-oriented nearest neighbor classifier are the most effective approaches to monitoring mangrove deforestation. Principal component bands 1, 2, and 3 generated

from a composite of all bands of start and end dates with the object-oriented nearest neighbor classifier and the same composite band with the unsupervised method can be expected to be similarly effective in detecting mangrove loss.

7.6 Global Mangrove Mapping

Even with the availability of >40 years of moderate-resolution satellite data (i.e., Landsat) and with a distinct signature of mangrove forests in the visible and near-infrared portion of the electromagnetic spectrum, mapping of mangrove forests at this resolution at the global scale was never attempted. Cost and computing facilities have been the primary limitations to using Landsat data for global studies. However, with the availability of free Global Land Survey (GLS), advent of making the Landsat archive freely available, and improvements in computing facilities, global mapping at Landsat scale became possible. Global data on the extent and conditions of mangrove forests could provide critical information needed for policy making and resource management. Giri et al. (2011b) used state-of-the-science remote sensing to prepare a wall-to-wall map of the mangrove forests of the world at 30 m resolution (Figure 7.4).

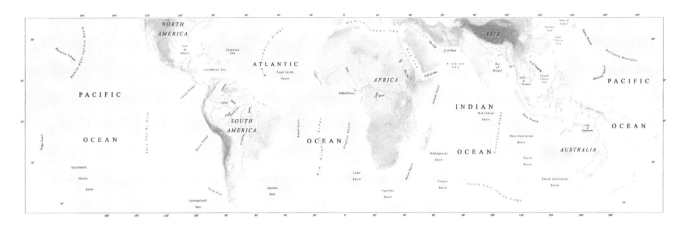

FIGURE 7.4 Mangrove forest distributions of the world for circa 2000 based on Landsat 30 m data. The study established that the total area of mangroves (shown in green) is 137,760 km² spread across 118 countries and territories in the tropical and subtropical regions of the world. Approximately 75% of mangroves worldwide are located in just 15 countries.

TABLE 7.2 Fifteen Most Mangrove-Rich Countries and Their Cumulative Percentages

S. No.	Country	Area (m²)	Cumulative (%)	Region
1	Indonesia	3112989.48	22.60	Asia
2	Australia	977975.46	29.70	Oceania
3	Brazil	962683	36.68	South America
4	Mexico	741917.00	42.07	North and Central Americas
5	Nigeria	653669.10	46.82	Africa
6	Malaysia	505386.00	50.48	Asia
7	Myanmar	494584.00	54.07	Asia
8	Papua New Guinea	480121.00	57.56	Oceania
9	Bangladesh	436570.00	60.73	Asia
10	Cuba	421538.00	63.79	North and Central Americas
11	India	368276.00	66.46	Asia
12	Guinea Bissau	338652.09	68.92	Africa
13	Mozambique	318851.10	71.23	Africa
14	Madagascar	278078.13	73.25	Africa
15	Philippines	263137.41	75.16	Asia

This database is the first most comprehensive, globally consistent, and highest-resolution (30 m) global mangrove database. Results from the Giri et al. (2011b) study showed that the remaining mangrove forest of the world is less than previously thought. This new estimate is ~12.3% smaller than the most recent estimate by the FAO of the United Nations. The present extent is 137,760 km² spread across 118 countries and territories in the tropical and subtropical regions of the world. The database includes mangrove vegetation excluding water bodies and barren lands necessary for global carbon accounting. The largest extent of mangroves is found in Asia (42%) followed by Africa (20%), North and Central Americas (15%), Oceania (12%), and South America (11%). Approximately 75% of mangroves are concentrated in just 15 countries worldwide (Table 7.2).

Mangrove area decreases with an increase in latitude, except between 20°N and 25°N latitude (Figure 7.5), which is where the Sundarbans are located (at the confluence of Ganges, Brahmaputra, and Meghna Rivers in Bangladesh and India on the Bay of Bengal). The Sundarbans are the largest tract of mangrove forests in the world (~10,000 km²). The study confirms earlier findings that biogeographic distribution of mangroves is generally confined to the tropical and subtropical regions of the world but can be found as far as 32°20′ N in Bermuda and 38°45′ S in Australia (Giri et al., 2011b). The largest percentage of mangrove is found in +5° to −5° latitudes (Figure 7.5).

7.7 Mangrove Monitoring

Time-series remote sensing data can be used to regularly monitor mangrove forests. Giri et al. (2008) monitored the regional (e.g., tsunami-impacted countries of Indonesia, Malaysia, Thailand, Myanmar, Bangladesh, India, and Sri Lanka) mangrove forest cover dynamics and identified their rates and causes of change from 1975 to 2005 using Landsat data. Results were derived using multitemporal satellite data and field observations. The repetitive coverage of satellite data provides an up-to-date and

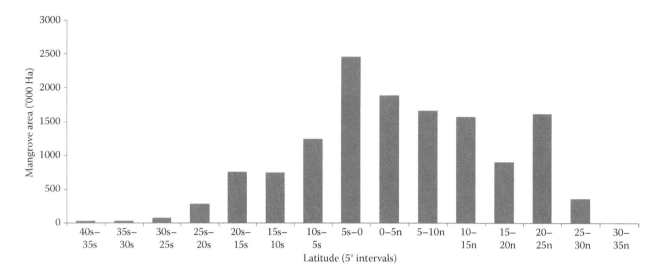

FIGURE 7.5 Latitudinal distribution of mangrove forests of the world.

consistent overview of the extent, distribution, and dynamics of mangrove forests with better spatial and thematic details than the existing coarse resolution and field inventory data. The analysis addressed the following research questions:

- How much mangrove forests remain?
- Where are these mangrove forests located?
- What are the spatial and temporal rate of change?
- What are the main reasons for the change?
- What are potential areas for rehabilitation/regeneration?

In this research, Landsat satellite data were preprocessed and classified for 1975, 1990, 2000, and 2005. Postclassification change analysis was performed subtracting the classification maps, 1975s–1990s, 1975s–2000s, 1975s–2005s, 1990s–2000s, 1990s–2005s, and 2000s–2005s. The change areas were interpreted visually with the help of secondary data and ground data information to identify the factors responsible for the change. Once the mangrove/nonmangrove areas were calculated for each period, the annual rate of change for the region and for each country was calculated. The time-series analysis revealed a net loss of 12% of mangrove forests in the region from 1975 to 2005 (Giri et al., 2008). Major hot-spot areas were also identified (Figure 7.6).

The rate of deforestation was not uniform in both spatial and temporal domains. The annual rate of deforestation during 1975–2005 was highest (~1%) in Myanmar compared to Thailand (0.73%), Indonesia (0.33%), Malaysia (0.2%), and Sri Lanka (0.08%). Mangrove areas in India and Bangladesh remained unchanged or increased slightly. Giri et al. (2008) identified the major deforestation fronts in Ayeyarwady Delta (Figure 7.7), Rakhine, and Tanintharyi in Myanmar; Sweetenham and Bagan in Malaysia; Belawan, Pangkalanbrandan, and Langsa in Indonesia; and southern Krabi and Ranong in Thailand. Major reforestation and afforestation areas are located on the southeastern coast of Bangladesh and Pichavaram, Devi Mouth, and Godavari in India.

Similarly, Giri et al. (2007), Giri and Muhlhausen (2008), Long et al. (2013), and Giri et al. (2015) performed change studies in the Sundarbans (Bangladesh and India), Madagascar, Philippines, and South Asia, respectively. In the Sundarbans, multitemporal satellite data from the 1970s, 1990s, and 2000s were used to monitor deforestation and degradation of mangrove forests. The spatiotemporal analysis showed that despite having the highest population density in the world in its immediate periphery, the areal extent of the mangrove forest of the Sundarbans has not changed substantially n the last ~25 years. However, the forest is constantly changing due to erosion, aggradation, deforestation, and mangrove rehabilitation programs. The net forest area increased by 1.4% from the 1970s to 1990 and decreased by 2.5% from 1990 to 2000. The change is insignificant in the context of classification errors and the dynamic nature of mangrove forests. The Sundarbans is an excellent example of the coexistence of humans with terrestrial and aquatic plant and animal life. The strong commitment of governments under various protection measures such as forest reserves, wildlife sanctuaries, national parks, and international designations is believed to be responsible for keeping this forest relatively intact (at least in terms of area). Nevertheless, the forest is under threat from natural and anthropogenic forces that could lead to forest degradation, primarily due to top-dying and overexploitation of forest resources.

Time-series analysis revealed that the mangrove forests of Madagascar are declining, albeit at a much slower rate (~1.5% per year) than the global average. The forests are declining due to logging, overexploitation, clear cutting, degradation, and conversion to other land uses. In this research, Giri and Muhlhausen (2008) interpreted time-series Landsat data from 1975, 1990, 2000, and 2005 using a hybrid supervised and unsupervised classification approach. Landsat data were geometrically corrected to an accuracy of plus-or-minus one-half pixel, an accuracy necessary for change analysis. The results showed that Madagascar lost 7% of mangrove forests from 1975 to 2005, to a present extent of ~27.97 km². Deforestation rates and causes

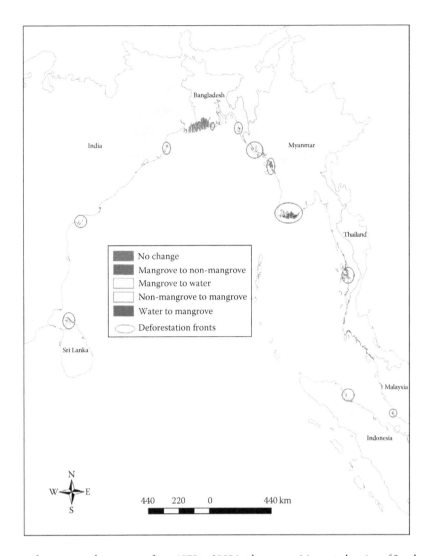

FIGURE 7.6 Major mangrove forest cover change areas from 1975 to 2005 in the tsunami-impacted region of South and Southeast Asia.

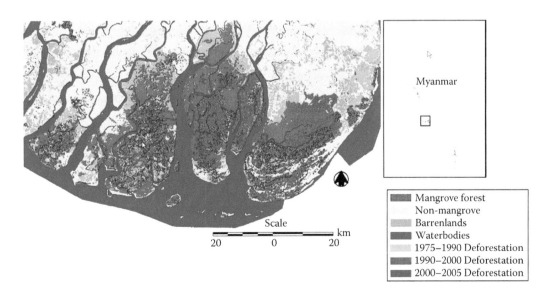

FIGURE 7.7 Spatial distribution of mangrove deforestation in Ayeyarwady Delta, Myanmar, from 1975 to 1990 (cyan), 1990 to 2000 (red), and 2000 to 2005 (purple).

varied both spatially and temporally. The forests increased by 5.6% (212 km²) from 1975 to 1990, decreased by 14.3% (455 km²) from 1990 to 2000, and decreased by 2.6% (73 km²) from 2000 to 2005. Major changes occurred in Bombetoka Bay, Mahajamba Bay, the coast of Ambanja, the Tsiribihina River, and Cap St. Vincent. The main factors responsible for mangrove deforestation include conversion to agriculture (35%), logging (16%), conversion to aquaculture (3%), and urban development (1%).

In the Philippines, several national mangrove estimates existed; however, information was unavailable at sufficient spatial and thematic detail for change analysis. Long et al. (2013) prepared a historical and contemporary mangrove distribution database of the Philippines for 1990 and 2010 at nominal 30 m spatial resolution using Landsat data. Image classification was performed using a supervised decision-tree classification approach. Additionally, decadal land cover change maps from 1990 to 2010 were prepared to depict changes in mangrove area using a postclassification technique. Total mangrove area decreased 10.5% from 1990 to 2010. A comparison of estimates produced from this study with selected historical mangrove area estimates revealed that total mangrove area in the Philippines decreased by approximately half (51.8%) from 1918 to 2010.

Mangrove forests in South Asia occur along the tidal sea edge of Bangladesh, India, Pakistan, and Sri Lanka. These forests provide important ecosystem goods and services to the region's dense coastal populations and support important functions of the biosphere. Mangroves are under threat from both natural and anthropogenic stressors; however, the current status and dynamics of the region's mangroves is poorly understood. Giri et al. (2015) mapped the current extent of mangrove forests in South Asia and identified mangrove forest cover change (gain and loss) from 2000 to 2012 using Landsat data. Three case studies were also conducted in the Indus Delta (Pakistan), Goa (India), and Sundarbans (Bangladesh and India) to identify rates, patterns, and causes of change in greater spatial and thematic detail than a regional assessment of mangrove forests.

Giri et al. (2015) found that the areal extent of mangrove forests in South Asia is approximately 11,874.76 km² representing ~7% of the global total. Approximately 921.35 km² of mangroves was deforested and 804.61 km² was reforested with a net loss of 116.73 km² from 2000 to 2012. In all three case studies, mangrove areas have remained unchanged or increased slightly; however, the turnover was greater than the net change. Both natural and anthropogenic factors are responsible for the change.

Giri et al. (2015) found that although the major causes of forest cover change are similar throughout the region, specific factors are dominant in specific areas. The major causes of deforestation in South Asia include (1) conversion to other land use (e.g., conversion to agriculture, shrimp farms, development, and human settlement), (2) overharvesting (e.g., grazing, browsing and lopping, and fishing), (3) pollution, (4) decline in freshwater availability, (5) flooding, (6) reduction of silt deposition, (7) coastal erosion, and (8) disturbances from tropical cyclones and tsunamis. The forests are changing due to distinct reasons in some locations, including sea salt extraction in the Indus Delta

in Pakistan, overharvesting of fruits in the Sundarbans, and garbage disposal in Mumbai, India. Conversely, mangrove areas are increasing in some regions because of aggradation, plantation efforts, and natural regrowth. Protection of existing mangrove areas is facilitating regrowth. The region's diverse socioeconomic and environmental conditions highlight complex patterns of mangrove distribution and change. Results from this study provide important insight to the conservation and management of the important and threatened South Asian mangrove ecosystem.

7.8 Species Discrimination

Accurate and reliable information on the spatial distribution of mangrove species is needed for a wide variety of applications, including sustainable management of mangrove forests, conservation and reserve planning, ecological and biogeographical studies, and invasive species management. Remotely sensed data have been used for such purposes with mixed results. Myint et al. (2008) employed an object-oriented approach with the use of a lacunarity technique to identify different mangrove species and their surrounding land use and land cover classes in southern Thailand using Landsat data. Results from the study showed that the mangrove zonation could be mapped using Landsat (Figure 7.8). It was also found that the object-oriented approach with lacunarity-transformed bands is more accurate (overall accuracy 94.2%; kappa coefficient = 0.91) than traditional per-pixel classifiers (overall accuracy 62.8%; and kappa coefficient = 0.57). Besides multispectral images, hyperspectral data have been used to discriminate mangrove species (Chakravortty et al., 2014).

7.9 Impact/Damage Assessment from Natural Disasters

Information regarding the present condition, historical status, and dynamics of mangrove forests was needed to study the impacts of the Gulf of Mexico oil spill of 2010. Such information was unavailable for Louisiana at sufficient spatial and thematic detail. Giri et al. (2011a) prepared mangrove forest distribution maps of Louisiana (before and after the oil spill) at 1 and 30 m spatial resolution using aerial photographs and Landsat data, respectively. Image classification was performed using a decision-tree classification approach. Maps were prepared of mangrove forest cover change pairs for 1983, 1984, and every 2 years from 1984 to 2010 depicting "ecosystem shifts" (e.g., expansion, retraction, and disappearance).

Direct damage to mangroves from the oil spill was minimal, but long-term impacts need to be monitored. This new spatiotemporal information can be used to assess long-term impacts of the oil spill on mangroves. The study also proposed an operational methodology based on remote sensing (Landsat, Advanced Spaceborne Thermal Emission and Reflection Radiometer [ASTER], hyperspectral, light detection and ranging [lidar], aerial photographs, and field inventory data) to monitor the existing and emerging mangrove areas and their disturbance and regrowth patterns. Several parameters such as

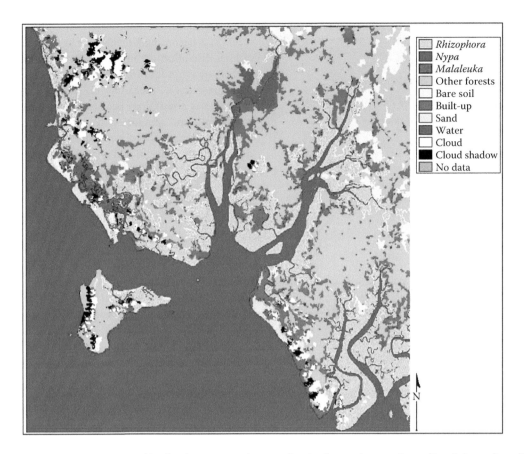

FIGURE 7.8   Species zonation map generated by the object-oriented approach using lacunarity-transformed bands in southern Thailand.

spatial distribution, ecosystem shifts, species composition, and tree height/biomass can be measured to assess the impact of the oil spill and mangrove recovery and restoration. Future research priorities will be to quantify the impacts and recovery of mangroves considering multiple stressors and perturbations, including the oil spill, winter freeze, sea-level rise, land subsidence, and land use/land cover change for the entire U.S. Gulf Coast.

Similarly, Landsat imagery is being used to map mangrove damage caused by Typhoon Haiyan in November 8, 2013, in the Philippines (Long et al. manuscript underdevelopment). The Normalized Difference Vegetation Index (NDVI) was used as a standardized measure. NDVI is one of the most widely used vegetation indexes (Tucker, 1979) to measure and monitor plant growth, vegetation cover, and biomass production. NDVI values range from 1.0 to −1.0 with dense vegetated areas (e.g., closed canopy tropical forest) generally yielding high NDVI values (0.6–0.8), sparsely vegetated areas (e.g., open shrub and grasslands) yielding moderate values (0.2–0.3), and nonvegetated (e.g., rock, sand, and snow) yielding low NDVI values (0.1 and below). Numerous studies have employed repeated measures of NDVI to monitor mangrove vegetation response from varying disturbances (Giri et al., 2011a), but few studies have applied this approach to monitor mangrove disturbance from typhoons (Wang, 2012) and still fewer at a 30 m spatial resolution.

Typhoon Haiyan transected five Landsat path/row footprints in the Philippines. Landsat 7 and Landsat 8 (ETM +

and OLI) imagery was used for the study. Image preprocessing for all imagery included converting from digital numbers to top-of-atmosphere (TOA) reflectance, stacking, masking for atmospheric contamination, and NDVI transformation. Mangrove areas were mapped prior to Haiyan for the year 2013 using a supervised decision-tree classification approach applied to the Landsat imagery captured before November 8, 2013 (Long et al., 2013). Independent variables used for mangrove classification included Landsat TOA imagery captured before November 8, 2013, a 30 m Shuttle Radar Topography Mission (SRTM) Digital Elevation Model (DEM), a slope index derived from the 30 m DEM, NDVI index transformed from Landsat TOA imagery, and a 2010 mangrove land cover map. Next, the 2013 mangrove land cover map was applied to mask values from all NDVI indexes (i.e., before and after) for mangrove areas. The NDVI images before and after the storm were differences to create NDVI change maps for all five Landsat footprints. These NDVI change maps illustrate where substantial changes occurred in NDVI following Typhoon Haiyan. Larger decreases in NDVI values indicate greater damage in mangrove canopy. Last, mangrove NDVI change maps were quality checked with field photos and field notes. Field data collection was independent from mapping and useful for quality checking mangrove damage maps.

Differing spatial patterns of mangrove damage resulted from variations in storm surge intensity, bathymetry and topography,

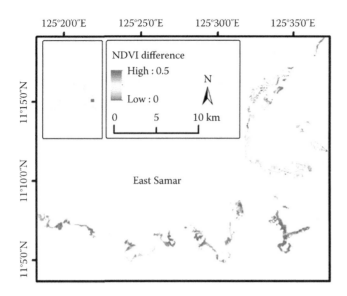

FIGURE 7.9   NDVI difference before and after Typhoon Haiyan in the Philippines.

wind speed and direction, and type and condition of the existing mangrove vegetation (Figure 7.9). Overall, mangrove damage with greatest decreases in NDVI values were observed along the Typhoon Haiyan's eye transect (east to west). As expected, mangrove damage decreased with increasing distance north and south from the eye path. Mangrove damage was also generally highest in the eastern Philippines and generally lessened westward corresponding with decreasing storm intensity from east to west. Average NDVI mangrove values prior to Haiyan were approximately 0.8, indicating vigorous vegetation growth. In this analysis, mangrove damage was considered significant if NDVI values reduced more than 0.5.

The total mangrove area affected (i.e., experiencing some decrease in NDVI) by Haiyan was estimated to be 214.45 km^2 about 9% of the Philippines' total mangrove area. Not all mangrove areas were included in the analysis because of insufficient data resulting from cloud cover in a few areas. Of the total affected mangrove area, only 6.53 km^2 of mangrove experienced a decrease in NDVI of 0.5 or greater, indicating substantial damage to mangrove health. Areas of substantial damage are considerably smaller than the mangrove areas experiencing minor damage.

7.10 High-Resolution Mangrove Mapping

As mentioned in the previous section, Landsat-based monitoring can capture the dynamic nature of mangrove forests. However, higher-resolution (<5 m spatial resolution) remotely sensed data are required to capture newly colonized individual stands or relatively small patches of mangrove stands that cannot be captured with 30 m Landsat imagery. At this resolution, it is possible to create precise information on individual land cover features such as species composition and alliances, ecosystem shifts, water courses, infrastructure, and developed areas.

High-resolution monitoring, deemed impossible until recently, is becoming a reality due to increasing the availability of high-resolution (<5 m) multispectral satellite data, improvement in image processing algorithms, and advancement in computing resources. Processing "big geo data" under this emerging paradigm is a challenge because of the data volume, software constraints, and computing needs. It is difficult, time consuming, and expensive to store, manage, visualize, and analyze these datasets using conventional image processing tools and computing resources. The USGS mangrove research team proposes to develop a cloud computing environment to visualize and analyze high-resolution satellite data for land change monitoring focusing on mangrove mapping and monitoring of Florida. This new methodology will develop the next generation of tools to analyze and visualize "big geo data" and improve our scientific understanding of mangrove forest cover change needed for decision making. Through the National Geospatial-Intelligence Agency (NGA), USGS, and commercial vendors, the research team has access to all available (unclassified) high-resolution satellite data (e.g., WorldView-2, GeoEye, IKONOS) from 2000 onward.

The USGS research team proposes a web-based mapping and monitoring platform and application to preprocess, analyze, and visualize the data. This application environment, based mainly on open-source tools, will be deployed in a cloud-based system for improved performance, effectively unlimited storage capacity, reduced software costs, scalability, elasticity, device independence, and increased data reliability. Using this platform, the team plans to map and monitor mangrove distribution and change (area, density, and species composition) and species expansion or squeeze due to climate change. The team also plans to quantify the impact and recovery from natural disasters such as hurricanes. High-resolution data are suitable to capture newly colonized individual mangrove stands, small island mangroves, or relatively small, fragmented, and linear patches of mangrove stands that cannot be captured with 30 m Landsat imagery (Figure 7.8).

The prototype system will improve the availability and usability of high-resolution data, algorithms, analysis tools, and scientific results through a centralized environment that fosters knowledge sharing, collaboration, innovation, and direct access to computing resources. Additional data such as Landsat and lidar can be added for similar analysis. The information generated from this study will improve our scientific understanding of distribution and dynamics of mangroves in Florida. The innovative tools and methods developed from this study will pave the way for future land change monitoring at high spatial resolution from local to national scales.

7.11 Conclusions and Recommendations

This chapter demonstrates how remote sensing data (Landsat and high-resolution satellite data) can be used for mangrove forest cover mapping and monitoring. These studies show that global mangroves can be mapped and monitored using Landsat; however, for specific use in localized areas, high-resolution satellite data or aerial photographs are needed. High-resolution data are useful to detect early colonization, small patches of mangrove forests, and changes. With the availability of high-resolution satellite data, advancement in computer technology, and improvement in classification algorithms, it will be possible to map and monitor global mangroves using high-resolution satellite data in the future. Remote sensing can help answer the following five major research topics:

1. Observed and predicted changes in RSLR, temperature, precipitation, atmospheric CO_2 concentration, hydrology, more frequent and destructive tropical storms, land use change in neighboring ecosystems, and human response to climate change are major components of environmental change that threaten the distribution and health of mangrove ecosystems (Alongi, 2008; Gilman et al., 2008; Saintilan et al., 2014). Monitoring mangrove forests using remote sensing could serve as an indicator of global climate change. For example, poleward migration beyond their historical range could indicate a changing climate.

2. Our scientific understanding of the impact of climate change on mangrove ecosystems and adaptation options is limited. RSLR is expected to be the major climate change component affecting mangrove ecosystems globally (Krauss et al., 2013; McKee et al., 2007). Mangroves may respond to RSLR by migrating upslope or maintaining their surface elevation through vertical accretion or sedimentation. However, several case studies around the world show that mangrove systems do not have the ability to keep pace with predicted RSLR rates (McKee et al., 2007), resulting in contraction or squeeze in extent at the shoreline from erosion and/or submergence, suffocation, and eventual death. RSLR could also impact the species composition of mangrove stands and related floral and faunal biodiversity. In contrast, "mangroves could be able to keep pace with sea-level rise in some places" (McIvor et al., 2013). Therefore, global impact will likely be site-specific, driven by the rate and degree of RSLR and other biophysical factors. Remote sensing observations could be used to quantify these changes spatially and temporally.

3. Mangrove forests are among the most carbon-rich habitats on the planet (Alongi, 2014; Donato et al., 2011) compared to other terrestrial forests, owing to the ecosystems' high productivity, rich organic input in long-term soil carbon storage, and low emissions of methane in the saltwater environment. These intertidal ecosystems are strong carbon sinks. Therefore, the forests could contribute to climate change mitigation efforts through carbon financing mechanisms such as reduced emissions from deforestation and degradation (REDD+) and other coastal marine conservation and mitigation initiatives (a.k.a. "Blue Carbon"). Remote sensing could be used to quantify carbon stock and carbon stock change in support of monitoring, reporting, and verification (MRV) of REDD+.

4. It has been suggested that mangrove ecosystems have expanded poleward in the last three decades primarily due to an increase in temperature or decrease in the frequency of winter freeze (Cavanaugh et al., 2014a; Saintilan et al., 2014). However, other reports contradict these findings stating that the observation period is inadequate (Giri et al., 2011a). For example, Cavanaugh et al. (2014b) used 1984–2011 Landsat data coupled with climate data and concluded that mangrove extent was expanding poleward along the northeast coast of Florida beyond its northern historical range. Giri and Long (2014) and (Giri et al. 2011a) conducted a similar study but included one additional year of Landsat data (i.e., 1983–2011) and concluded that although mangroves expanded from 1984 to 2010, the area has not reached the 1983 extent. The dramatic decrease in mangrove extent was caused by a severe winter freeze in December 1983 that reduced mangrove extent in the northern boundary of Florida and Louisiana by approximately 90%. Historic Landsat data dating back to 1972 and high-resolution satellite data could be used to answer whether mangrove is expanding poleward or not?

5. The Indian Ocean tsunami of December 2004, Hurricane Katrina along the U.S. Gulf Coast in 2005, Cyclone Nargis in Myanmar in 2008, and similar natural disasters in the past decade have highlighted the importance of the mangrove ecosystem as a "bioshield" or "natural barrier" in protecting vulnerable coastal communities from natural disasters. Several studies after the 2004 tsunami concluded that mangrove ecosystems provided protection to life and property (Barbier, 2006; Dahdouh-Guebas et al., 2005; Danielsen et al., 2005; Kathiresan and Rajendran, 2005; UNEP, 2005). Subsequent publications argued against this claim and concluded that mangrove ecosystems did not provide protection during the tsunami (Kerr and Baird, 2007; Vermaat and Thampanya, 2006). Studies conducted in India showed that villages located behind mangrove forests suffered less damage than those directly exposed to the coast (Kathiresan and Rajendran, 2005). In addition, Kathiresan (2012) suggested that the destructive power of the storm surge was exacerbated during Cyclone Nargis by recent loss of mangroves in Myanmar, although no primary evidence to support these statements was presented. Some researchers who are skeptical about the ability of mangroves to protect against tsunamis have noted that mangroves might be more capable of protecting against tropical storm surges. Storm surges differ from tsunamis in having shorter wavelengths and relatively more of their energy near the water surface. The current consensus is seemingly that mangroves provide protection to a certain extent, but proving it scientifically is a challenge. Increased frequency and intensity of extreme events could have profound impacts to coastal development and human safety in the future. Remote sensing data can provide unbiased and transparent information to help address some of these issues.

Acknowledgments

I thank the two anonymous reviewers who helped improve the manuscript. Any use of trade, firm, or product names is for descriptive purposes only and does not imply endorsement by the U.S. Government.

References

Alongi, D.M., 2002. Present state and future of the world's mangrove forests. *Environmental Conservation*, 29, 331–349.

Alongi, D.M., 2008. Mangrove forests: Resilience, protection from tsunamis, and responses to global climate change. *Estuarine, Coastal and Shelf Science*, 76, 1–13.

Alongi, D.M., 2014. Carbon cycling and storage in mangrove forests. *Annual Review of Marine Science*, 6, 195–219.

Barbier, E.B., 2006. Natural barriers to natural disasters: Replanting mangroves after the tsunami. *Frontiers in Ecology and the Environment*, 4, 124–131.

Caldeira, K., 2012. Avoiding mangrove destruction by avoiding carbon dioxide emissions. *Proceedings of the National Academy of Sciences*, 109, 14287–14288.

Cavanaugh, K.C., Kellner, J.R., Forde, A.J., Gruner, D.S., Parker, J.D., Rodriguez, W., Feller, I.C., 2014a. Poleward expansion of mangroves is a threshold response to decreased frequency of extreme cold events. *Proceedings of the National Academy of Sciences*, 111, 723–727.

Cavanaugh, K.C., Kellner, J.R., Forde, A.J., Gruner, D.S., Parker, J.D., Rodriguez, W., Feller, I.C., 2014b. Reply to Giri and Long: Freeze-mediated expansion of mangroves does not depend on whether expansion is emergence or reemergence. *Proceedings of the National Academy of Sciences*, 111, E1449.

Chakravortty, S., Ekta S., Arpita S.C. 2014. Application of spectral unmixing algorithm on hyperspectral data for mangrove species classification. In Applied Algorithms, Springer International Publishing, pp. 223–236.

Chakravortty, S., Shah, E., Chowdhury, A.S., 2014. Application of spectral unmixing algorithm on hyperspectral data for mangrove species classification. In *Applied Algorithms*. Springer, pp. 223–236.

Civco, D., Hurd, J.D., Wilson, E.H., Song, M., Zhang, Z., 2002. A comparison of land use and land cover change detection methods. *ASPRS-ACSM Annual Conference and FIG XXII Congress*, St. Louis, MO, April 22–26, 2002.

Dahdouh-Guebas, F., Jayatissa, L.P., Di Nitto, D., Bosire, J.O., Lo Seen, D., Koedam, N., 2005. How effective were mangroves as a defence against the recent tsunami? *Current Biology*, 15, R443–R447.

Danielsen, F., Sørensen, M.K., Olwig, M.F., Selvam, V., Parish, F., Burgess, N.D., Hiraishi, T., Karunagaran, V.M., Rasmussen, M.S., Hansen, L.B., 2005. The Asian tsunami: A protective role for coastal vegetation. *Science* (*Washington*), 310, 643.

Dittmar, T., Hertkorn, N., Kattner, G., Lara, R.J., 2006. Mangroves, a major source of dissolved organic carbon to the oceans. *Global Biogeochemical Cycles*, 20, GB1012.

Donato, D.C., Kauffman, J.B., Murdiyarso, D., Kurnianto, S., Stidham, M., Kanninen, M., 2011. Mangroves among the most carbon-rich forests in the tropics. *Nature Geoscience*, 4, 293–297.

Duke, N.C., Meynecke, J.-O., Dittmann, S., Ellison, A.M., Anger, K., Berger, U., Cannicci, S., Diele, K., Ewel, K.C., Field, C.D., 2007. A world without mangroves? *Science*, 317, 41–42.

Everitt, J., Escobar, D., Judd, F., 1991. Evaluation of airborne video imagery for distinguishing black mangrove (*Avicennia germinans*) on the lower Texas Gulf Coast. *Journal of Coastal Research*, 7, 1169–1173.

Gebhardt, S., Lam D.N., and Claudia, K., 2012. Mangrove ecosystems in the Mekong delta–overcoming uncertainties in inventory mapping using satellite remote sensing data. In *The Mekong Delta System*, Springer, Netherlands, pp. 315–330

Gebhardt, S., Nguyen, L.D., Kuenzer, C., 2012. Mangrove ecosystems in the Mekong delta—overcoming uncertainties in inventory mapping using satellite remote sensing data. In *The Mekong Delta System*. Springer, pp. 315–330.

Gilman, E.L., Ellison, J., Duke, N.C., Field, C., 2008. Threats to mangroves from climate change and adaptation options: A review. *Aquatic Botany*, 89, 237–250.

Giri, C.P., Long, J., 2014. Mangrove reemergence in the northernmost range limit of eastern Florida. *Proceedings of the National Academy of Sciences*, 111(15), E1447–E1448.

Giri, C., Long, J., Abbas, S., Murali, R.M., Qamer, F.M., Pengra, B., Thau, D., 2015. Distribution and dynamics of mangrove forests of South Asia. *Journal of Environmental Management*, 148, 101–111.

Giri, C., Long, J., Tieszen, L., 2011a. Mapping and monitoring Louisiana's mangroves in the aftermath of the 2010 Gulf of Mexico oil spill. *Journal of Coastal Research*, 27, 1059–1064.

Giri, C., Muhlhausen, J., 2008. Mangrove forest distributions and dynamics in Madagascar (1975–2005). *Sensors*, 8, 2104–2117.

Giri, C., Ochieng, E., Tieszen, L., Zhu, Z., Singh, A., Loveland, T., Masek, J., Duke, N., 2011b. Status and distribution of mangrove forests of the world using earth observation satellite data. *Global Ecology and Biogeography*, 20, 154–159.

Giri, C., Pengra, B., Zhu, Z.L., Singh, A., Tieszen, L.L., 2007. Monitoring mangrove forest dynamics of the Sundarbans in Bangladesh and India using multi-temporal satellite data from 1973 to 2000. *Estuarine, Coastal and Shelf Science*, 73, 91–100.

Giri, C., Zhu, Z., Tieszen, L.L., Singh, A., Gillette, S., Kelmelis, J.A., 2008. Mangrove forest distributions and dynamics (1975–2005) of the tsunami-affected region of Asia. *Journal of Biogeography*, 35, 519–528.

Hamdan, O., Khali Aziz, H., Mohd Hasmadi, I., 2014. L-band ALOS PALSAR for biomass estimation of Matang mangroves, Malaysia. *Remote Sensing of Environment*, 155, 69.

Heenkenda, M.K., Joyce, K.E., Maier, S.W., Bartolo, R., 2014. Mangrove species identification: Comparing WorldView-2 with aerial photographs. *Remote Sensing*, 6, 6064–6088.

IPCC, 2007. Climate change 2007: The physical science basis. In Solomon, S., D. Qin, M. Manning, Z. Chen, M. Marquis, K.B. Averyt, M. Tignor, and H.L. Miller (eds.), Contribution of Working Group I to the Fourth Assessment Report of the Intergovernmental Panel on Climate Change. Cambridge University Press, Cambridge, U.K.

Jeanson, M., Anthony, E.J., Dolique, F., Cremades, C., 2014. Mangrove evolution in Mayotte Island, Indian Ocean: A 60-year synopsis based on aerial photographs. *Wetlands*, 34, 459–468.

Jennerjahn, T.C., Ittekkot, V., 2002. Relevance of mangroves for the production and deposition of organic matter along tropical continental margins. *Naturwissenschaften*, 89, 23–30.

Kalubarme, M.H., 2014. Mapping and monitoring of mangroves in the coastal districts of Gujarat State using remote sensing and geo-informatics. *Asian Journal of Geoinformatics*, 14, 15–26.

Kathiresan, K., 2012. Importance of mangrove ecosystem. *International Journal of Marine Science*, 2, 70–89.

Kathiresan, K., Bingham, B.L., 2001. Biology of mangroves and mangrove ecosystems. *Advances in Marine Biology*, 40, 81–251.

Kathiresan, K., Rajendran, N., 2005. Coastal mangrove forests mitigated tsunami. *Estuarine, Coastal and Shelf Science*, 65, 601–606.

Kerr, A.M., Baird, A.H., 2007. Natural barriers to natural disasters. *BioScience*, 57, 102–103.

Krauss, K.W., McKee, K.L., Lovelock, C.E., Cahoon, D.R., Saintilan, N., Reef, R., Chen, L., 2013. How mangrove forests adjust to rising sea level. *New Phytologist*, 202, 19–34.

Kuenzer, C., Bluemel, A., Gebhardt, S., Quoc, T.V., Dech, S., 2011. Remote sensing of mangrove ecosystems: A review. *Remote Sensing*, 3, 878–928.

Leempoel, K., Bourgeois, C., Zhang, J., Wang, J., Chen, M., Satyaranayana, B., Bogaert, J., Dahdouh-Guebas, F., 2013. Spatial heterogeneity in mangroves assessed by GeoEye-1 satellite data: A case-study in Zhanjiang Mangrove National Nature Reserve (ZMNNR), China. *Biogeosciences Discussions*, 10, 2591–2615.

Long, J., Giri, C., Primavera, J., Trivedi, M., 2015. Assessing the Impact of Typhoon Haiyan on the Philippines' Mangrove Forests, U.S. Geological Survey, Sioux Falls.

Long, J., Napton, D., Giri, C., Graesser, J., 2013. A mapping and monitoring assessment of the Philippines' Mangrove Forests from 1990 to 2010. *Journal of Coastal Research*, 30, 260–271.

Lucas, R.M., Mitchell, A.L., Rosenqvist, A., Proisy, C., Melius, A., Ticehurst, C., 2007. The potential of L-band SAR for quantifying mangrove characteristics and change: Case studies from the tropics. *Aquatic Conservation: Marine and Freshwater Ecosystems*, 17, 245–264.

McIvor, A., Spencer, T., Möller, I., Spalding, M., 2013. The response of mangrove soil surface elevation to sea level rise. The Nature Conservancy and Wetlands International, Natural Coastal Protection Series: Report 3. Cambridge Coastal Research Unit, Cambridge, U.K.

McKee, K.L., Cahoon, D.R., Feller, I.C., 2007. Caribbean mangroves adjust to rising sea level through biotic controls on change in soil elevation. *Global Ecology and Biogeography*, 16, 545–556.

Mitchell, A., Lucas, R., 2001. Integration of aerial photography, hyperspectral and SAR data for mangrove characterization. *Geoscience and Remote Sensing Symposium, 2001 (IGARSS'01)*. IEEE 2001 International. IEEE, Piscataway, NJ, pp. 2193–2195.

Myint, S.W., Franklin, J., Buenemann, M., Kim, W.K., Giri, C.P., 2014. Examining change detection approaches for tropical mangrove monitoring. *Photogrammetric Engineering & Remote Sensing*, 80, 983–993.

Myint, S.W., Giri, C.P., Le, W., Zhu, Z.L., Gillette, S.C., 2008. Identifying mangrove species and their surrounding land use and land cover classes using an object-oriented approach with a lacunarity spatial measure. *GIScience and Remote Sensing*, 45, 188–208.

Rahman, A.F., Dragoni, D., Didan, K., Barreto-Munoz, A., Hutabarat, J.A., 2013. Detecting large scale conversion of mangroves to aquaculture with change point and mixed-pixel analyses of high-fidelity MODIS data. *Remote Sensing of Environment*, 130, 96–107.

Saintilan, N., Wilson, N.C., Rogers, K., Rajkaran, A., Krauss, K.W., 2014. Mangrove expansion and salt marsh decline at mangrove poleward limits. *Global Change Biology*, 20, 147–157.

Santos, L.C.M., Matos, H.R., Schaeffer-Novelli, Y., Cunha-Lignon, M., Bitencourt, M.D., Koedam, N., Dahdouh-Guebas, F., 2014. Anthropogenic activities on mangrove areas (São Francisco River Estuary, Brazil Northeast): A GIS-based analysis of CBERS and SPOT images to aid in local management. *Ocean & Coastal Management*, 89, 39–50.

Singh, A., 1989. Review article digital change detection techniques using remotely-sensed data. *International Journal of Remote Sensing*, 10, 989–1003.

Tomlinson, P., 1986. *The Botany of Mangroves*. Cambridge Tropical Biology Series. Cambridge University Press, Cambridge, U.K.

Tucker, C.J., 1979. Red and photographic infrared linear combinations for monitoring vegetation. *Remote Sensing of Environment*, 8, 127–150.

UNEP, 2005. After the tsunami: rapid environmental assessment report, February 22, 2005. UNEP, Nairobi, Kenya.

Valiela, I., Bowen, J.L., York, J.K., 2001. Mangrove Forests: One of the World's Threatened Major Tropical Environments At least 35% of the area of mangrove forests has been lost in the past two decades, losses that exceed those for tropical rain forests and coral reefs, two other well-known threatened environments. *Bioscience*, 51, 807–815.

Vermaat, J.E., Thampanya, U., 2006. Mangroves mitigate tsunami damage: A further response. *Estuarine, Coastal and Shelf Science*, 69, 1–3.

Wan, H., Wang, Q., Jiang, D., Fu, J., Yang, Y., Liu, X., 2014. Monitoring the invasion of *Spartina alterniflora* using very high resolution unmanned aerial vehicle imagery in Beihai, Guangxi (China). *The Scientific World Journal*, 2014, 638296.

Wang, Y., 2012. Detecting vegetation recovery patterns after hurricanes in South Florida using NDVI Time Series. Electronic Theses and Dissertations, University of Miami, pp. 61.

Wilkie, M., Fortune, S., 2003. Status and trends of mangrove extent worldwide. Forest Resources Assessment Working Paper No. 63. Forest Resources Division, FAO, Rome, Italy.

8

Wetland Mapping Methods and Techniques Using Multisensor, Multiresolution Remote Sensing: Successes and Challenges

D.R. Mishra
University of Georgia

Shuvankar Ghosh
University of Georgia

C. Hladik
Georgia Southern University

Jessica L. O'Connell
University of Georgia

H.J. Cho
Bethune-Cookman University

Acronyms and Definitions

GBM	Green biomass
ALVQ	Adaptive learning vector quantization
ALI	Advanced Land Imager
ALOS	Advanced Land Observing Satellite
ASAR	Advanced Synthetic Aperture Radar
AISA	Airborne Imaging Spectrometer for Applications
ALTM	Airborne Laser Terrain Mapper
AVIRIS	Airborne visible/infrared imaging spectrometer
ASPRS	American Society for Photogrammetry and Remote Sensing
ASD	Analytical Spectral Devices
BM	*Batis maritima*
BF	*Borrichia frutescens*
CHL	Canopy chlorophyll content
CSP	Carbon sequestration potential
CALMIT	Center for Advanced Land Management Information Technologies
CI_{green}	Chlorophyll index green
CI_{red}	Chlorophyll index red
CDR	Climate Data Record
CRD	Coastal Resources Division
R^2	Coefficient of determination
DGPS	Differential global positioning system
DEMs	Digital elevation models
EO-1	Earth Observation-1
ETM	Enhanced Thematic Mapper
ENVI	Environment for Visualizing Images
ENVISAT	Environmental Satellite
EVI2	Enhanced vegetation index 2
FODIS	Fiber-optic downwelling irradiance sensor
FOV	Field of view
N	Foliar nitrogen
FVA	Fundamental vertical accuracy
GA	Georgia
GWP	Global warming potential
GLAI	Green leaf area index
GPP	Gross primary productivity
GCPs	Ground control points

IFOV	Instantaneous field of view
MUD	Intertidal mud
JR	*Juncus roemerianus*
LEDAPS	Landsat Ecosystem Disturbance Adaptive Processing System
LAI	Leaf area index
LCC	Leaf chlorophyll content
LVQ	Learning vector quantization
LiDAR	Light Detection and Ranging
LUE	Light use efficiency
MLC	Maximum likelihood classification
MLLW	Mean lower low water
MODIS	Moderate Resolution Imaging Spectroradiometer
MSS	Multispectral scanning
NASA	National Aeronautics and Space Administration
NCALM	National Center for Airborne Laser Mapping
NOAA	National Oceanic Atmospheric Administration
NWI	National Wetlands Inventory
NIR	Near infrared
NDVI	Normalized difference vegetation index
nRMSEP	Normalized root mean squared error of prediction
OBIA	Object-based image analysis
PDSI	Palmer drought severity index
PLS	Partial least squares
%NRMSE	Percent normalized root mean squared error
PALSAR	Phased Array type L-band Synthetic Aperture Radar
PFT	Plant functional types
PRF	Pulse rate frequency
RTK	Real-time kinematic
ROIs	Regions of interest
RuBisCO	Ribulose-1,5-bisphosphate carboxylase–oxygenase
RMSE	Root mean squared error
RMSEP	Root mean squared error of prediction
SV	*Salicornia virginica*
SALT	Salt pan
SAVI	Soil-adjusted vegetation index
SA	*Spartina alterniflora*
SM	*Spartina* medium
SS	*Spartina* short
ST	*Spartina* tall
TM	Thematic Mapper
TOC	Top of canopy
U.S.	United States
USGS	United States Geological Survey
UTM	Universal Transverse Mercator
VF	Vegetation fraction
VARI	Visible atmospheric resistant index
WDRVI	Wide dynamic range vegetation index

8.1 Introduction

Wetlands occur as ecotones between terrestrial and deep-water aquatic systems. These ecosystems are defined by wide variations in location and size within landscapes, hydrology,

presence/absence and plant functional types (PFTs), productivity, and biogeochemical characteristics (Mitsch and Gosselink 2007; Keddy 2010). Wetlands have variable flood regimes over time (dry to wet) and usually are not considered to be waters >2 m deep (Mitcsch and Gosselink 2007). Wetlands tend to have anaerobic soils during some portion of the hydroperiod. Wetland types vary from freshwater to saltwater flood. Hydrology of wetlands is further determined by landscape elevation and geology, such that inundation frequency and length depend on wetland basin geomorphology (Brinson 1993). A common wetland definition is the one by Cowardin et al. (1979), subsequently adapted by U.S. Fish and Wildlife Service. According to this definition, wetlands share three common distinguishing features: presence of water within the plant root zone, unique hydric soil conditions, and presence of hydrophytes during some portion of the hydroperiod (Cowardin et al. 1979; Mitsch and Gosselink 2007). Because of unique and stressful environmental conditions within wetland types, wetland classes often are defined by distinctive vegetation communities adapted to patterns of inundation frequency, depth, and water chemistry within specific wetland classes (Cronk and Fennessy 2001). Thus, wetlands can be unvegetated barrens, moss-dominated acidic bogs and fens, and inland herbaceous wetlands including wet meadows, ephemeral ponds, prairie potholes, and semiarid playas. In coastal zones, wetland types include freshwater, brackish water, and saltwater herbaceous marshes. Woody wetlands include both freshwater bottomland hardwood swamps and saltwater mangroves (Mitsch and Gosselink 2007; Keddy 2010). Therefore, inundation, water chemistry, and dominant vegetation patterns vary considerably within and among wetland classes. Remote sensing helps in the inventory and monitoring of wetland distribution, productivity, and ecological status and often relies on an assessment of wetland vegetation (Ozesmi and Bauer 2002; Dahl 2004; Mayer and Lopez 2011). In this chapter, a review of the history of remote sensing techniques in wetland studies is presented and discussed. Three case studies highlight the new and advanced methodological aspects in remote sensing of wetlands. These new approaches have promise for solving ubiquitous problems in mapping wetland classes and understanding ecosystem dynamics with wetland habitats.

8.2 Evolution of Wetland Remote Sensing

8.2.1 Evolution

The history of remote sensing within wetlands mirrors that for uplands environments. Early remote sensing approaches were for delineating wetlands as habitats distinct from adjacent nonwetlands and were primarily dependent on visual interpretation of aerial photography. As in other systems, wetland remote sensing also took advantage of the development of other passive sensors, including multispectral and hyperspectral platforms. Passive remote sensors vary in terms of cost of acquisition, spatial coverage, and spatial, spectral, and temporal resolution. Although aerial photography has high spatial resolution and has been used frequently in wetland mapping, the acquisition cost and availability make them generally impractical for regional monitoring. The eventual deployment of multispectral satellite sensors has increased the spatial coverage, spectral range, and temporal frequency of data collection. However, multispectral satellite sensors generally have coarser spatial resolution than aerial photography. More recently, hyperspectral sensors on both airborne and space-borne platforms have significantly improved wetland remote sensing by providing narrowband spectra. High spectral resolution can aid in distinguishing certain biophysical and biochemical data. Both hyperspectral and multispectral data can estimate multiple vegetation parameters and identify broadscale patterns in wetland emergent properties. Suitability of each sensor for wetland remote sensing applications requires matching sensors' strengths and limitations to study goals.

8.2.2 Techniques and Challenges

Techniques used with passive remote sensors include unsupervised and supervised classification to map vegetation patterns (Hirano et al. 2003; Silvestri et al. 2003; Belluco et al. 2006; Wang et al. 2007). The main uses of aerial photography include the use as reference data for unsupervised classification of wetland vegetation and to collect training sites in supervised classification of satellite data. Applications of vegetation classification include monitoring invasive wetland vegetation (Albright et al. 2004; Rosso et al. 2006; Gilmore et al. 2008). Ramsey and Rangoonwala (2006) used classification to detect and categorize marsh dieback, onset, and progression. In their study, Ramsey and Rangoonwala (2006) documented enhanced performance of Earth Observation-1 (EO-1) Hyperion hyperspectral data compared to data from broadband multispectral sensors (EO-1 Advanced Land Imager and Landsat Enhanced Thematic Mapper). Airborne hyperspectral data have also been used to develop spectral libraries for classifying wetlands with greater accuracy than those derived from field spectroscopic data (Zomer et al. 2008).

In addition to vegetation classification, multispectral and hyperspectral sensors can be used to develop vegetation indices (VIs) of vegetation parameters. For example, hyperspectral imagery has been used extensively in salt marshes to document erosion and vegetation succession (Thomson et al. 2004), measure biomass and species richness and diversity (Wang et al. 2007; Lucas and Carter 2008), and detect vegetation change (Klemas 2011; Mishra et al. 2012), among other applications. In the early 1980s, multispectral sensing with handheld devices provided the first remote sensing of wetland biomass via the normalized difference vegetation index (NDVI) (Hardisky et al. 1983, 1984). More recently, region-wide collection of hyperspectral data allowed estimation of other vegetation parameters, such as foliar chlorophyll content. In this chapter, Case Study 2 discusses methodology for using Moderate Resolution Imaging

Spectroradiometer (MODIS) data to estimate key parameters for monitoring wetland health and restoration progress.

Remote sensing within wetlands has special challenges. For example, wetlands are patchy, and individual patches can be smaller than some sensor's spatial resolution. It is still a challenge to estimate species diversity from remotely sensed data particularly in freshwater wetlands with high species diversity. (Phinn 1998). Water inundation also influences spectral reflectance, reducing near-infrared (NIR) spectra and shifting the red-edge position, altering the effectiveness of NDVI and red-edge-type indices (Kearney et al. 2009; Turpie 2013). As such, wetland-specific indices are needed to account for the wetland optical properties. A method for development of such indices is described in Section 8.3.

A different approach for estimating vegetation patterns is multivariate analysis, such as partial least squares (PLS) regression. While VIs use either data mining or expert knowledge to select appropriate bands for indexing vegetation parameters, PLS regression uses eigenvector-based techniques to reduce full-spectrum data to components that maximize vegetation parameter prediction (Mevik and Wehrens 2007) and factors out nontarget cover-type influence from estimates (Chen et al. 2009). The use of PLS regression in wetland studies is illustrated in Study 3.

Recently, object-based image analysis (OBIA) has been used to solve classification and parameter estimation challenges within wetlands. Unlike pixel-based analysis, OBIA treats landscapes as clusters of objects representing surface entities and cover patches (Arbiol et al. 2006; Benz et al. 2004). Some advantages of OBIA are that object-shaped and neighborhood characteristics are included in the analyses (Yu et al. 2008). In wetlands, OBIA context information can aid classification, especially when high spatial heterogeneity, soil moisture, and inundation confound spectral analysis by other methods (Wright and Gallant 2007). Applications of OBIA include mapping PFTs (Dronova et al. 2012). Dronova et al. (2012) used a combination of OBIA and machine learning classifiers applied to Landsat 5 Thematic Mapper (TM) images in Lake Poyang, China. They established candidate PFTs by performing cluster analysis of physiological and phenological traits of 51 wetland species, resulting in generalized and specific PFTs. This method accurately captured the belt-shaped zonation of wetland vegetation types around water features, with the highest accuracy achieved for specific PFTs (91.1%). In another study, a combination of neural network and object-based texture measures was applied for mapping the vegetation in the Everglades, South Florida, using hyperspectral imageries (Zhang and Xie 2012). Fifteen land cover classes in the complex wetland ecosystem were identified, and results showed that adaptive learning vector quantization, a learning vector quantization classifier modified by adding a competitive layer, was the best neural network classifier, producing 94% overall accuracy in classification.

While conventional passive airborne and satellite sensors measure data in the visible and infrared spectral region, the advent of active sensors such as imaging radars provides fundamentally different information by acquiring signals at much longer wavelengths. One common imaging radar is synthetic-aperture radar (SAR), which provides the fine spatial resolution necessary for regional wetland mapping (Novo et al. 2002; Lang and McCarty 2008). Imaging radars have many advantages over sensors operating in the visible and infrared regions for wetland mapping. Microwave energy is sensitive to soil moisture variations and inundation and is least attenuated by wetland canopies (Kasischke et al. 1997; Townsend and Walsh 1998; Townsend 2000; Baghdadi et al. 2001; Townsend 2002; Rosenqvist et al. 2007; Lang and Kasischke 2008). Therefore, researchers often argue that SAR is the ideal sensor for studying wetlands (Hess et al. 1995; Hall 1996; Kasischke and Bourgeau-Chavez 1997; Kasischke et al. 1997; Dwivedi et al. 1999; Phinn et al. 1999; Rao et al. 1999; Toyra et al. 2002; Wilson and Rashid 2005; Costa and Telmer 2007). However, the radar signal is often reduced because of specular reflectance in flooded wetlands with low biomass (Kasischke et al. 1997). On the other hand, the radar signal is enhanced in flooded forested wetlands because of the double-bounce effect (Harris and Digby-Argus 1986; Dwivedi et al. 1999).

Advanced techniques have been proposed for mapping wetland types and plant species (Schmidt et al. 2004; Jensen et al. 2007; Yang et al. 2009; Klemas 2011). For example, many studies now use data fusion techniques to combine light detection and ranging (LiDAR), hyperspectral, multispectral, and/ or radar imageries to increase differentiation of wetland species and estimation of biophysical and biochemical parameters (Ozesmi and Bauer 2002; Schmidt et al. 2004; Artigas and Yang 2005; Filippi and Jensen 2006; Pengra et al. 2007; Adam 1990; Gilmore et al. 2010; Simard et al. 2010; Wang 2010). For example, combining hyperspectral imageries and LiDAR-derived digital elevation models (DEMs) has significantly improved the accuracy of mapping wetland vegetation (Geerling et al. 2007; Yang and Artigas 2010; Hladik et al. 2013). An application of multisensor data fusion is provided by Silva et al. (2010). They used MODIS and RADARSAT-1 to map temporal changes of macrophytes on the Amazon floodplain. Results revealed that radar image signatures for different cover types considerably overlapped, yet the method discriminated macrophytes with an overall accuracy of 93.5%. Wetland mapping in Brazilian Pantanal used Advanced Land Observing Satellite/Phased Array type L-band Synthetic Aperture Radar, RADARSAT-2, and Environmental Satellite/ Advanced Synthetic Aperture Radar imagery. This analysis employed a hierarchical OBIA approach under the eCognition Developer environment (Evans and Costa 2013). Backscattering analysis indicated a high degree of similarity (85%) between the woody classes and herbaceous classes. In Study 1, a case study demonstrating data fusion of hyperspectral and LiDAR is presented. Study 1 takes advantage of species sorting along elevation gradients within wetlands, i.e., lower elevations are flooded more often and have more

flood-tolerant species. Highly accurate LiDAR-derived elevation measures can both improve mapping of species groups and help resolve within-species productivity patterns along environmental gradients.

8.3 Proximal Sensing of Wetlands: Addressing Special Challenges

Remote sensing of terrestrial vegetation using medium-resolution satellite data has been successful because of the plant pigments, chlorophyll *a* and *b*, that absorb energy in the blue (centered at 450 nm) and the red (centered at 670 nm) wavelengths, as well as the leaf internal spongy mesophyll structure that is responsible for the high reflectance in the NIR region (700–1300 nm) (Lillesand et al. 2008). For terrestrial vegetation, simple spectral indices including simple vegetation index and NDVI have been successfully used to correlate with diverse plant characteristics at a broad span of scales from individual leaf areas to global vegetation dynamics (Graets 1990; Goward and Huemmich 1992; du Plessis 1999; Fang et al. 2001; Jackson et al. 2004; Geerken et al. 2005). These simple multispectral indices have also been used on wetland and aquatic vegetation, especially floating and emergent aquatics (Penuelas et al. 1993; Cho et al. 2008; Shekede et al. 2008; Kiage and Walker 2009). For example, NDVI calculated with Landsat multispectral scanning and TM data has been used to map the types and spatial extent of invasion by freshwater emergent and floating weeds (Shekede et al. 2008). However, surface water film or background water can affect the utility of these indices for monitoring wetland vegetation. Cho et al. (2008) found that the slight submergence of water hyacinth (*Eichhornia crassipes*) reduced the mean NIR reflectance from 76% to 6.3% during the close-range spectroscopy study in experimental outdoor tanks. In order to improve the classification accuracy of wetland plants, conventionally used VIs have been modified to include wetness factors in addition to plant greenness factors, to create vegetation water index (Wang et al. 2012). Vegetation ratios were used to incorporate moisture levels in the background mud when mapping of salt marsh. Vegetation ratios helped discriminate areas with dead marsh plants from areas with sparse live marsh plants (Ramsey and Rangoonwala 2006). In order to reparameterize an existing VI for wetland vegetation, proximal spectroscopy studies are needed to gain further understanding of wetland reflectance patterns and effect of water background on the NIR reflectance. The following is an example of top-of-canopy (TOC) reflectance data acquisition from tidal salt marsh vegetation using an in situ hyperspectral spectroradiometer (Mishra et al. 2012).

A dual-fiber system, with two intercalibrated Ocean Optics USB4000 hyperspectral radiometers (Ocean Optics Inc., Dunedin, FL, USA), mounted on a sturdy frame was used to acquire the TOC spectral percent reflectance (%*R*) data in the range of

FIGURE 8.1 Spectral reflectance data acquisition in salt marsh environments: (a) in situ spectral reflectance acquisition using Ocean Optics sensor; (b, c) sensor's IFOV; (d) sensor calibration using 99% Spectralon reflectance panel.

200–1100 nm with a sampling interval of 0.3 nm (Rundquist et al. 2004). The first radiometer with a field of view (FOV) of 25° pointed downward to acquire upwelling radiance (L; W m²/sr), while the second radiometer equipped with a cosine corrector pointed upward to acquire downwelling irradiance (E; W/m²) simultaneously. Based on the FOV and the height of the frame (5 m), the spatial resolution (instantaneous field of view [IFOV]) of the sensor was calculated to be 1.83 m (Figure 8.1), using the following relationship:

$$d = 2 \left\{ h \times \left(\tan \frac{\alpha}{2} \right) \right\} \tag{8.1}$$

where
  d is the diameter of the IFOV
  h is the height of the sensor from the target
  α is the FOV of the sensor

The radiometers were intercalibrated by comparing incident irradiance to measured upwelling radiance of a 99% white Spectralon reflectance panel (Labsphere, Inc., North Sutton, NH, USA), made of barium sulfate. In the case of changing sky conditions, the sensor was recalibrated at regular intervals (Figure 8.1). Noise removal in the raw hyperspectral data was accomplished by smoothing using a moving window average of 7 nm, after which the smoothed data were further interpolated at 1 nm intervals. Four scans of radiance and irradiance acquired per study plot converted to four percent reflectance (%R) readings. Mean %R was estimated from the four individual scans to obtain composite spectra of the study plot.

To examine inundation effects on NIR reflectance, %R of four salt marsh species were acquired based on the aforementioned experimental setup (Figure 8.2). The overall trend in %R showed a less prominent red edge in all species as compared to terrestrial vegetation, which might be attributed to soil moisture and standing water background. Moisture/water and vegetation have contrasting spectral response in the NIR region of the spectrum; while moisture/water shows more absorption and less/no scattering, vegetation scatters considerably, owing to foliar content and cellular structure. The differences in the NIR spectral response at the species level were very much evident and understandable, due to substantial difference in the foliar

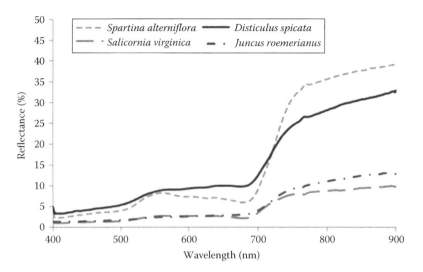

FIGURE 8.2   Sample hyperspectral spectral reflectance of four different salt marsh species as acquired by the Ocean Optics sensor.

structure and canopy architecture. In addition to %*R* spectra such as these, to calibrate and validate VI–based models using proximal or satellite data, plot level data on wetland biophysical properties are required.

8.4 Advanced Methods of Wetland Mapping

Three case studies have been incorporated in this chapter as examples of new methods in wetland remote sensing. These studies cover a broad geographic area and wetland diversity from freshwater wetlands to tidal salt marshes. The first study was conducted at Sapelo Island along the Georgia (GA) coast on salt marsh habitats. This study presents a method to overcome the respective limitations of LiDAR and hyperspectral imagery through the use of data fusion for an accurate elevation and habitat mapping in salt marsh environments. The second study was conducted on the salt marsh habitats of the entire GA coast. It discusses a MODIS-based mapping protocol to accurately map the biophysical characteristics of salt marsh vegetation. The third study was conducted in a low-diversity freshwater marsh environment to estimate and map belowground (BG) biomass and foliar nitrogen (N) using satellite data and hybrid modeling. These methods represent some of the recent developments in wetland remote sensing techniques. This chapter is aimed at providing an efficient and nondestructive mapping protocol for emergent wetlands. The techniques can be used for mapping spatial distribution of wetland species and their biophysical properties, delineating critical hot spots of wetland stress, assessing the success of previous marsh restoration projects, identifying areas of degradation as candidates for future restoration actions, and evaluating the overall productivity trend of wetlands at any geographic location.

8.5 Case Study 1: Accurate Mapping of Salt Marshes through Multisensor Integration of LiDAR and Hyperspectral Data

8.5.1 Introduction

Many coastal researchers use LiDAR to produce DEMs of salt marshes, as LiDAR provides broad coverage for areas that are large and sometimes difficult to access on the ground. However, there are several drawbacks to this approach. First, LiDAR tends to overestimate salt marsh elevations due to poor laser penetration of dense canopies (Montane and Torres 2006; Rosso et al. 2006; Sadro et al. 2007; Schmid et al. 2011). The majority of prior studies have focused on improving techniques to separate LiDAR returns (Wang et al. 2009) and optimizing DEM interpolation methods (Toyra et al. 2003; Schmid et al. 2011), both of which can help to reduce errors. These corrections have been applied without taking plant species into account and have had vertical accuracies (mean error) ranging from −0.02 to 0.12 m. In an earlier study, Hladik and Alber (2012) found that LiDAR-derived DEM mean error varied with vegetation cover in salt marshes. Of particular importance was the finding that tall, medium, and short height classes of the dominant macrophyte, *Spartina alterniflora*, the most common plant within eastern and southeastern salt marshes, required significantly different correction factors, ranging from 0.05 to 0.25 m. However, overall mean error could be reduced to −0.01 m by applying cover-class-specific correction factors to four test areas. Using these correction factors requires information on the distribution of cover classes throughout the study area, which was not available in the earlier study. A second, related limitation of topographic LiDAR is that it only receives spectral information at one wavelength in the NIR. It therefore cannot be used to distinguish among plant species, which requires information from the visible portion of the electromagnetic spectrum (Campbell 2006).

Hyperspectral imagery in the visible and NIR portion of the electromagnetic spectrum has been shown to be suitable for the separation of salt marsh vegetation species based on their spectral signatures (Schmidt and Skidmore 2003; Artigas and Yang 2005). Hyperspectral sensors are ideal for this purpose as they are able to collect a high number of contiguous spectral bands (sometimes greater than 200 bands) with narrow bandwidths and at a fine spatial resolution. There are several challenges in using hyperspectral imagery in salt marshes, particularly with respect to accurately classifying *Spartina* species. First, the different height classes of *S. alterniflora* (short, medium, and tall), which represent user-defined classes along a height continuum, are commonly confused in hyperspectral imagery classifications due to their spectral similarity in both the visible and NIR portions of the spectrum (Schmidt and Skidmore 2003; Artigas and Yang 2005). Another source of error results from mixed pixels that include either more than one class of vegetation and/or mud. Both of these complications are observed with *S. alterniflora*: the different height classes can be found adjacent to one another, and *S. alterniflora*'s erect structure and often sparse density mean that mud is spectrally mixed with vegetation (Silvestri et al. 2003; Thomson et al. 2003; Belluco et al. 2006). Silvestri et al. (2003) found that *Spartina maritima* is often misclassified because it is found in low-lying areas where mud and water interfere with its spectral signature. Thomson et al. (2003) hypothesized that microphytobenthos on mud may also cause mud to spectrally resemble *Spartina*. The inability to accurately classify the three height classes of *S. alterniflora*, compounded by the presence of mud in mixed pixels, is what we term the *Spartina* problem. A solution to the *Spartina* problem is especially important for ecological studies as the *S. alterniflora* height classes can have significantly different biomass and productivity values (Turner 1979; Morris and Haskin 1990; Schalles et al. 2013).

One way to potentially overcome the individual limitations of LiDAR-derived DEMs and hyperspectral imagery, and to address the *Spartina* problem, is through the use of multisensor data. Multisensor data integration combines data from different sources to improve classification performance and can include spectral, textural, and/or ancillary data such as DEMs (Pohl and van Genderen 1998; Lu and Weng 2007). LiDAR-derived DEMs have been included as a component band with multispectral and hyperspectral imagery to classify coastal habitats (Sadro et al. 2007; Chust et al. 2008; Yang and Artigas 2010) and as data layers in object-orientated classifications of salt marsh habitats (Brennan and Webster 2006; Gilmore et al. 2008), resulting in improved classification accuracies. LiDAR-derived DEMs have also been combined with land cover classifications post hoc to refine and improve classification products for urban areas (Pal and Mather 2003; Lu and Weng 2004), extract salt marsh species elevation ranges and distributions (Morris et al. 2005; Sadro et al. 2007), monitor the spread of invasive species (Rosso et al.

2006), model species habitat (Sellars and Jolls 2007; Moeslund et al. 2011), and predict the effects of sea level rise on coastal wetland inundation (Webster et al. 2006). These studies have all used multisensor data for classification purposes or for extracting additional elevation information. However, none have used elevation data to refine their existing classification of salt marshes.

The following example describes an approach for combining hyperspectral imagery of the salt marshes surrounding Sapelo Island, GA, USA, with a LiDAR-derived DEM through a decision tree, based on work by Hladik and Alber (2012). A decision tree is a nonparametric multistage or hierarchical classifier that can be applied to a single image or multiple coregistered images (Breiman et al. 1984). Using a multistage approach, a decision tree breaks down a complex decision into a series of nodes, or branches, where binary decisions are made to sequentially subdivide the data into predetermined classes. Data sources that can be used in decision trees include classified images, DEMs, and VIs. This example illustrates (1) the use of hyperspectral imagery to initially classify nine salt marsh cover classes, (2) the use of a decision tree to improve vegetation classification accuracy and address the *Spartina* problem by incorporating elevation information and NDVI, and (3) a methodology for combining the final vegetation classification with a LiDAR-derived DEM to produce corrected DEM elevations. These methods can produce both an accurate habitat classification and DEM for study areas. This approach will be of specific use to those interested in developing accurate maps of salt marshes, but will also be more broadly applicable as a demonstration of the combined power of LiDAR and hyperspectral imagery through the iterative use of multisensor data.

8.5.2 Methods

8.5.2.1 Study Site

This study included a total of 13.82 km² of salt marsh habitat in and around the Duplin River, a 13 km long tidal inlet that flows into Doboy Sound and forms the western boundary of Sapelo Island, GA, USA (Universal Transverse Mercator [UTM] Zone 17 N, 471480 E 3473972 N; Figure 8.3). The inlet is surrounded by a complex of salt marshes, tidal creeks, and back barrier islands. *S. alterniflora* is the dominant macrophyte in these marshes. For this study, *S. alterniflora* that was taller than 1 m was considered "tall." Tall *S. alterniflora* can grow to 2 m tall and is the dominant plant found along the regularly flooded creek banks in the low marsh. *S. alterniflora* that ranged from 0.5 to 1 m tall was considered "medium" and plants < 0.5 m were considered "short" (Reimold et al. 1973). Medium *S. alterniflora* dominates the midmarsh and short *S. alterniflora* is found in the irregularly flooded high marsh. The high marsh contains a mixed community of *Salicornia virginica* (more recently reclassified as *Sarcocornia* sp. [USDA 2010]), *Batis maritima*, *Distichlis spicata*, and short *S. alterniflora*, collectively termed marsh meadow or salt meadow

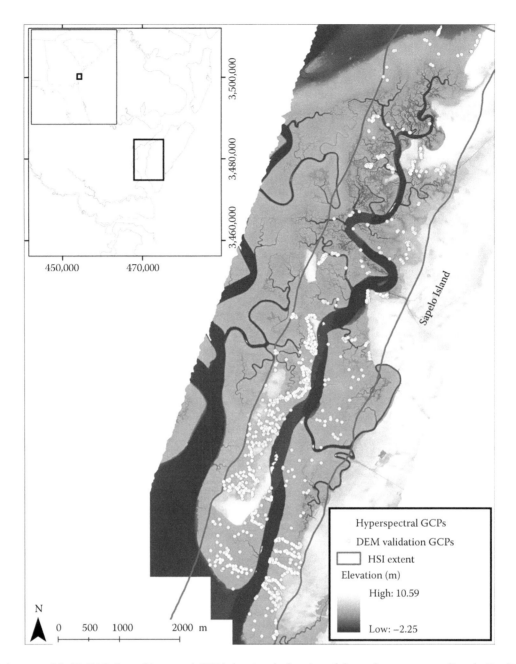

FIGURE 8.3 The unmodified LiDAR-derived bare earth DEM showing the location of the study area surrounding the Duplin River adjacent to Sapelo Island, GA, and the extent of the hyperspectral imagery evaluated for this study (red outline). Yellow dots indicate RTK ground control point (GCP) sampling locations used to validate DEM elevation accuracy. White dots indicate hyperspectral ground reference sampling locations.

(Wiegert and Freeman 1990). Marsh meadow has dense canopies generally less than 0.5 m tall. At the highest elevations along the upland fringe, *Juncus roemerianus* and *Borrichia frutescens* become the dominant species. Canopy heights of *J. roemerianus* and *B. frutescens* range from 0.5 m to over 2 m tall.

8.5.2.2 Hyperspectral Imagery

Airborne Imaging Spectrometer for Applications (AISA, Specim Imaging LTD, www.specim.fi) Eagle hyperspectral imagery of the site was acquired on June 20, 2006, by the Center for Advanced Land Management Information Technologies

(CALMIT) (Figure 8.3, Table 8.1). The flight lines were acquired in midmorning, to coincide with low tide (−0.11 m relative to mean lower low water [MLLW], National Oceanic Atmospheric Administration [NOAA, tidesandcurrents.noaa.gov]) and captured extensive areas of exposed intertidal mud (MUD). The AISA sensor was mounted in a Piper Saratoga plane flown at 1650 m above ground level. It collected spectral data in 63 bands in the visible and NIR portion of the electromagnetic spectrum, from 400 to 980 nm (9 nm full width half maximum spectral resolution, 2.3 nm spectral sampling, 2.9 nm spectral resolution). The high spectral resolution was selected

TABLE 8.1 LiDAR and Hyperspectral Sensor System Specifications Used to Acquire the Remote Sensing Data for This Study

	LiDAR	Hyperspectral
Sensor	Optech ALTM Gemini	AISA Eagle
Flight date	March 2009	June 2006
Altitude (m)	800	1650
Swath width (m)	370	1000
Overlap (%)	50	50
Number of spectral bands	1	63
Wavelengths (nm)	1047	400–980
Bandwidth	0.8 mrad	9 nm
Field of view (°)	16	68
Laser PRF (kHz)	125	—
Scan Freq (Hz)	40	—
Scan angle (°)	16	—
Scan cutoff (°)	3	—
Footprint (cm)	60	—
Pulse length (ns)	7	—
Pixel resolution (m)	1	1

to obtain contiguous spectral data for better discrimination among the dominant salt marsh species, and the fine spatial resolution (1 m) was selected to minimize the number of mixed pixels (patches of the various target cover classes range from a few square meters to tens of square meters). Initial postprocessing of data was performed by CALMIT using the CaliGeo (Specim Imaging LTD, www.specim.fi) and Environment for Visualizing Images (ENVI, Exelis, www.exelisvis.com) software programs. Processing included radiometric, geometric, and atmospheric correction procedures, and image data were produced with pixel values as normalized %R using a Specim fiber-optic downwelling irradiance sensor diffuse light collector. Hyperspectral processing routines are described in Hladik et al. (2013). An initial product of the hyperspectral imagery was an NDVI [NDVI = (NIR_{799} nm − RED_{675} nm)/(NIR_{799} nm + RED_{675} nm)] image (Rouse et al. 1974). Note that a number of VIs can be calculated using hyperspectral data to derive a number of wetland biophysical parameters. In this study, we tested a number of indices and found that their use had little effect on the overall final classification. NDVI has been used successfully in salt marsh vegetation mapping to discriminate vegetation from mud (Yang and Artigas 2010) and, therefore, was selected for use in this study. The NDVI image was used as one of the inputs in the decision tree classification.

8.5.2.3 LiDAR Data

The National Center for Airborne Laser Mapping acquired 35 km^2 of LiDAR data for the study site on March 9 and 10, 2009. Data were acquired when plant growth and biomass were seasonally low and during a spring low tide (−0.33 m relative to MLLW [NOAA, tidesandcurrents.noaa.gov]) to maximize laser penetration of the vegetation canopy and minimize standing water on the salt marsh surface. Data were collected with an Optech Gemini Airborne Laser Terrain Mapper (ALTM) mounted in a

twin-engine Cessna Skymaster flown at an altitude of 800 m above ground level. The survey was conducted with a laser pulse rate frequency (PRF) of 125 kHz and up to 4 returns. The high PRF was used to obtain a target point density of 9 hits/m^2. Reported vertical and horizontal accuracies (root mean squared error) (RMSE) for the sensor are 0.05–0.10 m and 0.10–0.20 m, respectively (Optech 2011). Absolute calibration was done using 662 ground control points (GCPs) surveyed with a vehicle-mounted kinematic GPS over paved roads near the Brunswick, GA, airport. These same road sections were surveyed with crossing flight lines using the ALTM, and the heights of the checkpoints were compared to the heights of the nearest neighbor LiDAR points within a radius of 20 cm. The RMSE of height differences was 0.11 m. LiDAR processing routines are described in Hladik and Alber (2012); sensor and overflight details can be found in Table 8.1.

A bare earth LiDAR-derived DEM was produced in SURFER Version 8 (Golden Software, http://www.goldensoftware.com) at 1.0 × 1.0 m resolution using a kriging algorithm that calculated the mean elevation value of all laser hits within each grid cell with a maximum standard deviation (SD) of 0.15 m. Elevations were all positioned in the NAD 83 reference frame and projected into UTM coordinate zone 17 N. Elevations are NAVD 88 orthometric heights (in meters) computed using the National Geodetic Survey GEOID 03.

8.5.2.4 Supporting Field Surveys

8.5.2.4.1 2006 Field Survey

We carried out an extensive field survey of 373 plots with horizontal submeter differential GPS (DGPS) positions (Trimble Geo-XH DGPS) from June to August 2006 to collect ground reference data for the AISA hyperspectral imagery. An additional 468 plots were surveyed in June and July 2007. Plots were positioned along 24 transects that spanned the salt marsh elevation gradient from low to high marsh with approximately 15–20 sampling locations per transect. Within each plot, all species present within a 1 × 1 m quadrat were documented and a dominant habitat type was assigned based on percent cover (Schalles et al. 2013). Additionally, vegetation stand polygons were delineated in the field using DGPS. Seven vegetation cover classes were included in the classification: *S. alterniflora* (short [SS], medium [SM], and tall [ST] height classes), *J. roemerianus* (JR), *B. maritima* (BM), *S. virginica* (SV), and *B. frutescens* (BF) and two nonvegetated classes, MUD and salt pan (SALT). The *S. virginica* class generally represented a mixture of high marsh plants, including *B. maritima*, *D. spicata*, and short *S. alterniflora*, and as such was rarely composed of only *S. virginica*. MUD consisted of patches of mud ≥ 1 m^2 within the vegetated portion of the salt marsh as well as mud on creek banks at elevations below tall *S. alterniflora*. SALTs were high marsh areas with hypersaline sediments and less than 25% vegetation cover. All data, that is, plot points, field-based vegetation polygons, and user-defined regions of interest, were randomly divided into training and validation datasets using the ArcGIS version 9.3 software program (http://www.esri.com), with approximately 75% of the data (33,923 m^2 pixels) used for supervised classifier training and 25% (10,701 m² pixels) reserved for validation of the classification results.

8.5.2.4.2 2009 Field Survey

To collect reference data for the LiDAR-derived DEM, a high-accuracy real-time kinematic (RTK) survey was conducted using a Trimble R6 RTK GPS receiver (Trimble 2009) with an observed vertical RMSE of 0.0037 m, a mean vertical error of 0.010 and mean horizontal error of 0.012 m. RTK elevations are NAVD 88 orthometric heights (in meters) computed using the National Geodetic Survey GEOID 03. As reported previously (Hladik and Alber 2012), RTK GCPs were collected throughout the Duplin River salt marshes in a survey that encompassed all of the vegetation cover classes considered in the hyperspectral imagery classification. At each GCP location, the RTK rover foot was placed flush with the marsh surface without disturbing the sediment and vegetation. The number of RTK points sampled per cover class ranged from 53 (MUD) to 267 (medium *S. alterniflora*) (Table 8.2). This range in sampling was primarily due to the relative dominance of the various cover classes in the salt marsh.

In total, 1830 RTK GCPs were acquired and were used for two purposes: to determine the elevation range of each salt marsh cover type for use in the decision tree analysis and to derive correction factors for DEM modification. The data were randomly divided into training and validation datasets using ArcGIS. Seventy-five percent (N = 1380) of the RTK GCPs were used to calibrate the elevation ranges and correction factors (Table 8.2), and 25% (N = 450, Figure 8.3) of the points were reserved as validation data for the modified DEM accuracy assessment.

8.5.2.5 Overview of Workflow

The workflow employed here used a two-step classification routine for vegetation classification, followed by DEM modification (Figure 8.4). To classify vegetation (Figure 8.4a), the maximum likelihood classification (MLC) algorithm was used

to produce an initial classification of the hyperspectral imagery based on spectral reflectance characteristics in the visible and NIR. The initial vegetation classification was then used in a decision tree in combination with the uncorrected elevations from the LiDAR-derived DEM and hyperspectral imagery-derived NDVI. This second classification iteration was used to correct misclassifications in the initial habitat map, especially those that occurred between the three height classes of *S. alterniflora* and/or mud pixels (the *Spartina* problem), as the height classes occupy distinct elevation zones (Wiegert and Freeman 1990; Hladik and Alber 2012) and NDVI has been shown to aid in the separation of vegetated (*S. alterniflora*) and nonvegetated (mud) pixels (Yang and Artigas 2010). Following this second step, the accuracy of the final vegetation classification was assessed. To correct elevations, the final vegetation classification was merged with the LiDAR-derived DEM and cover-class-specific correction factors were applied (Figure 8.4b). The accuracy of the final DEM was then assessed in comparison to RTK validation data.

8.5.2.6 Initial Habitat Classification

All image analyses were conducted using ENVI version 4.8. To carry out the supervised classifications, the mean spectrum for each cover class was calculated from training pixels. The MLC algorithm was used to classify salt marsh habitats using all 63 hyperspectral image bands. MLC is a commonly used parametric classifier that assumes that spectral values within each band for each class are normally distributed and calculates the probability that a given pixel belongs to a specific class based on variance and covariance measures (Hoffbeck 1995). A class membership probability threshold of 0.95 was used.

8.5.2.7 Final Habitat Classification

The initial MLC was combined with the LiDAR-derived DEM and the classification was refined through a decision tree that evaluated both spectral and elevation information. We used three input bands in the decision tree classification: the initial MLC; the DEM image, which provided information about the elevation of each pixel; and the NDVI image derived from the hyperspectral imagery. The decision tree nodes were straightforward for all classes except the three height classes of *S. alterniflora*. For all vegetation classes and areas initially classified as mud, the decision tree reassigned all pixels with an elevation less than −1.2 m as unclassified. This was done because −1.2 m was the lowest elevation for exposed creek bank areas based on RTK survey data, and, as such, any areas lower than −1.2 m would have been inundated at the time of LiDAR data acquisition. For areas initially classified as SALT, the decision tree reassigned pixels with an elevation less than 0.8 m into a new shell class. This was done because SALT and shell have similar spectral characteristics, but 0.8 m was the minimum observed RTK elevation for SALTs. Our knowledge of habitat distributions in the salt marsh supports this separation as oyster reefs are found in low-lying areas in the study site. More complicated decision tree nodes were created for areas initially classified as tall, medium, or short *S. alterniflora* (Figure 8.5).

TABLE 8.2   Cover-Class-Specific Correction Factors Used to Modify the LiDAR-Derived DEM When Combined with the Hyperspectral Classification

Cover Class	Correction Factor (m)	N	SD (m)	SE (m)
Tall *S. alterniflora*	0.25	152	0.17	0.01
Medium *S. alterniflora*	0.11	267	0.07	0.00
Short *S. alterniflora*	0.05	214	0.05	0.00
Intertidal mud	0.04	53	0.06	0.01
S. virginica	0.04	227	0.05	0.00
B. maritima	0.04	160	0.04	0.00
Salt pan	0.03	62	0.04	0.01
J. roemerianus	0.17	117	0.09	0.01
B. frutescens	0.12	78	0.07	0.01

The sample size (N), standard deviation (SD, 1 sigma), and standard error (SE, 1 sigma) for the derivation of the correction factors are also shown. Please refer to the text section for details regarding correction factors and Hladik and Alber (2012) for derivation of correction factors. All units (except N) are in meters (m).

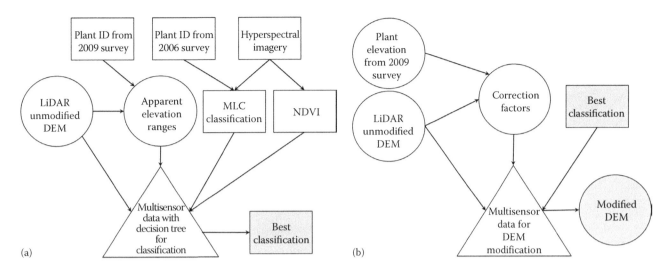

FIGURE 8.4   (a) Hyperspectral habitat classification and (b) LiDAR-derived digital elevation model correction workflows. Squares represent plant identification and habitat information; circles represent elevation information; triangles represent the use of multisensor data; and shaded shapes represent final data products.

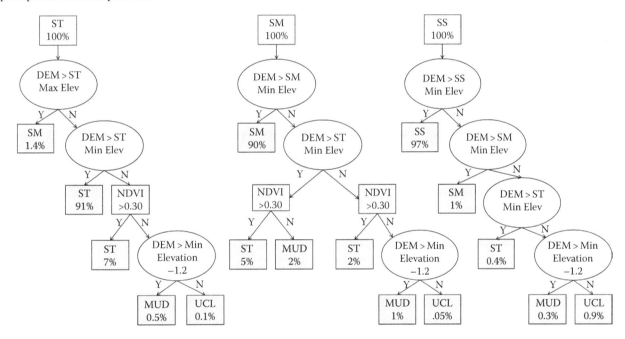

FIGURE 8.5   Decision tree nodes for evaluating pixels classified as tall, medium, and short *S. alterniflora*, combining hyperspectral imagery and LiDAR-derived DEM data sources. Squares represent habitat information (MLC initial classification and NDVI); circles represent elevation information (unmodified DEM); and shaded squares represent the cover class products of multisensor data integration following the application of the decision tree. Percentages represent the percent of pixels reassigned to each of the respective classes based on the decision tree rules. Percentages are rounded and do not sum to 100% for each tree.

In these cases, pixels were assessed based on the expected elevation range for each cover class obtained in our RTK ground survey data, extended using ±1 SD to account for the possibility that the RTK survey did not sample the entire range of habitat class elevations, as well as their NDVI value.

Classification accuracy of the decision tree output was evaluated by constructing a confusion matrix using the reserved validation data and calculating the overall accuracy and kappa coefficient (Congalton 1991), as well as the quantity disagreement and allocation disagreement (Pontius and Millones 2011). The producer's and user's accuracies for individual classes were also used to evaluate classifier performance.

8.5.2.8 DEM Modification

As described in Hladik and Alber (2012), class-specific correction factors were derived for the study area by subtracting the surveyed RTK elevation from the DEM elevation at the corresponding *x/y* coordinate of each GCP in the RTK training

TABLE 8.3   Maximum Likelihood Classification Confusion Matrix after Application of the Decision Tree for the Nine Marsh Cover Classes

Cover Class	Tall *S. alterniflora*	Medium *S. alterniflora*	Short *S. alterniflora*	Intertidal Mud	*S. virginica*	*B. maritima*	Salt Pan	*J. roemerianus*	*B. frutescens*
Unclassified	0	0	0	0	0	0	0	0	0
Tall *S. alterniflora*	94.50	2.71	0	1.76	0	0	0	2.10	0
Medium *S. alterniflora*	5.16	88.2	12.12	0.29	6.71	0	0	2.34	2.04
Short *S. alterniflora*	0	6.43	86.32	11.14	1.47	0	0	0	0
Intertidal mud	0	0.15	0.68	86.80	0	0	0	0	0
S. virginica	0	1.07	0.88	0	79.87	2.83	1.19	0	0
B. maritima	0	0	0	0	7.36	91.30	0	1.05	2.35
Salt pan	0	0	0	0	4.58	0	98.81	0	0
J. roemerianus	0	1.43	0	0	0	0	0	94.51	0
B. frutescens	0.34	0	0	0	0	5.87	0	0	95.61
Total	100	100	100	100	100	100	100	100	100

Columns represent the reference data (what the pixel actually was based on validation data) and rows represent the image data (what the pixel was classified as). Shaded cells are those where the classification was accurate. Percentages are rounded to the nearest decimal place and may not sum to 100% for each cover class.

dataset (N = 1380). Corrections ranged from 0.03 to 0.25 m and represent the mean error for each cover class (Table 8.2). To modify the DEM, the final vegetation classification of the hyperspectral imagery was brought into ArcGIS, converted to raster format by assigning a cover-class-specific correction factor to each polygon, and was then subtracted from the original, "unmodified" DEM to produce a "modified" DEM.

An accuracy assessment was performed on both the "modified" and "unmodified" DEMs using the reserved RTK survey validation data (N = 450), which had not been used to derive the correction factors. To examine the accuracy of the DEM elevation, the mean error, vertical RMSE, the fundamental vertical accuracy (FVA) with a 95% confidence level, and the 95th percentile errors for each cover class were calculated following the American Society for Photogrammetry and Remote Sensing (ASPRS) guidelines (ASPRS LiDAR Committee 2004). Modified DEM elevations from each cover class were compared to RTK elevations using paired t-tests. The calculated RMSE for each cover class in the DEM was also compared to the reported vertical RMSE of the LiDAR sensor (0.11 m based on vehicle-mounted GPS absolute calibration) to determine whether it was within the range of instrument error. Statistical results for all analyses in this study were considered significant when $p \leq 0.05$. All statistical analyses were done using the open-source program R version 2.10.1 (http://cran.r-project.org/).

8.5.3 Results and Discussion

8.5.3.1 Habitat Classification

This study tested a hybrid approach for habitat classification that used a decision tree to combine spectral and elevation information. The integration of a LiDAR-derived DEM with an initial MLC of the hyperspectral imagery through the application of the decision tree resulted in a clear improvement in the identification and separation of mud and tall *S. alterniflora* at creek edges as well as better separation of mud and medium *S. alterniflora* pixels. The final vegetation classification performed well, as seen in the confusion matrix and class accuracy results

TABLE 8.4   Maximum Likelihood Classification Errors of Commission, Errors of Omission, Producer's Accuracies and User's Accuracies for Each Cover Class Following Application of the Decision Tree for the Nine Cover Classes

Cover Class	Errors of Commission	Errors of Omission	Producer's Accuracy	User's Accuracy
Tall *S. alterniflora*	7.90	5.50	94.50	92.10
Medium *S. alterniflora*	12.41	11.80	88.20	87.59
Short *S. alterniflora*	12.70	13.68	86.32	87.30
Intertidal mud	6.03	13.20	86.80	93.97
S. virginica	13.01	20.13	79.87	86.99
B. maritima	13.93	8.70	91.30	86.07
Salt pan	5.32	1.19	98.81	94.68
J. roemerianus	3.86	5.49	94.51	96.14
B. frutescens	5.28	4.39	95.61	94.72

Percentages are rounded to the nearest decimal place and may not sum to 100% for each cover class.

(Tables 8.3 and 8.4). Overall accuracy was 90%, the kappa coefficient was 0.88, quantity disagreement was 1%, and allocation disagreement was 9% following application of the decision tree. The quantity and allocation disagreement errors indicate that the classifications were correctly assigning the total number of pixels to each cover class (lower quantity disagreement error), but showing them in the wrong location on the map (higher allocation error). Producer's and user's accuracies ranged from 80% to 99% for all cover classes. Producer's accuracies were highest in areas classified as SALT (99%) and *B. frutescens* (96%) and lowest in areas classified as *S. virginica* (80%) and short *S. alterniflora* (86%). The largest error of commission was *B. maritima* (14%) and the greatest error of omission was *S. virginica* (20%), indicating that *B. maritima* was overrepresented, whereas *S. virginica* was underrepresented in the final vegetation classification.

These high accuracies suggest that our approach is robust. Although it is difficult to compare the accuracies of separate studies due to differences in data and methods, previous studies that classified salt marsh habitats have attained a range of accuracies, from 59% to 99% (Rosso et al. 2006; Sadro et al. 2007;

Wang et al. 2007). The highest reported accuracies were in the lagoon of Venice, where Belluco et al. (2006) and Marani et al. (2006) had class producer's accuracies ranging from 75% to 99% using MLC and spectral angle mapper classifiers. Those studies classified four salt marsh vegetation classes (*S. maritima, Limonium, Salicornia,* and *J. roemerianus*), in addition to mud and water. Belluco et al. (2006) used data from a variety of airborne sensors to test numerous classification algorithms and achieved the highest class accuracies for *S. maritima* (98%) and *Limonium* (98%) using MLC. Marani et al. (2006) found that spectral angle mapper classified *Limonium* (98.5%) and *S. maritima* (97.9%) with the greatest accuracies.

The greatest challenge in classifying salt marsh vegetation based on MLC of hyperspectral imagery alone was to separate the different height classes of a single species, *S. alterniflora,* both from each other and from mud, which is something that previous studies have not done. These distinctions are important for a number of reasons. First, the different height classes of *S. alterniflora* vary in a number of ways, including biomass, productivity, carbon storage, and palatability to herbivores (Turner 1979; Morris and Haskin 1990; Goranson et al. 2004). Distinguishing among them in a habitat map is therefore important for deriving estimates of these parameters for the whole salt marsh. Second, accurate delineations of the interface between mud and tall *S. alterniflora* at the creek bank are critical for evaluating the tidally driven exchange of carbon, nutrients, and groundwater between the salt marsh and the estuary (Zedler 2001; Bouma et al. 2005; Townend et al. 2011). Additionally, tall *S. alterniflora* stands are vulnerable to creek bank erosion and wrack deposition, making high-accuracy habitat delineations useful for identifying these disturbances. Finally, a high-accuracy vegetation classification map was essential for performing the final DEM corrections as the classification determined the boundaries by which correction factors were applied to each pixel of the map. This is of particular importance for tall *S. alterniflora,* which has a correction factor that is significantly greater than that for the other vegetation classes.

Although the decision tree did not appreciably improve overall classification accuracies as compared to the initial MLC (overall accuracy improved from 89% to 90% and allocation disagreement was reduced from 10% to 9% following application of the decision tree [Tables 8.3 and 8.4]), it did produce small gains in separating the three height classes of *S. alterniflora* and mud (Table 8.3). The decision tree nodes for tall, medium, and short *S. alterniflora* presented in Figure 8.5 show the percent of pixels that were reclassified based on consideration of elevation and NDVI. Of the pixels initially classified as tall *S. alterniflora,* 1.4% were reassigned to medium *S. alterniflora,* 0.5% to mud and less than 0.1% to unclassified. Of the pixels initially classified as medium *S. alterniflora,* 7% were reclassified as tall *S. alterniflora,* 3% as mud, and 0.05% as unclassified. Only 1% of short *S. alterniflora* pixels were reassigned as medium *S. alterniflora,* 0.4% as tall *S. alterniflora,* 0.3% as mud, and 0.9% as unclassified.

The majority of the pixels reclassified by the decision tree occurred in the upper portion of the Duplin River, and it is instructive to focus on this area, where there is a high density of small creeks. The area is predominately classified as tall or medium *S. alterniflora* or mud. Three locations were selected within this portion of the salt marsh, each with an area of 0.15 km², which exemplify the effects of the decision tree on class distributions (Figures 8.6 and 8.7). In these example areas, there were 7%–14% gains in tall *S. alterniflora* pixels, 12%–22% losses in medium *S. alterniflora* pixels, and 0%–6% changes in pixels classified as mud following the use of the decision tree (Table 8.5, "Change in Proportion per Class" column). The majority of reclassifications involved medium *S. alterniflora* being reclassified as tall because the pixels occurred in low elevation areas where medium *S. alterniflora* is not observed, and mud pixels changed to unclassified because they occurred in areas less than −1.2 m. When expressed as a proportionate change per class, there were substantial increases in areas classified as both tall *S. alterniflora* (51%–319%) and mud (2%–29%) and corresponding losses in areas classified as short (32%–70%) and medium *S. alterniflora* (20%–33%) (Table 8.5, "Change in Proportion per Class" column). In these areas, the decision tree produced a more accurate representation of the salt marsh: low-lying areas on the edges of creeks, which were classified as medium *S. alterniflora* in the initial classification (darker blue areas in Figure 8.7), were reassigned to mud with a border of tall *S. alterniflora,* which meets our expectations. Although the reclassifications of *S. alterniflora* and mud affected 10% of all pixels in the map, this did not have a large effect on overall accuracy. This may be because we did not have many GCPs in the areas that showed the most change due to the difficulty of assessing these low-lying parts of the salt marsh. We therefore do not have validation pixels to quantify accuracy in these areas specifically.

Using multisensor data to classify salt marsh vegetation has generally resulted in improved accuracies. Using a LiDAR-derived DEM and multispectral imagery, Chust et al. (2008) classified a coastal wetland with 88% accuracy. Similarly, Yang and Artigas (2010), using a LiDAR-derived DEM and hyperspectral imagery, attained an accuracy of 68%. Geerling et al. (2007) mapped floodplain vegetation with 81% accuracy when they used LiDAR-derived DEMs in combination with hyperspectral imagery. These studies have all focused on the merging of visible imagery with DEM elevations prior to classification. Our approach is unique in that we combined the two data sources via a decision tree after the visible imagery was initially classified. To our knowledge, no prior studies have addressed or ameliorated the *Spartina* problem as effectively as our approach of multisensor data integration using a decision tree.

8.5.3.2 DEM Modification

The second use of multisensor data, which combined the hyperspectral classification and LiDAR-derived DEM for application of class-specific correction factors, greatly improved the accuracy of the DEM. An accurate DEM is crucial for hydrodynamic models used to simulate tidal flooding or to project the effects of storms and sea level rise on coastal habitats (Sanders 2007; Poulter and Halpin 2008; Gesch 2009). The overall mean DEM

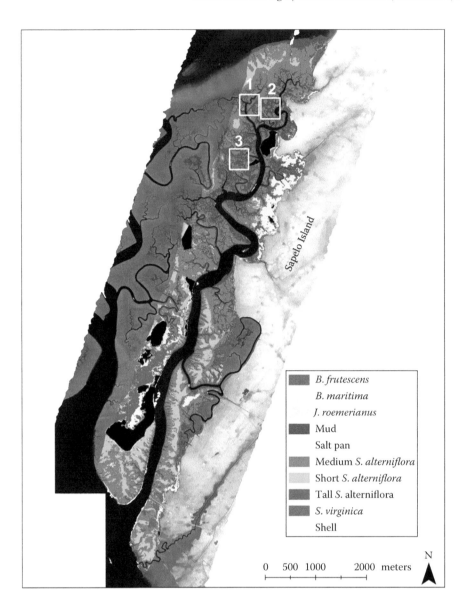

FIGURE 8.6 The final hyperspectral classification product (after application of decision tree). Areas 1, 2, and 3 represent locations of examples shown in Figures 8.7 and 8.8.

error was reduced from 0.10 to −0.003 m and the RMSE from 0.15 to 0.10 m (Table 8.6). Elevations in the unmodified DEM were significantly higher than the RTK ground elevations for all cover classes except for mud and SALT (Table 8.6). The overall RMSE of the unmodified DEM was 0.15 m, which is greater than the reported vertical accuracy of the LiDAR sensor (0.11 m). The RMSE for elevations in tall and medium *S. alterniflora*, *J. roemerianus*, and *B. frutescens* all exceeded instrument error, with the greatest value for tall *S. alterniflora* (0.34 m). In contrast, the values for short *S. alterniflora*, MUD, *B. maritima*, *S. virginica*, and SALT were all within the range sensor error.

The overall mean error in the modified DEM was substantially reduced to −0.003 ± 0.10 m (SD) (Table 8.6). The largest error reduction was for tall *S. alterniflora*, where mean error decreased from 0.28 to 0.08 m. *J. roemerianus* error was reduced from 0.16 to −0.005 m in the modified DEM. Applying the

cover-class-specific correction factors brought all DEM elevations in line with their true RTK elevations (Table 8.6) and resulted in elevations that were lower than those in the unmodified image (Figure 8.8). Overall RMSE (0.10 m) and RMSE for all cover classes, except tall *S. alterniflora* (0.22 m), fell within the reported instrument vertical RMSE in the modified DEM. The effect of modifying the DEM can be seen in Figure 8.8, which shows the same three locations featured in Figure 8.7. The overall elevation in these test areas decreased by an average of 110% (from 0.11 to −0.01 m), primarily as the result of reclassifying areas as tall *S. alterniflora* through the decision tree (Table 8.5). The modified DEM also shows fewer patches of high ground in the creeks, which appear as islands in the unclassified image.

These findings are in agreement with the results of DEM modification at four test sites within the Duplin River salt marshes using the same derived correction factors (Hladik and

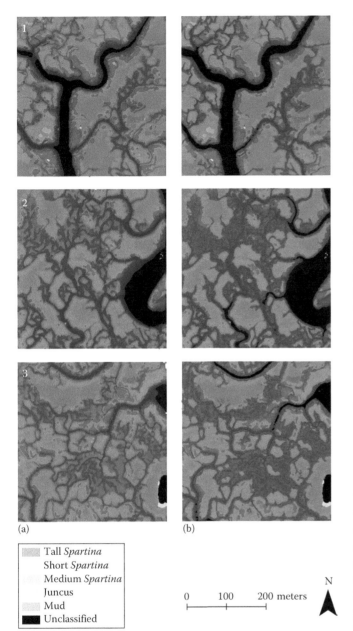

(a)

(b)

Tall *Spartina*
Short *Spartina*
Medium *Spartina*
Juncus
Mud
Unclassified

0   100   200 meters

N

FIGURE 8.7 Habitat classification (a) before and (b) after application of the decision tree on the MLC, for example, locations 1, 2, and 3 shown in Figure 8.6. Note the reassignment of medium *S. alterniflora* pixels (blue) to tall *S. alterniflora* (red) and mud (brown). Black areas are unclassified and are either upland areas or water pixels that gave unreliable LiDAR water returns. Each location is 0.15 km² and was selected to demonstrate the effect of the decision tree on class distributions.

Alber 2012). The reduced errors in the modified DEM are less than the elevation differences between vegetation classes, making the corrected DEM appropriate for use in salt marsh studies where small differences in elevation can have important ecological effects.

The DEM correction method illustrated here combined a hyperspectral classification with the DEM to modify DEM elevations. Sadro et al. (2007) used a similar approach and reported

reduced LiDAR-derived DEM errors for extracted elevation values following the application of cover-class-specific corrections in combination with an airborne visible/infrared imaging spectrometer (AVIRIS) classification of a California salt marsh composed mostly of high marsh plants. Following correction, they found no mean difference between survey and extracted LiDAR-derived DEM elevations, with an overall RMSE of 0.06 m. The Sadro et al. (2007) study did not modify the actual DEM surface, but rather modified extracted elevations according to cover-class-specific offsets. In contrast, we applied correction factors to modify the DEM and then performed a rigorous accuracy assessment.

Another approach to correcting LiDAR without multisensor data is the analysis of LiDAR las point clouds and generation of a new LiDAR-derived bare earth DEM surface. Wang et al. (2009) used statistical techniques to better differentiate ground and canopy returns in salt marsh vegetation, and other authors (Toyra et al. 2003; Schmid et al. 2011) have experimented with various DEM interpolation algorithms to produce the most representative ground surface. The errors in these efforts, however, were generally greater than, or comparable to, those reported here. The current study shows that the use of multisensor data is an accurate and viable alternative for DEM correction.

8.6 Case Study 2: Characterization of Salt Marsh Biophysical Properties for Georgia Coast Using MODIS Data

8.6.1 Introduction

The methodology provided in Case Study 1 demonstrates the utility of advanced remote sensing methods for habitat delineation and species-/community-level mapping, with potential applications for wetland loss/change estimation. These analyses provide valuable qualitative information on the presence/absence of marshes and their change in extent. However, if the goal is to assess previous restoration efforts and prioritize future actions, different methodologies and additional information on marsh condition are required. Biophysical properties such as canopy chlorophyll content (CHL), leaf area index (LAI) (a ratio of green foliage area vs. ground area), green vegetation fraction (VF) (percent green canopy cover), and aboveground (AG) green biomass (GBM) are necessary parameters to provide quantitative insight into the health and physiological status of vegetation. Further, these properties can serve as direct proxies of primary productivity and nitrogen content. Monitoring these characteristics through remotely sensed data can help infer the overall health and productivity of these valuable natural resources on a larger scale so that effective management strategies can be implemented to high-priority areas.

CHL is one of the most important foliar biochemicals and the content within a vegetation canopy is related closely to both vegetation productivity and absorbed photosynthetically active radiation. The importance of studying CHL in vegetation has

TABLE 8.5 Cover Class Areas Based on the Initial Maximum Likelihood Classification and Following the Application of the Decision Tree for Three Areas in the Upper Duplin River (Corresponding to Locations 1, 2, and 3 in Figures 8.6 and 8.7) and for the Entire Study Domain

Cover Class	MLC	Decision Tree	Change in Proportion per Class	Change in Total Proportion
Location 1				
Tall _S. alterniflora_ (ST)	0.15	0.22	0.51	0.07
Medium _S. alterniflora_ (SM)	0.60	0.48	−0.20	−0.12
Short _S. alterniflora_ (SS)	0.01	0.00	−0.32	−0.01
Intertidal mud (MUD)	0.17	0.17	0.02	0.00
Unclassified	0.07	0.12	0.63	0.05
Location 2				
Tall _S. alterniflora_ (ST)	0.03	0.14	3.19	0.11
Medium _S. alterniflora_ (SM)	0.58	0.39	−0.33	−0.19
Short _S. alterniflora_ (SS)	0.00	0.00	−0.70	0.00
MUD	0.31	0.35	0.13	0.04
Unclassified	0.08	0.12	0.63	0.04
Location 3				
Tall _S. alterniflora_ (ST)	0.08	0.22	1.64	0.14
Medium _S. alterniflora_ (SM)	0.69	0.47	−0.31	−0.22
Short _S. alterniflora_ (SS)	0.00	0.00	−0.36	0.00
MUD	0.22	0.28	0.29	0.06
Unclassified	0.01	0.03	1.88	0.02
Whole classification				
Tall _S. alterniflora_ (ST)	0.09 (0.13)	0.11 (0.17)	0.02	0.24
Medium _S. alterniflora_ (SM)	0.33 (0.50)	0.30 (0.47)	−0.03	−0.10
Short _S. alterniflora_ (SS)	0.08 (0.12)	0.08 (0.12)	0.00	−0.03
MUD	0.07 (0.11)	0.06 (0.09)	−0.02	−0.23
S. virginica (SV)	0.03 (0.05)	0.03 (0.05)	0.00	0.00
B. maritima (BM)	0.002 (0.01)	0.002 (0.01)	0.00	0.00
Salt pan	0.003 (0.02)	0.003 (0.01)	0.00	−0.17
J. roemerianus (JR)	0.03 (0.05)	0.03 (0.05)	0.00	0.00
B. frutescens (BF)	0.003 (0.01)	0.003 (0.01)	0.00	0.00
Shell	0.00	0.0005 (0.003)	0.00	1.00
Unclassified	0.34	0.37	0.03	0.01

The "MLC" column contains the area assigned for each class expressed as a proportion of the total number of pixels in each location prior to application of the decision tree. The "Decision Tree" column contains the proportion of the area for each class following the application of the decision tree. "Change in Proportion per Class" represents the change in cover class area, calculated on a per class basis, whereas "Change in Total Proportion" is calculated based on the total number of pixels in the area following the application of the decision tree ("Decision Tree" minus "MLC" column values). The values in parentheses for the Whole Classification are the proportion of the areas when unclassified pixels are excluded.

been recognized for decades (e.g., Danks et al. 1984). Changes in CHL are related to photosynthetic capacity (thus, productivity), developmental stage, and canopy stresses (e.g., Ustin et al. 1998). It has been suggested that CHL may appear to be the one community property most directly related to the prediction of productivity (Lieth and Whittaker 1975).

The synoptic view provided by airborne and space-borne sensors has the potential for estimating CHL on a regional and global basis. Variations in leaf CHL produce large differences in leaf reflectance and transmittance spectra; however, canopy reflectance is also strongly affected by other factors (e.g., canopy architecture, canopy CHL distribution, LAI, soil background) that mask and confound changes in canopy reflectance caused

by leaf CHL, thus making retrieval of total CHL in a vegetative canopy both complicated and challenging. Several remote sensing techniques using reflectance in the red and NIR spectral regions have been proposed to estimate CHL in leaves and canopies. Tucker (1977) reported red reflectance is very sensitive to low amounts of both biomass and CHL and might be the best spectral region to estimate low amounts of CHL. However, saturation of red reflectance at intermediate to high levels of CHL (e.g., Kanemasu 1974; Tucker 1977; Buschmann and Nagel 1993; Myneni et al. 1997) limits the applicability of the techniques that use red and NIR bands for CHL assessment, primarily due to choices of band location and width (e.g., Sellers 1985; Baret and Guyot 1991; Yoder and Waring 1994; Gitelson et al. 1996a;

TABLE 8.6 Summary of LiDAR-Derived DEM Accuracies

Cover Class	Mean Error (m)	N	SD (m)	SE (m)	RMSE (m)	FVA (m)	95th Percentile (m)	p-Value
Unmodified DEM								
Tall *S. alterniflora* (ST)	0.28	51	0.20	0.03	0.34	0.67	0.56	<0.001
Medium *S. alterniflora* (SM)	0.11	89	0.07	0.01	0.13	0.26	0.22	<0.001
Short *S. alterniflora* (SS)	0.05	71	0.05	0.01	0.07	0.14	0.13	0.006
Intertidal mud (MUD)	0.06	17	0.06	0.01	0.08	0.16	0.14	0.431
S. virginica (SV)	0.04	73	0.04	0.00	0.06	0.11	0.10	0.003
B. maritima (BM)	0.05	53	0.04	0.01	0.06	0.11	0.11	0.003
Salt pan	0.03	21	0.04	0.01	0.05	0.09	0.08	0.250
J. roemerianus (JR)	0.16	37	0.09	0.01	0.18	0.35	0.27	<0.001
B. frutescens (BF)	0.11	21	0.08	0.02	0.13	0.26	0.22	0.003
Overall	0.10	450	0.12	0.01	0.15	0.30	0.34	<0.001
Modified DEM								
Tall *S. alterniflora* (ST)	0.08	51	0.20	0.03	0.22	0.42	0.41	0.129
Medium *S. alterniflora* (SM)	−0.01	89	0.08	0.01	0.08	0.16	0.12	0.506
Short *S. alterniflora* (SS)	−0.03	71	0.06	0.01	0.07	0.14	0.06	0.081
MUD	−0.05	17	0.10	0.02	0.11	0.22	0.08	0.495
S. virginica (SV)	−0.01	73	0.05	0.01	0.05	0.09	0.06	0.522
B. maritima (BM)	0.00	53	0.04	0.01	0.04	0.08	0.07	0.911
Salt pan	−0.01	21	0.04	0.01	0.04	0.08	0.05	0.535
J. roemerianus (JR)	0.00	37	0.08	0.01	0.08	0.16	0.12	0.914
B. frutescens (BF)	0.00	21	0.08	0.02	0.08	0.15	0.08	0.985
Overall	0.00	450	0.10	0.00	0.10	0.19	0.17	0.870

Accuracies for each cover class are presented for both the unmodified and modified DEM relative to the RTK ground survey elevation. The table lists mean error (Mean Error), number of observations (N), standard deviation (SD), standard error (SE), root mean square error (RMSE), fundamental vertical accuracy (FVA), and 95th percentile error (95th Percentile). p-Values are from a paired t-test between the RTK elevations and the predicted DEM elevations for each cover class. All error units are in meters.

Jenkins et al. 2002). Further, the mathematical formulation of the NDVI limits this VI in CHL studies because the normalization procedure makes the NDVI insensitive to variation in R_{red} when $R_{NIR} \gg R_{red}$ (Gitelson and Merzylak 2004). These limitations prevent accurate estimation of CHL in moderate to high vegetation densities.

Other key variables required for estimating coast salt marsh status are green LAI, VF, and GBM. Remote estimations of these characteristics can be performed using transformations of spectral reflectance, called VIs (e.g., Rouse et al. 1974). A physically based algorithm for estimating LAI from NDVI observations has been developed (e.g., Myneni et al. 1997). However, the relationship between NDVI and LAI is essentially nonlinear and suffers a rapid decrease of sensitivity at moderate to high densities of photosynthetic GBM (Myneni et al. 1997; Gitelson et al. 2003a; Gitelson 2004). Alternative methods have been proposed that yield more linear relationships between remotely sensed data and VF, LAI, and GBM (e.g., Chen and Cihlar 1996; Gao et al. 2000; Gitelson et al. 2003a,b).

Monitoring these biophysical parameters not only helps in assessing overall wetland dynamics but also facilitates prioritization of restoration in areas that require immediate restoration and conservation measures. A robust biophysical mapping protocol is also critical in assessing the success/failure of previous restoration efforts (Hinkle and Mitsch 2005; Friess et al. 2012). The MODIS instrument holds considerable potential for

advancing our capabilities to estimate and monitor the biophysical characteristics of salt marshes across large geographic areas. MODIS provides a near-daily global coverage of moderate-resolution data in the red and NIR spectral regions (250 m) and in the green range (500 m) that are well calibrated and atmospherically corrected and has relatively high geolocational accuracy (Wolfe et al. 2002). However, the utility of MODIS 250 and 500 m data for quantitatively estimating biophysical characteristics of wetland/marshland has not been investigated and used to date. In Case Study 2, a method for reparameterization of existing VIs is presented. This method uses field data and MODIS surface reflectance data to map the biophysical characteristics of salt marsh habitats along the GA coast.

8.6.2 Methods

8.6.2.1 Field Data Collection of Wetland Biophysical Properties

8.6.2.1.1 Leaf Chlorophyll Content

Minolta 502 SPAD Chlorophyll Meter (Spectrum Technologies Inc., East Plainfield, IL, USA) was used to measure the in situ leaf level CHL (Figure 8.9). A total of twenty stratified random SPAD readings were acquired from each sampling location across varying chlorophyll levels inside the IFOV of the sensor. The twenty readings were averaged and converted to absolute chlorophyll

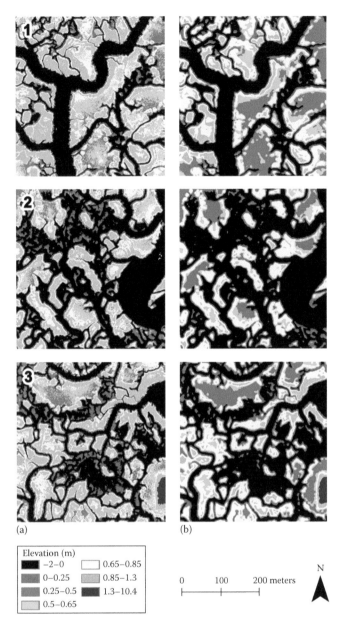

Elevation (m)

■ −2–0	□ 0.65–0.85
■ 0–0.25	■ 0.85–1.3
■ 0.25–0.5	■ 1.3–10.4
■ 0.5–0.65	

0   100   200 meters

N

FIGURE 8.8 (a) Unmodified and (b) modified DEMs, for example, locations 1, 2, and 3 shown in Figures 8.6 and 8.7. Note the overall decrease in elevation in the modified DEM as well as the reduction in elevation associated with tall *S. alterniflora* patches along creek margins.

values (mg/m²) by using coefficients derived from a calibration experiment conducted at the beginning of the field season. The calibration procedure involved laboratory-based analytical extraction of CHL from few leaf samples and development of a linear statistical relationship between the analytical CHL and the corresponding SPAD readings (Gitelson et al. 2005).

8.6.2.1.2 *Green Biomass*

AG GBM data were collected by destructive sampling from a 0.09 m² (1 ft²) subplot within each study plot using a PVC frame and clippers (Figure 8.9). Biomass samples were sorted to separate the live biomass from dead and oven dried at 65°C overnight (~24 h) to get rid of moisture. The dry weight was recorded using a standard measuring balance. Precautions were taken to avoid moisture absorption by the dried GBM during dry weight measurement. The dry GBM weights (g/ft²) were then rescaled to g/m².

8.6.2.1.3 *Vegetation Fraction*

Percent green VF was estimated from a circular crop of vertical digital photographs of the study plots acquired by OLYMPUS E – 400 digital SLR camera (Olympus America Inc., Center Valley, PA, USA). The camera was installed on the frame along with a laser pointer next to the hyperspectral radiometer. The laser pointer marked the center of the digital photograph and the IFOV. The digital photograph was cropped to match the IFOV of the hyperspectral radiometer, and the VF was estimated by the ratio of the number of green pixels to the total number of pixels in each photograph (Figure 8.9).

8.6.2.1.4 *Leaf Area Index*

LAI Plant Canopy Analyzer 2000 (LI-COR Biosciences Inc., Lincoln, NE, USA) (Gitelson 2004) and AccuPAR LP-80 Ceptometer (Decagon Devices Inc., Pullman, WA, USA) (Delalieux et al. 2008; Kovacs et al. 2009) were used to estimate LAI (Figure 8.9). The median of four LAI readings taken in each study plot was used as the LAI of the study plot. Each LAI measurement involved one above-canopy and four below-canopy readings. We estimated green leaf area index (GLAI) as the product of LAI and VF. Further, CHL (mg/m²) was calculated as the product of LAI and Leaf Level Chlorophyll (LLC) based on Gitelson et al. (2005).

The biophysical parameters were acquired from roughly 200 study plots during multiple field trips spanning over 2010–2011 for MODIS-based model calibration and validation.

8.6.2.2 Satellite Data

Multitemporal 8-day Level 1B atmospherically corrected surface reflectance composites for the GA coast were acquired from the National Aeronautics and Space Administration (http://modis-land.gsfc.nasa.gov) for the growing seasons (April–October) from 2000 through 2013. Both 250 and 500 m scenes from the MODIS sensor were downloaded and mosaicked. Salt marsh subsets were prepared using digital boundaries acquired from National Wetlands Inventory. (http://www.fws.gov/wetlands/index.html).

8.6.2.3 Model Calibration and Validation

Following the initial preprocessing of satellite data, in situ sampling locations were used to extract pixel values from MODIS images. MODIS scenes were chosen based on the proximity of the dates between the image acquisition and field data collection. Preexisting and well-established VIs were derived from the extracted pixel values for model calibration for each biophysical parameter (Table 8.7). For pixels containing multiple sampling locations, the average value of the individual biophysical parameter was calculated and used in model calibration. During this process, roughly 69 sampling plots acquired in 2010 were reduced to 10–15 MODIS 250 m and 7–10 MODIS 500 m

FIGURE 8.9 Biophysical data collection: (a, b) biomass collection from study plot; (c, d) leaf chlorophyll content measurement using SPAD 502 chlorophyll meter; (e, f) vegetation fraction measured from the IFOV of the sensor; and (g, h) LAI measurement using LICOR LAI Plant Canopy Analyzer 2000 and AccuPAR LP-80 Ceptometer.

TABLE 8.7   List of Satellite Image–Derived Vegetation Indices Used for Calibration and Validation of Models to Estimate Marsh Biophysical Parameters

Vegetation Index	Formula	Reference
Normalized difference vegetation index (NDVI)	$(R_{NIR} - R_{Red})/(R_{NIR} + R_{Red})$	Rouse et al. (1974)
Enhanced vegetation index 2 (EVI2)	$\{2.5 \times (R_{NIR} - R_{Red})/(R_{NIR} + 2.4 \times R_{Red} + 1)\}$	Huete et al. (2002)
Chlorophyll index red (CI$_{red}$)[a]	$(R_{NIR} - R_{Red})/R_{Red}$	Gitelson et al. (2006)
Wide dynamic range vegetation index (WDRVI)	$(\alpha \times R_{NIR} - R_{Red})/(\alpha \times R_{NIR} + R_{Red})$	Gitelson (2004)
Soil-adjusted vegetation index (SAVI)	$(R_{NIR} - R_{Red}) \times (I + L)/(R_{NIR} + R_{Red} + L)$	Huete (1988)
Chlorophyll index green (CI$_{green}$)	$(R_{NIR} - R_{Green})/R_{Green}$	Gitelson et al. (2006)
Visible atmospherically resistant index (VARI)	$(R_{Green} - R_{Red})/(R_{Green} + R_{Red})$	Gitelson et al. (2002)

[a] R_{red} was used instead of $R_{red\text{-}edge}$ as described in Mishra et al. (2012).

TABLE 8.8   Coefficients of Determination (R^2) and Percent Root Mean Square Error Values for MODIS 250 m– and 500 m–Derived Best Fit Models

		GLAI	VF (%)	GBM (g/m^2)	CHL (mg/m^2)	Best Fit Model
250 m	R^2	0.821	0.845	0.85	0.837	WDRVI ($\alpha = 0.1$)
	NRMSE (%)	14.967	22.055	19.616	25.817	
500 m	R^2	0.91	0.98	0.938	0.864	VARI
	NRMSE (%)	32.474	24.344	17.34	21.998	

pixels. Further, an independent field dataset was acquired during the field campaigns in 2011 containing another 121 sampling plots, subsequently reduced to 10–12 MODIS 250 m pixels and ~10 MODIS 500 m pixels, and were used for model validation (Table 8.8). Performance uncertainties were analyzed based on coefficient of determinations (R^2) and percent normalized root mean squared error (%NRMSE; RMSE/range) (Table 8.8).

8.6.2.4 Monthly Composite Products and Phenology Extraction

Following successful calibration and validation, 8-day time series composites were generated in ERDAS Imagine 2011 (Leica Geosystems, Heerbrugg, Canton of St. Gallen, Switzerland) for GLAI, VF, CHL, and GBM for the growing seasons (March–November) from 2000 to 2013, using the best fit models. Composites were generated for both 250 and 500 m resolution images. For each growing season month, four composites per parameter per resolution were created; therefore, over 9 months, 144 composites were created. As such, for all the four parameters, composites generated for 14 growing seasons amounted to almost 4032 composites (combining both 250 and 500 m composites for coastal GA). These composites were used for qualitative assessments of salt marsh habitats pre- and post significant natural and anthropogenic events such as hurricanes and droughts and to evaluate the performance of the best fit models chosen for the mapping of biophysical parameters. In addition, phenology charts for site-specific salt marsh patches were derived from these time series composites, using the spatial analysis module in ArcGIS 10.0 desktop for the entire fourteen-year growing season dataset. The phenology charts were examined for effects of discrete natural and anthropogenic events on the health of marsh patches, as well as for long-term trends in the biophysical values over the past 14 years.

8.6.3 Results and Discussion

8.6.3.1 Model Cal/Val

After an extensive testing of numerous VIs on MODIS data using the aforementioned methods for calibration and validation, the wide dynamic range vegetation index (WDRVI, $\alpha = 0.1$) (Gitelson 2004) was selected for estimating the biophysical properties (GLAI, CHL, VF, GBM) for the 250 m dataset, whereas the visible atmospheric resistant index (VARI) (Gitelson et al. 1996b) was selected for predicting the biophysical parameters for the 500 m dataset (Table 8.8). The indices were chosen based on a combination of R^2, %NRMSE, and residual trends. WDRVI used on the 250 m MODIS data for model calibration produced a R^2 of 0.821, 0.845, 0.85, and 0.837 and for model validation produced %NRMSE of 14.967%, 22.055%, 19.616%, and 25.815% for GLAI, VF, GBM, and CHL respectively (Table 8.8). Whereas VARI used on the 500 m MODIS data for model calibration produced a R^2 of 0.91, 0.98, 0.938, and 0.864 and for model validation produced %NRMSE of 32.474%, 24.344%, 17.34%, and 21.998% for GLAI, VF, GBM, and CHL respectively (Table 8.8).

8.6.3.2 Model Output: Biophysical Composites

Time series composites were generated for each biophysical parameter and for both resolutions separately, using the best fit models for growing seasons from 2000 to 2013 (Figure 8.10). The time series composites provided a qualitative estimation of the physiological health and productive capacity of the salt marshes, both at the site specific and landscape level. These composites elucidate seasonal variation in biophysical parameters and help restoration managers and conservationists to identify critical patches of marsh stress (Figure 8.10).

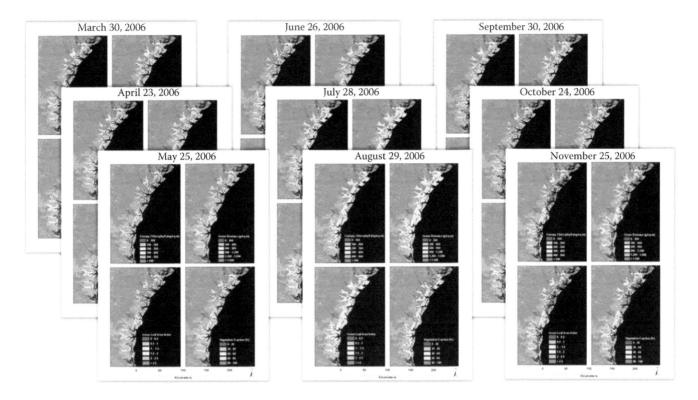

FIGURE 8.10 Examples of weekly map composites comprising of the four biophysical parameters produced using MODIS 250 m data–based models for the 2006 growing season.

8.6.3.3 Model Output: Time Series Phenology

Phenology charts for the last decade were extracted for selected wetland patch locations using a 3×3 or higher window size, depending on the extent of marsh patch (i.e., the patch size of marsh undergoing marsh dieback). The monthly averages of biophysical parameters were extracted to construct a 14-year phenology chart for wetland patches (Figure 8.11). For example, Figure 8.11 shows the 14-year phenology of a marsh patch along the Jericho River that has been repeatedly affected by dieback events since 2000 (J. Mackinnon, Coastal Resources Division [CRD], GA Department of Natural Resources (DNR), personal communication). A similar analysis was performed on dieback affected patches throughout coastal GA. This analysis was restricted to patches covering an area of at least 4 ha, ensuring study patches were detectable using the MODIS 250 m data. The yellow boxes indicate years when dieback was reported for these sites by CRD, GA. MODIS-250 m data isolated the dieback signal in all four biophysical parameters for this site (Figure 8.11). We also noticed a decreasing trend in marsh productivity at the site, indicating that it is experiencing increased stress over the last 15 years.

Since marsh diebacks have been linked to persistent drought in GA, we correlated the trends and fluctuations in biophysical parameters with monthly Palmer drought severity index (PDSI) estimates retrieved for each study area (Figure 8.11). As a measure of dryness based on precipitation and temperature, PDSI shows drought in terms of negative numbers and wet conditions in terms of positive numbers (Heim 2002). Correlations among PDSI and mean biophysical values were analyzed using Pearson's r, and linear regression was conducted using ordinary least squares regression during the peak of the growing season (July, August, and September) (results not shown). The phenological charts demonstrated that PDSI and biophysical variables covaried together over the growing seasons of the last 14 years; low PDSI for the years 2002–2003, 2007–2008, and 2011–2012 (major years of drought) coincided with low biophysical values as predicted by the model. Further, for the wet growing seasons (2004–2006 and 2009–2010), high PDSI values concurred with high levels of marsh productivity. Linear regression results suggested that PDSI has significant positive effects on each biophysical parameter during the peak growing season (July, August, and September) (results not shown). As PDSI decreased (increased drought conditions), there was a decreasing trend in all four biophysical parameters, suggesting increased stress in salt marshes. The time series composites along with phenological charts provide the end user with both qualitative and quantitative information of the wetland health and physiological status.

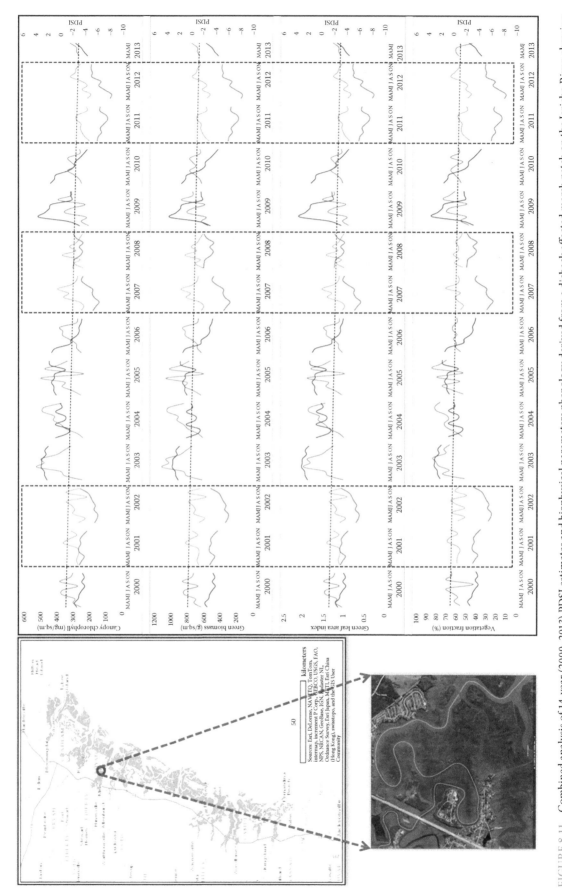

FIGURE 8.11 Combined analysis of 14-year (2000–2013) PDSI estimates and biophysical parameters phenology derived from a dieback affected marsh patch on the Jericho River, showing the overall spatiotemporal growth trend of the patch. Years with observed marsh dieback events at the location are highlighted in yellow.

8.7 Case Study 3: Above- and Belowground Biomass Mapping Using Spectroradiometer and Landsat Data–Based Hybrid Modeling

8.7.1 Introduction

Remote sensing models of above-ground (AG) biomass are commonly derived from optical data (Smith et al. 2002); however, monitoring BG biomass to understand whole plant productivity remains challenging. BG biomass estimates are needed because roots and rhizomes are the dominant source of soil organic carbon in wetlands (Moore 1987; Rasse et al. 2005). In addition, wetland BG biomass contributes to soil stability and accretion by adding organic matter volume (Nyman et al. 1993; Miller et al. 2008). BG biomass may consequently help maintain or build wetland elevation to prevent subsidence below sea level (Nyman et al. 2006). Therefore, tools for predicting and promoting wetland plant BG biomass can support wetland management for resilience and provision of ecosystem benefits.

Remote sensing may be very useful for whole plant biomass studies by fusing together remote sensing algorithms with ecological theories. Site-wide estimates of BG biomass from field measures are difficult to obtain because of wide fluctuations in AG vs. BG growth within and among species (Mokany et al. 2006). Thus, AG biomass often is a poor predictor of BG conditions. To reliably estimate BG production, we need models that describe plant allocation among AG and BG vegetation. Two such growth models exist. The first has been called balanced growth (Shipley and Meziane 2002) or optimal partitioning (McCarthy and Enquist 2007; Kobe et al. 2010) and suggests plants allocate growth toward growth limiting resources. In this view, new growth is directed belowground when water and nutrients are limiting and aboveground otherwise. Conversely, the isometric allocation growth model (McCarthy and Enquist 2007) suggests AG and BG growth scale together as a result of physiological constraints. Consequently, nutrient addition stimulates total net plant production, fertilizing AG and BG biomass equivalently. For either balanced or isometric growth, leaf N concentration, a measure of plant nutrient uptake, can inform biomass estimates. In the balanced growth case, additional nutrients, such as N and P, increase shoot growth without equally increasing BG biomass. Therefore, leaf N scales with root/shoot (RS) ratios and can be used to estimate BG biomass (O'Connell et al. 2014). Alternatively, plants responding isometrically to nutrient addition have stimulated BG and AG biomass and constant RS ratios with respect to N. Reliably modeling wetland fates requires studies that resolve relationships among plant nutrient uptake and biomass allocation across a range of environmental conditions.

Remote sensing may assist with whole plant biomass estimation by providing site-wide spectral reflectance estimates of plant nutrient uptake and AG biomass. For example, foliar N may be a good indicator of long-term plant available nutrients (Boyer et al. 2001; Cohen and Fong 2005) and can be estimated with remote sensing approaches (Curran 1989; Townsend et al. 2003). Several analytical techniques relate spectral data to canopy N and plant biomass, such as band-depth analysis (Kokaly and Clark 1999), spectral matching techniques (Gao and Goetz 1995), and PLS regression (Smith et al. 2002; Byrd et al. 2014). In particular, PLS regression, a full-spectrum multivariate method, is ideal for maximizing explained variance (EV) among plant responses and spectral reflectance signals (Geladi and Kowalski 1986; Mevik and Cederkvist 2004). PLS uses an eigenvector-based approach to reduce many correlated predictors, such as spectral data, to a few independent components. These components maximize predictive value and minimize background effects introduced by nontarget cover types such as water and dead vegetation (Chen et al. 2009). Such background effects commonly reduce the predictive value of spectral reflectance signals in emergent wetlands, as discussed earlier. Therefore, PLS regression may be an ideal analytical technique for wetland remote sensing. PLS regression analysis might be used to estimate site-wide measures of AG biomass and foliar N and then statistically related to BG biomass.

In early studies, Byrd et al. (2014) evaluated PLS regression with hyperspectral and multispectral data to estimate AG biomass within freshwater peat marshes of the Sacramento–San Joaquin Delta. Both hyper- and multispectral data had similar performance for AG biomass mapping within this system. In a follow-up study, O'Connell et al. (2014) explored the predictive capacity of spectral reflectance estimates of foliar N to model BG biomass. They compared the whole plant productivity of *Schoenoplectus acutus*, a common freshwater macrophyte, across a simple experimental N addition and control treatment. Results suggested *S. acutus* exhibited balanced growth, reducing BG biomass in fertilized plots (O'Connell et al. 2014). Additionally, spectral reflectance differed among high and low N plants. Particular differences were at 550 nm (peak greenness), 1250 nm (corresponding to leaf water content), 1750 nm, and 2100 nm (both corresponding to spectral N absorption features) (Curran 1989; Thenkabail et al. 2002, 2012). Thus spectral reflectance estimates of N-driven biomass allocation might correlate with BG biomass within *S. acutus* marshes, but evidences from field studies still were needed.

This study presents a method for identifying within site variability in BG biomass and RS ratios in a low-diversity coastal freshwater marsh. The effectiveness of multispectral remote sensing (Landsat 7) for estimating AG biomass and % foliar N was evaluated and compared to optimal remote sensing data, hyperspectral field spectroradiometer data. Ultimately, a hybrid modeling approach was used to linearly relate reflectance-based PLS regression estimates of foliar N and AG biomass to BG biomass and RS ratios. Initially, species-specific vs. mixed species models were compared (O'Connell et al. in review). Here, only the best fit satellite reflectance models are presented: species-specific models based on Landsat 7 satellite reflectance data.

8.7.2 Methods

8.7.2.1 Study Sites

Three impounded freshwater marsh units in the Sacramento–San Joaquin Delta, CA, USA, were sampled: Twitchell Island east unit (38.1073°, −121.6483°), Twitchell Island west unit (38.1069°, −121.6449°), and Mayberry Slough southeast unit on Sherman Island (38.0490°, −121.7660°). All units historically were part of extensive freshwater perennial peat marshes of the Sacramento–San Joaquin Delta. Emergent vegetation in all three units was mostly *S. acutus* and *Typha domingensis*, *T. latifolia*, *T. angustifolia*, and *Typha* spp. hybrids.

8.7.2.2 Experimental Design

AG and BG emergent wetland plant biomass and % foliar N were estimated using field methods during July, August, and September during one growing season (2012). These data were used to build relationships with spectral reflectance. AG biomass was estimated within 1 m² plots using validated species-specific allometric equations (Byrd et al. 2014). For BG biomass, ingrowth root cores were used to estimate relative variation in BG biomass within 30 cm of the wetland surface, where the majority of root production occurs (McKee et al. 2007; McKee 2011). At foliar N plots, leaf samples were collected for estimation of % foliar N. Sample locations were from permanent plots along transects spanning 4 Landsat 7 pixels per wetland. Landsat 7 pixels were used as the sample unit, and a single 1 m² estimate of vegetation parameters (AG and BG biomass, and foliar N) was calculated by averaging all values within each pixel for each vegetation survey. In total, 36 multitemporal Landsat 7 pixel-scale estimates were derived (3 wetland impoundments X 4 Landsat pixels per wetland X 3 sample dates).

8.7.2.3 Spectral Reflectance Data Collection

To build spectral models, in situ hyperspectral reflectance was collected using an Analytical Spectral Devices field spectroradiometer (FieldSpec Pro FR, Analytical Spectral Devices, Inc, Boulder, CO, USA). Field spectroradiometer collection was conducted for plots within 2 m of a boardwalk or wetland edge, allowing access with a spectroradiometer. Spectral readings were sampled every 1.4 nm over 350–1000 nm and 2 nm over 1000–2500 nm using a 25° FOV fore optics at nadir 1 m above the vegetation canopy. For plots inaccessible by portable spectroradiometer, spectral measurements from satellite platforms were relied on. These in situ hyperspectral data were used to build empirical models of % foliar N and AG biomass. Field spectroradiometer–based hyperspectral models were compared with the satellite-based multispectral models to calculate differences in model explanatory power. Landsat 7 Climate Data Record surface reflectance images (available via earthexplorer.usgs.gov) were acquired, which are atmospherically corrected surface reflectance data products processed by the United States Geological Survey using the Landsat Ecosystem Disturbance Adaptive Processing System algorithm (Masek et al. 2012).

8.7.2.4 Spectral Reflectance Model Development

To develop models predicting BG biomass and RS ratio, first two spectral reflectance model sets were generated to estimate % foliar N and AG biomass (model sets 1–2, Table 8.9). Spectral reflectance–derived predicted outcomes from models 1 and 2 were used to develop linear regressions for estimating BG biomass and RS ratio, that is, hybrid reflectance models. These hybrid models for predicting BG biomass included a balanced growth model (model set 3, Table 8.9), an isometric model (model set 4, Table 8.9), and, for completeness, a combined balanced and isometric growth model (model set 5, Table 8.9). A similar set of models were built for RS ratio (model sets 6–7, Table 8.9). Isometric growth was identified as equivalency in scale for AG and BG biomass, where RS ratios are roughly constant. For balanced growth, % foliar N should be negatively related to RS ratio. The combined balanced model uses both foliar N and AG biomass together to inform BG biomass.

To generate spectral reflectance estimates, the PLS package in R (Mevik and Wehrens 2007) was used to build PLS regression models that used full-spectrum reflectance as predictors for estimating % foliar N and AG biomass (each separately). We minimized model overfitting by selecting components corresponding to the first local minima for root mean squared error of prediction (RMSEP; e.g., averaged difference in predicted vs. measured values). RMSEP was estimated using leave-one-out cross-validation (Mevik and Wehrens 2007). We used a random subset as training data (70% of samples) and compared R^2 or EV and RMSEP with a testing dataset (30% of samples). We also calculated normalized RMSEP (nRMSEP = RMSEP/[response maximum − response minimum]), which approaches 0 for better models and 1 for poorer models. For hyperspectral models, loading plots for the first 3 PLS regression components (i.e., those explaining the majority of variance) were examined to determine wavelengths most associated with foliar N and AG biomass. These loadings were used to guide selection of an appropriate multispectral sensor. Landsat 7 ultimately was

TABLE 8.9　Spectral Reflectance and Hybrid Spectral Reflectance Model Set Development

Model Set	Model
Spectral models	
1	Predicted% foliar N ~ spectral reflectance
2	Predicted aboveground biomass ~ spectral reflectance
Hybrid spectral models	
3	Field belowground biomass ~ predicted N (model set 1)
4	Field belowground biomass ~ predicted aboveground biomass (model set 2)
5	Field belowground biomass ~ predicted aboveground biomass (model set 2) + predicted N (model set 1)
6	Field root/shoot ratio ~ predicted N (model set1)
7	Field root/shoot ratio ~ predicted belowground biomass (model set 3) / predicted aboveground biomass (model set 2)

Hybrid model 3 and 6 represent tests of balanced growth models, 4 isometric model tests, and 5 and 7 combined balanced and isometric growth models.

chosen because its measurement capabilities best corresponded to wavelengths with high absolute PLS loading scores. Landsat prediction accuracy (RMSE and nRMSE) also was greater than for other sensors tested (Hyperion and WorldView-2) (data not presented). To compare the performance of Landsat 7 under a variety of conditions, models were run with all sites included and then again with the shallowest and most spatially heterogeneous site excluded (Twitchell Island West Unit). The purpose was to generate recommendations for conditions where freely available, but spatially and spectrally coarse Landsat 7 might be utilized for productivity and N monitoring.

8.7.3 Results and Discussion

8.7.3.1 Estimating Foliar N and Aboveground Biomass with Hyper- and Multispectral Reflectance

Spectral reflectance of *Typha* spp. and *S. acutus* species were plotted and had subtle differences in visible (500–600 nm), NIR (800–1000 nm), and shortwave infrared (>1500 nm) wavelengths (Figure 8.12). PLS regression of in situ hyperspectral reflectance correlated well with % foliar N (Table 8.10). Models for *Typha* spp. explained the greatest variance in % foliar N (56%). Loading plots from these analyses suggested bands in the visible spectra (500–700 nm), in the NIR spectra (1100–1300 nm), and in the shortwave infrared spectra (>1700 nm) were important for associating % foliar N with reflectance. Landsat 7 data collection matched some PLS regression loading peaks for % foliar N, particularly bands 5 (1550–1750 nm) and 7 (2080–2350 nm) (Figure 8.13). However, other spectra with high loadings were missed, particularly 1250 nm (Figure 8.13). Multispectral Landsat 7 satellite reflectance was related to % foliar N, but for *Typha* spp., EV was reduced 25% from hyperspectral spectroradiometer reflectance data (Table 8.10; Figure 8.14). Conversely, *S. acutus* Landsat 7 foliar N models were roughly equivalent to hyperspectral models (EV 43%–46% each) (Table 8.10, Figure 8.14). Landsat 7 PLS regression also estimated AG biomass (EV > 60%, Table 8.10).

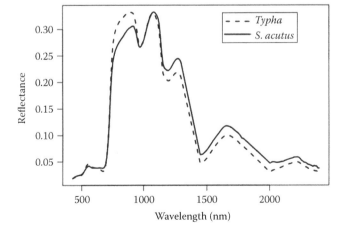

FIGURE 8.12 Spectral signature of *Typha* spp. and *S. acutus* from freshwater marsh impoundments in the Sacramento–San Joaquin Delta, CA, USA.

8.7.3.2 Predicting Belowground Biomass Using Satellite-Derived Multispectral Hybrid Models

A linear relationship was observed for reflectance derived estimates of % foliar N and BG biomass (Table 8.11; Figure 8.15), with strongest relationships observed for *Typha* spp. (29% RMSEP 56.6 g, nRMSEP 17%). Excluding the Twitchell west unit, where the greatest interspersion of cover types within pixels was observed, increased explanatory power for *S. acutus* models (34%, RMSEP 184.7 g, nRMSEP 23%). Spectral estimators of AG biomass had no detectable relationship to field-measured BG biomass (*P* ranged from 0.23 to 0.48 for all models, Table 8.11). Further, combined liner models of % foliar N and AG biomass spectral estimators did not improve model fit over models relying only on % foliar N (e.g. model set 5 from Table 8.9).

8.7.3.3 Predicting Root/Shoot Ratio Using Satellite-Derived Multispectral Hybrid Models

Hybrid modeling of RS ratio had greater correspondence to measured values than modeling of BG biomass (Table 8.11, Figure 8.16). Spectrally estimated % foliar N had a significant linear relationship with field-measured RS ratio, but best fit models combined spectral estimates of AG biomass with spectral estimates of BG biomass (predicted values from model set 3 derived from spectral estimates of % foliar N). As with previous *S. acutus* models, excluding the Twitchell west unit improved linear fit (EV with west unit 32%, without west unit 80.4%; Table 8.11, Figure 8.16).

8.7.3.4 Overall Performance

Remote sensing measures were useful for biomass estimation. Field-measured (data not presented) and spectrally estimated % foliar N both were associated with BG biomass and biomass allocation in *S. acutus* and *Typha* spp., while measures of AG biomass alone were poor predictors. For RS ratio, AG biomass helped parameterize models and improved model fit. Field spectroradiometer–based reflectance measurements explained greater than 45% of variation in foliar N concentration. Landsat 7 bands overlapped with some hyperspectral wavelengths associated with % foliar N. However, predictive power was lost by transitioning from hyperspectral, high spatial resolution field reflectance data to multispectral, moderate-resolution satellite data. Nonetheless, spectral estimators derived from Landsat 7 had relationships with foliar N, suggesting this public data source might be utilized for coarse canopy N estimation. Further, estimates of % foliar N from Landsat 7 bands had significant relationships with BG biomass.

Biomass allocation ratios were estimated with greater precision than BG biomass, benefiting from spectral estimates of AG biomass and % foliar N together. Emergent wetland macrophytes in these sites appeared to conform to a balanced growth model, as RS ratios were variable and negatively related to environmental nutrients (O'Connell et al. in review). Plants may increase allocation of photosynthate to BG biomass when nutrients are limited, both to increase absorptive area to

TABLE 8.10   PLS Regression Models of % Foliar N and Aboveground Biomass (g) from Spectroradiometer-Based Hyperspectral and Satellite-Based Multispectral Reflectance

Model			Training C	EV	RMSEP	nRMSEP	Testing RMSEP	nRMSEP
Field spectroradiometer hyperspectral reflectance models								
1a.	% foliar N *Typha* sp.	= Hyperspectral reflectance	7	56	0.56	0.18	0.66	0.22
1b.	% foliar N *S. acutus*	= Hyperspectral reflectance	5	46	0.53	0.26	0.43	0.26
Landsat 7 multispectral reflectance models								
1a.	% foliar N *Typha* sp.	= Landsat reflectance	3	27.0	0.67	0.32	0.77	0.47
1b.	% foliar N *S. acutus*	= Landsat reflectance	3	43.6	0.57	0.28	0.47	0.35
2a.	AG *Typha* sp.	= Landsat reflectance	3	59	448.2	0.24	309.3	0.25
2b.	AG *S. acutus*	= Landsat reflectance	2	67	155.5	0.20	262.7	0.34

Model set numbers refer to Table 8.9. C, the component number; EV, % explained variance; RMSEP, root mean square error of prediction; and nRMSEP, normalized RMSEP. Hyperspectral models used 1 m² plot level data, while Landsat models used pixel-scale data (30 m × 30 m). Training and testing data were 70% and 30% of samples.

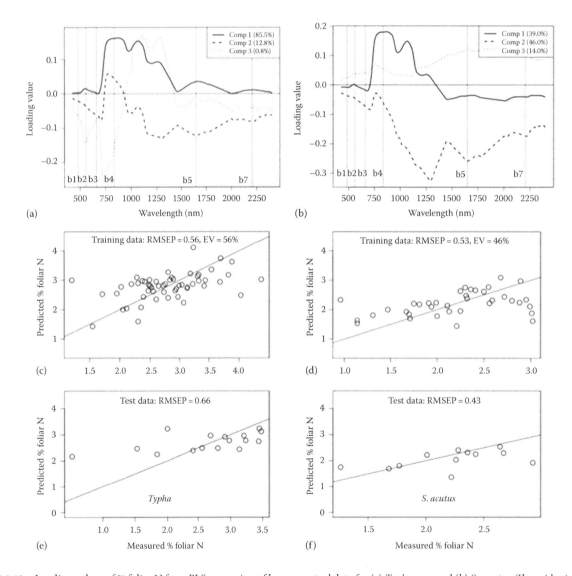

FIGURE 8.13   Loading values of % foliar N from PLS regression of hyperspectral data for (a) *Typha* spp. and (b) *S. acutus*. The midpoint of bands measured by Landsat 7 are superimposed and labeled with band number (b1–7). Measured vs. predicted values for training and testing datasets are presented in panels (c) and (e) for *Typha* and in panels (d) and (f) for *S. acutus*. EV is explained variance and RMSEP is root mean squared error of prediction.

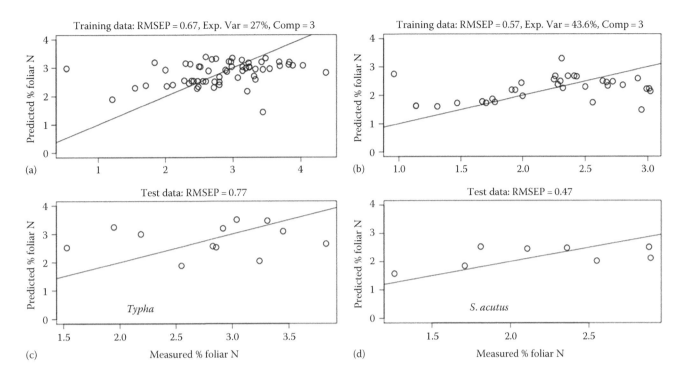

FIGURE 8.14 Measured vs. predicted % foliar N developed from PLS regression of Landsat 7. Training and testing data are plotted in panels (a) and (c) for *Typha* spp. and (b) and (d) for *S. acutus*.

TABLE 8.11 Hybrid Models

Growth Model Test Predictors		Training						Testing				
		F	df	P	β_0	β_1	β_2	EV	RMSEP	nRMSEP	RMSEP	nRMSEP
Typha spp. models												
3b. BG balanced	% foliar N (1b)	10.7	1,26	0.003	162	−45.7	—	29.1	56.6	0.17	48.8	0.52
4b. BG isometric	AG (2b)	0.73	1,26	0.4	—	—	—	2.7	—	—	—	—
6b. RS balanced	% foliar N (1b)	9.6	1,23	0.005	0.6	−0.2	—	29.5	0.23	0.21	0.09	0.28
7b. RS isometric	BG (3b)/AG (2b)	28.9	1,19	<0.001	0.0	2.7	—	60.3	0.15	0.14	0.26	0.31
S. acutus models												
3c. BG balanced	% foliar N (1c)	5.5	1,26	0.027	686	−253	—	17.5	184.7	0.24	310	0.32
4c. BG isometric	AG (2c)	0.5	1,26	0.48	—	—	—	1.9	—	—	—	—
6c. RS balanced	% foliar N (1c)	2.3	1,23	0.145	3.8	−1.4	—	9	1.7	0.23	0.37	0.22
7c. RS isometric	BG (3c)/AG (2c)	9.9	1,21	0.005	−0.6	3.1	—	32.1	0.72	0.24	2.27	0.3
3c. BG balanced[a]	% foliar N (1c)	10.5	1,20	0.004	1122	−438	—	34.4	176.5	0.23	309.3	0.32
4c. BG isometric[a]	AG (2c)	1.5	1,22	0.23	—	—	—	6.4	—	—	—	—
6c. RS balanced[a]	% foliar N (1c)	5.7	1,16	0.029	8.0	−3.3	—	26.4	1.8	0.23	0.36	0.21
7c. RS isometric[a]	BG (3c)/AG (2c)	65.7	1,16	<0.001	−0.2	3.1	—	80.4	0.89	0.12	0.68	0.22

Predictors are predicted outcomes from the models in Table 8.10, with source model indicated in parentheses. Legend: belowground (BG) biomass (g), root/shoot (RS) ratio, aboveground (AG) biomass (g), EV (% explained variance), and β_{0-2} (coefficients for regression parameter [intercept and slopes]). RMSEP and nRMSEP are presented for significant models. Training and testing data were 70% and 30% of samples.

[a] West unit excluded.

maximize nutrient capture and to increase rhizomatic carbohydrate storage as a buffer against future scarcity (Kobe et al. 2010). While plants in our study exhibited balanced growth, the model development process described here also would detect isometric growth and might assist with whole plant biomass estimates.

8.7.3.5 Correlations among Foliar N, Aboveground Biomass, and Spectral Reflectance

Remote sensing studies often estimate leaf N concentration (Tian et al. 2011; Stroppiana et al. 2012; Abdel-Rahman et al. 2013). Where full-spectrum field spectroradiometer reflectance data were available, estimation of foliar N also was possible.

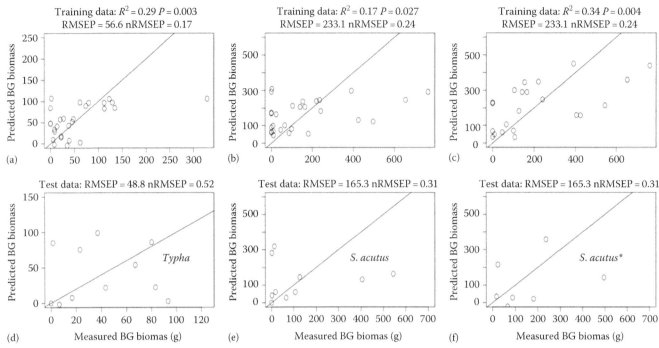

FIGURE 8.15 Measured vs. predicted BG biomass (g) developed from hybrid model set 3. Training and testing data are plotted in panels (a) and (d) for *Typha* spp., (b) and (e) for *S. acutus*, and (c) and (f) for *S. acutus* excluding the west unit.

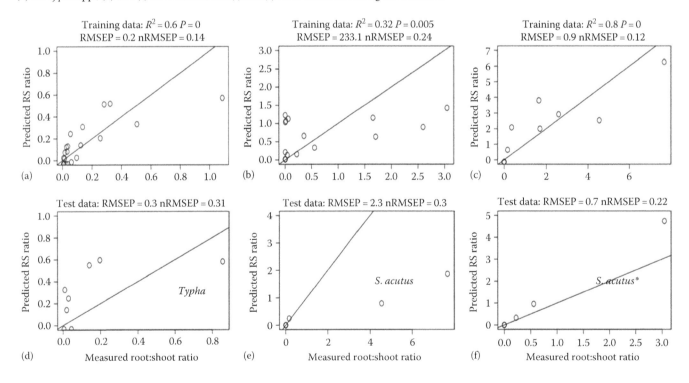

FIGURE 8.16 Measured vs. predicted RS ratio developed from hybrid model set 7. Training and testing data are plotted in panels (a) and (d) for *Typha* spp., (b) and (e) for *S. acutus*, and (c) and (f) for *S. acutus* excluding the west unit.

Reflectance data over 500–700 nm, 1100–1300 nm, and greater than 1700 nm were particularly informative for both *Typha* and *S. acutus*. Reflectance over 500–700 nm encompasses peak greenness and the red edge. Reflectance at 550 nm, peak greenness, is associated with chlorophyll (Thenkabail et al. 2002; Townsend et al. 2003), which contains 6.5% N by weight: 30%–50% of leaf N is within ribulose-1,5-bisphosphate carboxylase–oxygenase inside chloroplasts (Kokaly et al. 2009). Chlorophyll and foliar N are thus associated with photosynthesis and plant productivity. Red-edge reflectance also has been associated with plant stress and N accumulation (Thenkabail et al. 2012). For *S. acutus*, 1250 nm was the highest loading value estimating % foliar N from hyperspectral reflectance. This wavelength also had the greatest correspondence to % foliar N in earlier N addition experimental investigations of *S. acutus* (O'Connell et al. 2014). Reflectance at 1250 nm has been related to leaf water content (Thenkabail et al. 2002, 2012) and is also associated with plant turgor pressure, thermoregulation, and vigor (Kokaly et al. 2009).

Reducing hyperspectral to multispectral data resulted in some precision loss. Landsat 7 bands corresponded to some high loading values from hyperspectral analysis, particularly within shortwave infrared regions. Others demonstrated % foliar N was associated with shortwave infrared absorption at 1730, 2180, and 2240 nm (Curran 1989). In particular, increased absorption around 2050–2170 was associated with amide bond vibration within proteins (Kokaly et al. 2009). Landsat 7 bands 5 (1550–1750 nm) and 7 (2080–2350 nm) (landsat.usgs.gov; Figure 8.12) capture these important N spectra, suggesting Landsat 7 may be useful for regional N monitoring, particularly for detecting hot spot where foliar N differences may be very high. Landsat 8 (launched February 2013 [landsat.usgs.gov]) measures similar spectra and may prove useful in future studies.

8.7.3.6 Predicting Belowground Biomass and Root/ Shoot Ratio Using Multispectral Hybrid Models

Percent foliar N had negative relationships with both species-specific BG biomass and RS ratios. In combination with an earlier experimental study (O'Connell et al. 2014), this methodology presents the first use of spectral data to indirectly estimate BG biomass and biomass allocation. This effort links several relationships commonly observed in literature: remote sensing can be related to AG biomass (Smith et al. 2002; Kokaly et al. 2009) and foliar N concentration (Ryu et al. 2009; Stroppiana et al. 2009, 2012; Mokhele and Ahmed 2010; Tian et al. 2011; Abdel-Rahman et al. 2013), foliar N is linked to environmental N (Bedford et al. 1999; Sartoris et al. 1999; Kao et al. 2003; Larkin et al. 2012), and AG biomass and environmental N have relationships with RS ratios (Darby and Turner 2008a; Turner et al. 2009; Deegan et al. 2012; Morris et al. 2013).

Within the west unit, *S. acutus* was poorly predicted. Cover of *S. acutus* averaged 12% within west unit pixels. The west unit also was shallowest (<20 cm) and dominated by *Typha* spp., which are competitively dominate over *S. acutus* in shallower water. The west unit additionally had greater interspersion of dead thatch with live emergent vegetation. For *S. acutus*, thatch accumulation over lower portions of vertically oriented stems can mask biomass reflectance. In contrast, *Typha* spp. have broad horizontally oriented leaves. Horizontal broad canopies covers thatch accumulation at plant bases, reducing the influence of thatch in spectral signals. PLS regression using multispectral data has less capacity than hyperspectral analyses to minimize background effects from nontarget cover types. Areas with high heterogeneity might be better measured by hyperspectral sensors, such as AVIRIS. AVIRIS measures 400–2500 nm at 4–20 m spatial resolution depending on

platform (aviris.jpl.nasa.gov). However, Landsat is promising for measuring foliar N and whole plant biomass trends, particularly for pixels with high cover of target species.

8.8 Conclusions and Future Research

Accurate habitat and elevation maps of salt marsh habitats are important for many applications, but existing methodologies described in earlier sections have significant errors. Case Study 1 demonstrated the utility of combining multisensor data for both salt marsh habitat delineation and LiDAR-derived DEM correction. The hyperspectral classification, when combined with the LiDAR-derived DEM through a decision tree, produced a habitat map with a 90% overall accuracy, a quantity disagreement of 1%, and an allocation disagreement of 9%. Prior studies have noted the difficulty in correctly classifying *Spartina* areas but have not provided a solution to reduce class confusion. This study was not only comparable to previous efforts in terms of accuracy (*S. alterniflora* accuracy of 94.5%), but also provided a method for minimizing the *Spartina* problem. Combining the classification with the DEM to apply cover-class-specific elevation correction factors resulted in large reductions in mean error for nine cover classes with an overall mean error of −0.003 m. The multisensor data approach used here resulted in improved habitat and elevation maps and represents a significant advance over evaluating hyperspectral and LiDAR data independently. The results and future applications highlight the value of the remote sensing approach and show great promise for landscape level analyses of salt marsh plant habitats. Future work will focus on using remote sensing–derived metrics to classify imagery for other areas and incorporate these models to predict habitat change.

The biophysical mapping methodology developed through Case Study 2 can be used for identifying critical "hot spots" of wetland degradation due to factors other than natural disasters (e.g., developmental activities, localized drought, and urban pollutant runoff). These products have the potential to facilitate prioritization of restoration efforts, through identification of areas in need of immediate attention, and also help in comparing the health and growth of wetland patches/habitats pre- and postimplementation of restoration activities. Therefore, biophysical mapping will provide government regulators and restoration managers with important information to develop restoration and restoration plans. Time series composites can be further used to detect progressive change in the wetland habitats both inter- and intraseasonally. Further, long-term phenology plots provide documentation of the progressive improvement or decline in wetland health following natural or human-induced disasters. In addition, MODIS-derived long-term time series biophysical products can be used to develop light use efficiency models in order to map carbon sequestration potential (CSP) of the salt marsh habitats. Future research will include use of MODIS-derived biophysical products to map the gross primary productivity or CSP of salt marsh habitats.

Case Study 3 showed that the PLS regression technique on hyperspectral and multispectral data is promising for estimating % foliar N in wetlands and also for indirectly estimating variation in BG biomass, particularly where large differences exist as hot spots. Species-specific investigations are necessary to parameterize new modeling efforts at other sites. Currently, measures of BG biomass are difficult to obtain. Publicly available Landsat data provide global and long-term estimates of landscape change (Wulder et al. 2012). When properly supported by field and experimental studies, a hybrid modeling approach may reduce time and cost of broad-scale studies of BG biomass and RS ratios. Future studies should focus on Landsat 8 and on hyperspectral sensors, where available.

When working with multispectral data, our approach may have the greatest utility in sites with fairly homogeneous cover. Examples include salt and brackish marshes, typically dominated by one to a few species. Such areas also are vulnerable to sea level rise and there is an urgent need for understanding whole plant productivity patterns. For example, the approach outlined might help resolve controversy concerning the influence of eutrophication within coastal marshes. Some suggest elevated environmental nutrients results in loss of BG biomass and accelerated vegetated mash loss (Darby and Turner 2008a,b; Turner et al. 2009; Deegan et al. 2012). Others argue that eutrophication fertilizes BG production (Morris et al. 2013). Hybrid reflectance modeling may help estimate correlations among eutrophication, plant uptake as foliar N, resulting productivity responses, helping to resolve these issues and reduce uncertainty in managing wetlands for resiliency and provisioning of ecosystem services and benefits.

The potential broader impact of these studies is in the realm of CSP mapping for wetland vegetation species using satellite data. Mapping CSP directly from satellite data is not possible; however, it can be indirectly modeled via the biophysical properties. Therefore, accurate mapping of wetland species and biophysical properties is essential to developing a CSP model in the future. High CSP of wetlands make them critical habitats considering the exponential increment of greenhouse gases in the atmosphere since the last few decades. Often, their productive capacity has been compared to that of tropical evergreen rain forests. Interestingly, wetlands account for a third of global CH_4 flux; however, their contribution to global warming potential (GWP) is significantly offset by their carbon utilization capabilities and minimal N_2O emissions. Monitoring these wetland ecosystems in the United States and around the world, for managing and minimizing their GWP, requires precise information regarding their actual productivity, especially when global warming threatens the C_4 salt marsh vegetation with relatively higher CSP to be naturally succeeded by C_3 mangroves. The techniques and hybrid models presented in this chapter can be readily applied to similar ecosystems and would not require recalibration.

References

Abdel-Rahman, E.M., F.B. Ahmed, and R. Ismail. 2013. Random forest regression and spectral band selection for estimating sugarcane leaf nitrogen concentration using EO-1 Hyperion hyperspectral data. *Int J Remote Sensing* 34:712–728. doi: 10.1080/01431161.2012.713142.

Adam, P. 1990. *Saltmarsh Ecology*. Cambridge, U.K.: Cambridge University Press.

Albright, T.P., T.G. Moorhouse, and T.J. McNabb. 2004. The rise and fall of water hyacinth in Lake Victoria and the Kagera River Basin, 1989–2001. *J Aquat Plant Manage* 42:73–84.

Arbiol, R., Y. Zhang, and V. Palà. 2006. Advanced classification techniques: A review. In: ISPRS Commission VII Mid-term Symposium Remote Sensing: From Pixels to Processes. Enschede, NL. http://www.icc.cat/index.php/cat/layout/set/print/content/download/3831/12800/file/advanced_classification_techniques.pdf.

Artigas, F.J. and J.S. Yang. 2005. Hyperspectral remote sensing of marsh species and plant vigour gradient in the New Jersey Meadowlands. *Int J Remote Sensing* 26:5209–5220.

ASPRS LiDAR Committee. 2004. ASPRS LiDAR guidelines: Vertical accuracy reporting for LiDAR data. http://www.asprs.org/a/society/committees/LiDAR/Downloads/Vertical_Accuracy_Reporting_for_LiDAR_Data.pdf.

Baghdadi, N., M. Bernier, R. Gauthier, and I. Neeson. 2001. Evaluation of C-band SAR data for wetlands mapping. *Int J Remote Sensing* 22:71–88.

Baret, F. and G. Guyot. 1991. Potentials and limits of vegetation indices for LAI and PAR assessment. *Remote Sensing Environ* 35:161–173.

Bedford, B.L., M.R. Walbridge, and A. Aldous. 1999. Patterns in nutrient availability and plant diversity of temperate North American wetlands. *Ecology* 80:2151–2169. doi: 10.1890/0012-9658(1999)080[2151:PINAAP]2.0.CO;2.

Belluco, E., M. Camuffo, S. Ferrari, L. Modenese, S. Silvestri, A. Marani, and M. Marani. 2006. Mapping salt-marsh vegetation by multispectral and hyperspectral remote sensing. *Remote Sensing Environ* 105:54–67. doi: 10.1016/j.rse.2006.06.006.

Benz, U.C., P. Hofmann, G. Willhauck, I. Lingenfelder, and M. Heynen. 2004. Multi-resolution, object-oriented fuzzy analysis of remote sensing data for GIS-ready information. *ISPRS J Photogramm Remote Sensing* 58:239–258. doi: 10.1016/j.isprsjprs.2003.10.002.

Bouma, T.J., M.B. De Vries, E. Low, L. Kusters, P.M.J. Herman, I.C. Tanczos, S. Temmerman, A. Hesselink, P. Meire, and S. van Regenmortel. 2005. Flow hydrodynamics on a mudflat and in salt marsh vegetation: Identifying general relationships for habitat characterisations. *Hydrobiologia* 540:259–274. doi: 10.1007/s10750-004-7149-0.

Boyer, K.E., P. Fong, R.R. Vance, and R.F. Ambrose. 2001. *Salicornia virginica* in a southern California salt marsh: Seasonal patterns and a nutrient-enrichment experiment. *Wetlands* 21:315–326. doi: 10.1672/0277-5212(2001)021[0315:SVIASC]2.0.CO;2.

Breiman, L., J. Feidman, R. Olshen, and C. Stone. 1984. *Classification and Regression Trees*. Belmont, CA: Wadsworth, Inc.

Brennan, R. and T.L. Webster. 2006. Object-oriented land cover classification of LiDAR-derived surfaces. *Can J Remote Sensing* 32:162–172.

Brinson, M.M. 1993. A hydrogeomorphic classification for wetlands. US Army Corps of Engineers, Washington, DC. http://www.cpe.rutgers.edu/Wetlands/hydrogeomorphic-classification-for-wetlands.pdf.

Buschmann, C. and E. Nagel. 1993. In vivo spectroscopy and internal optics of leaves as basis for remote sensing of vegetation. *Int J Remote Sensing* 14: 711–722.

Byrd, K.B., J.L. O'Connell, S. Di Tommaso, and M. Kelly. 2014. Evaluation of sensor types and environmental controls on mapping biomass of coastal marsh emergent vegetation. *Remote Sensing Environ* 149:166–180.

Campbell, J.B. 2006. *Introduction to Remote Sensing*, 4th edn. New York: The Guilford Press.

Chen, J., S. Gu, M. Shen, Y. Tang, and B. Matsushita. 2009. Estimating aboveground biomass of grassland having a high canopy cover: An exploratory analysis of in situ hyperspectral data. *Int J Remote Sensing* 30:6497–6517.

Chen, J.M. and J. Cihlar. 1996. Retrieving leaf area index of boreal conifer forests using Landsat TM images. *Remote Sensing Environ* 55:153–162.

Cho, H.J., P. Kirui, and H. Natarajan. 2008. Test of multispectral vegetation index for floating and canopy-forming submerged vegetation. *Int J Environ Res Public Health* 5:289–295.

Chust, G., I. Galparsoro, A. Borja, J. Franco, and A. Uriarte. 2008. Coastal and estuarine habitat mapping, using LiDAR height and intensity and multi-spectral imagery. *Estuar Coast Shelf Sci* 78:633–643. doi: 10.1016/j.ecss.2008.02.003.

Cohen, R.A. and P. Fong. 2005. Experimental evidence supports the use of δ 15 N content of the opportunistic green macroalga *Enteromorpha intestinalis* (chlorophyta) to determine nitrogen sources to estuaries. *J Phycol* 41:287–293. doi: 10.1111/j.1529-8817.2005.04022.x.

Congalton, R.G. 1991. A review of assessing the accuracy of classifications of remotely sensed data. *Remote Sensing Environ* 37:35–46.

Costa, M.P.F. and K.H. Telmer. 2007. Mapping and monitoring lakes in the Brazilian Pantanal wetland using synthetic aperture radar imagery. *Aquat Conserv* 17:277–288.

Cowardin, L.M., V. Carter, F.C. Golet, and E.T. LaRoe. 1979. Classification of wetlands and deepwater habitats of the United States. FWS/OBS-79/31. U.S. Fish and Wildlife Service, Washington, DC.

Cronk, J.K. and M.S. Fennessy. 2001. *Wetland Plants*. Boca Raton, FL: Lewis Publication/ CRC Press LLC.

Curran, P.J. 1989. Remote sensing of foliar chemistry. *Remote Sensing Environ* 30:271–278. doi: 10.1016/0034-4257(89)90069-2.

Dahl, T.E. 2004. Remote sensing as a tool for monitoring wetland habitat change. U.S. Fish and Wildlife Service, Washington, DC, pp. 1–7.

Danks, S.M., E.H. Evans, and P.A. Whittaker. 1984. Photosynthetic systems: structure, function and assembly. New York: John Wiley, 162pp.

Darby, F.A. and R.E. Turner. 2008a. Below- and aboveground biomass of *Spartina alterniflora*: Response to nutrient addition in a Louisiana salt marsh. *Estuaries Coasts* 31:326–334.

Darby, F.A. and R.E. Turner. 2008b. Effects of eutrophication on salt marsh root and rhizome biomass accumulation. *Mar Ecol Prog Ser* 363:63–70. doi: 10.3354/meps07423.

Deegan, L.A., D.S. Johnson, R.S. Warren, B.J. Peterson, J.W. Fleeger, S. Fagherazzi, and W.M. Wollheim. 2012. Coastal eutrophication as a driver of salt marsh loss. *Nature* 490:388–392. doi: 10.1038/nature11533.

Delalieux, S., B. Somers, S. Hereijgers, W.W. Verstraeten, W. Keulemans, and P. Coppin. 2008. A near-infrared narrow-waveband ratio to determine Leaf Area Index in orchards. *Remote Sensing Environ* 112:3762–3772.

Dronova, I., P. Gong, N.E. Clinton, L, Wang, W. Fu, S. Qi, and Y. Liu. 2012. Landscape analysis of wetland plant functional types: The effects of image segmentation scale, vegetation classes and classification methods. *Remote Sensing Environ* 127:357–369.

du Plessis W.P. 1999. Linear regression relationships between NDVI, vegetation and rainfall in Etosha National Park, Namibia. *J Arid Environ* 42: 235–260.

Dwivedi, R., B. Rao, and S. Bhattacharya. 1999. Mapping wetlands of the Sundarban Delta and its environs using ERS-1 SAR data. *Int J Remote Sensing* 20:2235–2247.

Evans, T.L. and M. Costa. 2013. Landcover classification of the Lower Nhecolândia subregion of the Brazilian Pantanal wetlands using ALOS/PALSAR, RADARSAT-2 and ENVISAT/ASAR imagery. *Remote Sensing Environ* 128:118–137. doi:10.1016/j.rse.2012.09.022.

Fang, J., S. Piao, and Z. Tang. 2001. Interannual variability in net primary production and precipitation. *Science* 293:1723.

Filippi, A.M. and J.R. Jensen. 2006. Fuzzy learning vector quantization for hyperspectral coastal vegetation classification. *Remote Sensing Environ* 100:512–530.

Friess, D.A., T. Spencer, G.M. Smith, I. Möller, S.M. Brooks, and A.G. Thomson. 2012. Remote sensing of geomorphological and ecological change in response to saltmarsh managed realignment, The Wash, UK. *Int J Appl Earth Observ* 18:57–68.

Gao, B.C. and A.F.H. Goetz. 1995. Retrieval of equivalent water thickness and information related to biochemical components of vegetation canopies from AVIRIS data. *Remote Sensing Environ* 52:155–162.

Gao, X., A.R. Huete, W. Ni, and T. Miura. 2000. Optical-biophysical relationships of vegetation spectra without background contamination. *Remote Sensing Environ* 74:609–620.

Geerken, R., B. Zaitchick, and J.P. Evans. 2005. Classifying rangeland vegetation type and coverage from NDVI time series using Fourier filtered cycle similarity. *Int J Remote Sensing* 26:5535–5554.

Geerling, G.W., M. Labrador-Garcia, J.G.P.W. Clevers, A.M.J. Ragas, and A.J.M. Smits. 2007. Classification of floodplain vegetation by data fusion of spectral (CASI) and LiDAR data. *Int J Remote Sensing* 28:4263–4284. doi 10.1080/01431160701241720.

Geladi, P. and B.R. Kowalski. 1986. Partial least-squares regression: A tutorial. *Anal Chim Acta* 185:1–17. doi: 10.1016/0003-2670(86)80028-9.

Gesch, D.B. 2009. Analysis of LiDAR elevation data for improved identification and delineation of lands vulnerable to sea-level rise. *J Coastal Res* 25:49–58. doi: 10.2112/SI53-006.1

Gilmore, M.S., E.H. Wilson, N. Barrett, D.L. Civco, S. Prisloe, J.D. Hurd, and C. Chadwick. 2008. Integrating multi-temporal spectral and structural information to map wetland vegetation in a lower Connecticut River tidal marsh. *Remote Sensing Environ* 112:4048–4060. doi: 10.1016/j.rse.2008.05.020.

Gilmore, M.S., D.L. Civco, E.H. Wilson, N. Barrett, S. Prisloe, J.D. Hurd, and C. Chadwick. 2010. Remote sensing and in situ measurements for delineation and assessment of coastal marshes and their constituent species. In *Remote Sensing of Coastal Environments*, J. Wang (ed.). Boca Raton, FL: Springer.

Gitelson, A.A. 2004. Wide Dynamic Range Vegetation Index for remote quantification of crop biophysical characteristics. *J Plant Physiol* 161:165–173.

Gitelson, A., M. Merzlyak, and H. Lichtenthaler. 1996a. Detection of red edge position and chlorophyll content by reflectance measurements near 700 nm. *J Plant Physiol* 148:501–508.

Gitelson, A.A., Y.J. Kaufman, and M.N. Merzlyak. 1996b. Use of green channel in remote sensing of global vegetation from EOS-MODIS. *Remote Sensing Environ* 58:289–298.

Gitelson A.A., Y.J. Kaufman, R. Stark, and D. Rundquist. 2002. Novel algorithms for 830 remote estimation of vegetation fraction. *Remote Sens Environ* 80:76–87.

Gitelson, A.A., G.P. Keydan, and M.N. Merzlyak. 2006. Three-band model for noninvasive estimation of chlorophyll, carotenoids, and anthocyanin contents in higher plant leaves. *Geophys Res Lett* 33:L11402.

Gitelson, A.A. and M.N. Merzlyak. 2004. Non-destructive assessment of chlorophyll, carotenoid and anthocyanin content in higher plant leaves: Principles and algorithms. In *Remote Sensing for Agriculture and the Environment*, S. Stamatiadis, J.M. Lynch, J.S. Schepers (eds.). Larissa, Greece: Ella, pp. 78–94.

Gitelson, A.A., A. Vina, T.J. Arkebauer, D.C. Rundquist, G. Keydan, and B. Leavitt. 2003a. Remote estimation of leaf area index and green leaf biomass in maize canopies. *Geophys Res Lett* 30:1248. doi:10.1029/2002GL016450.

Gitelson, A.A., A. Vina, V. Ciganda, and D.C. Rundquist. 2005. Remote estimation of canopy chlorophyll content in crops. *Geophys Res Lett* 32:L08403.

Gitelson, A.A., S.B. Verma, A.Vina, D.C. Rundquist, G. Keydan, B. Leavitt, T.J. Arkebauer, G.G. Burba, and A.E. Suyker. 2003b. Novel technique for remote estimation of CO_2 flux in maize. *Geophys Res Lett* 30:1486. doi: 10.1029/2002GL016543.

Goranson, C.E., C.K. Ho, and S.C. Pennings. 2004. Environmental gradients and herbivore feeding preferences in coastal salt marshes. *Oecologia* 140:591–600. doi: 10.1007/s00442-004-1615-2.

Goward, S.N. and K.F. Huemmrich. 1992. Vegetation canopy PAR absorptance and the normalized difference vegetation index: An assessment using the SAIL model. *Remote Sensing Environ* 39:119–140.

Graets, D. 1990. Remote sensing of terrestrial ecosystem structure: An ecologist's pragmatic view. *Eco Studies* 79:5–30.

Hall, D.K. 1996. Remote sensing applications to hydrology: Imaging radar. *Hydrol Sci J* 41:609–624.

Hardisky, M.A., F.C. Daiber, C.T. Roman, and V. Klemas. 1984. Remote sensing of biomass and annual net aerial primary productivity of a salt marsh. *Remote Sensing Environ* 16:91–106. doi: 10.1016/0034-4257(84)90055-5.

Hardisky, M.A., R. Michael Smart, and V. Klemas. 1983. Growth response and spectral characteristics of a short *Spartina alterniflora* salt marsh irrigated with freshwater and sewage effluent. *Remote Sensing Environ* 13:57–67. doi: 10.1016/0034-4257(83)90027-5.

Harris, J. and S. Digby-Argus. 1986. The detection of wetlands on Radar imagery. *Proceedings of the Tenth Canadian Symposium on Remote Sensing*, Edmonton, Alberta, Canada.

Heim, R.R. 2002. A review of twentieth-century drought indices used in the United States. *Bull Am Meteorol Soc* 83:1149–1165.

Hess, L., J. Melack, S. Filoso, and Y. Wang. 1995. Delineation of inundated area and vegetation along the Amazon floodplain with the SIR-C synthetic aperture radar. *IEEE Trans Geosci Remote* 33:896–904.

Hinkle, R.L. and W.J. Mitsch. 2005. Salt marsh vegetation recovery at salt hay farm wetland restoration sites on Delaware Bay. *Ecol Eng* 25:240–251.

Hirano, A., M. Madden, and R. Welch. 2003. Hyperspectral image data for mapping wetland vegetation. *Wetlands* 23:436–448.

Hladik, C. and M. Alber. 2012. Accuracy assessment and correction of a LiDAR-derived salt marsh digital elevation model. *Remote Sensing Environ* 121:224–235. 10.1016/j.rse.2012.01.018.

Hladik, C., J. Schalles, and M. Alber. 2013. Salt marsh elevation and habitat mapping using hyperspectral and LiDAR data. *Remote Sensing Environ* 139:318–330. doi: 10.1016/j.rse.2013.08.003.

Hoffbeck, J.P. 1995. *Classification of High Dimensional Multispectral Data*. West Lafayette, IN: Purdue University. http://www.state.nj.us/dep/gis/digidownload/metadata/lulc02/anderson2002.html.

Huete, A.R. 1998. A soil–adjusted vegetation index (SAVI). *Remote Sens Environ* 25:295–309.

Huete, A., K. Didan, T. Miura, E.P. Rodriguez, X. Gao, and L.G. Ferreira. 2002. Overview of the radiometric and biophysical performance of the MODIS vegetation indices. *Remote Sens Environ* 83:195–213.

Jackson, T., D. Chen, M. Cosh, F. Li, M. Anderson, C. Walthall, P. Doraiswamy, and E.R. Hunt. 2004. Vegetation water content mapping using Landsat data normalized difference water index (NDWI) for corn and soybean. *Remote Sensing Environ* 92:475–482.

Jenkins, J.P., B.H. Braswell, S.E. Frolking, and J.D. Aber. 2002. Detecting and predicting spatial and interannual patterns of temperate forest springtime phenology in the eastern U.S. *Geophys Res Lett* 29:2201. doi:10.1029/2001GL014008.

Jensen, R.R., P. Mausel, N. Dias, R. Gonser, C. Yang, J. Everitt, and R. Fletcher. 2007. Spectral analysis of coastal vegetation and land cover using AISA+ hyperspectral data. *Geocarto Int* 22:17–28.

Kanemasu, E.T. 1974. Seasonal canopy reflectance patterns of wheat, sorghum, and soybean. *Remote Sensing Environ* 3:43–47.

Kao, J.T., J.E. Titus, and W.-X. Zhu. 2003. Differential nitrogen and phosphorus retention by five wetland plant species. *Wetlands* 23:979–987. doi:10.1672/0277-5212(2003)023[0979:DNAPRB]2.0.CO;2.

Kasischke, E. and L. Bourgeau-Chavez. 1997. Monitoring South Florida wetlands using ERS-1 SAR imagery. *Photogramm Eng Remote Sensing* 63:281–291.

Kasischke, E., J. Melack, and M. Dobson, 1997. The use of imaging radars for ecological applications—A review. *Remote Sensing Environ* 59:141–156.

Kearney, M.S., D. Stutzer, K. Turpie, and J.C. Stevenson. 2009. The effects of tidal inundation on the reflectance characteristics of coastal marsh vegetation. *J Coastal Res* 256:1177–1186. doi: 10.2112/08-1080.1.

Keddy, P.A. 2010. *Wetland Ecology: Principles and Conservation*. New York: Cambridge University Press.

Kiage, L.M. and N.D. Walker. 2009. Using NDVI from MODIS to monitor Duckweed bloom in Lake Maracaibo, Venezuela. *Water Res Manage* 23:1125–1135. doi:10.1007/s11269-008-9318-9.

Klemas, V. 2011. Remote sensing of wetlands: Case studies comparing practical techniques. *J Coastal Res* 27:418–427. doi:10.2112/Jcoastres-D-10-00174.1

Kobe, R.K., M. Iyer, and M.B. Walters. 2010. Optimal partitioning theory revisited: Nonstructural carbohydrates dominate root mass responses to nitrogen. *Ecology* 91:166–179. doi:10.1890/09-0027.1.

Kokaly, R.F., G.P. Asner, S.V. Ollinger, M.E. Martin, and C.A. Wessman. 2009. Characterizing canopy biochemistry from imaging spectroscopy and its application to ecosystem studies. *Remote Sensing Environ.* 113(Suppl. 1):S78–S91. doi:10.1016/j.rse.2008.10.018.

Kokaly, R.F. and R.N. Clark. 1999. Spectroscopic determination of leaf biochemistry using band depth analysis of absorption features and stepwise multiple linear regression. *Remote Sensing Environ* 67:267–287. doi:10.1016/S0034-4257(98)00084-4.

Kovacs, J.M., J.M.L. King, F.F. de Santiago, and F. Flores-Verdugo. 2009. Evaluating the condition of a mangrove forest of the Mexican Pacific based on an estimated leaf area index mapping approach. *Environ Monit Assess* 157:137–149.

Lang, M. and G. McCarty. 2008. Wetland mapping: History and trends. In *Wetlands: Ecology, Conservation and Management*. New York: Nova Publishers, pp. 74–112.

Lang, M.W. and E.S. Kasischke. 2008. Using C–band synthetic aperture radar data to monitor forested wetland hydrology in Maryland's coastal plain, USA. *IEEE Trans Geosci Remote Sens* 46:535–546.

Larkin, D.J., S.C. Lishawa, and N.C. Tuchman. 2012. Appropriation of nitrogen by the invasive cattail *Typha × glauca*. *Aquat Bot* 100:62–66. doi:10.1016/j.aquabot.2012.03.001.

Lieth, H. and R.H. Whittaker. 1975. *Primary Production of the Biosphere*. New York: Springer-Verlag, 339pp.

Lillesand, T.M., R.W. Kiefer, and J.W. Chipman. 2008. *Remote Sensing and Image Interpretation*, 6th edn. Hoboken, NJ: Wiley, 768pp.

Lu, D. and Q. Weng. 2004. Spectral mixture analysis of the urban landscape in Indianapolis with landsat ETM plus imagery. *Photogramm Eng Remote Sensing* 70:1053–1062.

Lu, D. and Q. Weng. 2007. A survey of image classification methods and techniques for improving classification performance. *Int J Remote Sensing* 28:823–870. doi:10.1080/01431160600746456.

Lucas, K.L. and G.A. Carter. 2008. The use of hyperspectral remote sensing to assess vascular plant species richness on Horn Island, Mississippi. *Remote Sensing Environ* 112:3908–3915. doi:10.1016/j.rse.2008.06.009.

Marani, M., E. Belluco, S. Ferrari, S. Silvestri, A. D'Alpaos, S. Lanzoni, A. Feola, and A. Rinaldo. 2006. Analysis, synthesis and modelling of high-resolution observations of salt-marsh eco-geomorphological patterns in the Venice lagoon. *Estuar Coast Shelf Sci* 69:414–426. doi:10.1016/j.ecss.2006.05.021.

Masek, J.G., E.F. Vermote, N. Saleous, R. Wolfe, F.G. Hall, F. Huemmrich, F. Gao, et al. 2012. LEDAPS Landsat calibration, reflectance, atmospheric correction preprocessing code. Oak Ridge National Laboratory Distributed Active Archive Center, Oak Ridge, TN. doi:10.3334/ORNLDAAC/1080.

Mayer, A.L. and R.D. Lopez. 2011. Use of remote sensing to support forest and wetlands policies in the USA. *Remote Sensing* 3:1211–1233.

Mccarthy, M.C. and B.J. Enquist. 2007. Consistency between an allometric approach and optimal partitioning theory in global patterns of plant biomass allocation. *Funct Ecol* 21:713–720. doi:10.1111/j.1365-2435.2007.01276.x.

McKee, K.L. 2011. Biophysical controls on accretion and elevation change in Caribbean mangrove ecosystems. *Estuar Coast Shelf Sci* 91:475–483. doi:10.1016/j.ecss.2010.05.001.

McKee, K.L., D.R. Cahoon, and I.C. Feller. 2007. Caribbean mangroves adjust to rising sea level through biotic controls on change in soil elevation. *Global Ecol Biogeogr* 16:545–556. doi:10.1111/j.1466-8238.2007.00317.x.

Mevik, B.H. and H.R. Cederkvist. 2004. Mean squared error of prediction (MSEP) estimates for principal component regression (PCR) and partial least squares regression (PLSR). *J Chemometrics* 18:422–429. doi:10.1002/cem.887.

Mevik, B.H. and R. Wehrens. 2007. The pls package: Principal component and partial least-squares regression in R. *J Stat Softw* 18:1–24.

Miller, R.L., M. Fram, R. Fujii, and G. Wheeler. 2008. Subsidence reversal in a re-established wetland in the Sacramento-San Joaquin Delta, California, USA. *San Franc Estuary Watershed Sci* 6:1–20. http://escholarship.org/uc/item/5j76502x.

Mishra, D.R., H.J. Cho, S. Ghosh, A. Fox, C. Downs, P.B.T. Merani, P. Kirui, N. Jackson, and S. Mishra. 2012. Post-spill state of the marsh: Remote estimation of the ecological impact of the Gulf of Mexico oil spill on Louisiana Salt Marshes. *Remote Sensing Environ* 118:176–185.

Mitsch, W.J. and J.G. Gosselink. 2007. *Wetlands*. Hoboken, NJ: John Wiley & Sons, Inc.

Moeslund, J.E., L. Arge, P.K. Bocher, B. Nygaard, and J.C. Svenning. 2011. Geographically comprehensive assessment of salt-meadow vegetation-elevation relations using LiDAR. *Wetlands* 31:471–482. doi: 10.1007/s13157-011-0179-2.

Mokany, K., R.J. Raison, and A.S. Prokushkin. 2006. Critical analysis of root: Shoot ratios in terrestrial biomes. *Global Change Biol* 12:84–96. doi:10.1111/j.1365-2486.2005.001043.x.

Mokhele, T.A. and F.B. Ahmed. 2010. Estimation of leaf nitrogen and silicon using hyperspectral remote sensing. *J Appl Remote Sensing* 4:1–18. doi:10.1117/1.3525241.

Montane, J.M. and R. Torres. 2006. Accuracy assessment of LiDAR saltmarsh topographic data using RTK GPS. *Photogramm Eng Remote Sensing* 72:961–967.

Moore, P.D. 1987. Ecological and hydrological aspects of peat formation. *Geol Soc Spec Publ* 32:7–15. doi:10.1144/GSL. SP.1987.032.01.02.

Morris, J.T. and B. Haskin. 1990. A 5-yr record of aerial primary production and stand characteristics of *Spartina- Alterniflora*. *Ecology* 71:2209–2217.

Morris, J.T., D. Porter, M. Neet, P.A. Noble, L. Schmidt, L.A. Lapine, and J.R. Jensen. 2005. Integrating LiDAR elevation data, multi-spectral imagery and neural network modelling for marsh characterization. *Int J Remote Sensing* 26:5221–5234.

Morris, J.T., G.P. Shaffer, and J.A. Nyman. 2013. Brinson review: Perspectives on the influence of nutrients on the sustainability of coastal wetlands. *Wetlands* 33(6):975–988. doi:10.1007/s13157-013-0480-3.

Myneni, R.B., R.R. Nemani, and S.W. Running. 1997 Estimation of global leaf area index and absorbed par using radiative transfer models. *IEEE Trans Geosci Remote* 35:1380–1393.

Novo, E.M.L.M., M.P.F. Costa, J.E. Mantovani, and I.B.T. Lima. 2002. Relationship between macrophyte stand variables and radar backscatter at L and C band, Tucurui reservoir, Brazil. *Int J Remote Sensing* 23:1241–1260.

Nyman, J.A., R.D. Delaune, H.H. Roberts, and W.H. Patrick. 1993. Relationship between vegetation and soil formation in a rapidly submerging coastal marsh. *Mar Ecol Prog Ser* 96:269–279.

Nyman, J.A., R.J. Walters, R.D. Delaune, and J.W.H. Patrick. 2006. Marsh vertical accretion via vegetative growth. *Estuar Coast Shelf Sci* 69:370–380.

O'Connell, J.L., K.B. Byrd, and M. Kelly. 2014. Remotely-sensed indicators of N-related biomass allocation in *Schoenoplectus acutus*. *PLoS One* 9:e90870. doi:10.1371/journal.pone.0090870.

O'Connell, J.L., K.B. Byrd, and M. Kelly. In review. A hybrid modeling approach for mapping relative differences in belowground biomass using spectral reflectance, foliar N and plant biophysical data. *Remote Sensing Environ*.

Optech. 2011. Gemini summary specification sheet. http://www. geo-konzept.de/data/downloads/altm_gemini.pdf.

Ozesmi, S.L. and M.E. Bauer. 2002. Satellite remote sensing of wetlands. *Wet Ecol Manage* 10:381–402.

Pal, M. and P.M. Mather. 2003. An assessment of the effectiveness of decision tree methods for land cover classification. *Remote Sensing Environ* 86:554–565. doi:10.1016/ S0034-4257(03)00132-9.

Pengra, B.W., C.A. Johnston, and T.R. Loveland. 2007. Mapping an invasive plant, *Phragmites australis*, in coastal wetlands using the EO-1 Hyperion hyperspectral sensor. *Remote Sensing Environ* 108:74–81.

Penuelas, J., J.A. Gamon, K.L. Griffin, and C.B. Field. 1993. Assessing community type, plant biomass, pigment composition, and photosynthetic efficiency of aquatic vegetation from spectral reflectance. *Remote Sensing Environ* 46:110–118.

Phinn, S.R. 1998. A framework for selecting appropriate remotely sensed data dimensions for environmental monitoring and management. *Int J Remote Sensing* 19:3457–3463. doi:10.1080/014311698214136.

Phinn, S., L. Hess, and C.M. Finalyson. 1999. An assessment of the usefulness of remote sensing for coastal wetland inventory and monitoring in Australia. In *Techniques for Enhanced Wetland Inventory and Monitoring*, C.M. Finlayson and A.G. Spiers (eds.). Canberra, Australian Capital Territory, Australia: Supervising Scientist Environment.

Phinn, S.R., D.A. Stow, and J.B. Zedler. 1996. Monitoring wetland habitat restoration in Southern California using airborne multi spectral video data. *Restoration Ecol* 4:412–422. doi:10.1111/j.1526-100X.1996.tb00194.x.

Pohl, C. and J.L. van Genderen. 1998. Multisensor image fusion in remote sensing: concepts, methods and applications. *Int J Remote Sensing* 19:823–854.

Pontius, R.G. and M. Millones. 2011. Death to Kappa: Birth of quantity disagreement and allocation disagreement for accuracy assessment. *Int J Remote Sensing* 32:4407–4429. doi:10.1080/01431161.2011.552923.

Poulter, B. and P.N. Halpin. 2008. Raster modelling of coastal flooding from sea-level rise. *Int J Geogr Inf Sci* 22:167–182. doi:10.1080/13658810701371858.

Ramsey, E. and A. Rangoonwala. 2006. Canopy reflectance related to marsh dieback onset and progression in coastal Louisiana. *Photogramm Eng Remote Sensing* 72:641–652.

Rao, B.R.M., R.S. Dwivedi, S.P.S. Kushwaha, S.N. Bhattacharya, J.B. Anand, and S. Dasgupta. Monitoring the spatial extent of coastal wetlands using ERS-1 SAR data. *Int J Remote Sens* 20:2509–2517.

Rasse, D.P., C. Rumpel, and M.F. Dignac. 2005. Is soil carbon mostly root carbon? Mechanisms for a specific stabilisation. *Plant Soil* 269:341–356.

Reimold, R.J., J.L. Gallagher, and D.E. Thompson. 1973. Remote sensing of tidal marsh. *Photogramm Eng Remote Sensing* 39:477–488.

Rosenqvist, A.K.E., C.M. Finlayson, J. Lowry, and D. Taylor. 2007. The potential of long-wavelength satellite-borne radar to support implementation of the Ramsar Wetlands Convention. *Aquat Conserv* 17:229–244.

Rosso, P.H., S.L. Ustin, and A. Hastings. 2006. Use of LiDAR to study changes associated with *Spartina* invasion in San Francisco Bay marshes. *Remote Sensing Environ* 100:295–306. doi:10.1016/j.rse.2005.10.012.

Rouse, J.W., R.H. Haas Jr., J.A. Schell, and D.W. Deering. 1974. Monitoring vegetation systems in the Great Plains with ERTS, NASA SP-351. *Third ERTS-1 Symposium*, NASA, Washington, DC, Vol. 1, pp. 309–317.

Rundquist, D., R. Perk, B. Leavitt, G. Keydan, and A. Gitelson. 2004. Collecting spectral data over cropland vegetation using machine-positioning versus hand-positioning of the sensor. *Comput Electron Agron* 43:173–178.

Ryu, C., M. Suguri, and M. Umeda. 2009. Model for predicting the nitrogen content of rice at panicle initiation stage using data from airborne hyperspectral remote sensing. *Biosyst Eng* 104:465–475. doi:10.1016/j.biosystemseng.2009.09.002.

Sadro, S., M. Gastil-Buhl, and J. Melack. 2007. Characterizing patterns of plant distribution in a southern California salt marsh using remotely sensed topographic and hyperspectral data and local tidal fluctuations. *Remote Sensing Environ* 110:226–239. doi:10.1016/j.rse.2007.02.024.

Sanders, B.F. 2007. Evaluation of on-line DEMs for flood inundation modeling. *Adv Water Resour* 30:1831–1843. doi:10.1016/j.advwatres.2007.02.005.

Sartoris, J.J., J.S. Thullen, L.B. Barber, and D.E. Salas. 1999. Investigation of nitrogen transformations in a southern California constructed wastewater treatment wetland. *Ecol Eng* 14:49–65. doi:10.1016/S0925-8574(99)00019-1.

Schalles, J.F., C.M. Hladik, A.A. Lynes, and S.C. Pennings. 2013. Landscape estimates of habitat types, plant biomass, and invertebrate densities in a Georgia Salt Marsh. *Oceanography* 26:88–97.

Schmid, K.A., B.C. Hadley, and N. Wijekoon. 2011. Vertical accuracy and use of topographic LIDAR data in coastal marshes. *J Coastal Res* 27(6A):116–132.

Schmidt, K.S. and A.K. Skidmore. 2003. Spectral discrimination of vegetation types in a coastal wetland. *Remote Sensing Environ* 85:92–108. doi:10.1016/S0034-4257(02)00196-7.

Schmidt, K.S., A.K. Skidmore, E.H. Kloosterman, H. Van Oosten, L. Kumar, and J.A.M. Janssen. 2004. Mapping coastal vegetation using an expert system and hyperspectral imagery. *Photogramm Eng Remote Sensing* 70:703–716.

Sellars, J.D. and C.L. Jolls. 2007. Habitat modeling for *Amaranthus pumilus*: An application of light detection and ranging (LiDAR) data. *J Coastal Res* 23:1193–1202. doi:10.2112/04-0334.1.

Sellers, P.J. 1985. Canopy reflectance, photosynthesis and transpiration. *Int J Remote Sensing* 6:1335–1372.

Shekede, M.D., S. Kusangaya, and K. Schmidt. 2008. Spatiotemporal variations of aquatic weeds abundance and coverage in Lake Chivero, Zimbabwe. *Phys Chem Earth* 33:714–721.

Shipley, B. and D. Meziane. 2002. The balanced-growth hypothesis and the allometry of leaf and root biomass allocation. *Funct Ecol* 16:326–331. doi:10.2307/826585.

Silva, T.S.F., M.P. Costa, and J.M. Melack. 2010. Spatial and temporal variability of macrophyte cover and productivity in the eastern Amazon floodplain: A remote sensing approach. *Remote Sensing Environ* 114:1998–2010.

Silvestri, S., M. Marani, and A. Marani. 2003. Hyperspectral remote sensing of salt marsh vegetation, morphology and soil topography. *Phys Chem Earth* 28:15–25. doi:10.1016/S1474-7065(03)00004-4.

Simard, M., L.E. Fatoyinbo, and N. Pinto. 2010. Mangrove canopy 3D structure and ecosystem productivity using active remote sensing. In *Remote Sensing of Coastal Environments*, J. Wang (ed.). Boca Raton, FL: CRC Press.

Smith, M.-L., S.V. Ollinger, M.E. Martin, J.D. Aber, R.A. Hallett, and C.L. Goodale. 2002. Direct estimation of aboveground forest productivity through hyperspectral remote sensing of canopy nitrogen. *Ecol Appl* 12:1286–1302. doi:10.1890/1051-0761(2002)012[1286:DEOAFP]2.0.CO;2.

Stroppiana, D., M. Boschetti, P.A. Brivio, and S. Bacchi. 2009. Plant nitrogen concentration in paddy rice from field canopy hyperspectral radiometry. *Field Crops Res* 111:119–129.

Stroppiana, D., F. Fava, M. Baschetti, and P.A. Brivio. 2012. Estimation of nitrogen content in crops and pastures using hyperspectral vegetation indices. In *Hyperspectral Remote Sensing of Vegetation*, P.S. Thenkabail, J.G. Lyon, and A. Huete (eds.). Boca Raton, FL: CRC Press, pp. 245–262. http://www.crcnetbase.com/doi/pdf/10.1201/b11222-16.

Thenkabail, P.S., J.G. Lyon, and A. Huete. 2012. Advances in hyperspectral remote sensing of vegetation and agricultural croplands. In *Hyperspectral Remote Sensing of Vegetation*, P.S. Thenkabail, J.G. Lyon, and A. Huete (eds.). Boca Raton, FL: Taylor & Francis Group, pp. 28–29.

Thenkabail, P.S., R.B. Smith, and E. De-Pauw. 2002. Evaluation of narrowband and broadband vegetation indices for determining optimal hyperspectral wavebands for agricultural crop characteristics. *Photogramm Eng Remote Sensing* 68:607–621.

Thomson, A.G., R.M. Fuller, M.G. Yates, S.L. Brown, R. Cox, and R.A. Wadsworth. 2003. The use of airborne remote sensing for extensive mapping of intertidal sediments and saltmarshes in eastern England. *Int J Remote Sensing* 24:2717–2737. doi:10.1080/0143116031000066918.

Thomson, A.G., A. Huiskes, R. Cox, R.A. Wadsworth, and L.A. Boorman. 2004. Short-term vegetation succession and erosion identified by airborne remote sensing of Westerschelde salt marshes, the Netherlands. *Int J Remote Sensing* 25:4151–4176. doi:10.1080/01431160310001647688.

Tian, Y.C., X. Yao, J. Yang, W.X. Cao, D.B. Hannaway, and Y. Zhu. 2011. Assessing newly developed and published vegetation indices for estimating rice leaf nitrogen concentration with ground- and space-based hyperspectral reflectance. *Field Crops Res* 120:299–310. doi:10.1016/j.fcr.2010.11.002.

Townend, I., C. Fletcher, M. Knappen, and K. Rossington. 2011. A review of salt marsh dynamics. *Water Environ J* 25:477–488. doi:10.1111/j.1747-6593.2010.00243.x.

Townsend, P. and S. Walsh. 1998. Modeling floodplain inundation using an integrated GIS with radar and optical remote sensing. *Geomorphology* 21: 295–312.

Townsend, P.A. 2000. A quantitative fuzzy approach to assess mapped vegetation classifications for ecological applications. *Remote Sensing Environ* 72:253–267.

Townsend, P.A. 2002. Relationships between forest structure and the detection of flood inundation in forested wetlands using C-band SAR. *Int J Remote Sensing* 23:443–460.

Townsend, P.A., J.R. Foster, R.A. Chastain, and W.S. Currie. 2003. Application of imaging spectroscopy to mapping canopy nitrogen in the forests of the central Appalachian Mountains using Hyperion and AVIRIS. *IEEE Trans Geosci Remote* 41:1347–1354. doi:10.1109/tgrs.2003.813205.

Toyra, J., A. Pietroniro, C. Hopkinson, and W. Kalbfleisch. 2003. Assessment of airborne scanning laser altimetry (LiDAR) in a deltaic wetland environment. *Can J Remote Sensing* 29:718–728.

Toyra, J., A. Pietroniro, L.W. Martz, and T.D. Prowse, 2002. A multi-sensor approach to wetland flood monitoring. *Hydrol Process* 16:1569–1581.

Trimble. 2009. Trimble R6 GPS receiver datasheet. http://trl.trimble.com/docushare/dsweb/Get/Document-333155/022543-259H_TrimbleR6GNSS_DS_0413_LR.pdf.

Tucker, C.J. 1977. Asymptotic nature of grass canopy spectral reflectance. *Appl Opt* 16:1151–1156.

Turner, R.E. 1979. Geographic variations in salt marsh macrophyte productions: A review. *Contrib Mar Sci* 20:47–68.

Turner, R.E., B.L. Howes, J.M. Teal, C.S. Milan, E.M. Swenson, and D.D.G. Toner. 2009. Salt marshes and eutrophication: An unsustainable outcome. *Limnol Oceanogr* 54:1634–1642. doi:10.4319/lo.2009.54.5.1634.

Turpie, K.R. 2013. Explaining the spectral red-edge features of inundated marsh vegetation. *J Coastal Res* 290:1111–1117. doi:10.2112/JCOASTRES-D-12-00209.1.

USDA. 2010. *The PLANTS Database*. Baton Rouge, LA: National Plant Data Center. http://plants.usda.gov.

Ustin, S.L., D.A. Roberts, S. Jacquemoud, J. Pinzón, M. Gardner, G.J. Scheer, C.M. Castaneda, and A. Palacios. 1998. Estimating canopy water content of chaparral shrubs using optical methods. *Remote Sensing Environ* 65:280–291.

Wang, C., M. Menenti, M.P. Stoll, E. Belluco, and M. Marani. 2007. Mapping mixed vegetation communities in salt marshes using airborne spectral data. *Remote Sensing Environ* 107:559–570. doi:10.1016/j.rse.2006.10.007.

Wang, C., M. Menenti, M.P. Stoll, A. Feola, E. Belluco, and M. Marani. 2009. Separation of ground and low vegetation signatures in LiDAR measurements of salt-marsh environments. *IEEE Trans Geosci Remote* 47:2014–2023. doi:10.1109/Tgrs.2008.2010490.

Wang, L., I. Dronova, P. Gong, W. Yang, Y. Li, and Q. Liu. 2012. A new time series vegetation-water index of phonological-hydrological trait across species and functional types for Poyang Lake wetland ecosystem. *Remote Sensing Environ* 125:49–63.

Wang, Y. 2010. Remote sensing of coastal environments: An overview. In *Remote Sensing of Coastal Environments*, J. Wang (ed.). Boca Raton, FL: CRC Press.

Webster, T.L., D.L. Forbes, E. MacKinnon, and D. Roberts. 2006. Flood-risk mapping for storm-surge events and sea-level rise using LiDAR for southeast New Brunswick. *Can J Remote Sensing* 32:194–211.

Wiegert, R.G. and B.J. Freeman. 1990. Tidal marshes of the southeast Atlantic coast: A community profile. U.S. Department of Interior, Fish and Wildlife Service, Biological Report 85(7.29), Washington, DC.

Wilson, B.A. and H. Rashid. 2005. Monitoring the 1997 flood in the Red River Valley using hydrologic regimes and RADARSAT imagery. *Can Geogr* 49:100–109.

Wolfe, R., M. Nishihama, A. Fleig, J. Kuyper, D. Roy, J. Storey, and F.S. Patt. 2002. Achieving sub-pixel geolocation accuracy in support of MODIS land science. *Remote Sensing Environ* 83:31–49.

Wright, C. and A. Gallant. 2007. Improved wetland remote sensing in Yellowstone National Park using classification trees to combine TM imagery and ancillary environmental data. *Remote Sensing Environ* 107:582–605. doi:10.1016/j.rse.2006.10.019.

Wulder, M.A., J.G. Masek, W.B. Cohen, T.R. Loveland, and C.E. Woodcock. 2012. Opening the archive: How free data has enabled the science and monitoring promise of Landsat. *Remote Sensing Environ* 122:2–10. doi:10.1016/j.rse.2012.01.010.

Yang, C., J.H. Everitt, R.S. Fletcher, J.R. Jensen, and P.W. Mausel. 2009. Mapping black mangrove along the south Texas gulf coast using AISA+ hyperspectral imagery. *Photogramm Eng Remote Sensing* 75:425–436.

Yang, J. and F.J. Artigas. 2010. Mapping salt marsh vegetation by integrating hyperspectral and LiDAR remote sensing. In *Remote Sensing of Coastal Environments*, Y. Wang (ed.). Boca Raton, FL: CRC Press.

Yoder, B.J. and R.H. Waring. 1994. The normalized difference vegetation index of small Douglas-fir canopies with varying chlorophyll concentrations. *Remote Sens Environ* 49:81–91.

Yu, Q., P. Gong, Y.Q. Tian, R. Pu, and J. Yang. 2008. Factors affecting spatial variation of classification uncertainty in an image object-based vegetation mapping. *Photogramm Eng Remote Sensing* 74:1007–1018. doi:10.14358/PERS.74.8.1007.

Zedler, J.B. 2001. *Handbook for Restoring Tidal Wetlands*. Boca Raton, FL: CRC Press.

Zhang, C. and Z. Xie. 2012. Combining object-based texture measures with a neural network for vegetation mapping in the Everglades from hyperspectral imagery. *Remote Sensing Environ* 124:310–320.

Zomer, R.J., A. Trabucco, and S.L. Ustin. 2008. Building spectral libraries for wetlands land cover classification and hyperspectral remote sensing. *J Environ Manage* 90:2170–2177.

9

Inland Valley Wetland Cultivation and Preservation for Africa's Green and Blue Revolution Using Multisensor Remote Sensing

Murali Krishna Gumma
International Crops Research Institute for the Semi Arid Tropics

Prasad S. Thenkabail
United States Geological Survey (USGS)

Irshad A. Mohammed
International Crops Research Institute for the Semi Arid Tropics

Pardhasaradhi Teluguntla
United States Geological Survey and Bay Area Environmental Research Institute

Venkateswarlu Dheeravath
World Food Program

Acronyms and Definitions

AEZ	Agroecological zones
AGRA	Alliance for a Green Revolution in Africa
ALI	Advanced Land Imager
ASTER	Advanced Spaceborne Thermal Emission and Reflection Radiometer
AVHRR	Advanced Very High Resolution Radiometer
CBERS-2	China-Brazil Earth Resources Satellite
CGIAR	Consultative Group on International Agricultural Research
CSI	Consortium for Spatial Information
DSS	Decision support system
EO	Earth observation
ERDAS	Earth Resources Data Analysis System
FAO	Food and Agriculture Organization of the United Nations
FORMOSAT Data	Taiwanis Satellite Operated by Taiwanis National Space Organization NSPO Marketed by SPOT
Hyperion	First Spaceborne Hyperspectral Sensor Onboard Earth Observing-1 (EO-1)
IITA	International Institute of Tropical Agriculture
IKONOS	High-Resolution Satellite Operated by GeoEye
IRS-1C/D-LISS	Indian Remote Sensing Satellite/Linear Imaging Self-Scanner
IRS-P6-AWiFS	Indian Remote Sensing Satellite/Advanced Wide Field Sensor

IVs	Inland Valleys
KOMFOSAT	Korean Multipurpose Satellite. Data Marketed by SPOT Image
Landsat-1, 2, 3 MSS	Multi spectral scanner
Landsat-4, 5 TM	Thematic Mapper
Landsat-7 ETM+	Enhanced Thematic Mapper Plus
MODIS	Moderate Imaging Spectral Radio Meter
NGO	Nongovernmental organization
QUICKBIRD	Satellite from DigitalGlobe, a private company in the United States
RAPID EYE—A/E	Satellite constellation from Rapideye, a German company
RESOURSESAT	Satellite launched by India
SPOT	Satellites Pour l'Observation de la Terre or Earth-observing Satellites
SWIR	Shortwave Infrared Sensor
VNIR	Visible Near-Infrared Sensor
WCA	West and Central Africa
WORLDVIEW	
USGS	United States Geological Survey

9.1 Introduction

Africa is the second largest continent after Asia with a total area of 30.22 million km² (including the adjacent islands). It has great rivers such as the River Nile, which is the longest in the world and flows a distance of 6650 km, and the River Congo, which is the deepest in the world, as well as the second largest in the world in terms of water availability. Yet, Africa also has vast stretches of arid, semiarid, and desert lands with little or no water. Further, Africa's population is projected to increase by four times by the year 2100, reaching about four billion from the current population of little over one billion. Food insecurity and malnutrition are already highest in Africa (Heidhues et al., 2004) and the challenge of meeting the food security needs of the fastest-growing continent in the twenty-first century is daunting. So, many solutions are thought of to ensure food security in Africa. These ideas include such measures as increasing irrigation in a continent that currently has just about 2% of the global irrigated areas (Thenkabail et al., 2009a, 2010), improving crop productivity (kg m⁻²), and increasing water productivity (kg m⁻³). However, an overwhelming proportion of Africa's agriculture now takes place on uplands that have poor soil fertility and water availability (Scholes, 1990). Thereby, the interest in developing sustainable agriculture in Africa's lowland wetlands, considered by some as the "new frontier" in agriculture, has swiftly increased in recent years. The lowland wetland systems include the big wetland systems that are prominent and widely recognized (Figure 9.1) as well as the less prominent, but more widespread, inland valley (IV) wetlands (Figures 9.2 through 9.8) that are all along the first to highest order river systems.

Africa's bigwetland ecosystems (Figure 9.1; MAW, 2014) are estimated to cover more than 131 million ha (4.33% of total geographic area of the continent) that vary in type from saline coastal lagoons in West Africa to fresh and brackish water lakes in East Africa. They deliver a wide range of ecosystem services that contribute to human well-being such as nutrition, water supply and purification, climate and flood regulation, coastal protection, feeding and nesting sites, recreational opportunities and increasingly, tourism (ESA, 2014). In contrast, the IV wetland systems (Figures 9.2 through 9.5) occupy roughly 6%–20% of various agroecosystems with higher percentage areas in the wetter agroecosystems and the lower percentage areas in the drier agroecosystems (Thenkabail et al., 2000b). Wetlands, with their abundant supply of fresh water, generally fertile soils, and high productivity, therefore play a central role in the economy of all river basins and coastal zones. They provide fish, water for agriculture, household uses, and transport. Additionally, many distant communities as well as entire cities and regions benefit from wetlands.

In this chapter, we will provide a focused study of wetlands of West and Central Africa (WCA) and demonstrate the richness and importance of wetlands in ensuring the food security of Africa. Throughout WCA, there is increasing pressure for agricultural development as a result of population growth and efforts to increase food security. The IV wetlands have high potential for growing agricultural crops due to (1) easy access to the river water, (2) significantly longer duration of adequate soil moisture to grow crops when compared with adjoining uplands, and (3) rich soils (depth and fertility) (FAO, 2005; WARDA, 2006; Tiner, 2009). However, 90% of WCA's current agriculture is concentrated in uplands, which have very poor soils and scarce water resources. In spite of such huge advantages over uplands, IV wetlands in WCA are highly underutilized mainly as a result of (1) waterborne diseases such as *Malaria, Bilharzias, Trypanosomiasis* (sleeping sickness), *Onchocerciasis* (river blindness), and *Dracontiasis* (guinea worm); and (2) difficulty in accessing them from roads–settlements–markets (WARDA, 2003; Lafferty, 2009). But these difficulties can be overcome with modern health care (Hetzel et al., 2007) and infrastructure (Woodhouse, 2009).

Given this background, it is increasingly felt that the best way to expedite WCA's green revolution (more crop per unit area) and blue revolution (more crop per unit of water) is to focus on its soil-water-rich and hitherto highly underutilized IV wetlands, which roughly constitute about 80% of WCA's total wetlands with the rest being river flood plains (12%) and coastal wetlands (8%) (Lyon, 2001; Mitsch and Gosselink, 2007; Thenkabail et al., 2009b). The WCA is yet to see a green revolution, so badly needed for the food security and economic progress of these countries, specifically for its subsistence farmers who constitute the overwhelming proportion of WCA's population of 350 million. The green revolution technologies developed in Asia in terms of improved agronomic, genetic traits,

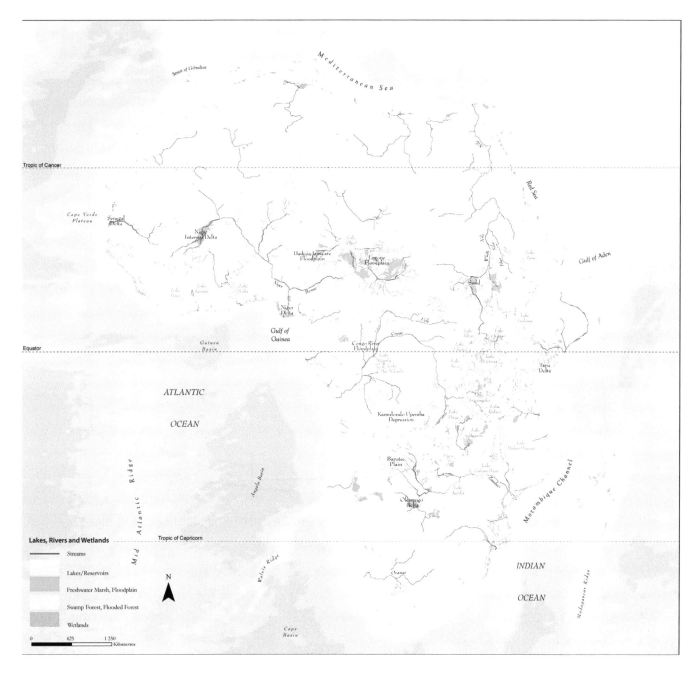

FIGURE 9.1 African wetlands (MAW, 2014). These are: "Areas of marsh, fen, peatland or water, whether natural or artificial, permanent or temporary, with water that is static or flowing…" (RAMSAR, 2004). But, these *do not* include inland valley wetlands.

and better water management can be adopted with minor modifications to WCA's own green revolution. The importance of IV wetlands is particularly high for rice cultivation as it is becoming a major staple in WCA. Records show a rapid increase of rice consumption in West Africa from 1 million tons in 1964 to 8.6 million tons in 2004 (WARDA, 2003; FAO, 2005). IV wetlands have higher crop yields than the equivalent upland areas. For example, potential yields of rice in IVs were estimated at 2.5–4.0 ton ha^{-1} compared to 1.5–2.0 ton ha^{-1} on uplands

(WARDA, 2006). Also, an important link in achieving food security is transportation; in these rural areas, fields nearest to the population have great value for supplying food needs and enhancing food security.

Balancing the need to bring in more land for agriculture by releasing land from other uses or natural cover are the ecological concerns about the environmental impacts of land cover such as wetland development (and the catchments that surround them) and the profound social and economic

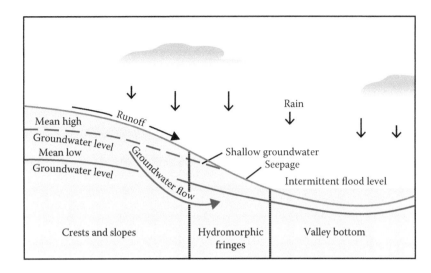

FIGURE 9.2   Depiction of wetlands. (From WARDA, Medium Term Plan 2007–2009, Charting the Future of Rice in Africa. Africa Rice Center (WARDA), Cotonou, Republic of Benin, 2006.)

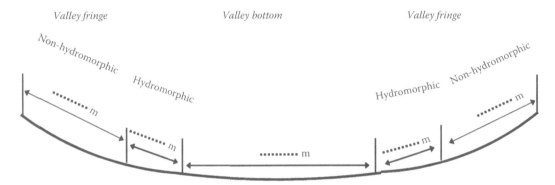

FIGURE 9.3   Inland valley wetlands consist of valley bottoms, hydromorphic valley fringes, and non-hydromorphic valley fringes.

FIGURE 9.4   Inland valley wetland illustration. The photos show valley bottoms. (From Gumma, M.K. et al., *J. Appl. Remote Sens.*, 3, 033537, 2009b.)

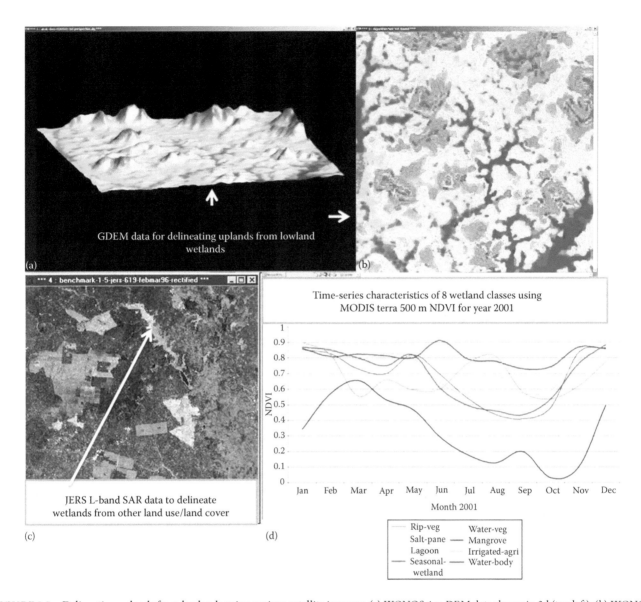

FIGURE 9.5 Delineating uplands from lowlands using various satellite imagery: (a) IKONOS 4 m DEM data shown in 3d (top left); (b) IKONOS 4 m DEM (top right); (c) JRTS SAR data (bottom left); and (d) MODIS temporal NDVI signatures of wetland classes. Inland valleys (IV) are seen in blue color in top two images. In the bottom left (JERS SAR), very high backscatter are areas of oil palm plantations. High backscatter (see arrow pointer) shows IV wetlands.

repercussions for people dependent on their natural resources and ecosystem functions. IV wetlands play an important role in bio-geochemical cycling, flood control, and recharging of aquifers. They are considered to be one of the richest and most productive biomes, serving as cradles of biological diversity that support unique flora and fauna (RAMSAR, 2004). They serve as potential sites for breeding waterfowl and significant carbon sinks in soils and plants (Lal et al., 2002; Mitsch and Gosselink, 2007).

Clearly, it is essential to incorporate wetlands explicitly within a natural resource management framework. There is the need to not only develop technologies that are adapted to farmers' economic needs to facilitate Africa's much awaited Green Revolution and supporting its Blue Revolution, but also sustain the integrity of the globally valuable WCA ecosystems. At present, the basis for making decisions relating to wetland utilization is weak (Gliessman, 2007). Given the fact that the characteristics of wetlands are known to vary dramatically within and across agroecosystems (Andriesse et al., 1994), it is important to map, characterize, and model different wetland systems (Gumma et al., 2009a, 2011b). This will provide impetus and enable the development of appropriate technologies for maximizing food production along with transportation (food security) with minimum ecological and environmental disturbance. A pre-requisite for sustainable management of IV wetlands is greater understanding of the

FIGURE 9.6 Inland valley wetland study areas across West and Central Africa (WCA). *Note:* background image is GTOPO30 1 km DEM data. Red dots are study areas. Sand color shows ground data points. Photos on right show typical IV wetlands.

interaction between climate, soil, topography, water, biophysical, health, and socioeconomic factors that influence both wetland utilization and the impacts that result including societal benefits.

Given the discussion, the three key action research goals presented in this chapter are:

First, identify, delineate, map, classify, and characterize wetlands of the entire WCA region using data fusion involving satellite multisensor data (e.g., Landsat ETM+, JERS SAR, ALOS PALSAR, MODIS, IKONOS/Quickbird; see Tables 9.1 and 9.2), secondary data (SRTM, FAO soils, precipitation), and in situ data (e.g., Fujii et al., 2010). IV wetlands are too small to appear on most maps and therefore the wetland surveys of the world have been mostly localized (Gilmore et al., 2008; Wdowinski et al., 2008) and limit themselves to large flood plains, swamps, and water bodies with or without irrigated areas. However, recent studies (Thenkabail and Nolte, 2000;

Lan and Zhang, 2006; Becker et al., 2007; Islam et al., 2008) have identified the potential of satellite remote sensing data and techniques for mapping different types of wetlands. None, however, has done so over very large areas such as nations, continents, and the world. Thereby, we propose to use multi-data fusion to best identify, map, classify, and characterize IV wetlands at high resolution (nominal 30 m) over entire WCA rapidly and accurately using automated and semiautomated methods.

Second, develop a decision support system (DSS) through spatial modeling to perform land suitability analysis in order to determine which of the IV wetland areas are best suited for: (1) agricultural development or (2) preservation. The goal is to balance food security-economic development with environmental conservation. Since the need is to maximize crop yields sustainably with minimal ecological and environmental impacts for the IV wetland ecosystems, we need to take into consideration climatic, soil, topographic, water, biophysical,

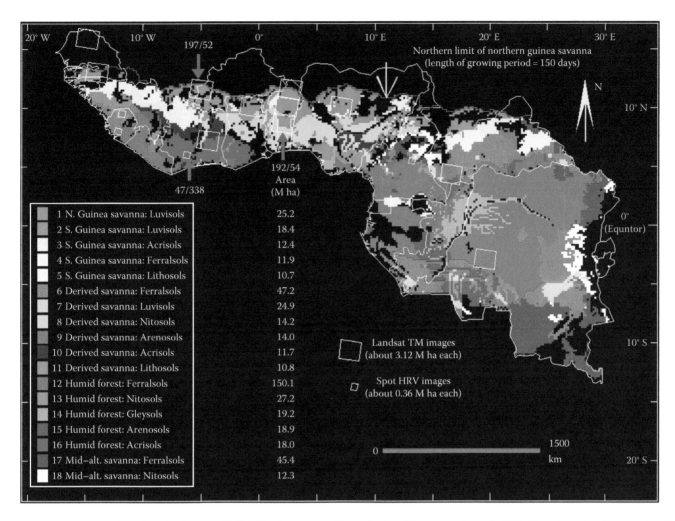

FIGURE 9.7 Agroecological and soil zones of WCA. The datasets used in producing this map are shown in Table 9.3 and consist of International Institute of Tropical Agriculture's (IITA) agroecological zones defined by the length of growing period (LGP), and FAO soils.

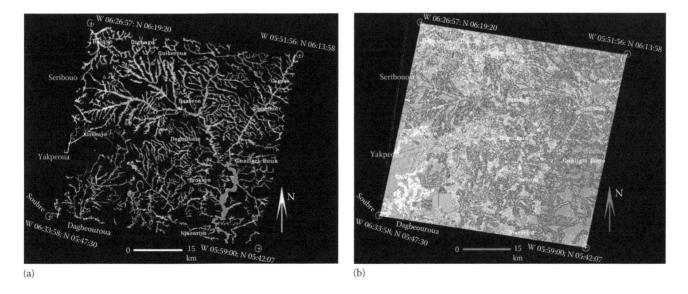

(a)   (b)

FIGURE 9.8 (a) Delineated inland valley wetlands using SPOT HRV data based on semi-automated methods (see Section 9.6.3) described in this chapter. (b) Land-use/land-cover classification of inland valley wetlands in Gaganoa, Côte d'Voire, using SPOT HRV images and semiautomated methods. *(Continued)*

1 Significant farmlands, uplands
2 Scattered farmlands, uplands
3 Savanna vegetation, uplands
4 Wetland/marshland, uplands
5 Dense forest, uplands
6 V. dense forest, uplands
7 Significant farmlands, fringes
8 Scattered farmlands, fringes
9 Insignificant farmlands, bottoms
10 Significant farmlands, bottoms
11 Scattered farmlands, bottoms
12 Insignificant farmlands, bottoms
13 Water
14 Built-up area or settlements
15 Roads
16 Barren land or desert

(c)

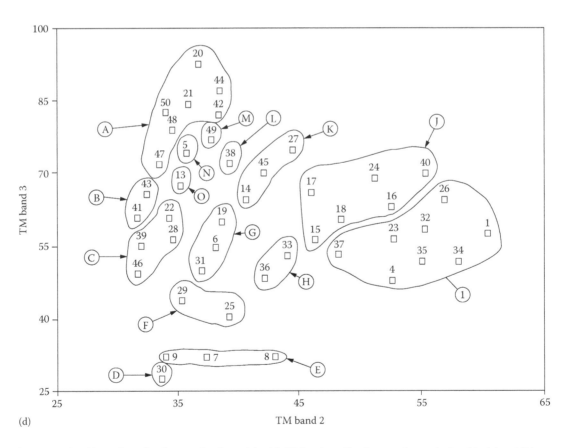

(d)

FIGURE 9.8 (*Continued*)   (c) Land-use/land-cover class legend for (a). (d) Land-use/land-cover classes depicted in (a) and (b).

TABLE 9.1 Wetland Delineation, Mapping, and Characterization Using Sensor Data Fusion

Sensor	Spatial (Meters)	Spectral (#)	Radiometric (Bit)	Band Range (µm)	Band Widths (µm)	Irradiance (W m^{-2} sr^{-1} µm^{-1})	Data Points (# per Hectares)	Frequency of Revisit (Days) Data Period
A. Moderate resolution								
1. MODIS terra/aqua	250, 500	2/7	12	0.62–0.67	0.05	1528.2	0.16, 0.04	8-day reflectance
				0.84–0.876	0.036	974.3	0.16, 0.04	2000–present
				0.459–0.479	0.02	2053		(wall to wall—Figure 9.1)
				0.545–0.565	0.02	1719.8		
				1.23–1.25	0.02	447.4		
				1.63–1.65	0.02	227.4		
				2.11–2.16	0.05	86.7		
B. High resolution on optical								
2. Landsat-TM/ ETM+	30	7	8	0.45–0.52	0.07	1970	11.1	16
				0.52–0.60	0.80	1843		GLS2005
				0.63–0.69	0.60	1555		(wall to wall—Figure 9.1)
				0.76–0.90	0.14	1047		
				1.55–1.74	0.19	227.1		
				10.4–12.5	2.10	0		
				2.08–2.35	0.25	80.53		
C. Radar								
3a. JERS/SAR	100, 500	L band	8	23.5 cm	L band	—	1, 0.04	Consolidated 1996
								Two periods
and/or								(Wall to wall—Figure 9.1)
3b. ALOS PALSAR	9–157	L band	8	23.5 cm	14–28 MHz	—	123, 0.4	2006–present
								For benchmark areas
								(See Figure 9.2)
D. Very high resolution optical								
4a. IKONOS	1–4	4	11	0.445–0.516	0.71	1930.9	10,000, 625	5
				0.506–0.595	0.89	1854.8		For benchmark areas
				0.632–0.698	0.66	1156.5		(See Figure 9.2)
and/or				0.757–0.853	0.96	1156.9		
4b. QUICKBIRD	0.61–2.44	4	11	0.45–0.52	0.07	1381.79	14,872, 625	5
				0.52–0.60	0.08	1924.59		For benchmark areas
				0.63–0.69	0.06	1843.08		(See Figure 9.2)
				0.76–0.89	0.13	1574.77		

Characteristics of data to be used in the study are listed.

health, and socioeconomic factors and potential societal benefits from the IV wetland ecosystem and use them in decision support systems. Stakeholders (e.g., Coalition for African Rice Development CARD/ Alliance for a Green Revolution in Africa [AGRA] network, Consultative Group on International Agricultural Research [CGIAR] network, International Institute for Tropical Agriculture [IITA]) will be involved in assigning weights to various spatial data layers used in the models of the DSS and hence will represent the collective knowledge of experts.

Third, provide access to data and products through USGS/NASA as well as stakeholder (e.g., CARD/ AGRA network, CGIAR Consortium of Spatial Information [CSI] network, IITA) through public domain web/data portals. This will help stakeholders to provide farmers and policy makers with sound science-based information that enables them to identify the best sites that could be developed to promote sustainable farming systems. The products will include (1) IV wetland maps, (2) wetland characteristics (e.g., phenology, land cover), (3) DSS, and (4) model outputs showing IV wetlands that are most suitable for (1) development as agricultural land and (2) conservation of biological diversity (outputs of goal 2).

TABLE 9.2 Satellite Sensor Data That Can Potentially Used in Wetland Studies

Sensor	Spatial (Meters)	Spectral (#)	Radiometric (Bit)	Band Range (μm)	Band Widths (μm)	Irradiance (W m⁻² sr⁻¹ μm⁻¹)	Data Points (# per Hectares)	Frequency of Revisit (Days)
A. Coarse resolution sensors								
1. AVHRR	1000	4	11	0.58–0.68	0.10	1390	0.01	Daily
				0.725–1.1	0.375	1410		
				3.55–3.93	0.38	1510		
				10.30–10.95	0.65	0		
				10.95–11.65	0.7	0		
2. MODIS	250, 500, 1000	36/7	12	0.62–0.67	0.05	1528.2	0.16, 0.04, 0.01	Daily
				0.84–0.876	0.036	974.3	0.16, 0.04, 0.01	
				0.459–0.479	0.02	2053		
				0.545–0.565	0.02	1719.8		
				1.23–1.25	0.02	447.4		
				1.63–1.65	0.02	227.4		
				2.11–2.16	0.05	86.7		
B. Multi spectral sensors								
3. Landsat-1, 2, 3 MSS	56 × 79	4	6	0.5–0.6	0.1	1970	2.26	16
				0.6–0.7	0.1	1843		
				0.7–0.8	0.1	1555		
				0.8–1.1	0.3	1047		
4. Landsat-4, 5 TM	30	7	8	0.45–0.52	0.07	1970	11.1	16
				0.52–0.60	0.80	1843		
				0.63–0.69	0.60	1555		
				0.76–0.90	0.14	1047		
				1.55–1.74	0.19	227.1		
				10.4–12.5	2.10	0		
				2.08–2.35	0.25	80.53		
5. Landsat-7 ETM+	30	8	8	0.45–0.52	0.65	1970	44.4, 11.1	16
				0.52–0.60	0.80	1843		
				0.63–0.69	0.60	1555		
				0.50–0.75	0.150	1047		
				0.75–0.90	0.200	227.1		
				10.0–12.5	2.5	0		
				1.75–1.55	0.2	1368		
				0.52–0.90(p)	0.38	1352.71		
5b. Landsat-8	30	11	8	0.433–0.453	0.02	1970	44.4, 11.1	16
				0.45–0.515	0.065	1843		
				0.53–0.60	0.07	1555		
				0.63–0.68	0.05	1047		
				0.845–0.885	0.04	227.1		
				1.56–1.66	0.1	0		
				2.10–2.30	0.2	1368		
				0.50–0.68	0.18	1352.71		
				1.360–1.390	0.03	1368		
				10.6–11.2	0.6	1352.71		
				11.5–12.5	1.0	1368		
6. ASTER	15, 30, 90	15	8	0.52–0.63	0.11	1846.9	44.4, 11.1, 1.23	16
				0.63–0.69	0.06	1546.0		

(*Continued*)

TABLE 9.2 (*Continued*) Satellite Sensor Data That Can Potentially Used in Wetland Studies

Sensor	Spatial (Meters)	Spectral (#)	Radiometric (Bit)	Band Range (μm)	Band Widths (μm)	Irradiance (W m⁻² sr⁻¹ μm⁻¹)	Data Points (# per Hectares)	Frequency of Revisit (Days)
				0.76–0.86	0.1	1117.6		
				0.76–0.86	0.1	1117.6		
				1.60–1.70	0.1	232.5		
				2.145–2.185	0.04	80.32		
				2.185–2.225	0.04	74.96		
				2.235–2.285	0.05	69.20		
				2.295–2.365	0.07	59.82		
				2.360–2.430	0.07	57.32		
			12	8.125–8.475	0.35	0		
				8.475–8.825	0.35	0		
				8.925–9.275	0.35	0		
				10.25–10.95	0.7	0		
				10.95–11.65	0.7	0		
7. ALI	30	10	12	0.048–0.69(p)	0.64	1747.8600		
				0.433–0.453	0.20	1849.5	11.1	16
				0.450–0.515	0.65	1985.0714		
				0.425–0.605	0.80	1732.1765		
				0.633–0.690	0.57	1485.2308		
				0.775–0.805	0.30	1134.2857		
				0.845–0.890	0.45	948.36364		
				1.200–1.300	1.00	439.61905		
				1.550–1.750	2.00	223.39024		
				2.080–2.350	2.70	78.072727		
8. SPOT-1	2.5–20	15	16	0.50–0.59	0.09	1858	1,600, 25	3–5
SPOT-2				0.61–0.68	0.07	1575		
SPOT-3				0.79–0.89	0.1	1047		
SPOT-4				1.5–1.75	0.25	234		
				0.51–0.73(p)	0.22	1773		
9. IRS-1C	23.5	15	8	0.52–0.59	0.07	1851.1	18.1	16
				0.62–0.68	0.06	1583.8		
				0.77–0.86	0.09	1102.5		
				1.55–1.70	0.15	240.4		
				0.5–0.75(p)	0.25	1627.1		
10. IRS-1	23.5	15	8	0.52–0.59	0.07	1852.1	18.1	16
				0.62–0.68	0.06	1577.38		
				0.77–0.86	0.09	1096.7		
				1.55–1.70	0.15	240.4		
				0.5–0.75(p)	0.25	1603.9		
11. IRS-P6-AWiFS	56	4	10	0.52–0.59	0.07	1857.7	3.19	16
				0.62–0.68	0.06	1556.4		
				0.77–0.86	0.09	1082.4		
				1.55–1.70	0.15	239.84		
12. CBERS-2	20 m pan		11	0.51–0.73	0.22	1934.03	25, 25	
CBERS-3B	20 m MS			0.45–0.52	0.07	1787.10		
CBERS-3	5 m pan			0.52–0.59	0.07	1587.97	400, 25	
CBERS-4	20 m MS			0.63–0.69	0.06	1069.21		
				0.77–0.89	0.12	1664.3		

(Continued)

TABLE 9.2 (*Continued*)   Satellite Sensor Data That Can Potentially Used in Wetland Studies

Sensor	Spatial (Meters)	Spectral (#)	Radiometric (Bit)	Band Range (μm)	Band Widths (μm)	Irradiance (W m⁻² sr⁻¹ μm⁻¹)	Data Points (# per Hectares)	Frequency of Revisit (Days)
C. Hyper-spectral sensor								
1. Hyperion	30	196[a]	16	196 effective calibrated bands VNIR (band 8–57) 427.55–925.85 nm SWIR (band 79–224) 932.72–2395.53 nm	10 nm wide (approx.) for all 196 bands	See data in Neckel and Labs (1984). Plot it and obtain values for Hyperion bands	11.1	16
D. Hyperspatial sensor								
1. World view-2	0.46–1.84	8	11	0.4–0.45	0.05	1758.2229	10,000, 625	3.7
				0.45–0.51	0.06	1974.2416		
				0.51–0.58	0.07	1856.4104		
				0.585–0.625	0.035	1738.4791		
				0.63–0.69	0.06	1559.4555		
				0.705–0.745	0.04	1342.0695		
				0.770–0.895	0.125	1069.7302		
				0.860–0.900	0.0.4	861.2866		
			PAN	0.860–0.900	0.0.4	1580.814		
2. IKONOS	1–4	4	11	0.445–0.516	0.71	1930.9	10,000, 625	5
				0.506–0.595	0.89	1854.8		
				0.632–0.698	0.66	1156.5		
				0.757–0.853	0.96	1156.9		
3. QUICKBIRD	0.61–2.44	4	11	0.45–0.52	0.07	1381.79	14,872, 625	5
				0.52–0.60	0.08	1924.59		
				0.63–0.69	0.06	1843.08		
				0.76–0.89	0.13	1574.77		
4. RESOURSESAT	5.8	3	10	0.52–0.59	0.07	1853.6	33.64	24
				0.62–0.68	0.06	1581.6		
				0.77–0.86	0.09	1114.3		
5. RAPID EYE-A RAPID EYE-E	6.5	5	12	0.44–0.51	0.07	1979.33	236.7	1–2
				0.52–0.59	0.07	1752.33		
				0.63–0.68	0.05	1499.18		
				0.69–0.73	0.04	1343.67		
				0.77–0.89	0.12	1039.88		
6. WORLDVIEW	0.55	1	11	0.45–0.51	0.06	1996.77	40,000	1.7–5.9
7. FORMOSAT-2	2–8	5	11	0.45–0.52	0.07	1974.93	2,500, 156.25	Daily
				0.52–0.60	0.08	1743.12		
				0.63–0.69	0.06	1485.23		
				0.76–0.90	0.14	1041.28		
				0.45–0.90(p)	0.45	1450		
8. KOMPSAT-2	1–4	5	10	0.5–0.9	0.4	1379.46	10,000, 625	3–28
				0.45–0.52	0.07	1974.93		
				0.52–0.6	0.08	1743.12		
				0.63–0.59	0.04	1485.23		
				0.76–0.90	0.14	1041.28		

Source: Adapted from Thenkabail, P., Lyon, G., Huete, A., Advances in hyperspectral remote sensing of vegetation and agricultural croplands. CRC Press/Taylor & Francis Group, Boca Raton, FL, 2011.

Note: First band is panchromatic, rest multi-spectral.

[a] Of the 242 bands, 196 are unique and calibrated. These are: (A) band 8 (427.55 nm) to band 57 (925.85 nm) that are acquired by visible and near-infrared (VNIR) sensor; and (B) band 79 (932.72 nm) to band 224 (2395.53 nm) that are acquired by short wave infrared (SWIR) sensor.

9.1.1 Carbon Budget of Wetlands

Wetlands, globally, contain about 771 billion tons of carbon? (20% of all the carbon on earth) (Lal et al., 2002; Pelley, 2008; Tiner, 2009). This is about the same amount of carbon as is now in the atmosphere. However, they also release methane, a greenhouse gas (Pelley, 2008) which is 22 times more potent than CO_2, on a per-unit-mass basis, in absorbing long-wave radiation on a 100-year time horizon (Zhuang et al., 2009). Nearly 60% of the planet's wetlands have been destroyed in the past 100 years, mostly for agriculture.

In Africa, since most wetlands are still intact, there is immense pressure to develop them to ensure African food security. Indeed, many consider wetlands as the best hope for Africa's green and blue revolution (WARDA, 2006) and a far better option for food security than the alternative of building large dams that will result in greater destruction of pristine rainforests (FAO, 2005). *Given the discussions*, WCA represents an unparalleled opportunity to guide agricultural expansion while being mindful of critical conservation goals and curtail the need for future remediation.

9.2 Definitions and Study Areas

9.2.1 Definition Used for Mapping Wetlands

Wetlands are (1) "Areas of marsh, fen, peatland or water, whether natural or artificial, permanent or temporary, with water that is static or flowing..." (RAMSAR, 2004), and (2) "...Seasonally or permanently waterlogged, including lakes, rivers, estuaries, and freshwater marshes; an area of low-lying land submerged or inundated periodically..." (USGS). In this study, we will map wetlands including irrigated agriculture, fresh water bodies, salt pans, lagoons, mangroves, riparian vegetation, permanent marshes, water bodies with or without aquatic plants, and seasonal wetlands. However, we will clearly demarcate IV wetlands that occur overwhelmingly on first- to fourth-order streams and roughly constitute about 80% of all wetlands in WCA (Andriesse et al., 1994). Hydromorphism is considered as a permanent or temporary state of water saturation in the soil associated with conditions of reduction (Figure 9.3). This condition is created easily in the soil each time the water stagnates in it and is not renewed. This is, for instance, the case in clayey soils with a slow internal drainage (Aguilar et al., 2003).

9.3 Remote Sensing Data for IV Wetland Characterization

The availability of multiple sensors at different resolution spatially and temporally and access to the scientific community being very easy, it is now the scientists who are exploiting such data for multiple applications. The critical ecosystems services and agroeconomic services provided by the wetlands makes them more important and crucial for conservation

and restoration. In this context, the identification and characterization of IV wetlands becomes a priority to sustain food production to the growing population where cultivable land is becoming scarce and water use is competed by many sectors of the society. Thenkabail and Nolte (1995a,b, 1996) and Thenkabail et al. (2000b) have used different sensors and also new techniques to map and characterize IV ecosystems in West Africa. Gumma et al. (2009) have modeled different layers of information derived from satellite imagery to identify suitable areas for cultivation of rice in the IV wetlands of Ghana. The use of remotely sensed data for such ecosystems also depends on the bio-physical characteristics of the IV wetlands, like the extent of the ecosystem. Morphometric characteristics of the river basin such as drainage network, drainage density, which in turn is dictated by the lithology and soils are also as important in the selection of remotely sensed imagery. Spatial resolution plays an important role in the IV wetland mapping, characterization, and modeling. The level of LULC classification that can be extracted is also dictated by the spatial resolution of the sensor. Especially, spatial resolution of elevation in the form of a DEM will dictate the extraction of stream order in different-sized IV wetlands. Even though water absorption bands like MIR and FIR are also useful to map such wetlands, specific sensors (Rebelo et al., 2009) have been designed to detect wetland areas like the ASTER (VNIR, SWIR and TIR subsystems).

9.4 Study Area and Ecoregional Approach

The 24 WCA nations are a perfect site for IV wetlands mapping and studied at nominal resolution of 30 m for the entire area (Figure 9.7, Table 9.3). The results are reported on an eco-regional basis across the WCA using the climate-length of growing period (LGP) method, FAO/UNESCO soils, and elevation (Figure 9.7). The 18 large ecoregions of 10 million ha or more (Figure 9.7) cover >90% of WCA's geographic area and are identified and mapped based on the definitions provided in Section 9.2 and Figure 9.4. Then, IV wetlands are categorized and characterized using time-series MODIS Terra/Aqua data (Figure 9.5), other temporal and spatial measures, including texture derivatives from very high resolution imagery (e.g., IKONOS, Quickbird, GeoEye; available to us from USGS sources—see data plan), along with other environmental variables derived from topography, soils, and other existing datasets. Information on habitat mapping of the species of flora and fauna that are identified for conservation is also generated. Finally, spatial models are developed to determine IV wetlands most suited for cultivation and conservation. For example, IV wetlands that form an isolated patch may be best to preserve, especially if they are part of a wildlife migration corridor, whereas wetlands near a population center, close to transportation, and with less-developed overstory vegetation may be best to cultivate.

TABLE 9.3 Parameters Describing the Level I Agroecological and Soil Zones

Level IAESZ[a]	Agroecological Zone According to IITA's Definition	LGP[b] (Days)	Major FAO Soil Grouping[c]	Area[d] (Million ha)
1	Northern Guinea savanna	151–180	Luvisols	25.2
2	Southern Guinea savanna	181–210	Luvisols	18.4
3	Southern Guinea savanna	181–210	Acrisols	12.4
4	Southern Guinea savanna	181–210	Ferralsols	11.9
5	Southern Guinea savanna	181–210	Lithosols	10.7
6	Derived savanna	211–270	Ferralsols	47.2
7	Derived savanna	211–270	Luvisols	24.9
8	Derived savanna	211–270	Nitosols	14.2
9	Derived savanna	211–270	Arenosols	14.0
10	Derived savanna	211–270	Acrisols	11.7
11	Derived savanna	211–270	Lithosols	10.8
12	Humid forest	>270	Ferralsols	150.1
13	Humid forest	>270	Nitosols	27.2
14	Humid forest	>270	Gleysols	19.2
15	Humid forest	>270	Arenosols	18.9
16	Humid forest	>270	Acrisols	18.0
17	Midaltitudesavanna[e]		Ferralsols	45.4
18	Midaltitudesavanna[f]		Nitosols	12.3

[a] AESZ, level I agroecological and soil zones.

[b] LGP, length of growing period.

[c] Names refer to the soil classification scheme of FAO/UNESCO (1974).

[d] The area figures are for West and Central Africa and were determined using the "AREA" procedure of IDRISI (Eastman, 1992).

[e] Area distribution of LGP in AEZ 17 is: 151–180 days 11%, 181–210 days 9%, 211–270 days 59%, >270 days 21%.

[f] Area distribution of LGP in AEZ 18 is: 151–180 days 2%, 181–210 days 5%, 211–270 days 53%, >270 days 40%.

9.5 Field Plot Data

We adopted multiple strategies to collect field plot data. *First*, we used a large and rich collection (1023 points) of field plot data on I) wetlands spread across WCA (see distribution and source of these points in Figure 9.6). For each point, we have data on (1) type of wetlands (e.g., hydromorphic, nonhydromorphic), (2) wetland order (e.g., first, second), (3) wetland bottom width, (4) land-use type (e.g., natural or cultivated), (5) moisture level, (6) land-cover percentages (e.g., trees, shrubs, grasses, water body, cultivated), and (7) digital photos. *Second*, through collaboration with CARD/AGRA, CGIAR/CSI, and other African networks of national and international institutes that are actively involved in Africa's wetland issues. These data will be collected during the year 1 project workshop in Africa (jointly hosted with CARD/AGRA, CGIAR/CSI). These data will include IV wetland point data as well as spatial data on socioeconomics and numerous other datasets (e.g., Figure 9.6). *Third*, we will source data from our previous projects in West Africa

(Gumma et al., 2009a; Fujii et al., 2010; Krishna et al., 2010). *Fourth*, very high resolution data (e.g., quickbird, IKONOS) are used as "groundtruth."

9.6 Methods of Rapid and Accurate IV Wetland Mapping of WCA

9.6.1 Existing Methods of Wetland Mapping

There are several studies that discuss methods of wetland mapping using remote sensing (Lyon and McCarthy, 1995; Lunetta and Balogh, 1999; Thenkabail et al., 2000a; Harvey and Hill, 2001; Lyon, 2001; Ozesmi and Bauer, 2002; Hirano et al., 2003; May et al., 2003; Töyrä and Pietroniro, 2005; Wagner et al., 2007; Wright and Gallant, 2007; Gumma et al., 2009; Jones et al., 2009). High levels of accuracy in delineating and mapping wetlands are feasible when multidate, multisensor, very high spatial resolution imagery are used (e.g., Lan and Zhang, 2006; Becker et al., 2007; Gilmore et al., 2008). Ramsey et al.

(1998) found an integrated ERS SAR-optical (TM and CIR) improved the accuracy of wetland classes by up to 20%. The SAR data are sensitive to soil moisture and are quite ideal for delineating lowlands (with high moisture) and uplands (with lower moisture) (Wagner et al., 2007). Recent research (Thenkabail and Nolte, 2000; Kulawardhana et al., 2007; Islam et al., 2008; Jones et al., 2009) demonstrated the ability to attain high levels of accuracy in delineating and mapping wetlands using multiple data. These data include (Table 9.4) (1) Global Land Survey 2005 (GLS 2005) Landsat 30 m, (2) Japanese Earth Resources Satellite Synthetic Aperture Radar (JERS SAR) 100 m, (3) MODIS 250–500 m, (4) Space Shuttle Topographic Mission (SRTM) 90 m, and (5) secondary datasets (e.g., soils).

9.6.2 Automated Methods of Wetland Delineation and Mapping

Automated methods of wetland delineation involve (Table 9.2; Lan and Zhang, 2006; Islam et al., 2008; Jones et al., 2009): (1) algorithms to rapidly delineate wetland streams using SRTM DEM data, (2) thresholds of SRTM-derived slopes, (3) thresholds of spectral indices and wavebands, and (4) automated classification techniques. *First*, wetlands are topographical lowlands and hence the DEM data offer a significant opportunity to delineate lowlands from uplands. Automated methods involving the SRTM-derived wetland boundaries have four known limitations (Islam et al., 2008): (1) generating non-existent or spurious wetlands, (2) providing nonsmooth alignment, (3) resulting in

TABLE 9.4 Automated Methods to Separate Wetlands, including Inland Valley Wetlands, from Non-Wetlands

Index or Parameter	Definition	Range (−1.0 to 1.0 Dimensionless or 0%–100%)	Threshold Values That Best Delineated Wetlands
a. Slope derived from SRTM DEM	This is the percentage slope derived using spatial analyst tools available in ArcGIS	0 to 100	<0.5%
b. Normalized difference vegetation index (NDVI) (Rouse et al., 1974)	$NDVI = \dfrac{\rho_4 - \rho_3}{\rho_4 + \rho_3}$ where ρ_3 and ρ_4 are the reflectance values derived from the bands 3 (red) and 4 (NIR) of Landsat ETM+ data respectively.	−1.0 to +1.0	−0.25 to 0.10
c. Tasseled-cap Wetness Index (TWI) (Crist and Cicone, 1984)	TWI = ([B1] * 0.1509 + [B2] * 0.1973 + [B3] * 0.3279 + [B4] * 0.3406 + [B5] * −0.7112 + [B7] * −0.4572) Where B1 to B7 are the DN values of the respective bands of Landsat ETM+ data. This index represents the overall degree of wetness over the area as reflected by the image data.	0 to 100	0 to 30
d. Normalized difference water index (NDWI) (McFeeters, 1996)	$NDWI = \dfrac{\rho_2 - \rho_4}{\rho_2 + \rho_4}$ where, ρ_2 and ρ_4 are the reflectance values derived from the bands 2 (Green) and 4 (NIR) of Landsat ETM+ data respectively.	−1.0 to +1.0	−0.15 to 0
e. Mid-infrared ratio (MIR) (Coppin and Bauer, 1994)	$MIR = \dfrac{Band\ 4}{Band\ 5}$ where bands 4 and 5 are NIR and mid infrared bands of Landsat ETM+ data respectively.	0 to 4	>0.25
f. Ratio vegetation index (RVI) (Tucker, 1979)	$RVI = \dfrac{Band\ 4}{Band\ 3}$ where bands 4 and 3 are NIR and red bands of Landsat ETM+ data respectively	0 to 6	<0.6
g. Green ratio (GR) (Lo, 1986)	$GR = \dfrac{Band\ 4}{Band\ 2}$ where bands 4 and 2 are NIR and green bands of Landsat ETM+ data, respectively.	0 to 4	0.5 to 0.8
h. Ratio of indices (this study)	RoI = B4/B7 * B4/B3 * B4/B2	0–240	12.5–20
i. Reflectance of SWIR 1 band (this study)	Band 5 where band 5 is the shortwave infrared band 1 of Landsat ETM + data.	0 to 47	<1

spatial dislocation of streams, and (4) absence of stream width. *Second*, the SRTM DEM data are used to derive local slope maps in degrees using the slope function of ArcInfo Workstation GIS. A threshold (Table 9.2) of degree slope provides areas of wetlands or low lying areas and nonwetlands. *Third*, the wetlands in the images can be highlighted by enhancing images (Lyon and McCarthy, 1995; Lunetta and Balogh, 1999). The thresholds of indices and wavebands will automatically delineate wetlands from nonwetlands (Kulawardhana et al., 2007; Schowengerdt, 2007). Numerous researchers have also attempted wetland separation through automated classification techniques on various remotely sensed data (Jensen et al., 1995; Fuller et al., 2006; Lan and Zhang, 2006) without first identifying and separating wetland areas from other land units based on their location in the toposequence. However, as Ozesmi and Bauer (2002) point out, this leads to difficulties of wetland categorization because of spectral confusion (Lan and Zhang, 2006). This is because the automated classification techniques are applied on entire image areas that include wetlands and other land units that often have significantly similar spectral properties. Classification accuracies improve when multitemporal data are used along with ancillary data such as soils and topography (Ozesmi and Bauer, 2002) in GIS modeling framework (Sader et al., 1995; Lyon, 2001; Fuller et al., 2006). Automated methods are rapid, but needs to be supplemented by semiautomated methods to increase accuracies and decrease errors of omissions and commissions.

9.6.3 SemiAutomated Methods of IV Wetland Delineation and Mapping

The semi-automated methods: (1) check any omissions or commissions of IV wetlands derived using automated methods, and (2) apply appropriate corrections to improve the mapping accuracies. The semi-automated methods involve (Thenkabail et al., 2000a): (1) image enhancement techniques involving ratio indices and applying simple thresholds were investigated for delineating wetlands automatically (Table 9.4; Lyon and McCarthy, 1995; Thenkabail and Nolte, 2000; Kulawardhana et al., 2007); (2) enhanced displays in red, green, blue (RGB) false color composites (FCCs) in different combinations of the ETM+ bands were also able to highlight wetland boundaries. The RGB FCCs that best highlight wetlands from other areas (Thenkabail et al., 2000a) were (a) ETM + 4/ETM + 7, ETM + 4/ ETM + 3, ETM + 4/ETM + 2; (b) ETM + 4, ETM + 3, ETM + 5; (c) ETM + 7, ETM + 4, ETM + 2; and (d) ETM + 3, ETM + 2, ETM + 1; and (3) once the images are enhanced (Section 3.2.1) and displayed (Section 3.2.2), they are subjected to object-oriented image analysis using eCognition software and delineate wetlands and nonwetlands (Bock et al., 2005) and then compare the results with the IV wetland maps derived using automated methods. Studies (Kulawardhana et al., 2007; Islam et al., 2008) have established that accuracies between 88% and 97% are attainable using ETM+ and SRTM data and the automated and semiautomated methods.

9.7 Characterization and Classification of IV Wetlands

The IV wetland areas are highlighted using various types of remote sensing and ancillary data (e.g., Figure 9.5). Any of the images with 30 m spatial resolution or better (see Tables 9.1 and 9.2) can be used to delineate, characterize, and map IV wetlands based on methods and approaches described in Section 9.3 and its subsection (Table 9.4).

9.7.1 Case Studies of a Location in Côte d'Voire and Entire Ghana

The IV wetland maps using SPOT HRV 20 m resolution image illustrated in Figure 9.8a for a location in Gagnoa, Côte d'Voire (see Figure 9.7 for the location), for the entire country of Ghana (Figure 9.9a) using Landsat ETM+ 30 m data in Figure 9.9b, and for a selected area within Ghana showing comparison between ETM+ derived versus IKONOS-derived IV wetlands (Figure 9.9c) are derived using the different methodologies explained in this chapter. These wetlands are then classified using optimized layered classification for monitoring wetland vegetation dynamics (Lan and Zhang, 2006; Wright and Gallant, 2007) using standard classification scheme such as the USGS Anderson (Table 9.5). The land-use categories derived from the imagery in this study are uplands, valley fringes, valley bottoms, and others. An equivalent level 1 class of the USGS classification systems is also compared. Since the classification systems used in this study is within the IVs and focused on agriculture as of the USGS system at different levels, it appropriately matches with the present study. It can also be seen that the toposequence followed in the classification system clearly shows the type of land-use/land-cover in the IVs. If we compare the class "significant farmland" in the uplands, it is agricultural land in the USGS system, in the valley fringes it is either agricultural land or range land due to the slope condition. Similarly in the valley bottoms, they are classified as wetlands in the USGS system, which can be potential rice cropland. A comparison to a standard classification system always helps in relating the different systems at different levels but also connects across scales. A glance at the statistics (Table 9.6) reveals the distribution of LULC in the study area. Even though the uplands occupy around 40% of the total area, the valley fringes and valley bottoms total to 58%, which can be potential rice croplands. The resulting outcome is shown for the Gagnoa, Côte d'Voire study area, in Figure 9.8b (with legend in Figure 9.8b and class bispectral plots in Figure 9.8c). Figure 9.10 shows the approach of using the tassel cap bispectral plots of the

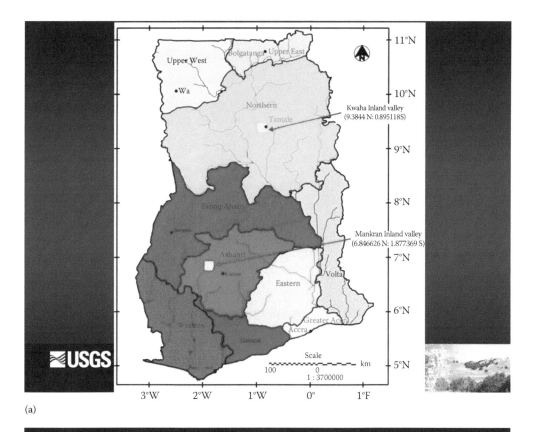

(a)

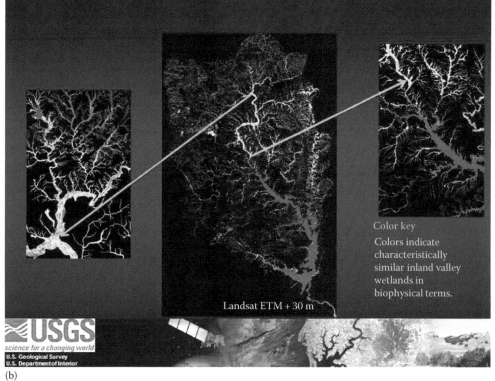

(b)

FIGURE 9.9 (a) Map of the Ghana. Inland valley (IV) wetlands were mapped for the entire country using Landsat ETM+ images. The Mankran and Kwaha study areas, greater details of IV wetlands, were studied using Qucikbird imagery. (b) Inland valley wetlands were delineated using Landsat ETM+ imagery based on semiautomated methods described in this chapter. Results showed that 11.4% (2,714,946 ha) of the total geographic area (23,853,300 ha) of Ghana was IV wetlands. Only 5% (130,000 ha) of IV wetlands is currently cultivated. *(Continued)*

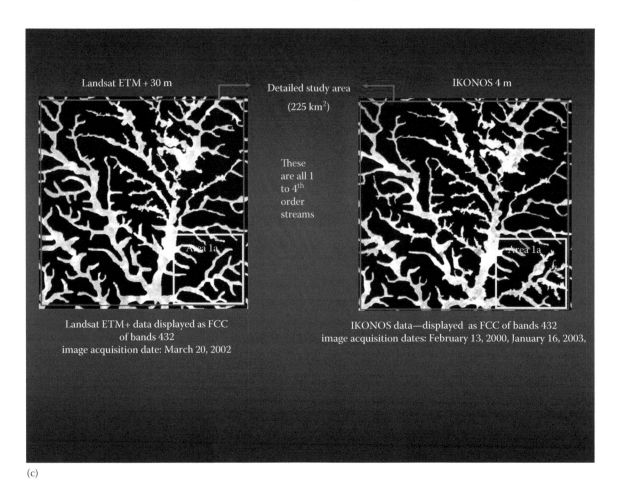

(c)

FIGURE 9.9 *(Continued)*   (c) Inland valley wetlands mapped for the Mankran, Kumasi, Ghana study area (225 km²). The left image is derived from Landsat ETM+ 30 m and the right image using IKONOS 4 m. Total area of IV wetlands was determined as 27.72% using Landsat ETM+ and 28.50% using IKONOS.

land-use classes to define and separate various distinct classes. Table 9.7 provides the percentage land-cover types in each of the 16 land-use classes.

Other classifications, not presented here, can include rule-based wetland mapping using fused MODIS, Landsat, secondary data like GDEM (e.g., Figure 9.5), and wetland change probability mapping (Nielsen et al., 2008; Wdowinski et al., 2008). Incorporating geostatistical evaluation of fine-scale spatial structure (e.g., Wallace and Marsh, 2005) will stratify wetlands based on overall canopy characteristics. Clustering algorithms, such as canonical correlation, will be used to group the wetlands into similar types based on various suites of environmental variables and their derivatives. The goal is to quantify various characteristics of the wetlands so that they can be compared for suitability given a set of criteria. For example, if two wetlands differ in total canopy cover but are otherwise similar, it may be preferable to develop the wetland with less canopy since the cost for clearing the land would be lower. It is also to be noted that even though the canopy cover

decides the type of action taken on it, the amount biological diversity in that wetland needs to be considered before any action is taken.

9.8 Spatial Data-Weights-Models for Identifying Areas for Agriculture versus Conservation

The goal of the spatial modeling (e.g., Figure 9.10a through c) is to pin-point IV wetland areas most suited for (1) cultivation and (2) conservation using spatial data layers (Figure 9.10a) and their relative weights (Table 9.8). For example, as a result of our extensive knowledge of the wetlands of WCA (see Fujii et al., 2010), a total of 29 *biophysical, technical, socioeconomic, and eco-environmental* factors (e.g., Table 9.8, Figure 9.10b) are considered important. In this project, weights will be assigned to these spatial data layers (e.g., Table 9.8) using expert knowledge solicited from stakeholder networks (e.g., CARD/AGRA,

TABLE 9.5 Comparison of the Land-Use/Land-Cover Classification System Used in This Study with the USGS Classification System

Classification System Used in This Study		This Classification of USGS			
		Level I		Level II	
Upland					
1	Significant farmlands	2	Agricultural land	21	Cropland and pasture
2	Scattered farmlands	2	Agricultural land, or		
		3	Rangeland		
3	Insignificant farmlands	3	Rangeland	32	Herbaceous rangeland
				33	Mixed rangeland
4	Wetland/marshland	6	Wetland		
5	Dense forest	4	Forest land	43	Mixed forest land
6	Very dense forest	4	Forest land	42	Evergreen forest land
Valley fringe					
7	Significant farmlands	2	Agricultural land, or		
		3	Rangeland, or	33	Mixed range land
		4	Forest land	43	Mixed forest land
8	Scattered farmlands	3	Rangeland, or	33	Mixed rangeland
		2	Agricultural land, or		
		4	Forest land	43	Mixed forest land
9	Insignificant farmlands	4	Forest land, or	43	Mixed forest land
		2	Agricultural land, or		
		3	Rangeland	33	Mixed rangeland
Valley bottom					
10	Significant farmlands	6	Wetland		
11	Scattered farmlands	6	Wetland		
12	Insignificant farmlands	6	Wetland	61	Forested land
Others					
13	Water	5	Water		
14	Built-up area/settlements	1	Urban or built-up land		
15	Roads	1	Urban or built-up land	14	Transportation Communication and utilities
16	Barren land or desert land	7	Barren land		

Source: Anderson, J.R., *A Land Use and Land Cover Classification System for use with Remote Sensor Data* (US Government Printing Office), 1976.

CGIAR CSI, IITA).The *socioeconomic factors* will include accessibility of settlements, road networks, markets, land tenure, labor force, credit systems, extension systems, social customs, gender, rice policy tariff, rice policy subsidy, and farmer's incentives. The models used algebra (e.g., coded in ERDAS modeler; Figure 9.10b) to arrive at the outputs that determined their suitability for cultivation and/or conservation. Two sets of data and four scenarios were considered to arrive at suitable areas in the IV wetlands. A 10 variable dataset where equal weights were assigned to the layers and varying weights for classes within the layers, varying weights for layers and varying weights for classes within layers produced 2 outputs, showing relatively lower area under "suitable" class. A nine-variable dataset with similar scenarios produced higher area under "suitable" class (Figure 9.10c). For example, if two wetlands differ only in their closeness to transportation and markets, it might be preferable to develop the wetland

nearest to the markets. As another example, if two wetlands are similar, but one forms an isolated patch of habitat important for migratory wildlife, that wetland may be prioritized for conservation.

9.9 Accuracies, Errors, and Uncertainties

Thematic accuracy of the wetland maps is assessed through an error matrix analysis and a regression analysis. A number of statistical considerations including appropriate sampling scheme, sample size, and sample unit are considered (Congalton and Green, 2008). Error matrix including overall, producers,' and users' accuracies (Congalton, 2009) are reported. The study used 1023 wetland data points already available with us (e.g., Figure 9.7), as well as data sourced through our African network partners during the project (Figure 9.10c).

TABLE 9.6   Land-Use Distribution in the Study Area[a,b]

| No. | Land-Use Category | Color | Full Study Area | | |
			Area (ha)	Study Area (Percent of Total)	Mean NDVI
	Uplands		157,601	40.1	
1	Significant farmlands	Gray	22,589	5.8	0.29
2	Scattered farmlands	Seafoam	31,992	8.1	0.34
3	Savanna vegetation[c]	Violet	0	0	—
4	Wetlands/marshland	Mocha	7,024	1.8	0.25
5	Dense vegetation	Rose	54,619	13.9	0.34
6	Very dense vegetation	Red-orange	41,377	10.5	0.39
	Valley fringes		158,606	40.3	
7	Significant farmlands	White	26,299	6.7	0.31
8	Scattered farmlands	Pine-green	39,376	10.0	0.32
9	Insignificant farmlands[d]	Red	92,931	23.6	0.38
	Valley bottom		70,638	18.0	
10	Significant farmlands	Cyan	11,490	2.9	0.29
11	Scattered farmlands	Yellow	19,058	4.9	0.33
12	Insignificant farmlands[e]	Magenta	40,090	10.2	0.35
	Others		6,268	1.6	
13	Water	Blue	358	0.1	−0.07
14	Built-up area/settlements	Tan	2,703	0.7	0.11
15	Roads	Navy	2,194	0.5	0.09
16	Barren land or desert lands	Sand	1,013	0.3	0.13

[a] The study area falls entirely into agroecological zone 16 of the level I map (Figure 9.1 and Table 9.1).

[b] For the composition of land-cover types and their distribution in each land-use class see Tables 9.2 and 9.3.

[c] Class 3 occurs only in Guinea savanna zones.

[d] Spectral characteristic of vegetation in class 9 is similar to that of classes 5, 6, and 12; the difference is mainly in the toposequence position.

[e] Mainly riparian vegetation; spectral characteristics of vegetation similar to classes 5, 6, and 9; the difference is mainly in the topo sequence position.

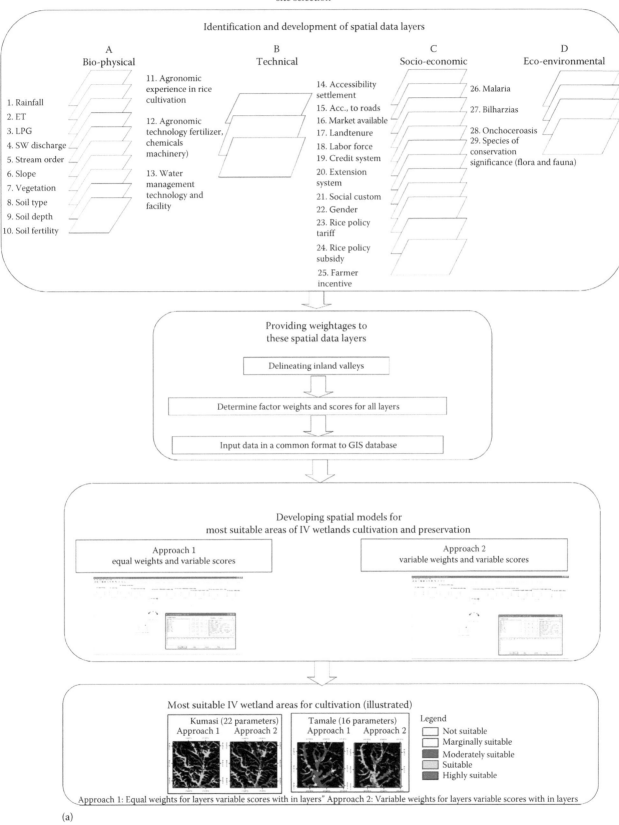

Data sets and spatial modeling framework for best
site selection

Identification and development of spatial data layers

A
Bio-physical

1. Rainfall
2. ET
3. LPG
4. SW discharge
5. Stream order
6. Slope
7. Vegetation
8. Soil type
9. Soil depth
10. Soil fertility

11. Agronomic
experience in rice
cultivation

12. Agronomic
technology fertilizer,
chemicals
machinery)

13. Water
management
technology and
facility

B
Technical

C
Socio-economic

14. Accessibility
settlement
15. Acc., to roads
16. Market available
17. Landtenure
18. Labor force
19. Credit system
20. Extension
system
21. Social custom
22. Gender
23. Rice policy
tariff
24. Rice policy
subsidy
25. Farmer
incentive

D
Eco-environmental

26. Malaria
27. Bilharzias
28. Onchoceroasis
29. Species of
conservation
significance (flora and fauna)

Providing weightages to
these spatial data layers

Delineating inland valleys

Determine factor weights and scores for all layers

Input data in a common format to GIS database

Developing spatial models for
most suitable areas of IV wetlands cultivation and preservation

Approach 1
equal weights and variable scores

Approach 2
variable weights and variable scores

Most suitable IV wetland areas for cultivation (illustrated)

Kumasi (22 parameters)
Approach 1   Approach 2

Tamale (16 parameters)
Approach 1   Approach 2

Legend
☐ Not suitable
☐ Marginally suitable
■ Moderately suitable
☐ Suitable
■ Highly suitable

Approach 1: Equal weights for layers variable scores with in layers" Approach 2: Variable weights for layers variable scores with in layers

(a)

FIGURE 9.10   (a) Spatial model steps involved in selecting the most suitable areas for rice cultivation in IV wetlands.   *(Continued)*

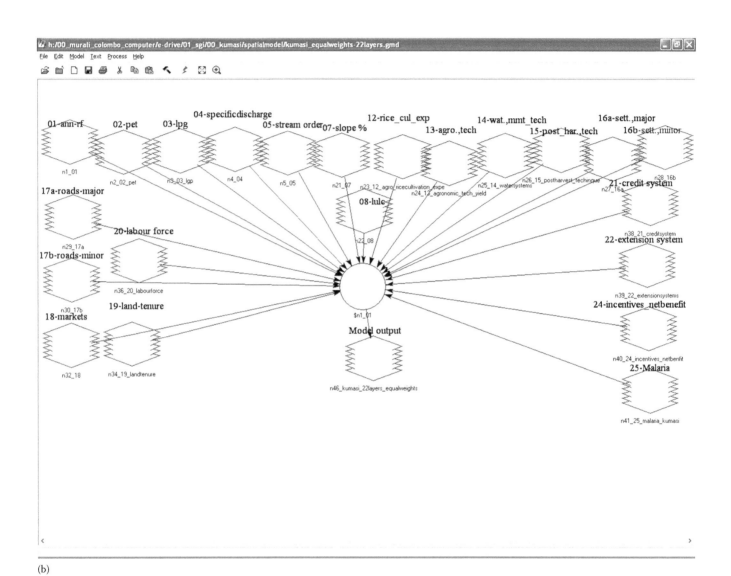

(b)

FIGURE 9.10 (*Continued*)   (b) Illustration of a typical spatial model built in ERDAS.

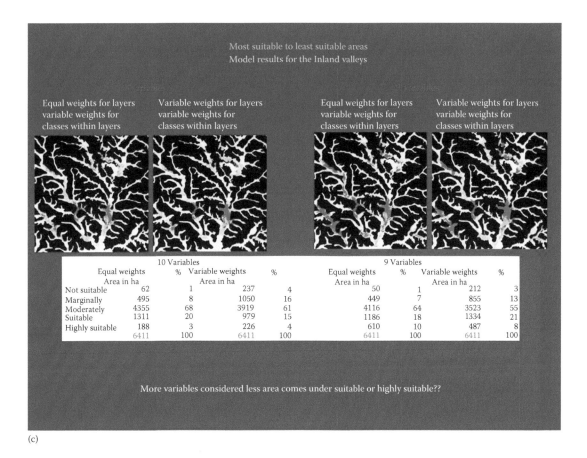

(c)

FIGURE 9.10 (*Continued*)   (c) Most suitable sites for IVs rice cultivation in (A) Kumasi (left) and (B) Tamale (right). For each location, the results and statistics are provided considering 16 variables and 2 approaches: (1) equal weight for layer, variable weight for classes within the layer; and (2) variable weight for layer, variable weight for classes within layer.

TABLE 9.7   Percentage Distribution of Land-Cover Types in the 16 Land-Use Classes for SPOT HRV K:44, J:338 Covering the Region of Gagnoa, Côte d'Ivoire (see the Area in Figure 9.8)

Code of Land-Use Classes[a]	Code of Land-Cover Types									
	1	2	3	4	5	6	7	8	9	10
1		4	14	12	58	12	0			0
2		20	30	25	10	15	0			0
3[b]										
4		21	31	27	4	7	1			9
5		48	25	0	0	0	0			27
6		83	17	0	0	0	0			0
7		10	19	4	57	6	3			1
8		19	39	6	13	6	2			15
9		30	55	5	2	1	1			6
10		7	6	6	60	21	0			0
11		17	6	5	17	0	0			0
12		32	52	11	5	0	0			0
13	100									
14								100		
15									100	
16							100			

[a] See land-use class names in Table 9.5.
[b] Class 3 (savanna vegetation) does not exist in this study area.

TABLE 9.8 Process of Assigning Weightage to Spatial Data Layers and Classes within Each Spatial Data Layer Based on Expert Opinion from Stakeholders

		Weight	Scores
A. Biophysical variables			
1	Rainfall	1	
	<700		1
	700–1,000		2
	1,000–1,300		3
	1,300–1,600		4
	>1,600		5
2	ET	1	
	<700		1
	700–1,000		2
	1,000–1,300		3
	1,300–1,600		4
	>1,600		5
3	LGP	1	
	90–120		1
	120–150		1
	150–180		2
	180–210		3
	210–240		4
	240–270		5
	>270		5
4	Water resources: surface water unit discharge	2	
	Very high		5
	High		4
	Moderate		3
	Low		2
	Very low		1
5	Water resources: stream order	5	
	1		1
	2		2
	3		3
	4		4
	5		5
6	Slope	1	
	<0.5		5
	0.5–1		5
	1–1.5		4
	1.5–2		3
	2.0–3.0		2
	3.0–5		1
	>5		1
7	Vegetation	4	
	Dense forest natural vegetation		1
	Fragmented natural vegetation		2
	Moderate natural vegetation		3
	Sparse natural vegetation		4
	Fallow lands and farmlands		5
8	Soil type	3	

(Continued)

TABLE 9.8 (*Continued*)   Process of Assigning Weightage to Spatial Data Layers and Classes within Each Spatial Data Layer Based on Expert Opinion from Stakeholders

		Weight	Scores
	Type 1		5
	Type 2		4
	Type 3		3
	Type 4		2
	Type 5		1
9	Soil depth	4	
	<10		1
	10–20		2
	20–30		3
	30–40		4
	>40		5
10	Soil fertility	3	
	Type 1		5
	Type 2		4
	Type 3		3
	Type 4		2
	Type 5		1
B. Technical factors	Water quality		
11	Agronomic experience in rice cultivation	2	
	<2 years experience		1
	2–5 years experience		2
	5–10 years experience		3
	10–15 years experience		4
	>15 years experience		5
12	Agronomic technology (fertilizer, chemicals, machinery)	3	
	Very high tech		5
	High tech		4
	Moderate		3
	Low tech		2
	Very low tech		1
13	Water management technology and facility	3	
	Major irrigation canal systems		5
	Minor canal systems		4
	Pump and lift irrigation		3
	Dug well and manual		2
	Rainfed		1
C. Socio-economic factors	Postharvest		
14a	Accessibility settlements: major (>500 people)	5	
	<500 m		5
	500 m–1,000 m		4
	1,000–2,000		3
	2,000–4,000		2
	>4,000		1
14b	Accessibility settlements: minor (<500 people)	3	
	<500 m		3
	500–1,000 m		2
	1,000–2,000		1
	2,000–4,000		1
	>4,000		1

(*Continued*)

TABLE 9.8 (*Continued*)   Process of Assigning Weightage to Spatial Data Layers and Classes within Each Spatial
Data Layer Based on Expert Opinion from Stakeholders

		Weight	Scores
15a	Accessibility roads: major	3	
	<500 m		5
	500–1,000 m		4
	1,000–2,000		3
	2,000–4,000		2
	>4,000		1
15b	Accessibility roads: minor	1	
	<500 m		3
	500–1,000 m		2
	1,000–2,000		1
	2,000–4,000		1
	>4,000		1
16a	Market: major (>50,000 people): define by size of settlement	3	
	<500 m		5
	500–1,000 m		4
	1,000–2,000		3
	2,000–4,000		2
	>4,000		1
16b	Market: moderate (10,000–50,000 people) define by size of settlement	2	
	<500 m		4
	500–1,000 m		3
	1,000–2,000		2
	2,000–4,000		1
	>4,000		1
16c	Market: minor (2,000–10,000 people) define by size of settlement	1	
	<500 m		3
	500–1,000 m		2
	1,000–2,000		1
	2,000–4,000		1
	>4,000		1
17	Land tenure	3	
	Ownership individual		5
	Ownership community/family		4
	Lease < 80 GHC per ha		3
	Lease 80–100		2
	Lease > 100		1
18	Labor force	3	
	Labor force enough		5
	Labor force OK		3
	Labor force shortage		2
	Labor force extremely short		1
19	Credit systems	3	
	Credit fully available		5
	Credit available		4
	Credit difficult		3
	Credit very difficult		2
	Credit not available		1

(*Continued*)

TABLE 9.8 (*Continued*)   Process of Assigning Weightage to Spatial Data Layers and Classes within Each Spatial Data Layer Based on Expert Opinion from Stakeholders

		Weight	Scores
20	Extension system	1	
	Available		5
	Inadequate		3
	Not available		1
21	Social customs	1	
22	Gender	3	
	Female gender obstacle		1
	Female gender not obstacle		3
	Male gender obstacle		1
	Male gender not obstacle		3
23	Rice policy tariff	3	
	No tariff		1
	Tariff 10%		2
	Tariff 10%–20%		3
	Tariff 21%–30%		4
	Tariff > 30%		5
24	Rice policy subsidy	4	
	No subsidy		1
	Low subsidy		2
	Moderate subsidy		3
	High subsidy		5
25	Farmers' incentive		
D. Ecoenvironmental factors			
26	Malaria	2	
	Very high incidence		1
	High incidence		2
	Moderate incidence		3
	Low incidence		4
	Negligible incidence		5
27	Bilhazias	1	
	Very high incidence		1
	High incidence		2
	Moderate incidence		3
	Low incidence		4
	Negligible incidence		5
28	Onchocercasis	3	
	Very high incidence		1
	High incidence		2
	Moderate incidence		3
	Low incidence		4
	Negligible incidence		5
29	Species of conservation significance flora and fauna		
	Critically endangered		1
	Endangered species		2
	Vulnerable		3
	Not endangered		5

Illustrated for IV wetlands of Ghana.

9.10 Conclusions

The chapter provides a comprehensive overview of mapping inland valley (IV) wetlands of Africa using remote sensing and GIS. Wetlands are in cusp of development *versus* preservation debate in Africa. Africa's food security, especially given that its population is projected to be four times (reaching about four billion) by the year 2100 relative to its present population of little over one billion, calls for urgent need to utilize inland valley wetlands for agriculture. At the same time, preserving the unique flora and fauna and the carbon sequestered in the wetlands is of utmost importance.

First, the chapter provides a roadmap for consistent IV wetland characterization and mapping at various spatial resolutions using a multitude of remote sensing data. For this, the chapter uses West and Central African (WCA) nations as case studies. *Second*, the chapter demonstrates wetland land-use/land-cover classification and study of their time-series phenological characteristics (Gumma et al., 2011a, 2014). *Third*, the remote sensing-derived products along with secondary data (e.g., length of growing period, soils, slope, elevation, temperature, agroecological zones), as well as a number of other data such as the biophysical data, socioeconomic data were assigned weights by experts for their importance and then harmonized, standardized, and built into a decision support spatial model that pin-pointed IV wetland areas that are (1) best suited for cultivation and (2) prioritized for conservation.

The chapter shows approaches and methods of utilizing EO for the purposes of (1) understanding inland valley wetlands as land units for Africa's green and blue revolution, and (2) balancing inevitable developmental activities with environmental/ecological solutions that inform which areas to preserve and which areas to develop. The outputs and outcomes of such a study is expected to benefit: (1) farmers to make decisions on where to focus their IV wetland agriculture based on pin-pointed areas most suitable for cultivation; (2) national governments to make decisions on promoting IV wetland cultivation and conservation; (3) financial institutions (e.g., African Development Bank) to make educated decisions on where to invest to fast forward Africa's green and blue revolution; and (4) researchers and NGOs working in Africa.

References

Aguilar, J., Fernandez, J., Dorronsoro, C., Stoops, G., Dorronsoro, B. 2003. Hydromorphy in soils. http://edafologia.ugr.es/hidro/indexw.

Anderson, J.R. 1976. *A land use and land cover classification system for use with remote sensor data* (US Government Printing Office).

Andriesse, W., Fresco, L., Van Duivenbooden, N., Windmeijer, P. 1994. Multi-scale characterization of inland valley agroecosystems in West Africa. *NJAS Wageningen Journal of Life Sciences* 42, 159–179.

Becker, B.L., Lusch, D.P., Qi, J. 2007. A classification-based assessment of the optimal spectral and spatial resolutions for Great Lakes coastal wetland imagery. *Remote Sensing of Environment* 108, 111–120.

Bock, M., Xofis, P., Mitchley, J., Rossner, G., Wissen, M. 2005. Object-oriented methods for habitat mapping at multiple scales—Case studies from Northern Germany and Wye Downs, UK. *Journal for Nature Conservation* 13, 75–89.

Congalton, R.G. 2009. Accuracy and error analysis of global and local maps: Lessons learned and future considerations. *Remote Sensing of Global Croplands for Food Security*, 441. In Chapter 19, *Remote Sensing of Global Croplands for Food Security*. CRC Press, 441p.

Congalton, R.G., Green, K. 2008. *Assessing the Accuracy of Remotely Sensed Data: Principles and Practices.* CRC Press, Boca Raton, FL.

Eastman, J.R. 1992. IDRISI: Version 4.0, March 1992. Clark University, Graduate School of Geography, Worcester, MA.

ESA. 2014. Europian Space Agency. http://www.esa.int/Our_Activities/Observing_the_Earth/World_Wetlands_Day_focuses_on_agriculture (accessed in May 6, 2014).

FAO. 2005. Limpopo Basin Profile report. http://www.arc.agric.za/limpopo/profile.htm (accessed in May 1, 2014).

Fujii, H., Gumma, M., Thenkabail, P., Namara, R. 2010 August. Suitability evaluation for lowland rice in inland valleys in West Africa. *Transactions of the Japanese Society of Irrigation, Drainage and Rural Engineering* 78(4), 47–55.

Fuller, L.M., Morgan, T.R., Aichele, S.S. 2006. Wetland delineation with IKONOS high-resolution satellite imagery, Fort Custer Training Center, Battle Creek, Michigan, 2005 (accessed in March 3, 2006). Scientific Investigations Report. United States Geological Survey, Reston, VA, p. 20.

Gilmore, M.S., Wilson, E.H., Barrett, N., Civco, D.L., Prisloe, S., Hurd, J.D., Chadwick, C. 2008. Integrating multi-temporal spectral and structural information to map wetland vegetation in a lower Connecticut River tidal marsh. *Remote Sensing of Environment* 112, 4048–4060.

Gliessman, S.R. 2007. *Agroecology: The Ecology of Sustainable Food Systems.* CRC Press, Boca Raton, FL.

Gumma, M.K., Nelson, A., Thenkabail, P.S., Singh, A.N. 2011a. Mapping rice areas of South Asia using MODIS multitemporal data. *Journal of Applied Remote Sensing* 5, 053547.

Gumma, M.K., Thenkabail, P.S., Fujii, H., Namara, R. 2009. Spatial models for selecting the most suitable areas of rice cultivation in the Inland Valley Wetlands of Ghana using remote sensing and geographic information systems. *Journal of Applied Remote Sensing* 3, 033537.

Gumma, M.K., Thenkabail, P.S., Hideto, F., Nelson, A., Dheeravath, V., Busia, D., Rala, A. 2011b. Mapping irrigated areas of Ghana using fusion of 30 m and 250 m resolution remote-sensing data. *Remote Sensing* 3, 816–835.

Gumma, M.K., Thenkabail, P.S., Maunahan, A., Islam, S., Nelson, A. 2014. Mapping seasonal rice cropland extent and area in the high cropping intensity environment of Bangladesh using MODIS 500 m data for the year 2010. *ISPRS Journal of Photogrammetry and Remote Sensing* 91, 98–113.

Harvey, K., Hill, G. 2001. Vegetation mapping of a tropical freshwater swamp in the Northern Territory, Australia: A comparison of aerial photography, Landsat TM and SPOT satellite imagery. *International Journal of Remote Sensing* 22, 2911–2925.

Heidhues, F., Atsain, A., Nyangito, H., Padilla, M., Ghersi, G., Vallée, L. 2004. Development strategies and food and nutrition security in Africa: An assessment. International Food Policy Research Institute, Washington, DC.

Hetzel, M.W., Iteba, N., Makemba, A., Mshana, C., Lengeler, C., Obrist, B., Schulze, A., Nathan, R., Dillip, A., Alba, S. 2007. Understanding and improving access to prompt and effective malaria treatment and care in rural Tanzania: The ACCESS Programme. *Malaria Journal* 6, 83.

Hirano, A., Madden, M., Welch, R. 2003. Hyperspectral image data for mapping wetland vegetation. *Wetlands* 23, 436–448.

Islam, M.A., Thenkabail, P., Kulawardhana, R., Alankara, R., Gunasinghe, S., Edussriya, C., Gunawardana, A. 2008. Semi-automated methods for mapping wetlands using Landsat ETM+ and SRTM data. *International Journal of Remote Sensing* 29, 7077–7106.

Jensen, J.R., Rutchey, K., Koch, M.S., Narumalani, S. 1995. Inland wetland change detection in the Everglades Water Conservation Area 2A using a time series of normalized remotely sensed data. *Photogrammetric Engineering and Remote Sensing* 61, 199–209.

Jones, K., Lanthier, Y., van der Voet, P., van Valkengoed, E., Taylor, D., Fernández-Prieto, D. 2009. Monitoring and assessment of wetlands using Earth Observation: The GlobWetland project. *Journal of Environmental Management* 90, 2154–2169.

Krishna, G.M., Prasad, T.S., Bubacar, B. 2010. Delineating shallow ground water irrigated areas in the Atankwidi Watershed (Northern Ghana, Burkina Faso) using Quickbird 0.61–2.44 meter data. *African Journal of Environmental Science and Technology* 4, 455–464.

Kulawardhana, R.W., Thenkabail, P.S., Vithanage, J., Biradar, C., Islam, M.A., Gunasinghe, S., Alankara, R. 2007. Evaluation of the Wetland Mapping Methods using Landsat ETM+ and SRTM Data. *Journal of Spatial Hydrology* 7(2), 62–96.

Lafferty, K.D. 2009. The ecology of climate change and infectious diseases. *Ecology* 90, 888–900.

Lal, R., Hansen, D.O., Uphoff, N. 2002. *Food Security and Environmental Quality in the Developing World*. CRC Press, Boca Raton, FL.

Lan, Z., Zhang, D. 2006. Study on optimization-based layered classification for separation of wetlands. *International Journal of Remote Sensing* 27, 1511–1520.

Lo, T.H.C., Scarpace, F.L., Lillesand, T.M. 1986. Use of multitemporal spectral profiles in agricultural land-cover classification. *Photogrammetric Engineering and Remote Sensing* 52, 535–544.

Lunetta, R.S., Balogh, M.E. 1999. Application of multitemporal Landsat 5 TM imagery for wetland identification. *Photogrammetric Engineering and Remote Sensing* 65, 1303–1310.

Lyon, J. 2001. *Wetland Landscape Characterization: Techniques and Applications for GIS Mapping, Remote Sensing, and Image Analysis*. CRC Press, Boca Raton, FL, 135pp.

Lyon, J., McCarthy, J. 1995. *Wetland and Environmental Applications of GIS*. CRC/Lewis Publishers, Boca Raton, FL, 368pp.

MAW. 2014. Map of African wetlands, African wetlands. http://africa.wetlands.org/Africanwetlands/ (accessed in May 6, 2014).

May, D., Wang, J., Kovacs, J., Muter, M. 2003. Mapping wetland extent using IKONOS satellite imagery of the O'Donnell point region, Georgian Bay, Ontario. *Proceedings, 25th Canadian Symposium on Remote Sensing, Canadian Aeronautics and Space Institute*, Kanata, Ontario, Canada, CD-9pp.

Mitsch, W.J., Gosselink, J.G. 2007. *Wetlands*. John Wiley & Sons, New York, 582pp.

Nielsen, E.M., Prince, S.D., Koeln, G.T. 2008. Wetland change mapping for the US mid-Atlantic region using an outlier detection technique. *Remote Sensing of Environment* 112, 4061–4074.

Ozesmi, S.L., Bauer, M.E. 2002. Satellite remote sensing of wetlands. *Wetlands Ecology and Management* 10, 381–402.

Pelley, J. 2008. Can wetland restoration cool the planet? *Environmental Science and Technology* 42, 8994–8994.

RAMSAR. 2004. RAMSAR convention. www.ramsar.org (accessed in May 1, 2014).

Ramsey III, E.W., Nelson, G.A., Sapkota, S.K. 1998. Classifying coastal resources by integrating optical and radar imagery and color infrared photography. *Mangroves and Salt Marshes* 2, 109–119.

Rebelo, L.-M., Finlayson, C.M., Nagabhatla, N. 2009. Remote sensing and GIS for wetland inventory, mapping and change analysis. *Journal of Environmental Management* 90, 2144–2153.

Sader, S.A., Ahl, D., Liou, W.-S. 1995. Accuracy of Landsat-TM and GIS rule-based methods for forest wetland classification in Maine. *Remote Sensing of Environment* 53, 133–144.

Scholes, R. 1990. The influence of soil fertility on the ecology of southern African dry savannas. *Journal of Biogeography* 17, 415–419.

Schowengerdt, R. 2007. *Remote Sensing: Models and Models for Image Processing*. Elsevier, New York.

Thenkabail, P., Biradar, C., Noojipady, P., Dheeravath, V., Li, Y., Velpuri, M., Gumma, M. et al. 2009a. Global irrigated area map (GIAM) for the end of the last millennium derived from remote sensing. *International Journal of Remote Sensing* 30(14), 3679–3733.

Thenkabail, P., Hanjra, M., Dheeravath, V., Gumma, M. 2010. A holistic view of global croplands and their water use for ensuring global food security in the 21st century through advanced remote sensing and non-remote sensing approaches. *Remote Sensing* 2, 211–261.

Thenkabail, P., Lyon, G., Huete, A. 2011. Advances in hyperspectral remote sensing of vegetation and agricultural croplands. CRC Press/Taylor & Francis Group, Boca Raton, FL.

Thenkabail, P., Lyon, J.G., Turral, H., Biradar, C. 2009b. *Remote Sensing of Global Croplands for Food Security*. CRC Press, Boca Raton, FL.

Thenkabail, P.S., Nolte, C. 1995a. Regional characterization of inland valley agroecosystems in Save, Bante, Bassila, and Parakou Regions in south-central Republic of Benin. Inland Valley Characterization Report 1. Resource and Crop Management Division. International Institute of tropical Agriculture, Ibadan, Nigeria, 60pp.

Thenkabail, P.S., Nolte, C. 1995b. Mapping and characterizing inland valley agroecosystems of west and central Africa: A methodology integrating remote sensing, global positioning system, and ground-truth data in a geographic information systems framework. RCMD Monograph No. 16. International Institute of Tropical Agriculture, Ibadan, Nigeria, 62pp.

Thenkabail, P.S., Nolte, C. 1996. Capabilities of Landsat-5 Thematic Mapper (TM) data in regional mapping and characterization of inland valley agroecosystems in West Africa. *International Journal of Remote Sensing* 17, 1505–1538.

Thenkabail, P.S., Nolte, C. 2000. Regional characterisation of inland valley agroecosystems in West and central Africa using high-resolution remotely sensed data. In: *GIS Applications for Water Resources and Watershed Management* (John G. Lyon, eds.). Taylor & Francis, New York, Book Chapter #8, pp. 77–100, 266.

Thenkabail, P.S., Nolte, C., Lyon, J.G. 2000a. Remote sensing and GIS modeling for selection of a benchmark research area in the inland valley agroecosystems of West and Central Africa. *Photogrammetric Engineering and Remote Sensing* 66, 755–768.

Thenkabail, P.S., Smith, R.B., De Pauw, E. 2000b. Hyperspectral vegetation indices and their relationships with agricultural crop characteristics. *Remote Sensing of Environment* 71, 158–182.

Tiner, R.W. 2009. Global distribution of wetlands. *Encyclopedia of Inland Waters*. pp. 526–530.

Töyrä, J., Pietroniro, A. 2005. Towards operational monitoring of a northern wetland using geomatics-based techniques. *Remote Sensing of Environment* 97, 174–191.

Tucker, C.J. 1979. Red and photographic infrared linear combinations for monitoring vegetation, *Remote sensing of Environment* 8(2), 127–150.

Wagner, W., Bloschl, G., Pampaloni, P., Calvet, J.-C., Bizzarri, B., Wigneron, J.-P., Kerr, Y. 2007. Operational readiness of microwave remote sensing of soil moisture for hydrologic applications. *Nordic Hydrology* 38, 1–20.

Wallace, C.S., Marsh, S. 2005. Characterizing the spatial structure of endangered species habitat using geostatistical analysis of IKONOS imagery. *International Journal of Remote Sensing* 26, 2607–2629.

WARDA. 2003. Strategic Plan 2003–2012. WARDA—The Africa Rice Center, Bouaké, Côte d'Ivoire, 56pp.

WARDA. 2006. Medium Term Plan 2007–2009. Charting the Future of Rice in Africa.Africa Rice Center (WARDA), Cotonou, Republic of Benin.

Wdowinski, S., Kim, S.-W., Amelung, F., Dixon, T.H., Miralles-Wilhelm, F., Sonenshein, R. 2008. Space-based detection of wetlands' surface water level changes from L-band SAR interferometry. *Remote Sensing of Environment* 112, 681–696.

Woodhouse, P. 2009. Technology, environment and the productivity problem in African agriculture: Comment on the World Development Report 2008. *Journal of Agrarian Change* 9, 263–276.

Wright, C., Gallant, A. 2007. Improved wetland remote sensing in Yellowstone National Park using classification trees to combine TM imagery and ancillary environmental data. *Remote Sensing of Environment* 107, 582–605.

Zhuang, Q., Melack, J.M., Zimov, S., Walter, K.M., Butenhoff, C.L., Khalil, M.A.K. 2009. Correction to "Global methane emissions from wetlands, rice paddies, and lakes". *EOS Transactions* 90, 92.

Snow and Ice

10

Remote Sensing Mapping and Modeling of Snow Cover Parameters and Applications

Hongjie Xie
University of Texas at San Antonio

Tiangang Liang
Lanzhou University

Xianwei Wang
Sun Yat-sen University

Guoqing Zhang
Chinese Academy of Sciences

Acronyms and Definitions

AMSR-E	Advanced Microwave Scanning Radiometer–Earth Observing System
APHRODITE	Asian Precipitation Highly Resolved Observational Data Integration Toward the Evaluation of Water Resources
ATR	Adjacent temporal reduction
AVHRR	Advanced very high resolution radiometer
DDF	Degree-day factor
E_{pan}	Pan evaporation
ENSO	El Nino Southern Oscillation
EOS	Earth Observing System
F_{SCA}	Fractional snow cover
GOES	Geostationary observational environmental satellite
IMS	Interactive multisensor snow and ice mapping system
MODIS	Moderate-resolution imaging spectroradiometer
MODSCAG	MODIS Snow Covered-Area and Grain size retrieval
NDSI	Normalized difference of snow index
NDVI	Normalized difference of vegetation index
NOHRSC	National Weather Service at the National Operational Hydrologic Remote Sensing Center
NS	Nash–Sutcliffe
SCA	Snow-covered area
SCD	Snow-covered duration
SCED	Snow cover end date
SCI	Snow cover index
SCOD	Snow cover onset date
SSM/I	Special sensor microwave/imager
SWE	Snow water equivalent
TP	Tibetan Plateau

10.1 Introduction

Snowpack or snow-covered area (SCA) is an important component of the hydrologic cycle, especially in mountainous basins where the majority of water originates from snowmelt. Basins with snowmelt water, such as the Western United States and the Tibetan Plateau (TP) and its surrounding areas, are a source of major water resources for agriculture, residential, industry, and many other needs. Due to its high albedo, snow cover (SC) is one of the key variables impacting the Earth's energy balance. Snow is also an excellent insulator because of its low thermal conductivity. Therefore, accurate mapping SC and/or snow water equivalent (SWE) have been a major task of the remote sensing community. Using its unique high spectral signature, snow can be relatively easy to be separated from other types of land cover (Brubaker et al., 2005; Simic et al., 2004). However, the accuracy of SC mapping techniques is affected by variables such as the sensor's spectral resolution, snow depth on the ground, cloud cover, and forest canopy (Hall et al., 1998; Vikhamar and Solberg, 2002). The importance of SC products as a substitute for ground stations/sensors is especially underscored in large areas where the ground station density is sparse or inaccessibility of the terrain due to the remoteness of the region and large spatial extent.

In situ measurements of snow depth and SWE can provide good snow depth observations at localized points and can be used as ground-truth data to validate remote sensing products for snow area extent (Wang et al., 2008; Zhou et al., 2005). However, these stations are limited in density and are usually located near open urban areas or at lower elevations, which can often bias the SC conditions at the mountainous areas or higher-elevation regions. Furthermore, these in situ stations provide point measurements of snow depth rather than the extent of SC. In some instances, airborne measurements are used for SC mapping, but these measurements are limited in time and space. The aforementioned requirements highlight the need for satellite-based SC products that provide a near-continuous mapping of snow, both in space and time.

SC has been derived from many remote sensors such as the Landsat imagery, special sensor microwave/imager, Advanced Microwave Scanning Radiometer–Earth Observing System (AMSR-E), geostationary observational environmental satellite, and advanced very high resolution radiometer. The last two sensors are used by the U.S. National Weather Service at the National Operational Hydrologic Remote Sensing Center in a physically based snow model to produce a daily gridded SC maps for the continental United States at the spatial resolution of 1 km. This product is used as a reference for SC mapping in the United States and has been used to validate other satellite SCA products, such as the daily moderate-resolution imaging spectroradiometer (MODIS) SC at 500 m resolution (Klein and Barnett, 2003; Maurer et al., 2003). Another satellite-based SCA product is the interactive multisensor snow and ice mapping system (IMS), which is produced globally at a daily temporal resolution and at two spatial resolutions (4 and 24 km). This product is cloud-free but has better spatial resolution than passive microwave remote sensing (such as AMSR-E) (Brubaker et al., 2005) and has shown compatible snow mapping accuracy as compared with MODIS (Mazari et al., 2013).

In this chapter, we primarily introduce MODIS, the mostly accurate, relatively high resolution, and widely used SC products in the world, for its application and modeling. MODIS flown on the Earth Observing System (EOS) Terra and Aqua platforms provides the most recent and most advanced SCA products for the global coverage (Hall et al., 2002).

10.2 Principles of MODIS Snow Cover Mapping and Standard Products

Snow has high reflectance in the visible wavelength (0.3–0.7 μm) and low reflectance in the mid-infrared wavelength (1.4 μm and longer) (Figure 10.1), with an albedo up to 0.9. This outstanding high albedo makes it possible to be mapped using optical remote sensing technology, separating objects based on radiative response to solar radiation. However, challenging does exist. First, optical remote sensing cannot map the earth surface during polar nights or nights. Second, electromagnetic radiation in the visible range cannot pass through thick cloud and thus cannot see the land surface under cloud; another problem is that thin cloud has similar high reflectance with snow in the visible range, frequently causing confusion between thin cloud and patchy snow, although most cloud has higher reflectance in the near and mid-infrared wavelength range, which is used to distinguish snow from cloud (Frei et al., 2012). Third, vegetations, particularly the dense forest canopies, obstruct visible and near-infrared signal to reach the snow surface under canopies and reduce the snow surface albedo, make it difficult to accurately detect snow depth and extent (Klein et al., 1998; Nolin, 2004). Lastly, surface heterogeneity, such as in mountainous areas or in the polar regions with wetlands and lakes, makes it difficult to map snow using moderate resolution visible and near-infrared imagery without high-resolution (tens of meters) land surface data (Frei and Lee, 2010).

10.2.1 Binary/Fractional Snow Cover Mapping

Similar as the normalized difference of vegetation index (NDVI) for vegetation mapping, normalized difference of snow index (NDSI) is used to map snow from no snow (Dozier, 1989; Hall et al., 1995). It was first tested using Landsat images: band 2 (0.52–0.60 μm) and band 5 (1.55–1.75 μm), with compatible wavelengths from MODIS, band 4 (0.545–0.565 μm) and band 6 (1.628–1.652 μm) for Terra MODIS (Equation 10.1), and band 7 (2.105–2.155 μm) for Aqua MODIS, due to the failure of band 6 sensor in Aqua MODIS. The snow mapping algorithm (SNOMAP) tested by Landsat images showed

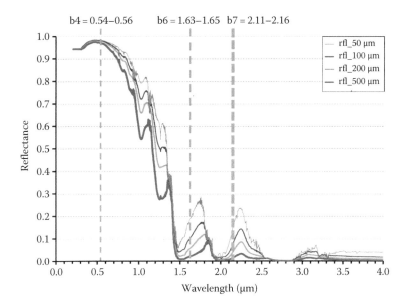

FIGURE 10.1 Snow model–predicted spectral reflectance of dry snow surface under clear sky at the Summit station (72.5794°N, 38.5042°W) of Greenland. Snow density is 250.0 kg m^{-3}, and snow grain radii are 50, 100, 200, and 500 μm from top to bottom, respectively. The three dashed lines from left to right are the wavelength positions of MODIS band 4, band 6, and band 7, respectively, and the width of the line is proportional to the width of wavelength. (Adapted from Wang, X. et al., Spatiotemporal variation of snow cover from space in Northern Xinjiang, in: Chen, Y. (ed.), *Water Resources Research in Northwest China*, Springer, the Netherlands, doi:10.1007/978-94-017-8016-2, 2014b.)

98% accuracy in identifying snow pixels that snow covered by 60% or more (Hall et al., 1995):

$$NDSI = \frac{band\ 4 - band\ 6}{band\ 4 + band\ 6} \quad (10.1)$$

The binary classification or hard classification, which only tells weather a pixel is covered by snow or not, while a pixel with less than 50% of SC might not be mapped as snow. The fractional SC tells the fractional snow (0%–100%) in a pixel. The MODIS' fractional SC (F_{SCA}) is derived based on the empirical relationships Equations 10.2 and 10.3 (Rittger et al., 2013; Salomonson and Appel, 2004, 2006):

$$Terra\ MODIS : F_{SCA} = -0.01 + 1.45 \times NDSI \quad (10.2)$$

$$Aqua\ MODIS : F_{SCA} = -0.64 + 1.91 \times NDSI \quad (10.3)$$

In the standard MODIS SC products, four masks are also used in classifying a certain NDSI value as snow or not snow, including a dense forest stands mask, a thermal mask, a cloud mask, and an ocean and inland water mask. A pixel in a nondensely forested region is mapped as snow if its NDSI is larger than or equal to 0.4 and reflectance in MODIS band 2 (0.841–0.876 μm) is larger than 0.11 and reflectance in MODIS band 4 is larger than 0.1. The latter prevents pixels containing very dark targets such as black spruce forests from being flagged as snow. The detailed algorithm and processing steps have been documented in several sources (Hall et al., 2002; Riggs et al., 2006).

Besides the standard MODIS algorithm and its SC products, the MODIS Snow Covered-Area and Grain size retrieval (MODSCAG) algorithm was developed to retrieve fractional SC and snow grain size based on spectral mixing analysis with MODIS spectral reflectance data (Dozier and Painter, 2004; Nolin et al., 1993; Painter et al., 2009; Sirguey et al., 2009). The algorithm uses the relative shape of the snow spectrum, which is sensitive to the spectral reflectance of the snow fraction. Thus, MODSCAG allows the snow's spectral reflectance to vary pixel by pixel and can address the spatial heterogeneity that characterizes snow and its albedo in mountainous and patchy snow regions. Validation and comparison study shows that the MODSCAG algorithm is more accurate to characterize fractional SC than the MODIS standard fractional SC algorithm Equations 10.2 and 10.3 based on NDSI (Rittger et al., 2013).

10.2.2 MODIS Standard Snow Cover Products and Accuracy

Two MODIS sensors on board the Terra and Aqua satellites as a part of NASA's EOS were launched on December 18, 1999, and May 4, 2002, respectively, with the aim of providing global monitoring of atmospheric, land, and ocean processes. Terra overpasses the equator at around 10:30 a.m. (10:30 p.m.) local time and Aqua at around 1:30 p.m. (1:30 a.m.). MODIS standard SC products are produced as a series (see Figure 10.2), beginning with a swath (scene) product at a nominal pixel spatial resolution of 500 m with nominal swath coverage of 2330 km × 2030 km. The multiple swath observations at 500 m resolution of SC (MOD10_L2) are then projected onto a sinusoidal gridded

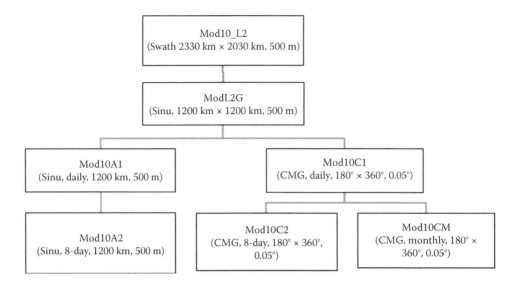

FIGURE 10.2 Terra/Aqua moderate-resolution imaging spectroradiometer (MODIS) standard snow cover products. Aqua MODIS snow cover product's names begin with MYD. (Adapted from Wang, X. et al., Spatiotemporal variation of snow cover from space in Northern Xinjiang, in: Chen, Y. (ed.), *Water Resources Research in Northwest China*, Springer, the Netherlands, doi:10.1007/978-94-017-8016-2, 2014b, Chapter 6.)

tile (1200 km × 1200 km) of MOD10L2G, which is further processed as a sinusoidal 500 m grid of daily (MOD10A1) and 8-day (MOD10A2) composite tile (1200 km × 1200 km) products, or 0.05° global climate modeling grid daily product (MOD10C1), 8-day product (MOD10C2), and monthly product (MOD10CM) (Hall et al., 2002; Riggs et al., 2006).

MODIS standard SC products (both daily and 8-day) have high snow accuracy (over 90%) in clear-sky conditions (Hall et al., 1998, 2007; Klein and Barnett, 2003; Liang et al., 2008b; Riggs et al., 2006; Wang et al., 2008, 2009; Wang and Xie, 2009; Zhou et al., 2005) and only 30%–50% in all-sky conditions (Gao et al., 2010a; Xie et al., 2009). In spite of the replacement of band 6 with the band 7 in Aqua MODIS due to the failure of band 6 sensor, both SC products (MOD10A1 and MYD10A1) have high agreement for cloud-free SC classification (Wang et al., 2009).

10.3 Improved Daily and Flexible Multiday Combinations of Snow Cover Mapping

As mentioned earlier, the MODIS' SC mapping suffers cloud contaminations. Obtaining cloud-free or even cloud percentage less than 10% of MODIS images remains a challenge. Various approaches were proposed to reduce cloud obscuration by improving the cloud mask (Riggs and Hall, 2003) or separating cloud-masked pixels into snow or land, through spatial–temporal or multisensor combinations. The improving the cloud mask method has its limitation since cloud existence is an unchangeable reality. Therefore, the replacing cloud-covered pixels through spatial–temporal or multisensor combinations has been the hot topic for the last 10 years, since the cloud movement is another unchangeable reality. Spatial approaches aim at replacing cloud pixels by the majority of noncloud pixels in

an eight-pixel neighborhood (Parajka and Blöschl, 2008; Tong et al., 2009; Zhao and Fernandes, 2009). This can effectively reduce cloud blockage of daily MODIS snow products by ~7%. But the method is not effective for massive area of cloud cover (i.e., all the neighboring pixels are cloud covered).

Temporal approaches merge daily or multiday MODIS SC products to minimize cloud coverage and maximize snow coverage, by the sacrifice of temporal resolution. Daily combination of Terra and Aqua (TAC) MODIS, which have a 3 h difference, can reduce ~10%–20% cloud cover (Parajka and Blöschl, 2008; Wang et al., 2009; Xie et al., 2009; Yang et al., 2006). The MODIS 8-day SC product (MOD10A2 or MYD10A2) is the representative of fixed-day combined products. Using the same algorithm, some user-defined multiday SC products are produced in fixed temporal windows (Liang et al., 2008b; Parajka and Blöschl, 2008; Yang et al., 2006; Yu et al., 2012). Based on a predefined maximum cloud coverage threshold such as 10%, other multiday composite products with flexible starting and ending dates are produced (Wang et al., 2009; Xie et al., 2009). These products have higher temporal resolution (average 2–3 days/image) and relatively low cloud coverage, but ignore some special situations. For example, when weather conditions remain overcast for over a week, this method may result in a composite product of over 8 days or even several weeks. Hall et al. (2010) developed a new method to fill the cloud gap based on the most recent cloud-free observations for each cloud pixel. This gap-filling strategy is a useful and dynamic method that uses all the most nearest noncloud observations. One disadvantage of this method is that it does not control the cloud percentage for the entire image area.

Multisensor approach can generate cloud-free SC maps through the merging of MODIS and AMSR-E snow products at the expense of spatial resolution. This approach takes advantage of both high spatial resolution of optical sensors and cloud penetration of passive microwave sensors (Foster et al., 2007; Gao et al., 2010a; Hall and Riggs, 2007; Liang et al., 2008a). The snow accuracy of the MODIS/Terra and AMSR-E blended daily SC products is 85.6%, which is much higher than the 30.7% of MODIS daily products in all-sky conditions (Liang et al., 2008a).

Here, we present three unique and improved SC products, based on daily TAC MODIS, that are currently widely used in the snow remote sensing community for various applications.

10.3.1 Flexible Multiday Combination of Snow Cover Mapping

Flexible multiday combination approach is controlled by two thresholds, namely, maximum cloud percentage (P) and maximum composite days (N), user-defined parameters that can be assigned depending on the application (Gao et al., 2010b). It is usually

defined the P as 10% and N as 8 days. The approach first calculates the cloud percentage of the first input image, a daily TAC MODIS. If it is less than the threshold P, the image directly outputs; else, it is combined with the second input image, and the cloud percentage of the study area in the combined image is updated. The process does not stop until either the cloud percentage is less than or equal to the threshold P, or the number of composite days equals to the threshold N. The resultant image from this process is a flexible multiday combination of SC map.

Figure 10.3 shows an example of cloud cover decrease and SC increase for daily TAC on October 10, 2002, and flexible multiday combination from October 7 to 10, 2002, in the Nam Co drainage basin in the TP. The cloud cover is decreased greatly from 27.19% in the origin MODIS/Terra and 21.67% in the MODIS/Aqua SC products to 1.87% in the flexible multiday combined SC mapping, while the SC percentage is improved from the ~45% to ~65%.

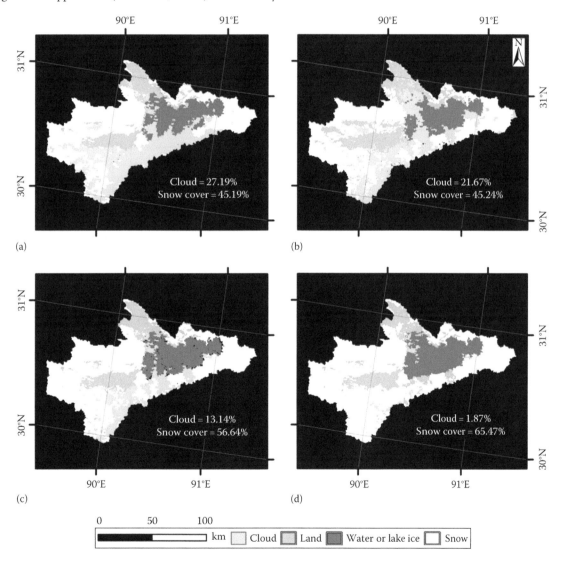

FIGURE 10.3 Cloud and snow cover changes after daily and multiday combinations of moderate-resolution imaging spectroradiometer (MODIS) images in the basin Nam Co of the Tibetan Plateau: (a) MODIS Terra, (b) MODIS Aqua, (c) daily Terra–Aqua combination, and (d) flexible multiday combination of October 07–10, 2002. (Adapted from Zhang, G. et al., *Water Resour. Res.*, 48(10), W10529, 2012.)

10.3.2 Daily Cloud-Free Snow Cover and Snow Water Equivalent Mapping

Although the cloud cover is significantly reduced by combination of daily Terra and Aqua MODIS, it still presents in the SC mapping even after flexible multiday combination. Gao et al. (2010a) presented a method that combines the daily AMSR-E SWE data with the daily TAC MODIS SC to produce cloud-free SC (MODIS Terra and Aqua MODIS combination snow cover [TAC-SC] and AMSR-E combined SC [MAC-SC]) and SWE (MODIS TAC-SC and AMSR-E combined SWE [MAC-SWE]) maps (at 500 m cell size) (Figure 10.4), through redistributing the 25 km grid cell of AMSR-E SWE product into those snow-covered pixels only (500 m). Equation 10.4 is used to calculate the SWE of every subpixel within one AMSR-E pixel:

$$SWE_n = \frac{SWE_o \times 2 \times 2500}{N_{snow}} \tag{10.4}$$

where

SWE_n is the SWE value (mm) of the new MAC-SWE product
SWE_o is the SWE value (mm) of the AMSR-E product
N_{snow} is the number of snow subpixels (500 m) within one AMSR-E snow pixel (25 km)
2 is the scaling factor

For example, the SWE_o value of one AMSR-E pixel is 3 mm, and it actually should be 6 mm because SWE values archived by the NSIDC are scaled down by a factor of 2 for storing in the hierarchical data format–EOS. Within the 2500 subpixels of one AMSR-E pixel, if we suppose 625 subpixels are snow, the SWE_n of the new MAC product should come from those 625 snow subpixels and thus should be 24 mm. This means that the rest of 1875 subpixels have neither SC nor SWE.

Figure 10.5 is a test example to demonstrate the outstanding capability in suppressing cloud obscuration of MODIS SC and increasing spatial resolution of AMSR-E SWE products.

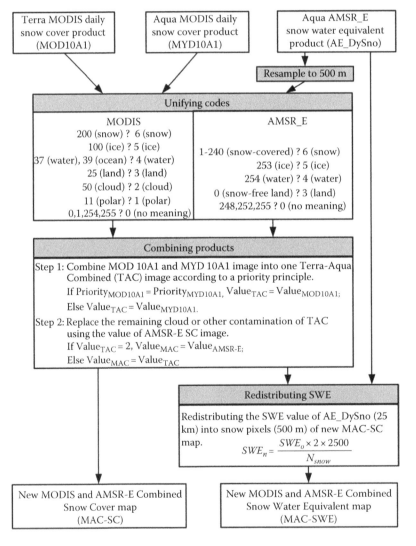

FIGURE 10.4 Flowchart of MODIS and AMSR-E combination (MAC) method for producing daily cloud-free snow cover (MAC-SC) and snow water equivalent (MAC-SWE) at 500 m cell size. (Adapted from Gao, Y. et al., *J. Hydrol.*, 385(1–4), 23, 2010a.)

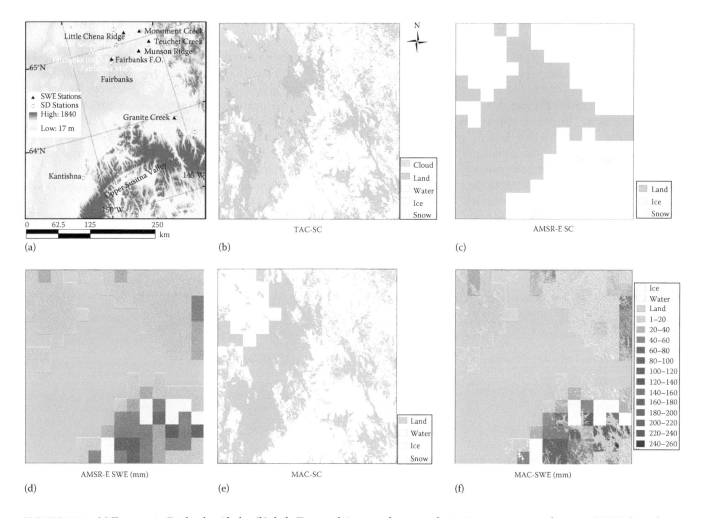

FIGURE 10.5 (a) Test area in Fairbanks, Alaska, (b) daily Terra and Aqua moderate-resolution imaging spectroradiometer (MODIS) combination snow cover (TAC-SC), (c) Advanced Microwave Scanning Radiometer–Earth Observing System (AMSR-E) snow cover, (d) AMSR-E snow water equivalent (SWE), (e) MODIS TAC-SC and AMSR-E combined snow cover (MAC-SC), and (f) redistributed SWE, for October 15, 2006. (Adapted from Gao, Y. et al., *J. Hydrol.*, 385(1–4), 23, 2010a.)

Image B (TAC-SC) is the combination of MOD10A1 and MYD10A1 of October 15, 2006, in which still significant cloud presents. Image E (MAC-SC) is the combination of TAC-SC and AMSR-E SC, in which the cloud cover is completely removed. However, there are some important points to mention. (1) Scattered cloud covers (less than 25 km × 25 km) in Image B are removed in E and the resolution of snow or land indeed increased to 500 m in those regions. (2) Massive and continuous cloud covers in Image B are simply replaced by the snow or no snow of AMSR-E SC product. Although those cloud coverage are removed, the real resolution of the snow or no snow pixels keeps 25 km. For example, the cloud covers in the upper left corner of Image B are now simply AMSR-E SC in the Image E, while the actual resolution of SC or land cover does not change to 500 m. (3) Ice cover (25 km in pixel size) in AMSR-E (Images C and D) is actually nicely increased to 500 m resolution in the Image E.

Image F (MAC-SWE) is the redistribution of D (AMSR-E SWE) based on E (MAC-SC), in which the 25 km SWE in D is now 500 m resolution in F. However, this is true only for those

pixels that have a real spatial resolution of 500 m in E; for those pixels such as in the upper left corner of E still keep the 25 km spatial resolution in F. For example, it is very clear that the upper right portion of the F has true 500 m SWE due to the true 500 m SC in E, while the most of the upper left portion of F has the false 500 m SWE due to the false 500 m SC in E.

The evaluation against in situ observations (Gao et al., 2010a) showed that snow accuracy in all-sky conditions increase from 31% (MYD10A1), 45% (MOD10A1), to 49% (TAC-SC), and 86% (MAC-SC). A similar combination of Terra MODIS and AMSR-E SC product was tested by Liang et al. (2008a) in the north Xinjiang in China against 20 in situ observations. They reported the snow accuracy of Terra MODIS (no Aqua MODIS involved) and AMSR-E combined SC product is 75%, which is much higher than the 34% of MOD10A1 alone, but lower than 86% found in Gao et al. (2010a), since the Aqua MODIS was involved in the latter study to remove cloud. The monthly comparison between MODIS and AMSR-E SC retrievals indicated that good correspondences were found in the land or snow stable period.

However, the AMSR-E maps were not as accurate as the MODIS maps in the snow accumulating and melting period (the early fall and the later spring) due to its coarse resolution. It was also found that for pixels with scattered cloud cover (less than 25 km in size) in the MODIS TAC-SC, the new daily 500 m MAC-SC and SWE products indeed improve the spatial resolution of those pixels to 500 m. For massive cloud cover (larger than 25 km in size), the real resolution of those pixels in the MAC products are actually 25 km, even in 500 m pixel size. Although with these limitations mentioned above, the resultant new daily and cloud-free MAC-SC and SWE maps are great additions to the current NASA's standard MODIS and AMSR-E snow products and are suitable for hydrological and meteorological modeling on a daily basis.

10.3.3 Improved Daily Cloud-Free Snow Cover Mapping

As indicated in Section 10.3.2, directly replacing the massive clouds in daily TAC image with the AMSR-E snow or no snow does remove those clouds. But it does not really increase the resolution, since there might be scattered snow within an AMSR-E pixel but was mapped as no snow due to the 25 km pixel size. Wang et al. (2015) developed an improved algorithm that greatly removes clouds, before combining with AMSR-E. The algorithm is shown in Figure 10.6, through four steps: (1) combination of MOD10A1/Terra and MYD10A1/Aqua (MOYD) (Xie et al., 2009), (2) adjacent temporal reduction based on 3-day (previous, current, and next days) images, (3) snow line method to further separate the remaining cloud into snow or cloud based on elevation and snow information (Parajka et al., 2010), and (4) MAC, for the remaining cloud pixel to replace with AMSR-E snow or land information. The final cloud-free products achieve high snow and overall accuracies (~85% and ~98%, respectively), much higher than those of existing daily SC products in all-sky conditions, and are very closer to or even slightly higher than those in clear-sky condition of the daily MODIS products based on the validation over the TP (Table 10.1) (Wang et al., 2015).

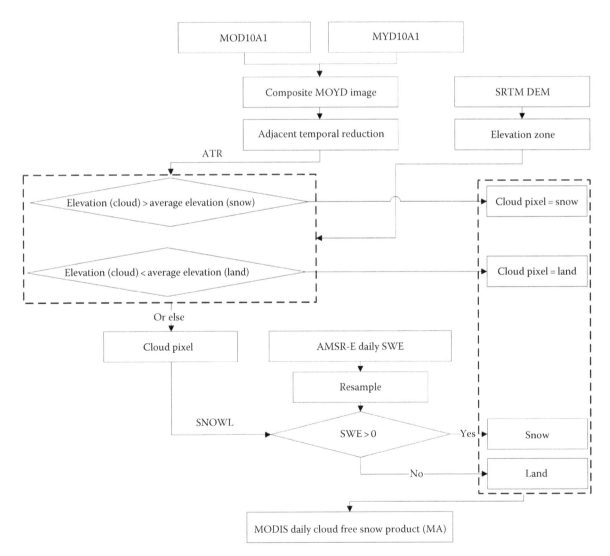

FIGURE 10.6 Flowchart of daily cloud-free snow cover composite image. (Adapted from Wang, W. et al., *Remote Sensing*, 7(1), 169, 2015.)

TABLE 10.1 Snow Classification Accuracy and Errors for Snow-Covered Area Images from 2003 to 2010

							Clear Sky		All Sky	
SCA Image	S-S	S-L	S-C	L-L	L-S	L-C	Snow Accuracy (%)	Overall Accuracy (%)	Snow Accuracy (%)	Overall Accuracy (%)
MOD10A1	1936	448	3855	139735	797	102290	81.21	99.13	31.03	56.88
MYD10A1	1452	613	4326	122914	1486	114580	70.31	98.34	22.72	50.68
MOYD	2552	546	2733	179275	2372	63011	82.38	98.42	43.77	72.59
ATR	3710	663	2183	200816	2461	41204	84.84	98.50	56.59	81.47
SNOWL	4670	781	1736	215291	2558	26068	85.67	98.50	64.98	87.60
MA	4979	875	0	245364	3199	0	85.05	98.40	85.05	98.40

Source: Adapted from Wang, W. et al., *Remote Sensing*, 7(1), 169, 2015.

Note: S, L, and C are, respectively, for snow, land, and cloud; the first letter is for climate station and the second letter for satellite. For example, S-S means that "snow" seen from climate stations is also seen as "snow" from satellite; L-S means that "land" seen from climate stations is seen as snow from satellite.

10.4 Snow Cover Parameters (SCOD, SCMD, SCD, and SCI)

Based on flexible multiday combination of SC or daily cloud-free SC maps, four SC parameters, that is, snow cover index (SCI), snow-covered duration/days (SCD) map, snow cover onset date (SCOD) map, and snow cover end dates (SCED) map, one each per hydrological year, can be further derived to examine the spatiotemporal variations of SC (Gao et al., 2011; Wang and Xie, 2009). The SCI (km² day) contains both SC extent and duration for one hydrological year and can be used to study the SC condition in a yearly basis. To intercompare SC condition for the multiple basins, a normalized SCI (NSCI, unit km² day/km²) can be used (Zhang et al., 2012). The NSCI is defined as the SCI divided by the corresponding basin area, which indicates the mean snow days in each basin. Figure 10.7 shows an example of SCOD, SCED, and SCD maps for the Pacific Northwestern United States for the 2006–2008 hydrological years (Gao et al., 2011). The three SCD maps have very similar patterns even though their corresponding SCOD and SCED maps change a lot from year to year or from area to area. When compared the SCD maps with elevation distribution (Figure 10.7d), it is clear that SCD has high correlation with elevation. SCD are less than 90 days (greenish) in the areas where elevations are lower than

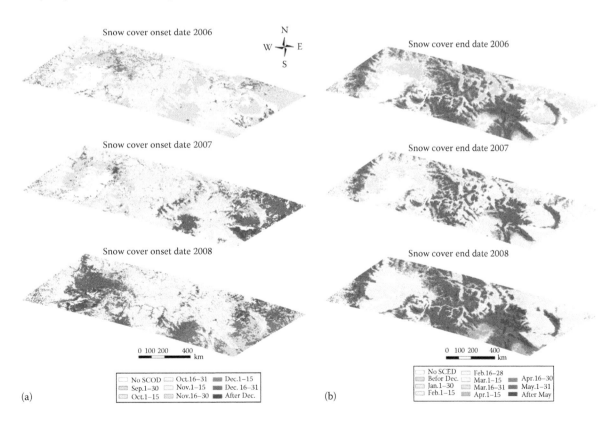

(a)

(b)

FIGURE 10.7 Flexible multiple combination of moderate-resolution imaging spectroradiometer snow cover–derived (a) snow cover onset date, (b) snow cover end date. *(Continued)*

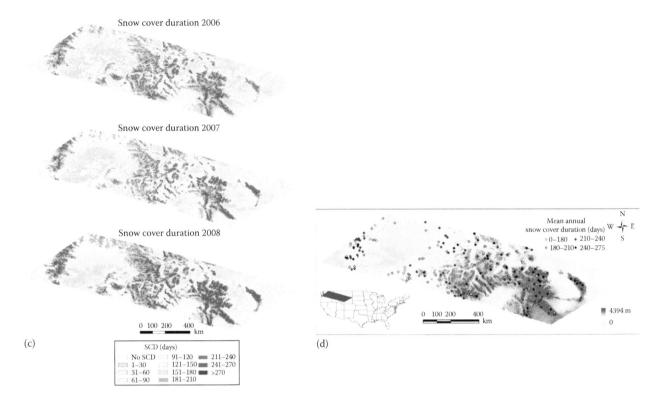

(c)

(d)

FIGURE 10.7 (*Continued*) Flexible multiple combination of moderate-resolution imaging spectroradiometer snow cover–derived (c) snow-covered duration (*SCD*) of (d) the Pacific Northwestern United States for 2006 to 2008 hydrological years. Dots on the digital elevation model map (d) are 244 snow telemetry stations with the mean annual SCDs in four different color groups. (Adapted from Gao, Y. et al., *Photogramm. Eng. Remote Sens.*, 77(4), 351, 2011.)

1000 m. In the middle 1000–2000 m areas, most of SCD is in the range of 90–150 days (yellowish). At the high-elevation areas, SCD is larger than 150 days and even larger than 240 days in mountaintop areas (reddish). At a few mountainous peak areas where elevation is higher than 3000 m, SCD even exceeded 270 days (deep red) especially in 2008.

Figure 10.8 shows an example of the NSCI variations for the four basins (Nam Co, Selin Co, Cedo Caka, Yamzhog Yumco) in the TP

(Zhang et al., 2012). Generally, the Nam Co basin shows the highest NSCI (i.e., the greatest SC condition) among the four basins in all years. The HY2003 shows the highest NSCI for the Nam Co basin and among the four basins. The smallest NSCI for the Nam Co basin occurs in HY2010. For Cedo Caka, Selin Co, and Yamzhog Yumco, the largest NSCI was in 2007. The HY2001 and 2010 show similar small NSCI for Cedo Caka, Selin Co, and Yamzhog Yumco basins, while the smallest NSCI for the Cedo Caka basin was in HY2004.

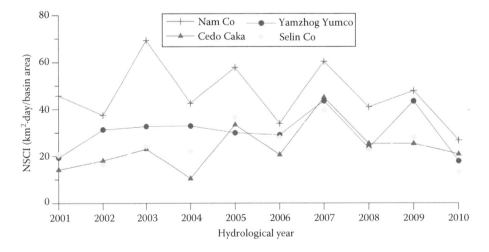

FIGURE 10.8 Normalized snow cover index for the basins Cedo Caka, Selin Co, Nam Co, and Yamzhog Yumco in the Tibetan Plateau during HY2001–2010. (Adapted from Zhang, G. et al., *Water Resour. Res.*, 48(10), W10529, 2012.)

10.5 Snow Cover as a Water Resource for Lake Level and Watershed Analyses

Snow plays an important role in the energy and water balance of drainage basin in alpine regions. Contribution of snowmelt to runoff is one of the important water resources in mountainous regions in addition to rainfall and glaciers melting. With the improved SC mapping based on the MODIS daily SC products, we can now examine the time series SC change as a water resource change for basin and watershed managements. Here, we present two application examples: (1) SC as a water resource for lake level change and (2) snowmelt and runoff for watershed analysis.

10.5.1 Snow Cover over Lake Basin for Lake Level Analysis

Snow over the TP greatly influences water availability of several major Asian rivers such as Yellow, Yangtze, Indus, Ganges,

Brahmaputra, Irrawaddy, Salween, and Mekong (Barnett et al., 2005; Immerzeel et al., 2009). Discharge from these rivers sustains the lives of more than 1 billion people living both in the region and downstream (Barnett et al., 2005). The TP has undergone warming in the past three decades (Liu and Chen, 2000). The glaciers over the TP showed an accelerated retreating because of the warmer climate, except for in the Karakoram (Bolch et al., 2010, 2012; Ding et al., 2006; Gardelle et al., 2012; Yao et al., 2007, 2010, 2012). Lake's water level studies using the NASA's ICES at laser altimetry data have shown over 70% of the lakes in the TP showing lake level increase in the past years (2003–2009) (Phan et al., 2012; Song et al., 2013; Zhang et al., 2011b, 2013). Did SC change play a role to the lake level increase?

Based on the flexible multiday combination of MODIS SC products, Zhang et al. (2012) examined four lake basins over the TP for SC dynamic change (Figure 10.9). The relation between lake level change and combined SC, precipitation, and pan evaporation during hydrological years 2001–2010 (September to

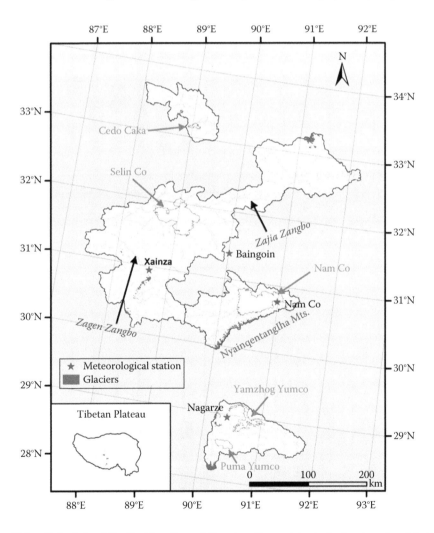

FIGURE 10.9 Location of lakes Cedo Caka, Selin Co, Nam Co, and Yamzhog Yumco (also in the inset map) and their corresponding drainage basins over Tibetan Plateau. Streams and boundaries of lake basins are delineated from Shuttle Radar Topography Mission DEM and glacier coverage from Global Land Ice Measurements from Space at www.glims.org. Four available meteorological stations are also denoted, with no station nearby the Cedo Caka basin. (Adapted from Zhang, G. et al., *Water Resour. Res.*, 48(10), W10529, 2012.)

August) were examined (Figure 10.10). The lake Cedo Caka had the largest lake level increase (+0.80 m/year) among all examined lakes in Zhang et al. (2011b). The lake Selin Co had the second largest lake level increase (+0.69 m/year). The lake Yamzhog Yumco, however, had the fastest drop in water level (−0.40 m/year) among all examined lakes in Zhang et al. (2011b).

As shown in Figures 10.9 and 10.10, except the Cedo Caka lake basin, all three lake basins have nearby weather stations. However, for the Cedo Caka lake, the correlation coefficient between SCA in the basin and lake level change is 0.94, indicating that SC variation played an important role for the lake level increase. This is saying that SC alone can explain 88%

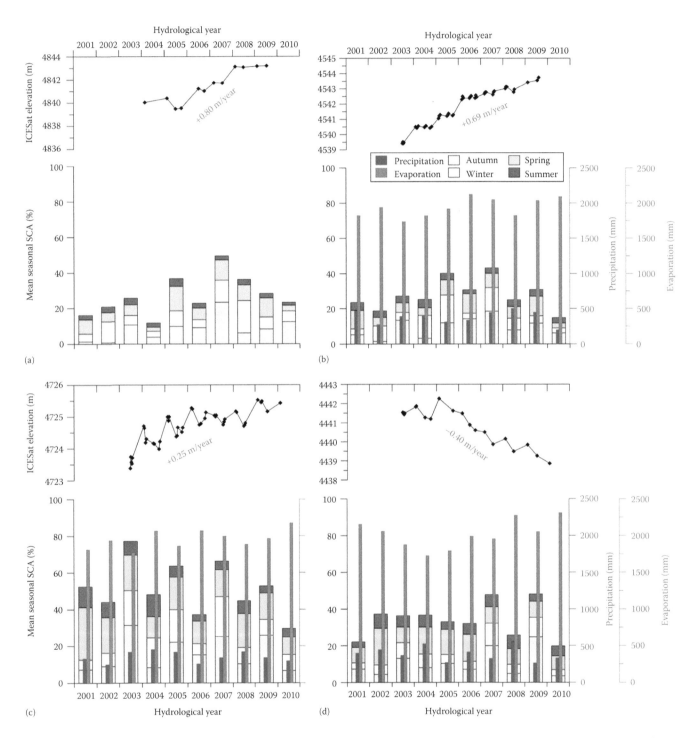

FIGURE 10.10 Mean seasonal snow-covered area (%), precipitation, and pan evaporation (observed at nearby stations) in the (a) Cedo Caka, (b) Selin Co, (c) Nam Co, and (d) Yamzhog Yumco basins during HY2001–2010; lake level changes derived from ICESat data for each basin during HY2003–2009 are also included. (Adapted from Zhang, G. et al., *Water Resour. Res.*, 48(10), W10529, 2012.)

lake level change, although it is not significant at the 95% confidence level ($p = 0.063$). This means other parameters such as precipitation, evaporation, and glaciers also played certain role for the lake level change. For the Selin Co lake, the precipitation, evaporation, and SC together can explain 98% of lake level change, and it is significant at the 95% ($p = 0.029$). This means glacier melts in the basin had almost no impact to the lake level change. For the Namco lake, the combined precipitation, evaporation, and SC can explain 76% of lake level change, although it is statistically insignificant at the 95% confidence level ($p > 0.05$). This suggests that the glacier melts from the Nyainqentanglha Mts. also played additional role for the lake level increase. For the Yamzhog Yumco lake, the combined precipitation, evaporation, and SC can explain 76% of the lake level decrease, although it is statistically insignificant at the 95% confidence level ($p > 0.05$). It is noticed that the operation of a power station nearby the Yamzhog Yumco lake would have played additional role for the lake's water decrease (Zhang et al., 2012).

10.5.2 Quantitative Water Resource Assessment Using Snowmelt Runoff Model

SC, in particular seasonal snowpack in mountainous areas, plays an important role in maintaining a global water balance (Goodison et al., 1999; Immerzeel et al., 2010). The monitoring of snowmelt runoff in the snow- and/or glacier-fed basins

is important for efficient flood forecasting and water resources management practices such as effective irrigation, controlled reservoir levels, and the generation of hydropower.

Several models have been developed to simulate snowmelt discharge, such as the degree-day snowmelt runoff model (SRM) (Martinec, 1975) and energy balance models (Blöschl et al., 1991a,b). However, physically based distributed models are often impractical to use in data-sparse mountainous regions such as the TP. Owing to the requirement of only minimal temperature, precipitation, and SC data as inputs, and only nominal calibration procedures, the SRM appears to be the most ideal model for use in data-sparse regions.

Zhang et al. (2014) did a SRM simulation on the Lake Qinghai (Figure 10.11) and compared the simulation performance of two different SC products: MODIS standard 8-day SC product (MOD10A2) and the flexible multiday combined MODIS SC product (MODISMC). Lake Qinghai is situated near the northeastern margin of the TP. The water level of this lake decreased overall by ~3.7 m from 1959 to 2004, but then increased by ~1 m between 2004 and 2009 (Zhang et al., 2011a). There is a long record of gauge runoff observations from the Buha watershed (the largest catchment area within the Lake Qinghai basin) in comparison with the other lake basins within the TP.

10.5.2.1 Datasets and Methodology

Besides the two MODIS SC datasets, MOD10A2 and MODISMC, for the period of HY2003–2009, daily observed data were also used, including temperature and precipitation

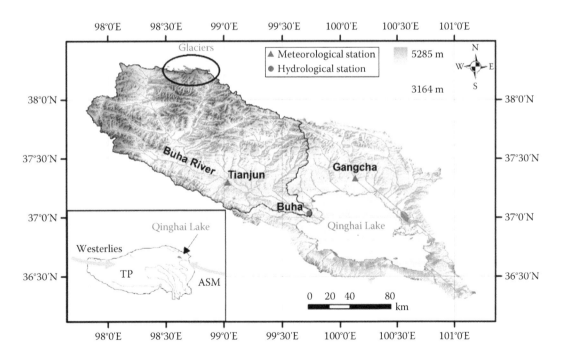

FIGURE 10.11 Lake Qinghai basin and Buha watershed delineated from SRTM DEM, including glacier coverage within this basin and beyond. The locations of meteorological stations Tianjin and Gangcha and hydrologic station Buha are denoted. The map inset shows the outline of Lake Qinghai and the large rivers on the Tibetan Plateau and the direction of the Westerlies and Asian summer monsoon that influence the lake. (Adapted from Zhang, G. et al., *J. Hydrol.*, 519, 976, 2014.)

at the Tianjin meteorological station and streamflow of the Buha watershed at the Buha hydrologic station and pan evaporation (E_{pan}) at the Gangcha meteorological station (the closest station to Lake Qinghai). The recently released Asian Precipitation Highly Resolved Observational Data Integration (APHRODITE) APHRO_MA_V1101 product (available at http://www.chikyu.ac.jp/precip/) for the period HY2003–2007 was also used. The APHRODITE-precipitation product is created by the APHRODITE project (Yatagai et al., 2012).

To derive the depletion curves of the SRM, short-lived SC in summer should not be used (Martinec et al., 2008) and were replaced from a linear interpolation based on adjacent days.

The daily simulated streamflows, as well as the discriminated runoff from snowmelt and rainfall in the Buha watershed, were calculated according to Equation 10.5:

$$Q_{n+1} = \underbrace{C_{sn} \cdot a_n(T_n + \Delta T_n)S_n \cdot A \cdot 0.116(1 - k_{n+1})}_{\text{snowmelt runoff}}$$

$$+ \underbrace{C_{rn}P_n.A \cdot 0.116(1 - k_{n+1})}_{\text{rain runoff}} + \underbrace{(Qs_n + Qr_n)k_{n+1}}_{\text{runoff contribution from the previous day}}$$

$$(10.5)$$

where

Q is the mean daily discharge (m³/s)
C is the snow (C_s) or rain (C_r) runoff coefficient
a is the degree-day factor (cm/°C/day)
$T + \Delta T$ are the degree days (°C days)
S is the ratio of the SCA to the total area
P is the measured precipitation on that day (cm)
A is the area of basin or elevation zone (km²)
k is the recession coefficient
n is the sequence of days during the simulation period

The factor of 0.116 converts data from cm km²/day to m³/s.

The accuracy of the SRM was evaluated using the Nash–Sutcliffe (NS) determination coefficient (R^2) and the volume difference (D_v). The model was run using both basin-wide simulations and zone-wise simulations for the Buha watershed during the period of HY2003–2009. The parameters were calibrated with data from HY2003 to 2005 and validated with data from HY2006 to 2007.

10.5.2.2 Results

The runoff simulations were conducted using measured precipitation and snow products for basin-wide applications, MODISMC-MB and MOD10A2-MB, and also with APHRODITE-precipitation and snow products, MODISMC-AB and MOD10A2-AB. The average NS determination coefficients are very similar between 0.74 and 0.75 for MODISMC-MB (Figure 10.12a), MOD10A2-MB (Figure 10.12b), MODISMC-AB (Figure 10.12c), and MOD10A2-AB (Figure 10.12d). The mean absolute value of volume difference (D_v%), however, is lower

when used MODISMC versus MODIS10A2, that is, 10.12% versus 16.55% (measured precipitation used), or 12.20% versus 19.65% (APHRODITE-precipitation used). These results indicate that MODISMC performs better in basin-wide simulation than MOD10A2. For all the different precipitation and SC used for basin-wide simulations, the correlation of determination is ~0.76 (Figure 10.12a through d).

Zone-wise runoff simulations were further conducted in the Buha watershed. The mean NS coefficient and absolute value of volume difference were 0.70% and 11.59%, respectively, for MODISMC-AZ data (Figure 10.12e). For the MOD10A2-AZ simulation, the mean NS coefficient and absolute values are 0.71% and 19.02%, respectively (Figure 10.12f). The results show that simulations with MODISMC data achieved lower volume differences than MOD10A2 data, although with smaller NS values. Figure 10.12g shows that there is a high correlation between computed runoff using measured precipitation and runoff using APHRODITE-precipitation for basin-wide simulations with the MODISMC snow product during the periods HY2003–2007 ($R^2 = 0.82$). In addition, runoff modeling from basin-wide applications has a high correlation with runoff from zone-wise simulations ($R^2 = 0.93$) (Figure 10.12h).

Figure 10.13 shows the daily runoff contribution fraction from snowmelt and rainfall in the Buha watershed. The average annual estimated runoff from snowmelt is 13% (range 10%–17%) of the total runoff, which indicates that rainfall is a major source of discharge to this lake.

A quantitative assessment (Table 10.2) showed that the lake level changes are highly correlated with single variables, R_m ($R^2 = 0.95$), R_s ($R^2 = 0.77$), R_s-rainfall ($R^2 = 0.75$), and SCA (zone B of Buha watershed, $R^2 = 0.68$) at the 95% confidence level, and multivariates, R_s-rain + P_L ($R^2 = 0.91$) and $R_m + P_L + E_{pan}$ ($R^2 = 0.98$) at the 95% confidence level. These results suggest that rainfall-derived runoff and precipitation over the lake's surface are major contributors to lake level changes in comparison to that of snowmelt-derived runoff. This differs from the four lake basins at inner TP examined that SC played important role for lake level change (Section 10.5.1).

10.5.2.3 Summary

Accurate snowmelt streamflow simulations and forecasts are critical for solving water resources management issues pertaining to irrigation, recreation, flood control, and hydroelectric power generation. The flexible multiday MODISMC product performs better than the standard 8-day product (MOD10A2), with a slightly higher NS coefficient for basin-wide streamflow simulation and overall smaller volume differences for basin- and zone-wide simulations. The 2–3 days combination seems give better SC product and then better runoff simulation. This study demonstrated that the SRM is a suitable hydrologic model for use in estimating snowmelt runoff in the TP and that precipitation in the Qinghai Lake basin plays a dominant role in lake level variations.

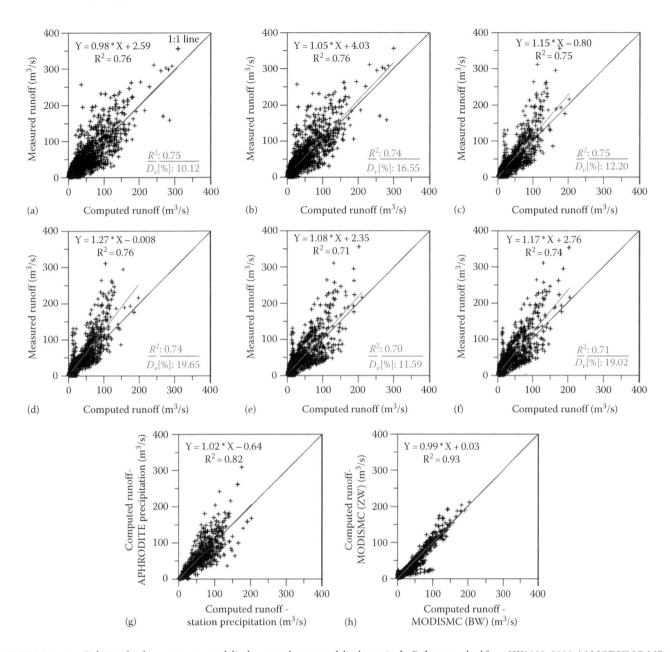

FIGURE 10.12 Relationship between measured discharge and computed discharge in the Buha watershed from HY2003–2009: (a) MODISMC-MB (Measured precipitation for basin-wide simulation), (b) MOD10A2-MB, (c) MODISMC-AB (Asian Precipitation Highly Resolved Observational Data Integration [APHRODITE]-precipitation for basin-wide simulation), (d) MOD10A2-AB, (e) MODISMC-AZ (APHRODITE-precipitation for zone-wise simulation), and (f) MOD10A2-AZ. Two computed discharges: (g) MODISMC-AB and MODISMC-MB and (h) MODISMC-AB and MODISMC-AZ. The blue number in the figures is the mean Nash–Sutcliffe coefficient of determination (R^2) (range) and the mean absolute value of volume difference (D_v) between the simulated and measured runoff. (Adapted from Zhang, G. et al., *J. Hydrol.*, 519, 976, 2014.)

10.6 Snow Cover Change in Responding to Climate Change

Using the daily cloud-free SC product (from Figure 10.6), Wang et al. 2015 derived the SCD maps for the entire TP (Figure 10.14) and examined for its correlation with temperature, precipitation under the current global warming condition. As shown in Figure 10.14 and Table 10.3, the overall spatial distribution pattern of SCD from 2003 to 2010 was actually very similar, with areas of SCD between 60 and 120 days accounting for the most, followed by the areas of SCD less than 60 days. Together, these two categories account for 71.26% of the total plateau area in 2003 and 63.44% in 2010, a clear decreasing trend for each category and for the two together. The areas with SCD between 121 and 180 and between 180 and 350 account for 14.82% of the total plateau

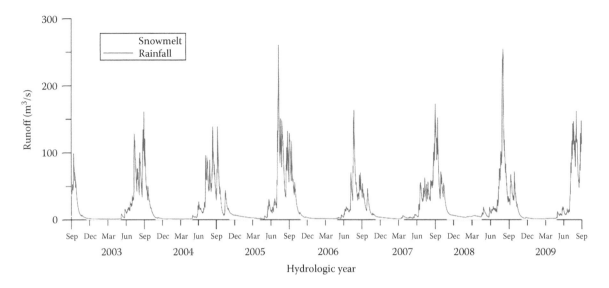

FIGURE 10.13 Daily runoff contribution fraction from snowmelt and rainfall from HY2003–2009 using MODISMC-MB. (Adapted from Zhang, G. et al., *J. Hydrol.*, 519, 976, 2014.)

TABLE 10.2 Coefficient of Determination (R^2) between Annual Differences of In Situ Lake Level Changes, Measured Runoff (R_m), and Simulated Runoff (R_s), for MODISMC-MB Simulation in the Buha Watershed, Precipitation Changes over the Lake Surface (P_L), Pan Evaporation (E_{pan}) Changes Measured at the Gangcha Station, and Snow-Covered Area Changes in the Buha Watershed (SCA_B and Zones A, B, C, D) and Qinghai Lake Basin (SCA_Q) from HY2003 to 2009

	Lake Level Change	
Variable	R^2	P Value
R_m	0.95	0.001
R_s	0.77	0.02
R_s-rain	0.75	0.03
P_L	0.64	0.06
R_s-rain + P_L	0.91	0.03
R_s-snow + P_L	0.64	0.22
$R_m + P_L + E_{pan}$	0.98	0.02
$R_s + P_L + E_{pan}$	0.95	0.08
R_s-rain + $P_L + E_{pan}$	0.95	0.08
SCA_B (Zone A, B, C, D)	0.60 (0.48, 0.68, 0.59, 0.38)	0.07 (0.13, 0.04, 0.07, 0.19)
SCA_Q	0.60	0.07
$SCA_B + P_L + E_{pan}$	0.94	0.09
$SCA_Q + P_L + E_{pan}$	0.85	0.22

Source: Adapted from Zhang, G. et al., *J. Hydrol.*, 519, 976, 2014.

area in 2003 and 23.54% in 2010, a clear increasing trend for each category and for the two together. The area with SCD > 350 can be treated as persistent SC or glacier, showing a slight decreasing tendency from 2003 to 2010.

Further analysis between SCD, precipitation, and temperature was done based on elevation zones. As expected, the maximum SCA percentage was in the elevation zone higher than 4500 m, while the minimum SCA percentage was in the elevation lower than 2000 m. The elevation between 4000 and 4500 m showed the fastest increase in SCA. There is a clear increase trend for

both the temperature (0.09°C/year or 0.75°C in 8 years) and precipitation (0.26 mm/year or 1.8 mm in 8 years). A clear trend of temperature decreases as elevation increase from low-elevation zone (<2000 m) to high-elevation zone (>4500 m), with the largest increase rate (0.27°C/year) in the elevation zone (<2000 m) and the smallest increase rate (0.06°C/year) in the elevation zone (between 3000 and 3500 m). Precipitation showed increase in all elevation zones lower than 4500 m, while decrease in the zone higher than 4500 m. The fastest increase rate (0.5 mm/year) is found in elevation zone 2000–2500 m.

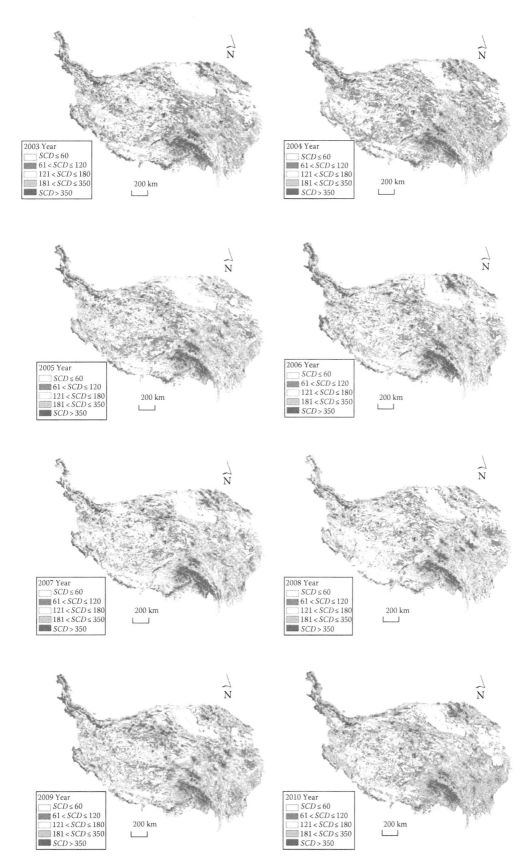

FIGURE 10.14 The snow cover duration maps of Tibetan Plateau based on the daily cloudless snow cover images for the years 2003 to 2010. (Adapted from Wang, W. et al., *Remote Sensing*, 7(1), 169, 2015.)

TABLE 10.3   Area Ratio (%) under Different Snow-Covered Durations over the Tibetan Plateau for the 2003–2010 Period

Year	Area of SCD (%)				
	$SCD \leq 60$	$61 < SCD \leq 120$	$121 < SCD \leq 180$	$181 < SCD \leq 350$	$SCD > 350$
2003	30.03	41.13	7.45	7.37	14.02
2004	30.95	40.51	7.6	7.6	13.34
2005	29.88	40.88	7.94	7.61	13.69
2006	29.83	39.7	8.09	8.94	13.44
2007	28.53	39.43	9.15	9.71	13.18
2008	27.49	38.82	8.68	11.92	13.09
2009	26.32	38.28	9.74	13.04	12.62
2010	25.78	37.66	10.19	13.35	13.02

Source: Adapted from Wang, W. et al., *Remote Sensing*, 7(1), 169, 2015.

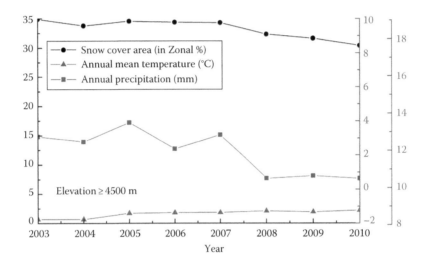

FIGURE 10.15   Mean annual snow cover area, temperature, and precipitation over the elevation higher than 4500 m in the Tibetan Plateau from 2003 to 2010. (Adapted from Wang, W. et al., *Remote Sensing*, 7(1), 169, 2015.)

The only elevation zone showing SCA decrease and precipitation decrease is the area higher than 4500 m, while temperature showing gradual increase (Figure 10.15). Note that the persistent SC and glaciers (SCD > 350 days) are sitting on elevation higher than 4500 m and showing decrease. Therefore, increased temperature and decreased precipitation in higher-elevation zone (>4500 m) are the two reasons for the decrease in persistent SC and glaciers.

Figure 10.16 further illustrates the change tendency of SCD and its significance level in pixel scale over the 8 years. It is clear, majority of the pixels (55.3%) show SCD decrease ($S < 0$), although only 5.8% were statistically significant ($p < 0.05$) and mostly decreased more than 6 days and mostly located at central and southern TP. In contrast, only 14.9% of pixels show increasing trend ($S > 0$), with only 2.1% statistically significant ($p < 0.05$) in the northwestern TP. About 39.8% pixels show no change of the SCD ($S = 0$) and scattered over the entire plateau. Overall, Majority pixels (55.3%) show a reducing tendency of SCD.

10.7 Snow-Caused Livestock Disasters in Pastoral Area: Early Warning

This section we present a case study of using the daily MODIS cloud-free SC products (the MAC-SC of Figure 10.4) and other parameters for application on snow-caused livestock disasters. In pastoral areas of China, the natural disaster resulting in a large amount of livestock deaths caused by continuous snowfall in winter and spring is called snow-caused livestock disaster or snow disaster (Liang et al., 2007). Snow has less impact on animal husbandry in developed countries due to better infrastructure in grassland and livestock industry. However, in pastoral areas of China, a great deal of snowfall often leads to grassland being buried and transportation disruption, resulting in a large number of deaths of livestock due to lower temperature and lack of forage stock. This has a severe influence on the sustainable development of grassland animal husbandry.

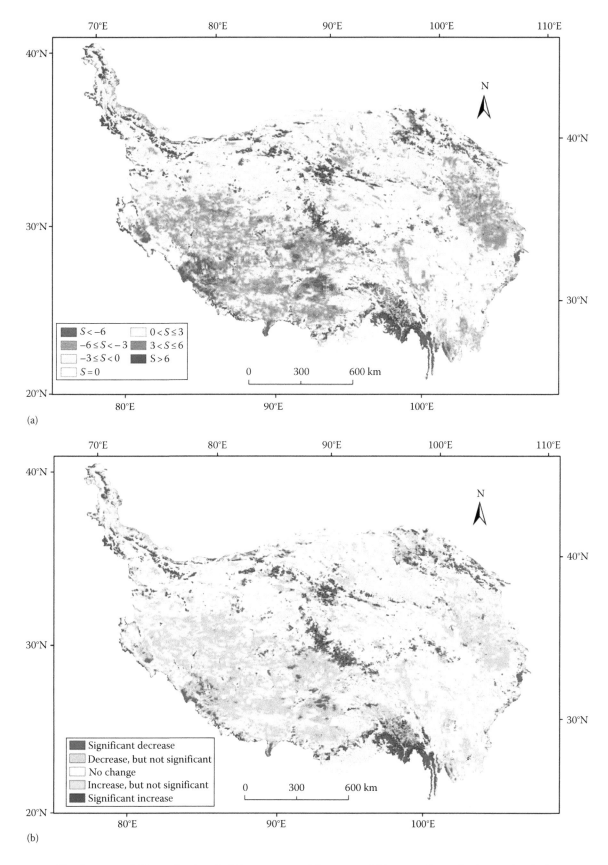

(a)

(b)

FIGURE 10.16 The change trend (*S*, unit days/year) of (a) snow-covered duration and (b) its significance level in pixel scale during 8 years in the TP region. (Adapted from Wang, W. et al., *Remote Sensing*, 7(1), 169, 2015.)

Early warning and risk assessment are the two upmost important yet difficult issues in the study of snow-caused livestock disasters. Early warning of snow disasters involves many factors such as grassland, snow, weather, livestock, society, and economy, of which many have strong temporal and spatial heterogeneity. The accuracy of warning results is not only closely related to weather forecasting information but also connected with regional environment conditions and capability of disaster prevention. Therefore, the premise for an operational snow disaster warning system is to establish snow disaster warning and risk assessment models based on a long-term series of regional snow disaster monitoring databases using remote sensing, ground-based observation data, and weather prediction data in pastoral areas. Furthermore, studying early warning mechanisms of snow disasters is extremely important in both theory and operational practice for improving the ability to prevent disaster and minimize loss in pastoral areas.

10.7.1 Factors of Snow Disaster Early Warning and Risk Assessment

Snow disaster warning and risk assessment mainly involve factors such as terrain, grassland, snow, weather, livestock, and social economy and are detailed in the following:

1. Terrain and grassland factors. These factors mainly include the topography conditions (e.g., slope, aspect) and grassland distribution and herbage growing status (e.g., herbage yield, cover and height of different grassland types).
2. Snow and weather factors. They mainly include (a) snow disaster probability based on multiple-year data (e.g., 50 years) on an administrative unit basis (e.g., township, county); (b) snow depth and number of snow-covered days, the rate of SCA, the rate of snow-covered grassland area, grassland burial index (i.e., the ratio of snow depth to grass height) in the unit; and (c) observed and forecasted meteorological conditions during a warning period (e.g., daily minimum, maximum, and average temperatures, precipitation, and snow depth observed and predicted at climatic stations) and continuous days of low temperature (e.g., below 0°C, below −10°C, or below −20°C).
3. Livestock factors. Livestock is the main hazard-affected body. The related factors mainly include number of livestock at the beginning of a year, number of livestock by the end of a year, actual livestock stocking rate, and fraction of small livestock.
4. Social economy factors. These factors mainly include (a) social and economic factors consisting of population, per capita gross domestic product (GDP), per capita farming income, density of road network, and spatial distribution of residential areas; (b) forage stock and

shed rate of livestock; and (c) per livestock GDP (i.e., the ratio of regional GDP to the number of livestock by the end of a year).

10.7.2 Models for Early Warning and Qualitative Risk Assessment of Snow Disasters

Snow disaster warning and risk assessment have been studied for a long time. Many studies have been focused on snow-caused disaster risk assessment (Liang et al., 2007; Romanov et al., 2002), loss evaluation of postdisaster (i.e., the research for the pastoral areas after snow disasters have occurred and have resulted in loss of livestock) (Nakamura and Shindo, 2001), and snow disaster and avalanche mapping, as well as their relations to climate change (Bocchiola et al., 2008; Delparte et al., 2008; Hendrikx et al., 2005; Hirashima et al., 2008; Jones and Jamieson, 2001; Lato et al., 2012; Martelloni et al., 2013; Mercogliano et al., 2013; Williamson et al., 2002). Tachiiri et al. (2008) evaluated the economic loss of snow disasters in arid inland pastoral areas of Mongolia using a tree-based regression model with parameters including livestock mortality rate of current year, grassland NDVI, SWE, and livestock numbers and mortality rates of previous years. In addition, Tominaga et al. (2011) predicted SC area and potential extent of snow disasters in buildup environments by combining a meteorological model and a computational fluid dynamics model. Nakai et al. (2012) established a snow disaster early warning system using meteorological factors (i.e., precipitation, wind speed, temperature), which would predict avalanche potential, visibility in blowing snow, and snow conditions on roads.

Due to lacking operational models and information system for real-time warning of snow disasters in pastoral areas in developing countries (Liang et al., 2007), there has been very difficult in snow disaster warning and risk assessments. As for snow disasters in pastoral areas in China, the emphasis has been placed on monitoring the change of snow distribution and livestock loss evaluation of postdisasters. Early warning of snow disaster requires developing a model that can be used to quantify the potential damages caused by heavy snowfall events in a particular period and area. In 2006, the Chinese government issued a national standard for grading snow disasters in pastoral area (GB/T 20482–2006) (General Administration of Quality Supervision of China, 2006). According to the standard, Zhou et al. (2006) analyzed the potential conditions of snow disasters based on evaluating vulnerability of hazard-affected bodies and dynamic change of precipitation in the natural environment. Liu et al. (2008) did a preliminary study on an indicating system for early warning and hazard assessment models of snow disasters in pastoral areas of northern Xinjiang utilizing livestock mortality rate as a factor of disaster assessment. Zhang et al. (2008) proposed several indicators and methods for quantification of

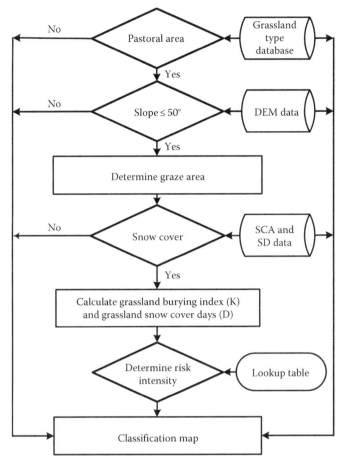

FIGURE 10.17 Workflow for risk intensity classification of snow-caused disaster. (Adapted from Wang, W. et al., *Nat. Haz. Earth Syst. Sci.*, 13, 1411, 2013.)

those indicators for early warning of snow disaster in the pastoral areas of northern Qinghai province. Wang et al. (2013) developed a model and a warning standard for early warning on county basis using the chosen 45 typical cases of snow disasters (Equation 10.6, Table 10.4) and proposed a method of qualitative risk assessment of snow disasters (Figure 10.17, Table 10.5) at 500 m resolution for pastoral areas in the TP. They chose 411 cases from 2008 to 2010 to validate the predicting results from the developed early warning model, with an overall accuracy of 85.64% in predicting snow disasters and no disasters on TP:

$$Z_{ij} = \frac{57.850 \times a_{ij} + 64.891 \times b_{ij} + 8496 \times \prod_{k=1}^{n}(c_{k,ij})}{d_{ij}} \quad (10.6)$$

where

Z_{ij} is the number of livestock death on county basis for a warning period

i is the number of counties, $i = 1, 2, 3, \ldots, n$

j is the number of warning periods, $j = 1, 2, 3, \ldots, m$ and $m = 18$, total warning periods in each snow season from October 1 to March 31 (10 days per warning period), for example, warning period 1 or 2 (i.e., $j = 1$ or 2) is from October 1 to October 10 or from October 11 to October 20

a is the mean annual probability of snow disaster

b is the continuous days of mean daily temperature below $-10°C$

c_k (where $k = 1-4$) are, respectively, the rate of snow-covered grassland, number of snow-covered days, grassland burial index, and livestock stocking rate

d is the per livestock GDP

TABLE 10.4 Warning Standard of Snow Disaster Grades

Snow Disaster Grade	Death Amount of Livestock (×1000)	Influence of Snow on Animal Husbandry
No disaster	0	There is snow in pastoral area, but grassland burial index is less than 35% or the rate of snow-covered grassland is less than 30%. It has no obvious effect on grazing.
Light disaster	50	It has an effect on yak grazing, less effect on sheep, and no effect on horse.
Moderate disaster	50–100	It has an effect on yak and sheep grazing, and no effect on horse.
Severe disaster	100–200	It has an effect on animal husbandry and has greater loss for yak and sheep.
Extremely severe disaster	>200	It has an effect on all animals husbandry and has the death of livestock on a massive scale.

Source: Adapted from Wang, W. et al., *Nat. Haz. Earth Syst. Sci.*, 13, 1411, 2013.

TABLE 10.5 Lookup Table for Defining Pixel-Based Risk Intensity of Snow Disaster

Indicator and Threshold		Number of Snow-Covered Days (K)		
		K < 10 Days	10 Days ≤ K < 20 Days	K ≥ 20 Days
Grassland burial index (D)	D < 30%	Low risk	Low risk	Moderate risk
	30% ≤ D < 60%	Low risk	Moderate risk	High risk
	D ≥ 60%	Moderate risk	High risk	High risk

Source: Adapted from Wang, W. et al., *Nat. Haz. Earth Syst. Sci.*, 13, 1411, 2013.

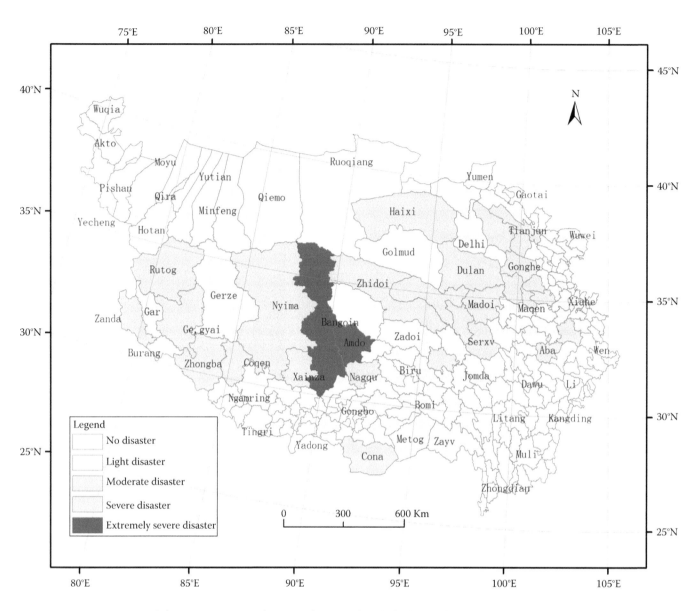

FIGURE 10.18 Snow-caused disaster warning simulation result (county basis) of Tibetan Plateau in late January 2008 (*J* = 12). (Adapted from Wang, W. et al., *Nat. Haz. Earth Syst. Sci.*, 13, 1411, 2013.)

According to the methodology for early warning and risk evaluation of snow disaster, the map of snow disaster grades on a county basis (such as Figure 10.18) and the potential risk intensity map at 500 m spatial resolution (such as Figure 10.19) can be made for decision making of relieving and preventing a snow disaster (Wang et al., 2013).

It is well known that the degree and extent of snow disasters are not only related to the infrastructure (e.g., roads, communications and sheds for livestock) of disaster resistance and forage stock in a pastoral area but also closely related to disaster rescue efforts. The occurrence of snow disasters was often not just the result of one heavy snowfall or temperature drop but also affected by continuous snowfall and other factors, climatic indicators, and SC condition being the main factors. However, the developed models for early warning of snow disasters and risk assessment in pastoral areas in China are completely based on existing and available datasets, without considering any interference or assistance of human beings such as disaster relief by supplying a great amount of forage from other regions. Uncertainties of the warning model require to be verified, and several indicators and parameters in the models and classification standards for early warning of snow disaster and risk assessment still need to be improved further in the future.

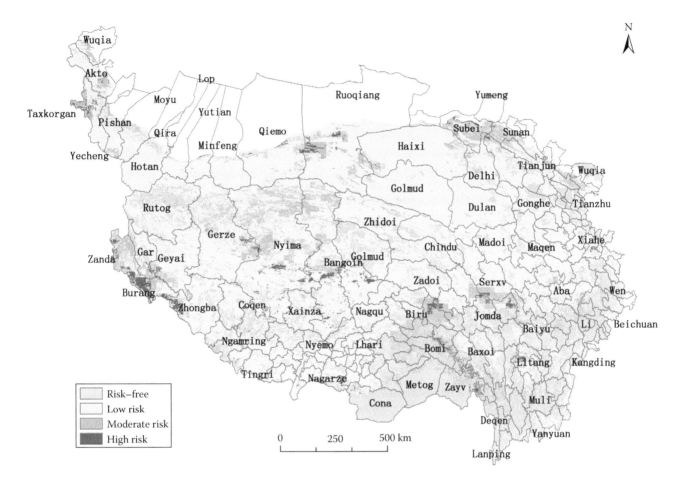

FIGURE 10.19 Simulated risk intensity of snow-caused disaster map at 500 m pixel scale in late January 2008 (*J* = 12). (Adapted from Wang, W. et al., *Nat. Haz. Earth Syst. Sci.*, 13, 1411, 2013.)

10.8 Conclusions

SC is an important fact for regional climate change studies, agriculture, and water source management. Satellite-based snow measurements have revolutionized the monitoring of spatiotemporal variation of SC in complex natural conditions at regional and global scales. This chapter introduces the principle of optical remote sensing SC detection and the MODIS standard SC products, summarizes the recent efforts and improvements on reducing the cloud contamination and increasing snow mapping accuracy in all-sky condition, and finally illustrates the spatiotemporal variations of SC and its applications in the TP and a few other places.

Overall, MODIS standard SC products have high accuracy in clear-sky observations and are widely applied in all kinds of studies around the world. In contrast, MODSCAG uses the relative shape of the snow's spectrum and is more accurate to characterize fractional SC than the MODIS standard SC products (Rittger et al., 2013). However, the impact of this improvement for fractional SC on the total water resource budget is limited. Meanwhile, cloud blockage severely obstructs the wide

application of MODIS daily SC products. Many efforts have been conduced to mitigate the negative effects of cloud cover.

Benefiting from daily twice observations of Terra and Aqua MODIS sensors and the cloud-free microwave AMSR-E/Aqua observations, the major improvements presented here include (1) the daily Terra and Aqua combination, (2) the flexible multiday combination, (3) the daily cloud-free SC processing, and (4) the daily redistribution of 25 km SWE to 500 m pixel size.

Based on the improved SC mapping, four specific SC parameters are introduced, namely, SCI, SCD map, SCOD map, and SCED map, one each per year or per hydrological year. While an SCD map gives the overall spatial distribution of SC duration, both SCOD and SCED maps provide the spatial distribution of the specific dates when the SC starts and when the SC melts away at the pixel scale of 500 m. The SCI (km² day) contains both SC extent and duration for a year and can be used to study the SC condition in a yearly basis. Together, SCD, SCI, SCOD, and SCED provide important information on the SC conditions and can be applied in any region of interests.

Based on those SC parameters, we present four examples of application: (1) SC as a water resource for lake level change study,

(2) SC as an input for snowmelt and runoff modeling to separate snow and rainfall contribution to total runoff, (3) SC change in response to global warming, and (4) SC as an input for developing model for assessing and predicting snow-caused disasters in pastoral area.

Overall, remotely sensed SC and parameters could be critical information for local government, such as land use planning, agriculture, livestock, and water resource management, for example, to mitigate snow-caused disasters and to plan for agriculture and industry water use. Long-term availability of MODIS type of SC data for producing such datasets is key to studying the connection between SC variations and climate change. For example, possible application for these data in the future is to studying the relation between SCI and El Nino Southern Oscillation. Additional questions can also be addressed through the type of data presented. Does global warming reduce the mean SC days for a region? Does global warming have potential to forward shift or backward shift the mean SC onset date and/or mean SC melting date? How does the spring sandstorm or dust soil from desert impact the melting of SC?

Acknowledgments

This work was in part supported by the National Natural Science Foundation of China (31228021, 31372367, and 41301063) and the University of Texas at San Antonio. Many figures and tables are adapted from published journal papers coauthored with many of our colleagues and graduate students, for which they are sincerely acknowledged. We also thank Dr. Prasad Thenkabail, editor in chief of the book volumes, for making this handbook possible and for his review comments to improve the chapter.

References

Barnett, T.P., Adam, J.C., Lettenmaier, D.P., 2005. Potential impacts of a warming climate on water availability in snow-dominated regions. *Nature*, 438 (7066): 303–309.

Blöschl, G., Gutknecht, D., Kirnbauer, R., 1991a. Distributed snowmelt simulations in an Alpine catchment: 2. Parameter study and model predictions. *Water Resour. Res.*, 27 (12): 3181–3188.

Blöschl, G., Kirnbauer, R., Gutknecht, D., 1991b. Distributed snowmelt simulations in an Alpine catchment: 1. Model evaluation on the basis of snow cover patterns. *Water Resour. Res.*, 27 (12): 3171–3179.

Bocchiola, D., Janetti, E.B., Gorni, E., Marty, C., Sovilla, B., 2008. Regional evaluation of three day snow depth for avalanche hazard mapping in Switzerland. *Nat. Haz. Earth Syst. Sci.*, 8 (4): 685–705.

Bolch, T. et al., 2010. A glacier inventory for the western Nyainqentanglha Range and Nam Co Basin, Tibet, and glacier changes 1976–2009. *Cryosphere*, 4 (3): 419–433.

Bolch, T. et al., 2012. The state and fate of Himalayan glaciers. *Science*, 336 (6079): 310–314.

Brubaker, K.L., Pinker, R.T., Deviatova, E., 2005. Evaluation and comparison of MODIS and IMS snow-cover estimates for the continental United States using station data. *J. Hydrometeorol.*, 6 (6): 1002–1017.

Delparte, D., Jamieson, B., Waters, N., 2008. Statistical runout modeling of snow avalanches using GIS in Glacier National Park, Canada. *Cold Reg. Sci. Technol.*, 54 (3): 183–192.

Ding, Y., Liu, S., Li, J., Shangguan, D., 2006. The retreat of glaciers in response to recent climate warming in western China. *Ann. Glaciol.*, 43: 97–105.

Dozier, J., 1989. Spectral signature of alpine snow cover from the landsat thematic mapper. *Remote Sens. Environ.*, 28: 9–22.

Dozier, J., Painter, T.H., 2004. Multispectral and hyperspectral remote sensing of alpine snow properties. *Annu. Rev. Earth Plan. Sci.*, 32: 465–494.

Foster, J. et al., 2007. Blended visible, passive microwave and scatterometer global snow products. *Proceedings of the 64th Eastern Snow Conference*, May 29 to June 1, 2007, St. John's, Newfoundland, Canada.

Frei, A. et al., 2012. A review of global satellite-derived snow products. *Adv. Space Res.*, 50 (8): 1007–1029.

Frei, A., Lee, S., 2010. A comparison of optical-band based snow extent products during spring over North America. *Remote Sens. Environ.*, 114 (9): 1940–1948.

Gao, Y., Xie, H., Lu, N., Yao, T., Liang, T., 2010a. Toward advanced daily cloud-free snow cover and snow water equivalent products from Terra-Aqua MODIS and Aqua AMSR-E measurements. *J. Hydrol.*, 385 (1–4): 23–35.

Gao, Y., Xie, H., Yao, T., 2011. Developing snow cover parameters maps from MODIS, AMSR-E, and blended snow products. *Photogramm. Eng. Remote Sens.*, 77 (4): 351–361.

Gao, Y., Xie, H., Yao, T., Xue, C., 2010b. Integrated assessment on multi-temporal and multi-sensor combinations for reducing cloud obscuration of MODIS snow cover products of the Pacific Northwest USA. *Remote Sens. Environ.*, 114 (8): 1662–1675.

Gardelle, J., Berthier, E., Arnaud, Y., 2012. Slight mass gain of Karakoram glaciers in the early twenty-first century. *Nature Geosci.*, 5: 322–325.

General Administration of Quality Supervision of China, Inspection and Quarantine of the People's Republic of China, Standardization Administration of the People's Republic of China, 2006. Chinese national standard. In: *Grade of Pastoral Area Snow Disaster*. China Standards Press, Beijing, China, pp. 1–7 (in Chinese).

Goodison, B.E., Brown, R.D., Crane, R.G., 1999. Cryospheric systems. In: King, M.D. (ed.), *EOS Science Plan: The State of Science in the EOS Program*, NASA, Washington, DC, pp. 261–307.

Hall, D.K., Foster, J.L., Verbyla, D.L., Klein, A.G., Benson, C.S., 1998. Assessment of snow-cover mapping accuracy in a variety of vegetation-cover densities in central Alaska. *Remote Sens. Environ.*, 66 (2): 129–137.

Hall, D.K., Riggs, G.A., 2007. Accuracy assessment of the MODIS snow products. *Hydrol. Processes*, 21 (12): 1534–1547.

Hall, D.K., Riggs, G.A., Foster, J.L., Kumar, S.V., 2010. Development and evaluation of a cloud-gap-filled MODIS daily snow-cover product. *Remote Sens. Environ.*, 114 (3): 496–503.

Hall, D.K., Riggs, G.A., Salomonson, V.V., 1995. Development of methods for mapping global snow cover using moderate resolution imaging spectroradiometer data. *Remote Sens. Environ.*, 54 (2): 127–140.

Hall, D.K., Riggs, G.A., Salomonson, V.V., DiGirolamo, N.E., Bayr, K.J., 2002. MODIS snow-cover products. *Remote Sens. Environ.*, 83 (1–2): 181–194.

Hendrikx, J., Owens, I., Carran, W., Carran, A., 2005. Avalanche activity in an extreme maritime climate: The application of classification trees for forecasting. *Cold Reg. Sci. Technol.*, 43 (1): 104–116.

Hirashima, H., Nishimura, K., Yamaguchi, S., Sato, A., Lehning, M., 2008. Avalanche forecasting in a heavy snowfall area using the snowpack model. *Cold Reg. Sci. Technol.*, 51 (2): 191–203.

Immerzeel, W.W., Beek, L.P.H., Bierkens, M.F.P., 2010. Climate change will affect the Asian water towers. *Science*, 328 (5984): 1382–1385.

Immerzeel, W.W., Droogers, P., De Jong, S.M., Bierkens, M.F.P., 2009. Large-scale monitoring of snow cover and runoff simulation in Himalayan river basins using remote sensing. *Remote Sens. Environ.*, 113 (1): 40–49.

Jones, A.S., Jamieson, B., 2001. Meteorological forecasting variables associated with skier-triggered dry slab avalanches. *Cold Reg. Sci. Technol.*, 33 (2): 223–236.

Klein, A.G., Barnett, A.C., 2003. Validation of daily MODIS snow cover maps of the Upper Rio Grande River Basin for the 2000–2001 snow year. *Remote Sens. Environ.*, 86 (2): 162–176.

Klein, A.G., Hall, D.K., Riggs, G.A., 1998. Improving snow cover mapping in forests through the use of a canopy reflectance model. *Hydrol. Processes*, 12 (10–11): 1723–1744.

Lato, M., Frauenfelder, R., Bühler, Y., Gauthier, D., Lanorte, A., 2012. Automated detection of snow avalanche deposits: Segmentation and classification of optical remote sensing imagery. *Nat. Haz. Earth Syst. Sci.*, 12 (9): 2893–2906.

Liang, T. et al., 2008a. Toward improved daily snow cover mapping with advanced combination of MODIS and AMSR-E measurements. *Remote Sens. Environ.*, 112 (10): 3750–3761.

Liang, T. et al., 2008b. An application of MODIS data to snow cover monitoring in a pastoral area: A case study in Northern Xinjiang, China. *Remote Sens. Environ.*, 112 (4): 1514–1526.

Liang, T., Liu, X., Wu, C., Guo, Z., Huang, X., 2007. An evaluation approach for snow disasters in the pastoral areas of northern Xinjiang, PR China. *N.Z. J. Agric. Res.*, 50 (3): 369–380.

Liu, X., Liang, T., Guo, Z., Zhang, X., 2008. Early warning and risk assessment of snow disaster in pastoral area of northern Xinjiang. *Chinese Journal of Applied Ecology,* 19 (1): 133–138.

Liu, X.D., Chen, B.D., 2000. Climatic warming in the Tibetan Plateau during recent decades. *Int. J. Climatol.*, 20 (14): 1729–1742.

Martelloni, G., Segoni, S., Lagomarsino, D., Fanti, R., Catani, F., 2013. Snow accumulation/melting model (SAMM) for integrated use in regional scale landslide early warning systems. *Hydrol. Earth Syst. Sci.*, 17: 1229–1240. doi:10.5194/hess-17-1229-2013.

Martinec, J., 1975. Snowmelt runoff model for river flow forecasts. *Nordic Hydrol.*, 6 (3): 145–154.

Martinec, J., Rango, A., Roberts, R., 2008. *Snowmelt Runoff Model (SRM) User's Manual, USDA Jornada Experimental Range*, New Mexico State University, Las Cruces, NM.

Maurer, E.P., Rhoads, J.D., Dubayah, R.O., Lettenmaier, D.P., 2003. Evaluation of the snow-covered area data product from MODIS. *Hydrol. Processes*, 17 (1): 59–71.

Mazari, N., Tekeli, A.E., Xie, H., Sharif, H.O., El Hassan, A.A., 2013. Assessment of ice mapping system and moderate resolution imaging spectroradiometer snow cover maps over Colorado Plateau. *J. Appl. Remote Sens.*, 7(1): 073540–073540.

Mercogliano, P., Segoni, S., Rossi, G., Sikorsky, B., Tofani, V., Schiano, P., Catani, F., Casagli, N., 2013. Brief communication "A prototype forecasting chain for rainfall induced shallow landslides". *Nat. Haz. Earth Syst. Sci.*, 13: 771–777. doi:10.5194/nhess-13-771-2013.

Nakai, S. et al., 2012. A Snow Disaster Forecasting System (SDFS) constructed from field observations and laboratory experiments. *Cold Reg. Sci. Technol.*, 70: 53–61.

Nakamura, M., Shindo, N., 2001. Effects of snow cover on the social and foraging behavior of the great tit Parus major. *Ecol. Res.*, 16 (2): 301–308.

Nolin, A.W., 2004. Towards retrieval of forest cover density over snow from the Multi-angle Imaging SpectroRadiometer (MISR). *Hydrol. Processes*, 18 (18): 3623–3636.

Nolin, A.W., Dozier, J., Mertes, L.A., 1993. Mapping alpine snow using a spectral mixture modelling technique. *Ann. Glaciol.*, 17: 121–124.

Painter, T.H. et al., 2009. Retrieval of subpixel snow covered area, grain size, and albedo from MODIS. *Remote Sens. Environ.*, 113 (4): 868–879.

Parajka, J., Blöschl, G., 2008. Spatio-temporal combination of MODIS images-potential for snow cover mapping. *Water Resour. Res.*, 44 (3): W03406.

Parajka, J., Pepe, M., Rampini, A., Rossi, S., Blöschl, G., 2010. A regional snow-line method for estimating snow cover from MODIS during cloud cover. *J. Hydrol.*, 381 (3–4): 203–212.

Phan, V.H., Lindenbergh, R., Menenti, M., 2012. ICESat derived elevation changes of Tibetan lakes between 2003 and 2009. *Int. J. Appl. Earth Obs. Geoinf.*, 17: 12–22.

Riggs, G.A., Hall, D.K., 2003. Reduction of cloud obscuration in the MODIS snow data product. *Proceedings of the 60th Eastern Snow Conference*, Sherbrooke, Québec, Canada, June 4–6, 2003, pp. 205–212.

Riggs, G.A., Hall, D.K., Salomonson, V.V., 2006. MODIS snow products user guide to collection 5. Online article, https://www.modis-snow-ice.gsfc.nasa.gov/userguides.html (Retrieved on January 2, 2007).

Rittger, K., Painter, T.H., Dozier, J., 2013. Assessment of methods for mapping snow cover from MODIS. *Adv. Water Res.*, 51: 367–380.

Romanov, P., Gutman, G., Csiszar, I., 2002. Satellite-derived snow cover maps for North America: Accuracy assessment. *Adv. Space Res.*, 30 (11): 2455–2460.

Salomonson, V., Appel, I., 2004. Estimating fractional snow cover from MODIS using the normalized difference snow index. *Remote Sens. Environ.*, 89 (3): 351–360.

Salomonson, V., Appel, I., 2006. Development of the Aqua MODIS NDSI fractional snow cover algorithm and validation results. *IEEE Trans. Geosci. Remote Sens.*, 44 (7): 1747–1756.

Simic, A., Fernandes, R., Brown, R., Romanov, P., Park, W., 2004. Validation of VEGETATION, MODIS, and GOES + SSM/I snow-cover products over Canada based on surface snow depth observations. *Hydrol. Processes*, 18 (6): 1089–1104.

Sirguey, P., Mathieu, R., Arnaud, Y., 2009. Subpixel monitoring of the seasonal snow cover with MODIS at 250 m spatial resolution in the Southern Alps of New Zealand: Methodology and accuracy assessment. *Remote Sens. Environ.*, 113 (1): 160–181.

Song, C., Huang, B., Ke, L., 2013. Modeling and analysis of lake water storage changes on the Tibetan Plateau using multimission satellite data. *Remote Sens. Environ.*, 135: 25–35.

Tachiiri, K., Shinoda, M., Klinkenberg, B., Morinaga, Y., 2008. Assessing Mongolian snow disaster risk using livestock and satellite data. *J. Arid. Environ.*, 72 (12): 2251–2263.

Tominaga, Y. et al., 2011. Development of a system for predicting snow distribution in built-up environments: Combining a mesoscale meteorological model and a CFD model. *J. Wind Eng. Ind. Aerodyn.*, 99 (4): 460–468.

Tong, J., Déry, S.J., Jackson, P.L., 2009. Interrelationships between MODIS/Terra remotely sensed snow cover and the hydrometeorology of the Quesnel River Basin, British Columbia, Canada. *Hydrol. Earth Syst. Sci. Discuss.*, 6 (3): 3687–3723.

Vikhamar, D., Solberg, R., 2002. Subpixel mapping of snow cover in forests by optical remote sensing. *Remote Sens. Environ.*, 84 (1): 69–82.

Wang, W., Huang, X., Deng, J., Xie, H., Liang, T., 2015. Spatiotemporal change of snow cover and its response to climate over the Tibetan plateau based on an improved daily cloud-free snow cover product[J]. *Remote Sensing*, 2015, 7(1): 169–194.

Wang, W., Liang, T., Huang, X., Feng, Q., Xie, H., Liu, X., Chen, M., Wang X., 2013. Early warning of snow-caused disasters in pastoral areas on the Tibetan Plateau. *Nat. Haz. Earth Syst. Sci.*, 13: 1411–1425.

Wang, X., Xie, H., 2009. New methods for studying the spatiotemporal variation of snow cover based on combination products of MODIS Terra and Aqua. *J. Hydrol.*, 371 (1–4): 192–200.

Wang, X., Xie, H., Liang, T., 2008. Evaluation of MODIS snow cover and cloud mask and its application in Northern Xinjiang, China. *Remote Sens. Environ.*, 112 (4): 1497–1513.

Wang, X., Xie, H., Liang, T., 2014b. Spatiotemporal variation of snow cover from space in Northern Xinjiang. In: Y. Chen (ed.), Book Chapter 6 of *Water Resources Research in Northwest China*, Springer, the Netherlands. doi:10.1007/978-94-017-8016-2.

Wang, X., Xie, H., Liang, T., Huang, X., 2009. Comparison and validation of MODIS standard and new combination of Terra and Aqua snow cover products in northern Xinjiang, China. *Hydrol. Processes*, 23 (3): 419–429.

Williamson, R.A., Hertzfeld, H.R., Cordes, J., Logsdon, J.M., 2002. The socioeconomic benefits of Earth science and applications research: Reducing the risks and costs of natural disasters in the USA. *Space Policy*, 18: 57–65.

Xie, H., Wang, X., Liang, T., 2009. Development and assessment of combined Terra and Aqua snow cover products in Colorado Plateau, USA and northern Xinjiang, China. *J. Appl. Remote Sens.*, 3: 033559.

Yang, W. et al., 2006. Analysis of leaf area index products from combination of MODIS Terra and Aqua data. *Remote Sens. Environ.*, 104 (3): 297–312.

Yao, T. et al., 2010. Glacial distribution and mass balance in the Yarlung Zangbo River and its influence on lakes. *Chin. Sci. Bull.*, 55 (20): 2072–2078.

Yao, T. et al., 2012. Different glacier status with atmospheric circulations in Tibetan Plateau and surroundings. *Nature Clim. Change*, 2 (9): 663–667.

Yao, T., Pu, J., Lu, A., Wang, Y., Yu, W., 2007. Recent glacial retreat and its impact on hydrological processes on the Tibetan Plateau, China, and surrounding regions. *Arctic Antarctic Alpine Res.*, 39 (4): 642–650.

Yatagai, A. et al., 2012. APHRODITE: Constructing a long-term daily gridded precipitation dataset for Asia based on a dense network of rain gauges. *Bull. Am. Meteorol. Soc.*, 93 (9): 1401–1415.

Yu, H. et al., 2012. A new approach of dynamic monitoring of 5-day snow cover extent and snow depth based on MODIS and AMSR-E data from Northern Xinjiang region. *Hydrol. Processes*, 26: 3052–3061.

Zhang, G. et al., 2008. Study on warning indicator system of snow disaster and risk management in headwaters region. *Pratacultural Sci.*, 26: 144–150.

Zhang, G., Xie, H., Duan, S., Tian, M., Yi, D., 2011a. Water level variation of Lake Qinghai from satellite and in situ measurements under climate change. *J. Appl. Remote Sens.*, 5: 053532.

Zhang, G., Xie, H., Kang, S., Yi, D., Ackley, S.F., 2011b. Monitoring lake level changes on the Tibetan Plateau using ICESat altimetry data (2003–2009). *Remote Sens. Environ.*, 115 (7): 1733–1742.

Zhang, G., Xie, H., Yao, T., Li, H., Duan, S., 2014. Quantitative water resources assessment of Qinghai Lake basin using Snowmelt Runoff Model (SRM). *J. Hydrol.*, 519: 976–987.

Zhang, G., Xie, H., Yao, T., Liang, T., Kang, S., 2012. Snow cover dynamics of four lake basins over Tibetan Plateau using time series MODIS data (2001–2010). *Water Resour. Res.*, 48 (10): W10529.

Zhang, G., Yao, T., Xie, H., Kang, S., Lei, Y., 2013. Increased mass over the Tibetan Plateau: From lakes or glaciers? *Geophys. Res. Lett.*, 40 (10): 2125–2130.

Zhao, H., Fernandes, R., 2009. Daily snow cover estimation from Advanced Very High Resolution Radiometer Polar Pathfinder data over Northern Hemisphere land surfaces during 1982–2004. *J. Geophys. Res.*, 114 (D5): D05113.

Zhou, B., Shen, F., Li, S., 2006. A synthetical forecasting model of snow disaster in Qinghai-Tibet Plateau. *Meteorological*, 9: 106–110.

Zhou, X., Xie, H., Hendrickx, J.M.H., 2005. Statistical evaluation of remotely sensed snow-cover products with constraints from streamflow and SNOTEL measurements. *Remote Sens. Environ.*, 94 (2): 214–231.

VI

Nightlights

287

11

Nighttime Light Remote Sensing: Monitoring Human Societies from Outer Space

Qingling Zhang
Yale University
and
Shenzhen Institutes of Advanced Technology

Noam Levin
The Hebrew University
of Jerusalem

Christos Chalkias
Harokopio University

Husi Letu
Research and Information Center
Tokai University

Acronyms and Definitions

AVHRR	Advanced Very High Resolution Radiometer
CIESIN	Center for International Earth Science Information Network
DMSP	Defense Meteorological Satellite Program
DN	Digital number
DNB	Day/night band
ETM+	Enhanced Thematic Mapper Plus
GDP	Gross domestic product
GIS	Geographic information system
HSC	High-sensitivity camera
HSI	Human settlement index
HSTC	High-sensitivity technological camera
ISS	International Space Station
MISR	Multiangle Imaging Spectroradiometer
MODIS	Moderate Resolution Imaging Spectroradiometer
NASA	National Aeronautics and Space Administration
NCEP	National Centers for Environmental Prediction
NDVI	Normalized difference vegetation index
NGDC	National Geophysical Data Center
NLP	Night-light pollution
NOAA	National Oceanic and Atmospheric Administration
NPOESS	National Polar-orbiting Operational Environmental Satellite System
NPP	Preparatory Project Satellite
NTL	Nighttime light
OLS	Operational Linescan System
SQM	Sky Quality Meter
TIR	Thermal infrared
TM	Thematic Mapper
VANUI	Vegetation-adjusted normalized urban index
VIIRS	Visible Infrared Imaging Radiometer Suite
VNIR	Visible–near infrared

11.1 Introduction

The land surface of our home planet Earth is a finite resource that is central to human welfare and the functioning of Earth systems. Human population growth is one of the major global-scale forcing that underlie the most recent global-scale state shift in Earth's biosphere (Barnosky et al. 2012). Due to industrialization, economic growth, technology advances, and population explosion, human activities worldwide are transforming the terrestrial environment at unparalleled rates and scales. Prime grasslands and forests have been converted to croplands and pastures, which cover about 40% of the global land surface (Foley et al. 2005), to support the rising need for food by more than 7 billion people (UNFPA 2011). Among anthropogenic activities, urbanization is the most irreversible and human-dominated form of land use, modifying land cover, hydrological systems, biogeochemistry, climate, and biodiversity worldwide (Grimm et al. 2008). Worldwide, urban expansion is one of the primary drivers of habitat loss and species extinction (Hahs et al. 2009). Urban areas also affect their local climates through the modification of surface albedo, evapotranspiration, and increased aerosols and anthropogenic heat sources, thereby creating elevated urban temperatures (Arnfield 2003) and changes in regional precipitation patterns (Marshall Shepherd et al. 2002; Rosenfeld 2000; Seto and Shepherd 2009). In many developing countries, urban expansion takes place on prime agricultural lands (del Mar López et al. 2001; Seto et al. 2000). Although occupying about only 2% of the global land surface (Akbari et al. 2009), cities worldwide are now hosting more than 50% of the world population (Heilig 2012), producing more than 90% of the world gross domestic product (GDP) (Gutman 2007), consuming more than 70% of energy (Nakićenović 2012), and generating more than 71% of anthropogenic greenhouse gas emissions (Hoornweg et al. 2010). However, cities also show themselves as a potential solution to climate change through efficient resource use. Compact urban development coupled with high residential and employment densities can reduce energy consumption, vehicle miles traveled, and carbon dioxide emissions (Gomez-Ibanez et al. 2009). Per capita energy use and greenhouse emissions are often lower in cities than national averages (Dodman 2009). Furthermore, increasing urban albedo could offset greenhouse gas emissions (Akbari et al. 2009).

At the same time, human societies are vulnerable to climate change. Low-lying human settlements in the coastal zones and on islands are increasingly threatened by sea level rise (Nicholls and Cazenave 2010) and severe weather events, such as extremely powerful hurricanes (Abramson and Redlener 2012; Brinkley 2006). People living in mountains are at risk of landslides (Galli and Guzzetti 2007) and flash floods (e.g., the 2013 North India floods, http://www.foxnews.com/world/2013/07/15/india-says-5748-missing-in-floods-now-presumed-dead/, retrieved on February 27, 2014). The two and a half billion people living in drylands worldwide are vulnerable to desertification and land degradation (Reynolds et al. 2007).

Monitoring and understanding the human dimensions of global change, including economic, political, cultural, and sociotechnical systems, and their interactions with the environmental systems is thus vital to the understanding of global environmental change and its consequences (Stern et al. 1991). However, monitoring human societies at the global scale can be very challenging. Social scientists have to rely on census and field survey to collect essential variables about humans, including government policies, land-tenure rules, GDP, population, population

density, energy consumption, and carbon emissions. Census and field survey can be very labor intensive, time consuming, and cost intensive and in many cases are just not impossible. Remote sensing data have long been used to monitor the global environment and its change due to its capability of taking measurements without contacting the objects of interest in a repeatable and relatively cheap way. However, conventional remote sensing techniques are limited to detecting only visible human artifacts such as buildings, crop fields, and roads, which are of less interest to many social scientists than the abstract variables that explain their appearance and transformation (Liverman et al. 1998). The variables of greatest interest to these social scientists, such as land use (e.g., residential, commercial, industrial), people's movement and focal locations of social interactions, population characteristics (e.g., size, poverty), as well as government policies, are not readily measured from space. Thus, there has been a deep doubt that remote sensing can measure anything considered important in social sciences.

We are currently living in an era of global change in hydrology, climatology, and biology that significantly differs from previous episodes of global change in the extent to which it is human in origin (Stern et al. 1991). A complex of social, political, economic, technological, and cultural variables are believed to be driving forces that influence the human activities that proximately cause global change (Stern et al. 1991). Understanding the linkages among all the driving forces is a major scientific challenge that will require developing new interdisciplinary teams, including social scientists and physical scientists.

This chapter focuses on a unique type of remote sensing technology: observing nighttime lights (NTLs) from outer space to monitor human societies. Contrasting to conventional environmental remote sensing, NTL remote sensing aims at measuring human activities from outer space. The Operational Linescan System on board the Defense Meteorological Satellite Program (DMSP/OLS) has been collecting routine NTL observations globally since the early 1970s and have been proved as a close proxy to human activities at night (Croft 1978). NTL remote sensing provides the great potential to fill the gap between social sciences and remote sensing. Furthermore, NTL analysis has its own environmental value: assessing the impacts of anthropogenic activities on the environmental systems.

Given the aforementioned discussions, this chapter will focus on the three areas of NTL remote sensing. First, a review of the history of NTL remote sensing and its evolution is provided. Second, major applications of NTL remote sensing related to human societies are illustrated. Third, challenges in future directions of NTL remote sensing are discussed and a number of suggestions are made.

11.1.1 Rationale That Underlies NTL Remote Sensing

Diurnality is a common phenomenon in the animal kingdom. Many animals' daily activities strictly depend on sunlight, active during the day and sleeping at night. Human beings used to be a member of the diurnal club. During its early history, human's activity at night was also greatly confined until they learned how to use fire to light up their living spaces. Technical advances in artificial lighting have given people greater flexibilities as to where, when, and how long their activities can take place. The use of street lighting was first recorded in the city of Antioch from the fourth century (Luckiesh 1920); later in the Arab Empire from the ninth to tenth centuries, especially in Cordova; and then in London from 1417 when Henry Barton (Lovatt and O'Connor 1995), the mayor, ordered "lanterns with lights to be hanged out on the winter evenings between Hallowtide and Candlemasse." At that time, candles were used for street lighting. After the invention of gaslight by William Murdoch in 1792, cities in Britain began to light their streets using gas. After Edison pioneered electric use, lightbulbs were developed for streetlights as well. On March 31, 1880, Wabash, Indiana, became the first electrically lighted city in the world (http://www.cityofwabash.com/city-information/history/electric-light-article/). Open arc lamps were used in the late nineteenth and early twentieth centuries by many large cities for street lighting. Some 20 years later, incandescent lightbulbs were introduced. In the late 1930s, the fluorescent lamp first became common, followed by mercury vapor streetlight assembly in 1948. Sodium vapor streetlights were put into service around 1970. In recent years, new types of streetlights have been introduced, including metal halide, ceramic discharge metal halide lamp, induction lamp, compact fluorescent lamp, and light-emitting diodes. The vivid nightscape dominated by artificial illumination of human settlements is the other side of our blue marble Earth (known as *black marble*, Kohrs et al. 2014) and can be seen from space by astronauts at night.

Artificial illumination allows human beings to break through its natural diurnality so their social, cultural, and economic activities can extend into night, creating the so-called nighttime economy (Lovatt and O'Connor 1995). At the first locations with gas and electricity lighting, the light was there to illuminate the goods on display and the people attracted to them (Baudelaire 1973; Geist 1983). Whether it is for nightclub, for bar, or for cinema, lighting becomes necessary. In modern days, lighting is first for convenience and then for safety. Night-light is thus an important and reliable indicator of human activities immediately at night and indirectly during daytime. Artificial illumination of buildings, transportation corridors, parking lots, and other elements of the built environment has become the hallmark of many contemporary urban settlements and urban activity. Our world is now a highly domesticated nature, with no place on our planet is truly free of human beings' footprints (Kareiva et al. 2007). This can be well illustrated by night-light satellite images (Figure 11.1). It is a very obvious fact that night on our planet is also highly domesticated. At the beginning of the millennium, more than two-thirds of the world's population (99% of the U.S. and the European Union population) and almost 20% of world terrain is under skies influenced by anthropogenic lights (Cinzano et al. 2001).

11.1.2 History of NTL Remote Sensing

Experiments in nighttime aerial photos were already done toward the end of World War I over East Anglia (Kingslake 1942), and during World War II (starting in 1940), nighttime aerial photos were often acquired using flash bombs for reconnaissance purposes (Katz 1948; Schulman and Rader 2012). However, it was not until the age of spaceborne imagery that the potential of this data source can contribute to our understanding of human activity has been realized. The OLS on board the U.S. Air Force DMSP was originally designed to detect clouds illuminated by moonlight for military bombing operations at nights. The first DMSP satellite carrying OLS began flying in 1976, soon after the pioneering program Sensor Aerospace Vehicle Electronics Package (1970–1976) ended. Since then a series of DMSP satellites have been launched into orbits and the latest four (F15–F18) are still in operation today. In most years, there are two or more satellites operating in space to form a constellation: one passes at dawn and the other at dusk. The DMSP program was declassified in 1972. A digital archive for the DMSP/OLS data was established in mid-1992 at the National Oceanic and Atmospheric Administration (NOAA) National Geophysical Data Center (NGDC) and made available to public access through the Internet. Digital DMSP/OLS data from 1972 to 1992 were not archived, and all scientific access to them was through films archived at the National Snow and Ice Data Center, University of Colorado.

OLS is an oscillating scan radiometer with a swath of about 3000 km. With 14 orbits per day, each OLS is capable of providing global nighttime coverage every 24 h. For cloud detection purposes, OLS has two spectral bands: visible–near infrared (VNIR) (0.5–0.9 μm) and thermal infrared (10.5–12.5 μm). Since moonlight is a weak illumination source compared with daytime solar illumination, the OLS VNIR band was designed to allow detecting radiances down to 10^{-9} W/sr/μm, which is more than five orders of magnitude lower than comparable bands of other daytime sensors, such as the NOAA Advanced Very High Resolution Radiometer and the Landsat Thematic Mapper (TM). OLS can acquire data in two spatial resolution modes: "fine" and "smoothed." In the "fine" mode, OLS captures full-resolution data with a nominal spatial resolution of 0.5 km. Onboard averaging of 5×5 pixels of fine data produces "smoothed" data with a nominal spatial resolution of 2.7 km. Most of the data received by the NOAA-NGDC is in the smooth spatial resolution mode.

Shortly after DMSP's declassification, the scientific research community found out that due to its low-light imaging capability, OLS can also detect nocturnal artificial lighting in clear night conditions without moonlight. Croft (1978) described the detection of cities, fires, fishing boats, and gas flares from the DMSP program, showing the potential of NTL data as an indicator of human activity. Since then, studies have shown strong relationships between NTL data and key socioeconomic variables such as urban population estimates (Amaral et al. 2006; Balk et al. 2006; Elvidge et al. 1997a; Sutton et al. 2001), population density (Sutton et al. 2003; Zhuo et al. 2009), economic activity (Doll et al. 2006), energy use, carbon emissions (Doll et al. 2000), impervious surfaces

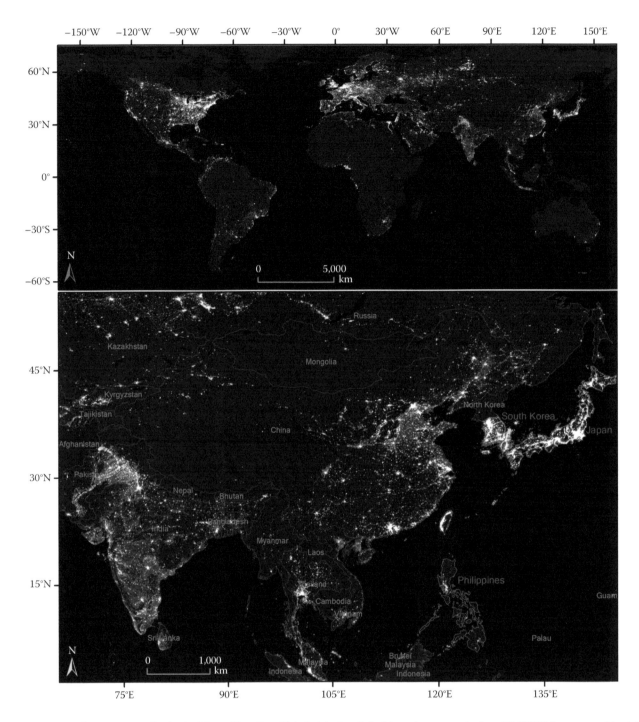

FIGURE 11.1 Nightscape as seen by the DMSP satellites. Top: Global nighttime light false color composite from DMSP/OLS average stable visible images (red, F162009; green, F152000; blue, F101992). Bottom: Close up to Asia. Urban intensity is indicated by the color intensity and change in urban intensity is indicated by the color tune. Red and yellow suggest recent growth, while blue means decline and white means stable dense urban.

(Elvidge et al. 2007b), and subnational estimates of GDP (Sutton et al. 2007). As such, they are often used as proxies for urban settlements and have been increasingly used to map aggregate measures of urban areas such as total area extent (Small et al. 2011) and regional and global urbanization dynamics (Zhang and Seto 2011).

There is a pressing need in the economic and policy community to map urban areas at finer spatial scales than those offered by the DMSP-OLS imagery. New detailed spatial resolution NTL imagery

can help to bridge the gap between social sciences (interested in explaining social processes) and remote sensing (offering the means to map spatial patterns and processes) (Rindfuss and Stern 1998). The recent launching of new sensors now allows us, for the first time, to study NTL at higher spatial resolution than before and thus will enable us to gain new insights on urban dynamics. The new high-sensitivity camera on board the Aquarius/Satelite de Aplicaciones Cientificas (SAC)-D, launched in June 2011, offers NTL images at a spatial resolution of 200–300 m (Sen et al. 2006). Since the launch of the National Polar-orbiting Operational Environmental Satellite System Preparatory Project Satellite (NPP), which was later renamed as Suomi NPP, on October 28, 2011, the onboard Visible Infrared Imaging Radiometer Suite (VIIRS) has been collecting high-quality nighttime images at a spatial resolution of 750 m in the day/night band (DNB) spanning the visible and infrared regions (Lee et al. 2006; Miller et al. 2005, 2006, 2012). With its dynamic radiometric range and advanced onboard calibration facilities, VIIRS takes continuous and consistent measurements of daytime Earth surface reflectivity and NTLs free of saturation, becoming the second- and next-generation sensor to provide daily global NTL observations from outer space. Primary evaluation shows that the VIIRS DNB is performing pretty well on orbit (Liao et al. 2013), and its capability to estimate GDP is much better than DMSP/OLS at various scales evaluated (Li et al. 2013d; Shi et al. 2014).

Other new sensors that can be used for mapping night-lights include astronaut photography from the International Space Station (ISS) (spatial resolution 10–100) (Anderson et al. 2010; Dawson et al. 2012; Doll 2008), dedicated airborne campaigns (Elvidge and Green 2005; Tardà et al. 2011), and high-sensitivity technological camera on board SAC-C launched in 2000 offering NTL images at a spatial resolution of 300 m (Colomb et al. 2003). The most accessible of these sources for medium spatial resolution images of night-lights are the astronaut photographs. Available since mission 6 of the ISS, using a handheld camera Kodak DCS 760 camera, there are now thousands of images showing night-lights of hundreds of cities around the world (Doll 2008), available for downloading through the National Aeronautics and Space Administration (NASA) Gateway to Astronaut Photography (http://eol.jsc.nasa.gov/). These images have been successful not only to capture public attention to light pollution through some visually stunning videos (as in Michael König's collection; http://vimeo.com/32001208), but can also be used to analyze spatial inequality in streetlights between cities and across borders with night color imagery (Levin and Duke 2012). However, ISS night photography suffers from some shortcomings, including imprecise geolocation, inconsistency in the spatial resolution and in the clarity of the image, and as the images are captured by relatively simple digital cameras, they cannot be quantitatively calibrated into photometric units (Doll 2008).

Imaging spectrometry of NTLs may enable a nearly full characterization of lighting type (Kruse and Elvidge 2011); however, this may also be accomplished using a multispectral sensor (Elvidge et al. 2010), and lamp classes were successfully identified from aerial color night-light images (Hale et al. 2013). The potential use of information from high spatial resolution NTL sensors has been summarized in a proposal submitted to NASA, to launch an NTL dedicated sensor to be termed Nightsat, with a suggested spatial resolution between 50 and 100 m (Elvidge et al. 2007c; Hipskinda et al. 2011).

The methodology for acquiring an aerial high spatial resolution night-lights image has been demonstrated over Berlin (1 m) by Küchly et al. (2012). This has been followed by an even finer (10 cm) aerial high spatial resolution night-lights orthophoto

acquired over Birmingham, United Kingdom (Hale et al. 2013). High spatial resolution NTL brightness can reflect political and socioeconomic factors at the city level (Levin and Duke 2012) and explain the nesting of sea turtles (Mazor et al. 2013). Additional research using higher spatial and temporal resolution night-light imagery will undoubtedly expand and "demonstrate the real potential and utility of satellite remote sensing for regional science in urban environments" (Patino and Duque 2013).

The full list of available NTL sensors is summarized in Table 11.1.

11.1.3 DMSP/OLS Data Archive

Given the long historical archive of DMSP/OLS NTL data and the increasing interest of using them, we give a short description of the process to generate the version 4 annual composites, based on the DMSP/OLS user guide (http://ngdc.noaa.gov/eog/gcv4_readme.txt). OLS has a wide view over the Earth surface—scanning as wide as 3000 km land surface below across the sub-satellite track (orbital swath) in every single pass. The NTL data collected by the DMSP/OLS measure light on Earth's surface such as those generated by human settlements, gas flares, fires, and illuminated marine vessels. The current image processing practiced by NOAA is to generate annual global composites through extracting best quality observations only. There are a number of sources that can contaminate NTL observations, including clouds, solar illumination, lunar illumination, aurora in the northern hemisphere, and lightning. Clouds are first excluded with the help of the OLS thermal band and National Centers for Environmental Prediction (NCEP, NOAA) surface temperature grids. Sunlit and moonlit data are excluded based on solar elevation angle and moon phase. Lighting from aurora is screened and excluded in the northern hemisphere on an orbit-by-orbit manner using visual inspection. Finally, only data from the center half of each 3000 km wide OLS swath go into the final composites to ensure high-quality observations in terms of geolocation, pixel size, and the consistency of radiometry. The version 4 annual composites contain three products in a nominal 30 arc second resolution (about 1 km):

1. Total number of cloud-free observations. Data quality in areas with low number of cloud-free observations is often reduced. In some years, there are areas with zero cloud-free observations.
2. Average visible band value with no further filtering. This product is generated by averaging the visible band digital number of all cloud-free observations and valid data range from 0 to 63. Areas with zero cloud-free observations are flagged by the value 255.
3. Stable average visible band value after cleaning. Ephemeral events, such as wildfires, have been screened and discarded. The final product contains lights from cities, towns, and other sites with persistent lighting year round. This is the commonly used DMSP/OLS NTL product.

The similar procedure is also carried out to generate a 2012 annual mosaic from VIIRS DNB imagery (Baugh et al. 2013) and

TABLE 11.1   Comparison of Available Spaceborne Sensors for Night-Lights Mapping, Sorted by Spatial Resolution

Sensor	Spatial Resolution (m)	Operational Years	Temporal Resolution	Products	Radiometric Range	Spectral Bands	Main References
DMSP-OLS	3000	Declassified in 1972, digital archive available from 1992 onward	Global coverage can be obtained every 24 h.	Stable lights, radiance calibrated, average DN	10^{-6}–10^{-9} W/ cm^2/sr/μm 6 bit	Panchromatic 400–1100 nm	Doll (2008), Elvidge et al. (1997b)
NPP-VIIRS	740	Launched in Oct 2011	Daily images can be downloaded.	Near real time from http:// ngdc.noaa.gov/eog/viirs/ download_dbs.html In development, including fire and power outage detection. Cloud-free composites available in radiance units of nW/(cm^2/sr)	14 bit	Panchromatic 505–890 nm	Miller et al. (2012)
SAC-C HSTC	300	Launched in Nov 2000	Sporadic.	N/A	8 bit	Panchromatic 450–850 nm	Colomb et al. (2003)
SAC-D HSC	200–300	Launched in June 2011	Sporadic.	N/A	10 bit	Panchromatic 450–900 nm	Sen et al. (2006)
Astronaut photographs on board the International Space Station (ISS)	10–200	From 2003 onward (since mission ISS006)	Photos taken irregularly.	Photos can be searched and downloaded from http://eol.jsc.nasa.gov/	8 bit	RGB	Doll (2008), Levin and Duke (2012b)
EROS-B	0.7	Night-light images offered since mid-2013	Commercial satellite, acquires images on demand.	N/A	16 bit	Panchromatic	Levin et al. (2014)

is also available for download from the website of NOAA NGDC. Recently, some monthly mosaics from VIIRS DNB have become available for downloading as well from the same website.

11.2 Major Applications of NTL

Since Croft (1978) first identified the potential of DMSP/OLS NTL imagery as an indicator of human activities, numerous applications have been reported in the literature to use them. A thorough review of all of them is not practical here due to limit of space (audiences are advised to the Center for International Earth Science Information Network thematic guide to NTL remote sensing (Doll 2008) for a detailed review of major applications before 2008). We place our current effort on new trends of applications emerging in recent years. Here we summarize them into several categories and focus on introducing the most representative cases within each category.

The majority of NTL remote sensing applications use only one single annual DMSP/OLS composite in a specific year. They can be further grouped into urban-related applications, such as urban extent, urban population, urban energy use, urban carbon emissions, urban population, urban GDP, conflict, and in-use stocks, and environment-related applications, such as gas flare, wildfires, and light pollution.

11.2.1 Urban Extent and Socioeconomic Variables

As one looks at the NTL images (Figure 11.1), the first impression one immediately gets is that NTL intensity varies with urban built-up intensity. It is often true that cities are well lit while rural areas and remote areas are not, which indicates that it is possible to find a threshold to delineate urban extents from DMSP/OLS NTL. Numerous efforts have been tried to find that optimal threshold (Imhoff et al. 1997; Small et al. 2005, 2011). However, the conclusion is that there is no single optimal threshold that can accurately delineate both large cities and small cities simultaneously. A larger threshold might be good for delineating large cities but will underestimate small cities. A smaller threshold can bring back small cities but tends to overestimate the extents of large cities. This dilemma mainly comes from the overglow effect in DMSP/OLS NTL but may also be related to the type of lighting as well as differing street lighting standards in different countries (Small 2005).

Empirical analyses find strong correlations between NTL and socioeconomic variables, such as population, GDP, carbon emissions, and energy use as mentioned earlier. Aggregated urban extent and later stable NTL intensity are used to build regression models. These models can then be used to disaggregate

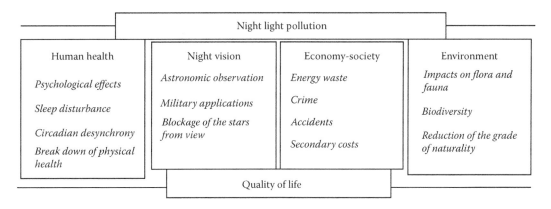

FIGURE 11.2 The main consequences of night-light pollution.

population (Zhuo et al. 2009), carbon emission (Oda and Maksyutov 2011), GDP (Chen and Nordhaus 2010), and in-use stock (Rauch 2009), which are often reported by administrative units, to a higher resolution or at the pixel level. However, it is found that the overglow and saturation problems cause biases in the derived models, and there might be a discrepancy between lighting and living when mapping the distribution of population at the pixel level. For example, industrial zones, airports, and commercial strips can have high NTL intensity but with few residents living there. This leads to efforts trying to figure out the relationship between lighting and land use and land cover with finer-resolution NTL images (Hale et al. 2013).

11.2.2 NTL Pollution

There is a dark side to NTL (Hölker et al. 2010a). Light pollution, the disturbance of the natural dark night sky due to night-light emissions, has been identified as one factor of environmental pollution. Night-light pollution described as "one of the most rapidly increasing alterations to the natural environment" is a problem whereby "mankind is proceeding to envelop itself in a luminous fog" (Cinzano et al. 2001). Recently, light pollution is considered not only as an alteration of the night sky for an observer but as a real environmental pollution (Cinzano and Falchi 2013).

The rapid growth of nighttime light pollution (NLP)— mainly above areas with intense human activities—is not only damaging the human's capability to perceive the universe (Marin 2009) but also has significant impact to all kinds of environmental, health, and socioeconomic factors related with the disturbance of the nighttime environment (Navara and Nelson 2007). Astronomers are among the worst affected by NLP (Falchi and Cinzano 2000; IDSA 1996), because of its direct impact on astronomical observation, but environmentalists and physicians are also worried about the direct and indirect effects on wildlife, as well as the reduction in the overall "quality of life" for the people exposed to significant night-light emissions (Gaston et al. 2013). In the last few decades negative

consequences of NLP have been reported to night vision, socioeconomic procedures, and environment, as well as to human health (Navara and Nelson, 2007; Figure 11.2).

11.2.2.1 Night Vision

In the document of Proclamation of 2009 as International Year of Astronomy (presented in 2005 at the 33rd Session of the UNESCO General Conference), the sky observation is presented as following: "Humankind has always observed the sky either to interpret it or to understand the physical laws that govern the universe. This interest in astronomy has had profound implications for science, philosophy, religion, culture and our general conception of the universe." Thus, sky is a significant component of our common universal heritage. The consequences of artificial night lighting to night vision and sky observation are profound. Many contemporary megacities produce a glow in the night sky that can be seen from more than 150 km away. At the beginning of the millennium, more than two-thirds of the world's population (99% of the United States and the European Union population) and almost 20% of world terrain are under light polluted skies (Cinzano et al. 2001). Furthermore, light pollution is considered an important driver for the loss of esthetic values such as the visibility of the Milky Way (Smith 2008).

11.2.2.2 Human Health: Medicine

Recently, many researchers argue that the light at night may lead to circadian desynchrony (among others; Pauley 2004; Salgado-Delgado et al. 2011). Accordingly, this desynchrony leads individuals to the loss of internal temporal order and to wide psychological disorders including impulsivity, mania, and depression. Moreover, many other health problems such as sleep and metabolic disorders are associated with internal desynchronization and potentially with the night-light pollution. Exposure to NLP suppresses the production of the pineal hormone melatonin, and since melatonin is an anticarcinogenic agent, lower levels in blood may encourage tumorigenesis (Bullough et al. 2006; Kloog et al. 2009).

11.2.2.3 Environment

The ecologic effects of NLP have been well documented. Light pollution has been shown to affect both plants and animals. The duration of darkness controls the metabolism as well as the growth of plants. Thus, NLP can disrupt plants by distorting their natural day–night cycle (Longcore and Rich 2004). Research on many wildlife animals shows that light pollution can alter their habitats, orientation, foraging areas, and even their physiology, not only in urban centers but in rural areas as well (Salmon 2003). Gaston et al. (2013) proposed a framework that focuses on the ways in which NLP influences biological systems by making the distinction between light as a resource and light as an information source. In their research, they argue that NLP has downstream effects to the structure and function of biological organizations from cell to ecosystem. They also underline that even low levels of NLP can have significant ecological impacts. Nowadays, NLP is treated as a biodiversity threat (Hölker et al. 2010b).

Artificial lighting also disturbs the "tranquility" (grade of naturality) of an area. This kind of pollution is directly correlated to the presence of human activities and for this reason is considered of high interest (see Levin et al. 2015). Tranquility maps are a valuable tool for the classification of parts of the countryside, as well as for the classification of areas that are relatively undisturbed by noise and visual intrusion, areas representative of "unspoilt" countryside. Tranquility can be defined as "the sense of peace, quiet and natural pureness of the countryside." While tranquility disturbance is profound in modern urban areas, suburban and rural areas also face the same problems due to urban growth, intense cultivation activities, transportation network expansion, etc. Thus, research has been addressed in order to assess the potential use of nighttime remotely sensed images for modeling light pollution, as well as to estimate the grade of light pollution in suburban areas (Chalkias et al. 2006).

11.2.2.4 Economy: Society

The increasing prevalence of exposure to light at night has significant social, behavioral, and economic consequences—among others—that are only recently becoming apparent (Navara and Nelson 2007). At the same time, there is an urgent need for light pollution policies given the dramatic increase in artificial light at night. This increase varies (from 0% to 20% per year—mean value, 6%) depending on the geographic region (Hölker et al. 2010a). Light pollution generates significant costs related not only to the profound waste of energy but also to negative impacts on wildlife, health, and the astronomy. According to Gallaway et al. (2010), the amount of this loss in the United States is approximately $7 billion annually. Current scientific models of light pollution are purely population based. Fractional logit models in their research showed that both population and GDP are significant explanatory variables of NLP.

Major applications using NTL time series are summarized in Table 11.2.

TABLE 11.2 Summary of Major Applications Using Nighttime Lights

Application	Strengths/Limitations	Data Source	References (Selected)
Urban extent	Quick/simple estimations at global scale; many differences according to the size of city and type of lighting; overglow effect.	DMSP/OLS; Suomi NPP/VIIRS DNB	Imhoff et al. (1997), Small et al. (2005, 2011)
Population/population density	Efficient representations of the spatial heterogeneity; low accuracy of the final outputs; differences across NLE data from various satellites.	DMSP/OLS; Suomi NPP/VIIRS DNB; ISS	Levin and Duke (2012), Zhuo et al. (2009)
GDP	Beneficial in countries with poor statistical systems; limited use to provide accurate indicators of economic activity, or for assessing regional GDP.	DMSP/OLS; Suomi NPP/VIIRS DNB	Bickenbach et al. (2013), Chen and Nordhaus (2010), Elvidge et al. (2007c), Shi et al. (2014), Sutton et al. (2007)
Carbon emissions	Strong correlation only in developed countries; coarse estimations. Stable lights are less appropriate than radiance calibrated products due to pixel saturation.	DMSP/OLS; Suomi NPP/VIIRS DNB	Ghosh et al. (2010), Letu et al. (2014), Oda and Maksyutov (2011)
Energy consumption	Suitable for global studies; limited availability of energy data time series in order to evaluate statistical analysis; underestimations caused by saturated pixels. Advantage of VIIRS over DMSP due to improved spatial resolution and the availability of data in physical units (W/cm²/sr).	DMSP/OLS; Suomi NPP/VIIRS DNB	Amaral et al. (2005), Coscieme et al. (2013), Shi et al. (2014)
In-use stock	Useful estimations of the in-use stock deposit dynamics; evaluation of the in-use stock in areas where statistical data are incomplete; not suitable for precise estimations. Improvements in methodology expected by using the new VIIRS sensor.	DMSP/OLS	Hattori et al. (2013), Rauch (2009), Takahashi et al. (2010)
Light pollution	Suitable for modeling light pollution as well as estimating light pollution in suburban areas; whereas DMSP/OLS data are limited by its coarse resolution, fine spatial resolution airborne and spaceborne sensors are now becoming available.	DMSP/OLS; emerging fine spatial resolution sensors (airborne, EROS-B)	Chalkias et al. (2006), Falchi et al. (2011), Hale et al. (2013), Kuechly et al. (2012), Kyba et al. (2011), Levin et al. (2014), Liu et al. (2013), Miller (2006), Navara and Nelson (2007), Olsen et al. (2013), Simpson et al. (2006)

11.3 Study of Urban Dynamics with NTL Time Series

11.3.1 Mapping Urban Extent Dynamics

The long historical archive of DMSP/OLS NTL data has the high potential in characterizing urbanization dynamics at regional and global scales. However, such a potential was not well explored until very recently due to differences in satellite orbits (dawn pass versus dusk pass) and sensor degradation, which causes NTL data collected by sensors on board different DMSP satellites to be significantly different, even when there are no real changes occurring on the ground. The version 4 DMSP/OLS NTL annual composites include data from five satellites: F10, F12, F14, F15, and F16 (Figure 11.3). In theory, a time series of NTL data could capture the dynamics of urbanization despite possible errors from sensor differences. This would require the signal of change to be larger than the error signal and also large enough to render the error signal (noise) unimportant.

We can derive five stylized facts from Figure 11.3:

1. The time series are very noisy. The shift between satellites brings in very large biases.
2. The biases show up in all countries in a similar way, which indicates systematic processes may have been in effect.

3. However, within the span of each satellite (except for F15, which has two distinct segments), there exists a clear urban development pattern in each country.
4. The overlapping sections of F12 and F14 as well as between F14 and F15 show very similar patterns, although there are differences in magnitude.
5. Each satellite can capture an urban development trend, that is, the noise is not large enough to overwhelm the signal within the span of each satellite.

These facts strongly suggest that the signal of change captured by the DMSP/OLS sensors is larger than the error signals and can be used as an indicator of urban change. This is further confirmed by validation results using population data at the country level as well as those using land cover data at state and regional levels in the United States (Zhang and Seto 2011). Zhang and Seto (2011) first explored the potential of utilizing time series NTL to map urbanization dynamics at regional and global scales. Using time series constructed with 17 NTL images from 4 satellites, F10 (1992–1994), F12 (1995, 1996), F14 (1997–1999), and F15 (2000–2008), they successfully identified five different types of urbanization dynamics in various regions (Figure 11.4 shows results in the Pearl River Delta region) with an iterative unsupervised classification procedure.

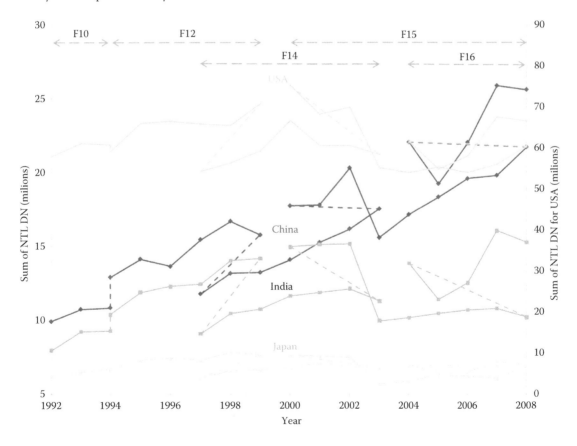

FIGURE 11.3 DMSP/OLS NTL time series composed of data from several satellites (F10, F12, F14, F15, and F16). Dotted line segments in the time series indicate satellite shifting. (Adapted from Zhang, Q. and Seto, K.C., *Remote Sens. Environ.*, 115, 2320, 2011.)

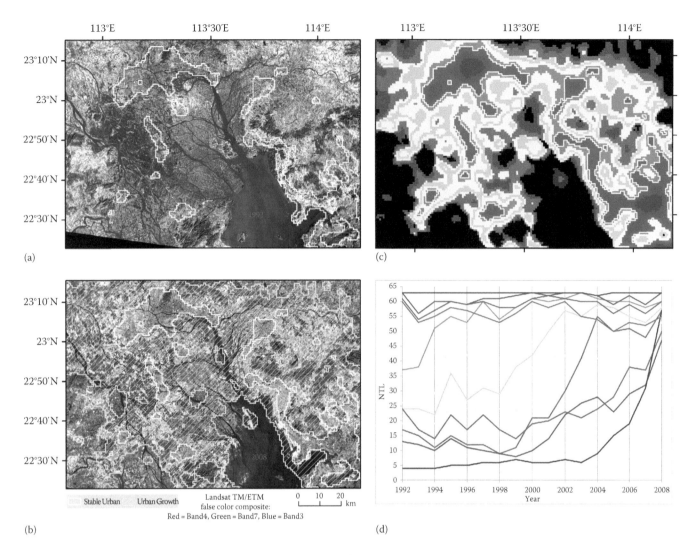

FIGURE 11.4 Urban dynamics in the Pearl River Delta region, China, from 1992 to 2008: (a) Stable urban areas in 1992, (b) urban growth areas in 2008, (c) multitemporal urbanization dynamics as detected from NTL, and (d) temporal NTL profiles of different types of urbanization dynamics. (Adapted from Zhang, Q. and Seto, K.C., *Remote Sens. Environ.*, 115, 2320, 2011.)

A number of applications have emerged in recent years to utilize multitemporal DMSP/OLS NTL annual composites for characterizing urbanization dynamics in different regions, such as China (Liu et al. 2012; Ma et al. 2012), Northeast China (Yi et al. 2014), South America (Álvarez-Berríos et al. 2013), India (Pandey et al. 2013), the conterminous United States (Zhang et al. 2014), and Hanoi, Vietnam (Castrence et al. 2014). Recently, Bennie et al. (2014a) examined the trends of light pollution across Europe from 1995 to 2010 using time series of DMSP/OLS NTL.

Liu et al. (2012) reported that urban growth patterns from 1992 to 2008 revealed by DMSP/OLS NTL time series in three cities of China are very similar to those revealed by Landsat TM/Enhanced Thematic Mapper Plus (ETM+) time series, with an average overall accuracy of 82.74%. However, mainly due to its relatively coarse spatial resolution and low radiometric resolution, urban expansion detected with DMSP/OLS are much larger

than that with Landsat TM/ETM+. During that period, urban areas expanded 107,593 ha in Beijing, 52,838 ha in Chengdu, and 41,004 ha in Zhengzhou, as detected with NTL time series. However, those numbers are only 49,733 ha in Beijing, 48,583 ha in Chengdu, and 25,992 ha in Zhengzhou, as detected with Landsat TM/ETM+ time series.

11.3.2 Detecting Socioeconomic Changes

Characterizing urbanization dynamics can help understand the relationship between urban changes and socioeconomic changes. Jiang et al. (2012) modeled urban expansion and cultivated land conversion for hot spot counties in China. Frolking et al. examined macroscale changes in urban structure in a number of global cities from 1999 to 2009 through the combination of DMSP/OLS NTL and NASA's SeaWinds microwave scatterometer data (Frolking et al. 2013). Research also finds that armed conflict events have significant impacts on NTL through an analysis of the relationship between armed conflicts and DMSP/OLS NTL variation from 1992 to 2010 in 159 countries (Li et al. 2013a). Shortland et al. (2013) discover that NTLs provide striking illustrations of economic decline and recovery and clearly show the contrast between the stable regions of Northern Somalia and the chaos and anarchy of Southern Somalia. Similarly, decadal economic decline in Zimbabwe is successfully detected using DMSP/OLS NTL time series (Li et al. 2013c).

11.3.3 Tracking Social Events with High Temporal Frequency NTL Time Series

The majority of change detection applications mentioned earlier depend on the time series constructed from the DMSP/OLS NTL annual composites. However, considering that DMSP/OLS can cover the entire globe every day, its temporal resolution is greatly reduced after the construction of annual NTL composites. Bharti et al. (2011) utilized a time series constructed from daily DMSP/OLS imagery and found that high temporal frequency DMSP/OLS NTL data offer a very powerful tool to measure seasonal variation in population density for studying measles epidemics. Measles epidemics in West Africa cause a significant proportion of vaccine-preventable childhood mortality. Epidemics are strongly seasonal, but the drivers of these

fluctuations are poorly understood, which limits the predictability of outbreaks and the dynamic response to immunization. Spatiotemporal changes in population density as measured by DMSP/OLS NTL are useful to explain measles seasonality. With dynamic epidemic models, measures of population density are essential for predicting epidemic progression at the city level and for improving intervention strategies. The ability to measure fine-scale changes in population density with high temporal frequency NTL data has implications for public health, crisis management, and socioeconomic development.

Such a potential can be further expanded by the VIIRS DNB, from which well-calibrated and well-normalized time series with daily coverage can be relatively easier to obtain. A preliminary analysis of the 2012 Ramadan, the ninth month of the Islamic calendar, in Cairo, Egypt, illustrates the improved potential of VIIRS DNB (Figure 11.5). Muslims worldwide observe Ramadan as a month of fasting. While fasting from dawn until sunset during Ramadan, Muslim people have food and drink after sunset and before sunrise every day, which means increased activities during night compared with non-Ramadan periods. Although still quite noisy, the unfiltered VIRRS DNB time series shows significant increase in NTL intensity during Ramadan in Cairo, Egypt, to reflect increased human activities during that special period (Figure 11.5).

Major applications using NTL time series are summarized in Table 11.3.

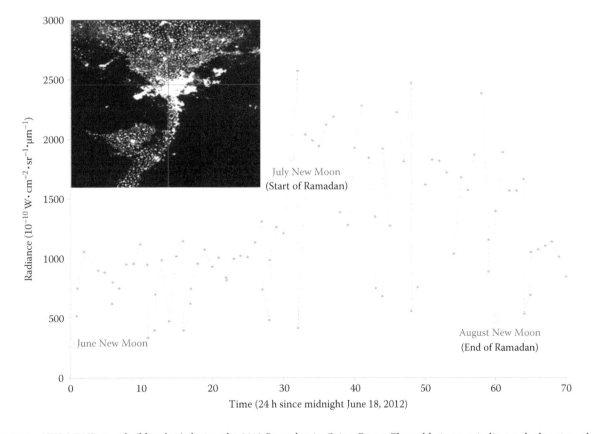

FIGURE 11.5 VIIRS DNB signals (blue dots) during the 2012 Ramadan in Cairo, Egypt. The red hair cross indicates the location where the VIIRS DNB time series was extracted.

TABLE 11.3   Summary of Major Applications Using Nighttime Lights Time Series

Application	Strengths/Limitations	Data Source	References (Selected)
Urban extent dynamics	Simple and fast for regional and global urban growth pattern detection; accuracy is limited due to its coarse spatial resolution and low radiometric resolution.	DMSP/OLS	Bennie et al. (2014a), Liu et al. (2012), Ma et al. (2012), Small and Elvidge (2013), Zhang et al. (2014), Zhang and Seto (2011)
Armed conflict	Night-light variation provides a good indicator for armed conflicts; can be confused with natural disasters.	DMSP/OLS	Li et al. (2013a)
Economic decline and recovery	Valuable in regions with poor economic data; reliability is limited by its coarse resolution and intersatellite variation.	DMSP/OLS	Li et al. (2013c), Shortland et al. (2013)
Seasonal population density variation	Useful for understanding the relationship between population density and disease transmission; high temporal resolution data are not readily available.	DMSP/OLS	Bharti et al. (2011)
Festival of Muslims' Ramadan	Simple and fast for tracking social events; raw data still contain a large amount of noise, which requires huge amount of efforts for image preprocessing.	Suomi NPP/ VIIRS DNB	The current manuscript

11.4 Challenges

The DMSP/OLS instrument was initially designed to observe moonlit clouds (Elvidge et al. 1997c). Due to its low-light imaging capability, the instrument can also detect artificial lighting at night in clear conditions without moonlight. The NTL data collected by the DMSP/OLS uniquely measure light on Earth's surface such as that generated by human settlements, gas flares, fires, and illuminated marine vessels. Although NTL does not directly measure human settlements or urban land cover, the data are strongly correlated with many characteristics of urban settlements, including carbon emissions (Doll et al. 2000) and economic activity (Sutton et al. 2007). However, a number of issues, including sensor degradation, limited radiometric range, and satellite orbit difference (dawn pass vs. dusk pass), limit the utility of NTL data for many applications. Chief among these is the well-documented saturation of data values in core urban areas (Elvidge et al. 2007a). Data values for the core of urban centers—where there are many lights—are much brighter entities than moonlit clouds. However, due to the limited radiometric range of DMSP/OLS, data values in these regions tend to be truncated. This problem of saturation is a significant challenge in using NTL for characterizations of interurban areas.

11.4.1 Correcting Saturation in DMSP/OLS NTL Imagery

In recent years, several methods have been proposed to correct or reduce NTL saturation. The correction algorithms vary in the spatial scale at which they are implemented, their complexity, and their success. In general, the correction methods fall into two categories: those that utilize only NTL data and those that use other satellite data to correct the NTL data. Of those that only use NTL data for the correction, Ziskin et al. (2010) generate nonsaturated NTL data by adding additional data taken at a "low gain" setting in dense urban areas to the operational NTL data that are taken at a "high gain" setting. It is the operational "high gain" setting of DMSP/OLS that causes bright urban centers to be saturated, whereas the "low gain" setting allows bright urban centers

to be captured with finer details. In terms of accuracy and data quality, Ziskin's method, utilizing the dynamic gain settings, is ideal. However, this method as currently applied with DMSP/OLS is very labor and cost intensive, and therefore the corrected NTL data are only available for a very limited number of years (Elvidge et al. 1999; Ziskin et al. 2010), and the method is unlikely to be used to correct the entire historical NTL archive. Letu et al. (2010) apply cubic regression models to correct saturation within an administrative unit. However, this method cannot be applied at the pixel scale. Letu et al. (2012) later propose a method to correct NTL saturation at the pixel scale based on 1999 nonsaturated NTL data. A major assumption of this approach is that the NTL intensity in saturated areas did not change from 1996 to 1999. Based on this assumption, they built a linear regression model based on data from the nonsaturated (in the 1996–1997 stable light image) regions and applied the derived model to correct saturation in NTL for the 1996–1997 image.

The major drawback of this method comes from the assumption on which it is based. While that assumption may hold in countries or regions with high urbanization levels (e.g., Japan, United States), it is not valid in many developing countries such as India and China where rapid urban expansion is the norm rather than the exception. Thus, their method has limited applicability for correcting NTL saturation globally.

It has been shown that vegetation abundance is closely and inversely correlated with impervious surfaces, a characteristic of many urban areas (Bauer et al. 2004; Pozzi and Small 2005; Small 2001; Weng et al. 2004, 2006). Recent investigations have attempted to combine information from normalized difference vegetation index (NDVI) and NTL to greatly enhance urban features (Cao et al. 2009; Lu et al. 2008), to examine the impact of urbanization on net primary productivity (Milesi et al. 2003), and to develop more accurate maps of urban areas (Lu et al. 2008; Schneider et al. 2003). This second category of NTL correction methods is based on the rationale that key urban features are inversely correlated with vegetation health and abundance. The human settlement index normalizes the NTL with Moderate Resolution Imaging Spectroradiometer (MODIS) NDVI, measures of vegetation health and quantity

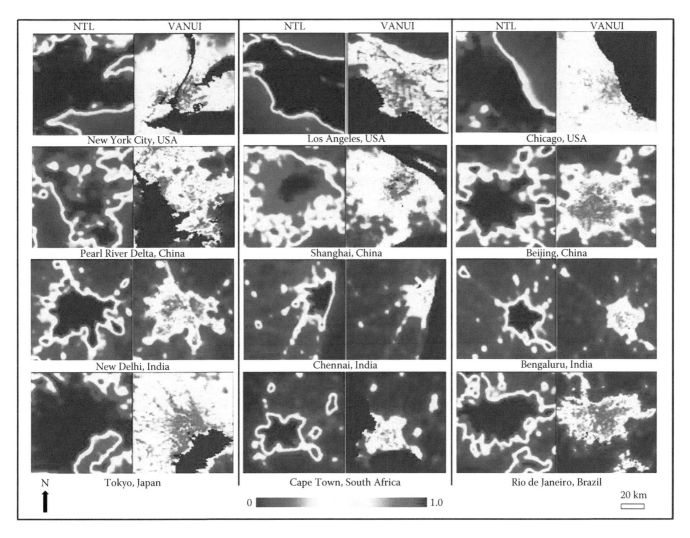

FIGURE 11.6 2006 normalized NTL with saturation and VANUI for select cities and urban regions around the world. Note that NTL and VANUI are normalized and unitless and are both scaled 0–1.0. (Adapted from Zhang, Q. et al., *Remote Sens. Environ.*, 129, 32, 2013.)

(Rouse Jr et al. 1974; Tucker 1979), to mitigate the saturation effects (Lu et al. 2008).

Zhang et al. (2013) proposed a simple method, vegetation-adjusted normalized urban index (VANUI), to correct NTL saturation with MODIS NDVI:

$$VANUI = (1 - NDVI) * NTL \qquad (11.1)$$

where NDVI is NDVI derived from MODIS. As negative MODIS NDVI values are usually associated with water and glacier, NDVI values here are constrained to the range of non-negative values between 0 and 1.0. The parenthetical expression 1 − NDVI inverses the shape of NDVI transects. This simple calculation assigns larger nonvegetative weights (as evidenced by 1 − NDVI) to core urban areas than to periurban areas, which results in an increased variability of data values within urban cores. Therefore, we expect core urban areas to have positive VANUI values close to 1, while nonurban, nonilluminated area—usually those with an abundance of vegetation—to have

low VANUI values close to 0. We also expect periurban areas and regions affected by overglow to exhibit lower VANUI values than the city core. Figure 11.6 shows VANUI for select urban regions around the world, chosen to represent a wide variation in city size, geography, and economic base. For all urban areas examined, the normalized NTL values were saturated in the urban cores, and within urban NTL, variations are not detectable. In contrast, VANUI captures the fine spatial details in and around urban areas.

11.4.2 Intercalibrating the DMSP/OLS NTL Time Series

Although signal of change in the DMSP/OLS NTL time series is larger than the error signal and also large enough to render the error signal (noise) unimportant, to facilitate accurate change analysis with NTL time series, it is necessary to calibrate first to minimize differences caused by satellite shift. The challenge to achieve successful radiometric calibration of remote sensing imagery obtained at different times is to find invariant ground targets that can be used as references for reliable comparison over time. Elvidge et al. (2009) chose Sicily, Italy, as the reference site and derive second-order regression models between the reference image F121999 and other images individually with all data points in Sicily. These models were then applied to calibrate the entire time series from 1992 to 2008. This method successfully reduces differences caused by satellite shift to some degree. However, it is questionable that models derived in Sicily can be generalized to cover the entire globe, since noises introduced by various sources might not be geographically consistent. For this reason, regional urbanization dynamic researchers derived their models that can suit closely to their specific regions by choosing local reference sites (Liu et al. 2011, 2012; Nagendra et al. 2012; Pandey et al. 2013). In an attempt to produce more generalized models for the entire globe, Wu et al. (2013) extended the Elvidge et al. (2009) method by selecting more reference sites, including Mauritius, Puerto Rico, and Okinawa, Japan. Despite that the Wu et al. (2013) method achieves improvement, the way they look for invariant regions is not essentially different than that applied by Elvidge et al. (2009) and also suffers the limitation of subjective. Li et al. (2013b) designed an automatic method to find invariant pixels in Beijing, China. They assumed that there exist stable pixels in the region and run a regression with all pixels without any screening. Based on the resulted regression model, they look for outliers, which are considered pixels that experienced change, and discard them iteratively. A final model can be built with only stable pixels identified. This automatic method can minimize bias introduced by subjective selection of invariant regions and has the potential to be extended to the entire globe. However, since the region of Beijing experienced dramatic change in the past decades, this method might lead to overcorrection to the NTL time series. Furthermore, the iterative procedure to identify stable pixels is very computation intensive thus cannot be directly implemented at the global scale, considering the gigantic amount of pixels.

11.5 Outlook: Fine-Resolution NTL Remote Sensing

The choice of the appropriate sensor to use is a function of the spatial extent of the area that needs to be covered, the temporal dynamics of the studied phenomena, its spatial heterogeneity, and the purpose for which this mapping will be used. The same general principles apply for mapping night-lights. While DMSP-OLS and VIIRS are the sensors of choice for mapping global patterns of night-lights and trends over time, medium spatial resolution sensors are required to examine urban patterns. Using ISS imagery, the development of new built-up areas and of lighting infrastructure can be mapped at spatial resolutions equivalent to those available from Landsat-type satellites. One of the most rapid developing regions in the world is that of the United Arab Emirates, where oil and gas revenues fuel the

economy and where the emirates try to position their countries on the global stage (Bagaeen 2007; Figure 11.7—urban changes in Abu Dhabi between 2003 and 2013). When even finer details are of interest, for example, to examine lighting at the street level or even for mapping individual streetlights, fine spatial resolution images (covering small areas) are required. While in the past such fine spatial resolution was only available from dedicated aerial campaigns, since mid-2013 ImageSat is offering high spatial resolution (<1 m) night-light images from its Earth Remote Observation System-B satellite (Levin et al. 2014). Medium and fine spatial resolution nighttime images can also assist to calibrate and validate models that aim at predicting light pollution. Such models use geographic information system layers of streetlights (location and type of streetlights), the topographic layout of terrain, buildings' heights (and vegetation if available), and visibility analysis algorithms so that we can better understand the extent of artificial light pollution (Bennie et al. 2014b; Chalkias et al. 2006; Gaston et al. 2012; Teikari 2007).

11.5.1 Applications of Fine Spatial Resolution Night-Lights

One of the basic needs for moderate and fine spatial resolution night-light images is to form a better understanding of the spatial distribution of point sources emitting night-lights and of the land use associated with light pollution. Using nighttime aerial photos, Küchly et al. (2012) estimated the contribution of streetlights to directed light pollution in Berlin to be 31.6%, whereas Hale et al. (2013) estimated that streetlights in Birmingham (United Kingdom) constituted about 38% of lit area using aerial photos. Based on these two studies, land use types that contribute most of the emitted night-lights are manufacturing, commercial, and housing. Capturing nighttime aerial photos of the University of Arizona campus at night, Kim and Hong (2013) suggested that highly reflective materials used in the built environment increase light pollution, even when full cutoff lights are used. However, additional studies from different cities around the world are needed to generalize such conclusions.

Once the spatial distribution of night-lights is mapped, it can contribute to different aspects of urban planning. Lit areas were found to be safer for traffic (Jackett and Frith 2013) and are often perceived as safer areas from crime (Loewen et al. 1993). Thus, by mapping and planning streetlights, planners can have an impact on the spatial distribution of crime and of accidents, which, as known, is not randomly distributed in the city (Weisburd and Amram 2014). Another important application using night-lights mapping for urban planning is related to the monitoring of energy consumption in the city, so that a city's carbon footprint can be reduced. Quantifying urban carbon footprint is not an easy task (Sovacool and Brown 2010), and night-lights mapping can contribute to this end. Urban planning can also use night-lights mapping to examine issues of spatial equality, examining the types and amount of streetlights in different neighborhoods, so as to determine whether all citizens are given similar standards of streetlights (Coulter 1983).

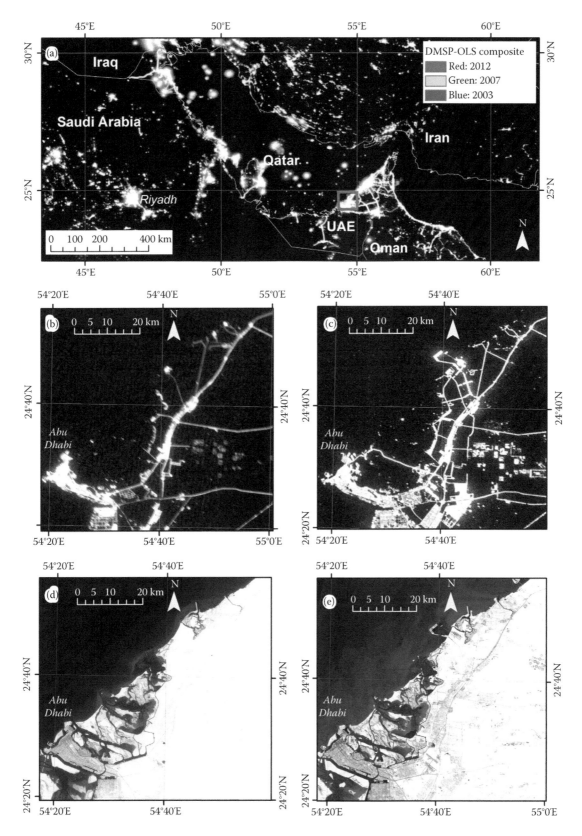

FIGURE 11.7 The reflection of rapid land use changes in Abu Dhabi in its night-lights: (a) a multitemporal composite of DMSP-OLS stable light images around the Persian Gulf (R = 2012, G = 2007, B = 2003); (b) ISS photo of Abu Dhabi, February 4, 2003, spatial resolution of 90 m; (c) ISS photo of Abu Dhabi, December 11, 2013, spatial resolution of 120 m; (d) Landsat 7 false color composite of Abu Dhabi, May 28, 2003; and (e) Landsat 8 false color composite of Abu Dhabi, December 9, 2013.

Last but not least, light pollution has been found to have negative impacts not only on astronomers that need dark sky but also on humans, animals, and plant species, due to the alteration of natural light regimes affecting our circadian clock (Chepesiuk 2009; Longcore and Rich 2004; Pauley 2004). In order to reduce our exposure to light pollution, mapping it is a basic requirement.

11.5.2 Calibrating with Ground Measurements

While fine spatial resolution mapping of night-lights is expected to benefit various applications, there are certain research gaps that need to be overcome in order to transform these data to be more quantitative. While in traditional optical remote sensing satellite images are atmospherically corrected to derive their reflectance values, it is not so clear which units should be used in night-light imagery. The DMSP-OLS imagery products are distributed as stable lights or average lights x percent (digital number values between 0 and 63). Often these products are used to calculate the total lit area or the total lights; however, these data are not in luminance units. Photometry is the measurement of the intensity of electromagnetic radiation in photometric units, like lumen and lux, or magnitudes. Radiometry is the measurement of optical radiation, with some of the many typical units encountered are W/m^2 and photons/s sr. The main difference between photometry and radiometry is that photometry is limited to the visible spectra as defined by the response of the human eye (Teikari 2007).

In recent years, there have been some attempts to calibrate fine spatial resolution images to photometric units. Hale et al. (2013) used ground measurements of incident lux along linear transects to calibrate their aerial night-light images into illuminance units. Another approach for field mapping of night-lights that can be used for calibrating aerial or spaceborne night-light imagery is using ground networks of instruments such as the Sky Quality Meter (manufactured by Unihedron, measuring the brightness of the night sky in magnitudes per square arc second; http://www.unihedron.com/projects/darksky/); however, ground networks aimed at monitoring light pollution are fairly recent (denOuter et al. 2011; Pun and So 2012; Pun et al. 2014). In an interesting study using Extech EasyView 30 light meters to map night brightness along a 10 m sampling grid on the Virginia Tech campus, brightness was measured twice: first, with the light meter pointing upward to catch direct light from the light fixtures at 30 cm from the ground and then with the light meter pointing downward to measure reflected light (Kim 2012). Thus, in addition to the inconsistency in the photometric units used for calibrating aerial night-light images, there is a gap with regard to how one should measure light on the ground so that it best corresponds with what an airborne capture or a spaceborne capture.

11.6 Summary and Future Directions

NTL remote sensing beginning with DMSP/OLS ushered in a range of important applications and has proved its value in social science research, which can help to bridge the gap between social sciences (interested in explaining social processes) and remote sensing (offering the means to map spatial patterns and processes) (Rindfuss and Stern 1998). NTL remote sensing provides a powerful tool to social scientists for estimating population size, economic activity, carbon emissions, and additional human activities, providing a bridge between physical scientists, social scientists, and remote sensing scientists.

Although almost as old as daytime optical remote sensing, such as Landsat, NTL remote sensing is still in its infancy stage and basically qualitative. Many applications are still empirical. To advance NTL remote sensing, there still exist various challenges, with regard to sensor availability, imagery acquisition, calibration between sensors and ground measurements, and the development of operational products.

There is a lack of understanding of the mechanism behind NTL remote sensing, due to the lack of studies at the ground level. The studies by researchers in the light pollution field (Chalkias et al. 2006; Gaston et al. 2012; Teikari 2007) aim to understand light emit from lighting source and how they are transferred and then make predictions of light pollution through modeling. This kind of work is highly required to better interpret NTL remote sensing images, just as radiative transfer models to daytime optical remote sensing. Understanding NTL transfer from lighting sources through the air to the sensor is critical for designing future NTL remote sensing sensors.

Another important issue is to generate high-quality data from existing NTL sensors, such as DMSP/OLS and Suomi NPP/VIIRS. The release of digital archive of DMSP/OLS and the annual composites have made remote sensing data easy to use among social scientists that often might have little training in remote sensing image processing and may feel uncomfortable of intense image processing. Numerous researches have benefited from this great remote sensing resource. However, during the compositing process, a lot of valuable information is discarded. The selection of high-quality observations is subjective. The high temporal frequency of DMSP/OLS and VIIRS DNB are valuable for intra-annual change analysis. But daily images of scientific quality are not ready available. New algorithms to generate better quality images from the long historical DMSP/OLS archives and the relatively new VIIRS data are highly desired.

Another point to make is that both DMSP/OLS and Suomi NPP/VIIRS are not dedicated to observing city lights at night, thus bringing numerous obstacles to NTL remote sensing. Sensors that are designed with a specific purpose of lighting sources on the ground are highly required in the future.

Currently, both DMSP/OLS and VIIRSS DNB are single-band sensors (the thermoinfrared band is only for cloud screening and provides no information about NTL). Multispectral band sensors are needed in order to discriminate different types of lighting techniques, which can convey very useful information.

Furthermore, both DMSP/OLS and Suomi NPP/VIIRS DNB are wide-view sensors, with swath widths greater than 3000 km, which means they can accumulate angular observations varying in a large range. Angular observations sometimes are not preferred, because they often cause variation across geography thus making mosaicking a big challenge. However, angular information is proved carrying valuable structural information and ironically is critical to normalize observations to the standard viewing–illuminating geometry, as seen in MODIS (Schaaf et al. 2002) and Multiangle Imaging Spectroradiometer (Diner et al. 1998). Due to the variation in street layout and building height, NTL is also expected to vary accordingly (Kyba et al. 2013). Angular observations from both DMSP/OLS and Suomi NPP/VIIRS DNB thus will release structural information about urban areas. Such information still remains unmined up to date. Future research is highly required to extract this invaluable information from both DMSP/OLS and Suomi NPP/VIIRS DNB.

Finally, we may suggest several promising research avenues in the field of fine (<10 m) and medium (10–100 m) spatial resolution night-light remote sensing:

1. Calibration—protocols and experience need to be gained about how to perform ground measurement of downward emitted light and upward reflected light that can be used to calibrate aerial and spaceborne images of night-lights and which units are best to use.
2. Mapping—the next step after the 2D mapping of night-scapes can be their mapping in three dimensions (e.g., also light escaping from windows), similar to 3D thermal studies, for example, using video cameras (as in Chudnovsky et al. 2004).
3. So far, almost all remote sensing studies of NTL (that we are aware of) were using either an image from a few single dates or annual composites. It is time to explore the temporal dynamics, both by creating monthly time series and by exploring the hourly changes in night-lights through the night.
4. Once the technical and methodological aspects of fine and medium spatial resolution imagery of night-lights are solved, thematic issues can be explored. Two examples include the correspondence between lit areas and hot areas (measured using a thermal sensor) and the correspondence between night-lights and human activity patterns (e.g., using traffic counts).

References

Abramson, D.M. and Redlener, I. (2012). Hurricane Sandy: Lessons learned, again. *Disaster Medicine and Public Health Preparedness, 6*, 328–329.

Akbari, H., Menon, S., and Rosenfeld, A. (2009). Global cooling: Increasing world-wide urban albedos to offset CO_2. *Climatic Change, 94*, 275–286.

Álvarez-Berríos, N.L., Parés-Ramos, I.K., and Aide, T.M. (2013). Contrasting patterns of urban expansion in Colombia, Ecuador, Peru, and Bolivia Between 1992 and 2009. *AMBIO, 42*, 29–40.

Amaral, S., Câmara, G., Monteiro, A.M.V., Quintanilha, J.A., and Elvidge, C.D. (2005). Estimating population and energy consumption in Brazilian Amazonia using DMSP night-time satellite data. *Computers, Environment and Urban Systems, 29*, 179–195.

Amaral, S., Monteiro, A., Camara, G., and Quintanilha, J. (2006). DMSP/OLS night-time light imagery for urban population estimates in the Brazilian Amazon. *International Journal of Remote Sensing, 27*, 855–870.

Anderson, S.J., Tuttle, B.T., Powell, R.L., and Sutton, P.C. (2010). Characterizing relationships between population density and nighttime imagery for Denver, Colorado: Issues of scale and representation. *International Journal of Remote Sensing, 31*, 5733–5746.

Arnfield, A.J. (2003). Two decades of urban climate research: A review of turbulence, exchanges of energy and water, and the urban heat island. *International Journal of Climatology, 23*, 1–26.

Bagaeen, S. (2007). Brand Dubai: The instant city; or the instantly recognizable city. *International Planning Studies, 12*, 173–197.

Balk, D., Deichmann, U., Yetman, G., Pozzi, F., Hay, S., and Nelson, A. (2006). Determining global population distribution: Methods, applications and data. *Advances in Parasitology, 62*, 119–156.

Barnosky, A.D., Hadly, E.A., Bascompte, J., Berlow, E.L., Brown, J.H., Fortelius, M., Getz, W.M., Harte, J., Hastings, A., and Marquet, P.A. (2012). Approaching a state shift in Earth/'s biosphere. *Nature, 486*, 52–58.

Baudelaire, C. (1973). A lyric poet in the era of high capitalism, trans. *Harry Zohn (London, 1973)*, 27–34.

Bauer, M.E., Heinert, N.J., Doyle, J.K., and Yuan, F. (2004). Impervious surface mapping and change monitoring using Landsat remote sensing. In *ASPRS Annual Conference Proceedings*, May 23–28, 2004, Denver, CO.

Baugh, K., Hsu, F.-C., Elvidge, C.D., and Zhizhin, M. (2013). Nighttime lights compositing using the VIIRS day-night band: Preliminary results. *Proceedings of the Asia-Pacific Advanced Network, 35*, 70–86.

Bennie, J., Davies, T.W., Duffy, J.P., Inger, R., and Gaston, K.J. (2014a). Contrasting trends in light pollution across Europe based on satellite observed night time lights. *Scientific Reports, 4*.

Bennie, J., Davies, T.W., Inger, R., and Gaston, K.J. (2014b). Mapping artificial lightscapes for ecological studies. *Methods in Ecology and Evolution*.

Bharti, N., Tatem, A.J., Ferrari, M.J., Grais, R.F., Djibo, A., and Grenfell, B.T. (2011). Explaining seasonal fluctuations of measles in Niger using nighttime lights imagery. *Science, 334*, 1424–1427.

Bickenbach, F., Bode, E., Lange, M., and Nunnenkamp, P. (2013). Night lights and regional GDP (No. 1888). In *Kiel Working Paper*.

Brinkley, D. (2006). *The Great Deluge: Hurricane Katrina, New Orleans, and the Mississippi Gulf Coast*: William Morrow.

Bullough, J.D., Rea, M.S., and Figueiro, M.G. (2006). Of mice and women: Light as a circadian stimulus in breast cancer research. *Cancer Causes & Control, 17*, 375–383.

Cao, X., Chen, J., Imura, H., and Higashi, O. (2009). A SVM-based method to extract urban areas from DMSP-OLS and SPOT VGT data. *Remote Sensing of Environment, 113*, 2205–2209.

Castrence, M., Nong, D.H., Tran, C.C., Young, L., and Fox, J. (2014). Mapping urban transitions using multi-temporal landsat and DMSP-OLS night-time lights imagery of the Red River Delta in Vietnam. *Land*, *3*, 148–166.

Chalkias, C., Petrakis, M., Psiloglou, B., and Lianou, M. (2006). Modelling of light pollution in suburban areas using remotely sensed imagery and GIS. *Journal of Environmental Management*, *79*, 57–63.

Chen, X. and Nordhaus, W.D. (2010). The value of luminosity data as a proxy for economic statistics. National Bureau of Economic Research.

Chepesiuk, R. (2009). Missing the dark: Health effects of light pollution. *Environmental Health Perspectives*, *117*, A20.

Chudnovsky, A., Ben-Dor, E., and Saaroni, H. (2004). Diurnal thermal behavior of selected urban objects using remote sensing measurements. *Energy and Buildings*, *36*, 1063–1074.

Cinzano, P. and Falchi, F. (2013). Quantifying light pollution. *Journal of Quantitative Spectroscopy and Radiative Transfer*.

Cinzano, P., Falchi, F., and Elvidge, C.D. (2001). Naked-eye star visibility and limiting magnitude mapped from DMSP-OLS satellite data. *Monthly Notices of the Royal Astronomical Society*, *323*, 34–46.

Colomb, R., Alonso, C., and Nollmann, I. (2003). SAC-C mission and the international AM constellation for Earth observation. *Acta Astronautica*, *52*, 995–1005.

Coscieme, L., Pulselli, F.M., Bastianoni, S., Elvidge, C.D., Anderson, S., and Sutton, P.C. (2013). A thermodynamic geography: Night-time satellite imagery as a proxy measure of energy. *AMBIO*, 1–11.

Coulter, P.B. (1983). Inferring the distributional effects of bureaucratic decision rules. *Policy Studies Journal*, *12*, 347–355.

Croft, T. (1978). Nighttime images of the earth from space. *Scientific American*, *239*, 86–98.

Dawson, M., Evans, C., Stefanov, W., Wilkinson, M.J., Willis, K., and Runco, S. (2012). Human settlements in the South-Central US, Viewed at Night from the International Space Station.

Del Mar López, T., Aide, T.M., and Thomlinson, J.R. (2001). Urban expansion and the loss of prime agricultural lands in Puerto Rico. *AMBIO: A Journal of the Human environment*, *30*, 49–54.

denOuter, P., Lolkema, D., Haaima, M., Hoff, R.v.d., Spoelstra, H., and Schmidt, W. (2011). Intercomparisons of nine sky brightness detectors. *Sensors*, *11*, 9603–9612.

Diner, D.J., Beckert, J.C., Reilly, T.H., Bruegge, C.J., Conel, J.E., Kahn, R.A., Martonchik, J.V., Ackerman, T.P., Davies, R., and Gerstl, S.A. (1998). Multi-angle imaging spectro radiometer (MISR) instrument description and experiment overview. *IEEE Transactions on Geoscience and Remote Sensing*, *36*, 1072–1087.

Dodman, D. (2009). Blaming cities for climate change? An analysis of urban greenhouse gas emissions inventories. *Environment and Urbanization*, *21*, 185–201.

Doll, C., Muller, J., and Elvidge, C. (2000). Night-time imagery as a tool for global mapping of socioeconomic parameters and greenhouse gas emissions. *AMBIO*, *29*, 157–162.

Doll, C., Muller, J., and Morley, J. (2006). Mapping regional economic activity from night-time light satellite imagery. *Ecological Economics*, *57*, 75–92.

Doll, C.N. (2008). CIESIN thematic guide to night-time light remote sensing and its applications. Center for International Earth Science Information Network of Columbia University, Palisades, New York.

Elvidge, C., Baugh, K., Dietz, J., Bland, T., Sutton, P., and Kroehl, H. (1999). Radiance calibration of DMSP-OLS low-light imaging data of human settlements. *Remote Sensing of Environment*, *68*, 77–88.

Elvidge, C., Baugh, K., Kihn, E., Kroehl, H., Davis, E., and Davis, C. (1997a). Relation between satellite observed visible-near infrared emissions, population, economic activity and electric power consumption. *International Journal of Remote Sensing*, *18*, 1373–1379.

Elvidge, C., Safran, J., Tuttle, B., Sutton, P., Cinzano, P., Pettit, D., Arvesen, J., and Small, C. (2007a). Potential for global mapping of development via a nightsat mission. *GeoJournal*, *69*, 45–53.

Elvidge, C., Tuttle, B., Sutton, P., Baugh, K., Howard, A., Milesi, C., Bhaduri, B., and Nemani, R. (2007b). Global distribution and density of constructed impervious surfaces. *Sensors*, *7*, 1962–1979.

Elvidge, C.D., Baugh, K.E., Kihn, E.A., Kroehl, H.W., and Davis, E.R. (1997b). Mapping city lights with night-time data from the DMSP Operational Linescan System. *Photogrammetric Engineering and Remote Sensing*, *63*, 727–734.

Elvidge, C.D., Baugh, K.E., Kihn, E.A., Kroehl, H.W., Davis, E.R., and Davis, C.W. (1997c). Relation between satellite observed visible-near infrared emissions, population, economic activity and electric power consumption. *International Journal of Remote Sensing*, *18*, 1373–1379.

Elvidge, C.D., Cinzano, P., Pettit, D., Arvesen, J., Sutton, P., Small, C., Nemani, R., Longcore, T., Rich, C., and Safran, J. (2007c). The Nightsat mission concept. *International Journal of Remote Sensing*, *28*, 2645–2670.

Elvidge, C.D. and Green, R.O. (2005). High-and low-altitude AVIRIS observations of nocturnal lighting. In *13th JPL Airborne Earth Science Workshop*, May 24–27, 2005, Pasadena, CA. Jet Propulsion Laboratory, National Aeronautics and Space Administration, Pasadena, CA.

Elvidge, C.D., Keith, D.M., Tuttle, B.T., and Baugh, K.E. (2010). Spectral identification of lighting type and character. *Sensors*, *10*, 3961–3988.

Elvidge, C.D., Ziskin, D., Baugh, K.E., Tuttle, B.T., Ghosh, T., Pack, D.W., Erwin, E.H., and Zhizhin, M. (2009). A fifteen year record of global natural gas flaring derived from satellite data. *Energies*, *2*, 595–622.

Falchi, F., Cinzano, P., Elvidge, C.D., Keith, D.M., and Haim, A. (2011). Limiting the impact of light pollution on human health, environment and stellar visibility. *Journal of Environmental Management*, *92*, 2714–2722.

Falchi, F. and Cinzano, P. (2000). Measuring and modeling light pollution. *Cinzano, P. ed., Mem. Soc. Astron. Ital. 71*, 139.

Foley, J.A., DeFries, R., Asner, G.P., Barford, C., Bonan, G., Carpenter, S.R., Chapin, F.S., Coe, M.T., Daily, G.C., and Gibbs, H.K. (2005). Global consequences of land use. *Science, 309*, 570–574.

Frolking, S., Milliman, T., Seto, K.C., and Friedl, M.A. (2013). A global fingerprint of macro-scale changes in urban structure from 1999 to 2009. *Environmental Research Letters, 8*, 024004.

Gallaway, T., Olsen, R.N., and Mitchell, D.M. (2010). The economics of global light pollution. *Ecological Economics, 69*, 658–665.

Galli, M. and Guzzetti, F. (2007). Landslide vulnerability criteria: A case study from Umbria, Central Italy. *Environmental Management, 40*, 649–665.

Gaston, K.J., Bennie, J., Davies, T.W., and Hopkins, J. (2013). The ecological impacts of nighttime light pollution: A mechanistic appraisal. *Biological Reviews, 88*, 912–927.

Gaston, K.J., Davies, T.W., Bennie, J., and Hopkins, J. (2012). REVIEW: Reducing the ecological consequences of nighttime light pollution: Options and developments. *Journal of Applied Ecology, 49*, 1256–1266.

Geist, J.F. (1983). *Arcades: The History of a Building Type.* MIT Press: Cambridge, MA.

Ghosh, T., Elvidge, C.D., Sutton, P.C., Baugh, K.E., Ziskin, D., and Tuttle, B.T. (2010). Creating a global grid of distributed fossil fuel CO_2 emissions from nighttime satellite imagery. *Energies, 3*, 1895–1913.

Gomez-Ibanez, D.J., Boarnet, M.G., Brake, D.R., Cervero, R.B., Cotugno, A., Downs, A., Hanson, S., Kockelman, K.M., Mokhtarian, P.L., and Pendall, R.J. (2009). Driving and the built environment: The effects of compact development on motorized travel, energy use, and CO_2 emissions. Oak Ridge National Laboratory (ORNL): Oak Ridge, TN.

Grimm, N.B., Faeth, S.H., Golubiewski, N.E., Redman, C.L., Wu, J., Bai, X., and Briggs, J.M. (2008). Global change and the ecology of cities. *Science, 319*, 756–760.

Gutman, P. (2007). Ecosystem services: Foundations for a new rural–urban compact. *Ecological Economics, 62*, 383–387.

Hahs, A.K., McDonnell, M.J., McCarthy, M.A., Vesk, P.A., Corlett, R.T., Norton, B.A., Clemants, S.E., Duncan, R.P., Thompson, K., and Schwartz, M.W. (2009). A global synthesis of plant extinction rates in urban areas. *Ecology Letters, 12*, 1165–1173.

Hale, J.D., Davies, G., Fairbrass, A.J., Matthews, T.J., Rogers, C.D., and Sadler, J.P. (2013). Mapping lightscapes: Spatial patterning of artificial lighting in an urban landscape. *PloS One, 8*, e61460.

Hattori, R., Horie, S., Hsu, F.-C., Elvidge, C.D., and Matsuno, Y. (2013). Estimation of in-use steel stock for civil engineering and building using nighttime light images. *Resources, Conservation and Recycling*.

Heilig, G.K. (2012). World Urbanization Prospects: The 2011 Revision. United Nations, Department of Economic and Social Affairs (DESA), Population Division, Population Estimates and Projections Section, New York.

Hipskinda, S., Elvidgeb, C., Gurneyc, K., Imhoffd, M., Bounouad, L., Sheffnera, E., Nemania, R., Pettite, D., and Fischerf, M. (2011). Global night-time lights for observing human activity.

Hölker, F., Moss, T., Griefahn, B., Kloas, W., Voigt, C.C., Henckel, D., Hänel, A., Kappeler, P.M., Völker, S., and Schwope, A. (2010a). The dark side of light: A transdisciplinary research agenda for light pollution policy. *Ecology & Society, 15*.

Hölker, F., Wolter, C., Perkin, E.K., and Tockner, K. (2010b). Light pollution as a biodiversity threat. *Trends in Ecology & Evolution, 25*, 681–682.

Hoornweg, D., Bhada, P., Freire, M., Trejos, C., and Sugar, L. (2010). Cities and climate change: An urgent agenda. World Bank, Washington, DC.

IDSA (1996). Astronomy's problem with light pollution. ISDA Information Sheet, No. 1, May 1996, International Dark-Sky Association: Tuscon, AZ.

Imhoff, M.L., Lawrence, W.T., Stutzer, D.C., and Elvidge, C.D. (1997). A technique for using composite DMSP/OLS "city lights" satellite data to map urban area. *Remote Sensing of Environment, 61*, 361–370.

Jackett, M. and Frith, W. (2013). Quantifying the impact of road lighting on road safety—A New Zealand Study. *IATSS Research, 36*, 139–145.

Jiang, L., Deng, X., and Seto, K.C. (2012). Multi-level modeling of urban expansion and cultivated land conversion for urban hotspot counties in China. *Landscape and Urban Planning, 108*, 131–139.

Kareiva, P., Watts, S., McDonald, R., and Boucher, T. (2007). Domesticated nature: Shaping landscapes and ecosystems for human welfare. *Science, 316*, 1866–1869.

Katz, A.H. (1948). Aerial photographic equipment and applications to reconnaissance. *JOSA, 38*, 604–605.

Kim, M. (2012). Modeling nightscapes of designed spaces—Case studies of the University of Arizona and Virginia Tech Campuses. In *13th International Conference on Information Technology in Landscape Architecture Proceedings*, pp. 455–463.

Kim, M. and Hong, S.-H. (2013). Relationship between the reflected brightness of artificial lighting and land-use types: A case study of the University of Arizona campus. *Landscape and Ecological Engineering*, 1–7.

Kingslake, R. (1942). Lenses for aerial photography. *Journal of the Optical Society of America, 32*, 129–134.

Kloog, I., Haim, A., Stevens, R.G., and Portnov, B.A. (2009). Global co-distribution of light at night (LAN) and cancers of prostate, colon, and lung in men. *Chronobiology International, 26*, 108–125.

Kohrs, R.A., Lazzara, M.A., Robaidek, J.O., Santek, D.A., and Knuth, S.L. (2014). Global satellite composites—20 Years of evolution. *Atmospheric Research, 135*, 8–34.

Kruse, F.A. and Elvidge, C.D. (2011). Characterizing urban light sources using imaging spectrometry. In *Urban Remote Sensing Event (JURSE), 2011 Joint* (pp. 149–152). IEEE.

Küchly, H.U., Kyba, C., Ruhtz, T., Lindemann, C., Wolter, C., Fischer, J., and Hölker, F. (2012). Aerial survey and spatial analysis of sources of light pollution in Berlin, Germany. *Remote Sensing of Environment, 126*, 39–50.

Kyba, C.C., Ruhtz, T., Fischer, J., and Hölker, F. (2011). Cloud coverage acts as an amplifier for ecological light pollution in urban ecosystems. *PloS One, 6*, e17307.

Kyba, C.C., Ruhtza, T., Lindemanna, C., Fischera, J., and Hölkerb, F. (2013). Two camera system for measurement of urban uplight angular distribution. In *Radiation Processes in the Atmosphere and Ocean (IRS2012): Proceedings of the International Radiation Symposium (IRC/IAMAS)* (pp. 568–571). AIP Publishing.

Lee, T.E., Miller, S.D., Turk, F.J., Schueler, C., Julian, R., Deyo, S., Dills, P., and Wang, S. (2006). The NPOESS VIIRS Day/Night Visible Sensor. *Bulletin of the American Meteorological Society, 87*.

Letu, H., Hara, M., Tana, G., and Nishio, F. (2012). A saturated light correction method for DMSP/OLS nighttime satellite imagery. *IEEE Transactions on Geoscience and Remote Sensing, 50*, 389–396.

Letu, H., Hara, M., Yagi, H., Naoki, K., Tana, G., Nishio, F., and Shuhei, O. (2010). Estimating energy consumption from night-time DMPS/OLS imagery after correcting for saturation effects. *International Journal of Remote Sensing, 31*, 4443–4458.

Letu, H., Nakajima, T.Y., and Nishio, F. (2014). Regional-scale estimation of electric power and power plant CO2 emissions using DMSP/OLS nighttime satellite data. *Environmental Science and Technology Letters*.

Levin, N. and Duke, Y. (2012). High spatial resolution night-time light images for demographic and socio-economic studies. *Remote Sensing of Environment, 119*, 1–10.

Levin, N., Kark, S., and Crandall, D.J. (2015). Where have all the people gone? Enhancing global conservation using night lights and social media. *Ecological Applications*.

Levin, N., Johansen, K., Hacker, J.M., and Phinn, S. (2014). A new source for high spatial resolution night time images—The EROS-B commercial satellite. *Remote Sensing of Environment*.

Li, X., Chen, F.R., and Chen, X.L. (2013a). Satellite-observed nighttime light variation as evidence for global armed conflicts. *IEEE Journal of Selected Topics in Applied Earth Observations and Remote Sensing, 6*, 2302–2315.

Li, X., Chen, X., Zhao, Y., Xu, J., Chen, F., and Li, H. (2013b). Automatic intercalibration of night-time light imagery using robust regression. *Remote Sensing Letters, 4*, 45–54.

Li, X., Ge, L., and Chen, X. (2013c). Detecting Zimbabwe's decadal economic decline using nighttime light imagery. *Remote Sensing, 5*, 4551–4570.

Li, X., Xu, H., Chen, X., and Li, C. (2013d). Potential of NPP-VIIRS nighttime light imagery for modeling the regional economy of China. *Remote Sensing, 5*, 3057–3081.

Liao, L., Weiss, S., Mills, S., and Hauss, B. (2013). Suomi NPP VIIRS day-night band on-orbit performance. *Journal of Geophysical Research: Atmospheres, 118*(12),7 05–712, 718.

Liu, J., Peng, X.C., Zhong, Q.J., Lin, K., Feng, H.L., Hong, H.J., Dong, J.H., Fang, Q.L., and Wu, Y.Y. (2013). Case study of investigation and evaluation on the urban light pollution in Macau. *Applied Mechanics and Materials, 295*, 678–687.

Liu, Z., He, C., and Yang, Y. (2011). Mapping urban areas by performing systematic correction for DMSP/OLS nighttime lights time series in China from 1992 to 2008. In *Geoscience and Remote Sensing Symposium (IGARSS), 2011 IEEE International* (pp. 1858–1861). IEEE.

Liu, Z., He, C., Zhang, Q., Huang, Q., and Yang, Y. (2012). Extracting the dynamics of urban expansion in China using DMSP-OLS nighttime light data from 1992 to 2008. *Landscape and Urban Planning, 106*, 62–72.

Liverman, D., Moran, E.F., Rindfuss, R.R., and Stern, P.C. (1998). People and pixels: Linking remote sensing and social science.

Loewen, L.J., Steel, G.D., and Suedfeld, P. (1993). Perceived safety from crime in the urban environment. *Journal of Environmental Psychology, 13*, 323–331.

Longcore, T. and Rich, C. (2004). Ecological light pollution. *Frontiers in Ecology and the Environment, 2*, 191–198.

Lovatt, A. and O'Connor, J. (1995). Cities and the night-time economy. *Planning Practice and Research, 10*, 127–134.

Lu, D., Tian, H., Zhou, G., and Ge, H. (2008). Regional mapping of human settlements in southeastern China with multisensor remotely sensed data. *Remote Sensing of Environment, 112*, 3668–3679.

Luckiesh, M. (1920). *Artificial Light: Its Influence Upon Civilization.* University of London Press: London, U.K.

Ma, T., Zhou, C., Pei, T., Haynie, S., and Fan, J. (2012). Quantitative estimation of urbanization dynamics using time series of DMSP/OLS nighttime light data: A comparative case study from China's cities. *Remote Sensing of Environment, 124*, 99–107.

Marin, C. (2009). StarLight: A common heritage. *Proceedings of the International Astronomical Union, 5*, 449–456.

Marshall Shepherd, J., Pierce, H., and Negri, A.J. (2002). Rainfall modification by major urban areas: Observations from spaceborne rain radar on the TRMM satellite. *Journal of Applied Meteorology, 41*.

Mazor, T., Levin, N., Possingham, H.P., Levy, Y., Rocchini, D., Richardson, A.J., and Kark, S. (2013). Can satellite-based night lights be used for conservation? The case of nesting sea turtles in the Mediterranean. *Biological Conservation, 159*, 63–72.

Milesi, C., Elvidge, C.D., Nemani, R.R., and Running, S.W. (2003). Assessing the impact of urban land development on net primary productivity in the southeastern United States. *Remote Sensing of Environment, 86*, 401–410.

Miller, M.W. (2006). Apparent effects of light pollution on singing behavior of American robins. *The Condor, 108*, 130–139.

Miller, S.D., Hawkins, J.D., Kent, J., Turk, F.J., Lee, T.F., Kuciauskas, A.P., Richardson, K., Wade, R., and Hoffman, C. (2006). NexSat: Previewing NPOESS/VIIRS imagery capabilities. *Bulletin of the American Meteorological Society, 87*.

Miller, S.D., Lee, T.F., Turk, F.J., Kuciauskas, A.P., and Hawkins, J.D. (2005). Shedding new light on nocturnal monitoring of the environment with the VIIRS day/night band. In *Optics and Photonics 2005* (pp. 58900W-58900W-58909). International Society for Optics and Photonics.

Miller, S.D., Mills, S.P., Elvidge, C.D., Lindsey, D.T., Lee, T.F., and Hawkins, J.D. (2012). Suomi satellite brings to light a unique frontier of nighttime environmental sensing capabilities. *Proceedings of the National Academy of Sciences*, 109, 15706–15711.

Nagendra, H., Lucas, R., Honrado, J.P., Jongman, R.H.G., Tarantino, C., Adamo, M., and Mairota, P. (2012). Remote sensing for conservation monitoring: Assessing protected areas, habitat extent, habitat condition, species diversity, and threats. *Ecological Indicators.*

Nakićenović, N. (2012). *Global Energy Assessment: Toward a Sustainable Future*. Cambridge University Press: Cambridge, U.K.

Navara, K.J. and Nelson, R.J. (2007). The dark side of light at night: Physiological, epidemiological, and ecological consequences. *Journal of Pineal Research*, 43, 215–224.

Nicholls, R.J. and Cazenave, A. (2010). Sea-level rise and its impact on coastal zones. *Science*, 328, 1517–1520.

Oda, T. and Maksyutov, S. (2011). A very high-resolution (1 km × 1 km) global fossil fuel CO_2 emission inventory derived using a point source database and satellite observations of nighttime lights. *Atmospheric Chemistry and Physics*, 11, 543–556.

Olsen, R.N., Gallaway, T., and Mitchell, D. (2013). Modelling US light pollution. *Journal of Environmental Planning and Management*, 1–21.

Pandey, B., Joshi, P., and Seto, K.C. (2013). Monitoring urbanization dynamics in India using DMSP/OLS night time lights and SPOT-VGT data. *International Journal of Applied Earth Observation and Geoinformation*, 23, 49–61.

Patino, J.E. and Duque, J.C. (2013). A review of regional science applications of satellite remote sensing in urban settings. *Computers, Environment and Urban Systems*, 37, 1–17.

Pauley, S.M. (2004). Lighting for the human circadian clock: Recent research indicates that lighting has become a public health issue. *Medical Hypotheses*, 63, 588–596.

Pozzi, F. and Small, C. (2005). Analysis of urban land cover and population density in the United States. *Photogrammetric Engineering and Remote Sensing*, 71, 719–726.

Pun, C.S.J. and So, C.W. (2012). Night-sky brightness monitoring in Hong Kong. *Environmental Monitoring and Assessment*, 184, 2537–2557.

Pun, C.S.J., So, C.W., Leung, W.Y., Wong, C.F. (in press). Contributions of artificial lighting sources on light pollution in Hong Kong measured through a night sky brightness monitoring network. *Journal of Quantitative Spectroscopy and Radiative Transfer.*

Rauch, J.N. (2009). Global mapping of Al, Cu, Fe, and Zn in-use stocks and in-ground resources. *Proceedings of the National Academy of Sciences*, 106, 18920–18925.

Reynolds, J.F., Smith, D.M.S., Lambin, E.F., Turner, B., Mortimore, M., Batterbury, S.P., Downing, T.E., Dowlatabadi, H., Fernández, R.J., and Herrick, J.E. (2007). Global desertification: Building a science for dryland development. *Science*, 316, 847–851.

Rindfuss, R.R. and Stern, P.C. (1998). Linking remote sensing and social science: The need and the challenges. *People and Pixels: Linking Remote Sensing and Social Science*, 1–27.

Rosenfeld, D. (2000). Suppression of rain and snow by urban and industrial air pollution. *Science*, 287, 1793–1796.

Rouse Jr, J., Haas, R., Schell, J., and Deering, D. (1974). Monitoring vegetation systems in the Great Plains with ERTS. *NASA Special Publication*, 351, 309.

Salgado-Delgado, R., Tapia Osorio, A., Saderi, N., and Escobar, C. (2011). Disruption of circadian rhythms: A crucial factor in the etiology of depression. *Depression Research and Treatment*, 2011.

Salmon, M. (2003). Artificial night lighting and sea turtles. *Biologist*, 50, 163–168.

Schaaf, C.B., Gao, F., Strahler, A.H., Lucht, W., Li, X., Tsang, T., Strugnell, N.C., Zhang, X., Jin, Y., and Muller, J.-P. (2002). First operational BRDF, albedo nadir reflectance products from MODIS. *Remote Sensing of Environment*, 83, 135–148.

Schneider, A., Friedl, M.A., and Woodcock, C.E. (2003). Mapping urban areas by fusing multiple sources of coarse resolution remotely sensed data. In *Proceedings of 2003 IEEE International Geoscience and Remote Sensing Symposium, 2003. IGARSS'03.* (pp. 2623–2625). IEEE.

Schulman, S. and Rader, A.C. (2012). RAF night photography. *The RUSI Journal.*

Sen, A., Kim, Y., Caruso, D., Lagerloef, G., Colomb, R., and Le Vine, D. (2006). Aquarius/SAC-D mission overview. In *Remote Sensing* (pp. 63610I-63610I-63610). International Society for Optics and Photonics.

Seto, K.C., Kaufmann, R.K., and Woodcock, C.E. (2000). Landsat reveals China's farmland reserves, but they're vanishing fast. *Nature*, 406, 121–121.

Seto, K.C. and Shepherd, J.M. (2009). Global urban land-use trends and climate impacts. *Current Opinion in Environmental Sustainability*, 1, 89–95.

Shi, K., Yu, B., Huang, Y., Hu, Y., Yin, B., Chen, Z., Chen, L., and Wu, J. (2014). Evaluating the ability of NPP-VIIRS nighttime light data to estimate the gross domestic product and the electric power consumption of china at multiple scales: A comparison with DMSP-OLS data. *Remote Sensing*, 6, 1705–1724.

Shortland, A., Christopoulou, K., and Makatsoris, C. (2013). War and famine, peace and light? The economic dynamics of conflict in Somalia 1993–2009. *Journal of Peace Research*, 50, 545–561.

Simpson, S., Winebrake, J., and Noel-Storr, J. (2006). Willingness to pay for a clear night sky: Use of the contingent valuation method. *Bulletin of the American Astronomical Society*, 1195pp.

Small, C. (2001). Estimation of urban vegetation abundance by spectral mixture analysis. *International Journal of Remote Sensing, 22*, 1305–1334.

Small, C. (2005). A global analysis of urban reflectance. *International Journal of Remote Sensing, 26*, 661–682.

Small, C. and Elvidge, C.D. (2013). Night on Earth: Mapping decadal changes of anthropogenic night light in Asia. *International Journal of Applied Earth Observation and Geoinformation, 22*, 40–52.

Small, C., Elvidge, C.D., Balk, D., and Montgomery, M. (2011). Spatial scaling of stable night lights. *Remote Sensing of Environment, 115*, 269–280.

Small, C., Pozzi, F., and Elvidge, C.D. (2005). Spatial analysis of global urban extent from DMSP-OLS night lights. *Remote Sensing of Environment, 96*, 277–291.

Smith, M. (2008). Time to turn off the lights. *Nature, 457*, 27–27.

Sovacool, B.K. and Brown, M.A. (2010). Twelve metropolitan carbon footprints: A preliminary comparative global assessment. *Energy Policy, 38*, 4856–4869.

Stern, P.C., Young, O.R., and Druckman, D. (1991). *Global Environmental Change: Understanding the Human Dimensions*. National Academies Press: Washington, DC.

Sutton, P., Elvidge, C., and Obremski, T. (2003). Building and evaluating models to estimate ambient population density. *Photogrammetric Engineering and Remote Sensing, 69*, 545–554.

Sutton, P., Roberts, D., Elvidge, C., and Baugh, K. (2001). Census from Heaven: An estimate of the global human population using night-time satellite imagery. *International Journal of Remote Sensing, 22*, 3061–3076.

Sutton, P.C., Elvidge, C.D., and Ghosh, T. (2007). Estimation of gross domestic product at sub-national scales using night-time satellite imagery. *International Journal of Ecological Economics & Statistics, 8*, 5–21.

Takahashi, K.I., Terakado, R., Nakamura, J., Adachi, Y., Elvidge, C.D., and Matsuno, Y. (2010). In-use stock analysis using satellite nighttime light observation data. *Resources, Conservation and Recycling, 55*, 196–200.

Tardà, A., Palà, V., Arbiol, R., Pérez, F., Viñas, O., Pipia, L., and Martínez, L. (2011). Detección de la iluminación exterior urbana nocturna con el sensor aerotransportado CASI 550. In *Proceedings of the International Geomatic Week*, March 2011, Barcelona, Spain.

Teikari, P. (2007). Light pollution: Definition, legislation, measurement, modeling and environmental effects. Universitat politécnica de Catalunya. Barcelona, Catalunya, 10.

Tucker, C.J. (1979). Red and photographic infrared linear combinations for monitoring vegetation. *Remote Sensing of Environment, 8*, 127–150.

UNFPA. (2011). State of world population 2011: People and possibilities in a world of 7 billion. The United Nations Population Fund.

Weisburd, D. and Amram, S. (2014). The law of concentrations of crime at place: The case of Tel Aviv-Jaffa. *Police Practice and Research*, 1–14.

Weng, Q., Lu, D., and Liang, B. (2006). Urban surface biophysical descriptors and land surface temperature variations. *Photogrammetric Engineering and Remote Sensing, 72*, 1275–1286.

Weng, Q., Lu, D., and Schubring, J. (2004). Estimation of land surface temperature–vegetation abundance relationship for urban heat island studies. *Remote Sensing of Environment, 89*, 467–483.

Wu, J., He, S., Peng, J., Li, W., and Zhong, X. (2013). Intercalibration of DMSP-OLS night-time light data by the invariant region method. *International Journal of Remote Sensing, 34*, 7356–7368.

Yi, K., Tani, H., Li, Q., Zhang, J., Guo, M., Bao, Y., Wang, X., and Li, J. (2014). Mapping and evaluating the urbanization process in Northeast China using DMSP/OLS nighttime light data. *Sensors, 14*, 3207–3226.

Zhang, Q., He, C., and Liu, Z. (2014). Studying urban development and change in the contiguous United States using two scaled measures derived from nighttime lights data and population census. *GIScience and Remote Sensing*, 1–20.

Zhang, Q., Schaaf, C., and Seto, K.C. (2013). The vegetation adjusted NTL urban index: A new approach to reduce saturation and increase variation in nighttime luminosity. *Remote Sensing of Environment, 129*, 32–41.

Zhang, Q. and Seto, K.C. (2011). Mapping urbanization dynamics at regional and global scales using multi-temporal DMSP/OLS nighttime light data. *Remote Sensing of Environment, 115*, 2320–2329.

Zhuo, L., Ichinose, T., Zheng, J., Chen, J., Shi, P., and Li, X. (2009). Modelling the population density of China at the pixel level based on DMSP/OLS non-radiance-calibrated nighttime light images. *International Journal of Remote Sensing, 30*, 1003–1018.

Ziskin, D., Baugh, K., Hsu, F.C., Ghosh, T., and Elvidge, C. (2010). Methods used for the 2006 radiance lights. In *Proceedings of the Asia Pacific Advanced Network* (pp. 131–142).

VII

Geomorphology

12

Geomorphological Studies from Remote Sensing

James B. Campbell
Virginia Tech

Lynn M. Resler
Virginia Tech

Acronyms and Definitions

ADAR	Airborne Data Acquisition and Registration
ALB	Airborne LiDAR Bathymetry
ASTER	Advanced Spaceborne Thermal Emission and Reflection Radiometer
CORONA	Coordinating Research on the North Atlantic; early U.S. national security satellite observation system
CSC	Coastal Service Center (South Carolina)
DEM	Digital elevation model
DOQQ	Digital orthophoto quarter quad
DTM	Digital terrain model
ERS	European Remote Sensing (earth observation satellites)
GIS	Geographic information system
GISc	Geographic Information Science
GPS	Global Positioning System
InSAR	Interferometric SAR
JPL	Jet Propulsion Laboratory
Landsat	U.S. land observation system originated in the early 1970s
LiDAR	Light detection and ranging
LOS	Line of sight
MLS	Mobile laser scanning
NAS	National Academy of Sciences
NASA	National Aeronautics and Space Administration
NOAA	National Oceanographic and Atmospheric Administration
NRC	National Research Council
OBIA	Object-based image analysis
PDTD	Photogrammetrically derived topographic data
PSI	Persistent/permanent scatterer interferometry
RADARSAT	Canadian imaging radar satellite remote sensing system
RGA	Rapid geomorphic assessment
SAR	Synthetic aperture radar
SONAR	Sound navigation and ranging
SRTM	Shuttle Radar Topography Mission
SSC	Suspended sediment concentration
TS	Total station
TanDEM-X	German radar satellite system
TLS	Terrestrial laser scanning
TM	Thematic Mapper
TopoSAR	A specific NASA aircraft SAR system
USGS	United States Geological Survey

12.1 Introduction

Remote sensing (RS) analysis has provided important insights into geomorphic systems and landforms through its analytical capabilities and data integration capacities. This chapter reviews the role of remotely sensed imagery in geomorphic inquiry, initially through an historical overview of contributions of aerial imagery to understanding geomorphological systems at varied spatial and temporal scales.

With historical background as context, the discussion following documents the significance of RS' role in developing current geomorphic understanding through (1) enhancing the ability to conduct inquiries on established topics and (2) developing technologies that open new lines of inquiry that were previously unavailable. Thus, the remaining portions of the chapter will address the following questions as they apply to a selection of topics (Table 12.1), with a focus on selected subfields of geomorphology that represent peri-

glacial, glacial, mass-wasting, fluvial, coastal, aeolian, and biogeomorphological processes:

- In what ways has RS been used in various subfields of geomorphology?
- How has RS analysis provided unique perspectives and advanced understanding in each subfield?

Although applications of RS already form a significant dimension to geomorphic inquiry, their impact remains largely dispersed among various subfields and is seldom discussed in an overview. We intend our sketch of these developments to form the basis for further efforts to examine their contributions.

12.1.1 Historical Perspective

The practice of RS has advanced the field of geomorphology for almost 100 years (Vitek et al. 1996). The history of RS in geomorphology begins with the role of aerial imagery in providing insights into landscape evolution at local to global scales.

Aerial photography played a pivotal role in the early adoption of RS technologies for examination of physiography, and later, of geomorphic systems. Aerial photography, as the prevailing aerial survey technology of the day, already had a long history by the beginning of World War I. During that conflict, aerial photography began its transformation from a status as a curiosity into a practical and scientific tool for examining soils, geological features, coastlines, and river systems. Previously, the airplane and camera existed as two separate, largely incompatible, technologies that could be used together only with great inconvenience. By the time of the 1918 armistice, the groundwork was set for design of photographic systems that could systemically collect aerial imagery to support analytical objectives, together with organizational structures to interpret and report results of aerial survey, which form forerunners of today's practices.

After military demobilization, in the 1920s, such developments formed foundations for the institutionalization of aerial survey in both military and civil enterprises. Lee's (1922) survey of aerial photography's potential for civil applications outlines its future role in urban planning, land use planning, and examination of geomorphic features, such as coastal landforms, overwash fans, tidal deltas, coastal marshes, and spits.

Progress in development of aerial cameras, aircraft, and supporting technologies during the interwar decades increased interest in applications of aerial photography. By 1939 (i.e., Melton 1939), aerial photographs were used as teaching resources for introductory geology courses, especially for illustrating basics of landforms shaped by steams, wind, waves, and glaciers.

Although aerial photography contributed to geomorphological inquiry during the interwar years, such knowledge was fragmented. By the beginning of World War II, however, methods and techniques were organized systematically into military handbooks, field manuals, and training programs. Examples include the detailed knowledge of coastal landforms, especially beach profiles, initially in the Pacific Theatre, but later in

TABLE 12.1 Summary of Selected Geomorphic Topics

Selected Topics	Selected References

Alpine and periglacial
  Active layer thickness
  Thermokarst
  Soil organic layer
  Ground ice
	French and Thorn (2006), Kääb (2005), Kääb (2008), Polar Research Board

Glacial landforms
  Glacial extent
  Changes in glacial extent
  Ice movement
	Kääb et al. (2005)

Mass-wasting
  Landslides
  Debris flows
  Soil creep
	Metternicht et al. (2005), Scaioni (2013)

Fluvial landforms
  Channel migration
  Flooding and floodplain analysis
  Stream bank retreat
	Carbonneau and Piegay (2012)

Coastal landforms
  Coastal form and change
  Suspended sediment concentrations
  Bathymetry
	Allen and Wang (2010), Gao (2011)

Aeolian landforms
  Dune patterns
  Rates of movement
  Dune volume
	Mitasova et al. (2005), Livingston et al. (2007)

Biogeomorphology
  Feedbacks between process and pattern
  Tree line position and dynamics
  Channel migration
	Walsh et al. (1998), Bryant and Gilvear (1999), Hudson et al. (2006)

European campaigns. What formerly was considered esoteric geographic knowledge of remote locations became essential for planning successful operations, since many of World War II's key campaigns were conducted in locations lacking systematic geomorphological knowledge.

In this context, aerial photography, as the prevailing form of RS imagery of the time, formed an essential component of geomorphological knowledge supporting (1) combat intelligence, supporting studies of terrain, trafficability (assessment of terrain suitability for motorized vehicles), relief, and intervisibility (assessing ability to observe landscapes from a given location) and (2) engineering intelligence, supporting construction of airfields and roads in remote, unmapped regions. Successful military operations required development, often without systematic field data, of detailed understanding of, for example, subsurface beach configurations, river systems, and wetlands. In this context, the role of aerial photography grew from a source of qualitative data to form a more precise quantitative tool for assessing, for example, topographic slope and depths of offshore topography. World War II saw some of the first attempts to integrate earlier piecemeal practices into systematized photointerpretation to support geologic inquiry (Smith 1942, 1943).

In the years after World War II, many of these capabilities found applications in civil society and simultaneously in the national security domain, as the Cold War era began. Many of the successful uses of aerial photography in World War II confirmed and validated its use for civil applications in government and commerce, which in turn contributed to a growing archive of imagery and to investments in new sensor technologies.

For the earth sciences, many of these initial applications can be seen as providing descriptive content, basically recording locations of labeled features—an important mapping contribution, but one that did little to advance the cause of geomorphology's primary mission—investigation of geomorphic processes. During the post–World War II era, aerial imagery contributed to the emerging paradigm of landscape quantification; analysis of aerial imagery became an increasingly important tool for extraction of quantitative landscape metrics (Chorley 2008). The "midcentury revolution" in fluvial geomorphology (Chorley 2008) sought to replace a descriptive, unidimensional approach to geomorphology with a quantitative/analytical strategy, attributed in part to the impact of World War II, especially in facilitating application of concepts derived from engineering and the physical sciences. For example, Horton (1945) and Strahler (1957) proposed drainage density as a strategy for quantitative assessment of watersheds and drainage systems and more generally for introducing quantitative analysis in the earth sciences. Such measures, initially proposed in the context of map analysis, were later applied to aerial photography and other remotely sensed imagery (Rudraiah et al. 2008). Thus, although the midcentury revolution began before the field of RS was formalized, it set the context for uses of aerial imagery as a source of quantitative analysis.

Ray's (1960) *Aerial Photographs in Geologic Interpretation and Mapping*, a compilation of annotated aerial photographs illustrating landforms and geologic structures, in addition promoted aerial imagery as a source of quantitative information to support geomorphic and geological analysis. This work was later reinforced by Scheidegger's (1961) survey of quantitative strategies for geomorphological analysis.

Parallels between the advancement of RS techniques, and in geomorphic knowledge, have emerged more recently. Increasingly sophisticated RS data acquisition and analysis have extended our understanding of geomorphic processes and patterns at varied spatial and temporal scales (e.g., Walsh et al. 1998). For example, by the 1960s and 1970s, the availability of new RS systems, such as imaging radars and later Landsat imagery, provided the ability to examine geomorphic features at broader scales. Early applications of imaging radars focused upon its ability to provide a synoptic view, with clear, crisp, and clearly defined representations of relief, drainage, open water, and coastlines (Simpson 1966).

By 1972, the availability of Landsat imagery provided additional capabilities for broader-scale examination of geomorphic features (Short and Blair 1986). Beginning in the 1980s and continuing into subsequent decades, Landsat's systematic multispectral coverage has provided a comprehensive sequential record of the earth's surface and changes to topography, vegetation, and land use.

RS technologies increased to include light detection and ranging (LiDAR) and specialized applications of imaging radars, to be discussed in the succeeding text. In a geomorphological context, their value has been to provide data that extend imagery's capabilities beyond providing a basic maplike view of the terrain to providing metric data that record precise detail and changes in terrain elevation and motion. These innovations have moved the field of geomorphology beyond description and labeling of landforms to providing an improved understanding of process through modeling and forecasting.

Current geomorphological research has employed several forms of remotely sensed imagery, including aerial photography, satellite imagery, imaging radars, and LiDAR imagery. Aircraft and satellite systems contribute to geomorphological analysis through their ability to provide data with maplike views of the terrain surface and represent landscape features in relation to drainage, relief, and land use patterns. LiDAR and imaging radars are *active systems* that illuminate terrain using their own energy, then compare transmitted and received signals to derive detailed information about terrain surfaces. Aerial photography collects imagery using panchromatic (a single channel in the visible spectrum) imagery, natural color (separate channels for the blue, green, and red primaries), and color infrared (using green, red, and near-infrared channels). Such imagery often provides spatial detail at about 1 m, or finer, but coarser detail if acquired at higher altitudes. Aerial photography is widely used for a range of geomorphic topics, but often is limited in function to providing the spatial context for field reconnaissance and planning fieldwork.

Satellite imagery provides systematic coverage using standardized formats and technical specifications. For present purposes, we can think of satellite imagery as either (1) fine-scaled, detailed data (at a resolution of perhaps 1 meter or less), with three to four spectral channels, (in the optical region of the spectrum), typically acquired by commercial organizations, or (2) broad-scale data with coarser levels of detail (perhaps ≥30 m).

Commercial satellite corporations collect imagery at a variety of resolutions and spectral regions; the most common specifications and formats resemble those of those of aerial photography—a selection of panchromatic, natural color, and color infrared, with spatial detail at several meters or at submeter detail for some systems.

For regional coverage, land resource satellites provide coarse detail for broader regions using multispectral imagery (perhaps 8–12 spectral channels) at spatial detail in the range of about 10–30 m. Such systems employ systematic coverage, consistent in format, and in coverage pattern. For example, there is special value in the ability of satellite imagery, especially Landsat imagery, to provide sequential coverage. Landsat has an image archive of many decades, and imagery with consistent calibration permits reliable change detection, including coastline changes, fluvial systems, and mass-wasting events. Sequential imagery provides the ability to assess before/after changes for events, including landslides, hurricane beach erosion, or flooding, to permit diagnostic analysis. Equally valuable is the capability provided by sequential imagery to examine incremental processes, such as coastal erosion, fluvial processes, and retreating periglacial landforms. In each instance, it is the ability of RS to illuminate an understanding of geomorphic process that forms its value to geomorphic inquiry.

Two families of RS systems that have special significance for geomorphic inquiry, *LiDAR* and *synthetic aperture radars* (*SARs*), are increasingly finding important roles in geomorphic inquiry. Although other RS systems will continue to be valuable, we can expect LiDAR and SAR imagery to form the core of RS analysis through their ability to acquire accurate and detailed terrain data, as outlined in the succeeding text.

In the 1980s, LiDAR imagery began to find a role in observing landscapes using instruments carried by aircraft to systematically observe a swath of terrain at fine levels of detail. LiDARs observe the earth's surface by transmitting pulses of light at wavelengths in or near the visible spectrum (approximately, within the interval 0.25 μm [250 nm] to 10 μm [10,000 nm], depending upon the specific instrument) toward the earth and recording the time delay of the returned radiation. Although basic LiDAR technology had been developed in earlier decades, by the mid-1980s, they were integrated with navigation systems and Global Positioning System (GPS) to form scanning systems that observe the earth at fine detail and high accuracy. Such systems can detect differences in time delays of returned pulses to separate those reflected from forest canopy and those that have penetrated the canopy to reach the terrain below. Thus, LiDAR

offers geomorphologists the ability to observe the terrain surface free of vegetation cover at levels of detail and accuracy not previously available. Although LiDAR imagery is often thought of as a resource best suited to observe relatively small regions at fine spatial detail, we note that, in the United States, states such as Iowa and North Carolina have collected statewide LiDAR coverage and that North Carolina is now covered by sequential LiDAR coverage—a likely precursor to future developments that will routinely provide broader coverage of the fine-scale LiDAR data. Because of these capabilities, LiDAR is rapidly becoming one of geomorphology's principal research tools, a growth that will likely accelerate as acquisitions create larger archives that can be applied for sequential analyses of landscapes and landscape changes.

Likewise, SAR, developed over several decades beginning in the 1970s, plays a significant, and growing, role in geomorphological, geophysical, and geologic inquiry. SAR, a specific form of imaging radar, collects imagery by broadcasting a beam of microwave energy (wavelengths in the range approximately 1 cm to 1 m, as selected for specific systems), then recording the energy backscattered from the earth's surface to form images. As active systems, SAR can measure the time delay between transmission of the original signal and receipt of its echo, as well as differences in phase, and polarization. These differences provide a basis for characterization of terrain surfaces, to include roughness, moisture status, subsurface features, vegetation cover and structure, and drainage systems.

Special techniques based upon comparison of signal phase, known as *interferometric* SAR (*InSAR*), can assess changes in surface elevation using dual views of the same terrain from different tracks to acquire two views of the same region (Klees and Massonnet 1999) (Figure 12.1). A related implementation, known as *along-track interferometry*, examines two views illuminated at different times as the satellite or aircraft follows its trajectory along a single path. If the echo of this signal is generated by a moving object, phase differences between transmitted and received signals will reveal the velocity of the lateral motion (e.g., in ocean currents, volcanic flows, or displacement of glacial ice). Another InSAR variation, *persistent/permanent scatterer interferometry* (*PSI*) Crosetto et al. (2010) (Section 12.4.1), can assess a collection of sequential interferograms to detect lateral movement of surface debris, which may signal increasing danger of mass-wasting events.

InSAR, therefore, forms a valuable tool for assessing subsidence, displacement, and lateral motion of terrain surfaces associated with geomorphological analysis. Reported accuracies of InSAR analyses vary with specifics of terrain, sensor systems (including wavelength), atmospheric conditions, and validation strategies, among others. The European Space Agency's TanDEM-X SAR system is specifically designed for terrain assessment at levels of detail and accuracies that support assessment of mass-wasting hazards (Bamler et al. 2009; Zebker et al. 2010). (In general, tabulation of summary information regarding accuracies of geomorphic RS applications represents

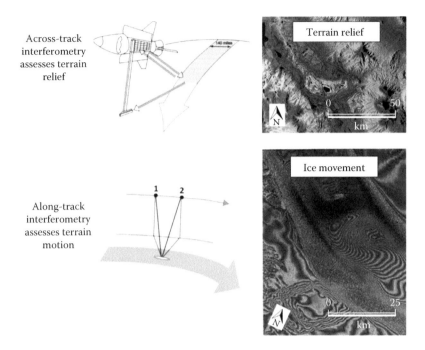

Across-track interferometry assesses terrain relief

Terrain relief

Along-track interferometry assesses terrain motion

Ice movement

FIGURE 12.1   Schematic diagrams illustrating uses of interferometric synthetic aperture radar to illuminate the terrain from two separate views (shown here with the two antennas of the Shuttle Imaging Radar Topography Mission), or alternatively from two passes of the same instrument, known as *cross-track interferometry*, and analysis of phase differences in the returned signal to derive terrain elevation. *Along-track interferometry* broadcasts microwave signals from differing positions to illuminate a surface and then analyzes phase differences in the reflected signals to assess surface movement, such as glacial ice (shown here), ocean currents, or lava flows. The colored contours (known as *banded color* or *wrapped interferograms*) indicate elevation differences, terrain subsidence, uplift, or lateral motion, depending upon the application. (Terrain relief image credit: USGS; Ice movement image credit: Jet Propulsion Laboratory, California Institute of Technology; SRTM diagram credit: Campbell, J.B. and Wynne, R.H., *Introduction to Remote Sensing*, 2011, Copyright Guilford Press, New York. Reprinted with permission of The Guilford Press.)

so many different measures, methodologies, and variations in experimental design, which the data are often not comparable.)

LiDAR and SAR, therefore, play important roles in current progress in geomorphological RS, for their ability to provide cogent geomorphological data and their inherent capability to provide metric data at high positional detail and accuracy. Their significance will increase in future decades, as analytical capabilities advance, as numbers of systems increase, and as the scope of archives expand. Although these systems may be seen as displacing conventional imagery, it seem likely that future applications will use multiple forms of imagery together to exploit their synergistic capabilities.

Despite active investigation of RS applications to geomorphology, relatively few such applications are employed operationally, in routine usage, to meet societal needs. One such example of operational usage is North Carolina's application of LiDAR imagery for flood zone mapping. Such applications require an accurate understanding of the accuracy and precision in mapping flood plain topography, for its use in meeting societal needs in land use planning and insurance underwriting. However, basic geomorphic research is usually conducted in a much different context, to investigate specific research objectives with findings that may be difficult to generalize across topics or disciplines.

Figure 12.2 illustrates relationships between spatial and temporal scales as they apply to uses of remotely sensed data for geomorphic analysis (Smith and Paine 2009). This diagram, based upon a previous analysis by Millington and Townshend (1987), depicts a data space defined by spatial resolution (X-axis) and temporal resolution (Y-axis). The dashed lines illustrate the analytical space for selected geomorphic analyses, defined by the capabilities of archives and the limits of spatial detail, as available at the time of the 1987 study. The red lines represent current thresholds, which define a larger space created by improvements in spatial detail of sensor systems and the increased capabilities of archives systems, and sensor systems and data processing capabilities (Smith and Pain 2009). This increased analytical space offers opportunities for future RS analysis to contribute to geomorphic inquiry.

12.2 Alpine and Polar Periglacial Environments

In both alpine and polar periglacial environments, geomorphological processes are tied to seasonal or perennial freezing and thermal changes in the ground. Furthermore, such processes are influenced by climatic variation that may occur over decadal or millennial temporal scales (Humlum 2008). Many processes in arctic permafrost

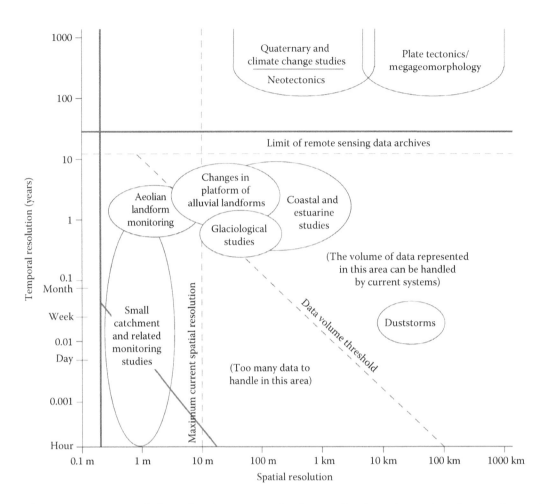

FIGURE 12.2 Schematic representation of relationships between spatial and temporal scales as they apply to the uses of remotely sensed data for geomorphic analysis. Dashed lines represent thresholds as reported in 1987; red lines signify current thresholds, indicating expansion of the capabilities of archives, improved resolutions, and improved data handling capabilities. (From Smith, M.J. and Paine, C.F., *Prog. Phys. Geogr.*, 33, 568, 2009. Used with permission.)

regions have impacts of global relevance, such as the carbon cycle of wet tundra lowlands (Semiletov et al. 1996; Grosse et al. 2006) and their rapid responses to climate change (ACIA 2004).

In periglacial geomorphology, field measurements have a long and rich history. Manual measurements prevailed prior to the 1980s (Humlum 2008). Arctic periglacial geomorphology has been associated with the use of high-resolution aerial imagery (Cabot 1947; Frost et al. 1966), but spatial and temporal coverage of such imagery is limiting. Furthermore, political (in addition to physical) remoteness prevails in broad regions of the Russian Arctic, where aerial imagery and detailed topographic maps are still classified or unavailable.

Adoption of RS techniques in remote arctic and mountain environments has increased our ability to analyze locations that are difficult to access and lack detailed topographic data. Accessibility is often cited as an obstacle to obtaining understanding of slope instability processes on periglacial high-mountain landscapes (Fisher et al. 2011). Applications of RS data for the monitoring of these remote and extensive areas are

cost-effective and, therefore, a fast-growing research area in the field of periglacial geomorphology.

12.2.1 Overview

Periglacial geomorphologists often combine a historical perspective coupled with interests in contemporary processes and patterns (Humlum 2008). In recent years, RS techniques have been employed to detect surface change (Grosse et al. 2005), and map and classify general permafrost properties (Morrissey et al. 1986; Leverington and Duguay 1997; Plug et al. 2008) and distributions (Ulrich et al. 2009) (Figure 12.3). These techniques often combine data from multiple sensors to enhance the landscape view, analytical power, and temporal perspective. Kääb (2008) provides an excellent review of spaceborne, airborne, and ground-based RS techniques applied in periglacial studies, especially as applied to periglacial hazards.

Degradation of permafrost is a widespread, striking, and important process interrelated with global change dynamics.

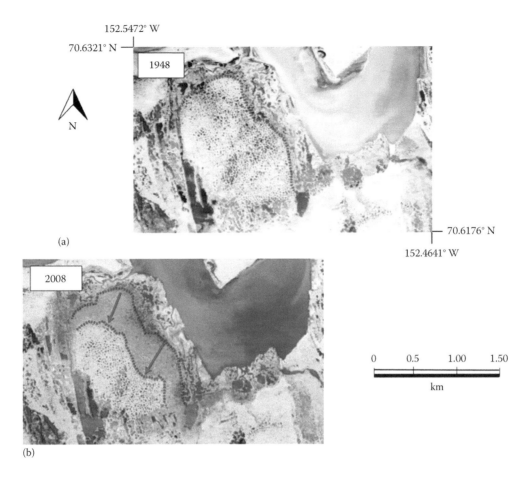

FIGURE 12.3 Changes in patterned ground, 1948–2008, Gary Creek, Alaska. Sequential aerial photography, 1948 (a) and 2008 (b), illustrating subsiding ice wedge polygonal ground near the mouth of Garry Creek, North Slope of Alaska. Arrows in the 2008 image indicate retreat of the polygonal ground during the interval 1948–2008. Field observations indicate that this and similar near-coastal areas of subsided tundra, has been covered with brackish sediments and colonized by saline vegetation. (From USGS 2011.)

The value of RS for permafrost studies has been demonstrated (e.g., Morrissey et al. 1986) (Figure 12.3). Analysis of permafrost degradation has involved both identification of periglacial terrain and change detection and in recent years has integrated RS methods that optimize the spatial, temporal, and spectral resolution of different sensors their using them in combination. For example, Grosse et al. (2005) present an integrated use of optical RS digital elevation model (DEM) and geographic information system (GIS)-based image stratification for the identification of periglacial terrain surfaces with a focus on thermokarst terrain.

In periglacial environments, practical obstacles to acquiring airborne data have led to reliance upon satellite imagery, which provide economical coverage of remote regions. As a result, Landsat-7 Enhanced Thematic Mapper Plus image, DEM data, and CORONA images offer options for coverage and, in some instances, may be useful as reference data to supplement field data. Hjort and Luoto's (2006) integrative approach incorporating topographical, ground, and remote sensing in predictive geomorphological mapping used generalized additive modeling. Plug et al. (2008) used Landsat scenes from 1978 to

2001 to classify and determine changes in lake coverage on the Tuktoyaktuk Peninsula in northwestern Canada.

12.2.2 Change Analysis

Change in mountainous periglacial geomorphic features has also incorporated integrated RS techniques (Walsh et al. 2003; Fischer et al. 2011). For example, Fischer et al. (2011) developed a time series of high-resolution digital terrain models (DTMs) with a 2 m resolution from digital aerial photogrammetry for 1956, 1988, and 2001 and from airborne LiDAR for 2005 and 2007 to characterize topographic change in steep periglacial high-mountain faces in the Swiss Alps associated with increased rock and ice avalanche activity. Their study necessitated spatial resolution and accuracy assessment at a meter scale and, therefore, used digital photogrammetry based on aerial images and combined LiDAR for multitemporal topographic analysis of fine-scale change in periglacial processes. A time series of high-resolution DTMs from high-precision digital aerial photogrammetry and airborne LiDAR was developed to study topographic changes in steep periglacial rockwalls in detail over a long time. Walsh et al. (2003)

used remotely sensed imagery in concert with spatial analysis to investigate geomorphic elements hypothesized to be direct or indirect influences on alpine tree line in Glacier National Park. Relict solifluction terraces were investigated and mapped using data fusion techniques combining airborne Airborne Data Acquisition and Registration (ADAR) data from 1999, with a spatial resolution of 50 cm to 3 m and Landsat Thematic Mapper (TM). In their study, ADAR data provided greater spatial resolution and Landsat greater spectral resolution (seven spectral channels). Walsh et al. (2003) suggested that broad-scale imagery, such as Landsat TM, can be used to describe signatures of periglacial processes or patterns, whereas finer-resolution data can be used to assess the nature of the process or pattern.

12.2.3 Emergent Opportunities and Challenges

Many RS applications in periglacial geomorphology, despite availability of high-resolution data, remain focused on characterization (description) of periglacial features, and analytical research to date still remains largely focused on change detection analysis. Relatively little emphasis has been placed upon analysis and prediction at this point. Furthermore, in dynamic environments, such as rapidly occurring processes and hazards, airborne and spaceborne RS may offer support for assessing hazard susceptibility (Kääb 2008).

However, the use of RS imagery into studies of periglacial processes is likely to enable analysis of spatial patterns a more fundamental part of analyses in the development, parameterization, and validation of spatially explicit models (Walsh et al. 2003). Humlum (2008) claims that a practical priority for periglacial geomorphology should be to improve understanding of climatic controls on geomorphic processes and to predict hazards and hazard risks in the context of climate change. Additionally, modeling landscape evolution in cold climates (French and Thorn 2006) will likely be aided as they increasingly integrate RS into their development and validation.

12.3 Glacial Geomorphology

Glacial geomorphology is the study of geomorphological processes and resultant landforms and relief tied to glacier dynamics. Glacial dynamics have global relevance through their rapid response to climate change and resultant geomorphic impacts. Mountain glaciers are especially sensitive to climate variation, thus are considered to be a high-priority climate indicator (Kääb et al. 2005). Glacial landforms are often studied as proxies for past glacial activities (Smith et al. 2006) (Figure 12.3).

12.3.1 Overview

RS techniques have supported analysis of glacial landforms and sediment in remote locations lacking detailed topographic data. Mapping of previously glaciated terrain inherently involves application of an historical perspective to glacial landscapes, (e.g., Humlum 2000 for rock glaciers) given that

glaciated landforms are landscape manifestations of past processes. Reconstruction of glacier dynamics through landforms originally involved contour mapping derived from field observations. Analysis using remotely sensed data, DEMs (e.g., Clark and Meehan 2001), and/or aerial photography, to a large extent, has replaced, or corroborates, this methodology (Smith et al. 2006).

The study of glaciated terrain from the air has its roots, and still carries a strong tradition, in aerial photography. Aerial photography analysis remains as a valuable methodology because it enables fine-resolution mapping of glacial landforms that may not be assessed with coarser-scale data (Smith et al. 2006). Humlum (2000), for example, estimated weathering rates, debris volume, and Holocene rockwall retreat rates at 400 locations on Disko Island, Greenland, using a mixed method approach incorporating aerial photography.

Digital procedures to extract terrain models from aerial photography have enhanced the collection of topographic data. Use of DEMs to outline topography and to study surface and erosional processes is often used in conjunction with or to corroborate other analyses. For example, Schiefer and Gilbert (2007) prepared high-resolution (submeter) DEMs from archived photography to map changes since the mid-1900s in proglacial terrain. Their study exemplifies a unique application of historical aerial photography for generation of historical DEMs that provide information on landscape response to environmental change.

Satellite imagery, such as low- to moderate-resolution Landsat and Advanced Spaceborne Thermal Emission and Reflection Radiometer data, has broader aerial coverage and relatively low costs compared to aerial photography that has been used effectively to uncover coarse-scale patterns created by scale glacier processes (Smith et al. 2006). For example, Clark (1993) used Landsat images to unveil previously unsuspected large-scale pattern of streamlining within drift that is assumed to reflect former phases of ice flow. Use of Landsat, as opposed to aerial photography, was essential in documenting a previously undocumented pattern of an ice-molded landscape. They have been applied to document form and topology of features such as glacial lineations (large-scale ice-molded landform assemblages), drumlins, glacial megaflutes, and previously undocumented ice-molded landforms. These features have such broad scales that they are not effectively examined using the narrower coverage of airborne sensors. In general, a pattern cannot be recognized as such unless a large enough area is viewed, to reveal its continuity or the repetition of its internal elements.

12.3.2 Emergent Opportunities and Challenges

Despite the availability of high-resolution data, RS applications in glacial geomorphology still emphasize description of glacial features. Analytical research to date largely remains focused on analysis of change over time, with little emphasis on predictive modeling. However, with the recognition of the importance of surface erosion in mountain building (Raymo and Ruddiman 1992; Pinter and Brandon 2005), one promising application of RS technologies for glacial geomorphologists is in mapping and

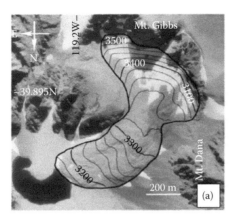
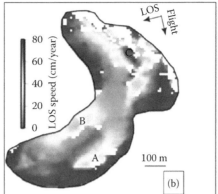

FIGURE 12.4 Rock glacier movement as assessed by InSAR. Liu et al. (2013) applied InSAR analysis to examine the movement of rock glaciers in California's Sierra Nevada mountain range. (a) Aerial photography, September 25, 1993, showing the Mt. Gibbs rock glacier outlined by the heavy black line and 20 m contours as narrower black lines. (b) InSAR analysis of rock glacier motion (along the LOS axis) with letters and red tones indicating zones of rapid motion (maximum = 48 cm/year). (From Liu et al., 2013. Used with permission.) (c) Aerial photography, September 14, 2013, accessed from Google Earth, June 23, 2014.

assessing glacial erosion and feedbacks with climate as they relate to mountain building. Bishop and Shroder (2000) and Bishop et al. (2003) used satellite imagery and digital elevation data to examine erosion and depositional features at Nanga Parbat to identify geomorphic events and to better understand the role of surface processes in the denudation cascade. Furthermore, in dynamic glacial environments, airborne and spaceborne RS may offer increased support for assessing hazard susceptibility, such as glacial lake outbursts (Huggel et al. 2002). For example, Liu et al. (2013) examined movement of rock glaciers in the California's Sierra Nevada mountains, reporting local flow rates of 48 cm/year (Figure 12.4), finding that their motion exhibited high seasonal variation and high year-to-year variation, indicating the significance of high frequency of observations over long time intervals to understand dynamics of rock glaciers.

Important limitations to our understanding of mountains through RS are imposed by the challenges associated with complex terrain and environmental factors such as atmosphere and land cover, which control irradiant and radiant flux. Landforms with forested cover may be represented inaccurately in locations with forest cover, even using LiDAR imagery (Smith et al. 2006). Reliable information depends upon reducing spectral variation

caused by topography and land cover, often by application of band ratios to reduce atmosphere and topographic effects. Such constraints lead to development of new and innovative approaches in the combined area of RS and geomorphometry, as shown in Bishop and Schroder (2004). Furthermore, more advanced forms of modeling may be needed to accurately apply RS to investigate landform development.

12.4 Mass-Wasting

Concisely defined, mass-wasting is the bulk transport of regolith downslope, chiefly under the force of gravity, often facilitated by saturation and/or tectonic and weather events. Typical mass-wasting processes include landslides, soil creep, debris flows, and solifluction. Contributing factors often include interactions between local slope, surface materials, geologic structure, hydrology, freeze–thaw, shrink–swell, and vegetative cover, which, in specific circumstances, can assume dominant or causative roles. Mass-wasting events can be incremental or episodic; in some cases, discrete events are preceded by long intervals of incremental processes that set the stage for events often observed as distinct, unexpected events.

Alos-band [mm/year]

| 60 | 30 | 20 | 10 | 5 | 0 |

FIGURE 12.5 Mapped landslide in Estellencs, Italy, showing terrain displacement as assessed by PSI (colored arrows) with an inset illustrating a subsequent rotational landslide event in March 2010. (From Bianchini, S. et al., *Remote Sens.*, 5, 6198, 2013. Used with permission.)

The study of mass-wasting includes, among others, geotechnical engineering, hydrology, and geography. These disciplines all include, to varying degrees, impacts of including human behavior, especially prevailing and historic land use practices, providing the historical context (e.g., previous events that signal the prevailing dynamics), as well as an understanding of local lithology and geological structure, climate, vegetation, and hydrology.

12.4.1 Role of Synthetic Aperture Radar in Landslide Analysis

Research has long used aerial photography, optical satellite imagery, and supporting topographic, geologic, and vegetation data to assess landslide and mass movement risk. These analyses basically target risk by identifying the geomorphic context for increased hazard. However, in recent decades, development of analytical strategies using SAR imagery increases opportunities to extract more direct information specifying locations and nature of landslides and risks they may present. Because these techniques offer the potential to extract immediate, site-specific information about landslide risk, they form an important resource for research addressing mass-wasting processes.

As an active RS system, SAR illuminates terrain with microwave energy of known wavelengths and polarizations, so, as it receives backscatter, it can detect changes in wavelength, phase, and polarization. Analysis of phase change provides the basis for detection of terrain displacement and, therefore, the monitoring of subsidence and assessing hazard.

A variation on the basic InSAR strategy, known as PSI, relies upon analysis of SAR pixels that maintain stable (known as "coherent") backscatter over a sequence of SAR images of the same region. A critical, and often difficult, step in implementation of PSI is identification of these stable pixels (i.e., those easily identifiable in sequential images) within a sequence of SAR images of the same region. Tantianuparp et al. (2013) and Del Ventisette et al. (2013) offer specifics.

Whereas basic InSAR extracts changes in vertical terrain displacement, PSI seeks to define horizontal displacement, which often forms a precursor to landslide events. Because SAR satellites observe terrain from an (almost) overhead perspective, assessment of vertical displacement is easier than assessment of horizontal displacement (Figure 12.5). Nonetheless, using both processing algorithms and ground-based networks of ground observations, projects are underway in Europe to monitor lateral motion to provide warnings of developing landslide hazards (e.g., Ghuffar 2013). Because urban regions usually have larger and more reliable populations of stable backscatters, they are well positioned to benefit from such systems. In contrast, for rural regions, the number of radar benchmarks can be low, limiting application of the PSI strategy.

12.4.2 Landslide Studies

Landslides are like people: Every one is different. They occur in settings ranging from jungles to deserts. The materials involved range from mud to rock to ice, including mixtures of all three. Some slides are wet; others are dry. Some roar down steep mountainsides; others creep along at barely perceptible rates. Some are so rigid that they are truly "slides"; others, so fluid that they are best described as "flows." But the motion of landslides, small or large, is always governed by the conservation laws of physics.

Kieffer (2014, p. 298)

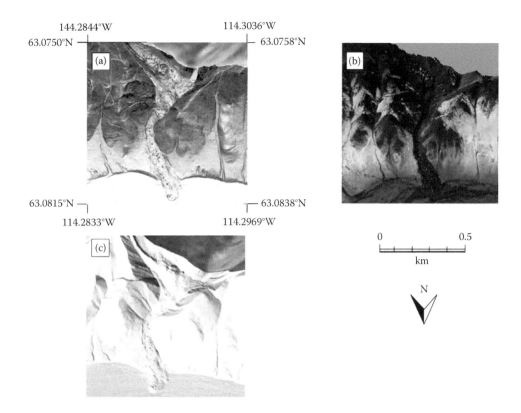

FIGURE 12.6 LiDAR image of a landslide in the Alaska Range along Denali–Totschunda fault system, Alaska. Pictured here are the oblique views of a landslide adjacent to the Denali fault. The last major rupture of this fault system was a magnitude 7.9 earthquake on November 3, 2002, which could have caused this landslide. (a) A slope-shade model, which darkens steeper slopes and illuminates gentle slopes. Notice that there are dark, round spots in the landslide patch, which are likely large boulders displaced by the event. (b) The lineament of a fault is visible in the blue shading extending from higher (light blue) to lower (dark blue) elevations. (c) A hillshaded relief display of the same region. (Image courtesy of Emily Kleber, OpenTopography. Dataset: EarthScope (2008). EarthScope Alaska Denali Totschunda LiDAR Project. Plate Boundary Observatory & National Center for Airborne Laser Mapping. Distributed by OpenTopography. http://dx.doi.org/10.5069/G9QN64NF, date accessed: February 25, 2014. Used with permission. Google Earth imagery accessed May 15 2014.)

As is clear from Kieffer's statement, landslides are among the most dangerous of geomorphic phenomena and among the most difficult to study. RS, in partnership with other research methods, provides an important resource for understanding landslides and their behavior. Figure 12.6 illustrates the value of the detail provided by LiDAR imagery to record the configuration of landslide events and permit examination of changes. Scaioni (2013) identified several themes in applications of RS to landslide studies: (1) inventories of past landslide events, (2) site-by-site examination and monitoring of potential mass-wasting sites, largely using SAR imagery, and (3) application of ground-based lasers to examine active sites and monitor stability.

1. *Inventory*: The inventory strategy seeks to apply geospatial data to identify areas of increased risk through convergence of contributing factors. Sankar and Kanungo (2004) developed integrated RS and GIS strategy through their studies of a region of the Darjeeling Himalaya. They used Indian Remote Sensing Satellite data, field data, and topographic maps integrated using GIS to assess topographic, hydrologic, and geological conditions within their study area. Their numerical rating strategy that defined four susceptibility classes, provided close agreement with existing field instability reports and with event of observed landslides.

   Strozzi et al. (2013) combined interpretations of aerial photographs and surface displacement derived from InSAR to examine landslides and their intensities. Such approaches provide valuable contributions toward analysis of landslide hazards in areas where traditional monitoring techniques are sparse or unavailable.

2. *Monitoring critical sites*: Othman and Gloaguen (2013) examined satellite imagery of river channels at the Iraq/Iran borderlands to examine displacement of river channels by landslides that have blocked river channels, thereby diverting river channels. They quantified river offsets using two geomorphic indices that assess the displacement with respect to basin midline and with respect to the river's principal orientation. Their analysis used these indices to assess intensities of landslides and to assess risks of future events.

   Tofani et al. (2013) examined landslides in the northern Apennines, Italy, to assess surficial displacement as measure of surface behavior and its relationships with triggering factors. They applied PSI to estimate the yearly deformation velocity of the Santo Stefano d'Aveto landslide. Their study compared PSI results with data from in situ data to confirm the value of the PSI strategy. Tantianuparp et al. (2013) applied InSAR and PSI to examine multiple SAR datasets of the Three Gorges area in China to study slow moving landslides over long periods. Ghuffar (2013) examined an active landslide in Doren, Austria, using multitemporal airborne and terrestrial laser scanning (TLS), 2003–2012. They applied 3D motion vectors for the time series to identify displacements up to 10 m, whereas other regions often did not change for several years. Akbarimehr et al. (2013) applied observations from both InSAR (2004–2006) and GPS (2010–2012) to assess slope stability of the Sarcheshmeh landslide in northeast Iran.

3. *Applying ground-based laser scans to examine active sites and monitor stability*: At local scales, RS can assess site-specific risks, warn of impending hazards, and provide diagnostic analysis of historic events. At broader perspectives, RS can identify the significance of latent causative factors and link to contributing factors (e.g., biogeography, hydrology, geology, tectonics, terrain, and land use). Band et al. (2012) linked landslides in mountainous terrain to forest regulation of hydrologic dynamics at the Coweeta Hydrologic Laboratory, a U.S. Forest Service experimental watershed in North Carolina. Their ecohydrological approach examined impacts of hydrologic and canopy patterns upon slope stability and development of landslide risk in steep forested catchments. LiDAR data provided the basis for deriving the flow path structure and the pattern of pore pressures, in relation to elevation, slope, and aspect. This strategy revealed ecological and hydrological dimensions of mass-wasting processes at this site.

12.4.3 Summary and Future Outlook

Recent developments in RS technologies and analytical strategies have advanced studies of mass-wasting events from one that basically focuses upon descriptive before and after observations of landforms to one that can deploy RS systems to observe dynamic processes in progress, both to examine geomorphic processes and to assess risk with temporal and spatial detail not previously feasible. Key technologies include LiDARs, InSAR, and PSI as mentioned earlier. Also significant are application of these technologies with established image analytical strategies, including multitemporal LiDAR and object-based image analysis, which provide the detail and accuracy necessary for volumetric analyses.

Future applications will see increased applications of LiDAR, especially sequential LiDAR analyses, as the archive of repeat coverage increases as acquisitions increase. Research to apply RS mass-wasting will apply multisource RS; continued use to integrate RS with other geospatial data, especially DEMs; increased use of in situ data, including in situ GPS; and applications of fine-resolution InSAR to study geomorphological dynamics and features at finer scales. We note also that some of these applications require systematic and sequential coverage at resolutions, revisit times, and spatial coverage not always routinely available using current systems.

12.5 Fluvial Landforms

12.5.1 Overview

Fluvial landforms are formed by channelized flow within rivers and streams, as well as associated deposits and landforms, including terraces, sedimentary deposits, erosional features, stream beds, and river valleys that accompany channelized flow. Fluvial systems are characterized by their dynamic behavior (flooding, channel migration, alternate episodes of deposition and erosion) and their multiscale character (processes that operate simultaneously at varied spatial and temporal scales). Temporal variation in stream systems combines with incremental processes and those that act in an episodic manner. Although these characteristics present numerous obstacles for systematic inquiry, the field of RS, with its capabilities for multitemporal (sequential) and multiscale imagery, provides insight into many of these dimensions of fluvial systems that otherwise would be less well understood.

Aerial imagery of fluvial landscapes was among the earliest images used for geomorphic analysis; Lee (1922), for example, provides representations of complex landscapes that were previously difficult to depict from ground-level views. Likewise, early imaging radar imagery (Simpson 1966) and, later, Landsat images (Short and Blair 1986; El-Baz 2000; Sarkar et al. 2012) provided synoptic views of landscapes to highlight regional variations in drainage and fluvial systems and in active geomorphological processes. (Figure 12.7 provides a more recent example.) Currently, LiDAR and sequential aerial photography form two of the most valuable resources for the study of fluvial systems. The detail and accuracy of LiDAR (Hohenthal et al. 2011) form an especially valuable resource for study of fluvial systems and the length of the aerial photography archive, especially in the United States. The ability of SAR imagery, because of its all-weather capabilities and its ability to detect flooded areas, forms an important resource for understanding flood events and flood plain delineation, especially when cloud cover prevents observation using other sensors. For broad-scale inquiry, the length and consistency of the Landsat archive provide an invaluable resource for geomorphic inquiry. National Academy of Sciences (2007) provides a useful overview of floodplain mapping technologies. Overall, with respect to RS of fluvial systems, Marcus and Fonstad (2010, p. 1868) note that the "... recent proliferation of review papers on the topic is one indicator that this is a rapidly emerging field (Mertes 2002; Gilvear and Bryant 2003; Handcock et al. 2006; Feurer et al. 2008; Marcus and Fonstad 2008), as are science agency reports advocating increased use of remote sensing to systematically monitor, manage, and understand river systems (Environmental Protection Agency 2006; NRC 2007)."

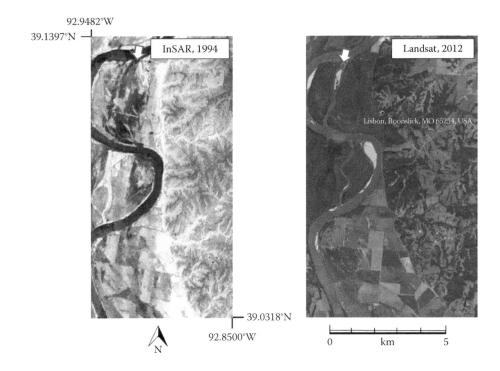

FIGURE 12.7   Missouri River floodplain, August 1994, following a major flooding during the summer of 1993, observed using airborne InSAR (TOPOSAR). Here, the colors represent elevations, with the darkest blues indicating low elevations created by recent scour channels as flood waters cut incipient channels. The right-hand side image depicts the same region as imaged by Landsat in 2012. The two arrows identify the location of the incipient scour channel identifiable in the 1994 image as a linear blue streak, but in 2012 presenting as a major diversion of the river flow. (Courtesy of NASA/JPL-Caltech/ PIA01806.)

12.5.2 Floodplain Analysis

Defining and understanding the nature and dynamics of floodplain form an important dimension of fluvial geomorphology, to include understanding sediment budgets, fluvial erosion, and deposition, with significant implications for land use policy, hazardous waste, property values, and insurance policy. RS technologies offer capabilities to examine the fundamental characteristics of fluvial systems (e.g., Figure 12.8).

Administrative floodplain mapping applied conventional methods based upon analysis of topographic maps and aerial photography to define floodplain limits and to inventory floodplain land use. In recent decades, despite limitations in LiDAR's horizontal accuracy, its superior vertical accuracy, fine spatial detail, (Chen 2007), and ability to measure ground elevation in vegetated areas have demonstrated its advantages relative to conventional aerial survey (Sasaki et al. 2008). As a result, LiDAR data form a principal source of topographic information for hydrologic/hydraulic analysis and floodplain delineation (Deshpande 2013).

Analytical and research investigations often require broader spatial scope and a broader range of spectral observations, so a range of RS technologies still form significant resources for floodplain analyses. For example, Ghoshal et al. (2010) examined a 100-year record of stream bank erosion in the Yuba River, California, an interval that encompasses a history of hydraulic mining, 1853–1884. Their study focused on the interval

1906–1999, recording a period of recovery from the disruption from mining, using historical maps, aerial photographs, digital orthophoto quarter quads, sound navigation and ranging, and photogrammetrically derived topographic data. Their analysis, using aerial photograph for the period 1947–2006, documented flooding, erosion, sedimentation, reworking and redistribution of sediments, calculation of volumes, and reworking of sediment throughout the 100-year interval, 1906–2006. Their study reveals the long period of hydraulic adjustment to episodic sedimentation events—a long but declining reaction to a major disturbance.

Townsend and Foster (2002) employed SAR imagery as part of a simulation of flood extent and duration for the Roanoke River floodplain (the United States). Their results suggest that the intermediate sector of the hydrologic gradient has been compressed to emphasize extremes of either wetter or drier conditions. Their model represents a simple method to simulate hydroperiod regimes at landscape scales in situations where data necessary for more complex physical models are not available.

12.5.3 Channel Migration

Channel migration describes the lateral movement of a river channel across the alluvial floodplain, through incremental erosion and deposition, and especially through *avulsion*—the sudden translocation of a river channel, typically in response

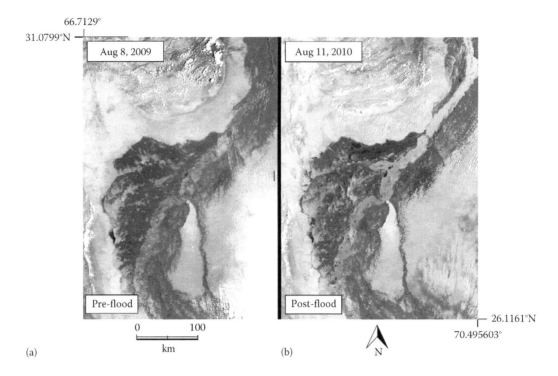

FIGURE 12.8   Flooding of the Indus River as recorded by sequential satellite imagery, Sindh Province, South Pakistan, August 2010 (a: August 8, b: August 11). The Indus River crosses the image from upper right to lower left. In late July 2010, heavy monsoon rains caused flooding in several regions of Pakistan, including the Khyber Pakhtunkhwa, Sindh, Punjab, and parts of Baluchistan. These false-color views display the near-infrared, red, and green regions of the spectrum in a combination that highlights contrasts between water and vegetation on the river banks. Such imagery provides the capability to record the extent, duration, and impacts of such broad-scale flood events with precision and accuracy. (Courtesy of NASA/GSFC/LaRC/JPL, MISR Team. PIA13337.)

to flood events. Understanding channel migration is critical dimension of addressing geomorphological and river management problems. Because of the large magnitudes and episodic and rapid rates of change, special surveillance systems are needed to efficiently measure and monitor channel migration. Because channel migration must be observed at rather broad scales, at infrequent intervals, RS forms an important tool for recording and analyzing channel migration.

Yang et al. (1999) examined sequential Landsat imagery of the Yellow River delta, China, encompassing an interval of approximately 19 years. They used GIS to map channel position and systematically examined changes of river banks and channel centerlines, to relate these to natural and human processes. Likewise, Bhaskar and Kumar (2011) examined satellite imagery to study geomorphic processes that control river channel migration in the Thengapattanam coastal tract bordering the Arabian Sea in the Kanyakumari District, Tamil Nadu, southern India. Their strategies permitted examination of geomorphic units, recording positions of river channels and, over time, documenting migration of the Kuzhithurai river channel in coordination with field data revealing entrenching of the channel, possibly relative uplift relative to mean sea level.

Sequential multispectral satellite imagery of alluvial rivers provides specifics of channel characteristics and positions over time and provides the geomorphic and land use context for

understanding the nature of changes and interrelationships with human occupation of the landscape of Amazon floodplain (Mertes et al. 1995).

12.5.4 Stream Bank Retreat

While considerable effort has been directed toward reducing erosion from agricultural and urban lands, the significance of stream bank degradation has only recently been acknowledged. Studies have shown that sediment loss from stream bank retreat (SR) can account for as much as 85% of watershed sediment yields, and bank retreat rates as great as 1.5 m to 1100 m/year have been documented (Simon et al. 2000). In addition to water quality impairment, SR impacts floodplain residents, riparian ecosystems, bridges, and other structures (ASCE 1998).

SR typically occurs by a combination of subaerial erosion, fluvial erosion, and stream bank failure, all related to local soil conditions, land use, streamflow regime, and drainage basin hydrology. Heeren et al. (2012) describe their procedure for evaluating their rapid geomorphic assessments for assessing SR using sequential aerial photography, 2003–2008.

Recent research has developed strategies for applying ground-based LiDAR. Resop and Hession (2010) surveyed an 11 m stream bank at Stroubles Creek, Blacksburg, Virginia, six times over a 2-year period using both conventional surveying with total station and TLS. Wang et al. (2013) employed mobile laser scanning to record high-precision data (including grain size, sphericity, and orientation) of coarse sediment in a gravel bar of a subarctic stream, Utsjoki River, northern Finland.

12.5.5 Summary

Relative to other topics examined here, RS applications to examine fluvial processes have employed a wide variety of technologies over a broad range of spatial scales and temporal scales. Because of the broad spatial and temporal scales require to monitor channel migration, satellite systems have often been employed to examine channel migration. At finer scales, LiDAR imagery provides the spatial detail and the ability to record bare-earth terrain data necessary for modern floodplain maps. Stream bank erosion research has employed sequential aerial photography and LiDAR imagery, but has also used sequential ground-based LiDAR, with other instruments, to examine local changes in stream banks. It seems likely that

LiDAR will continue to increase in its significance but that this mix of different technologies will continue to be important in fluvial studies.

12.6 Coastal Geomorphology

12.6.1 Overview

Coastal geomorphology encompasses the dynamic interface between oceans and land surfaces—a zone subject to constant wave action, which forms the principal driver among the numerous processes that erode, transport, and deposit sediments. Such processes are significant now and in the past, but have special significance in the current context of rising sea levels and increases in populations inhabiting coastal zones (Figure 12.9).

RS' capabilities to support understanding of the dynamic interplay of processes at local and regional scales, and over time, forms an important resource for the geomorphic study of coastal environments. RS is especially valuable for the study of dynamic

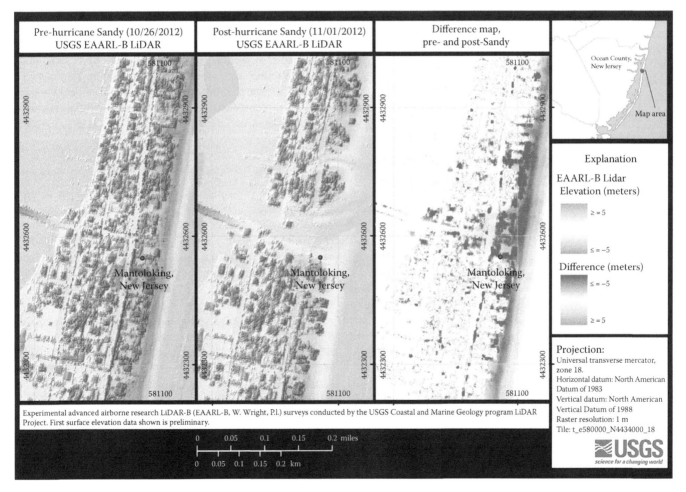

FIGURE 12.9 Before and after LiDAR surveys of coastal elevations at Mantoloking, New Jersey, documenting impacts of Hurricane Sandy (a, October 26, and b, November 1, 2012). (c) The difference map shows, in red, losses in elevation of as much as 5 m (16 ft.). Sequential imagery with high levels of precision and detail of LiDAR form valuable geomorphological information in dynamic coastal environments. Figure by Cindy Thatcher, USGS. (Courtesty of U.S. Geological Survey.)

phenomena, such as dunes, coastal erosion, and deposition. Geomorphic dimensions of current concerns such as changing sea levels, protection of coastlines from wave action, formation of land use policies in the littoral zone, and nearshore bathymetry, all can benefit from applications of remote-sensed imagery. Allen and Wang (2010) provide a review of RS contributions to studies of coastal regions.

12.6.2 Multispectral Bathymetry

Water depth and subsurface topography are important dimensions of coastal geomorphology, especially in monitoring changes in coastal processes and landforms. Multispectral bathymetry techniques examine brightnesses in several spectral channels to assess differences in brightness against a reference brightness, in the same channels, recorded in a region known to have deep water (e.g., deeper than 50 m), assuming a clear, calm, water body. Thus, for example, values of the brightnesses in the blue and green regions of the spectrum are observed to increase in shallower water due to scattering of solar radiation from the subsurface terrain. From such observations, water depth can be estimated from reference data that records maximum penetration of the several spectral channels (again, assuming a clear, calm, water surface). As the number of spectral channels available increases, and radiometric resolution increases, depth estimates can attain greater detail and reliability. For example, Gao (2011) derived bathymetric information over clear waters at depths up to 70 m. Gao's (2011) review of RS-based bathymetric methods reports that detectable depths are usually limited to about 20 m, with accuracy decreasing with increasing depth, especially at a depths below 12 m. Subsurface topography includes the same irregularities in surface materials, configuration, and vegetation cover, observed in terrestrial RS, which all further contribute to inaccuracies.

12.6.3 Airborne LiDAR Bathymetry

Airborne LiDAR Bathymetry uses LiDAR technology to assess water depth directly by illuminating the water surface with a LiDAR beam (Irish 2000). As each LiDAR pulse reaches the air/water interface, a portion of the transmitted energy is returned from the surface back to the sensor (surface return). The remaining portion of the energy propagates through the water column, to be reflected from the subsurface topography (bottom return), assuming water depth is shallow enough for the pulse to reach the bottom; the time delay between the two returns permits assessment of water depth. An important capability for bathymetric LiDARs is the ability to use a single instrument to collect, in a single pass, high-resolution topographic data, as well as reliable water surface elevation, a capability especially useful at the interface between land and shallow water. Bathymetric LiDAR instruments specifically tailored for bathymetric surveys include green bands designed for penetration of the water body to permit subsurface mapping at varied depths.

12.6.4 Suspended Sediment Concentration

RS has been applied to observe sediment transport, provide information for sediment budgets, monitor sediment sources, and identify sources of pollution. Multispectral imagery and, in some instances, panchromatic imagery provide the ability to observe distinctions between clear and turbid water. For example, Brakel (1984) examined Landsat Multispectral Scanner data to track sediments plumes from rivers discharging into the Indian Ocean. They found seasonal patterns related to seasonal rain events that carry sediment to the coastline, alternately northward and southward along the coastline, as monsoon winds and currents change seasonally.

RS has also been applied to derive quantitative estimates of suspended sediment concentration (SSC) using both airborne and spaceborne sensors. Such surveys offer synoptic overviews of large water bodies that can, in principle, be coordinated with simultaneous, or near-simultaneous, on-site data collection. These methods exploit the positive associations between spectral radiance measured at the instrument and SSC as assessed on-site, usually by boat survey.

Curran and Novo (1988) reviewed the ability of multispectral observation to estimate SSC. Their review highlights difficulties of defining robust relationships between RS observations and on-site SSC because of effects of atmosphere, solar angle, sensor altitude, sensor altitude, and wave height. They conclude that future research should focus upon improved field sampling of SSC and compensation of environmental influences upon observed radiances.

12.6.5 Coastal Change and Retreat

The ability of sequential imagery to record changes in positions and configurations of coastal landforms provides the opportunity to observe and analyze changes at varied scales (Figure 12.10). Moore and Griggs (2002) applied aerial photography, GIS, and soft-copy photogrammetry to examine long-term coastal cliff retreat and erosion (1953–1994), estimating an average rate of retreat of 7–15 cm/year, but identified episodic hot spots with rates as high as 20–63 cm/year. Gibson et al. (2009) examined sequential aerial photography of Gulf Shores, Alabama, 1955–2006. At coarser scales, they only identified small coastline changes, but at finer scales, their analysis revealed a variety of effects arising from interactions between residential development, hurricane overwash fans, and artificially engineered tidal inlets.

Several investigators have devised image analysis strategies to track shoreline changes. Shu et al. (2010) devised thresholding algorithms for semiautomated analyses of RADARSAT-2 for shoreline extraction. Marghany et al. (2010) examined multitemporal SAR (European Remote Sensing Satellite-1 and RADARSAT-1) to examine spit migration along the Malaysian coastline, applying filters to define the spit and to track migration of the spit boundary, 1993–2003. Their analysis concluded that their filtering strategy was able to accurately track

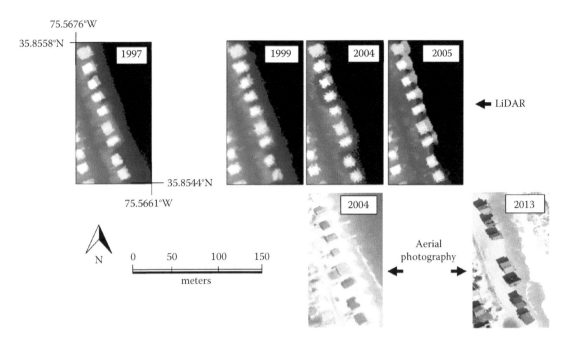

FIGURE 12.10 Coastal erosion as recorded by sequential LiDAR, North Carolina Coast, 1997–2005. (a) Sequential LiDAR imagery; 1997–2005. Here, black represents open water, the white rectangular shapes are beachfront homes, and the gray background represents the sand surfaces of the beach. Progress of coastal retreat is revealed by the advance of the coastline toward the residential structures; 1997–2005. (b) Aerial photography of the same region, with the 2013 image depicting the loss of one of the structures visible in the earlier imagery. LiDAR point data collected during 1997–2005 by NASA/NOAA coastal mapping programs, hosted by the Coastal Service Center (CSC) of NOAA, Charleston, SC, and processed at East Carolina University. Ground reference data were collected between 2006 and 2008. (From White, S.A. and Wang, Y., *Remote Sens. Environ.*, 113, 39, 2003; Allen, T.R. and Wang, Y., Selected scientific analyses and practical applications of remote sensing: Examples from the coast, in *Manual of Geospatial Science and Technology*, 2nd edn., eds. J.D. Bossler, J.B. Campbell, R.B. McMaster, and C. Rizos, Taylor & Francis, London, U.K., 2010, pp. 467–485. Used with permission. GoogleEarth imagery accessed May 15, 2014.)

accretionary processes over this interval. Brock et al. (2002) describe applications of LiDAR surveys to examine bare-earth topography, vegetation structure, coastal dunes, barrier islands, shoreline change, landslides coastal cliffs, subsidence, storm surge, and tsunami inundation.

Kawakuboa et al. (2011) examined TM data of coastlines of southeastern Brazil, applying spectral linear mixing and contextual classification using a segmentation technique, and resolved differences between vegetated zones, clear water (a proxy for shade), and soil. Comparison of these classification results revealed a transformation process active in coastal environments, and that erosive and depositional features are highly dynamic over short periods of time.

12.7 Aeolian Landforms

Aeolian landforms are formed by the action of wind, either by erosion or deposition of sand-, silt-, or clay-sized particles. Erosional features include lag deposits, formed by selective erosion of finer particles by wind, leaving behind coarser sediments as distinctive geomorphic surfaces, and deflation hollows (blowouts). Such features are typically formed in arid or semi-arid regions where local vegetation cover is disturbed, allowing wind to remove finer particles to create shallow depressions. Here, our discussion focuses chiefly upon depositional forms,

including both coastal and desert landscapes. Aeolian entrainment occurs by rolling or sliding (*creep*), the Bernoulli effect of winds (*lift*), bouncing (*saltation*), and the impact of one particle upon another. Entrained particles ultimately form dunes, notable for the distinctive forms and their mobility. Dunes are commonly associated with sand, although notable dunes formed from gypsum (White Sands National Monument, the United States), and loess deposits throughout the world from silt-sized sediments.

Dune formation depends upon nearby sources of sediment, such as playas, washes, and fans; some dunes active others relict from previous episodes when conditions favored dune formation. Dunes can stabilized by natural vegetation, perhaps associated with changes in local climate, or by stabilization programs managed to prevent or minimize dune movement, especially in coastal zones. Because of the dynamic character of active dunes, and their potential threats to agriculture in some regions and their role in protecting coastal zones, RS has become a valuable tool for recording movement of dunes and assessing their volume and changes in volume.

Livingston et al. (2007) report that "…in the past many of the single-dune studies described a largely inductive approach: that is, they collected field data, generally about wind flow and sand flux, and then tried to make sense of those data" (p. 254). RS imagery, especially LiDAR imagery, provides the ability to

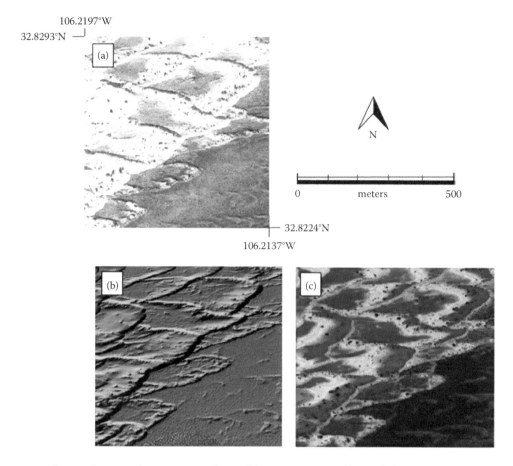

FIGURE 12.11   Dunes, White Sands National Monument, as observed by LiDAR imagery. (a) Aerial photography (Google Earth) of the crescent-shaped gypsum barchan dunes formed by SW–NE winds. (b) The same region as represented by LiDAR imagery, shown as hillshaded relief display. (c) Color-coded relief display of LiDAR data (red represents the highest elevations). LiDAR's detail and accuracy permit calculation of sand volume and, in the case of sequential imagery, measurement of the rates of movement. (Image courtesy of Emily Kleber, OpenTopography. Dataset: Ewing, R. (2010). White Sands National Monument, NM: LiDAR Survey of Dune Fields. National Center for Airborne Laser Mapping. Distributed by OpenTopography. GoogleEarth imagery accessed May 2014. http://dx.doi.org/10.5069/G97D2S2D date accessed: February 25, 2014. Used with permission. GoogleEarth imagery accessed May 15, 2014.)

support analysis of volumetric data and, in sequential studies, to examine dune movement and provide a new paradigm opening new arenas for inquiry (Figure 12.11).

Aerial photography has long been a resource for assessing dunes and their movement (Brown and Arbogast 1999; Mitasova et al. 2005). Other technologies have provided the synoptic view to understand dune systems in their broader geologic and ecological contexts. Blumberg (2006) examined dunes using Shuttle Radar Topography Mission, and Short and Blair (1986) discuss applications of Landsat imagery to the study of dune fields. Al-Masrahy and Mountney (2013) examined the broad-scale context for dune fields of Saudi Arabia. However, at fine scales, the detail and accuracy of LiDAR imagery establish it as the premier sensor system for RS of aeolian landforms.

Hugenholtz et al. (2012) reviewed applications of RS to examine aeolian dunes, focusing upon (1) dune activity, (2) dune patterns and hierarchies, and (3) extraterrestrial dunes. LiDAR's ability to assess dune volume has been studied under varied circumstances (Woolard and Colby 2002; Grohmann and

Sawakuchi 2013). Integration of RS data with field-based measurements of vegetation cover, structure, and aeolian transport rate in order to develop predictive models of dune field activity; expanding observational evidence of dune evolution at temporal and spatial scales that can support validation and refinement of simulation models.

12.8 Biogeomorphology

Biogeomorphology is the subfield of geomorphology that explicitly recognizes interdependencies between ecologic and geomorphic systems (Viles 1988; Stallins 2006). Different methodological and theoretical approaches employed by both biogeographers and geomorphologists can confound formation of unified theories in biogeomorphic research (Fonstad 2006); however, RS may make important contributions in this area. Through synthesis at different scales, analysis of remotely sensed imagery may activate scale-independent perspectives of biogeomorphic analyses to integrate the process-centered focus of geomorphology with the pattern-centered emphasis of biogeography. Furthermore, RS analysis may contribute to a fundamental goal in biogeomorphic research to determine the types and geographic extent of landscape patterns that emerge from interactions among vegetation dynamics and geomorphic processes (Phillips 1999).

Biogeomorphological inquiry that may profit from RS analysis include fluvial geomorphology and hydrology, and corresponding vegetation cover and/or land use and land cover change (e.g., Mertes et al. 1995; Bryant and Gilvear 1999; Hudson et al. 2006; Zhou et al. 2008) Mertes et al. (1995) used Landsat TM data to investigate relationships between hydrogeomorphology and vegetation communities in the Amazon. Using Landsat images and semivariogram analysis, they identified important constraints on the transfer and storage of water, sediment, and other materials during high flood levels, as revealed on three Landsat images. Similar questions have also been examined through the use of SAR (e.g., Martinez and Le Toan 2007).

Characterization of pattern gives important clues to process, especially if examining pattern over time. Archival aerial photography provides the best temporal coverage for such analyses and are still applied in innovative ways to examine pattern–process relationships over time. For example, Meitzen (2009) measured lateral channel migrations and their effects on structure and function of the riparian forests of the Congaree River, South Carolina, using aerial photography (1938–2006) to measure migration rates. She found that lateral migration of channels produced a directional control on riparian forests, with forest types and successional stage dependent upon controls related to frequency of floods, elevation, and migration rates. They note that recent advances in geospatial technologies have provided "efficient and accurate" methods for examining such dynamics using temporally sequenced aerial photographs.

Innovative uses of RS technologies have also been applied to biogeographic studies in mountain environments at multiple spatial scales (e.g., Butler et al. 2003; Crowley et al. 2003; Walsh et al. 2003). Walsh et al. (2003), for example, studied how biogeographic and geomorphic processes are constrained at alpine tree line by geomorphic features. They use a combined Geographic Information Science approach that incorporates multiresolution RS systems, geospatial information, and tree line simulations. Data fusion linking ADAR and Landsat TM data enhanced their visualizations and analytical power by linking datasets with "different biological sensitivities." ADAR provided greater spatial resolution and thus could discern smaller objects and fine-scale landscape pattern. Landscape TM corroborates the ADAR data by providing greater spectral resolution. Using this integrated strategy, they discerned previously undetected vegetation patterns related to fine-scale periglacial features.

Emerging areas include development of simulation models for uncovering feedbacks between process and pattern. Fonstad (2006), for example, argues that cellular automata models that use remotely sensed data in raster grid formats are highly flexible, enabling them to link ecology and geomorphology that involve competing schema and different process rates of change. Such an approach enables detection of feedbacks and emergent phenomena. Geodiversity mapping—an emerging area of research that examines the close relationship between geomorphic and topographic diversity with biological diversity—also forms a promising and emerging research area where RS is likely to make important contributions (e.g., Hjort and Luoto 2012).

12.9 Remote Sensing and Geomorphic Inquiry

RS' contributions to the field of geomorphology may be perceived to result primarily from enhancement of data acquisition capabilities with advances in imaging technologies. However, our examples here clearly show that these imaging capabilities have contributed to the development of the discipline and its epistemology in profound ways. Geomorphic knowledge has been expanded to encompass increasingly broader ranges of landscapes by enabling extraction of information with detail and precision not previously feasible. Examples mentioned previously include LiDAR imagery and InSAR, which each convey data of a nature, quality, and precision that were beyond any reasonable expectations just a few decades ago. Applications of these newly emerging technologies contribute to increase the focus upon the role of temporal and spatial change within geomorphic inquiry and to advance the role of analytical strategies in relation to descriptive methodologies.

However, these capabilities form only a part of the broader contributions of RS to studies of geomorphology, which also include the following:

1. Archives and data resources. Remotely sensed data, as well as related data in cognate fields, are often available at low cost, in convenient formats, with minimal administrative overhead.

2. Ancillary benefits of such archives include access to sequential data, available often with the quality, continuity, and consistency, to provide the ability to examine changes over time with the confidence to know that observed changes indicate changes within the landscape rather than changes in the sensor systems.

3. Capabilities within the broader field of RS have been applied to good effect to support research in geomorphology. For example, Band et al. (2012) applied the normalized difference vegetation index ratio, originally developed in agriculture and ecology, to examine forest canopies to evaluate hydrologic processes and their contributions to mass-wasting.

4. Applications of RS open a gateway into the broader realm of geospatial data. Figure 12.12 represents (as a schematic approximation) each of the three principal components of geospatial data (GIS, RS, and GPS), initially developed within their own realms largely isolated from external interactions, as they have increasingly interact with each other to form an integrated system. (In Figure 12.12, status as an independent technology is represented by the

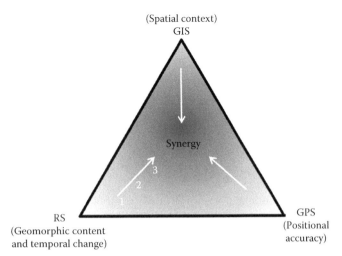

(Spatial context)
GIS

Synergy

3
2
1

RS   GPS
(Geomorphic content   (Positional
and temporal change)   accuracy)

FIGURE 12.12 Schematic diagram depicting interrelationships between the three key geospatial technologies illustrating evolution over time from independent, isolated development toward mutually supportive, synergistic contributions. The three dimensions are intended to approximate the contributions of remote sensing, GPS, and GIS in providing thematic content, precise location, and spatial context, respectively, to the geospatial enterprise. The arrows, and the numbers, indicate progressive degrees of interactions among the three dimensions. Within this framework, we can visualize how geomorphic analysis using these technologies has migrated from the periphery of this diagram toward the central zone of multifaceted applications of geospatial technologies.

apexes of the triangle.) RS is devoted to acquisition of the informational (geomorphic, in this context), and the temporal dimensions of an analysis, whereas GPS provides real-time positional data, and GIS contributes a locational framework, to permit analysis in a common spatial context. Thus, much of the power of RS derives from the ability of these technologies to integrate informational, temporal, positional, and contextual dimensions of a geomorphic problem into a common framework.

5. Development of new subfields of geomorphology that rely primarily on remotely sensed data, such as the emerging field of planetary geomorphology. In the last decade, improvements in spatial, spectral, and temporal resolution of remotely sensed data has allowed for identification of a range of active and relict geomorphic processes, many of which are similar to those on earth, including impact cratering, volcanism, aeolian, fluvial, mass-wasting, rock breakdown, periglacial, coastal, and meteoritic gardening. This has sparked renewed activity within this subfield, as evidenced through new publications (e.g., Melosh 2011; Greeley 2013) and the emergence of professional groups. Earth field analogs are often used to better understand geomorphic processes on other planets and new ways to examine equifinality (e.g., Burr et al. 2009).

6. Initially, such technologies develop in their own realms, largely isolated from each other. As they mature, these technologies interact with each other through improved

design, data formats, and instrumentation. For example, the capabilities of LiDAR systems reside in its ability to integrate its sensor systems with the real-time GPS data and orientation data from the aircraft's navigation system.

Figure 12.12 illustrates transitions of geospatial systems as they evolve from an initial narrowly defined technological focus (represented by 1 in Figure 12.12) toward a broader integration with cognate technologies (2 and 3 in Figure 12.12). Such integration and interactions exploit mutual synergies, represented by positions occupying the central regions of the diagram (2 and 3), distant from the apex positions at the periphery of the diagram.

Likewise, we can envision the applications of RS to geomorphic inquiry within this diagram, some following a trajectory from the edges toward more central positions. Geomorphic applications of imaging radars (SAR), considered over the decades, can be seen to have made this transition as it changed from use as a static image to a multitemporal resource using InSAR and PSI capabilities that provide quantitative data at fine spatial and temporal detail (Figure 12.5). Likewise, LiDAR systems, operating near the central portion of Figure 12.12, acquire accurate site-specific data, which when deployed to acquire sequential imagery, provide precise information that documents geomorphic processes at a detail not previously feasible.

As such technologies find their niches within geographic inquiry, they offer the opportunity to expand and strengthen our methodologies and increase the caliber and robustness of its contributions to scientific inquiry, to informing the public of geomorphic and environmental hazards and to addressing current and emerging societal problems.

References

ACIA. 2004. *Impacts of a Warming Arctic: Arctic Climate Impact Assessment*. Cambridge, U.K.: Cambridge University Press.

Akbarimehr, M., M. Motagh, and M. Haghshenas-Haghighi. 2013. Slope stability assessment of the Sarcheshmeh Landslide, Northeast Iran, investigated using InSAR and GPS observations. *Remote Sens* 5:3681–3700.

Allen, T. R. and Y. Wang. 2010. Selected scientific analyses and practical applications of remote sensing: Examples from the coast. In: *Manual of Geospatial Science and Technology*, 2nd edn., eds. J. D. Bossler, J. B. Campbell, R. B. McMaster, and C. Rizos. London, U.K.: Taylor & Francis, pp. 467–485.

Al-Masrahy, M. A. and N. P. Mountney. 2013. Remote sensing of spatial variability in aeolian dune and interdune morphology in the Rub' Al-Khali, Saudi Arabia. *Aeolian Res* 1:155–170, doi:10.1016/j.aeolia.2013.06.004.

ASCE. 1998. River width adjustment. I: Processes and mechanisms. *Journal of Hydraulic Engineering* 124(9):881–902.

Bamler, R., M. Eineder, N. Adam, X. Zhu, and S. Gerhardt. 2009. Interferometric potential of high-resolution spaceborne SAR. *Photogramm Fernerkund Geoinform* 5:407–419.

Band, L. E., T. Hwang, T. C. Hales, J. Vose, and C. Ford. 2012. Ecosystem processes at the watershed scale: Mapping and ecohydrological control of landslides. *Geomorphology* 137:159–167.

Bhaskar, A. S. and R. B. Kumar. 2011. Remote sensing of coastal geomorphology to understand river migration in the Thengapatnam area, southern India. *Int J Remote Sens* 32:5287–5301.

Bianchini, S., G. Herrera, R. M. Mateos, D. Notti, I. Garcia, O. Mora, and S. Moretti. 2013. Landslide activity maps generation by means of persistent scatterer interferometry. *Remote Sens* 5:6198–6222, http://www.mdpi.com/2072–4292/5/12/6198.

Bishop, M. P. and J. F. Shroder, Jr. 2000. Remote sensing and geomorphometric assessment of topographic complexity and erosion dynamics in the Nanga Parbat massif. *Geol Soc London Spec Publ* 170:181–200.

Bishop, M. P. and J. F. Shroder, Jr., eds. 2004. *Geographic Information Science and Mountain Geomorphology*. Berlin, Germany: Springer-Praxis.

Bishop, M. P., J. F. Shroder, Jr., and J. D. Colby. 2003. Remote sensing and geomorphometry for studying relief production in high mountains. *Geomorphology* 55:345–361.

Blumberg, D. G. 2006. Analysis of large aeolian (wind-blown) bedforms using the Shuttle Radar Topography Mission (SRTM) digital elevation data. *Remote Sens Environ* 100:179–189, http://dx.doi.org/10.1016/j.rse.2005.10.011.

Brakel, W. H. 1984. Seasonal dynamics of suspended-sediment plumes from the Tana and the Sabaki Rivers, Kenya: Analysis of Landsat data. *Remote Sens Environ* 16:165–173.

Brock, J. C., C. W. Wright, A. H. Salenger, and W. N. Swift. 2002. Basis and methods of NASA Airborne Topographic Mapper LiDAR surveys for coastal studies. *J Coastal Res* 18:1–13.

Brown, D. G. and A. F. Arbogast. 1999. Digital photogrammetric change analysis as applied to active coastal dunes in Michigan. *Photogramm Eng Remote Sens* 65:467–474.

Bryant, R. G. and D. J. Gilvear. 1999. Quantifying geomorphic and riparian land cover changes either side of a large flood event using airborne remote sensing: River Tay, Scotland. *Geomorphology* 29:307–321.

Burr, D. M., K. L. Tanaka, and K. Yoshikawa. 2009. Pingos on Earth and Mars. *Planet Space Sci* 57:541–555.

Butler, D. R., G. P. Malanson, M. F. Bekker, and L. M. Resler. 2003. Lithologic, structural, and geomorphic controls on ribbon forest patterns in a glaciated mountain environment. *Geomorphology* 55:203–217.

Cabot, E. C. 1947. The northern Alaskan coastal plain interpreted from aerial photographs. *Geogr Rev* 37:639–648.

Campbell, J. B. and R.H. Wynne. 2011. *Introduction to Remote Sensing*. The Guilford Press, New York.

Carbonneau, P. and H. Piegay. 2012. Fluvial Remote Sensing for Science and Management. Hoboken, New Jersey. Wiley-Blackwell. 458 p.

Chen, Q. 2007. Airborne LiDAR data processing and information extraction. *Photogramm Eng Remote Sens* 73:109–112.

Chorley, R. J. 2008. The mid-century revolution in fluvial geomorphology. Chapter 19. In: *The History of the Study of Landforms*, eds. T. P. Burt, R. J. Chorley, D. Brunsden, and A. S. Goudie. London, U.K.: The Geological Society, pp. 925–960.

Clark, C. D. 1993. Mega-scale glacial lineations and cross-cutting ice-flow landforms. *Earth Surf Proc Land* 18:1–29.

Clark, C. D. and R. T. Meehan. 2001. Subglacial bedform geomorphology of the Irish Ice Sheet reveals major configuration changes during growth and decay. *J Quatern Sci* 16:483–496.

Committee on Floodplain Mapping Technologies, National Research Council. 2007. Assessment of floodplain mapping technologies. In: *Elevation Data for Floodplain Mapping* (http://www.nap.edu/catalog/11829.html), Washington, DC: National Academies Press, pp. 89–114.

Crosetto, M., O. Monserrat, A. Jungner, and B. Crippa. 2010. Persistent scatterer interferometry: Potential and limitations. *Photogramm Eng Remote Sens* 76:1061–1069.

Crowley, J. K., B. E. Hubbard, and J. C. Mars. 2003. Analysis of potential debris flow source areas on Mount Shasta, California, by using airborne and satellite remote sensing data. *Remote Sens Environ* 87:345–358.

Curran, P. J. and E. M. M. Novo. 1988. The relationship between suspended sediment concentration and remotely sensed spectral radiance: A review. *J Coast Res* 4:351–368.

Del Ventisette, C., A. Ciampalini, M. Manunta, F. Calò, L. Paglia, F. Ardizzone, A. C. Mondini et al. 2013. Exploitation of large archives of ERS and ENVISAT C-Band SAR data to characterize ground deformations. *Remote Sens* 5:3896–3917.

Deshpande, S. S. 2013. Improved floodplain delineation method using high-density LiDAR data. *Comp Aided Civil Infrastruct Eng* 28:68–79.

Feuirer, D., J.-S. Bail, C. Puech, Y. Le Coarer, and Alain. A. Viau. 2008. Very highresolution mapping of river-immersed topography by remote sensing. *Progress in Physical Geography* 32, 403–419.

Fischer, L., H. Eisenbeiss, A. Kääb, C. Huggel, and W. Haeberli. 2011. Monitoring topographic changes in a periglacial high-mountain face using high-resolution DTMs, Monte Rosa East Face, Italian Alps. *Permafrost Periglacial Process* 22:140–152.

Farouk, E.-B. 2000. Satellite observations of the interplay between wind and water processes in the Great Sahara. *Photogrammetric Engineering and Remote Sensing* 66(6):777–782.

Fonstad, M. A. 2006. Cellular automata as analysis and synthesis engines at the geomorphology–ecology interface. *Geomorphology* 77:217–234.

French, H. and C. E. Thorn. 2006. The changing nature of periglacial geomorphology. *Géomorphologie Relief Process Environ* 3: 165–174.

Frost, R. E., J. H. Mclerran, and R. D. Leighty. 1966. Photointerpretation in the arctic and subarctic. In: *Proceedings of the First International Conference on Permafrost*, West Lafayette, IN. National Academy of Sciences, National Research Council of Canada, Publication 1287, pp. 343–348.

Gao, J. 2011. Bathymetric mapping by means of remote sensing: Methods, accuracy and limitations. *Prog Phys Geogr* 35:782–809.

Ghoshal, S., L. A. James, M. B. Singer, and R. Aalto. 2010. Channel and floodplain change analysis over a 100-year period: Lower Yuba River, California. *Remote Sens* 2:1797–1825.

Ghuffar, S., B. Székely, A. Roncat, and N. Pfeifer. 2013. Landslide displacement monitoring using 3D range flow on airborne and terrestrial LiDAR data. *Remote Sens* 5:2720–2745.

Gibson, G., J. B. Campbell, and L. Kennedy. 2009. Fifty-one years of shoreline change at Little Lagoon, Alabama. *Southeast Geog* 49:67–83.

Greeley, R. 2013. *Introduction to Planetary Geomorphology.* Cambridge University Press, Cambridge.

Grohmann, C. H. and A. O. Sawakuchi. 2013. Influence of cell size on volume calculation using digital terrain models: A case of coastal dune fields. *Geomorphology* 180–181:130–136.

Grosse, G., L. Schirrmeister, V. V. Kunitsky, and H. W. Hubberten. 2005. The use of CORONA images in remote sensing of periglacial geomorphology: An illustration from the NE Siberian coast. *Permafrost Periglacial Process* 16:163–172.

Grosse, G., L. Schirrmeister, and T. J. Malthus. 2006. Application of Landsat-7 satellite data and a DEM for the quantification of thermokarst-affected terrain types in the periglacial Lena–Anabar coastal lowland. *Polar Res* 25:51–67.

Handcock R. N., A. R. Gillespie, K. A. Cherkauer, J. E. Kay, S. J. Burges, and S. K. Kampf. 2006. Accuracy and uncertainty of thermal-infrared remote sensing of stream temperatures at multiple spatial scales. *Remote Sensing of Environment* 100:457–473.

Heeren, D. M., A. R. Mittelstet, G. A. Fox, D. E. Storm, A. T. Al-Madhhachi, T. L. Midgley, A. F. Stringer, K. B. Stunkel, and R. D. Tejral. 2012. Using rapid geomorphic assessments to assess streambank stability in Oklahoma Ozark Streams. *Trans Am Soc Agric Biol Eng* 55:957–968.

Hjort, J. and M. Luoto. 2006. Modelling patterned ground distribution in Finnish Lapland: An integration of topographical, ground and remote sensing information. *Geografiska Annaler: Series A, Phys Geog* 88:19–29.

Hjort, J. and M. Luoto. 2012. Can geodiversity be predicted from space? *Geomorphology* 153:74–80.

Hohenthal, J., P. Alho, J. Hyyppä, and H. Hyyppä. 2011. Laser scanning applications in fluvial studies. *Prog Phys Geog* 35:782–809.

Horton, R. E. 1945. Erosional development of streams and their drainage basins: Hydrophysical approach to quantitative morphology. *Bull Geol Soc Am* 56:275–370.

Hudson, P. F., R. R. Colditz, and M. Aguilar-Robledo. 2006. Spatial relations between floodplain environments and land use–land cover of a large lowland tropical river valley: Panuco basin, Mexico. *Environ Manag* 38:487–503.

Hugenholtz, C., H. Noam Levin, T. E. Barchyn, and M. C. Baddock. 2012. Remote sensing and spatial analysis of aeolian sand dunes: A review and outlook. *Earth Sci Rev* 111:319–334.

Huggel, C., A. Kääb, W. Haeberli, P. Teysseire, and F. Paul. 2002. Remote sensing based assessment of hazards from glacier lake outbursts: A case study in the Swiss Alps. *Can Geotech J* 39:316–330.

Humlum, O. 2000. The geomorphic significance of rock glaciers: Estimates of rock glacier debris volumes and headwall recession rates in West Greenland. *Geomorphology* 35:41–67.

Humlum, O. 2008. Alpine and polar periglacial processes: The current state of knowledge. Plenary paper, *Ninth International Conference on Permafrost,* pp. 743–759.

Irish, J. L. 2000. An introduction to coastal zone mapping with airborne LiDAR: The SHOALS System. Technical report: Corp of Engineers, Mobile, AL. Joint Airborne LiDAR Bathymetry Technical Center of Expertise.

Kääb, A. 2008. Remote sensing of permafrost-related problems and hazards. *Permafrost Periglacial Process* 19:107–136.

Kääb, A., C. Huggel, L. Fischer, S. Guex, F. Paul, I. Roer, N. Salzmann et al. 2005. Remote sensing of glacier-and permafrost-related hazards in high mountains: An overview. *Nat Hazards Earth System Sci* 5:527–554.

Kawakuboa, F. S., R. G. Moratoa, R. S. Naderb, and A. Luchiari. 2011. Mapping changes in coastline geomorphic features using Landsat TM and ETM+ imagery: Examples in southeastern Brazil. *Int J Rem Sens* 32:2547–2562, doi:10.1080/01431161003698419

Kieffer, S. W. 2014. The deadly dynamics of landslides. *Am Sci* 102:298–303.

Klees, R. and D. Massonnet. 1999. Deformation measurements using SAR interferometry: Potential and limitations. *Geologie en Mijinbouw* 77:161–176.

Lee, W. T. 1922. *The Face of the Earth as Seen from the Air.* Special Publication No. 4. New York: American Geographical Society.

Liu, L., C. I. Millar, R. D. Westfall, and H. A. Zebke. 2013. Surface motion of active rock glaciers in the Sierra Nevada, California, USA: Inventory and a case study using InSAR. *The Cryosphere* 7:1109–1119.

Livingston, I., G. F. S. Wiggs, and C. M. Weaver. 2007. Geomorphology of desert sand dunes: A review of recent progress. *Earth Sci Rev* 80:239–257.

Leverington, D. W. and C. R. Duguay. 1997. A neural network method to determine the presence or absence of permafrost near Mayo, Yukon Territory, Canada. *Permafrost Periglacial Process* 8:205–215.

Marcus, W. A. and Fonstad, M. A. 2008. Optical remote mapping of rivers at sub-meter resolutions and watershed extents. *Earth Surf. Process. Landforms* 33:4–24.

Marcus, A. and M. A. Fonstad. 2010. Remote sensing of rivers: The emergence of a subdiscipline in the river sciences. *Earth Surf Proc Land* 35:1867–1872.

Marghany, M., Z. Sabu, and M. Hashim. 2010. Mapping coastal geomorphology changes using synthetic aperture radar data. *Int J Phys Sci* 5:1890–1896.

Martinez, J.-M. and T. Le Toan. 2007. Mapping of flood dynamics and spatial distribution of vegetation in the Amazon floodplain using multitemporal SAR data. *Remote Sens Environ* 108:209–223.

Meitzen, K. M. 2009. Lateral channel migration effects on riparian forest structure and composition, Congaree River, South Carolina, USA. *Wetlands* 29:465–475.

Melosh, H. J. 2011. *Planetary Surface Processes* (No. 13). Cambridge University Press, Cambridge.

Melton, F. A. 1939. Aerial photographs and the first course in geology. *Photogramm Eng* 5:74–77.

Mertes, L. A. K. 2002. Remote sensing of riverine landscapes. *Freshwater Biology* 47(4):799–816.

Mertes, L. A. K., D. L. Daniel, J. M. Melack, B. Nelson, L. A. Martinelli, and B. R. Forsberg. 1995. Spatial patterns of hydrology, geomorphology, and vegetation on the floodplain of the Amazon River in Brazil from a remote sensing perspective. *Geomorphology* 13:215–232.

Metternicht, G., L. Hurni, and R. Gogu. 2005. Remote sensing of landslides: An analysis of the potential contribution to geo-spatial systems for hazard assessment in mountainous environments. *Rem Sens Environ* 98:284–303.

Millington, A. C. and Townshend, J. R. G. 1987. The potential of satellite remote sensing for geomorphological investigations: An overview. In: *International Geomorphology*, ed., V. Gardiner. Chichester, U.K.: Wiley, pp. 331–342.

Mitasova, H., M. Overton, and R. S. Harmon. 2005. Geospatial analysis of a coastal sand dune field evolution: Jockey's Ridge, North Carolina. *Geomorphology* 72:204–221.

Moore, L. J. and G. B. Griggs. 2002. Long-term cliff retreat and erosion hotspots along the central shores of the Monterey Bay National Marine Sanctuary. *Marine Geol* 181:265–283.

Morrissey, L. A., L. Strong, and D. H. Card. 1986. Mapping permafrost in the boreal forest with thematic mapper satellite data. *Photogramm Eng Remote Sens* 52:1513–1520.

National Academy of Sciences Committee on Floodplain Mapping Technologies, National Research Council. 2007. Elevation Data for Floodplain Mapping. Washington, D.C., 168 p.

Othman, A. and R. Gloaguen. 2013. River courses affected by landslides and implications for hazard assessment: A high resolution remote sensing case study in NE Iraq–W Iran. *Remote Sens* 5:1024–1044, doi:10.3390/rs5031024.

Phillips, J. D. 1999. Divergence, convergence, and self-organization in landscapes. *Ann Assoc Am Geogr* 89:466–488.

Pinter, N. and M. T. Brandon. 2005. How erosion builds mountains. *Sci Am* 15:74–81.

Plug, L. J., C. Walls, and B. M. Scott. 2008. Tundra lake changes from 1978 to 2001 on the Tuktoyaktuk Peninsula, western Canadian Arctic. *Geophys Res Lett* 35(3):L0352.

Ray, R. G. 1960. Aerial photographs in geologic interpretation and mapping. USGS professional paper. GPO, Washington, DC.

Raymo, M. E. and W. F. Ruddiman. 1992. Tectonic forcing of late Cenozoic climate. *Nature* 359:117–122.

Resop, J. P. and W. C. Hession. 2010. Terrestrial laser scanning for monitoring streambank retreat: Comparison with traditional surveying techniques. *J Hydraulic Eng* 136:794–798.

Ritter, D. F. 1988. Landscape analysis and the search of geomorphic unity. *Bull Geol Soc Am* 100:160–171.

Rudraiah, M., S. Govindaiah, and S. Srinivas Vittala. 2008. Morphometry using remote sensing and GIS techniques in the sub-basins of Kagna River Basin, Gulburga District, Karnataka, India. *J India Soc Remote Sens* 36:351–360.

Sankar, S. and D. P. Kanungo. 2004. An integrated approach for landslide mapping using remote sensing and GIS. *Photogramm Eng Remote Sens* 70:617–625.

Sarkar, A., R. D. Garg, and N. Sharma. 2012. RS-GIS based assessment of river dynamics of Brahmaputra River in India. *J Water Resource Protect* 4:63–72, http://dx.doi.org/10.4236/jwarp.2012.42008, published Online February 2012 (http://www.SciRP.org/journal/jwarp).

Sasaki, T., J. Imanishi, K. Ioki, Y. Morimoto, and K. Kitada. 2008. Estimation of leaf area index and canopy openness in broad-leaved forest sing airborne laser scanner in comparison with high-resolution near-infrared digital photography. *Land Ecol Eng* 4:47–55.

Scaioni, M. 2013. Remote sensing for landslide investigations: From research into practice. *Remote Sens* 5:5488–5492.

Scheidegger, A. E. 1961. *Theoretical Geomorphology*. New York: Springer, 333pp.

Schiefer, E. and R. Gilbert. 2007. Reconstructing morphometric change in a proglacial landscape using historical aerial photography and automated DEM generation. *Geomorphology* 88:167–178.

Semiletov, I. P., I. I. Pipko, N. Ya Pivovarov, V. V. Popov, S. A. Zimov, Yu V. Voropaev, and S. P. Daviodov. 1991. Atmospheric carbon emission from North Asian Lakes: A factor of global significance. *Atmospheric Environ* 30:1657–1671.

Short, N. and R. W. Blair (eds.). 1986. *Geomorphology from Space: A Global Overview of Regional Landforms*. NASA SP-486. Washington, DC: GPO.

Shu, Y., J. Li, and G. Gomes. 2010. Shoreline extraction from RADARSAT-2 intensity imagery using a narrow band level set segmentation approach. *Marine Geodesy* 33:187–203, doi:10.1080/01490419.2010.496681.

Simon, A., A. Curini, S. E. Darby, and E. J. Langendoen. 2000. Bank and near-bank processes in an incised channel. *Geomorphology* 35(3–4):193–217.

Simpson, R. 1966. Radar-geographic tools. *Ann Assoc Am Geog* 56:80–96, doi:10.1111/j.1467-8306.1966.tb00545.x.

Smith, H. T. U. 1942. Aerial photographs in geomorphic studies. *Photogramm Eng* 8:129–155.

Smith, H. T. U. 1943. *Aerial Photographs and Their Applications*. New York: Century Crofts, Inc.

Smith, M. J. and C. F. Pain. 2009. Applications of remote sensing in geomorphology. *Prog Phys Geogr* 33:568–582, doi:10.1177/0309133309346648.

Smith, M. J., J. Rose, and S. Booth. 2006. Geomorphological mapping of glacial landforms from remotely sensed data: An evaluation of the principal data sources and an assessment of their quality. *Geomorphology* 76:148–165.

Stallins, J. A. 2006. Geomorphology and ecology: Unifying themes for complex systems in biogeomorphology. *Geomorphology* 77:207–216.

Strahler, A. N. 1957. Quantitative analysis of watershed geomorphology. *Trans Am Geophysical Union* 38:913–920.

Strozzi, T., C. Ambrosi, and H. Raetzo. 2013. Interpretation of aerial photographs and satellite SAR interferometry for the inventory of landslides. *Remote Sens* 5:2554–2570.

Tantianuparp, P., X. Shi, L. Zhang, T. Balz, and M. Liao. 2013. Characterization of landslide deformations in Three Gorges area using multiple InSAR data stacks. *Remote Sens* 5:2704–2719.

Tofani, V., F. Raspini, F. Catani, and N. Casagli. 2013. Persistent scatter interferometry (PSI) technique for landslide characterization and monitoring. *Remote Sens* 5:1045–1065.

Townsend, P. A. and J. R. Foster. 2002. A synthetic aperture radar–based model to assess historical changes in lowland floodplain hydroperiod. *Water Resour Res* 38:20–21.

Ulrich, M., G. Grosse, S. Chabrillat, and L. Schirrmeister. 2009. Spectral characterization of periglacial surfaces and geomorphological units in the Arctic Lena Delta using field spectrometry and remote sensing. *Remote Sens Environ* 113:1220–1235.

Viles, H. A. 1988. *Biogeomorphology*. Oxford, UK: Blackwell.

Vitek, J. D., J. R. Giardino, and J. W. Fitzgerald. 1996. Mapping geomorphology: A journey from paper maps, through computer mapping to GIS and virtual reality. *Geomorphology* 16:233–249.

Walsh, S. J., D. R. Butler, and G. P. Malanson. 1998. An overview of scale, pattern, process relationships in geomorphology: A remote sensing and GIS perspective. *Geomorphology* 21:183–205.

Walsh, S. J., D. R. Butler, and G. P. Malanson, K. A. Crews-Meyer, J. P. Messina, and N. Xiao. 2003. Mapping, modeling, and visualization of the influences of geomorphic processes on the alpine treeline ecotone, Glacier National Park, MT, USA. *Geomorphology* 53:129–145.

Wang, Y., X. Liang, C. Flener, A. Kukko, H. Kaartinen, M. Kurkela, M. Vaaja, H. Hyyppä, and P. Alho. 2013. 3D Modeling of coarse fluvial sediments based on mobile laser scanning data. *Remote Sens* 5:4571–4592.

Woolard, J. W. and J. D. Colby. 2002. Spatial characterization, resolution, and volumetric change of coastal dunes using airborne LIDAR: Cape Hatteras, North Carolina. *Geomorphology* 48:269–287.

Yang, X., C. J. Michiel, R. A. Damen, and R. A van Zuidam. 1999. Satellite remote sensing and GIS for the analysis of channel migration changes in the active Yellow River Delta, China. *Int J Appl Earth Obs Geoinf* 1:146–157, dx.doi.org/10.1016/S0303-2434(99)85007-7.

Zebker, H. A., Hensley, S., Shanker, P., and Wortham, C. 2010. Geodetically accurate InSAR data processor. *IEEE Transactions on Geoscience and Remote Sensing*, 48(12):4309–4321.

Zhou, D., H. Gong, and Z. Liu. 2008. Integrated ecological assessment of biophysical wetland habitat in water catchments: Linking hydro-ecological modelling with geo-information techniques. *Ecol Model* 214:411–420.

VIII

Droughts and Drylands

13

Agricultural Drought Detection and Monitoring Using Vegetation Health Methods

Felix Kogan
Center for Satellite Applications and Research

Wei Guo
University Research Court

Acronyms and Definitions

BT	Brightness temperature
BT_{max}	32-Year absolute maximum BT
BT_{min}	32-Year absolute minimum BT
dY	Yield deviation from technological trend
E	Exceptional drought
EE	Extreme-to-exceptional drought
ENSO	El Nino Southern Oscillation
FAO	Food and Agricultural Organization
GAC	Global area coverage
GVH	Global vegetation health
GVI	Global Vegetation Index
IMSG	I.M. System Group, Inc.
IPCC	Intergovernmental Panel on Climate Change
NDVI	Normalized Difference Vegetation Index
$NDVI_{max}$	32-Year absolute maximum NDVI
$NDVI_{min}$	32-Year absolute minimum NDVI
NIR	Near infrared reflectance
NOAA	National Oceanic and Atmospheric Administration
NPOESS	National Polar-Orbiting Operational Satellite System
NPP	NPOES Preparatory Platform
LOM	Law of minimum
LOT	Law of tolerance
PCC	Principle of carrying capacity
SE	Severe-to-exceptional drought
TCI	Temperature Condition Index
VCI	Vegetation Condition Index
VIIRS	Visible Infrared Imager Radiometer Suite
VIS	Visible reflectance
VH	Vegetation health
VHI	Vegetation Health Index

13.1 Introduction

Drought is a part of Earth's climate, which occurs every year without warning, recognizing borders and political and economic differences. Drought has wide-ranging impacts on water resources, ecosystems, energy, agriculture, forestry, human health, recreation, transportation, food supply and demands, and other resources and activities. Drought affects the largest number of people on Earth and is a very costly disaster. In the United States, a country of high technology, drought is considered a "14 billion dollar" annual event in terms of incurred losses (NCDC 2013). In the developing world, drought leads to food shortages, famine, population displacement, and death of people. Since drought is a very complex and the least understood phenomenon, drought prediction, what would be an effective way to fight with drought consequences, is a very challenging task. Therefore, an early detection and monitoring is currently an important way to deal with drought in assessing its impact and developing mitigation measures.

Weather data have been used traditionally as a tool for drought monitoring. However, weather station network is sparse, especially in climate, ecosystem, and population marginal areas.

In the last 30 years, satellite technology has successfully filled this gap due to comprehensive coverage of high-quality data, its availability, and access as well as the ability to use the same data for multiple applications. Besides, satellite indices and products provide cumulative approximation of weather impacts on the environment and economies, including estimation of drought-related losses in agriculture. Since the launch of the first operational weather satellites in the late 1960s, space-based remote sensing has proven to be a perfect method for operational drought management complementing weather data. In the past 20 years, operational remote sensing was improved considerably since the introduction of vegetation health (VH) method (Kogan 1997, 2002, Kogan and Guo 2014a, Kogan et al. 2012, 2013). This chapter discusses the principle of the VH method and its products and how they are applied for monitoring global droughts and their impacts on agricultural losses.

13.2 Drought: Unusual Weather Disaster

Drought is quite different than other weather and geophysical disasters because it starts unnoticeably, develops cumulatively and slow, and produces cumulative impacts, especially on agriculture, and by the time damages are visible, it is too late to mitigate the consequences (Kogan 1997). Besides, drought is a very complex and least understood phenomenon because in addition to be triggered by weather, other factors such as soils, ecosystems, climate, and human activities also contribute to drought initiation, development, and impacts. Although it is known that drought is a period with dry and hot weather, even a few months with such weather do not necessarily indicate drought, if the preceding to dryness period was wet. Owning to the creeping nature of drought, its effect often takes weeks, months, seasons, and even years to appear. Since drought is a multidimensional phenomenon by its appearance, properties, origination, and impacts, weather-based parameters and indices are insufficient to characterize special and temporal drought features, especially such characteristic as drought start, which is important for timely initiation of mitigating measures. Another specific drought feature is its cumulative development with the following cumulative impacts that are not immediately observable (Wilhite 2000). By the time when damages are visible, it is too late to mitigate the consequences. The latest is very important in regions and countries with limited and/or variable water supplies and with economies that are highly dependent on agriculture.

13.3 How to Measure Drought

13.3.1 Traditional Approach

Traditionally, drought was measured with weather, soil moisture data, and their combination in the form of indices. The most widely used weather and soil parameters are precipitation, temperature, air humidity, and soil moisture. From weather indices, the most widely used are Palmer Drought Severity Index (Palmer 1965), Standardized Precipitation Index (McKee et al. 1993), U.S. Drought Monitor index (Svoboda et al. 2002), and a few others, specific to a country environment and economic resources. Weather parameter and indices have many useful features in their applications. However, the most important shortcomings are calculated from a limited weather station network, especially in marginal areas and developing countries.

13.3.2 Remote Sensing Approach

Satellite indices were introduced in the late 1970s since the launch of operational weather satellites and started to be widely used from the 1980s, with accumulation of knowledge and satellite data. The indices were developed from the Earth surface reflectance measured by the operational environmental satellites' sensors. The most popular for land monitoring was the normalized difference vegetation index (NDVI) introduced by Deering (1978). The NDVI calculated as a ratio of visible (VIS) and near infrared (NIR) sun's reflectance (NDVI = (NIR − VIS)/(NIR + VIS)) is a dimensionless measure, which is used as a proxy to characterize land surface greenness and vigor (Cracknel 1997). However, NDVI alone was not sufficient to monitor vegetation condition, drought, and variation in agricultural production. Therefore, in the 1990s, the new VH method and indices were introduced. The new numerical method combined NDVI with thermal emission converted to brightness temperature (BT) of land surface (Kogan 1990). This method was built on the three basic environmental laws, law of the minimum, law of tolerance, and the principle of carrying capacity, which provide a theoretical basis for calculation of biophysical climatology (expressed as the lowest and the highest ecosystem level that the environmental resources can support weekly inside a 4 km² pixel). The satellite reflectance was pre- and postlaunch calibrated, NDVI and BT were calculated, high-frequency noise was completely removed from NDVI and BT time series, ecosystems were stratified, and medium-to-low frequency vegetation fluctuations associated with weather variations were singled out (Kogan 1997). Finally, three indices characterizing moisture (VCI), thermal (TCI), and VH index (VHI) conditions were derived by the following equations:

$$VCI = 100 * \frac{NDVI - NDVI_{min}}{NDVI_{max} - NDVI_{min}} \qquad (13.1)$$

$$TCI = 100 * \frac{BT_{max} - BT}{BT_{max} - BT_{min}} \qquad (13.2)$$

$$VHI = a * VCI + (1 - a) * TCI \qquad (13.3)$$

where
NDVI, NDVI$_{max}$, and NDVI$_{min}$ are no noise weekly NDVI and 32-year absolute maximum and minimum, respectively
BT, BT$_{max}$, and BT$_{min}$ are similar data characterizing BT
a is a coefficient that quantifies a share of VCI and TCI contribution to the total VH

The three indices were scaled from zero indicating extreme vegetation stress to 100 indicating optimal condition. A comparison of VH data with crops showed that 5%–10% reduction in crop production signals about the beginning of drought, which corresponds to a reduction in index (VCI, TCI, and VHI) values below 40. The new method was first applied to the NOAA Global Vegetation Index data set (16 km²), issued routinely since 1985 (Kidwell 1995). Currently, the method, called Global VH (GVH), is applied to the NOAA's Global Area Coverage data set produced from 1981 through the present (34 years) issued weekly for each 4 km² land surface between 75° N and 55° S. The data and products are delivered every week to the following NOAA web (http://www.star.nesdis.noaa.gov/smcd/emb/vci/VH/index.php).

13.4 Vegetation Health Interpretation

The VH method stems from the properties of green vegetation to reflect and emit incoming sunlight, which is converted to greenness (NDVI) and thermal (BT) indices. In drought-free years, vegetation is greener and cooler than in climate-normal and below-normal years stimulating NDVI increase and BT decrease compared to climatology (expressed by the denominators in Equations 13.1 and 13.2, respectively). This resulted in VCI, TCI, and VHI increase above 60. Drought normally depresses vegetation greenness and increases canopy temperature, resulting in NDVI decrease, BT increase, and decrease in VCI, TCI, and VHI below 40 (Kogan 1997, Kogan et al. 2012).

Figure 13.1 demonstrates color-coded map of VHI on October 9, 2012. Stressful, fair, and favorable conditions are represented by red, green/yellow, and blue colors, respectively. Large-area VHI-related intensive vegetation stress serves normally as an indicator of crop yield and pasture biomass reduction, intensive fire activities, and reduction of water level in reservoirs and rivers (see the United States, Kazakhstan, Ukraine, Brazil, and Argentina on the map). On the contrary, optimal VH triggered by a cool and wet weather indicates favorable conditions for above average crop production, water availability, and no fire activities, but is an indication for development of mosquito-borne diseases because they require moist and cool conditions (see blue area in sub-Sahara and Southern Africa, northwestern India, and eastern China).

13.5 Unusual Droughts of the Twenty-First Century

The twenty-first century began with a series of widespread, long, and intensive droughts around the world, continuing the previous two- to three-decade tendencies. Up to 16% of the 2000–2013 world land was affected by droughts of severe and stronger intensity (Kogan et al. 2013). They reduced grain production in the important agricultural regions, leading to a negative balance between global food supplies and demands (PotashCorpo 2012). The world experienced the most dreadful droughts during the recent 3 years (2010–2012). The United States was affected twice, in 2011 and 2012 (Figure 13.2). The worst drought of 2012 hit primary agricultural areas of the Great Plains sharply reducing corn, soybeans, hay, and pasture production, which resulted in price increase for food and farmland (U.S. Drought 2012). This drought has also affected grain production in other parts of the world, Kazakhstan and southern Russia, resulting in 20% grain losses in Russia and almost 50% in Kazakhstan (UNDP 2012). In the Southern Hemisphere, the 2012 drought affected agricultural regions of Brazil, northern Argentina, and southern Australia.

The 2011 U.S. drought was much stronger than in 2012, but it covered only Texas, Oklahoma, New Mexico, and parts of the neighboring states (Figure 13.2). In the hardest hit states,

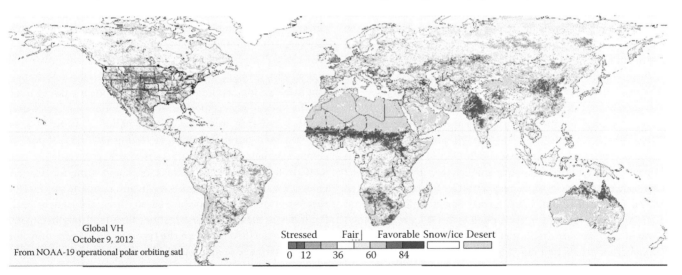

Global VH
October 9, 2012
From NOAA-19 operational polar orbiting satl

Stressed Fair | Favorable Snow/ice Desert
0 12 36 60 84

FIGURE 13.1 Color-coded global VH index map for October 9, 2012. An intensive October 2012 vegetation stress in the United States, southern Europe, east Siberia, and a part of Amazon indicated potential for fire activities. In the United States, this stress started in April and continued through the entire summer leading to considerable agricultural losses.

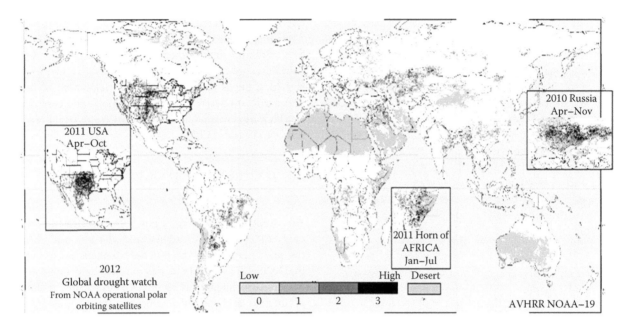

FIGURE 13.2 Drought area and intensity (derived from VHI < 40) in the major agricultural areas of the world during 2010–2012.

moderate-to-exceptional drought covered up to 100% of their territories (AgriLife 2011). Another extremely strong drought developed in 2010, covering huge agricultural areas of Russia, Ukraine, and Kazakhstan (Figure 13.2). Russian grain production dropped to 75 million metric tons (versus 97 in 2009, FAO 2012) forcing the Russian government to impose a grain embargo, which triggered a sharp increase in global wheat prices. This drought affected also the 2011 Russian harvest, since winter wheat was planted in dry soil. In addition to agriculture and water resources, the 2010 drought has also triggered hundreds of fires and heavy smoke, deteriorating human health and increasing the death rate in Russia.

Compared to 2010–2012, the 2013/2014 agricultural year was favorable for global grain production, which is forecasted to exceed 8% of the 2012 level due to recovery of corn in the United States and wheat in the former Soviet Union (FAO 2013). Meanwhile, as seen in Figure 13.3, the area affected by stronger than severe drought in 2013/2014 is still large, especially in the Southern Hemisphere. The VHI estimated that 16% of the globe was affected by drought (stronger than severe) in 2013/2014, which is typical for the twenty-first century (Kogan and Guo 2014a), except for 2012, when the drought area reached 21%. It is important to emphasize that the 2013/2014 global drought impact on agriculture was much smaller than in the previous years since large grain-producing countries (China, the United States, European Union, India, former Soviet Union) were not affected (severe drought in the United States [Figure 13.3, July] covered only western states, causing an intensive fire activity). Global grain production might have been even larger than projected for 2013/2014 if severe drought had not affected grain-producing areas of Argentina, Australia, and eastern Brazil (Figure 13.3, February).

13.6 Drought Products

VH-based drought products include VH, moisture and thermal stress, drought start/end, intensity, duration, magnitude, area, season, origination, and impacts. Drought start and end are identified when VCI, TCI, and VHI decline below 40 thresholds, which are experimentally determined from in situ observations (crop yield reduction). Drought intensity (commonly referred as the severity) was graded by a percent of yield reduction (dY) below the technological trend. Mild drought is identified if dY is 2%–9% below trend, yield reduction of 10%–14% indicates moderate drought, 15%–24% extreme, and below 24% exceptional. Drought duration is measured in the number of days with drought in different VH intensities. Magnitude accounts for the combination of a drought's intensity and duration (e.g., the number of days with severe drought). Drought area can be measured as a percentage of an administrative region, with drought of different intensity: this is a very important measure for developing mitigation strategies and evaluating the budget (e.g., imposing some limitations for water use). Another criterion might be a combination of drought area, intensity, and duration. Drought timing is a very important component determining if urgent measures are needed to mitigate drought impacts or if there is enough time for drought recovery with minimal losses to incur. Drought origination specifies if drought is moisture based, thermal based, or both; the latter is the worst combination. Finally, drought impacts should be specified based on the type of affected economy and extent of damage (e.g., crop yield reduction and water depletion). Each drought is unique, but common features of the most severe droughts that have far-reaching human and environmental impacts include time of drought start, intensity of

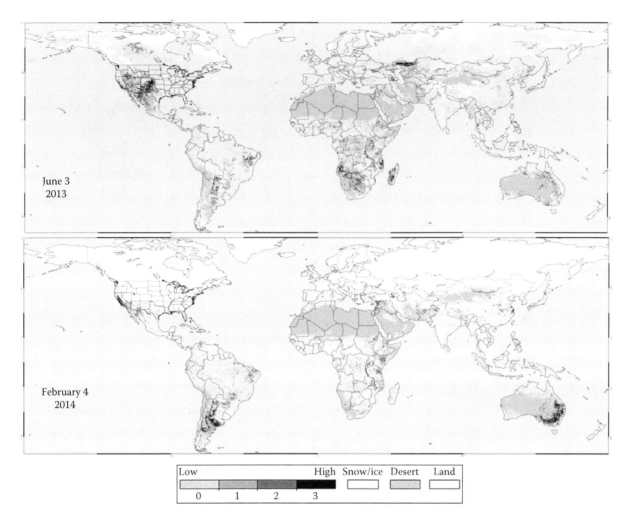

FIGURE 13.3 Drought area and intensity in the 2013/2014 growing season. Drought is derived from VHI < 40; intensity is indicated by a darker color. For example, central valley in California, the Pampas in Argentina, and southeast Australia are facing severe drought relative to long-term climatology.

progress, long duration, large moisture deficits, severe thermal stress, large areal extent, and severe impact.

The following VH-based drought products are discussed below (Section 13.6): moisture stress from VCI, thermal stress from TCI, drought area, drought intensity, drought duration, fire risk, yield, and biomass reduction (Figures 13.4 and 13.5).

Figure 13.4 demonstrates moisture and thermal stress in the principal agricultural areas of southern and eastern Australia (states Queensland, New South Wales, and Victoria). A very dramatic event occurred in 2006–2007, when severe drought reached the apogee slashing water supplies, crops, and rangeland production severely. This was one of the worst droughts for agriculture due to the combined impacts of extreme moisture and thermal stress. Following FAO (2012), Australian wheat yield dropped 46% in 2006 and 37% in 2007 (below the 1960–2010 yield's trend level). An example of localized VHI and thermal (TCI) stress by different intensities and percent of the affected area is shown in Figure 13.5a and b. Drought intensity is demonstrated during a 5-year (2007–2011) period in Kenya (Horn of Africa) when

the country was affected for 4 years as shown in Figure 13.5c. Following moisture and thermal stress estimation, moderate-to-severe intensity droughts occurred in 2009 and 2011.

Thousands of acres of vegetative land are burnt every year worldwide leaving huge scars on the land, polluting the atmosphere, and affecting human health. An early assessment of fire risk can help to mitigate fire consequences. Drought is the principal factor creating fire risk conditions. In drought years, the burnt area increases twofold. VH fire risk monitoring is based on assessment of severity and duration of vegetation stress (Kogan 2002). Figure 13.5d shows maximum fire risk area during the 2007 and 2010 seasons in Russia and Ukraine. These two droughts caused huge fires in both years. Based on VH fire risk product, Russia was the most affected in 2010 and Ukraine in 2007, although partial effects were observed in northern Ukraine as well. Although the fire risk area in the north was not large, it covered the region of the 1986 nuclear accident in Chernobyl. Soil and vegetation in that area are still having radioactive remnants that were thrown into the air by wildfires.

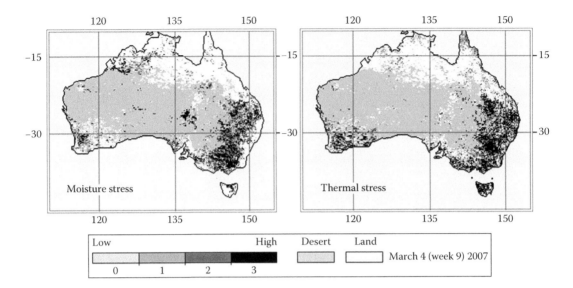

FIGURE 13.4 Moisture (VCI < 40) and thermal (TCI < 40) stress in agricultural areas of Australia, March 4, 2007. Numerical values of drought intensity are indicated by a darker color. Southern and eastern Australia are affected by both extreme moisture and thermal stress, which is the worst combination of drought severity, and consequently expected considerable losses of crop production.

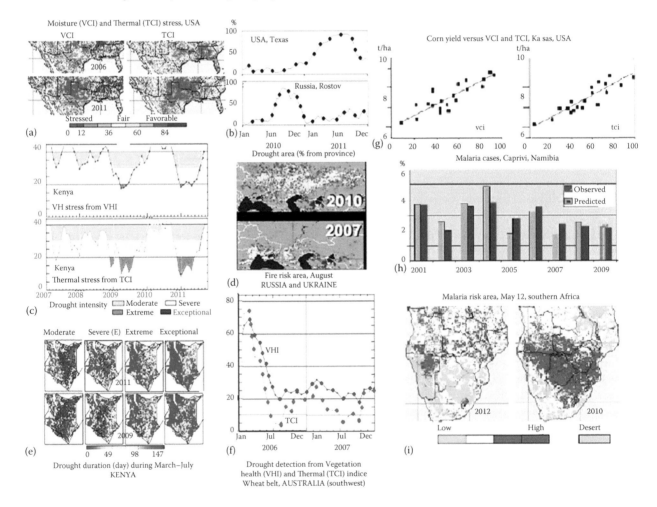

FIGURE 13.5 Drought products from VH available at the NOAA website. (a) Moisture (VCI) and Thermal (TCI) stress; (b) Drought area; (c) Drought intensity; (d) Fire risk area; (e) Drought duration; (f) Drought detection; (g) Corn yield versus VCI and TCI; (h) Malaria cases; (i) Malaria risk area. (http://www.star.nesdis.noaa.gov/smcd/emb/vci/VH/index.php).

Figure 13.5e shows duration of droughts of different intensity in Kenya during the two most drought-affected years, 2009 and 2011. During March–July, the minor and major growing seasons were affected by moderate and severe drought. Some differences between the years are observed for the extreme and exceptional drought intensities, which affected southern Kenya in 2009 and central Kenya in 2011. Early drought detection is the most important part of drought management and mitigation strategy development. Since there is no reliable drought prediction method, drought can be detected prior to its appearance from VH dynamics. As seen in Figure 13.5f, the first signs of the approaching 2006–2007 drought in southern Australia (for both VHI and TCI) appeared in March–April 2006, 2 months prior the time series crossed down the 40 threshold (indicating drought start) and vegetation was still in good health (VHI > 50).

Over the years of VH application for drought monitoring and verification, it has been revealed that VH methodology can be successfully applied for monitoring and predicting crop yield losses and risk of malaria (Kogan et al. 2012, Rahman et al. 2011). Testing VH indices in 34 countries of North America, South America, Africa, Europe, and Asia revealed that yield of such crops as wheat (both winter and spring), corn, sorghum, rice, and soybeans has strong correlation with the VH indices during the critical period of crops' growth, development, and reproduction. Figure 13.5g demonstrates some of these results. As seen, mean

Kansas (USA) corn yield correlates strongly with VCI and TCI during the critical period of corn development in July. During drought years (the indices <40), up to 40% corn production could be lost (depending on drought severity) compared to normal and wet years. For example, in 2011, Texas drought slashed corn yield by 28% compared to the average for the last 10 years and 36% relative to the highest yield received in the 2010 season (Taxes 2012).

Malaria risk area and intensity are estimated from VCI and TCI values, which correlated with the number of malaria cases in some regions of India, Bangladesh, Colombia, and Namibia (Kogan 2002, Nizamuddin et al. 2013, Rahman et al. 2011). An example from Caprivi province of Namibia indicating high correlation between observed and independent VH-simulated number of malaria cases is shown in Figure 13.5h. Finally, Figure 13.5i shows two images of VH-estimated malaria risk with a smaller area in 2012 (drought year) versus larger area in 2010 (wet year). This information is renewed at http://www.star.nesdis.noaa.gov/smcd/emb/vci/VH/index.php every week.

13.7 Droughts in a Warmer World

The 2012 Intergovernmental Panel on Climate Change (IPCC) report stated that the average Earth surface temperature in the past 100 years increased 0.85° (IPCC 2012). The warming trend

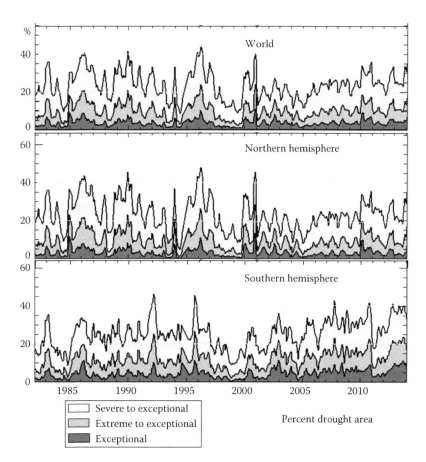

FIGURE 13.6 Percent drought area (from each region) for three intensities.

has continued until 1998, after which the world temperature leveled off and for the past 15 years was remaining at that level (although in the Northern Hemisphere the warming trend is continuing). Based on IPCC (2012), it is anticipated that the risk of droughts in a warmer world will increase and they expand their area and intensify. Therefore, it is quite possible that such new tendencies have already started. We used a 34-year VHI to estimate global and hemispheric drought dynamics (percent of affected area) by three intensities, severe to exceptional (SE), extreme to exceptional (EE), and exceptional (E), following the classification by Svoboda et al. (2002). Figure 13.6 presents these results for the entire world, Northern and Southern Hemispheres. The statistical assessment of drought dynamics presented in this figure has not supported upward trend in either of these regions for all three intensities. Meanwhile, visually, there is a very small reduction in the global and Northern Hemisphere percent drought area in SE and EE intensities during the twenty-first century and slight increase in Southern Hemisphere's percent drought area in all three intensities since 2005. These changes are so negligible that can be either incidental or related to local features of the regions. No upward trend conclusion is in line with the most recent revised analysis of the U.S. Palmer Drought Severity Index over the past 60 years and drought trend analysis in the United States, Ukraine, Horn of Africa, and southern Australia (Kogan and Guo 2014a, Kogan et al. 2013, Sheffield et al. 2013). Figure 13.6 indicates also that droughts of SE, EE, and E intensities covered 25%–35%, 15%–20%, and up to 10%, respectively, of these regions.

13.8 Conclusion

Drought is a part of Earth's climate, occurring every year without warning and recognizing boarders and political and economic differences. Drought affects the largest number of people on Earth and is a very costly disaster affecting water resources, ecosystems, agriculture, forestry, energy, human health, recreation, transportation, food supply and demands, and other resources and activities. Unlike other natural disasters, drought starts unnoticeably, develops cumulatively and slow, and produces cumulative impacts, and by the time damages are visible, it is too late to mitigate the consequences. Drought is characterized by the days of its start/end, intensity, duration, magnitude, area, season, origination, and impacts.

One of the main challenges in drought detection and monitoring is sparse weather station network. Satellite data avoid this problem. Since the introduction of the VH method, early drought detection and assessment of drought intensity, duration, origination, and impacts become the most widely used by global community. The VH method is accurate, simple to understand and interpret, available in real time, estimates drought every week for each 4 km² land surface, and was validated comprehensively against in situ data (climate, weather, agriculture, health, etc.) in 34 countries, including all major agricultural producers. A few drought products have been developed based on VH methodology. The VH-based products include moisture index, thermal index, drought area, drought

intensity, drought duration, soil wetness, and fire risk. The VH method and data set are used for prediction of climate and land cover trends, crop yield, risk of malaria, climate forcing (ENSO) impacts on vegetation productivities, and others. The products and data are available on the NOAA website (Kogan and Guo, 2014b), which is accessed by 3000–4500 users every month. The web provides assessments for the entire globe, continents, 194 countries, and around 4000 first-order countries' administrative divisions. Finally, it is important to emphasize that drought detection and monitoring method will be considerably improved with the new generation of satellite technology that started in 2011. The new Visible Infrared Imager Radiometer Suite on the National Polar-orbiting Operational Environmental Satellite System Preparatory Platform will provide observations for each 375 m of the globe, with much sharper view at the edge of the scan and four times more spectral bands (compared to its predecessor).

References

AgriLife. 2011. Research News. http://agrilife.org/today/2011/08/17/texas-agricultural-drought-losses-reach-record-5-2-billion/. Accessed August 17.

Cracknel, A.P. 1997. *The Advanced Very High Resolution Radiometer.* Taylor & Francis, Great Britain, England, 350pp.

Deering, D.W. 1978. Rangeland reflectance characteristics measured by aircraft and spacecraft sensors. PhD dissertation, Texas A&M University, College Station, TX, 388pp.

FAO. 2012. Crop statistics. http://faostat.fao.org/?PageID=567#ancor. (Accessed June 8, 2015.)

FAO. 2013. http://en.mercopress.com/2013/10/07/world-2013-14-cereal-production-forecasted-to-surpass-2012-level-by-8-says-fao. Accessed March 10.

IPCC. 2012. Summary for policymakers. Twelfth session of Working Group 1, WGI AR5. http://www.climatechange2013.org/images/uploads/WGIAR5-SPM_Approved27Sep2013.pdf. Accessed October 5, 2013, 36pp.

Kidwell, K.B. (ed.). 1995. NOAA polar orbiting data user's guide. U.S. Department of Commerce Technical Report, Washington, DC, 92pp.

Kogan, F. 1997. Global drought watch from space. *Bulletin of the American Meteorological Society,* **78,** 621–636.

Kogan, F. 2002. World droughts from AVHRR-based vegetation health indices. *Eos, Transactions, American Geophysical Union,* **83**(48), 557–564.

Kogan, F., T. Adamenko, and W. Guo. 2013. Global and regional drought dynamics in the climate warming era. *Remote Sensing Letters,* **4,** 364–372.

Kogan, F. and W. Guo. 2014a. Early twenty-first-century droughts during the warmest climate. *Geomatics, Natural Hazards and Risk.* http://dx.doi.org/10.1080/19475705.2013.878399. (Accessed July 21, 2014.)

Kogan, F. and W. Guo. 2014b. Satellite-based Vegetation health for drought, climate, land cover, agriculture, malaria, food security. http://www.star.nesdis.noaa.gov/smcd/emb/vci/VH/index.php.

Kogan, F., L. Salazar, and L. Roytman. 2012. Forecasting crop production using satellite based vegetation health indices in Kansas, United States. *International Journal of Remote Sensing*, 3, 2798–2814. DOI: 10.1080/01431161.2011.621464.

Kogan, F.N. 1990. Remote sensing of weather impacts on vegetation in nonhomogeneous areas. *International Journal of Remote Sensing*, 11, 1405–1419.

McKee, T.B., N.J. Doesken, and J. Kleist. 1993. The relationship of drought frequency and duration to time scale. In: *Preprints Eighth Conference on Applied Climatology*, Anaheim, CA.

NCDC (National Climatic Data Center). 2013. Billion dollar U.S. weather disasters. http://www.ncdc.noaa.gov/oa/reports/billionz.html. Accessed July 25, 2011.

Nizamuddin, M., F. Kogan, R. Dihman, W. Guo, and L. Roytman. 2013. Modeling and forecasting malaria in Tripura, India using NOAA/AVHRR-based vegetation health indices. *International Journal of Remote Sensing Applications*, 3(3), 108–116.

Palmer, W.C. 1965. Meteorological drought. U.S. Weather Bureau, Washington, DC, Research Paper No. 45, 58pp.

PotashCorpo. 2012. Agriculture: Crop overview. http://www.potashcorp.com/industry_overview/2011/agriculture/16. Accessed November 22, 2012.

Rahman, A., F. Kogan, L. Roytman, M. Goldberg, and W. Guo. 2011. Modeling and prediction of malaria vector distribution in Bangladesh from remote sensing data. *International Journal of Remote Sensing*, 32(5), 1233–1251.

Sheffield, J., E.F. Wood, and M.L. Roderick. 2013. Little change in global drought over the past 60 years. *Nature*, 491, 435–438.

Svoboda, M., D. LeComte, M. Hayes, R. Heim, K. Gleason, J. Angel, B. Rippey et al. 2002. The drought monitor. *Bulletin of the American Meteorological Society*, 83(8), 1181–1190.

Taxes. 2012. USAD Census of Agriculture. http://www.nass.usda.gov/Statistics_by_State/Texas/Charts_&_Maps/zcorn_y.htm. Accessed November 26.

UNDP. 2012. Drought in Russia and Kazakhstan. http://europe-andcis.undp.org/aboutus/show/. Accessed September 10, 2012.

U.S. Drought. 2012. Science. *New York Times*. http://topics.nytimes.com/top/news/science/topics/drought/index.html. Accessed December 10, 2012.

Wilhite, D.A. 2000. Drought as a natural disaster. In: *Drought* (D. Wilhite, Ed.). Routledge Hazards and Disasters Series. Routledge, Taylor & Francis Group, New York, pp. 3–19.

Agricultural Drought Monitoring Using Space-Derived Vegetation and Biophysical Products: A Global Perspective

Felix Rembold
Joint Research Centre

Michele Meroni
Joint Research Centre

Oscar Rojas
Food and Agriculture Organization of the United Nations

Clement Atzberger
University of Natural Resources and Life Sciences (BOKU)

Frederic Ham
Action Contre la Faim—International

Erwann Fillol
Action Contre la Faim—International

Acronyms and Definitions

ACF	Action Contre la Faim
ASIS	Agricultural Stress Index System
AVHRR	Advanced Very High Resolution Radiometer
BOKU	University of Natural Resources and Life Sciences
CART	Classification and regression tree
CDI	Combined Drought Index
CFAPAR	Cumulated FAPAR
CMI	Crop Moisture Index
CNDVI	Crop-specific NDVI
DMP	Dry matter productivity
ECMWF	European Centre for Medium-Range Weather Forecasts
EOS	Earth Observing System (operated by NASA)
ESA	European Space Agency
EUMETSAT	European Organisation for the Exploitation of Meteorological Satellites
FAO	Food and Agriculture Organization of the United Nations
FAPAR	Fraction of absorbed photosynthetically active radiation
FEWSNET	Famine Early Warning System
GIEWS	Global Information and Early Warning System
GIS	Geographic information system
GLC	Global land cover
IPCC	Intergovernmental Panel on Climate Change
JRC	Joint Research Centre
LAI	Leaf area index
MARS	Monitoring Agricultural ResourceS
MERIS	MEdium Resolution Imaging Spectrometer
METOP	Small satellite series operated by EUMETSAT

MODIS	Moderate Resolution Imaging Spectroradiometer
MSG	Meteosat Second Generation
MVC	Maximum-value composite
NASA	National Atmospheric and Space Administration
NDVI	Normalized Difference Vegetation Index
NIR	Near infrared
NOAA	National Oceanic and Atmospheric Administration
NRT	Near real-time
PDSI	Palmer Drought Severity Index
POES	Polar Operational Environmental Satellite
PRI	Photochemical Reflectance Index
PROBA_V	Satellite succeeding the VEGETATION instruments
RDI	Reclamation Drought Index
RFE	Rainfall estimates of the Climate Prediction Centre of the National Oceanic and Atmospheric Administration
SEAWIFS	Sea-Viewing Wide Field-of-View Sensor
SENTINEL	Multisatellite project of the European Space Agency that started in 2014
SPI	Standardized Precipitation Index
SPOT	Satellite Pour l'Observation de la Terre
SWB	Small water bodies
SWIR	Shortwave infrared
SWSI	Surface Water Supply Index
TAMSAT	Tropical Applications of Meteorology using SATellite of the Reading University
TCI	Temperature Condition Index
UN	United Nations
U.S.	United States
USAID	United States Agency for International Development
VCI	Vegetation Condition Index
VDI	Vegetation Drought Index
VHI	Vegetation Health Index
VI	Vegetation index
VIS	Visible (refers to the electromagnetic spectrum)
VITO	Flemish Institute for Technological Research
WFP	World Food Programme
WMO	World Meteorological Organization
ZVI	Standard Score Vegetation Index

14.1 Introduction

For a long time, agricultural monitoring systems have been using space remote sensing (RS) instruments to provide timely and synoptic information about drought. A variety of approaches are currently being used and most of them are based on the analysis of RS data in the optical domain. They permit the mapping of vegetation vigor, as well as hydrological variables such as rainfall and evapotranspiration when using imagery in the thermal domain. Sensors operating in the microwave domain provide additional and valuable information regarding soil moisture. In this chapter, after providing background information about drought-monitoring indices and systems in general, we focus on the current use of satellite-derived biophysical indicators of vegetation status from RS in the optical domain.

14.1.1 What Is Drought?

Depending on the nature of drought and its impact, one can define meteorological, hydrological, or agricultural droughts, which all have different physical and socioeconomic impacts. A meteorological drought is an extreme climate event over land characterized by below-normal precipitation over a period of time. This event may lead to what is generally defined as agricultural drought, a period with declining soil moisture and consequent crop failure (Mishra and Singh, 2010). In this review we focus on agricultural drought, a phenomenon that is characterized by a severe reduction of actual evapotranspiration of crops as compared to the potential one. Besides prevailing weather conditions, this imbalance is affected by other agriculture-specific characteristics such as stage of growth and the soil's physical and biological properties, among others.

Drought is part of the climate variability in arid regions, but it is different from aridity itself, which is a permanent climate characteristic, mainly defined by low average precipitation. Drought is a complex phenomenon that originates from anomalous rainfall deficiency. It results in low runoff, groundwater and soil moisture, and finally the shortage of available water for plants, animals, and humans. However, drought does not only depend on precipitation but also on other factors such as air temperature, humidity, wind speed, and soil properties. All these factors can substantially contribute to exacerbate drought severity.

14.1.2 How Does Drought Affect Agricultural Production and Food Security?

Drought, with its negative effects on agricultural production, is one of the main causes of food insecurity worldwide. Extreme droughts like those that hit the Sahel region in the 1970s and 1980s, the Ethiopian drought in 1984, and the recent Horn of Africa drought in 2010/2011 have received extensive media attention because they directly caused hunger and death of hundreds of thousands of people (Checchi and Courtland Robinson, 2013). With the recent trend of persistently high food prices and a continuously increasing demand for agricultural production to satisfy the food needs and dietary preferences of an increasing world population, drought is one of the climate-related factors with the highest potential of negative impact on food availability and societal development. Droughts aggravate the competition and conflicts for natural resources in those areas where water is already a limiting factor for agriculture, pastoralism, and human health. Climate change may further deteriorate this picture by increasing drought frequency and extent in many regions of the world due to the projected increased aridity in the next decades (IPCC AR5, 2013).

For drought to negatively affect agricultural production in a region, one has to consider both its spatial and temporal dimensions (Rojas et al., 2011). Drought is usually a slow-onset problem that does negatively impact crop production and ultimately food security only if it persists for a period long enough to seriously reduce plant growth and health and if the area concerned is large enough to substantially reduce food production in a region. To estimate the drought impact on food security, it is also important to take into account the level of vulnerability and coping capabilities of the exposed population as shown once again by the recent famine in Somalia (Maxwell and Fitzpatrick, 2012). This example shows clearly that for a crisis to evolve from a prolonged agricultural drought into a famine, many other factors are at play, such as high international food prices, limited access to the drought-affected area, and civil conflicts and political difficulties in organizing humanitarian interventions.

Crop failures and pasture biomass production losses are the primary direct impact of drought on the agricultural sector productivity. Drought-induced production losses cause negative supply shocks, but the amount of incurred economic impacts and distribution of losses depends on the market structure and interaction between the supply and demand of agricultural products (Ding et al., 2011). These adverse shocks affect households in a variety of ways, but typically the key consequences are on assets (UN, 2009). First, households' incomes are affected, as returns to assets (e.g., land, livestock, and human capital) tend to collapse, which may lead to or exacerbate poverty. Assets themselves may be lost directly due to the adverse shocks (e.g., loss of cash, live animals, and impacts on health or social networks) or may be used or sold in attempts to buffer income fluctuations, affecting the ability to generate income in the future.

Droughts are by their nature covariate phenomena, with many people affected at the same time, making their consequences even harder. For instance, rural population affected by drought may be forced to opt for limiting or even disrupting copying strategies including selling their production assets (e.g., livestock, land and tools). Short-term impacts can last a few weeks or months. If response action (e.g., food relief or cash transfer) is taken to the household or community level right after the disaster, the consumption drops or income losses can be softened. On the contrary, if households have few assets to protect themselves during hardships and public protective measures come too late, the negative transitory effects on their members can deteriorate into more permanent disadvantages, for instance, migration or nutrition shortfalls in children that in turn could affect their development later in life.

Drought as a climate-related disaster is hard to prevent, and main efforts towards reducing drought impacts traditionally focus on mitigation and on strengthening the resilience of drought-exposed livelihoods by efficient drought management. One way of mitigating drought impacts is by improving early warning and monitoring systems fed by objective and reliable drought indices (vegetation and precipitation) and near real-time (NRT) weather information (Wilhite et al., 2007; Sheffield and Wood, 2011; Boyd et al., 2013).

Obviously, even if the impact of a drought can be timely assessed, having an operational early warning system in place is only a first step toward ensuring rapid and efficient response (Hillbrunner and Moloney, 2012).

For the implementation of programs that aim to increase food security, the identification of drought-prone areas and the estimation of the probability of drought occurrence are also fundamental. For example, knowing the probability of drought occurrence is of basic importance for risk management programs (e.g., crop and livestock insurances) as well as for planning efficient food-aid delivery. Furthermore, drought information is very important for a better interpretation of potential effects of climate change in Africa (Rojas et al., 2011) and elsewhere.

14.1.3 What Can Remote Sensing Do?

Most drought-monitoring methods have focused mainly on rainfall and rainfall anomalies and many indices have been developed over time. A significant number of indices for agricultural drought monitoring are based on a water balance approach, where several climatic and physical variables are observed over a certain period in order to estimate the soil moisture deficit. This family of indices includes the Palmer Drought Severity Index (PDSI), the Crop Moisture Index (CMI), the Surface Water Supply Index (SWSI), and the Reclamation Drought Index (RDI) (Palmer, 1965; Shafer and Dezman, 1982).

Other indices are of more statistical nature and look at the time series of precipitation. This is, for example, the case of the Standardized Precipitation Index (SPI) (Wu et al., 2007; WMO, 2012). To the same group belongs the more simple precipitation decile index (Gibbs and Maher, 1967) and the percent of normal precipitation method. The latter index simply indicates the relative difference of current rainfall as compared to a long-term average. A good overview of agrometeorological drought indices is made available by M.G Hayes (http://www.civil.utah.edu/~cv5450/swsi/indices.htm).

At regional to continental scale, drought monitoring and drought risk assessment based on any of the aforementioned indices are often hampered by the scarcity of reliable rainfall data. In particular, the coverage of operational weather stations in many drought-prone countries shows large spatial gaps and individual stations often provide discontinuous data. Additionally, the spatial representativeness of in situ measurements is often very restricted, and a continuous spatial description of precipitation is difficult to be achieved because of the known limitation in spatial interpolation of rainfall data from meteorological stations. Due to these reasons, rainfall measurements are commonly replaced by estimates generated by atmospheric circulation models or derived from meteorological satellite observations. Commonly used rainfall estimate datasets for drought monitoring are reported in Table 14.1.

Another approach to drought monitoring is to evaluate "vegetation health status" by using optical RS. The large spatial coverage and high temporal revisit frequency of low spatial resolution satellite instruments such as the Moderate Resolution Imaging

TABLE 14.1   Rainfall Estimate Datasets Used for Drought Monitoring in Africa

Dataset	References	Website
European Centre for Medium-Range Weather Forecasts (ECMWF)	ECMWF (2013)	http://www.ecmwf.int/products/forecasts/d/charts
Rainfall estimates (RFEs) of the Climate Prediction Centre of the National Oceanic and Atmospheric Administration	NOAA (2001)	http://www.cpc.ncep.noaa.gov/products/fews/ AFR_CLIM/afr_clim.html
Tropical Applications of Meteorology using SATellite (TAMSAT) of Reading University	Grimes (2003)	http://www.met.reading.ac.uk/tamsat/about/
Rainfall estimates of the Food and Agriculture Organization (FAO-RFE)	Alessandrini and Evangelisti (2011)	http://geonetwork3.fao.org/climpag/FAO-RFE.php
Tropical Rainfall Measuring Mission (TRMM) of National Aeronautics and Space Administration	Zhong et al. (2012)	http://trmm.gsfc.nasa.gov/

Spectroradiometer (MODIS) or the Satellite Pour l'Observation de la Terre (SPOT)-VEGETATION make them particularly useful for NRT information collection at the regional and global scale (Rembold et al., 2013). Thanks to their large swath width, low-resolution systems have currently a much better synoptic view and temporal revisit frequency compared to high spatial resolution sensors (e.g., Landsat instruments). The individual scenes span a width of up to 3000 km, so that the entire Earth surface is scanned every day and the specific costs per ground area unit are very low.

A pragmatic and widespread approach to extract the relevant information from the various spectral bands of such satellite sensors relies on the computation of vegetation indices (VIs). Among the different VIs, the Normalized Difference Vegetation Index (NDVI; Rouse et al., 1974) based on the red and near-infrared reflectances has become the most popular indicator for studying vegetation health and crop production. Research in vegetation monitoring has shown that NDVI is nonlinearly related to the leaf area index (LAI) and linearly related to the fraction of absorbed photosynthetically active radiation (FAPAR) and hence the vegetation's photosynthetic activity. VIs are subject to intrinsic limitations (e.g., saturation of the signal) and contaminations from different sources (e.g., illumination and observation geometry, 3D structure of the vegetated medium, and background reflectance)

(Gobron et al., 1997). Alternative approaches make use of canopy radiative transfer models to derive key vegetation variables such as LAI and FAPAR from canopy reflectances (e.g., Myneni et al., 2002; Gobron et al., 2005; Baret et al., 2013). The advantage of these methods is that they provide access to inherent vegetation properties largely decontaminated of external factors (Pinty et al., 2009). In particular, FAPAR acts as an integrated indicator of the status and health of vegetation and plays a major role in driving gross primary productivity (Prince and Goward, 1995). These biophysical variables are also independent with regard to the exact spectral band location and width of the satellite instrument used, thus potentially offering the possibility of building long-term datasets composed by multiple generation of instruments.

Time series of up to 30 years are nowadays available for low-resolution satellite sensors, which makes time series analysis of biophysical indicators a common tool for drought monitoring (Table 14.2). Repeated acquisition of the same area is performed with hourly to daily frequency using coarse- to moderate-resolution instruments. Daily images can be used for specific purposes, but most of the monitoring systems described in this chapter make use of temporal synthesis of daily images (so-called composites), typically aggregating 10 daily acquisitions into a 10-day composite image (also called dekadal composite). The maximum-value

TABLE 14.2   Properties of the Most Common Optical Low- and Medium-Resolution Operational and Planned Sensors Relevant for Vegetation Monitoring

Sensor	Platform	Spectral Range	Number of Bands	Resolution	Swath Width	Repeat Coverage	Launch
AVHHR	NOAA POES 6–19	VIS, NIR	5	1100 m	2400 km	12 h	1978
AVHRR	METOP	VIS, NIR, SWIR	5	1100 m	2400 km	12 h	2007
SEAWIFS	OrbView-2	VIS, NIR	8	1100 m 4500 m	1500 km 2800 km	1 day	1997
VEGETATION	SPOT 4, 5	VIS, NIR, SWIR	4	1100 m	2200 km	1 day	1998
MODIS	EOS AM1/PM1	VIS, NIR, SWIR	36	250–1000 m	2330 km	<2 days	1999
MERIS	ENVISAT	VIS, NIR	15	300 m (1200 m)	1150 km	<3 days	2000
PROBA-V	PROBA-V	VIS, NIR, SWIR	4	300 m (1000 m)	2250 km	1 day	2013
SENTINEL 3	SENTINEL	VIS, NIR	21	300 m	1270 km	<2 days	Foreseen 2015

The following abbreviations are used for different intervals of the electromagnetic spectrum used for optical drought monitoring: VIS, visible (350–750 nm); NIR, near infrared (750–1400 nm); SWIR, shortwave infrared (1400–3000 nm).

NB: The ENVISAT mission stopped officially in May 2012. SPOT-VEGETATION mission stopped in May 2014 with PROBA-V ensuring the continuity of product generation.

composite procedure (MVC; Holben, 1986) that retains the highest NDVI value within the compositing period is often used for the purpose although several compositing techniques have been proposed (for an overview, see Chuvieco et al., 2005). Compositing also reduces considerably the amount of noise that is present in the time series of NDVI or any other RS-derived biophysical indicator (due to different sources, including atmospheric perturbation, undetected clouds, and anisotropy of the surface). In addition, temporal smoothing techniques are usually employed to further reduce the residual noise in the data (e.g., Chen et al., 2004; Atzberger and Eilers, 2011).

Table 14.2 is not taking into consideration low spatial resolution geostationary satellites that belong primarily to the meteorological domain like Meteosat and Meteosat Second Generation (MSG). Nevertheless, optical and thermal measurements offered by these satellites can be used for drought monitoring too (Fensholt et al., 2011). In addition, we acknowledge the use of radar RS for soil moisture estimation. For an introduction on the use of radar RS in drought monitoring, see Nghiem et al. (2012).

14.2 Operational Methods and Techniques

In this section we describe four approaches/systems operationally used for drought monitoring with satellite-derived VIs and biophysical products:

1. NDVI-based vegetation anomaly indicators for drought and crop monitoring bulletins
2. FAO Agricultural Stress Index System (ASIS) approach targeting agriculture
3. Action Contre la Faim (ACF) approach targeting pastures
4. Joint Research Centre (JRC) approach for the early detection of biomass production–deficit hot spots

We thus focus on agricultural drought-monitoring systems (including pastoral systems) working in NRT at global to regional scale. Several other approaches and systems exist and have been extensively described in recent books such as *Famine Early Warning Systems and Remote Sensing Data* by Molly Brown in 2008 or *Remote Sensing of Drought: Innovative Monitoring Approaches* by Brian D. Wardlow et al. in 2012 as well as in other chapters of this book. The first book focuses mainly on the use of RS for food security early warning. It includes also a number of drought and hydrological stress monitoring applications. The second book explores a broad range of applications for monitoring and estimating vegetation health, soil moisture, precipitation, and evapotranspiration.

Table 14.3 summarizes the main characteristics of the four drought-monitoring approaches described in this chapter and mentions main strengths and limitations for each of them.

14.2.1 NDVI-Based Vegetation Anomaly Indicators Used Mainly for Drought or Crop Monitoring Bulletins

Vegetation anomaly methods are mostly based on the qualitative (or semiquantitative) interpretation of RS-derived indicators (often NDVI). They generally compare the actual crop status to previous seasons or to what can be assumed to be the average or *normal* situation. Detected anomalies are then used to draw conclusions on possible vegetation *health* or yield limitations.

Simple, but timely and accurate, vegetation monitoring systems working both at the national and regional scale are particularly necessary in arid and semiarid countries, where temporal and geographic rainfall variability leads to high interannual fluctuations in primary production and to a large risk of famines (Hutchinson, 1991). These environmental situations, along with the wide extent of the areas to monitor and the generally poor availability of efficient ground data collection systems,

TABLE 14.3 Summary of the Main Aspects of the Four Drought-Monitoring Approaches Described in This Chapter

	Input Data	Methods	Strengths	Limitations
NDVI-based vegetation anomaly indicators for drought and crop monitoring	NDVI derived from medium- and low-resolution data. Usually weekly or 10-day composites	Temporal comparisons and identification of anomalies for ongoing vegetative season	Relatively simple and straightforward to be produced	NDVI is only a proxy of vegetation vigor and health. Phenology not accounted for
FAO Agriculture Stress Index System approach targeting agriculture	VHI derived from low-resolution NDVI and temperature data. 10-day composites	Computation of cumulative value over time (crop season) and space (administrative areas) and use of thresholds to classify a drought event	Provides information directly in terms of percentage of agricultural area affected	Depends on accuracy of crop mask
Action Contre la Faim approach targeting pastures	DMP based on Monteith approach. Small water bodies (SWB) product	Computation of seasonal cumulative values	Simple calculation and easy to understand. Results expressed in biomass physical units	DMP is not water limited reducing sensitivity to drought
Joint Research Centre approach for the early detection of biomass production–deficit hot spots	FAPAR 10-day composites	Computation of seasonal cumulative values and computation of probability of final seasonal deficit	Provides a forecast for the end of the growing season adjusted for the uncertainty of the estimation	More complex than the others

Region: The GLOBE
Period: August, 2012
Theme: Normalized Difference Vegetation Index (NDVI)
       difference w.r.t. historical mean (Act. - Hist.)
Source: SPOT-VEGETATION

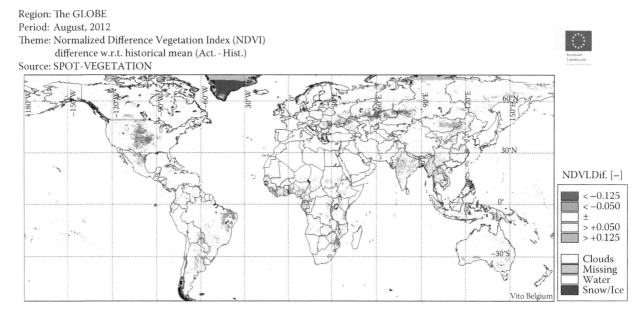

FIGURE 14.1 Global map of NDVI anomalies during the 2012 boreal hemisphere summer crop growing season (August). Negative anomalies are visible mainly in central United States, central Asia, and northern Brazil, while a positive situation is evident in eastern China and southern Brazil. Data are from SPOT-VEGETATION. Anomalies are expressed as deviation in absolute NDVI units from the 1999–2011 average of NDVI for the month of August.

represent a scenario where qualitative monitoring can quickly produce valid information for releasing early warnings about possible stress of crops and pastures. Such systems are typically used in many food-insecure countries by the Food and Agriculture Organization of the United Nations (FAO), Famine Early Warning System (FEWSNET) of the United States Agency for International Development (USAID), and the Monitoring of Agricultural ReSources (MARS) project of the JRC of the European Commission. However, qualitative monitoring is not necessarily linked to an early warning context in arid areas but can also be very useful to get a quick overview of vegetation stress for large areas in different climatic zones of the world. An example is given in Figure 14.1, which depicts NDVI anomalies during the 2012 northern hemisphere crop growing season. The NDVI anomaly is computed as the difference between the mean NDVI value for the month of August 2012 and the mean NDVI value computed for the same month over the historical NDVI archive (in this case from 1998 to 2011 for a total of 14 years). A clear stress for summer crops is visible in central parts of the United States and in southern parts of Russia due to poor rainfall distribution (data not shown), whereas favorable conditions are observed in large parts of China. In the southern hemisphere, negative vegetation anomalies are visible in Northeastern Brazil and Southern Africa, whereas favorable conditions can be observed in the southern part of Brazil.

In addition to analyzing anomaly images for qualitative vegetation growth monitoring, useful information can be derived from temporal profiles of remotely derived VIs. These temporal profiles are extracted for the administrative area of interest by averaging all pixel values within the area or, to better focus on

agricultural land, by considering only those pixels where crops are dominant. The profiles give a complete picture of the vegetation development during the seasonal cycle and can be compared with other (e.g., previous) crop seasons and the long-term average vegetation profile. If available, information regarding the major phenological development stages can be considered during the evaluation as stress effects on crop growth differ for different phenological stages.

Signatures from coarse-resolution pixels usually represent mixtures of different land use classes (so-called mixed pixel). Several approaches have been elaborated for extracting land use–specific signatures from the low-resolution pixel. A simple and common one is the crop-specific NDVI (CNDVI) method (Genovese et al., 2001). When computing the average NDVI value of an area of interest (e.g., a region, a department), the CNDVI approach adds proportional weights to the NDVI values based on the fraction of area covered by crop within each low-resolution pixel. More sophisticated methods are based on so-called unmixing models that consider the NDVI of a given low-resolution pixel as a linear mixture of so-called end-member spectral signatures (Busetto et al., 2008). Both CNDVI and unmixing approaches require a land cover map at a spatial resolution higher than that of the coarse-resolution time series. This land cover map is used to compute the relative presence of the classes of interest (e.g., crops). Relative advantages and limitations of CNDVI and unmixing approaches have been recently studied in Atzberger et al. (2014).

For comparing recent VI images to *normal* conditions and its historical distribution, a variety of statistical indices have been proposed beyond the simple difference described so far.

Following classical statistical theory, one approach is to calculate a so-called standard score (Standard Score Vegetation, ZVI) by subtracting, for a given dekad (10-day compositing period) d, the historical mean ($NDVI_{AVG}$) from the observed value and divide it by its standard deviation ($NDVI_{SD}$, also obtained from historical data):

$$ZVI(d) = \frac{NDVI(d) - NDVI_{AVG}(d)}{NDVI_{SD}(d)} \quad (14.1)$$

ZVI thus indicates how many standard deviations an observed NDVI value is below (or above) the historical average. To avoid the assumption of normal distribution of Equation 14.1, a nonparametric version of the index can be computed using, for instance, the median and the interquartile distance of the observed distribution.

Another example of a drought index is the Vegetation Condition Index (VCI) of Kogan (1995). The VCI locates the current VI value in the historical range of all preceding images acquired at the same time of the year:

$$VCI(d) = 100 \frac{NDVI(d) - NDVI_{MIN}(d)}{NDVI_{MAX}(d) - NDVI_{MIN}(d)} \quad (14.2)$$

where d refers to the dekad of the year at which the VCI is computed, subscripts MIN and MAX refer the minimum and maximum value observed for dekad d in the historical archive, respectively.

VCI should therefore be interpreted as a percentage expressing the vegetation status of a given pixel in relation to its historic range of variability represented by the minimum (worst conditions) and maximum (best conditions) NDVI over the years. VCI is, by definition, extremely sensitive to outliers in the series that would affect the maximum and minimum value of NDVI, which makes the use of temporal smoothing a basic requirement of the method.

The VCI can be combined with an analogous index built with remotely sensed surface temperature (i.e., the Temperature Condition Index [TCI]) into the Vegetation Health Index (VHI) as proposed by Kogan (2001). Spatial and temporal aggregations of the VHI have been successfully used as an agricultural drought indicator by Rojas et al. (2011).

In a similar way, Balint et al. (2011) proposed the Combined Drought Index (CDI) that, besides NDVI and temperature, takes into account precipitations. In the CDI framework, drought (or vegetation stress) is conceived as a combination of magnitude and time persistence of the following factors: rainfall deficit, temperature excess, and soil moisture deficit. Because of limited availability of soil moisture observations at 1 km resolution, Balint et al. (2011) proposed to approximate the soil moisture component by NDVI deficits and deficit persistence. The three individual drought indices, that is, *Precipitation Drought Index* (PDI), *Temperature Drought Index* (TDI), and *Vegetation*

Drought Index (VDI—as a substitute for the *Soil Moisture Drought Index*), are computed by taking into account the magnitude of the anomaly and its duration and then combined as a weighted sum to yield the CDI.

14.2.2 Agricultural Stress Index System Approach Targeting Agriculture

FAO is developing the ASIS to detect agricultural areas with a high likelihood of water stress (drought) at the global level. Based on RS data, ASIS will support the vegetation monitoring activities of the FAO Global Information and Early Warning System (GIEWS). The idea behind ASIS is to setup an agricultural drought-monitoring system where the RS data processing part is highly automated and the end user can concentrate the analysis on the results of the system. ASIS provides maps of "drought hot spots" updated every 10 days to the GIEWS officers and external users. To ensure that the system will not produce false alerts due to external factors such as atmospheric perturbations affecting low-resolution optical images, the officers then verify the *hot spots* with auxiliary information, for example, by contacting the Ministry of Agriculture of the affected country or by monitoring prices of the commodities. ASIS uses the VHI as derived from the NDVI and surface brightness temperature (see Section 14.2.1) dekadal product from the METOP-Advanced Very High Resolution Radiometer (AVHRR) sensor at 1 km resolution. The first step consists in computing the temporal average of the VHI assessing the intensity and duration of the dry period(s) occurring during the crop cycle at pixel level. The second step consists in the calculation of the Agricultural Stress Index (ASI) as the percentage of agricultural area affected by drought (pixels with VHI < 35%—a value identified as critical in previous studies) to assess the spatial extent of the drought. Finally, each administrative area is classified into one of several drought severity classes according to the percentage of the area affected.

By definition, VHI can potentially detect drought conditions at any time of the year. For agriculture, however, the analysis is restricted to the period between the start and end of the crop season. ASIS assesses the severity (intensity, duration, and spatial extent) of agricultural drought and provides summary statistics at the selected administrative level, allowing the comparison with the official agricultural statistics, where available. The full methodology is described in Rojas et al. (2011).

For the operational implementation, the Flemish Institute for Technological Research (VITO) intercalibrated the METOP-AVHRR data (available only since 2007) with the NOAA-AVHRR time series (since 1984) to produce a consistent long-term historical archive. The ASIS database thus allows the analysis of 30 years of potential agricultural hot spots, starting from year 1984 when the Sahel was severely affected by drought. As an example of global hot spot maps produced, Figure 14.2 shows the ASI map for the year 1989, when a large fraction of the global agricultural land suffered from water scarcity (see also Figure 14.3).

FIGURE 14.2   ASI map for the first crop season of 1989. The ASI is an indicator developed by FAO that highlights anomalous vegetation growth and potential drought in arable land during a given cropping season. ASI assesses the temporal intensity and duration of dry periods and calculates the percentage of arable land affected by drought as pixels with a VHI value below 35% (reference threshold taken from literature). In 1989 large portions of agricultural land worldwide were affected by drought, with more than 85% of agricultural areas affected by drought in the United States, in Argentina, and in central Asia. Figure 14.3 provides an overview of the percentage of the agricultural land affected by drought in the time period 1984–2012, as derived from the ASIS method.

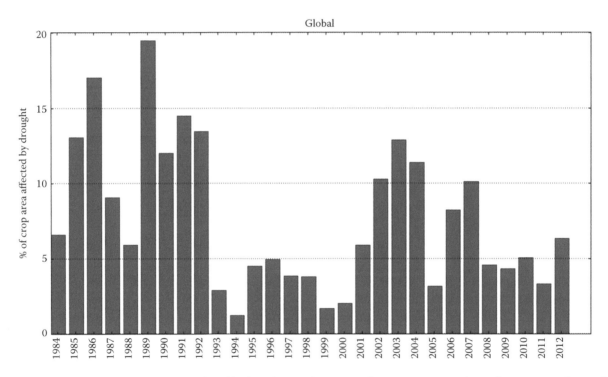

FIGURE 14.3   Percentage of agricultural land affected by drought according to ASIS by year 1984–2012. The total crop area used as a reference amounts to 23,440,622 km² and was derived from the crop mask shown in Figure 14.2 and obtained using the crop zones developed by FAO in the 1990, and the Global Land Cover (GLC2000). (From Bartholomé, E. and Belward, A.S., *Int. J. Remote Sens.*, 26(9), 1959, 2005.)

14.2.3 Action Contre la Faim Approach Targeting Pastures

In Sahelian pastoral and agropastoral areas, livestock systems are fully dependent on the spatial repartition of pastoral resources, namely, pasture and water. Their availability in space and time is frequently affected by drought, which can lead to livestock movements and locally tuned adaptation strategies. In this semiarid region, pastoral systems rely on rangelands and crop residues and are mainly nomadic or transhumant in the northern Sahel and more or less seasonally mobile in the southern part (rainfed and irrigated mixed crop–livestock systems). The unconditional need for mobility of livestock and families results from the scarce natural forage resources, while feed is unavailable or unaffordable owing to geographical isolation, high transportation costs, and low purchasing power (Ickowicz et al., 2012).

To address these specificities, ACF's approach consists of using a geographic information system (GIS)- and RS-based system designed to monitor pastoral resources on the one hand and livestock spatial adaptation strategies on the other hand. The full system—currently under development—ultimately aims at producing early alert and surveillance indicators in order to prevent and mitigate the impacts of drought events, targeting most affected areas and communities, and anticipating tenses and possible conflicts for the access to the resources (Ham et al., 2011). The two main pastoral resources (pastures and surface water) are monitored through RS indicators.

14.2.3.1 Pasture Monitoring

This module is handled using the software package BioGenerator (Fillol, 2011a) used to calculate dry matter biomass produced during the growing season by cumulating the dry matter productivity (DMP) dekadal product at 1 km resolution produced by VITO from SPOT-VEGETATION data using a modified light-use efficiency approach (Eerens et al., 2004). The cumulative value obtained at the end of the growth period is used to produce maps of biomass production (Figure 14.4) and biomass production anomalies (expressed as percent of average value) over the Sahel region (Figure 14.5). The output data of BioGenerator have been validated through several field studies, integrating field observations from Mali and Niger.

As compared to classical VIs, the use of DMP product directly provide amounts of production in physical units (kg/ha). The DMP calculation approach is derived from the Monteith (1972) model and uses the FAPAR, solar radiation, and temperature information retrieved from the European Centre for Medium-Range Weather Forecasts (ECMWF) global climate model. The DMP model does not directly take into consideration water available to the plants and is therefore not extremely drought

sensitive; also it only provides total biomass production estimates and does not give any information on the actual pasture usability nor accessibility. Nevertheless, experience shows that the data are suitable and consistent for early warning purposes. GIS operations are then used to aggregate biomass production per geographical units (administrative, agroecological) in order to produce zonal balances. The output data of this module have already been used to assess drought impact at national and regional level during known drought events in the Sahel region (2004–2005, 2009–2010, and 2011–2012) and are currently used operationally to analyze food security in pastoral areas (FAO-WFP, 2013). As an example, Figure 14.5 shows the biomass anomaly for year 2013 compared to the 1999–2013 average, which appears to be an average year except in certain localized areas scattered along the pastoral zones bordering the desert to the South.

14.2.3.2 Surface Water Monitoring

The methodology is based on the use of the small water bodies (SWB) dekadal product available at a resolution of 1 km and providing information on the presence of surface water (Haas et al., 2009). SWB is produced by VITO from SPOT-VEGETATION data since 1999. This product has been validated using high-resolution satellite data and showed an overall accuracy of 95.4% considering only commission errors. Omission errors were significantly higher (about 30%) mainly due to the low spatial resolution of the product that hampers the detection of water bodies occupying only a minor fraction of the pixel. Integrating this information, the software package HydroGenerator (Fillol, 2011b) compiles annual statistical information on surface water. For each pixel and year, an index of water accessibility is calculated, using an integration of detections of all ponds within a 30 km circular buffer. The integration is weighted by the distance to these ponds. This monitoring of surface water should ideally be coupled with a consistent inventory of functioning underground water points (wells and boreholes). Having these two complementary

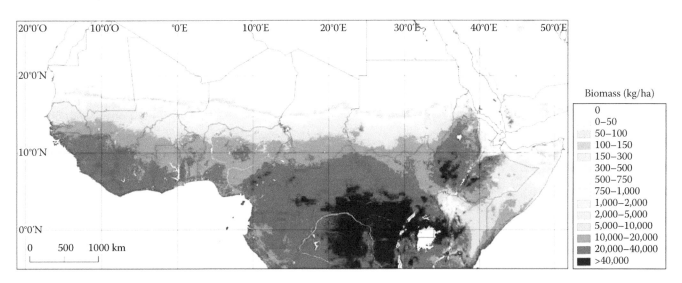

FIGURE 14.4 Average biomass annual productivity over the sub-Saharan area (1999–2013).

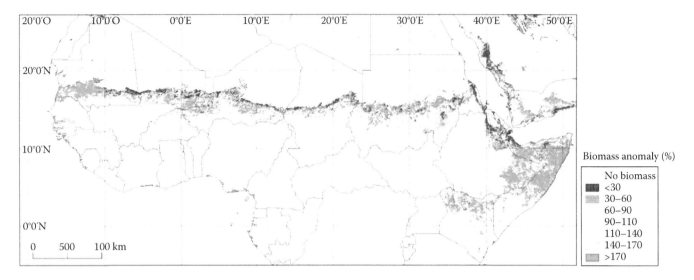

FIGURE 14.5 Biomass annual productivity anomaly for year 2013 in comparison with average (1999–2013).

information, it would be possible to make a realistic assessment of water availability and infer a differential accessibility of pastures. Nevertheless, the underground water inventory is not yet completed across the Sahel countries and therefore the current product does not strictly reflect the actual situation of water availability. The current version of the product also does not account for security issues, social habits, livestock movements, or even financial constraints herders might face to access the water points.

As an example, Figure 14.6 shows the position and occurrence of surface water expressed in percentage of the normal dekadal occurrence calculated over the 1999–2013 period and the position of boreholes and artificial water points in the North Mali

(Source: Direction Nationale Hydraulique Mali). Using this information it is possible to compute the accessible biomass over the area for each year as well as the inaccessible biomass defined as the difference between the total and the accessible biomasses (data not shown). This quantification of inaccessible or *lost* biomass could also support the expression of needed boreholes or water points in a way to optimize the biomass availability.

14.2.3.3 Integration with Livestock Movements

To estimate the evolution of the pastoral situation and assess the breeders' spatial strategy up to the next rainy season, it is necessary to get consistent knowledge of livestock movements.

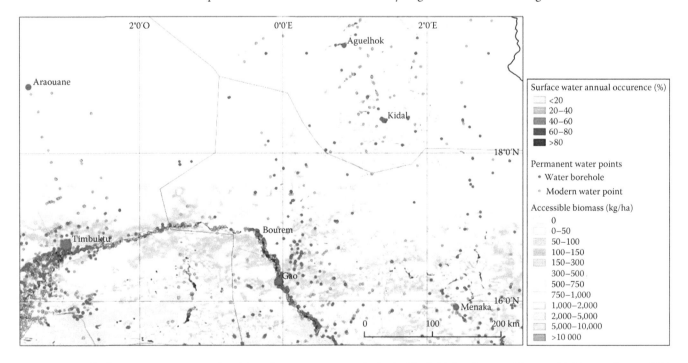

FIGURE 14.6 Surface water occurrence and position of permanent water point in the North Mali and areas with accessible biomass for year 2013.

By combining pastoral resources monitoring data with grazing areas and projected movements, it would be possible to draw a differential scenario in order to target the most vulnerable areas and identify the plausible zones of tension and/or conflict for the access to the resources. This module is currently the least developed even if it is a crucial one. Research activities for the implementation of the module are currently focused on delineating *grazing shed* (i.e., the space used by a given herders group during a full pastoral cycle) and subzones used by animals during different periods of time, estimating the pastoral importance of each zone and calculating their respective load capacity, identifying and mapping transhumance routes in favorable and critical periods, and identifying changing movement patterns (difference in precocity or direction) as early warning indicators.

All these indicators will be part of the pastoral vulnerability model under construction. While the first module of the model is quite deeply used and diffused (pasture monitoring), the water and the movement modules still need further research and require both technical developments and field data collection.

14.2.4 JRC Approach for the Early Detection of Biomass Production–Deficit Hot Spots

Timely information on vegetation growth at regional scale are needed in arid and semiarid regions where rainfall variability leads to high interannual fluctuations in crop and pasture productivity and to high risk of food crisis in the presence of severe drought events (FAO, 2011). Monitoring systems should provide information on impending risks to national and international food security stakeholders as early as possible during the growing season. A regular update throughout the season is necessary to allow effective decision making.

As mentioned in Section 14.2.1, current RS monitoring methods are based on the computation of the anomaly of a vegetation status indicator (typically NDVI) with respect to a reference value (e.g., the historical mean). The interpretation of such anomalies is not straightforward as the comparison is made at a fixed time of the year regardless of the actual plant development stage. Neglecting the actual development stage of the crop leads to a nonstandardized spatial information. This is because anomalies at different locations may refer to different stages of development. For example, in one country the growing cycle could be nearly completed (high reliability of the information provided), while in another country it has just started (low reliability).

In addition, the traditional approach captures only a single-time snapshot of vegetation status but misses an overall view of the entire seasonal development. To overcome such potential problems, a probabilistic approach has been recently proposed to estimate the probability of experiencing an end-of-season critical biomass production deficit during the ongoing growing season on the basis of the statistical analysis of long-term time series (from year 1998 to today) of moderate-resolution SPOT-VEGETATION FAPAR observations (Meroni et al., 2014). The cumulative value of FAPAR during the growing season (CFAPAR) is used as a proxy of vegetation gross primary production (e.g., Fensholt et al., 2006; Jung et al., 2008; Dardel et al., 2014) and of crop yield (e.g., Lobell et al., 2003; Funk and Budde, 2009; Meroni et al., 2013).

The method is applicable at the regional to continental scale and can be updated regularly during the season to provide a synoptic view of the hot spots of likely production deficit. The specific objective of the procedure is to deliver to the food security analyst, as early as possible during the season, only the relevant information (e.g., masking out areas without active vegetation at the time of analysis), expressed through a reliable and easily interpretable measure of impending risk.

Within-season forecasts of the final biomass production, expressed in terms of probability of experiencing a critical deficit, are based on a statistical approach taking into account the similarity between the current CFAPAR profile and past profiles observed in the time series and the uncertainty of past predictions of seasonal outcome (derived using jackknifing technique). Processing is pixel based and proceeds in five main steps: (1) retrieval of key phenological parameters (start and end of growing season), (2) computation of historical CFAPAR values and definition of a critical deficit based on the historical distribution, (3) definition of a metric estimating the likelihood of deficit occurrence, (4) assessment of its uncertainty in detecting the deficit occurrence, and (5) estimation of a formal deficit probability. Details about the processing algorithm can be found in Meroni et al. (2014).

The procedure is applied in near-real time whenever a new satellite observation becomes available to compute the probability of ending with a critical deficit. A critical production deficit is defined as the occurrence of a seasonal CFAPAR value below a certain threshold, conveniently set in relative terms as the *i*th percentile of the distribution of the observed CFAPAR seasonal values. The method is thus aimed at detecting extremes that occur less than or equal to *i*% of the time. In addition, one can interpret the percentile in terms of return period (inverse of the frequency of occurrence) of an event of that magnitude or lower (e.g., the return period of a final CFAPAR falling in the first quartile is 4 years). In this way, the threshold value in CFAPAR units changes pixel by pixel as it is the ranking with respect to the historical distribution that defines it.

As an example, Figure 14.7 shows the probability of deficit for the Sahel region as derived from the satellite observations available as of the first of September 2009, halfway through the season, a timing that can be considered appropriate for early warning analysis.

During 2009, poor rains were reported for a large fraction of the region shown in Figure 14.7. This is well captured by the estimated probability of deficit. For example, Figure 14.7 shows a widespread presence of hot spots of high deficit probability ranging from Mali, Niger, north Nigeria, Chad, and South Sudan. At the end of the 2009 season, significant reductions in

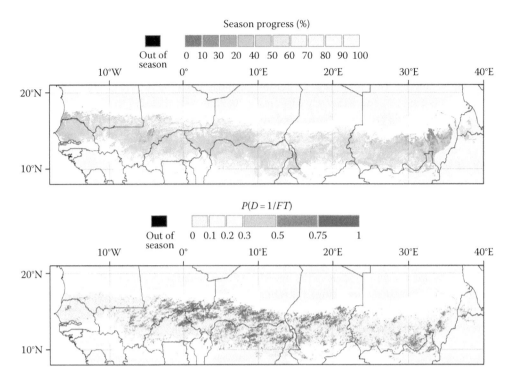

FIGURE 14.7 Progress of the season (upper panel) and probability of deficit (lower panel) as of September 1, 2009. Deficit threshold set to the first quartile of seasonal CFAPAR distribution. The analysis is limited to the herbaceous and cropland land covers (other classes in white) and the five main ecoregions of the Sahel (other regions in gray).

grain harvest and pasture biomass production were reported in the Sahel, particularly in eastern Mali, Niger, Chad, northern Nigeria, and Burkina Faso (Fewsnet, 2009, 2010).

The method has been designed to timely provide an easily interpretable information: the probability of experiencing a critical deficit, a pragmatic indicator that can be easily understood despite the relative complexity involved in estimating it. Differently from standard anomaly products, the procedure maps the probability only where a growing season is actually ongoing, so that the analyst can focus on the relevant information only. Finally, the delivered information at a given time is already *weighted* for its reliability, as assessed using past estimations performed at that specific time.

In NRT operations, in the case a problem may be emerging, the method would allow identifying its geographic dimensions, and identifying where and how quickly it is developing. The appearance of deficit hot spots may also guide the decision on where to concentrate in-depth field assessments. In addition, starting from the probability map, country- or district-level key summary statistics can be extracted in tabular form, as, for example, the fraction of the cropland or pasture area exposed to different levels of deficit risk (e.g., the fraction with a risk greater than the business-as-usual scenario, or greater than a selected threshold). This would allow assigning, for example, a risk score to each administrative unit taking into account the magnitude of the deficit probability and, at the same time, its spatial extent.

14.3 Other Recent Methodological Approaches

In this section we highlight a nonexhaustive number of other and complementary recent advances made in drought monitoring using RS-based biophysical parameters.

A first approach (Tadesse et al., 2005; Brown et al., 2008, Wu et al., 2013) makes use of data mining techniques to establish a relationship between a variety of inputs (including RS-derived variables) and observed historical droughts. The novelty of this approach is that the link between observed variable of different nature (RS, meteorological, and ancillary) and the occurrence of a drought is established using a multilevel decision tree (classification and regression tree, CART). This allows to classify and stratify the input variables before applying a linear regression scheme defining the proposed drought index. As a result, the index is defined by a set of rules that control the application of a specific regression involving one or more input variables. The regression tree nicely accommodates the fact that one input variable, for instance, the surface temperature, may contribute to explain the occurrence of a drought only when other conditions apply, for instance, if the precipitation was below a certain threshold. When such conditions do not apply, other variables may be picked up by the model. Rules and regression parameters are established in an iterative and automatic process that is finally cross-validated to ensure that the highly parameterized final model can provide the required predictive power. A drawback of the method may stand in the fact that it is not easy to interpret and gain physical insight from the generated rules.

A second approach developed by Sepulcre-Canto et al. (2012) at the European Drought Observatory (http://edo.jrc.ec.europa.eu) differs from the ones described so far as it is aimed to a qualitative classification of drought severity based on different data sources (including RS-based FAPAR) and a kind of "convergence of evidences" approach. The authors proposed a Combined Drought Indicator (CDI) built on an idealized cause–effect relationship for agricultural drought: a shortage of precipitation (formalized by the 3-month Standardized Precipitation Index, SPI-3) leads to a soil moisture deficit (estimated through the LISFLOOD hydrological precipitation–runoff model) that in turn results in a reduction of vegetation productivity (monitored with FAPAR anomalies). By setting thresholds indicating extreme values for the three indicators and after having empirically defined a time lag between the three events, they combine the three time-lagged indicators (anomaly of FAPAR at dekad n, anomaly of soil moisture at n-1, and SPI-3 computed over the interval [n-9, n]) and classify such triplets into four CDI categories: *watch* when there is a precipitation deficit (SPI-3 is below the threshold), *warning* when there is also a soil moisture deficit (both SPI-3 and anomaly of soil moisture are below their thresholds), "alert 1" when there is vegetation stress following a precipitation deficit (both SPI-3 and anomaly of FAPAR are below their thresholds), and finally "alert 2" when vegetation stress follows a precipitation and soil moisture deficit (all indicators below their thresholds). This approach was applied in Europe to assess its reliability in spotting recent historical droughts and it is currently being used operationally by the European Drought Observatory.

Another inspiring example of possible approach to drought detection is the work of Zscheischler et al. (2013). In this study they explore how to detect and quantify extreme events as spatiotemporal phenomena spanning over both the spatial and temporal dimensions. Working with a 30-year FAPAR time series constructed with different satellite sensors, they first define an extreme event in terms of FAPAR anomalies and then compute the contiguous volume occupied by the extreme event in the 3D space made up by latitude, longitude, and time. In this way, the severity of a drought is determined by its spatial magnitude and its persistency over time.

14.4 Discussion and Future Developments

Concerning coverage and quality in RS data, the continuous trend in increasing spatial and temporal resolution of new satellite sensors (e.g., the recently launched PROBA-V and Sentinel 2 satellites and the forthcoming Sentinel 3 satellite of the European Space Agency, ESA) is expected to allow improvements of the RS methods described. However, the availability of a long-term time series will still remain a relevant issue as it

was shown to be a precondition in many of the drought-monitoring methods described. In fact, in the absence of reliable ground measurements of crop yield and rangeland production in most of the food-insecure regions of the world, most monitoring systems are based on the use of long-time series for comparison with previous years or with the average situation. Such approaches will not be able to profit immediately from the upcoming availability of higher spatial resolution sensors. This means that even when the next generation of Earth observing satellites with smaller spatial ground sampling distance will be launched (e.g., Sentinel-2), a number of years will pass until the benefits of the increased spatial resolution will deploy their full impact on improving the quality of drought-monitoring products. This time lapse could only be reduced by unmixing the signal recorded by sensors of different resolution at administrative level to a common *end member* signature. Increased research efforts on sensor intercalibration and on methods for exploiting the current long-term archive for supporting the analysis of higher spatial resolution instruments are needed to simplify access to long-time series of remotely sensed data from different sensors.

Ideally, one would have a suitable land process model at hand, describing the main processes with sufficient detail, and assimilate the remotely sensed observations to constrain the simulation outcomes. The climate community is actively working on such models (mainly for the purpose of improving weather forecasts and climate predictions), but nowadays such dynamic models are still too data demanding (e.g., relating to soil and vegetation properties) for being useful in operational programs.

Another critical issue in the use of RS time series that deserves further consideration refers to the trade-off between timeliness and noise reduction. Typically, noise reduction is obtained using temporal smoothing of the time series. Effective smoothing is in turn achieved when temporal observations before and after the data point being smoothed are available. As a result one has to choose between the exploitation of less reliable but more recent observation and the more reliable but not updated observations. An option to overcome this impasse may be represented by the use of a more or less simple vegetation growth model and data assimilation techniques (for a review see Dorigo et al., 2007) for ingesting RS data. In such a way, the increasing uncertainty of the most recent observations may be taken into account.

Finally, it is worth mentioning that the tremendous advances made in vegetation stress detection using RS techniques, developed mainly for precision farming applications using high spectral and spatial resolution data, have not been translated yet into operational and regional drought monitoring. Employed techniques span from the use of narrowband VIs to detected leaf pigment concentration (e.g., Haboudane et al., 2002; Rossini et al., 2013) and leaf water content (Colombo et al., 2011 for a recent review), to the use of advanced optical indices such as the Photochemical Reflectance Index (PRI), chlorophyll fluorescence, and thermal imagery (e.g., Meroni et al., 2009; Suárez et al., 2010, Zarco-Tejada et al., 2012) to characterize leaf physiology and detect vegetation stress in its early stages. There are many reasons for such a gap: technical constrains related to the unavailability of the required spectral and spatial resolution observations at the regional to global scale, relative novelty of such approaches as compared to more traditional ones based on well-known VIs such as NDVI, complex effects of the 3D canopy structure and illumination and view angles on the advanced indices, difficulties in applying empirical methods tuned at the landscape scale to the regional scale, and lack of calibration/validation data in drought-prone countries (e.g., Africa), among others. However, successful examples of application in carbon flux estimation over different biomes (see recent reviews and a discussions see Grace, 2007; Garbulsky et al., 2011; Penuelas et al., 2011) offer promising prospect for enhanced capability in regional to global drought monitoring, for instance, through the use of the PRI available globally and with the required temporal frequency from MODIS instruments Terra and Aqua.

14.5 Conclusions

The operational agricultural drought-monitoring systems using space observations and the recent methodological developments show that satellite data play a key role in drought surveillance and early warning at global level. The examples of monitoring systems described in this chapter also show that the information derived from RS is directly used for drought management approaches that aim at designing response actions improving food and water availability in dry areas. At the same time, monitoring and early warning systems are also becoming important components of more comprehensive (drought) risk management systems. These include crop and livestock insurance schemes that can be based on remotely sensed indices and national risk management systems using RS-based drought information for early design of contingency plans for response actions.

It is also important to remember that RS-based methods observe primarily the physical signals of drought, while decision makers and the general public are usually more interested in drought impact on local communities, which depend on many other factors such as the vulnerability of the population and the economic and political situation at national level. Therefore it is important to combine RS information with other sources of data, including socioeconomic aspects addressing coping and adaptation strategies, to strengthen community's resilience while securing the production systems that are most exposed to uncertainty.

References

Alessandrini, S. and Evangelisti, M. 2011. FAO RFE the FAO African rainfall estimate. *Workshop: Rainfall Estimates for Crop Monitoring and Food Security*, October 22–23, 2008, Ispra (VA), Italy. http://geonetwork3.fao.org/climpag/FAO-RFE-ISPRA2008.pdf. (Accessed April 27, 2015.)

Atzberger, C. and Eilers, P.H.C. 2011. Evaluating the effectiveness of smoothing algorithms in the absence of ground reference measurements. *International Journal of Remote Sensing*, 32, 13, 3689–3709.

Atzberger, C., Formaggio, A.R., Shimabukuro, Y.E., Udelhoven, T., Mattiuzzi, M., Sanchez, G.A, and Arai, E. 2014. Obtaining crop-specific time profiles of NDVI: The use of unmixing approaches for serving the continuity between SPOT-VGT and PROBA-V time series. *International Journal of Remote Sensing*, 35, 7, 2615–2638.

Balint, Z., Mutua, F.M., and Muchiri, P. 2011. *Drought Monitoring with the Combined Drought Index*. FAO-SWALIM, Nairobi, Kenya, 32pp.

Baret, F., Weiss, M., Lacaze, R., Camacho, F., Makhmara, H., Pacholcyzk, P., and Smets, B. 2013. GEOV1: LAI and FAPAR essential climate variables and FCOVER global time series capitalizing over existing products. Part 1: Principles of development and production. *Remote Sensing of Environment*, 137, 299–309.

Bartholomé, E. and Belward, A.S. 2005. GLC2000: A new approach to global land cover mapping from earth observation data. *International Journal of Remote Sensing*, 26, 9, 1959–1977.

Boyd, E., Cornforth, R.J., Lamb, P.J., Tarhule, A., Lélé, M.I., and Brouder, A. 2013. Building resilience to face recurring environmental crisis in African Sahel. *Nature Climate Change*, 3, 631–637.

Brown, J., Wardlow, B., Tadesse, T., Hayes, M., and Reed, B. 2008. The Vegetation Drought Response Index (VegDRI): A new integrated approach for monitoring drought stress in vegetation. *GIScience Remote Sensing*, 45, 1, 16–46.

Brown, M.E. 2008. Famine early warning systems and remote sensing data. Springer Verlag, Heidelberg, Germany.

Busetto, L., Meroni, M., and Colombo, R. 2008. Combining medium and coarse spatial resolution satellite data to improve the estimation of sub-pixel NDVI time series. *Remote Sensing of Environment*, 112, 118–131.

Checchi, F. and Courtland Robinson, W. 2013. *Mortality Among Populations of Southern and Central Somalia Affected by Severe Food Insecurity and Famine During 2010–2012*. FAO/FSNAU; Fewsnet, Rome, Italy; Washington, DC. www.fsnau.org/downloads/Somalia_Mortality_Estimates_Final_Report_8May2013_upload.pdf. (Accessed April 27, 2015.)

Chen, J., Jonsson, P., Tamura, M., Gu, Z., Matsushita, B., and Eklundh, L. 2004. A simple method for reconstructing a high-quality NDVI time-series data set based on the Savitzky–Golay filter. *Remote Sensing of Environment*, 91, 332–344.

Chuvieco, E., Ventura, G., Pilar Martin, P., and Gomez, I. 2005. Assessment of multitemporal compositing techniques of MODIS and AVHRR images for burned land mapping. *Remote Sensing of Environment*, 94, 450–462.

Colombo, R., Meroni, M., Busetto, L., Rossini, M., and Panigada, C. 2011. Optical remote sensing of vegetation water content. In: Thenkabail, P.S., Lyon, J.G., and Huete, A. (Eds.), *Hyperspectral Remote Sensing of Vegetation*. CRC Press, Taylor & Francis Group, Boca Raton, FL, pp. 227–244.

Dardel, C., Kergoat, L., Hiernaux, P., Mougine, E., Grippa, M., and Tucker, C.J. 2014. Re-greening of Sahel: 30 year of remote sensing data and field observations (Mali, Niger). *Remote Sensing of Environment*, 140, 350–364.

Ding, Y., Hayes, M., and Widhalm, M. 2011. Measuring economic impact of drought: A review and discussion. *Disaster Prevention and Management*, 20, 4, 434–446.

Dorigo, W.A., Zurita-Milla, R., de Wit, A.J.W., Brazile, J., Singh, R., and Schaepman, M.E. 2007. A review on reflective remote sensing and data assimilation techniques for enhanced agroecosystem modelling. *International Journal of Applied Earth Observation and Geoinformation*, 9, 2, 165–193.

ECMWF. 2013. User guide to ECMWF forecast products. Technical report, 121pp. http://old.ecmwf.int/products/forecasts/guide/user_guide.pdf. (Accessed April 27, 2015.)

Eerens, H., Piccard, I., Royer, A., and Orlandi, S. 2004. *Methodology of the MARS Crop Yield Forecasting System*. Remote Sensing Information, Data Processing and Analysis. Vol. 3. Joint Research Centre European Commission, Ispra, Italy, 76pp.

FAO, Word Bank, and UNSC. 2011. Global strategy to improve agricultural and rural statistics. Report No. 56719-GLB, World Bank, Washington, DC, 40pp.

FAO and WFP. 2013. Sécurité alimentaire et implications humanitaires en Afrique de l'ouest et au Sahel, N°51—Novembre/Décembre 2013, 7pp. http://reliefweb.int/sites/reliefweb.int/files/resources/Note%20conjointe%20FAO%20PAM%20%20N%2051%20Novembre%202013.pdf. (Accessed April 27, 2015.)

Fensholt, R., Anyamba, A., Huber Gharib, S., Proud, S.R., Tucker, C., Small, J., Pak, E., Rasmussen, M.O., Sandholt, I., and Shisanya, C. 2011. Analysing the advantages of high temporal resolution geostationary MSG SEVIRI data compared to Polar operational environmental satellite data for land surface monitoring in Africa. *International Journal of Applied Earth Observation and Geoinformation*, 13, 5, 721–729.

Fensholt, R., Sandholt, I., Rasmussen, M.S., Stisen, S., and Diouf, A. 2006. Evaluation of satellite based primary production modelling in the semi-arid Sahel. *Remote Sensing of Environment*, 105, 173–188.

Fewsnet. 2009. Sahel and West Africa, food security update, November 2009. Technical report, 4pp. http://www.fews.net/sites/default/files/documents/reports/West_Africa_FSU_2009_11_en.pdf. (Accessed April 27, 2015.)

Fewsnet. 2010. Sahel and West Africa, food security update, February 2010. Technical report, 7pp. http://www.fews.net/sites/default/files/documents/reports/West_FSU_2010_02_en.pdf.

Fillol, E. 2011a. BioGenerator (v2.1), Guide de l'utilisateur. Action Contre la Faim International, 12pp. http://www.accioncontraelhambre.org/publicaciones_biblioteca.php?sec=4#4, under Section "Manuales y Guias". (Accessed April 27, 2015.)

Fillol, E. 2011b. HydroGenerator (v1.1), Guide de l'utilisateur. Action Contre la Faim International, 15pp. http://www.accioncontraelhambre.org/publicaciones_biblioteca.php?sec=4#4, under Section "Manuales y Guias".

Funk, C.C. and Budde, M.E. 2009. Phenologically-tuned MODIS NDVI-based production anomaly for Zimbabwe. *Remote Sensing of Environment*, 113, 115–125.

Garbulsky, M.F., Penuelas, J., Gamon, J., Inoue, Y., and Filella, I. 2011. The photochemical reflectance index (PRI) and the remote sensing of leaf, canopy, and ecosystem radiation use efficiencies. *Remote Sensing of Environment*, 115, 281–297.

Genovese, G., Vignolles, C., Nègre, T., and Passera, G. 2001. A methodology for a combined use of normalised difference vegetation index and CORINE land cover data for crop yield monitoring and forecasting. A case study on Spain. *Agronomie*, 21, 91–111.

Gibbs, W.J. and Maher, J.V. 1967. Rainfall deciles as drought indicators. Bureau of Meteorology Bulletin, No. 48, Commonwealth of Australia, Melbourne, Victoria, Australia.

Gobron, N., Pinty, B., Taberner, M., Mélin, F., Verstraete, M., and Widlowski, J.-L. 2005. Monitoring the photosynthetic activity of vegetation from remote sensing data. *Advances in Space Research*, 38, 2196–2202.

Gobron, N., Pinty, B., and Verstraete, M.M. 1997. Theoretical limits to the estimation of the leaf area index on the basis of visible and near-infrared remote sensing data. *IEEE Transaction on Geoscience and Remote Sensing*, 35, 1438–1445.

Grace, J., Nichol, C., Disney, M., Lewis, P., Quiafe, T., and Bowyer, P. 2007. Can we measure terrestrial photosynthesis from space, using spectral reflectance and fluorescence? *Global Change Biology*, 13, 1484–1497.

Grimes, D.I.F. 2003. Satellite-based rainfall monitoring for food security in Africa. In: Rijks, D., Rembold, F., Negre, T, Gommes, and R., Cherlet, M. (Eds.), *"Crop and Rangeland Monitoring in Eastern Africa—For Early Warning and Food Security," Proceedings of the International Workshop on Crop and Rangeland Monitoring in East Africa*, Nairobi, Kenya, January 2003, European Commission (pub).

Haas, E.M., Bartholome, E., and Combal, B. 2009. Time series analysis of optical remote sensing data for the mapping of temporary surface water bodies in sub-Saharan western Africa. *Journal of Hydrology*, 370, 1–4, 52–63.

Haboudane, D., Miller, J.R., Tremblay, N., Zarco-Tejada, P.J., and Dextraze, L. 2002. Integrated narrow-band vegetation indices for prediction of crop chlorophyll content for application to precision agriculture. *Remote Sensing of Environment*, 81, 416–426.

Ham, F., Métais, T., Hoorelbeke, P., Fillol, E., Gómez, A., and Crahay, P. 2011. One horn of the cow, an innovative GIS-based surveillance and early warning system in pastoral areas of Sahel. In: *Risk Returns*. Rose, T. (Ed.). UNISDR, Geneva, pp. 127–131.

Hillbrunner, C. and Moloney, G. 2012. When early warning is not enough—Lessons learnt from the 2011 Somalia Famine. *Global Food Security*, 1, 1, 20–28.

Holben, B. 1986. Characteristics of maximum-value composite images from temporal AVHRR data. *International Journal of Remote Sensing*, 7, 11, 1417–1434.

Hutchinson, C.F. 1991. Uses of satellite data for famine early warning in sub-Saharan Africa. *International Journal of Remote Sensing*, 12, 1405–1421.

Ickowicz, A., Ancey, V., Corniaux, C., Duteurtre, G., Poccard-Chappuis, R., Touré, I., Vall, E., and Wane, A. 2012. Crop–livestock production systems in the Sahel—Increasing resilience for adaptation to climate change and preserving food security. In: *Proceedings of the Joint FAO/OECD Workshop on Building Resilience for Adaptation to Climate Change in the Agriculture Sector*, Rome, Italy, April 23–25, 2012, pp. 261–294. http://www.fao.org/docrep/017/i3084e/i3084e.pdf. (Accessed April 27, 2015.)

Intergovernmental Panel on Climate Change. 2013. Climate change 2013: The physical science basis. Contribution of Working Group I to the Fifth Assessment Report of the Intergovernmental Panel on Climate Change. Stocker, T.F., Qin, D., Plattner, G.-K., Tignor, M., Allen, S.K., Boschung, J., Nauels, A., Xia, Y., Bex, V., and Midgley, P.M. (Eds.), Cambridge University Press, Cambridge, U.K., 1535pp.

Jung, M., Verstraete, M., Gobron, N., Reichstein, M., Papale, D., Bondeau, A., Robustelli, M., and Pinty, B. 2008. Diagnostic assessment of European gross primary production. *Global Change Biology*, 14, 10, 2349–2364.

Kogan, F.N. 1995. Droughts of the late 1980s in the United States as derived from NOAA polar orbiting satellite data. *Bulletin of the American Meteorological Society*, 76, 655–668.

Kogan, F.N. 2001. Operational space technology for global vegetation assessment. *Bulletin of the American Meteorological Society*, 89, 1949–1964.

Lobell, D.B., Asner, G.P., Ortiz-Monasterio, J.I., and Benning, T.L. 2003. Remote sensing of regional crop production in the Yaqui Valley, Mexico: Estimates and uncertainties. *Agriculture, Ecosystems and Environment*, 94, 205–220.

Maxwell, D. and Fitzpatrick, M. 2012. The 2011 Somalia famine: Context, causes and complications. *Global Food Security*, 1, 1, 5–12.

Meroni, M., Fasbender, D., Kayitakire, F., Pini, G., Rembold, U., and Verstraete, M.M. 2014. Early detection of biomass production deficit hot-spots in semi-arid environment using FAPAR time series and a probabilistic approach. *Remote Sensing of Environment*, 142, 57–68.

Meroni, M., Rossini, M., Guanter, L., Alonso, L., Rascher, U., Colombo, R., and Moreno, J. 2009. Remote sensing of solar-induced chlorophyll fluorescence: Review of methods and applications. *Remote Sensing of Environment*, 113, 2037–2051.

Meroni, M., Verstraete, M., Marinho, M., Sghaier, N., and Leo, O. 2013. Remote sensing based yield estimation in a stochastic framework—Case study of Tunisia. *Remote Sensing*, 5, 2,539–557.

Mishra, A.K. and Singh, V.P. 2010. A review of drought concepts. *Journal of Hydrology*, 391, 202–216.

Monteith, J.L. 1972. Solar radiation and productivity in tropical ecosystems, *Journal of Applied Ecology*, 9, 747–766.

Myneni, R.B., Hoffman, S., Knyazikhin, Y., Privette, J.L., Glassy, J., Tian, Y., Wang, Y. et al. 2002. Global products of vegetation leaf area and fraction absorbed PAR from year one of MODIS data. *Remote Sensing of Environment*, 83, 214–231.

Nghiem, S.V., Wardlow, B.D., Allurer, D., Svoboda, M.D., LeComte, D., Rosencrans, M., Chan, S.K., and Neumann, G. 2012. Microwave remote sensing of soil moisture. In: Wralow, B.D., Anderson, M.C., and Verdin, J.P. (Eds.), *Remote Sensing of Drought: Innovative Monitoring Approaches*. CRC Press, Boca Raton, FL, pp. 179–226.

NOAA. 2001. The NOAA climate prediction center African rainfall estimation algorithm, Version 2.0. Technical report, 4pp. http://www.cpc.ncep.noaa.gov/products/fews/RFE2.0_tech.pdf. (Accessed April 27, 2015.)

Palmer, W.C. 1965. Meteorological drought. Research Paper No. 45, U.S. Department of Commerce Weather Bureau, Washington, DC.

Penuelas, J., Garbulsky, M., and Filella, I. 2011. Photochemical reflectance index (PRI) and remote sensing of plant CO_2 uptake. *New Phytologist*, 191, 596–599.

Pinty, B., Lavergne, T., Widlowsky, J.L., Gobron, N., and Verstraete, M.M. 2009. On the need to observe vegetation canopies in the near-infrared to estimate visible light absorption. *Remote Sensing of Environment*, 113, 10–23.

Prince, S.D. and Goward, S.N. 1995. Global primary production: A remote sensing approach. *Journal of Biogeography*, 22, 815–835.

Rembold, F., Atzberger, C., Savin, I., and Rojas, O. 2013. Using low resolution satellite imagery for yield prediction and yield anomaly detection. *Remote Sensing*, 5, 4, 1704–1733.

Rojas, O., Vrieling, A., and Rembold, F. 2011. Assessing drought probability for agricultural areas in Africa with coarse resolution remote sensing imagery. *Remote Sensing of Environment*, 115, 343–352.

Rossini, M., Fava, F., Cogliati, S., Meroni, M., Marchesi, A., Panigada, C., Giardino, C. et al. 2013. Assessing canopy PRI from airborne imagery to map water stress in maize. *ISPRS Journal of Photogrammetry and Remote Sensing*, 86, 168–177.

Rouse, J.W., Haas, R.H., Schell, J.A., Deering, D.W., and Harlan, J.C. 1974. Monitoring the vernal advancement of retrogradation of natural vegetation. Final Report, Type III, NASA/GSFC, Greenbelt, MD, 371pp.

Sepulcre-Canto, G., Horion, S., Singleton, A., Carrao, H., and Vogt, J. 2012. Development of a Combined Drought Indicator to detect agricultural drought in Europe, *Natural Hazards and Earth System Sciences*, 12, 3519–3531.

Shafer, B.A. and Dezman, L.E. 1982. Development of a Surface Water Supply Index (SWSI) to assess the severity of drought conditions in snowpack runoff areas. In: *Proceedings of the Western Snow Conference*, pp. 164–175. http://www.westernsnowconference.org/sites/westernsnowconference.org/PDFs/1982Shafer.pdf. (Accessed April 27, 2015.)

Sheffield, J. and Wood, E.F. 2011. *Drought: Past Problems and Future Scenarios*. Earthscan, London, U.K., 192pp.

Suárez, L., Zarco-Tejada, P.J., Gonzalez-Dugo, V., Berni, J.A.J., Sagardoy, R., Morales, F., and Fereres, E. 2010. Detecting water stress effects on fruit quality in orchards with time-series PRI airborne imagery. *Remote Sensing of Environment*, 114, 286–298.

Tadesse, T., Brown, J.F., and Hayes, M.J. 2005. A new approach for predicting drought-related vegetation stress: Integrating satellite, climate, and biophysical data over the U.S. central plains. *ISPRS Journal of Photogrammetry and Remote Sensing*, 59, 4, 244–253.

United Nations. 2009. *Global Assessment Report on Disaster Risk Reduction*. United Nations, Geneva, Switzerland. http://www.preventionweb.net/english/hyogo/gar/report/index.php?id = 9413. (Accessed April 27, 2015.)

Wardlow, B.D., Anderson, M.C., and Verdin, J.P. 2012. Remote sensing of drought: Innovative monitoring approaches. CRC Press, Boca Raton, FL.

Wilhite, D.A., Svoboda, M.D., and Hayes, M.J. 2007. Understanding the complex impacts of drought: A key to enhancing drought mitigation and preparedness. *Water Resources Management*, 21, 763–774.

WMO. 2012. Standardized precipitation index user guide. Technical document no. 1090, Geneva, Switzerland. http://www.wamis.org/agm/pubs/SPI/WMO_1090_EN.pdf. (Accessed April 27, 2015.)

Wu, H., Svodoba, M.D., Hayes, M.J., Wilhite, D.A., and Wen, F. 2007. Appropriate application of the standardized precipitation index in arid locations and dry seasons, *International Journal of Climatology*, 27, 65–79.

Wu, J., Zhou, L., Liu, M., Leng, S., and Diao, C. 2013. Establishing and assessing the Integrated Surface Drought Index (ISDI) for agricultural drought monitoring in mid-eastern China. *International Journal of Applied Earth Observation and Geoinformation*, 23, 397–410.

Zarco-Tejada, P.J., Gonzalez-Dugo, V., and Berni, J.A.J. 2012. Fluorescence, temperature and narrow-band indices acquired from a UAV platform for water stress detection using a micro-hyperspectral imager and a thermal camera. *Remote Sensing of Environment*, 117, 322–337.

Zhong, L., Ostrenga, D., Teng, W., and Kempler, S. 2012. Tropical rainfall measuring mission (TRMM) precipitation data and services for research and applications. *Bulletin of the American Meteorological Society*, 93, 1317–1325.

Zscheischler, J., Mahecha, M.D., Harmeling, S., and Reichstein, M. 2013. Detection and attribution of large spatiotemporal extreme events in Earth observation data. *Ecological Informatics*, 15, 66–73.

15

Remote Sensing of Drought: Emergence of a Satellite-Based Monitoring Toolkit for the United States

Brian Wardlow
University of Nebraska-Lincoln

Martha Anderson
United States Department of Agriculture

Tsegaye Tadesse
University of Nebraska-Lincoln

Chris Hain
University of Maryland

Wade T. Crow
United States Department of Agriculture

Matt Rodell
Goddard Space Flight Center

Acronyms and Definitions

AAU	Addis Ababa University
ACIS	Applied Climate Information System
ALEXI	Atmosphere-Land Exchange Inverse
AMO	Atlantic Multidecadal Oscillation Index
AMSR-E	Advanced Microwave Scanning Radiometer Earth Observing System (EOS)
ASCAT	Advanced Scatterometer
AVHRR	Advanced Very High Resolution Radiometer
CART	Classification and regression tree
CDF	Cumulative distribution function
CLSM	Catchment land surface model
CMI	Crop Moisture Index
CMORPH	Climate Prediction Center (CPC) MORPHing
CONUS	Continental United States
DEM	Digital elevation model
DisALEXI	Disaggregated Atmosphere-Land Exchange Inverse
EnKF	Ensemble Kalman filter
ENSO	El Niño and Southern Oscillation
EOS	Earth Observing System
ES	End of Season
EROS	Earth Resources Observation Science
ERS	European Remote Sensing
ESA	European Space Agency
ESI	Evaporative Stress Index
ET	Evapotranspiration
FEWS	Famine Early Warning System
FSA	Farm Service Agency
GOES	Geostationary Operational Environmental Satellites
GRACE	Gravity Recovery and Climate Experiment
HPRCC	High Plains Regional Climate Center
LSM	Land surface model
LST	Land surface temperature
LULC	Land use/land cover
MEI	Multivariate ENSO Index
MERRA	Modern-Era Retrospective Analysis for Research and Applications
MetOp	Meteorological Operational

MIR	Middle infrared
MIrAD-US	MODIS Irrigated Agriculture Dataset for the United States
MJO	Madden–Julian oscillation
MODIS	Moderate Resolution Imaging Spectroradiometer
MW	Microwave
NADM	North American Drought Monitor
NAO	North Atlantic Oscillation
NARR	North American Regional Reanalysis
NDDI	Normalized Difference Drought Index
NDMC	National Drought Mitigation Center
NDVI	Normalized difference vegetation index
NDWI	Normalized difference water index
NIR	Near infrared
NLCD	National Land Cover Dataset
NLDAS	North American Land Data Assimilation System
NMDI	Normalized multiband drought index
NOAA	National Oceanic and Atmospheric Administration
OS	Out of season
PASG	Percent Annual Seasonal Greenness
PDO	Pacific Decadal Oscillation
PDSI	Palmer Drought Severity Index
PERSIANN	Precipitation Estimation from Remotely Sensed Information using Artificial Neural Networks
PNA	Pacific North American Index
QuikSCAT	Quick Scatterometer
RFI	Radio-frequency interference
SAR	Synthetic aperture radar
SAWC	Soil available water capacity
SCAN	Soil Climate Analysis Network
SC-PDSI	Self-calibrated Palmer Drought Severity Index
SDNDVI	Standard Deviation of Normalized Difference Vegetation Index
SG	Seasonal greenness
SMAP	Soil Moisture Active Passive
SMOS	Soil Moisture and Ocean Salinity
SOI	Southern Oscillation Index
SOS	Start of season
SOSA	Start of season anomaly
SPI	Standardized Precipitation Index
SSG	Standardized seasonal greenness
SSM/I	Special sensor microwave/imager
SMMR	Special sensor microwave imager sounder
STATSGO	State Soil Geographic
SVI	Standardized Vegetation Index
SWOT	Surface Water and Ocean Topography
SWSI	Surface Water Supply Index
TCI	Temperature Condition Index
TIR	Thermal infrared
TMI	TRMM Microwave Imager
TRMM	Tropical Rainfall Measuring Mission
TSEB	Two-source energy balance
TWS	Terrestrial water storage
UNK	University of Nebraska–Kearney
USDA	U.S. Department of Agriculture

USDM	U.S. Drought Monitor
USGS	U.S. Geological Survey
VegDRI	Vegetation Drought Response Index
VegOut	Vegetation outlook
VIIRS	Visible/Infrared Imager Radiometer Suite
VHI	Vegetation Health Index
VI	Vegetation index
WFAS	Wildland Fire Assessment System

15.1 Introduction

Drought is a naturally recurring climatic feature of most regions of the world that can negatively impact many sectors of society, including agriculture, water resources, energy, ecosystem services, and economic conditions. In the United States, drought has been a common natural hazard, often impacting large regions of the country. Figure 15.1 shows that since the late 1880s, an average of more than 20% of the country has been affected by drought at any given time, including key drought episodes such as the 1930s Dust Bowl and pronounced periods in the 1950s, late 1980s, and 2000s (NCDC, 2013). As a result, drought ranks as one of the most costly natural hazards in the United States, accounting for $6–$8 billion in losses per year (NCDC, 2014). The impact and significance of drought is expected to increase as the climate changes and climatic extremes increase (Dai, 2012), and as demands on finite water supplies continue to increase to support competing sectoral demands (e.g., agricultural, municipal, industrial, and ecological).

15.1.1 Drought Definitions

Monitoring and early warning is a critical component of effective and proactive actions by decision makers to mitigate the impacts of drought. However, drought monitoring is complex and challenging because it lacks a single universal definition, which makes the identification and assessment of key drought characteristics such as duration, intensity (or severity), and geographic extent difficult (Mishra and Singh, 2010). In response, three physically based operational drought definitions have been established to characterize different types of drought (meteorological, agricultural, and hydrologic) (Wilhite and Glantz, 1985). The primary factor distinguishing these categories is the temporal length of dryness needed to initiate and recover from a drought, with the time period increasing in descending order of the three types of drought listed. As a result, a defined period of dryness may result in the occurrence of one type of drought (e.g., meteorological) but not the others, whereas in the instance of longer, more severe periods, several types of drought may be occurring concurrently. Other defining factors of these three types of drought are the specific impacts and appropriate measures to monitor each drought type. For example, agricultural drought is commonly manifested through soil moisture depletion and observed plant stress on crops, whereas hydrologic drought

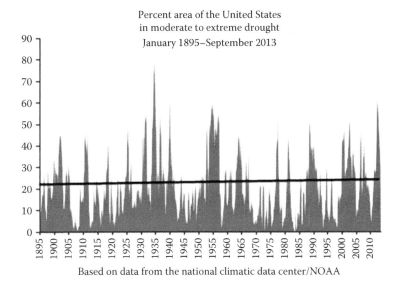

Percent area of the United States
in moderate to extreme drought
January 1895–September 2013

Based on data from the national climatic data center/NOAA

FIGURE 15.1   Percent area of the United States in moderate to severe drought as defined by the Palmer Drought Severity Index (PSDI) for 1985 to 2013 (modified from NDMC, 2014), with the horizontal line demarcating the 20.5% historical average area of the United States under this severity range of drought conditions.

may be best observed through declining stream flows and reservoir and groundwater levels.

15.1.2   Traditional Drought Monitoring Approaches

Traditional drought monitoring has relied heavily upon in situ–based meteorological (i.e., precipitation and temperature) and hydrologic (e.g., lake/reservoir level, stream flow, soil moisture, and groundwater elevation) observations that represent discrete point-based information. Typically, an extended record of historical observations is used to develop *anomaly* measures to detect deviations (i.e., type and magnitude) from some historical average condition for drought detection. The in situ–based observations are commonly spatially interpolated or areally aggregated to an administrative unit (e.g., county or crop reporting district) for decision-making purposes. Figure 15.2a shows an example of the Palmer Drought Severity Index (PDSI; Palmer, 1965) areally aggregated from weather station locations to the National Oceanic and Atmospheric Administration (NOAA) multicounty climate divisions, and Figure 15.2b shows a gridded Standardized Precipitation Index (SPI, McKee et al., 1995) map spatially interpolated from station-based index values across the continental United States (CONUS) in August 2011. The PDSI has been used since the 1960s to characterize agricultural drought and is calculated from precipitation, temperature, and soil available water content data using a supply-and-demand concept of a water balance equation. This index considers past conditions on the scale of 9–12 months, and the intensity of drought for a specific time period and location is determined through estimates of evapotranspiration (ET), soil moisture recharge, runoff, and moisture loss from the soil surface layer through the analysis of the preceding precipitation

and temperature observations in the water balance model. The primary limitation of the PDSI is the inherent lag in identifying the start of a drought because of the longer time scale of prior precipitation and temperature conditions incorporated into the model. The SPI is a precipitation-based index, which is calculated from a long-term record of precipitation observations (typically station-based, in situ measures) by fitting the data to a normal distribution and allowing the index to be normalized and calculated over various time scales (e.g., 3, 6, or 1 month(s)) for a given location. The strength of the SPI is its flexibility to monitor both short- and long-term drought conditions, but it does not include a temperature component in its calculation so temperature-related influences are not considered in this index. Both of these indices have been commonly used for operational drought monitoring with these types of map products and datasets. Products based on in situ observations are highly accurate for specific locations and local areas, but often do not adequately characterize spatial variations in drought conditions across large areas, particularly in regions with sparse observation networks such as parts of the western United States (e.g., Great Basin). In addition to the PSDI and SPI, several other indices such as the Crop Moisture Index (CMI; Palmer, 1968), Keetch–Byram Drought Index (Keetch and Byram, 1968), and Surface Water Supply Index (SWSI, Shafer and Dezman, 1982) have been developed from station-based observations to monitor specific types of drought.

15.1.3   U.S. Drought Monitor

The development of the U.S. Drought Monitor (USDM; Svoboda et al., 2002) in 1999 marked a major advance in drought monitoring capabilities for the United States. The USDM represents a composite index tool that incorporates several commonly used

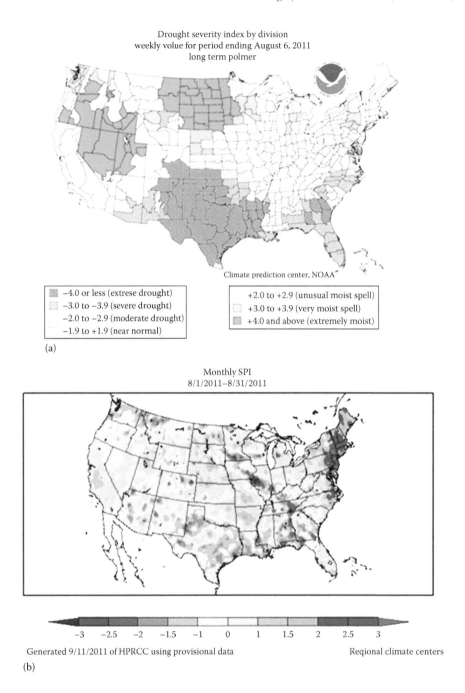

FIGURE 15.2   Examples from August 2011 of the PDSI map (NOAA, 2011) of station-based index values aggregated to the multicounty NOAA climate division (a) and a spatially interpolated, monthly SPI map (NOAA/HPRCC, 2011) from station-based precipitation observations (b).

drought indices such as PDSI, SPI, and SWSI along with other in situ measurements (e.g., stream flow and soil moisture) and guidance from a group of experts (e.g., climatologists, water and natural resource managers, and agriculturalists) to depict both short- and long-term drought conditions across the CONUS. Figure 15.3 presents the USDM map for August 9, 2011, when much of the southern United States was experiencing extreme to exceptional drought conditions.

Until recently, the use of satellite remote sensing in the development of the USDM had been very limited, with the Vegetation

Health Index (VHI; Kogan, 1995) being the sole remote sensing–based input. The VHI incorporates both the normalized difference vegetation index (NDVI)–based Vegetation Condition Index (VCI) and thermal-based Temperature Condition Index (TCI) to produce an indicator suitable for agricultural drought monitoring. The VHI concept built upon well-established value of satellite-based NDVI data from the NOAA Advanced Very High Resolution Radiometer (AVHRR) for vegetation condition assessments, which was first demonstrated for agricultural drought applications in sub-Saharan Africa in

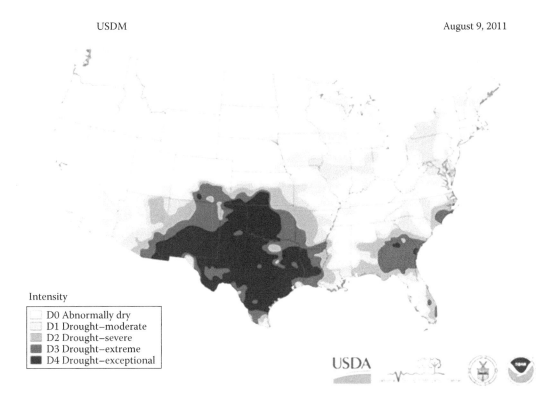

FIGURE 15.3 USDM map drought conditions over the CONUS for August 16, 2011. (*Note*: White areas on USDM map represent areas that are not experiencing drought conditions on that date.)

the 1980s by Tucker et al. (1986) and is currently used in several current vegetation monitoring programs in the United States such as the U.S. Forest Service Wildland Fire Assessment System (WFAS; http://www.wfas.net/index.php/avhrr-ndvi-moisture—drought-47) and the Green Report (http://kars.ku.edu/products/greenreport/greenreport.shtml). The VHI also incorporated a thermal component through the TCI to represent changes in land surface temperature (LST) conditions related to plant stress-induced changes in ET rates and subsequent sensible heat fluxes. Although valuable, the VHI only characterizes agricultural drought conditions and has been shown to have limited utility in energy-limited environments such as high-elevation or high-latitude locations (Karnieli et al., 2010).

There is an increasing demand for more timely and spatially detailed drought information by the USDM and many other drought monitoring activities both in the United States (e.g., state disaster management planning and federal relief programs) and internationally (e.g., Famine Early Warning System [FEWS]) to address this natural hazard at a more localized scale. The USDM represents a prime example of this need as several major decision-making activities have designated the USDM as trigger for disaster response actions at a county-level scale. For example, the U.S. Department of Agriculture (USDA) Farm Service Agency (FSA) uses the USDM to establish the county-level eligibility of agricultural producers for drought-related assistance. As a result, the USDM and other drought monitoring tools for the United States in general are increasingly being looked upon to provide drought information at a county to subcounty spatial

scale where many administrative decisions are made. The majority of traditional drought monitoring tools, which are primarily based on in situ observational data, lack the necessary spatial resolution and scale to meet this demand. This is illustrated in the USDM map in Figure 15.3, where large groupings of counties within many states were assigned to a single drought category by the USDM when in reality there were likely local-scale variations in the severity of those conditions that could not be captured with the spatial resolution of the majority of data inputs into the USDM. Satellite-based remote sensing holds considerable potential to improve drought information at local scales not only for the USDM but also for other drought monitoring activities in the United States and internationally, as the tools and products they provide are of a higher resolution, spatially continuous nature across large areas that can fill in data gaps and complement traditional in situ measurements and existing drought assessment tools.

15.1.4 Traditional Remote Sensing of Drought

Historically, the use of remote sensing for operational drought monitoring has involved the application of NDVI-based products from the NOAA AVHRR. The lineage of AVHRR NDVI-based measures for this application can be traced back more than 20 years to the early work of Hutchinson (1991), Tucker et al. (1986), Kogan (1990), Burgan et al. (1996), and Unganai and Kogan (1998). The NDVI is a simple mathematical transformation developed in the early 1970s (Rouse et al., 1974) that incorporates spectral data from both the visible red and near-infrared (NIR) regions, which are sensitive to changes in plant chlorophyll content and intercellular spaces of the spongy mesophyll layers of the plants' leaves, respectively. Considerable research has shown that NDVI has a strong relationship with several biophysical vegetation characteristics (e.g., green leaf area and biomass) (Asrar et al., 1989; Baret and Guyot, 1991) and temporal changes in index values are highly correlated with interannual climate variations (Peters et al., 1991; Yang et al., 1998; McVicar and Bierwith, 2001; Ji and Peters, 2003). The value of using the AVHRR NDVI (and similar vegetation indices [VIs]) data for drought monitoring is the ability to calculate anomaly measures representing a departure from longer-term historical average conditions from the extended AVHRR NDVI data time series dating back to the 1980s. As a result, AVHRR NDVI-based tools have been used routinely in several drought-related monitoring efforts in the United States (highlighted earlier), as well as internationally as part of systems like FEWS.

Building on the NDVI, researchers have developed new indices that extend the initial NDVI concept with new types of mathematical transformations and/or additional data inputs from other spectral regions. A prime example is the VCI, which is a precursor to the VHI now routinely used for global drought assessments. The VCI (Kogan and Sullivan, 1993) is based on the assumption that for a given location, the historical maximum and minimum NDVI values represent the upper and lower bounds of possible vegetation conditions, with abnormally low NDVI values on given data being indicative of drought-stressed vegetation. A companion thermal-based index called the TCI was developed by Kogan (1995) based on data from the AVHRR thermal bands using the historical maximum (unfavorable vegetation conditions) and minimum (favorable conditions) to establish similar upper and lower bounds for vegetation conditions with drought being manifested as unusually high thermal values near the maximum. Kogan (1995) unified the VCI and TCI into a single index through the creation of the VHI. Other efforts such as the Standardized Vegetation Index (SVI; Peters et al., 2002) describe the probability of NDVI variation from a normal (or average) NDVI value for a weekly time period of the extended AVHRR NDVI time series, with much lower-than-normal NDVI values indicative of drought. Others further refined the integration of thermal and NDVI data into indices such as the temperature/NDVI ratio (McVicar and Bierwirth, 2001) and 2D geometric expressions (Karnieli and Dall'Olmo, 2003). As time-series middle infrared (MIR) spectral data became available from the Moderate Resolution Imaging Spectroradiometer (MODIS), a number of efforts tried to incorporate data from the MIR region, which is sensitive to plant water content, with other MODIS spectral bands to develop new indices representative of changes in moisture status of vegetation (and thus agricultural drought conditions). The normalized difference water index (NDWI), which integrated MIR and NIR data in a similar fashion to NDVI, was tested for drought monitoring by Gu et al. (2007). The NDWI was extended further by Gu et al. (2008) into the Normalized Difference Drought Index

(NDDI), which incorporates NDVI and NDWI in an attempt to leverage the relative strengths of both indices. Wang et al. (2007) also built upon the original NDWI concept by developing a three-band index called the normalized multiband drought index (NMDI) that uses the NDVI in combination with the MIR bands available from MODIS, which are designed to be sensitive to soil and plant water content, respectively.

Collectively, these more traditional remote sensing efforts, utilizing primarily remotely sensed observations in the visible, NIR, MIR, and thermal infrared (TIR) regions, have attempted to develop improved and more effective drought assessment tools over the years. Some of these remote sensing–based tools (such as the NDVI and VHI) have been adopted for operational use in key monitoring systems highlighted earlier, while others were research efforts to test new approaches that have yet to be operationalized. This work has provided a valuable foundation for advancing the use of satellite-based tools for drought monitoring, but has primarily focused on agricultural drought.

15.1.5 New Era of Remote Sensing Observations

Since 2000, a number of new satellite-based earth observing systems have been launched that collect many new types of remotely sensed data, which can be used to provide new perspectives for drought monitoring. The launch of MODIS onboard NASA's Terra and Aqua platforms allowed the collection of spectral observations in the visible and NIR regions to extend the global time series of NDVI data that was established with the AVHRR. The near-daily global coverage of MODIS spectral observations also extends into the MIR region, allowing indices such as the NDWI to be calculated routinely over large areas, and the TIR region, which can be used to develop thermal-based tools for ET estimation. Microwave (MW) sensors such as the Advanced Microwave Scanning Radiometer Earth Observing System (EOS) (AMSR-E) and Quick Scatterometer (QuikSCAT) collect key observations that can be used to estimate soil moisture (Bolten et al., 2010). Gravity field observations from NASA's Gravity Recovery and Climate Experiment (GRACE) also provide new insights into water cycle variables including soil moisture and groundwater (Rodell and Famiglietti, 2002). Collectively, this suite of new remote sensing observations and datasets that has become available over the past 10+ years, coupled with advances in environmental models and algorithms as well as computing capabilities, has resulted in the rapid emergence of many new tools that monitor different aspects of the hydrologic cycle that influences drought conditions.

This chapter will highlight several satellite-based remote sensing tools that have been developed to support and enhance operational drought monitoring in the United States. The focus is on tools that characterize terrestrial components of the hydrologic cycle related to drought, including vegetation, health, ET, soil moisture, and groundwater. These tools either are also currently operational or have the potential to be operational in the near future. For each tool, the objectives and methods will be summarized, informational drought products highlighted, and current/future work and efforts to extend them internationally discussed. Several upcoming satellite missions that hold considerable potential to further advance our drought monitoring will also be presented.

15.2 New Drought Monitoring Tools

15.2.1 Vegetation Drought Response Index

The Vegetation Drought Response Index (VegDRI) is a hybrid VI designed to detect drought-related vegetation stress through integration of satellite, climate, and biophysical data (Figure 15.4; Brown et al., 2008). For the satellite input, VegDRI uses an NDVI-based measure to detect vegetation stress. Although NDVI has proven useful for this application, analysis of only NDVI information for drought monitoring is problematic because many types of environmental events (e.g., fire, flooding, pest infestation, plant disease, and land use/land cover [LULC]

change) can produce a decline in NDVI values that mimics drought stress. As a result, VegDRI incorporates climatic information to distinguish drought-impacted areas experiencing abnormally dry conditions from other areas impacted by these nondrought events. A biophysical component is added to VegDRI to include other environmental factors that can influence climate–vegetation interactions such as LULC type, soil characteristics, elevation, and rainfed versus irrigated systems. Table 15.1 lists the specific satellite, climate, and biophysical variables. An empirical-based modeling approach is used to analyze the historical behavior of these data inputs over a 20-year period (1989–2008) to produce the VegDRI models. VegDRI (http://vegdri.unl.edu) has been operationally produced over the CONUS at a nominal 1 km spatial resolution since 2009 (with a complete history dating back to 1989) in a collaborative effort between the U.S. Geological Survey (USGS) Earth Resources Observation Science (EROS) Center and the National Drought Mitigation Center (NDMC) at the University of Nebraska–Lincoln.

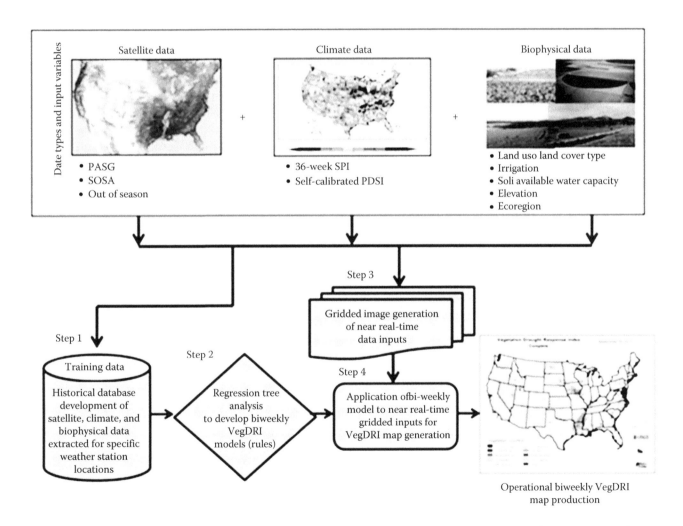

FIGURE 15.4 Overview of the VegDRI methodology. (Adapted from Wardlow, B.D. et al., The Vegetation Drought Response Index (VegDRI): An integration of satellite, climate, and biophysical data, in *Remote Sensing of Drought: Innovative Monitoring Approaches*, eds. B.D. Wardlow, M.A. Anderson, and J. Verdin, CRC Press, Boca Raton, FL, pp. 51–74.)

TABLE 15.1   Attributes Used to Develop the VegOut-Ethiopia Model

No.	Attribute	Acronym	Type	Format	Source
1	Standardized Deviation of NDVI	SDNDVI	Satellite	Raster	NOAA AVHRR
2	Digital elevation model	DEM	Biophysical	Raster	USGS
3	Soil water holding capacity	WHC	Biophysical	Raster	USGS
4	Ecological type	Eco_Ethiopia	Biophysical	Vector	Ecodiv.org
5	Land cover	Landcover	Biophysical	Raster	ESA
6	Three-month SPI	SPI_3month	Climate	Raster	IRI
7	Pacific Decadal Oscillation	PDO	Oceanic/atmospheric	Point data	CPC/NOAA
8	Atlantic Multidecadal Oscillation Index	AMO	Oceanic/Atmospheric	Point data	CPC/NOAA
9	North Atlantic Oscillation	NAO	Oceanic/atmospheric	Point data	CPC/NOAA
10	Pacific North American Index	PNA	Oceanic/atmospheric	Point data	CPC/NOAA
11	Multivariate ENSO Index	MEI	Oceanic/atmospheric	Point data	CPC/NOAA

Source: Adapted from Tadesse, T. et al., *Water Resour. Res.*, doi:10.1002/2013WR0142812014, 2014.

15.2.1.1 Input Data

15.2.1.1.1 Satellite Variables

Time-series AVHRR NDVI data were used as the data source to derive three satellite-based variables for VegDRI. The first NDVI-derived variable is Percent Annual Seasonal Greenness (PASG), which relates how vegetation conditions for a specific period during the growing season compare to the historical average conditions for the same period in the 20-year AVHRR NDVI record. A moving-window averaging technique (Reed et al., 1994) is used to determine the pixel-level start and end of season (SOS and EOS) day of year (DOY), and the NDVI date is summarized within these beginning and end dates to calculate the PASG, which is updated on a biweekly time step in concert with the temporal compositing period of the AVHRR NDVI dataset (Eidenshink, 2006). A second NDVI-based measure, the start of season anomaly (SOSA), is included in VegDRI to represent the departure in the pixel-level SOS for a specific year from the historical SOS date. SOSA was included to discriminate between areas that have a normal SOS with a low PASG because of interannual climate variations (e.g., late freeze from cold temperatures or drought) and areas that have experienced nonclimatic factors (e.g., LULC change such as a crop rotation) resulting in a comparably low PASG. A third NDVI-derived metric, out of season (OS), is included to represent the nongrowing season period of the year when the vegetation is dormant and agricultural drought monitoring is not necessary. OS is calculated for each pixel using the SOS and EOS DOYs and is defined as the period for the EOS (e.g., December 1) to the SOS (e.g., March 1). The OS measure is used to exclude VegDRI calculations during this time to avoid anomalous agricultural drought from being detected during periods when there is no actively growing vegetation and the NDVI signal is driven by nonvegetation factors (e.g., soils).

15.2.1.1.2 Climate Variables

VegDRI includes two commonly used climate indices for drought monitoring: the self-calibrated PDSI (SC-PDSI) and the SPI. Historical data for both indices were taken from 2417 weather station locations across the CONUS from the Applied Climate

Information System (ACIS; http://rcc-acis.unl.edu/) with sufficiently long and relatively complete data records (i.e., 30-year length and less than 20% missing observations).

The SC-PDSI is a modified version of the original PDSI calculation developed by Palmer (1965), which was based on a simple supply-and-demand water balance model that considers precipitation, temperature, and the available water holding capacity of the soil at a given location. The SC-PDSI developed by Wells et al. (2004) calibrates the constants and duration factors in the original PDSI calculation to the local soil characteristics of each station. The PDSI is a well-established index for agricultural drought assessment in the United States, and a modified version of its drought severity classification scheme was adopted for VegDRI. The VegDRI classification scheme includes four drought classes (extreme, severe, moderate, and predrought), one normal class, and three classes of better-than-normal conditions (unusually, very, and extremely moist) as defined by the PDSI. An OS class derived from the OS metric calculated from the AVHRR NDVI data is also included to mask areas where there is no actively growing vegetation and agricultural drought monitoring is not required at that time.

The SPI is designed to quantify precipitation anomalies over varying time periods (e.g., 1-, 3-, or 6-month periods) based on fitting a long-term precipitation record for a station location over a specific time interval to a probability function, which is then transformed into a gamma distribution to standardize the mean SPI for all locations and time periods (McKee et al., 1985). A 36-week (or 9-month) SPI was selected to represent seasonal precipitation conditions in the VegDRI models.

15.2.1.1.3 Biophysical Variables

The dominant LULC type for each 1 km pixel across the CONUS is considered in VegDRI to account for the varying interactions of climate and vegetation response by different cover types. The LULC information is derived from the national-level 30 m National Land Cover Dataset (NLCD, circa 2001; Homer et al., 2004), which was aggregated to 1 km using a majority zonal function calculation. Further LULC information related to the irrigation status of each pixel is included through the incorporation of a national-level irrigated/rainfed agricultural land dataset called the MODIS Irrigated Agriculture Dataset for the United States (MIrAD-US; Pervez and Brown, 2010). The irrigation variable was included to represent the different effects that drought can have on crops and other land cover under irrigated and rainfed conditions. The soil available water holding capacity (SAWC) variable was derived from the State Soil Geographic (STATSGO) database (USDA, 1994) and incorporated into VegDRI to represent the potential of the soil to hold moisture that is available to plants, which is an indicator of the susceptibility of vegetation to drought stress. An ecoregion variable derived from the Omernik Level II ecoregion data (Omernik, 1987) was added to the VegDRI model to represent a geographic framework across the CONUS that accounts for the considerable environmental variability encountered across the nation by grouping areas into regional ecosystems where vegetation has adapted to the local abiotic and biotic conditions and resources. Elevation is the final biophysical variable included in VegDRI to account for the elevational influence on vegetation, and the different sensitivity to drought a vegetation type may experience across a vertical gradient. It should be noted that all biophysical variables were *static* across the historical record and do not reflect changes such as LULC conversion or crop rotations because CONUS-scale datasets with sufficient temporal resolution are not available.

15.2.1.2 Model Development and Implementation

The VegDRI methodology is presented in Figure 15.4. The initial step is the development of a training database of historical data for each input variable, which were extracted from 2417 weather station locations from across the CONUS. Climate data were in a tabular format and could be incorporated directly into the database. For the gridded satellite and biophysical datasets, a zonal calculation within a 3×3 pixel window (i.e., 9 km² area) centered on each station location was used to populate the database for each of these variables. The zonal average was used for continuous variables (e.g., SAWC) and the zonal majority for thematic variables (e.g., LULC type). The final station locations used in model development were sited in nonurban areas and away from large water bodies to minimize the potential influence of urban areas (e.g., built structures) and water within the 1 km AVHRR NDVI data, which both could result in NDVI signals that are not representative of rainfed vegetation. The NLCD dataset was used as the LULC reference and the majority of LULC type within the 3-by-3 pixel window surrounding each station was analyzed to screen urban or water locations. All extracted historical data were incorporated into an SQL Server database to prepare the input data files for model development. The data were organized into 26 biweekly periods spanning the calendar year, and for each period, the complete set of historical data for all stations was extracted for analysis. For the satellite- and climate-related variables, 20 years of observations for each specific biweekly period was included in this training dataset along with the biophysical variables, which remain static across the historical record.

A commercial regression tree analysis algorithm called Cubist (Quinlan, 1993) was used to analyze the historical data for each

specific biweekly period and generate a rule-based, piecewise linear regression VegDRI model for that period. The SC-PDSI was the dependent variable for these empirical models, providing a well-established classification scheme of agricultural drought severity recognized within the drought community. The model development stage yielded 26 period-specific VegDRI models. Each Cubist-derived model consists of an unordered set of rules with each rule having the syntax "if x conditions are satisfied, then use the associated linear multiple regression equation" to calculate the VegDRI value. The number of rule sets can vary by period and often more than one rule set may apply, resulting in multiple linear regression equations being applied and the average of those products used as the final VegDRI value.

For the mapping phase, period-specific VegDRI models are applied to the corresponding gridded data inputs using MapCubist software developed by the USGS EROS. An inverse distance weighted interpolation method is used to convert the point-based SPI data into a 1 km grid over the CONUS to match the other gridded satellite and biophysical datasets. During model implementation to the gridded data using MapCubist, the values of all input variables associated with a pixel location are considered to determine the specific rule set (s) and associated linear regression equation(s) to apply to calculate a VegDRI value. This process is repeated for all pixels across the CONUS, producing a 1 km resolution VegDRI map.

These VegDRI models were moved into operational production over the CONUS in 2009, producing national-level VegDRI maps of agricultural drought conditions in near real time (i.e., less than 24 h from the last satellite and climate data observation) that are updated on a biweekly time step consistent with the AVHRR NDVI data production schedule. The operational VegDRI production system resides at USGS EROS, where the appropriate VegDRI model is applied to up-to-date AVHRR NDVI-derived data inputs generated at USGS EROS, current SPI data acquired from the NOAA ACIS via the High Plains Regional Climate Center (HPRCC), and static biophysical information. A variety of value-added products (e.g., maps, tables, animations, and summaries) are staged for the general public on the VegDRI webpage (http://vegdri.unl.edu/), as well as within a dynamic USGS VegDRI Viewer (http://vegdri.cr.usgs.gov/viewer/viewer.htm) that enables visualization of VegDRI maps in combination with other spatial data (e.g., USDM, geopolitical boundaries, and precipitation anomalies). More recently, the VegDRI models were extended to use customized MODIS-based NDVI inputs from the USGS eMODIS system (Jenkerson et al., 2010), enabling weekly operational updates of VegDRI to complement the initial biweekly AVHRR-based product. From an operational standpoint, transitioning VegDRI to MODIS NDVI inputs was important for several reasons: (1) a new remote sensing data source for long-term VegDRI data products as the final AVHRR sensor is in operation and (2) flexibility to update VegDRI maps on a customized production schedule (e.g., updated each Monday) tailored to decision makers' needs rather than the inflexible biweekly production cycle of AVHRR NDVI data inputs for the original VegDRI implementation.

A similar crosswalk exercise will be needed in the near future to extend VegDRI production from MODIS to the new operational NOAA Visible/Infrared Imager Radiometer Suite (VIIRS), which is intended to extend the satellite-based NDVI data record established by AVHRR and MODIS into the future.

15.2.1.3 Examples of VegDRI

An example of a VegDRI map over the CONUS is presented in Figure 15.5 for August 11, 2011, when many parts of the southern United States were experiencing severe-to-extreme drought conditions. The most extreme drought conditions were located in the south-central United States, centered on western Texas where many locations had the driest (or one of the driest) years in the instrumental data record for precipitation, which often spanned more than 100 years. The inset VegDRI map of the south-central U.S. states shows that extreme drought was being detected by this mid-July period and conditions continued to worsen and expand over a larger geographic area in subsequent VegDRI maps as the growing season progressed, which was consistent

with record crop and forage losses across this area. The severity and intensification of drought conditions over this area were further exacerbated by a prolonged period of days with air temperature exceeding 100°F, with many locations within the core area of extreme drought characterized by VegDRI experiencing 70+ days of triple-digit temperatures (NCDC, 2013).

A comparison of the general drought patterns depicted in VegDRI (Figure 15.5) and the USDM (Figure 15.3) reveals that similar patterns and severity levels of drought were detected by both tools over the CONUS. The severe-to-extreme droughts that were widespread across much of New Mexico, Oklahoma, and Texas are represented, as well as the extension of these drought conditions along the southern edge of the Gulf Coast region into the Atlantic Coast region. However, the higher spatial resolution of VegDRI compared to the USDM is quite apparent as subtle within-state variations in drought conditions are captured in VegDRI over areas where the USDM has assigned the same area to a single drought severity designation. A closer view of the VegDRI output for the southern Great

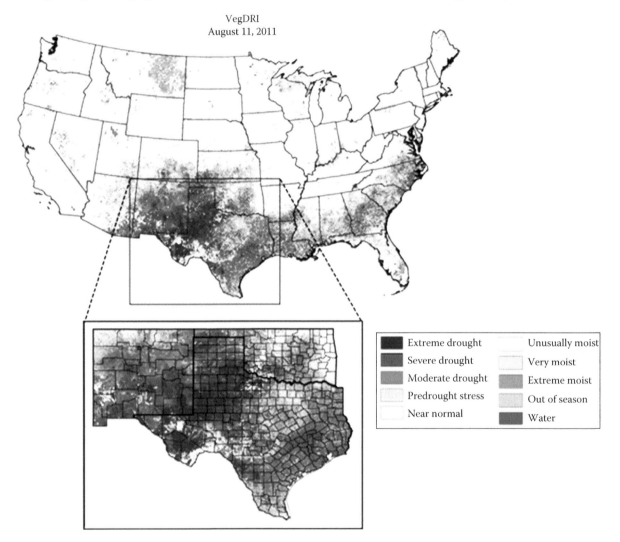

FIGURE 15.5 VegDRI map for August 11, 2011, over the CONUS (top) and the southern Great Plains region comprising New Mexico, Oklahoma, and Texas.

Plains in Figure 15.5 illustrates the subtle spatial variations in drought conditions detected at a localized county to subcountry scale.

15.2.1.4 Applications of VegDRI

Information from VegDRI is increasingly being used for a range of drought monitoring applications and activities. The authors of the USDM routinely consult VegDRI in combination with other data sources to develop the USDM map. VegDRI is used to provide more local-scale information on spatial variations in agricultural drought conditions in an attempt to provide more accurate spatial depictions of drought within the USDM. The Bureau of Land Management has also adopted VegDRI as a response trigger and field visit priority guide within the bureau's drought management plans (DOI, 2013). The National Weather Service field offices are incorporating VegDRI information into drought impact statements. Several states such as Arizona, Montana, New Mexico, and Kansas have added VegDRI as a tool in their respective statewide drought monitoring systems and/or assessment reports. Currently, several countries (including Canada, China, Czech Republic, India, and Mexico) are attempting to develop modified versions of the VegDRI methodology for drought monitoring systems that are customized to their available data resources and specific information needs. The VegDRI approach provides a flexible framework within which various types of remote sensing, climatic, and biophysical data can be integrated to develop an agricultural drought-based indicator. For example, Canada is working to develop a VegDRI approach identical to the U.S. version so the datasets produced in both countries can be spatially merged seamlessly to the multinational North American Drought Monitor (NADM; http://www.ncdc.noaa.gov/temp-and-precip/drought/nadm/). In comparison, a country such as India may have a different set of climate, satellite, and/or biophysical data inputs that can be integrated in a similar fashion to develop a customized agricultural drought monitoring tool for that specific area.

15.2.2 Vegetation Outlook

The vegetation outlook (VegOut) is an experimental hybrid drought monitoring tool that provides outlooks of general vegetation conditions in a series of 1 km resolution maps at multiple time steps (Tadesse et al., 2010). This tool implements a similar modeling methodology, as well as several of the same data inputs as VegDRI, to provide complementary information about projected vegetation conditions within a 1–2-month period. The VegOut tool is being developed for the CONUS by researchers at the NDMC and the University of Nebraska at Kearney (UNK). A modified version of the predictive VegOut tool is also being developed for Ethiopia by the NDMC in collaboration with researchers at Addis Ababa University (AAU), which is tailored to the data resources available for that region. VegOut is designed to predict vegetation conditions based on the analysis of "historical patterns" of the satellite-based vegetation condition observations, climate-based drought indices, general biophysical

information about the environment (e.g., elevation, land cover type, and soil available water capacity), and several oceanic indices (e.g., sea surface temperature and Southern Oscillation Index [SOI]). The primary difference in the data inputs between the U.S. and Africa VegOut tools is the climate data. For the United States, observed measures from weather station locations were used, while for Ethiopia, a merged product of satellite rainfall estimate and weather stations records was included. In addition, the biophysical irrigated land variable was excluded in the VegOut-Ethiopia model because no comparable dataset exists for this area. However, the lack of an irrigated lands input is expected to have little impact on the VegOut-Ethiopia models since there is no significant irrigation coverage in Ethiopia.

15.2.2.1 Input Data

Similar to the VegDRI, the underlying models of the VegOut tool include satellite, climate, and biophysical input variables and use an empirical regression tree modeling approach to collectively analyze these historical data inputs for model development. However, VegOut also incorporates a fourth teleconnection component in the models by integrating oceanic indices from the Atlantic and Pacific oceans to drive regional-scale climate patterns and intra- and intervariations. Figure 15.6 shows the VegOut historical database of all input variables and a general schematic of the VegOut model and outlook map development process. The four components and their associated input variables for VegOut are briefly discussed as follows.

15.2.2.1.1 Satellite Data

The VegOut models are built upon a derivative of AVHRR-based NDVI observations called seasonal greenness (SG), which is the time-integrated (or accumulated) NDVI above a baseline NDVI value (termed "latent" NDVI, representing the nonvegetated background signal from soil and plant litter) from the start of the growing season to a specified time during the growing season (Brown et al., 2008). In the VegOut model, the SG is standardized (z-index) to a measure called the standardized seasonal greenness (SSG) to make all values directly comparable across both space and time (Tadesse et al., 2010). The VegOut models are built upon historical SSG observations along with the other variables discussed below to predict SSG values, which are representative of future general vegetation conditions. Thus, the dependent variable (the prediction value) is the SSG. The SSG predictions (i.e., VegOut values) are based on historical time-lag relationships between satellite-observed vegetation conditions and the preceding climatic, oceanic, and satellite observations. The other satellite-derived variable used in the VegOut model is the SOSA measure, which was discussed earlier for VegDRI. The SOSA was included in the VegOut model to account for the different interannual timings of emergence of natural and agricultural vegetation that determines the initiation of vegetated activity and influences the seasonal vegetation performance in a given year. This historical (1989–2008) dataset of SSG and SOSA values is used to develop the empirical VegOut models (Figure 15.6).

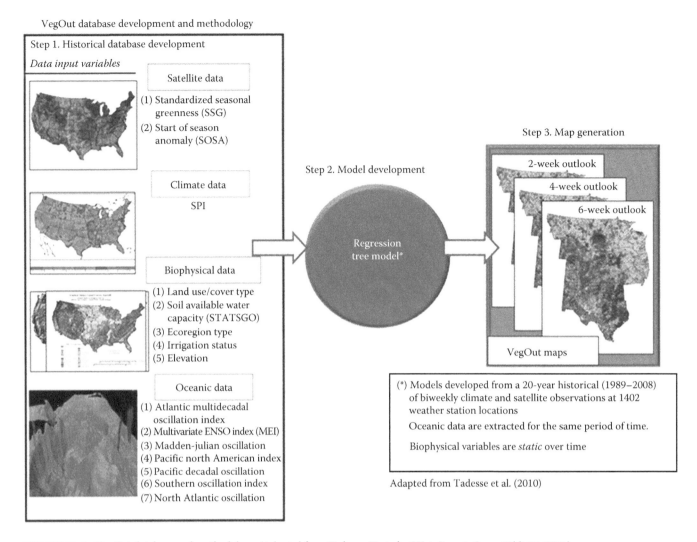

VegOut database development and methodology

Step 1. Historical database development

Data input variables

Satellite data

(1) Standardized seasonal
    greenness (SSG)
(2) Start of season
    anomaly (SOSA)

Climate data

SPI

Biophysical data

(1) Land use/cover type
(2) Soil available water
    capacity (STATSGO)
(3) Ecoregion type
(4) Irrigation status
(5) Elevation

Oceanic data

(1) Atlantic multidecadal
    oscillation index
(2) Multivariate ENSO index (MEI)
(3) Madden-julian oscillation
(4) Pacific north American index
(5) Pacific decadal oscillation
(6) Southern oscillation index
(7) North Atlantic oscillation

Step 2. Model development

Regression
tree model*

Step 3. Map generation

2-week outlook

4-week outlook

6-week outlook

VegOut maps

(*) Models developed from a 20-year historical (1989–2008)
    of biweekly climate and satellite observations at 1402
    weather station locations

    Oceanic data are extracted for the same period of time.

    Biophysical variables are *static* over time

Adapted from Tadesse et al. (2010)

FIGURE 15.6　VegOut database and methodology. (Adapted from Tadesse, T. et al., *GISci. Remote Sens.*, 47(1), 25, 2010.)

15.2.2.1.2 Climate Data

The SPI is integrated in the VegOut model to represent degree of dryness at a location and represents the seasonal climate impact on vegetation conditions. Similar to the VegDRI models, a 36-week SPI was selected to represent seasonal precipitation conditions in the VegOut models.

15.2.2.1.3 Biophysical Data

The biophysical data used in VegOut include land use and LULC-type, Omernik Level III ecoregion data (Eco), percent of irrigated agricultural land (Percent_Irrig), SAWC, and digital elevation model (DEM) data. Each variable is described earlier in the VegDRI section of this chapter.

15.2.2.1.4 Oceanic/Atmospheric Indices

Oceanic/atmospheric indices constitute the final component of the VegOut models and are integrated into the predictions because of the indirect temporal and spatial relationships

between ocean–atmosphere dynamics and climate–vegetation interactions (teleconnection patterns such as El Niño and La Niña) that have been observed (Verdin et al., 1999; Asner et al., 2000; Los et al., 2001; Lyon, 2004; Barnston et al., 2005; Tadesse et al., 2005). The oceans play a significant role in shaping the complex nature of weather and climate patterns over land through interactions with the atmosphere resulting in climate patterns (e.g., drought or excessive rainfall and warm or cold temperatures) that affect vegetation conditions (Boyd et al., 2002; Tadesse et al., 2010). Thus, to capitalize on historical climate–vegetation interactions and ocean–climate teleconnections expressed over longer periods of time, selected oceanic/atmospheric indices are incorporated into the VegOut model that influences the general climatic patterns that impact general vegetation conditions from year to year. Seven oceanic indices are integrated into the predictions to account for the temporal and spatial relationships between ocean–atmosphere dynamics and climate–vegetation interactions (i.e., teleconnection patterns): the Atlantic Multidecadal Oscillation Index (AMO), Multivariate El Niño and Southern Oscillation (ENSO) Index (MEI), Madden–Julian oscillation (MJO), Pacific North American Index (PNA), Pacific Decadal Oscillation (PDO), SOI, and North Atlantic Oscillation (NAO).

15.2.2.2 Examples of VegOut Models Development and Implementation

15.2.2.2.1 *VegOut for the Central United States*

The basic algorithm underlying the VegOut model to predict SSG is based on a series of multiple linear regression equations defined by the classification and regression tree (CART)–based Cubist software through the analysis of the historical data discussed earlier. The model calculates the SSG value for future biweekly period $t = i$ (e.g., $t = 2$ weeks into the future) by applying a set of linear regression equations associated with historical periods in the database that exhibited similar patterns (or behavior) among the set of independent variables. To calculate the predicted value, the regression equation(s) is (are) applied using the conditions of the current week ($t = 0$). The following is the general form of the linear regression equation defined by Cubist that is applied to calculate the SSG for a future biweekly time period $t = i$:

$$\mathbf{VegOut}(i) = f_{1,i}(SSG, SOSA)_{t=0} + f_{2,i}(SPI)_{t=0} + f_{3,i}(LULC, Eco,$$
$$Percent\_Irrig, SAWC, DEM)_{t=0}$$
$$+ f_{4,i}(MEI, MJO, NAO, PDO, SOI, AMO, PNA)_{t=0}$$
$$(15.1)$$

where VegOut(i) is the predicted SSG at future biweekly time period i as a function of the current ($t = 0$) values of the input variables. The equation shows that the VegOut is defined as four functions (f_1, f_2, f_3, and f_4) of the current (i.e., the date on which the SG prediction is made) climate, environmental, and satellite variables and the values of the oceanic indices, respectively.

A set of biweekly VegOut models was developed from a historical 20-year record of climatic, oceanic, and satellite observations for each 14-day period across the year to calculate 2-, 4-, and 6-week SSG outlooks. Longer outlooks over periods up to a maximum of 52 weeks (i.e., 1 year) are possible, but most of the VegOut work to date has focused on these shorter outlooks because consistently higher predictive accuracies have been attained for 2–6-week time intervals across the growing season (Tadesse et al., 2010). For a given biweekly period, each outlook model is applied to the current observations to predict future SSG values at the 2-, 4-, and 6-week time steps based on events in the historical record that had a similar pattern among the climate, oceanic, and satellite variables.

Experimental VegOut maps were produced for a 15-state region of the central United States. Figure 15.7 shows an example of a series of VegOut maps produced for mid-July and early August 2009 over the central United States. The predicted SSG

patterns of the 2-, 4-, and 6-week VegOut maps calculated on June 29 had strong spatial agreement with the observed SSG patterns on the corresponding July and August dates of the three outlooks. This example illustrates that VegOut performed well over the region's diverse climatic conditions (ranging from arid to humid) and land cover types (crops, forest, grassland, and shrubland). The only notable discrepancy between the predicted and observed SSGs was the slight underprediction of SSG values at the lower end of the SSG value range over the far northern part of this area in the 6-week VegOut map (Figure 15.7d and g). In statistical testing of the VegOut models, Tadesse et al. (2010) found relatively high predictive accuracy ($r^2 > 0.8$) for the three outlook intervals across the growing season. Specifically, the correlation coefficient (r^2) of observed and predicted values ranged from 0.94 to 0.98 for 2-week, 0.86 to 0.96 for 4-week, and 0.79 to 0.94 for 6-week predictions. Although the predicted SSG changes depicted in the VegOut maps represent a composite of both normal seasonal vegetation changes (e.g., green up or senescence) and weather-related conditions (e.g., surplus or deficit of soil moisture for plants), this outlook information could be used as an early indicator of future drought stress for drought monitoring and highlight geographic areas of interest where drought conditions might be expected to emerge or intensify. Work is planned to improve the VegOut model by including other important variables for vegetation (e.g., ET data) and expand the coverage of VegOut across the CONUS. In addition, the VegOut concept has raised great interest internationally to develop similar but country-specific VegOut models by modifying the VegOut inputs to the specific datasets that are available for a given country or region. The following section highlights this research effort in Ethiopia as an example.

15.2.2.2.2 *VegOut for Ethiopia*

In a recent study, Tadesse et al. (2014) used the VegOut approach to develop and evaluate another experimental drought monitoring tool called VegOut for Ethiopia (VegOut-Ethiopia) to predict the vegetation condition in Ethiopia and other eastern African countries. For this VegOut-Ethiopia model, a 24-year time series (1983–2006) of 10-day (dekadal) composited 8 km AVHRR NDVI data was used. The monthly NDVI data were calculated by summing the three-dekadal data in each month. Then, a standardized monthly NDVI (Standard Deviation of Normalized Difference Vegetation Index [SDNDVI]) was calculated and used as a dependent variable to reflect the vegetation conditions (SG) at the end of each month during the growing season as compared to the historical average conditions for that same period over the historical record. Attributes that are used in the VegOut-Ethiopia models are listed in Table 15.1.

The VegOut-Ethiopia models are composed of an unordered set of rules, with each rule having the syntax "if x conditions are met, then use the associated linear regression tree model."

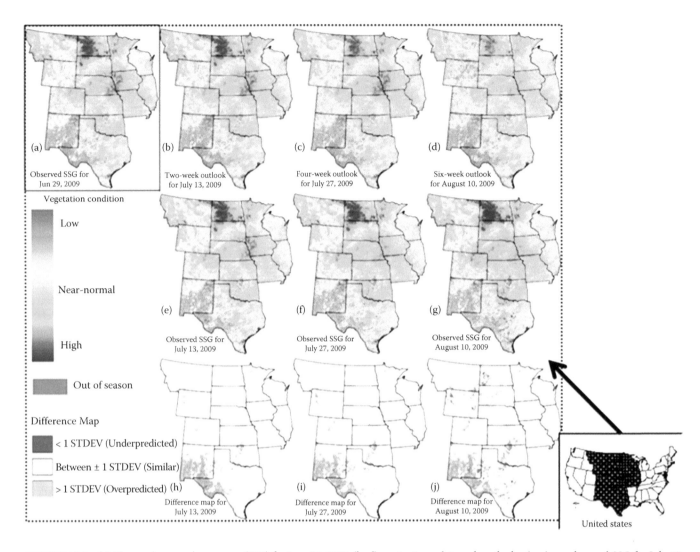

FIGURE 15.7 (a) Observed seasonal greenness (SSG) for June 29, 2009; (b–d) are 2-, 4-, and 6-week outlooks; (e–g) are observed SSG for July 13, July 27, and August 10, corresponding to the 2-, 4-, and 6-week outlooks, respectively; and (h–j) show the difference between the predicted and observed greenness for the corresponding 2-, 4-, and 6-week outlooks, respectively. (Adapted from Tadesse, T. et al., *GISci. Remote Sens.*, 47(1), 25, 2010.)

The following provides an example of the rules generated by the Cubist algorithm for VegOut-Ethiopia:

Rule 1(Example):

If:

Land cover is {mosaic cropland-vegetation grassland shrubland forest}
Ecoregion is {Afroalpine Ericacous Belt}
$SPI \leq -1.5$
$WHC \leq 7.5$
$MEI > 0.0$
$AMO < 0.75$
$PDO < -2.5$

Then:

VegOut-Ethiopia $= -1.25 + 0.6SDNDVI$ (previous month) $+ 0.8SPI - 0.2WHC + 0.25AMO = 0.4MEI + 0.35PDO$

According to this example, if the data associated with a specific case meet the conditional statement for the ecoregion (i.e., Afroalpine Ericaceous Belt) and the threshold criteria for the five continuous variables (i.e., SPI, AWC, AMO, PDO, and MEI) and are represented by the specific land cover classes (i.e., Mosaic cropland [50%–70%], vegetation [grassland/shrubland/ forest]) identified by the Cubist regression algorithm, then the aforementioned multivariate linear regression equation is used to calculate a VegOut-Ethiopia value (i.e., future SDNDVI value) for a specific outlook period interval. Notice that the rule generated in this example includes 50% of the SDNDVI values for the previous month, which helps establish the initial condition of the vegetation for the prediction. The percentage changes for each model depend on the patterns of the historical records in the database. If two or more rules in the Cubist model apply to the case, then the predicted values from each regression equation will be averaged to arrive at the final predicted VegOut value. For more information about the VegOut modeling approach in Ethiopia, readers are referred to Tadesse et al. (2014) and Berhan et al. (2014).

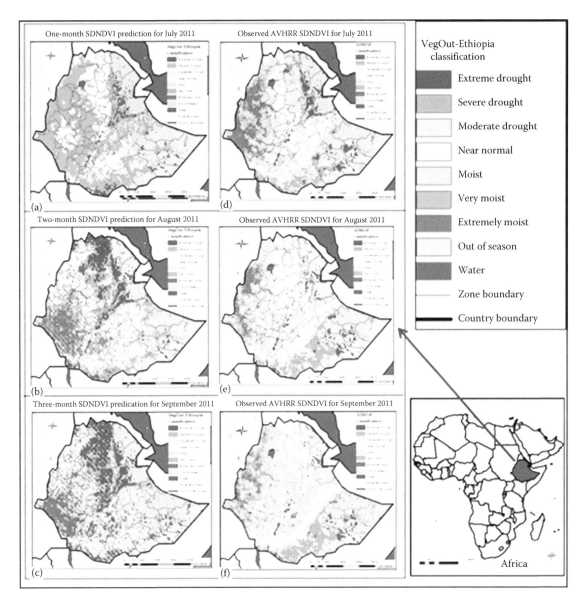

FIGURE 15.8 VegOut-Ethiopia models maps for June 2011: (a) 1-month, (b) 2-month, and (c) 3-month prediction (i.e., predicting SDNDVI for July, August, and September, respectively); (d–f) the corresponding observed satellite-derived SDNDVI for each month. (Adapted from Tadesse, T. et al., *Water Resour. Res.*, 2014, doi:10.1002/2013WR0142812014.)

Figure 15.8 shows an example of the VegOut-Ethiopia models maps for June 2011. Figure 15.8a through c shows the 1-, 2-, and 3-month prediction maps applying these three VegOut-Ethiopia models developed for June to predict SDNDVI in July, August, and September 2011, respectively. These predicted maps were compared with the satellite-derived observed AVHRR SDNDVI maps (Figure 15.8d through f) for the corresponding months. The 1-month prediction of June (Figure 15.8a) showed predrought to severe drought over the eastern parts of the northern highlands of Ethiopia in July, with moderate drought in the southern and eastern highlands of Ethiopia. In other words, in July 2011, various parts of the country, including the north, northeast, and central and eastern highlands of Ethiopia, were predicted to have highly stressed vegetation conditions as compared to the

historical mean of the 24 years of data used in the model. When the predicted SDNVI in Figure 15.8a is compared to the observed SDNVI in Figure 15.8d, the model predicted the spatial patterns and intensity of the drought in July 2011 with a relatively good accuracy ($r^2 = 0.72$). The June 2-month prediction (for August) showed extreme drought conditions to be expanded across the northern highlands (Figure 15.8b). The observed SDNDVI for the corresponding period in August (Figure 15.8e) also showed the drought patterns, but the prediction overestimated the intensity, especially over the northeastern highlands. The overall accuracy of the model as compared to the observed values of SDNDVI for the 2-month prediction was 0.59 r^2. The 3-month prediction of the June model (for September) showed that there was moderate-to-severe drought in most parts of the country

(Figure 15.8c). However, compared to the observed SDNDVI for September (Figure 15.8f), the prediction was not good over the central and northern highlands of Ethiopia, which recovered from drought in September 2011. In this study, the northern rift valley (Afar region) and the southeastern lowlands of Ethiopia that normally have less than 15% of vegetation and are out of growing season during the study period (June–September) were masked to avoid the potential model bias of predicting long-term means for these areas (gray color in Figure 15.8a through f).

A visual comparison of the observed and predicted maps in Figure 15.8a through f shows general agreement in their SDNDVI patterns and intensity. However, further investigation is needed to improve this model and make it an operational tool so that it can provide support for food security management decisions. In addition, VegOut must be integrated with other monitoring and prediction tools to be used efficiently.

15.2.3 Evaporative Stress Index

Of the various components of the hydrologic budget, ET is perhaps most directly tied to vegetation health. Canopy transpiration rates are regulated by root uptake and stomatal conductance and are strongly correlated with plant photosynthetic functioning, while the soil evaporation component of ET is directly related to the surface soil moisture status. The Evaporative Stress Index (ESI) reflects standardized anomalies in a ratio of actual-to-reference ET, with actual ET retrieved using the remote sensing–based Atmosphere-Land Exchange Inverse (ALEXI) model of surface energy balance (Norman et al., 1995). Negative anomalies in ESI signify lower-than-normal consumptive water use, typically indicating depletion of root-zone soil moisture reserves (i.e., agricultural drought) and/or poor vegetation condition.

The primary diagnostic inputs to ALEXI are maps of LST retrieved from satellite imagery collected in the TIR wavebands. Using principles of energy balance, ALEXI estimates the land evaporative flux required to keep the surface at the temperature observed from the satellite platform. One advantage of this diagnostic approach to estimating ET is that it does not require a priori information about precipitation or other ancillary sources of moisture—a necessity in prognostic land surface models (LSMs) governed by water balance. In the energy balance approach, LST conveys proxy information about the surface moisture status, with moisture deficiencies reflected in elevated canopy and soil temperatures. Therefore, the LST-based ESI provides an independent assessment of drought conditions in comparison with standard precipitation-based indices and is arguably more directly related to actual stress in the vegetative canopy.

15.2.3.1 Model Development and Implementation

15.2.3.1.1 Two-Source Energy Balance Model

The ALEXI energy balance model is built on the two-source energy balance (TSEB) LSM of Norman et al. (1995), with subsequent revisions by Kustas and Norman (1999, 2000). The TSEB treats the land surface as imaged by the remote TIR sensor as a composite of soil and canopy components, each with a characteristic temperature (T_S and T_C, respectively) that is constrained by the bulk radiometric surface temperature observation, $T_{RAD}(\theta)$, and the local fraction of green vegetation cover, $f(\theta)$, both apparent at the view zenith angle θ:

$$T_{RAD}(\theta) \approx \left[f(\theta) T_C{}^4 + \left[1 - f(\theta) \right] T_S{}^4 \right]^{1/4} \quad (15.2)$$

These component temperatures in turn constrain fluxes of sensible heat from the soil and canopy components, H_S and H_C, computed using a series of resistance network regulating temperature gradients between the surface and above-canopy airspace, as described schematically in Figure 15.9.

Resistances factors diagrammed in Figure 15.9 are estimated based on local wind, stability, and surface roughness conditions following Norman et al. (1995). The net radiation flux above the canopy (RN) is likewise partitioned between energy divergence within the canopy (RN_C) and energy available at the soil surface (RN_S) based on a two-stream model of radiative transfer within the canopy (Anderson et al., 2000), with a diurnally varying fraction of RN_S conducted into the soil (G) following Santanello and Friedl (2003).

To solve the system of equations describing the energy balance associated with the two flux sources, an initial estimate of the canopy component of latent heat, λE_C, describing transpiration expected for a well-watered canopy is defined using either a modified Priestley–Taylor approximation or an analytical canopy conductance model (Anderson et al., 2008). The soil evaporation rate, λE_S, is then computed as an overall residual to the system energy balance:

$$\lambda E_S = (RN_S + RN_C) - (H_S + H_C) - G - \lambda E_C \quad (15.3)$$

If the canopy transpiration is less than potential (e.g., because of soil moisture restrictions and stress-induced stomatal closure), λE_C will be overestimated and Equation 15.2 will lead to negative soil evaporation (i.e., condensation), which is not expected midday when polar-orbiting thermal sensors typically overpass. In this case, λE_C is throttled back until reasonable λE_S is obtained. ET (E) is merely the total latent heat flux, $\lambda E = \lambda E_C + \lambda E_S$ (W m^{-2}), expressed in units of mass flux (kg s^{-1} m^{-2} or mm s^{-1}), where λ is the latent heat of vaporization (J kg^{-1}).

15.2.3.1.2 Regional Application of the TSEB Model (ALEXI and DisALEXI)

Although the TSEB has been demonstrated to work well in local applications using in situ measurements of T_{RAD} and air temperature (T_A in Figure 15.9), direct regional implementation is confounded primarily by (1) errors in LST retrieval due to atmospheric and emissivity corrections and (2) errors in gridded boundary conditions in near-surface air temperature, T_A. These lead to errors in H and therefore in λE (ET) by residual. The ALEXI model (Anderson et al., 1997, 2007a) was designed to

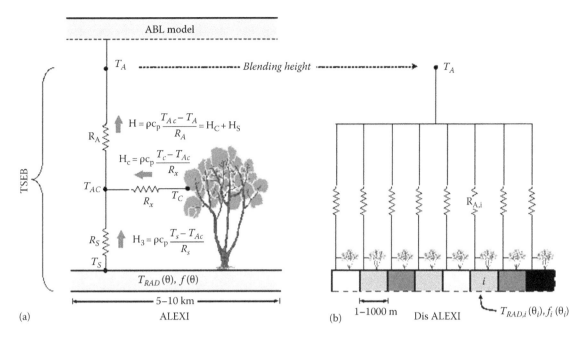

FIGURE 15.9 Schematic diagram representing the ALEXI (a) and DisALEXI (b) modeling schemes, highlighting fluxes of sensible heat (*H*) from the soil and canopy (subscripts "*s*" and "*c*") along gradients in temperature (*T*), and regulated by transport resistances R_A (aerodynamic), R_x (bulk leaf boundary layer), and R_S (soil surface boundary layer). DisALEXI uses the air temperature predicted by ALEXI near the blending height (T_A) to disaggregate 10-km ALEXI fluxes, given vegetation cover ($f(q)$) and directional surface radiometric temperature ($T_{RAD}(q)$) information derived from high-resolution remote-sensing imagery at look angle *q*.

provide a more robust framework for applying the TSEB regionally. This is accomplished by (1) applying TSEB in time-differential mode, rendering it sensitive primarily to time changes in LST that can be retrieved much more accurately than absolute (instantaneous) LST and (2) driving the time-differential TSEB with a modeled air temperature field governed by a simple slab model of morning atmospheric boundary layer (ABL) development. In (2), the air temperature boundary conditions are consistent with both the surface temperatures and modeled sensible heat flux, which serves to heat the air within the ABL over the morning hours.

ALEXI therefore requires at least two measurements of LST acquired during the morning, typically obtained from thermal imagers or sounders on geostationary satellite platforms with typical spatial resolution on the order of 3–10 km. For higher resolution applications, requiring field or sub-field-scale sampling, the Disaggregated Atmosphere-Land Exchange Inverse (DisALEXI) algorithm for spatially disaggregating ALEXI fluxes using TIR data from polar-orbiting sensors or airborne systems was developed (Norman et al., 2003). DisALEXI constitutes a single time-of-day application of TSEB in gridded mode, but using the ALEXI-modeled air temperature field T_A as a first guess (Figure 15.9). These air temperatures are iteratively modified until the DisALEXI sensible heat flux field aggregates to the ALEXI-derived field (Anderson et al., 2012). Again, no measurements of near-surface air temperature are required. Combined, ALEXI/DisALEXI provides multiscale ET modeling capabilities from field to continental to global scales.

15.2.3.1.3 ESI Algorithm

The ESI represents standardized anomalies in a normalized clear-sky ET ratio, f_{RET} = ET/ET$_{ref}$, where ET is the actual ET retrieved at midday using ALEXI or DisALEXI and ET$_{ref}$ is a reference ET scaling flux used to minimize impacts of nonmoisture-related drivers on ET (e.g., seasonal variations in radiation load) (Anderson et al., 2007b, 2011, 2013). Here, we use the FAO-96 Penman–Monteith reference ET for grass, as described by Allen et al. (1998). Anderson et al. (2013) compared several different scaling fluxes over the CONUS and found the FAO PM equation provided best agreement with drought classifications in the USDM and with soil moisture–based drought indices. It is hypothesized that use of clear-sky ET retrievals (as opposed to all-sky estimates) results in better separation of soil moisture–induced controls on ET from drivers related to variable radiation load such as cloud cover.

To compute ESI, daily values of clear-sky f_{RET} are composited over moving windows of 2-, 4-, 8-, and 12-week lengths, advancing in 7-day increments. Normal conditions (mean and standard deviation) are computed over the period of record (currently 2000–present) for each 7-day period and compositing interval. Finally, ESI is computed as a z-index, indicating the number of standard deviations (σ) the current composited f_{RET} value deviates from the normal moisture conditions for that period of time, with typical values ±3σ.

Daily maps of ESI over the CONUS are distributed throughout the growing season at http://hrsl.arsusda.gov/drought. The U.S. product currently is at 10 km resolution using TIR data from the Geostationary Operational

Environmental Satellites (GOES) Sounder instrument but will be upgraded to 4 km using GOES Imager data by mid-2014. Expanded domains covering North and South America are under development and will be made web accessible as they become routinely available. Also distributed at this site are maps of ESI change anomalies, indicating areas where ESI is rapidly decreasing (drought onset) or increasing (recovery) (Anderson et al., 2013). Studies by Otkin et al. (2013a,b) suggest that these change anomalies may have value as an early indicator of rapid drought development—capturing early thermal signals of increasing canopy stress during the so-called "flash drought" events.

15.2.3.2 Input Data

As noted earlier, ALEXI uses time-differential measurements of surface temperature change during the morning hours (between sunrise and local noon), typically obtained from geostationary platforms (ALEXI_GEO). This poses a challenge for international or global applications, requiring access to and integration of time-series thermal information from multiple satellites operated by different countries (e.g., the GOES [covering North and South America] and the Meteosat satellites [Europe and Africa]). Additionally, GEO satellites lose effectiveness for ET retrieval at latitudes beyond approximately ±60° where the view angle becomes significantly oblique. To circumvent these problems, a version of ALEXI (ALEXI_POLAR) based on day–night LST observations from polar-orbiting systems, such as MODIS on Terra or Aqua, was developed. This version conveniently provides global land coverage from a single satellite system, including the near-polar regions, at workable view angles.

Using DisALEXI, ALEXI_GEO fluxes can be spatially disaggregated to 1 km using MODIS or AVHRR LST products (approximately daily depending on cloud cover) and to 30 m resolution using sharpened TIR imagery from Landsat (approximately biweekly to monthly). Data fusion algorithms have been developed to combine GEO, MODIS, and Landsat ET retrievals to approximate daily coverage at 30 m resolution—the spatiotemporal requirements for many field-scale water management applications and for early detection of developing crop stress (Cammalleri et al., 2013, 2014).

In addition to LST, gridded information on fraction cover ($f(\theta)$ in Equation 15.2) is obtained from shortwave VIs and/or standard leaf area index products (e.g., from MODIS or Meteosat). Meteorological inputs of wind speed (required for resistances) and atmospheric temperature profile (required for the ABL growth component of ALEXI) are obtained from specialized mesoscale analyses or standard regional or global reanalysis datasets (e.g., North American Regional Reanalysis [NARR], Modern-Era Retrospective Analysis for Research and Applications [MERRA]). Land cover class in conjunction with cover fraction is used to specify canopy characteristics such as height and clumping factor, needed for roughness and canopy radiation transfer computations.

15.2.3.3 Examples of ESI

Figure 15.10 shows a 3-month ESI composite ending August 5, 2011, generated at 4 km resolution using LST from the GOES-East and GOES-West Imager systems over the CONUS. This can be compared to the USDM and VegDRI maps for mid-August shown in Figures 15.3 and 15.5. A full monthly time series and comparison with USDM maps is shown in Figure 15.12 in Anderson et al. (2013), with a zoom-in on the southern United States in Figure 15.13 in Otkin et al. (2013a). In addition to the severe drought that encompassed Texas for much of 2011, rapid changes in drought severity classifications in the USDM, indicative of flash drought onset, began in June over Arkansas and progressed to Missouri and northward in July. This drought was a result of not only lower-than-normal precipitation but also high winds and air temperature driving rapid depletion of remaining soil moisture reserves. The ALEXI energy balance model used to construct the ESI considers each of these ancillary factors, and it captured the expansion of the drought-affected area with good time fidelity (Otkin et al., 2013a).

A comparable map of global ESI for this same time period, created using ALEXI_MODIS, is shown in Figure 15.11. From a global perspective, we see that the 2011 drought in Texas extended well into Mexico. The severe drought leading to famine conditions in the Horn of Africa is also apparent in this image.

15.2.3.4 Applications of ESI

Given the strong physical relationship between the transpiration component of ET and vegetation health, ESI development to date has focused primarily on applications in the agricultural sector—in detecting crop stress, scheduling irrigation, estimating yield, and informing crop insurance programs. Utility of 4–10 km GEO-derived ESI products as an early predictor of yield is under investigation using data collected in the United States, Brazil, and Africa. In addition, 30 m ESI datasets generated using fusion of GEO/MODIS/Landsat imagery are being used to forensically dissect coarse-scale drought signals apparent in maps such as those in Figures 15.10 and 15.11. At this scale, stress signals can be segregated by crop type, which may improve yield predictability, particularly when combined with field-scale remotely derived phenology information that can help to identify when different crops are in stages that are most susceptible to moisture stress. The ultimate goal is a multiscale integrated drought assessment tool, with full global and regional coverage and ability to map at high spatial resolution in drought-impacted areas.

The primary limitation on the TIR-based ET estimates currently used in the ESI products is temporal sampling—imposed by the inability to retrieve LST through clouds using thermal band techniques. This can result in long periods of time where f_{RET} cannot be updated, particularly within persistently cloudy regions such as the Amazon and Equatorial Africa. This limitation may be addressed by combining TIR-derived moisture estimates with MW SM moisture retrievals within a data assimilation framework (as discussed in Section 15.2.4.3).

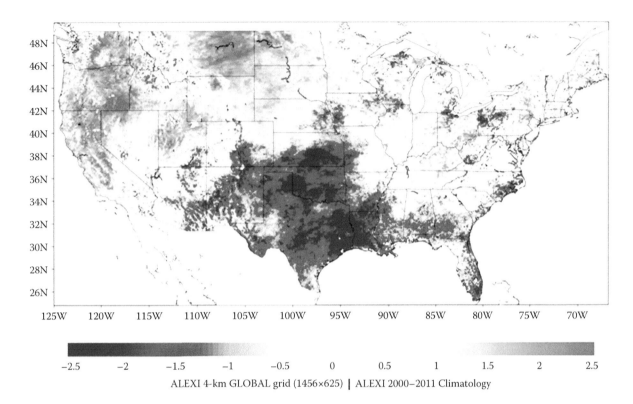

ALEXI 4-km GLOBAL grid (1456×625) | ALEXI 2000–2011 Climatology

FIGURE 15.10   Three-month ESI composite over the continental United States ending August 5, 2011, generated using ALEXI-GEO applied to LST retrievals from the GOES East and West imager instruments at 4-km spatial resolution.

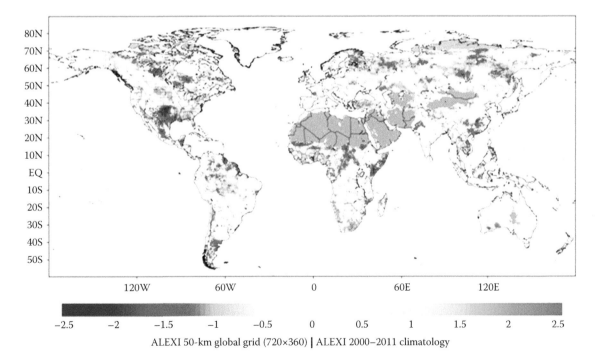

ALEXI 50-km global grid (720×360) | ALEXI 2000–2011 climatology

FIGURE 15.11   Three-month global ESI composite ending August 5, 2011, generated using ALEXI-MODIS applied to MODIS day/night LST and aggregated to 50-km spatial resolution.

15.2.4 Microwave-Based Surface Soil Moisture Retrievals

Soil moisture is a key parameter in drought monitoring because it strongly influences water availability to plants (reflecting agricultural drought conditions) and is also tied to longer-term hydrologic drought conditions. The retrieval of soil moisture conditions and the detection of negative anomalies in these conditions representative of drought are important for drought monitoring and early warning. In situ observations of soil moisture conditions are challenging, and existing networks are limited, particularly internationally. Satellite MW remote sensing is being looked upon to help complement in situ observations and fill in the informational gaps on spatial variations in soil moisture conditions.

Because of the strong contrast between the dielectric properties of water versus dry soil, land surface emissivity in the 1–10 GHz MW frequency range is strongly dependent on near-surface soil water content (Shutko, 1986). This sensitivity forms the physical basis of remote sensing algorithms that retrieve surface soil moisture content from satellite-based MW measurements. In contrast to visible and TIR-based sources of drought information discussed earlier, MW-based observations provide a relatively direct measurement of soil moisture. In addition, satellite-based MW retrievals can be acquired at relatively dense temporal frequencies (on the order of 3–5 retrievals per week) because of the development of wide-swath satellite measurement techniques and the lack of interaction between 1 and 10 GHz MW radiation and clouds. However, these advantages must be weighed against the poor spatial resolution of satellite-based surface soil moisture retrievals (typically >30 km), the relatively shallow vertical support of MW observations within the soil column (typically estimated to be on the order to 1–5 cm), and the tendency for decreased retrieval sensitivity under moderate and dense biomass conditions. Despite these challenges, the past few years have seen the first successful applications of MW-based surface soil moisture within operational drought monitoring systems (Bolten et al., 2010; Bolten and Crow, 2012).

15.2.4.1 Input Data

Surface soil moisture retrieval algorithms have been developed for both active and passive MW measurements of the land surface. Active measurements are based on measuring the fraction of transmitted energy scattered back toward a satellite-based radar emitter (and a co-located antennae). The spatial resolution of observed backscatter is frequently enhanced via synthetic aperture radar (SAR) processing whereby the motion of the satellite (or airborne) platform is used to increase the effective size of the receiving antennae. In contrast, passive MW techniques are based on the measurement of the natural gray-body emission of the land surface in the MW spectral region.

Input requirements vary slightly between active- and passive-based surface soil moisture retrieval techniques. Active retrieval techniques are typically based on simplified forward backscatter modeling, and they attempt to model the scattering of an incident MW energy pulse at the land surface as a function of land surface roughness, vegetation structure, and soil emissivity (i.e., soil wetness). The successful inversion of these models to retrieve absolute surface soil moisture values typically requires relatively detailed ancillary information concerning the statistical structure of surface roughness and the spatial distribution of water in the vegetation canopy (Wagner et al., 2007). However, the temporal contrast between the seasonal variation of soil roughness and vegetation properties and relatively shorter time scales of soil moisture variability have enabled the development of effective active MW-based change detection approaches with modest input requirements (Naeimi et al., 2009).

In contrast, passive MW observations are generally considered to be less sensitive to variations in surface roughness and vegetation structure. For example, soil moisture retrieval from passive brightness temperature observations is commonly performed using a so-called zero-order "tau-omega" model for surface MW emission, which reduces the parameterization of soil/vegetation to the specification of bulk vegetation water content, a single-scattering albedo, and a root-mean-square (rms) indicator of surface roughness (Jackson et al., 1982). However, unlike active measurements, the retrieval of surface emissivity from passive MW observations also requires an accurate estimate for the physical temperature of the land surface. Such temperatures are commonly obtained from either higher frequency 37 GHz MW observations or numerical weather prediction model output (Holmes et al., 2009, 2012). Finally, after land surface emissivity is estimated, it is typically converted into a soil moisture value based on knowledge of soil texture and the application of a soil moisture mixing model relating soil emissivity to volumetric soil water content.

It is also worth noting that a number of soil moisture retrieval algorithms can utilize multipolarization and/or multi-incident angle observations to simultaneously estimate both soil moisture and vegetation canopy water content (Owe et al., 2008). While less frequently utilized than soil moisture retrievals, MW-based canopy water content estimates have also demonstrated the ability to detect agricultural drought through variations in vegetation canopy water content (Liu et al., 2013).

15.2.4.2 Algorithm Development and Implementation

Early research on MW remote sensing revealed the optimality of relative low-frequency L-band (1–2 GHz) observations for soil moisture retrieval (Jackson, 1993). Higher frequency 2–20 GHz MW observations can also be used, but resulting soil moisture retrievals are subject to reduced sensitivity under moderate-to-dense vegetation cover (Jackson, 1993). Examples of higher-frequency passive MW retrievals include the use of 19 GHz special sensor microwave/imager (SSM/I) (Wen et al., 2005), 6.9 GHz special sensor microwave imager sounder (SMMR) (Owe et al., 2008), and 10.7 GHz Tropical Rainfall Measuring Mission (TRMM) Microwave Imager (TMI) (Gao et al., 2006) brightness temperature observations for the generation of retrospective surface soil moisture products. However, it was not until the 2002 launch of AMSR-E observations that such observations were used to generate an operational soil moisture product (Njoku et al., 2003). One disappointing development was the discovery of widespread radio-frequency interference (RFI) in the AMSR-E 6.9 GHz (C-band) channel over the United States and Japan (Li et al., 2004). This discovery necessitated the more widespread use of the 10.7 (X-band) AMSR-E channel for soil moisture retrievals.

Despite the relative success of these higher frequency-based products, the first dedicated L-band satellite sensor was not launched until the European Space Agency (ESA) Soil Moisture and Ocean Salinity (SMOS) mission in late 2009 (Kerr and Levine, 2008). SMOS has been successfully generating an L-band surface soil moisture product since early 2010, although it has been challenged by the presence of significant L-band RFI over areas of Europe and East Asia (Oliva et al., 2012).

For active retrievals, the development of global data products has primarily been limited to the use of scatterometer (as opposed to SAR) observing systems. The first global-scale products were based on the use of retrospective 5.3 GHz European Remote Sensing (ERS)-1/2 scatterometer observations (Wagner et al., 1999). This system was later upgraded into a real-time operational system applied to 5.2 GHz Meteorological Operational (MetOp) Advanced Scatterometer (ASCAT) retrievals (Naeimi et al., 2009). Despite theoretical concerns about the sensitivity of active remote sensing retrievals over moderate-to-dense vegetation areas, validation results suggest that scatterometer-based retrievals are surprisingly accurate over moderately vegetated areas (Crow and Zhan, 2007). However, the reliance of both active (non-SAR) scatterometer and passive MW soil moisture products on low-frequency MW observations has—to date—limited the ground spatial resolution satellite-derived surface soil moisture retrievals to >30 km.

A natural goal has been the design of satellite instrumentation capable of simultaneous active/passive MW observations. Although specifically designed for ocean salinity measurements, the Aquarius satellite mission provided the first example of a satellite-based dual/active L-band MW system and is currently being used to generate a soil moisture product (T.J. Jackson, personal communication). One intriguing possibility is integrating (relatively) high-sensitivity—but low-resolution—passive MW observations with lower-sensitivity but higher resolution active MW observations that have been subjected to SAR processing (Entekhabi et al., 2004, 2010). This strategy allows for retrieval of high-quality soil moisture data products down to resolutions as fine as 10 km and represents a primary motivation for the upcoming 2014 launch of the NASA Soil Moisture Active Passive (SMAP) mission. As a result, the SMAP mission will represent the first satellite-based sensor to leverage both the accuracy of L-band MW radiometry and the spatial resolution advantages of SAR processing.

15.2.4.3 Applications of Microwave-Based Surface Soil Moisture Retrieval Products

The shallow vertical penetration depth of MW surface moisture retrievals is typically assumed to represent a significant barrier to their successful integration into agricultural drought

monitoring activities (Albergel et al., 2008). The assumed vertical support of these measurements (less than 5 cm) samples only a small fraction of the 30–150 cm total column rooting depth commonly assigned to vegetation. As a result, the application of MW-based soil moisture retrievals to agricultural drought has generally been confined to their use in land data assimilation systems. Such systems are designed to optimally integrate continuous (but error-prone) dynamic models with (presumably) more accurate, but less temporally/spatially complete, observations that can be related to the states of the model. Ideally, the relative weighting of both sources of information is based on the relative uncertainty of the continuous forecast model versus the diagnostic observations. By calculating error covariance information between multiple model states and background error states (typically via a Monte Carlo ensemble) and applying Kalman filtering concepts, such systems can also be used to update (unobserved) profile soil moisture states beyond the near surface using only surface soil moisture retrievals.

Using ground-based soil moisture observations in the United States as source of validation, past results have demonstrated the ability of such land data assimilation systems to improve the accuracy of LSM surface and root-zone soil moisture predictions via the assimilation of a passive-based surface soil moisture retrieval product (Reichle and Koster, 2005).

Such systems are also easily adapted to integrate multiple types of satellite-based soil moisture products. For example, the added value of simultaneously assimilating both active- and passive-based soil moisture retrievals has been demonstrated (Draper et al., 2012). Data assimilation has also proven to be an effective technique for merging MW-based soil moisture retrieval with TIR-based drought monitoring products (discussed earlier in Section 15.2.3). In general, the two methods have been shown to be highly complementary (Li et al., 2011). TIR methods (such as ESI) provide relatively high spatial resolution on the order of the resolution of the particular TIR sensor (~100 m–10 km) and lower temporal resolution due to constraints on TIR-based retrievals being limited to clear-sky conditions (typical repeat cycles of 2–7 days for geostationary satellites [Hain et al., 2011]). In contrast, MW methods provide soil moisture retrievals at relatively low spatial resolution (25–60 km) and high temporal resolution (retrievals possible through nonprecipitating cloud cover with typical repeat cycles of 1–2 days). The two techniques also differ in their sensitivity to vegetation cover; TIR methods provide the opportunity to attain a soil moisture signal over a wide range of cover regimes, while MW methods can suffer significant degradation in accuracy as the density of vegetation cover increases.

Hain et al. (2012a,b) examined the efficacy of assimilating TIR-based ESI products and passive MW-based soil moisture retrievals using an ensemble Kalman filter (EnKF) in the Noah LSM to improve modeled soil moisture predictions. The methodology employed a data denial framework to quantify the ability of the two different soil moisture retrieval datasets to correct for errors in precipitation forcing (using a satellite-based precipitation product) in an open-loop simulation as compared to a control simulation forced with a higher quality, gauge-based precipitation dataset. The study found that the joint assimilation of TIR and MW soil moisture into the Noah LSM provided better soil moisture estimates than did retrieval method (either TIR or MW) in isolation. As expected, added value of TIR assimilation over MW alone is most significant in areas of moderate-to-dense vegetation cover, where MW retrievals have limited sensitivity to soil moisture at any depth. Ongoing research is aimed at implementing this joint assimilation technique in a near-real-time uncoupled framework, where the assimilation of TIR and MW soil moisture has the potential to provide improvements to the representation of soil moisture anomalies, particularly in the case of fast-developing "flash drought" scenarios when signals of vegetation stress are apparent in the ESI signal (through elevated canopy temperatures) before a degradation in vegetation health occurs. Additionally, efforts are being made to assess the impact of the SM assimilation framework in a coupled numerical weather prediction framework. Improvements in the representation of the initial soil moisture states have the potential to lead to improvements in temperature and precipitation forecasts produced by the National Weather Prediction models.

The ability of land data assimilation systems to improve the characterization of soil moisture anomalies appears to carry over to crop yield predictions. For example, Ines et al. (2013) documented improvements in corn yield predictions associated with the assimilation of MW-based soil moisture retrievals into a crop system model. Finally, for cases in which a water balance approach is not utilized, land data assimilation concepts can also be applied to correct satellite-based rainfall accumulations using a time series of MW surface soil moisture retrievals (Pellarin et al., 2008; Crow et al., 2011; Brocca et al., 2013).

In general, the magnitude of improvements associated with the application of a land data assimilation system increases as the quality of input data into the baseline model is degraded (Qing et al., 2011). For instance, using NDVI anomalies as a proxy for root-zone soil moisture availability in water-limited ecosystems, Bolten and Crow (2012) were able to map the global impact of assimilating passive MW-based soil moisture retrievals into a water balance model. In general, improvements were modest in areas of the world with high-quality operational rain gauge networks (which provide key forcing data for the water balance model), but much larger added value was noted in data-poor regions (e.g., Central Asia and the Horn of Africa) where operational rainfall information is unavailable (Figure 15.12). The USDA Foreign Agricultural Service has recently begun the

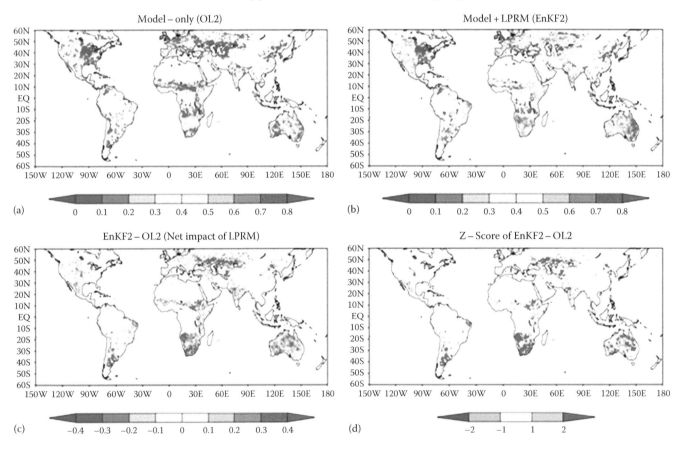

FIGURE 15.12 The lag +1 month correlation between future NDVI and current root-zone soil moisture predictions obtained from both: (a) a land surface model alone and (b) the land surface model enhanced by the assimilation of AMSR-E LPRM passive microwave surface soil moisture retrievals. (c) Plots the difference between parts (b) and (a) (i.e., the increase in correlation associated with the assimilation of the retrievals). (d) Relates the Z-score of the differences in part (c). (Taken from Bolten, J.D. and Crow, W.T., *Geophys. Res. Lett.*, 39, L19406, 2012, doi:10.1029/2012GL053470.)

operational generation of an SMOS-based root-zone data assimilation product to monitor soil moisture conditions in data-poor (and frequently food insecure) regions of the world (http://www.pecad.fas.usda.gov/cropexplorer/).

15.2.5 GRACE Satellite–Based Soil Moisture and Groundwater

Although surface and root-zone soil moisture play an important role in the characterization of agricultural drought, a complete description of hydrologic drought requires more vertically integrated information. The NASA/German GRACE satellite mission is uniquely able to provide global maps of anomalies (deviations from the long-term mean) of terrestrial water storage (TWS; the sum of groundwater, soil moisture, surface waters, snow, and ice). Thus, after accounting for natural seasonal variations in TWS, GRACE is well suited for drought assessment. Several studies have made direct use of GRACE data to quantify drought (e.g., Yirdaw et al., 2008; Chen et al., 2009; LeBlanc et al., 2009). In particular, Thomas et al. (2014) described how GRACE TWS data could be used to quantify the volume of water required by a region to return to normal water storage conditions and combined that storage deficit with event duration to assess drought severity.

However, the spatial and temporal resolutions of GRACE are low relative to other measurement types, and the data latency (typically 2–4 months) makes GRACE difficult to apply directly for operational drought monitoring. A solution is to integrate GRACE TWS data with other hydrologic observables that have higher resolutions and lower latency. A numerical LSM, which can be described as a sophisticated version of the land component of a weather or climate prediction model, is a useful vehicle for this sort of data integration. An added benefit is that data assimilation within an LSM allows vertical disaggregation of the GRACE TWS data into the TWS component contributions. As described in Section 15.2.5.3, such an approach has been successfully demonstrated and the results applied for operational drought monitoring in the United States.

15.2.5.1 Input Data

15.2.5.1.1 *GRACE Terrestrial Water Storage*

GRACE, a joint mission of NASA and the German Space Agency, launched in 2002. It comprises two satellites following each other around the earth in a near-polar orbit, roughly 200 km apart, at an initial altitude of about 500 km. Unlike most earth observing satellites, which measure light emitted or reflected from earth's surface, the GRACE satellites continuously and precisely measure the distance (and its rate of change) between the two using a K-band MW ranging system. Heterogeneities in earth's gravity field perturb the satellite orbits in a predictable manner, such that the intersatellite range measurements can be used to construct a new numerical model of earth's gravity field nominally once per month (Tapley et al., 2004). Global representations of earth's gravity field are produced as sets of spherical harmonic coefficients. These can be manipulated using numerical methods

such as Gaussian averaging functions in order to isolate temporal mass anomalies (deviations from the mean) over regions of interest (Wahr et al., 1998). The primary sources of temporal variations in earth's gravity field are mass variations in the oceans, atmosphere, and terrestrial water. By accounting for the first two using oceanic and atmospheric analysis models, it is thus possible to quantify monthly TWS anomalies (Wahr et al., 1998; Rodell and Famiglietti, 1999).

Challenges to using GRACE TWS anomaly data include their coarse spatial and temporal resolutions, the vertically integrated nature of the measurements, and data processing issues that include accounting for *leakage* of gravity signals from adjacent regions. There is a trade-off between spatial resolution and accuracy, such that 150,000 km^2 is the approximate minimum area that can be resolved with a reasonable degree of certainty at mid-latitudes (e.g., Swenson et al., 2006). While GRACE-based TWS fields are available from multiple sources, the examples below make use of gridded GRACE data products from the NASA Jet Propulsion Laboratory's Tellus website (http://grace.jpl.nasa.gov; Swenson and Wahr, 2006; Landerer and Swenson, 2012).

15.2.5.1.2 *Meteorological Data*

LSMs are not prognostic and therefore rely on meteorological inputs to drive them forward in time. The required forcing variables are precipitation, downward shortwave and longwave radiation, near-surface air temperature, specific humidity, wind speed, and surface pressure. In the example presented in the below in this paragraph, these forcing fields were obtained from two sources. One source is the Princeton Global Meteorological Forcing Dataset (Sheffield et al., 2006). The Princeton dataset spans from 1948 to 2010 with 3 h temporal and 1° spatial resolution. It was produced by combining observation-based datasets with a reanalysis product. The second source is the North American Land Data Assimilation System Phase 2 (NLDAS-2; Xia et al., 2012). The NLDAS-2 forcing dataset covers central North America from 1979 to the present and incorporates precipitation gauge observations, bias-corrected shortwave radiation, and surface meteorology reanalyses. The precipitation data have been temporally disaggregated using Stage II Doppler radar data, and the resulting 0.125° resolution dataset is considered to be one of the finest available anywhere.

15.2.5.2 Algorithm Development and Implementation

Zaitchik et al. (2008) developed and tested an algorithm for assimilating GRACE TWS data into the catchment LSM (CLSM; Koster et al., 2000). CLSM is well suited for GRACE data assimilation because it is one of the first LSMs to simulate groundwater storage, which is requisite for properly representing TWS as a variable that can be updated using the GRACE observations. The algorithm is a form of ensemble Kalman smoother (Evensen and van Leeuwen, 2000) that merges basin-scale monthly GRACE-derived TWS anomalies into CLSM using information on the uncertainty in both the observation and the model. Whereas Kalman filter data assimilation integrates observations as they become available and updates only the most recent model states, smoothers use information from a series of observations to update model states over a window of time. Smoothers are therefore well suited for GRACE observations, which are noninstantaneous. Prior to assimilation, the GRACE TWS anomalies are converted to absolute TWS values by adding the corresponding regional GRACE-period mean TWS from the open-loop (no assimilation) CLSM simulation.

Data assimilation enhances the value of GRACE observations in three ways. First, the spatial resolution improves from the scales of midsized river basins to the scales of counties or small catchments. Most practical applications benefit from data at the finest possible scales. Second, the resulting time series have subdaily temporal resolution, as opposed to monthly, and the latest fields can be generated within a day of real time. Standard GRACE hydrology products normally become available after a 2–5 month lag, which is unacceptable for operational applications. Third, GRACE assimilation output includes gridded maps of groundwater, soil moisture, and snow variations, which in most cases are easier to interpret than aggregate TWS. Hence, GRACE assimilation output combines the veracity of an observation with the fine spatial and temporal resolutions of a model. Uncertainty in GRACE data assimilation–based groundwater and soil moisture change estimates is highly variable and depends on a number of factors. Data assimilation finds the optimal estimate based on two or more independent estimates (one from the model and the other from the observation) and knowledge of the errors in each. Hence, the error in the result should be smaller than that in either of the input values. At the river basin scale (roughly 500,000 km² or larger), uncertainty in the monthly GRACE TWS anomalies is on the order of 1–2 cm equivalent height of water (Wahr et al., 2006). Uncertainty in the model estimates, which is determined based on ensemble spread, is similar in magnitude. At the fine spatial and temporal scales of the model output, the errors are likely to be larger than at the basin scale. Further, there is uncertainty associated with disaggregating TWS into its components, although error in a given component is expected to be smaller than the error in the total. Determining actual errors in assimilated groundwater or soil moisture output in a specific region and at a certain scale requires comparison with independent data. For example, Zaitchik et al. (2008) showed that modeled groundwater agreed better with independent in situ measurements after the assimilation of GRACE TWS observations (Figure 15.13), with rms errors of 1.9–4.0 cm equivalent height of water in the Mississippi River basin and its four major subbasins, compared with 2.4–6.3 cm for the open-loop simulation.

15.2.5.3 Application to Drought Monitoring

The ability to observe changes in deep soil moisture and groundwater from space is what makes GRACE unique as a tool for hydrology and drought monitoring. Unlike surface moisture conditions, which fluctuate rapidly with the weather, deep soil moisture and groundwater integrate meteorological conditions over weeks and months, making them natural gauges of drought

severity (Rodell, 2012). Downscaling and extension to near real time via data assimilation, as described in the previous subsection, make possible the application of GRACE data for operational drought monitoring.

Premised on this potential, Houborg et al. (2012) developed surface and root-zone soil moisture and groundwater drought indicators based on GRACE data assimilation results. They first evaluated improvements in the output fields of soil moisture and groundwater storage that resulted from assimilation of GRACE TWS data using in situ groundwater observations from the USGS and soil moisture observations from the Soil Climate Analysis Network (SCAN). Houborg et al. (2012) then introduced a process for deriving drought indicators, which has since been refined and is described below in the following paragraphs. These indicators are now being produced weekly at NASA Goddard Space Flight Center and distributed by the NDMC (http://drought.unl.edu/MonitoringTools/ NASAGRACEDataAssimilation.aspx). They are also used as inputs to the USDMs and NADMs.

The soil moisture and groundwater indicators are generated as follows. First, an open-loop simulation of CLSM is executed for the period 1948–2010 using as input a meteorological forcing dataset developed at Princeton University (Sheffield et al., 2006). The climatology of the open-loop simulation is the basis for quantification of wetness/dryness conditions as a probability of occurrence during that 63-year period. The monthly GRACE TWS anomalies are converted to absolute TWS by adding the temporal mean TWS from the open-loop LSM output. Thus, the mean field of the assimilated TWS output is assured to be nearly identical to that of the open loop (i.e., GRACE data assimilation does not introduce a bias).

Data assimilation begins in a separate branch of the CLSM simulation in 2002, when the GRACE data become available. At the same time in that branch simulation, NLDAS-2 meteorological forcing data replace the Princeton forcing data. NLDAS-2 data are of higher resolution and higher quality and extend to near-real time. However, while bias due to GRACE data assimilation is avoided as previously described, biases between the two forcing sources are significant enough that they must be accounted for and removed. Furthermore, the GRACE data (and therefore the assimilation results), although unbiased, could still have a larger or smaller range of variability than the open-loop LSM results at any given location. That is an important consideration because drought monitoring concerns the extremes. Therefore, two statistical adjustments are applied based on the approach of Reichle and Koster (2004). The first one matches the cumulative distribution functions (CDFs) of the NLDAS-2 forcing variables to the CDFs of the Princeton forcing variables. The same adjustment method is applied to the output variables to correct for biases and differences in amplitude between the assimilation-mode NLDAS2-forced model output and the open-loop Princeton-forced model output, using the results from the overlapping period (2002–2010).

Finally, weekly drought indicator fields for surface (top several centimeters) soil moisture, root-zone soil moisture, and

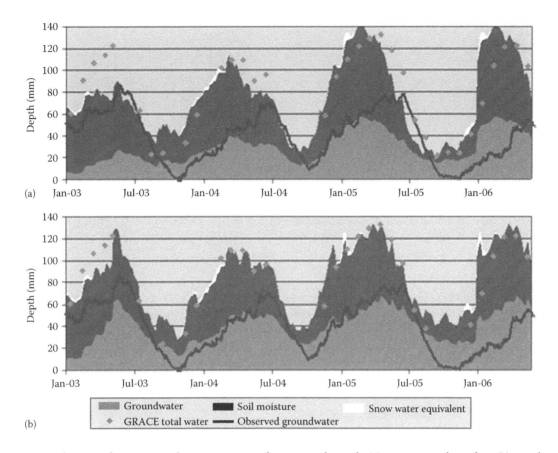

FIGURE 15.13 Groundwater, soil moisture, and snow water equivalent averaged over the Mississippi river basin from (a) open loop CLSM and (b) GRACE-DAS. Also shown are daily observation-based groundwater and monthly GRACE-derived TWS anomalies. GRACE and modeled TWS were adjusted to a common mean, as were observed and modeled groundwater. The correlation coefficient between simulated and observed groundwater improved from 0.59 (open loop) to 0.69 (GRACE-DAS). Unlike GRACE alone, the assimilated product is 3-h and vertically distributed. (From Zaitchik, B.F. et al., *J. Hydrometeorol.*, 9(3), 535, 2008.)

groundwater are computed based on probability of occurrence at the same location (grid pixel) during the same time of year (i.e., eliminating seasonal variations) in the 1948–2010 record, using CDFs as before. For ease of comparison with the USDM, wetness (dryness) conditions are characterized from D0 (abnormally dry) to D4 (exceptional), corresponding to cumulative probability percentiles of 20%–30%, 10%–20%, 5%–10%, 2%–5%, and 0%–2%.

Figure 15.14 compares June 2007 GRACE data assimilation–based drought indicators for surface and root-zone soil moisture and groundwater with the original GRACE TWS anomalies (seasonal cycle intact) and the USDM. Note that the GRACE-based indicators did not begin to be incorporated into the USDM (as a reference for the USDM authors) until 2012, and hence they are completely independent in this comparison. The level of agreement and disagreement typifies that displayed by the various GRACE products and USDM, though the patterns of correlation are variable. In this example, all five maps generally showed drought afflicting the west (save for the Pacific northwest), the southeastern states, and the area around Lake Superior. The western drought was more widespread in the four drought indicator maps than it was in the GRACE TWS anomaly

map, which highlights the importance of both accounting for the seasonal cycle and combining GRACE TWS with precipitation and other data. GRACE and the GRACE-based indicators showed the eastern drought extending north into Michigan and the northeastern states, while USDM did not. These discrepancies may have been due to errors in the GRACE-based maps or the fact that the USDM is conservatively adjusted each week by its authors and thus typically lags other indicators. In this case, by mid-July, the USDM did show drought in Michigan and part of Maine, and by September, it showed drought in New York and southern New England, although that is not definitive evidence that the GRACE-based maps were accurate leading indicators in June.

The three GRACE-based drought indicator maps are not identical because the deeper the water, the more slowly it responds to atmospheric conditions. Here, it can be seen that the surface soil was wetter than the root zone in Texas and northern Montana and North Dakota on June 25, 2007, possibly owing to recent rainfall, while the surface soil was drier than the root zone along the southwestern coast of California. The patterns of groundwater wetness percentiles were noticeably different from those of soil moisture in the central and western United States.

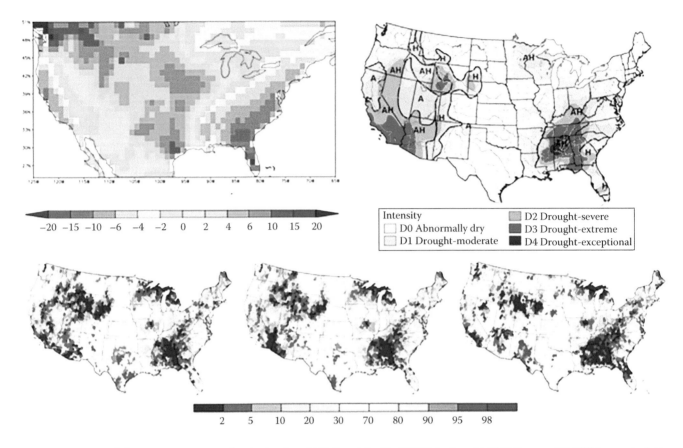

FIGURE 15.14   Top left: GRACE TWS anomalies (cm) in June 2007. Top right: The USDM for June 26, 2007. Bottom: GRACE data assimilation based drought indicators for surface soil moisture (left), root zone soil moisture (middle), and groundwater (right) on June 25, 2007.

In general, based on several years of comparisons, the USDM is more closely correlated with the root-zone soil moisture drought indicator than the other two. That does not suggest that the root-zone indicator is the best of the three; rather, it is indicative of the type of information incorporated into the USDM and the needs of the end users that it serves. Part of the value of the GRACE-based products is that they provide a type of drought indicator that previously didn't exist (maps of groundwater percentiles) and they help the USDM authors to distinguish agricultural (short-term) drought from hydrologic (long-term) drought (A and H on the USDM map).

GRACE launched in 2002 and has survived well past its original 5-year mission plan. At the time of writing, the satellites were suffering from aging battery cells, which caused the GRACE mission managers to shut down the instruments approximately 2–3 months per year (depending on the orbital cycle and the amount of time the satellite solar panels spend in the earth's shadow). The GRACE Follow-On Mission, which will by very similar to GRACE but with some technological improvements, is expected to launch in 2017. It is likely that a data gap will occur between the two missions, but the drought indicators can continue to be produced in the interim by running CLSM with the same atmospheric inputs but without the benefit of GRACE data assimilation.

15.3 Concluding Thoughts

Since the early 2000s, there has been a tremendous increase in the number and types of satellite-based earth observations being collected that can be used to monitor many of the major components of the hydrologic cycle. In parallel, the methods and tools to analyze and retrieve estimates of key hydrologic parameters such as ET, soil moisture, and groundwater have developed and matured over this same time period. As a result, many innovative remote sensing–based tools have been developed for drought monitoring in the United States and elsewhere. This chapter has reviewed many of the recent efforts to enhance drought monitoring capabilities for the United States using several types of satellite-based observations and modeling approaches that collectively provide insights into various aspects of the terrestrial hydrologic cycle relevant to the onset and intensification of drought conditions.

Further enhancements of these tools are expected as methods improve, new satellite observations become available, and the current satellite records are extended to provide longer climatologies for more meaningful anomaly detection. Upcoming missions such as SMAP and the GRACE Follow-On Mission (highlighted earlier) are prime examples of new observations that hold potential to continue to improve the characterization of soil moisture and other TWS conditions, particularly at higher spatial resolutions.

Other remote sensing efforts not summarized in this chapter provide additional opportunities to analyze other hydrologic parameters related to drought, such as snow cover and precipitation. Work is ongoing using both satellite-based thermal and passive MW observations to provide rainfall estimates globally, which can be used to better spatially define areas of negative rainfall anomalies that are indicative of drought conditions. Prime examples include extended historical records of rainfall estimates from these types of satellite observations using established rainfall-retrieval algorithms such as the TRMM 3B42 (Huffman et al., 2007), the Climate Prediction Center (CPC) MORPHing (CMORPH; Joyce et al., 2004), and the Precipitation Estimation from Remotely Sensed Information using Artificial Neural Networks (PERSIANN; Sorooshian et al., 2000). Other innovative work to improve the spatial resolution and precision of gridded climate datasets related to temperature and precipitation has been conducted using ground-based meteorological station measurements in concert with remotely sensed rainfall estimates to produce precipitation grids and satellite LST data to generate air temperature grids, as demonstrated by Funk et al. (2012) over eastern Africa and the Sahel region. Considerable progress has also been made in the use of both optical and MW satellite remote sensing for characterizing snow cover characteristics such as snow cover area and depth, as well as snow water equivalent (e.g., Kongoli et al., 2012). Snow cover information is an often overlooked but important parameter in drought assessments, as snow cover provides a critical water source for soil moisture and ground and surface water recharge. A lack of snow cover and water produced from snow melt and runoff can cause antecedent soil moisture and/or surface and groundwater conditions that make an area more susceptible for both agricultural and hydrologic droughts if dry conditions persist into the growing season or sustain the multiyear drought from the previous year. Additional new perspectives on surface water conditions will be provided by the NASA Surface Water and Ocean Topography (SWOT) instrument, which is being designed to estimate and monitor the water surface elevation of the ocean, rivers, and other inland water bodies (e.g., lakes and reservoirs). The SWOT mission, which is scheduled for launch in 2020, will use an interferometric altimeter that includes a Ka-band SAR interferometer within two antennas to estimate water surface elevation and associated change over time (Durand et al., 2010). This new capability to monitor water elevation changes of rivers and other surface water bodies will generate critical new information for drought monitoring and other water management activities by providing insights into locations and rates of surface water depletion that are associated with hydrologic drought.

These new satellite technologies are increasingly being paired with data assimilation systems to create value-added data products. These products are now being generated within the official product stream of satellite missions. For example, the SMAP Level 4 surface and root-zone soil moisture product will merge SMAP surface (0–5 cm) soil moisture retrievals with a continuous LSM to vertically extrapolate surface moisture information and constrain estimates of deeper root-zone soil moisture variations (Reichle et al., 2012). Comparable Level 4 data products are envisioned for the SWOT and GRACE Follow-On Missions.

Long-term continuity of key satellite observations required to sustain key drought monitoring variables such as the NDVI, thermal-based ET, and TWS is critical. Given that drought monitoring primarily relies upon the detection of anomalous conditions relative to a longer-term historical baseline condition, extended multidecade records of consistent satellite observations and data products from these various tools are needed to detect accurate and meaningful drought severity information. A notable example of such an effort is the development of a 20+-year historical record of 1 km AVHRR NDVI observations (1989 to the present) over the CONUS, which has represented the core remote sensing dataset for operational agricultural drought monitoring for many years. Similar but coarser resolution AVHRR NDVI datasets are also available globally and have often been used for drought assessments. However, the creation of such an extended dataset required several considerations, including strategies to compile an NDVI dataset from a series of NOAA AVHRR instruments that included both morning and afternoon data acquisition times. Other recent efforts have utilized NDVI data from the more recent NASA MODIS instruments, which have a shorter observational record dating back to 2000. Given that MODIS is a research rather than operational sensor and the series of AVHRR instruments available is now exhausted, the extension of NDVI datasets to observations from the new operational NOAA VIIRS instrument is crucial to maintain effective NDVI-based drought monitoring tools. The development of a long intersensor NDVI dataset that extends AVHRR and/or MODIS NDVI to the VIIRS era of observations will require careful consideration of the varying spectral resolutions of the input bands among these sensors, as well as sensor degradation within the data record, which could introduce anomalous trends/variations in the NDVI record that mimic a drought signal. Several efforts have attempted to intercalibrate the spectral data from the series of NOAA AVHRR sensors (Tucker et al., 2005; Eidenshink, 2006) and merge NDVI observations from AVHRR with MODIS (van Leeuwen et al., 2006) to create extended multisensor NDVI data records. Similar intersensor, crosswalking work of spectral and NDVI data among these sensors will be required in the new VIIRS era to provide useful information for drought monitoring.

Similar multisensor retrospective analysis can be applied to other drought variables. For example, the ESA Climate Change Initiative has recently released a 30-year surface (1979–2009) soil moisture retrospective data product based on the merger of active and passive MW observations obtained from a variety of satellite sensors (Wagner et al., 2012). Analogous efforts to extend the observational record of other remote sensing products such as ESI and TWS are also possible with the launch of new space-borne sensors collecting the required data inputs (e.g., TIR and gravity, respectively) for these tools. However, similar intersensor calibrations and/or model modifications will likely be necessary to ensure that a temporally consistent historical data record is maintained.

Remote sensing of drought has rapidly evolved and tremendous progress has been made over the past decade in the development and use of satellite-based tools to improve operational drought monitoring in the United States. The complexity and multiple dimensions of this complex natural hazard require a set of tools rather than a single tool to effectively map and monitor drought conditions. The remote sensing products highlighted in this chapter represent an emerging tool kit that will be capable of monitoring many key components of the hydrologic cycle related to drought across a vertical continuum ranging from subsurface groundwater and soil moisture to general plant health and surface-boundary layer conditions (i.e., ET). The development of these tools and products has been of tremendous benefit to improving the drought characterizations by the USDM, which is the primary national-level drought assessment tool for the United States. The benefit of many of these tools is now beginning to be realized in other parts of the world as these remote sensing–based methods are transferred internationally to other drought monitoring activities. The role and contribution of satellite-based remote sensing tools will continue to be important as the need for more local-scale drought information across multiple time scales (e.g., weeks, months, and seasons) increases for decision making at local, state, regional, and national scales across multiple sectors (i.e., agriculture, water resources, energy, insurance, and public health).

References

Albergel, C., C. Rüdiger, T. Pellarin, J.C. Calvet, N. Fritz, F. Froissard, D. Suquia, A. Petitpa, B. Piguet, and E. Martin. 2008. From near-surface to root-zone soil moisture using an exponential filter: An assessment of the method based on in situ observations and model simulations. *Hydrologic Earth System Science* 12:1323–1337.

Allen, R.G., L.S. Pereira, D. Raes, and M. Smith. 1998. Crop evapotranspiration: Guidelines for computing crop water requirements. United Nations FAO, Irrigation and Drainage Paper 56. Rome, Italy.

Anderson, M.C., C.R. Hain, J.A. Otkin, X. Zhan, K.C. Mo, M. Svoboda, B. Wardlow, and A. Pimstein. 2013. An intercomparison of drought indicators based on thermal remote sensing and NLDAS-2 simulations with U.S. Drought Monitor classifications. *Journal of Hydrometeorology* 14:1035–1056.

Anderson, M.C., C.R. Hain, B. Wardlow, J.R. Mecikalski, and W.P. Kustas. 2011. Evaluation of drought indices based on thermal remote sensing of evapotranspiration over the continental U.S. *Journal of Climate* 24:2025–2044.

Anderson, M.C., W.P. Kustas, J.G. Alfieri, C.R. Hain, J.H. Prueger, S.R. Evett, P.D. Colaizzi, T.A. Howell, and J.L. Chavez. 2012. Mapping daily evapotranspiration at Landsat spatial scales during the BEAREX'08 field campaign. *Advances in Water Resources* 50:162–177.

Anderson, M.C., J.M. Norman, G.R. Diak, W.P. Kustas, and J.R. Mecikalski. 1997. A two-source time-integrated model for estimating surface fluxes using thermal infrared remote sensing. *Remote Sensing of Environment* 60:195–216.

Anderson, M.C., J.M. Norman, W.P. Kustas, R. Houborg, P.J. Starks, and N. Agam. 2008. A thermal-based remote sensing technique for routine mapping of land-surface carbon, water and energy fluxes from field to regional scales. *Remote Sensing of Environment* 112:4227–4241.

Anderson, M.C., J.M. Norman, J.R. Mecikalski, J.A. Otkin, and W.P. Kustas. 2007a. A climatological study of evapotranspiration and moisture stress across the continental U.S. based on thermal remote sensing: I. Model formulation. *Journal of Geophysical Research* 112, D10117, doi:10110.11029/12006JD007506.

Anderson, M.C., J.M. Norman, J.R. Mecikalski, J.A. Otkin, J.A., and W.P. Kustas. 2007b. A climatological study of evapotranspiration and moisture stress across the continental U.S. based on thermal remote sensing: II. Surface moisture climatology. *Journal of Geophysical Research* 112, D11112, doi:11110.11029/12006JD007507.

Anderson, M.C., J.M. Norman, T.P. Meyers, and G.R. Diak, G.R. 2000. An analytical model for estimating canopy transpiration and carbon assimilation fluxes based on canopy light-use efficiency. *Agricultural and Forest Meteorology* 101:265–289.

Asner, G.P., A.R. Townsend, and B.H. Braswell. 2000. Satellite observation of El Niño effects on Amazon forest phenology and productivity. *Geophysical Research Letters* 27:981–984.

Asrar, G., R.B. Myneni, and E.T. Kanemasu. 1989. Estimation of plant canopy attributes from spectral reflectance measurements. In G. Asrar (ed.), *Theory and Applications of Optical Remote Sensing* (pp. 252–296). New York: Wiley Publishers.

Baret, F. and G. Guyot. 1991. Potentials and limits to vegetation indices for LAI and APAR assessments. *Remote Sensing of Environment* 35:161–173.

Barnston, A.G., A. Kumar, L. Goddard, and M.P. Hoerling. 2005. Improving seasonal prediction practices through attribution of climate variability. *Bulletin of the American Meteorological Society* 86:59–72.

Berhan, G., S. Hill, T. Tadesse, and S. Atnafu. 2014. Drought prediction system for improved climate change mitigation. *IEEE Transactions on Geoscience and Remote Sensing* 52(7):4032–4037.

Bolten, J.D. and W.T. Crow. 2012. Improved prediction of quasi-global vegetation conditions using remotely-sensed surface soil moisture. *Geophysical Research Letters* 39, L19406, doi:10.1029/2012GL053470.

Bolten, J.D., W.T. Crow, X. Zhan, T.J. Jackson, and C.A. Reynolds. 2010. Evaluating the utility of remotely sensed soil moisture retrievals for operational agricultural drought monitoring. *IEEE Journal of Selected Topics in Applied Earth Observations and Remote Sensing* 3(1):57–66.

Boyd, D.S., P.C. Philips, G.M. Foody, and R.P.D. Walsh. 2002. Exploring the utility of NOAA AVHRR middle infrared reflectance to monitor the impacts of ENSO-induced drought stress on Sabah rainforest. *International Journal of Remote Sensing* 23:5141–5147.

Brocca, L., F. Melone, T. Moramarco, and W. Wagner. 2013. A new method for rainfall estimation through soil moisture observations. *Geophysical Research Letters* 40(5):853–858, doi:10.1002/grl.50173.

Brown, J.F., B.D. Wardlow, T. Tadesse, M.J. Hayes, and B.C. Reed. 2008. The Vegetation Drought Response Index (VegDRI): A new integrated approach for monitoring drought stress in vegetation. *GIScience and Remote Sensing* 45(1):16–46.

Burgan, R.E., R.A. Hartford, and J.C. Eidenshink. 1996. Using NDVI to assess departure from average greenness and its relation to fire business. General Technical Report. INT-GTR-333, U.S. Department of Agriculture, Forest Service, Intermountain Research Station, Ogden, UT, 8pp.

Cammalleri, C., M.C. Anderson, F. Gao, C.R. Hain, and W.P. Kustas. 2013. A data fusion approach for mapping daily evapotranspiration at field scale. *Water Resources Research* 49:1–15, doi:10.1002/wrcr.20349.

Cammalleri, C., M.C. Anderson, F. Gao, C.R. Hain, and W.P. Kustas. 2014. Mapping daily evapotranspiration at field scales over rainfed and irrigated agricultural areas using remote sensing data fusion. *Agricultural and Forest Meteorology* 186:1–11.

Chen, J.L., C.R. Wilson, B.D. Tapley, Z.L. Yang, and G.Y. Niu. 2009. 2005 drought event in the Amazon River basin as measured by GRACE and estimated by climate models. *Journal of Geophysical Research* 114:B05404, doi:10.1029/2008JB006056.

Crow, W.T., M.J. van Den Berg, G.F. Huffman, and T. Pellarin. 2011. Correcting rainfall using satellite-based surface soil moisture retrievals: The Soil Moisture Analysis Rainfall Tool (SMART). *Water Resources Research* 47, W08521, doi:10.1029/2011WR010576.

Crow, W.T. and X. Zhan. 2007. Continental-scale evaluation of remotely-sensed soil moisture products. *IEEE Geoscience and Remote Sensing Letters* 4(3):451–455.

Dai, A. 2012. Increasing drought under global warming in observations and models. *Nature Climate Change* 3(1):52–58.

Department of the Interior (DOI). 2013. Final Environmental Assessment—Carson City District Drought Management. DOI-BLM-NV_C000-2013-0001-EA report, 236 pp.

Draper, C.S., R.H. Reichle, G.J.M. De Lannoy, and Q. Liu. 2012. Assimilation of passive and active microwave soil moisture retrievals. *Geophysical Research Letters* 39:L04401.

Durand, M., L.L. Fu, D.P. Lettenmaier, D. Alsdorf, E. Rodriguez, and D. Esteban-Fernandez. 2010. The Surface Water and Ocean Topography mission: Observing terrestrial surface water and oceanic submesoscale eddies. *Proceedings of IEEE* 98(5):766–779.

Eidenshink, J.C. 2006. A 16-year time series of 1 km AVHRR satellite data of the conterminous United States and Alaska. *Photogrammetric Engineering and Remote Sensing* 72:1027–1035.

Entekhabi, D., E. Njoku, P. Houser, M. Spencer, T. Doiron, J. Smith, R. Girard et al. 2004. The Hydrosphere State (HYDROS) mission concept: An earth system pathfinder for global mapping of soil moisture and land freeze/thaw. *IEEE Transactions on Geoscience and Remote Sensing* 42(10):2184–2195.

Entekhabi, D., E. Njoku, P. O'Neill, K. Kellogg, W. Crow, W. Edelstein, J. Entin et al. 2010. The Soil Moisture Active and Passive (SMAP) mission. *Proceedings of the IEEE* 98(5):704–716.

Evensen, G. and P.J. van Leeuwen. 2000. An Ensemble Kalman smoother for nonlinear dynamics. *Monthly Weather Review* 128:1852–1867.

Funk, C., J. Michaelsen, and M.T. Marshall. 2012. Mapping recent decadal climate variations in precipitation and temperature across eastern Africa. In B.D. Wardlow, M.A. Anderson, and J. Verdin (eds.), *Remote Sensing of Drought: Innovative Monitoring Approaches* (pp. 331–358). Boca Raton, FL: CRC Press/Taylor & Francis.

Gao, H., E.F. Wood, T.J. Jackson, M. Drusch, and R. Bindlish. 2006. Using TRMM/TMI to retrieve surface soil moisture over the southern United States from 1998 to 2002. *Journal of Hydrometeorology* 7:23–38.

Gu, Y., J.F. Brown, J.P. Verdin, and B. Wardlow. 2007. A five-year analysis of MODIS NDVI and NDWI for grassland drought assessment over the central Great Plains of the United States. *Geophysical Research Letters* 34, 6pp. doi:10.1029/2006GL029127.

Gu, Y., E. Hunt, B. Wardlow, J.B. Basara, J.F. Brown, and J.P. Verdin. 2008. Evaluation of MODIS NDVI and NDWI for vegetation drought monitoring using Oklahoma Mesonet soil moisture data. *Geophysical Research Letters* 35, 5pp. doi:10.1029/2008GL035772.

Hain, C.R., W.T. Crow, M.C. Anderson, and J.R. Mecikalski. 2012a. Developing a dual assimilation approach for thermal infrared and passive microwave soil moisture retrievals. *Water Resources Research* 48:W11517, doi:11510.11029/12011WR011268.

Hain, C.R., W.T. Crow, M.C. Anderson, and J.R. Mecikalski. 2012b. An EnKF dual assimilation of thermal-infrared and microwave satellite observations of soil moisture into the Noah land surface model. *Water Resources Research* 48:W11517, doi:10.1029/2011WR011268.

Hain, C.R., W.T. Crow, J.R. Mecikalski, M.C. Anderson, and T. Holmes. 2011. An intercomparison of available soil moisture estimates from thermal-infrared and passive microwave remote sensing and land-surface modeling. *Journal of Geophysical Research* 116:D15107, doi:15110.11029/12011JD015633.

Holmes, T., R. de Jeu, M. Owe, and A. Dolman. 2009. Land surface temperature from Ka band passive microwave observations. *Journal of Geophysical Research of the Atmosphere* 114:D04113, doi:10.1029/2008JD010257.

Holmes, T., J. Jackson, R. Reichle, and J. Basara. 2012. An assessment of surface soil temperature products from numerical weather prediction models using ground-based measurements. *Water Resources Research* 48:W02531, doi:10.1029/2011WR010538.

Homer, C., C. Huang, L. Yang, B. Wylie, and M. Coan. 2004. Development of a 2001 national land cover database for the United States. *Photogrammetric Engineering and Remote Sensing* 70:829–840.

Houborg, R., M. Rodell, B. Li, R. Reichle, and B. Zaitchik. 2012. Drought indicators based on model assimilated GRACE terrestrial water storage observations. *Water Resources Research* 48:W07525, doi:10.1029/2011WR011291.

Huffman, G., R. Alder, D. Bolvin, G. Gu, E. Nelkin, K. Bowman, Y. Hong, E.F. Stocker, and D.B. Wolff. 2007. The TRMM multiscale precipitation analysis: Quasi-global, multiyear combined-sensor precipitation estimates at fine scale. *Journal of Hydrometeorology* 8:38–55.

Hutchinson, C.F. 1991. Use of satellite data for famine early warning in sub-Saharan Africa. *International Journal of Remote Sensing* 12:1405–1421.

Ines, A.V.M., N.N. Das, J.W. Hansen, and E.G. Njoku. 2013. Assimilation of remotely sensed soil moisture and vegetation with a crop simulation model for maize yield prediction. *Remote Sensing of Environment* 138:149–164.

Jackson, T.J. 1993. Measuring surface soil moisture using passive microwave remote sensing. *Hydrological Processes* 7(2):139–152.

Jackson, T.J., T.J. Schmugge, and J.R. Wang. 1982. Passive microwave remote sensing of soil moisture under vegetation canopies. *Water Resources Research* 18:1137–1142.

Jenkerson, C., T. Maiersperger, and G. Schmidt. 2010. eMODIS—A user-friendly data source. U.S. Geological Survey Open-File Report 2010–1055. http://pubs.usgs.gov/of/2010/1055/ (accessed March 28, 2014).

Ji, L. and A.J. Peters. 2003. Assessing vegetation response to drought in the northern great plains using vegetation and drought indices. *Remote Sensing of Environment* 87:85–98.

Joyce, R., J. Janowiak, P. Arkin, and P. Xie. 2004. CMORPH: A method that produces global precipitation estimates from passive microwave and infrared data at high spatial and temporal resolution. *Journal of Hydrometeorology* 5:487–503.

Karnieli, A., N. Agam, R.T. Pinker, M. Anderson, M.L. Imhoff, G.G. Gutman, N. Panov, and A. Goldberg. 2010. Use of NDVI and land surface temperature for drought assessment: Merits and limitations. *Journal of Climate* 23(3):618–633.

Karnieli, A. and G. Dall'Olmo. 2003. Remote-sensing monitoring of desertification, phenology, and drought. *Management of Environmental Quality: An International Journal* 41(1):22–38.

Keetch, J.J. and G.M. Byram. 1968. A drought index for forest fire control. Research Paper SE-38. Asheville, NC: U.S. Department of Agriculture, Forest Service, Southeastern Forest Experiment Station, 32pp. (Revised 1988).

Kerr, Y.H. and D. Levine. 2008. Forward to the special issue on the Soil Moisture and Ocean Salinity (SMOS) mission. *IEEE Transaction in Geoscience and Remote Sensing* 46(3):583–585.

Kogan, E.N. 1990. Remote sensing of weather impacts on vegetation. *International Journal of Remote Sensing* 11:1405–1419.

Kogan, F. 1995. Application of vegetation index and brightness temperature for drought detection. *Advances in Space Research* 15:91–100.

Kogan, F. and J. Sullivan. 1993. Development of global drought-watch system using NOAA/AVHRR data. *Advances in Space Research* 13:219–222.

Kongoli, C., P. Romanov, and R. Ferraro. 2012. Snow cover monitoring from remote-sensing satellites: Possibilities for drought assessment. In B.D. Wardlow, M.A. Anderson, and J. Verdin (eds.), *Remote Sensing of Drought: Innovative Monitoring Approaches* (pp. 359–386). Boca Raton, FL: CRC Press/Taylor & Francis.

Koster, R.D., M.J. Suarez, A. Duchame, M. Stieglitz, and P. Kumar. 2000. A catchment-based approach to modeling land surface processes in a general circulation model: 1. Model structure. *Journal of Geophysical Research: Atmospheres* 105(D20): 24809–24822.

Kustas, W.P. and J.M. Norman. 1999. Evaluation of soil and vegetation heat flux predictions using a simple two-source model with radiometric temperatures for partial canopy cover. *Agricultural and Forest Meteorology* 94:13–29.

Kustas, W.P. and J.M. Norman. 2000. A two-source energy balance approach using directional radiometric temperature observations for sparse canopy covered surfaces. *Agronomy Journal* 92:847–854.

Landerer, F.W. and S.C. Swenson. 2012. Accuracy of scaled GRACE terrestrial water storage estimates. *Water Resources Research* 48:W04531, 11 PP, doi:10.1029/2011WR011453.

Leblanc, M.J., P. Tregoning, G. Ramillien, S.O. Tweed, and A. Fakes. 2009. Basin-scale, integrated observations of the early 21st century multiyear drought in southeast Australia. *Water Resources Research* 45:W04408, doi:10.1029/2008WR007333.

Li, L., E.G. Njoku, E. Im, P.S. Chang, and K.S. Germain. 2004. A preliminary survey of radio-frequency interference over the US in Aqua AMSR-E data. *IEEE Transactions in Geoscience and Remote Sensing* 42:380–390.

Liu, Q., R.H. Reichle, R. Bindish, M.H. Cosh, W.T. Crow, R. de Jeu, G. De Lannoy, G.J. Huffman, and T.J. Jackson. 2011. The contributions of precipitation and soil moisture observations to the skill of soil moisture estimates in a land data assimilation system. *Journal of Hydrometeorology* 12(5):750–765.

Liu, Y.Y., A.I.J.M. van Dijk, M.F. McCabe, J.P. Evans, and R.A.M. de Jeu. 2013. Drivers of global vegetation biomass trends (1988–2008) and attribution to environmental and human drivers. *Global Ecology and Biogeography*, 22(6):692–705.

Los, S.O., G.J. Collatz, L. Bounoua, P.J. Sellers, and C.J. Tucker. 2001. Global interannual variations in sea surface temperature and land surface vegetation, air temperature, and precipitation. *Journal of Climate* 14:1535–1549.

Lyon, B. 2004. The strength of El Niño and the spatial extent of tropical drought. *Geophysical Research Letters* 31:L21204, doi:10.1029/2004GL020901.

McKee, T.B., N.J. Doesken, and J. Kleist. 1995. Drought monitoring with multiple time scales. *Ninth Conference on Applied Climatology*, Dallas, TX, January 15–20.

McVicar, T.R. and P.B. Bierwirth. 2001. Rapidly assessing the 1997 drought in Papua New Guinea using composite AVHRR imagery. *International Journal of Remote Sensing* 22:2109–2128.

Mishra, A.K. and V.P. Singh. 2010. A review of drought concepts. *Journal of Hydrology* 391(1):202–216.

Naeimi, V., K. Scipal, Z. Bartalis, S. Hasenauer, and W. Wagner. 2009. An improved soil moisture retrieval algorithm for ERS and METOP scatterometer observations. *IEEE Transactions on Geoscience and Remote Sensing* 47(7):1999–2013.

National Climatic Data Center (NCDC). 2013. Climate Monitoring: Temperature, Precipitation, and Drought. http://www.ncdc.noaa.gov/climate-monitoring/ (last accessed February 10, 2014).

National Climatic Data Center (NCDC). 2014. Billion Dollar U.S. Weather Disasters. http://www.ncdc.noaa.gov/oa/reports/billionz.html (last accessed February 7, 2014).

Njoku, E.G., T.J. Jackson, V. Lakshmi, T.K. Chan, and S.V. Nghiem. 2003. Soil moisture retrieval from AMSR-E. *IEEE Transactions on Geoscience and Remote Sensing* 41(2):215–229.

Norman, J.M., M.C. Anderson, W.P. Kustas, A.N. French, J.R. Mecikalski, R.D. Torn, G.R. Diak, T.J. Schmugge, and B.C.W. Tanner. 2003. Remote sensing of surface energy fluxes at 10^1-m pixel resolutions. *Water Resources Research* 39, 18pp. doi:10.1029/2002WR001775.

Norman, J.M., W.P. Kustas, and K.S. Humes. 1995. A two-source approach for estimating soil and vegetation energy fluxes from observations of directional radiometric surface temperature. *Agricultural and Forest Meteorology* 77:263–293.

Oliva, R., E. Daganzo, Y.H. Kerr, S. Mecklenburg, S. Nieto, P. Richaume, and C. Gruhier. 2012. SMOS radio frequency interference scenario: Status and actions taken to improve the RFI environment in the 1400–1427-MHz passive band. *IEEE Transactions on Geoscience and Remote Sensing* 50(5):1427–1439.

Omernik, J.M. 1987. Ecoregions of the conterminous United States. *Annals of the Association of American Geographers* 77(1):118–125.

Otkin, J.A., M.C. Anderson, C.R. Hain, L.E. Mladenova, J.B. Basara, and M. Svoboda. 2013a. Examining rapid onset drought development using the thermal infrared based Evaporative Stress Index. *Journal of Hydrometeorology* 14:1057–1074.

Otkin, J.A., M.C. Anderson, C.R. Hain, and M. Svoboda. 2013b. Examining the relationship between drought development and rapid changes in the Evaporative Stress Index. *Journal of Hydrometeorology*, 15, 19 pp.

Owe, M., R. de Jeu, and T. Holmes. 2008. Multisensor historical climatology of satellite-derived global land surface moisture. *Journal of Geophysical Research* 113:F01002, doi:10.1029/2007JF000769.

Palmer, W.C. 1965. Meteorological drought. Research Paper Number 4, Office of Climatology, U.S. Weather Bureau, Washington, DC, 58pp.

Palmer, W.C. 1968. Keeping track of crop moisture conditions, nationwide: The new crop moisture index. *Weatherwise* 21(4):156–161.

Pellarin, T., A. Ali, F. Chopin, I. Jobard, and J.-C. Bergs. 2008. Using spaceborne surface soil moisture to constrain satellite precipitation estimates over West Africa. *Geophysical Research Lett*ers 35:L02813, doi:10.1029/2007GL032243.

Pervez, M.S. and J.F. Brown. 2010. Mapping irrigated lands at 250-m scale by merging MODIS data and national agricultural statistics. *Remote Sensing* 2:2388–2412.

Peters, A.J., D.C. Rundquist, and D.A. Wilhite. 1991. Satellite detection of the geographic core of the 1988 Nebraska drought. *Agricultural and Forest Meteorology* 57:35–47.

Peters, A.J., E.A. Walter-Shea, L. Ji, A. Vina, M. Hayes, and M.D. Svoboda. 2002. Drought monitoring with NDVI-based Standardized Vegetation Index. *Photogrammetric Engineering and Remote Sensing* 68(1):71–75.

Qing, L., R. Reichle, R. Bindlish, M.H. Cosh, W.T. Crow, R. de Jeu, G. de Lannoy, G.J. Huffman, and T.J. Jackson. 2011. The contributions of precipitation and soil moisture observations to the skill of soil moisture estimates in a land data assimilation system. *Journal of Hydrometeorology* 12(5):750–765.

Quinlan, J.R. 1993. *C4.5 Programs for Machine Learning*. San Mateo, CA: Morgan Kaufmann Publishers.

Reed, B.C., J.F. Brown, D. VanderZee, T.R. Loveland, J.W. Merchant, and D.O. Ohlen. 1994. Measuring phenological variability from satellite imagery. *Journal of Vegetation Science* 5:703–714.

Reichle, R.H., W.T. Crow, R.D. Koster, J. Kimball, and G. De Lannoy. 2012. Algorithm theoretical basis document: SMAP Surface and Root-zone Soil Moisture (L4_SM) (version 1). *NASA SMAP Mission Documentation.* 32(2), http://smap.jpl.nasa.gov/files/smap2/L4_SM_InitRel_v1.pdf.

Reichle, R.H. and R.D. Koster. 2004. Bias reduction in short records of satellite soil moisture. *Geophysical Research Letters* 31:L19501, doi:10.1029/2004GL020938.

Reichle, R.H. and R.D. Koster. 2005. Global assimilation of satellite surface soil moisture retrievals into the NASA Catchment land surface model. *Geophysical Research Letters* 32:L02404, doi:10.1029/2004GL021700.

Rodell, M. 2012. Satellite gravimetry applied to drought monitoring. In B. Wardlow, M. Anderson, and J. Verdin (eds.), *Remote Sensing of Drought: Innovative Monitoring Approaches* (pp. 261–280). Boca Raton, FL: CRC Press/Taylor & Francis.

Rodell, M. and J.S. Famiglietti. 1999. Detectability of variations in continental water storage from satellite observations of the time dependent gravity field. *Water Resources Research* 35:2705–2723.

Rodell, M. and J.S. Famiglietti. 2002. The potential of satellite-based monitoring of groundwater storage changes using GRACE: The High Plains aquifer, Central US. *Journal of Hydrology* 263(1–4):245–256.

Rouse, J.W. Jr., R.H. Haas, J.A. Schell, D.W. Deering, and J.C. Harlan. 1974. Monitoring the vernal advancement and retrogradation (green wave effect) of natural vegetation. *NASA/GSFC Type III Final Report*, Greenbelt, MD.

Santanello, J.A. and M.A. Friedl. 2003. Diurnal variation in soil heat flux and net radiation. *Journal of Applied Meteorology* 42:851–862.

Shafer, B.A. and L.E. Dezman. 1982. Development of a surface water supply index (SWSI) to assess the severity of drought conditions in snowpack runoff areas. *Preprints, Water Snow Conference*, Reno, NV, pp. 164–175.

Sheffield, J., G. Goteti, and E.F. Wood. 2006. Development of a 50-yr high-resolution global dataset of meteorological forcings for land surface modeling. *Journal of Climate* 19(13):3088–3111.

Shutko, A.M. 1986. *Microwave Radiometry of Water and Terrain Surfaces.* Moscow, Russia: Nauka Publisher.

Sorooshian, S., K.L. Hsu, X. Guo, H.V. Gupta, B. Imam, and D. Braithwaite. 2000. Evaluation of PERSIANN system satellite-based estimates of tropical rainfall. *Bulletin of the American Meteorological Society* 81(9):2035–2046.

Svoboda, M., D. LeComte, M. Hayes, R. Heim, K. Gleason, J. Angel, B. Rippey et al. 2002. The Drought Monitor. *Bulletin of the American Meteorological Society* 83(8):1181–1190.

Swenson, S. and J. Wahr. 2006. Post-processing removal of correlated errors in GRACE data. *Geophysical Research Letters* 33:L08402, doi:10.1029/2005GL025285.

Swenson, S., P.J.-F. Yeh, J. Wahr, and J. Famiglietti. 2006. A comparison of terrestrial water storage variations from GRACE with in situ measurements from Illinois. *Geophysical Research Letters* 33:L16401, doi:10.1029/2006GL026962.

Tadesse, T., G.B. Demisse, B. Zaitchik, and T. Dinku. 2014. Satellite-based hybrid drought monitoring tool for prediction of vegetation condition in Eastern Africa: A case study for Ethiopia. *Water Resources Research*, 50(3):2176–2190.

Tadesse, T., B.D. Wardlow, M.J. Hayes, M., Svoboda, and J.F. Brown. 2010. The vegetation outlook (VegOut): A new method for predicting vegetation seasonal greenness. *GIScience and Remote Sensing* 47(1):25–52.

Tadesse, T., D.A. Wilhite, M.J. Hayes, S.K. Harms, and S. Goddard. 2005. Discovering associations between climatic and oceanic parameters to monitor drought in Nebraska using data-mining techniques. *Journal of Climate* 18(10):1541–1550.

Tapley, B.D., S. Bettadpur, J.C. Ries, P.F. Thompson, and M.M. Watkins. 2004. GRACE measurements of mass variability in the Earth system. *Science* 305:503–505.

Thomas, A.C., J.T. Reager, J.S. Famiglietti, and M. Rodell. 2014. A GRACE-based water storage deficit approach for hydrological drought characterization. *Geophysical Research Letters*, 41(5):1537–1545.

Tucker, C.J., C.O. Justice, and S.D. Prince. 1986. Monitoring the grasslands of the Sahel 1984–1985. *International Journal of Remote Sensing* 7:1571–1581.

Tucker, C.J., J.E. Pinzon, M.E. Brown, D.A. Slayback, E.W. Pak, R. Mahoney, E.F. Vermote, and N. El Saleous. 2005. An extended AVHRR 8-km NDVI dataset compatible with MODIS and SPOT Vegetation NDVI data. *International Journal of Remote Sensing* 26(20):4485–4498.

Unganai, L.S. and F.N. Kogan. 1998. Drought monitoring and corn yield estimation in southern Africa from AVHRR data. *Remote Sensing of Environment* 63:219–232.

USDA (United States Department of Agriculture). 1994. State Soil Geographic (STATSGO) Data Base: Data Use Information. USDA Miscellaneous Publication 1492:1–113.

Van Leeuwen, W.J.D., B.J. Orr, S.E. Marsh, and S.M. Hermann. 2006. Multi-sensor NDVI data continuity: Uncertainties and implications for vegetation monitoring applications. *Remote Sensing of Environment* 100(1):67–81.

Verdin, J., C. Funk, R. Klaver, and D. Robert. 1999. Exploring the correlation between Southern Africa NDVI and Pacific sea surface temperatures: Results for the 1998 maize growing season. *International Journal of Remote Sensing* 20:2117–2124.

Wagner, W., G. Blöschl, P. Pampaloni, J.-C. Calvet, B. Bizzarri, J.-P. Wigneron, and Y. Kerr. 2007. Operational readiness of microwave remote sensing of soil moisture for hydrologic applications. *Nordic Hydrology* 38(1)1–20.

Wagner, W., W. Dorigo, R. de Jeu, D. Fernandez, J. Benveniste, E. Haas, and M. Ertl. 2012. Fusion of active and passive microwave observations to create an Essential Climate Variable data record on soil moisture. *ISPRS Annuals of the Photogrammetry, Remote Sensing and Spatial Information Sciences*, Volume I-7, XXII ISPRS Congress, Melbourne, Victoria, Australia, August 25–September 1, 2012, pp. 315–321.

Wagner, W., G. Lemoine, and H. Rott. 1999. A method for estimating soil moisture from ERS scatterometer and soil data. *Remote Sensing of Environment* 70(2):191–207.

Wahr, J., M. Molenaar, and F. Bryan. 1998. Time-variability of the Earth's gravity field: Hydrological and oceanic effects and their possible detection using GRACE. *Journal of Geophysical Research* 103(30):205–230.

Wahr, J., S. Swenson, and I. Velicogna. 2006. Accuracy of GRACE mass estimates. *Geophysical Research Letters* 33, L06401, doi:10.1029/2005GL025305.

Wang, L., J.J. Qu, and X. Hao. 2007. Forest fire detection using the normalized multi-band drought index (NMDI) with satellite measurements. *Agricultural and Forest Meteorology* 148:1767–1776.

Wardlow, B.D., T. Tadesse, J.F. Brown, K. Callahan, S. Swain, and E. Hunt. 2012. The Vegetation Drought Response Index (VegDRI): An integration of satellite, climate, and biophysical data. In *Remote Sensing of Drought: Innovative Monitoring Approaches*, eds. B.D. Wardlow, M.A. Anderson, and J. Verdin. Boca Raton, FL:CRC Press, pp. 51–74.

Wells, N., S. Goddard, and M.J. Hayes. 2004. A self-calibrating Palmer Drought Severity Index. *Journal of Climate* 17(12):2335–2351.

Wen, J., T.J. Jackson, R. Bindlish, A.Y. Hsu, and Z.B. Su. 2005. Retrieval of soil moisture and vegetation water content using SSM/I data over a corn and soybean region. *Journal of Hydrometeorology* 6:854–863.

Wilhite, D.A. and M.H. Glantz. 1985. Understanding the drought phenomenon: The role of definitions. *Water International* 10:111–120.

Xia, Y., K. Mitchell, M. Ek, J. Sheffield, B. Cosgrove, E. Wood, L. Lifeng et al. 2012. Continental-scale water and energy flux analysis and validation for the North American Land Data Assimilation System project phase 2 (NLDAS-2): 1. Intercomparison and application of model products. *Journal of Geophysical Research* 117:D03109, doi:10.1029/2011JD016048.

Yang, L., B. Wylie, L.L. Tieszen, and B.C. Reed. 1998. An analysis of relationships among climatic forcing and time-integrated NDVI of grasslands over the U.S. northern and central Great Plains. *Remote Sensing of Environment* 65:25–37.

Yirdaw, S Z., K.R. Snelgrove, and C.O. Agboma. 2008. GRACE satellite observations of terrestrial moisture changes for drought characterization in the Canadian Prairie. *Journal of Hydrology* 56:84–92.

Zaitchik, B.F., M. Rodell, and R.H. Reichle. 2008. Assimilation of GRACE terrestrial water storage data into a land surface model: Results for the Mississippi River Basin. *Journal of Hydrometeorology* 9(3):535–548.

16

Regional Drought Monitoring Based on Multisensor Remote Sensing

Jinyoung Rhee
APEC Climate Center

Jungho Im
Ulsan National Institute of Science and Technology

Seonyoung Park
Ulsan National Institute of Science and Technology

Acronyms and Definitions

AET	Actual evapotranspiration
AMSR-E	Advanced Microwave Scanning Radiometer for Earth Observing System
ARIMA	Autoregressive integrated moving average
AVHRR	Advanced Very High Resolution Radiometer
AWC	Available water capacity
CART	Classification and regression tree
CMI	Crop moisture index
CPC	Climate Prediction Center
DSI	Drought severity index
ESI	Evaporative stress index
EVI	Enhanced vegetation index
IPAD	International Production Assessment Division
KBDI	Keetch–Bryam drought index
LAI	Leaf area index
LST	Land surface temperature
MIDI	Microwave integrated drought index
MODIS	Moderate Resolution Imaging Spectroradiometer
NASS	National Agricultural Statistics Service
NCDC	National Climatic Data Center
NDDI	Normalized difference drought index
NDII	Normalized difference infrared index
NDMC	National Drought Mitigation Center
NDVI	Normalized difference vegetation index
NDWI	Normalized difference water index
NLCD	National Land Cover Database
NMDI	Normalized multiband drought index
NRCS	Natural Resources Conservation Service
OBDI	Objective blend drought index
PALSAR	Phased Array type L-band Synthetic Aperture Radar
PASG	Percent of average seasonal greenness
PDSI	Palmer drought severity index
PET	Potential evapotranspiration
SDCI	Scaled drought condition index
SEBS	Surface Energy Balance System
SM	Soil moisture
SMOS	Soil Moisture and Ocean Salinity
SNOTEL	Snowpack Telemetry
SOSA	Start of season anomaly
SPEI	Standardized precipitation and evapotranspiration index
SPI	Standardized precipitation index
SWD	Soil water deficit
SWDI	Soil wetness deficit index
SWI	Soil wetness index
TRMM	Tropical Rainfall Measuring Mission

TVDI	Temperature–vegetation dryness index
USDA	United Stated Department of Agriculture
USDM	United States Drought Monitor
USGS	United States Geological Survey
VegDRI	Vegetation drought response index
VHI	Vegetation health index
VSDI	Visible and shortwave infrared drought index
VTCI	Vegetation temperature condition index
WWAI	Wetland water area index

16.1 Introduction

There are numerous definitions of drought since it has been recognized as a very costly natural disaster. The recent special report of the IPCC (2012) defines drought as "a period of abnormally dry weather long enough to cause a serious hydrological imbalance." Drought can also be defined for three types: a meteorological drought refers to "a period with an abnormal precipitation deficit," soil moisture (SM) drought or agricultural drought is "a deficit of SM," and hydrological drought refers to "negative anomalies in streamflow, lake, and/or groundwater levels."

In order to monitor drought conditions, hydrometeorological variables need to be estimated for each type of drought. Precipitation is an appropriate variable for monitoring meteorological drought, and hydrological variables such as streamflow and reservoir levels can be used to detect hydrological drought. SM drought or agricultural drought can be measured using variables of SM condition and crop yield (Mishra and Singh, 2011). Variables such as precipitation and evapotranspiration can be considered as the drivers of drought, while variables such as streamflow and reservoir levels are the response variables impacted by drought. Both types of variables may be used for efficient drought monitoring.

The type of drought is determined as a function of the spatial location of the drought, timescale of it, as well as stakeholders involved. Different types of drought are related, as one type of drought may develop into another, and the variables for quantifying drought or those affected by drought are interconnected through land–atmosphere interactions. There exist couplings and feedbacks between variables, which may appear differently according to the regions of interests. For example, the coupling between SM and evapotranspiration is controlled by SM in dry regions, while it appears otherwise in wet regions (Seneviratne et al., 2010). The couplings and feedbacks of SM temperature and SM precipitation are more complex and intertwined between variables (Seneviratne et al., 2010; Figure 16.1).

Remote sensing techniques have been used to estimate variables related to each type of drought (Jensen, 2000). In some cases, however, the estimation of each variable may not fully represent the actual target when it is derived from algorithms using many other variables. The use of multiple variables in combined forms can improve the quantification of drought since the variables are related through the couplings and feedbacks. However, multicollinearity among the variables should be kept in mind when linear combination of variables is employed. Drought types and variables required to monitor each type of drought are listed in Table 16.1; socioeconomic drought is not included because it should be explained by many other factors than biophysical variables. It needs to be noted that the variables in Table 16.1 include both the drivers and impacts of drought, there exist overlaps of variables between drought types, and the list is not exhaustive. Generally, drought indices or combinations of the variables are useful for drought monitoring compared to single variables. Most variables can be estimated using remote sensing, which are well covered by Wardlow et al. (2012).

There are examples of using a variety of remote sensing–derived variables for drought monitoring, including the United States Drought Monitor (USDM), objective blend drought index (OBDI) of Climate Prediction Center (CPC), and vegetation drought response index (VegDRI; Brown et al., 2008).

The USDM is a drought monitoring system that provides maps with the spatial extent and severity of drought on a weekly basis for the conterminous United States. Drought information are

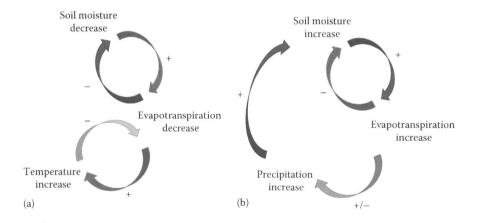

FIGURE 16.1 Processes contributing to (a) SM temperature and (b) SM precipitation coupling and feedback loops. (Adapted from Seneviratne, S.I. et al., *Earth Sci. Rev.*, 99, 125, 2010.)

TABLE 16.1 Drought Types and Variables Required to Monitor Each Type of Drought

Drought Type	Climatic/Hydrological Variables Required to Monitor Drought
Meteorological drought	Precipitation
	Evapotranspiration
Soil moisture drought or agricultural drought	Precipitation
	Evapotranspiration
	Soil moisture
	Vegetation activity
	Crop yields
Hydrological drought	Precipitation
	River discharge (streamflow)
	Reservoir storage
	Groundwater level

derived mainly from six key physical indicators, which include observation-based indicators of the Palmer drought severity index (PDSI; Palmer, 1965), Percent of Normal Precipitation (Willeke et al., 1994), standardized precipitation index (SPI; McKee et al., 1993), U.S. Geological Survey Daily Streamflow Percentiles, model-based CPC Soil Moisture Model Percentiles (Huang et al., 1996), and satellite-based vegetation health index (VHI; Kogan, 1995a,b). Ancillary data are also used in the USDM including, but not limited to, drought indices of the Palmer crop moisture index (CMI; Palmer 1968) and Keetch–Bryam drought index (KBDI; Keetch and Byram, 1968), as well as reservoir and lake levels, groundwater levels, and SM field observations. For the western United States, the Natural Resources Conservation Service's Snowpack Telemetry observations to measure snow water equivalent for mountain sites are also additionally used (Svoboda et al., 2002).

The USDM integrates subject judgments of the *authors* who are drought experts in the National Drought Mitigation Center, United Stated Department of Agriculture (USDA), CPC, and National Climatic Data Center. Since the indicators in use and their weights are determined by the authors, the OBDI is also not entirely objective (Svoboda et al., 2002). The produced drought information from OBDI, however, is replicable because they are based on the fixed weights of determined indicators. The type of indicators and their weights are selected considering the timescale of drought of interest.

The VegDRI measures drought-induced vegetation stress by combining climate-based drought indices of SPI, self-calibrated PDSI (Wells et al., 2004) and USDM, and remote sensing–based vegetation index metrics, which are normalized difference vegetation index, percent of average seasonal greenness, and start of season anomaly. It also utilizes biophysical information, which includes land cover/land use, soil available water capacity, irrigated agriculture, and ecoregions. The VegDRI adopted the supervised classification and regression tree (CART) analysis to determine vegetation stress.

The examples listed earlier are based on various available data sources without limiting the indicators to remotely sensed ones. The combination of multiple variables can be applied

for drought monitoring solely based on remote sensing. The combination of multiple indicators enables the production of customized drought indices that can be applied to specific regions of interests by determining the types and weights of the indicators.

This chapter investigates the regional drought monitoring methodology that utilizes multisensor remote sensing data. Recent trends in studies based on remote sensing–based drought monitoring are presented in the following section, and the development of the scaled drought condition index (SDCI) and its advances are introduced. Case studies of the Korean Peninsula and the United States are provided by applying the advanced SDCI for regional drought monitoring.

16.2 Recent Trends of Remote Sensing–Based Drought Monitoring

Since a multitude of factors are related with drought conditions, scientists have tried to incorporate such factors in remote sensing–based drought monitoring. Recent studies since 2010 show that the trends in remote sensing–based drought monitoring include (1) development of customized monitoring approaches to different types of drought, (2) development of new drought indices especially focusing on the incorporation of SM and evapotranspiration, and (3) data assimilation of remotely sensed products in process-based models for drought monitoring and prediction.

16.2.1 Customized Drought Monitoring Approaches

Since drought is a slow-onset event, it is hard to identify the exact start and end dates as well as types of drought. In particular, as drought has somewhat different characteristics by type, scientists have proposed customized monitoring approaches that are appropriately applicable to each type of drought. Because many drought indices are based on precipitation such as SPI, studies on monitoring meteorological drought typically investigated the relationship between remote sensing–derived products and in situ drought indices. For example, Caccamo et al. (2011) assessed the Moderate Resolution Imaging Spectroradiometer (MODIS)–based drought indices through comparison with SPI. They found that normalized difference infrared index (NDII) using band 6 ([band 2 – band6]/[band2 + band6]) outperformed the other remote sensing–derived indices showing strong correlation with SPI, which implied that NDII can be operationally used for monitoring meteorological drought. Scaini et al. (2014) used Soil Moisture and Ocean Salinity (SMOS) SM product to determine drought conditions by examining the relationships between SM anomalies and SPI as well as standardized precipitation and evapotranspiration index (SPEI). They found that the short-term remotely sensed anomalies had a high response to precipitation events, and thus the optimal timescale for the SMOS values was 1 month.

Since drought directly affects crop yields, efforts have been made to monitor agricultural drought using remote sensing–derived products including vegetation indices and SM. Shaheen and Baig (2011) assessed drought conditions in arid areas using SPI, NDVI, crop yield anomaly, rainfall anomaly, and evapotranspiration focusing on meteorological and agricultural droughts. Remote sensing–based vegetation indices such as NDVI and VHI have been successfully used for agricultural drought monitoring (Anderson et al., 2010; Rojas et al., 2011; Gao et al., 2014). SM is one of the key indicators for monitoring agricultural drought as it is closely related with vegetation growth. Bolten et al. (2010) incorporated the Advanced Microwave Scanning Radiometer for Earth Observing System (AMSR-E) SM into the International Production Assessment Division (IPAD) Water Balance Model to operationally monitor agricultural drought. Crop yield data have been commonly used to assess agricultural drought (Rhee et al., 2010; Choudhary et al., 2012). However, correlation between crop yield and drought severity may not be always high especially for irrigated agricultural lands (Park et al., 2014).

Vegetation phenology has been examined with drought conditions because drought during growing seasons can significantly affect crop yields. Ivits et al. (2014) investigated spatiotemporal patterns of drought conditions focusing on vegetation phenology using SPEI and NDVI based on major European bioclimatic zones. Garcia et al. (2014) evaluated the temperature–vegetation dryness index (TVDI) to estimate water deficits using MODIS data and found that the best conditions for TVDI performance agreed with the growing season that typically showed higher soil water content and lower vapor pressure deficit.

Hydrologic variables such as streamflow, SM, *evapotranspiration*, and water level have been used as reference indicators to identify hydrologic drought. For example, Choi et al. (2013) compared remote sensing–based drought indices with standard indices using streamflow and SM measurements. They found that evaporative stress index (ESI) successfully captured severe drought conditions to be a promising drought index for characterizing streamflow and SM anomalies. Since ESI calculation uses a thermal remote sensing energy balance framework, it is closely related with evapotranspiration deficits, which can be used as a diagnostic fast-response indicator. The combination of ESI and SM can be used as a valuable early warning tool for rapidly evolving flash drought conditions (Anderson et al., 2013). Seitz et al. (2014) examined water mass from Gravity Recovery and Climate Experiment, water stage from Envisat, and water extent from Phased Array type L-band Synthetic Aperture Radar to identify their spatial and temporal variability. They found that water extent and water level measurements in heavy drought conditions could be used to identify water volume changes that are closely related with hydrological drought.

16.2.2 Development of New Remote Sensing–Based Drought Indices

Many studies have recently proposed new drought indices that use multisensor satellite products related with potential drought indicators such as precipitation, temperature, vegetation healthiness, SM, and evapotranspiration. Table 16.2 summarizes recently introduced drought indices from multisensor satellite data. For example, Mu et al. (2013) proposed a new drought severity index (DSI) that uses MODIS evapotranspiration, potential evapotranspiration (PET), and NDVI products. Huang et al. (2011) developed a wetland water area index (WWAI) from PDSI, NDVI, and normalized difference water index (NDWI) to predict water surface area, and they were able to identify intra- and interannual water change between 1910 and 2009. They also proposed a water allocation model to simulate spatial distribution of water bodies, which can be used to simulate major changes in wetland water surface for ecosystem service. Wang et al. (2013) utilized passive microwave AMSR-E data to develop drought indices in Huaihe river basin focusing on SM.

As drought has different characteristics by type, drought indices have been developed for a specific type of drought. Rhee et al. (2010) proposed the SDCI that uses MODIS land surface temperature (LST), NDVI, and Tropical Rainfall Measuring Mission (TRMM) precipitation through linear combination and found that the index could be useful for agricultural drought monitoring in both arid and humid regions. Zhang and Jia (2013) proposed a microwave integrated drought index (MIDI) to monitor short-term drought (i.e., meteorological drought) over semiarid regions using TRMM-derived precipitation and AMSR-E-derived SM and LST. Keshavarz et al. (2014) proposed soil wetness deficit index based on soil wetness index from MODIS satellite data to examine agricultural drought conditions. Zhang et al. (2013) proposed a visible and shortwave infrared drought index (VSDI) for monitoring moisture in soil and vegetation. VSDI is expected to be efficient for agricultural drought monitoring over different land cover types during the plant-growing season.

16.2.3 Data Assimilation for Drought Monitoring and Prediction

Process-based physical models have been used to quantify various drought-related factors such as SM and evapotranspiration. Products generated from such models can be combined with remote sensing–derived drought indices to better document drought conditions. For example, Anderson et al. (2011) investigated drought conditions using ESI calculated based on evapotranspiration and PET produced from the ALEXI model. They found that ESI performed similarly to short-term precipitation-based indices such as SPI at higher spatial resolution without using any precipitation data. Anderson et al. (2012) used Mapping EvapoTranspiration at high Resolution with Internalized Calibration (METRIC), Atmospheric Land Exchange Inverse

TABLE 16.2 Summary of the Recently Proposed Satellite-Based Drought Indices

Drought Index	Description	Strength	Weakness	References
Normalized difference drought index (NDDI)	(NDVI – NDWI)/(NDVI + NDWI)	Contains strength of both NDVI and NDWI. Works well for identifying agricultural drought	Less reliable for short-term drought monitoring	Gu et al. (2007)
Normalized multiband drought index (NMDI)	(NIR – (1640 – 2130 nm)/ (NIR – (1640 + 2130 nm)	Works well for identifying agricultural drought in areas with a high vegetation rate	Less reliable for short-term drought monitoring	Wang and Qu (2007)
Normalized difference water index (NDWI)	(NIR – SWIR)/(NIR + SWIR)	Responds to drought faster than NDVI		Gao (1996)
Scaled drought condition index (SDCI)	0.25*scaledLST + 0.5*scaledTRMM + 0.25*scaledNDVI	Works for both meteorological and agricultural drought. Works for both arid and humid regions	Needs refinement to optimize weights of variables	Rhee et al. (2010)
Microwave integrated drought index (MIDI)	a*PCI + b*SMCI + (1 – a – b)*TCI (PCI = scaled TRMM SMCI = scaled soil moisture)	Appropriate for short-term drought monitoring. Provides high temporal resolution products based on microwave data	Less reliable for long-term drought monitoring	Zhang and Jia (2013)
Visible and shortwave infrared drought index (VSDI)	1 – [(SWIR-Blue) + (Red-Blue)]	Appropriate for use as a real-time drought indicator. Can be applied to various land covers.	Considers the very limited number of drought factors	Zhang et al. (2013)
Evaporative stress index (ESI)	Based on the ratio of ET and PET and anomalies	Monitors drought without using antecedent precipitation and subsurface soil characteristics	Can be affected by local factors such as groundwater	Anderson et al. (2010)
Vegetation drought response index (VegDRI)	Calculate index values using climatic, satellite, and biophysical components in rule-based models	Considers various drought factors. Provides high spatial resolution products	Does not consider some important variables such as land surface temperature	Brown et al. (2008)
Vegetation outlook (VegOut)	Combine satellite, climatic, and oceanic data as well as biophysical data to monitor drought biweekly	Able to predict drought using historical data	Focus only on agricultural drought based on vegetation greenness	Tadesse et al. (2005, 2010)

(ALEXI), and Surface Energy Balance Algorithm for Land (SEBAL) models with Landsat 7 Enhanced Thematic Mapper Plus (ETM+), MODIS, and Geostationary Operational Environmental Satellite (GOES) satellite data to explore the utility of moderate-resolution thermal satellite imagery in water resources management. They found that the fusion of the multisensor data at different scales could be effective for drought monitoring.

Zhong et al. (2014) investigated soil water deficit (SWD) from the Surface Energy Balance System model using Advanced Very High Resolution Radiometer (AVHRR) and MODIS data. They validated the SWD index using AMSR-E SM data and found that SM may have diurnal variations. Sakamoto et al. (2014) examined daily AMSR-E SM data to derive SM anomalies based on four spatial aggregation approaches. They found the spatial aggregation approaches could provide useful information on SM anomalies for a relatively short period of remote sensing data available.

Remote sensing–derived products can be combined with forecasting models to predict future drought conditions. Han et al. (2010) used an autoregressive integrated moving average model with vegetation temperature condition index data produced between 1999 and 2006 to examine the feasibility of drought forecasting based on the proposed approach. Remote sensing–based approaches are yet limited in drought forecasting. Remote sensing data have been used to provide input values as initial conditions for drought forecasting models through data assimilation, as well as to produce ancillary data such as land cover. As remote sensing–derived drought information documents past and present drought conditions, it can be used to regularly update modeling results to increase reliability of drought forecasting models.

16.2.4 Issues of Multisensor Data Combination

The integration of multiple variables from different sensors for drought monitoring can be viewed as a special case of data fusion. Data fusion combines data from multiple sensors to produce improved information and is known to have diverse challenges including conflicting data, data modality, data correlation, and data association (Khaleghi et al., 2013). The combination of multisensor-derived variables does not share all the challenges of data fusion since it is based on the *use of best available data* for the detection of drought of the areas of interests and uses multiple variables to extract useful drought information rather than dealing with single variables that may be conflicting or inconsistent. However, the combination of multiple variables still has some issues such as remote sensing data imperfection and outliers and different spatial resolutions of data.

Remote sensing data have impreciseness (Khaleghi et al., 2013) compared to observation data since remote sensing data rely heavily on estimation algorithms. Outliers also exist due to many factors interfering the acquisition of high-quality data such as atmospheric conditions. These issues become the sources of uncertainty. Use of artificial intelligence techniques such as neural networks or machine learning can help relieving such problems (Mishra and Desai, 2006). Such techniques can recognize the relationships between variables without considering the physics explicitly and perform well even with outliers (Mishra and Desai, 2006). The issue regarding different spatial resolutions is commonly ignored based on the *use of best available data* for the detection of drought of the areas of interests as previously mentioned. Downscaling approaches can relieve the spatial resolution issue though.

16.3 Development of the SDCI

16.3.1 Drought in Humid Regions

Droughts in arid regions have drawn much attention historically compared to humid regions due to the arid climatological characteristic. Since drought is a natural phenomenon and relative in nature, drought also occurs in humid regions. Many existing studies using remote sensing–based drought indices were mostly performed for arid/semiarid regions (e.g., Ji and Peters, 2003; Wan et al., 2004), and only a limited number of studies considered humid/subhumid regions (e.g., Kogan, 1995a,b).

There are historical documentations of drought in humid/subhumid regions; in the southeastern United States, it is known that drought occurred almost every decade since the 1920s (Knutson and Hayes, 2002; Weaver, 2005). In 1986, South Carolina, United States, experienced a drought with the return period of over 100 years (Karl and Young, 1987; Cook et al., 1988), and the economic loss by the 4-year drought 1998–2002 in the Carolinas was enormous (Carbone et al., 2007).

16.3.2 Introduction of SDCI, Solely Based on Remote Sensing Data

The SDCI has been proposed for use in humid regions as well as arid regions proving the performance of multisensor data fusion mainly for agricultural drought monitoring (Rhee et al., 2010). It is composed of three components based on remote sensing data: a temperature component using LST, a vegetation component using one of the vegetation-related indices, and a precipitation component using remotely sensed rainfall data. The Terra MODIS LST data and the TRMM monthly rainfall data were used for remotely sensed LST and precipitation data, respectively. Compared were various vegetation indices including the Terra MODIS NDVI and relatively new indices including the NDWI, the normalized difference drought index (NDDI), and the normalized multiband drought index (NMDI) calculated based on the Terra MODIS surface reflectance data. The land cover from the National Land Cover Database (NLCD)

2001 product was used as ancillary data and only locations with grassland or cropland areas were used for the analyses. All analysis was performed for the period 2000–2009.

Additive linear combinations of three variables as scaled from zero to one, each from one of the three components, were tested with three sets of weights against 3- and 6-month SPI as well as PDSI values obtained from observation data. Among the tested sets, the combination of scaled LST, scaled TRMM, and scaled NDVI where the weight for the precipitation component is twice of the others was selected as an optimum index that performed great in both arid and humid regions.

16.3.3 Validation of the SDCI

The developed SDCI was validated in two ways—the drought conditions derived by the SDCI were compared to USDM maps for well-known drought events (Figures 16.2 and 16.3), and the gridded values were merged for each county to be compared to crop yield.

The states of Arizona and New Mexico, United States (95°W–122°W longitude/31°N–37°N latitude), were selected as an arid region, while the states of North Carolina and South Carolina, United States (the 71°W–87.5°W longitude/32°N–37°N latitude), were chosen as a humid region. The year-to-year changes of drought conditions of the region for the period 2000–2009 were examined and compared to USDM maps.

Drought conditions based on SDCI for all land cover types were compared to USDM in Figures 16.2 and 16.3, despite that SDCI is optimized only for grassland and cropland land cover types. It is not appropriate to compare SDCI and USDM for this reason, and also because the USDM includes subject judgments of drought authors and utilizes numerous inputs.

The areas marked with "A (agricultural drought)" and "AH (agricultural and hydrological drought)" in USDM maps are examined more closely for comparisons. The months of May and September for arid and humid region, respectively, were selected since those are when SDCI showed the highest correlations with observational drought indices. The year-to-year changes of spatial distribution and severity of drought generally agreed between SDCI and USDM (Figures 16.2 and 16.3).

The SDCI was also tested against crop yield data. The main cultivated crops for the areas are cotton wheat, corn, and sorghum for arid region and soybean, corn, and cotton for humid region based on 2007 census of agriculture. Crop yield data were obtained from the National Agricultural Statistics Service (NASS) of USDA, and Curry County of New Mexico, Lee County of South Carolina, and Edgecombe, Greene, Lenoir, Pitt, Robeson, Sampson, Wayne, and Wilson County of North Carolina were chosen with larger than 33% total area of croplands.

Regression analyses were performed between monthly SDCI and yearly crop yield data. The remote sensing index values with 2001 NLCD cultivated crops land cover type for each county were averaged and correlated with the crop yield data for the county. In the arid region, only May SDCI showed statistically significant correlation with yearly cotton yield. In the humid region,

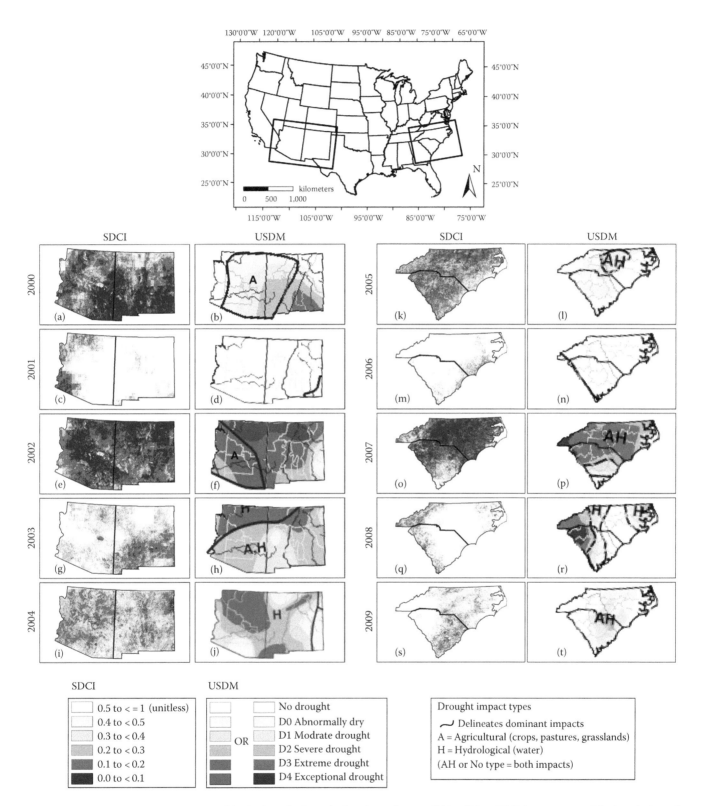

FIGURE 16.2 Year-to-year comparisons of the SDCI and USDM (a–j) in the arid region (AZ and NM, USA) during May and (k–t) in the humid region (NC and SC, USA) during September for 2000–2009. (Adapted from Rhee, J. et al., *Remote Sensing Environ.*, 114, 2875, 2010.)

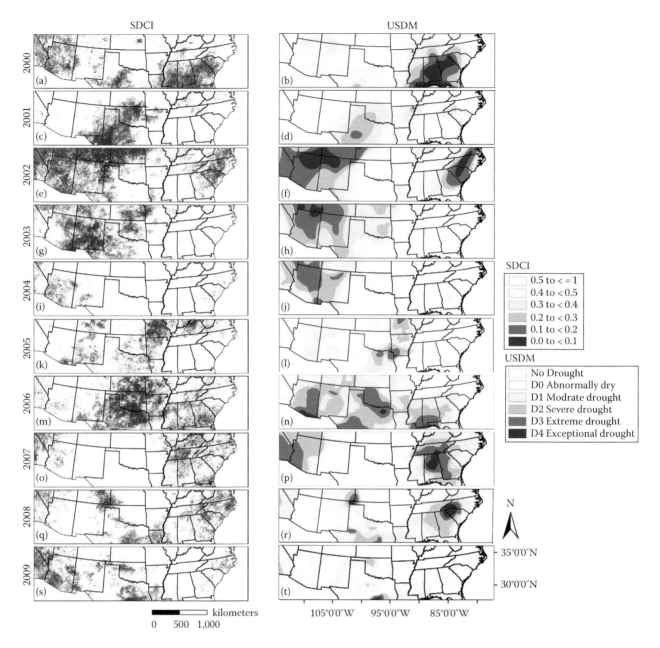

FIGURE 16.3   Year-to-year comparisons of the SDCI and USDM in the southern part of the United States during July for 2000–2009.

June and July SDCI showed high correlations with corn yield for some counties, August SDCI with cotton and soybean yield, and September SDCI with soybean yield. The months showing high correlations with yearly crop yield are the active growing periods during the phonological cycles of the crops (Savitsky, 1986; Jensen, 2007). Scatterplots of some examples are shown in Figure 16.4.

16.3.4 Development of the Advanced SDCI

Although SDCI with linear combinations of three components can be successfully used to monitor agricultural drought, it has some limitations: the weights are arbitrarily determined, and it includes only limited types of variables. These issues can be solved by integrating more variables related to drought and combining artificial intelligence, such as neural networks or machine learning techniques.

As previously mentioned, SM is the most important variable for agricultural drought monitoring. Since there exist couplings and feedbacks between SM and precipitation, temperature, and evapotranspiration, the inclusion of the three components may explain much portion of other variables that are not considered. In order to more fully represent the responses of the variables during drought, more variables are added: actual evapotranspiration and PET, enhanced vegetation index (EVI), and leaf area index (LAI) estimated from MODIS and SM from AMSR-E sensor. The SDCI was improved by providing multiple blends of multisensor indices for different types of drought.

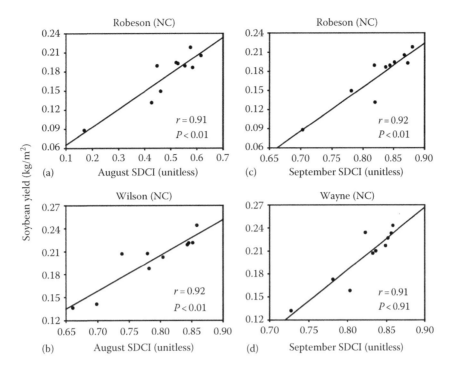

FIGURE 16.4 Scatterplots of the SDCI versus soybean yield data in NC, United States, for 2000–2009. The SDCI values with 2001 NLCD cultivated crops land cover type for each county were averaged and correlated with the crop yield data for the county. (Adapted from Rhee, J. et al., *Remote Sensing Environ.*, 114, 2875, 2010.)

TABLE 16.3 Strengths and Limitations of SDCI Compared to USDM

	SDCI	USDM
Data source	Utilizes only remote sensing data	Combines observation-based indicators, satellite-based indicators, model-based data, ancillary data, and expert judgments
Strengths	Can be applied to any region, especially to areas without observation data	Maximizes the use of available data/information
		Currently provides optimized information for conterminous United States
		Provides information for various types of drought
Limitations	Currently optimized for agricultural drought monitoring	Requires long-term in situ data to derive observation-based indicators
		Needs subjective inputs from experts when applied to additional regions other than conterminous United States

The blends of multisensor indices can be used in two ways—linear combinations of variables can be used as SDCI but with weights determined using a machine learning technique, or nonlinear combinations of variables can be used as trained using the machine learning technique though a semi-*blackbox*. Strengths and limitations of SDCI compared to USDM are shown in Table 16.3. Two multisensor blending examples are described in the following two sections.

16.4 Case Study: Linearly Combined SDCI in Korean Peninsula

The advanced SDCI with a linear combination was applied to Korean peninsula including North and South Korea (Figure 16.5) for agricultural drought monitoring. The study area has four distinct seasons over the course of the year and contains complex topography with a variety of land cover types. Remote sensing data

were obtained for the period 2003–2011, and observation data from 39 out of 106 weather stations with grassland or cropland land cover types were used for training to obtain weights (Figure 16.5).

Weights for variables (LST, EVI, LAI, PET, SM, and precipitation [PRCP]) were determined using a random forest machine learning approach. Random forest uses an ensemble approach to predict a target variable through combining predictions returned by multiple CART using the Gini index for selecting an attribute at a node (Han and Kamber, 2011; Li et al., 2014). Random forest introduces two levels of randomness into CART to improve the weaknesses of CART such as overfitting to samples and dependency to training data configuration: (1) a random subset of training samples and (2) a random subset of candidate variables at each node. The randomness allows to generate numerous decorrelated trees, which can deliver robustness to noise, outliers, and overfitting. Random forest provides relative variable importance using out-of-bag data (i.e., accuracy change when a

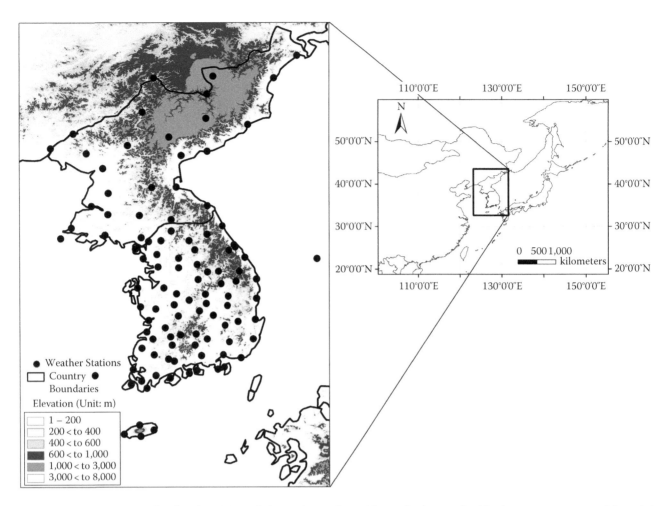

FIGURE 16.5 Korean peninsula—data from 39 out of 106 weather stations with grassland or cropland land cover types were used for training to derive weights.

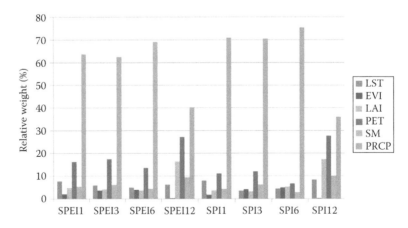

FIGURE 16.6 Weights for each variable were obtained to fit the widely used drought indices; the SPI and SPEI with 1–12 months of timescales.

variable is out of bag). The training was performed for SPI and SPEI with 1–12 months of timescales (Figure 16.6).

The advanced SDCI and a couple of remote sensing–based indices including NMDI and NDWI were tested using the do (province)-level yield data of Highland Radish for two weather stations locations of Jangsu, Jeollabuk-do, and Daegwallyeong,

Kangwon-do, located above 400 m level (Figure 16.7). The June SDCI showed statistically significant high correlations with SPEI12 (Figure 16.7a; r = 0.71, p = 0.031) and SPI12 (Figure 16.7b; r = 0.70, p = 0.035) in Jangsu, Jeollabuk-do, as well as July NMDI (Figure 16.7d; r = 0.68, p = 0.042). The May NDWI (Figure 16.7c, r = 0.82, p = 0.007) showed good correlations to yield data in

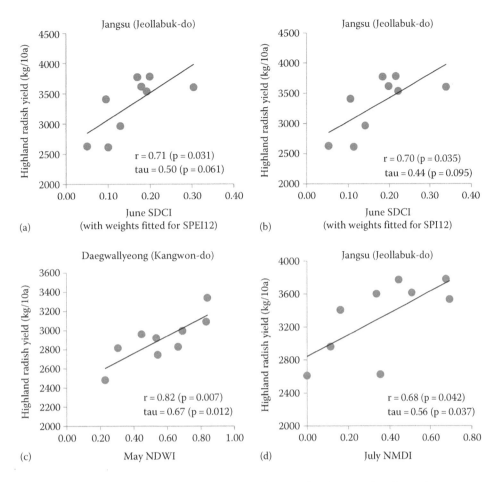

FIGURE 16.7 Scatterplots with correlation coefficient r as well as Kendall's tau values between the advanced SDCI, NDWI, and NMDI at two weather stations locations versus do (province)-level Highland Radish yield.

Daegwallyeong, Kangwon-do. The Kendall's tau values were also examined since the number of samples is small; only the May NDWI and July NMDI showed statistical significance with the significance level of 0.05 (tau = 0.67, p = 0.012, and tau = 0.56, p = 0.037, respectively).

The scatterplots in Figure 16.7 are for the comparisons between the do (province)-level yearly Highland Radish yield and the SDCI values for weather stations locations, not averaged value for the highland areas of the corresponding province. It explains somewhat smaller correlation coefficient values compared to Figure 16.4. The comparisons between the averaged grid values and do (province)-level yield data are being performed in the following study. Since the case study did not use the full range of variables, a further study is required to explore the effect of the use of each variable.

16.5 Case Study: Nonlinearly Combined SDCI in the United States

The drought indicators that are used to produce SDCI can be nonlinearly combined to increase the predictability of drought. Machine learning approaches were evaluated to estimate SPI using SDCI drought indicators, including MODIS LST, NDVI,

NDWI, NDMI, NDDI, evapotranspiration, and TRMM-derived precipitation over both arid and humid regions in the United States, the same study regions in Rhee et al. (2010) and Park et al. (2014). The arid region includes Arizona and New Mexico states and the humid region includes North and South Carolina. MODIS and TRMM products (i.e., a total of 15 variables) from 2000 and 2012 were scaled from 0 to 1 to temporally normalize the data (Rhee et al., 2010) (Table 16.4). SPI was calculated for 54 weather stations (28 stations in the arid region and 26 in the humid region) at the accumulated 1-, 3-, 6-, 9-, and 12-month timescales (i.e., SPI1, SPI3, SPI6, SPI9, and SPI12).

Three machine learning techniques were adopted in estimating SPI at each timescale, including boosted regression trees, random forest, and Cubist. Boosted regression trees are similar with random forest in that an ensemble approach based on CART to predict a target variable is used. Boosted regression trees produce a series of weighted predictions from individual trees. Weights assigned to each sample are updated at every iteration based on how the sample was explained in the previous iteration. That way, misinterpreted samples will get more attention in the next iteration, which can increase overall accuracy, but risks overfitting the model to such samples (Han and Kamber, 2011). Boosted regression trees and random forest were

TABLE 16.4   Variables Used for Nonlinear Combination of SDCI

Satellite Sensor	Variable	Description
MODIS	Land surface temperature (LST)	
	Evapotranspiration (ET)	
	Normalized difference vegetation index (NDVI)	
	NDVI500	Using 500 m data
	Normalized difference water index (NDWI5)	Using band 5
	NDWI6	Using band 6
	NDWI7	Using band 7
	NDDI5	Using band 5
	NDDI6	Using band 6
	NDDI7	Using band 7
	NMDI	
TRMM	TRMM1	For 1 month
	TRMM3	For 3 months
	TRMM6	For 6 months
	TRMM9	For 9 months
	TRMM12	For 12 months

Source: Adapted from Park, S. et al., *Agric. Forest Meteorol.*, 2014.

implemented in R (Robert Gentleman and Ross Ihaka, version 2.7.2) and its contributed packages (R Core Development Team, 2008) (*gbm* and *randomForest* packages, respectively) with default settings, except that 1000 trees were used instead of the default tree numbers.

Cubist uses a modified regression tree system developed by Quinlan (1993) to build rule-based predictive models. Cubist is a commercial product and has been proved useful in various remote sensing–based regression tasks (Im et al., 2009, 2012; Li et al., 2014). Each rule generated from Cubist is associated with a multivariate regression model to estimate a target variable. When multiple rules are applied to a sample, the output is averaged from the associated regression models. Cubist was applied using the Cubist software package (Rulequest Research, 2012).

Results show that random forest among the three machine learning approaches outperformed the other two methods

regardless of the target variable and region used (Figure 16.8). Prediction of SPI in the arid region was slightly better than the humid region for all three approaches used. All three machine learning approaches provide relative variable importance: boosted regression trees and random forest document change in accuracy when a variable is out of bag. Cubist documents how many times a variable is used in rules and multivariate regression models.

Relative variable importance by machine learning approach when all variables were used against SPI1, SPI6, and SPI12 for the arid region is summarized in Table 16.5. For the arid region, precipitation was the most important variable when random forest and boosted regression trees were used. Vegetation indices were also very useful for estimating SPI12 when Cubist was used (not shown). For both arid and humid regions, precipitation was dominantly important for short-term drought (i.e., SPI1, SPI3),

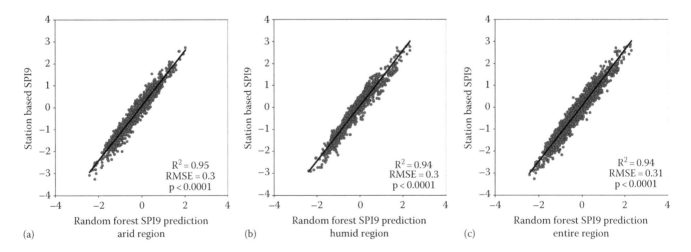

FIGURE 16.8   Scatterplots between random forest-based and station-based SPI9 values for (a) the arid region, (b) the humid region, and (c) the entire region. (Adapted from Park, S. et al., *Agric. Forest Meteorol.*, 2014.)

TABLE 16.5 Relative Variable Importance in Percentage by Random Forest (RF) and Boosted Regression Trees (BRT) against SPI1, SPI6, and SPI12 for the Arid Region

Variable	RF	BRT	RF	BRT	RF	BRT
LST	**6.4**	**8.5**	5.4	4.8	4.0	2.5
ET	**8.7**	**6.3**	6.0	4.0	6.4	2.8
NDVI	5.8	4.7	**7.8**	5.5	7.8	**5.9**
NDVI 500	4.0	1.3	6.9	1.0	7.5	2.5
NDWI5	2.0	1.2	3.4	1.8	4.3	1.9
NDWI6	5.0	2.7	**9.6**	**7.8**	9.0	3.1
NDWI7	5.2	1.4	**9.3**	**10.5**	9.5	**8.6**
NDDI5	2.9	1.7	4.6	1.0	5.9	1.3
NDDI6	4.5	3.2	**9.1**	**7.3**	10.8	**7.9**
NDDI7	2.6	1.2	5.2	2.6	7.1	3.6
NMDI	1.2	0.6	4.2	0.6	5.2	1.3
TRMM1	**21.3**	**38.0**	3.5	1.2	1.8	0.9
TRMM3	**9.6**	**12.0**	5.3	2.7	5.8	2.5
TRMM6	**10.3**	**11.6**	**17.8**	**36.6**	14.2	13.3
TRMM9	4.8	1.9	7.2	**7.5**	7.4	1.8
TRMM12	5.5	3.7	6.0	5.0	**20.0**	**40.2**

Top five important variables are in bold for each case.

while vegetation indices became important for long-term drought (i.e., SPI9, SPI12). It should be considered that some variables have high multi-collinearity, and thus further analysis is required to determine weights for drought indicators to identify each type of drought.

16.6 Conclusions

In this chapter, we examined recent trends in remote sensing–based drought monitoring and investigated the regional drought monitoring methodology based on the use of multisensor remote sensing data. The development of the SDCI, which integrates three components of temperature, precipitation, and vegetation-related stress, was introduced along with its further improvements by integrating more pivotal variables including SM and evapotranspiration and utilizing machine learning techniques. Case studies of the Korean Peninsula and the United States were presented.

In the case study for the Korean Peninsula, the advanced SDCI with a linear combination of variables showed fairly good correlations with the yield of Highland Radish, which is generally cultivated without irrigation, proving the usability for monitoring and assessment of agricultural drought. The June SDCI showed statistically significant high correlations with Highland Radish yield data with weights fitted for SPEI12 (r = 0.71, p = 0.031) and SPI12 (r = 0.70, p = 0.035) in Jangsu, Jeollabuk-do. In the case study for the arid and humid regions in the United States, machine learning approaches were successfully applied to estimate SPI at different timescales and to examine the variable importance for documenting drought. Such variable importance can be used to determine weights for key variables to monitor drought conditions by drought type.

The development and advancement of SDCI follows the trend of recent studies of providing customized monitoring approaches to different types of drought and developing new indices focusing on the consideration of more variables including SM and evapotranspiration. Currently, the integration of remote sensing data–based SDCI with process–based land surface models is under development, differentiating the use of the SDCI from existing ones emphasizing its use for regions with limited observation data.

Since the coupling and feedback between variables are intertwined, it is not appropriate to investigate only limited variables to monitor each type of drought. Multiple indices need to be examined together for better drought monitoring and assessment. The use of drought indicators solely from globally available satellite data in obtaining SDCI enables the proposed index to be used in areas with limited regional in situ data. Drought nowcast and forecast information is especially important for those areas, since it helps stakeholders to make informed decisions on the allocation of water resources and the determination of planting dates. Great cost may be avoided by the introduction of multisensor remote sensing drought indicators, which are based on globally available data and can be easily customized.

Acknowledgments

This research was supported by the National Space Lab Program through the National Research Foundation of Korea (NRF) funded by the Ministry of Science, ICT, and Future Planning (Grant: NRF-2013M1A3A3A02042391).

References

Anderson, L., Y. Malhi, L. Aragão, R. Ladle, E. Arai, N. Barbier, and O. Phillips. 2010. Remote sensing detection of droughts in Amazonian forest canopies. *New Phytol* 187:733–750.

Anderson, M., R. Allen, A. Morse, and W. Kustas. 2012. Use of Landsat thermal imagery in monitoring evapotranspiration and managing water resources. *Remote Sensing Environ* 122:50–65.

Anderson, M., C. Hain, J. Otkin, X. Zhan, K. Mo, M. Svoboda et al. 2013. An intercomparison of drought indicators based on thermal remote sensing and NLDAS-2 simulations with US drought monitor classifications. *J Hydrometeorol* 14:1035–1056.

Anderson, M., C. Hain, B. Wardlow, A. Pimstein, J. Mecikalski, and W. Kustas. 2011. Evaluation of drought indices based on thermal remote sensing of evapotranspiration over the continental United States. *J Climate* 24:2024–2044.

Bolten, J., W. Crow, X. Zhan, T. Jackson, and C. Reynolds. 2010. Evaluating the utility of remotely sensed soil moisture retrievals for operational agricultural drought monitoring. *IEEE J Sel Top Appl Earth Observ Remote Sensing* 3:57–66.

Brown, J. F., B. D. Wardlow, T. Tadesse, M. J. Hayes, and B. C. Reed. 2008. The vegetation drought response index (VegDRI): A new integrated approach for monitoring drought stress in vegetation. *GISci Remote Sensing* 45:16–46.

Caccamo, G., L. Chisholm, R. Bradstock, and M. Puotinen. 2011. Assessing the sensitivity of MODIS to monitor drought in high biomass ecosystems. *Remote Sensing Environ* 115:2626–2639.

Carbone, G., J. Rhee, H. Mizzell, and R. Boyles. 2007. A regional-scale drought monitoring tool for the Carolinas. *Bull Am Meteorol Soc* 89:20–28.

Choi, M., J. Jacobs, M. Anderson, and D. Bosch. 2013. Evaluation of drought indices via remotely sensed data with hydrological variables. *J Hydrol* 476:265–273.

Choudhary, S. S., P. K. Garg, and S. K. Ghosh. 2012. Mapping of agriculture drought using remote sensing and GIS. *Int. J. Sci. Eng. Technol.*, 1(4):149–157.

Cook, E., M. Kablack, and G. Jacoby. 1988. The 1986 drought in the southeastern United States: How rare an event was it? *J Geophys Res* 93:14257–14260.

Gao, B.-C. 1996. NDWI—A normalized difference water index for remote sensing of vegetation liquid water from space. *Remote Sensing of Environment*, 58(3):257–266.

Gao, Z., Q. Wang, X. Cao, and W. Gao. 2014. The responses of vegetation water content (EWT) and assessment of drought monitoring along a coastal region using remote sensing. *GISci Remote Sensing* 51:1–16.

Garcia, M., N. Fernández, L. Villagarcía, F. Domingo, J. Puigdefábregas, and I. Sandholt. 2014. Accuracy of the Temperature–Vegetation Dryness Index using MODIS under water-limited vs. energy-limited evapotranspiration conditions. *Remote Sensing Environ* 149:100–117.

Gu, Y., J. F. Brown, J. P. Verdin, and B. Wardlow. 2007. A five-year analysis of MODIS NDVI and NDWI for grassland drought assessment over the central Great Plains of the United States. *Geophysical Research Letters*, 34(6). doi: 10.1029/2006GL029127.

Han, J. and M. Kamber. 2011. *Data Mining: Concepts and Techniques*. Burlington, MA: Elsevier.

Han, P., P. Wang, S. Zhang, and D. Zhu. 2010. Drought forecasting based on the remote sensing data using ARIMA models. *Math Comput Model* 51:1398–1403.

Huang, J., H. Van den Dool, and K. P. Georgakakos. 1996. Analysis of model-calibrated soil moisture over the United States (1931–93) and application to long-range temperature forecasts. *J Climate* 9:1350–1362.

Huang, S., D. Dahal, C. Young, G. Chander, and S. Liu. 2011. Integration of Palmer Drought Severity Index and remote sensing data to simulate wetland water surface from 1910 to 2009 in Cottonwood Lake area, North Dakota. *Remote Sensing Environ* 115:3377–3389.

Im, J., J. R. Jensen, M. Coleman, and E. Nelson. 2009. Hyperspectral remote sensing analysis of short rotation woody crops grown with controlled nutrient and irrigation treatments. *Geocarto Int* 24:293–312.

Im, J., Z. Lu, J. Rhee, and L. J. Quackenbush. 2012. Impervious surface quantification using a synthesis of artificial immune networks and decision/regression trees from multi-sensor data. *Remote Sensing Environ* 117:102–113.

IPCC. 2012. Managing the risks of extreme events and disasters to advance climate change adaptation. A special report of Working Groups I and II of the Intergovernmental Panel on Climate Change [Field, C. B., V. Balrros, T. F. Stocker, D. Qin, D. J. Dokken, K. L. Ebi, M. D. Mastrandrea, K. J. Mach, G.-K. Plattner, S. K. Allen, M. Tignor, and P. M. Midgley (eds.)]. Cambridge University Press, Cambridge, UK/New York, 582pp.

Ivits, E., S. Horion, R. Fensholt, and M. Cherlet. 2014. Drought footprint on European ecosystems between 1999 and 2010 assessed by remotely sensed vegetation phenology and productivity. *Glob Chang Biol* 20:581–593.

Jensen, J. R. 2000. *Remote Sensing of the Environment: An Earth Resource Perspective*. Upper Saddle River, NJ: Prentice Hall, Chapter 10.

Ji, L. and A. Peters. 2003. Assessing vegetation response to drought in the northern Great Plains using vegetation and drought indices. *Remote Sensing Environ* 87:85–98.

Karl, T. and P. Young. 1987. The 1986 Southeast drought in historical perspective. *Bull Am Meteorol Soc* 68:773–778.

Keetch, J. and G. Byram. 1968. A drought index for forest fire control. Forest Service Research Paper SE-38. U.S. Department of Agriculture, Washington, DC, 32pp.

Keshavarz, M., M. Vazifedoust, and A. Alizadeh. 2014. Drought monitoring using a Soil Wetness Deficit Index (SWDI) derived from MODIS satellite data. *Agric Water Manage* 132:37–45.

Khaleghi, B., A. Khamis, F. O. Karray, and S. N. Razavi. 2013. Multisensor data fusion: A review of the state-of-the-art. *Inform Fusion* 14(1): 28–44.

Kogan, F. N. 1995a. Drought of the late 1980s in the United States as derived from NOAA polar–orbiting satellite data. *Bull Am Meteorol Soc* 76:655–668.

Kogan, F. N. 1995b. Application of vegetation index and brightness temperature for drought detection. *Adv Space Res* 15:91–100.

Knutson, C. and M. Hayes. 2002. South Carolina drought mitigation and response assessment: 1998–2000 drought. Quick Response Research Report #136. National Hazards Research and Applications Information Center, University of Colorado, Boulder, CO.

Li, M., J. Im, L. Quackenbush, and L. Tao. 2014. Forest biomass and carbon stock quantification using airborne LiDAR data: A case study over Huntington Wildlife Forest in the Adirondack Park. *IEEE J Sel Top Appl Earth Observ Remote Sensing* 7(7):3143–3156.

McKee, T. B., N. J. Doesken, and J. Kleist. 1993. The relationship of drought frequency and duration to time scales. Preprints, Eighth Conference on Applied Climatology, Dallas, TX. *Am Meteorol Soc* 179–184.

Mishra, A. K. and V. R. Desai. 2006. Drought forecasting using feed-forward recursive neural network. *Ecol Model* 198:127–138.

Mishra, A. K. and V. Singh. 2011. Drought modeling—A review. *J Hydrol* 403:157–175.

Mu, Q., M. Zhao, J. Kimball, N. McDowell, and S. Running. 2013. A remotely sensed global terrestrial drought severity index. *Bull Am Meteorol Soc* 94:83–98.

Palmer, W. C. 1965. Meteorological drought. Research Paper No. 45. U.S. Department of Commerce, Weather Bureau, Washington, DC, p. 58.

Palmer, W. C. 1968. Keeping track of crop moisture conditions, nationwide: The new crop moisture index. *Weatherwise* 21:156–161.

Park, S., J. Im, E. Jang, H. Yoon, and J. Rhee. 2015. Multi-sensor blending for regional drought monitoring using machine learning approaches. *Agric Forest Meteorol*, in review.

Quinlan, J. 1993. *C4.5: Programs for Machine Learning*. San Mateo, CA: Morgan Kaufman.

R Core Development Team. 2008. R: A language and environment for statistical computing. http://www.R-project.org/. (Accessed May 17, 2014.)

Rhee, J., J. Im, and G. J. Carbone. 2010. Monitoring agricultural drought for arid and humid regions using multisensor remote sensing data. *Remote Sensing Environ* 114:2875–2887.

Rojas, O., A. Vrieling, and F. Rembold. 2011. Assessing drought probability for agricultural areas in Africa with coarse resolution remote sensing imagery. *Remote Sensing Environ* 115:343–352.

Rulequest Research. 2012. Data mining with Cubist. http://www.rulequest.com/cubist-info.html. (Accessed May 17, 2014.)

Sakamoto, T., A. Gitelson, and T. Arkebauer. 2014. Near real-time prediction of US corn yields based on time-series MODIS data. *Remote Sensing Environ* 147:219–231.

Savitsky, B. G. 1986. Agricultural remote sensing in South Carolina: A study of crop identification capabilities utilizing Landsat multispectral scanner data. Unpublished master thesis. University of South Carolina Geography Department, Columbia, SC, 78pp.

Scaini, A., N. Sánchez, S. Vicente-Serrano, and J. Martínez-Fernández. 2014. SMOS-derived soil moisture anomalies and drought indices: A comparative analysis using in situ measurements. *Hydrol Processes* 29(3):373–383.

Seitz, F., K. Hedman, F. Meyer, and H. Lee. 2014. Multi-sensor space observation of heavy flood and drought conditions in the Amazon Region. In *Earth on the Edge: Science for a Sustainable Planet*. Berlin, Germany: Springer, pp. 311–317.

Seneviratne, S. I., T. Corti, E. L. Davin, M. Hirschi, E. B. Jaeger, I. Lehner, B. Orlowskly, and A. J. Teuling. 2010. Investigating soil moisture-climate interactions in a changing climate: A review. *Earth Sci Rev* 99:125–161.

Shaheen, A. and M. Baig. 2011. Drought severity assessment in arid area of Thal Doab using remote sensing and GIS. *Int J Water Resour Arid Environ* 1:92–101.

Svoboda, M., D. LeComte, M. Hayes, R. Heim, K. Gleason, J. Angel, B. Rippey et al. 2002. The drought monitor. *Bull Am Meteorol Soc* 83:1181–1190.

Tadesse, T., J. F. Brown, and M. J. Hayes. 2005. A new approach for predicting drought-related vegetation stress: Integrating satellite, climate, and biophysical data over the US central plains. *ISPRS Journal of Photogrammetry and Remote Sensing* 59(4):244–253.

Tadesse, T., B. D. Wardlow, M. J. Hayes, M. D. Svoboda, and J. F. Brown. 2010. The vegetation outlook (VegOut): A new method for predicting vegetation seasonal greenness. *GIScience & Remote Sensing* 47(1):25–52.

Wan, Z., P. Wang, and X. Li. 2004. Using MODIS Land Surface Temperature and Normalized Difference Vegetation Index products for monitoring drought in the southern Great Plains, USA. *Int J Remote Sensing* 25:61–72.

Wang, L. and J. J. Qu. 2007. NMDI: A normalized multi-band drought index for monitoring soil and vegetation moisture with satellite remote sensing. *Geophysical Research Letters*, 34(20). L20405.

Wang, R., J. Xu, D. Wang, X. Xie, and P. Wang. 2013. Construction of Drought Indices from passive microwave remote sensing AMSR-E data in Huaihe River Basin. *Appl Mech Mater* 397:2503–2506.

Wardlow, B. D., M. C. Anderson, and J. P. Verdin (Eds.). 2012. *Remote Sensing of Drought: Innovative Monitoring Approaches*. Boca Raton, FL: CRC Press, 422pp.

Weaver, J. C. 2005. The drought of 1998–2002 in North Carolina—Precipitation and hydrologic conditions. Scientific Investigations Report, 2005-5053. U.S. Geological Survey, U.S. Department of the Interior, Reston, VA, 88pp.

Wells, N., S. Goddard, and M. J. Hayes. 2004. A self-calibrating Palmer Drought Severity Index. *J Climate* 17:2235–2351.

Willeke, G., J. R. M. Hosking, J. R. Wallis, and N. B. Guttman. 1994. *The National Drought Atlas*. Institute for Water Resources Report 94-NDS-4. U.S. Army Corps of Engineers, Washington, DC, CD-ROM.

Zhang, A. and G. Jia. 2013. Monitoring meteorological drought in semiarid regions using multi-sensor microwave remote sensing data. *Remote Sensing Environ* 134:12–23.

Zhang, N., Y. Hong, Q. Qin, and L. Liu. 2013. VSDI: A visible and shortwave infrared drought index for monitoring soil and vegetation moisture based on optical remote sensing. *Int J Remote Sensing* 34:4585–4609.

Zhong, L., Y. Ma, Y. Fu, X. Pan, W. Hu, Z. Su et al. 2014. Assessment of soil water deficit for the middle reaches of Yarlung-Zangbo River from optical and passive microwave images. *Remote Sensing Environ* 142:1–8.

17

Land Degradation Assessment and Monitoring of Drylands

Marion Stellmes
Trier University

Ruth Sonnenschein
European Academy of Bozen/Bolzano

Achim Röder
Trier University

Thomas Udelhoven
Trier University

Stefan Sommer
Joint Research Centre

Joachim Hill
Trier University

Acronyms and Definitions

ACRIS	Australian Collaborative Rangeland Information System
ARC2	Africa Rainfall Estimate Climatology Version 2
AVHRR	Advanced Very High Resolution Radiometer
BFAST	Breaks For Additive Season and Trend
BRDF	Bidirectional reflectance distribution function
CAP	Common Agricultural Policy
CCDC	Continuous change detection and classification
CDR	Landsat surface reflectance climate data record
CMFDA	Continuous monitoring of forest disturbance algorithm
CNES	Centre National d'Études Spatiales
CSIRO	Commonwealth Scientific and Industrial Research Organisation
DDP	Drylands Development Paradigm
DLR	German Aerospace Center
DSITIA	Department of Science, Information Technology, Innovation and the Arts, Queensland, Australia
EC-JRC	European Commission Joint Research Centre
EnMAP	Environmental Mapping and Analysis Program
ENSO	El Niño-Southern Oscillation
ESA	European Space Agency
ETM+	Enhanced Thematic Mapper
EVI	Enhanced Vegetation Index
EU	European Union
FAPAR	Fraction of Absorbed Photosynthetically Active Radiation
FASIR	Fourier-adjusted, sensor and solar zenith angle-corrected, interpolated, and reconstructed AVHRR archive
GES-DAAC	GSFC Earth Sciences Distributed Active Archive Center
GIMMS	Global Inventory Monitoring and Modeling System

GLADIS	Global Land Degradation Information System
GLASOD	Global Assessment of Human-Induced Soil Degradation
GLFC	Global Land Cover Facility
GPCC	Global Precipitation Climatology Centre
GFSC	NASA's Goddard Space Flight Center
HANPP	Human appropriation of net primary production
HyspIRI	Hyperspectral Infrared Imager
ISLSCP II	International Satellite Land Surface Climatology Project, Initiative II
JRSRP	Joint Remote Sensing Research Program, University of Queensland, Australia
LADA	Land degradation assessment in drylands
LandTrendr	Landsat-based Detection of Trends in Disturbance and Recovery
LCC	Land capability class
LNS	Local net primary productivity scaling
LEDAPS	Landsat Ecosystem Disturbance Adaptive Processing System
LP DAAC	Land Processes Distributed Active Archive Center
LTDR	Land Long Term Data Record
LUCC	Land use/land cover changes
MEDOKADS	Mediterranean Extended Daily One-km AVHRR Data Set
MERIS	Medium Resolution Imaging Spectrometer
MODIS	Moderate Resolution Imaging Spectroradiometer
NASA	National Aeronautics and Space Administration
NOAA	National Oceanic and Atmospheric Administration
NDVI	Normalized difference vegetation index
NDVI3g	Third-generation GIMMS NDVI from AVHRR
NPP	Net primary productivity
OLI/TIRS	Operational Land Imager and Thermal Infrared Sensors
PAL	Pathfinder AVHRR land archive
RESTREND	Residual trend analysis
RUE	Rain-use efficiency
SAVI	Soil-adjusted vegetation index
SeaWIFS	Sea-Viewing Wide Field-of-View Sensor
SMA	Spectral mixture analysis
SSI	Soil stability index
SPOT	Satellite Pour l'Observation de la Terre
STARFM	*Spatial* and Temporal Adaptive Reflectance Fusion Model
TC	Tasselled Cap Transformation
Timesat	Software tool to derive phenological metrics
Timestats	Software tool for time series analysis
TM	Thematic Mapper
UNCCD	United Nations Convention to Combat Desertification
UNEP	United Nations Environment Programme

USGS	U.S. Geological Survey
USGS-ESPA	USGS EROS Science Processing Architecture
VCT	Vegetation Change Tracker
VITO	Company "Vision On Technology"
VOD	Vegetation optical depth
WAD	World Atlas of Desertification
WELD	Web-Enabled Landsat Data

17.1 Introduction

17.1.1 Drylands

Drylands cover about 41% of the earth's land surface, comprising hyperarid to dry subhumid climate zones that are defined by low mean annual precipitation amounts compared to potential evaporation, that is, a ratio of mean precipitation to potential evaporation less than 0.65 (Thomas and Middleton, 1994; Safriel et al., 2005; see Figure 17.1). They include a large number of ecosystems that belong to the four broad biomes, forests, Mediterranean, grasslands, and deserts (Safriel et al., 2005), and are home to about one-third of the global population, with many residents directly depending on dryland ecosystem services including the provision of food, forage, water, and other resources (Millennium Ecosystem Assessment, 2005a). Drylands also provide ecosystem services of global significance, such as climate regulation by sequestering and storing vast amounts of carbon due to the large areal extent (Lal, 2004) (Table 17.1).

Drylands are characterized by high variability in both rainfall amounts and intensities and the occurrence of cyclic and prolonged periods of drought. Most frequently, soils contain low nutritious reserves and have low contents of organic matter and nitrogen (Skujins, 1991). In addition, surface runoff events, soil-moisture storage, and groundwater recharge in drylands are generally more variable and less reliable than in more humid regions (Koofhafkan and Stewart, 2008).

Water availability and the tolerance to periods of water scarcity are key factors in dryland productivity (Stafford Smith et al., 2009). In response to water scarceness and prolonged drought periods, fauna and flora of dryland ecosystems have adapted to these conditions following manifold strategies (morphological, physical, chemical), such as the development of drought-avoiding (i.e., ephemeral annual grasses) or drought-enduring (i.e., xerophytes) plant species as well as plant adaptations such as xeromorphological leaf structures. Fire is a further important element in functioning and maintenance of dryland ecosystems (Bond and Keeley, 2005).

17.1.2 Land Use in Dryland Areas

For thousands of years humans developed strategies to use the goods and services provided by drylands in a sustainable way (Table 17.1), thereby responding to the level of aridity. Thus, land use systems in drylands are very diverse, including a variety of

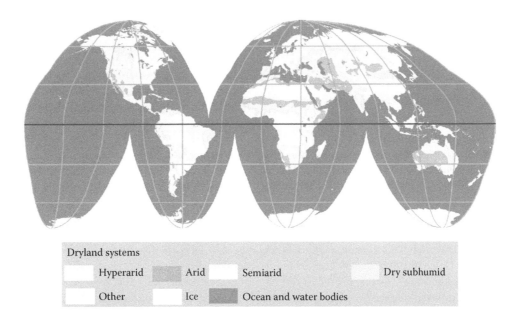

FIGURE 17.1 The spatial extent of drylands based on the aridity index (AI equals ratio of rainfall (P) and potential evapotranspiration (PET) for the period 1951–1980). Hyperarid: P/PET < 0.05; arid: 0.05 ≤ P/PET < 0.20; semiarid: 0.20 ≤ P/PET < 0.50; and dry subhumid: 0.50 ≤ P/PET < 0.65. Goode homolosine projection. (From UNEP (United Nations Environment Programme) The UNEP Environmental Data Explorer, as compiled from UNEP/DEWA/GRID, Geneva, Switzerland, 2014, http://geodata.grid.unep.ch, ESRI Data & Maps.)

TABLE 17.1   Key Dryland Ecosystem Services

Supporting Services: Services That Maintain the Conditions for Life on Earth		
	Soil development (conservation, formation)	
	Primary production	
	Nutrient cycling	
	Biodiversity	
Provisioning Services: Goods Produced or Provided by Ecosystems	*Regulating Services*: Benefits Obtained from Regulation of Ecosystem Processes	*Cultural Services*: Nonmaterial Benefits Obtained from Ecosystems
Provision derived from biological productivity: food, fiber, forage, fuelwood, and biochemicals	Water purification and regulation	Recreation and tourism
Fresh water	Pollination and seed dispersal	Cultural identity and diversity
	Climate regulation (local through vegetation cover and global through carbon sequestration)	Cultural landscapes and heritage values
		Indigenous knowledge systems
		Spiritual, aesthetic, and inspirational services

Source: After Millennium Ecosystem Assessment, Ecosystems and human well-being: Desertification synthesis, World Resources Institute, Washington, DC, 2005a.

shifting agriculture systems, annual croplands, home gardens, and mixed agriculture–livestock systems, including nomadic pastoral and transhuman systems (Koofhafkan and Stewart, 2008). The vast majority of drylands that support vegetation are used as rangelands (69%), which sustain about 50% of the world's total livestock population, whereas 25% of the dryland areas are used as croplands (Reid et al., 2004). However, land use varies largely among dryland climates. The proportion of rangeland increases with aridity, from 34% in subhumid regions to 97% in hyperarid areas (Millennium Ecosystem Assessment, 2005b), whereas arable cultivation is restricted to semiarid and dry subhumid regions (Koofhafkan and Stewart, 2008). Also the use of fire as a land use management tool has a history of millennia in drylands and includes the use of fire by pastoralists to improve rangeland conditions (Naveh, 1975), but also for slash and burn agriculture, honey collection, charcoal production, and

opening landscapes to facilitate hunting as practiced in African Savannahs (Mbow et al., 2000). Even though dryland ecosystems are adapted to fires, changing fire regimes may cause land degradation and loss of biodiversity as they impact species composition and vegetation and severely affect nutrient cycle (e.g., Trapnell, 1959, Anderson et al., 2003).

Countries with drylands differ in their socioeconomic development. Differences range from agrarian via industrialized to service-oriented societies, whereby at least 90% of the dryland population lives in developing countries (Safriel et al., 2005). The development stage defines to a large extent the land use systems and the corresponding process framework of land use/land cover changes (LUCCs) (DeFries et al., 2004). Even though land use changes are affecting almost all terrestrial ecosystems, drylands are considered as most vulnerable to degradation processes. Thus, water scarcity, overuse of resources, and climate change are a much greater threat for dryland ecosystems than for nondryland systems (Millennium Ecosystem Assessment, 2005a).

17.1.3 Land Degradation and Desertification

Degradation of terrestrial dryland ecosystems, also termed desertification, is recognized as one of the major threats to the global environment impacting directly on human well-being (Millennium Ecosystem Assessment, 2005a) and threatening to reverse the gains in human development in many parts of the world (UNU, 2006). The terms land degradation and desertification received worldwide attention following the prolonged Sahel drought during the 1970s and 1980s that caused a humanitarian catastrophe. As a result of the United Nations Conference on Desertification in 1977, a "Plan of Action to Combat Desertification" was approved. Limited progress in reducing the problem of desertification since then led the Rio Conference in 1992 to call on the United Nations General Assembly to prepare through intergovernmental negotiation a Convention to Combat Desertification. Thus, in 1994, the United Nations Convention to Combat Desertification (UNCCD) was adopted and brought into force in 1996 having received notification of the 50th ratification of the Convention, which by now has 193 signatory parties. The definition of both terms was subject to highly controversial debates (Hermann and Hutchinson, 2005).

A nowadays widely accepted definition of land degradation and desertification is provided by the UNCCD. According to the UNCCD (1994), land degradation is defined as

> "the reduction or loss, in arid, semi-arid and dry sub-humid areas, of the biological or economic productivity and complexity of rainfed cropland, irrigated cropland, or range, pasture, forest and woodlands resulting from land uses or from a process or combination of processes, including processes arising from human activities and habitation patterns". Desertification is defined as "land degradation in arid, semiarid and dry sub-humid areas, resulting from various factors, including climatic variations and human activities".

This definition aims to cover at large the broad range of complex processes that cause a sustained decrease of ecosystem services throughout all terrestrial ecosystems in drylands. Nevertheless, this also leaves room for interpretation and uncertainties concerning the terminology (Vogt et al., 2011) and, hence, also arouses different perceptions of the processes that lie behind these two terms.

17.1.4 Scientific Perception of Land Degradation

In the past decades, the scientific communities' understanding has undergone a shift concerning the key factors that are required to allow for adequate assessment and monitoring of land degradation. The assessment of land degradation changed from a mere biophysical perception to a more holistic approach where human-induced or climate-driven underlying forces as well as spatial and temporal scale issues have been recognized as factors that should be considered to understand and identify land degradation processes (Vogt et al., 2011).

The understanding of land degradation processes, including their causes and consequences on ecosystem functioning as well as the identification of affected areas and regions at risk, is a prerequisite to develop strategies to mitigate and avoid land degradation. Accordingly, over the past decades, many national and international research initiatives reviewed the status of land degradation sciences and identified gaps and developed strategies to assess and monitor land degradation and desertification.

This chapter provides an overview of important studies on remote sensing of land degradation in drylands. Section 17.2 presents general considerations regarding the assessment and monitoring of land degradation including suitable indicators as well as sensor systems. The following sections give a review of the state of the art on the assessment of land condition (Section 17.3), the monitoring of LUCCs to assess land degradation processes (Section 17.4), and the identification of human-induced drivers of land degradation using integrated concepts (Section 17.5), whereas Section 17.6 describes limits and uncertainties regarding dryland observation. This chapter concludes with a summary of land degradation assessment and monitoring by remote sensing techniques (Section 17.7).

17.2 Remote Sensing of Dryland Degradation Processes

Various scientific disciplines contribute valuable information that enhances the understanding of land degradation and desertification at different temporal and spatial scales. These include studies ranging from the plot scale to global assessments as well as the collection of biophysical or socioeconomic data and the implementation of models to predict land use changes in future decades.

Earth observation is a tool that essentially contributes to the assessment and monitoring of ecosystems from a local to a global scale. Hence, information extracted from remote sensing data can be employed to (1) assess the extent and condition of ecosystems and (2) monitor changes of ecosystem conditions and services over long time periods (Foley et al., 2005; Turner et al., 2007). The use of earth observation data fundamentally contributes to the understanding of dynamics and responses of vegetation to climate and human interactions (DeFries, 2008).

Monitoring drylands requires observation data that are able to observe long-term trends and short-term disturbances across large areas. For this reason, remote sensing data are important components of monitoring strategies, as they provide objective, repetitive, and synoptic observations across large areas (Graetz, 1996; Hill et al., 2004). Three major components are particularly important to (1) provide a comprehensive observation of dryland areas, (2) ensure their relevance for policy and management, and (3) help prevent unsustainable use of ecosystems goods and services:

1. Assessment of actual land condition, that is, the capacity of an ecosystem to provide goods and services (Section 17.3)
2. Monitoring of land cover changes and assessment of their implications for land condition separating natural processes, that is, climate variability and fire, from human-induced land use/land cover-related processes (Section 17.4)
3. Integrated concepts that link remotely sensed results to the human dimension in order to identify drivers of land degradation (Section 17.5)

Neither the condition of ecosystems nor the processes affecting them can directly be measured by earth observation data. Rather, suitable indicators have to be identified (Verstraete, 1994) that (1) can be related to the status and processes and (2) can be derived in standardized and replicable way.

17.2.1 Suitable Remote Sensing Indicators for Dryland Observation

A range of approaches and models has been developed allowing to derive a variety of biophysical parameters appropriate for the observation of drylands (Lacaze, 1996; Hill, 2008). Depending on the spatial and spectral characteristics of the remote sensing data, these qualitative and quantitative measures include vegetation indices related to greenness, vegetation cover, pigment and water content, soil organic matter of the topsoil, landscape metrics, etc. (e.g., Blaschke and Hay, 2001; Hill et al., 2004).

Even though land degradation indicators related to soil have proven to provide important information on land degradation, vegetation cover hampers the remotely sensed assessment of soil properties. Thus, soil properties can only be reliably assessed at low vegetation cover (Jarmer et al., 2009). Furthermore, many of the proposed indicators, for example, grain size distribution, mineral content, and soil organic carbon, require hyperspectral data. To date, these data are mostly acquired using airborne systems, making them costly and only available for small areas. As a result, only few studies exist that use hyperspectral imagery for land degradation assessment (e.g., Shrestha et al., 2005, De Jong and Epema, 2011). However, various hyperspectral, spaceborne missions are currently being developed (e.g., Environmental Mapping and Analysis Program under the lead of the German Aerospace Center or Hyperspectral Infrared Imager by the National Aeronautics and Space Administration [NASA]), and it is to be expected that the utilization of this hyperspectral imagery in the context of land degradation assessment will increase in the near future.

17.2.2 Biophysical Remote Sensing Indicators for Long-Term Dryland Observation

The biological productivity of ecosystems is one of the key factors that describe the functioning of an ecosystem, and it is also explicitly stated in the definition of desertification and land degradation of the UNCCD (Del Barrio et al., 2010). Parameters related to productivity such as greenness, vegetation cover, and biomass can therefore serve as proxies to assess and monitor land degradation. These parameters are especially suitable for earth observation methods due to the distinct spectral signature of vegetation.

A commonly used vegetation index calculated from the red and near-infrared spectral information is the normalized difference vegetation index (NDVI) (Rouse et al., 1974; Tucker, 1979). It was shown that the NDVI is a proxy for greenness and is linearly related to the fraction of absorbed photosynthetic active radiation (FAPAR) (Myneni and Williams, 1994; Fensholt et al., 2004), which in itself is an important factor of assessing the net primary productivity (NPP). However, the NDVI has well-known weaknesses due to its sensitivity to soil background, especially when vegetation cover is low (Price, 1993; Elmore et al., 2000). Advanced vegetation indices overcome these problems, like the enhanced vegetation index (EVI) (Huete et al., 2002) and the soil-adjusted vegetation index (SAVI) (Huete, 1988). More advanced methods to derive parameters that are related to vegetation are the Tasselled Cap Transformation (TC) (Kauth and Thomas, 1976) and spectral mixture analysis (SMA) (Adams et al., 1986; Smith et al., 1990). The latter directly provides vegetation cover if correctly parameterized and is often used for Landsat-based land degradation assessment in drylands (Sonnenschein et al., 2011). Nevertheless, for temporal analysis, it seems to be of more decisive importance to employ a robust and consistent measure (Udelhoven and Hill, 2009; Sonnenschein et al., 2011).

17.2.3 Earth Observation Platforms Used in Dryland Observation

High variability of precipitation amounts infers also a high variability of vegetation cover and its vitality. Moreover, disturbances like fires create abrupt changes of vegetation cover. Dryland observation requires to consider these variations by using long-term observation to separate gradual long-term trends from short-term variations. Among the numerous spaceborne sensors, only few satellites are fulfilling the two criteria of collecting data that (1) cover a long time period and (2) provide a systematic global coverage. These systems can be distinguished in two major groups: The first provides a medium spatial resolution, but has a limited temporal resolution, and the second provides a coarse-scale resolution, but has a high temporal resolution. Table 17.2 gives information of important sensors and derived archives, which are presented in more detail in the following sections.

TABLE 17.2   Selection of Important Sensors and Available Data Products Suitable for Dryland Monitoring

Sensor	Coverage	Name	Source	Observation Period	Spatial Resolution	Temporal Resolution	Indicator
Coarse resolution							
NOAA AVHRR	Global	PAL	GES-DAAC, NOAA/NASA James and Kalluri (1994), Smith et al. (1997)	1981–2001	8 km	10 daily	NDVI
		FASIR (ISLSCP II)	ISLSCP II Sietse (2010)	1981–1998	8 km	Monthly	NDVI
		GIMMS*	GLFC, University of Maryland Tucker et al. (2005)	1981–2006	8 km	Bimonthly	NDVI
		GIMMS3g*	GLFC, University of Maryland Pinzon and Tucker (2014)	1981–2012	8 km	Bimonthly	NDVI, LAI, FAPAR
		LTDR v4	NASA/GFSC; University of Maryland Pedelty et al. (2007)	1981–present	0.05°	Daily	NDVI
	Regional	NDVI (Australia)	Bureau of Meteorology; AusCover BOM (2014)	1992–present	0.01°/0.05°	10 daily/monthly	NDVI
		MEDOKADS (Mediterranean)	Freie Universität Berlin Koslowsky (1998)	1989–2005	1 km	10 daily	NDVI
SeaWiFS	Global	L3 Land—NDVI	NASA/GFSC OceanColor WEB (http://oceancolor.gsfc.nasa.gov/)	1997–2010	4 km/9 km	Daily, 8 daily, monthly, annual	NDVI
		FAPAR	EC-JRC Gobron et al. (2006)	1997–2006	0.01°	10 daily	FAPAR
SPOT Vegetation		VGT-S10	VITO; CNES Archard et al. (1995)	1998–present	1 km	10 daily	NDVI
Envisat-MERIS	Global	EM-10	VITO; ESA; Belspo; EC-JRC Gobron (2011)	2002–2012	1.2 km	10 daily	NDVI, FAPAR
		MGVI	ESA/JRC-EC Gobron et al. (1999)	2002–2012	1.2 km	10 daily	FAPAR
MODIS Terra/Aqua	Global	MOD/MYD13Q1*	NASA LP DAAC Huete et al. (1999)	2000–present	250 m–1 km	16 daily/monthly	EVI, NDVI
		MOD/MYD/MCD15A	NASA LP DAAC Knyazikhin et al. (1999), Myneni et al. (2002)	2000–present	1 km	4 daily/8 daily	FAPAR
		MOD/MYD17A	NASA LP DAAC Running et al. (2000)	2000–present	1 km	8 daily/annual	GPP/NPP
		TIP, FAPAR	EC-JRC Pinty et al. (2011)	2000–present	1 km	16 daily	FAPAR
Combined							
NOAA AHRR/MODIS	Global	LTDR v3	NASA/GFSC; University of Maryland Pedelty et al. (2007)	1991–2012	0.05°	Daily	NDVI
SeaWIFS/MERIS		FAPAR	EC-JRC Ceccherini et al. (2013)	1997–2012	1.2 km	10 daily	FAPAR
Moderate resolution							
Landsat	Local	Individual*	Individually; USGS Röder et al. (2008a), Sonnenschein et al. (2011)	1982–present	30 m	Multiseasonal to annual	Vegetation indices, SMA
		Surface Reflectance Climate Data Record	USGS-ESPA Masek et al. (2006)	1982–present	30 m	Multiseasonal to annual	Vegetation indices

(Continued)

TABLE 17.2 *(Continued)*   Selection of Important Sensors and Available Data Products Suitable for Dryland Monitoring

Sensor	Coverage	Name	Source	Observation Period	Spatial Resolution	Temporal Resolution	Indicator
	Regional	Seasonal fractional vegetation cover, Queensland (Australia)	JRSRP; DSITIA; AusCover Danaher et al. (2010), Flood et al. (2013), Muir et al. (2011)	1986–present	30 m	Multiseasonal	Fractional vegetation cover
	Global	WELD v1.5 product	NASA LP DAAC/USGS Roy et al. (2010)	2002–2012	30 m	Annual	Tree cover
SPOT	Local	Individual	Individually; CNES; ASTRIUM (http://www.astrium-geo.com/)	1986–present	6 m/20 m	Multiseasonal to annual	Vegetation indices

Archives marked with an asterisk are/were often used in dryland studies (as of 2014).

17.2.3.1 Medium Spatial Resolution Sensors

The Landsat program consists of a series of multispectral optical sensors that record the reflected radiance in the visible to middle infrared domain (complemented by band[s] in the thermal domain) that allows the derivation of several surrogates related to vegetation properties (Fang et al., 2005). Landsat Thematic Mapper (TM), Enhanced Thematic Mapper (ETM+), and Operational Land Imager and Thermal Infrared Sensors (OLI/TIRS), respectively, are providing data of the earth's surface with a spatial resolution of 30 m × 30 m since 1982 (Goward and Masek, 2001). The temporal revisit rate of the sensor is 16 days and could theoretically provide a time series of earth observations with similar density compared to those provided by coarse-scale sensors, but also in many dryland areas, cloud cover impedes the acquisition of utilizable images. Thus, often only few images of sufficient quality can be acquired per season.

The Satellite Pour l'Observation de la Terre (SPOT) satellites operated by Centre National d'Études Spatiales (CNES) provide multispectral data since 1986 with a spatial resolution of 6 m × 6 m up to 20 m × 20 m with a revisit rate of 26 days. The SPOT system is operated commercially, which offers the possibility to prioritize the observation of specific areas. Whereas the Landsat sensors are restricted to Nadir acquisition, the SPOT sensors are able to incline the sensor allowing for the acquisition of data for specific areas more often than these 26 days. At the same time, this means that other areas are not recorded on a regular basis.

17.2.3.2 Coarse Spatial Scale Satellite Sensors

Regional to global dryland studies are mostly based on coarse-scale imagery with higher temporal resolution. Due to the long legacy of the mission, the National Oceanic and Atmospheric Administration (NOAA) Advanced Very High Resolution Radiometer (AVHRR) sensor series is one of the most important sensors in this context. Within the Global Inventory Monitoring and Modeling System (GIMMS) project, the most commonly used global NOAA AVHRR time series are provided. The recent version, third-generation GIMMS NDVI from AVHRR (NDVI3g), spans the time period from 1981 to 2012 and consists of bimonthly measurements of the NDVI data at a pixel size of about 8 km × 8 km.

Higher resolution NOAA AVHRR archives data are available for some parts of the world, such as the Mediterranean Extended Daily One-km AVHRR Data Set (MEDOKADS). The archive consists of a 10-day maximum value composite of full-resolution NOAA AVHRR channel data covering the whole Mediterranean region from 1989 to 2004 with a spatial resolution of about 1 km^2 (Koslowsky, 1996). Another regional datasets is, for example, a 1 km^2 dataset covering Australia (BOM, 2014).

A prerequisite for long-term observation analyses are well-calibrated data archives. This is especially demanding in the case of the NOAA AVHRR data archives as preprocessing comprises the correction of effects caused by orbital drift of the sensor (i.e., changing overpass time) as well as the intercalibration of the spectral channels between the different AVHRR sensors employed to create the long-term archives. Due to the limited spectral properties of the NOAA AVHRR sensors, the derivation of biophysical parameters is limited and usually based on the NDVI.

The Moderate Resolution Imaging Spectroradiometer (MODIS) provides a better spatial and spectral resolution that allows to derive more enhanced biophysical surrogates. NDVI and EVI are provided as standard vegetation parameter products. Moreover, the sensor properties facilitate the provision of a consistent high-quality data archive including the possibility to derive bidirectional reflectance distribution function (BRDF) corrected data (Strahler et al., 1999). Other sensors delivering time series suitable for land degradation assessment are, for example, SPOT Vegetation, Sea-Viewing Wide Field-of-View Sensor (SeaWIFS), and Medium Resolution Imaging Spectrometer (MERIS). However, in comparison to the NOAA AVHRR datasets, these archives are still confined to rather short observation periods. Several studies aimed at combining different data archives to overcome the different spectral responses, differing observation characteristics including observation geometry and diverging spatial resolutions of the sensor systems (Ceccherini et al., 2013).

17.2.3.3 Recent Developments for Obtaining Medium Spatial and High Temporal Resolution Time Series

Although both coarse and medium sensor types provide data that allow for adequate dryland observation, there is a trade-off between geometric and spectral level of detail, areas covered, and temporal resolution that needs to be considered. With the planned launch of the European Space Agency (ESA) Sentinel-2 satellites in 2015 and 2016, two additional Landsat-type sensors will be available. Together with the Landsat OLI, the repetition rate of acquiring data from the entire globe will be much higher, augmenting also the probability of cloud-free observations. Another promising technique is the fusion of Landsat and MODIS images with the Spatial and Temporal Adaptive Reflectance Fusion Model (STARFM) (Gao et al., 2006) aiming at providing time series with a temporal resolution of MODIS but the spatial resolution of Landsat. The approach was applied successfully to dryland areas (Schmidt et al., 2012; Walker et al., 2012) and offers the possibility to monitor land degradation processes in more detail. One drawback of this procedure is that the fusion can only be performed after the launch of MODIS Terra in the year 2000.

17.2.3.4 Analysis Techniques

Long-term monitoring requires accurate geometric and radiometric correction of the data to reduce noise that originates from observational conditions including observation geometry, atmospheric conditions, and sensor degradation. A meaningful analysis necessitates a rigorous preprocessing scheme for all the time series images (Röder et al., 2008a).

The creation of a medium-resolution time series is challenging because images should originate from comparable phenological stages. Therefore, many of the early studies investigating time trajectories of vegetation based on Landsat time series are confined to only one observation per season (e.g., Hostert et al., 2003; Röder et al., 2008a).

The opening of the Landsat archives distributed by the U.S. Geological Survey (USGS) has enabled new opportunities to assess land cover changes based on the full range of available data from the archive, including images with high cloud cover. Thus, new approaches move from image-based analysis toward pixel-based analysis. This comes along with new methodologies that allow for preprocessing and analyzing the data in an automated way. It includes the provision of geometrically corrected Landsat L1T data by USGS, cloud detection via fmask (Zhu and Woodcock, 2012), and automated radiometric correction schemes like the Landsat Ecosystem Disturbance Adaptive Processing System (Masek et al., 2006) or the Australian BRDF correction scheme (Flood et al., 2013). Recently, the USGS has embarked to distribute higher level Landsat data products, for example, Landsat surface reflectance climate data record (CDR) and Landsat surface reflectance-derived spectral indices (USGS, 2015). For Queensland, Australia, a fractional vegetation cover product is available since 1986 providing seasonal images for the entire state (http://www.auscover.org.au).

With the changes in data policy and increases in data quality as well as computational improvements, time series approaches were developed that allow for the detection of gradual or abrupt changes, or both simultaneously. Several methodologies and tools were published, for example, Landsat-based Detection of Trends in Disturbance and Recovery (LandTrendr) (Kennedy et al., 2010),

the Vegetation Change Tracker (VCT) (Huang et al., 2010), Breaks for Additive Seasonal and Trend (BFAST) (Verbesselt et al., 2010), continuous monitoring of forest disturbance algorithm (Zhu et al., 2012), and continuous change detection and classification (Zhu and Woodcock, 2014). Many of these approaches were implemented and tested in boreal and temperate forest ecosystems (e.g., Griffiths et al., 2011; Schroeder et al., 2011). In such ecosystems, the vegetation signal is high and yearly variations are small compared to dryland areas. Moreover, vegetation communities in drylands are often very complex and the spatial arrangement of the landscape very heterogeneous. These factors plus the occurrence of fires hamper the detection of subtle modifications of vegetation cover due to land degradation processes. Therefore, enhanced time series analyses tools are gaining more and more importance as they allow for monitoring not only the overall increase or decrease of greenness but also more complex change patterns including its character, that is, gradual and abrupt changes (De Jong et al., 2012). This represents reality better as trends are rarely uniform during a long observation period, for example, due to droughts, fire events, and macro weather situations.

The techniques that are used to explore the coarse-scale data archives are very similar to the ones used to examine Landsat time series. Additionally, due to the dense temporal resolution, the phenology of vegetation and its changes can be portrayed by deriving phenological metrics using specialized software like Timesat (Jönsson and Eklundh, 2002) and Timestats (Udelhoven, 2011).

17.3 Assessing Land Condition

Land degradation may be defined as a long-term loss of an ecosystem's capacity to provide goods and services. Therefore, a major component of a comprehensive dryland observation is the assessment of land condition that can be linked to ecosystem status. Even though land degradation is recognized as a severe threat, only few global land degradation assessments have been carried out until today (Millennium Ecosystem Assessment, 2005a; Vogt et al., 2011).

The first global assessment of land quality was provided in the framework of the Global Assessment of Human-Induced Soil Degradation (GLASOD) project (Global Assessment of Human-Induced Soil Degradation, 1987–1990) where human-induced soil degradation (extent, type, and grade) was mapped at a scale of 1:10 million based on expert judgment (Oldeman et al., 1990). Another global assessment was provided by Dregne and Chou (1992) who also integrated information on vegetation status based on secondary sources. Whereas the map provided by GLASOD indicated that 20% of soils in drylands were degraded, Dregne and Chou estimated that 70% of dryland areas were affected either by degradation of soil or vegetation. A more recent study (Lepers, 2003) prepared for the Millennium Ecosystem Assessment covered over 60% of all dryland areas. Several data sources, including remote sensing data, were integrated in the analyses and indicated that 10% of the observed area was affected by land degradation. One of the major points of criticisms is related to the subjectivity of the studies that impede operational use or comparability (Millennium Ecosystem Assessment, 2005a). In recent years, different concepts were developed and implemented to assess land condition, which will be described in the following part. A selection of studies and the techniques used is summarized in Table 17.3.

17.3.1 Assessment of Land Condition Related to the Biological Productivity of Ecosystems

In recent years, the assessment of land condition has been primarily related to the biological productivity of ecosystems. The concept is based on the fact that land degradation, which might be caused by a wide variety of climate- and human-induced processes, results in a decline of the potential of the soil to sustain plant productivity (Del Barrio et al., 2010). Using the example of rangelands, Figure 17.2 clearly illustrates the dependence of biological productivity on grazing pressure, rainfall, and soil properties. In this respect, soil properties like water holding capacity and nutrient supply are essential factors that directly affect primary productivity. Ongoing overgrazing drives feedback loops between vegetation and soil, resulting in a degradation of these soil properties, and triggers a sustained decrease of the soil's capacity to sustain primary productivity. As a consequence, the ecosystem's capacity to utilize local resources (such as soil nutrients and water availability) in relation to its potential capacity may be defined as land condition. This in turn allows drawing conclusions on the degradation status of observed areas (Boer and Puigdefabregas, 2005). Hence, biological productivity is considered a suitable surrogate to assess land condition and surrogates derived from remote sensing are predestined to support this assessment.

At local scale, Boer and Puigdefabrégas (2005) conceptualized and implemented a spatial modeling framework to assess land condition based on climate data as well as on NDVI data derived from the Landsat sensor, which served as a proxy for primary productivity. The approach is based on the assumption that in arid and semiarid areas, water availability is the major limiting factor of productivity and, furthermore, that the water balance, which depends on rainfall, soil properties (evaporation), vegetation (interception and transpiration), and discharge, reflects land condition. Based on this theoretical concept, they proposed a long-term ratio of mean actual evapotranspiration and precipitation to assess land condition.

Prince (2004) and Prince et al. (2009) introduced the local net primary productivity scaling (LNS) method where the actual NPP is compared to the potential NPP of the corresponding land capability class (LCC). The LCCs are homogenous areas that are determined by climate, soils, land cover, and land use and are independent of actual NPP. The magnitude of the difference provides a measure of land degradation and at the same time the loss of carbon sequestration. The actual NPP is derived for each pixel from multitemporal earth observation data. The potential NPP, that is, the NPP that could be expected without human land use,

equals the maximum NPP found in the corresponding LCC and enables to implement this approach for large physical heterogeneous areas. The implementation of this method for Zimbabwe (Prince et al., 2009; see Figure 17.3) showed that only 16% of the land cover reached the level of the potential NPP, whereas over 80% was found to have an actual NPP far below the potential one suggesting a loss of carbon sequestration of 7.6 million tons C year^{-1}. Similar methodologies were developed by Bastin et al. (2012) and Reeves and Baggett (2014) to identify rangeland conditions in Queensland, Australia, and the southern and northern Great Plains, USA, respectively.

17.3.2 Assessment of Land Condition Including Climate and Its Variability

Wessels et al. (2007) used a residual trend analysis (RESTREND) to identify potentially degraded areas by decoupling the NDVI signal from rainfall variability based on NOAA AVHRR data. This methodology identifies areas where a reduction in productivity per unit rainfall has occurred by comparing modeled accumulated NDVI values based on rainfall data to the observed NDVI. While the method proved capable of identifying potentially degraded areas in South Africa, Wessels et al. (2007) stressed that the cause of the negative trend cannot be explained solely by this approach, but needs detailed investigation. Li et al. (2012) transferred the RESTREND methodology to a rangeland area in Inner Mongolia, China. Their results showed that until the year 2000, heavy overgrazing deteriorated rangelands in this area, but grasslands recovered afterward due to the implementation of new land use policies. The authors concluded that the methodology is useful to identify human-induced changes in drylands, but also underlined that the results need careful interpretation.

Other developed approaches make use of the concept of rain-use efficiency (RUE), which was introduced by Le Houérou (1984). RUE is defined as the ratio of NPP to precipitation over a given time period and may be interpreted as being "proportional to the fraction of precipitation released to the atmosphere" (Del Barrio et al., 2010). Several studies explored RUE in dryland areas based on remote sensing (e.g., Prince et al., 1998; Bai et al., 2008) causing debates between scientists due to supposed weaknesses in the rationale (Hein and de Ridder, 2006; Prince et al., 2007; Wessels, 2009). In the framework of the Land degradation assessment in drylands (LADA) project, Bai et al. (2008) proposed a methodology to assess and monitor land condition by deriving RUE based on the global NOAA AVHRR GIMMS dataset. The implemented methodology and the results were criticized (Wessels, 2009) because rainfall is not a limiting factor in more humid areas, and moreover, RUE values are dependent on precipitation amounts and thus impede the direct comparison of RUE values from regions of diverging aridity level. Fensholt et al. (2013) proposed to only use NPP proxies that are positively linearly correlated to precipitation and to only consider the rainy-season variation of NDVI for those areas where the correlation between RUE and annual precipitation is close to zero.

TABLE 17.3   Selection of Studies Evaluating Land Condition Using Remote Sensing Data

Extent	Study Area	Remote Sensing Data and Indicator	Methodology	Observation Period	Result	References
Local						
	Spain	Landsat NDVI	Water balance model	1993–1994	Long-term ratio of mean actual evapotranspiration and precipitation able to assess land condition, e.g., poor land condition due to soil erosion.	Boer and Puigdefabregas (2005)
Regional						
	Zimbabwe	MOD13Q1 NDVI	LNS.	2000–2005	Over 80% were found to have an actual NPP far below the potential one.	Prince et al. (2004, 2009)
	North east Queensland, Australia	Landsat persistent ground cover time series	Automated detection of rangeland condition based on reference areas.	1986–2008	Management-related change in ground cover in savanna woodlands at three spatial scales was detected.	Bastin et al. (2012)
Regional to global						
	Great Plains, USA	MOD13Q1 NDVI	Rangeland productive capacity is derived relative to reference conditions.	2000–2012	16% of the northern and 9% of the southern study area are degraded.	Reeves and Baggett (2014)
	Bishri Mountain, Syria	NOAA AVHRR NDVI	RESTREND	1981–1996	Areas showing a negative temporal trend in residuals of NDVImax and rainfall coincide with areas that are most heavily used by humans.	Geerken and Ilaiwi (2004)
	South Africa	NOAA AVHRR VI	RESTREND	1985–2003	Identification of potentially degraded areas in South Africa.	Wessels et al. (2007)
	Inner Mongolia, China	GIMMS NDVI	RESTREND	1981–2006	Heavy overgrazing deteriorated rangelands in this area but grasslands recovered afterward due to the implementation of new land use policies.	Li et al. (2012)
	Sahel, Africa	GIMMS NDVI(NPP)	RUE	1982–1990	Systematic increase of RUE in the Sahel; recovery of vegetation after the severe drought.	Prince et al. (1998)
	Spain	MEDOKADS green vegetation fraction	RUE	1989–2000	Ongoing land degradation appeared only in localized areas caused by current or recent intensive land use.	Del Barrio et al. (2010)
		GIMMS NDVI; MODIS MOD13C1 NDVI	RUE, RESTREND	1982–2007	Very limited anthropogenic land degradation in the Sahel–Sudanian zone could be observed by trend analyses.	Fensholt and Rasmussen (2011)
	Sahel, Africa	GIMMS3g NDVI (SPOT Vegetation NPP)	RUE	1982–2010	Only few areas (0.6%) were affected by land degradation processes.	Fensholt et al. (2013)
Global						
	60% of global drylands covered	Several (meta analysis)	Expert judgment	1980–2000	Sahel is not a hot spot of land degradation, Asia shows the largest area of degradation, but other drylands are not covered well by studies.	Lepers (2003)
		GIMMS NDVI (MOD17 NPP)	RUE	1981–2003	Declining RUE-adjusted NDVI on ca. 24% of the global land area.	Bai et al. (2008)

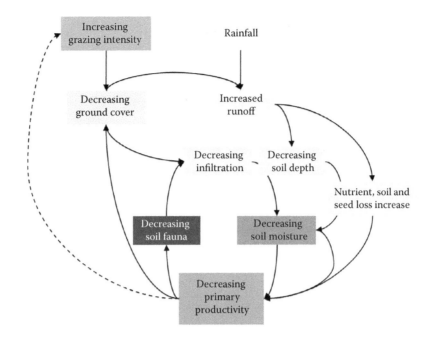

FIGURE 17.2 Aspects of landscape function using the example of grazing. Changes of ground cover, which are at short timescales mainly driven by rainfall variability and grazing pressure, can affect soil properties negatively. If thresholds are crossed, the cycle moves toward a new state that is characterized by degraded soil properties and a long-term loss in productivity. Even though a negative feedback exists to grazing intensity, management interventions often weaken this mechanism by maintaining constant stock numbers. (Modified from Stafford Smith, M. et al., Drylands: Coping with uncertainty, thresholds, and changes in state, in F.S. Chapin III, G.P. Kofinas, and C. Folke (eds.), *Principles of Ecosystem Stewardship*, Springer, New York, 2009, pp. 171–195.)

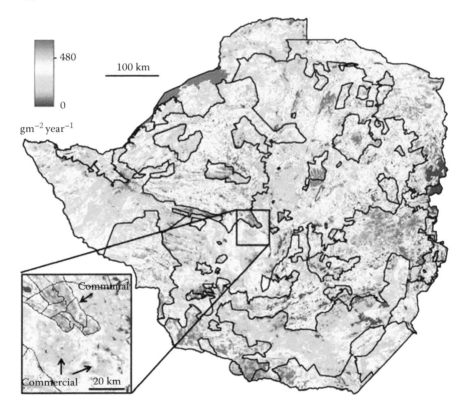

FIGURE 17.3 Local NPP scaling (LNS) of Zimbabwe, where LNS provides the NPP lost as a result of degradation. Communal and commercial area boundaries are in black. Inset—higher resolution segment southwest of Gweru showing communal area degradation (top left) and commercial area degradation (lower right). (From Prince, S.D. et al., *Remote Sens. Environ.*, 113, 1046, 2009.)

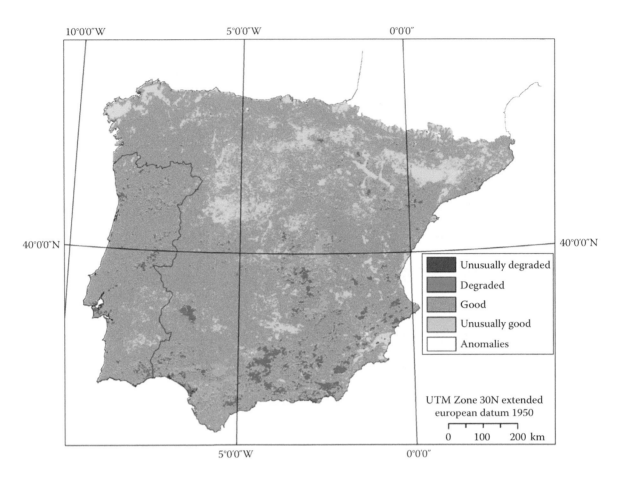

FIGURE 17.4 Assessment of land condition in the Iberian Peninsula (1989–2000). (From Del Barrio, G. et al., *Remote Sens. Environ.*, 114, 1817, 2010.)

Del Barrio et al. (2010) presented an approach that takes the dependency of RUE on aridity into account. The approach was implemented for the Iberian Peninsula based on 1 km² NOAA AVHRR NDVI data and spatially interpolated climate data. Due to the strong climatic gradient across the Iberian Peninsula, the derived RUE values were in a first step detrended for aridity to ensure the comparability of the derived data between different climatic zones. In a next step, statistically derived boundaries of minimum and maximum RUE were employed to calculate relative RUE values. Based on the assumption that healthy and undisturbed vegetation is characterized by a maximum RUE value, the relative RUE can be treated as a measure for land condition. Results of this study indicated that land condition of the Iberian Peninsula was better than expected with localized areas of ongoing land degradation caused by current or recent intensive land use (Del Barrio et al., 2010; see Figure 17.4). This study also focused on the monitoring of changes in primary productivity and considered effects of climatic variations. The results suggest that areas already in good conditions further improved, whereas degraded areas remained static. One disadvantage of this approach is the statistical determination of land condition lacking absolute references of vegetation performance comparable to the LNS method.

17.4 Monitoring of Land Use/Land Cover Changes to Assess Land Degradation Processes

Land cover is defined by the attributes of the land surface including all aspects such as flora, soil, rock, water, and anthropogenic surfaces, whereas land use has been defined as the purpose for which humans employ land cover (Lambin et al., 2006). Changes in land use are often accompanied by alterations in land cover that always imply changes in ecosystem functions, such as primary productivity, soil quality, water balance, and climatic regulation (e.g., Foley et al., 2005; Turner et al., 2007). The monitoring of landscape dynamics forms therefore an essential component for dryland observation as it provides information about the nature and extent of the changes and allows for the evaluation of the consequences for ecosystem functions.

Land cover changes can be distinguished in two major groups: (1) conversion and (2) modification (Lambin et al., 2006). Land use conversion commonly involves the replacement of one land use/land cover class by another (e.g., shrublands with arable land), whereas modification is usually related to gradual changes within one thematic class (e.g., shrub encroachment within natural ecosystems). The assessment of both conversion and modification is important to provide a comprehensive picture of LUCCs. The assessment of land use/land cover conversion is often based on land use change detection performed at defined years of interest. Several strategies and methods were developed to optimize the results of change detection analyses. A detailed overview of change detection techniques and their application, potentials, and limits is, for instance, given in Hecheltjen et al. (2014).

The assessment of modifications is a crucial element in dryland areas, because land cover changes related to land degradation are often associated with a modification of the landscape (Lambin and Geist, 2001; Lambin, et al. 2006). These include, for instance, vegetation cover loss due to overgrazing or primary or secondary succession on abandoned fields and rangelands. The detection and monitoring of a modification is often more challenging as changes of biophysical properties have to be observed and distinguished from interannual variability. This is especially important for dryland areas where primary productivity is dependent on the highly variable climatic conditions in terms of rainfall (Turner et al., 2007). Time series analysis of remote sensing archives is a suitable methodology to assess gradual changes of land cover (Udelhoven, 2011), providing means to delineate interannual variability from long-term trends. This requires consistent long-term data of biophysical parameters connected to surface properties, such as those provided by the broad remote sensing data sources described in the previous chapter.

The high geometric detail of Landsat data often matches the scale of land management decisions (Cohen and Goward, 2004; Lambin et al., 2006), whereas coarse-scale NOAA AVHRR data are more suitable to cover large areas providing a much higher temporal repetition rate. These data archives therefore permit the detection of changing parameters connected to vegetation cover as well as the deduction of changes in phenology (e.g., Andres et al., 1994; Brunsell and Gillies, 2003; Stellmes et al., 2013). Numerous studies exist that assess landscape dynamics in dryland areas based on these data sources, giving useful insights on process patterns from local to regional and global scales (see Table 17.4).

17.4.1 Local-Scale Studies to Detect Land Degradation–Related Modifications

At a local scale, many studies focused on monitoring both long-term and abrupt modifications using Landsat time series. In this context, the impact of grazing pressure on vegetation cover has been analyzed in different parts of the world, for example, in Bolivia (Washington-Allen et al., 2008), Greece (Hostert et al., 2003; Röder et al., 2008a; Sonnenschein et al., 2011), and Nepal (Paudel and Andersen, 2010). These studies used either vegetation indices like the NDVI or enhanced parameters such as proportional vegetation cover derived from SMA. Degradation processes were identified in all study areas and often additional information layers were used to explain these findings. Washington-Allen et al. (2008) assessed the effect of an El Niño

Southern Oscillation (ENSO)-induced drought on a rangeland system in Bolivia employing Landsat time series. This study showed that the decrease in vegetation cover of the rangelands resulted in an increased risk of soil erosion. In northern Greece (Röder et al., 2008a), patterns of over- and undergrazing were identified following changed rangeland management practices from transhumance to sedentary pastoralism (see Figure 17.5). Similar patterns were observed on the island of Crete, Greece (Hostert et al., 2003). Another important dimension in land degradation science is the understanding of impacts of LUCCs on ecosystems. Hill et al. (2014) used the ecosystem services concept (Millennium Ecosystem Assessment, 2005a) to estimate changes in ecosystem services introduced by LUCCs detected using a Landsat TM/ETM+ time series in Inner Mongolia, China, between 1987 and 2007.

Other studies were focusing on abrupt changes caused by fires, including studies on mapping fire patterns (Diaz-Delgado and Pons, 2001; Bastarrika et al., 2011) or postfire recovery (e.g., Viedma et al., 1997; Röder et al., 2008b). Yet assessments of the relationship of gradual and abrupt vegetation changes in the Mediterranean are largely missing (Sonnenschein et al., in review).

While many studies focused on local areas, that is, covering one Landsat scene, operational systems for the monitoring of rangeland areas have been set up in Australia during the last decades (Wallace et al., 2006, 2004). Several regional projects use parameters derived from Landsat time series to monitor land cover changes and land condition, which are integrated in the Australian Collaborative Rangeland Information System on a nationwide level. Furthermore, the software tool VegMachine was developed where satellite imagery and expert knowledge are combined to assess the health status of grazing grounds and to support pastoral producers as well as management decisions (CSIRO, 2009; Sonnenschein, 2013).

17.4.2 Regional- to Global-Scale Studies to Detect Land Degradation–Related Modifications

Most studies covering large dryland areas (including continental and global studies) are based on NOAA AVHRR archives or similar sensor systems. Many of these studies focused on ordinary least-square regression or nonparametric trend tests such as the Mann–Kendall test on NDVI values, which were often seasonally aggregated and served as a proxy for NPP (e.g., Eklundh and Olsson, 2003; Anyamba and Tucker, 2005) or other parameters related to greenness (e.g., Lambin and Ehrlich, 1997; Cook and Pau, 2013). Only in recent years, changes in phenological metrics were analyzed to monitor dryland areas (e.g., Heumann et al., 2007; Stellmes et al., 2013; Hilker et al., 2014), and change detection techniques were applied to also describe nonlinear trends (e.g., Jamali et al., 2014).

A major concern of monitoring dryland areas is the distinction between land cover changes driven by climatic fluctuations and those caused by human intervention. Various techniques were

TABLE 17.4   Selection of Studies Monitoring Land Degradation in Drylands

Extent	Study Area	RS Data and Indicator	Methodology	Time Range	Result	References
Local						
Grazing-induced vegetation loss/gain in rangelands						
	California, USA; Utah, USA	Landsat SAVI/SSI time series	Trend analysis based on SAVI/SSI	1982–1997	The landscape showed an increased susceptibility to soil erosion due to drought events and grazing.	Washington-Allen et al. (2006, 2010)
	Crete, Greece	Landsat SMA time series	Trend analysis based on SMA-derived vegetation abundances	1984–2000	Pattern of over- and undergrazing as a result of rangeland management practices from transhumance to sedentary pastoralism.	Hostert et al. (2003)
	Lagadas, Greece	Landsat SMA time series	Trend analysis based on SMA-derived vegetation abundances	1984–2000	Pattern of over- and undergrazing as a result of rangeland management practices from transhumance to sedentary pastoralism.	Röder et al. (2008a)
	Crete, Greece	Landsat vegetation proxy time series	Comparative trend analysis based on SMA, NDVI, and TC	1984–2006	Different vegetation estimates result in similar vegetation trend pattern.	Sonnenschein et al. (2011)
	Nepal	Landsat NDVI time series/GIMMS NDVI	Landsat: Trend analysis based on NDVI; GIMMS: RESTREND	1976–2008/ 1981–2006	Interannual vegetation variability driven by annual precipitation, degradation result of overgrazing or other processes.	Paudel and Andersen (2010)
Fire regime						
	Catalonia, Spain	Annual NDVI Landsat images	NDVI time series to generate a map series of fire history	1975–1993	Methodology to create maps of fire distribution.	Diaz-Delgado and Pons (2001)
	Portugal, Southern California	Multiseasonal Landsat imagery	Two-step approach to detect fires at medium resolution	1993	The algorithm showed a good agreement with the official burned area perimeters.	Bastarrika et al. (2011)
	Alicante, Spain	Landsat NDVI time series	Nonlinear regression analysis of NDVI values and time elapsed since the fire event	1984–1994	After fire events, two recovery trends were found that can be explained by species type.	Viedma et al. (1997)
	Ayora, Spain	Landsat SMA time series	Trend analysis and diachronic thresholding to procure a fire perimeter database and depict postfire dynamics	1975–1990	Typical recovery phases were described by exponential functions and were related to plot-based botanical information.	Röder et al. (2008b)
	Peloponnese, Greece	Landsat time series/ MODIS NBR time series	Analysis of the temporal dimension of assessing burn severity	2006–2008	Within the limitations of available Landsat imagery, caution is recommended for the temporal dimension when assessing postfire effects.	Veraverbeke et al. (2010)
Relationship of vegetation trends and climatic factors						
	Altiplano, Bolivia	Landsat time series	Mean-variance analysis	1972–1987	The landscape showed an increased susceptibility to soil erosion during ENSO-induced droughts.	Washington-Allen et al. (2008)
Regional to Global						
Change of vegetation cover						
	Sub-Saharan Africa	NOAA AHRR NDVI and surface temperature	Seasonal analysis of surface temperature–NDVI trajectories	1982–1991	Only 4% of the study area showed consistent trends (increase/decrease) of vegetation cover.	Lambin and Ehrlich (1997)
	Sahel, Africa	PAL NDVI	Trend analysis of NDVI	1982–1999	Increase in seasonal NDVI was observed over large areas in the Sahel.	Eklundh and Olsson (2003)

(Continued)

TABLE 17.4 (Continued) Selection of Studies Monitoring Land Degradation in Drylands

Extent	Study Area	RS Data and Indicator	Methodology	Time Range	Result	References
	Sahel, Africa	GIMMS NDVI	Trend analysis of NDVI/RESTREND with gridded rainfall data	1982–2003	Rainfall was found to be a major reason for the increase in vegetation greenness, and most of the Sahel does not show large-scale human-induced land degradation.	Herman et al. (2005)
	Sahel, Africa	GIMMS NDVI	Trend analysis of NDVI	1981–2003	NDVI data indicate a gradual and slow but persistent recovery from the peak drought conditions that affected the region in the early to mid-1980s.	Anyamba and Tucker (2005)
	Sahel, Africa	GIMMS NDVI3g, MOIS MOD12C2 NDVI	Trend analysis of NDVI	1981–2011	Recovery rate of vegetation is dependent on factors like soil type and soil depth.	Dardel et al. (2014)
	Sahel, Africa	GIMMS NDVI3g	Trend analysis of growing season averages of NDVI	1981–2012	NDVI behavior reflects the variability of rainfall condition such as the drought in the 1980s and the weather conditions starting in 1994; data might be used as a land surface CDR in a semiarid areas where detailed ground-based meteorological data are missing.	Anyamba et al. (2014)
	Sahel, Africa	GIMMS NDVI3	Automated mapping of vegetation trends with polynomials	1982–2006	Dominance of positive linear trends distributed in an east–west band across the Sahel. Regions of nonlinear change occur on the peripheries of larger regions of linear change.	Jamali et al. (2014)
	Global, pastures	GIMMS LAI3g	Trend analysis of maximum LAI; correlation analysis with rainfall and temperature	1982–2008	Degradation of pastures is not a globally widespread phenomenon, but an increase of greenness in many areas was observed; precipitation was the dominant climate control on interannual variability of LAImax in pastures.	Cook and Pau (2013)
	Global	GIMMS NDVI	BFAST used to map gradual and abrupt changes of NDVI and breakpoints	1982–2008	Abrupt greening prevailed in semiarid regions, probably due to their strong reactions to climatic variations. These abrupt greening events were often followed by periods of gradual browning.	De Jong et al. (2012)
Change in phenological characteristics						
	Sahel and Soudan, Africa	GIMMS NDVI	Trend analysis of phenological metrics derived with Timesat	1981–2005	Significant positive trends for the length and the end of the growing season for the Soudan and Guinean regions were detected but not in the Sahel; this can be attributed to two types of greening trends associated with rainfall change since the drought in the early 1980s.	Heumann et al. (2007)
			Trend analysis of phenological metrics for an integrated analysis		Listed in Table 17.5.	Hill et al. (2008), Stellmes et al. (2013), Hilker et al. (2014)
Relationship of vegetation trends and climatic factors						
		RUE			Listed in Table 17.3.	
		RESTREND			Listed in Table 17.3.	

(Continued)

TABLE 17.4 (Continued)   Selection of Studies Monitoring Land Degradation in Drylands

Extent	Study Area	RS Data and Indicator	Methodology	Time Range	Result	References
	Major areas of global drylands	GIMMS NDVI	Linear regression models	1981–2003	A strong general relationship between NDVI and rainfall over time characterizes large parts of the drylands; no large-scale land degradation was observed but rather an increase of vegetation cover.	Helldén and Tottrup (2008)
	Spain	MEDOKADS NDVI	Distributed lag models	1989–1999	Significant relationships between lagged NDVI and rainfall anomalies up to 3 months are confined to subhumid/semiarid areas; severe drought periods might have an enduring influence on biomass production in subsequent years.	Udelhoven et al. (2009)
	Global semiarid areas	GIMMS NDVI	Correlation analysis of NDVI and precipitation/air temperature	1981–2007	Semiarid areas, on average, experience an increase in greenness; similar increases in greenness may have widely different explanations.	Fensholt et al. (2012)
	Okavango, Kwando, upper Zambezi, Africa	GIMMS NDVI3g/MODIS MOD13A3 NDVI	Dynamic factor analysis	1982–2010/2001–2010	The spatial distribution of soil moisture and precipitation as determinants of NDVI is important in areas with mean annual precipitation under 750 mm.	Campo-Bescós et al. (2013)
	Global	GIMMS NDVI3g/MODIS MOD and MYD13C2	Multiple stepwise regression	1982–2010	Precipitation showed the highest correlation to temperate to tropical water-limited herbaceous systems where rainfall partially explains more than 40% of NDVI variability.	Zeng et al. (2013)
Teleconnections						
	Sahel, Africa	GIMMS NDVI	Correlations between NDVI and climate indices and global sea surface temperatures	1982–2007	Global SST anomalies and Sahelian NDVI showed strong correlations with different characteristics for western, central, and eastern Sahel.	Huber and Fensholt (2011)
	Africa	GIMMS NDVI/MODIS NDVI for correction	Land surface model driven by meteorological data and NDVI to analyze response of photosynthesis to macro weather situations	1982–2003	ENSO and IOD induce large seasonal anomalies of precipitation, vegetation, humidity, as well as photosynthesis across the main part of Africa.	Williams and Hanan (2011)
Fire regime						
	Central Asia	MODIS active fire and burned area product	Validation of MODIS products and mapping of fire occurrence	2001–2009	In average, about 15 million ha of land burns annually across Central Asia with the majority of the area burned in August and September in grassland areas.	Loboda et al. (2010)
	Southern Africa	MODIS burned area product	Random forest regression tree procedure to determine the factors of wildfires	2003	Areas were identified where fire is rare due to low-rainfall regions, regions where fire is under human control and higher rainfall regions where burnt area is determined by rainfall seasonality.	Archibald et al. (2009)
	Mediterranean biomes	MODIS active fire product	Statistical fire–climate models driven by ensembles of climate projections under the IPCC A2 emissions scenario	2001–2007	Fire activity was found to be sensitive to environmental changes, and productivity may be the key to future fire occurrence in this biome.	Battlori et al. (2013)

TABLE 17.5 Selection of Studies Evaluating Land Degradation Based on Integrated Concepts and Use of Remote Sensing Products

Extent	Study Area	RS Data	Methodology	Observation Period	Result	References
Local						
	Northwestern Spain	Time series of orthorectified aerial photographs	Species distribution modeling techniques (MaxEnt and BIOMOD)	1956–2004	Land use history primarily controlled forest expansion rates, as well as upward altitudinal shift.	Alvarez-Martinez et al. (2014)
	Northeastern Spain	Bitemporal analysis of aerial photographs	Logistic regression models	1956–2006	Effects of several topographic and socioeconomic variables were analyzed; patterns of observed forest expansion are highly related to patterns of farmland abandonment.	Améztegui et al. (2010)
	Lagadas, Greece	Landsat SMA image	Cost surface modeling to understand the influence of grazing management on vegetation cover.	2000	Uneven distribution of livestock causes both over- and undergrazing to occur in close proximity, which negatively affects the ecosystem through various feedback loops.	Roeder et al. (2007)
	Northeast Spain	Landsat MSS and TM land cover maps	Multiple logistic regressions (MLOR) combining biophysical and human variables	MSS, 1977–1993; TM, 1991–1997	EU subsidies were major drivers of LUCCs, e.g., intensification of subsidized herbaceous crops on the coastal agricultural plain.	Serra et al. (2008)
	Lagadas, Greece	Annual Landsat TM/ETM+ vegetation fraction time series	Combined use of household-level land use data, remote sensing products, and standardized socioeconomic data	1984–2000	Major driver of LUCCs were EU subsidies, e.g., low-profit farmers maintained extensive farming activities on the most erodible, steep-sloped land due to subsidies.	Lorent et al. (2008)
	Xilinhot, Inner Mongolia, China	Landsat TM/ETM+ land use and NDVI (three time steps)	Multinomial logistic regression model	1991–2005	Main drivers of observed trends in rangelands were altitude, slope, annual rainfall, distance to highway, soil organic matter, sheep unit density, and fencing policy.	Li et al. (2012)
	Lake Nakuru drainage basin, Kenya	Landsat TM/ETM+ land use maps (three time steps)	Logistic regression models	1985–2011	Major drivers of forest–shrubland conversions, grassland conversions, and cropland expansions were identified; significance of the influential factors varied depending on the time period observed and the land cover change type.	Were et al. (2014)

(Continued)

TABLE 17.5 (*Continued*) Selection of Studies Evaluating Land Degradation Based on Integrated Concepts and Use of Remote Sensing Products

Extent	Study Area	RS Data	Methodology	Observation Period	Result	References
Regional						
	Mongolia	MODIS daily 1B data (MYD021KM), NDVI (MAIAC correction)	Regression analysis between variables on a provincial level	2002–2012	About 80% of the decline in NDVI explained by increase in livestock; 30% of changes across the country by precipitation.	Hilker et al. (2014)
	Uzbekistan	MOD13Q1 NDVI	Spatial logistic regression modeling	2000–2010	One-third of the area was characterized by a decline of greenness. Groundwater table, land use intensity, low soil quality, slope, and salinity of the groundwater were identified as the main drivers of degradation.	Dubovyk et al. (2013)
	Spain	MEDOKADS NDVI	Syndrome approach	1989–2004	Only a few areas affected by land degradation in the sense of productivity loss; shrub and woody vegetation encroachment due to land abandonment of marginal areas, intensification, urbanization trends along the coastline caused by migration/increase of mass tourism.	Hill et al. (2008), Stellmes et al. (2013)
Global						
		SPOT Vegetation global land cover map 2000 (GLC2000)	HANPP	~2000	Annual loss of NPP due to land degradation at 4% to 10% of the potential NPP of drylands, ranging up to 55% in some degraded agricultural areas.	Zika and Erb (2009)
		NOAA AVHRR NDVI and MOD17A	GLADIS: NPP trend detrended via RUE and RESTREND; biomass, soil quality, water quantity, biodiversity, and economic and social services used as indicators to describe the status of land degradation	1981–2003 (with MODIS 1981–2006)	Degraded lands are found to be highly variable; degraded land occurs mostly in drylands and steep lands; the capacity to deliver ecosystem services is also generally less in developing countries as compared to industrial nations.	Nachtergaele et al. (2011)

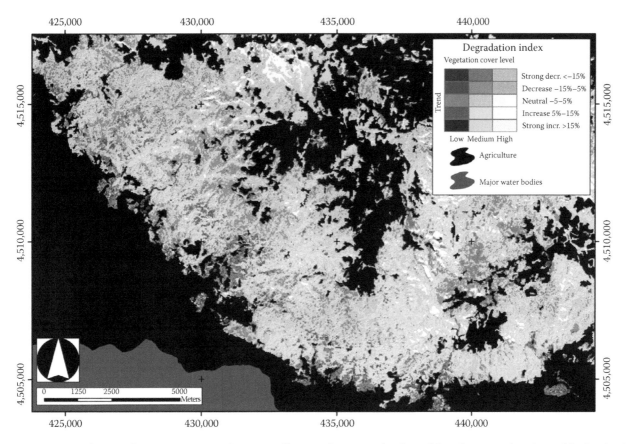

FIGURE 17.5 Degradation index map integrating the gain coefficient and average value derived from linear trend analysis of the Landsat TM/ETM+ time series for the rangelands of Lagadas, Greece; agricultural areas were masked out (black). (From Röder, A., et al. *Remote Sensing of Environment*, 112, 2863, 2008a.)

employed for assessing the effect of climatic variability such as the aforementioned RUE (Geerken and Ilaiwi, 2004; Fensholt et al., 2013) and RESTREND methodology (Wessels et al., 2007), but furthermore linear regression analysis (Helldén and Tottrup, 2008), distributed lag models (Udelhoven et al., 2009), multiple stepwise regression (Zeng et al., 2013), and dynamic factor analysis (Campo-Bescós et al., 2013) of teleconnections of macro weather situations (Williams and Hanan, 2011), and global sea surface temperature (Huber and Fensholt, 2011).

Many of the large-scale studies focused on Africa, especially on Sub-Saharan Africa including the Sahel region. Droughts in the first decades of the twentieth century as well as in the 1960s to 1980s caused disastrous famines in the Sahel zone and had a strong impact on vegetation cover. Yet resilience in these systems often led to recovery under more profitable climatic conditions, while the term desertification involves a permanent and irreversible reduction in vegetation productivity. In the 1990s, remote sensing studies started to support the analysis based on time series analyses. Recent studies dealing with greening trends in the Sahel found vegetation recovery in most parts of the Sahel (e.g., Eklundh and Olsson, 2003; Herman et al., 2005). Heumann et al. (2007) showed that both annual and perennial vegetation recovery processes drive the observed greening, and Dardel et al. (2014) demonstrated that soil type and soil

depth are important factors for recovery. Jamali et al. (2014) implemented an automated approach to account for nonlinear changes. Results showed a dominance of positive linear trends distributed in an east–west band across the Sahel, whereas regions of nonlinear change occur only in limited areas, mostly on the peripheries of larger regions of linear change (see Figure 17.6). These studies all implied that vegetation recovered after the severe droughts in the 1970s and 1980s and that land degradation not related to water availability/droughts is not a widespread phenomenon but is confined to smaller areas (Fensholt et al., 2013).

Also in other parts of the world, a large proportion of dryland areas showed "greening-up" trends (e.g., Helldén and Tottrup, 2008; Hill et al., 2008; De Jong et al., 2012; Fensholt et al., 2012; Stellmes et al., 2013). The global study of Cook and Pau (2013) focused on rangeland productivity between 1982 and 2008 and indicated that almost 25% of the rangelands were affected by significant trends. These trends were found to be mostly with increasing productivity, whereas decreasing productivity related to land degradation was found in rather isolated spots, mainly in China, Mongolia, and Australia. Whereas in many other regions rainfall was the dominant factor influencing NDVI, in Mongolia, 80% of the decline in greenness could be attributed to an increase in livestock (Hilker et al., 2014).

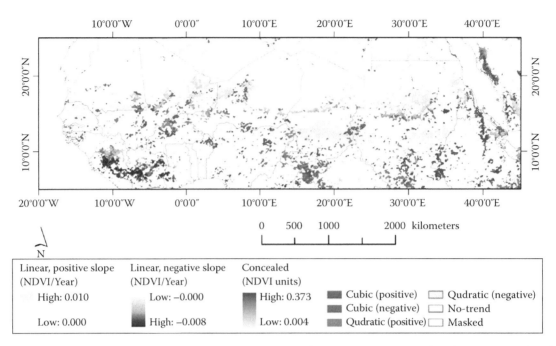

FIGURE 17.6   Results of a polynomial fitting–based approach to account also for nonlinear trends (Jamali et al., 2014). Trend slope for the linear trends, range of annual variations of NDVI for the concealed trends, and trend sign for the cubic and quadratic trends obtained by using the annual GIMMS–NDVI data series for the Sahel (1982–2006) in the trend classification scheme. Concealed trends indicate that no net change in vegetation productivity has occurred, but the curve exhibits at least one minimum or maximum. Areas with a mean yearly NDVI < 0.1 were masked out. (From Jamali, S. et al., *Remote Sens. Environ.*, 141, 79, 2014.)

Generally, a comprehensive analysis of land degradation needs to include, also at regional to global scale, the fire regime and possible interlinkages to land use and land cover, for example, by analyzing recovery after fire events (Katagis et al., 2014). The two MODIS fire products, active fire and burned area, allow monitoring of important variables of fire regimes (Justice et al., 2006; Loboda et al., 2012), such as fire frequency, fire seasonality, and fire intensity, and allow for identifying drivers (Archibald et al., 2009) and model potential changes (Batllori et al., 2013).

17.5 Integrated Concepts to Assess Land Degradation

The previous sections have illustrated that time series analysis allows to discriminate human-induced land cover changes and changes caused by interannual climatic variability. Beyond this, a crucial element of land degradation assessment is the identification of underlying and proximate causes of human-induced changes (e.g., Reynolds et al., 2007). Only in this manner the coupled human-natural character of land cover changes can be understood and an identification of the mechanisms that drive land degradation is possible. This knowledge provides the foundation to support the development of sustainable land management strategies.

A comprehensive framework designed to capture the complexity of land degradation and desertification was provided by Reynolds et al. (2007). They introduced the term "Drylands Development Paradigm" (DDP), which "represents a convergence of insights and key advances drawn from a diverse array of research on desertification, vulnerability, poverty alleviation,

and community development" (Reynolds et al., 2007). The DDP aims at identifying and synthesizing those dynamics central to research, management, and policy communities (Reynolds et al., 2007). The essence of this paradigm, which consists of five principles, builds on the assumption that desertification cannot be measured by solitary variables, but that it has to consider biophysical and socioeconomic data at the same time (Vogt et al., 2011) as well. A limited number of *slow* variables (e.g., soil fertility) are usually sufficient to explain the human–natural system dynamics. These slow variables possess thresholds, and if these thresholds are exceeded, the system moves to a new state. *Fast* variables, for instance, climatic variability, often mask the slow variables and, thus, aggravate the assessment of the slow variables, which is a prerequisite to understand the ecosystem behavior. Moreover, it is important to consider that human–natural systems are "hierarchical, nested, and networked across multiple scales" (Reynolds et al., 2007). Accordingly, both the human component, for example, stakeholders at different levels, and the biophysical component, for example, slow variables at one scale, can be affected by the change of slow variables operating at another scale (Reynolds et al., 2007).

Prior to the DDP, Geist and Lambin (2004) examined the main mechanisms that trigger land degradation processes and conclude that these processes, which often manifest in LUCCs, are governed by proximate causes (immediate human and biophysical actions) that depend on the underlying drivers (fundamental social and biophysical processes). Figure 17.7 illustrates the dependencies of LUCCs from proximate causes and underlying drivers. Furthermore, alterations of ecosystem services

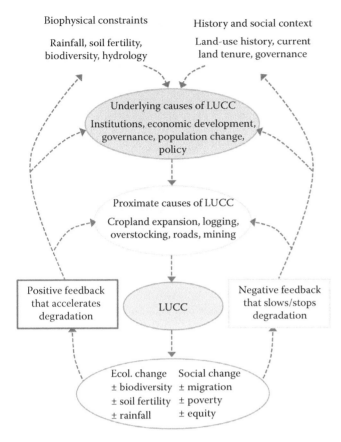

Biophysical constraints

Rainfall, soil fertility, biodiversity, hydrology

History and social context

Land-use history, current land tenure, governance

Underlying causes of LUCC

Institutions, economic development, governance, population change, policy

Proximate causes of LUCC

Cropland expansion, logging, overstocking, roads, mining

Positive feedback that accelerates degradation

LUCC

Negative feedback that slows/stops degradation

Ecol. change   Social change
± biodiversity   ± migration
± soil fertility   ± poverty
± rainfall   ± equity

FIGURE 17.7 Conceptual model illustrating the feedback loop of land use/land cover changes (LUCC), its consequences, and the underlying and proximate causes. (Modified from Reid, R.S. et al., Linking land-change science and policy: Current lessons and future integration, in E.F. Lambin and H. Geist (eds.), *Land-Use and Land-Cover Change: Local Processes and Global Impacts*, Springer, Berlin, Germany, 2006.)

caused by LUCCs can again alter underlying drivers, proximate causes, and even external constraints, hence resulting in a feedback loop. Policy plays an important role in avoiding positive feedback mechanisms that can accelerate unsustainable land use (Reid et al., 2006).

17.5.1 Integrated Studies at Local Scale

Several local studies linked the biophysical dimension of LUCCs to the human dimension in various dryland areas such as Spain (Serra et al. 2008; Améztegui et al., 2010; Alvarez-Martinez et al., 2014), Greece (Lorent et al., 2008), Kenya (Were et al., 2014), China (Li et al., 2012), Mongolia (Hilker et al., 2014), and Uzbekistan (Dubovyk et al., 2013). Regression-based models are the most widely used approach to identify the major drivers of change (Were et al., 2014) and mostly rely on land cover changes derived from land use/land cover classifications at several time steps. However, time series of remotely sensed data were only rarely used (Lorent et al., 2008; Dubovyk et al., 2013). The drivers of LUCC depend very much on the contextual framework of the study area including physical and socioeconomic characteristics.

Therefore, it is essential to first set up a hypothesis that identifies major underlying drivers of LUCC. For instance, in Spain and Greece, the Common Agricultural Policy subsidies of the European Union (EU) were identified as one of the important drivers. These largely influenced agricultural developments like intensification and land abandonment, where abandonment of marginal areas involved forest expansion and bush encroachment (Lorent et al., 2008; Serra et al., 2008; Améztegui et al., 2010). In the grasslands of Inner Mongolia/China, many factors explained observed grassland degradation between 1990 and 2000 and the reduced degradation rate between 2000 and 2005, which were altitude, slope, annual rainfall, distance to highway, soil organic matter, sheep unit density, and fencing policy. Fencing policy was negatively correlated suggesting that fencing of sensitive areas can reduce land degradation (Li et al., 2012). The analysis of cropland degradation in the Khorezm Region, Uzbekistan, based on MODIS time series (Dubovyk et al., 2013) revealed that one-third of the area was characterized by a decline of greenness between 2000 and 2010. Groundwater table, land use intensity, low soil quality, slope, and salinity of the groundwater were identified as the main drivers of degradation. These examples show that the combination of remote sensing supported LUCC, and underlying and proximate causes may reveal the most important drivers of land degradation. However, this analysis is often hampered by the fact that for each study area (1) all potential and relevant drivers have to be identified and (2) spatially explicit information of each driver or a proxy has to be available with a sufficient spatial resolution.

17.5.2 Integrated Studies at Regional to Global Scale

Another approach capable to support land degradation assessment is the syndrome approach that has been developed in the context of global change research (Petschel-Held et al., 1999; Cassel-Gintz and Petschel-Held, 2000). It aims at a place-based, integrated assessment by describing global change by archetypical, dynamic, coevolutionary patterns of human–nature interactions instead of regional or sectoral analyses. In this framework, syndromes (as a "combination of symptoms") describe bundles of interactive processes ("symptoms") that appear repeatedly and in many places in typical combinations and patterns. Sixteen global change syndromes were suggested and distinguished into utilization, development, and sink syndromes. Downing and Lüdeke (2002) applied the approach to land degradation. Based on the set of global change syndromes, they identified the syndromes that are of relevance in dryland areas and linked vulnerability concepts to degradation processes. The syndrome concept is considered a suitable interpretation framework that allows for an integrated assessment of land degradation (Sommer et al., 2011; Verstraete et al., 2011). This concept was transferred to earth observation–based studies and implemented for Spain based on NOAA AVHRR data between 1989 and 2004 (Hill et al., 2008; Stellmes et al., 2013), thus enabling to monitor changes in land cover after the accession of Spain

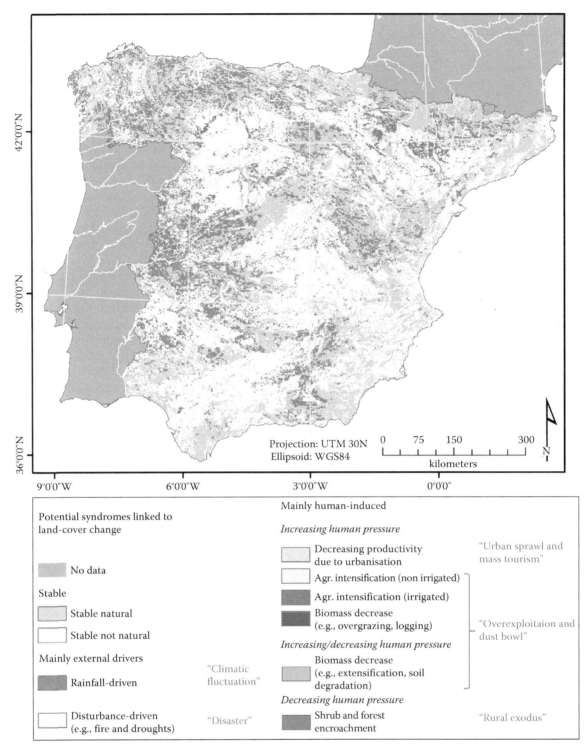

FIGURE 17.8 Syndromes and main drivers of the identified land cover changes in Spain derived from MEDOKADS NDVI data, 1989–2004. (From Stellmes, M. et al., *Land Use Policy*, 30, 685, 2013.)

to the EU (see Figure 17.8). In these studies, the focus was not only on the identification of land cover changes but also on the link of these findings to underlying causes enabling the designation of syndromes of land use change. The main findings of the two studies comprise three major land cover change processes caused by human interaction: shrub and woody vegetation

encroachment in the wake of land abandonment of marginal areas; intensification of nonirrigated and irrigated, intensively used fertile regions; and urbanization trends along the coastline caused by migration and the increase of mass tourism.

At a global scale, LADA has recently implemented a Global Land Degradation Information System (GLADIS), which provides information on land degradation with a spatial resolution of 8 km × 8 km. The interpretation of ecosystem changes in GLADIS includes RUE, NPP and climatic variables and is based on an integrated land use system map. This map entails information about the main proximate causes of LUCCs such as livestock pressure and irrigation. The major constraints of this approach concerns the derivation of the RUE and the NPP from the GIMMS NOAA AVHRR dataset (compare Section 17.3) (Wessels, 2009) and the coarse spatial resolution that hampers the detection of land cover changes (Vogt et al., 2011). Nevertheless, Vogt et al. (2011) emphasized that this assessment is a first step toward an integrated assessment.

Another spatially explicit assessment concept that was not specifically designed in the context of land degradation, but was adapted and implemented, is the human appropriation of net primary production (HANPP; Haberl et al., 2007; Erb et al., 2009). HANPP represents the aggregated impact of land use on biomass available each year in ecosystems as a measure of the human domination of the biosphere. Global maps of the parameter were prepared based on vegetation modeling, agricultural and forestry statistics, and geographical information systems data on land use, land cover, and soil degradation (Haberl et al., 2007; Erb et al., 2009). In a global study, Zika and Erb (2009) estimated the annual loss of NPP due to land degradation at 4%–10% of the potential NPP of drylands, ranging up to 55% in some degraded agricultural areas.

17.6 Uncertainties and Limits

Manifold methods were developed for assessing and monitoring land degradation ranging from detailed local to broad global studies. Nevertheless, until today, no comprehensive picture of the state of drylands is available. This results from different aspects some of which shall be discussed here.

17.6.1 Uncertainties Regarding the Definition of Land Degradation and Its Derivation

Monitoring of drylands is often based on analyzing indicators related to the productivity of vegetation. Thereby, the loss of productivity is considered to be linked to degradation processes. However, it should be stressed that the decrease of primary productivity does not necessarily imply land degradation processes. This was, for instance, illustrated by an example in Syria where unsustainable irrigation agriculture was transformed to near-natural rangelands in Syria (Udelhoven and Hill, 2009). In turn, a positive trend of productivity is not always an indicator for improving land condition, and a greening up of, for instance, rangelands does not necessarily imply an improvement of pastures (Miehe et al., 2010). In marginal areas of the European Mediterranean, greening up has been shown to be caused by bush encroachment due to land abandonment, and the consequences for ecosystems are heavily discussed (Stellmes et al., 2013). Thus, on the one hand, soils can be stabilized and soil erosion can be reduced (Thomas and Middleton, 1994), and more

carbon can be sequestered (Padilla et al., 2010), but on the other hand, runoff and groundwater recharge are reduced (Beguería et al., 2003), biodiversity is altered (Forman and Collinge, 1996), and the fire regime changes (Duguy et al., 2007). Thus, including additional information sources, for instance, on land use, is required to allow a meaningful interpretation of time series results (Vogt et al., 2011). The same is also true in the case of fires that strongly affect the time series signal, for example, induced short-term decreases in productivity and subsequent increase in productivity due to vegetation recovery.

17.6.2 Uncertainties Regarding Remote Sensing Data

17.6.2.1 Remote Sensing Archives and Their Analysis

Uncertainties in remote sensing observations pose a set of methodological and practical challenges for both the analysis of long-term trends and the comparison between different data archives. Creating consistent remote sensing time series is challenging and the prerequisite for a meaningful trend analysis. Using combined data from different sensors affording high temporal resolution such as AVHRR, MODIS and SPOT Vegetation in principle allow for the construction of time series in surface reflectance and related changes back to the early 1980s. However, this is hampered by several sources of uncertainties in the comparability between different sensor products (Yin et al., 2012). Comparison of the absolute NDVI values from different archives as well as the derived trends showed strong differences; where a good correspondence of derived NDVI trends was found at the global scale, spatial trends at the local to regional scale often showed remarkable discrepancies (Hall et al., 2006; Beck et al., 2011; Fensholt and Proud, 2012; Yin et al., 2012). Beck et al. (2011) found, among others, good agreements between GIMMS; Pathfinder AVHRR land archive (PAL); Fourier-adjusted, sensor and solar zenith angle-corrected, interpolated, and reconstructed AVHRR archive (FASIR); and Land Long Term Data Record version 3 (LTDR v3) (see Table 17.2) in Australia and tundra regions and moderate consistency for North America and China but inconsistent trends for Europe and Africa including the Sahel zone. A comparison with Landsat NDVI showed that MODIS data perform better than any of the NOAA AVHRR archives. Also, the trends of NDVI between GIMMS and SPOT Vegetation considerably disagreed for different land use systems across Northern China (Yin et al., 2012) indicating that trends have to be interpreted with caution and bearing in mind the limitations of the datasets. LTDR v3 showed apparent trends within the Sahara (Beck et al., 2011), which hints on calibration problems. A new and enhanced version of the GIMMS dataset was published in June 2014 (http://ltdr.nascom.nasa.gov/cgi-bin/ltdr/ltdrPage.cgi), but still regional inconsistencies with MODIS data appear (Figure 17.9).

As an example, Figure 17.9 shows trends derived from NOAA NDVI3g and MODIS MOD13Q1 NDVI data covering the same observation period (2001–2011) of the eastern Sahel. Even though the general picture is quite similar for the mean annual

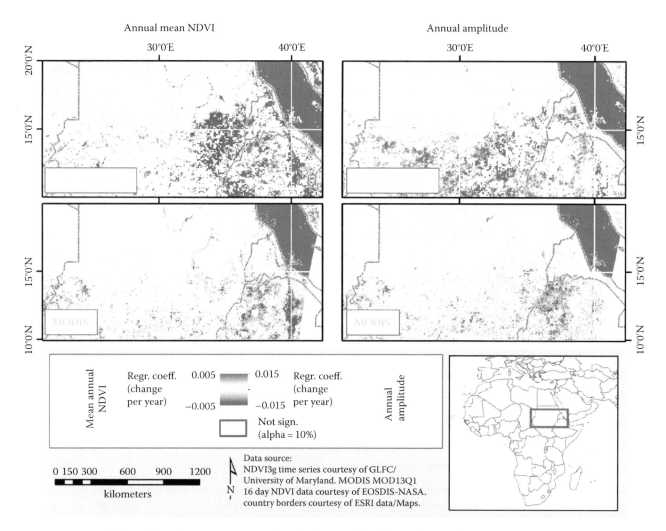

FIGURE 17.9   Trends derived from linear regression analysis for the annual total sum of NDVI based on the NOAA NDVI3g archive (a) and MODIS MOD13Q1 NDVI time series (b) for the eastern Sahel from 2001 to 2012. Time series analysis performed with Timestats. (From Udelhoven, T, *IEEE Journal of Selected Topics in Applied Earth Observations and Remote Sensing (J-STARS)*, 4(2), 310, 2011.)

NDVI trends, a more detailed analysis reveals a considerable disagreement between both datasets that also addresses the temporal trends for a phenological parameter (i.e., the amplitudes of the annual NDVI cycle). Possible explanations for these incoherencies include different data preprocessing schemes for different sensors. The effects of sensor degradation on the captured signal are different, and AVHRR data need to be additionally corrected for orbital drift effects that introduce systematic changes in the bidirectional characteristics of surfaces. Another factor is different spectral mixture effects in heterogeneous regions that arise from the different spatial resolutions of the GIMMS and MODIS data products.

The comparability of many studies is additionally hampered by the fact that the used methods and techniques, vegetation proxies, and thresholds to exploit the time series are very diverse, since the implemented methods are often adapted to specific objectives and certain study areas. This is often necessary as drylands are very diverse concerning the degradation processes and the environmental settings including climate, soil, geology, fauna, and flora.

17.6.2.2 Observation Period

As outlined before, rainfall variability is a key driver of variability of vegetation productivity within drylands. In consequence, the observation period will substantially influence the derived trends depending on the assembly of drier and wetter periods. Figure 17.10 illustrates the difference of trends for different observation periods derived from the NOAA NDVI3g archive.

This underlines that dryland monitoring should always consider rainfall variability, for example, implemented in the RESTREND method (Wessels et al., 2007). Hereby, similar to remote sensing archives, the homogeneity and reliability of the precipitation time series is of utmost importance. Even though some authors generated interpolated precipitation fields for their studies themselves (Wessels et al., 2007; Del Barrio, 2010), diverse global and regional gridded precipitation data are available, for example, Global Precipitation Climatology Centre (Meyer-Christoffer et al., 2011) and Africa Rainfall Estimate Climatology Version 2 (Novella and Thiaw, 2013). The choice of an appropriate dataset should be based on plausibility checks

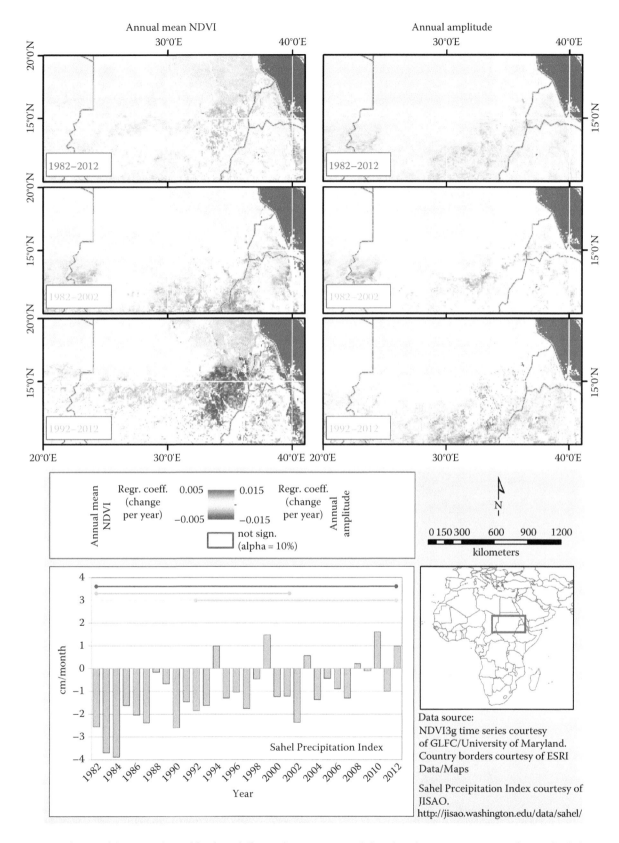

FIGURE 17.10 Change of the NDVI derived for three different observation periods based on the NOAA NDVI3g archive and Sahel Precipitation Index from 1982 to 2012 (Janowiak, 1988). Time series analysis performed with Timestats. (From Udelhoven, T, *IEEE Journal of Selected Topics in Applied Earth Observations and Remote Sensing (J-STARS)*, 4(2), 310, 2011.)

(e.g., Anyamba et al., 2014). Tozer et al. (2012) demonstrated for three monthly gridded Australian rainfall datasets that interpolated data are rather restricted as a useful proxy for observed point data, although these grids are *based* on observed data. Gridded datasets often significantly vary from gauged rainfall datasets, and they do not capture gauged extreme events. Apart from observation errors, these uncertainties are mainly introduced by the spatial interpolation algorithms, which always introduce some artificiality. Furthermore, it is difficult to verify the *ground truth* of the gridded data in areas or epochs with sparse observation gauges. Tozer et al. (2012) recommend always to acknowledge these uncertainties in using gridded rainfall data and to try to quantify and account for it in any study, if possible.

17.6.2.3 Spatial Scale

One drawback of regional to global studies is the coarse pixel resolution, which often impedes the monitoring of small-scale land degradation processes (e.g., Stellmes et al., 2010; Fensholt et al., 2013) as illustrated in Figure 17.11.

Moreover, species composition cannot be identified and vegetation structure is not resolved, and often the focus is put on green vegetation cover even though dry vegetation is an important component in drylands. Some approaches try to solve some of these gaps, for example, decomposition of time series to assess woody and herbaceous components (Lu et al., 2001) or using a clumping index to estimate woody cover from MODIS data (Hill

et al., 2011). Other methods make use of alternative sensor systems such as passive microwave radar to derive vegetation optical depth, which is sensitive to both photosynthetic active and nonactive biomass (Andela et al., 2013), or combine the analysis of optical and radar imagery (Bucini et al., 2010).

Methods like STARFM or the improved availability of Landsat-like medium-resolution data (e.g., Sentinel-II mission) will only partially solve the problem, since dryland observation requires long-term archives. However, these sensors and methods will improve the situation over the long term. The same is true for operational satellite-based hyperspectral data that will allow for the development and application of enhanced indicators for dryland observation.

17.7 Summary and Conclusions

The definition and perception of land degradation and desertification have undergone a substantial transformation within the past decades. While in the beginning the biophysical assessment of degradation processes, which often focused on soil degradation, was the primary objective of many research initiatives, in recent years, the necessity to investigate the mechanisms of human–environmental systems as a prerequisite to create a comprehensive understanding of land degradation processes has been recognized. This is considered essential to understand the impacts of land degradation on the provision of ecosystem goods and services and, thus, its impact on human well-being

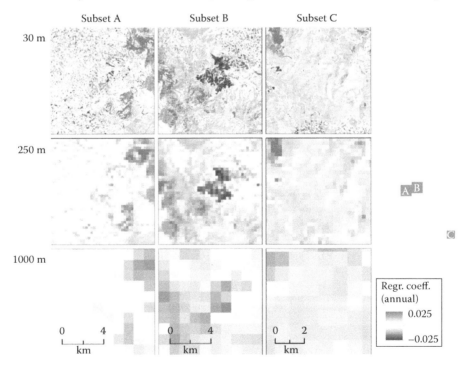

FIGURE 17.11 Effects of spatial degradation of Landsat TM/ETM+ time series (1990–2000) from a geometric resolution of 30 m × 30 m to 1000 m × 1000 m on the derived regression coefficient of a linear trend analysis. The three presented subsets represent different types and scales of land cover change. (Modified from Stellmes, M. et al., *Remote Sens. Environ.*, 114, 2111, 2010.)

(Millennium Ecosystem Assessment, 2005a). In recent years, great efforts were put into developing methodologies to enhance the understanding of coupled human–environmental systems and the influence of natural climatic variations.

One of the major challenges that remains is to link these observations with socioeconomic data, thus connecting biophysical and socioeconomic information to yield combined information of land change processes and their underlying causes. This proves especially crucial for large-scale assessments of land degradation from national to global scales. This intricacy even increases if degradation is not only defined as a loss of productivity of ecosystems but as the decline of important ecosystem services as suggested by the Millennium Ecosystem Assessment (2005a). Even though this definition further increases the complexity of dryland assessment, it might be more compliant with the needs of policy makers and land management to develop and establish sustainable land use practices.

This complexity might also explain that until today no comprehensive picture of dryland condition is available, even though manifold methods were developed for assessing and monitoring land degradation ranging from detailed local to broad global studies. Moreover, dryland studies differ in implemented techniques, indicators, observation periods, thresholds and significance levels as well as the spatial and temporal resolution and the spectral characteristics of the sensor, hampering a comparison of the studies to form a picture of global dryland condition. Therefore, it is of utmost interest to promote international cooperation in order to harmonize dryland studies, such as the initiative to compile a new World Atlas of Desertification under the lead of the United Nations Environment Programme and the European Commission Joint Research Centre (EC-JRC). In particular, it is not likely that one singular methodology will be sufficient to comprehensively analyze drylands; rather, depending on the respective physical and socioeconomic framework, complementary approaches have to be applied as introduced in the preceding sections.

The land degradation and desertification topic is part of the more broadly perceived debate on global change. Thereby, climate change and its environmental and economic consequences are major environmental issues of global interest. Human activities have transformed a major part of the earth's terrestrial ecosystems to meet rapidly growing demands for food, fresh water, timber, fiber, and fuel. Land use practices have not only affected global and regional climate due to the emission of relevant greenhouse gases but also altered energy fluxes and water balances (Foley et al., 2005). Additionally, even seemingly "unaffected" areas are also influenced and altered indirectly through pollutants and climate change (DeFries et al., 2004; Foley et al., 2005).

Whereas in the past, conservation of ecosystems was given priority to maintain ecosystem services, in the face of global change, it cannot be assumed that the future behavior of ecosystem responses to changes will be the same as in the past (Chapin et al., 2010). Instead, the challenge of future land use management

will include the assessment of trade-offs between acute human needs and the long-term capacity of ecosystems to provide goods and services (DeFries et al., 2004; Foley et al., 2005).

It is essential to consider that ecosystem responses to land use changes vary in time and space, and moreover, analysis should encompass larger areas with sufficient spatial resolution to ensure that on- and off-site ecosystem responses are detected. Sustainable management of ecosystems requires information concerning the actual conditions and, furthermore, alterations of ecosystems in relation to reference states. Such information allows for a thorough analysis of ecosystem functionality and enables rating trade-offs between ecosystem services in which policy decisions (where necessary considering climate change scenarios) could impose by inducing land use changes (DeFries et al., 2004). The understanding of the impact of LUCCs is even more urgent in the context of climate change, and the prospect land use will be further intensified to satisfy humanities' growing demand for resources (Foley et al., 2011). Especially when considering the expected rise to 10 billion people by the end of the twenty-first century (Lee, 2011), pressure on dryland ecosystems could further increase, making the development of integrated, multicomponent dryland observation and management even more important.

References

Achard, F., J.P. Malingreau, T. Phulpin, G. Saint, B. Saugier, B. Seguin, and D. Vidal Madjar. 1995. A mission for global monitoring of the continental biosphere. http://nieuw.vgt.vito.be/pdf/achard85.pdf (accessed July 2, 2014).

Adams, J.B., M.O. Smith, and P.E. Johnson. 1986. Spectral mixture modeling: A new analysis of rock and soil types at the Viking Lander 1 site. *J. Geophys. Res.* 91(B8): 8098–8112.

Álvarez-Martínez, J.M., S. Suárez-Seoane, J.J. Stoorvogel, and E. de Luis Calabuig. 2014. Influence of land use and climate on recent forest expansion: A case study in the Eurosiberian-Mediterranean limit of northwest Spain. *Journal of Ecology* 102: 905–919. doi: 10.1111/1365-2745.12257.

Améztegui, A., L. Brotons, and L. Coll. 2010. Land-use changes as major drivers of mountain pine (*Pinus uncinata* Ram.) expansion in the Pyrenees. *Global Ecology and Biogeography* 19: 632–641.

Andela, N., Y.Y. Liu, A.I.J.M. van Dijk, R.A.M. de Jeu, and T.R. McVicar. 2013. Global changes in dryland vegetation dynamics (1988–2008) assessed by satellite remote sensing: Comparing a new passive microwave vegetation density record with reflective greenness data. *Biogeosciences* 10: 6657–6676.

Anderson, A.N., G.D. Cook, and R.J. Williams. 2003. *Fire in Tropical Savannas.* New York: Springer.

Andres, L., W.A. Salas, and D. Skole. 1994. Fourier analysis of multi-temporal AVHRR data applied to land cover classification. *International Journal of Remote Sensing* 15: 1115–1121.

Anyamba, A., J.L. Small, C.J. Tucker, and E.W. Pak. 2014. Thirty-two years of Sahelian zone growing season non-stationary NDVI3g patterns and trends. *Remote Sensing* 6: 3101–3122.

Anyamba, A. and C.J. Tucker. 2005. Analysis of Sahelian vegetation dynamics using NOAA-AVHRR NDVI data from 1981–2003. *Journal of Arid Environments* 63: 596–614.

Archibald, S., D.P. Roy, B.W. van Wilgen, and R.J. Scholes. 2009. What limits fire? An examination of drivers of burnt area in Southern Africa. *Global Change Biology* 15(3):613–630.

Bai, Z.G., D.L. Dent, L. Olsson, and M.E. Schaepman. 2008. Proxy global assessment of land degradation. *Soil Use and Management* 24: 223–234.

Bastarrika, A., E. Chuvieco, and M.P. Martin. 2011. Mapping burned areas from Landsat TM/ETM+ data with a two-phase algorithm: Balancing omission and commission errors. *Remote Sensing of Environment* 115: 1003–1012.

Bastin, G., P. Scarth, V. Chewings, A. Sparrow, R. Denham, M. Schmidt, P. O'Reagain, R. Shepherd, and B. Abbott. 2012. Separating grazing and rainfall effects at regional scale using remote sensing imagery: A dynamic reference-cover method. *Remote Sensing of Environment* 121: 443–457.

Batllori, E., M. Parisien, M.A. Krawchuk, and M.A. Moritz. 2013. Climate change-induced shifts in fire for Mediterranean ecosystems. *Global Ecology and Biogeography* 22(10): 1118–1129.

Beck, H.E., T.R. McVicar, I.J.M. van Dijk, J. Schellekens, R.A.M. de Jeu, and L.A. Bruijnzeel. 2011. Global evaluation of four AVHRR–NDVI data sets: Intercomparison and assessment against Landsat imagery. *Remote Sensing of Environment* 115(10): 2547–2563.

Beguería, S., J.I. López-Moreno, A. Lorente, M. Seeger, and J.M. García-Ruiz. 2003. Assessing the effect of climate oscillations and land-use changes on streamflow in the Central Spanish Pyrenees. *Ambio* 32: 283–286.

Blaschke, T. and G.J. Hay. 2001. Object-oriented image analysis and scale-space: Theory and methods for modeling and evaluating multiscale landscape structure. *International Archives of Photogrammetry and Remote Sensing* 34: 22–29.

Boer, M.M. and J. Puigdefabregas. 2005. Assessment of dryland condition using spatial anomalies of vegetation index values. *International Journal of Remote Sensing* 26: 4045–4065.

BOM. 2014. Normalised difference vegetation index (NDVI) http://www.bom.gov.au/climate/austmaps/about-ndvi-maps.shtml (accessed July 2, 2014).

Bond, W.J. and J.E. Keeley. 2005. Fire as a global 'herbivore': The ecology and evolution of flammable ecosystems. *Trends in Ecology & Evolution* 20: 387–394.

Brunsell, N.A. and R.R. Gillies. 2003. Determination of scaling characteristics of AVHRR data with wavelets: Application to SGP97. *International Journal of Remote Sensing* 24: 2945–2957.

Bucini, G., N.P. Hanan, R.B. Boone, I.P.J. Smit, S. Saatchi, M.A. Lefsky, and G.P. Asner. 2010. Woody fractional cover in Kruger National Park, South Africa: Remote-sensing-based maps and ecological insights. In M.J. Hill and N.P. Hanan (eds.), *Ecosystem Function in Savannas: Measurement and Modeling at Landscape to Global Scales*. Boca Raton, FL: CRC Press, pp. 219–237.

Campo-Bescós, M.A., R. Muñoz-Carpena, J. Southworth, L. Zhu, P.R. Waylen, and E. Bunting. 2013. Combined spatial and temporal effects of environmental controls on long-term monthly NDVI in the Southern Africa Savanna. *Remote Sensing* 5(12): 6513–6538.

Cassel-Gintz, M. and G. Petschel-Held. 2000. GIS-based assessment of the threat to world forests by patterns of non-sustainable civilisation nature interaction. *Journal of Environmental Management* 59: 279–298.

Ceccherini, G., N. Gobron, and M. Robustelli. 2013. Harmonization of fraction of absorbed photosynthetically active radiation (FAPAR) from sea-viewing wide field-of-view sensor (SeaWiFS) and medium resolution imaging spectrometer instrument (MERIS). *Remote Sensing* 5(7): 3357–3376.

Chapin III, F.S., S.R. Carpenter, G.P. Kofinas, C. Folke, N. Abel, W.C. Clark, P. Olsson et al. 2010. Ecosystem stewardship: Sustainability strategies for a rapidly changing planet. *Trends in Ecology & Evolution* 25: 241–249.

Cohen, W.B. and S.N. Goward. 2004. Landsat's role in ecological applications of remote sensing. *BioScience* 54: 535–545.

Cook, B.I. and S. Pau. 2013. A global assessment of long-term greening and browning trends in pasture lands using the GIMMS LAI3g dataset. *Remote Sensing* 5: 2492–2512.

CSIRO. 2009. Property-scale solution for environmental monitoring and land management. http://www.csiro.au/solutions/Vegmachine.html. (Accessed July 31, 2014.)

Danaher, T., P. Scarth, J. Armston, L. Collet, J. Kitchen, and S. Gillingham. 2010. Remote sensing of tree–grass systems: The eastern Australian woodlands. In M.J. Hill and N.P. Hanan (eds.), *Ecosystem Function in Savannas: Measurement and Modelling at Landscape to Global Scales*. Boca Raton, FL: CRC Press.

Dardel C., L. Kergoat, P. Hiernaux, E. Mougin, M. Grippa, and C.J. Tucker. 2014. Re-greening Sahel: 30 years of remote sensing data and field observations (Mali, Niger). *Remote Sensing of Environment* 140: 350–364.

DeFries, R. 2008. Terrestrial vegetation in the coupled human-earth system: Contributions of remote sensing. *Annual Review of Environment and Resources* 33: 369–390.

DeFries, R.S., J.A. Foley, and G.P. Asner. 2004. Land-use choices: Balancing human needs and ecosystem function. *Frontiers in Ecology and the Environment* 2: 249–257.

De Jong, S. and G.F. Epema. 2011. Imaging spectrometry for surveying and modelling land degradation. In F.D. van der Meer and S.M. De Jong (eds.), *Imaging Spectrometry*. Dordrecht, the Netherlands: Springer.

De Jong, R., J. Verbesselt, M.E. Schaepman, and S. de Bruin. 2012. Trend changes in global greening and browning: Contribution of short-term trends to longer-term change. *Global Change Biology* 18: 642–655.

Del Barrio, G., J. Puigdefabregas, M.E. Sanjuan, M. Stellmes, and A. Ruiz. 2010. Assessment and monitoring of land condition in the Iberian Peninsula, 1989–2000. *Remote Sensing of Environment* 114: 1817–1832.

Diaz-Delgado, R. and X. Pons. 2001. Spatial patterns of forest fires in Catalonia (NE of Spain) along the period 1975–1995: Analysis of vegetation recovery after fire. *Forest Ecology and Management* 147: 67–74.

Downing, T.E. and M. Lüdeke. 2002. International desertification. Social geographies of vulnerability and adaptation. In J.F. Reynolds and D.M. Stafford-Smith (eds.), *Global Desertification. Do Humans Cause Deserts?* Berlin, Germany: Dahlem University Press, pp. 233–252.

Dregne, H.E. and N.-T. Chou. 1992. Global desertification dimensions and costs. In H.E. Dregne (ed.), *Degradation and Restoration of Arid Lands.* Lubbock, TX: Texas Tech University.

Dubovyk, O., G. Menz, C. Conrad, E. Kan, M. Machwitz, and A. Khamzina. 2013. Spatio-temporal analyses of cropland degradation in the irrigated lowlands of Uzbekistan using remote sensing and logistic regression modelling. *Environmental Monitoring and Assessment* 185(6): 4775–4790.

Duguy, B., J.A. Alloza, A. Röder, R. Vallejo, and F. Pastor. 2007. Modeling the effects of landscape fuel treatments on fire growth and behaviour in a Mediterranean landscape (eastern Spain). *International Journal of Wildland Fire* 16: 619–632.

Eklundh, L. and L. Olsson. 2003. Vegetation trends for the African Sahel in 1982–1999. *Geophysical Research Letters* 30: 1430–1434.

Elmore, A.J., J.F. Mustard, S.J. Manning, and D.B. Lobell. 2000. Quantifying vegetation change in semiarid environments: Precision and accuracy of spectral mixture analysis and the Normalized Difference Vegetation Index. *Remote Sensing of Environment* 73: 87–102.

Erb, K.-H., F. Krausmann, V. Gaube, S. Gingrich, A. Bondeau, M. Fischer-Kowalski, and H. Haberl. 2009. Analyzing the global human appropriation of net primary production—Processes, trajectories, implications. An introduction. *Ecological Economics* 69: 250–259.

Fang, H., S. Liang, M.P. McClaran, W. Van Leeuwen, S. Drake, S.E. Marsh, A.M. Thomson, R.C. Izaurralde, and N.J. Rosenberg. 2005. Biophysical characterization and management effects on semiarid rangeland observed from Landsat ETM+ data. *IEEE Transactions on Geoscience and Remote Sensing* 43: 125–134.

Fensholt, R., T. Langanke, K. Rasmussen, A. Reenberg, S.D. Prince, C. Tucker, R.J. Scholes et al. 2012. Greenness in semi-arid areas across the globe 1981–2007—An earth observing satellite based analysis of trends and drivers. *Remote Sensing of Environment* 121: 144–158.

Fensholt, R. and S.R. Proud. 2012. Evaluation of earth observation based global long term vegetation trends—Comparing GIMMS and MODIS global NDVI time series. *Remote Sensing of Environment* 119: 131–147.

Fensholt, R. and K. Rasmussen. 2011. Analysis of trends in the Sahelian 'rain-use efficiency' using GIMMS NDVI, RFE and GPCP rainfall data. *Remote Sensing of Environment* 115: 438–451.

Fensholt, R., K. Rasmussen, P. Kaspersen, S. Huber, S. Horion, and E. Swinnen. 2013. Assessing land degradation/recovery in the African Sahel from long-term earth observation based primary productivity and precipitation relationships. *Remote Sensing* 5(2): 664–686.

Fensholt, R., I. Sandholt, and M. Rasmussen. 2004. Evaluation of MODIS LAI, FAPAR and the relation between FAPAR and NDVI in a semi-arid environment using in situ measurements. *Remote Sensing of Environment* 91(3–4): 490–507.

Flood, N., T. Danaher, T. Gill, and S. Gillingham. 2013. An operational scheme for deriving standardised surface reflectance from Landsat TM/ETM+ and SPOT HRG Imagery for Eastern Australia. *Remote Sensing* 5: 83–109.

Foley, J.A., R. DeFries, G.P. Asner, C. Barford, G. Bonan, S.R. Carpenter, F.S. Chapin et al. 2005. Global consequences of land use. *Science* 309: 570–574.

Foley, J.A., N. Ramankutty, K.A. Brauman, E.S. Cassidy, J.S. Gerber, M. Johnston, N.D. Mueller et al. 2011. Solutions for a cultivated planet. *Nature* 478: 337–342. doi:10.1038/nature10452.

Forman, R.T.T. and S.K. Collinge. 1996. The "spatial solution" to conserving biodiversity in landscapes and regions. In R.M. DeGraaf and R.I. Miller (eds.), *Conservation of Faunal Diversity in Forested Landscapes.* London, U.K.: Chapman & Hall, pp. 537–568.

Gao, F., J. Masek, M. Schwaller, and F. Hall. 2006. On the blending of the Landsat and MODIS surface reflectance: Predicting daily Landsat surface reflectance. *IEEE Transactions on Geoscience and Remote Sensing* 44(8): 2207–2218.

Geerken, R. and M. Ilaiwi. 2004. Assessment of rangeland degradation and development of a strategy for rehabilitation. *Remote Sensing of Environment* 90: 490–504.

Geist, H.J. and E.F. Lambin. 2004. Dynamic causal patterns of desertification. *BioScience* 54: 817–829.

Gobron, N. 2011. Envisat's medium resolution imaging spectrometer (MERIS) algorithm theoretical basis document: FAPAR and rectified channels over terrestrial surfaces. JRC-EC, Ispra, Italy. https://earth.esa.int/instruments/meris/atbd/atbd_2.10.pdf (accessed July 2, 2014).

Gobron, N., B. Pinty, O. Aussedat, J.M. Chen, W.B. Cohen, R. Fensholt, K.F. Huemmrich et al. 2006. Evaluation of fraction of absorbed photosynthetically active radiation products for different canopy radiation transfer regimes: Methodology and results using joint research center products derived from SeaWIFS against ground-based estimations. *Journal of Geophysical Research* 111(D13): D13110.

Gobron, N., B. Pinty, M. Verstraete, and Y. Govaerts.1999. The MERIS Global Vegetation Index (MGVI): Description and preliminary application. *International Journal of Remote Sensing* 20: 1917–1927.

Goward, S.N. and J.G. Masek. 2001. Landsat—30 years and counting. *Remote Sensing of Environment* 78: 1–2.

Graetz, R.D. 1996. Empirical and practical approaches to land surface characterization and change detection. In J. Hill and D. Peter (eds.), *The Use of Remote Sensing for Land Degradation and Desertification Monitoring in the Mediterranean Basin—State of the Art and Future Research. Proceedings of a Workshop*, jointly organized by JRC/IRSA and DGXII/D-2/D-4, Valencia, Spain, June 13–15, 1994. Luxembourg: Office for Official Publications of the European Communities, pp. 9–23.

Griffiths, P., T. Kuemmerle, E.R. Kennedy, I.V. Abrudan, J. Knorn, and P. Hostert. 2011. Using annual time-series of Landsat images to assess the effects of forest restitution in post-socialist Romania. *Remote Sensing of Environment* 118: 199–214.

Haberl, H., K.H. Erb, F. Krausmann, V. Gaube, A. Bondeau, C. Plutzar, S. Gingrich, W. Lucht, and M. Fischer-Kowalski. 2007. Quantifying and mapping the human appropriation of net primary production in earth's terrestrial ecosystems. *PNAS* 104: 12942–12947.

Hall, F., J.G. Masek, and G.J. Collatz. 2006. Evaluation of ISLSCP Initiative II FASIR and GIMMS NDVI products and implications for carbon cycle science. *Journal of Geophysical Research* 111: D22S08. doi:10.1029/2006JD007438.

Hecheltjen, A., F. Thonfeld, and G. Menz. 2014. Recent advances in remote sensing change detection—A review. In I. Manakos and M. Braun (eds.), *Land Use and Land Cover Mapping in Europe. Practices and Trends.* Dordrecht, the Netherlands: Springer, pp. 145–178.

Hein, L. and N. de Ridder. 2006. Desertification in the Sahel: A reinterpretation. *Global Change Biology* 12: 751–758.

Helldén, U. and C. Tottrup. 2008. Regional desertification: A global synthesis. *Global and Planetary Change* 64(3–4): 169–176.

Herman, S.M., A. Anyamba, and C.J. Tucker. 2005. Recent trends in vegetation dynamics in the African Sahel and their relationship to climate. *Global Environmental Change* 15: 394–404.

Hermann, S.M. and C.F. Hutchinson. 2005. The changing contexts of the desertification debate. *Journal of Arid Environments* 63: 538–555.

Heumann, B.W., J.W. Seaquist, L. Eklundh, and P. Jönnson. 2007. AVHRR derived phenological change in the Sahel and Soudan, Africa, 1982–2005. *Remote Sensing of Environment* 108: 385–392.

Hilker, T., E. Natsagdorj, R.H. Waring, A. Lyapustin, and Y. Wang. 2014. Satellite observed widespread decline in Mongolian grasslands largely due to overgrazing. *Global Change Biology* 20: 418–428.

Hill, J. 2008. Remote sensing techniques for monitoring desertification. LUCINDA Booklet Series A, Number 3. http://geografia.fcsh.unl.pt/lucinda/booklets/Booklet%20A3%2-0EN.pdf. (Accessed July 31, 2014.)

Hill, J., P. Hostert, and A. Röder. 2004. Long-term observation of Mediterranean ecosystems with satellite remote sensing. In S. Mazzoleni, G. di Pasquale, M. Mulligan, P. di Martino, and F. Rego (eds.), *Recent Dynamics of the Mediterranean Vegetation and Landscape.* Chichester, U.K.: John Wiley & Sons Ltd., pp. 33–43.

Hill, J., M. Stellmes, and C. Wang. 2014. Land transformation processes in NE China: Tracking trade-offs in ecosystem services across several decades with Landsat-TM/ETM+ time series. In I. Manakos and M. Braun (eds.), *Land Use and Land Cover Mapping in Europe. Practices and Trends.* Heidelberg, Germany: Springer, pp. 383–410.

Hill, J., M. Stellmes, T. Udelhoven, A. Röder, and S. Sommer. 2008. Mediterranean desertification and land degradation mapping related land use change syndromes based on satellite observations. *Global and Planetary Change* 64: 146–157.

Hill, M.J., M.O. Roman, C.B. Schaaf, L. Hutley, C. Brannstrom, A. Etter, and N.P. Hanan. 2011. Characterizing vegetation cover in global savannas with an annual foliage clumping index derived from the MODIS BRDF product. *Remote Sensing of Environment* 115: 2008–2024.

Hostert, P., A. Röder, J. Hill, T. Udelhoven, and G. Tsiourlis. 2003. Retrospective studies of grazing-induced land degradation: A case study in central Crete, Greece. *International Journal of Remote Sensing* 24: 4019–4034.

Huang, C.Q., S.N. Coward, J.G. Masek, N. Thomas, Z.L. Zhu, and J.E. Vogelmann. 2010. An automated approach for reconstructing recent forest disturbance history using dense Landsat time series stacks. *Remote Sensing of Environment* 114: 183–198.

Huber, S. and R. Fensholt. 2011. Analysis of teleconnections between AVHRR-based sea surface temperature and vegetation productivity in the semi-arid Sahel. *Remote Sensing of Environment* 115: 3276–3285.

Huete, A., C. Justice, and W. Leeuwen. 1999. MODIS vegetation index (Mod13)—Algorithm theoretical basis document—v3. http://modis.gsfc.nasa.gov/data/atbd/atbd_mod13.pdf(accessed July 2, 2014).

Huete A., K. Didan, T. Miura, E.P. Rodriguez, X. Gao, and L.G. Ferreira. 2002. Overview of the radiometric and biophysical performance of the MODIS vegetation indices. *Remote Sensing of Environment* 83: 195–213.

Huete, A.R. 1988. A soil-adjusted vegetation index (SAVI). *Remote Sensing of Environment* 25: 295–309.

Jamali, S., J. Seaquist, L. Eklundh, and J. Ardö. 2014. Automated mapping of vegetation trends with polynomials using NDVI imagery over the Sahel. *Remote Sensing of Environment* 141: 79–89.

James, M.E. and S.N.V. Kalluri. 1994. The pathfinder AVHRR land data set: An improved coarse resolution data set for terrestrial monitoring. *International Journal of Remote Sensing* 15: 3347–3363.

Janowiak, J.E.1988. An investigation of interannual rainfall variability in Africa. *Journal of Climate* 1: 240–255. doi:10.6069/H5MW2F2Q.

Jarmer, T., H. Lavée, P. Sarah, and J. Hill. 2009. Using reflectance spectroscopy and Landsat data to assess soil inorganic carbon in the Judean Desert (Israel). In A. Röder and J. Hill (eds.), *Recent Advances in Remote Sensing and Geoinformation Processing for Land Degradation Assessment*. London, U.K.: CRC Press/Balkema (Taylor & Francis Group), pp. 227–241.

Jönsson, P. and L. Eklundh. 2002. Seasonality extraction by function fitting to time-series of satellite sensor data. *IEEE Transactions on Geoscience and Remote Sensing* 40: 1824–1832.

Justice, C., L. Giglio, L. Boschetti, D. Roy, I. Csiszar, J. Morisette, and Y. Kaufman. 2006. MODIS fire products—Version 2.3. Algorithm technical background document.

Katagis, T., I.Z. Gitas, P. Toukiloglou, S. Veraverbeke, and R. Goossens. 2014. Trend analysis of medium- and coarse-resolution time series image data for burned area mapping in a Mediterranean ecosystem. *International Journal of Wildland Fire* 23: 668. http://dx.doi.org/10.1071/WF12055.

Kauth, R.J. and G.S. Thomas. 1976. The Tasselled cap—A graphic description of the spectral-temporal development of agricultural crops as seen by LANDSAT. In *LARS Symposia*, Purdue University of West Lafayette, Indiana, Paper 159.

Kennedy, R.E., Z. Yang, and W.B. Cohen. 2010. Detecting trends in forest disturbance and recovery using yearly Landsat time series: 1. LandTrendr—Temporal segmentation algorithms. *Remote Sensing of Environment* 114: 2897–2910.

Knyazikhin, Y., J. Glassy, J. L. Privette, Y. Tian, A. Lotsch, Y. Zhang, Y. Wang et al. 1999. MODIS leaf area index (LAI) and fraction of photosynthetically active radiation absorbed by vegetation (FPAR) product (MOD15) algorithm theoretical basis document, http://eospso.gsfc.nasa.gov/atbd/modistables.html (accessed July 2, 2014).

Koohafkan, P. and B.A. Stewart. 2008. Water and cereals in drylands. FAO Publications/EarthScan, London, U.K./Sterling VA. http://www.fao.org/docrep/012/i0372e/i0372e00.htm (accessed March 10, 2014).

Koslowsky, D. 1996. Mehrjährige validierte und homogenisierte Reihen des Reflexionsgrades und des Vegetationsindexes von Landoberflächen aus täglichen AVHRR-Daten hoher Auflösung. Berlin, Germany: Institute for Meterology, Free University Berlin.

Lacaze, B. 1996. Spectral characterisation of vegetation communities and practical approaches to vegetation cover changes monitoring. In J. Hill and D. Peter (eds.), *The Use of Remote Sensing for Land Degradation and Desertification Monitoring in the Mediterranean Basin—State of the Art and Future Research. Proceedings of a Workshop*, Jointly organized by JRC/IRSA and DGXII/D-2/D-4, Valencia, Spain, June 13–15, 1994, pp. 149–166. Valencia, Spain: Office for Official Publications of the European Communities.

Lal, R. 2004. Carbon sequestration in dryland ecosystems. *Environmental Management* 33: 528–544.

Lambin, E.F. and D. Ehrlich. 1997. Land-cover changes in sub-Saharan Africa (1982–1991): Application of a change index based on remotely sensed surface temperature and vegetation indices at a continental scale. *Remote Sensing of Environment* 61: 181–200.

Lambin, E.F. and H.J. Geist. 2001. Global land-use and land-cover change: What have we learned so far? *LUCC Newsletter* 46: 27–30.

Lambin, E.F., H. Geist, and R.R. Rindfuss. 2006. Introduction: Local processes with global impacts. In E.F. Lambin and H. Geist (eds.), *Land-Use and Land-Cover Change. Local Processes and Global Impacts*. Berlin, Germany: Springer.

Le Houérou, H.N. 1984. Rain use efficiency: A unifying concept in arid-land ecology. *Journal of Arid Environments* 7: 213–247.

Lee, R. 2011. The outlook for population growth. *Science* 333: 569–573.

Lepers, E. 2003. Synthesis of the main areas of land-cover and land-use change. Millennium ecosystem assessment, Final report. http://www.geo.ucl.ac.be/LUCC/lucc.html.

Li, S., P.H. Verburg, S. Lv, J. Wu, and X. Li. 2012. Spatial analysis of the driving factors of grassland degradation under the conditions of climate change and intensive use in Inner Mongolia, China. *Regional Environmental Change* 12: 461–474.

Loboda, T.V., L. Giglio, L. Boschetti, and C.O. Justice. 2012. Regional fire monitoring and characterization using global NASA MODIS fire products in drylands of Central Asia. *Frontiers of Earth Science* 6(2): 196–205.

Lorent, H., C. Evangelou, M. Stellmes, J. Hill, V. Papanastasis, G. Tsiourlis, A. Roeder, A., and E.F. Lambin. 2008. Land degradation and economic conditions of agricultural households in a marginal region of northern Greece. *Global and Planetary Change* 64: 198–209.

Lu, H., M.R. Raupach, and T.R. McVicar. 2001. Decomposition of vegetation cover into woody and herbaceous components using AVHRR NDVI time series. CSIRO Land and Water Technical Report 35/01. CSIRO Land and Water, Canberra, Australia. http://www.clw.csiro.au/publications/technical2001/tr35-01.pdf.

Masek, J.G., E.F. Vermote, N. Saleous, R. Wolfe, F.G. Hall, K.F. Huemmrich, F. Gao, J. Kutler, and T.K. Lim. 2006. A Landsat surface reflectance data set for North America, 1990–2000. *Geoscience and Remote Sensing Letters* 3: 68–72.

Mbow, C., T.T. Nielsen, and K. Rasmussen. 2000. Savanna fires in east-central Senegal: Distribution patterns, resource management and perceptions. *Human Ecology* 28(4): 561–583.

Meyer-Christoffer, A., A. Becker, P. Finger, B. Rudolf, U. Schneider, and M. Ziese. 2011. GPCC climatology version 2011 at 0.25°: Monthly land-surface precipitation climatology for every month and the total year from rain-gauges built on GTS-based and historic data. Global Precipitation Climatology Centre at Deutscher Wetterdienst, Germany. doi:10.5676/DWD_GPCC/CLIM_M_V2011_025. (Accessed July 31, 2014.)

Miehe, S., J. Kluge, H. von Wehrden, and V. Retzer. 2010. Long-term degradation of Sahelian rangeland detected by 27 years of field study in Senegal. *Journal of Applied Ecology* 47: 692–700.

Millennium Ecosystem Assessment. 2005a. Ecosystems and human well-being: Desertification synthesis. Washington, DC: World Resources Institute.

Millennium Ecosystem Assessment. 2005b. Ecosystems and human well-being: Synthesis. Washington, DC: Island Press.

Muir, J., M. Schmidt, D. Tindall, R. Trevithick, P. Scarth, and J. Stewart. 2011. Guidelines for Field measurement of fractional ground cover: a technical handbook supporting the Australian collaborative land use and management program. Tech. rep., Queensland Department of Environment and Resource Management for the Australian Bureau of Agricultural and Resource Economics and Sciences, Canberra.

Myneni, R.B., S. Hoffman, Y. Knyazikhin, J.L. Privette, J. Glassy, Y. Tian, Y. Wang et al. 2002. Global products of vegetation leaf area and fraction absorbed PAR from year one of MODIS data. *Remote Sensing of Environment* 83(1): 214–231.

Myneni, R.B. and D.L. Williams. 1994. On the relationship between FAPAR and NDVI. *Remote Sensing of Environment* 49(3): 200–211.

Nachtergaele, F.O., M. Petri, R. Biancalani, G. van Lynden, H. van Velthuizen, and M. Bloise. 2011. An information database for land degradation assessment at global level. LADA Technical Report n. 17. Global Land Degradation Information System (GLADIS), version 1.0, Food and Agriculture Organization of the United Nations (FAO). http://www.fao.org/nr/lada/index.php?option=com_docman&~task=doc_download&gid=773&lang=en. (Accessed July 31, 2014.)

Naveh, Z. 1975. The evolutionary significance of fire in the Mediterranean region. *Vegetati* 29: 199–208.

Novella, N.S. and W.M. Thiaw. 2013. African rainfall climatology version 2 for famine early warning systems. *Journal of Applied Meteorology and Climatology* 52: 588–606.

Oldeman, L.R., R.T.A. Hakkeling, and W.G. Sombroek. 1990. World map on status of human-induced soil degradation (GLASOD). UNEP, Nairobi, Kenya.

Padilla, F.M., B. Vidal, J. Sánchez, and F.I. Pugnaire. 2010. Land-use changes and carbon sequestration through the twentieth century in a Mediterranean mountain ecosystem: Implications for land management. *Journal of Environmental Management* 91: 2688–2695

Paudel, K.P. and P. Andersen. 2010. Assessing rangeland degradation using multi temporal satellite images and grazing pressure surface model in Upper Mustang, Trans Himalaya, Nepal. *Remote Sensing of Environment* 114: 1845–1855.

Pedelty, J., S. Devadiga, E. Masuoka, M. Brown, J. Pinzon, C. Tucker, D. Roy et al. 2007. Generating a long-term land data record from the AVHRR and MODIS instruments. In *Geoscience and Remote Sensing Symposium, IGARSS 2007. IEEE International*, New York, pp. 1021–1025.

Petschel-Held, G., M.K.B. Lüdeke, and F. Reusswig. 1999. Actors, structures and environment: A comparative and transdisciplinary view on regional case studies of global environmental change. In B. Lohnert and H. Geist (eds.), *Coping with Changing Environments: Social Dimensions of Endangered Ecosystems in the Developing World*. London, U.K.: Ashgate, pp. 255–291.

Pinty, B., M. Clerici, I. Andredakis, T. Kaminski, M. Taberner, M.M. Verstraete, N. Gobron, S. Plummer, and J.-L. Widlowski. 2011. Exploiting the MODIS albedos with the two-stream inversion package (JRC-TIP): 2. Fractions of transmitted and absorbed fluxes in the vegetation and soil layers. *Journal of Geophysical Research* 116(D9), D09106.

Pinzon, J.E. and C.J. Tucker. 2014. A Non-Stationary 1981–2012 AVHRR NDVI3g time series. *Remote Sensing* 6: 6929–6960.

Price, J.C. 1993. Estimating of leaf area index from satellite data. *IEEE Transactions on Geoscience and Remote Sensing* 31: 727–734.

Prince, S.D. 2004. Mapping desertification in southern Africa. In G. Gutman, A. Janetos, C.O. Justice, E.F. Moran, J.F. Mustard, R.R. Rindfuss, D. Skole, and B.L. Turner II (eds.), *Land Change Science: Observing, Monitoring, and Understanding Trajectories of Change on the Earth's Surface*. Dordrecht, the Netherlands: Kluwer, pp. 163–184.

Prince, S.D., I. Becker-Reshef, and K. Rishmawi. 2009. Detection and mapping of long-term land degradation using local net production scaling: Application to Zimbabwe. *Remote Sensing of Environment* 113: 1046–1057.

Prince, S.D., E.B. De Colstoun, and L.L. Kravitz. 1998. Evidence from rain-use efficiencies does not indicate extensive Sahelian desertification. *Global Change Biology* 4: 359–374.

Prince, S.D., K.J. Wessels, C.J. Tucker, and S.E. Nicholson. 2007. Desertification in the Sahel: A reinterpretation of a reinterpretation. *Global Change Biology* 13: 1308–1313.

Reeves, M.C. and L.S. Baggett. 2014. A remote sensing protocol for identifying rangelands with degraded productive capacity. *Ecological Indicators* 43: 172–182.

Reid, R.S., P.K. Thornton, G.J. McCRabb, R.L. Kruska, F. Atieno, and P.G. Jones. 2004. Is it possible to mitigate greenhouse gas emissions in pastoral ecosystems of the tropics? *Development and Sustainability* 6: 91–109.

Reid, R.S., T.P. Tomich, J. Xu, H. Geist, A. Mather, R. DeFries, J. Liu, D. Alves et al. 2006. Linking land-change science and policy: Current lessons and future integration. In E.F. Lambin and H. Geist (eds.), *Land-Use and Land-Cover Change. Local Processes and Global Impacts*. Berlin, Germany: Springer.

Reynolds, J.F., D.M. Stafford-Smith, E.F. Lambin, B.L. Turner II, M. Mortimore, S.P.J. Batterbury, T.E. Downing et al. 2007. Global desertification: Building a science for dryland development. *Science* 316: 847–851.

Röder, A., J. Hill, B. Duguy, J.A. Alloza, and R. Vallejo. 2008b. Using long time series of Landsat data to monitor fire events and post-fire dynamics and identify driving factors. A case study in the Ayora region (eastern Spain). *Remote Sensing of Environment* 112: 259–273.

Röder, A., T. Kuemmerle, J. Hill, V.P. Papanastasis, and G.M. Tsiourlis. 2007. Adaptation of a grazing gradient concept to heterogeneous Mediterranean rangelands using cost surface modelling. *Ecological Modelling* 204: 387–398.

Röder, A., T. Udelhoven, J. Hill, G. Del Barrio, and G.M. Tsiourlis. 2008a. Trend analysis of Landsat-TM and -ETM+ imagery to monitor grazing impact in a rangeland ecosystem in Northern Greece. *Remote Sensing of Environment* 112: 2863–2875.

Rouse, J.W., R.H. Haas, J.A. Schell, and D.W. Deering. 1974. Monitoring vegetation systems in the Great Plains with ERTS. In *Third ERTS Symposium*, NASA, Washington, D.C., SP-351 I, pp. 309–317.

Roy, D.P., J. Ju, K. Kline, P.L. Scaramuzza, V. Kovalskyy, M.C. Hansen, T.R. Loveland, E.F. Vermote, and C. Zhang. 2010. Web-enabled Landsat Data (WELD): Landsat ETM+ composited mosaics of the conterminous United States. *Remote Sensing of Environment* 114: 35–49.

Running, S.W., P.E. Thornton, R. Nemani, and J.M. Glassy. 2000. Global terrestrial gross and net primary productivity from the Earth Observing System. In O. Sala, R. Jackson, and H. Mooney (eds.), *Methods in Ecosystem Science*. New York: Springer Verlag, pp. 44–57.

Safriel, U., Z. Adeel, D. Niemeijer, J. Puigdefabregas, R. White, R. Lal, M. Winslow et al. 2005. Dryland systems. In R.M. Hassan, R.J. Scholes, and N. Ash (eds.), *Millennium Ecosystem Assessment: Ecosystems and Human Well-being: Current State and Trends: Findings of the Condition and Trends Working Group*. Washington, DC: Island Press, pp. 623–662.

Schmidt, M., T. Udelhoven, T. Gill, and A. Röder. 2012. Long term data fusion for a dense time series analysis with modis and landsat imagery in an Australian savanna. *Journal of Applied Remote Sensing* 6(1): 063512-1–063512-18.

Schroeder, T.A., M.A. Wulder, S.P. Healey, and G.G. Moisen. 2011. Mapping wildfire and clearcut harvest disturbances in boreal forests with Landsat time series data. *Remote Sensing of Environment* 115: 1421–1433.

Serra, P., X. Pons, and D. Saurí. 2008. Land-cover and land-use change in a Mediterranean landscape: A spatial analysis of driving forces integrating biophysical and human factors. *Applied Geography* 28: 189–209.

Shrestha, D.P., D.E. Margate, F. van der Meer, and H.V. Anh. 2005. Analysis and classification of hyperspectral data for mapping land degradation: An application in southern Spain. *International Journal of Applied Earth Observation and Geoinformation* 7: 85–96.

Sietse, O.L. 2010. ISLSCP II FASIR-adjusted NDVI, 1982–1998. In Hall, Forrest G., G. Collatz, B. Meeson, S. Los, E. Brown de Colstoun, and D. Landis (eds.), *ISLSCP Initiative II Collection. Data Set*. Oak Ridge National Laboratory Distributed Active Archive Center, Oak Ridge, Tennessee.

Skujins, J. (1991). *Semiarid Lands and Deserts: Soil Resource and Reclamation*. Boca Raton, FL: CRC Press.

Smith, M.O., S.L. Ustin, J.B. Adams, and A.R. Gillespie. 1990. Vegetation in deserts: I. A regional measure of abundance from multispectral images. *Remote Sensing of Environment* 31: 1–26.

Smith, P.M., S.N.V. Kalluri, S.D. Prince, and R.S. DeFries, 1997. The NOAA/NASA Pathfinder AVHRR 8-km land data set. *Photogrammetric Engineering and Remote Sensing* 63: 12–31.

Sommer, S., C. Zucca, A. Grainger, M. Cherlet, R. Zougmore, Y. Sokona, J. Hill, R.D. Peruta, J. Roehrig, and G. Wang. 2011. Application of indicator systems for monitoring and assessment of desertification from national to global scales. *Land Degradation & Development* 22: 184–197. doi: 10.1002/ldr.1084.

Sonnenschein, R. 2013. Land use change and its effects on vegetation trends and fire patterns in Mediterranean rangelands. Doctoral Thesis. Geomatics Series, No. 7, Berlin, Geography Department Humboldt-Universität zu Berlin.

Sonnenschein, R., T. Kuemmerle, T. Udelhoven, M. Stellmes, and P. Hostert. 2011. Differences in Landsat-based trend analyses in drylands due to the choice of vegetation estimate. *Remote Sensing of Environment* 115(6): 1408–1420.

Stafford Smith, M., N. Abel, B. Walker, and F.S. Chapin III. 2009. Drylands: Coping with uncertainty, thresholds, and changes in state. In F.S. Chapin III, G.P. Kofinas, and C. Folke (eds.), *Principles of Ecosystem Stewardship*. New York: Springer, pp. 171–195.

Stellmes, M., A. Röder, T. Udelhoven, and J. Hill. 2013. Mapping syndromes of land change in Spain with remote sensing time series, demographic and climatic data. *Land Use Policy* 30: 685–702.

Stellmes, M., T. Udelhoven, A. Röder, R. Sonnenschein, and J. Hill. 2010. Dryland observation at local and regional scale—Comparison of Landsat TM/ETM+ and NOAA AVHRR time series. *Remote Sensing of Environment* 114(10): 2111–2125.

Strahler, A.H., W. Lucht, C.B. Schaaf, T. Tsang, F. Gao, X. Li, J.-P. Muller, P. Lewis, and M.J. Barnsley. 1999. MODIS BRDF/Albedo product: Algorithm theoretical basis document version 5.0. NASA, Washington, DC. http://modis.gsfc.nasa.gov/data/atbd/atbd_mod09.pdf (accessed March 5, 2014).

Thomas, D.S.G. and N.J. Middleton. 1994. *Desertification: Exploding the Myth*. Chichester, U.K.: Wiley.

Tozer, C.R., A.S. Kiem, and D.C. Verdon-Kidd. 2012. On the uncertainties associated with using gridded rainfall data as a proxy for observed. *Hydrology and Earth System Science* 16: 1481–1499. doi:10.5194/hess-16-1481-2012.

Trapnell, C.G. 1959. Ecological results of woodland burning experiments in Northern Rhodesia. *Journal of Ecology* 47: 129–168.

Tucker, C.J. 1979. Red and photographic infrared linear combinations for monitoring vegetation. *Remote Sensing of Environment* 8: 127–150.

Tucker, C.J., J.E. Pinzon, M.E. Brown, D.A. Slayback, E.W. Pak, R. Mahoney, E.F. Vermote, and N. El Saleous. 2005. An extended AVHRR 8-km NDVI dataset compatible with MODIS and SPOT vegetation NDVI data. *International Journal of Remote Sensing* 26: 4485–4498.

Turner II, B.L., E.F. Lambin, and A. Reenberg. 2007. The emergence of land change science for global environmental change and sustainability. *PNAS* 104: 20666–20671.

Udelhoven, T. 2011. TimeStats: A software tool for the retrieval of temporal patterns from global satellite archives. *IEEE Journal of Selected Topics in Applied Earth Observations and Remote Sensing (J-STARS)* 4(2), 310–317. doi: 10.1109/JSTARS.2010.2051942.

Udelhoven, T. and J. Hill. 2009. Change detection in Syria's rangelands using long-term AVHRR data (1982–2004). In A. Röder and J. Hill (eds.), *Recent Advances in Remote Sensing and Geoinformation Processing for Land Degradation Assessment*. London, U.K.: Taylor & Francis.

Udelhoven, T., M. Stellmes, G. del Barrio, and J. Hill. 2009. Assessment of rainfall and NDVI anomalies in Spain (1989–1999) using distributed lag models. *International Journal of Remote Sensing* 30: 1961–1976.

UNCCD. 1994. United Nations convention to combat desertification in countries experiencing serious drought and/or desertification, particularly in Africa. http://www.unccd.int/en/about-the-convention/Pages/Text-overview.aspx.

UNU. 2006. International year of deserts and desertification. United Nations University Institute for Water, Environment and Health, Hamilton, Ontario, Canada. http://inweh.unu.edu/desertification06/ (accessed March 15, 2014).

USGS. 2015. Landsat Higher Level Science Data Products, 05.02.2015.

Veraverbeke, S., S. Lhermitte, W.W. Verstraeten, and R. Goossens. 2010. The temporal dimension of differenced Normalized Burn Ratio (dNBR) fire/burn severity studies: The case of the large 2007 Peloponnese wildfires in Greece. *Remote Sensing of Environment* 114: 2548–2563.

Verbesselt, J., R. Hyndman, G. Newnham, and D. Culvenor. 2010. Detecting trend and seasonal changes in satellite image time series. *Remote Sensing of Environment* 114: 106–115.

Verstraete, M. 1994. The contribution of remote sensing to monitor vegetation and to evaluate its dynamic aspects. In F. Veroustrate and R. Ceulemans (eds.), *Vegetation, Modeling and Climate Change Effects*. The Hague, the Netherlands: SPB Academic Publishing, pp. 207–212.

Verstraete, M.M., C.F. Hutchinson, A. Grainger, M. Stafford-Smith, R.J. Scholes, J.F. Reynolds, P., Barbosa, A. Léon, and C. Mbow. 2011. Towards a global drylands observing system: Observational requirements and institutional solutions. *Land Degradation & Development* 22: 198–213.

Viedma, O., J. Melia, D. Segarra, and J. GarciaHaro. 1997. Modeling rates of ecosystem recovery after fires by using Landsat TM data. *Remote Sensing of Environment* 61: 383–398.

Vogt, J.V., U. Safriel, G. Von Maltitz, Y. Sokona, R. Zougmore, G. Bastin, and J. Hill. 2011. Monitoring and assessment of land degradation and desertification: Towards new conceptual and integrated approaches. *Land Degradation & Development* 22: 145–149. doi: 10.1002/ldr.1075.

Walker, J.J., K.M. de Beurs, R.H. Wynne, and F. Gao. 2012. Evaluation of Landsat and MODIS data fusion products for analysis of dryland forest phenology. *Remote Sensing of Environment* 117: 381–393.

Wallace, J., G. Behn, and S. Furby. 2006. Vegetation condition assessment and monitoring from sequences of satellite imagery. *Ecological Management and Restoration* 7: 31–36.

Wallace, J., P.A. Caccetta, and H.T. Kiiveri. 2004. Recent developments in analysis of spatial and temporal data for landscape qualities and monitoring. *Australian Ecology* 29: 100–107.

Washington-Allen, R. A., R. D. Ramsey, N. E. West, and B. E. Norton. 2008. Quantification of the ecological resilience of drylands using digital remote sensing. *Ecology and Society* 13(1): 33.

Washington-Allen, R.A., N.E. West, R.D. Ramsey, and R.A. Efroymson. 2006. A protocol for retrospective remote sensing-based ecological monitoring of rangelands. *Rangeland Ecology & Management* 59: 19–29.

Washington-Allen, R.A., N.E. West, R.D. Ramsey, D.H., Phillips, and H.H. Shugart. 2010. Retrospective assessment of dryland soil stability in relation to grazing and climate change. *Environmental Monitoring and Assessment* 160: 101–121.

Were, K., Ø.B. Dick, and B.R. Singh. 2014. Exploring the geophysical and socio-economic determinants of land cover changes in Eastern Mau forest reserve and Lake Nakuru drainage basin, Kenya. *Geojournal* 79(6): 775–790.

Wessels, K. 2009. Letter to the editor: Comments on 'Proxy global assessment of land degradation' by Bai et al. (2008). *Soil Use and Management* 25: 91–92.

Wessels, K.J., S.D. Prince, J. Malherbe, J. Small, P.E. Frost, and D. van Zyl. 2007. Can human-induced land degradation be distinguished from the effects of rainfall variability? A case study in South Africa. *Journal of Arid Environment* 68(2): 271–297.

Williams, C.A. and N.P. Hanan. 2011. ENSO and IOD teleconnections for African ecosystems: Evidence of destructive interference between climate oscillations. *Biogeosciences* 8: 27–40.

Yin, H., T. Udelhoven, R. Fensholt, D. Pflugmacher, and P. Hostert. 2012. How Normalized Difference Vegetation Index (NDVI) trends from advanced very high resolution radiometer (AVHRR) and Système Probatoire d'Observation de la Terre VEGETATION (SPOT VGT) time series differ in agricultural areas: An inner Mongolian case study. *Remote Sensing* 4(11): 3364–3389. doi:10.3390/rs4113364.

Zeng, F.-W., G.J. Collatz, J.E. Pinzon, and A. Ivanoff. 2013. Evaluating and quantifying the climate-driven interannual variability in global inventory modeling and mapping studies (GIMMS) Normalized Difference Vegetation Index (NDVI3g) at global scales. *Remote Sensing* 5: 3918–3950.

Zhu, Z. and C.E. Woodcock. 2012. Object-based cloud and cloud shadow detection in Landsat imagery. *Remote Sensing of Environment* 118: 83–94.

Zhu, Z. and C.E. Woodcock. 2014. Continuous change detection and classification of land cover using all available Landsat data. *Remote Sensing of Environment* 144: 152–171.

Zhu, Z., C.E. Woodcock, and P. Olofsson. 2012. Continuous monitoring of forest disturbance using all available Landsat imagery. *Remote Sensing of Environment* 122: 75–91.

Zika, M. and K.-H. Erb. 2009. The global loss of net primary production resulting from human-induced soil degradation in drylands. *Ecological Economics* 69: 310–318.

Disasters

Disasters:
Risk Assessment, Management, and Post-Disaster Studies Using Remote Sensing

Norman Kerle
University of Twente

Acronyms and Definitions

ALOS	Advanced Land Observing Satellite
ASAR	Advanced Synthetic Aperture Radar (flying on-board ENVISAT)
AVHRR	Advanced Very High Resolution Radiometer
COSMO-SkyMed	Constellation of Small Satellites for the Mediterranean Basin Observation
DEM	Digital Elevation Model
DMC	Disaster Monitoring Constellation
DNSS	Defense Navigation Satellite System
DRM	Disaster Risk Management
EaR	Elements at Risk
EMS	European Macroseismic Scale
EnMAP	Environmental Mapping and Analysis Program
ENVISAT	Environmental Satellite
EO	Earth Observation
EO-1	Earth Observing-1 Mission
ERS-1/2	European Remote Sensing Satellite (1 and 2)
ERTS	Earth Resources Technology Satellites
FEWS	Famine Early Warning System Network
GDEM	Global Digital Elevation Model
GEO-CAN	Global Earth Observation-Catastrophe Assessment Network
GEOS	Geostationary Operational Environmental Satellite
GMES	Global Monitoring for Environment and Security (now called Copernicus)
GPS	Global Positioning System
HyspIRI	Hyperspectral Infrared Imager
JAXA	Japan Aerospace Exploration Agency
JERS-1	Japanese Earth Resources Satellite (also known as Fuyo-1)
LiDAR	Light Detection and Ranging
HALE	High Elevation–Long Endurance
MODIS	Moderate Resolution Imaging Spectroradiometer

MSS	Multispectral Scanner
NOAA	National Oceanic and Atmospheric Administration
OBIA/OOA	Object-Based Image Analysis/Object-Oriented Analysis
PALSAR	Phased Array Type L-Band Synthetic Aperture Radar
PRISMA	Prototype Research Instruments and Space Mission Technology Advancement
PROBA	Project for On-Board Autonomy (carrying the CHRIS hyperspectral sensor)
PSInSAR	Permanent Scatterer SAR Interferometry
RISAT-1	Radar Imaging Satellite 1
SAR	Synthetic Aperture Radar
SERTIT	Service Régional de Traitement d'Image et de Télédétection
SPOT	Satellite Pour l'Observation de la Terre
SRTM	Shuttle Radar Topography Mission
SWIR	Short Wave Infrared
TanDEM-X	TerraSAR-X Add-On for Digital Elevation Measurement
TIR	Thermal Infrared
TIROS	Television Infrared Observation Satellite
TOMS	Total Ozone Mapping Spectrometer
UAV	Unmanned Aerial Vehicle
UN	United Nations
UNSPIDER	United Nations Platform for Space-Based Information for Disaster Management and Emergency Response
VHRR	Very High Resolution Radiometer
VNIR	Visible and Near-Infrared

18.1 Introduction

The number and consequences of natural disasters have shown dramatic developments in past decades. The number of disaster events rose from annually less than 10 in the first decade of the nineteenth century, to about 30 per year by the early 1960s. From then onward, in the 50-year timeframe that is the focus of this book, numbers first rose rapidly to more than 500 events by the year 2000, but in the past 10 years showed a declining trend (Figure 18.1). The annual total economic damage figures are more erratic, on one hand also showing a strong increase since about 1970, but with increasing high inter-annual variation owing to individual, exceptionally costly events, such as the Hurricane Katrina in 2005, or the 2011 Tohoku (Japan) earthquake and tsunami event that caused damage in excess of 300 bn US$. All damage statistics must be considered with care. For one, to be included in the CRED EM-DAT database (CRED, 2014) that is the source of the preceding numbers, events either need to cause at least 10 fatalities, affect at least 100 people, or lead to a state of emergency or a call for international assistance, an arbitrary definition. Furthermore, with reporting, especially of smaller, more local events getting increasingly sketchy the further we look back (or indeed still today in more remote places, or those suffering from weak governments or from conflicts), the preceding numbers suffer from variable reliability and completeness. It is nevertheless interesting to note that the timeframe of this book appears to coincide with a new, more intense, epoch of natural disaster occurrence. The increase clearly reflects global population numbers that also started to rise rapidly around 1960 (Kerle and Alkema, 2011), and a strong growth in wealth and thus the basis for economic damage. The trend also owes to growing transparency and interest in global affairs, better

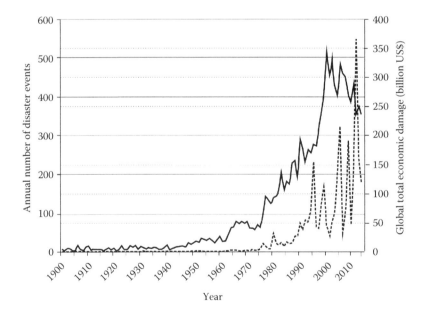

FIGURE 18.1 Number of annual natural disasters between 1900 and 2014 in bold (primary axis), and total annual economic disaster damage for the same period in hatched line (secondary axis). (From CRED: Center for Research on the Epidemiology of Disasters (2014) EMDAT: The OFDA/ CRED International Disasters Data Base. Brussels, Belgium: School of Public Health, Université Catholique de Louvain, Belgium, EM-DAT: The OFDA/CRED international disaster database, 2014.)

reporting, and clearly also remote sensing has played a role. While disasters have been a focus area as long as remote sensing has existed, in particular the advent of space-based remote sensing in the 1960s (military) and 1970s (civilian) has led to better technical means to detect and assess disaster events, including events national rulers would have preferred not to be seen by the world (e.g., earthquake or flooding disasters in China in the 1970s, or the Chernobyl nuclear power plant disaster in 1986). Hence, remote sensing has been playing a pivotal role in allowing increasingly informed disaster risk management (DRM), and in achieving a global inventory of disaster events.

There has been enormous progress in remote sensing in the last 50 years that has vastly benefitted DRM, the summary term used here to address risk and management of, and response to disasters. These developments are the focus of this chapter. On one hand this includes the many technical advances made—new and better sensor types and platforms, the move from analogue to digital data capture and distribution, as well as better computing resources and software. But better data have also led to an equally fundamental shift in our understanding of what causes disasters and has led to new ways to assess and quantify risk, to anticipate events with disaster potential and provide early warning, and also to respond rapidly to damaging events. In each of the preceding categories, a host of conceptual and methodological advances can be identified. The preceding, mostly technical progress, is flanked by significant organizational developments that include better international cooperation (leading to global monitoring and response systems), but also a growing involvement of civil society and lay persons.

This chapter first introduces several key DRM terms and concepts, before discussing selected domain developments, and the role remote sensing has played in those. The second part of the chapter details a range of relevant trends and developments, as well as remaining gaps and limitations where more work is needed. The chapter only addresses natural risks and disasters, excluding technological events or industrial accidents, as well as complex humanitarian emergencies linked to political or ethnic conflicts.

18.2 From Hazards to Disaster Risk: Terms and Concepts

The very use of the term *disaster risk management*, and that it has largely replaced the more historic term *disaster management*, is evidence of profound developments in our understanding of the nature of disasters and our ability to deal with them. Until well into the twentieth century, disasters were widely seen as the result of a violent nature that somehow had to be tamed, increasingly with engineering measures (Smith, 2004, a mindset that still frequently appears today). Events that did occur were responded to and managed. The middle of the century saw a fundamental shift in perspective, one marked by political ecology views (Kerle and Alkema, 2011). Those not only considered nature as a hazard, but added social, economic, and political elements to the equation, in an attempt to understand why nature and society intersected

disastrously at ever shorter intervals. This led to the concept of vulnerability, meaning the capacity for loss or damage to people, systems, or material assets (Blaikie et al., 1994). Hence, it became clear that, for a disaster to occur, a number of factors had to be in place: a hazardous event or process, and so-called elements at risk (EaR) that are located in the area exposed to the hazard, and that are vulnerable to the hazard in question, at a given magnitude. This can be summed up in a risk equation that typically takes the form of: Risk = Hazard × Value × Vulnerability, the latter two terms referring to all EaR present. This provides the basis for a quantitative assessment of risk in the form of expected losses per time period considered, typically per year. The hazard is thus treated as a process or phenomenon of a certain type, magnitude, and probability of occurrence, while the value and vulnerability are, respectively, the total economic value of a given EaR, and the fraction that is expected to be lost in case of the hazard event being considered. For example, if an event has a 10% probability of occurrence in a given year and the capacity to damage an EaR by 10%, the expected loss, or risk, in that year is 1% of the total value of the EaR (for more background on risk assessment see van Westen et al., 2011).

This conceptualization contains a number of aspects that are directly relevant from a remote sensing perspective: (1) all elements are spatial in nature, meaning that they have a geographic location and characteristics such as extent, proximity, or adjacency to other risk elements, and can thus be associated with attributes that are linked to a geographic place or area; (2) for each EaR a number of characteristics must be known (e.g., location, size, and occurrence interval of a hazard, or the location, type, and vulnerability of an EaR), and for the assessment of many of those remote sensing-based approaches have been developed; and (3) a careful match of the specific risk element with a suitable remote sensing data type or processing method is needed. At the same time, though, remote sensing is versatile and can address various needs. DRM is often conceptualized as a cycle that includes mitigation, preparedness, and the disaster event, followed by response, recovery, and rehabilitation, which then leads again to mitigation and preparedness (Figure 18.2). Remote sensing-based methods can often benefit

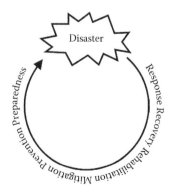

FIGURE 18.2 DRM cycle.

different phases in the cycle. For example, methods designed to map EaR (buildings, infrastructure, etc.) can be adapted to map post-event structural damage.

18.3 Domain Developments and Selected Technical Advances

18.3.1 Early Disaster Mapping with Remote Sensing

The unique perspective afforded by remote sensing quickly led to many application areas to be identified. While initially still too limited in terms of available and suitable instruments and their deployability to provide a useful tool to respond to disasters, this soon changed. Early DRM was largely limited to providing information following disaster events. Not counting balloon-based visual assessment of damage caused by military action during the US Civil War in the 1860s, the first documented deployment of an airborne sensing instrument following a natural disaster occurred in 1906, when George R. Lawrence launched a 20 kg camera on a series of kites some 600 m above San Francisco to map the damage caused by a severe earthquake only days earlier (see also O'Rourke et al., 2006). Developments were subsequently spurred primarily by World War I, leading to a situation where aerial imagery would be routinely acquired, including for environmental damage assessment. As early as 1919, an image-based forestry mapping program was established in Canada, with the first airborne thermal infrared images being acquired in the same year. Methods and scope of operations continuously improved until the field of remote sensing changed fundamentally in 1960, when the first satellite-based image was successfully taken. This marks the beginning of the period reviewed in this chapter.

18.3.2 Dawn of the Satellite Era

Technological developments for military purposes, which continue to be a key driver in Earth observation (EO), and geoinformation science in general, led to the acquisition of the first satellite-based images in 1960. At the height of the Cold War, the Corona program, declassification of which finally began in 1995 and only finished in 2002, saw the launch of nearly 150 satellite missions aimed at image capture over areas relevant for the national security of the United States. For DRM purposes Corona as well as sister programs Argon and Lanyard were primarily relevant insofar as they advanced satellite and sensor developments, since the image acquisition was aimed at areas of military relevance, and data were not made available outside the intelligence community. However, following declassification the data acquired by Corona over its 12-year mission duration have been used for a variety of studies related to DRM. Those range from geological mapping to population growth studies (Dashora et al., 2007), time series analysis of glacier retreat (Narama et al., 2010), volcanic hazard assessment (Karakhanian et al., 2003), or deforestation mapping for landslide hazard assessment (Kerle et al., 2003). The majority of the images were taken with black and white film (with a few being taken in color or infrared), with later missions delivering images with spatial resolutions of up to 1.8 m (Galiatsatos et al., 2008). Only three decades later would satellite images of comparable resolution become available for civilian users. The data had two additional advantages. The film strips covered a very narrow (14 km) but also very long (188 km) corridor (Figure 18.3), explaining why often also areas that do not seem to be of much military relevance were imaged. Some of the Corona missions also carried stereo cameras, the data from which, following declassification, have been used to create digital elevation models (DEMs). Especially prior to the more widespread availability of detailed Light Detection and Ranging

FIGURE 18.3   Example of a Corona satellite photograph from 1967 of parts of Casita volcano, Nicaragua, covering an area of 14 × 188 km, and part of a stereo pair (a). Close-up of the two stereo scenes to illustrate the amount of detail visible (b, c). Approximate location shown by box in (a).

(LiDAR) data or images from modern high resolution stereo satellites, the resulting DEMs with resolutions approximately ranging between 9 m (Schmidt et al., 2001) and 17 m (Galiatsatos et al., 2008) were a valuable data source, for example, to study glacier hazards (Lamsal et al., 2011).

The study of atmospheric hazards, such as tropical storms, benefits tremendously from EO. First experiments with spaceborne meteorological observations and measurements also date back to 1960, the beginning of the TIROS (Television Infrared Observation Satellite) program. The data from these missions were unclassified, making TIROS the first civilian EO satellite, though providing images with a rather low spatial resolution of less than 1 km. This continued when the satellites started to be named NOAA in 1970, which, after initially operating the Very High Resolution Radiometer (VHRR), since 1978 have been flying the Advanced Very High Resolution Radiometer (AVHRR). The series continues to this day, providing images with a spatial resolution ranging from 1.1 km (at nadir) up to a maximum of 3.8 km for peripheral pixels. TIROS III was the first satellite to capture images of a developing hurricane (Esther in September 1961; Perlroth, 1962), and such data have since then been of vital importance both to understand atmospheric hazards and to provide early warning. Despite their modest spatial resolution the NOAA satellites became valuable due to their high temporal resolution, broad swath width (nearly 2400 km), as well as additional provision of infrared information (Hastings and Emery, 1992). The daily coverage and moderate spatial detail started to allow the monitoring of hazardous events such as large floods (e.g., the Mississippi flood of 1973), tropical storms, as well as of sea ice dynamics (Space Science Board-National Research Council, 1974). One of the NOAA satellites also provided first repeat coverage of a large volcanic eruption (St. Augustine, Alaska, in 1976).

Spatial resolution of the NOAA satellites was improved by an order of magnitude by the launch of ERTS-1 (Earth Resources Technology Satellites) in 1972, which marked the beginning of the Landsat series, the longest and most significant civilian EO program. The spatial resolution was initially limited to 100 m for national security reasons (Ramachandran et al., 2010). The principal instrument on ERTS-1 was the four-channel multispectral scanner (MSS) that had a nominal spatial resolution of 80 m (Mika, 1997). The same Mississippi flooding observed by the NOAA-2 satellite, and that lasted approximately 3 months, was mapped in 16 day intervals, and with considerably more detail, by ERTS-1. The satellite operated for nearly 6 years, and the successful MSS sensor continued to be deployed on the follow-up missions of Landsat-2 and -3 (1978–1981 and 1978–1983, respectively). While the daily coverage of the VHRR system allowed for monitoring of dynamic events, ERTS-1's better spatial resolution started to allow the more detailed study of spatial phenomena related to natural hazards. This included the mapping of volcanic areas and other major geological units (Space Science Board-National Research Council, 1974), as well as geological evidence of seismic activity, such as fault lines (Gedney and VanWormer, 1973).

The lifetime of ERTS-1 also coincided with two major earthquakes, the 1972 magnitude 5.6 event that destroyed much of Managua, Nicaragua, and the magnitude 7.0 earthquake in August 1973 near Mexico City. The ERTS-1 satellite was rapidly programmed to obtain data of both events (Carter and Rinker, 1976), marking the beginning of dedicated post-disaster surveillance with satellite technology.

The year the lifespan of ERTS-1 ended, 1978, saw another remote sensing milestone with the launch of SeaSat, the first synthetic aperture radar (SAR) instrument. Though operating for less than 6 months, it laid the technical and application foundations for subsequent radar missions. The high spatial resolution images (ca. 25 m) were primarily aimed at answering oceanographic questions. However, they were also used for terrestrial studies, including those related to DRM, such as changes in forest cover (Hoffer and Lee, 1989), or to study volcanic features (Mackenzie and Ringrose, 1986), and also its potential to address agricultural disasters was explored (Howard et al., 1978). The ability of satellite radar data to map inundated areas was also realized (National Research Council (U.S.) Committee on Remote Sensing Programs for Earth Resource Surveys, 1977), which was to become the most commonly used application area for all later SAR instruments.

Several other technical developments of relevance to DRM occurred in the 1970s. In 1975, the first geostationary weather satellite (GOES-1) was launched, providing hemispheric meteorological data. The GOES satellites were later joined by comparable instruments from Japan, Russia, Europe, and India, to provide global coverage and collectively proving data that are critical for the study of atmospheric hazards. In 1978, the polar orbiting Nimbus 7 satellite was launched that carried the Total Ozone Mapping Spectrometer (TOMS), which collected global ozone data until 1994 and allowed the detection of the ozone hole (Herman et al., 1993), a global health hazard. The 1970s also saw the groundwork being laid for the Defense Navigation Satellite System (DNSS), later renamed NAVSTAR and subsequently Global Positioning System (GPS). Though not a remote sensing system, GPS has nevertheless been playing a critical supporting role in the use of EO data in nearly all DRM application fields (Kerle, 2013a).

The 1960s and 1970s were thus the pioneering decades in spaceborne remote sensing, and included many milestones in optical and radar EO, events that fundamentally and lastingly defined the principles of monitoring and mapping of the Earth's land, oceans, and atmosphere. All subsequent spaceborne missions can be characterized by continuity and refinement of the early missions detailed earlier. The growing amount of instruments in orbit, and more rapidly available data of a range of spatial, spectral, and temporal resolutions, led to hazard and disaster research making increasing use of such data. Two additional developments advanced this process: the launch and operation of EO satellites by countries other than the United States, and later also by private companies, both developments with profound effects for EO and DRM. Already in 1979 India launched the polar orbiting Bhaskara-I experimental satellite, a first step that later morphed into one of the most extensive and successful EO programs in the world. In 1986, SPOT-1 (Satellite Pour l'Observation de la Terre) was launched by France, providing images with a ground-breaking 10 m (pan-chromatic) resolution. This was followed by Japan's first EO satellite (MOS-1) in 1987. The U.S. Land Remote Sensing Policy Act of 1992 provided the basis for a development that has changed the remote sensing field in equal measure. For the first time private companies in the United States were allowed to build and operate civilian remote sensing satellites, leading to the launch of OrbView-1 in 1995, the world's first commercial EO satellite. This led to a race to provide ever more sophisticated imagery, but also undermined governmental EO efforts, and hence hurt data acquisition continuity. It has also been shifting the focus to targets that are commercially valuable, which carries the risk that fewer data are acquired of areas for which no commercial market exists.

18.3.3 Disaster Risk as a Multifaceted Spatial Phenomenon, and the Role of Operational Remote Sensing

As explained earlier, disaster risk is understood to be a function of a hazardous process or phenomenon that spatially intersects with vulnerable EaR. With increasing availability of remote sensing data of various types, as well as better geographic information system (GIS) tools and advanced spatial modeling, a fundamental shift has occurred from detection and mapping of hazards or disaster events, toward more comprehensive analysis of disaster risk. In the following sections, the significance of remote sensing for the different aspects of DRM is discussed in more detail. Table 18.1 provides definitions for major aspects of DRM, and aspects that determine the utility of remote sensing data.

18.3.3.1 Hazard Assessment with EO Data

Apart from the multitude of man-made hazards that are not the focus of this chapter, many processes in nature can be harmful to people, the infrastructure they build, and the systems (e.g., environmental, social, economic, political, transport) they create. Those hazards can be characterized by their origin and extent, as well as their rate of occurrence and intensity/magnitude. A common approach to assess those parameters is via a frequency–magnitude relationship that is based on the realization that low-magnitude events occur more frequently than larger events. A frequency–magnitude analysis rests on two assumptions: that the history of events of a given type in a given area is well known (e.g., all flooding events to affect a given area, and their magnitude as expressed by duration and spatial extent), and that the processes that led to the hazardous events remain largely unchanged, and that thus the past can be seen as a guide to the future. This type of analysis is an indicator of the probability of occurrence of a hazard of a given type in a given area, and thus critical for risk assessment. A frequency analysis requires knowledge of the existence of the hazard in the first place, something remote sensing frequently provides. For example, Landsat data were used to identify previously unknown volcanoes in the Andes (Francis and De Silva, 1989). Similarly, airborne laser scanning (LiDAR) data, with the ability to survey the ground beneath forest, have been used to identify landslides through the (revegetated) scars they leave (Razak et al., 2013; Van Den Eeckhaut et al., 2012), thus allowing a more complete hazard assessment than would otherwise be possible.

A remote sensing archive of some 40 years now exists for many parts of the world, useful in particular to create inventories of larger events, such as flooding. Flood frequency assessment has been carried out with optical images (e.g., Huang et al., 2012) as well as radar data (Hoque et al., 2011). Similar time-series analyses have been done to understand drought frequencies (e.g., Heumann et al., 2007) and trends in desertification (Dardel et al., 2014; Piao et al., 2005), though especially for studies related to climate change the extent of our archive, and how reliably it allows the detection of frequencies and trends, has been called into question (Loew, 2014). The record for LiDAR and other airborne data typically flown for dedicated missions also remains patchy for many parts of the worlds, and repeat datasets are rare.

Given its rich archive that spans more than 40 years, Landsat data have been employed in a range of hazard-related time-series studies, ranging from glacier-lake expansion (Nie et al., 2013) to multidecadal wild fire frequency analysis (Oliveira et al., 2012). For events of lower spatial extent, and thus requiring data with higher spatial resolution, a shorter record exists, and assessments into the occurrence of past hazard events are limited to more recent times. Nevertheless, remote sensing data have been instrumental in the assessment of smaller events, such as landslides, including the dynamics of continuously developing mass movements (e.g., Martha et al., 2012). Also the temporal behavior of phenomena that constitute no direct hazard to people but to their livelihood, such as erosion, has been effectively analyzed with multitemporal data (Shruthi et al., in press-a). Table 18.2 provides a comprehensive overview of the utility of different types of remote sensing data to assess a range of hazards, as well as other elements of the DRM cycle.

18.3.3.2 Hazard Assessment: A Multifaceted Problem

Talking of hazards in a broad sense—for example, *the* flood hazard, or *the* volcanic hazard—may be meaningful when describing general situations, or when communicating with decision makers or potentially exposed people. From a risk assessment and planning perspective, a more nuanced approach is needed, since those broad hazards, in a strict sense, do not exist. Flooding comes in many different types (riverine, coastal, flash flooding, among others), each with a unique behavior, and each requiring a different approach for their assessment and monitoring. Likewise, rock falls and debris flows are both gravity-driven mass movements, but both their genesis and behavior of movement differ dramatically. *The* volcanic hazard also does not exist; instead we can distinguish about 20 different hazardous processes that are associated with a volcano, some of which

TABLE 18.1   Definition of Key DRM Terms and Aspects That Determine the Utility of Remote Sensing

DRM Parameter	Definition	Aspects That Determine the Use of EO Data
Hazard	Two distinct ways of defining a hazard exist. It can be seen (1) as a potentially damaging physical event, phenomenon, or human activity that may cause loss of life or injury, property damage, social and economic disruption, or environmental degradation; or (2) the probability of occurrence within a specified period of time and within a given area of a potentially damaging phenomenon. The types of natural processes that can cause damage are manifold and can be broadly grouped into hydro-meteorological and geophysical. Hazards are typically assessed using one of three approaches: heuristically (based on expert knowledge), statistically (based on evaluation of past events and causative factors), and deterministically (based on physically based process modeling).	Hazards vary tremendously in terms of spatial extent, duration, frequency, the presence of direct and indirect indicator, as well as the nature of the indicators (e.g., physical, chemical), with profound consequences for the utility of EO data in terms of their spatial, spectral, and temporal resolutions. The value of image data also strongly depends on the type of hazard assessment method. The main contributions of EO data are the identification of past hazard events (e.g., floods, landslides), and the characterization of relevant parameters (e.g., slope angle, land cover).
Disaster	A serious disruption of the functioning of a community or a society causing widespread human, material, economic, or environmental losses that exceed the ability of the affected community or society to cope using its own resources, and necessitating external assistance. Thresholds for what constitutes a disaster vary. For inclusion in the EM-DAT database one of the following are required: at least 10 fatalities, at least 100 people affected, or declaration of a state of emergency or a call for international assistance. We can further distinguish slow- and fast-onset events, and while disasters are typically of limited duration, some event types (e.g., drought, desertification, or even climate change) that can also be considered to be disasters can last years or decades.	Disasters come in many forms and lead to highly variable consequences. Some are physical and may thus be directly detectable in image data (see damage). Damage to systems and processes may be detectable through proxies (e.g., reduction of ships in a harbor. It is important to distinguish between an area being affected by a hazard event (e.g., through flood extent) and actual damage.
Risk	The probability of harmful consequences, or expected losses (deaths, injuries, property, livelihoods, economic activity disrupted or environment damaged), resulting from interactions between (natural, human-induced, or man-made) hazards and vulnerable EaR.	Risk cannot be directly detected from remote sensing data, though the principal factors determining risk—hazard and vulnerability—can be assessed (see entries for hazard and vulnerability).
Element at risk	Population, properties, economic activities, including public services, or any other defined values exposed to hazards in a given area, also referred to as "assets." The amount of EaR can be quantified either in numbers (of buildings, people, etc.), area, in monetary value (replacement costs, market costs, etc.), or perception (importance of elements at risk EaR).	EO data are very well suited to assess and characterize EaR, in particular physical ones. This includes using high spatial resolution imagery (optical images, radar, LiDAR) to identify and characterize (e.g., type, size, height) infrastructure elements (e.g., buildings, bridges), but also natural entities (wetlands, national parks, etc.). Systemic elements such as networks (road/infrastructure) can be identified based on the detection of their constituting components. Some not directly visible elements (e.g., economic activity) can be detected using physical proxies.
Vulnerability	Capacity for loss or damage to people, systems, or material assets, typically ranging from 0 (no loss) to 1 (complete loss). Since damage can be inflicted on physical EaR as well as systems and processes, different types of vulnerability exist. Typically considered in DRM are physical, social, environmental, and economic. However, the vulnerability of other systems also, such as political or institutional, can be considered. Vulnerability should always be assessed as a function of a given hazard type and magnitude.	Assessing how a physical entity or system will fare in a specific hazard situation is challenging, in particular considering a range of scenarios, and the utility of EO data is often limited. For different building types vulnerability curves are created based on the relationship of hazard intensities and observed damage. Image data can then be used, within limits, to identify buildings of a given type and vulnerability. For non-physical or process vulnerabilities physical proxies can be extracted from image data (e.g., physical state of neighborhoods, road networks, or model bottlenecks).
Resilience	"Capacity of a system, community or society potentially exposed to hazards to adapt, by resisting or changing in order to reach and maintain an acceptable level of functioning and structure" UNISDR (2004). At times resilience is seen as the inverse of vulnerability.	Direct measurement with EO data is not possible. However, by measuring recovery (both of physical assets and systems/functions) from image parameters, resilience of an affected area can be assessed.

(Continued)

DRM Parameter	Definition	Aspects That Determine the Use of EO Data
Damage	Impairment or harm inflicted on an element at risk by a hazardous event or phenomenon, and can be either physical or functional. For different hazard types different damage scales have been created, such as the European Macroseismic Scale (EMS-98), that classes building damage from D1 (none or negligible damage) to D5 (total collapse). Alternatively, a percentage for system-related damage also a percentage of capacity or performance reduction can be used.	EO data of many different types have been used to detect and characterize disaster damage, in particular for physical assets. Focus has been on building damage mapping, in particular with very high resolution optical satellite images, LiDAR, high resolution SAR, and airborne oblique imagery. Detailed, per-building assessment, especially on intermediary damage, remains difficult. EO are very suitable for other damage types (e.g., forests, wetlands), while methods continue to be largely lacking for systemic damage assessment.
Recovery/ reconstruction	Decisions and actions taken after a disaster with a view to restoring or improving the predisaster living conditions of the affected community, while encouraging and facilitating necessary adjustments to reduce disaster risk.	Image data are well suited to assess physical post-disaster reconstruction, also using semantic or ontological analysis methods to account for reconstruction that can occur in different ways. Assessment of system or process recovery (e.g., economic activity, ecosystem performance) with image data can be done via physical proxies, though fewer methods exist.

relate to magmatic or eruptive activity, while others are typical for dormant or extinct structures. Some of those specific or sub-hazards tend to affect a small area on the volcanic edifice itself (e.g., volcanic bombs or fumaroles), while others, such as gas and ash injected into the stratosphere, have global consequences. Some, such as seismic activity related to ascending magma, may last hours or days, while degassing can continue over decades or longer. A clear understanding of those specific processes is thus always needed before deciding on a suitable remote sensing strategy.

Remote sensing research has made immense progress at this more detailed level that is the basis for actual decision making and planning. For example, while flood hazards are typically assessed using 1D or 2D flood models (Kerle and Alkema, 2011), remote sensing provides critical data on land cover (to estimate evapotranspiration and surface roughness; Straatsma and Baptist, 2008; van der Sande et al., 2003), as well as on soil type (to allow estimation of water infiltration capacity), and topography (e.g., Kraus and Pfeifer, 1998). In particular, the combination of multispectral information with airborne LiDAR data has allowed both a detailed characterization of the vegetation in flood-prone area (which has a strong effect on flow behavior), as well as the underlying topography (Geerling et al., 2009; Straatsma, 2008). The same multiparameter assessment is done for wild fire hazard assessment and monitoring, as was reviewed by Chuvieco (2003). Remote sensing data are being used to assess the amount of burnable vegetation matter, its dryness, and the topography of the area, but meteorological information is also used to monitor atmospheric parameters such as wind and temperature that directly affect wild fire potential. In addition, LiDAR data, whose use for detailed vegetation mapping was already mentioned, also allow a better wildfire hazard assessment (Newnham et al., 2012). Radar data have also been shown to allow a detailed characterization of biomass, canopy height, or forest types (Balzter, 2001). As something of an exception for hazards, a comparably comprehensive assessment for seismic hazard is less straightforward. Again, for detailed planning the seismic movement is only the starting point: The actual hazard to people and infrastructure is determined by geology, soil type and thickness, and the topography. Remote sensing data have allowed the mapping and characterization of fault lines (Ramasamy, 2006), regolith thickness (Shafique et al., 2011), as well as detailed topography that informs the amount of site-specific ground amplification (Shafique et al., 2012; Tronin, 2010). In particular, interferometric SAR data have allowed the assessment of fault line dynamics and crustal deformation (Hooper et al., 2012). In a similar fashion, the relevant parameters for other hazard types are being assessed with remote sensing (see, e.g., Joyce et al., 2009; Kerle, 2013b).

Both flooding and (some) volcanic hazards are good examples of a distance effect that can be well addressed with remote sensing data. In both cases, but also for other hazard types, large distances may lie between the hazard source and the area where damage may be caused. The melt or rain water may originate in mountains far from settlements, yet pose a downstream flood hazard, as do very mobile mass movements, such as volcanic lahars that can reach distances of 50–100 km from the point of origin. Similarly, a remote volcano has the potential to alter weather patterns elsewhere on the globe, at times leading to a global temperature drop of 0.5°C or more, effects that can last for years. Remote sensing constitutes a vital information source to detect and characterize those hazard sources, and to link them to EaR. To that effect the information derived from remote sensing is frequently combined with models or other forms of GIS analysis. Table 18.2 summarizes the utility of remote sensing for a different part of the DRM cycle for a range of natural hazards.

18.3.3.3 Mapping of Elements at Risk

Anything of value, be it monetary, historical, sentimental, cultural, or otherwise, can get adversely affected by hazardous events or processes. Such EaR can thus be highly diverse, and range from the obvious—buildings or infrastructure—to features less

TABLE 18.2 Utility of Remote Sensing for Different DRM Aspects of Different Natural Hazard Types

Hazard Type	Subtype	Hazard Assessment	Prevention/Mitigation[a]	Monitoring/Early Warning	Syn-Event Monitoring	Damage Assessment	Recovery/Rehabilitation
Flood	Riverine	++ HRV, DEM, LiDAR	+ HRV, DEM, LiDAR	++ Met	++ HRV, MRV	++ (V)HRV	++ (V)HRV, LiDAR
	Flash flood	+ HRV, DEM	+ HRV, DEM	+ Met	−	+ VHRV	+ VHRV
	Coastal	++ MRV, HRV, DEM	+ MRV, DEM	+ Met	+ MRV	++ MRV, HRV	++ MRV, HRV
Tsunami[b]		−	−	−	−	++ MRV, (V)HRV	++ MRV, (V)HRV
Storm[b]	Regional (cyclones)	++ Met	−	++ Met	++ Met	++	++ MRV, HRV
	Local (tornado)	+ Met	−	+ Met, DR	+ Met, DR	++ (V)HRV, Obl	+ (V)HRV
Earthquake		+ SAR	−	+[c] SAR	−	++ (V)HRV, Obl	++ (V)HRV
Drought		++ IR, Met	+ IR, Met	++ IR, Met	++ IR	++ IR	+ IR
Volcanic	Magmatic activity	+ HRV, SAR, TRS	−	++ TRS	++ TRS	+ (V)HRV	+ MRV
	Lahar (mudflow)	+ HRV, DEM	+ HRV, DEM	+[d] Met	−	++ MRV, HRV, IR	++ MRV, HRV, IR
	Gas emission (local)	+ Hyp[e]	−	+ TRS, Hyp	+ Hyp	+[f] IR	+ IR
	Gas/ash emission (stratospheric)	+ Hyp	−	+ TRS, Hyp	++ Hyp	+[g] Met	+ Met
Landslide	Crater lake breach	++ VHRV, LiDAR	−	+ Met, HRV	−	++ (V)HRV	++ MRV
	Fast slope failures	++ HRV, DEM, IR	+ HRV, DEM	+ Met	−	+ HRV	+ HRV
	Slow-moving slides	++ SAR, HRV, SD, LiDAR	+[h] HRV, IR	+ Met	+ SAR, SD	+ HRV	+ HRV
	Re-vegetated slides	+ LiDAR	−	−	−	−	−
Wild fire		++ IR, DEM, Met	+ IR, Met	++ TRS, Met	++ TRS, Met, MRV	++ MRV, HRV, IR	++ MRV, HRV, IR
Desertification		+ LRV, IR, Met	+ LRV, IR	+ IR	−	+[i] LRV, MRV	+ LRV, MRV
Erosion		+ VHRV, LiDAR	+ VHRV	−	−[j]	+ (V)HRV	+ (V)HRV

Key: −, limited or no use; +, moderate utility; ++, high utility.

Remote sensing data: VHRV, very high resolution visible (spatial resolution better than 1 m); Obl, oblique imagery; HRV, high resolution visible (spatial resolution >1–10 m); SD, stereo imagery; MRV, moderate resolution visible (>10–30 m); LiDAR, light detection and ranging; LRV, low resolution visible (>30 m); Hyp, hyperspectral imagery; TRS, thermal imagery; Met, meteorological[k]; SAR, imaging radar, interferometric; SAR IR, infrared (typically a NIR band in optical sensors); DR, Doppler radar.

[a] Mitigation or prevention in a sense of remote sensing data allowing strategies or activities to be identified that reduce the hazard or the probability of an event occurring.

[b] Storms come in a large variety of forms and area affected, ranging from small water spouts to winter storms and tropical cyclones. Here two main categories in terms of area affected are discussed: tornadoes (local) and cyclones (which include tropical and extra-tropical/mid-latitude storms) with more regional range.

[c] Through SAR interferometry revealing surface deformation indication seismic stress build-up; actual remote sensing-based detection of earthquake precursors continues to be investigated.

[d] For example, through monitoring of crater lake volumes, rainfall amounts, or remaining amount of mobilizable ash deposits.

[e] For example, mapping of SO_2 concentrations with the hyperspectral (ultraviolet–visible range) Ozone Monitoring Instrument (OMI) on NASA's EOS AURA mission, or similar approaches based on absorption spectroscopy.

[f] Through detection of vegetation damage caused by acid rain; health damage cannot be detection in image data.

[g] Large amounts of volcanic gases injected into the stratosphere can cause global cooling, which can be considered a form of damage, and which can be measured indirectly with weather satellites. Recovery would then be a return to pre-event temperatures.

[h] For example, through soil moisture data allowing a better understanding of slide dynamics and triggers, allowing intervention and stabilization measures to be defined.

[i] In the sense of desertified areas.

[j] Erosion progresses both continuously and through individual, high intensity rainfall events. Remote sensing can only monitor changes in erosion prevalence at different times.

[k] Including from geostationary meteorological satellites and from specific weather-related mission such as the Tropical Rainfall Measuring Mission (TRMM).

frequently considered in a risk analysis, such as places of cultural or natural heritage, national parks, sites of high biological diversity such as some tropical forests or coral reefs, or even beaches or other coastal assets, such as cliffs. In particular, natural sites of value to the tourism industry are frequently taken for granted, with their significance, and susceptibility to damage, only being appreciated after a disaster strikes and income is lost (see also Liu, 2014). All of the preceding are primarily physical features. However, in DRM we can also consider valuable non-physical systems, processes, or functions that may get affected by hazardous events, be it cultural diversity, political systems, or economic processes. Not surprisingly, remote sensing has been predominantly used to map and characterize physical EaR, in particular buildings and infrastructure (Ferro et al., 2013). Those include physical assets exposed to flooding (e.g., Müller, 2013), or specific building types. Vertical optical satellite images have been useful for building footprint detection. To provide extra information, such as the height of a structure, shadow analysis has been used (Lee and Kim, 2013). Alternatively, data from instruments with stereoscopic capabilities, such as Ikonos, Cartosat-1, or GeoEye, can be processed photogrammetrically (Poli and Caravaggi, 2013). In particular the growing availability of LiDAR data has been a great asset for building detection, since it not only allows effective discrimination of vegetation that plagues other image processing methods, but also because it very readily allows the ground surface to be determined, resulting in building height (Sithole and Vosselman, 2004). Since LiDAR data tend to be very detailed, with modern systems collecting dozen of points per square meter, they provide useful information that allows characterization of building types (Alexander et al., 2009), or of entire urban landscapes, for example, to support informed flood risk assessment and management (Talebi et al., 2014).

Detecting specific building types and similar forms of investigation go in the direction of semantic analysis. For decades image processing was primarily focused on using spectral information, which gives insight into the material exposed on the ground. However, many hazard parameters, as well as EaR, must be defined—or indeed can only be fully recognized—from their context, that is, spectral information needs to be combined with other forms of information, be it topographic, topological, or contextual. In particular, the detection of many EaR that are not clearly defined physical units must be approached in such a fashion. For example, parameters related to biodiversity in forests have been identified with remote sensing data in such a manner (Torontow and King, 2011). Ontological frameworks that encode feature knowledge have also been used as a basis for image-based detection of specific urban units such a slums, which can also be considered as EaR in specific hazard scenarios (Kohli et al., 2012).

The success and utility of remote sensing is always a function of the data, and how they are processed. Detection and characterization of EaR in particular has been benefitting from a shift away from pixel-based processing toward object-oriented analysis (OOA, also called object-based image analysis, OBIA). Fundamentally a two-stage technique that combines image segmentation with subsequent classification, it is far more than that, with dramatic consequences for DRM, including the detection and characterization of EaR. The strength of OOA is that it allows an effective incorporation of process and features knowledge, different data types, and its flexibility in handling 2D–4D data (the latter being multitemporal data cubes). This has become an asset in the detection of specific hazards, such as different landslide types (Martha et al., 2010; Figure 18.4) or forms of erosion (Shruthi et al., 2011), but also to characterize EaR. For example, Kohli et al. (2013) used the ontological framework mentioned earlier to provide the knowledge basis to detect slum units using OOA. Such knowledge-driven approaches are also the key to detecting more complex EaR, such as more complex processes (level of economic activity, or the state of a transport system), though research is only starting to address these questions. OOA, for all its strengths, also has limitations. Segments can be formed in a number of ways, and >100 features and characteristics are readily calculated for those units by modern software tools, such as eCognition. However, which features (and in which combination) best identify a given feature or object class, what thresholds to use in the classification, etc., remain significant challenges. Here, computer vision and machine learning are becoming more influential, evident, for example, in the use of Random Forest approaches (e.g., Shruthi et al., in press-b; Stumpf and Kerle, 2011).

18.3.3.4 Vulnerability Assessment

In simple terms, vulnerability is defined as the capacity for loss, that is, how much damage a given EaR will sustain in a specific hazard scenario (type, magnitude, and duration). In practical terms, vulnerability is a more subtle phenomenon, one where remote sensing can also help, but within limits. The concept of vulnerability has its roots in the social sciences of the 1970s and is a response to the purely hazard-oriented understanding of disaster risk at that time. Vulnerability assessment has been primarily applied to physical features—buildings and infrastructure—and how those may get affected by hazards, due to physical forces exerted by ground motion, water, wind, etc. Vulnerability ranges between 0 (no damage) and 1 (total loss). Within the scientific community definitions of vulnerability, and optimal ways to analyze and quantify it, continue to be discussed (see, e.g., Birkmann, 2007; Galderisi and Ferrara, 2013). A number of characteristics make vulnerability assessment particularly challenging, which need to be considered when assessing the utility of EO data. (1) Vulnerability has different facets. The ones commonly considered today are physical, social, economic, and environmental (United Nations Office for Disaster Risk Reduction (UNISDR), 2004), yet an inclusion of other types, such as political or institutional, is equally valid. (2) Vulnerability is dynamic, hence changes over time. For example, as a building ages its vulnerability to certain hazards can increase, while a human being, as it grows from child to adult, may become less vulnerable to certain environmental forces. (3) Vulnerability is strongly scale-dependent, meaning that it can, and often must, be assessed on a scale that ranges from the individual (building, person,

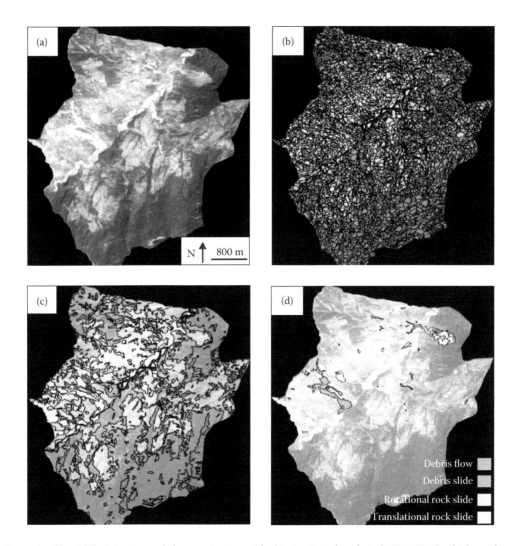

FIGURE 18.4 Example of landslide detection and characterization with object-oriented analysis (OOA/OBIA), which combines image segmentation and feature- or process knowledge-driven analysis of the resulting segments. Image (a) shows part of a Resourcesat-1 image (5.8 m multispectral resolution) of the Indian Himalayas (Okhimath) that experienced a variety of landslide types. The image was first finely segmented, with (b) and (c) showing the effect of different segmentation scales. Following the identification of all bare areas that may be landslides, a number of false positives (e.g., shadow, river sands, and clear-cuts) were sequentially eliminated based on image spectral and textural characteristics, as well as information from a DEM. The resulting landslide objects were classified, again using knowledge of their different spatial and contextual characteristics, into different landslide types (d). For more details see Martha et al. (2010).

road, etc.) up to a city/country, community, or network level. (4) Vulnerability is highly dependent on the hazard type. The same building that might withstand flooding well (low vulnerability) might suffer extensively during an earthquake (high vulnerability). Hence, in principle every EaR has as many vulnerabilities as the number of hazard types it is exposed to. Spatial Multi Criteria Evaluation is often used in vulnerability assessment, providing a way to include different vulnerability indicators. Those can be physical (e.g., building material), social (e.g., ethnic or age distribution), or related to capacity (e.g., distance to hospital, risk awareness levels). Those indicators are weighted, using a pairwise criteria comparison, arriving at a vulnerability index (for more details see van Westen et al., 2011).

To further complicate matters, the natural counter-weight to vulnerability is resilience, which must be known for a comprehensive analysis (Galderisi and Ferrara, 2013). For communities this relates to the strengths and resources available that can collectively reduce the effect of a hazardous event. Inherently, resilience relates primarily to social, political, and organizational parameters. For a more in-depth introduction to vulnerability, and its assessment with geospatial data, see van Westen et al. (2011).

The preceding explains why in geoinformation science, and in remote sensing in particular, the main focus to date has been on physical vulnerability. The key to its assessment lies in so-called vulnerability curves that express the relationship between hazard intensity and the resulting damage. (Note the focus on intensity; as stated in Section 18.3.3.2 the actual magnitude of an event, e.g., an earthquake, is only the starting point. What matters in risk assessment is the actual location-specific hazard intensity.) Vulnerability curves are constructed for different types of structures (e.g., adobe brick buildings vs. reinforce concrete structures of different heights). We also distinguish between relative curves, which indicate what percentage of the value of a structure will be lost, and absolute curves that show the actual losses, since the value of the assets being considered is already incorporated in the analysis. Hence, from a remote sensing perspective physical vulnerability assessment is largely reduced to identifying EaR and determining their respective category, in terms of type, material, etc. For example, see studies by Dell'Acqua et al. (2013) and Ehrlich et al. (2013). The value of OOA for EaR detection was already described. The concept is also useful to assess the physical vulnerability of the detected features (e.g., Wu et al., 2014).

Other types of vulnerability are no less important in comprehensive risk assessment, yet less progress has been made in terms of remote sensing-based methodologies. This is primarily due to the fact that a direct detection is typically not possible; instead, physical proxies must be employed. For example, social vulnerability can be defined as "people's differential incapacity to deal with hazards, based on the position of the groups and individuals within both the physical and social worlds" (Clark et al., 1998), which allows the definition of relevant physical indicators that can explain the social vulnerability of people living in a given place. The possibility of such an assessment based on satellite images, and also making use of OOA techniques, was shown by Ebert et al. (2009). Similar approaches can also be used for environmental vulnerability assessment (see, e.g., Petrosillo et al., 2010; Poompavai and Ramalingam, 2013). To assess to what extent complex phenomena, such as economic or political systems, can be affected by hazardous processes, remote sensing data will be of less use; instead, more domain-specific modeling must be employed (such as macro-economic agent-based modeling to understand linkages in economic systems, and how adverse effects might spread within them; e.g., Kromker et al., 2008).

18.3.3.5 Monitoring and Early Warning to Detect Potentially Hazardous Situations

Once hazards are understood in terms of their spatiotemporal characteristics, efforts can be made to mitigate them where possible (e.g., by giving rivers more flood plains to spread into when needed, re-establishing mangrove forests for coastal protection, or by removing dry vegetation close to buildings to create a wild fire buffer). Another strategy is to monitor hazards. Due to the continuity provided by EO systems—as well as the availability of an archive for reference and benchmarking purposes—their data are exceptionally well suited to keep a permanent eye on many potentially hazardous situations, and provide advance warning in case of a looming threat. For every hazard type monitoring, methodologies have been developed, and for nearly all of them remote sensing plays a major role. An exception is tsunamis, where buoy systems installed on the ocean floor are used. Also the image-based real-time monitoring of seismic activity remains limited. While interferometric SAR can be used

to monitor surface deformation that can be related to seismic potential (Hooper et al., 2012), research is still ongoing to identify precursor signals of earthquakes that can be extracted from remote sensing data (Tronin, 2010).

The principal challenge of hazard monitoring is to understand the natural variability of parameters linked to hazards, to be able to spot unusual developments. For example, keen process understanding is needed to pick up early drought indicators from within a potentially complex phenological cycle of a crop. Existing agriculture monitoring schemes based on remote sensing were recently reviewed by Atzberger (2013). Such data play a critical role in combating hunger in the severely drought-affected parts of the world, especially the region around the Horn of Africa. AVHRR data provide daily information, the use of which to link drought and crop yields was already shown many years ago by Unganai and Kogan (1998). Other systems that provide very frequent information of use for detailed vegetation state mapping are MODIS and SPOT vegetation. Therefore, remote sensing is a core ingredient in operational drought monitoring systems such as FEWS (Famine Early Warning System Network; http://www.fews.net/), that has been in operation already for 30 years. Remote sensing data are the key to successful monitoring of a variety of other atmospheric hazards, due to the long history of such instrumentation, the strength stemming from highly frequent geostationary observation (nearly 100 scenes per day of Africa and most of Europe from the current Meteosat satellite), coupled with better than daily and more detailed images from polar orbiters (e.g., AVHRR) and land-based instruments (Doppler radar), and rapid and wide availability of the data. As a consequence, in most parts of the world the developments of broad atmospheric hazards can now be monitored. Countries that also possess a detailed network of ground stations can go beyond that and also assess the developments of more local and short-lived phenomena such as tornados or flash floods. Indeed it was remote sensing—the study of Doppler radar data—that nearly 40 years ago allowed the signature fingerprint of developing tornadoes to be identified (Ray and Hane, 1976), paving the way for modern early warning systems.

The suitability of remote sensing is tightly coupled with the specific characteristics of a given hazard. In broad terms, volcanoes are probably the most rewarding hazard setting for remote sensing to be applied to. This is because nearly all of the many sub-hazards previously mentioned emit some form of measurable signal prior to the commencement of a threatening process. The consequent potential for EO technology was quickly noticed, and concepts for satellite-based volcano surveillance appeared soon after the launch of the first civilian satellite, ERTS-1 (Cochran and Pyle, 1978; Endo et al., 1974). The major detectable signs are (1) seismic activity that signals magma movement within the edifice, (2) thermal signals on flanks or in crater lakes that indicate magma approaching the surface, (3) topographic changes due to magma movement or structural deformation that may lead to mass movements, and (4) gas emissions. As such, virtually all types of ground-, air-, and spaceborne EO instruments at our disposal today have been successfully deployed on volcanoes (for a more detailed review see, e.g., Joyce et al., 2009; Tralli et al., 2005). In particular, satellites with hemispheric or global coverage are being used for operational volcano monitoring. For example, the MODVOLC system (http://modis.higp.hawaii.edu/) operated by the University of Hawaii has been providing global volcanic hotspot information based on MODIS imagery already since 2003. Other researchers have focused on the combined strength of different systems to optimize monitoring (Cochran and Pyle, 1978; Murphy et al., 2013). Satellite data are also a principal input to Volcanic Ash Advisories issued by the International Civil Aviation Organization to alert aircraft pilots to volcanic aviation hazards stemming from eruption clouds. Examples of recent remote sensing developments that have been benefitting the monitoring of volcanoes include Permanent Scatterer SAR Interferometry (PSInSAR) for ground deformation detection, and the deployment of unmanned aerial vehicles (UAVs/Drones; e.g., Williams, 2013).

18.3.3.6 Post-Disaster Response and Damage Assessment

Despite the many efforts to mitigate hazards, prepare for disasters and to provide early warning, the reality shown in Figure 18.1 is that we need to deal with several hundred disasters worldwide every year. Due to site accessibility problems, potentially widespread damage, and health hazards that damage sites themselves pose for first responders, remote sensing has become an indispensable tool to provide first intelligence on the nature and consequences of a disaster. Given the very long history of image-based damage assessment that stretches back to Lawrence's efforts in 1906 (e.g., Figure 18.5a), and the many potentially useful tools and methods available, a very rich body of literature exists on this topic. Without a doubt decades of research have led to significant developments. At the same time a more sober assessment would lead to the following conclusion: remote sensing-based damage mapping has seen the least progress of all the DRM aspects reviewed in this chapter. To phrase it more drastically: in some ways image-based damage mapping is done as it was in 1906 (Figure 18.5).

This statement requires qualification and explanation. Lawrence flew his kites at a time when no airplanes, let alone satellites, existed. In terms of technology of course there has been enormous progress, both on the sensor and platform side, and we have seen dozens of combinations of the two. Virtually any type of sensor in existence, be it optical, thermal or radar, mono or stereo, active or passive, has been fitted to satellites of different types, to airplanes large or small, or flying at low or stratospheric heights, or to balloons, kites, or UAVs, in short any platform that can be brought into the air or space, and aimed at a disaster site, the typical operational altitude range for some of which is shown in Figure 18.6. The use of air- and space-based remote sensing for emergency response and damage mapping was extensively reviewed by Kerle et al. (2008) and Zhang and Kerle (2008), respectively, showing a rich palette of possibilities, as well as technical advances. This technology provides us with spectrally diverse images of high spatial resolution. Indeed, very significant progress has been made in terms of temporal and

spatial resolution, and current delays to initial image acquisition after an event are measured in hours, instead of days or week. Nevertheless, those high quality images continue to be primarily analyzed manually.

Several issues require further explanation. First, we must be clear on what we mean by disaster response and damage mapping. Detecting disaster sites as anomalies, as a departure from the normal, tends to be quite straightforward, in particular when suitable pre-event reference data exist, and when the event leads to significant physical changes on the ground. However, caution must be used to separate the evidence of an event from evidence of damage. For example, water extent provides a clear delineation of the reach of a flooding event, and can be automatically detected in image data, yet it is not synonymous with damage. It makes sense here to distinguish between something being affected (by flooding, wind, etc.) and being damaged. From a damage assessment perspective, it is also important to know if damage is permanent or not. Crops that suffer short-term flooding may survive, experience partial loss, or be completely destroyed. Likewise, we can readily identify the extent of wild fires in remote sensing data. However, how specifically the vegetation fared, for example, if only brush was lost and trees largely survived, requires a more detailed analysis, or subsequent imagery to track recovery.

When we talk about disaster damage mapping, almost invariably we talk about structural damage assessment, which is precisely where the main challenges lie. Ground-based structural damage mapping frequently makes use of damage categorizations such as the European Macroseismic Scale of 1998 (EMS98; Grünthal, 1998), and image-based assessment tries to emulate that process. In terms of vagueness of the damage signal, the damage scale that ranges from D1 (no damage) to D5 (complete destruction) shows a Gaussian distribution: the extreme ends, D1 and D5, are comparatively easy to detect, provided the spatial resolution of the image is suitable with respect to the average size of the structural objects. Hence, for places that suffered blanket destruction, and where damage is quite well delimited from intact areas, such as caused by strong tornados (Figure 18.7), even automated damage mapping is possible with good accuracies (see, e.g., Brown et al., 2012).

However, for sites characterized by more intermediate damage states, automatic assessment methods quickly find their limits. This is true for virtually any image type that has been tried, be it airborne TV footage (Mitomi et al., 2000), radar data (see Arciniegas et al., 2007; Uprety et al., 2013 for results with coherence and intensity data, respectively), or various types of vertical optical images (e.g., Ehrlich et al., 2009). Image-based damage mapping has been a particular challenge for earthquake sites, and the state of the art was recently reviewed by Dell'Acqua and Gamba (2012), and also critically assessed by Kerle (2010). A widespread assumption in the remote sensing community has long been that better data, in particular with increasing spatial resolution, would allow more detailed and more accurate analysis, yet that has only been partly the case. Following the 2010 Haiti earthquake damage analysis was first based on Geoeye-1

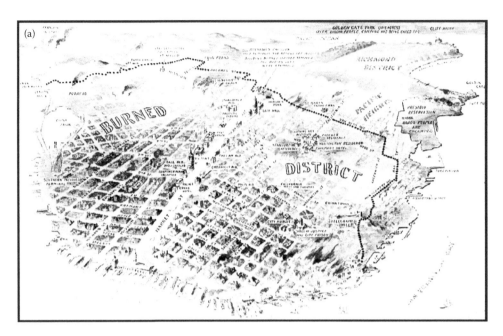

FIGURE 18.5 Damage map of San Francisco following the 1906 earthquake, based on kite-borne imagery by George R. Lawrence (a). Damage map of Port-au-Prince, Haiti, following the 2010 earthquake, prepared by SERTIT by visual interpretation of 50 cm resolution GeoEye imagery (b).

images (0.41 m panchromatic resolution) and repeated later using airborne images with approximately 15 cm resolution. This difference led to the amount of damage being identified to increase by a factor of 10 (see Gerke and Kerle, 2011; van Aardt et al., 2011). Overall, however, even such high quality vertical image data have turned out to be of limited utility for structural damage mapping, since they provide a very limited perspective (mostly only of the roof) of what is in reality a very complex 3D scene. Hence, analysis strongly relies on proxies such as blowout debris, or changes in shadow or building position, the latter being of particular use when reference images exist (Kerle and Hoffman, 2013). In recent years, high quality multiperspective oblique imaging solutions have become available, such as

the Pictometry© system. The camera system acquires images in five directions (nadir and the four cardinal directions at a 45° angle), in principle providing a very comprehensive view of the disaster scene (Figure 18.8). The images are also acquired in stereo, allowing for photogrammetric processing. Pictometry data of Port-au-Prince following the 2010 earthquake were used by Gerke and Kerle (2011) to map damage, in an approach based on OOA and machine learning. Damage indicators were classified with accuracies between 60% and 70%, values comparable with what studies using other remote sensing have found. However, the work has confirmed a problem: Most research efforts on image-based structural damage mapping are experimental and tend to work well on specific test data but have insufficiently

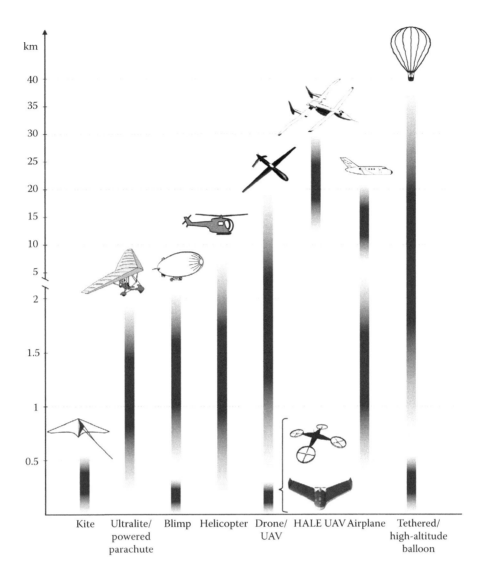

FIGURE 18.6 Overview of approximate operating altitude ranges of different airborne platforms (shaded bars). For a number of platform types, such as blimps or balloons, both low altitudes (typically tethered) and sophisticated polyurethane ones that reach stratospheric heights exit. Similarly, low-flying airplanes can carry out surveys at altitudes of approximately 1–5 km, while other planes, in particular military surveillance aircraft, can fly as high as about 21 km. Drones/UAVs have seen the most dynamic development in recent years. Those, too, include high-flying, mostly military, versions, and numerous low-flying types that can be classed into fixed-wing and multirotor devices. HALE UAVs focus on both high altitude and long-term autonomous deployment. (Modified from Kerle, N. et al., Real-time data collection and information generation using airborne sensors, in Zlatanova, S. and Li, J., eds., *Geospatial Information Technology for Emergency Response*, Taylor & Francis, London, U.K., 2008, pp. 43–74.)

proven ready transferability to other sites and data types (see also Booth et al., 2011). This explains why organizations charged with rapid post-disaster damage mapping, for example, by mandate of the International Charter "Space and Major Disasters" (www.disasterscharter.org), continue with manual damage mapping. While those efforts lead to maps of high cartographic quality, map styles and data accuracy continue to be quite variable (Kerle, 2010).

The ultimate reason for the limited progress in image-based damage mapping can be readily identified: images address damage as a physical entity; however, damage is better thought of as a complex concept with different facets. While damage can justifiably be seen as a physical state, where damage can be depicted in absolute or relative terms with respect to the entire structure or the pre-event state, this says little about internal damage and, more significantly, even less about its functional damage. A nuclear power station may appear to be structurally largely intact yet may leak radiation, while a hospital that suffered disruption to its electricity or water supply may be structurally sound, but severely incapacitated in functional terms. More work is needed to semantically link physical external damage indicators to functional capacity.

FIGURE 18.7   Subset of a GeoEye image of Washington, Illinois, that suffered severe damage during a series of tornados on November 18, 2013. Damage patterns are nearly binary (no/little damage versus complete destruction), typical for tornados.

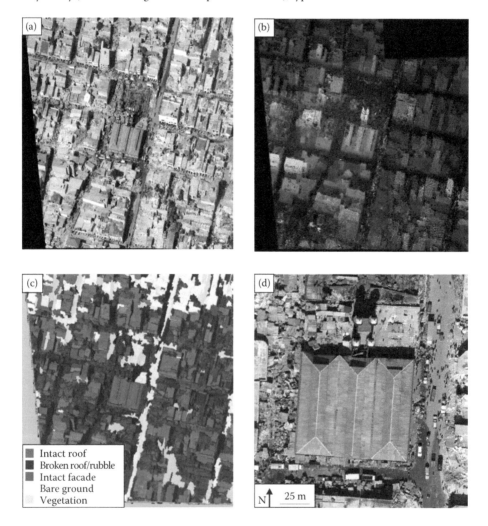

FIGURE 18.8   Oblique Pictometry image of part of Port-au-Prince, Haiti, following the 2010 Earthquake (a), derived depth information that shows geometric homogeneity (b), classified damage features based on segmentation-based supervised classification and machine learning (for details see Gerke and Kerle, 2011; c), and a vertical image of part of the area to show the limitations of only that perspective (d). (Images a and c ©Pictometry International Corp.)

18.4 Trends and Developments

In the final section of this chapter, a number of noteworthy developments with relevance to the use of remote sensing data in DRM will be assessed, covering both technical and organizational aspects.

18.4.1 New Platforms

Throughout most of its history, remote sensing has been a high-tech and costly endeavor, limiting it to a few, mostly government, players. When commercial opportunities became apparent, the private sector joined in, though costs of infrastructure and operation remained high, both in air- and spaceborne remote sensing. A number of significant changes have taken place. As mentioned before, many countries have launched their own satellites, owing in part to the development of micro- and nano-satellites that combine good performance with relative affordability, which also paves the way for novel intelligent networks (Sandau et al., 2010). Some existing satellites already form virtual constellations (e.g., the Disaster Monitoring Constellation, DMC, managed by DMCii; http://www.dmcii.com/), which increase the image temporal resolution the participating countries can take advantage of, also providing for more rapid response in case of a disaster. Indeed, one of the DMC satellites (NigeriaSat-1) was the first to provide high-resolution spaceborne images of the area affected by Hurricane Katrina in 2005 (Salami et al., 2010).

The other significant platform development is that of UAVs. On one hand this closes the circle back to Lawrence's kites, allowing nearly anyone do-it-yourself remote sensing. The other end of the UAV spectrum leads closer to the stratosphere, with the development of high elevation–long endurance (HALE) instruments. While some technical and legislation issues still need to be sorted out, the idea from a DRM perspective is that such instruments, as part of a network, can either provide continuous, high spatial resolution coverage of significant places (large metropolises, etc.) or can be relatively rapidly deployed to a disaster area to provide comprehensive and continuous coverage of the post-event situation. Also, the utility of small UAVs that fly at low elevations (a maximum of 100–200 m above the ground) continue to be hampered by regulations, or often rather the lack of them. In many countries it is simply not allowed to operate them, though pressure from industry and the many emerging potential users will likely force a chance to that (for an overview of current regulation in Europe with regards to deployment of UAVs see http://uvs-info.com). UAV-based imagery of areas of seismic damage has recently been processed with OOA methods that made use of detail-rich 3D point clouds, coupled with the oblique image data (Fernandez Galarreta et al., in press), demonstrating well the potential of this new class of platforms for damage mapping (Figures 18.9 and 18.10). Also, other DRM areas have started to see the benefit of UAV instruments, for example, for glacier change mapping and landslide monitoring (Niethammer et al., 2012). Ongoing developments in miniaturization, autonomous navigation, and swarm intelligence mean

that small UAVs will soon be deployable *en masse* for disaster site surveillance, including the ability to enter structures to assess damage and search for victims or survivors (Kuntze et al., 2012). Table 18.3 provides a detailed overview of recent and upcoming novel platforms useful for DRM.

18.4.2 New Sensors

A number of recent and upcoming sensors are of interest from a DRM perspective. While new and more advanced optical satellites continued to get launched with some regularity since Landsat 1, the situation for radar instruments was different. For one, their number has always been much lower, and after an initial phase in the early 1990s where several instruments were launched (e.g., ERS-1 [1991], JERS-1 [1992], ERS-2 [1995], RADARSAT-1 [1995]), follow-up missions were scarce. A noteworthy but quite isolated new radar mission was ASAR on ENVISAT (launch in 2002, and operating for 10 years). Only more recently has spaceborne radar seen a renaissance, starting with the launch of PALSAR on the ALOS satellite (2006, working until 2011), and followed by missions such as COSMO-SkyMed (constellation with 4 radar satellites; 2007), TerraSar-X (2007), RISAT-2 (2009), and TanDEM-X (2010). In particular, the high spatial resolution data from the TerraSAR-X and TanDEM-X couple that reaches 1 m in Spotlight mode have been shown to be useful for structural damage mapping (Ferro et al., 2013; Uprety et al., 2013). The data will also be used to create a new global digital elevation model (WorldDEM) that is expected to have a resolution of 12 m × 12 m, a substantial improvement on currently available ASTER-based GDEM and the SRTM dataset. Yet another radar instrument flies on a platform that is likely going to become part of a premier constellation: Sentinel-1, a satellite that carries a C-band SAR instrument, was launched in 2014 (Torres et al., 2012). It is to be followed by five additional Sentinel spacecraft. The constellation will form the core of the European GMES program (see Aschbacher and Milagro-Perez, 2012).

Hyperspectral remote sensing, in particular spaceborne, has always been a niche technology. Experimental missions such as Hyperion (flying on EO-1 since 2000), the CHRIS instrument on PROBA (launch in 2001), and the short-lived Indian Hyper Spectral Imager (2008) were later joined by the Chinese HJ-1A instrument, and demonstrated the utility of such data. Only in recent years have developments on more mature and operational hyperspectral systems started. In 2015, the Italian PRISMA mission is expected to be launched. The German-led EnMAP mission is to provide 30 m resolution data following a launch planned for about 2017. Around the same time (with the launch date also having been moved back several times) the HyspIRI mission is to be launched. Those data are useful for a wide range of DRM applications, such as volcanic hazard assessment (gases, ash, etc.; Abrams et al., 2013), or urban mapping (e.g., Bhaskaran et al., 2004).

An original new addition to the sensor-platform menu is real-time spaceborne video, offered by a company called Skybox. The micro-satellite acquires high-resolution optical images, but in addition also high-definition video clips of maximum 90 s, with pixel resolutions better than 1 m. Such video streams have the

FIGURE 18.9   UAV-based photo of a church in Maranello, Italy, destroyed in a 2012 earthquake (a; © Aibotix Italy), and detailed textured point cloud (b).

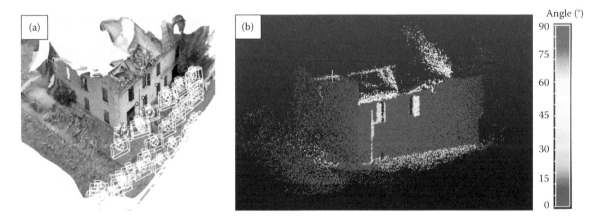

FIGURE 18.10   Textured 3D point cloud of a building damaged by an earthquake constructed from a multitude of UAV and ground-based images (Camera positions shown in white in (a)). The model provides detailed geometric information (b), such as angle information (vertical sections in blue), or rubble piles (orange).

TABLE 18.3 Selected Recent and Upcoming Novel Platforms of Use for DRM

Platform Type/Name		Maker/Operator	Description	Technical Parameters	Utility for DRM	Limitations	Notes	Example Literature
Airborne	Pictometry©	Pictometry, Inc.	Piloted aircraft with five cameras; since 2002	One nadir camera, four oblique at 45° in each cardinal direction). Typical flying height: 1000 m, resulting in stereo imagery of approximately 15 cm. Images georeferenced, visible range only.	Detailed vertical and oblique imagery of a structure from five perspectives, allowing structural EaR to be characterized including in 3D.	Expensive data; manned aircraft with limited survey flexibility; Limited spatial resolution for detail.		Booth et al. (2011), Gerke and Kerle (2011)
	HALE UAV	AeroVironment (Global Observer)	High-flying UAV; since 2010	Approx. 160 kg in weight; flying at max. 20 km for up to 7 days; image spatial resolution of ca. 30 cm); providing real-time image data downlink; solar-powered.	Highly valuable for sustained syn- and post-disaster surveillance of extensive and/or dynamic events, such as wild fires or flooding.	Expensive system, predominantly developed for military applications.	HALE UAVs have been in development since about 2000; legislation remains a challenge.	Van Achteren et al. (2013)
	UAVs	Various	Fixed-wing or multirotor platforms, ranging from ca. 3 g to several kg	Dozens of types available, including experimental devices (e.g. Delfly; at <4 g including a camera), inexpensive platforms with limited performance, and survey-grade devices with live-data downlink and carrying capacity of 1–2 kg. Stereo data can be processed photogrammetrically.	Excellent for detailed multiperspective building and infrastructure damage assessment; experiments on-going with UAVs entering damaged structures. Ready repeat acquisitions for change detection.	Legislation varies greatly among countries. Civilian UAVs often limited to a total weight of 5 kg and a maximum flying height of 200–300 m; battery capacity of multirotors limited, allowing only local surveys.	Most dynamic remote sensing segment, with many solutions available; rapidly growing maturity. Processing of uncalibrated stereo imagery readily possible.	Fernandez Galarreta et al. (in press), Niethammer et al. (2012), Hinkley and Zajkowski (2011)
Spaceborne	Micro-/nano-constellations	DMC	Satellites made by Surrey Satellite Technology Ltd (SSTL), owned by individual countries	Microsatellites owned by Algeria, China, Nigeria, Thailand, Turkey, United Kingdom, Vietnam	Different satellite generations have been produced, starting with AlSAT-1 (Algeria) in 2002 (32 m resolution), and similar instruments for Turkey and Nigeria. Beijing-1 (launched in 2005) added a 4 m pan band. Most recent satellite (NigeriaSAT-2; 2011) added a 2.5 m pan, and 5 m multispectral band. New constellation of 3 satellites (DMC-3; 1 m pan, 4 m ms) is being planned.	DMC satellites are noted for their moderate spatial resolution, but very extensive swath width (up to 650 km), allowing most parts on Earth to be imaged daily, ideal for extensive disasters. High-resolution instruments offer lower coverage but more details.	High data costs (marketed by DMC International Imaging Ltd. (DMCii)).	DMC satellites provided the first satellite data of areas devastated by the Asian tsunami (2004) and Hurricane Katrina (2005).	Sandau et al. (2010), Wu et al. (2010)

(*Continued*)

TABLE 18.3 (*Continued*)   Selected Recent and Upcoming Novel Platforms of Use for DRM

Platform Type/Name		Maker/ Operator	Description	Technical Parameters	Utility for DRM	Limitations	Notes	Example Literature
	Sentinel	European Space Agency	Six Sentinel missions with two satellites each planned (Copernicus EO program)	Different mission focusing on land, ocean, and atmosphere, to be launched successively until ca. 2021. Includes SAR instruments (Sentinel-1; launched in 2014), high resolution optical instruments (S-2), altimeters for ocean studies (S-3/6), spectrometers and sounders for atmospheric studies (S-4/5).	Sentinel was designed for comprehensive environmental monitoring purposes, with double-satellites ensuring revisits of 6 days (or better). Many risk and disaster-related problems can be addressed with Sentinel.	Sentinel is not specifically dedicated to DRM.	Copernicus is the new name for the Global Monitoring for Environment and Security (GMES) program that started in 1998. Sentinel data access will be free of charge.	Torres et al. (2012)
TerraSar-X and TanDEM-X		German Aerospace Center (DLR)	Radar twin-satellite system for high-resolution DEM generation	Satellites flying as close as 250 m to each other, working to create WorldDEM, a global DEM with up to 12 m resolution; satellites provide SAR data or up to 0.25 m resolution (in Staring SpotLight (ST) mode).	High resolution SAR imagery is useful for hazard assessment (e.g., EaR, floodplain characterization); Topographic information from WorldDEM will be very useful for process modeling (flooding, landslides).	Expensive data (15 km² 0.25 m resolution data for 7000 Euro).		Ferro et al. (2013), Rossi and Gemhardt (2013)
Skybox		Skybox Imaging, Inc.	Satellite-based video acquisition; since 2013	Video clips of up to 90 s duration, 30 fps, spatial resolution of up to 1.1 m, field of view of 1.1. × 2 km; constellation of 24 satellites planned; <1 m resolution images also acquired.	Potentially useful to provide coverage of dynamic post-disaster situations.	Only 90 second video clips; acquisition limited by orbital characteristics.	Partnership with Google.	(No scientific literature yet)

TABLE 18.4 Overview of Selected Recent and Upcoming Sensors of Relevance for DRM for Details on TerraSAR-X/ TanDEM-X, and Sentinel, Also Relevant from a Sensor Perspective, Table 18.3

Sensor Name	Operator, Launch	Sensor Type	Technical Specifications	Utility for DRM	Limitations	Example Literature
COSMO-SkyMed	Italian Space Agency; 4 missions, launched between 2007 and 2010); COSMO second generation being developed (two SAR satellites, launch from 2016)	SAR	X-band instruments (3.1 cm); constellation complete since 2011, resolution varying from 15 to 100 m, depending on mode; field of view ranging from 10 × 10 km to 200 × 200 km	Suitable for landslide and ground deformation studies, volume estimation (e.g., of landslides), damage mapping, volcanic hazard, and post-eruption studies (e.g., flow emplacement)	Latitudinal coverage of only ±20°–60° (Mediterranean focus)	Bovenga et al. (2012), Covello et al. (2010)
RISAT-2	Indian Space Research Organisation (ISRO); 2012	SAR	C-band (5.6 cm), spatial resolution up to 3 m, depending on mode, swath width ranging from 30 to 240 km	Data are useful for hazard assessment, element at risk mapping, flood mapping, estimation for near-shore bathymetry		Hegde et al. (2009), Mishra et al. (2014)
PRISMA	Italian Space Agency/ Temporary Industrial Group; 2015 (expected)	Hyper-spectral	Precursor mission; hyperspectral data with 20–30 m, pan data with 2.5–5 m resolution	Useful for hazard studies where detailed spectral information or surface materials is needed (e.g., volcanology, wild fire)	Only a precursor for future operational missions	Sacchetti et al. (2010)
EnMAP	German Space Agency; 2017 (expected)	Hyper-spectral	VNIR and SWIR coverage; 30 m resolution, 30 km swath width; steerable for off-nadir acquisition, 4 day temporal resolution	Same as for PRISMA; in addition more useful for post-event imaging due to higher temporal resolution		Stuffler et al. (2009)
HyspIRI	NASA; 2017 (expected)	Hyper-spectral	VNIR-SWIR sensor, as well as mid-TIR instrument, with temporal resolution of 19 and 5 days, respectively, and 60 m spatial resolution	Very suitable for studies of volcanic, wild fire and drought hazards		Abrams et al. (2013)

potential to become a valuable tool in disaster response scenarios, and the company plans to operate a constellation of 24 such satellites (see www.skyboximaging.com). Table 18.4 summarizes recent and upcoming new sensors of relevance for DRM.

18.4.3 Better Data Analysis Methods

Many methodological advances have already been mentioned in previous sections, such as OOA, photogrammetry with uncalibrated imagery, or machine learning, and a very long list of technical advances could be named here. Data analysis, however, has also been improved in non-technical ways. Just like multicore or distributed computing has accelerated data processing, people are also being increasingly made use of to support data analysis. Crowdsourcing, or citizen sensing, has become a broad area of many subtypes where people, typically, but not exclusively, lay persons, actively or passively provide geographical information. From a DRM perspective such Volunteered Geographic Information (VGI; e.g., Goodchild, 2007) offers many useful contributions, ranging from better base data (e.g., via Google Map Maker or OpenStreetMap) to better post-disaster response. The latter has been benefiting from people located in affected areas providing information, including imagery, of the disaster site. Other efforts have pooled remote sensing expertise by

getting analysts to perform rapidly a shared image-based structural damage assessment. Following the 2010 Haiti earthquake, more than 600 volunteers with remote sensing expertise participated in what became known as the Global Earth Observation-Catastrophe Assessment Network (GEO-CAN; see Ghosh et al., 2011). While this allowed an area of >1000 km^2 to be mapped in detail within a few weeks, the effort raised questions such as how best to instruct volunteers for this type of work, how to get them to map consistently as a group, and how best to process and validate their contributions (Kerle and Hoffman, 2013).

18.4.4 Organization

Disasters create empathy, especially with larger events creating much visibility and an international response. While at a practical level much has been done to allow fast response of search and rescue teams, higher organizational levels have seen strong improvements. The already mentioned Disaster Charter is the most prominent example, which since 2000 has been providing rapid post-disaster support based on satellite imagery. The protocols provide for data acquisition on a priority basis, and drawing on the space assets of virtually all space agencies, but also commercial entities such as Digital Globe and GeoEye. The data from the more than 300 Charter activations to date are primarily processed (and, as mentioned before, largely in a manual manner) by three main agencies, UNOSAT, DLR-ZKI, and SERTIT (for details see Kerle, 2013b). The Charter has been a successful example of a global effort to use collective resources to address disaster events. However, we are still faced with a situation where nearly a dozen organizations are now routinely engaging in image-based damage mapping, leading to duplication of efforts, and creating an overwhelming amount of damage map products that becomes counter-productive (see also Kerle, 2011; Voigt et al., 2011).

In addition to the Charter other global efforts are working toward a better use of EO assets for DRM. The United Nations Platform for Space-based Information for Disaster Management and Emergency Response (UN-SPIDER) operates a Knowledge Portal that aims at providing background information, as well as guidance and best practice information on how best to use (spaceborne) remote sensing for disasters. In that context training sessions are organized, and national governments advised on how best to work toward remote sensing-based DRM.

The Sentinel Asia program also focuses on coordinating the use of existing space infrastructure, together with auxiliary spatial data, for DRM, rather than promoting the development of new instruments. It is coordinated by JAXA, the Japan Aerospace Exploration Agency.

18.5 Gaps and Limitations

Since the early days of spaceborne remote sensing in the 1960s, we have witnessed a multitude of exciting developments that continue to provide us with more and better data that benefit every aspect of DRM, and ever more rapidly. The preceding sections are an attempt to show how technical advances, deeper insights in disaster risk as a complex phenomenon, but also better international cooperation, have led to a situation where remote sensing has morphed into a vital instrument in our efforts to understand and reduce risk, but also to warn of events and respond effectively where they do happen. In the process much has been achieved, and gaps have been reduced or closed. Nevertheless, a number of limitations remain that are briefly discussed in this section.

18.5.1 The Military Side

Military interests can safely be seen as the driving force behind all significant remote sensing developments, usually with military developments later leading to civil adaptations. Hence, the many attempts and efforts originating from the military/ national security community—both successes and failures— over decades have led to today's arsenal of remote sensing instruments and tools that benefit DRM in countless ways.

At the same time, the military has always kept a skeptical eye on the civilian use of their assets, and frequently imposed limitations and restrictions. Several of those have already been mentioned in this chapter. This has included limiting the detail that remote sensing instruments were allowed to acquire

(e.g., the 100 m maximum resolution of ERTS-1), or the artificial signal limitation of GPS accuracy until selective availability was abolished in 2000. Other datasets continue to suffer those restrictions. Until as recently as 2013, the U.S. military imposed a limit to the maximum spatial resolution (0.5 m) that data by commercial companies could provide, and only recently those rules have started to get relaxed. Likewise, DEM data based on the Shuttle Radar Topography Mission flown in 2000 continue to be available for non-U.S. territories only at a reduced resolution of 90 m (versus the 30 m available for the United States). Also, other countries have shown a very restrictive data sharing policy, among the major EO nations notably India and China.

In sum, military/national security interests have been both a blessing and a burden for the civil remote sensing community, and its efforts to use EO information.

18.5.2 Methodological Gaps in DRM

DRM, both in theoretical and practical terms, has seen tremendous advances. It has moved away from seeing disasters as a result of a violent nature toward a realization that, while nature indeed contains destructive forces and processes, it is usually human beings that move into hazardous terrain, or who change the probability of events occurring, be it through poor land management or excessive greenhouse gas emissions. This means both nature and human society with its assets and activities must be jointly considered if the aim is to reduce costs and damage. Modern risk assessment is based on a complex understanding of the role of the various parameters that influence the outcome of an event, and EO data have helped to understand those processes and connections. In practical terms, those data are also a critical asset to help manage the delicate balance of an ever growing society living in a closely coupled and complex manner with nature.

Our understanding and abilities have limitations, though. While risk assessment methods are mathematically flexible and extendable, they have been optimized for single-hazard situations. Dealing with multiple, or with cascading hazard situations, remains a challenge. For example, an area hit by a seismic event may suffer direct structural damage due to ground shaking. The same force may lead to the rupture of gas or water mains, potentially resulting in explosions/fires or flooding, respectively. At the same time the earthquake may lead to landslides, which can block waterways, subsequently resulting in dangerous breakout floods. The water left in the area quickly poses a health hazard due to waterborne diseases. Assessing the total risk in situations with this type of complexity remains a methodological challenge. Likewise, assessing the total risk, including to remote places, or accounting for secondary or tertiary consequences remain difficult. For example, the relatively minor 2010 eruption of the Eyjafjallajökull volcano in Iceland shut down production at several BMW plants in Bavaria (Jones and Mendoza Bolivar, 2011), since the closed airspace prevented needed parts to be brought in from Japan. A complete risk assessment would have to account for damage from the lost production, as well as financial consequences airlines and stranded passengers suffered.

We have a good understanding of the role vulnerability plays in risk, and that we need to distinguish between different vulnerability types. However, risk, as expected losses, has traditionally been quantified in monetary terms. How, for example, social or political vulnerability can be reconciled with those traditional approaches still requires a consensus. Some researchers (e.g., Adger and Kelly, 1999; Smit and Pilifosova, 2003) have thus focused more on understanding functional relationships.

18.5.3 Lack of Standards and Suitable Legislation

Both points have already been mentioned in earlier sections. Lack of standards relates primarily to post-disaster damage mapping that continues to see a largely redundant creation of many maps (more than 2000 damage map products were created following the 2010 Haiti earthquake), using a number of different styles and nomenclatures (Kerle, 2011). Also for damage maps created in a crowdsourcing manner, operational standards continue to be lacking. At this point it has not been conclusively shown that such a collaborative effort, even when recruiting volunteers only from a pool of people with remote sensing experience, can provide results that are sufficiently accurate. Also operating procedures for how to engage with volunteers, how to provide instructions for the specific job, including corrective feedback (Kerle and Hoffman, 2013), do not yet exist.

18.6 Summary

The purpose of this chapter is to assess the development of remote sensing over the last 50 years in terms of its effect on DRM, that is, for the entire spectrum from hazard and risk assessment to post-disaster response. Disaster risk with all its components is a spatial phenomenon, hence, for all its social and organizational facets, can only be effectively addressed with spatial data. Therefore, the main source that feeds DRM models and analysis tools is remote sensing information, and without such data meaningful risk assessment, early warning, or effective post-disaster response would simply not be possible. This chapter has highlighted that in DRM a vast number of methods and techniques have been developed that make use of virtually any remote sensing sensor or platform ever developed. At the same time, the analysis also showed that we often have very suitable or interesting data, but not yet adequate ways to use them, for example, in damage mapping. The chapter also made the point that DRM is an application field that can be seen as native to EO, in that its potential was one of the first uses of remote sensing data to be realized, and where essentially continuous developments to maximize the utility of EO data have been taking place.

The process of making EO data a critical pillar in DRM has been a very dynamic one. It reflects as much the race for military supremacy during the two World Wars and the subsequent Cold War, as the more recent, more concerted international efforts aimed at finding global solutions, such as the Disaster Charter or UN efforts mentioned. Despite being strongly driven by military spending, remote sensing has benefitted in large measure from the effort of individuals, be it pioneers who showed what can be done (such as Lawrence in 1906), or the people who transitioned remote sensing from a governmental activity into private businesses. In particular the latter has created a new race, a commercial one, which too has led to rapid advance in technology and use of EO data.

The chapter finished reviewing a number of remaining gaps and limitations, and further work for different communities has been identified. For a very long time disasters were seen, in a fatalistic fashion, as unavoidable acts of nature (or perhaps divine punishment), which later gave way to a more reasoned analysis of the role society plays. EO technology now provides us with all the data and analysis means we need to make informed decisions. This, however, means that this must be followed by an adequate socio-political decision-making process. Technology, in EO or otherwise, is not going to eliminate disasters. What is needed is a sound understanding—among all stakeholders—of the state of nature and societal processes, as provided in large part by EO instruments, but also how human activity relates with natural systems and processes. Hence, remote sensing is an effective means to reveal the relevant connections, but a sustained reduction of the number of annual disasters can only be achieved if the insights a risk assessment process provides are converted into effective risk reduction measures.

References

Abrams, M., Pieri, D., Realmuto, V., and Wright, R. 2013. Using EO-1 Hyperion data as HyspIRI preparatory data sets for volcanology applied to Mt Etna, Italy. *IEEE Journal of Selected Topics in Applied Earth Observations and Remote Sensing*, 6, 375–385.

Adger, W. N. and Kelly, P. M. 1999. Social vulnerability to climate change and the architecture of entitlements. *Mitigation and Adaptation Strategies for Global Change*, 4, 253–266.

Alexander, C., Smith-Voysey, S., Jarvis, C., and Tansey, K. 2009. Integrating building footprints and LiDAR elevation data to classify roof structures and visualise buildings. *Computers Environment and Urban Systems*, 33, 285–292.

Arciniegas, G., Bijker, W., Kerle, N., and Tolpekin, V. A. 2007. Coherence- and amplitude-based analysis of seismogenic damage in Bam, Iran, using Envisat ASAR data. *IEEE Transactions on Geoscience and Remote Sensing*, 45, 1571–1581.

Aschbacher, J. and Milagro-Perez, M. P. 2012. The European Earth monitoring (GMES) programme: Status and perspectives. *Remote Sensing of Environment*, 120, 3–8.

Atzberger, C. 2013. Advances in remote sensing of agriculture: Context description, existing operational monitoring systems and major information needs. *Remote Sensing*, 5, 949–981.

Balzter, H. 2001. Forest mapping and monitoring with interferometric synthetic aperture radar (InSAR). *Progress in Physical Geography*, 25, 159–177.

Bhaskaran, S., Datt, B., Forster, B., Neal, T., and Brown, M. 2004. Integrating imaging spectroscopy (445–2543 nm) and geographic information systems for post-disaster management: A case of hailstorm damage in Sydney. *International Journal of Remote Sensing*, 25, 2625–2639.

Birkmann, J. 2007. Risk and vulnerability indicators at different scales: Applicability, usefulness and policy implications. *Environmental Hazards*, 7, 20–31.

Blaikie, P., Cannon, T., Davis, I., and Wisner, B. 1994. *At Risk: Natural Hazards, People's Vulnerability, and Disasters.* London, U.K.: Routledge, 284pp.

Booth, E., Saito, K., Spence, R., Madabhushi, G., and Eguchi, R. T. 2011. Validating assessments of seismic damage made from remote sensing. *Earthquake Spectra*, 27, S157–S177.

Brown, T. M., Liang, D. A., and Womble, J. A. 2012. Predicting ground-based damage states from windstorms using remote-sensing imagery. *Wind and Structures*, 15, 369–383.

Carter, W. D. and Rinker, J. N. 1976. Structural features related to earthquakes in Managua, Nicaragua, and Cordoba. Mexico: U.S. Geological Survey.

Chuvieco, E. 2003. *Wildland Fire Danger Estimation and Mapping: The Role of Remote Sensing Data.* Series in Remote Sensing. River Edge, NJ: World Scientific Publishing, 280pp.

Clark, G. E., Moser, S. C., Ratick, S. J., Dow, K., Meyer, W. B., Emani, S., Jin, W., Kasperson, J. X., Kasperson, R. E., and Schwarz, H. E. 1998. Assessing the vulnerability of coastal communities to extreme storms: The case of Revere, MA, USA. *Mitigation and Adaptation Strategies for Global Change*, 3, 59–82.

Cochran, D. R. and Pyle, R. L. 1978. Volcanology via satellite. *Monthly Weather Review*, 106, 1373–1375.

Covello, F., Battazza, F., Coletta, A., Lopinto, E., Fiorentino, C., Pietranera, L., Valentini, G., and Zoffoli, S. 2010. COSMO-SkyMed an existing opportunity for observing the Earth. *Journal of Geodynamics*, 49, 171–180.

CRED. 2014. EM-DAT: The OFDA/CRED international disaster database.

Dardel, C., Kergoat, L., Hiernaux, P., Mougin, E., Grippa, M., and Tucker, C. J. 2014. Re-greening Sahel: 30 years of remote sensing data and field observations (Mali, Niger). *Remote Sensing of Environment*, 140, 350–364.

Dashora, A., Lohani, B., and Malik, J. N. 2007. A repository of earth resource information—CORONA satellite programme. *Current Science*, 92, 926–932.

Dell'Acqua, F. and Gamba, P. 2012. Remote sensing and earthquake damage assessment: Experiences, limits, and perspectives. *Proceedings of the IEEE*, 100, 2876–2890.

Dell'Acqua, F., Lanese, I., and Polli, D. A. 2013. Integration of EO-based vulnerability estimation into EO-based seismic damage assessment: A case study on L'Aquila, Italy, 2009 earthquake. *Natural Hazards*, 68, 165–180.

Ebert, A., Kerle, N., and Stein, A. 2009. Urban social vulnerability assessment with physical proxies and spatial metrics derived from air- and spaceborne imagery and GIS data. *Natural Hazards*, 48, 275–294.

Ehrlich, D., Guo, H. D., Molch, K., Ma, J. W., and Pesaresi, M. 2009. Identifying damage caused by the 2008 Wenchuan earthquake from VHR remote sensing data. *International Journal of Digital Earth*, 2, 309–326.

Ehrlich, D., Kemper, T., Blaes, X., and Soille, P. 2013. Extracting building stock information from optical satellite imagery for mapping earthquake exposure and its vulnerability. *Natural Hazards*, 68, 79–95.

Endo, E. T., Ward, P. L., Harlow, D. H., Allen, R. V., and Eaton, J. P. 1974. A prototype global volcano surveillance system monitoring seismic activity and tilt. *Bulletin Volcanologique*, 38, 315–344.

Fernandez Galarreta, J., Kerle, N., and Gerke, M. In press. UAV-based urban structural damage assessment using object-based image analysis and semantic reasoning. *Natural Hazards and Earth System Sciences*.

Ferro, A., Brunner, D., and Bruzzone, L. 2013. Automatic detection and reconstruction of building radar footprints from single VHR SAR images. *IEEE Transactions on Geoscience and Remote Sensing*, 51, 935–952.

Francis, P. W. and De Silva, S. L. 1989. Application of the Landsat Thematic Mapper to the identification of potentially active volcanoes in the Central Andes. *Remote Sensing of Environment*, 28, 245–255.

Galderisi, A. and Ferrara, F. F. 2013. Resilience, in Bobrowsky, P. T., ed., *Encyclopedia of Natural Hazards.* Dordrecht, the Netherlands: Springer, pp. 849–850.

Galiatsatos, N., Donoghue, D. N. M., and Philip, G. 2008. High resolution elevation data derived from stereoscopic CORONA imagery with minimal ground control: An approach using Ikonos and SRTM data. *Photogrammetric Engineering and Remote Sensing*, 74, 1093–1106.

Gedney, L. and VanWormer, J. 1973. ERTS-1, earthquakes, and tectonic evolution in Alaska. Washington, DC, December 10–14, 1973, Volume Vol. 1, Sect. A, pp. 745–756.

Geerling, G. W., Vreeken-Buijs, M. J., Jesse, P., Ragas, A. M. J., and Smits, A. J. M. 2009. Mapping river floodplain ecotopes by segmentation of spectral (CASI) and structural (LiDAR) remote sensing data. *River Research and Applications*, 25, 795–813.

Gerke, M. and Kerle, N. 2011. Automatic structural seismic damage assessment with airborne oblique Pictometry imagery. *Photogrammetric Engineering and Remote Sensing*, 77, 885–898.

Ghosh, S., Huyck, C. K., Greene, M., Gill, S. P., Bevington, J., Svekla, W., DesRoches, R., and Eguchi, R. T. 2011. Crowdsourcing for rapid damage assessment: The Global Earth Observation Catastrophe Assessment Network (GEO-CAN). *Earthquake Spectra*, 27, S179–S198.

Goodchild, M. 2007. Citizens as sensors: The world of volunteered geography. *GeoJournal*, 69, 211–221.

Grünthal, G. 1998. European Macroseismic Scale 1998 (EMS-98). *Cahiers du Centre Européen de Géodynamique et de Séismologie*, Vol. 15. Luxembourg: Centre Européen de Géodynamique et de Séismologie, p. 99.

Hastings, D. A. and Emery, W. J. 1992. The advanced very high-resolution radiometer (AVHRR)—A brief reference guide. *Photogrammetric Engineering and Remote Sensing*, 58, 1183–1188.

Herman, J. R., McPeters, R., and Larko, D. 1993. Ozone depletion at northern and southern latitudes derived from January 1979 to December 1991 Total Ozone Mapping Spectrometer data. *Journal of Geophysical Research-Atmospheres*, 98, 12783–12793.

Heumann, B. W., Seaquist, J. W., Eklundh, L., and Jonsson, P. 2007. AVHRR derived phenological change in the Sahel and Soudan, Africa, 1982–2005. *Remote Sensing of Environment*, 108, 385–392.

Hinkley, E. A. and Zajkowski, T. 2011. USDA forest service-NASA: Unmanned aerial systems demonstrations—Pushing the leading edge in fire mapping. *Geocarto International*, 26, 103–111.

Hoffer, R. M. and Lee, K. S. 1989. Forest change classification using Seasat and SIR-B satellite SAR data, in *Proceedings Geoscience and Remote Sensing Symposium (IGARSS '89), 12th Canadian Symposium on Remote Sensing*, Vancouver, British Columbia, Canada, July 10–14, 1989, Vol. 3, pp. 1372–1375.

Hooper, A., Bekaert, D., Spaans, K., and Arikan, M. 2012. Recent advances in SAR interferometry time series analysis for measuring crustal deformation. *Tectonophysics*, 514, 1–13.

Hoque, R., Nakayama, D., Matsuyama, H., and Matsumoto, J. 2011. Flood monitoring, mapping and assessing capabilities using RADARSAT remote sensing, GIS and ground data for Bangladesh. *Natural Hazards*, 57, 525–548.

Howard, J. A., Barrett, E. C., and Heilkema, J. U. 1978. The application of satellite remote sensing to monitoring of agricultural disasters. *Disasters*, 2, 231–240.

Huang, C., Wu, J. P., Chen, Y., and Yu, J. 2012. Detecting floodplain inundation frequency using MODIS time-series imagery, in *Proceedings 2012 First International Conference on Agro-Geoinformatics (Agro-Geoinformatics)*, Shanghai, PR China, August 2–4, 2012, pp. 349–354.

Jones, S. and Mendoza Bolivar, E. 2011. Natural disasters and business: The impact of the Icelandic volcano of April 2010 on European logistics and distribution—A case study of Malta. Maastricht School of Management, Maastricht, the Netherlands.

Joyce, K. E., Belliss, S. E., Samsonov, S. V., McNeill, S. J., and Glassey, P. J. 2009. A review of the status of satellite remote sensing and image processing techniques for mapping natural hazards and disasters. *Progress in Physical Geography*, 33, 183–207.

Karakhanian, A., Jrbashyan, R., Trifonov, V., Philip, H., Arakelian, S., Avagyan, A., Baghdassaryan, H., Davtian, V., and Ghoukassyan, Y. 2003. Volcanic hazards in the region of the Armenian Nuclear Power Plant. *Journal of Volcanology and Geothermal Research*, 126, 31–62.

Kerle, N. 2010. Satellite-based damage mapping following the 2006 Indonesia earthquake—How accurate was it? *International Journal of Applied Earth Observation and Geoinformation*, 12, 466–476.

Kerle, N. 2011. Remote sensing based post—Disaster damage mapping: Ready for a collaborative approach?

Kerle, N. 2013a. Global positioning systems (GPS) and natural hazards, in Bobrowsky, P. T., ed., *Encyclopedia of Natural Hazards*. Dordrecht, the Netherlands: Springer, pp. 416–417.

Kerle, N. 2013b. Remote sensing of natural hazards and disasters, in Bobrowsky, P. T., ed., *Encyclopedia of Natural Hazards*. Dordrecht, the Netherlands: Springer, pp. 837–847.

Kerle, N. and Alkema, D. 2011. Multiscale flood risk assessment in urban areas—A geoinformatics approach, in Richter, M. and Weiland, U., eds., *Applied Urban Ecology*. John Wiley & Sons, Ltd., pp. 93–105.

Kerle, N., de Vries, B. V., and Oppenheimer, C. 2003. New insight into the factors leading to the 1998 flank collapse and lahar disaster at Casita volcano, Nicaragua. *Bulletin of Volcanology*, 65, 331–345.

Kerle, N., Heuel, S., and Pfeifer, N. 2008. Real-time data collection and information generation using airborne sensors, in Zlatanova, S. and Li, J., eds., *Geospatial Information Technology for Emergency Response*. London, U.K.: Taylor & Francis, pp. 43–74.

Kerle, N. and Hoffman, R. R. 2013. Collaborative damage mapping for emergency response: The role of Cognitive Systems Engineering. *Natural Hazards and Earth System Sciences (NHESS)*, 13, 97–113.

Kohli, D., Sliuzas, R., Kerle, N., and Stein, A. 2012. An ontology of slums for image-based classification. *Computers Environment and Urban Systems*, 36, 154–163.

Kohli, D., Warwadekar, P., Kerle, N., Sliuzas, R., and Stein, A. 2013. Transferability of object-oriented image analysis methods for slum identification. *Remote Sensing*, 5, 4209–4228.

Kraus, K. and Pfeifer, N. 1998. Determination of terrain models in wooded areas with airborne laser scanner data. *ISPRS Journal of Photogrammetry and Remote Sensing*, 53, 193–203.

Kromker, D., Eierdanz, F., and Stolberg, A. 2008. Who is susceptible and why? An agent-based approach to assessing vulnerability to drought. *Regional Environmental Change*, 8, 173–185.

Kuntze, H., Frey, C. W., Tchouchenkov, I., Staehle, B., Rome, E., Pfeiffer, K., Wenzel, A., and Wollenstein, J. 2012. SENEKA—Sensor network with mobile robots for disaster management, in *Proceedings IEEE Conference on Technologies for Homeland Security (HST)*, November 13–15, 2012, pp. 406–410.

Lamsal, D., Sawagaki, T., and Watanabe, T. 2011. Digital terrain modelling using Corona and ALOS PRISM data to investigate the distal part of Imja Glacier, Khumbu Himal, Nepal. *Journal of Mountain Science*, 8, 390–402.

Lee, T. and Kim, T. 2013. Automatic building height extraction by volumetric shadow analysis of monoscopic imagery. *International Journal of Remote Sensing*, 34, 5834–5850.

Liu, T. M. 2014. Analysis of the economic impact of meteorological disasters on tourism: The case of typhoon Morakot's impact on the Maolin National Scenic Area in Taiwan. *Tourism Economics*, 20, 143–156.

Loew, A. 2014. Terrestrial satellite records for climate studies: How long is long enough? A test case for the Sahel. *Theoretical and Applied Climatology*, 115, 427–440.

Mackenzie, J. S. and Ringrose, P. S. 1986. Use of SeaSat SAR imagery for geological mapping in a volcanic terrain—Askja-caldera, Iceland. *International Journal of Remote Sensing*, 7, 181–194.

Martha, T. R., Kerle, N., Jetten, V. G., van Westen, C. J., and Vinod Kumar, K. 2010. Characterising spectral, spatial and morphometric properties of landslides for semi-automatic detection using object-oriented methods. *Geomorphology*, 116, 24–36.

Martha, T. R., Kerle, N., van Westen, C. J., Jetten, V., and Kumar, K. V. 2012. Object-oriented analysis of multi-temporal panchromatic images for creation of historical landslide inventories. *ISPRS Journal of Photogrammetry and Remote Sensing*, 67, 105–119.

Mika, A. M. 1997. Three decades of Landsat instruments. *Photogrammetric Engineering and Remote Sensing*, 63, 839–852.

Mitomi, H., Yamzaki, F., and Matsuoka, M. 2000. Automated detection of building damage due to recent earthquakes using aerial television images, in *Proceedings 21st Asian Conference on Remote Sensing*, Taipei, Taiwan, 2000. GIS Development, pp. 401–406.

Müller, A. 2013. Flood risks in a dynamic urban agglomeration: A conceptual and methodological assessment framework. *Natural Hazards*, 65, 1931–1950.

Murphy, S. W., Wright, R., Oppenheimer, C., and Souza, C. R. 2013. MODIS and ASTER synergy for characterizing thermal volcanic activity. *Remote Sensing of Environment*, 131, 195–205.

Narama, C., Kääb, A., Duishonakunov, M., and Abdrakhmatov, K. 2010. Spatial variability of recent glacier area changes in the Tien Shan Mountains, Central Asia, using Corona (~1970), Landsat (~2000), and ALOS (~2007) satellite data. *Global and Planetary Change*, 71, 42–54.

National Research Council (U.S.) Committee on Remote Sensing Programs for Earth Resource Surveys. 1977. Microwave remote sensing from space for earth resource surveys. Washington, DC: National Academy of Sciences.

Newnham, G. J., Siggins, A. S., Blanchi, R. M., Culvenor, D. S., Leonard, J. E., and Mashford, J. S. 2012. Exploiting three dimensional vegetation structure to map wildland extent. *Remote Sensing of Environment*, 123, 155–162.

Nie, Y., Liu, Q., and Liu, S. Y. 2013. Glacial lake expansion in the Central Himalayas by Landsat images, 1990–2010. *PloS One*, 8, 8.

Niethammer, U., James, M. R., Rothmund, S., Travelletti, J., and Joswig, M. 2012. UAV-based remote sensing of the Super-Sauze landslide: Evaluation and results. *Engineering Geology*, 128, 2–11.

O'Rourke, T. D., Bonneau, A. L., Pease, J. W., Shi, P., and Wang, Y. 2006. Liquefaction and ground failures in San Francisco. *Earthquake Spectra*, 22, 91–112.

Oliveira, S. L. J., Pereira, J. M. C., and Carreiras, J. M. B. 2012. Fire frequency analysis in Portugal (1975–2005), using Landsat-based burnt area maps. *International Journal of Wildland Fire*, 21, 48–60.

Perlroth, I. 1962. Relationship of central pressure of hurricane Esther (1961) and the sea surface temperature field. *Tellus*, 14, 403–408.

Petrosillo, I., Zaccarelli, N., and Zurlini, G. 2010. Multi-scale vulnerability of natural capital in a panarchy of social-ecological landscapes. *Ecological Complexity*, 7, 359–367.

Piao, S. L., Fang, J. Y., Liu, H. Y., and Zhu, B. 2005. NDVI-indicated decline in desertification in China in the past two decades. *Geophysical Research Letters*, 32, 4.

Poli, D. and Caravaggi, I. 2013. 3D modeling of large urban areas with stereo VHR satellite imagery: Lessons learned. *Natural Hazards*, 68, 53–78.

Poompavai, V. and Ramalingam, M. 2013. Geospatial analysis for coastal risk assessment to cyclones. *Journal of the Indian Society of Remote Sensing*, 41, 157–176.

Ramachandran, B., Justice, C. O., and Abrams, M. J. 2010. *Land Remote Sensing and Global Environmental Change: NASA's Earth Observing System and the Science of ASTER and MODIS*. New York: Springer.

Ramasamy, S. M. 2006. Remote sensing and active tectonics of South India. *International Journal of Remote Sensing*, 27, 4397–4431.

Ray, P. S. and Hane, C. E. 1976. Tornado-parent storm relationship deduced from a dual-Doppler radar analysis. *Geophysical Research Letters*, 3, 721–723.

Razak, K. A., Santangelo, M., Van Westen, C. J., Straatsma, M. W., and de Jong, S. M. 2013. Generating an optimal DTM from airborne laser scanning data for landslide mapping in a tropical forest environment. *Geomorphology*, 190, 112–125.

Sacchetti, A., Cisbani, A., Babini, G., and Galeazzi, C. 2010. The Italian precursor of an operational hyperspectral imaging mission, in Sandu, R. et al., eds., *Small Satellite Missions for Earth Observation: New Developments and Trends*. Berlin, Germany: Springer-Verlag, pp. 73–81.

Salami, A. T., Akinyede, J., and de Gier, A. 2010. A preliminary assessment of NigeriaSat-1 for sustainable mangrove forest monitoring. *International Journal of Applied Earth Observation and Geoinformation*, 12, S18–S22.

Sandau, R., Briess, K., and D'Errico, M. 2010. Small satellites for global coverage: Potential and limits. *ISPRS Journal of Photogrammetry and Remote Sensing*, 65, 492–504.

Schmidt, M., Goossens, R., and Menz, G. 2001. Processing techniques for CORONA satellite images in order to generate high-resolution digital elevation models, in Bégni, G., ed., *Observing Our Environment from Space: New Solutions for a New Millennium*. Lisse, the Netherlands: A.A. Balkema Publishers, pp. 191–196.

Shafique, M., van der Meijde, M., and Rossiter, D. G. 2011. Geophysical and remote sensing-based approach to model regolith thickness in a data-sparse environment. *Catena*, 87, 11–19.

Shafique, M., van der Meijde, M., and van der Werff, H. M. A. 2012. Evaluation of remote sensing-based seismic site characterization using earthquake damage data. *Terra Nova*, 24, 123–129.

Shruthi, R. B. V., Kerle, N., and Jetten, V. 2011. Object-based gully feature extraction using high spatial resolution imagery. *Geomorphology*, 134, 260–268.

Shruthi, R. B. V., Kerle, N., Jetten, V., Abdellah, L., and Machmach, I. In press-a. Quantifying temporal changes in gully erosion areas with object oriented analysis. *Catena*.

Shruthi, R. B. V., Kerle, N., Jetten, V., and Stein, A. In press-b. Object-based gully system prediction from medium resolution imagery using Random Forests. *Geomorphology*.

Sithole, G. and Vosselman, M. G. 2004. Experimental comparison of filter algorithms for bare-earth extraction from airborne laser scanning point clouds. *ISPRS Journal of Photogrammetry & Remote Sensing*, 59, 85–101.

Smit, B. and Pilifosova, O. 2003. From adaptation to adaptive capacity and vulnerability reduction, in Huq, S., Smith, J., and Klein, R. T. J., eds., *Enhancing the Capacity of Developing Countries to Adapt to Climate Change*. London, U.K.: Imperial College Press, pp. 9–25.

Smith, K. 2004. *Environmental Hazards*. London, U.K./New York: Routledge, 432pp.

Space Science Board-National Research Council. 1974. United States space science program: Report to COSPAR National Academy of Sciences. Washington, DC: National Research Council.

Straatsma, M. W. 2008. Quantitative mapping of hydrodynamic vegetation density of floodplain forests under leaf-off conditions using airborne laser scanning. *Photogrammetric Engineering and Remote Sensing*, 74, 987–998.

Straatsma, M. W. and Baptist, M. J. 2008. Floodplain roughness parameterization using airborne laser scanning and spectral remote sensing. *Remote Sensing of Environment*, 112, 1062–1080.

Stuffler, T., Forster, K., Hofer, S., Leipold, M., Sang, B., Kaufmann, H., Penne, B., Mueller, A., and Chlebek, C. 2009. Hyperspectral imaging—An advanced instrument concept for the EnMAP mission (Environmental Mapping and Analysis Programme). *Acta Astronautica*, 65, 1107–1112.

Stumpf, A. and Kerle, N. 2011. Object-oriented mapping of landslides using Random Forests. *Remote Sensing of Environment*, 115, 2564–2577.

Talebi, L., Kuczynski, A., Graettinger, A. J., and Pitt, R. 2014. Automated classification of urban areas for storm water management using aerial photography and LiDAR. *Journal of Hydrologic Engineering*, 19, 887–895.

Torontow, V., and King, D. 2011. Forest complexity modelling and mapping with remote sensing and topographic data: A comparison of three methods. *Canadian Journal of Remote Sensing*, 37, 387–402.

Torres, R., Snoeij, P., Geudtner, D., Bibby, D., Davidson, M., Attema, E., Potin, P. et al. 2012. GMES Sentinel-1 mission. *Remote Sensing of Environment*, 120, 9–24.

Tralli, D. M., Blom, R. G., Zlotnicki, V., Donnellan, A., and Evans, D. L. 2005. Satellite remote sensing of earthquake, volcano, flood, landslide and coastal inundation hazards. *ISPRS Journal of Photogrammetry and Remote Sensing*, 59, 185–198.

Tronin, A. A. 2010. Satellite remote sensing in seismology. A review. *Remote Sensing*, 2, 124–150.

Unganai, L. S. and Kogan, F. N. 1998. Drought monitoring and corn yield estimation in Southern Africa from AVHRR data. *Remote Sensing of Environment*, 63, 219–232.

UN-ISDR (United Nations International Strategy for Disaster Reduction). 2004. Living with risk: A global review of disaster reduction initiatives. Geneva, Switzerland: UN/ISDR.

United Nations Office for Disaster Risk Reduction (UNISDR). 2004. Living with risk: A global review of disaster reduction initiatives.

Uprety, P., Yamazaki, F., and Dell'Acqua, F. 2013. Damage detection using high-resolution SAR imagery in the 2009 L'Aquila, Italy, earthquake. *Earthquake Spectra*, 29, 1521–1535.

van Aardt, J. A. N., McKeown, D., Faulring, J., Raqueno, N., Casterline, M., Renschler, C., Eguchi, R. et al. 2011. Geospatial disaster response during the Haiti earthquake: A case study spanning airborne deployment, data collection, transfer, processing, and dissemination. *Photogrammetric Engineering and Remote Sensing*, 77, 943–952.

Van Den Eeckhaut, M., Kerle, N., Poesen, J., and Hervas, J. 2012. Object-oriented identification of forested landslides with derivatives of single pulse LiDAR data. *Geomorphology*, 173, 30–42.

van der Sande, C. J., de Jong, S. M., and de Roo, A. P. J. 2003. A segmentation and classification approach of IKONOS-2 imagery for land cover mapping to assist flood risk and flood damage assessment. *International Journal of Applied Earth Observation and Geoinformation*, 4, 217–229.

van Westen, C. J., Alkema, D., Damen, M. C. J., Kerle, N., and Kingma, N. C. 2011. *Multi-Hazard Risk Assessment*. Enschede, the Netherlands: UNU-ITC DGIM.

Voigt, S., Schneiderhan, T., Twele, A., Gahler, M., Stein, E., and Mehl, H. 2011. Rapid damage assessment and situation mapping: Learning from the 2010 Haiti earthquake. *Photogrammetric Engineering and Remote Sensing*, 77, 923–931.

Williams, S. C. P. 2013. Studying volcanic eruptions with aerial drones. *Proceedings of the National Academy of Sciences of the United States of America*, 110, 10881–10881.

Wu, H., Cheng, Z. P., Shi, W. Z., Miao, Z. L., and Xu, C. C. 2014. An object-based image analysis for building seismic vulnerability assessment using high-resolution remote sensing imagery. *Natural Hazards*, 71, 151–174.

Zhang, Y. and Kerle, N. 2008. Satellite remote sensing for near-real time data collection, in Zlatanova, S. and Li, J., eds., *Geospatial Information Technology for Emergency Response*. London, U.K.: Taylor & Francis, pp. 75–102.

Humanitarian Emergencies: Causes, Traits, and Impacts as Observed by Remote Sensing

Stefan Lang
University of Salzburg

Petra Füreder
University of Salzburg

Olaf Kranz
Helmholtz-Association

Brittany Card
Harvard Humanitarian Initiative

Shadrock Roberts
University of Georgia

Andreas Papp
*Médecins Sans Frontières
(MSF) Austria*

Acronyms and Definitions

ASM	Artisanal and small-scale mining techniques	LSHTM	London School of Hygiene and Tropical Medicine
CNL	Cognition Network Language	MODIS	Moderate Resolution Imaging Spectroradiometer
DRC	Democratic Republic of the Congo	MSF	Médecins Sans Frontières
EO	Earth Observation	NASA	National Aeronautics and Space Administration
EMS	Emergency Management Service	NGO	Nongovernmental Organization
ESA	European Space Agency	OBIA	Object-based image analysis
GIO	GMES Initial Operations	OSM	OpenStreetMap
GIS	Geographic information system	SAF	Sudan Armed Forces
HCS	Hyperspherical color sharpening	SPLA	Sudan People's Liberation Army
HOT	Humanitarian OpenStreetMap Team	UAV	Unmanned aerial vehicle
HR	High (spatial) resolution (1: 4 m <resolution< = 10 m, 2: 10 m <resolution< = 30 m)	UN	United Nations
		UNEP	United Nations Environment Programme
ICCM	International Network of Crisis Mappers	UNHCR	United Nations High Commissioner for Refugees
IDP	Internally displaced person	UNICEF	United Nations International Children's Fund
		UNITAR	United Nations Institute for Training and Research
		UNMIS	United Nations Mission in Sudan

UNOOSA   United Nations Office for Outer Space Affairs
VHR   Very high (spatial) resolution (1: resolution < = 1 m;
          2: 1 m < resolution < = 4 m)
VGI   Volunteered geographic information
V&TC   Volunteer and Technical Communities
WFP   World Food Program

19.1 Introduction

19.1.1 Humanitarian Disasters: A Particular Case?

Drawing a sharp conceptual line between natural and humanitarian disasters is difficult because of the mutual relationships among them that all too often lead to humanitarian crises. In fact, the notion of any *disaster* has a human component, not necessarily in a causal, but always is an affected sense. Without human reference or any impact on the anthropogenic sphere, a natural event like an earthquake, a landslide, or a river flood would rather be considered a disturbance in the sense of an episodic event inherent to an (eco-)system's integrity. Moreover, due to climate change and other large-scale anthropogenic effects, natural disasters apparently increase both in terms of occurrence and severity (IPCC 2001); thus the term "natural" is even more deceptive. In order to find a (pragmatic) borderline to literature dealing with natural disasters in a stricter sense, we shall concentrate on disasters that are either caused or reinforced by crises or conflicts, whether they ultimately root in natural (e.g., a drought spanning over several years) or societal causes (any type of aggression, fights over resources, etc.). This corresponds, again with many transitions and uncertainties, to the field of humanitarian action. Unlike natural disaster response, which operates in fairly distinct phases in a rather distinct disaster management cycle (Joyce et al. 2009), humanitarian action faces

more gradual, at times protracted, response phases. In particular, humanitarian conflict situations often lack a distinct peak situation (as compared to catastrophic events such as floods, earthquakes, or wildfires); thus, it is hard to pinpoint the exact point in time when a man-made conflict leads to a humanitarian disaster. While conceptually disconnected here, natural disasters or resource scarcity/abundance may overlay and reinforce conflict situations or contribute to secondary risks through the outbreak of a disease or other calamities (e.g., as in the case of the Haiti earthquake in 2010).

19.1.2 Forced Migrations and Regional Conflicts

The most obvious, and concurrently also the most adverse, impact of humanitarian emergencies is the forced displacement of large numbers of people. Next to natural disasters and changing environmental conditions (land degradation, desertification, large-scale land investments, etc.), violent regional conflicts are among the main drivers that make people flee their homes. As indicated earlier, the causes are often multi layered (Humanitarian Coalition n.d.), with systemic and reinforcing cycles, and ultimately lead to high amounts of population displacements (see Figure 19.1).

Depending on whether persons are displaced within their country of origin or crossed an international border, we differentiate between internally displaced persons (IDPs) and refugees. Currently, there are about 51.2 million people displaced, among them 33.3 million IDPs, 16.7 million refugees, and 1.2 million asylum seekers (UNHCR 2013, IDMC 2014). Most of these people gather in camps or informal settlements. The prevalence of man-made crises in the shadow of conflicts and wars has led to an ever-increasing number of displaced people in the last decade. The highest amounts of IDPs in 2013 were found in

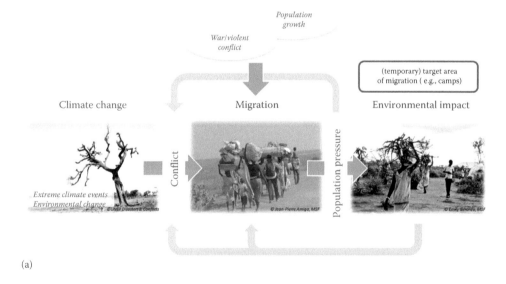

(a)

FIGURE 19.1   (a) Conflict-related migrations and environmental impact of large-scale spontaneous settlements influencing each other in a reinforcing feedback loop potentially intensified by climate change.   *(Continued)*

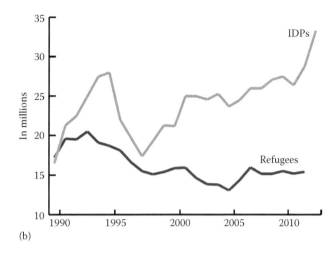

(b)

FIGURE 19.1 (*Continued*) (b) Numbers of IDPs and refugees since the 1990s. (From IDMC, *Global Overview* 2014. People internally displaced by conflict and violence. I. D. M. Centre, 2014.)

Syria where their number has reached more than 6.5 million that is almost five times the figure of the last decade. This country is followed by Colombia, Nigeria, Democratic Republic of the Congo (DRC), and Sudan (in that order, IDMC 2014; Table 19.1).

Forced migration, the rise of large spontaneous settlements, most often in areas scarce in resources anyway, is a global problem with strong local implications. In turn, increasing pressure on scare resources is a common factor for intensification or even the outbreak of new local conflicts. The United Nations Environment Programme (UNEP) Expert Advisory Group on Environment, Conflict and Peacebuilding stated in 2009 that "there is significant potential for conflict over natural resources to intensify in the coming decades" (UNEP 2009) (p. 5, see also Walker et al. [2012] for recent facts and figures on this issue). Conflicts over natural resources are one of the most frequent conflict items in Sub-Saharan Africa (HIIK 2013), but not

TABLE 19.1 Important Refugee Figures and Causes of Displacement during the Last Two Decades Due to Armed Conflict, Human Rights Violations, or Generalized Violence

Country of Origin	Duration	IDPs	IDPs (Peak)	Refugees	Refugees (Peak)
Syria	2011–ongoing	6.5 million (May 2014)	6.5 million (2013)	2.8 million (January 2014)	2.9 million (2014)
Afghanistan	1978[a]–ongoing (start IDPs: 2001)	670,000 (April 2014)	1.2 million (2002)	2.6 million (January 2014)	6.3 million (1990)
Colombia	1960–ongoing	5.7 million (December 2013)	5.7 million (2013)	400,000 (January 2014)	550,000 (2007)
Mozambique	1981–1995	—	4–4.5 million[b]	—	1.4 million (1992)
Iraq	1968–ongoing	2.3 million (January 2014)	2.8 million (2008)	400,000 (January 2014)	2.3 million (2007)
Democratic Republic of the Congo	1996–ongoing	2.6 million (March 2014)	3.4 million (2003)	500,000 (January 2014)	510,000 (2012)
Nigeria	1999–ongoing	3.3 million (March 2014)	3.3 million (2013)	20,000 (January 2014)	30,000 (2013)
Sudan	1984–ongoing	2.9 million (June 2014)	4.7 million (2008)	650,000 (January 2014)	730,000 (2004)
South Sudan[c]	2011–ongoing	1.1 million (July 2014)	1.1 million (July 2014)	560,000 (July 2014)	560,000 (July 2014)
Central African Republic	2005–ongoing	530,000 (July 2014)	940,000 (2013)	380,000 (July 2014)	380,000 (July 2014)
Rwanda	1994–1996[d]	—	650,000[e] (1998–1999)	84,000 (2013)	2.3 million (1994)
Somalia	1991–ongoing	1.1 million (March 2014)	1.5 million (2007[f])	1.1 million (January 2014)	1.1 million (2013)
Burundi	1972–ongoing[g]	80,000 (December 2013)	800,000 (1999)	70,000 (January 2014)	870,000 (1993)

Sources: IDMC, *Global Overview* 2014. People internally displaced by conflict and violence. I. D. M. Centre, 2014, UNHCR Population Statistics (http://popstats.unhcr.org), UNHCR historical refugee data, and Information Portals (http://data.unhcr.org), UNHCR Global Trends 2013 (numbers > 1 million rounded to the nearest 100,000, numbers < 1 million rounded to the nearest 10,000).

[a] Forced migration review, University of Oxford, Oxford, UK, http://www.fmreview.org/FMRpdfs/FMR13/fmr13.3.pdf.

[b] United Nations, ONUMOZ, http://www.un.org/en/peacekeeping/missions/past/onumozFT.htm.

[c] South Sudan gained independence from Sudan in 2011.

[d] War 1990–1994.

[e] Norwegian Refugee Council (2005): Profile of Internal Displacement: Rwanda.

[f] UNHCR figures 2009.

[g] Conflict 1993–2005.

only occur in countries that are scarce in natural resources. Resource abundance might also lead to violent conflicts—the so-called resource curse (Le Billon 2001, Mildner et al. 2011), as, for example, related to the financing of weapons, supplies, and corruption through revenues from the exploitation of natural resources. Extraction and trade of minerals, timber, oil, and other resources are often controlled or even conducted by military groups. As a result, the conflict situation intensifies eventually leading to a complex emergency.

19.1.3 Role of Satellite Remote Sensing in Humanitarian Action

Due to its observational power, ubiquitous usage, and availability, remote sensing in combination with advanced analysis techniques has become a decision-supporting tool for humanitarian professionals (Bjorgo 2001). Remote sensing and related derived information products are complementary data sources to field-based surveys enriching the pool of spatially aware technologies (Cowan 2011, Verjee 2011) for humanitarian relief support. This chapter focuses on the use of remotely sensed data from Earth observation (EO) satellites, that is, orbital sensors in support to civilian applications in various societal benefit areas.

19.1.3.1 Objectivity, Sensitivity, and the Crowd

Above all, remote sensing has the advantage of an objective imaging device that captures data over large areas under equal conditions. This is much like a *neutral* observer's camera, just covering a much larger extent, and taken from an orbital view. Thus, remote sensing entails a trend for a "democratizing tool" (Lang et al. 2013), designed to reveal the situation "as is," nondistorted, nonmanipulated, and potentially accessible to everyone.

The potential for this accessibility of remote sensing data to act as a convening point among humanitarian actors was noted as commercial satellite imagery was becoming available in the early part of the 2000s, with one significant concern: it was thought that the high cost and limited availability could mitigate its widespread use (Bjorgo 2001). However, there is a large variety of different optical as well as radar sensor types available today that can provide information at all kinds of spatial, temporal, and spectral resolution applicable for many different humanitarian disaster scenarios. For slow-onset disasters such as droughts, floods, or diseases, frequent coverage of large areas is ensured through high- to medium-resolution satellites (e.g., Moderate Resolution Imaging Spectroradiometer [MODIS], Landsat). Analyses of soil moisture in support of the monitoring of emerging droughts could be done with medium-resolution radar data (e.g., ScanSAR modes of TerraSAR-X, Radarsat, Cosmo-SkyMed). The same is true for slow-onset flood situations as they, for example, frequently occur in the north of Namibia. The derived information is crucial for early warning, for better response to and mitigation (or even prevention) of humanitarian disasters in this context.

Since very-high-resolution (VHR) imagery became commercially available more than a decade ago, it has become more common as an input to humanitarian cartographic products

and is increasingly explored as a point of collaboration between humanitarian agencies and the larger public: specifically through crowdsourcing — the distribution of a task to a large group of undefined people.* Advancements in Internet technology and the greater availability of remote sensing data now allow a wide range of nontraditional actors to engage during a humanitarian response. Indeed, new forms of online volunteers—often called "volunteer and technical communities (V&TC)"—have organized with the specific goal of supporting humanitarian response; see Capelo et al. (2012). Because many of the biophysical features found in VHR imagery are recognizable by nonspecialists, there have been several cases of humanitarian organizations experimenting with "volunteered geographic information (VGI)" derived by manual image interpretation.

This phenomenon was emphasized by the creation of vector data, manually digitized from satellite imagery, in Port-au-Prince, Haiti, in the hours following the devastating earthquake in 2010. Using OpenStreetMap (OSM) (an online mapping platform and database that allows users to trace satellite imagery, include GPS data or use their local knowledge to create open data that is often abbreviated as OSM), several hundred online volunteers created more than 1.4 million edits in the weeks following the quake: making the OSM data of Haiti the primary source of cartographic data for humanitarian responders. Since then, a "Humanitarian OpenStreetMap Team" (HOT) has formed to act as a bridge between the humanitarian responders and the OSM community. Other online platforms that make VHR imagery available for interpretation, such as Tomnod, have experimented with the enumeration of IDP dwelling units in Somalia (SBTF 2011).

This type of public engagement, whether solicited or not by humanitarian agencies, presents both opportunities and challenges for the humanitarian sector. The data quality of any type of VGI is a long-standing research focus (Elwood 2008), and ensuring that the fundamentals of imagery interpretation, such as color interpretation, minimum mapping unit, or other functions of scale, are understood and consistently applied by the general public is a challenge. However, research concerning the spatial accuracy and precision of data found within OSM shows that there is no reason, a priori, to dismiss these types of data since their accuracy can be quite good compared to *authoritative* data sets (Hakley 2010). However, a recent comparison conducted by the REACH initiative and the American Red Cross of a crowd-sourced damage assessment with field-based assessments after the Typhoon Haiyan (Yolanda) in the Philippines found that, while the OSM community did reasonably well at identifying affected buildings, they were less accurate when reporting the type of damage sustained. Overall, when compared to the field data, completely destroyed buildings were overestimated by 134%, while majorly damaged and partially/undamaged buildings were underestimated by 25% and 18%, respectively.

---

* While not yet fully explored and applied by NGOs, it should be noted that the authors see large and unexplored potential in various ways of support from the *crowd*. We see similar acceptance and proliferation patterns as with remote sensing data, even more rapid and wide ranging.

The highest uncertainty revealed the class "destroyed building," where only 16% were actually destroyed and around 70% were actually majorly or partially damaged. Conversely, buildings tagged as "undamaged" actually had major damage or were destroyed 50% of the time. Overall, the proportion of buildings that were accurately tagged by OSM contributors was only 36%. The assessment concluded that the OSM data were not reliable enough to utilize for damage analysis and recovery planning, but still sees OSM as strong platform for damage assessments in the future, if modest technological investments and better coordination mechanisms are provided (ARC 2014).

The potential surge of unsolicited, crowd-sourced data that may be of a type or format that does not fit easily into existing humanitarian data pipelines or workflows has created challenges for humanitarian responders, and the simple fact that nonprofessionals are engaged in any aspect of humanitarian response may cause consternation* (HHI 2011). However, some humanitarian institutions are paving the way for the public use of VHR imagery to support humanitarian operations. The Humanitarian Information Unit of the U.S. Department of State, together with representatives from the National Geospatial-Intelligence Agency and the U.S. Agency for International Development, met with the HOT to establish the necessary policies and protocols for federally purchased VHR imagery to be served via OSM precisely for the engagement of the V&TC. This process, known as "Imagery to the Crowd" has been used to map refugee camps in Kenya and Ethiopia (Humanitarian Information Unit). The increasing availability of refugee camp data available in OSM together with emerging disaster risk reduction projects that rely heavily upon it (Soden et al. 2014) and provide guidance for its use suggests that leveraging the crowd to interpret VHR imagery will become increasingly common. One attempt in this direction is the Missing Maps Project which supports the OSM HOT team and aims to "map the most vulnerable places in the developing world, in order that international and local NGOs, and individuals can use the maps and data to better respond to crises affecting the areas" (Missing Maps Project)

While the use of remote sensing has great advantages for direct observations, it is not undisputable from an ethical point of view. This particularly applies to the use of VHR sensors (Slonecker et al. 1998). VHR imagery, below 1 m resolution (see technical details in Table 19.3), provides a level of detail that can detect features relating to individual privacy (e.g., housing, cars). While individual persons cannot be identified on current satellite imagery, larger groups of people such as refugee trails can be traced (Ehrlich et al. 2009). The sensitivity of remotely sensed image data is a greater issue when resolution increases to the level of standard digital cameras, in the case of any kind of "in situ," that is, near-ground, remote sensing devices. The usage of unmanned aerial vehicles (UAVs, also known as "drones") is currently discussed from this perspective. UAVs, while often resembling

small military drones and thereby potentially mistaken,[†] offer a wide range of options limited not only to imaging but also in support to logistical tasks or even medical treatment, all still in experimental stage though (e.g., as recently tested by MSF to support tuberculosis treatment in Papua New Guinea).

19.1.3.2 Indication Based on Time Steps

Critical information can be provided via remote sensing technology in each phase of humanitarian crisis response (see Section 19.1.4). Before entering into more technical details regarding the actual observable parameters in the following section, we broadly reflect on the general capacity of EO systems in response to information needs by humanitarian actors. Satellite imagery captures a "snapshot in time" capable to represent the dynamic of a given phenomenon in steps only. To different but increasing degree, satellite sensors provide sequential imagery under comparable conditions, a crucial prerequisite to any kind of monitoring activity (Aschbacher 2002). As in any other application domain, the time span between two captured scenes is critical to understanding changes or directions of nonstatic events accordingly. Thus, the frequency at which imagery is captured needs to be commensurate to a situation's dynamics (see Section 19.3).

Just like other remote sensing applications, EO-based crisis response relies on suitable indicators, as usually the phenomenon under concern is not directly observable. Depending on the information need (see Section 19.1.4) and the respective type spatial/spectral resolution, passive/active, etc.), such indicators can be derived in different scales and extent, as follows: (1) The prevalence of a specific dwelling type (e.g., a tent) derived within a refugee camp area from submeter WorldView-2 data may represent vivid temporary living conditions for an estimated average occupation of people; (2) a specific pattern of linear structures in vegetation-scarce areas observed on Landsat imagery may indicate geological fault zones likely to store groundwater; and (3) high vegetation index values derived on district level from medium-resolution MODIS data may point to intensified agricultural activities suggesting increasing human presence, or in contrast a decreasing vegetation index over time might indicate extraction of wood resources due to locally increasing population in the camps.

19.1.3.3 Remote Sensing vs. Field Mapping

In comparison to conventional terrestrial field mapping and observations on the ground, EO-based humanitarian response benefits from general assets of remotely sensed data:

- *From a distance*: The core principle of remote sensing data, its obtainment from indirect contact with the object of concern, is critical to crisis-related applications. Often the area affected by the crisis is inaccessible or difficult to reach or, from a security point of view, too dangerous

---

* One of the reasons why NGOs are sometimes "consternated" is that EO data may include sensitive (i.e., political, military) information and NGOs might lose control on how it is used or abused—with adverse consequences, such as accusations of spying and not being impartial.

[†] Practically, NGOs have just started to look into the options of applying UAVs, and the problem is the hallmarking as a military tool, in fact a weapon to assassinate the enemy. At this stage, UAVs are hardly appropriate for humanitarian activities within complex emergencies. Instead, other types of natural disasters, chemical and nuclear accidents, etc., may greatly benefit from this right now.

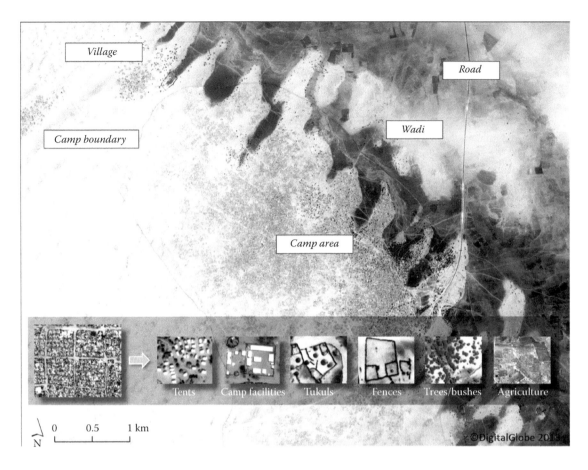

FIGURE 19.2 Detail and overview of a camp area provided by VHR data (here: Zam Zam IDP camp, Darfur, Sudan, WorldView-2). Depending on the type of camp, different dwelling structures can be identified besides camp infrastructure and surrounding land cover. The number of single dwellings, derived by visual interpretation or automated feature extraction, can be used for population estimations.

to enter (Aschbacher 2002). This means that information derived from remote sensing is the only information available (see Figure 19.2).

- *Area-wide coverage*: Depending on the granularity (spatial resolution), areas can be covered with variable extent, under the same imaging conditions and characteristics. The trade-off between resolution and extent is thereby a limiting factor that is also reflected in costs and timeliness of data provision (see Figure 19.6).
- *Global availability*: Satellite data are globally available with a theoretical cover rate of some 95% of the inhabitable space of the globe and factually 100% of the permanent settlement area. Note that cloud cover (e.g., in tropical latitudes) is a limiting factor for optical acquisition and analysis (see Section 19.3.2).
- *Retrospective view*: Time series enable not only constant monitoring in future time steps but also ex post assessments by past sequences. This is a key factor for estimating detected trend patterns in a more reliable matter (see Figure 19.4).

19.1.3.4 Space Policy and Regulations

In support to disaster management using information derived from remotely sensed data, there are several global, international,

and national mechanisms in place. The *International Charter on Space and Major Disasters* is an agreement among space agencies around the globe to provide satellite-based data and information in support of relief operations during emergencies and crises caused by major disasters (Voigt et al. 2007). Such disasters can either be natural or man-made, while the latter is limited to oil spills and industrial accidents and does not include conflict or crisis situations. Several national space agencies as well as the European Space Agency (ESA) participate in this endeavor and provide critical information resources to mitigate the effects of disasters. The Charter is activated by dedicated authorized users such as disaster management authorities from countries of Charter members. Up to now, through this mechanism, imagery free of charge has been provided in more than 400 cases since the first activation in the year 2000 (cf. International Charter).

Next to the provision of imagery, there are organizations or institutionalized services to provide dedicated information products derived from satellite data: Since April 2012, the European Copernicus* *Emergency Management Service* (EMS) is operational. This service provides timely geospatial information

---

* Formerly Global Monitoring for Environment and Security (GMES): the first operational services were called GMES Initial Operations ("GIO").

generated from satellite images and in situ or open data sources to actors involved in the management of natural disasters, man-made emergencies, and humanitarian crises. Besides a so-called "rush mode" providing information rapidly for immediate response for emergency management, a "nonrush mode" is dedicated to post disaster needs assessments and IDP/refugee camp monitoring. In addition to the EMS, the Copernicus program of the European Union (EU) has been funding research activities toward operationally providing geospatial intelligence information in support of EU external action. The aim is to fulfill a key objective of the EU that is to prevent global and transregional threats leading to destabilization. Consequently, information needs are—among others—related to conflict prevention and mitigation also including support to humanitarian crisis management.

The *REACH Initiative* is a joint initiative of international nongovernmental organizations (NGOs) to "facilitate the development of information tools and products that enhance the capacity of aid actors to make evidence-based decisions in emergency, recovery and development contexts" (REACH Initiative). One of the key players within REACH, United Nations Institute for Training and Research (UNITAR)'s operational satellite application program (UNOSAT), delivers satellite solutions and analysis to relief and development organizations within and outside the United Nations (UN). The aim is to provide reliable information to those "who work at reducing the impact of crises and disasters and help nations plan for sustainable development" (www.unitar.org).

A critical issue during disasters is the collection of and access to timely and reliable information. When it comes to this exchange of disaster-related information, two networks need to be mentioned: *ReliefWeb* (www.reliefweb.int) and *UN-SPIDER* (www.un-spider.org). Both organizations focus on providing fast and efficient access to space-based information to involved actors by compiling links and information on major disasters. As stated with respect to the Charter all the mentioned mechanisms can be activated for natural disasters; some include man-made emergencies and humanitarian crises, but there is no mechanism as yet in place dedicated to conflict situations.

Along with the aforementioned institutions/initiatives and programmes, new networks of humanitarian actors have formed. The *International Network of Crisis Mappers (ICCM)* was founded in 2009 and is composed of more than 6500 members worldwide whose experience with remote sensing in the humanitarian sector ranges from professional to volunteer (see http://crisismappers.net). ICCM maintains a "Google Group" that has become an important forum for ad hoc coordination of geospatial information and remotely sensed data. Its members point one another to existing sources of remotely sensed information, and the various V&TC that are part of the network have, in some cases, engaged directly with satellite imagery providers to acquire imagery. These providers alert ICCM if they are releasing satellite imagery without charge, as was the case in Haiti when DigitalGlobe made a significant number of VHR images publicly available at no cost. The annual conference of the network, now in its 6th year, is regularly attended by industry representatives, and companies such as DigitalGlobe and ESRI have been corporate sponsors.

The question often arises: can anyone (including NGOs) have access to satellite imagery? The answer is simple: yes–provided the technical and financial means are available for data handling and licensing costs). Generally speaking, neither the operation of satellites nor the acquisition of satellite data is bound to territorial sovereignty, as in airspace control. The international space law and related principles pursued by the *United Nations Office for Outer Space Affairs* (UNOOSA) underlines that remote sensing is "for the benefit and in the interests of all countries, irrespective of their degree of economic, social or scientific and technological development, and taking into particular consideration the needs of the developing countries. [It shall promote] the protection of the Earth's natural environment [and] the protection of mankind from natural disasters." This was phrased in a time (in the 1980s), where VHR data did not play a role yet in civil applications. Today, with submeter resolutions available, there are ethical issues arising whose regulations are yet to come (Slonecker et al. 1998; Table 19.2).

19.1.4 Information Needs for Humanitarian Action

In general, humanitarian action relies on firsthand, reliable information about the development of the situation within a certain region of interest. These regions of interest are often too remote or too insecure, or the situation is too dynamic to gather the required information timely in the field. Through satellite-based assessments, information can be provided for—among others—supporting the strategic planning of humanitarian relief missions (Bjorgo 2001, Kranz et al. 2010, Tiede et al. 2013). This kind of information is generally (highly) sensitive, as it might be used for help and support, but likewise misused for any strategic movements of actors in the conflict itself.

With respect to the phases of humanitarian crisis response, remote sensing can provide the following critical information:

1. Information in support to *early warning* or *disaster preparedness* (see Section 19.2.1). Prior to the outbreak of a humanitarian disaster, remote sensing can provide the necessary synoptic viewpoint that is strategically critical to understand the severity, geographical focus, and characteristics of an emergency situation, thereby improving preparedness for a disaster (e.g., drought). Likewise, the dynamics and directions of conflict-related destabilizing trends can be identified* to support planning of appropriate countermeasures. Situational awareness also includes the maintenance and cohesion of geospatial data layers for ensuring an adequate preparedness level.

---

* For example, a negative trend in vegetation cover as monitored by remote sensing may signalize increasing pressure on the local food production that may cause migration or conflicts about this increasingly scarcer resource. Even a hunger crisis might be predicted using indices derived from remote sensing data.

TABLE 19.2   Overview of Disaster Initiatives (Selection)

Disaster Initiative by Agency	Type of Service or Data	Specificities
International Charter on Space and Major Disasters (24/7 operational service)	Satellite-based data and information during major natural or man-made disasters	• Satellite data free of charge for authorized users. • Only for major natural disasters and limited to accidents with respect to man-made disasters. • Not for conflict regions or humanitarian crisis situations.
Copernicus Emergency Management Service (Operational service) *Rush mode*: within hours or days *Nonrush mode*: weeks/months	Geospatial information derived from satellite remote sensing images and in situ or open data sources in the course of natural disasters, man-made emergencies, and humanitarian crises *Rush mode*: standardized products—reference maps, delineation maps (providing an assessment of the event extent), and grading maps (providing an assessment of the damage grade and its spatial distribution) *Nonrush mode*: prevention, preparedness, disaster risk reduction, and recovery phases (reference maps, predisaster situation maps, and postdisaster situation maps)	• Free of charge for users. • Not for conflict regions and limited with respect to humanitarian crisis situations (depending on future strategic direction). • Results publicly available at http://emergency.copernicus.eu/mapping/.
Copernicus "security" service (Preoperational service) *Rush mode*: within hours or days *Nonrush mode*: weeks/months	Geospatial information in support of EU External Action (conflict prevention and mitigation, support to humanitarian crisis management), based on preoperational research projects	• The combination of preoperational service provision with research activities ensures further important developments. • Potentially toward services for conflict and humanitarian situations (depending on future strategic direction).
REACH Initiative	Information products for aid actors during emergency, recovery, and development phase; combines fieldwork and satellite imagery analysis	• Explicitly incorporates humanitarian disasters; products are open to all aid actors.
UN-SPIDER	Fast and efficient access to space-based information by compiling links and information on major disasters	• Collection of links and information provides an overview about different activities during certain disasters. • No own mapping and analysis capabilities.
ReliefWeb	Largest humanitarian information portal, compiling links, reports, maps, guidelines, assessments, info graphics, etc., on global crises and disasters	• Public access. • Overview about different activities during certain disasters; provision of several background and additional information, reports, and publications about certain regions. • No own mapping and analysis capabilities.
International Network of Crisis Mappers	Community platform	• At the intersection of humanitarian crises, new technology, crowdsourcing, and crisis mapping.

2. Information about the current situation (see Section 19.2.2) in support to *crisis monitoring* and *humanitarian action*. In the course of a conflict or during the peak of a disastrous event, remote sensing can be used to direct humanitarian response activities,* including indications (and verifications) of the affected population and settlements, destroyed infrastructure and other assets, or large-scale displacements and potential secondary crises—scenarios due to emergence of spontaneous settlements and pressure on local resources.

3. Information on the *mid- and long-term effects* of humanitarian disasters in support to potential integration and rehabilitation. Remote sensing data and image analysis help understand the effects of displacements and the impact on environmental conditions and resources, livelihoods, and land use practices, as well as resettlement and repatriation scenarios with (potentially) recursive consequences.

In the following, the example of refugee/IDP camp monitoring (Bjorgo 2000, Giada et al. 2003, Kranz et al. 2010, Lang et al. 2010, Kemper and Heinzel 2014) illustrates the information needs according to the underlying questions of humanitarian relief during the different phases of disaster

---

* As the timing of aid delivery is always crucial, EO-based information can help identify the spots where humanitarian needs are biggest. It does not replace missing movement clearances from other actors.

management. In order to prepare for a humanitarian crisis situation in a certain region, *indicators* need to be investigated that are suitable for risk assessment and early warning. The monitoring of large population movements and respective agglomeration might be a sign for increasing potential of local conflicts. The same is true for decreasing availability of natural resources in the vicinity of refugee/IDP camps due to increased extraction of water and firewood. During humanitarian disasters, information about the development of the situation in and around certain refugee/IDP camps is required focusing on an effective camp management or even mission planning. Generated information includes population dynamics, camp development and structure, and the impact on the environment, including potential pressure on natural resources. Underlying questions are related to the treatment of the inhabitants with food, water, medicine, and shelter and where the supply is most needed. In addition, it is important to gather information about the security situation in the area. Related to this issue, it might be important to get information about arising local conflicts through certain military or other violent attacks on camps and settlements in the region. The phase of *integration* and *rehabilitation* requires information suitable for supporting progressive stabilization of the situation. Besides a comprehensive picture about the effects of displacement and related environmental impact, this includes also reliable information about the options of long-term integration of migrants. Local governments and relief organizations require information that allow for estimating or even modeling the sustainability of the entire region with respect to natural resources (water, firewood, and building timber), setup of infrastructure, and development of socioeconomic parameters.

19.2 Crisis-Related Earth-Observable Indicators

EO satellite data is said to be *ubiquitous*, which relates to the general capacity to cover any point on Earth under orbit, but unfortunately not at the same time. Remote sensing data are limited to snapshots in time, even though the time spans in between can be quite low, up to the range of a few days or even hours. Here we have to distinguish between two cases: EO satellites of moderate (medium to high) spatial resolution record permanently while orbiting the Earth. For example, the USGS Landsat 8 program as well as the MODIS aboard National Aeronautics and Space Administration's Terra satellite, deliver data in a fixed, sequential tracking mode, covering any particular location on the Earth in a regular interval, say every 16 days. On the other hand, VHR sensors, like WorldView-2 or TerraSAR-X, capture and deliver data on demand. Depending on the technical setup (skewing sensor, sensor constellation, etc.), a higher frequency of coverage can be achieved as the orbital revisiting rate allows. The hypothetical (yet not factual) availability makes remote sensing a responsive, reactive tool

that either requires searching in the archives, waiting, or tasking. In other words, whether or not remotely sensed imagery is available is a function of ideal capturing conditions ("chance") and a certain trigger ("action"). Again, this is very much like a journalist's camera capturing a certain event, while on a different scale. But even if the sensor would be ready and in place, with an ideal temporal matching scenario, other constraints may hamper the availability of a *perfect scene* in the end: atmospheric conditions, natural conditions (e.g., seasonal vegetation cycles), shutter controls, and, last but not least, costs involved.

19.2.1 "Early Warning"

Any capability of remote sensing to act as an "early warning tool" that prevents conflict from escalating or a natural event from turning into a catastrophe is highly desirable. There are many encouraging examples toward this potential – e.g., the inventory of FEWSNET (www.fews.net). On the other hand, remote sensing data have limited predictive power due to their indirect measurements, availability, and continuity and also to the unpredictability inherent to the systemic effects of such events, in particular of the human behavior. To consider the limited foresight potential, the term "prediction" will be replaced with "indication" in this context, as it is difficult to tell whether any observed parameter is an indication of something already going on or yet to come. Examples of the indicative power of remote sensing that can affect levels of *sensitivity* and *preparedness* include the following:

- Anomalies in time series of soil-moisture data as compared to normal seasonal variability may indicate a forthcoming drought threatening food security (Wagner et al. 2003, Kuenzer et al. 2008, Rhee et al. 2010).
- Rapid growth of a refugee/IDP camp may indicate the rise of additional security and safety issues within the camp, as well as the conflicts related to the diminishing surrounding resource supply (Hagenlocher et al. 2012) (see Section 19.5.2).
- A damage density map showing different magnitudes of building damages and their distribution caused by an earthquake, flood, or conflict-related destruction may indicate areas most affected and in most urgent need to response (Pesaresi et al. 2007, Tiede et al. 2011).
- Detected areas where illegal or informal activities (logging, mining, cropping, etc.) are carried out in an increasing scale can be an indicator for potentially upcoming regional instabilities and conflicts (Schöpfer and Kranz 2010, Lüthje et al. 2014).
- Decreasing vegetation cover in the vicinity of IDP camps might indicate increasing pressure on natural wood resources and longer distances for the local population to walk for collecting firewood (and consequently growing insecurity especially for women), potentially leading to regional conflicts.

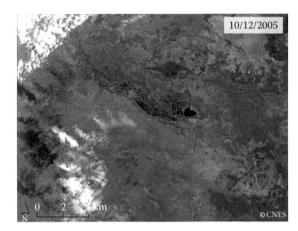
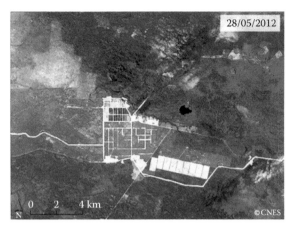

FIGURE 19.3   Conversion of a piece of land from forest, shrubland, and grassland to large-scale agricultural fields including construction of roads and irrigation systems in Gambella, Ethiopia, visible on VHR data from SPOT-5 between 2005 and 2012.

- Rapid conversion of a stretch of land from unmanaged savannah or forest into a large agricultural complex may represent large-scale land investments that potentially imply the displacement of smallholders and pastorals, destruction of villages, higher pressure on local resources due to intensification of production, and the like (FAO 2013) (see Figure 19.3).

The above statement on the limited predictive power of remote sensing may sound a bit discouraging. This is indeed the most challenging and most research-intensive aspect of this technology. This application domain competes, as many others, with the benchmark of operational weather forecast. The key ingredient of any indication that shall gain more robustness, in terms of *evidence* (backward looking) or *prediction* (forward looking), is repeated analysis based on *time series*. Time series capture the dynamics of a phenomenon and allow filtering out anomalies from the regular case or a higher degree of change than usual. Note that all of the examples mentioned earlier contain some changing conditions, amplified by the word "rapid." A single snapshot in time needs either subsequent, recursive information or reference information from a time slot in the past. In addition, as in many other technological fields, the complementarity of tools and available data collection devices may be the key to trigger efficient action "ahead of time."

19.2.2 Crisis Monitoring

As mentioned earlier, the exact stage when a humanitarian crisis begins is often difficult to determine. The period prior, during and after a disaster is often a gradual transition and not a fixed point in time. Here we examine the immediate effects that accompany such disasters and how the crisis can be monitored and further escalation mitigated:

- The provision of drinking water is the first and foremost key prerequisite when maintaining a refugee or IDP camp. Remote sensing can help to narrow down the range of (if not allocate) potential groundwater supply (Drury and Deller 2002).

- Understanding the population dynamics within a camp including the general growth of the camp, its extent camp structure, densification of camp sections, and partly de- or reconstruction. Remote sensing can help estimate the overall number of people present in a camp and get an overview on camp management facilities.

- Monitoring the impact of refugee or IDP camps on the surrounding environment, for example, based on changes in wood resources or agricultural activities, helps assessing the increasing pressure on natural resources and might support decisions for mitigation and even prevention of raising conflicts (transitions to midterm impact).

19.2.3 Mid- to Long-Term Impact

Humanitarian crises, in particular complex or protracted crises, may have a long-term impact on the societal and environmental integrity of the affected area. Remote sensing can help assess this impact through the use of time series, including a retrospective view (Lang et al. 2010, Hagenlocher et al. 2012) (see Figure 19.4):

- The prolonged existence of a temporary camp* runs the risk of a de facto transition to a semipermanent settlement with critical effects on both the societal integrity of the hosting community(ies) and the carrying capacity of the environment (Martin 2005). Remote sensing can help analyze the spatial impact of such large-scale, long-lasting settlements in terms of infrastructure and food and resource supply.

---

* What sounds like a contradiction in terms relates to the fact that politically speaking there is no such thing as a semipermanent camp or any such transition. While (especially IDP) camps may de facto exist over time spans of more than 10 years, such development is neither intended nor supported by NGOs, the UN, or the host communities. According to UNHCR, the term "protracted refugee situation" is used for long(er)-term camps that are defined as "one where more than 25,000 refugees have been in exile for more than five years." According to this definition, around 6.3 million refugees were in a protracted situation by the end of 2013 (UNHCR, Global Trends, 2013).

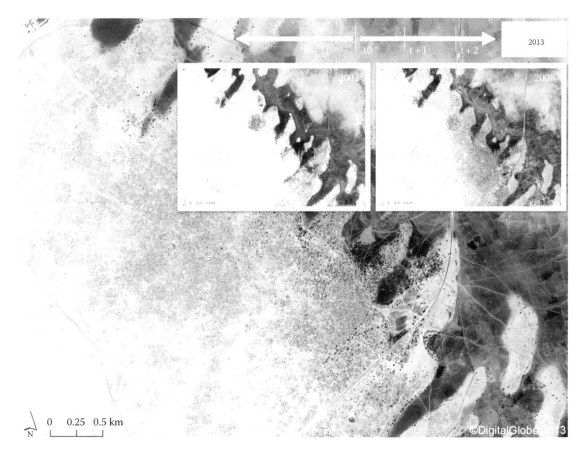

FIGURE 19.4   Retrospective view using VHR satellite time series over Zam Zam IDP camp, Darfur, Sudan.

- The impact of resource extraction, such as logging and exploitation of minerals on the environment, is critical as well. A sustainable management of natural resources could contribute to long-term stabilization of a certain conflict region.
- The mid- to long-term monitoring of natural resources in conflict regions provides important information for international transparency initiatives to be set in place. This is also contributing to stabilization and peacebuilding in the affected region.

19.3 (Satellite) EO Capacities

This section provides a broad synopsis of satellite sensors that are currently used in the context of humanitarian action and disaster response (see also section on case studies, where the actual usage is demonstrated in various examples). As in other application domains of remote sensing, there are evolving dynamics at play: (1) technical developments including new sensor technology, image retrieval and analysis, and mobile communication devices and (2) user developments including changes in acceptance and debate of such technological assets

over recent years. Only when we can find a match between both, the technical capacity and the trust and willingness to use it, the former may turn out to be a clear asset for the latter.

19.3.1 Usage of EO Data: A Matter of Reliable Supply and Acceptance

While talking about assets and advantages of this technology we need to accept that there is still a gap between the capacity and the actual usage. To understand this, we need to address what the main concerns are that users from the humanitarian domain bring about remote sensing. We hardly find concerns clearly reflected in scientific publications; they are rather expressed informally during workshops and discussion rounds. For example, at the annual Humanitarian Congress in Berlin, in 2012 and 2013, two technical sessions focused on the use of satellite and UAV technology in the context of humanitarian action. The aim was to demonstrate use cases and the general potential of geographic information systems (GIS), satellite and near-field remote sensing, and some other technological achievements of recent years to the user community (field workers, practitioners, decision makers, other scientists). While the community of practice in general seems to be eager to use, adapt, or at least test such new technologies, there is still reluctance from individuals or institutional barriers, due to either bad own experiences or the latent concern of a potential dual use of satellites and drones.

19.3.2 Different Tasks: Different Sensors

In general, most of the available sensor types can be used in the humanitarian context as well, depending on the nature of the task. Sensors can roughly be divided into two broad types: optical (passive) and microwave (active) sensors. Both can be subdivided according to their spatial resolution into low (>300 m), medium (30 to <300 m), high (HR2 10 to <30 m, HR1 4 to <10 m), and very high (VHR2 1 to 4 m, VHR1 <1 m) spatial resolution sensor types. Besides the spatial resolution, sensors are characterized by different specifications in terms of spectral bands, repetition rate, etc. Since data acquisition is a matter of these characteristics in conjunction with costs, acquisition time, and effort, the decision whether satellite data are applied needs to be taken carefully. For example, if it is about wider impact of a large-scale resettlement of people in the course of a conflict, a medium- to high-resolution sensor like Landsat or MODIS may be sufficient. If instead the issue is about population monitoring and dynamics, the extraction of individual housing may require VHR imagery such as WorldView-2. In the first example, the temporal resolution might be a key factor for the analysis when frequent availability over a longer time period allows conclusions about the causes and the direction of certain trend patterns. Another factor influencing the decision are the costs involved, especially if long-time monitoring for large areas is required. In the case of monitoring environmental impact,

Landsat data may suffice, balancing out the level of detail and costs involved. When it comes to retrospective analyses, the decision between different sensors within the same family (HR1, HR2, VHR1, VHR2), is mostly determined by available archive data (Figure 19.5).

19.3.2.1 Optical Sensors

Optical satellite sensors are most commonly used for *direct mapping* of geographical features relevant to a humanitarian crisis scenario.

Direct mapping means that visual cues on optical images are directly comprehensible to human vision, thus can be identified and mapped. Examples include (see also Section 19.2)

- The recent conversion and current use of a piece of land as aggregates of agricultural fields as compared to a previous state (e.g., in the course of large-scale land investments sensu "land grabbing" Figure 19.3).
- The prevalence of unusual color (i.e., spectral) characteristics of cropland during a vegetative drought.
- The emergence of a temporary settlement (e.g., IDP camp) on previously agricultural land.
- The presence of different dwelling types and infrastructure inside a refugee camp (e.g., tents, huts, camp management facilities, or pathways).
- The appearance of a quarry indicating quarry, indicating illegal mining activities or a clear-cut suggesting illegal logging.
- The inconsistent shape of buildings and the presence of clutter around their footprints assuming heavy damage or collapse.

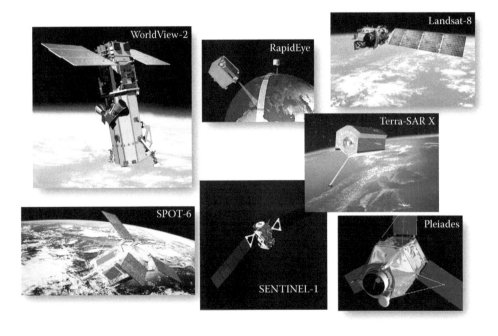

FIGURE 19.5 Selection of HR and VHR optical and SAR satellite systems in support of humanitarian action and disaster response.

These features appear intuitively to a skilled interpreter, able to abstract features from different scale of representation (*remote sensing*), different perspective (*top* view), and (most often) different spectral characteristics as when viewed from the ground. Optical imagery can thus be interpreted visually (some say *manually* as usually there is some kind of tool, a pencil or mouse, that is operated manually) or (semi-)automatically by the use of dedicated algorithms. Computer algorithms prove to be strong in discerning different spectral behavior when working on a pixel-by-pixel basis. On the other hand, the human eye is superior in identifying specific structures or patterns in multiple scales across the image, but it is highly subjective and unrepeatable (Jensen 2005). Nevertheless, some experiences in interpretation are required especially when it comes to false-color interpretations. The trait of human perception is difficult to automate in particular on a reasonable level of efficiency and reliability. Recently, great efforts went into the automation of visual processes, for example, using object-based image analysis (OBIA) (Blaschke 2010). This approach is likewise promising for operational tasks but also requires care and operator's responsibility considering the complexity and high level of sensitivity of information extraction from VHR EO data (Lang 2008) (see also Section 19.4; Figure 19.6).

Table 19.3 gives an overview of optical sensor characteristics, including spatial, spectral, radiometric, and temporal resolution as well as indicative costs and some of the key features and application areas in the context of humanitarian crisis response.

The dependence on atmospheric conditions, even more than the match of sensor specifics to extractable features in terms of image resolution (spatial, spectral, etc.) and revision time, constrains the usability of optical sensors. *Optical sensors* record sunlight reflected by the Earth's surface and thereby work in a *passive* mode, that is, not producing their own radiation. Atmospheric conditions, in particular the presence of water vapor and other gas molecules, interact with the wavelengths, which to optical sensors are sensitive. These influences may hamper the quality of images when clouds partly or fully obscure features on the ground. When critical features (e.g., an IDP camp) are not visible, an entire satellite scene may be rendered useless, as the required information simply cannot be obtained. A class of sensors not affected by atmospheric conditions, but with other constraints, are radar sensors.

19.3.2.2 Radar Sensors

As opposed to optical imagery, *radar sensors* are *active* devices, recording microwave radiation that they emit. Due to their specific wavelength range, microwaves (or radio waves) do not interact with water vapor or other atmospheric particles and thereby also penetrate clouds. The generated signal is independent of any external energy source, allowing radar image acquisition under bad weather conditions; over clouded, tropical forests; and at night. While this independence from weather and daylight is an undoubted pro for their usability, radar data are more complex to process and analyze and much less (some say not at all) intuitive to interpret. In operational application scenarios in the disaster response domain, radar data are mostly limited to the detection and extraction of specific, well-defined aereal features, such as a water mask in the aftermath of a flood or the trafficability of a road network after the peak of the flood has receded. Other applications try to utilize standardized measurements of biophysical parameters such as soil moisture to better understand drought occurrence or even predict them (Wagner et al. 2003). Other current research looks into the transfer of radar technology to optical image analysis domains such as dwelling extraction in refugee camps (Bernhard 2013), by combining the extractability of features using higher spatial resolutions with the ubiquity of light-independent radar technology.

In Table 19.4, some of the most prominent radar sensors are listed and characterized, again with potential (or actual)

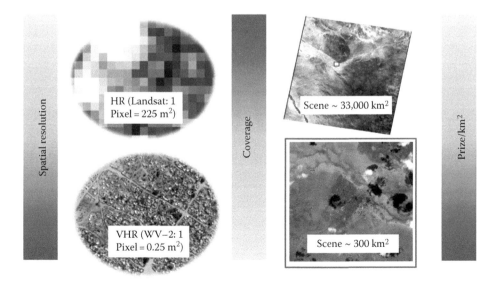

FIGURE 19.6 Acquiring satellite data is a trade-off between resolution (here, pan sharpened), extent, and prize.

TABLE 19.3   Selection of Optical Sensors: Key Characteristics and Application-Relevant Features

Platform/Sensor	Resolution (Spatial/Spectral/Radiometric/Temporal)	Humanitarian Application Domain (Including Examples of Detectable Features)	Indicative Costs (Archive/Tasking) (EUR per km²)[a]	Minimum Order Size (km²)
MODIS	250 m (ground sample distance)/36 bands/16 bit (quantization)/16 days (to revisit)	The NDVI product allows the analysis of time series for vegetation trend patterns	0	1 scene
Landsat-7	15 m/8 bands/8 bit/16 days	Large-scale land cover change analysis	0	1 scene
Landsat-8	15 m/11 bands/16 bit/16 days	Vegetation trend patterns	0	1 scene
DMCii	22 m/3 bands/8 or 10 bit/daily	Logging activities, clearings, road infrastructure built-up	0.01–0.14 (minimum price for archive, 1,260)	25,600
RapidEye	6.5 m/5 band/16 bit/daily	Logging activities, clearings, general infrastructure built-up Bare soil detection (as indication for mining sites)	1	500 for archive, 3,500 for tasking
SPOT-4	10 m/5 bands/8 bit/2–3 days	Land cover change analysis Vegetation trend patterns	1–2.5 (depending on size)	400
DMCii	2.5 m/4 bands/16 bit/2 days	Road network, detecting larger buildings, detailed land cover information	1.9–2.5	1600
SPOT-5	2.5 m/5 bands/8 bit/2–3 days		2–8 (depending on size)	400
FORMOSAT-2	2 m/5 bands/8 bit/daily		3–4.5	576
SPOT-6/7	1.5 m/5 bands/12 bit/daily if using both satellites		4–4.5	250 for archive, 1000 for normal tasking, 100 for emergency tasking
IKONOS	1 m/5 bands/11 bit/3 days	Feature extraction	7–14	25 for archive, 100 for tasking
QuickBird	0.6 m/5 bands/11 bit/1–3.5 days	Single dwellings/huts/tents	12–18	
GeoEye-1	0.5 m/5 bands/11 bit/2–8 days	Single trees Fences, walls		
WorldView-1	0.5 m/1 band (pan)/11 bit/1.7 days	Groups of people	10–16	
WorldView-2	0.5 m/8 bands/11 bit/1.1 days	Mining site detection Stereo capability: DEM, 3D structures	4 bands: 12–18 8 bands: 21–28	
Pléiades 1/2	0.5 m/5 bands/12 bit/daily		10–17	
WorldView-3	0.31 m/17 bands/11 bit/daily		4 bands: 32–58 (for 0.3 m) 8 bands: 40–73 (for 0.3 m)	25 for archive, 100 for tasking
SkySat	0.9 m/5 bands/11 bit/3 times per day, constellation of 24 satellites	Same as above + monitor population movements and infrastructure development		50

[a] The information on costs is only a rough indication. Actual costs depend on priority and preprocessing levels, whether from archive or tasked and other parameters. Details to be obtained from data provider catalogues; handling or administrative fees are not considered. Prices refer to bundle (panchromatic and multispectral bands), where applicable.

application domains. Note that the table also includes the recently launched ESA satellite Sentinel-1, which provides data in operational mode since September 2014.

19.3.2.3 Nano-/Microsatellites

There is a substantial growing market of small satellites that are designed to reduce costs by minimizing mass. Several categories are distinguished: small satellites (100–500 kg), microsatellites (10–100 kg), and nanosatellites (1–10 kg). Beyond that there are picosatellites (<1 kg) and femtosatellites (10–100 g) in production. In 2013, almost 100 micro- and nanosatellites were launched. Many of them are built in the CubeSat standard format, with a volume of exactly 1 L (10 cm cube) and a mass of no more than 1.33 kg. The spatial resolution is up to 1 m with revisiting times of up to several hours. Companies like Planet Labs, Spire (formerly Nanosatisfi), Surrey Satellite Technology, Dauria Aerospace, or Skybox

Imaging are planning to launch many more of their nano- and microsatellites in the upcoming years.

19.3.2.4 Unmanned Aerial Vehicles

Intuitive in usage ("toy factor") and potentially unlimited in their usability, *unmanned aerial vehicles (UAV)* are still highly disputed in the humanitarian response domain. In fact the Humanitarian UAV Network has been established with the aim to bridge humanitarian and UAV communities and to establish clear standards for humanitarian use. The advantage over satellite data may be seen in the flexibility and controllability of the tool, and the lack of additional costs once a device is purchased. However, legal and factual constraints may limit the usage of UAVs due to accessibility or risk. The operator needs to be physical on-site: in the course of natural disasters, this is more feasible as NGOs like MapAction may share their immediate information needs with first responders who have their own UAVs. But

TABLE 19.4 Selection of Radar Sensors and Their Specifics

	Band	Polarization	Spatial Resolution/Repeat Cycle	Application Domains
Sentinel-1	C-band	HH, VV, HV, VH		Flood detection, oil spills, sea ice monitoring, ship detection, forest monitoring (forest cover, vertical structure, biomass), surface movement monitoring
(1) Interferometric wide swath mode			5 × 20 m	
(2) Wave mode			5 m	
(3) Strip map			5 m	
(4) Extrawide swath			20 × 40 m	
TerraSAR-X:	X-band	HH, VV, HV, VH		
(1) ScanSAR mode			18 m/2.5 days	
(2) Stripmap mode			3 m/2.5 days	
(3) Spotlight mode			1 m/2.5 days	
Radarsat-1	C-band	HH		
(1) Fine res.			8 m	
(2) Standard			30 m	
(3) ScanSAR			50–100 m	
Radarsat-2	C-band	HH, VV, HV, VH		
(1) Spotlight			3 × 1 m	
(2) Ultrafine			3 × 3 m	
(3) Fine res.			10 × 9 m	
(4) Standard			25 × 28 m	
(5) ScanSAR			50 × 50 m	
(6) Fine quad-pol			25 × 28 m	
ALOS-PALSAR	L-band	HH, VV, HV, VH		
(1) Fine beam single			10 m	
(2) Fine beam dual			20 m	
(3) Direct downlink			20 m	
(4) ScanSAR wide beam			100 m	
(10) Poliametric			30 m	
COSMO-SkyMed	X-band	HH, VV, HV, VH		
(1) ScanSAR mode			30 m	
(2) Stripmap mode			3–15 m	
(3) Spotlight mode			1 m	

for (protracted) humanitarian disasters, it turns out to be difficult, when restricted access and security issues do not allow acquisitions or proper devices are simply not available. While the general strength of UAVs may be seen in their universality, the need for careful operation and handling and the limited coverage make it even questionable in direct comparison to VHR data, from the group of micro-/nanosatellites in particular.

19.4 Image Analysis Techniques

19.4.1 General Workflow: Example Population Monitoring

Image analysis does not start at the interpretation stage; the workflow starts much earlier (Lang et al. 2006). Figure 19.7 shows the principal workflow, as pursued by an information service run by the Department of Geoinformatics-Z_GIS, University of Salzburg for Médecins Sans Frontières (MSF) on population monitoring. Situational awareness is a crucial first step to understand the information needs and the required products. This can be achieved by a standardized questionnaire or user request form. Accordingly,

imagery is acquired—from archive if less time critical or tasked when the actual situation is to be represented. In addition to image acquisition, any available auxiliary data (such as local reports, data on camp infrastructure) need to be collated and integrated. The image preprocessing part (georeferencing, orthographic and atmospheric correction, calibration) is followed by semiautomated analyses of the satellite image and spatial analysis. The latter steps are described in more detail in the following chapters. The final step of the workflow is the user-targeted information delivery by choosing the most appropriate communication medium and performing validation in both technical (accuracy) and usability aspects.

19.4.2 Visual Image Interpretation

The majority of EO-based information in the context of disaster management is currently retrieved by visual image interpretation. In humanitarian disasters, the focus is often on detailed analyses, requiring interpretation of specific features (e.g., buildings/dwellings, roads, infrastructure facilities). Visual interpretation is seen as most reliable for such complex interpretations, as human vision usually outperforms algorithmic approaches

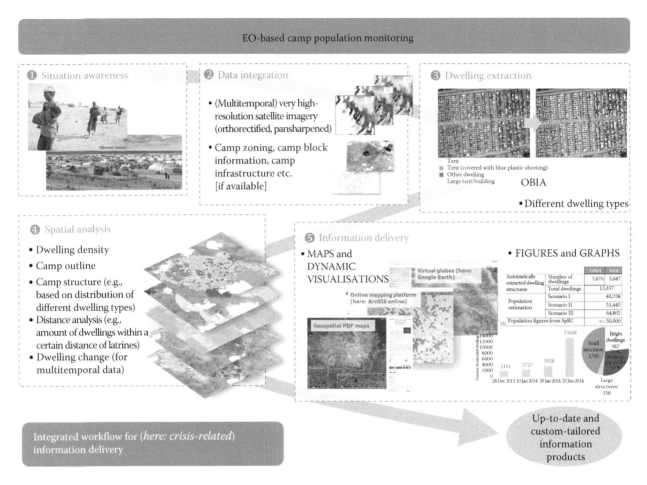

FIGURE 19.7   Workflow for EO-based information delivery in the context of camp population monitoring (Z_GIS).

in identifying complex features and structures, even at different scales, image contexts, image quality, or seasonal effects (Checchi et al. 2013). Visual perception can be trained by experience and thus adapt to various situations. There is an "in-built" capacity to delineate features in a given scale domain.

However, when it comes to repetitive, standardized tasks (e.g., the enumeration of single dwellings), visual inspection becomes increasingly inefficient and prone to errors or rash generalizations. Here, automated classification and feature extraction techniques (see Section 19.4.3) start getting more efficient and superior in operational, analogous settings (Tiede et al. 2013). Ideally, human vision and automated techniques are combined in a hybrid approach (Füreder et al. 2014) each contributing its specific strengths in a complementary manner. In fact, human vision is considered a benchmark for automated approaches. There is currently no such automated techniques that better *see* than humans. But often human vision faces similar challenges as automated approaches in differentiating complex structures.

When, for example, looking at a VHR satellite image of a refugee camp, the focus for the interpretation—after a first assessment of the situation including accessibility, water sources, and land use/land cover—is the camp itself. A human interpreter can quickly assess the extent of the settled areas and focuses

the subsequent interpretation on that area. False positives in the outskirts of the camp (e.g., dry vegetation, tree shadows, or rocks with similar spectral and geometric characteristics like traditional huts) can thereby be avoided. Depending on the type of camp (refugee/IDP camp, self-settled, unplanned or planned, managed camp, informal settlement at urban fringes and geographical area, the camp itself shows different dwelling structures (e.g., tents covered with white or blue plastic sheeting with rectangular or tunnel shape, traditional round huts, shelters made of local materials, dwellings with metal roof, makeshift shelters), fences to separate family compounds (more common in IDP camps), and larger camp facility areas. The human eye can quite easily adapt to such different camp structures and dwelling types. The visual interpretation of single dwellings is, however, limited in complex situations, for example, connected dwellings, dense complex pattern of roofs, dwellings with little contrast to bare soil or dwellings covered by vegetation (Checchi et al. 2013).

Before starting the visual interpretation, a set of features has to be defined on which to focus. This predefined interpretation strategy is less flexible to adapt during or after interpretation (e.g., introducing new classes or dividing one class into two depending on specific criteria). In some cases automated techniques may provide additional information not directly visible from one single scene. Humanitarian disasters caused by droughts, famine, or slow-onset floods are examples where the automated analysis of frequently acquired satellite images over a certain period of time provides important information about the dynamic of the situation in the region of interest. In the context of slow-onset floods, it is important to get information not only about flooded areas but also about intact infrastructure to serve as humanitarian aid, potential evacuation routes, or gathering areas. Furthermore, such analyses can be used for an effective early warning system to prevent humanitarian disasters. While less visible, droughts have also significant impact on both human lives and economic development and frequently cause devastating humanitarian crisis situations, especially in sub-Saharan Africa. Indications on the onset of drought conditions provide a reliable basis for an effective early warning system.

19.4.3 Automated Feature Extraction and Image Classification

The use of automated feature extraction methods in an operational mode is still limited to targeted analyses (e.g., extraction of flood mask, extraction of distinct dwelling structures). More complex feature extraction is still considered "experimental."

Automated methods have their advantages in being scalable and transferable. This makes it predestined for monitoring analyses, analyses of large areas or large amount of features, respectively (Figure 19.8).

Several methods for automated feature extraction have been recently reported (Tiede et al. 2010, Kemper and Heinzel 2014). OBIA techniques (Lang 2008) are flexible to adapt during classification (e.g., reclassifying classes) and can to some degree deliver more detailed information (e.g., size of buildings). In complex ground situations, where automated feature extraction methods reach their limits, it makes sense to combine automated and manual interpretations. Automated methods can concentrate on the simpler structures; more complex structures can be manually refined. This would reduce analysis time compared to purely manual interpretation and still deliver reliable information (Figure 19.9).

19.4.4 Spatial Analysis and Modeling

The results of image interpretation and classification can be enhanced by GIS analysis. Aggregated information often provides a better overview of a situation, wherever there is a high amount of features or classes that cannot be grasped easily. Such aggregated maps can be provided by density calculations

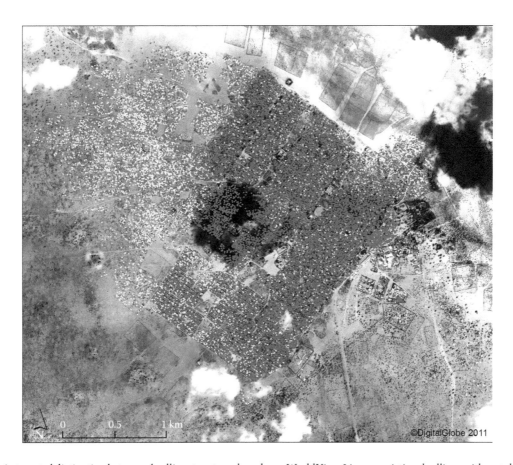

FIGURE 19.8 Automated distinction between dwelling structures based on a WorldView-2 image: existing dwellings with metal roof (violet) and new dwellings (yellow, white plastic sheeting; red, makeshift shelter).

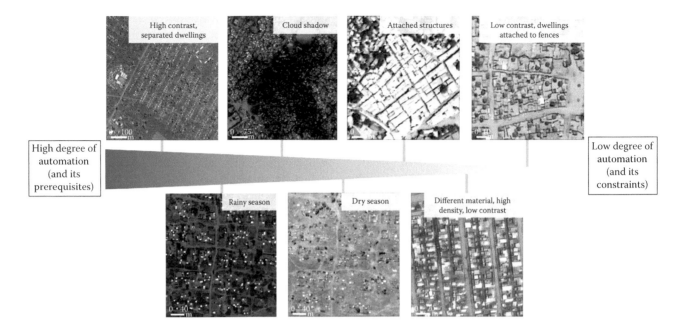

FIGURE 19.9   Automation level for dwelling extraction in relation to degree of complexity of image data. (Adapted from: Füreder, P. et al., *South-Eastern European Journal of Earth Observation and Geomatic*, 3(2s), 539, 2014.)

from extracted point data (e.g., dwelling density to see where population in a refugee camp is concentrated or density of damaged buildings for a first indication of the most affected areas) or aggregating classification results to larger reporting units (administrative units, regular grids, hexagons, etc.) to show predominant classes (e.g., regarding land cover or dwelling types).

The extracted information from EO data can also be enriched with other data. To keep with the example of spatial-explicit information of single dwellings, this information can be used to calculate distance maps for camp planning, for example, how many dwellings are within a specific distance to boreholes, hospitals, latrines, etc. (see Figure 19.10).

For monitoring purposes, multitemporal analyses provide changes between two or more time stamps. A post classification comparison can either look at every change, but for a large number of classes (e.g., land cover), such multidirectional changes are difficult to interpret, or changes for selected classes. Also here the interpretation can be significantly improved if relevant changes are aggregated to larger reporting units.

19.4.5   Validation and Ground Reference Information

The technical part of validation consists of an assessment of the accuracy of an image classification. A classification error is defined as a discrepancy between the classification and reality (Foody 2002). The results of an accuracy assessment, however, relies on accurate ground data to (Lang et al. 2010): ground data may also contain errors and are often dependent on subjective interpretations. In most of the presented case studies, often

spatially explicit ground data were not available due to the unsafe situation. In a similar study, the London School of Hygiene and Tropical Medicine and the Department of Geoinformatics-Z_ GIS at Salzburg University assessed the estimation of population sizes using two different image analysis methods in Am Timan City, Chad (unpublished). The comparison of automated object-based dwelling extraction and manual count was benchmarked at a standard population survey (quadrat method), carried out in January 2012. It turned out that the mean total structure count for the automated method was some 10% lower than the manual method, mainly due to the fact that the automated method could not discern between close-by buildings that were identified as one larger building. Manual postprocessing led to very similar results, with a difference of some 2% at a total amount of some 12,000 dwelling structures. In terms of time required for the analysis, the manual method did not require more time as compared to the automated one. Once the algorithms are developed, however, the transfer to similar tasks will significantly lower the performance time of the automated method as compared to the manual one, which rather operates linearly from case to case.

19.5   Case Studies

In this chapter, we want to illustrate the discussed material on selected case studies. They cover each of the three phases discussed earlier (early warning, crisis monitoring, and mid- to long-term impact) and also demonstrate the variety of data usage including free data sources. All case studies refer to *real cases*, scenarios and analyses that have been performed on real data, extensive time series, etc.

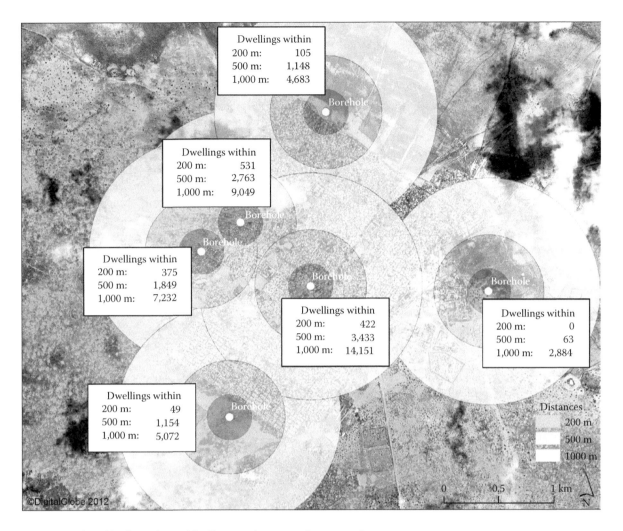

Dwellings within
200 m: 105
500 m: 1,148
1,000 m: 4,683

Dwellings within
200 m: 531
500 m: 2,763
1,000 m: 9,049

Dwellings within
200 m: 375
500 m: 1,849
1,000 m: 7,232

Dwellings within
200 m: 422
500 m: 3,433
1,000 m: 14,151

Dwellings within
200 m: 0
500 m: 63
1,000 m: 2,884

Dwellings within
200 m: 49
500 m: 1,154
1,000 m: 5,072

Distances
200 m
500 m
1000 m

0   0.5   1 km

©DigitalGlobe 2012

FIGURE 19.10   GIS-based buffer analysis of dwelling numbers around water supply posts.

19.5.1 Monitoring Population Dynamics in the Refugee Camp Dagahaley in Kenya

19.5.1.1 Background

The Dagahaley refugee camp is located in the Garissa district in Kenya, around 100 km from the border to Somalia. The camp was established by the UN refugee agency (United Nations High Commissioner for Refugees [UNHCR]) in 1992 around the town Dadaab together with the camps Ifo and Hagadera. Initially planned to host up to 90,000 people, mainly refugees from the civil war in Somalia, it has been extended by two additional camps in 2011 (Ifo 2 East and West and Kambioos) and is now the world's largest refugee camp complex with about 350,000 people. Until 2006, the population of the Dagahaley camp was quite stable with about 30,000 inhabitants. New refugees arrived in 2006, 2008, 2010, and 2011 (UNHCR 2014). The large influx of refugees in 2011 was caused by a combination of a severe drought and famine at the Horn of Africa and ongoing conflicts and violence in Somalia. New arrivals settled in the outskirts of Dagahaley, where in July 2011 around 25,000 people lived (MSF 2011). The enormous influx of people to the Dagahaley refugee

camp brought the camp registration to a halt and revealed the need for a more efficient camp monitoring. Moreover, emerging violence against Kenyan security forces but also humanitarian workers at the end of 2011 hampered the ability of aid agencies to provide relief assistance, protection, and essential services (UNHCR 2012).

19.5.1.2 Data Used

For monitoring the development of the Dagahaley camp, WorldView-2 images from July 2011 and December 2011 with eight multispectral bands (2 m spatial resolution) and a panchromatic band (0.5 m spatial resolution) were available. Due to cloud cover in the December image, the eastern part of the image was replaced by an additional WorldView-2 image from January 2012.

The images have been pan-sharpened using the hyperspherical color sharpening (HCS) method, designed for sharpening WorldView-2 images. This method can handle any input bands and produces acceptable color and spatial recovery (Padwick et al. 2010). The December and January images have been georeferenced based on the July image to ensure spatial match between the images for change analysis. Due to different recording conditions (off-nadir angle, target azimuths), a small offset between the images occured.

19.5.1.3 Methods Applied

Dwellings were semiautomatically extracted using a core algorithm for dwelling extraction that relies on OBIA. This ruleset, coded in Cognition Network Language within eCognition (Trimble Geospatial Imaging), is based on adapted segmentation using edge filtering to extract settlement areas in a first step. For dwelling extraction, relative spectral differences and spatial characteristics are used to increase transferability (Lang et al. 2010). This *master ruleset* (Tiede et al. 2010) is transferable to other time intervals or other geographical areas with similar camp setting. For the Dadaab region, the ruleset was adapted regarding dwelling types and the increased spectral information of WorldView-2.

For the analysis of the July image, three dwelling types were distinguished: tents, huts, and dwellings with metal roof. The analysis was conducted in several steps. A rapid assessment of the newly settled areas in the western outskirts of Dagahaley concentrated on tents, the prevailing dwelling type there. Makeshift huts were digitized manually due to small and complex structures and similarity to dry vegetation. The first results were produced within one day after data preprocessing. Subsequently, the algorithm was adjusted to the whole camp area, where also dwellings with metal roof, the predominant dwelling type within the camp, were extracted. The ruleset applied to the July image was transferred to the December image. However, clearly distinctive indicators of newly settled areas were missing and only a few makeshift huts were still present. Many dwellings with metal roof, which were present in the July image, were covered with white plastic sheeting in the December/January images due to the rainy season, which made distinction to white tents difficult. Therefore only one class dwelling was extracted for the second timeslot. Finally, minor manual refinement was performed to eliminate obvious classification errors. The rulesets of both images were slightly adapted to camp areas covered by cloud shadows. Thereby, the classification of shaded and unshaded areas and the combination of the results were integrated within one ruleset to increase the degree of automation (see Figure 19.11).

19.5.1.4 Delivered Information

The base information is the amount of single dwellings distinguished between different dwelling types. For July 2011, about 23,400 dwellings were extracted: 13,950 dwellings with metal roof, 6,650 tents, and 2,800 huts. The combined results

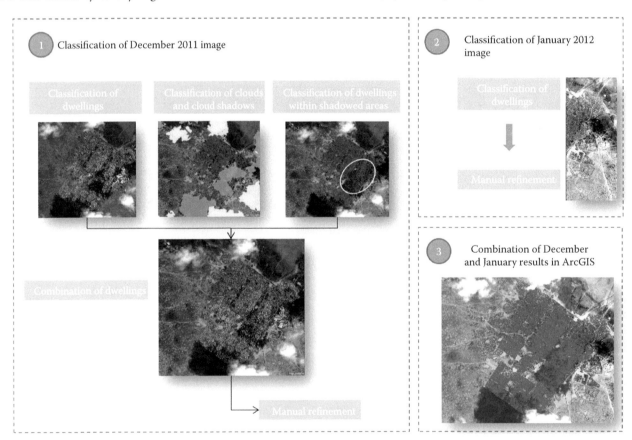

FIGURE 19.11 Workflow of dwelling extraction (here, an example with cloud shadows and different images). (From Füreder, P. et al., Monitoring Refugee Camp Evolution and Population Dynamics in Dagahaley, Kenya, based on VHSR satellite data, *Ninth International Conference African Association of Remote Sensing of the Environment (AARSE)*, El Jadida, Morocco, 2012.)

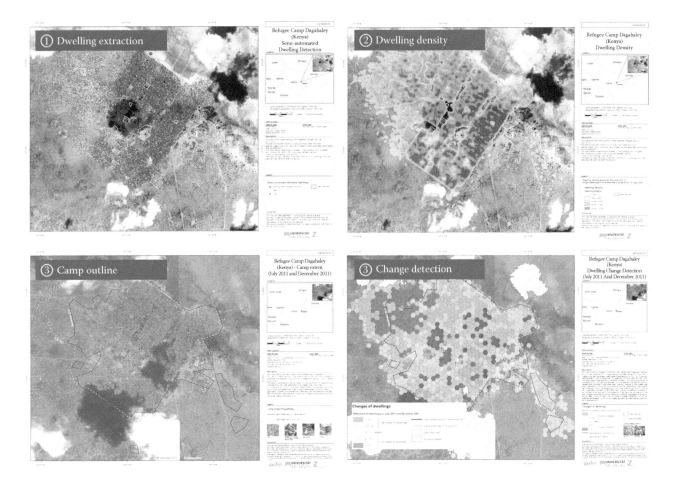

FIGURE 19.12 Delivered information products based on single extracted dwellings.

of the December and January images revealed 21,950 dwellings. An additional information product comprises dwelling density (dwellings/km²) calculated using kernel density methods (Lang et al. 2010) to provide a better overview of the spatial distribution of dwellings. A camp outline can be derived from dwelling density, which is important for rapidly expanding camps, where the extent can hardly be observed on the ground. A change analysis of dwellings between July 2011 and December 2011/January 2012 aggregated to hexagonal units provided an overview of major changes within the camp. It showed a decrease of dwellings in the western outskirts of the camp and a slight increase of dwellings in the main part of the camp. Areas covered by clouds in either of the two images were excluded from the change analysis (Figure 19.12).

19.5.1.5 Impact

In comparison to IDP camps, in UNHCR-managed refugee camps there is usually good knowledge about the camp population. However, in extreme events, with enormous influx of refugees, reliable information on population can often not be assessed on the ground. In the case of Dagahaley, there was limited access to the outskirts of the camp due to security reasons. The information provided could therefore help to understand where new arrivals were settling and estimate the amount of new arrivals in need for assistance. Monitoring changes of dwelling structures showed that population moved away from the outskirts. They have been mainly relocated to Ifo 2, but also found place within the camp.

Methodologically, it has to be taken into account that some areas were affected by cloud shadows in both images. Due to the high radiometric resolution of the WorldView-2 images, these shadow areas indeed can be analyzed, but the expected accuracy is lower than in well-illuminated regions. Therefore, the figures on dwelling numbers within cloud shadows have likely been underestimated.

19.5.2 Evolution and Impact of the IDP Camp Zam Zam, Sudan, at Local and Regional Scale

19.5.2.1 Background

The IDP camp Zam Zam is located to the southwest of Wadi El Ko and Wadi Golo, approximately 15 km south of El Fasher, the capital city of Northern Darfur, between longitudes 25°17′ and 25°19′ east and latitudes 13°28′ and 13°30′ north. The region is characterized by a semiarid environment with very low annual rainfall but a high seasonal variability. The surrounding land is used for cultivating crops and vegetables. When the first IDPs came to Zam Zam in 2003, they settled around existing villages; since then, the number of IDPs rose from approximately 18,900 in 2004 to over 50,000 in 2008 with continuing trend. In 2013, UN agencies and NGOs working in the camp estimated a number of 164,000 IDPs living in Zam Zam (UNOCHA 2013).

19.5.2.2 Data Used

For detailed analysis (Lang et al. 2010), two QuickBird scenes were ordered from DigitalGlobe image archive (June 18, 2002, and December 20, 2004). A third one was tasked and recorded on May 8, 2008. QuickBird has a spatial resolution of up to 0.6 m and multispectral images are collected at a resolution of 2.4 m. The three QuickBird scenes have three optical bands and a near infrared.

For the analysis of the entire region around Zam Zam, MODIS vegetation indices (MOD13Q1) over a time period of more than one decade have been acquired starting on February 18, 2000, and ending on May 25, 2010. The MODIS image data sets are provided every 16 days as a gridded level-3 product at a spatial resolution of 250 m, and thus the entire time series contains 237 data sets.

19.5.2.3 Methods Applied

The workflow is a sequence of two subsequent analysis steps: (1) analyzing camp extent and population dynamics in several time slices and (2) assessing likely impacts on the surrounding environmental conditions—at local and regional scale. The first part is methodologically similar to the case study Dagahaley, with the particularity to perform an ex-post analysis, applying a retrospective view (Lang et al. 2010). The study was performed in 2008, looking at the evolvement of the Zam Zam IDP camp over a few years back in time. Since QuickBird imagery is available from 2001 onward, authors were able to cover the entire time span of camp evolution from 2002 (before) until 2008. Currently, the analysis of camp evolution was extended with two additional time slices from 2010 to 2013.

For an integrated assessment of the changes in the immediate surroundings of the camp, Hagenlocher et al. (2012) used a weighted natural resource depletion index. This index reflects the relative importance of single land use classes for both human security and ecosystem integrity, as judged by experts. The calculation was done using weighted overlay techniques. Extending the view to the entire region around Zam Zam, trend patterns resulting from time series analyses of MODIS vegetation indices between 2000 and 2010 were generated. The trend analysis was based on the Seasonal Kendall test, deriving significance and slope values of detected trend patterns. This variation of Mann-Kendall was applied to incorporate the high seasonality of the climate in North Darfur (Kranz et al. 2010). Finally, based on the derived significance and slope values, a trend classification has been set up that allows for a more detailed separation of positive and negative changes.

19.5.2.4 Delivered Information

The results of the medium-resolution analysis can be described as a detection of hot spots of significant positive and negative trends in vegetation cover over the period of time considered. Due to their impact on the surrounding environment, significant negative trend patterns are located in the vicinity of IDP/refugee camps such as Zam Zam, Abu Shouk, and Al Salaam (Figure 19.13). This information on a large scale gives indications on the influence of human activities on the vegetation cover for large regions. This indicates increasing pressure on natural resources (e.g., firewood) and consequently is one indicator showing increasing potential for local conflict.

19.5.2.5 Impact

Population displacement often leads to large-scale settlements with a potential impact on the local environment. The dynamics apply both to the structural changes of the camp area (extent and population density) and the wider impacts on the surroundings. Environmental deterioration can result in violent conflicts over access to, or control of, scarce natural resources, or in renewed migration of the population (Hagenlocher et al. 2012). Remote sensing can capture actual population dynamics and hint to environmental changes in the surrounding areas. While the analysis of VHR satellite imagery provides detailed information about camp structure and local changes of the environment, medium-resolution data have the capability to provide trend patterns of environmental impact on a regional scale.

19.5.3 Indications of Destructions: The Case of Abyei, Disputed Border Region between Sudan and South Sudan

19.5.3.1 Background

On May 21, 2011, Sudan Armed Forces (SAF) invaded Abyei Town, located in the Abyei administrative area. The Abyei area is a contested region on the border of Sudan and South Sudan, then Republic of South Sudan. SAF's invasion of Abyei town followed multiple skirmishes between SAF, SAF-aligned forces, and the Sudan People's Liberation Army (SPLA). As fighting broke out between SAF-aligned and SPLA-aligned forces, civilians fled the area, and NGO personnel were evacuated to South Sudan. The United Nations Mission in Sudan (UNMIS) was also present in Abyei Town at the time of the attack, although their freedom of movement was restricted. During fighting, civilian dwellings, infrastructure, and humanitarian compounds were attacked. Figures estimate that between 30,000 and 80,000 civilians, primarily Dinka Ngok, were forcefully displaced from their homes (AlAchkar et al. 2013).

19.5.3.2 Data Used

Two data sets were used to visually confirm reported events that impacted the civilian and humanitarian communities in Abyei Town, VHR satellite imagery and open-source data

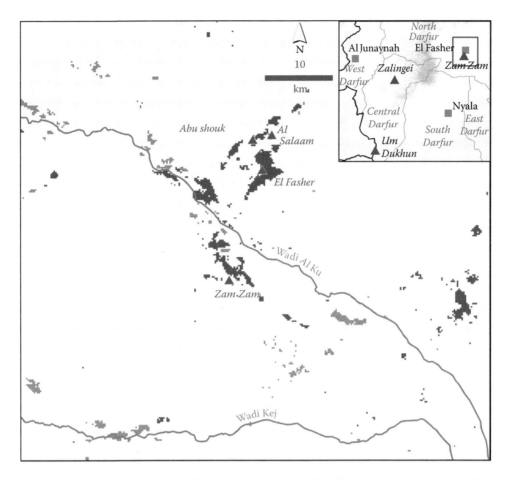

FIGURE 19.13 Negative (red) and positive (green) trend pattern of NDVI values based on a MODIS time series analysis with 237 data sets between 2000 and 2010 in the area of El Fasher/Zam Zam. Trend patterns have been derived using Seasonal Kendall test.

from news reports and UN reports. This case study examines the Harvard Humanitarian Initiative's use of satellite imagery to confirm events reported by UNMIS Human Rights. An UNMIS report dated May 29, 2011, claims that "The Abyei market and all the shops have been looted and burned. The World Food Programme [WFP] and UNICEF warehouse, which were fully stocked before the SAF attack, as well as the MSF (Médecines Sans Frontières), AECOM and Humanitarian Affairs compounds have also been completely plundered" (UNMIS 2011).

To confirm these reported events, analysis was conducted using two archive VHR images. The pre-event image, captured on February 12, 2011, is from WorldView-2 with a resolution of 50 cm (commercial use). The post-event image, from QuickBird with a resolution of 61 cm, was acquired 5 days after the events on May 26, 2011. Images were obtained from DigitalGlobe in georeferenced and pan-sharpened mode.

19.5.3.3 Methods Applied

Two adjustments were made to the raw images in ERDAS Imagine before analysis was performed. Cubic convolution resampling was conducted to refine the pixels image to allow for better analysis. Min–max stretch rescaling was also applied to

balance the color and contrast in the image. The areas of interest, the Abyei Market, and WFP and MSF compounds, were then located in the satellite image using open-source mapping sources, like OSM and Wikimapia.

Manual analysis was conducted on both images. Damage to and the destruction of structures were identified by conducting change detection between the two images in geographically synced windows in ERDAS Imagine. Burned structures were identified due to their darker color, similar appearance to other burned structures, and the apparent presence of ash. Razed structures appeared to be damaged, with their original structure size undetectable due to apparent dismantling. Ground photographs of looting were used to help assess what the results of this act look like in imagery.

19.5.3.4 Delivered Information

Analysis of the post-event satellite imagery corroborates UNMIS reports that the WFP and MSF compounds were looted and that the Abyei Market was looted and razed. Additional indications of destructions are also apparent in the imagery (Figure 19.14).

At both the WFP and MSF compounds, scattered debris is visible both inside and outside the perimeter. This debris is indicative of looting. A cited UN media report states that "800 metric

Looting at WFP and UN Patrol
26 May 2011
Abyei Town, Abyei Area

(a)

Looting and Razing at MSF
26 May 2011
Abyei Town, Abyei Area

(b)

FIGURE 19.14   WFP and MSF compounds after the attacks.

Abyei Market Razed - Before
12 February 2011
Abyei Town, Abyei Area

Abyei Market Razed - After
26 May 2011
Abyei Town, Abyei Area

FIGURE 19.15   Abyei Market before (left) and after (right) the attacks.

tons of food, enough to feed 50,000 people for three months" and other supplies were looted from the WFP compound (UNRadio 2011). Imagery analysis also reveals 12 razed structures at the MSF compound, 10 of which appear to have been burned. Debris and razed structures are also apparent at the Abyei Market (Figure 19.15).

19.5.3.5 Impact

Visually documenting the results of the attacks on the Abyei Market and WFP and MSF compounds is critical for many reasons. First, these attacks are documented as part of a larger effort to understand the scale of destruction throughout Abyei Town. The results of these acts not only signify events that immediately impacted the civilian and humanitarian communities but also presented long-term challenges. Documenting the destruction of critical civilian and humanitarian infrastructure provides insights into impediments that may prohibit the local population from returning to their community. This point is exemplified and amplified by the report that the WFP compound, prior to the looting, held enough food to feed 50,000 for three months.

The proliferation of this technology has also raised questions concerning the possible role that remote sensing may play in documenting violations of international humanitarian law. Current widespread standards for the use of remote sensing in this capacity do not exist. Current research reflects that if this particular use is aimed at, then imagery acquisition, analysis, and documentation must occur within a proper chain of custody due to potential perceived politicization of the analysis.

19.5.4 Logging and Mining Activities in Relation to the Conflict Situation in the Democratic Republic of the Congo

19.5.4.1 Background

The Democratic Republic of the Congo (DRC) is one of the most resource-rich countries in the world but economically one of the poorest and on the edge of a failed state (UNDP 2011, Haken et al. 2013). Natural resources locally and abundantly available do not only cover minerals but also wood—especially high-value tree species. These resources are known to fuel the conflict in the country either by directly investing revenues into weapons or by rivalry for the resources itself (USAID 2005, Le Billon 2006, Global Witness 2013). Furthermore, armed groups are taking the control over wood and mineral resources as well as their trade. The insecure and instable situation especially in the eastern provinces of the DRC (North and South Kivu) caused significant numbers of IDPs. With 2.6 million IDPs and another 500,000 refugees, the displaced population is one of the largest in the world (IDMC 2014).

Mining in the DRC is mainly carried out by civilians using artisanal and small-scale mining techniques, which in fact means working with shovel, pickaxe, and hammers or even using bare hands (Garrett 2008). With respect to logging, one can distinguish three types: industrial logging, informal logging, and slash-and-burn activities. In general, the fragmentation of rainforests by logging results in well-analyzed secondary effects, such as increasing agricultural land use alongside the logging roads (UNEP 2006), charcoal production, poaching, and mining activities (Potapov et al. 2012).

Detecting land cover changes related to mining and logging activities provides information about the situation on the ground in the insecure and often remote areas of interest.

19.5.4.2 Data Used

With respect to the monitoring of logging and slash-and-burn activities, DMCii images of three timeslots (January, May, and September 2010) with a spatial resolution of 22 m were analyzed covering an area of 50 × 50 km in the Oriental province in DRC. As a reference for the identification of the defined land cover classes, VHR satellite scenes from IKONOS (March 2011) and GeoEye-1 (June 2009) have been used.

Data of the same VHR sensors have been applied for monitoring mining activities in North and South Kivu provinces of the DRC. Of particular interest was the most well-known cassiterite mining site of Bisie located in North Kivu. For this area, three scenes were available comprising a period of April and September 2010 (GeoEye-1) to March 2011 (IKONOS). For the aim of covering a large region to identify hot spots of mining activities, RapidEye data with a spatial resolution of 6.5 m have been acquired. For preprocessing, orthorectification and atmospheric correction have been applied to all images with "Atmospheric/Topographic Correction for Satellite Imagery" (Richter 1997). The GeoEye-1 and IKONOS data were pan-sharpened, and additional thematic layers were used as ancillary data.

19.5.4.3 Methods Applied

The overall methodological procedure follows a multiscale approach developed by Schöpfer and Kranz (2010). The initial step is the analysis of HR satellite data with the aim of identifying hot spots of mining/logging activities on large survey extends. In order to reach this goal, a transferable feature extraction scheme is generated building upon OBIA concepts. In the case of the analysis of DMCii imagery, pixel-based approaches have proven to be very efficient for the detection of logging and shifting cultivation. These activities as well as the extraction of minerals imply complete clearing of the sites, a fact that can be used for focusing the detection of corresponding activities on bare soil areas only.

With respect to mining, the following assumptions for further investigations can be made: All exploitation activities are related to small-scale artisanal mining. Thus, major interventions into the environment through large excavation, soil accumulation, or pollution are not expected. Within the RapidEye data, a further distinction between bare soil and potential mining site is only possible through secondary indicators, such as distance to settlements/dwellings, roads, and water bodies/rivers. The identified hot spots of resource extraction activities are visually assessed to decide whether VHR data should be acquired in order to investigate the most interesting areas in more detail. Subsequently, for the selected sites, a multitemporal change detection analysis is conducted. In the case of the mining site at Bisie, a monitoring has been conducted on GeoEye-1/IKONOS data to evaluate the evolution of the detected mining area.

The outlined methodology has several advantages: (1) the coverage of large survey extents for the identification of hot spots of resource exploitation activities, (2) highly detailed analyses of the detected hot spot areas, and (3) the multitemporal monitoring resulting in land cover changes for more qualitative statements about developments and trends in relation to the conflict situation in the region.

19.5.4.4 Delivered Information

The result is twofold referring to the different scales the analyses are based on. First, hot spots of mining and logging activities are detected on HR data that allow covering a large area and indicating the most relevant sites of interest. As such, areas are identified that are characterized by significant land cover changes potentially related to the conflict situation on the ground. The information resulting from the VHR data–based monitoring provides complementary information to reports and statistics about the extraction of conflict resources. These documents frequently give only a limited insight about ongoing activities in a certain area of interest that is often too remote or too insecure or even both to be investigated by field surveys. As an example, a brief summary of a monitoring conducted during a mining ban (September 11, 2010, to March 10, 2011) for the mining site at Bisie in the North Kivu province, DRC, is given in the following.

The analysis is based on two VHR satellite images acquired three days prior to the announcement of a mining ban by President Joseph Kabila and exactly at the date the ban has been lifted. As a result, an expansion of the mining site indicates a continuation of exploitation activities during the mining ban although this does not provide confirming evidence (see Figure 19.16).

Linking these results with statistical data and information gathered from reports, higher reliability have been reached by identification of further indications underlining the hypothesis of continued mining during the ban (Zingg-Wimmer and Hilgert 2011). As a conclusion, one can highlight the added value of the integrated assessment of satellite-based information and socioeconomic data, especially in unstable regions where only the merging of different information sources may facilitate a comprehensive picture of the situation in the region. For future analysis, the estimation of population through an assessment of the settlement structure at the mining site has been identified as an important indicator for mining activities through the monitoring of workers at the mine.

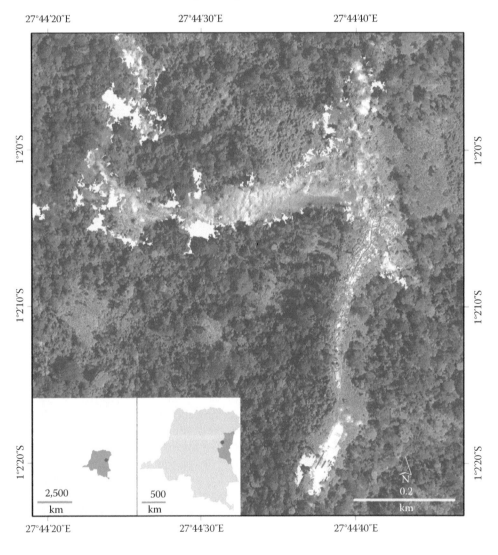

FIGURE 19.16 Expansion of the mining area at Bisie between September 2010 and March 2011. (Modified from Lüthje, F. et al., *Rem Sens*, in press, 2014.)

19.5.4.5 Impact

The expansion detected trough the analysis of satellite imagery has been said to be "the clearest sign of continued mining activities at Bisie" (Zingg Wimmer and Hilgert 2011) by the International Peace Information Service (IPIS)—a research center supporting governmental, nongovernmental, and intergovernmental development actors. In an additional study, IPIS compared official statistics about cassiterite trade in the region with the expansion of the mining area. The overall aim of this analysis was to raise the awareness of the real consequences of the mining ban that are (a) mining continued, (b) production volumes went down considerably, and (c) the military increased their control over mineral resources in the region (which is contrary to the objectives of setting up the ban). As a major result, recommendations may be given, namely, strengthening the demilitarization of mines and trading routes, deployment of disciplined regiments for securing the territory, and strengthening the mining police. The implementation of these recommendations would provide a stable basis for international transparency initiatives to continue to be set in place. Finally, the international electronic industry would restart buying minerals from the DRC, which has been identified as an additional stimulus for the economic development and a stabilizing factor of the entire region.

19.6 Conclusion

This chapter has shown the potential of remote sensing and EO for humanitarian emergency response, but also the challenges faced by those who want to utilize it. Many restrictions are to a lesser degree of technical nature, but more related to institutional, operational, or political constraints. The authors discussed in a critical, yet supportive and forward-looking way, the areas where remotely sensed data along with the appropriate analysis techniques can be supportive to existing workflows. Only if there is mutual trust among those who deliver and those who receive, thus an approximation from both technological and operational side, remote sensing will help where it is needed most—to ultimately safe lives.

References

Al Achkar, Z. et al. 2013. *Sudan: Anatomy of a Conflict.* H. University, Cambridge, MA, pp. 31–41.

ARC. 2014. OpenStreetMap Damage Assessment Review: Typhoon Haiyan (Yolanda) Interim Report. A. R. C. A. a. REACH, available online at http://americanredcross.github. io/OSM-Assessment/#intro (accessed Aug 2014).

Aschbacher, J. 2002. Monitoring environmental treaties using earth observation. *VERTIC Verification Yearbook* 171–186.

Bernhard, E. 2013. Refugee camp mapping in Jordan using TerraSAR-X data.

Bjorgo, E. 2000. Refugee camp mapping using very high spatial resolution satellite sensor images. *Geocarto International* 15 (2): 79–88.

Bjorgo, E. 2001. Supporting humanitarian relief operations. In: *Commercial Observation Satellites: At the Leading Edge of Global Transparency*, J. C. Baker, K. M. O'Connell, and R. A. Williamson (eds). Santa Monica, CA: ASPRS RAND, pp 403–427.

Blaschke, T. 2010. Object based image analysis for remote sensing. *Int J Photogram Rem Sens* 65(1): 2–16.

Capelo, L., N. Chang, and A. Verity. 2012. *Guidance for Collaborating with Volunteer and Technical Communities.* D. H. Network, available online at http://digitalhumanitarians.com/content/guidance-collaborating-volunteer-technical-communities (accessed Aug 2014).

Checchi, F., B. T. Stewart, J. J. Palmer, and C. Grundy. 2013. Validity and feasibility of a satellite imagery-based method for rapid estimation of displaced populations. *Int J Health Geogr* 12(4): 1–12.

Cowan, N. M. 2011. A geospatial data management framework for humanitarian response. *Eighth International ISCRAM Conference*, Lisbon, Portugal, pp. 1–5.

Drury, S. A. and M. E. A. Deller. 2002. Remote sensing and locating new water sources. The Use of Space Technology for Disaster Management for Africa, Addis Ababa.

Ehrlich, D., S. Lang, G. Laneve, S. Mubareka, S. Schneiderbauer, and D. Tiede. 2009. Can Earth observation help to improve information on population? Indirect population estimations from EO derived geospatial data. In: *Remote Sensing from Space Supporting International Peace and Security*, B. Jasani, M. Pesaresi, S. Schneiderbauer and G. Zeug (eds). Berlin, Germany, Springer, pp. 211–237.

Elwood, S. 2008. Volunteered geographic information: Key questions, concepts and methods to guide emerging research and practice. *GeoJournal* 72: 133–135.

European Union, Copernicus emergency mapping service, http://emergency.copernicus.eu (Accessed August, 2014).

FAO. 2013. *Trends and Impacts of Foreign Investment in Developing Country Agriculture: Evidence from Case Studies*. Rome, Food and Agricultural Organization of the United Nations (FAO).

FEWS NET team, FEWS NET–famine early warning systems network, http://www.fews.net (Accessed August, 2014).

Foody, G. M. 2002. Status of land cover classification accuracy assessment. *Rem Sens Environ* 80(1): 185–201.

Forced migration review, University of Oxford, Oxford, UK, http://www.fmreview.org/FMRpdfs/FMR13/fmr13.3.pdf. (Accessed August, 2014).

Füreder, P., D. Hölbling, D. Tiede, S. Lang, and P. Zeil. 2012. Monitoring refugee camp evolution and population dynamics in Dagahaley, Kenya, based on VHSR satellite data. *Ninth International Conference African Association of Remote Sensing of the Environment* (AARSE), El Jadida, Morocco.

Füreder, P., D. Tiede, F. Lüthje, and S. Lang. 2014. Object-based dwelling extraction in refugee/IDP camps – challenges in an operational mode. *South-Eastern Eur J Earth Observ Geom* 3(2s): 539–544.

Garrett, N. W. 2008. Artisanal Cassiterite mining and trade in North Kivu—Implications for poverty reduction and security. C. a. S.-s. M. (CASM). Communities and Artisanal & Small-scale Mining initiative (CASM).

Giada, S., T. de Groeve, and D. Ehrlich. 2003. Information extraction from very high resolution satellite imagery over Lukole refugee camp, Tanzania. *Int J Remote Sens* 24(22): 4251–4266.

Global Witness. 2013. Breaking the links between natural resources and conflict: The case for EU regulation. A civil society position paper. G. W. Publishers, available online at https://www.globalwitness.org/sites/default/files/breakingthelinks%28eng%29.pdf (accessed Aug 2014).

Hagenlocher, M., S. Lang, and D. Tiede. 2012. Integrated assessment of the environmental impact of an IDP camp in Sudan based on very high resolution multi-temporal satellite imagery. *Rem Sens Environ* 126: 27–38.

Haken, N., J. J. Messner, K. Hendry, P. Taft, K. Lawrence, and F. Umaa. 2013. Failed State Index IX 2013. Technical report. F. f. Peace.

Hakley, M. 2010. How good is volunteered geographical information? A comparative study of OpenStreetMap and ordnance survey datasets. *Environ Plan B: Plan Des* 37(4): 682–703.

HHI. 2011. *Disaster Relief 2.0: The Future of Information Sharing in Humanitarian Emergencies*. Washington, DC and Berkshire, U.K., UN Foundation and Vodafone Foundation Technology Partnership.

HIIK. 2013. Conflict Barometer 2013. H. I. f. I. C. R. (HIIK). Heidelberg, Germany.

Humanitarian Coalition. n.d. Humanitarian Coalition factsheets, http://humanitariancoalition.ca/info-portal/factsheets/what-is-a-humanitarian-crisis (Retrieved June 1, 2014).

Humanitarian Information Unit, U.S. Department of State, https://hiu.state.gov/ittc/ittc.aspx. (Accessed August, 2014).

IDMC. 2014. Global Overview 2014. People internally displaced by conflict and violence. I. D. M. Centre.

International Charter, International Charter on Space and Major Disasters, HYPERLINK "http://www.disasterscharter.org" www.disasterscharter.org. (Accessed August, 2014).

IPCC. 2001. *Climate Change 2001: The Scientific Basis. Contribution of Working Group I to the Third Assessment Report of the Intergovernmental Panel on Climate Change.* Cambridge, U.K., Cambridge University Press.

Jensen, J. R. 2005. *Introduction to Digital Image Processing. A Remote Sensing Perspective.* Upper Saddle River, NJ, Prentice-Hall.

Joyce, K. E., S. E. Belliss, S. V. Samsonov, S. J. McNeill, and P. J. Glassey. 2009. A review of the status of satellite remote sensing and image processing techniques for mapping natural hazards and disasters. *Prog Phys Geogr* 33(2): 183–207.

Kemper, T. and J. Heinzel. 2014. Mapping and monitoring of refugees and internally displaced people using EO data. In: *Global Urban Monitoring and Assessment through Earth Observation*, Weng, Qihao, CRC Press, Boca Raton, pp. 195–216.

Kranz, O. et al. 2010. Monitoring refugee/IDP camps to support international relief action. In: *Geoinformation for Disaster and Risk Management—Examples and Best Practices. Joint Board of Geospatial Information Societies (JB GIS)*, O. Altan, R. Backhaus, P. Piero Boccardo, and S. Zlatanova, United Nations Office for Outer Space Affairs (UNOOSA), Vienna, pp. 51–56.

Kuenzer, C., M. Bartalis, M. Schmidt, D. Zhaoa, and W. Wagner. 2008. Trend analyses of a global soil moisture time series derived from ERS1/2 scatterometer data: floods, droughts and long-term changes. *The International Archives of the Photogrammetry, Remote Sensing and Spatial Information Sciences* XXXVII, 1363–1368.

Lang, S. 2008. Object-based image analysis for remote sensing applications: Modeling reality—Dealing with complexity. In: *Object-Based Image Analysis—Spatial Concepts for Knowledge-Driven Remote Sensing Applications,* T. Blaschke, S. Lang, and G. J. Hay (eds). Berlin, Germany, Springer, pp. 3–28.

Lang, S., C. Corbane, and L. Pernkopf. 2013. Earth Observation for Habitat and Biodiversity Monitoring. In: *Ecosystem and Biodiversity Monitoring: Best Practice in Europe and Globally* [section title], S. Lang and L. Pernkopf (eds). Heidelberg, Germany, Wichmann, pp. 478–486.

Lang, S., D. Tiede, D. Hölbling, P. Füreder, and P. Zeil. 2010. EO-based ex-post assessment of IDP camp evolution and population dynamics in Zam Zam, Darfur. *Int J Rem Sens* 31(21): 5709–5731.

Lang, S., D. Tiede, and F. Hofer. 2006. Modeling ephemeral settlements using VHSR image data and 3D visualization—The example of Goz Amer refugee camp in Chad. *Photogrammetrie, Fernerkundung, Geoinformatik* 4: 327–337.

Le Billon, P. 2001. The political ecology of war: Natural resources and armed conflicts. *Pol Geogr* 20: 561–584.

Le Billon, P. 2006. *Fuelling War: Natural Resources and Armed Conflict*. Abingdon, U.K., Routledge.

Lüthje, F., E. Schöpfer, and O. Kranz. 2014. Geographic object-based image analysis using optical satellite imagery and GIS data for the detection of mining sites in the Democratic Republic of the Congo. *Rem Sens* 6: 6636–6661.

Martin, A. 2005. Environmental conflict between refugee and host communities. *J Peace Res* 42(3): 329–347.

Mildner, S. A., G. Lauster, and W. Wodni. 2011. Scarcity and abundance revisited: a literature review on natural resources and conflict. *Int J Conf Viol* 5(1): 155–172.

Missing Maps Project, http://wiki.openstreetmap.org/wiki/Missing_Maps_Project (Accessed April 28, 2015).

MSF. 2011. Kenya: Humanitarian Crisis on the Outskirts of Overcrowded Dadaab Camp.

MSF USA, Innovating to Fight Tuberculosis in Papua New Guinea, http://www.doctorswithoutborders.org/article/innovating-fight-tuberculosis-papua-new-guinea. (Accessed April 28, 2015).

Padwick, C., M. Deskevich, F. Pacifiä, and S. Smallwood. 2010. WorldView-2 pan-sharpening. *Proc Am Soc Photogram Rem Sens* 13, online article available at http://info.asprs.org/publications/proceedings/sandiego2010/sandiego10/Padwick.pdf (accessed Aug 2014).

Pesaresi, M., A. Gerhardinger, and F. Haag. 2007. Rapid damage assessment of built-up structures using VHR satellite data in tsunami-affected areas. *Int J Remote Sens* 28(13): 3013–3036.

Potapov, P. V. et al. 2012. Quantifying forest cover loss in Democratic Republic of the Congo, 2000–2010, with Landsat ETMC data. *Rem Sens Environ* 122: 106–116.

REACH Initiative, REACH—Informing more effective humanitarian action, Geneva, Switzerland, http://www.reachinitiative.org (Accessed August, 2014).

ReliefWeb, ReliefWeb–Informing humanitarians worldwide, http://www.reliefweb.int. (Accessed August, 2014).

Rhee, J., I. Jungho, and G. J. Carbone. 2010. Monitoring agricultural drought for arid and humid regions using multisensor remote sensing data. *Rem Sens Environ* 114(12): 2875–2887.

Richter, R. 1997. Correction of atmospheric and topographic effects for high spatial resolution satellite imagery. *Int J Remote Sens* 18: 1099–1111.

SBTF. 2011. Crowdsourcing Satellite Imagery Analysis for UNHCR-Somalia: Latest Results. S. T. F. (SBTF).

Schöpfer, E. and O. Kranz. 2010. Monitoring natural resources in conflict using an object-based multiscale image analysis approach. *International Archives of Photogrammetry, Remote Sensing, and Spatial Information Sciences* XXXVIII-4(C7).

Slonecker, E. T., D. M. Shaw, and T. M. Lillesand. 1998. Emerging legal and ethical issues in advanced remote sensing technology. *Photogramm Eng Rem Sens* 64(6): 589–595.

Soden, R., N. Budhathoki, and L. Palen. 2014. Resilience-building and the crisis informatics agenda: Lessons learned from Open Cities Kathmandu. *11th International ISCRAM Conference*, S. R. Hiltz, M. S. Pfaff, L. Plotnick and A. C. Robinson (eds). Penn State University, Pennsylvania, PA, 339–348.

Tiede, D., S. Lang, P. Füreder, D. Hölbling, C. Hoffmann, and P. Zeil. 2011. Automated damage indication for rapid geospatial reporting. An operational object-based approach to damage density mapping following the 2010 Haiti earthquake. *Photogramm Eng Rem Sens* 9: 933–942.

Tiede, D., S. Lang, D. Hölbling, and P. Füreder. 2010. Transferability of OBIA rule sets for IDP camp analysis in Darfur. *Int Arch Photogram, Rem Sens Spatial Inform Sci* XXXVIII-4(C-7), online available at http://www.isprs.org/proceedings/XXXVIII/4-C7/pdf/Tiede_137.pdf (accessed Aug 2014).

Tiede, D., P. Füreder, S. Lang, D. Hölbling, and P. Zeil. 2013. Automated analysis of satellite imagery to provide information products for humanitarian relief operations in refugee camps—From scientific development towards operational services. *Photogrammetrie—Fernerkundung—Geoinformation* 3: 185–195.

UNDP. 2011. Human Development Report 2011. Technical report. U. N. D. P. (UNDP), available online at http://hdr.undp.org/sites/default/files/hdr_2013_en_technotes.pdf (accessed Aug 2014).

UNEP. 2006. Africa Environment Outlook 2. Our Environment, our Wealth. D. o. E. W. a. A. United Nations Environment Programme (UNEP). Nairobi, Kenya.

UNEP. 2009. From Conflict to Peacebuilding. The Role of Natural Resources and the Environment. U. N. E. P. (UNEP).

UNHCR. 2012. Global Trends 2011. A year of crisis.

UNHCR. 2013. Global Trends 2013. War's human cost.

UNHCR. 2014. Refugees in the Horn of Africa: Somali displacement crisis.

UNHCR, UNHCR population statistics, http://popstats.unhcr.org. (Accessed August, 2014).

UNHCR, UNHCR—the UN refugee agency, http://data.unhcr.org. (Accessed August, 2014).

United Nations, ONUMOZ, http://www.un.org/en/peacekeeping/missions/past/onumozFT.htm. (Accessed August, 2014).

UNITAR, UNITAR–knowledge to lead, http://www.unitar.org. (Accessed August, 2014).

UNMIS. 2011. Update on the attack and occupation of Abyei by SAF. U. N. M. i. Sudan.

UNOCHA. 2013. Sudan: Zamzam IDP Camp Profile. http://reliefweb.int/sites/reliefweb.int/files/resources/sud13_North%20Darfur_Zamzam%20IDP%20Camp%20Profile_a3_09may13.pdf (Retrieved May 29, 2014).

UNOOSA, UN SPIDER knowledge portal, http://www.un-spider.org. (Accessed August, 2014).

UNRadio. 2011. Humanitarian supplies looted in Abyei town of Sudan. U.N. Radio.

USAID. 2005. Forest and Conflict. A toolkit for Intervention. O. o. C. M. a. M. U.S. Agency for International Development, Washington, DC.

Verjee, F. 2011. *GIS Tutorial for Humanitarian Assistance*. Redlands, CA, ESRI Press.

Voigt, S., T. Kemper, T. Riedlinger, R. Kiefl, K. Scholte, and H. Mehl. 2007. Satellite Image Analysis for Disaster and Crisis-Management Support. *IEEE Trans Geosci Rem Sens* 45(6): 1520–1528.

Wagner, W., K. Scipal, C. Pathe, D. Gerten, W. Lucht, and B. Rudolf. 2003. Evaluation of the agreement between the first global remotely sensed soil moisture data with model and precipitation data. *J Geophys Res* 108: 4611.

Walker, P., J. Glasser, and S. Kambli. 2012 Climate change as a driver of humanitarian crises and response.

Ziemke, J. and Meier, P. Crisis mappers–the humanitarian technology network, http://crisismappers.net. (Accessed August, 2014).

Zingg Wimmer, S. and F. B. Hilgert. 2011. A one-year snapshot of the DRCs principal cassiterite mine. Technical report. I. P. I. S. (IPIS), online available at: http://ipisresearch.be/publication/bisie-one-year-snapshot-drcs-principal-cassiterite-mine/ (accessed Aug 2014).

Volcanoes

Remote Sensing of Volcanoes

Robert Wright
University of Hawaii at Mānoa

Acronyms and Definitions

ADEOS	Advanced Earth Observing Satellite
AIRS	Atmospheric Infrared Sounder
ALOS	Advanced Land Observing Satellite
ASTER	Advanced Spaceborne Thermal Emission and Reflection Radiometer
ATSR-1	Along-Track Scanning Radiometer-1
AVHRR	Advanced Very High Resolution Radiometer
BTD	Brightness temperature difference
DEM	Digital elevation model
DInSAR	Differential interferometric SAR
DU	Dobson unit
ENVISAT	Environmental Satellite
EO-1	Earth Observing-1
ERS-1 (and ERS-2)	European Remote Sensing satellites
ETM+	Enhanced Thematic Mapper Plus
GOES	Geostationary Operational Environmental Satellite
GOME-2	Global Ozone Monitoring Experiment-2
GOSAT	Greenhouse Gas Observing Satellite
GPS	Global Positioning System
IASI	Infrared atmospheric sounding interferometer
IFOV	Instantaneous field of view
InSAR (and IFSAR)	Interferometric synthetic aperture radar

MIR	Middle infrared
MODIS	Moderate Resolution Imaging Spectroradiometer
MTSAT	Multifunctional Transport Satellite
NASA	National Aeronautics and Space Administration
NPP	National Polar-orbiting Partnership
OMI	Ozone mapping instrument
OMPS	Ozone Profiler and Mapping Suite
PS	Permanent scatterer
SCIAMACHY	Scanning Imaging Absorption Spectrometer for Atmospheric Chartography
SEVERI	Spinning Enhanced Visible and Infrared Imager
SWIR	Shortwave infrared
TANSO-FTS	Thermal and near-infrared sounder for carbon observation-Fourier transform spectrometer
TIR	Thermal infrared
TM	Thematic Mapper
TOMS	Total Ozone Mapping Spectrometer
TOVS-HIRS	TIROS Operational Vertical Sounder-High Resolution Infrared Radiation Sounder
UV	Ultraviolet
VAAC	Volcanic Ash Advisory Center
VNIR	Visible and near infrared

20.1 Introduction

There are about 1500 active and potentially active volcanoes on Earth, of which an average of 70 erupt in any given year (Siebert et al., 2010). This subareal volcanism (i.e., that which is not hidden from us beneath the surface of the ocean) is the most obvious and dramatic manifestation of energy loss from Earth's interior. The impact of volcanism on the Earth and humanity is profound; volcanism provided us with an atmosphere that it has modified ever since (Condie, 2005) and may have influenced the evolution of life on Earth through the role eruptions have played in mass extinctions (Rampino, 2010). Volcanism has been a source of myth (Sigurdsson, 1999), subject for art and literature (Sigurdsson and Lopes-Gautier, 2000), and yields resources that society can exploit (Wohletz and Heiken, 1992).

This chapter will provide an overview of how satellite remote sensing can provide quantitative information before, during, and after a volcanic eruption. After an overview of the physical processes that take place at active volcanoes that we might be interested in measuring, the manner in which remote sensing can be used to provide quantitative information about them will be discussed, along with some contextual information about how these

measurements can be made in situ. Although by no means exhaustive, a list of citations for significant papers will be provided, so that the reader can trace the development of remote sensing as a tool for studying volcanism on Earth over the last several decades.

20.2 What Do Volcanoes Do That We Might Be Interested in Measuring?

Before describing how remote sensing can (and cannot) be used to aid in the study of Earth's volcanoes, it will be helpful to describe what it is that an erupting volcano, or one that is about to erupt, does, which scientists might be interested in measuring (Figure 20.1), in order to place the remote sensing measurements in context.

Consider a dormant volcano that is about to reawaken. Melting of the mantle takes place at depths of between ~20 and ~150 km, depending on tectonic setting. The melt is buoyant and begins to rise from the source region. As it approaches the surface, it must displace crustal rocks. This results in seismicity, as the ground shakes as the surrounding country rocks are fractured, and as gases and fluids associated with the rising magma pass through cracks and fractures. The displacement of rocks is

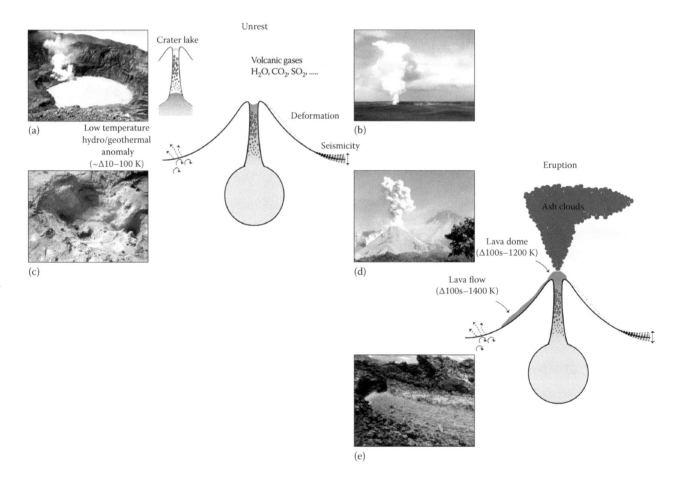

FIGURE 20.1 Cartoons illustrating the processes that occur before and during a volcanic eruption, which scientists find desirable to measure. (a) Hot, acidic, crater lake at Poas, Costa Rica; (b) hydrothermal mudpot in Iceland; (c) gas plume at Kīlauea volcano, Hawaii; (d) lava dome and ash plume at Santa Maria volcano, Guatemala; and (e) lava flow at Mount Etna, Sicily.

accompanied by the deformation of the surface of the volcano—the volcano swells and shrinks as magma arrives from depth and moves around beneath and within the edifice.

In addition to shaking and deforming the ground, this transfer of mass to shallow crustal levels also transports thermal energy and volatiles toward the surface. At depth, under pressure, species such as water, carbon dioxide, and sulfur dioxide (SO_2) are soluble in magmas. But as the magma rises, and the pressure decreases, these volatiles come out of solution promoting the growth of a vapor phase within the melt. The most insoluble volatiles (such as the noble elements and carbon dioxide) begin to exsolve at a depth of about 80 km, while at depths of about 10 km, confining pressures are reduced to the extent that water enters the gas phase. Other volatiles, such as SO_2 and HF, enter the mix when the magma reaches to within about 1 km of the surface. While some of these gases are common in the atmosphere (e.g., H_2O and CO_2), others are not (e.g., SO_2) and act as almost incontrovertible proof that fresh magma has been injected beneath the volcano from the primary melt region. Changes in the amount and type of gases exhaled by a volcano provide insights into the mass of magma involved and the nature of its movement between the shallow magma reservoir and the surface, via the conduit.

The presence of a shallow magma body heats meteoric water, resulting in convective cycling of water within the permeable zone between the heat source and the surface. This results in relatively low-temperature manifestations of volcanism at the surface (e.g., up to a few hundreds of °C), such as hot crater lakes, mud pools, geysers, and fumaroles (i.e., steam vents). Chemical reactions between this circulating groundwater and the magma also endow these fluids with chemical elements, which can be precipitated at the surface as the fluids cool. This gives rise to changes in surface composition, including sulfur, silica, and calcium deposition, as well as acid alteration of surface rocks to clays.

In many cases, the story ends here, and the ascending magma stalls at several kilometers depth in a crustal magma chamber—a failed eruption. But if magma arrives at the surface, an eruption begins. Eruptions take one of two end-member forms. In an effusive eruption, magma that arrives at the surface is erupted as lava (a contiguous body of silicate melt within which is suspended crystals and gas bubbles). Most commonly the lava extends downhill from the eruptive vent under the influence of gravity as one or more lava flows, which tend to be long, narrow, and thin, such as those commonly observed at Kilauea volcano, Hawaii, or Mount Etna, Sicily. Although dramatic, lava flows rarely pose a threat to human life as the speeds at which they travel downslope are little more than a fast walking speed. However, they do often have a substantial economic cost, due to the destruction of infrastructure.

In cases where the lava is more viscous, gas rich, and erupted at a lower volumetric flux rate (as happens in convergent tectonic environments, exemplified by the "Ring of Fire" that surrounds the Pacific Ocean), the lava does not flow downhill significantly but rather piles up atop the conduit, forming a relatively small, but thick, lava dome. Active lava domes are associated with the most famously destructive volcanic eruptions (e.g., Mount St. Helens, Washington (1980), and Montagne Pelee, Martinique (1902). This is a result of the higher gas content and viscosity of the lava and the fact that the domes can pile up to the extent that they become gravitationally unstable, collapse, and spawn fast-moving pyroclastic flows down the flanks of the volcano. Such pyroclastic flows were responsible for the deaths of ~28,000 people during the May 1902 eruption of Mount Pelee.

Finally, with regard to effusive eruptions, persistent lava lakes are much rarer. Here, hot, gas-rich, low-viscosity lava circulating in the conduit via convection can maintain a permanent lava *lake* at the top of the conduit, within a summit crater. Such lava lakes can persist for decades, perhaps even hundreds of years. Although they pose little or no threat to life, they are of interest to volcanologists partly because of their rarity and partly because they are sites where circulation of magma and volatiles between the surface and the shallow chamber can be directly studied. Whether it results in a lava flow, lava dome, or lava lake, an effusive eruption is characterized by the emplacement of lava, with temperatures as high as ~1150°C, onto the surface of the volcano.

At the other extreme, an eruption can be explosive rather than effusive. In an explosive eruption, ascending magma experiences a reduction in pressure as it approaches the surface, and rapid expansion of the gases contained in growing bubbles within it tears the magma apart, resulting in a mixture of solids (pyroclasts, fragmented magma) suspended in gas. This mixture accelerates rapidly up the conduit before being ejected at great speed into the atmosphere. Although nature provides a spectrum of exit velocities, masses, mass flux rates, and plume heights, it is the volcanic plumes generated in large explosive eruptions, such as the eruption of Mount Pinatubo, Philippines, in 1991, or Mount St. Helens in 1980, that have the most far-reaching effects. In these eruptions, the energy involved pulverizes the fragments into very small ash-sized particles, forming volcanic ash clouds. During the 1991 Pinatubo eruption, the cloud reached heights of 40 km and covered an area of 230,000 km². Collapse of these plumes (once they have entrained enough cold air to become denser than the surrounding atmosphere) generates dangerous pyroclastic flows. The ash can be hazardous to human health, as it may be fine enough to be respirable. Ash clouds are also a hazard to global aviation.

For the interested reader, introductions to volcanology are provided by Parfitt and Wilson (2008) and Francis and Oppenheimer (2004). A more advanced treatment of all aspects of the science of volcanology can be found in the volume edited by Sigurdsson et al. (2000).

20.3 Why Use Satellite Remote Sensing to Study Active Volcanism?

As subjects for study, active volcanism leverages the advantages afforded by an orbital perspective. Erupting volcanoes are temporally dynamic; active lava flows expand, volcanoes deform, volumetric flux rates vary; eruptions themselves last from days, to weeks, to months. For such targets, the acquisition of data of uniform quality, at repetitive time intervals, over extended periods of time, from the safety of space, is obviously beneficial. A synoptic view is of importance, given the size that some targets can attain; lava flows can easily attain length of several kilometers, while volcanic ash clouds are commonly tens of kilometers in diameter. Furthermore, these ash clouds can expand to cover large areas as they disperse; the ash cloud produced during the climactic eruption of June 15, 1991, at Mount Pinatubo extended more than 1000 km from the vent in less than 11 h (Holasek et al., 1996). The aerosol cloud produced by the 1982 eruption of El Chichon, Mexico, circled the globe within 2 weeks. Such large eruption products cannot be adequately sampled from the ground, or with a narrow field of view. Remote sensing of volcanism also, as will become apparent, leverages data acquired from submicron to centimeter wavelengths, in the ultraviolet (UV) (for the analysis of volcanic gas plumes), through the shortwave and midwave infrared (for retrieving the temperatures of active lavas) and the long-wave infrared (for detecting and tracking volcanic ash clouds) to the microwave (for quantifying volcano deformation).

The relative importance of a remote sensing in providing information about active volcanoes varies depending on which volcano is the subject. Very well monitored volcanoes (such as Kilauea volcano in Hawaii) do not rely on satellite remote sensing to provide information about volcanic processes, given the comprehensive in situ monitoring systems that exist (including seismometers, permanent Global Positioning System (GPS) arrays, tilt meters, ground-based gas-sensing spectrometers, thermal cameras, infrasound arrays). At this class of volcano, satellite remote sensing may provide only supplemental data. Alternatively, satellite remote sensing may be the only source of information regarding eruptions at poorly monitored volcanoes, such as the large number in

the remote, and unpopulated, Aleutian island chain. Between these end members, remote sensing plays a greater or lesser role in providing information about the volcanic processes described in the previous section.

20.4 Remote Sensing of Volcano Deformation

Volcanoes inflate as magma is intruded from beneath and deflate as magma is erupted onto the surface or redistributed within the edifice. In situ measurements of ground deformation have been shown to reliably document these movements of magma, at many volcanoes around the world (Figure 20.2).

Measurements of volcano deformation were traditionally made using conventional field surveying techniques including precise leveling and trilateration (e.g., Murray et al., 1995). Tilt measurements (using the wet tilt technique, Figure 20.2, or dry tilt) have also been a mainstay of deformation monitoring at volcanoes around the world. In recent years, GPS has somewhat supplanted these methods.

Although able to resolve very subtle changes in volcano topography, these methods suffer from the fact that they provide a very incomplete spatial sample. Volcanoes are large (Mount Etna in Sicily covers an area of approximately 1200 km², and leveling lines, surveying benchmarks sample at a relative handful of point locations, requiring interpolation between disparate measurements to yield a volcano-wide deformation field. In addition, with the exception of the permanently recording GPS networks installed at some volcanoes, temporal sampling also leaves much to be desired, as these surveying methods require boots on the ground, with temporal sampling limited by the expense that entails. This combination of factors meant that although the deformation behavior of some volcanoes was very well constrained (e.g., Mount Etna, Sicily, and Kilauea, Hawaii), many (most) others, either because of remoteness, lack of funds, or lack of interest, were unmonitored.

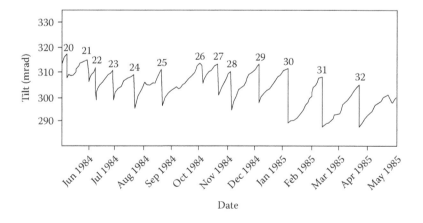

FIGURE 20.2 Tilt measurements obtained on the rim of the caldera at the summit of Kilauea volcano, Hawaii. Increases in tilt angle correspond to periods of volcano inflation. Decreases correspond to deflation during an eruption. Numbers denote eruptive episodes. (From http://volcanoes. usgs.gov/activity/methods/deformation/tilt/kilauea.php, courtesy of the USGS.)

20.4.1 Quantifying the Surface Deformation Field

The field of volcano deformation monitoring has been revolutionized by the technique of interferometric synthetic aperture radar (InSAR; sometimes referred to as IFSAR). Figure 20.3 illustrates the technique.

For those interested in more details, Burgamann et al. (2000) and Richards (2007) provide reviews of the technique. Side-looking spaceborne synthetic aperture radars sample the amplitude and phase of radar echoes from ground resolution elements. Amplitude images are familiar to many. Phase information less so, but it is the crucial component of the InSAR technique. The electromagnetic waves that the SAR emits travel to the target and are scattered back to the antenna. The distance is an integer number of complete wave cycles, plus a fraction of a cycle—this fraction is the phase, which combined yield the range from the antenna to each point on the ground within the image. Figure 20.3 shows two amplitude and phase images acquired by the first European

Remote Sensing satellite (ERS-1) SAR at Peulik volcano, Alaska. Here, spatial resolution is about 30 m. The images encode phase as colors varying between 0° and 360° (alternatively expressed as 0 to 2π radians). One cycle of interferometric phase is called a "fringe."

An idealized description of the processing follows, before a description of how nonideal conditions influence the applicability of the technique. Volcano deformation occurs over time, so a single image contains no deformation information. InSAR uses two images of the same target separated in time (Δt) and computes the phase for each pixel. If no deformation has occurred during the time interval, there will be no phase change ($\Delta\varphi = \varphi_1 - \varphi_2 = 0$). However, if deformation has occurred, then the same point on the ground will have moved to or away from the satellite, and the $\Delta\varphi$ will not be 0. First, the two SAR images are coregistered. This is necessary in order that the phase values retrieved from the first and second images that are to be differenced relate to the same physical area on the ground. Once this has been achieved, the phase difference that has accumulated

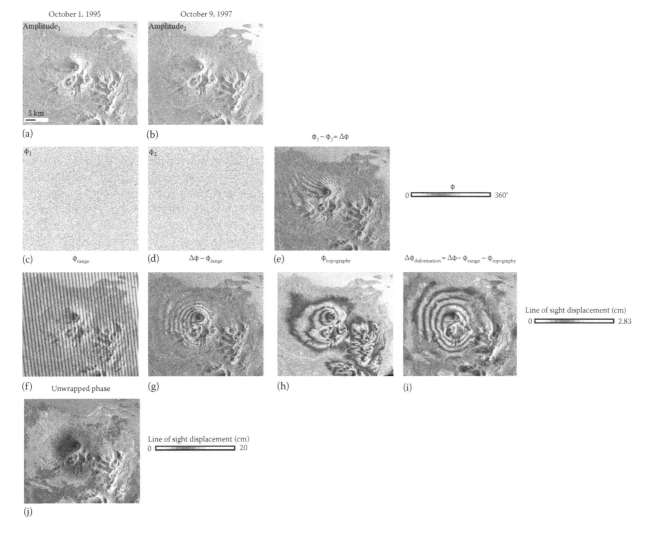

FIGURE 20.3 Overview of the InSAR data processing technique, using ERS-1 SAR images acquired at Mount Peulik volcano, Alaska. Images were processed by Zhong Lu, Southern Methodist University, TX, and the authors express their gratitude to Zhong for taking the time.

between observation time t_1 and t_2, for each point on the surface of the volcano, can be determined, and the interferogram generated (i.e., $\varphi_1 - \varphi_2 = \Delta\varphi$; Figure 20.3e).

If the two images were acquired by the same sensor at exactly the same position in space, then $\Delta\varphi$ would be the result of deformation alone. However, spacecraft orbital and attitude control does not allow for imaging from the exact same position, so in practice the images will be acquired from *about* the same position. This difference in position of the satellites at times t_1 and t_2 is referred to as the baseline and should be as small as possible for deformation monitoring. The limit depends on the sensor used; for ERS-1, a theoretical maximum of 1100 m has been calculated, but in practice a value of closer to 600 m was found to be more realistic. Imaging the same target from two different positions in space means that $\Delta\varphi$ is not purely the result of deformation. The change of viewing angle introduces a parallax effect, shifting the apparent position of ground points within the images and resulting in a phase difference for each ground target due to the topography ($\varphi_{topography}$). Even in the absence of any topography, a flat (or curved) Earth would still yield a gradual phase change across the image ($\varphi_{flat-earth}$) because of the gradient in look angle across the scene. Both the phase difference due to topography and the phase difference caused by the difference in satellite positions over a flat Earth must be removed from $\Delta\varphi$ to isolate the deformation component.

The flat-Earth phase (which is manifest as strong linear fringes running across the image; Figure 20.3f) can be predicted (i.e., based on knowledge of the satellites' geometry with respect to a reference ellipsoid) and removed using a geometry-based approach (e.g., Zebker et al., 1994). This is referred to as flattening the interferogram ($\Delta\varphi - \varphi_{flat-earth}$, Figure 20.3g) and is performed before removal of the topographic phase. Removal of the topographic phase information can be achieved by either using a digital elevation model (DEM) and solving for the phase difference that that topography would contribute to the observed phase (e.g., generate a synthetic interferogram from the DEM; Massonnet et al., 1993) or using a third radar image to generate a DEM from the radar data itself. In this case, the first pair of images (which contain the deformation signal) are spaced with a sufficient time interval (Δt) to capture the deformation signal of interest. The third image must be captured very close in time to either of these two images, such that it can be assumed that no deformation has had chance to occur, and the only phase change this pair records is that of the topography caused by the nonzero baseline (e.g., Gabriel et al., 1989).

The result (Figure 20.3i, $\varphi_{deformation} = \Delta\varphi - \varphi_{flat-earth} - \varphi_{topography}$) is the phase change due to deformation of the volcano surface between time t_1 and t_2. As phase is assumed to be proportional to distance at the scale of the wavelength, in these images one cycle/fringe corresponds to a distance of 2.8 cm, the wavelength of the C-band ERS-1 SAR being 5.6 cm (ground movement of $\lambda/2$ corresponds to a whole wavelength change in the round-trip range and is thus equal to one fringe). At this point, the actual amount of deformation at any point in the image is not known, as $\varphi_{deformation}$ merely cycles, or wraps, between 0 and 2π. The interferogram depicted in Figure 20.3i is said to be wrapped. As previously stated, the actual range between the antenna and the satellite for each

point in the image is given by $n \times 2\pi$ (where n is the complete number of wave cycles between the target and antenna—the integer ambiguity) *plus* the small fraction of interferometric phase measured, $\Delta\varphi$. In order to work out the unambiguous phase for each pixel in the image, a value of n has to be added to, or subtracted from, each $\varphi_{deformation}$ measurement. This process, determining the correct value of n for each pixel, is called phase unwrapping. A generalized method is to choose a seed pixel in areas of low noise and then add values of 2π to its phase value in expanding contours away from the seed while minimizing the change between adjacent points. There are many phase-unwrapping algorithms described in the literature (e.g., Zebker and Lu, 1998).

Phase unwrapping is an important part of the process, required to obtain a quantitative deformation field for the study area, detect subtle deformations, and also if the InSAR data are to be used to retrieve the properties of the subsurface source that caused the deformation, via numerical modeling. It is common that (a) the interferogram is unwrapped, (b) the unwrapped interferogram is then used for deformation source modeling, and (c) the observed interferogram, the modeled interferogram, and residual interferogram are then plotted in a *wrapped* fashion for publication and interpretation. Figure 20.3j shows the unwrapped interferogram for this example. As such, the relative deformation recorded in an wrapped interferogram such as the one depicted in Figure 20.3i can be determined by (1) choosing a point in the far field where no fringes occur (and hence there was no deformation) and then (2) counting the number of fringes inward from that point and multiplying by $\lambda/2$. In this example, approximately 18 cm of deformation occurred at the summit of the volcano over the 2-year period sampled by the interferogram. Whether this is subsidence (movement away from the satellite) or inflation (movement toward) depends on the direction of the change in phase (or order of the colors as shown in the color scale); when counting in from the edge of the fringe pattern toward the center of the deformation feature, if the interferometric phase decreases toward the center (i.e., trends from 2π to 0), then the distance between the antenna and the ground has become shorter toward the center, so the ground must have moved toward the satellite (i.e., inflation). The opposite is true for ground subsidence. In this case, the colors cycle from cyan–yellow–magenta as the center of the fringe pattern is approached—the volcano experienced a period of net inflation in the 2-year period encompassed by this interferogram.

The precision with which volcano deformation can be quantified is usually quoted at the sub-cm or mm scale, and the ability to quantify deformation at this scale over the entire volcanic edifice has revolutionized ground deformation monitoring. This mm sensitivity depends on several factors. The phase change due to deformation must be isolated from the interferometric phase measured, which is also composed of phase shifts due to several other factors. Although expressed differently by different authors, these components can be summarized as factors that introduce a change in phase due to uncertainty in the spacecraft orbit, errors in the DEM used to estimate topographic phase, the changing atmosphere, and the scattering properties of the surface (e.g., Massonnet and Feigl, 1998). Uncertainties in orbit (baseline uncertainties) impact the success of the flat-Earth correction. For Environmental Satellite (ENVISAT) SAR (part of the ERS series), this uncertainty has been estimated to be equivalent to a few tenths of a mm per km across the image (Wang et al., 2009) but this depends on which SAR is used, as this error can be eliminated if spacecraft orbit is known perfectly (and they are known more, or less, perfectly for different spacecraft). As previously discussed, isolating the deformation signal requires that the topographic phase be calculated, often using a DEM. The DEMs derived from the Shuttle Radar Topography Mission have a relative vertical accuracy of less than 10 m. It may seem counterintuitive that a DEM with ~10 m uncertainties can be used as part of a process that derives ground displacements on the mm scale, but the sensitivity of the phase measurement to deformation is several thousands of times greater than to the sensitivity to the topography (Equations 15 and 16 in Zebker et al., 2001). This explains why deformation can be estimated to sub-cm precision even if the DEM used to correct for topographic phase is inaccurate on the order of meters.

The at-satellite phase can also be influenced by the atmosphere (mainly water vapor concentration), which alters the propagation speed of the signal. Zebker et al. (1997) estimate that a 20% change in relative humidity can translate into 10 cm of perceived deformation. A common solution to this is to compute several interferograms of the same target spanning different time periods and then stack the interferograms, driving down the error due to the atmosphere by one over the square root of the number of interferograms (Zebker et al., 1997), although this does reduce the temporal resolution of the deformation measurement. Alternatively, meteorological measurements can also be used to estimate the magnitude of the atmospheric phase contribution (Delacourt et al., 1998), as can GPS measurements, which are also affected by the atmosphere (Webley et al., 2002), or other remote sensing data sets that allow atmospheric water vapor to be estimated independently (e.g., Li et al., 2005).

Finally, changes in the scattering properties of the surface (or changes in the apparent backscatter from that surface as viewing geometry changes) contribute to the interferometric phase (Zebker and Villasenor, 1992), often referred to as decorrelative noise. In short, the phase measurement from any given parcel of ground is unpredictable, being the sum of all interactions of the radar wave with all subpixel-sized scattering components, which means that the useful phase information that is proportional to target range cannot be extracted. But, if in the time between two SAR image acquisitions the target does not alter its scattering properties (at the scale of the radar wavelength), but merely moves toward or away from the antenna, then the difference in phase between these two observations times does tell us about this change in range, because the aforementioned random component of the phase cancels out. Coherence is a measure of how well correlated the phase measurements are between the two images. If they are perfectly correlated (high coherence), then the aforementioned assumption that the random component of the measured phase cancels out holds. If not, then differencing the two phase images (Figure 20.3) does not leave the phase change due to propagation delay, and deformation may not be quantifiable.

There are several sources of decorrelation, apparent on interferograms as areas without fringes (i.e., noisy areas, Figure 20.3). One source is poor coregistration of the image pair. Decorrelation takes place when the nature of the surface scatterers within each pixel changes substantially between SAR images acquisitions. This could be because a new lava flow has been emplaced over a previously lava-free area, ash has fallen on a previously ash-free surface, or snow has fallen (a common problem on volcanoes). Because such changes tend to be cumulative over time, long time periods between SAR image acquisition lead to increased incidence of decorrelation. Decorrelation can also occur because of overly large acquisition baselines. The signal received from the same subpixel arrangement of surface scatterers will change as viewing geometry changes (i.e., baseline increases) as the apparent relative positions of the scatterers change. As such, decorrelation can be ameliorated by using smaller temporal and spatial baselines (although obviously enough time must elapse for deformation to take place).

Given that decorrelation is wavelength dependent, different SAR instruments suffer to different extents. For example, as the size of leaves are of the order of the C-band wavelength (5.7 cm), interferograms derived from C-band SAR (such as ERS-1, ERS-2, ENVISAT, and the Canadian RADARSAT series) suffer from low coherence over vegetated surfaces, as the arrangement of the leaves of trees between acquisitions varies with the wind, or season. As vegetation is not uncommon on volcanoes, this has been a problem. Longer wavelength SARs (such as the Japanese ALOS L-band SAR; wavelength of 24 cm) are not affected by scatterers as small as leaves; they are affected by the tree trunks, but these do not change (as much) between SAR acquisitions. Consequently, interferograms derived from such L-band data exhibit greater coherence spatially than those derived from C-band data.

The technique has been widely applied to quantify volcano deformation since Massonnet et al. (1995) were the first to use ERS-1 SAR to calculate 11 cm of deflation at Mount Etna, coincident with 1991–1993 eruption, during which almost 500 million cubic meters of lava were erupted onto the surface. (It should be noted that a single interferogram resolves motion of the ground in the spacecraft's line-of-sight direction and does not return true vertical or horizontal ground motions, as in situ surveying methods do. However, by computing interferograms from SAR images acquired from different look angles [e.g., Fialko et al., 2001], then the true 3-D displacement field can be computed.) An obvious question is whether InSAR has detected deformation prior to eruptions, to which the answer is yes. Kizimen volcano, Kamchatka, erupted in 2010 for the first time since the late 1920s. Inflation of 6 cm over a near 2-year period prior to this eruption was detected using orbital InSAR, which also allowed the source of the deformation (propagation of a near vertical dike) to be constrained from the satellite data (Ji et al., 2013). Although most dramatic deformation usually occur at mafic volcanoes (because the erupted volumes are larger), substantial preeruptive deformation has also been observed at silicic volcanoes using InSAR. Jay et al. (2014) describe how three cycles of inflation preceded the 2011 explosive eruption of Cordon Caulle,

Chile. When deformation is not followed by an eruption, it is still a useful indicator of volcanic unrest. Subsidence of 84 mm was detected at Campi Flegrei (near Naples, Italy), a so-called "supervolcano" because of its potential to generate large explosive eruptions, between 1993 and 1996 (Avallone et al., 1999). During 2003, Mauna Loa volcano, Hawaii, received a great deal of media attention because of a renewed period of inflation (the volcano has not erupted since 1984). Although InSAR detected the inflation (Amelung et al., 2007), no eruption ensued.

Perhaps less common is the situation when an eruption occurs with no discernable deformation signature, as observed by Moran et al. (1996). Here two eruptions at Shishaldin volcano, Alaska, in 1996 and 1999 were accompanied by no apparent ground motion. These authors speculate that the reason for this was that magma rises so fast at this volcano, and from such depth, that significant deformation of the edifice does not occur on a resolvable time scale. At Lascar volcano, Chile, Pritchard and Simons (2002) advance similar reasons as to why no deformation was observed at that volcano during a period when several eruptions took place involving an amount of magma that should have produced resolvable deformation: either the magma source was too deep or the InSAR pair spanned a period of time before and after an inflation–deflation cycle over which time deformation was relatively short-lived and perfectly elastic. Sigmundsson et al. (1999) describe how although substantial deformation was measured by InSAR after the onset of the 1998 eruption of Piton de la Fournaise (Reunion Island) there was no deformation prior to the eruption, upon which the eruption could have been predicted. Again, very deep storage of the magma prior to eruption was advanced as the reason.

20.4.2 Quantifying Subsurface Magma Bodies

The deformation data that InSAR provides are very valuable in itself, and the technique has also been used to quantify thermal contraction of cooling lava flows (Stevens et al., 2001), lava flow volumes (Lu et al., 2002), and volcanic mass wasting (Ebmeier et al., 2010). However, its role in constraining the nature of the subsurface magma bodies responsible for the observed deformation is probably of greater importance. For a magma source at given depth, and geometry, deforming at a given rate, the change in surface topography with distance from the source can be predicted. A common source is a spherical point source of deformation, or a "Mogi source" (Mogi, 1958). Using numerical minimization techniques, the depth and volume change of a Mogi source can be stipulated, and the surface deformation field that combination would produce can be computed and compared with that obtained via the InSAR technique. The source parameters (location and volume change) are then iterated until the residuals between the observed and the predicted deformation fields are minimized, providing a best-fit estimate of the location of the magma chamber and the rate at which it is growing or shrinking as magma is added or removed, very important pieces of information for those wishing to assess the likelihood of future eruptions at volcanoes, by identifying which

volcanoes are deforming and which are not. The properties of the magma chamber, as derived from InSAR, also provide crucial constraints for modeling other volcanological processes and behaviors (e.g., Anderson and Segall, 2013). Although the Mogi point source of dilation is commonly used as the basis for this sort of geophysical inversion, other deformation models that reflect deformations resulting from intrusion of magma as dikes and sills, nonspherical magma chambers, or deformation due to faulting can also be used, if appropriate to the volcano in question (e.g., Okada, 1985; Davies, 1986; Amelung et al., 2000).

20.4.3 Quantifying Volcano Topography

Finally, although this section has focused on volcano deformation, it should be noted that volcano topography itself is an important product that the InSAR technique yields (see Rowland, 1996). Knowledge of volcano topography derived from InSAR is vital for modeling the downslope propagation of volcanic flows (lava or pyroclasts). As might be imagined, large spatial baselines (to maximize the stereoscopic effect that yields the topographic phase information) and short temporal baselines (to minimize phase changes due to topography) are preferable in this instance. (In fact, some restrict the use of the term InSAR to be solely associated with derivation of topography, the term DInSAR reserved for deformation studies.)

The issue of decorrelation has been mentioned before, as a factor that reduces the ability to resolve deformation across the entire surface of the volcano, for most volcanoes on Earth. The kind of long temporal baselines required to resolve subtle ground motions are precluded in most cases by the issue of temporal decorrelation. Although many SAR image pairs are often acquired of a target volcano, only a subset meet the critical baseline requirements, the remainder being subject to degradation via baseline decorrelation. Ferretti et al. (2001) proposed a technique that allows deformation measurements to be made even under these nonideal conditions, called the permanent scatterer (PS) technique. The technique identifies individual pixels in coregistered SAR images that display high coherence (correlation) over long periods of time. As such (because the phase scattering characteristics are dominated by one stable point scatterer), long temporal baselines can be employed, as well as large spatial baselines because of the high degree of correlation. These permanent scatter pixels in effect behave as a natural GPS network (Ferretti et al., 2001) allowing volcano deformation to be quantified even when no traditional InSAR fringes can be retrieved, as the stability of their scattering characteristics means that the phase information relating to changes in antenna-target range can still be retrieved for these scattered ground locations. Figure 20.4 shows a comparison of ground deformation derived from the InSAR permanent scatter technique with motions derived from traditional ground surveying.

Although initially applied on a volcano-by-volcano basis, increased availability of data (as more InSAR-capable missions have been launched), combined with the advent of more widely available and user-friendly InSAR processing software (and scientists who can use that software), and an ever lengthening

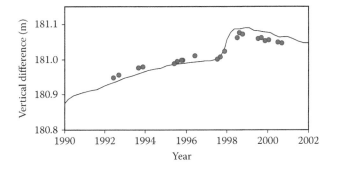

FIGURE 20.4 Comparison of ground deformation retrieved from PS InSAR with in situ measurements. Solid line shows height changes derived from electronic distance measurements; circles are the PS InSAR data. (Adapted from Hooper, A. et al., *Geophys. Res. Lett.*, 31, L23611, 2004, doi:10.1029/2004GL021737.)

archive of SAR data, has allowed the application of the technique to be expanded spatially and temporally. Local and regional analyses of volcanic deformation have been conducted (e.g., Pritchard and Simons, 2004; Lu and Dzuirsin, 2014). A recent paper by Biggs et al (2014) described a global analysis of deformation at 198 volcanoes over an 18-period finding that, of these, 54 exhibited deformation, while 25 of these went on the erupt. Analysis of 540 volcanoes over a shorter 3-year period allowed Biggs et al. to place this deformation–eruption/no eruption relationship in the context of wider petrotectonic factors.

20.5 Remote Sensing of Volcanic Degassing

The importance of gas measurements stems from the insights that they can yield into a wide range of volcanic phenomena. Gas measurements provide constraints on the composition, quantity, and origin of magmatic volatiles (e.g., Symonds et al., 1994), the hazards posed by an active volcano, insofar as fluxes of magmatic volatiles document changes in magma supply from depth (Roberge et al., 2009), and the deleterious effects, on both humans and landscapes, of the gases emitted (Williams-Jones and Rymer, 2000).

Volcanic gas emissions are comprised of a relatively restricted range of elements as molecular species. Although variable from volcano to volcano, H_2O, CO_2, and SO_2 are universally the most abundant volatiles emitted, in something like the proportions listed earlier (see Table 3 in Symonds et al., 1994, for much more data). Direct sampling, either using a "Giggenbach bottle" (flask partially filled with an absorbing solution) or filter packs close to gas vents (the resulting samples subsequently analyzed using traditional wet chemistry techniques, chromatography, or mass spectrometry), has been used for a long time. Most of the previously listed species can be measured using these techniques with high precision, although in situ gas measurements suffer from poor temporal resolution, spatial resolution, and an element of danger to the person doing the sampling. Direct sampling by flying through the gas plume (e.g., Gerlach et al., 1999) somewhat overcomes the safety issue without overcoming the spatiotemporal sampling

issues (and adding cost). This has led volcanologists to look toward remote sensing as a source of data on volcanic gas emissions.

The list of species amenable to characterization from low Earth orbit, using recently and currently operational sensors, is more restricted. The issue is one of the path length to space and the effect this has on isolating signal from noise (i.e., detecting a relatively small amount of volcanic gas in a long atmospheric path that may well contain that same gas naturally), the often coarse spatial resolution of the remote sensing instruments available, and/or the inability of these instruments to resolve the fine spectral resolution wavelength-to-wavelength contrasts that denotes the characteristic transitions of the relevant molecules, particularly given the masking effect that other species (such as H_2O) can have. Nevertheless, progress has been made.

From low Earth orbit, only volcanic emissions of SO_2 have consistently, and successfully, been mapped. SO_2 is a magmatic gas that is not present naturally in the atmosphere (although it is produced by burning fossil fuels), making it a relatively easy target. It is emitted by most volcanoes in prodigious amounts. The SO_2 molecule also has absorption features at wavelengths, in the UV and the long-wave thermal infrared (TIR), that are relatively free of obscuration by other gases and at which orbital remote sensing instruments commonly make measurements.

20.5.1 Quantifying Volcanic Sulfur Dioxide Emissions in the Ultraviolet

Kruger (1983) was the first to demonstrate that volcanic SO_2 could be measured from orbit, using data acquired by the Nimbus 7 Total Ozone Mapping Spectrometer (TOMS), during the 1982 eruption of El Chichon, Mexico. Figure 20.5 shows the extent of the SO_2 cloud on April 8, 1982, 4 days after the eruption occurred. TOMS-like volcanic SO_2 retrievals are based on the assumption that the reflectivity of the atmosphere in the UV is the sum of backscattering of UV light by air molecules, aerosols and clouds, and the ground, modulated by absorption by ozone and sulfur dioxide present in the column (although TOMS was designed for mapping ozone, O_3 and SO_2 have similar absorption spectra in the UV; Figure 20.5).

Dave and Mateer (1967) described the basis for ozone mapping from orbit, something that volcanologists have managed to exploit for SO_2 mapping. The amount of UV light available to measure at the top of the atmosphere is assumed to depend on the backscatter from the atmosphere (air molecules, aerosols, and clouds) and the ground, modulated by absorption by ozone in the atmospheric column, given the O_3 molecule's absorption cross section at these wavelengths. Clearly for the emergent signal to contain any information about O_3 concentrations (or SO_2), the effective scattering layer (i.e., the layer of the atmosphere from which most light is reflected back to the spacecraft) must be below the height in the atmosphere at which the O_3 (or SO_2) resides (if it is above, then the light never has a chance to interact with the target molecule).

By using radiative transfer modeling to predict the diffuse reflectance from the top of the atmosphere as a function of UV wavelength (for assumed solar zenith angles, O_3 profiles, and

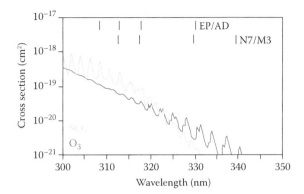

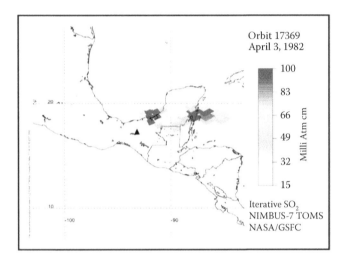

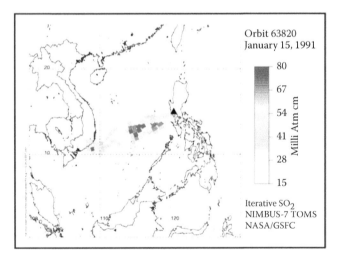

FIGURE 20.5 Top: absorption spectra of O_3 and SO_2 between 300 and 350 nm. Also indicated are the approximate wavelengths sampled by the Nimbus 7 and Meteor-3 (N and M3) and Earth Probe and ADEOS (EP and AD) TOMS instruments. OMI provides contiguous sampling across this region at 0.4 nm resolution. Middle: TOMS-derived estimate of the SO_2 placed into the 1982 eruption of El Chichon, Mexico. Bottom: the same data for the 1991 eruption of Mount Pinatubo, Philippines. The bottom two figures are courtesy of NASA/GSFC. (http://so2.gsfc.nasa.gov.)

the reflectivity from the physical layer of the model atmosphere above which scattering and absorption takes place), a measurement of the actual reflectance at one wavelength could be used to infer the total amount of O_3 in the column (i.e., interpolating between the reflectance predicted for each model O_3 distribution and that observed for an a priori unknown real O_3 distribution). Two measurements made at different UV wavelengths (one where O_3 attenuation is strong, the other weak) was found to reduce the error. Here the key parameter, N_λ, is given by

$$N_\lambda = -100\log_{10}\left(\frac{I_\lambda}{F_\lambda}\right) \qquad (20.1)$$

where

I_λ is the intensity of light reflected in the nadir direction (the observation)

F_λ is the exoatmospheric solar flux normally incident

λ corresponds to the wavelength at which the measurement is made

Figure 20.5 denotes the wavelengths at which Nimbus 7 TOMS and its successors acquired data. The model assumes no SO_2 in the atmosphere. But given the overlap in the spectra of SO_2 and O_3, presence of volcanic SO_2 would lead to an overestimate of total O_3; the issue for volcanologists is how to discriminate between attenuation by O_3 (always present) and attenuation by SO_2 (sometimes present; Krueger et al., 1995).

Several different algorithms have been used to process TOMS-like UV data for volcanic SO_2 retrievals. For example, the original work of Krueger (1983) analyzed regions within the volcanic SO_2 plume, as well as plume-free (and SO_2-free) regions adjacent to it. The apparent *excess ozone*, SO_2, in the plume region was estimated by determining that typical for the local atmosphere (determined from the plume-free regions), compared to the plume region. Krueger et al. (1995) used a least squares minimization approach based on multiple TOMS wavebands. The absorption by SO_2 and O_3 is retrieved via

$$N_\lambda = -\left[a + b(\lambda - \lambda_0) + s(\alpha_{1\lambda}\omega_1 + \alpha_{2\lambda}\omega_2)\right] \qquad (20.2)$$

where

a and b are coefficient that describe the relative radiance of a scattering atmosphere

s is optical path

ω_1 and ω_2 are the total column amount of O_3 and SO_2 from the ground to the top of the atmosphere

α_1 and α_2 are the O_3 and SO_2 absorption coefficients

N_λ is given by Equation 20.1, modified to include a term to account for changes in solar zenith angle

Here, the four unknowns (α_1, α_2, ω_1, ω_2) were determined by using four values for N_λ obtained at four TOMS wavelengths. Papers by Krotkov et al. (2006) and Yang et al. (2007) describe iterations that have allowed for improved retrievals of volcanic SO_2 using UV reflectance measurements.

These new algorithms have been developed as new sensors have been launched, and SO_2 detection limits have improved as a result of increased spatial resolution and changes in wavelengths sampled (see Table 1 in Carn et al., 2003). Nimbus 7 TOMS and Meteor-3 TOMS (operational 1978–1994) had spatial resolution of 50 and 63 km at nadir, with SO_2 detection limits of ~12,000–17,000 tonnes. The launch of Earth Probe TOMS (operational 1996–2005) and ADEOS TOMS (operational 1996–1997) provided data with a nadir resolution of 24–42 km, but an increased sensitivity down to ~1000–4000 tonnes of SO_2. Still, only large eruptions that injected large quantities of SO_2 into the upper troposphere and lower stratosphere could be quantified. The launch of the ozone mapping instrument (OMI) on board National Aeronautics and Space Administration's (NASA) Aura platform in 2004 led to a substantial improvement in our ability to detect volcanic SO_2 in the atmosphere, as a result of its smaller footprint (13×24 km at nadir), better spectral resolution, and improved detector noise characteristics. This combination of factors has lowered the sensitivity limit to <100 tonnes (using equivalent metrics) and allowed volcanologists to monitor nonexplosive passive degassing of Earth volcanoes, as well as low-intensity eruptive degassing that remains confined to the lower troposphere (see Carn et al., 2008).

Although the TOMS series of sensors (and now MI) have proven to be the workhorses for characterizing volcanic SO_2 emissions from space, other sensors have also been used, including the Scanning Imaging Absorption Spectrometer for Atmospheric Chartography (SCIAMACHY) launched in 2002 onboard ENVISAT (see Loyola et al., 2008) and the Global Ozone Monitoring Experiment 2 (GOME-2) launched on MetOp-A in 2006 (see Rix et al., 2012). The recently launched Ozone Profiler and Mapping Suite (OMPS) (SUOMI NPP) can also be used for quantifying SO_2 in the atmosphere (Yang et al., 2013). Once more, alternative algorithms have been developed for the data that these sensors provide. Although all of the algorithms cannot be described in this chapter (see cited papers for details), it is worth noting that regardless of sensor and algorithm, the unit universally used to quantify SO_2 in the remote sensing of volcanic SO_2 literature is the Dobson unit (DU), which is a vertical column abundance (1 DU = 1 milli-atm-cm = 2.691016 molecules cm^{-2}, = 0.0285 g of SO_2 per m^2), or the extinction per centimeter thickness of that gas under standard temperature and pressure. The total mass of SO_2 in a volcanic plume, such as that depicted in Figure 20.5, can then be easily estimated based on the instantaneous field of view (IFOV) of the sensor in question and the number of pixels comprising the plume, from which mass time series can be constructed using multitemporal observations. However, field volcanologists commonly report (or ultimately desire) an estimate of the *flux* of SO_2 being emitted from a volcano at a given time and how this changes (i.e., in kg s^{-1}, or t d^{-1}). Converting the satellite-derived SO_2 masses (sometimes referred to as "burden") to fluxes is not straightforward, and Theys et al. (2013) give a detailed overview of this issue and attempts to reconcile these quantities. For example, a simple approach is to assume that the flux of SO_2 from a volcano (M T^{-1}) is given by the mass of SO_2 determined from a UV satellite observation (M) divided by the residence time of SO_2 in the atmosphere (T).

Given that volcanic degassing is a temporally dynamic phenomenon, the question of temporal resolution at which satellite data are acquired is of importance. The aforementioned SCIAMACHY sensor provides complete global coverage once every 6 days; GOME-2 matches the temporal revisit of OMI (~24 h), but at much coarser spatial resolution (40×80 km). The early TOMS sensors did not provide complete global coverage in a 24 h period because of variations in their mission orbits. However, OMI and OMPS allow true global monitoring of volcanic SO_2 emissions (down to the quoted sensitivity limits) for most of Earth's volcanoes on a daily basis (see Krotkov, NASA Global Sulfur Dioxide Monitoring).

20.5.2 Quantifying Volcanic Sulfur Dioxide Emissions in the Thermal Infrared

SO_2 also exhibits absorption features at 7.3 and 8.6 μm, which means that TIR remote sensing also can be used to quantify its abundance in volcanic plumes (Realmuto et al., 1994; Watson et al., 2004). (For those who are interested, the paper by Guo et al. (2004) provides a direct comparison of the infrared and UV approaches for measuring volcanic SO_2, using the 1991 eruption of Mount Pinatubo as an example.) Figure 20.6 shows the absorption features in question, compared with the spectral response functions of the Terra Moderate Resolution Imaging Spectroradiometer (MODIS) and Advanced Spaceborne Thermal Emission and Reflection Radiometer (ASTER) TIR channels.

ASTER data only provide information regarding the less intense 8.6 μm feature, which will be discussed first. Adjacent is an ASTER TIR false color composite image of Miyake-jima volcano, Japan. The absorption of ground radiance by SO_2 in channel 11 reduces the amount of radiance reaching the sensor in this wavelength interval, causing the plume to appear in hues of yellow in this band triplet. Realmuto (1994) describes a technique for using thermal data of this kind to estimate the column abundance (g m^{-2}) of SO_2 in volcanic plumes, by modeling how much SO_2 must be present to produce the observed depression in at-satellite radiance in the 8.6 μm channel. The plume appears yellow because radiance from the scene is not substantially attenuated at wavelengths sampled by ASTER bands 14 and 13.

The at-satellite spectral radiance, $L_s(\lambda)$, for a vertical path through the atmosphere, can be written as (Realmuto et al., 1994)

$$L_s(\lambda) = \left\{ \varepsilon_\lambda L(\lambda, T_0) + [1 - \varepsilon_\lambda] L_d(\lambda) \right\} \times \tau_\lambda + L_u(\lambda) \quad (20.3)$$

where

ε_λ is spectral emissivity of the ground

$L(\lambda, T_0)$ is the spectral radiance from the ground surface radiating at temperature T_0

τ_λ is the spectral transmissivity of the atmosphere

L_d and L_u represent the downwelling and upwelling radiance produced by the atmosphere, respectively

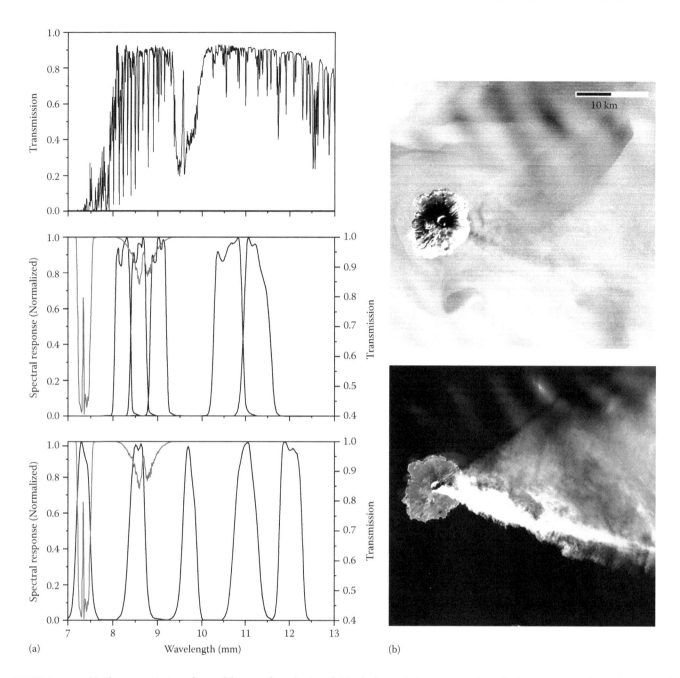

FIGURE 20.6   (a) The transmission of a model atmosphere (top) and SO_2 (red curve). Superimposed on the lower two panels are the spectral response of the Terra ASTER TIR subsystem (bands 10, 11, 12, 13, and 14 from left to right) and a selection of the MODIS emissive channel (bands 28, 29, 30, 31, and 32, left to right) sensors. (b) Terra ASTER false color composite (bands 14, 13, and 12, in red, green, blue) of Miyaka-jima volcano, Japan. The sulfur dioxide plume appears yellow in this band triplet. The lowermost image on the right is an ASTER near-infrared and visible wavelength false color composite (bands 3N, 2, and 1, in red, green, and blue). In reflected light, the plume appears white, as it is predominately composed of water.

The amount of radiance leaving the ground is a function of the radiance emitted by the ground ($\varepsilon_\lambda L(\lambda, T_0)$) plus the downwelling radiance reflected by the ground ($[1 - \varepsilon_\lambda]Ld(_\lambda)$). Upon passage from the ground to the sensor through the atmosphere, some of this radiance is attenuated (τ_λ) and some additional radiance is contributed ($L_u(\lambda)$).

In summary, the algorithm uses a radiative transfer model (e.g., MODerate resolution atmospheric TRANsmission [MODTRAN]; Berk et al., 1989) to predict how spectral radiance from a ground surface of temperature T_0 and emissivity ε_λ is attenuated as it passes through a model atmosphere from ground to sensor. The concentration of SO_2 at a particular altitude in the model atmosphere is varied until the predicted top-of-the-atmosphere radiance matches the observed at-satellite radiance for a range of TIR wavelengths centered on, and adjacent to, the 8.6 μm absorption feature. SO_2 column abundance is the product of SO_2 concentration and plume thickness. Integrating this for each plume pixel and multiplying by wind speed yields SO_2 flux (e.g., kg s⁻¹;

Realmuto et al., 1994). The temperature and emissivity of the ground beneath the volcanic plume can be obtained from image data. Although the height and thickness of the plume (which determines the layer of the plane-parallel model atmosphere in which SO_2 concentration will be varied) and its speed (which is used to compute the mass flux) could be constrained using stereoscopic or photoclinometric techniques, values for these parameters are often assumed, or obtained from in situ observations.

This technique has been used successfully to estimate SO_2 fluxes at a range of volcanoes using data acquired from space (e.g., Urai, 2004; Pugnaghi et al., 2006) and has the advantage of being able to image volcanic plumes by day or night (the UV approaches clearly cannot work without sunlight). The high spatial resolution of ASTER allows even low-intensity passive degassing to be detected and quantified. Comparisons between in situ measurements and ASTER-derived SO_2 abundances have shown an encouraging level of agreement (e.g., Henney et al., 2012). An alternative approach to retrieving SO_2 path concentrations that does not use the Realmuto et al. (1994) radiative-transfer-based approach is described by Campion et al. (2010). MODIS data can also be used as input to the algorithm (Watson et al., 2004; Novak et al., 2008), although the coarser spatial resolution (1 km, rather than 90 m) means that the volcanic emissions (e.g., plume size and SO_2 concentration) must be larger (see Realmuto, 2000 for sensitivity analysis).

The technique is limited by uncertainties regarding plume geometry and speed, atmospheric water vapor content, and the fact that the transmissivity of the atmosphere at 8.6 µm is also a function of absorption and scattering by other constituents of volcanic clouds, such as ash, sulfate aerosols, and ice particles (Watson et al., 2004). (The presence of ash mixed with SO_2 also confounds unambiguous SO_2 detection in the UV.) Furthermore, using ASTER and MODIS, the technique has not been operationally implemented to document global volcanic SO_2 fluxes in the manner seen with the previously described UV approaches. ASTER's low duty cycle (it is not a mapping mission) means that temporal coverage for most of Earth's volcanoes can be poor. The algorithm described by Realmuto (2000) also requires more user interaction than the UV approaches, making autonomous, operational implementation more challenging.

It has also been demonstrated that TIR data can also be used to quantify volcanic SO_2 using the 7.3 µm absorption feature (e.g., Prata et al., 2003; Prata and Bernardo, 2007). Although the atmospheric column as a whole is opaque at this wavelength (due to water absorption), the water vapor itself emits energy toward the sensor, which can be absorbed by SO_2 in the path (i.e., high altitude SO_2 clouds). Sensors such as MODIS, TIROS Operational Vertical Sounder-High Resolution Infrared Radiation Sounder (TOVS-HIRS), and the Aqua Atmospheric Infrared Sounder (AIRS) instrument all allow this absorption to be measured. In the case of HIRS (Prata et al., 2003), by making measurements of at-satellite radiance over an SO_2 cloud in wavebands adjacent to this absorption feature (6.7 and 11.1 µm), the radiance from an SO_2-free atmosphere at 7.3 µm ($L_{7.3pred}$) can be computed as radiance will vary linearly between these three wavelengths. Measured departures from this predicted radiance ($L_{7.3obs}$) can then be related to the SO_2 content of the cloud:

$$L_{7.3obs} - L_{7.3pred} = (1 - \tau_{SO2}) \times (L_{cloud, T} - L_{7.3pred}) \quad (20.4)$$

where

τ_{SO2} is the transmission of SO_2
$L_{cloud, T}$ is the spectral radiance from the cloud at temperature T

As sulfate aerosols and ash particles do not strongly affect extinction at this wavelength, the 7.3 µm region allows absorption by SO_2 to be measured more directly than is possible at 8.6 µm. However, successful detection of volcanic SO_2 at this wavelength is limited to plumes with altitudes greater than ~3 km (i.e., above the majority of the H_2O, Watson et al., 2004), for MODIS and HIRS. AIRS has the advantage that, as an imaging spectrometer, it has sufficient spectral resolution to isolate the SO_2 absorption feature and measure between the water absorption features, allowing SO_2 to be detected at lower altitudes. This must be offset against the fact that the poor spatial resolution of these sensors (HIRS resolution is ~19 km at nadir, increasing to ~32 km × 63 km at the edge of the scan; AIRS resolution is ~15 km^2 at nadir, increasing to 18 km × 40 km at the edge of the scan) equates to lower sensitivity.

20.5.3 Other Gases

Although SO_2 is the gas that volcanologists have most easily been able to measure from space, it is not necessarily the most desirable. Carbon dioxide is typically emitted in greater quantities than SO_2, exsolves at greater depth, and is relatively inert in shallow hydrothermal systems and the atmosphere (Burton et al., 2000), reducing its susceptibility to the scrubbing suffered by acid gases such as SO_2 (Edmonds, 2008). This means that the flux of CO_2 from a volcano at the surface is more a direct function of that exsolved at depth than SO_2, and given that this depth is greater an increase in CO_2 output can herald an increase in volcanic unrest before the subsequent shallow degassing of SO_2 (i.e., CO_2 may provide early warning of an eruption).

It is, however, not easy to detect volcanic CO_2 from space. Unlike SO_2 (which exists in nonvolcanic regions at the ppbv to pptv level; Seinfeld and Pandis, 1997), the atmosphere contains almost 400 ppmv of CO_2, against which volcanic CO_2 fluxes must be identified and then quantified. And the volcanic signal can be low. Gerlach et al. (2002) found volcanic CO_2 concentrations (volcanic = total − background) of up to 700 ppmv at the summit of Kilauea volcano, Hawaii (which is a very large CO_2 source), decreasing to approximately 300 ppmv within a horizontal distance of ~300 m and to <100 ppmv within a lateral distance of ~1500 m.

Although little has been published in the literature at the time of writing, there are some abstract volumes that report efforts to measure volcanic CO_2 degassing from space. Thus far, most (e.g., Schwandner et al., 2014) have focused on the potential for the Japanese Greenhouse Gas Observing Satellite (GOSAT) to serve this purpose. The Fourier transform spectrometer onboard GOSAT (TANSO-FTS) provides spot measurements at 10 km IFOV, acquiring spectra in the shortwave infrared (SWIR) and TIR with a spectral resolution of 0.2 wavenumbers. At the time of writing, little or no data are in the peer-reviewed literature to show whether these attempts have met with any success.

20.6 Remote Sensing of Geothermal and Hydrothermal Activity

The arrival of magma at shallow depths in the crust results in a range of interactions that manifest themselves as geothermal and hydrothermal phenomena that can be quantified from orbit, as these systems manifest themselves at the surface in the form of fumaroles, mudpots, geysers, and crater lakes, which exhibit either anomalous temperatures, surface compositions, or both. These phenomena yield insights into the nature of the magma body and can also act as precursory warning on impending eruptions. The famous 1902 eruption of Mount Pelée was preceded by 3 years of increasing fumarole temperatures (Chrétien and Brousse, 1989), with similar examples cited by Francis (1979).

Hydrothermal and geothermal activity tends to manifest itself as relatively low-temperature thermal anomalies, where low temperature means elevations in temperature up the order of a few hundred degrees kelvin above ambient. Planck's blackbody radiation law relates the spectral radiance emitted by a surface to its kinetic temperature, and so a space-based measurement of the spectral radiance emitted by such a volcanic target, at an appropriate wavelength, can be inverted to retrieve its temperature. Of course it is not just the temperature of the target that is of relevance but also its spatial abundance at the subpixel scale. Although the vapor vented through fumaroles can approach magmatic temperatures, the fumaroles themselves (fumaroles are cracks in rocks heated by passage of magmatic gases and steam from heated meteoric water) are very small compared to the size of the IFOV of remote sensing instruments. For example, at Vulcano, Aeolian Islands, Harris and Stevenson (1997) estimated that the fumaroles themselves occupied only 2.5% of the 16,000 m² area covered by the fumarole field. For a Landsat Thematic Mapper (TM) pixel, this amounts to only 22 m² out of 900 m². Thus, there is a strong dilution effect whereby such targets, while hot enough to "glow" at visible wavelengths at the resolution of the human eye, are so small that they do not emit sufficient spectral radiance, at the IFOV scale, to be apparent at visible and near infrared (VNIR), SWIR, or middle-infrared (MIR) wavelengths when observed from space. Because of this, and the fact that many of these volcanic features are water related, remote sensing of hydrothermal and geothermal phenomena has focused largely on the use of long-wave infrared (8–14 μm) data.

Figure 20.7 illustrates some of the issues. On the left are a series of Planck curves indicative of the amount of spectral radiance that a fumarole field might be expected to emit. The blue curve shows emittance for a surface at 300 K that fills the sensor IFOV. The red curve shows the emittance from a surface at 400 K that also fills the IFOV, some kind of geothermal feature (perhaps a large fumarole field, or a very hot acidic crater lake). Clearly, they are separable in the TIR and the MIR, but not sufficiently hot to radiate at all in the SWIR. However, it is usual that the emitted spectral radiance is a mixture of that from ground radiating at ambient temperature (e.g., 300 K) and the geothermal feature of interest, as the volcanic target is often subpixel in size. The green curve shows the emittance from a pixel that contains fumaroles at 400 K (occupying only 1% of the IFOV, e.g., Harris and Stevenson, 1997) surrounded by ambient temperature ground at 300 K (i.e., occupying the remainder of the pixel). Clearly, an image pixel containing this volcanic anomaly would barely be distinguishable from an adjacent pixel containing no such anomaly, making it very hard to detect and harder to quantify the excess volcanogenic radiance. When the size of the volcanic feature increases to cover 10% of the image pixel (orange curve), the anomalous pixel becomes somewhat distinct

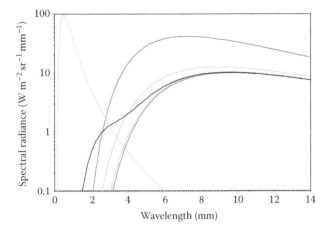
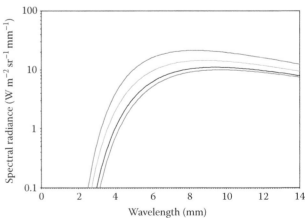

FIGURE 20.7 Simulated spectral radiance emitted from surfaces analogous to low-temperature hydrothermal and geothermal volcanic phenomena.

from the background (blue curve) more so at about 4 μm (the MIR) than the long-wave infrared.

Although fumaroles can reach higher temperatures (e.g., Harris and Maciejewski, 2000 measured some in the range of 800 K), these very hot fumaroles occupy an even smaller fraction of the total radiating area. The black curve illustrates the emitted radiance from a pixel containing fumaroles at 800 K occupying only 0.1% of the IFOV. A pixel containing such a volcanic target would be indistinct from its neighbors in the TIR but would radiate some energy in the MIR and SWIR. However, the preceding discussion has assumed that the observations are made at night. During the day, reflected sunlight will mask thermal anomalies in the SWIR and MIR. The gray curve shows the reflected sunlight from a 15% reflector. In summary, low-temperature hydrothermal and geothermal phenomena tend to radiate predominantly in the TIR. If they are sufficiently hot to radiate at shorter wavelengths, they are generally of such small subpixel size that the amount of energy radiated is in fact small at these wavelengths and easily masked by reflected sunlight during the day. It should also be noted that solar heating of surfaces can also mask volcanogenic sources of heat during the day and into the early part of the evening. These issues are obviously exacerbated as the spatial resolution of the data worsens (e.g., as one moves from Landsat TM-class resolution to MODIS-class resolution).

Despite the challenges, relatively high spatial resolution sensors (including the Landsat TM and its successor, the Enhanced Thematic Mapper Plus, ETM+) and Terra ASTER have been used to successfully quantify such low-temperature volcanism. Gaonac'h et al. (1994) used Landsat TM to estimate heat fluxes at Vulcano volcano, measuring excess radiant flux of up to ~50 W m² in the crater region, using nighttime data. Harris and Stevenson (1997) arrived at similar values using daytime data but noted the existence of solar heating anomalies of similar magnitude to the anomaly associated with the fumarole field itself. In both cases, no emittance was observed in the SWIR (Figure 20.7). Furthermore, Harris and Stevenson (1997) found that at 1 km spatial resolution of the Advanced Very High Resolution Radiometer (AVHRR) sensor, the temperature of the fumarole field was only 1 K higher than adjacent area, as a result of the dilution effect previously discussed. To be somewhat contrary, Kaneko and Wooster (1999) report thermal emission at SWIR wavelength from fumaroles at Unzen volcano Japan (albeit based on analysis of nighttime Landsat TM images), finding positive correlations between this thermal emission and SO_2 flux and magma discharge rate. Patrick et al. (2004) used TIR data acquired by the Landsat 7 ETM+ to document the areal expansion rate and surface heat flux from flows erupted from mud volcanoes with temperatures in the range of 10°C–40°C. Yellowstone National Park plays host to the full spectrum of hydrothermal and geothermal phenomena (including hot springs, geysers, fumaroles, and mudpots), and Vaughan et al. (2012) describe how ASTER TIR (and to a lesser extent MODIS TIR data) can be used to quantify geothermal heat fluxes at the scale of the park from all such sources, as well as monitor how these fluxes change over time.

Although substantial attention has been paid to measuring the thermal characteristics of low-temperature volcanism, this type of volcanic activity also gives rise to signatures that can be detected via either reflectance or emittance spectroscopy. The interaction of hydrothermal fluids alters surface rock compositions; minerals are also precipitated on the surface as warm, element-charged hydrothermal fluids equilibrate to surface temperatures and (and to a lesser extent pressures). Broadly speaking, carbonates, sulfates, clays, and silica-rich compositions result. For commercial reasons, much effort has been focused on detecting these targets from orbit and a large body of literature exists describing these exploration-centric activities (see Sabins, 1999, for a review, and Pour et al., 2013, for a recent example). Rather less has been reported on remote spectroscopic analysis for the purposes of studying active volcanic processes. Hellman and Ramsey (2004) investigated the use of ASTER data for studying active and fossil hot-spring deposits at Yellowstone. As hydrothermal alteration weakens rocks, mapping alteration on volcanoes can identify zones that are prone to catastrophic collapse and landslides. Crowley et al. (2003) describe the use of hyperspectral reflectance measurements acquired by NASA's Earth Observing-1 Hyperion sensor to do this.

In addition to the aforementioned manifestations of hydrothermal and geothermal volcanism are volcanic crater lakes. There are more than 100 volcanic crater lakes on Earth (Delmelle and Bernard, 2000). The volume, temperature, and bulk composition (which controls water color) of the water that resides in the lake are controlled by the influx and efflux of water, enthalpy, and chemical elements. The arrival of a fresh batch of magma beneath a crater lake will perturb the lake system, and the water (which acts as a calorimeter and chemical condenser for volcanic heat and chemical elements that rise from the underlying magmatic system) will respond.

The literature contains many examples of instances when changes in either crater lake temperature or color have preceded eruptions. For example, lake temperatures increased by 10°C in the 3 months preceding the 1965 eruption of Taal volcano, Philippines, an eruption that resulted in over 200 deaths (Moore et al., 1966). A similar 10°C rise (this time over a period of 4 months) preceded the 1990 eruption of Kelut volcano, Indonesia (Badrudin, 1994). Such temperature increases are clearly within the range that can be observed from space.

Changes in lake color have also been observed to precede eruptions (Oshawa et al., 2010). Coloration results from organic and inorganic materials dissolved and/or suspended in the water. At Ruapehu, New Zealand, a shift in color from blue–green to gray has been shown to indicate an increase in hydrothermal flux, the gray color due to the mobilization of bottom sediments by enhanced subaqueous fumarolic flow (Christenson, 1994). Dramatic color changes from gray–green to yellow–green at Poás, Costa Rica, have been attributed to increased fluxes of SO_2, which oxidizes dissolved iron, changing the spectral absorptance of the water (Delmelle and Bernard, 2000). Substantial volume changes (i.e., disappearance of a lake, something which would be relatively easy to detect from orbit) have been observed

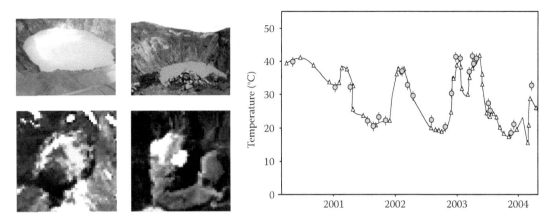

FIGURE 20.8 Left: Landsat ETM+ true color composites of the crater lakes at Maly Semyachik, Russia (left), and Irazu (Costa Rica, right) with corresponding photographs beneath. Right: ASTER-derived crater lake temperatures at Mount Ruapehu, New Zealand. Open triangles are in situ data; filled circles are ASTER measurements. (Adapted from Trunk, L. and Bernard, A., *J. Volcanol. Geotherm. Res.*, 178, 259, 2008.)

to precede eruptions at Poás in Costa Rica (Brown et al., 1989), as a result of increased magmatic energy input from depth.

The temperature and color of crater lakes has been measured from space (Figure 20.8). With regard to temperature, Oppenheimer (1996) presents a detailed heat budget analysis of several crater lakes based on temperatures derived from Landsat TM TIR data, noting the need to correct for the skin effect in order to force better agreement between the remote sensing data and in situ measurements, which typically report bulk water temperatures. Trunk and Bernard (2008) present an extensive analysis of ASTER-derived crater lake temperatures at four volcanoes and show how ASTER allows lake temperatures to be estimated to within 1.5°C of in situ measurements. Obviously, higher-resolution data are preferable, as the mixing effect previously discussed makes the same lake temperature anomaly harder to detect in lower-resolution data. In Figure 20.7 (right), the blue curve shows emitted spectral radiance from a 300 K blackbody, while the red curve shows the spectral radiance from a 350 K blackbody (i.e., the maximum crater lake temperature reported by Trunk and Bernard, 2008). The black curve shows the spectral radiance from a pixel containing a lake at 350 K covering 10% of a 1 km pixel (e.g., MODIS or AVHRR), assuming that 90% of the pixel radiates at the ambient background temperature of 300 K, approximately equal to the size of the lake at Ruapehu (dia. 400 m). The gray curve shows the same, but for a lake that covers 40% of the 1 km pixel (e.g., indicative of the dia. 700 km lake at Kawah Ijen, Indonesia). Regarding color, Figure 20.8 shows that the color of volcanic crater lakes can be retrieved provided true color data are acquired. Oppenheimer (1997) provides some examples of the use of Landsat TM to determine lake water color.

20.7 Remote Sensing of Active Lavas: Effusive Eruptions

Once magma breaches the surface, an eruption begins. Effusive eruptions involve the eruption of lava, which flows under the influence of gravity from the vent. At basaltic volcanoes (such

as Mount Etna, Sicily), this lava may be as hot as 1150°C and is sufficiently fluid to flow downhill, forming lava flows that are longer than they are wide and wider than they are thick. At felsic volcanoes (such as Mount St. Helens, United States), the lava is erupted at temperatures perhaps as few hundreds of degrees lower. This combined with its different chemistry means that felsic lava tend to be rheologically stiffer and piles up over the vent forming lava domes, which have higher thickness to area ratios.

In either case, the lava cools exponentially when exposed to the atmosphere. This causes rheological changes that influence the ability of the lava to expand from the vent. Ultimately, the areal expansion of a lava flow is halted because either these rheological changes cause the lava to become sufficiently stiff that it will not flow any further, or the supply of lava from the vent shuts off. In the case of lava domes, growth can cease because of either of these two factors or because the dome is disrupted in an explosive eruption, with a new dome taking its place (felsic lavas tend to have higher volatile contents, and higher viscosity, than mafic lavas, promoting generation of gas overpressures that may be released explosively).

The role of remote sensing in analyzing effusive eruption can be broadly divided into two fields. First, much attention has been diverted to developing autonomous systems that use orbiting satellites to detect the onset and cessation of effusive eruptions, based on their heat signatures observed from space. These studies have largely focused on the use of low spatial resolution data (such as AVHRR, MODIS, and even the geostationary missions, such as the Geostationary Operational Environmental Satellite series, GOES) because of the high temporal resolution at which such data are acquired, essential for timely detection and documentation of temporally dynamic eruptions that occur all over the globe. Second, there is a substantial body of literature targeted at analyzing the detailed thermophysical characteristics of lava flows and domes. These studies have tended to use higher spatial and spectral resolution data sets, such as Landsat TM, Terra ASTER, and to a lesser extent EO-1 Hyperion. It will become apparent that there is overlap between these two general fields. The book by Harris (2013) should be consulted for further details.

20.7.1 Detecting the Thermal Signature of Erupting Volcanoes

It is obvious that Earth orbiting satellites should be used to detect the heat signatures of erupting volcanoes, and many papers have been published describing approaches for doing this. The algorithms themselves have almost everything in common with those developed by remote sensors interested in detecting and mapping global wildfires (e.g., Prins and Menzel, 1992). The temporal revisit of Landsat-class instruments (~16 days, and often daytime acquisitions only) is of no use for monitoring eruptions that may last a matter of days, particularly as sensors such as TM, ETM+, and ASTER do not have 100% duty cycles. As a result, low spatial resolution but high temporal resolution sensors such as AVHRR, MODIS and GOES have been used extensively for monitoring eruptions in near-real-time. These provide global coverage at a temporal frequency of between 15 min and 12 h, with day and night imaging.

Although the spatial resolution of these sensors is coarse (about 1–4 km²), this is not a barrier to detecting lavas from space given the high temperature of the targets, and provided data are acquired at suitable wavelengths (Figure 20.9).

To detect an active lava body requires that the image pixel(s) that contain it are distinguishable from adjacent image pixels that do not contain active lava. In Figure 20.9, the blue curve shows the spectral radiance emitted from a blackbody at 300 K, representative of this ambient background. The red curve shows the spectral radiance emitted by a 1400 K blackbody, equivalent to the eruption

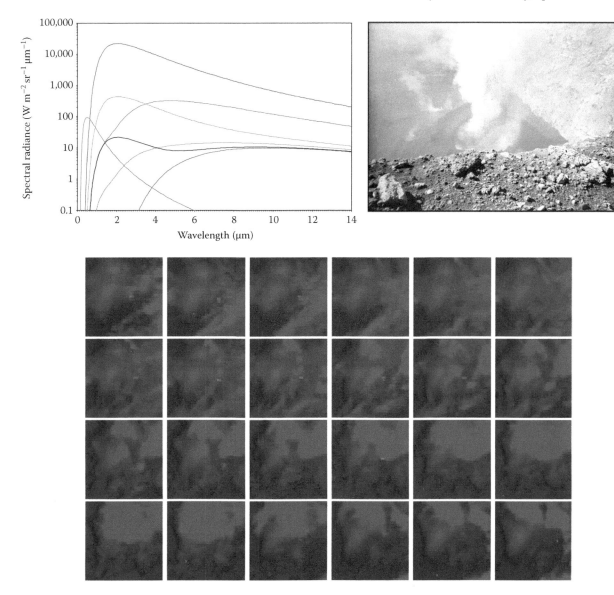

FIGURE 20.9   Left: Spectral radiance emitted from hypothetical surfaces analogous to high-temperature lavas. See text for details. Right: photograph of an active vent at Mount Etna, Sicily, September 1997. Vent is 80 m in diameter. Bottom: 6 h of GOES data acquired at Kilauea volcano, Hawaii. Images have been processed to highlight high-temperature material as magenta, making the lava flows from the volcano visible in the center of each image, cloud permitting.

temperature of an active lava flow. Clearly, they are distinguishable at all wavelengths. Note that given the high temperature of active lava, the wavelength of peak emission is in the SWIR, in contrast to the low-temperature hydrothermal phenomena described in the previous section for which only TIR data were useful. Note also the change in scale in ordinate between Figures 20.7 and 20.9; active lavas emit prodigious amounts of radiance at all wavelengths.

However, lava at eruption temperature never fills a 1–4 km² IFOV. Rather, the surface of an active lava is covered by a cool crust, within which cracks expose much hotter material from the flow interior (Figure 20.9), the total spectral radiance being areally weighted mixture of that from these components. Some active lava flows are sufficiently large that they fill several 1 km IFOVs. The orange illustrates the spectral radiance from such a pixel that contains an active lava flow, where the flow fills the pixel, and has a crust temperature of 600 K, within which cracks (occupying only 0.1% of the flow surface) expose lava at 1400 K. Such a curve is indicative of either a large lava body imaged at coarse spatial resolution data or a smaller lava body imaged at higher spatial resolution. However, in the present context (detection of volcanic thermal signatures using low spatial resolution data), the target of interest is often (perhaps usually) smaller than the pixel. The green curve shows the spectral radiance from an IFOV where the lava body still has a crust temperature of 600 K (and hot cracks still occupying 0.1% of the lava surface), but in this case, the lava body only fills 3% of the IFOV (e.g., a 100 m diameter lava dome). The gray curve shows another example, this time for a vent such as the one depicted in Figure 20.9 that, at 80 m diameter, would occupy 2% of a 1 km² IFOV. In this latter case, there is no cool crust—the 2% of the IFOV is occupied by material at 1400 K, with the remainder at ambient temperature (300 K). Finally, the black curve depicts the situation if that vent is made even smaller, occupying only 0.1% of the IFOV (i.e., a 30 m diameter vent radiating at a temperature of 1400 K). The magenta curve shows the reflected sunlight from a 15% reflector.

This graph conveys most of that which is required to understand the detection of active lavas from space using low spatial resolution satellite data (or indeed any data). First, and contrary to the situation for the low-temperature geo-/hydrothermal phenomena described previously, pixels containing active lava are most easy to detect if one looks in the MIR (~4 µm) and the SWIR (1.2–2.2 µm), as order of magnitude separation in the amount of emitted spectral radiance between target and background is apparent at these wavelengths. Second, coarse spatial resolution is no barrier to detection, as even vastly subpixel-sized active lava bodies emit enough radiance at SWIR and MIR wavelengths to influence the gross at-satellite signal, making pixels that contain these targets easily discriminable from the background. Third, the wavelength of peak emission for some of the hypothetical lava bodies depicted in this chart lies in the SWIR. At night, the SWIR signal is purely volcanogenic, with adjacent ground being too cool to radiate. Wooster and Rothery (1997) exploited this fact and used the 1.6 µm channel of the ERS-1 Along-Track Scanning Radiometer (ATSR-1) to document the SWIR spectral radiance emitted by the lava dome at

Lascar volcano, Chile, a feature vastly smaller than the 1 km² size of the ATSR IFOV. However, as shown in Figure 20.9, during the day, the SWIR signal from small active lava bodies is masked by reflected sunlight, rendering them undetectable. As a result, algorithms for the automated detection of volcanic hot spots have largely relied on the 4 µm atmospheric window. Inspection of Figure 20.9 reveals that all of the hypothetical lavas depicted are most easily distinguishable from the background (including solar heating and any reflected sunlight during the day) at this wavelength.

Several algorithms have been developed that exploit these principles, using low spatial but high temporal resolution data sources to monitor volcanoes in near real time at both the regional (e.g., Dehn et al., 2000; Harris et al., 2000) and global scales (e.g., Wright et al., 2002). Early work focused on the use of AVHRR (e.g., Harris et. Al., 1997a) and ATSR (e.g., Wooster and Rothery, 1997). Harris et al. (1997b) showed how even very-low-resolution (4 km) data acquired by the geostationary GOES satellite could be used to detect and track variations in thermal emission (but not map) from active lava flows at Kilauea volcano, in this case leveraging the geostationary vantage point to achieve 15 min temporal frequency (Figure 20.9).

The details of the algorithms are many and varied, and for the sake of brevity, all cannot be recounted here (for details of several techniques spanning the spectrum of approaches that have been adopted, see Higgins and Harris, 1997; Dehn et al., 2000; Wright et al., 2002; Pergola et al., 2004; Steffke and Harris, 2011). All interrogate large volumes of image data and seek to distinguish pixels that contain volcanic thermal anomalies from adjacent pixels that do not, by resolving the kinds of thermal emittance signatures depicted in Figure 20.9. Figure 20.10 shows, schematically, how a typical hot spot detection algorithm might work (in this example, the kind of spatiospectral-contextual algorithm employed by Higgins and Harris, 1997).

All algorithms involve comparing the radiant properties of a pixel against some kind of detection threshold (either absolute, based on the behavior of that pixel's neighboring pixels, or based on time-series analysis of the thermal history of a particular volcano). As such, near-real-time volcano monitoring algorithms must balance the desire to set thresholds low enough to detect the smallest and coolest hot spots possible against the need to minimize false positives, which erode confidence in algorithm performance. Substantial progress has been made in this arena, to the point that truly autonomous, global monitoring for volcanic thermal anomalies has been achieved, using data from the Terra and Aqua MODIS sensors (http://modis.higp.hawaii.edu; see Wright et al., 2002). At this website, the details of all volcanic thermal hot spots detected MODIS around the globe are made available within about 12 h of satellite overpass.

20.7.2 Quantifying the Thermophysical Characteristics of Active Lava Bodies

In addition to simple detection, these thermal data can also be used to analyze the nature of the volcanic activity responsible.

1. A kernal is passed over the image. It is assumed that each pixel in the image could potentially contain a volcanic "hot-spot". A pixel is reclassified as an actual volcanic hot-spot if its spectral radiance/temperature characteristics are found to differ significantly from its neighbors (background pixels).

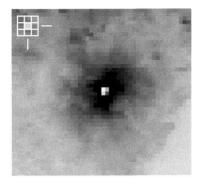

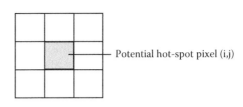

Potential hot-spot pixel (i,j)

2. The brightness temperature for each pixel in the image is computed at 4 and 12 mm(T_4 and T_{12}), in this example, ATSR-1 bands 2 and 4, respectively). Pixels that do not contain active lava will exhibit similar brightness temperatures. Pixels that are entirely filled with active lava will also have $T_4 \approx T_{12}$, but in this case the temperatures will be much higher than adjacent pixels in the kernal. Such a situation may arise when imaging an areally extensive lava flow. Pixels that contain sub-pixel-sized volcanic features are characterized by $T_4 \gg T_{12}$ (below). In this example, the elevated T_4 of the pixels in the center of the image is the result of small amounts of active lava present in Mount Etna's summit craters, vastly sub-pixel in scale, but sufficiently radiant at 3.7 mm to cause T_4 for these pixels to be significantly higher than adjacent pixels that do not contain lava. DT ($T_4 - T_{12}$) describes this difference and is effective at distinguishing pixels that contain small volcanic features (such as lava lakes, lava domes, and volcanic vents) from adjacent pixels that do not (background pixels).

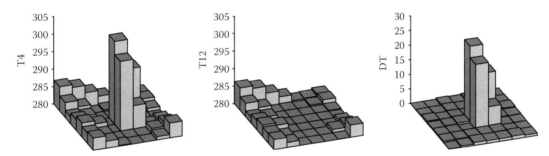

3. Comparing both T4 and DT for each pixel within the image with its immediate neighbors allows pixels containing active lava to be distinguished from those that do not. Hot-spot pixels are characterized by T_4 and DT which are more than n standard deviations (s) above the mean ($T_{4,bm}$; DT_{bm}) of the background pixels.

T4,b	T4,b	T4,b
T4,b	T4,(i,j)	T4,b
T4,b	T4,b	T4,b

DTb	DTb	DTb
DTb	DT(i,j)	DTb
DTb	DTb	DTb

Potential hot-spot pixel reclassified as an actual hot-spot pixel when:

$$T_{4,(i,j)} > T_{4,bm} + n\sigma T_{4,bm}$$

and

$$DT_{(i,j)} > DT_{bm} + n\sigma DT_{bm}$$

FIGURE 20.10   Overview of a volcanic hot spot detection algorithm.

Although initially this focused on the analysis of high-resolution Landsat-class data (e.g., Rothery et al., 1988), it quickly became apparent that volcanologically useful data could also be extracted from low spatial resolution data (e.g., Harris et al., 1997a). Determination of lava temperature was the early focus of this work, with the paper by Rothery et al. (1988), one of the most influential, in which they describe how SWIR data acquired by the Landsat TM can be used to constrain the subpixel thermal structure of lava flow surfaces (i.e., the temperature of the crust, the temperature of the hot cracks, and the area of the flow surface that the hot cracks cover) using a modified subpixel temperature unmixing approach based on the Dozier (1981) method. This work was subsequently expanded by others (e.g., Oppenheimer, 1991). Although in principle, this method works, it was in fact limited by the fact that (1) the Landsat TM only had two useable SWIR wavebands, forcing the assumption of an unrealistically simple thermal mixture model, and (2) the SWIR channels of TM readily saturated over highly radiant lava flow surfaces. Wright et al. (2010) show how the use of hyperspectral data acquired across the full 0.4–2.5 μm range alleviated these restrictions, allowing the full subpixel temperature distribution of active lavas to be retrieved from space, to the point that lava eruption temperature (a proxy for lava chemistry) can also be identified (Wright et al., 2011).

A particularly significant development was the realization that thermal measurements of active lava flows could be used to derive the volumetric flux of lava at the vent. This flux, the lava effusion rate, is of particular importance in volcanology as it is, after lava composition, the single most important variable that determines the final length that a lava flow can attain (Walker, 1973). Harris et al. (1997a) showed how AVHRR data acquired during the 1991–1993 eruption of Mount Etna could be used to estimate the lava effusion rate 27 times during the eruption. Not only were the individual AVHRR-derived estimates comparable with in situ estimates, but upon integration of the effusion rates ($m^3 \ s^{-1}$) over time (s), the final volume of the lava flow estimated from the AVHRR-derived fluxes was within the bounds identified from GPS surveying. In short the method exploits a simple physics-based proportionality between lava effusion rate and active flow area, which had been proposed earlier by Pieri and Baloga (1986), in which a higher effusion rate produces a larger lava flow. By making measurements of the area of active lava at the instant of satellite overpass, the remote sensing method allows the antecedent effusion rate that produced that amount of lava to be estimated, via an empirically determined constant (see Wright et al., 2001; Harris and Baloga, 2009 for details).

The importance of this development has been twofold. First, as mentioned previously, effusion rate is of paramount importance in determining how far a lava flow will advance. Wright et al. (2008) showed how satellite-derived effusion rates could be used to autonomously drive, and update, physics-based lava flow predictions. Second, the total amount of lava that cools to the atmosphere during an effusive eruption places constraints on how the volume of magma available at shallow depth is partitioned between the component that is erupted and the component that is intruded within the edifice. This has been used to constrain magma budgets at several volcanoes, including those hosting permanently active lava lakes, using both high and low spatial resolution data sets (e.g., Francis et al., 1993; Harris et al., 2000; Steffke et al., 2011).

It has also been found that simply plotting the amount of energy radiated by erupting volcanoes contains valuable information about volcanic processes. Trends in satellite-derived thermal emission correlate with other geophysical metrics (Figure 20.11).

Although many studies have sought to correlate satellite-measured thermal emission measurements with volcanic processes, two studies separated by two decades perhaps illustrate the best evidence that such satellite records contain signals that can be construed precursory to eruptions taking place. Oppenheimer et al. (1993) found that large decreases in SWIR spectral radiance as measured by Landsat TM at Lascar volcano, Chile, were

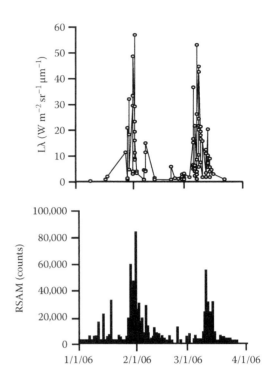

FIGURE 20.11 Top: 3.959 μm spectral radiance emitted during the 2006 eruption of Augustine volcano, Alaska, as measured by Terra and Aqua MODIS band 21. Bottom: real-time seismic amplitude measured in situ. (RSAM data from Power, J.A. et al., *Trans. Am. Geophys. Union*, 87, 373, 2006.)

followed by large explosive eruptions. Here, the hot fumaroles that gave rise to the excess SWIR radiance sealed, causing gas pressures within the dome to build up to the point of explosion. More recently, Van Manen et al. (2013) report how systematic increases in emitted MIR spectral radiance at Bezymianny volcano, Russia, as measured by AVHRR, were statistically likely to be followed by explosions.

20.8 Remote Sensing of Volcanic Ash Clouds: Explosive Eruptions

During an explosive volcanic eruption, the rapid rate at which volcanic ash plumes rise through the atmosphere and disperse within it makes near-real-time detection of their location, and how this changes over time, essential for aviation. There have been more than 100 inadvertent encounters between aircraft and volcanic ash clouds (Webley and Mastin, 2009). Flying through an eruption cloud causes abrasion and damage to an aircraft's exterior surfaces; ingestion of ash can cause engines to stall. There have been two well-documented cases where passenger jets have lost power to all four engines after inadvertently flying through ash clouds. Nine Volcanic Ash Advisory Centers around the world issue warnings about the location of volcanic ash clouds to the aviation community. Satellite remote sensing provides a significant amount of the information upon which these warnings are based.

Reliable detection of volcanic ash clouds depends upon the ability to effectively distinguish them from regular meteorological clouds. Prata (1989) suggested that TIR satellite data could be used to distinguish volcanic ash clouds from meteorological clouds based on their transmissive properties in this wavelength region. Following Prata (1989), the at-satellite spectral radiance ($L_s(\lambda)$) when viewing the ground through a partially transparent cloud (silicate ash or H_2O) can be written as

$$L_s(\lambda) = e^{-\tau_\lambda} L(\lambda, T_s) + (1 - e^{-\tau_\lambda}) L(\lambda, T_c) \qquad (20.5)$$

where

τ_λ is the cloud optical depth along the line of sight
T_c is the temperature of the cloud top
T_s is the temperature of the ground beneath the cloud

The method assumes a plume that is not opaque to the transmission of light from the ground beneath. Spectral radiance from the ground at temperature T_s passes up and encounters the cloud or plume. As the transmissivity of H_2O is higher at ~11 μm (λ_i) than at ~12 μm (λ_j), the difference in at-satellite radiance at these two wavelengths (when expressed as a difference in brightness temperature; BTD = $T_i - T_j$) will be positive for water clouds. Conversely, volcanic ash clouds are less transmissive at 11 μm than at 12 μm, and the equivalent brightness temperature difference (BTD) will be negative. This so-called "split window" technique has been widely used as the basis for detecting and tracking volcanic ash clouds for more than 20 years, using images provided by any sensor that acquires data at these wavelengths (e.g., AVHRR, GOES, MODIS). Data from geostationary spacecraft allow ash clouds to be detected and tracked in near real time (although this capability is reduced at high latitudes where the satellite zenith angle becomes large). Figure 20.12 illustrates the technique, using MODIS data acquired during the 2010 eruption of Eyjafjallajökull, Iceland. The ash cloud is clearly discriminated from other scene elements by virtue of its strongly negative BTD.

Although there are many sources of data regarding ash clouds that are used by the aviation authorities when issuing advisories (including visible wavelength satellite images and pilot reports themselves), the method of Prata (1989) remains an important means for detecting volcanic ash clouds using

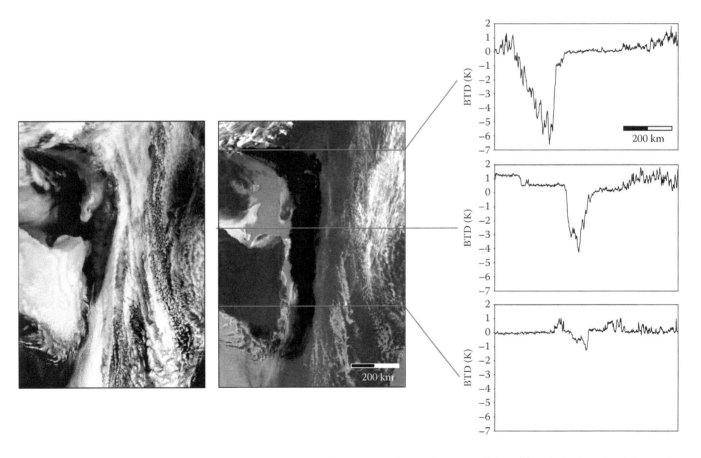

FIGURE 20.12   Left: Terra MODIS simulated true color composite (bands 1, 4, and 3 in red, green, and blue) of the ash cloud produced during the 2010 eruption of Eyjafjallajökull, Iceland. The ash cloud is clearly visible as the brown streak running approximately north–south. Center: MODIS band 31 (11.02 μm) minus MODIS band 32 (12.02 μm) brightness temperature difference (BTD) image. The black pixels running north–south correspond to strongly negative BTD values. Right: three BTD profiles across the image subset (red lines mark transects). The horizontal scale is the same as for the horizontal transect on the images displayed.

TIR satellite data, and there are many papers in the literature that have employed the technique (e.g., Mayberry et al., 2001; Dean et al., 2004; Dubuisson, et al., 2014). Although not perfect (see Prata et al., 2001), it is effective in great many instances provided that the results are interpreted correctly, and the conditions under which, what is actually a relatively simple algorithm that should not be expected to work flawlessly, are taken into account. For example, it is known (Prata et al., 2001) that negative BTDs can result over deserts (which themselves have a silicate absorption spectrum, like the ash), clear land surfaces at night (due to surface temperature and moisture inversions), and at ash cloud edges (due to misregistration of the two channels used to compute the BTD index), yielding false positives. It is also known that under certain circumstances, ash clouds can exhibit a positive BTD (opaque ash clouds; coatings of ice on ash particles; see Watson et al., 2004). The Prata (1989) method also has the distinct advantage that it can be exploited by any remote sensing mission that acquires data in the 11 and 12 μm passbands, including those acquired from high temporal resolution geostationary instruments such as GOES, Multifunctional Transport Satellite (MTSAT), and Spinning Enhanced Visible and Infrared Imager (SEVIRI). This allows ash clouds to be tracked with a resolution of minutes, rather than hours or days. Alternative techniques for volcanic ash detection have been published. The method of Filizzola et al. (2007) does not rely on any radiative transfer modeling of ash cloud physical properties but rather uses a change detection approach, largely equivalent to the approach described by Pergola et al. (2004).

Although this section has focused on ash detection in the TIR, volcanic ash can also be detected in the UV, using TOMS (Seftor et al., 1997), based on an aerosol index computed from measurements of backscatter at two UV wavelengths. Krotkov et al. (1999) show excellent agreement between the previously described TIR retrieval and this UV-derived metric for the same ash cloud imaged almost simultaneously by AVHRR and TOMS. Constantine et al. (2000) show less impressive convergence, for a different eruption. The issue of ash (and SO_2) detection in either the UV or the TIR is complicated by their frequent coexistence in the same cloud, in addition to the presence of other species (water, ice, sulfate aerosols) that also absorb light in the same spectral passbands (Watson et al., 2004; Figure 20.13).

Rose et al. (1995) showed how in one case the presence of ice in an ash cloud completely masked the negative BTD thermal signature described earlier. Clearly, higher spectral resolution provides a solution to this. Unfortunately, higher spectral resolution tends to come at the expense of spatial resolution (e.g., AIRS, IASI).

Detection and tracking of ash is obviously important. But providing quantitative information about the properties of the cloud is also of value. Following from the original work of Prata (1989), Wen and Rose (1994) developed a method for using TIR satellite data to retrieve the sizes and total mass of ash particles in volcanic clouds. Assuming a thin plane-parallel cloud

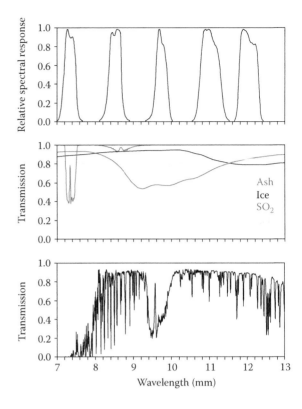

FIGURE 20.13 Top: relative spectral response of MODIS channels 28, 28, 30, 31, and 32, from left to right. Middle: transmission spectra of ice, silicate ash, SO_2, and SO_4^{2-}. Bottom: transmissivity of a model atmosphere, calculated using MODTRAN. (Figure simplified from Watson, I.M. et al., *J. Volcanol. Geotherm. Res.*, 135, 75, 2004.)

comprising spherical ash particles overlying a homogenous surface, they write at-satellite spectral radiance as

$$L_s(\lambda) = (1 - r_c(r_e, \tau_\lambda))L(\lambda, T_c) + t_c(r_e, \tau_\lambda)(L(\lambda, T_s) - L(\lambda, T_c))$$

(20.6)

where

r_c is the reflectivity of the cloud
r_e is the effective radius of the spherical ash particles
t_c is the transmissivity of the cloud

Assuming two measurements of emitted radiance at wavelengths λ_i (~11 μm) and λ_j (~12 μm), and converting to brightness temperature, allows the possible set of contours of r_e and τ_λ to be plotted in a $T_i - T_j$ versus T_i brightness temperature space. Comparing real values of $T_i - T_j$ versus T_i values obtained from analysis of AVHRR images of volcanic clouds (where T_i and T_j are the brightness temperature in AVHRR channels 4 and 5, respectively) allows the effective radius of the ash particles comprising the cloud and its optical depth to be determined on a pixel-by-pixel basis. Knowledge of the density of the ash allows the mass of ash present to be computed.

At a more rudimentary level, the planimetric shape of the ash cloud and its height can also be determined from satellite data. These are important boundary conditions for numerical models that seek to predict the dispersal of ash during an eruption, as well as providing validation data for the output of these models (e.g., Searcy et al., 1998). Ash cloud height is also an important constraint on the intensity of an eruption (i.e., the mass flux of magma from the vent) as the theoretical relationship between plume height and mass flux is well constrained (e.g., Wilson et al., 1978). Several authors have demonstrated how ash cloud height can be determined from satellite measurements of cloud-top temperature (e.g., Kienle and Shaw, 1979; Holasek et al., 1996). Here a simple measurement of cloud-top temperature can be used to estimate its height above sea level if the local lapse rate of the atmosphere is known and the cloud is in thermal equilibrium with the atmosphere (and the emissivity of the ash cloud is known or can safely be assumed). The cloud must also be opaque, to prevent the cloud-top temperature being an integral over a range of depths within the cloud. In addition, the momentum of a rising ash cloud can cause it to overshoot the level of neutral buoyancy, resulting in error in height estimation.

An ash cloud casts a shadow, and measurements of shadow length cast by an ash cloud on the ground below can also be used to estimate cloud height (Glaze et al., 1989). Clearly, this technique only works during the day but is reliable if the satellite and solar zenith and azimuth angles are known, and the plume is placed against a simple background (e.g., the ocean). If the shadow is cast upon a cloud deck at a lower altitude, then the shadow can still be used to estimate height but only in conjunction with thermal measurements of these clouds to constrain their height above sea level. The shapes of the tops of ash clouds add a level of complexity. For example, Holasek et al. (1996) found that the central region of the Pinatubo eruption cloud was ~15 km higher than the cloud edge, meaning multiple measurements were required, each with a different datum. Heights can also be retrieved from stereoscopic image pairs (see Prata and Turner, 1997). With regard to the shape of ash clouds, Glaze et al. (1999) present a method for determining their surface morphology using photoclinometric analysis of AVHRR images.

20.9 Conclusion

Although no remote sensing missions have been designed to study active volcanism, scientists have found imaginative ways to extract volcanologically useful data from sensors of varying spatial, spectral, and temporal resolution. This has been particularly true as the number of Earth observations missions has increased, making data more widely available, the cost of image processing facilities has come down, allowing more graduate students to find inventive ways to squeeze as much information as possible out of the datasets. Real- or near-real-time delivery of data via the internet (at no cost) has been a significant development, eruptions to be detected at any volcano on Earth within minutes to hours of the spacecraft passing over. It seems that increased capabilities will become apparent as the spatial and spectral resolution of geostationary imaging systems improves. As temporal archives expand, remote sensing has been used to provide a more useful longer terms perspective on volcanic behavior, based on

early proof of concept studies that, because of scarcity of data, largely ignored the temporal component. Although not the cure-all that some may have envisaged 30 years ago, remote sensing plays a significant role in monitoring Earth's active volcanoes.

References

Amelung, F., Jonsson, S., Zebker, H., and Segall, P. 2000. Widespread uplift and 'trapdoor' faulting on Galapagos volcanoes observed with radar interferometry. *Nature*, 407, 993–996.

Amelung, F., Yun, S., Walter, T.R., Segall, P., and Kim, S. 2007. Stress control of deep rift intrusion at Mauna Loa volcano, Hawaii. *Science*, 316, 1026–1030.

Anderson, K. and Segal, P. 2013. Bayesian inversion of data from effusive volcanic eruptions using physics-based models: Application to Mount St. Helens 2004–2008. *Journal of Geophysical Research*, 118, 2017–2037.

Avallone, A., Zollo, A., Briole, P., Delacourt, C., and Beauducel, F. 1999. Subsidence of Campi Flegrei (Italy) detected by SAR interferometry. *Geophysical Research Letters*, 26, 2303–2306.

Badrudin, M. 1994. Kelut volcano monitoring: Hazards, mitigation and changes in water chemistry prior to the 1990 eruption. *Geochemical Journal*, 28, 233–241.

Berk, A., Bernstein, L.S., and Robertson, D.C. 1989. MODTRAN: A moderate resolution model for LOWTRAN7. Final Report, GL-TR-0122, Airforce Geophysics Laboratory, Hanscom Airforce Base, Bedford, MD.

Biggs, J., Ebmeier, S.K., Aspinall, W.P., Lu, Z., Pritchard, M.E., Sparks, R.S.J., and Mather, T. 2014. Global link between deformation and volcanic eruption quantified by satellite imagery. *Nature Communications*, 5, 3471, doi:10.1038/ncomms4471.

Brown, G., Rymer, H., Dowden, J., Kapadia, P., Stevenson, D., Barquero, J., and Morales, L.D. 1989. Energy budget analysis for Poas crater lake: Implications for predicting volcanic activity. *Nature*, 339, 370–373.

Burgmann, R., Rosen, P.A., and Fielding, E.J. 2000. Synthetic aperture radar interferometry to measure Earth's surface topography and its deformation. *Annual Reviews of Earth and Planetary Sciences*, 28, 169–209.

Burton, M.R., Oppenheimer, C., Horrocks, L.A., and Francis, P.W. 2000. Remote sensing of CO$_2$ and H$_2$O emissions from Masaya volcano, Nicaragua. *Geology*, 28, 915–918.

Campion, R., Salerno, G.G., Coheur, P., Hurtmans, D., Clarisse, L., Kazahaya, K., Burton, M., Caltabiano, T., Clerbaux, C., and Bernard, A. 2010. Measuring volcanic degassing of SO$_2$ in the lower troposphere with ASTER band ratios. *Journal of Volcanology and Geothermal Research*, 194, 42–54.

Carn, S.A., Krueger, A.J., Bluth, G., Schaefer, S.J., Krotkov, N.A., Watson, I.M., and Datta, S. 2003. Volcanic eruption detection by the Total Ozone Mapping Spectrometer (TOMS) instruments: A 22-year record of sulphur dioxide and ash emissions. In: C. Oppenheimer et al. (eds) *Volcanic Degassing*. London, U.K.: Geological Society of London, pp. 177–202.

Carn, S.A., Krueger, A.J., Krotkov, N.A., Arellano, S., and K. Yang. 2008. Daily monitoring of Ecuadorian volcanic degassing from space. *Journal of Volcanology and Geothermal Research*, 176, doi:10.1016/j.jvolgeores.2008.01.029.

Chrétien, S. and Brousse, R. 1989. Events proceeding the great eruption of 8 May, 1902 at Mount Pelée, Martinique. *Journal of Volcanology and Geothermal Research*, 38, 67–75.

Christenson, B.W. 1994. Convection and stratification in Ruapehu crater lake, New Zealand: Implications for Lake Nyos–type gas release eruptions. *Geochemical Journal*, 28, 185–197.

Condie, K.C. 2005. *Earth as an Evolving Planetary System*. London, U.K.: Elsevier Academic Press.

Constantine, E.K., Bluth, G.J.S., and Rose, W.I. 2000. TOMS and AVHRR observations of drifting volcanic clouds from the August 1991 eruptions of Cerro Hudson. In: J. Mouginis-Mark et al. (eds) *Remote Sensing of Active Volcanism*. AGU Geophysical Monograph Series. Washington, DC: American Geophysical Union, pp. 45–64.

Crowley, J.K., Hubbard, B.C., and Mars, J.C. 2003. Analysis of potential debris flow source areas on Mount Shasta, California, by using airborne and satellite remote sensing data. *Remote Sensing of Environment*, 87, 345–358.

Dave, J.V. and Mateer, C.L. 1967. A preliminary study on the possibility of estimating total atmospheric ozone from satellite measurements. *Journal of Atmospheric Sciences*, 24, 414–427.

Davis, P.M. 1986. Surface deformation due to inflation of an arbitrarily oriented triaxial ellipsoidal cavity in an elastic half-space, with reference to Kilauea Volcano, Hawaii. *Journal of Geophysical Research*, 91, 7429–7438.

Dean, K., Dehn, J., Papp, K.R., Smith, S., Izbekov, P., Peterson, R., Kearney, C., and Steffke, A. 2004. Integrated satellite monitoring of the 2001 eruption of Mount Cleveland, Alaska. *Journal of Volcanology and Geothermal Research*, 135, 51–74.

Dehn, J., Dean, K., and Engle, K. 2000. Thermal monitoring of North Pacific volcanoes from space. *Geology*, 28, 755–758.

Delacourt, C., Briole, P., and Achache, J. 1998. Tropospheric corrections of SAR interferograms with strong topography: Application to Etna. *Geophysical Research Letters*, 25, 2849–2852.

Delmelle, P. and Bernard, A. 2000. Volcanic Lakes. In: H. Sigurdsson et al. (eds) *Encyclopedia of Volcanoes*. London, U.K.: Academic Press, pp. 877–895.

Dozier, J. 1981. A method for satellite identification of surface temperature fields of subpixel resolution. *Remote Sensing of Environment*, 11, 121–129.

Dubuisson, P., Herbin, H., Minvielle, F., Compiegne, M., Thieuleux, F., Parol, F., and Pelon, J. 2014. Remote sensing of volcanic ash plumes from thermal infrared: A case study analysis from SEVERI, MODIS and IASI instruments. *Atmospheric Measurement Techniques*, 7, 359–371.

Ebmeier, S.K., Biggs, J., Mather, T.A., Wadge, G., and Amelung, F. 2010. Steady downslope movement on the western flank of Arenal volcano, Costa Rica. *Geochemistry, Geophysics and Geosystems*, 11, Q12004.

Edmonds, M. 2008. New geochemical insights into volcanic degassing. *Philosophical Transactions of the Royal Society of London*, 366, 4559–4579.

Ferretti, A., Prati, C., and Rocca, F. 2001. Permanent scatterers in SAR interferometry. *IEEE Transactions on Geoscience and Remote Sensing*, 39, 8–20.

Fialko, Y., Simons, M., and Agnew, D. 2001. The complete (3-D) surface displacement field in the epicentral area of the 1999 Mw 7.1 Hector Mine earthquake, California, from space geodetic observations. *Geophysical Research Letters*, 28, 3063–3066.

Filizzola, C., Lacava, T., Marchese, F., Pergola, N., Scaffidi, I., and Tramutoli, V. 2007. Assessing RAT (Robust AVHRR Techniques) performances for volcanic ash cloud detection and monitoring in near real-time: The 2002 eruption of Mount Etna (Italy). *Remote Sensing of Environment*, 107, 440–454.

Francis, P.W. 1979. Infra-red techniques for volcano monitoring and prediction—A review. *Journal of the Geological Society of London*, 136, 355–359.

Francis, P.W. and Oppenheimer, C. 2004. *Volcanoes*, 2nd ed. Oxford, U.K.: Oxford University Press.

Francis, P.W., Oppenheimer, C., and Stevenson, D.S. 1993. Endogenous growth of persistently active volcanoes. *Nature*, 366, 544–557.

Gabriel, A.K., Goldstein, R.M., and Zebker, H.A. 1989. Mapping small elevation changes over large areas: Differential radar interferometry. *Journal of Geophysical Research*, 94, 9183–9191.

Gaonac'h, H., Vandemeulebrouck, J., Stix, J., and Halbwachs, M. 1994. Thermal infrared satellite measurements of volcanic activity at Stromboli and Vulcano. *Journal of Geophysical Research*, 99, 9477–9485.

Gerlach, T.M., Doukas, M.P., McGee, K.A., and Kessler, R. 1999. Airborne detection of diffuse carbon dioxide emissions at Mammoth Mountain, California. *Geophysical Research Letters*, 26, 3661–3664.

Gerlach, T.M., McGee, K.A., Elias, T., Sutton, A.J., and Doukas, M.P. 2002. Carbon dioxide emission rate of Kilauea volcano: Implications for primary magma and summit reservoir. *Journal of Geophysical Research*, 107, doi:10.1029/2001JB000407.

Glaze, L., Francis, P., Self, S., and Rothery, D. 1989. The 16 September 1986 eruption of Lascar volcano, north Chile: Satellite investigations. *Bulletin of Volcanology*, 51, 149–160

Glaze, L.S., Wilson, L., and Mouginis-Mark, P.J. 1999. Volcanic eruption plume top topography and heights determined from photoclinometric analysis of satellite data. *Journal of Geophysical Research*, 104, 2989–3001.

Guo, S., Bluth, G.J.S., Rose, W.I., Watson, I.M., and Prata, A.J. 2004. Re-evaluation of SO_2 release of the 15 June 1991 Pinatubo eruption using ultraviolet and infrared satellite sensors. *Geochemistry, Geophysics and Geosystems*, 5, Q04001.

Harris, A.J.L. 2013. *Thermal Remote Sensing of Active Volcanoes: A User's Manual*. Cambridge, U.K: Cambridge University Press.

Harris, A.J.L. and Baloga, S.M. 2009. Lava discharge rates from satellite measured heat flux. *Geophysical Research Letters*, 36, L19302, doi:10.1029/2009GL039717.

Harris, A.J.L., Blake, S., Rothery, D.A., and Stevens, N.F. 1997a. A chronology of the 1991 to 1993 Etna eruption using advanced very high resolution radiometer data: Implications for real–time thermal volcano monitoring. *Journal of Volcanology and Geothermal Research*, 102, 7985–8003.

Harris, A.J.L., Keszthelyi, L., Flynn, L.P., Mouginis-Mark, P.J., Thornber, C., Kauahikikaua, J., Sherrod, D., Trusdell, F., Sawyer, M.W., and Flament, P. 1997b. Chronology of the episode 54 eruption at Kilauea Volcano, Hawai'i, from GOES-9 satellite data. *Geophysical Research Letters*, 24, 3181–3184.

Harris, A.J.L. and Maciejewski, A.J.H. 2000. Thermal surveys of the Vulcano Fossa fumarole field 1994–1999: Evidence for fumarole migration and sealing. *Journal of Volcanology and Geothermal Research*, 102, 119–147.

Harris, A.J.L. and Stevenson, D.S. 1997. Thermal observations of degassing open conduits and fumaroles at Stromboli and Vulcano using remotely sensed data. *Journal of Volcanology and Geothermal Research*, 76, 175–198.

Hellman, M.J. and Ramsey, M.S. 2004. Analysis of hot springs and associated deposits in Yellowstone National Park using ASTER and AVIRIS remote sensing. *Journal of Volcanology and Geothermal Research*, 135, 195–219.

Henney, L.A., Rodriguez, L.A., and Watson, I.M. 2012. A comparison of SO_2 retrieval techniques using mini-UV spectrometers and ASTER imagery at Lascar volcano, Chile. *Bulletin of Volcanology*, 74, 589–594.

Higgins, J. and Harris, A.J.L. 1997. VAST: A program to locate and analyse volcanic thermal anomalies automatically from remotely sensed data. *Computers and Geoscience*, 23, 627–645.

Holasek, R., Self, S., and Woods, A.W. 1996. Satellite observations and interpretation of the 1991 Mount Pinatubo eruption plumes. *Journal of Geophysical Research*, 101, 27635–27655.

Hooper, A., Zebker, H., Segall, P., and Kampes, B. (2004). A new method for measuring deformation on volcanoes and other natural terrains using InSAR persistent scatterers. *Geophysical Research Letters*, 31, L23611, doi:10.1029/2004GL021737.

Jay, J., Costa, F., Pritchard, M., Lara, L., Singer, B., and Herrin, J. 2014. Locating magma reservoirs using InSAR and petrology before and during the 2011–2012 Cordon Caulle silicic eruption. *Earth and Planetary Science Letters*, 395, 254–266.

Ji, L., Lu, Z., Dzurisin, D., and Senyukov, S. 2013. Pre-eruption deformation caused by dike intrusion beneath Kizimen volcano, Kamchatka, Russia, observed by InSAR. *Journal of Volcanology and Geothermal Research*, 256, 87–95.

Kaneko, T. and Wooster, M.J. 1999. Landsat infrared analysis of fumarole activity at Unzen volcano: Time-series comparison with gas and magma fluxes. *Journal of Volcanology and Geothermal Research*, 89, 57–64.

Kienle, J. and Shaw, G.E. 1979. Plume dynamics, thermal energy and long distance transport of vulcanian eruption clouds from the Augustine volcano. *Journal of Volcanology and Geothermal Research*, 6, 139–164.

Krotkov, N., Torres, O., Seftor, C., Krueger, A.J., Kostinski, A., Rose, W.I., Bluth, G.J.S., Schneider, D., and Schaefer, S.J. 1999. Comparison of TOMS and AVHRR volcanic ash retrievals from the August 1992 eruption of Mt. Spurr. *Geophysical Research Letters*, 26, 455–458.

Krotkov, N.A. NASA Global Sulfur Dioxide Monitoring Home Page, http://so2.gsfc.nasa.gov.

Krotkov, N.A., Carn, S.A., Krueger, A.J., Bhartia, P.K., and Yang, K. 2006. Band residual difference algorithm for retrieval of SO_2 from the Aura Ozone Monitoring Instrument (OMI). *IEEE Transactions on Geoscience and Remote Sensing*, 44, 1259–1266.

Kruger, A.J. 1983. Sighting of El Chichon sulfur dioxide clouds with the Nimbus 7 Total Ozone Mapping Spectrometer. *Science*, 220, 1377–1379.

Krueger, A.J., Walter, L.S., Bhartia, P.K., Schnetzler, C.C., Krotkov, N.A., Spord, I., and Bluth, G.J.S. 1995. Volcanic sulfur dioxide measurements from the total ozone mapping spectrometer instruments. *Journal of Geophysical Research*, 100, 14057–14076.

Li, Z., Muller, J.P., Cross, P., and Fielding, E.J. 2005. Interferometric synthetic aperture radar (InSAR) atmospheric correction: GPS, Moderate Resolution Imaging Spectroradiometer (MODIS), and InSAR integration. *Journal of Geophysical Research*, 110, doi:10.1029/2004JB003446.

Loyola, D., van Geffen, J., Valks, P., Erbertseder, T., Van Roozendael, M., Thomas, W., Zimmer, W., and Wißkirchen, K. 2008. Satellite-based detection of volcanic sulphur dioxide from recent eruptions in Central and South America. *Advances in Geosciences*, 14, 35–40.

Lu, Z. and Dzurisin, D. 2014. *InSAR Imaging of Aleutian Volcanoes: Monitoring a Volcanic Arc from Space*. Heidelberg, Germany: Springer Praxis Books.

Lu, Z., Wicks, C., Dzurisin, D., Power, P., Moran, C., and Thatcher, W. 2002. Magmatic inflation at a dormant stratovolcano: 1996–1998 activity at Mount Peulik volcano, Alaska, revealed by satellite radar interferometry. *Journal of Geophysical Research*, 107, doi:10.1029/2001JB000471.

Massonnet, D., Briole, P., and Arnaud, A. 1995. Deflation of Mount Etna monitored by spaceborne radar interferometry. *Nature*, 375, 567–570.

Massonnet, D. and Feigl, K.L. 1998. Radar interferometry and its application to changes in the earth's surface. *Reviews in Geophysics*, 36, 441–500.

Massonnet, D., Rossi, M., Carmona, C., Adragna, F., Peltzer, G., Feigl, K., and Rabaute, T. 1993. The displacement field of the Landers earthquake mapped by radar interferometry. *Nature*, 364, 138–142.

Mayberry, G.C., Rose, W.I., and Bluth, G.J. 2001. Dynamics of the volcanic and meteorological clouds produced by the December 26 (Boxing Day) 1997 eruption of Soufriere Hills volcano, Montserrat, WI. In: T.H. Druitt et al. (eds) *The Eruption of Soufriere Hills Volcano, Montserrat, from 1995 to 1999*. London, U.K.: Geological Society of London, pp. 539–556.

Mogi, K. 1958. Relations between the eruptions of various volcanoes and the deformations of the ground surfaces around them. *Bulletin of the Earthquake Research Institute of Tokyo*, 36, 99–134.

Moore, J.G., Nakamura, K., and Alcarez, A. 1966. The eruption of Taal volcano, Philippines, September 28–30, 1965. *Science*, 151, 955–960.

Moran, S.C., Kwoun, O., Masterlark, T., and Lu, Z. 1996. On the absence of InSAR-detected volcano deformation spanning the 1995–1996 and 1999 eruptions of Shishaldin volcano, Alaska. *Journal of Volcanology and Geothermal Research*, 150, 119–131.

Murray, J.B., Pullen, A.D., and Saunders, S. 1995. Ground deformation surveying of active volcanoes. In: B. McGuire et al. (eds) *Monitoring Active Volcanoes*. London, U.K.: UCL Press, pp. 113–150.

Novak, M.A.M., Watson, I.M., Delgado-Granados, H., Rose, W.I., Cardenas-Gonzalez, L., and Realmuto, V.J. 2008. Volcanic emissions from Popocatepetl volcano, Mexico, quantified using moderate resolution imaging spectroradiometer (MODIS) infrared data: A case study of the December 2000–January 2001 emissions. *Journal of Volcanology and Geothermal Research*, 170, 76–85.

Okada, Y. 1985. Surface deformation due to shear and tensile faults in a half space. *Bulletin of the Seismological Society of America*, 75, 1135–1154.

Oppenheimer, C. 1991. Lava flow cooling estimated from Landsat Thematic Mapper infrared data: The Lonquimay eruption (Chile, 1989). *Journal of Geophysical Research*, 96, 21865–21878.

Oppenheimer, C. 1996. Crater lake heat losses estimated by remote sensing. *Geophysical Research Letters*, 23, 1793–1796.

Oppenheimer, C. 1997. Remote sensing of the color and temperature of volcanic crater lakes. *International Journal of Remote Sensing*, 18, 5–37.

Oppenheimer, C., Francis, P.W., Rothery, D.A., Carlton, R.W.T., and Glaze, L.S. 1993. Infrared image analysis of volcanic thermal features: Lascar Volcano, Chile, 1984–1992. *Journal of Geophysical Research*, 98, 4269–4286.

Oshawa, S. Saito, T., Yoshikawa, S., Mawatari, H., Yamada, M., Amita, K., Takamatsu, N., Sudo, Y., and Kagiyama, T. 2010. Color change of lake water at the active crater lake of Aso volcano: Is it in response to change in water quality induced by volcanic activity? *Limnology*, 11, 207–215.

Parfitt, E.A. and Wilson, L. 2008. *Fundamentals of Physical Volcanology*. Oxford, U.K.: Blackwell Publishing.

Patrick, M., Dean, K., and Dehn, J. (2004). Active mud volcanism observed with Landsat 7 ETM+. *Journal of Volcanology and Geothermal Research*, 131, 307–320.

Pergola, N., Marchese, F., and Tramutoli, V. 2004. Automated detection of thermal features of active volcanoes by means of infrared AVHRR records. *Remote Sensing of Environment*, 93, 311–327.

Pieri, D.C. and Baloga, S.M. 1986. Eruption rate, area, and length relationships for some Hawaiian lava flows. *Journal of Volcanology and Geothermal Research*, 30, 29–45.

Pour, A.B., Hashim, M., and Genderan, J. 2013. Detection of hydrothermal alteration zones in a tropical region using satellite remote sensing data: Bau goldfield, Sarawak, Malaysia. *Ore Geology Reviews*, 54, 181–196.

Power, J.A., Nye, C.J., Coombs, M.L., Wessels, R.L., Cervelli, P.F., Dehn, J., Wallace, K.L., Freymueller, J.T., and Doukas, M.P. 2006. The reawakening of Alaska's Augustine volcano, EOS, *Transactions of the American Geophysical Union*, 87, 373.

Prata, A.J. 1989. Infrared radiative transfer calculations for volcanic ash clouds. *Geophysical Research Letters*, 16, 1293–1296.

Prata, A.J. and Bernardo, C. 2007. Retrieval of volcanic SO_2 column abundance from Atmospheric Infrared Sounder data. *Journal of Geophysical Research*, 112, D20204, doi:10.1029/2006JD007955.

Prata, A.J., Bluth, G., Rose, W.I., Schneider, D., and Tupper, A. 2001. Comments on failures in detecting volcanic ash from a satellite-based technique. *Remote Sensing of Environment*, 78, 341–346.

Prata, A.J., Rose, W.I., Self, S., and O'Brien, D.M. 2003. Global, long-term sulphur dioxide measurements from TOVS data: A new tool for studying explosive volcanism and climate. In: A. Robock et al. (eds) *Volcanism and the Earth's Atmosphere*. Washington, DC, American Geophysical Union, pp. 75–92.

Prata, A.J. and Turner, P.J. 1997. Cloud top height determination from the ATSR. *Remote Sensing of Environment*, 59, 1–13.

Prins, E.M. and Menzel, W.P. 1992. Geostationary satellite detection of biomass burning in South America. *International Journal of Remote Sensing*, 13, 2783–2799.

Pritchard, M. and Simons, M. 2002. A satellite geodetic survey of large-scale deformation of volcanic centres in the central Andes. *Nature*, 418, 167–171.

Pritchard, M. and Simons, M. 2004. An InSAR-based survey of volcanic deformation in the southern Andes. *Geophysical Research Letters*, 31, L15610, doi:10.1029/2004GL020545.

Pugnaghi, S., Gangale, G., Corradini, S., and Buongiorno, M.F. 2006. Mt. Etna sulfur dioxide flux monitoring using ASTER–TIR data and atmospheric observations. *Journal of Volcanology and Geothermal Research*, 152, 74–90.

Rampino, M. 2010. Mass extinctions of life and catastrophic flood basalt volcanism. *Proceedings of the National Academy of Sciences of the United States of America*, 107, 6555–6556.

Realmuto, V.J. 2000. The potential use of Earth Observing System data to monitor the passive emission of sulfur dioxide from volcanoes. In: P.J. Mouginis-Mark et al. (eds) *Remote Sensing of Active Volcanism*. AGU Geophysical Monograph Series. Washington, DC: American Geophysical Union, pp. 101–115.

Realmuto, V.J., Abrams, M.J., Buongiorno, M.F., and Pieri, D.C. 1994. The use of multispectral thermal infrared image data to estimate the sulfur dioxide flux from volcanoes: A case study from Mount Etna, Sicily, July 29, 1986. *Journal of Geophysical Research*, 99, 481–488.

Richards, M.A. 2007. A beginner's guide to interferometric SAR concepts and signal processing. *IEEE Aerospace and Electronic Systems Magazine*, 22, 5–29.

Rix, M., Valks, P., Hao, N., Loyola, D., Schlager, H., Huntrieser, H., Flemming, J., Koehler, U., Schumann, U., and Inness, A. 2012. Volcanic SO_2, BrO and plume height estimations using GOME-2 satellite measurements during the eruption of Eyjafjallajökull in May 2010. *Journal of Geophysical Research*, 117, doi:10.1029/2011JD016718.

Roberge, J., Delgado-Grandos, H., and Wallace, P.J. 2009. Mafic magma recharge supplies high CO_2 and SO_2 gas fluxes at Popocatepetl volcano, Mexico. *Geology*, 37, 107–110.

Rose, W.I., Delene, D.J., Schneider, D.J., Bluth, G.J.S., Krueger, A.J., Sprod, I., McKee, C., Davies, H.L., and Ernst, G.J. 1995. Ice in the 1994 Rabaul eruption: Implications for volcanic hazard and atmospheric effects. *Nature*, 375, 477–479.

Rothery, D.A., Francis, P.W., and Wood, C.A. 1988. Volcano monitoring using short wavelength infrared data from satellites. *Journal of Geophysical Research*, 93, 7993–8008.

Rowland, S.K. 1996. Slopes, lava flow volumes and vent distributions on Volcan Fernandina, Galapagos Islands. *Journal of Geophysical Research*, 101, 27657–27672.

Sabins, F.F. 1999. Remote sensing for mineral exploration. *Ore Geology Reviews*, 14, 157–183.

Schwandner, F., Carn, S.A., Kuze, A., Kataoka, F., Shiomi, K., Goto, N., Popp, C. et al. 2014. Can satellite-based monitoring techniques be used to quantify volcanic CO_2 emissions? In: *Abstract Presented at 2014 General Assembly*, European Geophysical Union, Vienna, Austria, April 27 to May 2.

Searcy, C., Dean, K., and Stringer, W. 1998. PUFF: A high resolution volcanic ash tracking model. *Journal of Volcanology and Geothermal Research*, 80, 1–16.

Seftor, C.J., Hsu, N.C., Herman, J.R., Bhartia, P.K., Torres, O., Rose, W.I., Schneider, D.J., and Krotkov, N. 1997. Detection of volcanic ash clouds from Nimbus-7/total ozone mapping spectrometer. *Journal of Geophysical Research*, 102, 16749–16759.

Seinfeld, J.H. and Pandis, S.N. 1997. *Atmospheric Chemistry and Physics*. New York, John Wiley and Sons.

Siebert, L., Simkin, T., and Kimberly, P. 2010. *Volcanoes of the World*, 3rd ed. Los Angeles, CA, University of California Press.

Sigmundsson, F., Durand, P., and Massonnet, D. 1999. Opening of an eruptive fissure and seaward displacement at Piton de la Fournaise volcano measured by RADARSAT satellite radar interferometry. *Geophysical Research Letters*, 26, 533–536.

Sigurdsson, H. 1999. *Melting the Earth: The History of Ideas on Volcanic Eruptions*. Oxford, U.K.: Oxford University Press.

Sigurdsson, H., Houghton, B., McNutt, S.R., Rymer, H., and Stix, J. (eds). 2000. *Encyclopedia of Volcanoes*. London, U.K.: Academic Press.

Sigurdsson, H. and Lopes-Gautier, R. 2000. Volcanoes in literature and film. In: H. Sigurdsson et al. (eds) *Encyclopedia of Volcanoes*. London, U.K.: Academic Press, pp. 1339–1360.

Steffke, A.M. and Harris, A.J.L. 2011. A review of algorithms for detecting volcanic hot spots in satellite infrared data. *Bulletin of Volcanology*, 73, 1109–1137.

Steffke, A.M., Harris, A.J.L., Burton, M., Caltabiano, T., and Salerno, G. 2011. Coupled use of COSPEC and satellite measurements to define the volumetric balance during effusive eruptions at Mt. Etna, Italy. *Journal of Volcanology and Geothermal Research*, 205, 47–53.

Stevens, N.F., Wadge, G., Williams, C.A., Morley, J.G., Muller, J.-P., Murray, J.B., and Upton, M. 2001. Surface movements of emplaced lava flows measured by synthetic aperture radar interferometry, *Journal of Geophysical Research*, 106, doi:10.1029/2000JB900425.

Symonds, R.B., Rose, W.I., Bluth, G.J.S., and Gerlach, T.M. 1994. Volcanic-gas studies: Methods, results and applications. *Reviews in Mineralogy*, 30, 1–66.

Theys, N., Campion, R., Clarisse, L., Brenot, H., van Gent, J., Dils, B., Corradini, S. et al. 2013. Volcanic SO_2 fluxes derived from satellite data: A survey using OMI, GOME-2, IASI and MODIS. *Atmospheric Chemistry and Physics*, 13, 5945–5968. doi:10.5194/acp-13-5945-2013.

Trunk, L. and Bernard, A. 2008. Investigating crater lake warming using ASTER thermal imagery: Case studies at Ruapehu, Poás, Kawah Ijen, and Copahué Volcanoes. *Journal of Volcanology and Geothermal Research*, 178, 259–270.

Urai, M. 2004. Sulfur dioxide flux estimation from volcanoes using Advanced Spaceborne Thermal Emission and Reflection Radiometer—A case study of the Miyakejima volcano, Japan. *Journal of Volcanology and Geothermal Research*, 134, 1–13.

Van Manen, S., Blake, S., Dehn, J., and Valcic, L. 2013. Forecasting large explosions at Bezymianny Volcano using thermal satellite data. *Geological Society of London Special Publication*, 380, 187–201

Vaughan, R.G., Keszthelyi, L.P., Lowenstern, J.B., Jaworowski, C., and Heasler, H. 2012. Use of ASTER and MODIS thermal infrared data to quantify heat flow and hydrothermal change at Yellowstone National Park. *Journal of Volcanology and Geothermal Research*, 233–234, 72–89.

Walker, G.P.L. 1973. Lengths of lava flows. *Philosophical Transactions of the Royal Society of London*, 273, 107–118.

Wang, H., Wright, T., and Biggs, J. 2009. Interseismic slip rate of the northwestern Xianshuihe fault from InSAR data. *Geophysical Research Letters*, 36, L03302, doi:10.1029/2008GL036560.

Watson, I.M., Realmuto, V.J., Rose, W.I., Prata, A.J., Bluth, G.J.S., Gu, Y., Bader, C.E., and Yu, T. 2004. Thermal infrared remote sensing of volcanic emissions using the moderate resolution imaging spectroradiometer. *Journal of Volcanology and Geothermal Research*, 135, 75–89.

Webley, P., and Mastin, L. 2009. Improved prediction and tracking of volcanic ash clouds: *Journal of Volcanology and Geothermal Research*, 186, doi:10.1016/j.jvolgeores.2008.10.022.

Webley, P.W., Bingley, R.M., Dodson, A.H., Wadge, G., Waugh, S.J., and James, I.N. 2002. Atmospheric water vapor correction to InSAR surface motion measurements on mountains: Results from a dense GPS network on Mount Etna. *Physics and Chemistry of the Earth*, 27, 363–370.

Wen, S. and Rose, W.I. 1994. Retrieval of sizes and total mass of particles in volcanic ash clouds using AVHRR bands 4 and 5. *Journal of Geophysical Research*, 99, 5421–5431.

Williams-Jones, G. and Rymer, H. 2000. Hazards of volcanic gases. In: H. Sigurdsson et al. (eds) *Encyclopedia of Volcanoes*, London, U.K.: Academic Press, pp. 997–1004.

Wilson, L., Sparks, R.S.J., Huang, T.C., and Watkins, N.D. 1978. The control of volcanic column heights by eruption energetics and dynamics. *Journal of Geophysical Research*, 83, 1829–1836.

Wohletz, K. and Heiken, G. 1992. *Volcanology and Geothermal Energy*. Los Angeles, CA: University of California Press.

Wooster, M.J. and Rothery, D.A. 1997. Thermal monitoring of Lascar Volcano, Chile using infrared data from the Along Track Scanning Radiometer: A 1992–1995 time-series. *Bulletin of Volcanology*, 58, 566–579.

Wright, R., Blake, S., Rothery, D.A., and Harris, A.J.L. 2001. A simple explanation for the space–based calculation of lava eruption rates. *Earth and Planetary Science Letters*, 192, 223–233.

Wright, R., Flynn, L.P., Garbeil, H., Harris, A.J.L., and Pilger, E. 2002. Automated volcanic eruption detection using MODIS. *Remote Sensing of Environment*, 82, 135–155.

Wright, R., Garbeil, H., and Davies, A.G. 2010. Cooling rate of some active lavas determined using an orbital imaging spectrometer. *Journal of Geophysical Research*, 115, doi:10.1029/2009JB006536.

Wright, R., Garbeil, H., and Harris, A.J.L. 2008. Using infrared satellite data to drive a thermo–rheological/stochastic lava flow emplacement model: A method for near–real–time volcanic hazard assessment. *Geophysical Research Letters*, 35, L19307, doi:10.1029/2008GL035228.

Wright, R., Glaze, L., and Baloga, S.M. 2011. Constraints on determining the composition and eruption style of terrestrial lavas from space. *Geology*, 39, 1127–1130.

Yang, K., Dickerson, R.R., Carn, S.A., Ge, C., and Wang, J. 2013. First observations of SO_2 from the satellite Suomi NPP OMPS: Widespread air pollution events over China. *Geophysical Research Letters*, 40, 4957–4962, doi:10.1002/grl.50952.

Yang, K., Krotkov, N.A., Krueger, A.J., Carn, S.A., Bhartua, P.K., and Levelt, P.F. 2007. Retrieval of large volcanic SO_2 columns from the Aura Ozone Monitoring Instrument: Comparison and limitations. *Journal of Geophysical Research*, 112, D24S43, doi:10.1029/2007JD008825.

Zebker, H.A., Amelung, F., and Jonsson, S. 2001. Remote sensing of volcano surface and internal processes using radar interferometry. In: P.J. Mouginis-Mark et al. (eds) *Remote Sensing of Active Volcanism*. AGU Geophysical Monograph Series. Washington, DC: American Geophysical Union, pp. 179–205.

Zebker, H.A. and Lu, Y. 1998. Phase unwrapping algorithms for aradr interferometry: Residue cut, least squares and synthesis algorithms. *Journal of Optical Applications*, 15, 586–598.

Zebker, H.A., Rosen, P.A., Goldstein, R.M., Gabriel, A., and Werner, C.L. 1994. On the derivation of co-seismic displacement fields using differential radar interferometry: The Landers earthquake. *Journal of Geophysical Research*, 99, 19617–19634.

Zebker, H.A., Rosen, P.A., and Hensley, S. 1997. Atmospheric effects in interferometric synthetic aperture radar surface deformation and topographic maps. *Journal of Geophysical Research*, 102, 7547–7563.

Zebker, H.A. and Villasenor, J.A. 1992. Decorrelation in interferometric radar echos. *IEEE Transactions on Geoscience and Remote Sensing*, 30, 950–959.

Fires

21

Satellite-Derived Nitrogen Dioxide Variations from Biomass Burning in a Subtropical Evergreen Forest, Northeast India

Krishna Prasad Vadrevu
University of Maryland

Kristofer Lasko
University of Maryland

Acronyms and Definitions

GHGs   Greenhouse gases
GOME   Global Ozone Monitoring Experiment
SCIAMACHY   Scanning Imaging Absorption Spectrometer for Atmospheric Cartography
OMI   Ozone Monitoring Instrument
FRP   Fire radiative power
MODIS   Moderate Resolution Imaging Spectroradiometer
NASA   National Aeronautics and Space Administration
EOS   Earth Observing System
DOMINO   Dutch OMI-NO$_2$
TEMIS   Tropospheric Emission Monitoring Internet Service

21.1 Introduction

Biomass burning is an important source of greenhouse gas (GHG) emissions and aerosols including carbon dioxide (CO_2), methane (CH_4), carbon monoxide (CO), nitrogen oxide (NOx), ammonia (NH_4), and volatile organic compounds (Andreae and Merlet, 2001). Global annual areas burned for the years 1997 through 2011 vary from 301 to 377 Mha, with an average of 348 Mha (Giglio et al., 2013). Of the different regions, tropical Asia is considered a major source of biomass burning (Streets et al., 2003; Vadrevu and Justice, 2011). Important sources of biomass burning emissions in tropical Asia include deforestation (van Der Werf et al., 2008), slash-and-burn agriculture (Prasad et al., 2000; Langner et al., 2007), agricultural residue burning (Badarinath et al., 2009; Vadrevu et al., 2011; 2012; Cheewaphongphan and Garivait, 2013), management fires (Murdiyarso and Level, 2007), and peat land burning (Heil et al., 2007). Present estimates suggest that globally, wildfires contribute about 20% of the fossil fuel carbon emissions to the atmosphere and global fire emissions averaged over 1997–2009 amount to 2.0 Pg C year^{-1} (van der Werf et al., 2010). It is estimated that carbon monoxide (CO) and nitrogen dioxide (NOx) emissions from fires comprise approximately 30% and 15% of global total direct emissions, respectively (Jaeglé et al., 2005; Müller and Stavrakou, 2005; Arellano et al., 2006). Enhanced CO and NOx concentrations can impact tropospheric ozone formation and affect the oxidizing capacity of the atmosphere by regulating the hydroxide lifetime (Logan et al., 1981). Aerosols released from the biomass burning can be elevated by midlatitude wave cyclones and sometimes can travel long distances to possibly influence climate and weather patterns. Specific to climate impacts, Wang et al. (2014) have shown that Asian pollution invigorates winter cyclones over the northwest Pacific, increasing precipitation by 7% and net cloud radiative forcing by 1.0 W m^{-2} at the top of the atmosphere and by 1.7 W m^{-2} at the Earth's surface. No single system can provide all the necessary data, and to address air quality and climate impacts of GHGs, several studies infer the need to integrate both top-down and bottom-up approaches including modeling (Martin et al., 2002).

The science of Earth observation, more specifically, remote detection of tropospheric gases using satellite instruments has significantly improved over the past 20 years (Burrows et al., 2011). Several GHG concentrations such as O_3, CO, CO_2, CH_4, HCHO, NO_2, SO_2, and BrO can be measured from nadir-looking sensors that record these gases in the lower troposphere (<6 km) (Palmer et al., 2001; Boersma et al., 2004). Important sources of these GHGs include both fossil fuel combustion (automobile combustion, industrial production and heating) and natural sources (terrestrial vegetation, soils, lightning, wetlands, biomass burning). Of the different natural sources, biomass burning that is prevalent in tropical regions contributes significantly to GHG emissions including NO_2. The exposure to NO_2 pollutants can cause or worsen respiratory disease, such as emphysema and bronchitis (Schwartz, 2004).

NO_2 is an important trace gas in both the troposphere and stratosphere that exhibits high atmospheric variability. In the troposphere, NO_2 is a precursor for ozone formation. The photolysis of NO_2 in the presence of strong solar radiation releases atomic oxygen, which then combines with molecular oxygen to form ozone (Logan, 1983). In an unpolluted atmosphere, the natural sources of SO_2 and NO_2 provide a mechanism by which the pH of aerosols and rain are expected to be slightly acidic. However, large amounts of NO, NO_2 (NOx), and SO_2 produced during fossil fuel and biofuel combustion can result in acid rain (Burrows et al., 2011). NO_2 is not only involved in catalytic ozone depletion in the stratosphere but also reacts with halogen oxides to form reservoir substances and thereby reduces the ozone depletion potential of Cl and Br. NO_2 is removed from the atmosphere through chemical conversion to other nitrogen-contained species, nitrate aerosols, and uptake by vegetation and soils (Shaw, 1976). It also exhibits a distinct diurnal cycle (Thomas et al., 1998).

While the main sources and source regions of NO_2 are known, large uncertainties remain as to the individual source strengths and their latitudinal and seasonal variation (Richter and Burrows, 2002). NO_2 can be measured either by in situ chemical methods, on airborne or balloon platforms, or by remote sensing in the ultraviolet (UV)/visible (VIS) and infrared spectral regions. From satellite instruments, NO_2 can be measured as a column integral from solar backscatter instruments from space, since it absorbs light in the visible portion of the electromagnetic spectrum. Tropospheric NO_2 columns

are retrieved from the total columns by subtracting the stratospheric contribution, assuming zonal invariance of the stratospheric contribution. Because the satellite signal is closely related to the total amount of NO_2, there is a more direct link with area-averaged concentrations than with surface in situ observations, which depend strongly on local sources and local removal processes (Burrows et al., 2011). Tropospheric NO_2 columns retrieved from satellite measurements, for example, by the Global Ozone Monitoring Experiment (GOME), GOME-II, Scanning Imaging Absorption Spectrometer for Atmospheric Cartography (SCIAMACHY), and Ozone Monitoring Instrument (OMI), have contributed to mapping spatiotemporal variations in NOx sources (e.g., Burrows et al., 1999; Richter and Burrows, 2002; Martin et al., 2006; van der A et al., 2006; Boersma et al., 2007; Stavrakou et al., 2008; Kurokawa et al., 2009; Zhao and Wang, 2009; Lin et al., 2010; Russell et al., 2011; Miyazaki et al., 2012). For example, Richter and Burrows (2002) using GOME measurements observed enhanced tropospheric NO_2 from biomass burning above Africa during the fall of 1997. Similarly, Thomas et al. (1998) recorded a twofold increase in the vertical NO_2 content over large parts of the smoke cloud formed from rainforest biomass burning episodes in Java and Borneo during September 1997. Using the combined GOME and SCIAMACHY retrievals, van der A et al. (2006, 2008) mapped a significant increase in the NO_2 concentrations over China from 1996 to 2006 and attributed these emissions to the rapidly growing economy and the associated increase of fossil fuel consumption, industries, and traffic. Other examples of using satellite-derived NO_2 products for characterizing anthropogenic emissions can be found in Beirle et al. (2003), Bertram et al. (2005), Jaeglé et al. (2005), Kim et al. (2006), Martin et al. (2006), Boersma et al. (2007), Hudman et al. (2007), Ghude et al. (2008), Mijling et al. (2009), Bucsela (2010), Lin et al. (2010), Mebust et al. (2011), Valin et al. (2011), Zhou et al. (2012), and David and Nair (2013).

In contrast to these studies conducted all over the globe, evaluation of OMI and SCIAMACHY-NO_2 signals in relation to evergreen forest fires in India have not yet been attempted. Assessment of satellite observations of tropospheric NO_2 columns is needed over a range of environments to improve the validation efforts (Lamsal et al., 2010). In this study, the following questions relating to fire–NO_2 concentrations were addressed: (1) How does biomass burning due to slash-and-burn agriculture impact NO_2 concentrations for subtropical evergreen forests? (2) How are the Moderate Resolution Imaging Spectroradiometer (MODIS)-retrieved fire counts and fire radiative power (FRP) products related to NO_2 concentrations in evergreen forest burning? (3) Which satellite product, for example, SCIAMACHY-NO_2 or OMI-NO_2, best correlates with the MODIS fire products? (4) What is the correlation strength between OMI and SCIAMACHY-NO_2 products in a biomass burning landscape? (5) How do these products correlate with MODIS-retrieved aerosol optical depth (AOD)? (6) How much variance in NO_2 concentrations can be accounted by the MODIS Terra and Aqua fire counts and the FRP together?

We answered the aforementioned questions using MODIS active fires, SCIAMACHY-NO$_2$ (2003–2011), OMI-NO$_2$ (2005–2011), and MODIS Aqua AOD (2003–2011) datasets. In addition to the aforementioned questions, we hypothesized that the correlation strength between the FRP and the NO$_2$ concentrations will be much stronger than the correlation strength between the number of fire counts and NO$_2$ concentrations because the FRP has been previously associated with the strength of the fires (Wooster et al., 2005). From these questions and hypothesis testing, the results from this study are expected to provide robust information on fire–NO$_2$ relationships in the subtropical evergreen forests.

21.2 Study Area

To test the fire–NO$_2$ relationships, we selected the subtropical evergreen forests of Northeast India where biomass burning is prevalent due to slash-and-burn agriculture. Northeast India refers to the easternmost region of India consisting of the contiguous seven states of Arunachal Pradesh, Assam, Manipur, Meghalaya, Mizoram, Nagaland, and Tripura, occupying

~255,083 km^2 (see boxed area in Figure 21.1a). Seventy percent of the region is occupied by hills. Nearly 400,000 families belonging to 100 different indigenous tribes practice slash-and-burn agriculture, locally called *jhum*. Nearly 3863 sq.km is affected by slash-and-burn cultivation annually (Majumder et al., 2011). *Jhum* cultivation involves the clearing of forest vegetation and burning (March–May) just before the monsoon (June), followed by mixed cropping on steep slopes of 30°–40°. In the *jhum* plots, varieties of crops are grown including cereals (*Oryza sativa*, *Zea mays*), tuberous crops (*Manihot esculenta*, *Dioscorea* spp.), vegetable crops (*Cucurbita moschata*, *Solanum melongena*, etc.), and spices (*Zingiber officinale*, *Capsicum* sp., etc.). Due to an increasing population, overexploitation of forest resources, and loss of soil fertility, the *jhum* cycle has reduced to 3–5 years from the more traditional 20–30 years on land that had already been occupied for slash-and-burn agriculture (Majumdar et al., 2011). The forests are mainly subtropical evergreen type, interspersed by bamboo forest. Subtropical temperate climax forests of mixed broad-leaved forest with early successional species of pure pine or mixed pine were also reported from this region.

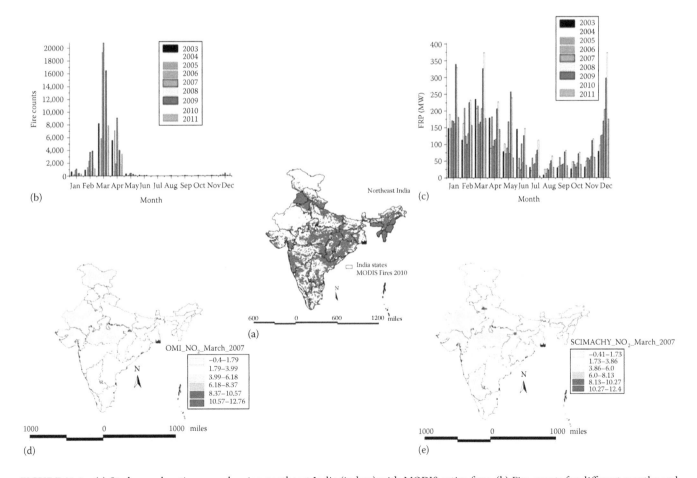

FIGURE 21.1 (a) Study area location map showing northeast India (in box) with MODIS active fires. (b) Fire counts for different months and years in the northeast India. (c) FRP (MW) for different months and years in northeast India. (d) OMI-NO$_2$ (x-axis values in 10^{15} molecules cm^{-2}) for the peak month of March (2007) in northeast India. (e) SCIAMACHY-NO$_2$ (x-axis values in 10^{15} molecules cm^{-2}) for the peak month of March (2007) in northeast India.

21.3 Datasets

21.3.1 Active Fires and Fire Radiative Power

We aggregated the monthly active fires and FRP data from MODIS for the northeastern states of India. The two MODIS sun-synchronous, polar-orbiting satellites pass over the Equator at approximately 10:30 am (Terra) and 1:30 pm (Aqua), and they have a revisit time of 1–2 days. The data collected by the sensors are processed by the MODIS Advanced Processing System, using the enhanced contextual fire detection algorithm (Giglio et al., 2003) into the Collection 5 Active Fire product. For this study, we analyzed the data from 2003 to 2011. The fire data are at 1 km nominal spatial resolution at nadir; however, under ideal conditions, flaming fires as small as 50 m² can be detected. FRP is the rate of fire energy released per unit time, measured in megawatts (Kaufman et al., 1998). The MODIS algorithm for FRP is calculated as the relationship between the brightness temperature of fire and background pixels in the middle infrared (band center near 4 μm). It is given as (Kaufman et al., 1998)

$$\text{FRP} = 4.34 \times 10^{-19} \left(T_{MIR}^8 - T_{bgMIR}^8 \right)$$

where

 FRP is the rate of radiative energy emitted per pixel

 4.34×10^{-19} (MW km^{-2} K^8) is the constant derived from the simulations

 T_{MIR} (Kelvin) is the radiative brightness temperature of the fire component

 T_{bgMIR} (Kelvin) is the neighboring nonfire background component

 MIR refers to the middle infrared wavelength here, 3.96 μm

In this study, we utilized the Collection 5 Terra and Aqua monthly climate modeling grid datasets (MOD14CMH/MYD14CMH) that represent cloud and overpass corrected fire pixels data along with the mean FRP data.

21.3.2 OMI-NO$_2$

OMI is one of four instruments on board National Aeronautics and Space Administration's (NASA) Earth Observing System-Aura satellite, launched on July 15, 2004. OMI is a nadir-viewing imaging spectrometer (Levelt et al., 2006; Boersma et al., 2007) (Table 21.1). Aura traces a sun-synchronous, polar orbit with a period of 100 min and has a local equator crossing time of about 13:45 pm. OMI provides measurements of both direct and atmospheric backscattered sunlight in the UV–VIS range from 2.7 to 5 μm useful to retrieve tropospheric NO$_2$ columns (Miyazaki et al., 2012). OMI pixels are 13 km × 24 km at nadir, increasing in size to 24 km × 135 km for the largest viewing angles. The instrument achieves near-daily coverage and the retrievals are sufficient to extract global NOx concentrations on a daily basis. These capabilities of OMI make it unique as compared to GOME and SCIAMACHY retrievals, which have relatively lower spatial and temporal resolutions and less frequent global coverage.

In this study, we specifically used the Dutch OMI-NO$_2$ (DOMINO) product (version 2.0) (Boersma et al., 2011). The DOMINO is a post-processing dataset based on the most complete set of OMI orbits, improved level-1b (ir)radiance data (Collection 3), analyzed meteorological fields, and actual spacecraft data. The better data coverage, the improved calibration of level-1b data, and the use of analyzed rather than forecast data make the DOMINO product superior to the near-real-time NO$_2$ data (Boersma et al., 2011). The differential optical absorption spectroscopy spectral analysis technique is used to determine NO$_2$ slant column densities, and then the stratospheric portion of the column is subtracted to yield a tropospheric slant column. The multiplicative air mass factor (AMF) is then used to convert the slant column to a vertical column based on the output from a radiative transfer model, which accounts for terrain, profile, cloud, and viewing parameters. The DOMINO data contain geolocated column integrated NO$_2$ concentrations, or NO$_2$ columns (in units of molecules cm^{-2}). DOMINO data constitute a pure level 2 product, that is, it provides geophysical information for every ground pixel observed by the instrument, without the additional binning, averaging, or gridding typically applied for level 3 data. In addition to vertical NO$_2$ columns, the product contains intermediate results, such as the result of the spectral fit, fitting diagnostics, assimilated stratospheric NO$_2$ columns, averaging kernel, cloud information, and error estimates. To reduce the errors resulting from cloud cover, we excluded those pixels with an effective cloud fraction exceeding 20%. Further, for the study area, there was no contamination due to row anomalies and pixels at the swath edges.

21.3.3 SCIAMACHY-NO$_2$

The SCIAMACHY instrument on board Envisat is an 8-channel UV/VIS/near-infrared grating spectrometer covering the wavelength region of 220–2400 nm with a 0.2–1.5 nm spectral resolution 10:00 am local time equator crossing and a global coverage of every 6 days (Table 21.1).

TABLE 21.1 SCIAMACHY and OMI Instrument Characteristics

Instrument	Satellite	Nadir View	Global Coverage	Wavebands	Spectral Resolution (at 440 nm)
SCIAMACHY	Envisat	30 × 60 km²	6 days	UV-SWIR: 240–314, 309–3405, 394–620, 604–805, 785–1050, 1000–1750, 1940–2040, and 2265–2380 nm	0.44 nm
OMI	EOS-Aura	13 × 24 km²	1 day	270–500 nm	0.63 nm

It measures trace gas constituents in nadir, limb, and occultation configuration. The UV/VIS nadir measurements of SCIAMACHY are very similar to those performed by GOME, the main difference being the better spatial resolution (30×30 to 30×240 km^2) as compared to 40×320 km^2 for GOME. Similar to OMI data, we used the data from Tropospheric Emission Monitoring Internet Service. These datasets have been validated against in situ and aircraft measurements and compared with regional air quality models (e.g., Schaub et al., 2006; Blond et al., 2007). As discussed by Boersma et al. (2007) and Lin et al. (2010), systematic errors in OMI (DOMINO v1) and SCIAMACHY retrievals are expected to correlate well with each other, since these retrievals are derived with a very similar algorithm.

21.3.4 MODIS Aerosol Optical Depth Variations

We used the MODIS Collection 5.1 (MYD08_M3.051) AOD at 550 nm (Remer et al., 2005; Levy et al., 2007) level 3 monthly product for characterizing the fire–AOD variations from 2003 to 2011. The aerosol properties are derived from the inversion of the MODIS-observed reflectance using precomputed radiative transfer lookup tables based on aerosol models (Remer et al., 2005; Levy et al., 2007).

21.4 Methods

21.4.1 Descriptive Statistics

Mean NO$_2$ concentrations from OMI (2005–2011) and SCIAMACHY (2003–2011) and corresponding active fire numbers and FRP (MW) values from the MODIS datasets were extracted for the Northeast India states. Both the interannual and the seasonal variations—winter (January–February), summer or premonsoon (March–May), rainy season (June–September), and postmonsoon season (October–December)—in NO$_2$ were analyzed. We also extracted MODIS-AOD values corresponding to NO$_2$ and fires. A yearly coefficient of variation in NO$_2$ was calculated to determine the variability and amplitude based on the temporal data. Time-series datasets of active fire numbers, FRP, NO$_2$, and AOD were plotted to assess fire and FRP-NO$_2$-AOD signal variations. Pearson correlation coefficients (two-tailed test of significance 0.05 level) were computed among the datasets. In addition to reporting descriptive statistics of fire counts, FRP, OMI-NO$_2$, SCIAMACHY-NO$_2$, and MODIS-AOD values, we also used cumulative relative frequency plots to assess the NO$_2$ concentrations from OMI and SCIAMACHY.

21.4.2 Time-Series Regression

We used the time-series regression to assess the combined contribution of fire counts and FRP affecting the NO$_2$ concentrations from OMI and SCIAMACHY independently. We used fire counts and FRP as predictor variables of NO$_2$.

When using the time-series data, ordinary least squares (OLS) estimates may not be reliable. A certain amount of smoothing is introduced in the time-series data, by averaging the data over months (or months from days, or quarters to years). Thus, some of the randomness inherent in the data is lost. The smoothing of data can lead to systematic patterns in the error terms, thus leading to the possibility of autocorrelation. As a result, the estimated variances of the OLS estimators are biased and tend to underestimate the true variances and standard errors, which may inflate the "t" values, thus potentially leading to erroneous conclusions. As a result, the usual F- and t-tests are not reliable, including the estimated R^2. To account for these errors, Prais–Winsten regression was used, which utilizes the generalized least-squares method to estimate the parameters in a linear regression model in which the errors are serially correlated. The errors are assumed to follow a first-order autoregressive process. Once a model is estimated from the time-series data, the Durbin–Watson (DW) statistic (also known as the d-statistic) is calculated for diagnosing whether the residuals are serially correlated (Durbin and Winsten, 1950). DW statistic is given as

$$DW = \frac{\sum_{i=1}^{n} (e_i - e_{i-1})^2}{\sum_{i=1}^{n} e_i^2}$$

Residual (e_i) is the difference between the observed value and the predicted value at a certain level of X:

$$e_i = y_i - \hat{y}_i$$

The DW statistic lies in the 0–4 range, with a value near two indicating no first-order autocorrelation. Positive serial correlation is associated with DW values below two and negative serial correlation with DW values above two. DW values close to two are desirable, suggesting no autocorrelation errors.

21.5 Results

21.5.1 Fires in the Northeast India

Active fires retrieved from the MODIS datasets for the extent of India are shown in Figure 21.1a. The aggregated yearly and monthly MODIS fire counts and FRP for Northeast India from 2003 to 2011 are shown in Figure 21.1b and c. MODIS recorded 21,417 fire counts per year from Northeast India, which mostly correspond to evergreen forest fires from slash-and-burn agriculture. Of the different years, 2009 recorded the highest number of fire counts followed by 2006 and 2010. Analysis of fire counts for monthly variations (Figure 21.1b) suggested that March had the highest fire counts with 63% of all fires occurring during that month followed by April (21%), February (9.15%), etc.

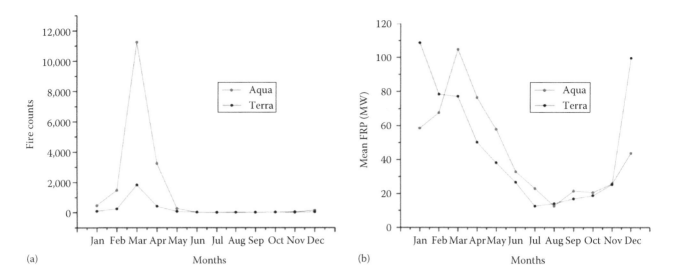

FIGURE 21.2 Nine-year data has been averaged to arrive at the monthly fire counts and FRP (MW): (a) aqua and terra fire counts for northeast India and (b) mean FRP for northeast India.

Monthly fire counts as well as the mean FRP (MW) averaged over a 9-year time period are shown in Figure 21.2a and b. Results suggested that fire counts in the early afternoon from Aqua were six times higher than the morning Terra detections. The higher fire detections from Aqua are justified as most of the clear felled slash is left till late afternoon for drying and then burnt by the locals (Majumdar et al., 2011). The averaged FRP for Terra was 12.5–108.7 MW, whereas Aqua was 12.6–104.8 MW, with the peak during March (91.02 MW averaged for Aqua and Terra) (Figure 21.2b).

21.5.2 NO$_2$ Temporal and Seasonal Variations

Monthly and seasonal variations in the mean columnar NO$_2$ concentrations obtained by averaging OMI (2005–2011) and SCIAMACHY (2003–2011) data are shown in Figure 21.3a and b and the peak concentration of NO$_2$ spatial patterns

during March in Figure 21.1d and e. The temporal mean of OMI data suggested a concentration of 1.4×10^{15} molecules cm^{-2} with 1.59×10^{15} molecules cm^{-2} ($+1\sigma$) and 0.70×10^{15} molecules cm^{-2} (-1σ). Relatively higher concentrations of NO$_2$ were observed during March (2.3×10^{15} molecules cm^{-2}), April (1.4×10^{15} molecules cm^{-2}), February (1.18×10^{15} molecules cm^{-2}), and the lowest concentration was observed during July (0.75×10^{15} molecules cm^{-2}). From the SCIAMACHY data, the 9-year temporal mean suggested an NO$_2$ concentration of 0.95×10^{15} molecules cm^{-2}, with 1.27×10^{15} molecules cm^{-2} ($+1\sigma$) and 0.63×10^{15} molecules cm^{-2} (-1σ). Although with lower concentrations, SCIAMACHY-NO$_2$ also exhibited a similar trend with the highest NO$_2$ values during March (1.5×10^{15} molecules cm^{-2}), April (1.20×10^{15} molecules cm^{-2}), and February (1.18×10^{15} molecules cm^{-2}), however, with the least NO$_2$ during October (0.65×10^{15} molecules cm^{-2}). Evaluation of

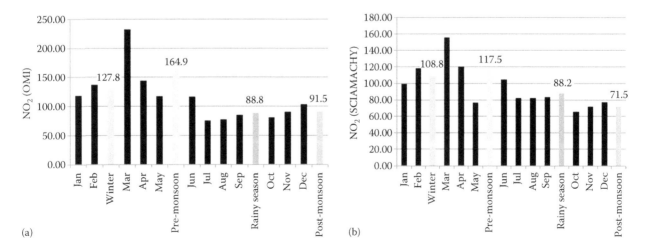

FIGURE 21.3 (a) Seasonal variations in OMI-NO$_2$ (y-axis values in 10^{13} molecules cm^{-2}) and (b) SCIAMACHY-NO$_2$ (y-axis values in 10^{13} molecules cm^{-2}) in the northeast India.

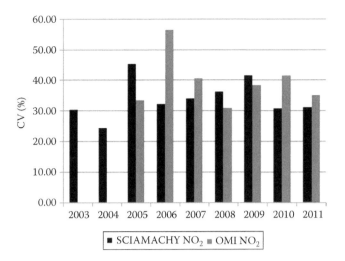

FIGURE 21.4 Coefficient of variation (CV) % in the OMI- and SCIAMACHY-NO_2 retrievals.

the seasonal patterns in OMI-NO_2 concentrations showed the highest concentrations during premonsoon, followed by winter, postmonsoon, and the lowest concentration during the rainy season (Figure 21.3a). Similar to OMI-NO_2, SCIAMACHY-NO_2 showed higher concentrations during premonsoon followed by winter; however, the rainy season concentrations were relatively higher than that of the postmonsoon (Figure 21.3b). Analysis for interannual variations in NO_2 (Figure 21.4) suggested that OMI showed relatively high variations during 2006, 2007, 2010, and 2009. The empirical cumulative distribution function plots with the cumulative relative frequency for OMI-NO_2 and SCIAMACHY-NO_2 data are shown in Figure 21.5a and b. The OMI-NO_2 data had the maximum of 3.0×10^{15} molecules cm^{-2}, mean of 1.1×10^{15} molecules cm^{-2}, and median of 1.05×10^{15}

molecules cm^{-2} compared to the SCIAMACHY maximum of 1.75×10^{15} molecules cm^{-2}, mean of 0.95×10^{15} molecules cm^{-2}, and median of 0.9×10^{15} molecules cm^{-2} values. The other statistics are shown in the quantile plots, where the 90th percentile value for the OMI-NO_2 data was 1.63×10^{15} molecules cm^{-2} in contrast to the SCIAMACHY data with 1.41×10^{15} molecules cm^{-2}. These plots (Figure 21.5a and b) clearly suggest relatively higher concentrations of NO_2 captured by OMI than by the SCIAMACHY data.

21.5.3 Correlations and Time-Series Regression

A clear increase in NO_2 signal corresponding with the fire counts during the peak biomass burning month of March can be seen in the OMI-NO_2 time-series data (Figure 21.6a) and with a lesser correspondence in the SCIAMACHY-NO_2 data (Figure 21.6b). OMI-NO_2 coincides more strongly with the AOD signal than SCIAMACHY-NO_2 (Figure 21.6c and d). Scatter plots with correlations for different datasets are shown in Figure 21.7a through i. The results suggested a stronger correlation of fire counts with OMI-NO_2 ($R^2 = 0.74$) than fire counts with SCIAMACHY-NO_2 ($R^2 = 0.35$) (Figure 21.7a and b). In addition, the FRP and NO_2 correlation was weak as compared to the sum of fire counts and NO_2 (Figure 21.7c and d). OMI-NO_2 showed a stronger correlation coefficient with MODIS-AOD ($R^2 = 0.54$) than SCIAMACHY-NO_2 and MODIS-AOD ($R^2 = 0.36$) (Figures 21.7e, f, and 21.6e). The results also suggested a relatively higher correlation of the sum of fire counts with AOD ($R^2 = 0.40$) than FRP with AOD (0.21) (Figure 21.7g and h). OMI-NO_2 and SCIAMACHY-NO_2 showed correlation strength of 57% (Figures 21.7i and 21.6f).

We used a time-series regression to address the total variance in the NO_2 observations contributed by the fire counts

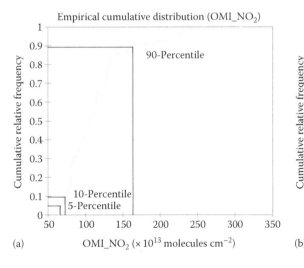

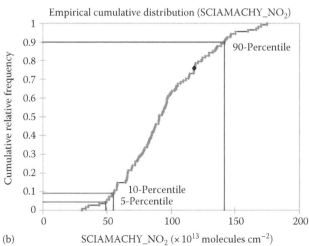

FIGURE 21.5 Empirical cumulative distribution plots for the OMI-NO_2 and the SCIAMACHY-NO_2 obtained from long-term data sets. For the OMI-NO_2, the value of the 5th percentile was 66.4 ($\times 10^{13}$ molecules cm^{-2}), the 10th percentile was 72.54 ($\times 10^{13}$ molecules cm^{-2}), and the 90th percentile was 163.6 ($\times 10^{13}$ molecules cm^{-2}); For SCIAMACHY-NO_2 the value of the 5th percentile was 48.96 ($\times 10^{13}$ molecules cm^{-2}), the 10th percentile was 55.05 ($\times 10^{13}$ molecules cm^{-2}), and the 90th percentile was 141.5 ($\times 10^{13}$ molecules cm^{-2}). Clearly, the OMI-NO_2 showed relatively high concentrations (90th percentile) than the SCIAMACHY-NO_2.

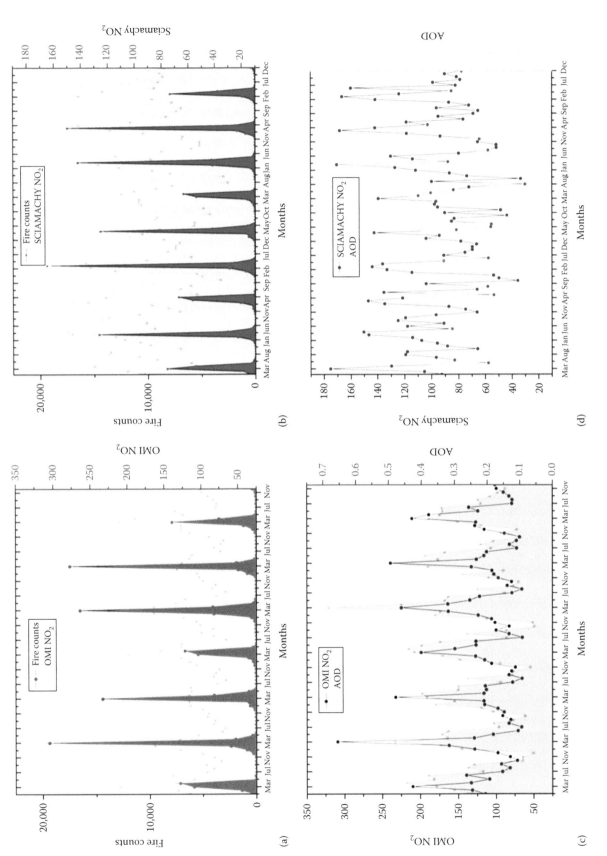

FIGURE 21.6 Time series plots between fire counts versus OMI-NO$_2$: (a) NO$_2$ values for all the graphs in 10^{13} molecules cm^{-2}, (b) fire counts versus SCIAMACHY-NO$_2$, (c) OMI-NO$_2$ and AOD, (d) SCIAMACHY-NO$_2$ and AOD.

(Continued)

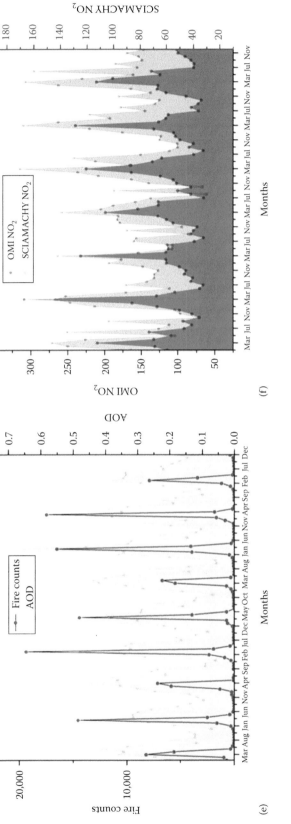

FIGURE 21.6 *(Continued)* Time series plots between fire counts versus OMI-NO₂: (e) fire counts and AOD, and (f) OMI-NO₂ and SCIAMACHY-NO₂.

and FRP. Results using the Prais–Winsten time-series regression for OMI-NO$_2$ and SCIAMACHY-NO$_2$ data are shown in Tables 21.2 and 21.3. For the Prais–Winsten regression, the regression coefficients, residual sum of squares, F-test, R^2, adjusted R^2, residual standard errors, t-statistic, p-values, 95% confidence intervals, "rho," and the DW statistic were

all reported. The adjusted R^2 value is useful for comparing the explanatory power of models with different numbers of predictors. A model with more terms may appear to have a better fit simply because it has more terms, thus adjustment is needed to account for all the predictors. As depicted in the adjusted R^2 values, fire counts and FRP together explained

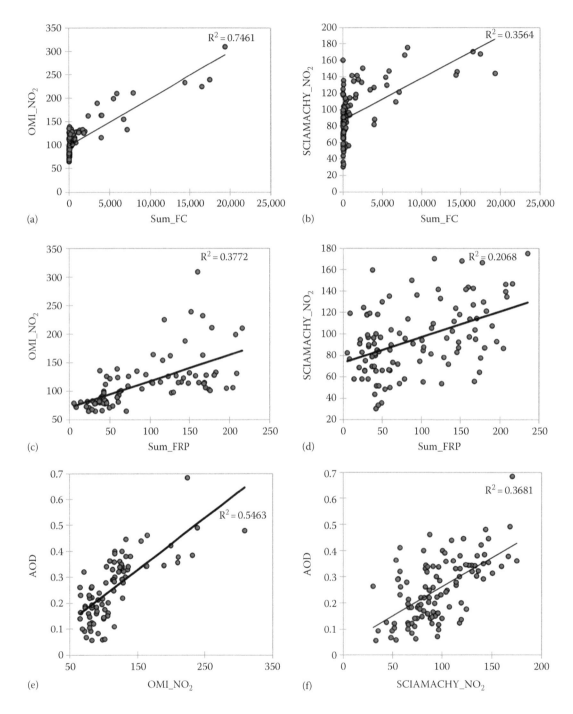

FIGURE 21.7   Scatter plots with Pearson correlation coefficients (two-tailed test of significance (0.05 level)): (a) sum of fire counts (FC) versus OMI-NO$_2$ values in 10^{13} molecules cm^{-2}, (b) sum of FC versus SCIAMACHY-NO$_2$ values in 10^{13} molecules cm^{-2}, (c) sum of fire radiative power (FRP in MW) versus OMI-NO$_2$ values in 10^{13} molecules cm^{-2}, (d) sum of FRP (MW) versus SCIAMACHY-NO$_2$ values in 10^{13} molecules cm^{-2}, (e) OMI-NO$_2$ (values in 10^{13} molecules cm^{-2}) versus AOD, (f) SCIAMACHY-NO$_2$ (values in 10^{13} molecules cm^{-2}) versus AOD.   *(Continued)*

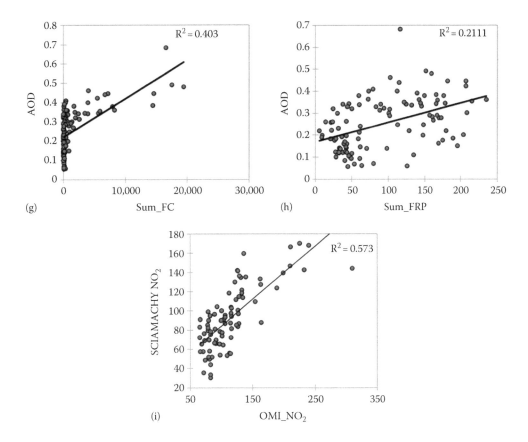

FIGURE 21.7 (*Continued*)   Scatter plots with Pearson correlation coefficients (two-tailed test of significance (0.05 level)): (g) sum of FC versus AOD, (h) sum of FRP versus AOD, and (i) OMI-NO₂ versus SCIAMACHY NO₂ (values in 10^{13} molecules cm^{-2}).

TABLE 21.2   Time-Series Prais–Winsten Regression of Fire Counts and FRP with OMI NO₂

Omi_NO$_2$	Coef.	Std. Err.	t	P > |t|
Sum_FC	0.0081107	0.0005831	13.91	0.000
Sum_FRP	0.2182609	0.0435501	5.01	0.000
Cons	81.55099	4.757556	17.14	0.000

R-square = 0.79; adjusted R-square = 0.78.

TABLE 21.3   Time-Series Prais–Winsten Regression of Fire Counts and FRP with SCIAMACHY-NO₂

SCIAMACHY_~2	Coef.	Std. Err.	t	P > |t|
Sum_FC	0.0038616	0.0007331	5.27	0.000
Sum_FRP	0.2318535	0.0969293	2.39	0.019
Cons	78.53321	5.036624	15.59	0.000

R-square = 0.34; adjusted R-square = 0.33.

78% of NO₂ variance in the OMI data compared to 33% in the SCIAMACHY data. The t-statistic and its corresponding p-value determine whether a regression coefficient is significantly different from zero. In our case, the p-values for the estimated coefficients for the sum of fire counts and mean FRP are both 0.000 at alpha = 0.05, indicating that they are significantly related to NO₂ concentrations in the OMI data. In the SCIAMACHY-NO₂ data, FRP seems to be less significant than the fire counts (Table 21.3). The standard errors reflect the variability and are a measure of the precision with which the regression coefficients are measured. In both the OMI-NO₂ and SCIAMACHY-NO₂ models, the standard errors seem to show high precision. An F-test assesses how well the set of independent variables, as a group, explains the variation in the dependent variable. The significant F-value (Tables 21.2 and 21.3) suggests that the calculation of R² in the model was best fit. The "rho" parameter in the model indicates the level of autocorrelation, and in the Prais–Winsten regression, it is calculated iteratively. The "rho" parameters converged at 0.24 in the OMI-NO₂ model and 0.15 in the SCIAMACHY-NO₂ model indicating that both of these datasets were more stationary time series having a clear structure and were not too far from the mean. Finally, the DW statistic clearly indicated that the Prais–Winsten transformation was helpful with the values close to 2 in both the models, indicating that the error term is serially independent, thus no autocorrelation.

21.6 Discussion

Accurate data on biomass burning emissions can help to assess land–atmosphere interactions effectively. In tropical regions, uncertainties still exist for biomass burning emissions, and thus there is a need to evaluate different approaches and datasets that are useful for quantifying the emissions. Intercomparison of OMI and SCIAMACHY-NO_2 in relation to fires clearly suggested stronger correlations with the OMI than SCIAMACHY. Comparison of seasonal trends clearly showed that the NO_2 patterns from both OMI and SCIAMACHY matched well with the peak biomass burning episodes (March, April, and February). However, OMI-NO_2 captured 1.5 times more NO_2 concentrations during the peak month of March than the SCIAMACHY-NO_2 as observed from the time-series data. Earlier, Martin et al. (2002), using the GOME data, reported seasonal enhancements in NO_2 during December, January, and February over Northern Africa; March, April, and May in India; and from June to October in Central Africa and South America. Further, comparison of NO_2 concentrations in the other biomass burning regions suggests that NO_2 concentrations from the evergreen forest fires of Northeast India are relatively lower (1.14×10^{15} molecules cm^{-2}) than the other biomass burning regions. For example, Thomas et al. (1998) reported total NO_2 concentrations of around 2.5×10^{15} (plus or minus 0.8×10^{15} molecules cm^{-2}) over southern Borneo, Indonesia, mainly due to rainforest burning. Similarly, Jaegle et al. (2005) reported NO_2 values in the range of $2.5–4.0 \times 10^{15}$ molecules cm^{-2} in the South African savanna burning areas using GOME data.

The relatively higher NO_2 column retrievals observed by OMI than SCIAMACHY during the evergreen forest fires are attributed to the satellite overpass time, that is, SCIAMACHY during morning (10.00 am) and OMI during afternoon (13:45 pm). This inference in our case is also supported by the relatively higher number of fire counts captured by the MODIS Aqua having later afternoon satellite overpass compared to Terra. For example, Aqua-MODIS captured six times more fires compared to Terra-MODIS in the study area. The differences in NO_2 from different vegetation types are attributed to the amount of biomass burnt and gas-phase oxidation of NO_2 with OH in different environments. In addition, comparison of urban NO_2 concentrations elsewhere with the biomass burning NO_2 concentrations clearly suggests relatively higher NO_2 in urban areas than biomass burning areas. For example, using the GOME NO_2 retrievals, Velders et al. (2001) noted the maximum NO_2 columns during January for the eastern United States, Western Europe, and Eastern China in the range of $33–43 \times 10^{15}$ molecules cm^{-2} compared to the other months in the range of $14–20 \times 10^{15}$ molecules cm^{-2}. Also, NO_2 from evergreen forest fires of Northeast India seem to be relatively low compared to other large point sources located in India such as from thermal power plants, steel plants, cement plants, industrial processes, and fossil fuel extractions, which exceed 4.0×10^{15} molecules cm^{-2} (Ghude et al., 2008; David and Nair, 2013). These inferences suggest that relative to the biomass burning, NO_2 emissions from urban sources should be given priority for mitigation efforts.

Boersma et al. (2007, 2011) noted that SCIAMACHY observes higher NO_2 than OMI (up to 40%) in most industrial regions of northern midlatitudes, while it observes lower NO_2 than OMI (up to 35%) in tropical biomass burning regions.

Further, Boersma et al. (2007) through the detailed analysis of slant columns and AMF concluded that differences between SCIAMACHY and OMI-NO_2 columns in source regions could not be ascribed to a retrieval artifact but to the underlying chemistry governing the NO_2 concentrations. The lifetime of NO_2 varies from 1 h at the surface to days in the upper atmosphere, depending on the season, latitude, altitude, and other local atmospheric properties. The lifetime variations during different seasons reflect gas-phase oxidation of NO_2 with OH in the summer with an increasing role for heterogeneous chemistry in the winter (Lamsal et al., 2010; Miyazaki et al., 2012). The seasonal differences highlight the temporal variability of NO_2 from different sources and regions.

Although we hypothesized that FRP would be a better predictor of NO_2 concentrations, we found that fire counts had a better correlation. We also found that use of fire counts in conjunction with FRP increased the correlation with NO_2 concentrations. Use of FRP values for characterizing the fire cycle may be constrained due to various other reasons such as smoke plumes, MODIS-viewing geometry, and FRP sampling (Ichoku et al., 2008; Vadrevu et al., 2011). Considering these discrepancies, we infer that more studies are needed on using FRP as a surrogate measure for quantifying GHG concentrations and emissions. Further, results from the time-series regression clearly suggested combined use of the sum of fire counts and mean FRP as predictors of NO_2 concentrations. Relating to the other uncertainties, although the errors in the tropospheric NO_2 columns caused due to cloud fraction (~30%), surface albedo (25%), AMF (10%), etc., are addressed in previous studies using the DOMINO product (Boersma et al., 2012), more detailed analysis including modeling might be needed in order to address the remaining uncertainties specific to Indian monsoon climate.

21.7 Conclusions

Biomass burning is an important source of GHG emissions and aerosols. We evaluated the nitrogen dioxide (NO_2) emissions from OMI and SCIAMACHY data from the biomass burning of subtropical evergreen forests in Northeast India. We also assessed the long-term (2003–2011) MODIS-AOD signal in relation to MODIS fire retrievals. Results suggested a stronger correlation of MODIS fire counts with OMI-NO_2 ($R^2 = 0.74$) than with SCIAMACHY-NO_2 ($R^2 = 0.35$). Also, OMI-NO_2 showed a stronger correlation with MODIS-AOD ($R^2 = 0.54$) than SCIAMACHY-NO_2 ($R^2 = 0.36$). Further, intercomparison of OMI and SCIAMACHY-NO_2 showed correlation of 57%. We attribute these results to the satellite overpass time, that is, SCIAMACHY during morning (10.00 am) and OMI during afternoon (13:45 pm), ecosystem variations, and oxidation reactions influencing NO_2 formation. Results from time-series modeling suggested combined use of fire counts and FRP for explaining NO_2 concentrations than either of those variables alone. Our results highlight the potential MODIS fires and OMI datasets in characterizing biomass burning episodes from subtropical evergreen forests. We infer the need for field campaigns that characterize the relationship between satellite retrievals of emissions and ground-based measurements to resolve satellite-based uncertainties.

Acknowledgments

Authors thank the MODIS fire, MODIS aerosol, SCIAMACHY, and OMI-NO$_2$ science teams for the datasets. This research was supported by NASA grant NNX10AU77G.

References

Andreae, M. O. and Merlet, P. 2001. Emission of trace gases and aerosols from biomass burning. *Global Biogeochemical Cycles*, 15(4), 955–966.

Arellano, A. F., Kasibhatla, P. S., Giglio, L., van der Werf, G. R., Randerson, J. T., and Collatz, G. J.2006. Time-dependent inversion estimates of global biomass-burning CO emissions using Measurement of Pollution in the Troposphere (MOPITT) measurements. *Journal of Geophysical Research*, 111, D09303, doi:10.1029/2005JD006613.

Badarinath, K. V. S., Kharol, S. K., Sharma, A. R., and Krishna Prasad, V. 2009. Analysis of aerosol and carbon monoxide characteristics over Arabian Sea during crop residue burning period in the Indo-Gangetic Plains using multi-satellite remote sensing datasets. *Journal of Atmospheric and Solar-Terrestrial Physics*, 71(12), 1267–1276.

Beirle, S., Platt, U., Wenig, M., and Wagner, T. 2003. Weekly cycle of NO$_2$ by GOME measurements: A signature of anthropogenic sources. *Atmospheric Chemistry and Physics*, 3, 2225–2232.

Bertram, T. H., Heckel, A., Richter, A., Burrows, J., and Cohen, R. C. 2005. Satellite measurements of daily variations in soil NOx emissions. *Geophysical Research Letters*, 32, L24812, doi:10.1029/2005GL024640.

Blond, N., Boersma, K. F., Eskes, H. J., van der R. J. A., Roozendael, M. V., De Smedt, I., Bergametti, G., and Vautard, R. 2007. Intercomparison of SCIAMACHY nitrogen dioxide observations, in-situ measurements, and air quality modeling results over Western Europe. *Journal of Geophysical Research*, 112, D10311, doi:10.1029/2006JD007277.

Boersma, K. F., Eskes, H. J., and Brinksma, E. J. 2004. Error analysis for tropospheric NO$_2$ retrieval from space. *Journal of Geophysical Research*, 109, D04311, doi:10.1029/2003JD003962.

Boersma, K. F., Eskes, H. J., Dirksen, R. J., van der, A. R. J., Veefkind, J. P., Stammes, P., Huijnen, V. et al. 2011. An improved retrieval of tropospheric NO$_2$ columns from the Ozone Monitoring Instrument. *Atmospheric Measurement Techniques*, 4, 1905–1928.

Boersma, K. F., Vinken, G. C. M., Castellanos, P., De Ruyter, M., and Eskes, H. J. 2012. Reductions in nitrogen oxides over Europe driven by environmental policy and economic recession. Tijdschrift Lucht, 8, 13–17.

Boersma, K. F., Eskes, H. J., Veefkind, J. P., Brinksma, E. J., van der, R. J., Sneep, M., van den Oord, G. H. Levelt, P. F., Stammes, P., Gleason, J. F., and Bucsela, E. J. 2007. Near-real time retrieval of tropospheric NO$_2$ from OMI. *Atmospheric Chemistry and Physics*, 7, 2103–2118.

Bucsela, E. J. 2010. Lightning-generated NOx seen by OMI during NASA's TC4 experiment. *Journal of Geophysical Research*, 115, D00J10, doi:10.1029/2009JD013118.

Burrows, J. P., Platt, U., and Borrell, P. 2011. Tropospheric remote sensing from space. In: Burrows, J. P., Platt, U., and Borell, P. (eds.), *The Remote Sensing of Tropospheric Composition from Space*. Physics of Earth and Space Environments. Springer Heidelberg, New York.

Burrows, J. P., Weber, M., Buchwitz, M., Rozanov, V., Weißenmayer, A. L., Richter, A., DeBeek, R. et al. 1999. The Global Ozone Monitoring Experiment (GOME): Mission concept and first scientific results. *Journal of Atmospheric Science*, 56, 151–175.

Cheewaphongphan, P. and Garivait, S. 2013. Bottom up approach to estimate air pollution of rice residue open burning in Thailand. *Asia-Pacific Journal of Atmospheric Sciences*, 49(2), 139–149.

David, L. M. and Nair, P. R. 2013. Tropospheric column O$_3$ and NO$_2$ over the Indian region observed by Ozone Monitoring Instrument (OMI): Seasonal changes and long-term trends. *Atmospheric Environment* 65(2013), 25–39.

Durbin, J. and Winsten, G. S. 1950. Testing for serial correlation in least-squares regression II. *Biometrika*, 37, 409–428.

Ghude, S. D., Fadnavis, S., Beig, G., Polade, S. D., and van der, R. J. A. 2008. Detection of surface emission hot spots, trends, and seasonal cycle from satellite-retrieved NO$_2$ over India. *Journal of Geophysical Research*, 113, D20305, doi:10.1029/2007JD009615.

Giglio, L., Descloitres, J., Justice, C. O., and Kaufman, Y. 2003. An enhanced contextual fire detection algorithm for MODIS. *Remote Sensing of Environment* 87, 273–282.

Giglio, L., Randerson, J. T., and van der Werf, G. R. 2013. Analysis of daily, monthly, and annual burned area using the fourth—Generation global fire emissions database (GFED4). *Journal of Geophysical Research: Biogeosciences* 118(1), 317–328.

Heil, A., Langmann, B., and Aldrian, E. 2007. Indonesian peat and vegetation fire emissions: Study on factors influencing large-scale smoke haze pollution using a regional atmospheric chemistry model. *Mitigation and Adaptation Strategies for Global Change*, 12(1), 113–133.

Hudman, R. C., Jacob, D. J., Turquety, S., Leibensperger, E. M., Murray, L. T., Wu, S., Gilliland, A. B. et al. 2007. Surface and lightning sources of nitrogen oxides over the United States: Magnitudes, chemical evolution, and outflow. *Journal of Geophysical Research*, 112, D12S05, doi:10.1029/2006JD007912.

Ichoku, C., Giglio, L., Wooster, M. J., and Remer, L. A. 2008. Global characterization of biomass-burning patterns using satellite measurements of fire radiative energy. *Remote Sensing of Environment*, 112, 2950–2962.

Jaeglé, L., Steinberger, L., Martin, R. V., and Chance, K. 2005. Global partitioning of NOx sources using satellite observations: Relative roles of fossil fuel combustion, biomass burning and soil emissions. *Faraday Discussion*, 130, 407–423, doi:10.1039/b502128f.

Kaufman, Y. J., Justice, C. O., Flynn, L., Kendall, J. D., Prins, E. M., Giglio, L., Ward, D., Menzel, W., and Setzer, A. 1998. Potential global fire monitoring from EOS-MODIS. *Journal of Geophysical Research* 103, 32215–32238.

Kim, S.-W., Heckel, A., McKeen, S. A., Frost, G. J., Hsie, E. Y., Trainer, M. K., Richter, A., Burrows, J. P., Peckham, S. E., and Grell, G. A. 2006. Satellite-observed U.S. power plant NOx emission reductions and their impact on air quality. *Geophysical Research Letters*, 33, L22812, doi:10.1029/2006GL027749.

Kurokawa, J. I., Yumimoto, K., Uno, I., and Ohara, T. 2009. Adjoint inverse modeling of NOx emissions over eastern China using satellite observations of NO$_2$ vertical column densities. *Atmospheric Environment*, 43, 1878–1887, doi:10.1016/2008.12.030.

Lamsal, L. N., Martin, R. V., van Donkelaar, A., Celarier, E. A., Bucsela, E. J., Boersma, K. F., Dirksen, R., Luo, C., and Wang, Y. 2010. Indirect validation of tropospheric nitrogen dioxide retrieved from the OMI satellite instrument: Insight into the seasonal variation of nitrogen oxides at northern midlatitudes. *Journal of Geophysical Research*, 115, D05302, doi:10.1029/2009JD013351.

Langner, A., Miettinen, J., and Siegert, F. 2007. Land cover change 2002–2005 in Borneo and the role of fire derived from MODIS imagery. *Global Change Biology*, 13(11), 2329–2340.

Levelt, P. F., van den Oord, G. H. J., Dobber, M. R., Malkki, A., Visser, H., de Vries, J., Stammes, P., Lundell, J. O. V., and Saari, H. 2006. The ozone monitoring instrument. *IEEE Transactions on Geoscience and Remote Sensing*, 44, 1093–1101.

Levy, R. C., Remer, L. A., and Dubovik, O. 2007. Global aerosol optical properties and application to Moderate Resolution Imaging Spectroradiometer aerosol retrieval over land. *Journal of Geophysical Research*, 112, D13210, doi:10.1029/2006JD007815.

Lin, J. T., McElroy, M. B., and Boersma, K. F. 2010. Constraint of anthropogenic NOx emissions in China from different sectors: A new methodology using multiple satellite retrievals. *Atmospheric Chemistry and Physics*, 10, 63–78, doi:10.5194/acp-10-63-2010.

Logan, J. A. 1983. Nitrogen oxides in the troposphere: Global and regional budgets. *Journal of Geophysical Research*, 88, 10785–10807.

Logan, J. A., Prather, M. J., Wofsy, S. C., and McElroy, M. B. 1981. Tropospheric chemistry: A global perspective. *Journal of Geophysical Research*, 86, 7210–7254, doi:10.1029/JC086iC08p07210.

Majumder, M., Shukla, A. K., and Arunachalam, A. 2011. Agricultural practices in Northeast India and options for sustainable management. In: E. Lichtfouse (ed.), *Biodiversity, Biofuels, Agroforestry and Conservation Agriculture*, pp. 287–315. Sustainable Agriculture Rev., 5, doi:10.1007/978-90-481-9513-8_10.

Martin, R. V., Chance, K., Jacob, D. J., Kurosu, T. P., Spurr, R. J. D., Bucsela, E., Gleason, J. F. et al. 2002. An improved retrieval of tropospheric nitrogen dioxide from GOME. *Journal of Geophysical Research*, 107(D20), 4437, doi:10.1029/2001JD001027.

Martin, R. V., Sioris, C. E., Chance, K., Ryerson, T. B., Bertram, T. H., Wooldridge, P. J., Cohen, R. C., Neuman, J. A., Swanson, A., and Flocke, F. M. 2006. Evaluation of space-based constraints on global nitrogen oxide emissions with regional aircraft measurements over and downwind of eastern North America. *Journal of Geophysical Research*, 111, D15308, doi:10.1029/2005JD006680.

Mebust, A. K., Russell, A. R., Hudman, R. C., Valin, L. C., and Cohen, R. C. 2011. Characterization of wildfire Nox emissions using MODIS fire radiative power and OMI tropospheric NO$_2$ columns. *Atmospheric Chemistry and Physics*, 11, 5839–5851.

Mijling, B., van der A. R. J., Boersma, K. F., Van Roozendael, M., De Smedt, I., and Kelder, H. M. 2009. Reductions of NO$_2$ detected from space during the 2008 Beijing Olympic Games, *Geophysical Research Letters*, 36, L13801, doi:10.1029/2009GL038943.

Miyazaki, K., Eskes, H. J., and Sudo, K. 2012. Global NOx emission estimates derived from an assimilation of OMI tropospheric NO$_2$ columns. *Atmospheric Chemistry and Physics*, 12, 2263–2288.

Müller, J.-F. and Stavrakou, T. 2005. Inversion of CO and NOx emissions using the adjoint of the IMAGES model. *Atmospheric Chemistry and Physics*, 5, 1157–1186, doi:10.5194/acp-5-1157-2005.

Murdiyarso, D. and Lebel, L. 2007. Local to global perspectives on forest and land fires in Southeast Asia. *Mitigation and Adaptation Strategies for Global Change*, 12(1), 3–11.

Palmer, P. I., Jacob, D. J., Chance, K., Martin, R. V., Spurr, R. J. D., Kurosu, T. P., Bey, I., Yantosca, R., Fiore, A., and Li, Q. 2001. Air-mass factor formulation for spectroscopic measurements from satellites: Application to formaldehyde retrievals from the Global Ozone Monitoring Experiment. *Journal of Geophysical Research*, 106, 539–550.

Prasad, K. V., Gupta, P. K., Sharma, C., Sarkar, A. K., Kant, Y., Badarinath, K. V. S., and Mitra, A. P. (2000). NO< sub> x emissions from biomass burning of shifting cultivation areas from tropical deciduous forests of India–estimates from ground-based measurements. *Atmospheric Environment*, 34(20), 3271–3280.

Remer, L. A., Kaufman, Y. J., Mattoo, S., Martins, J. V., Ichoku, C., Levy, R. C., Kleidman, R. G. et al. 2005. The MODIS aerosol algorithm, products and validation. *Journal of Atmospheric Science*, 62, 947–973, doi:10.1175/JAS3385.1

Richter, A. and Burrows, J. P. 2002. Tropospheric NO_2 from GOME measurements. *Advances in Space Research*, 29, 1673–1683.

Russell, A. R., Perring, A. E., Valin, L. C., Bucsela, E., Browne, E. C., Min, K. E., Wooldridge, P. J., and Cohen, R. C. 2011. A high spatial resolution retrieval of NO_2 column densities from OMI: Method and evaluation. *Atmospheric Chemistry and Physics*, 11, 12411–12440.

Schaub, D., Boersma, K. F., Kaiser, J. W., Weiss, A. K., Folini, D., Eskes, H. J., and Buchmann, B. 2006. Comparison of GOME tropospheric NO_2 columns with NO_2 profiles deduced from ground-based in situ measurements. *Atmospheric Chemistry and Physics*, 6, 3211–3229.

Schwartz, J. 2004. Air pollution and children's health. *Pediatrics*, 113, 1037–1043.

Shaw, G. E. 1976. Nitrogen dioxide-optical absorption in the visible. *Journal of Geophysical Research*, 81, 5791–5792.

Stavrakou, T., Muller, J. F., Boersma, K. F., De Smedt, I., and van der, A. R. J. 2008. Assessing the distribution and growth rates of NOx emission sources by inverting a 10-year record of NO_2 satellite columns. *Geophysical Research Letters*, 35, L10801, doi:10.1029/2008GL033521.

Streets, D. G., Bond, T. C., Carmichael, G. R., Fernandes, S. D., Fu, Q., He, D., Yarber, K. F. et al. 2003. An inventory of gaseous and primary aerosol emissions in Asia in the year 2000. *Journal of Geophysical Research: Atmospheres (1984–2012)*, 108, D21.

Thomas, W., Hegels, E., Slijkhuis, S., Spurr, R., and Chance, K. 1998. Detection of biomass burning combustion products in Southeast Asia from backscatter data taken by the GOME spectrometer. *Geophysical Research Letters*, 25, 1317–1320.

Vadrevu, K. P., Ellicott, E., Badarinath, K. V. S., and Vermote, E. 2011. MODIS derived fire characteristics and aerosol optical depth variations during the agricultural residue burning season, north India. *Environmental Pollution*, 159, 1560–1569.

Vadrevu, K. P., Ellicott, E., Giglio, L., Badarinath, K. V. S., Vermote, E., Justice, C., and Lau, W. K. 2012. Vegetation fires in the himalayan region–Aerosol load, black carbon emissions and smoke plume heights. *Atmospheric Environment*, 47, 241–251.

Vadrevu, K. P. and Justice, C. O. 2011. Vegetation fires in the Asian region: Satellite observational needs and priorities. *Global Environmental Research*, 15(1), 65–76.

Valin, L. C., Russell, A. R., Bucsela, E. J., Veefkind, J. P., and Cohen, R. C. 2011. Observation of slant column NO_2 using the super-zoom mode of AURA-OMI. *Atmospheric Measurement Techniques Discussions*, 20, 1989–2005, doi:10.5194/amtd-4-1989-2011.

van der, A. R. J., Eskes, H. J., Boersma, K. F., van Noije, T. P. C., van Roozendael, M., De Smedt, I., Peters, D. H. M. U., and Meijer, E. W. 2008. Trends, seasonal variability and dominant NOx source derived from a ten year record of NO_2 measured from space. *Journal of Geophysical Research*, 113, 1–12, doi:10.1029/2007JD009021.

van der, A. R. J., Peters, D. H. M. U., Eskes, H., Boersma, K. F., Van Roozendael, M., De Smedt, I., and Kelder, H. M. 2006. Detection of the trend and seasonal variation in tropospheric NO_2 over China. *Journal of Geophysical Research*, 111, D12317, doi:10.1029/2005JD006594.

Van der Werf, G. R., Dempewolf, J., Trigg, S. N., Randerson, J. T., Kasibhatla, P. S., Giglio, L., and DeFries, R. S. 2008. Climate regulation of fire emissions and deforestation in equatorial Asia. *Proceedings of the National Academy of Sciences*, 105(51), 20350–20355.

van der Werf, G. R., Randerson, J. T., Giglio, L., Collatz, G. J., Mu, M., Kasibhatla, P. S., Morton, D. C., deFries, R. S., Jin, Y., and van Leeuwen, T. T. 2010. Global fire emissions and the contribution of deforestation, savanna, forest, agricultural, and peat fires (1997–2009). *Atmospheric Chemistry and Physics*, 10, 11707–11735.

Velders, G. J. M., Granier, C., Portmann, R. W., Pfeilsticker, K., Wenig, M., Wagner, T., Platt, U., Richter, A., and Burrows, J. P. 2001. Global tropospheric NO_2 column distributions: Comparing three-dimensional model calculations with GOME measurements. *Journal of Geophysical Research*, 106, D12.

Wang, Y., Renyi, Z., and Saravanan, R. 2014. Asian pollution climatically modulates mid-latitude cyclones following hierarchical modelling and observational analysis. *Nature Communications*, 5.

Wooster, M. J., Roberts, G., Perry, G. L. W., and Kaufman, Y. J. 2005. Retrieval of biomass combustion rates and totals from fire radiative power observations: FRP derivation and calibration relationships between biomass consumption and fire radiative energy release. *Journal of Geophysical Research*, 110, D24311–D24311.

Zhao, C. and Wang, Y. 2009. Assimilated inversion of NOx emissions over east Asia using OMI NO_2 column measurements. *Geophysical Research Letters*, 36, L06805, doi:10.1029/2008GL037123.

Zhou, W., Cohan, D. S., Pinder, R. W., Neuman, J. A., Holloway, J. S., Peischl, J., Ryerson, T. B., Nowak, J. B., Flocke, F., and Zheng, W. G. 2012. Observations and modelling of the evolution of Texas plant plumes. *Atmospheric Chemistry and Physics*, 12, 455–468.

22

Remote Sensing–Based Mapping and Monitoring of Coal Fires

Anupma Prakash
University of Alaska Fairbanks

Claudia Kuenzer
Earth Observation Center (EOC),
German Aerospace Center (DLR)

Acronyms and Definitions

ASTER	Advanced spaceborne thermal emission spectrometer
ATCOR	Atmospheric correction and haze reduction
AVHRR	A very high resolution radiometer
BIRD	Bi-spectral infrared detection
CBERS	China-Brazil Earth Resources Satellite
DLR	Deutsches Zentrum für Luft- und Raumfahrt or German Space Agency
EM	Electromagnetic
ETM+	Enhanced thematic mapper
FLAASH	Fast line-of-sight atmospheric analysis of hypercubes
LST	Land surface temperature
MIR	Mid infrared
MODIS	Moderate resolution imaging spectrometer
MODTRAN	Moderate resolution atmospheric transmission
NIR	Near infrared
OLI	Operational land imager
SWIR	Shortwave infrared
TIR	Thermal infrared
TIRS	Thermal infrared sensor
TM	Thematic mapper
VIS	Visible

22.1 Introduction

Coal fires, or fires occurring in reserves of coal, are a phenomenon of global occurrence. Almost all large coal mining operations in the world have been threatened by coal fires at some time. An excellent compilation of case studies of fires in different parts of the world is presented in Stracher et al. (2013), which includes an online database and a scalable map showing locations of several reported coal fires worldwide (Gens, 2013). This section introduces commonly used terminology for different types of coal fires, why these fires are so common, what damage they can cause, and how remote sensing can be used to investigate coal fires.

22.1.1 Terminology

Depending on the location of coal fires, several different terms have been coined. A good first review of classification of coal fires is presented by Zhang et al. (2004a). A fire in a coal mine is generically referred to as coal mine fire. If the fire is in an in situ coal seam it is also referred to as coal seam fire. If the coal seam that is on fire is exposed on the surface, it is occasionally referred to as an outcrop fire. The bigger differentiation for remote sensing–based investigation is based on whether the fire is at the surface and exposed (then known as

561

surface coal fire) or whether it is buried at depth (then known as subsurface or underground coal fire). Surface coal fires can include coal seam fires or also fires that occur in mining-related waste dumps (overburden dumps), in which case these are commonly called overburden coal fires. A further in-depth classification of coal fires was also published by Kuenzer and Stracher (2011), who presented a detailed overview on coal fire induced geomorphologic features, and classified fires not only according to their spatial occurrence (surface, subsurface, within a seam or in an artificial coal storage of coal waste pile), but also differentiated them according to age (paleo coal fires versus recent coal fires), genesis (natural coal fires versus human-induced coal fires), and burning behavior (steadily burning, accelerating, burning out, extinct, reignited). In this chapter, we will restrict to using the terms surface and subsurface coal fires, the former being associated with very high temperature fires exposed on the surface and the latter being associated with fires underground that are associated with warm surfaces, much lower in temperature than surface fires (Figure 22.1). More attributes of these coal fires are discussed later in this chapter.

Domestic and industrial coal fires are not a subject of consideration in this study. This study presents only the use of thermal infrared (TIR) and shortwave infrared (SWIR) remote sensing (used for temperature mapping) that forms the foundation of remote sensing investigations of coal fires. Related work (see Table 22.1) such as use of geophysical techniques for identification of active or previously burned areas, interferometric techniques to map fire and mining-related subsidences,

and optical images to study environmental impacts and land use/land-cover change due to coal are beyond the scope of this chapter.

22.1.2 Causes

The most commonly reported cause of coal fires is spontaneous combustion, a process where in the presence of some moisture acting as a catalyst, carbon in the coal reacts with oxygen in an exothermic reaction, causing enough heat accumulation to set the coal on fire. For an excellent background on this topic, and on the propensity of different coal types to ignite by spontaneous combustion, the readers are referred to the book by Banerjee (1985). In the Jharia coalfield, India, the first fire started in the Bhowrah colliery in 1916 due to spontaneous combustion, and nearly 100 years later this fire is still reported to be active (BCCL, 2014). Other natural causes of coal fires are lightning strikes, forest fires, and fires occurring in peat lands or waste dumps—the latter being a cause for the well-known coal fire under the town of Centralia, Pennsylvania, USA. In many cases, forest fires and coal fires are closely and cyclically related with one being the cause of the other (Prakash et al., 2011; Waigl et al., 2014; Whitehouse and Mulyana, 2004).

Most other causes of coal fires are linked to mining or related human activities. These include mining induced subsidence, short circuits, frictional heat from mining equipment, dust explosions, or negligence of safety protocols in mining operations (e.g., Chen et al., 2012; Zheng et al., 2009). Occasionally, coal fires have been reported to start from people using abandoned mines as safe abode for illegal distilling of alcohol or to hide from cold harsh weather (Glover, 2011; Prakash, 2014). Fire in an overburden dump can come in contact with a coal seam and set the coal seam aflame. Fire in one coal seam can migrate and ignite the coal in another seam above or below it. Regardless of how a fire starts, it continues to propagate as long the three essential elements (1) coal (2) heat, and (3) oxygen are present (Kim, 2010; Kuenzer et al., 2007a; Stracher et al., 2010; van Dijk et al., 2011).

22.1.3 Hazards

Hazards from coal fires are widely documented in popular visual and printed media. This is especially true for some underground fires in the United States, India, and China, where these fires have caused substantial adverse impacts to the environment and the communities. In the formerly prosperous town of Centralia (PA, USA), an underground coal seam caught fire arguably in 1962 and several decades later it is still ablaze. Incidences of sudden land subsidence and constant gaseous emissions from the burning coal made the living conditions dangerous, requiring relocation of all residents. According to unconfirmed media reports, Centralia is now an uninhabited ghost town where the population dwindled from 2761 in 1890 to seven remaining long-time residents who were

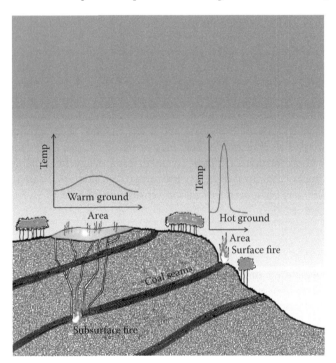

FIGURE 22.1 A conceptual diagram showing surface and subsurface fires in coal seams and the differences in temperature anomalies they cause at the land surface.

TABLE 22.1 Regions of the Electromagnetic Spectrum, Commonly Associated Wavelength Ranges, and the Main Coal Mining and Coal Fire Applications Associate with These Regions

Region of Spectrum	Wavelength Range	Main Coal Mining and Coal Fire Applications	Example References
Visible	0.4–0.7 μm	Land cover/Land use	Kumar and Pandey (2013)
			Kuenzer et al. (2012)
			Kuenzer et al. (2008a)
			Martha et al. (2010)
			Kuenzer et al. (2007d)
			Prakash and Gupta (1998)
		Geomorphology and faults	Kuenzer and Stracher (2011)
			Yang et al. (2008)
Near infrared (NIR)	0.7–1.3 μm	Vegetation	Kuenzer and Voigt (2003)
			Gupta and Prakash (1998)
			Lambin and Ehrlich (1996)
		Soil moisture	Hummel et al. (2001)
		Gas emissions	Engle et al. (2012a,b)
			van Dijk et al. (2011)
			Kuenzer et al. (2007c)
			Gangopadhyay (2007)
Shortwave infrared (SWIR) through mid infrared (MIR)	1.3–3.0 μm through 3.0–8.0 μm	Soil composition	Ben-Dor (2002)
		Associate forest fires	Waigl et al. (2014)
			Prakash et al. (2011)
			Whitehouse and Mulyana, (2004)
		High-temperature surface coal fires	**Kuenzer et al. (2013)**
			Kuenzer et al. (2008b–d)
			Hecker et al. (2007)
			Zhang and Kuenzer (2007)
			Zhang J. et al. (2007)
			Tetzlaff (2004)
			Voigt et al. (2004)
			Zhang et al. (2004b)
			Prakash and Gupta (1999)
Thermal infrared (TIR)	8.0–14 μm	**Underground coal fires**	**Martha et al. (2010)**
			Kuenzer et al. (2008b–d)
			Kuenzer et al. (2007a)
			Hecker et al. (2007)
			Zhang and Kuenzer (2007)
			Zhang et al. (2007)
			Tetzlaff (2004)
			Cassells (1998)
			Prakash et al. (1997, 1999)
			Zhang X et al. (1997)
		Fire depths	Berthelote et al. (2008)
			Wessling et al. (2008a,b)
			Prakash and Berthelote (2007)
			Peng et al. (1997)
			Prakash et al. (1995b)
			Panigrahi et al. (1995)
			Mukherjee et al. (1991)
		Gas emissions	van Dijk et al. (2011)
			Kolker et al. (2009)

(Continued)

TABLE 22.1 (*Continued*) Regions of the Electromagnetic Spectrum, Commonly Associated Wavelength Ranges, and the Main Coal Mining and Coal Fire Applications Associate with These Regions

Region of Spectrum	Wavelength Range	Main Coal Mining and Coal Fire Applications	Example References
Microwave	0.75 cm–1 m	Faults	Bonforte et al. (2013)
			Gens (2009)
		Subsidence	Jiang et al. (2011)
			Voigt et al. (2004)
			Prakash et al. (2001) Genderen et al. (2000)

Source: Adapted and updated from Prakash, A., Coal fire research: Heading from remote sensing to remote measurement, in *Second International Conference on Coal Fire Research*, Berlin, Germany, May 19–21, 2010.

Note: Selected references for each application area are also listed.

granted permission in 2013 to live out their lives there. The condition in the Jharia coal mining area in India, and larger coal fire areas in China are similar but the situation is more complicated due to the high population density in the mining areas.

The obvious issue posed by fires is the loss of coal, a precious nonrenewable energy resource. The coal burns down to ash reducing its economic value to nothing. As the coal burns, mining the unburned coal around it becomes dangerous and sometimes practically impossible causing further economic loss. Burning coal produces oxides and dioxides of carbon, nitrogen, and sulfur, methane, toxic compounds, along with steam and particulate matter (Carras et al., 2009; Engle et al., 2012a,b; Gangopadhyay, 2007; Kolker et al., 2009) that contribute to atmospheric pollution. The environmental pollution due to coal fires affects not only the atmosphere but also the surrounding land and water resources (Kuenzer et al., 2007b). Locally dry patches of land devoid of healthy vegetation, referred to as barren aureoles (Gupta and Prakash, 1998), are common in some coal fire areas. Such patches show anomalous signatures in both optical and TIR images (Lambin and Ehrlich, 1996). A larger issue associated with underground coal fires is the land stability. As the coal burns underground and turns to ash, the material volume decreases leaving voids underground that ultimately cause the overlying land surface

to sink in and subside (Figure 22.2). Massive mining and fire-related subsidences are reported from many parts of the world (Genderen et al., 2000; Jiang et al., 2011; Kuenzer and Stracher, 2011; Prakash et al., 2001; Voigt et al., 2004) where buildings, transportation network, and mining equipment have been completely engulfed or have suffered massive damage due to subsidence and coal fires. Environmental pollution leads to serious health hazards (Finkelman, 2004; Hower et al., 2009; Stracher and Taylor, 2004) and a general socioeconomic decline in the affected areas.

Early detection and containment of a coal fire, before it becomes uncontrollable, is therefore critical. Remote sensing offers a powerful tool for ongoing mapping and monitoring of coal fires to effectively target and manage fire-fighting efforts, which vary depending on the severity of the fire, its proximity to infrastructure and communities, availability of resources, and funding for containment efforts. In principle, all fire-fighting efforts are targeted at cutting off access to one or more of the three factors (oxygen, heat, and coal) required to sustain a coal fire. The most common practice for putting out surface and near-surface coal fires is to simply isolate the burning area by trenching or building a retaining structure, and then douse the flames and excavate the remaining coal. For subsurface fires, it is more effective to cut off the oxygen supply by either completely blanketing the area with sand and loess

FIGURE 22.2 A sink hole in a coalfield in India, created due to sudden subsidence of land as subsurface coal fires consumed material. Heat emanating from the underground fires has baked the rocks and even turned some to ash (see grey region on one side of the sink hole). At the surface of this sink hole is approximately 3 m by 3 m in this area. (Photo Credit: Anupma Prakash, 2006.)

(a) (b)

FIGURE 22.3 Field photo of coal fires in the Jharia coal field, India. (a) A surface fire showing a burning coal seam. Flames, red hot rock, and ash are clearly visible. The width shown in this photograph is approximately 3 m. (b) A subsurface fire area that was blanketed with top soil in efforts to fill in the cracks and contain the fire. The cracks resurfaced soon. In this photo, smoke is seen emitting from these cracks. Note the huts and trees in the background for scale. (Photo Credit: Anupma Prakash, 2006.)

(e.g., Figure 22.3b). In other cases, prior to sand blanketing, the voids underground are also filled with different materials such as a sand-slurry mixture, swelling clays, fire-resistant colloidal mixtures, or by injecting water, inert gases or foam. In more sophisticated and effective fire-fighting efforts, the entire fires areas were flushed with liquid nitrogen to bring down the temperature of coal seams so that the fire would not reignite (Ray and Singh, 2007).

22.1.4 Attributes of Surface and Subsurface Coal Fires

Surface coal fires in coal mining areas are much easier to locate in the field than subsurface fires. Sometimes they are associated with flames (Figure 22.3a). At other times, they become visible when the surface material collapses exposing the coal to a fresh gust of wind that fans the hot surface and makes it glow. Some surface fires are associated with dense smoke whereas the others just smolder and emanate small amounts of smoke. Large surface fires can extend for hundreds of meters when a long stretch of a coal seam outcrop is on fire, as in some coalfields in India and China. Sometimes the coal outcrop is on fire at several different locations causing a string of surface fires. However, the most common occurrence is an individual instance of surface fire that may persistently burn for a long time or appear and disappear intermittently due to seasonal factors or fire-fighting efforts.

Detecting subsurface fires in the field is hard (Zhang and Kuenzer, 2007) because one has to rely on indirect indicators such as the presence of heated ground, smoke emitting vents (Figure 22.3b), dry and barren soils, local subsidence pockets, mineral occurrences in cracks (Stracher et al., 2005) and sometimes steam effusions after a rainfall event. Such surface manifestations are subtle and span a large spatial extent making field-based identification and delineation more complex.

Remote sensing techniques have proven to be increasingly successful in recent years in more reliably delineating regions affected by both surface and subsurface fires as described in the following section.

22.2 Remote Sensing of Coal Fires

Coal fires are high-temperature phenomena that lend well to remote sensing–based detection as fire areas emit higher amounts of energy due to elevated temperatures. Use of remote sensing technology starting from the early 1970s to current day is presented along with a discussion on the future use of this technology for coal fire studies.

22.2.1 History and Recent Evolution

It is no surprise that remote sensing–based fire detections date as far back as the development and use of passive sensors that measure Earths' emitted energy. Coal fires were mapped in the early

1970s using broadband TIR cameras mounted on aircrafts by Ellyett and Fleming (1974). Satellite-based TIR remote sensing lagged behind due to the coarse spatial resolution of the TIR sensors. Mansor et al. (1994) demonstrated the utility of even coarse 1.1 km spatial resolution images from a very high resolution radiometer (AVHRR) TIR channel to map large regions of coal fires in the eastern part of the Jharia Coalfield in India. Kuenzer et al. (2008b) and Hecker et al. (2007) also successfully used the moderate resolution imaging spectrometer (MODIS) TIR bands to map and monitor coal fires in the Jharia Coalfield in India. However, the use of AVHRR and MODIS data is limited only to regional investigations of a few very large coal fire areas. TIR bands of the Landsat thematic mapper (TM) and enhanced thematic mapper (ETM+), and the advanced spaceborne thermal emission spectrometer (ASTER) are by far the most widely used satellite data for coal fire investigations in the last two decades (see Table 22.1). Data from some experimental satellite missions, such as the bi-spectral infrared detection (BIRD) operated by the German Aerospace Agency (DLR), have also successfully demonstrated the potential of sensors operating in the mid-infrared wavelength ranges for mapping surface coal fires (Tetzlaff, 2004; Zhukov et al., 2006).

22.2.2 Spatial and Temporal Resolution

The spatial and temporal resolution of popular remote sensing systems used for mapping and monitoring coal fires is shown in Figure 22.4. Earth-observing satellites such as Landsat and ASTER are particularly useful for coal fire studies as their spatial resolution is commensurate with the spatial extent of coal fires that range typically from a few meters to hundreds of meters. Their temporal resolution of 16–18 days also works well for long-term monitoring of large underground coal fires. Coal fires are not as dynamic as forest fires. Though some small surface coal fires can move fast in the orders of meters per day, most

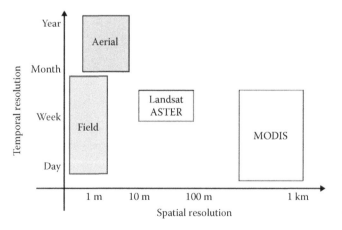

FIGURE 22.4 Spatial and temporal resolution of popular coal fire mapping and monitoring systems. Commonly used satellite images are shown in yellow boxes. The spatial and temporal resolutions of aerial and field surveys are not fixed as shown in this figure, but are dependent on resource availability.

subsurface coal fires migrate at a much slower rate with perceptible change on satellite images occurring over several months to years.

As indicated in Figure 22.4, the higher spatial resolution airborne TIR systems are ideal for detailed mapping and investigation of coal fires on a one-time basis, or in some cases for monitoring coal fires, contingent on whether the mining agency can afford to acquire frequent airborne coverage. The coarser resolution satellite images such as AVHRR and MODIS have limited potential and are useful only for regional scale studies with little practical use in decision support for targeted fire-fighting.

22.2.3 Spectral Resolution

Different spectral regions have different significance for remote sensing of coal fires and related phenomena (Table 22.1). As mentioned in Section 22.1.1, here we focus only on the bold fonts of the table that deals with characterizing the surface and subsurface coal fires. The SWIR region (1.3–3 µm) and the TIR region (8–14 µm) of the electromagnetic (EM) spectrum are by far the most important regions for remote sensing of coal fires as they allow for quantitative estimates of temperatures of underground and surface coal fires. The most popularly used satellites for coal fire studies that have sensors operating in these spectral regions are ASTER and the Landsat series of satellites. Table 22.2 provides details of the spatial and spectral resolution of the sensors operating on these satellites that are conducive for coal fire research.

To detect coal fires, we rely on the fact that fires have elevated temperatures and these can be derived from satellite images acquired in the TIR and SWIR regions. The principle and techniques of temperature estimation from remote sensing data are covered in detail in literature (e.g., Chander et al., 2009; Gupta, 2003; Kuenzer, 2015; Kuenzer and Dech, 2013; Quattrochi et al., 2009). Coal fire temperature estimation is based on the principle of inversing Planck's function to change an image-derived spectral radiance to temperature values (e.g., Kuenzer et al., 2007a; Prakash and Gupta, 1999; Prakash et al., 1995a; Saraf et al., 1995; Zhang et al., 1997). Figure 22.5 shows the typical emission curves of blackbodies at various temperatures as determined by Planck's function. The locations of the Landsat and ASTER spectral bands are also plotted on this graph to give an idea of the temperature sensitivity ranges of these spectral bands.

The key message from Figure 22.5 and previous studies is that data from the broadband TIR region (approximately 8–14 µm) are better suited for deriving surface temperatures associated with underground fires, whereas the data from SWIR regions (1.3–3 µm) are better suited for estimating temperature of surface fires and open flames.

22.3 Methods

Methods used for remote sensing of coal fires rely primarily on the accurate land surface temperature (LST) estimations. They can be divided into methods for (1) atmospheric correction

TABLE 22.2 Characteristics of Important Satellite Sensors and Their Applicability for Coal Fire Research

		Landsat 4/5 TM	Landsat 7 ETM+	Landsat 8 OLI & TIRS	ASTER	Application in Coal Fire Studies
Spectral Resolution [μm]	**VIS/NIR**	1: 0.45–0.52	1: 0.45–0.52	1: 0.43–0.45	1: 0.52–0.60	Primarily used for land use land cover mapping. Aster 3a (nadir) and 3b (backward looking) also used for elevation mapping
				2: 0.45–0.51	2: 0.63–0.69	
		2: 0.52–0.60	2: 0.52–0.60	3: 0.53–0.59	3a: 0.78–0.86	
				4: 0.64–0.67	3b: 0.78–0.86	
		3: 0.63–0.69	3: 0.63–0.69	5: 0.85–0.88		
		4: 0.76–0.90	4: 0.77–0.90			
	SWIR	5: 1.55–1.75	5: 1.55–1.75	6: 1.57–1.65	4: 1.600–1.700	Mapping and monitoring of high-temperature surface fires; subpixel temperature and area analysis
		7: 2.08–2.35	7: 2.09–2.35	7: 2.11–2.29	5: 2.145–2.185	
					6: 2.185–2.225	
					7: 2.235–2.285	
					8: 2.295–2.365	
					9: 2.360–2.430	
	TIR	6: 10.40–12.50	7: 10.40–12.50	10: 10.60–11.19	10: 8.125–8.475	Mapping/monitoring subsurface fires; temperature estimation; emissivity (ASTER only)
				11: 11.50–12.51	11: 8.475–8.825	
					12: 8.925–9.275	
					13: 10.25–10.95	
					14: 10.95–11.65	
	PAN		8: 0.52–0.90	8: 0.50–0.68		As high spatial resolution base image
				9: 1.36–1.38 (Cirrus band—not PAN)		
Spatial Resolution	**PAN**		15 m × 15 m	15 m × 15 m	15 m × 15 m	Thermal band of Landsat 7 is most useful for coal fire studies due to its superior spatial resolution
	VIS	30 m × 30 m	30 m × 30 m	30 m × 30 m	30 m × 30 m	
	TIR	120 m × 120 m	60 m × 60 m	100 m × 100 m	90 m × 90 m	
Temporal Resolution		16 days	16 days	16 days	16 days	Used for long-term monitoring of slow fires
Swath Width		185 km	185 km	185 km	60 km	No special implication for coal fire studies
Quantization		8 bit	8 bit	12 bit	TIR: 12 bit other: 8 bit	Landsat 8 TIRS and ASTER TIR bands have a better signal-to-noise ratio compared to earlier missions
Equatorial Overpass		10:00 a.m. ±15 min	10:00 a.m. ±15 min	10:00 a.m. ±15 min	10:30 a.m.	All capable of on-demand nighttime acquisition for TIR data that is better for coal fire studies

and emissivity compensation, (2) LST estimation, (3) thresholding for fire delineation, and (4) time series analysis for fire monitoring.

22.3.1 Atmospheric Correction and Land Surface Emissivity Variations

An important consideration in accurate temperature retrievals from remote sensing data is accounting for atmospheric correction and land surface emissivity variations (Gens and Cristobal, 2015).

The emitted spectral radiance measured at the sensor (L_s) is different from the spectral radiance emitted from the ground (L_g), as it includes the path radiance (L_p) or the upwelling spectral radiance emitted from the atmospheric particles. The signal reaching the sensor is also attenuated in complex ways that can be corrected by accounting for atmospheric

transmissivity (τ). The ground emitted spectral radiance can be computed using Equation 22.1:

$$L_{g\lambda} = \frac{L_{s\lambda} - L_{p\lambda}}{\tau_\lambda} \qquad (22.1)$$

where

λ is the central wavelength (in μm) of the spectral band in consideration

L_g, L_s, and L_p are all measured in W m^{-2} μm^{-1} sr^{-1}

τ is unitless

Water vapor and aerosols (smoke, dust, haze—all very common in coal fire areas) are the main atmospheric components influencing the at-sensor spectral radiance, and therefore are the most important inputs in most radiative transfer models, such as moderate resolution atmospheric transmission

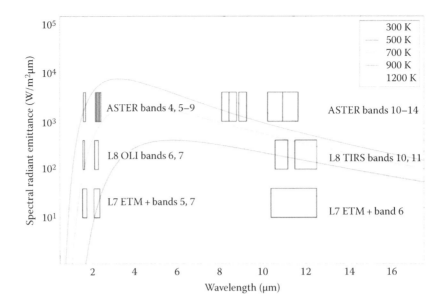

FIGURE 22.5 Positions of the SWIR and TIR bands of the Aster, Landsat 7, and Landsat 8 satellites, superimposed on Planck's emission curves for blackbodies at different temperatures. Note that only the spectral bandwidths are indicated and the saturation temperatures of individual spectral bands are not reflected in this figure. Generally the SWIR bands are used for surface fire mapping and the TIR bands are used for subsurface fire mapping.

(MODTRAN), used in atmospheric corrections of TIR data (Cristóbal et al., 2009; Jiménez-Muñoz, 2014; Schott et al., 2012). The water vapor values are typically either taken from field measurements where available, or from higher order image products from other satellites, for example, MODIS (MOD05 water vapor), and AVHRR (Sobrino et al., 1999). Transmissivity is estimated from general conditions of visibility reported from field sites. Commercial software packages such as atmospheric correction and haze reduction (ATCOR) and fast line-of-sight atmospheric analysis of hypercubes (FLAASH) provide a graphical user interface to input these variables (or use pre-set standard atmospheric conditions based on general parameters) and run a MODTRAN or MODTRAN-like atmospheric correction model in the background to atmospherically correct TIR data and retrieve spectral radiance that can be used further for temperature retrievals (see Section 22.3.2).

Emissivity is the ratio of the spectral radiance emitted by a target at a given temperature and the spectral radiance emitted by a blackbody at the same the same temperature. Being a ratio, emissivity has no unit. The emissivity of a blackbody (an ideal energy absorber) is 1 and that of a white body (an ideal reflector) is 0. All natural objects typically have an emissivity value between 0 and 1. Several approaches have been taken into account in coal fire studies to assign target emissivity values. These include (1) using a constant, uniform and high emissivity value usually ranging from 9.5 to 9.8; (2) assigning each land-cover class a general published emissivity value; (3) using NDVI to determine emissivity; or (4) using image-based emissivity products, such as the ASTER emissivity product.

Methods to retrieve emissivity from remote sensing images are well documented in literature (Gillespie et al., 1998; Hook, 1992; Hulley et al., 2012; Martha et al., 2010; Schmugge et al., 1998; Sobrino et al., 2008).

22.3.2 Fire Temperature Estimation

Spectral radiance can be related to LST using the Planck's function $B_{\lambda T}$:

$$L_{\lambda T} = \varepsilon_\lambda B_{\lambda T} = \varepsilon_\lambda \frac{C_1 \lambda^{-5}}{\pi \left[\exp\left(C_2 / \lambda T \right) - 1 \right]} \tag{22.2}$$

where

$L_{\lambda T}$ is the spectral emitted radiance in W m^{-2} μm^{-1} sr^{-1} at wavelength λ (μm) and temperature T (K)

ε_λ is the spectral emissivity (dimensionless)

$C_1 = 2\pi hc^2$ = a constant value of 3.7418×10^{-16}

$C_2 = hc/k$ = a constant value of 1.4388×10^{-2} mK

Other values used to derive C_1 and C_2 are Planck's constant (h = 6.626076×10^{-34} Js), speed of light (c = 2.998×10^8 m s^{-1}), and Boltzmann constant (k = 1.3806×10^{-23} J K^{-1}).

Equation 22.2 can be inverted to directly derive LST (T):

$$T = \frac{C_2}{\lambda \ln \left[\varepsilon_\lambda C_1 / \pi \lambda^5 L + 1 \right]} \tag{22.3}$$

To find warm areas associated with subsurface coal fires, TIR images acquired in the 8–14 μm are very useful. However, the temperature of the exposed coal fires and their flaming fronts may rise well over 1000°C saturating the 8–14 μm wavelength bands that cannot be used for temperature retrievals. This makes it necessary to rely on shorter 1.3–3 μm wavelength infrared channels for temperature retrievals.

Many surface fires are also smaller in size than the spatial resolution of the sensor that is being used to study them. In such cases, dual channel processing techniques need to be applied to retrieve subpixel area and subpixel temperature estimates of surface fires (Matson and Dozier, 1981; Prakash and Gupta, 1999; Tetzlaff, 2004). The main limitation in using Landsat SWIR images for subpixel level investigation is the limited availability of spectral bands in the SWIR region.

22.3.3 Thresholding to Delineate Fire Area

Once the LST have been computed for an image, it is important to set a threshold value (a temperature cut-off value) to delineate fire areas from cooler nonfire background areas. Threshold determination is an important step and researchers have approached it differently. An overview of some of the common thresholding techniques is presented by Prakash and Gens (2010), Kuenzer et al. (2013), and Raju et al. (2013).

The most common practice is to determine the threshold by trial-and-error guided by field knowledge (Gangopadhyay et al., 2005; Prakash et al., 1995a). The success of this practice stems from the fact that it works well locally for areas that are well studied in the field, as is the case with many important coal mining areas. Some researchers have simply used a fixed image standard deviation as a threshold criterion, for example, a threshold of two was deemed to be good for delineating fires in the Jharia coalfield in India by Prakash et al. (1999). However, more recent studies have shown that such low standard delineations may not work well in all conditions. Waigl et al. (2014) demonstrated that in automated processing of large scenes from high-latitude regions that show a great variation in LSTs, a standard deviation of four is more suitable for delineating hot spots related to near surface coal fires. A recent study Raju et al. (2013) determined that the highest spectral radiance that can be attributed to solar reflection on Landsat band-7 SWIR image can serve as a conservative threshold to segregate the surface fire pixels from nonfire pixels.

Other techniques rely on the derivation of scene-dependent relative thresholds based on the image histogram, for example, using the change in slope of the histogram of a small image window around a fire (Prakash and Vekerdy, 2004; Rosema et al., 1999; Zhang et al., 2004a), or the first local minimum after the main maximum digital value during several iterations of a moving window kernel (Kuenzer et al., 2007a), as a good indicator to separate fire and nonfire areas. This latter automatic method enables to detect thermal anomalies of different temperatures within an image: an anomalous cluster of 65°C in a surrounding of 50°C heated bedrock will likewise be detected, just as a cluster of 70°C in a surrounding background of 60°C heated bedrock. This method pays attribute to the fact that coal fire–related anomalies are (e.g., compared to forest fires) extremely subtle anomalies, which are hard to detect against the background. A dark shale surface or a coal dust covered surface can easily reach 60°C and more at 10:30 during local Landsat overpass time. Thus, natural surfaces heated by solar radiation easily mimic coal fire–related anomalies. Simple thresholding with a temperature threshold in a full thermal satellite scene will thus not only lead to the extraction of coal fire–related anomalies, but also to the extraction of heated bedrock surfaces, sun-exposed slopes, and other thermal anomalies. So trial-and-error-based thresholding only helps to delineate coal fire areas that are already known well from field survey. However, the main purpose of remote sensing should be to detect phenomena in hard to access and large area where the coal fire locations are not known. This challenge could be addressed in China (Kuenzer et al., 2007a) and in a coal fire in the boreal forest in Alaska (Prakash et al., 2011), where unknown coal fires were first detected in remote sensing imagery and later validated in the field.

22.3.4 Time Series Analysis for Fire Monitoring

Time series analysis implies analysis of many images acquired at different times over the same study area. One purpose of carrying out a time series analysis is to determine whether the thermal anomaly detected by thresholding TIR images is transient or persistent. Transient anomaly is one that appears infrequently on a TIR image of 1 day or a few days, but then is not detectable on other days. This may happen due to some unusual heating locally at a particular time or due to local weather conditions, especially fluctuating wind speeds, rain events, etc. On the other hand, persistent thermal anomalies tend to appear again and again at the same spot in a temporal stack of processed images. Waigl et al. (2014) classified a time series of 40 Landsat scenes and defined a thermal anomaly as persistent if it appeared in more than one-third of the scenes in which the target pixel was present (in some cases, the pixel was masked out due to cloud cover).

The second and more important purpose of a time series analysis is to monitor the movement or progression of a coal fire over a longer period of time based on multitemporal Earth observation data. Early efforts in coal fire monitoring were reported by Mansor et al. (1994), Saraf et al. (1995), and Prakash et al. (1999) with publications following from Zhang et al. (2004a), Hecker et al. (2007), Chatterjee et al. (2007), and Kuenzer et al. (2008a–d), all employing multitemporal data sets of daytime and night time data to extract fire-related anomalies based on manual-, semiautomated, and automated methods. The monitoring activities, among others, aimed to determine if fires shrink or increase in size, estimate how large is variability of the fire-related thermal anomaly, and monitor if extinguished fires remain extinguished.

22.4 Selected Results

To map the locations of surface and subsurface coal fires in a selected part of the Jharia coalfield, India, we used a recent Landsat image acquired in April 2014, and processed it using the approach outlined by Prakash et al. (1997). It is important to remember that in coal mining areas it is common to find instances where coal fires exist only at the surface, or only in the subsurface, or coexist on the surface and subsurface due to the geologic setting and fire history. In the subsurface, one or more coal fires may occur in one seam or in different coal seams at different depths. Figure 22.6 shows the processing result.

Figure 22.6a is a false color composite generated by coding the Landsat 8 spectral band 7 (SWIR 2), band 6 (SWIR 1), and band 4 (Red) in red green and blue, respectively. On this figure, surface fires that emit significantly in the SWIR regions stand out as yellow and red pixels. Figure 22.6b is a density-sliced color-coded TIR image (Landsat 8, band 10) of the same area. On this image red, orange and yellow represent successively warmer LST–associated with subsurface fires. The grey-scale image in the background shows variation in the background

LSTs, with darkest grey tones associated with the cooler temperatures of the Damodar River. Comparing Figure 22.6a with 22.6b, we clearly see a situation marked with a yellow arrow, where there is a large temperature anomaly likely due to a subsurface fire but there is no obvious surface fire present. In contrast, there is a small surface fire south of the river (marked with a red arrow on Figure 22.6a), which shows no corresponding thermal anomaly on the TIR image, possibly due to the fact that the thermal signature disappears when averaged over the coarser TIR pixels and because there is no anomalous heating from a coexisting subsurface fire. The blue and green arrows show more complex situations of coexisting fires at the same location or where a subsurface fire is shifted but in proximity to the surface fires, a situation comparable to the conceptual coal fire scenario shown in Figure 22.1 in which a coal seam is on fire where it is exposed at the surface, as well as at a location where it is buried deeper down.

The second example (Figure 22.7) shows results of a coal fire monitoring study in the Wuda coal field in China that was based on a combination of in situ mapping, and processing a Landsat 7 image acquired on December 10, 2002. The right panel shows

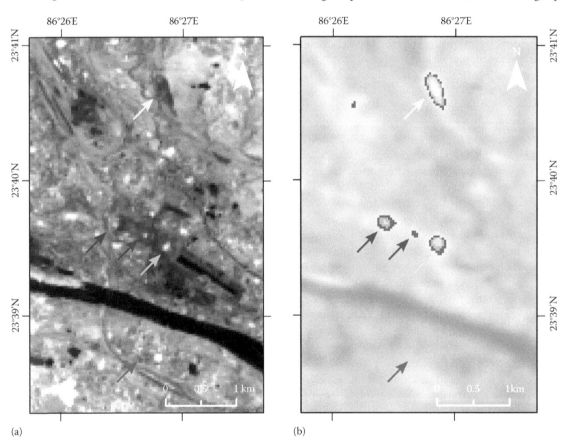

(a)   (b)

FIGURE 22.6   A comparative analysis of locations of surface and subsurface coal fires in the southeastern part of the Jharia coal field in India, using Landsat 8 satellite images from April 2014. (a) A false color composite generated by coding the SWIR bands in red and green colors so that high temperature surface fires show in shades of yellow and red. (b) A density-sliced color-coded TIR image of the same date showing successively higher LST due to subsurface fires in red (36°C–37°C), orange (37°C–39°C), and yellow (39°C–46°C). Fires may occur only on the surface (red arrow), only in the subsurface (yellow arrow), or both on the surface and underground at the same location (blue arrows) or adjacent location (green arrow).

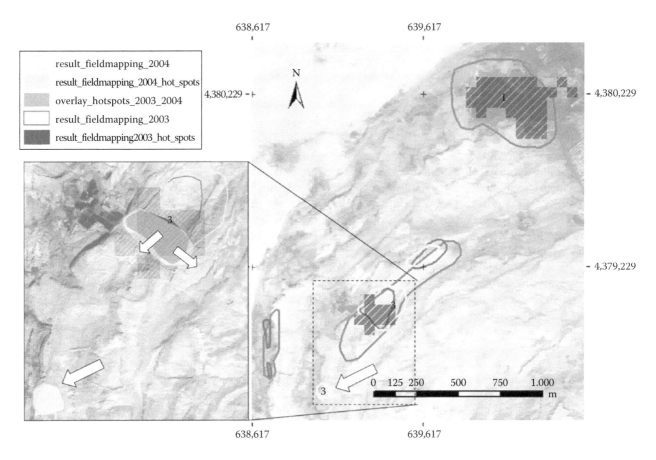

FIGURE 22.7 Coal fire monitoring for part of the Wuda coal field, China. A panchromatic QuickBird image from October 5, 2003, is used as a background. Coal fire areas mapped in situ in 2000, 2003, and 2004 are shown as orange, blue, and yellow outlined polygons, whereas the hottest areas are depicted as solid-filled polygons. The red striped areas are coal fire extents extracted from the TIR band of Landsat ETM+ acquired on December 10, 2002. Only the hottest fire areas could be detected and extracted from the remote sensing images.

the fire zones number 1, 2, and 3 in the Wuda coal fire area and the left panel shows an enlarged view of fire zone 3.

For in situ mapping, the field crew traversed the terrain in a predefined grid pattern, logged the GPS track, and recorded the LST using contact thermometers and handheld radiometers. This helped to map the surface thermal expression, or footprint, of the underground fire (Kuenzer et al., 2005). In situ mapping was carried out during 2000, 2003, and 2004 and fire boundaries from these surveys are shown as orange, blue, and yellow polygons, respectively. The boundaries of fires zones 1, 2, and 3 are clear for the 2000 survey that was carried out by mining engineers. Fire 1–related anomalies could not be found in 2003 and 2004 field campaigns. Fire 2 was still active in 2003, but had burned out by 2004. Fire 3 had formed new high-temperature clusters in 2004.

The 2002 Landsat scene was processed using the method presented by Kuenzer et al. (2007a), where an unbiased automatic moving window approach was implemented to extract regional thermal anomalies. As opposed to using a single predefined threshold for the entire image to delineate fire area, this technique defined variable region-specific threshold values, picking up *relative* hot spots. This implied that 50°C pixel in a 40°C background was treated the same as a 60°C pixel in a 50°C

surrounding, and the algorithm defined both cases as a thermal anomaly. This processing technique was successful in extracting only the hottest areas of fires 1 and 3. Large parts of fire 3 and the entire area of fire 2 remained undetected.

Monitoring studies, as exemplified earlier, are very important. So far no frequent long-term monitoring study has been published to our knowledge. The most extensive monitoring study published has been based on repeated in situ field mappings of the coal fires of Wuda, China (Kuenzer et al., 2008c) depicting that fires can move for several tenths to hundreds of meters per year, and that coal fire dynamics are quite accentuated. This publication also highlighted that remote sensing–based analyses miss many coal fire–related hot spots in a coal fire area, and can only detect the hottest and largest fires.

22.5 Discussion

Efforts in remote sensing–based coal fire mapping and monitoring have steadily grown since the mid-1990s. However, compared to the research community using remote sensing for studying other high-temperature events, such as volcanic eruptions and wild fires, the coal fire research community is relatively small and transient, in part due to the sporadic nature of funding in this area. Based on the body of existing research and our experience in coal fire studies, we outline the issues and limitations of current research, discuss future needs, and make recommendations for remote sensing of coal fires. The following sections are centered on some of the main issues.

22.5.1 Coal Fire Research Is Local in Nature

In the past decades, most scientists focused on specific coal fire localities and consequently the research publications have a very local, or at best regional, focus. A global, standardized approach to monitor coal fires does not exist to date. While it is true that each coal fire/coal fire area is different, with variability due to locality, climate, weather, bedrock, soils, type of coal, depth of the fire, mining methods, etc., it is important to establish a generally accepted, standardized approach to map and monitor the fires to generate comparable and continuous results. Just as there are established standard algorithms for atmospheric correction, or multispectral image classification, or time series analyses, it would be most useful to have a "coal fire toolbox" with a set of standardized algorithms and tools to analyze or monitor a coal fire area based on a certain type of thermal data. More refined remote sensing tools for estimating depth of underground coal fires, fire heat flux (e.g., Fischer et al., 2011; Haselwimmer et al., 2013), amount of coal burned, and material requirement for fire containment need to be developed and tested. This will require a larger dedicated team of researchers for developing and testing algorithms to work for varying conditions, and standardizing them as an operational tool-set for the global coal fire community.

Currently, the biggest growth in the number of coal fire remote sensing scientists is in China as this country is one that is most seriously impacted by coal fires. The challenge here is that much of their research is published only in Chinese and thus not accessible to researchers outside of China. However, it is likely that future coal fire research will be driven by the fleet of young researchers there who are investigating further in this area.

22.5.2 Coal Fires Are Dynamic and Baseline Data Does Not Exist

Coal fire–related parameters (such as the occurrence of thermal anomalies, the size of thermal anomalies, the temperature of anomalies, amount of emissions, etc.) fluctuate a lot with local climate, weather, and other influences. To get a reliable estimate of these parameters and quantify the average energy released (that helps to estimate the amount of coal burned) requires analysis of a large volume of thermal data. Most published coal fire research is based on the analysis of only a few images, or at best four thermal images in a given year. Therefore, a comprehensive baseline data for coal fires, comparable to the baseline data available for volcanoes, does not exist. This is a much required gap. It will be important to see future coal fire remote sensing researchers be involved in processing large volumes of thermal data (e.g., all available cloud-free Landsat and ASTER

TIR scenes for targeted coal fire areas), to establish and grow the baseline data on coal fires.

22.5.3 Timing of Data Acquisition Is Important

Coal fire monitoring literature indicates that there can be significant diurnal temperature variability of natural bedrock surfaces with daytime temperature easily mimicking coal fire–related anomalies. To limit the influence of solar heating, nighttime data is better suited for coal fire analyses. The *nighttime* thermal data of Landsat is not truly nighttime as the overpass time is around 10:00 p.m., and it still contains effects of uneven solar heating during the day. Additionally, as there is very limited availability of nighttime TIR images, most studies rely on daytime TIR images, but minimize the effect of solar heating by empirical methods or by using energy budget models (Deng et al., 2001; Haselwimmer and Prakash, 2013).

Predawn is best for thermal anomaly extraction data. MODIS has predawn acquisitions (e.g., images acquired around 4:00 a.m.), which are ideally timed but the images lack sufficient spatial resolution to be useful.

For coal fire studies, summer time data should be avoided. Snow-free winter data, or data from spring or fall are best suited, as here the contrast between the anomaly and the background temperatures will be most accentuated.

22.5.4 Spaceborne TIR Imagery Is Coarse and Discontinuous

The largest gap in coal fire research is indirectly a result of limited data continuity and changing sensor characteristics in the TIR region. While the free Landsat archive now enables optical data access for most spots on the Earth over the past 40 years, the situation is complicated in the thermal domain. Landsat MSS did not have a thermal band, so Landsat thermal data exist only since the launch of Landsat 4 in 1982. However, the thermal band of Landsat 4 and 5 have a coarse spatial resolution of 120 m, which only enables to pick up thermal anomalies induced by extremely hot and large surface or subsurface fires. Even gaseous emissions at 500° temperature are insufficient to make a detectable signature on the coarse 120 m thermal pixel, if such emissions come from vents that are only in the order of centimeters in magnitude.

The coal fire research experienced a strong push forward with the launch of Landsat ETM+ in 2001, which had a low gain and a high gain thermal band at 60 m spatial resolution. Unfortunately, Landsat 7 ETM+ experienced a scanner problem in May of 2003, and since then hardly any useful thermal data has been available. This was one reason that publications in the late 2000s focused on analyses of data acquired before May 2003, or exploited the potential of ASTER or other low resolution thermal data, such as MODIS thermal data. Although ASTER had 5 thermal bands enabling emissivity retrievals, it proved to be inferior to Landsat 7 ETM+, for coal fire research due to its coarser 90 m spatial resolution. Unfortunately, ASTER also encountered the failure of its SWIR bands in 2009 and had limited data availability that posed challenges in data analysis. Thus, little coal fire–related research has been published based on thermal data newer than 2009. The launch of Landsat 8 in February 2013 gave new hope to the coal fire community. However, unlike ETM+ at 60 m resolution, this sensor now has a thermal band at only 100 m spatial resolution—a step backward for coal fire–related research.

It is anticipated that the next years will bring out some repeated local studies in well-known coal fire areas, such as Jharia in India, Wuda in China, or selected fires in other localities based on such Landsat-8, as well as selected China–Brazil Earth Resources Satellite (CBERS) thermal data (similar resolution): however, ground-breaking developments will require a sustained availability of higher spatial resolution thermal data. The optimal sensor for coal fire–related thermal monitoring would have a band or two in the SWIR region and at the least a band in the TIR region with a spatial resolution of 45 m or better. It would have a repeat cycle of less than 15 days with a possibility of on-demand nighttime acquisitions. Furthermore, data would need to be free of charge to facilitate research and grow the community of scientists working in this important area.

In the absence of a reliable source of higher resolution spaceborne TIR imagery, in situ field mapping along with airborne thermal campaigns are the methods of choice to extract and deliver information for a comprehensive and stakeholder-relevant assessment of coal fires in a region.

22.6 Conclusions

Coal fires occur on the surface and in the subsurface in coal mines across the globe. They present a threat to the environment and also an opportunity in remote sensing research to characterize, map, and monitor these fires with the intent to support mining operations, firefighters, and local communities. Based on review of past research on remote sensing of coal fires, and near future plans of space agencies for new satellite launches, we conclude that

- Remote sensing of coal fires is an area of research that has witnessed slow progress due sporadic and limited funding, but has potential to grow in the era of heightened environmental concerns.
- Coarse spatial resolution images from AVHRR and MODIS type sensors are not useful for providing meaningful fire-specific information. Most research has relied on the use TIR and SWIR images from Landsat and ASTER, which are good for studying the larger and known fires, but often miss small individual fires.
- Predawn cloud-free and snow-free airborne images acquired in the nonsummer seasons, coupled with concurrent field investigations are ideal for coal fire studies.

- More research in remote sensing of coal fires will help to test and improve algorithms for fire detection and characterization, which could then be standardized for wide-scale implementation and global coal fire database generation.
- There is a pressing need in coal fire research for a global spaceborne TIR mission that could acquire TIR images at better than 45 m spatial resolution with at least a repeat coverage of twice in a given month.

Acknowledgments

We thank Derek Starkenburg, Dr. Rudiger Gens, and Chris Waigl for their assistance in generating Figures 22.1, 22.4, and 22.5, respectively.

References

Banerjee, S.C., 1985, *Spontaneous Combustion of Coal and Mine Fires*. A.A. Balkema, Rotterdam, The Netherlands, 168pp.

BCCL, 2014, Bharat Coking Coal Limited, Dhanbad. A subsidiary of Coal India Limited. Official Website: http://www.bccl.co.in. (Accessed October 2014.)

Ben-Dor, E., 2002, Quantitative remote sensing of soil properties. *Advances in Agronomy*, 75, 173–231.

Berthelote, A.R., Prakash, A., and Dehn, J., 2008, An empirical function to estimate the depths of linear hot sources: Applied to the Kuhio Lava tube, Hawaii. *Bulletin of Volcanology*, 70(7), 813–824.

Bonforte, A., Federico, C., Giammanco, S., Guglielmino, F., Liuzzo, M., and Neri, M., 2013, Soil gases and SAR measurements reveal hidden faults on the sliding flank of Mt. Etna (Italy). *Journal of Volcanology and Geothermal Research*, 251, 27–40.

Carras, J.N., Day, S.J., Saghafi, A, and Williams, D.J., 2009, Greenhouse gas emissions from low-temperature oxidation and spontaneous combustion at open-cut coal mines in Australia. *International Journal of Coal Geology*, 78(2), 161–168.

Cassells, C.J.S., 1998, Thermal modelling of underground coal fires in Northwest China, University of Dundee, Dundee, U.K. (ITC publication 51).

Chander, G., Markham, B.L. and Helder, D.L., 2009, Summary of current radiometric calibration coefficients for Landsat MSS, TM, ETM+, and EO-1 ALI sensors. *Remote Sensing of Environment*, 113, 893–903.

Chatterjee, R.S., Shah, A., Raju, E.V.R., Lakhera, R.C., and Dadhwal, V.K., 2007, Dynamics of coal fire in Jharia Coalfield, Jharkhand, India during the 1990s as observed from space. *Current Science*, 92(1), 61–68.

Chen, H., Qi, H., Long, H., and Zhang, M., 2012, Research on 10-year tendency of China coal mine accidents and the characteristics of human factors. *Safety Science*, 50(4), 745–750.

Cristóbal, J., Jiménez-Muñoz, J.C., Sobrino, J.A., Ninyerola, M., and Pons, X., 2009, Improvements in land surface temperature retrieval from the landsat series thermal band using water vapour and air temperature. *Journal of Geophysical Research*, 114(D08103), 1–16.

Deng, W., Wan, Y.Q., and Zhao, R.C., 2001, Detecting coal fires with a neural network to reduce the effect of solar radiation on Landsat Thematic Mapper thermal infrared images. *International Journal of Remote Sensing*, 22, 933–944.

Ellyett, C.D. and Fleming, A.W., 1974, Thermal infrared imagery of the Burning Mountain coal fire. *Remote Sensing of Environment*, 3, 79–86.

Engle, M.A., Radke, L.F, Heffern, E.L., O'Keefe, J., Hower, J.C., Smeltzer, C.D., Hower, J.M. et al., 2012a, Gas emissions, minerals, and tars associated with three coal fires, Powder River Basin, USA. *Science of Total Environment*, 420, 146–59.

Engle, M.A., Radke, L.F, Heffern E.L., O'Keefe, J., Smeltzer, C.D., Hower, J.C., Hower, J.M. et al., 2012b, Quantifying greenhouse gas emissions from coal fires using airborne and ground-based methods. *International Journal of Coal Geology*, 88 (2–3), 147–151.

Finkelman, R.B., 2004, Potential health impacts of burning coal beds and waste banks. *International Journal of Coal Geology*, 59(1–2), 19–24.

Fischer, C., Li, J., Ehrler, C., and Wu, J., 2011, Radiative energy release quantification of subsurface coal fires. In *Proceedings of the ISPRS Conference*, Munich, Germany. Available at http://www.isprs.org/proceedings/2011/isrse-34/211104015Final00345.pdf. (Accessed April 2015.)

Gangopadhyay, P.K., 2007, Application of remote sensing in coal-fire studies and coalfire related emissions. *Reviews in Engineering Geology*, 18, 239–248.

Gangopadhyay, P.K., Maathuis, B., and van Dijk, P., 2005, ASTER derived emissivity and coal fire related surface temperature anomaly: A case study in Wuda, North China. *International Journal of Remote Sensing*, 24, 5555–5751.

Gens, R., 2009, Spectral information content of remote sensing imagery. In Li et al. (eds.), *Geospatial Technology for Earth Observation*, Springer, New York, Dordrecht, Heidelberg, London. 558p.

Gens, R., 2013, Global distribution of coal and peat fires. In Stracher et al. (eds.), *Interactive World Map of Coal and Peat Fires Made To Accompany, Coal and Peat Fires: A Global Perspective, Volume 2, Photographs and Multimedia Tours*. Elsevier, Amsterdam, Netherlands.. Available at http://booksite.elsevier.com/brochures/coalpeatfires/interactive-map.html. (Accessed April, 2015.)

Gens, R., and Cristobal, J.R., 2015, Remote sensing data normalization. Chapter 5 this book.

Gillespie, A., Rokugawa, S., Matsunaga, T., Cothern, J.S., Hook, S., and Kahle, A.B., 1998, A temperature and emissivity separation algorithm for advanced spaceborne thermal emission and reflection radiometer (ASTER) images. *IEEE Transactions on Geoscience and Remote Sensing*, 36(4), 1113–1126.

Glover, S.S., 2011, *Coal Mining in Jefferson County*. Arcadia Publishing, Charleston, South Carolina, 128p.

Gupta, R.P., 2003, *Remote Sensing Geology*, 2nd edn. Springer, Berlin, Heidelberg, New York, 656p.

Gupta, R.P. and Prakash, A., 1998, Reflection aureoles associated with thermal anomalies due to subsurface mine fires in the Jharia Coalfield, India. *International Journal of Remote Sensing*, 19(14), 2619–2622.

Haselwimmer, C. and Prakash, A., 2013, Chapter 22—Thermal infrared remote sensing of geothermal systems, pp. 453–474. In Kuenzer, C. and Dech, S. (eds.), *Thermal Infrared Remote Sensing: Sensors, Methods, Applications*, 554p. Springer, Dordrecht, Heidelberg, New York, London.

Haselwimmer, C., Prakash, A., and Holdmann, G., 2013, Quantifying the heat flux and outflow rate of hot springs using airborne thermal imagery: Case study from Pilgrim Hot Springs, Alaska. *Remote Sensing of Environment*, 136, 37–46.

Hecker, C., Kuenzer, C., and Zhang, J., 2007, Remote sensing based coal fire detection with low resolution MODIS data. *Reviews in Engineering Geology*, 18, 229–239.

Hook, S.J., 1992, A comparison of techniques for extracting emissivity information from thermal infrared data for geologic studies. *Remote Sensing of Environment*, 42(2), 123–135.

Hower, J.C., Henke, K., O'Keefe, J.M.K., Engle, M.A., Blake, D.R., and Stracher, G.B., 2009, The Tiptop coal-mine fire, Kentucky: Preliminary investigation of the measurement of mercury and other hazardous gases from coal-fire gas vents. *International Journal of Coal Geology*, 80(1), 63–67.

Hulley, G.C., Hughes, C.G., and Hook, S.J., 2012, Quantifying uncertainties in land surface temperature and emissivity retrievals from ASTER and MODIS thermal infrared data, *Journal of Geophysical Research*, 117, D23113, doi:10.1029/2012JD018506.

Hummel, J.W., Suddutha, K.A., and Hollingerb, S.E., 2001, Soil moisture and organic matter prediction of surface and subsurface soils using an NIR soil sensor. *Computers and Electronics in Agriculture*, 32(2), 149–165.

Jiang, L., Lin, H., Ma, J., Kong, B., and Wang, Y, 2011, Potential of small-baseline SAR interferometry for monitoring land subsidence related to underground coal fires: Wuda (Northern China) case study. *Remote Sensing of Environment*, 115(2), 257–268.

Jiménez-Muñoz, J.C., Sobrino, J.A., Skoković, D., Mattar, C., and Cristóbal J., 2014, Land surface temperature retrieval methods from landsat-8 thermal infrared sensor data. *Geoscience and Remote Sensing Letters*, 11(10), 1840–1843.

Kim, A.J., 2010, Coal formation and the origin of coal fires. In Stracher, G.B., Prakash, A., and Sokol, E.V. (eds.), *Coal and Peat Fires: A Global Perspective, Volume 1, Coal—Combustion and Geology*. Elsevier: Oxford, U.K.

Kolker, A., Engle, M., Stracher, G., Hower, J., Prakash, A., Radke, L., ter Schure, A., and Heffern, E., 2009, Emissions from coal fires and their impact on the environment: U.S. Geological Survey Fact Sheet, 2009–3084, 4p.

Kuenzer, C., 2015, Chapter X, this book

Kuenzer, C., Bachmann M., Mueller, A., Lieckfeld, L., and Wagner, W. 2008a, Partial unmixing as a tool for single surface class detection and time series analysis. *International Journal of Remote Sensing*, 29(11), 1–23.

Kuenzer, C. and Dech, S., 2013, Theoretical background of thermal infrared remote sensing. In Kuenzer, C. and Dech, S. (eds.), *Thermal Infrared Remote Sensing—Sensors, Methods, Applications. Remote Sensing and Digital Image Processing Series*, Vol. 17, Springer, New York, Dordrecht, Heidelberg, London. 572pp., pp. 1–26.

Kuenzer, C., Hecker, C., Zhang, J., Wessling, S., Kuenzer, C., and Wagner, W., 2008b, The potential of multidiurnal MODIS thermal band data for coal fire detection. *International Journal of Remote Sensing*, 29(3), 923–944.

Kuenzer, C. and Stracher, G.B., 2011, Geomorphology of coal seam fires. *Geomorphology*, 138, 209–222.

Kuenzer, C. and Voigt, S., 2003, Vegetationsdichte als möglicher Indikator für Kohleflözbrände? Untersuchung mittels Fernerkundung und GIS. In: Strobl, J., Blaschke, T. and Griesebner, G., (ed.): Angewandte Geographische Informationsverarbeitung XV. Beiträge zum AGIT-Symposium Salzburg 2003. Heidelberg, Wichmann, pp. 256–261.

Kuenzer, C., Wessling, S., Zhang, J., Litschke, T., Schmidt, M., Schulz, J., Gielisch, H., and Wagner, W., 2007c, Concepts for green house gas emission estimating of underground coal seam fires. *Geophysical Research Abstracts*, 9, 11716, EGU 2007, April 16–20, 2007, Vienna.

Kuenzer, C., Zhang, J., Hirner, A., Bo, Y., Jia, Y., and Sun, Y., 2008c, Multitemporal in-situ mapping of the Wuda coal fires from 2000 to 2005—Assessing coal fire dynamics. In *UNESCO Beijing, 2008, Spontaneous Coal Seam Fires: Mitigating a Global Disaster*, Vol. 4, 602pp. ERSEC ecological book series, Tsinghua University Press, Beijing, China, pp. 132–148.

Kuenzer, C., Zhang, J., Li, J., Voigt, S., Mehl, H., and Wagner, W., 2007a, Detecting unknown coal fires: Synergy of automated coal fire risk area delineation and improved thermal anomaly extraction. *International Journal of Remote Sensing*, 28(20), 4561–4585.

Kuenzer, C., Zhang, J., Sun, Y., Jia, Y., and Dech, S., 2012, Coal fires revisited: The Wuda coal field in the aftermath of extensive coal fire research and accelerating extinguishing activities, *International Journal of Coal Geology*, 102, 75–86.

Kuenzer, C., Zhang, J., Tetzlaff, A., and S. Dech, 2013, Thermal Infrared Remote Sensing of Surface and underground Coal Fires. In Kuenzer, C. and Dech, S. (eds.), *Thermal Infrared Remote Sensing—Sensors, Methods, Applications. Remote Sensing and Digital Image Processing Series*, Vol. 17, 572pp., pp. 429–451.

Kuenzer, C., Zhang, J., Tetzlaff, A., van Dijk, P., Voigt, S., Mehl, H., and Wagner, W., 2007b, Uncontrolled coal fires and their environmental impacts: Investigating two arid mining regions in north-central China. *Applied Geography*, 27, 42–62.

Kuenzer, C., Zhang, J., Tetzlaff, A. Voigt, S., and Wagner, W., 2008d, Automated demarcation, detection and quantification of coal fires in China using remote sensing data. In *UNESCO Beijing, 2008, Spontaneous Coal Seam Fires: Mitigating a Global Disaster*, Vol. 4, 602pp. ERSEC ecological book series, pp. 362–380.

Kuenzer, C., Zhang, J., Voigt, S., and Wagner, W., 2007d, Remotely sensed land-cover changes in the Wuda and Ruqigou-Gulaben coal mining areas China. *Reviews in Engineering Geology*, Tsinghua University Press, Beijing, China, 18, 219–228

Kumar, A. and Pandey, A.C., 2013, Evaluating impact of coal mining activity on landuse/landcover using temporal satellite images in South Karanpura Coalfields and Environs, Jharkhand State, India. *International Journal of Advanced Remote Sensing and GIS*, 2(1), 183–197.

Lambin, E.F. and Ehrlich, D, 1996, The surface temperature-vegetation index space for land cover and land-cover change analysis. *International Journal of Remote Sensing*, 17(3), 463–487.

Mansor, S.B., Cracknell, A.P., and Shilin, B.V., 1994, Monitoring of underground coal fires using thermal infrared data. *International Journal of Remote Sensing*, 15, 1675–1685.

Martha, T.R., Guha, A., Kumar, K.V., Kumaraju, M.V.V., and Raju, E.V.R., 2010, Recent coal-fire and land-use status of Jharia Coalfield, India from satellite data. *International Journal of Remote Sensing*, 31(12), 3243–3262.

Matson, M. and Dozier, J., 1981, Identification of subresolution high temperature sources using a thermal IR sensor. *Photogrammetric Engineering and Remote Sensing*, 47(9), 1311–1318.

Mukherjee, T.K., Bandhopadhyay, T.K., and Pande, S.K., 1991, Detection and delineation of depth of subsurface coalmine fires based on an airborne multispectral scanner survey in part of Jharia Coalfield, India. *Photogrammetric Engineering and Remote Sensing*, 57, 1203–1207.

Panigrahi, D.C., Singh, M.K., and Singh, C., 1995, Predictions of depth of mine fire from the surface by using thermal infrared measurement. In *Proceedings of the National Seminar on Mine Fires, Varanasi*, India, February 24–25, 1995. Banaras Hindu University, Varanasi, India, pp. 122–134.

Peng, W.X., van Genderen, J.L., Kang, G.F., Guan, H.Y., and Yongjie, Tan, 1997, Estimating the depth of underground coal fires using data integration techniques. *Terra Nova*, 9(4), 180–183.

Prakash, A., 2010, Coal fire research: Heading from remote sensing to remote measurement. In *Second International Conference on Coal Fire Research*, May 19–21, Berlin, Germany.

Prakash, A., 2014, Coal fires, http://www2.gi.alaska.edu/~prakash/coalfires/causes_hazards.html (last accessed April, 2014).

Prakash, A. and Berthelote A.R., 2007, Subsurface coal mine fires: Laboratory simulation, numerical modeling and depth estimation. *Reviews in Engineering Geology*, 18, 211–218.

Prakash, A., Fielding, E.J., Gens, R., van Genderen, J.L., and Evans, D.L., 2001, Data fusion for investigating land subsidence and coalfire hazards in a coal mining area. *International Journal of Remote Sensing*, 22(6), 921–932.

Prakash, A. and Gens, R., 2010, Remote sensing of coal fires. In Stracher, G.B., Prakash, A., and Sokol, E.V. (eds.), *Coal and Peat Fires: A Global Perspective, Volume 1, Coal—Combustion and Geology*. Elsevier: Oxford, U.K.

Prakash, A., Gens, R., and Vekerdy Z., 1999, Monitoring coal fires using multi-temporal night-time thermal images in a coalfield in North-west China. *International Journal of Remote Sensing*, 20(14), 2883–2888.

Prakash, A. and Gupta, R.P., 1998, Land-use mapping and change detection in a coal mining area—A case study of the Jharia Coalfield, India. *International Journal of Remote Sensing*, 19(3), 391–410.

Prakash, A. and Gupta, R.P., 1999, Surface fires in Jharia Coalfield, India—Their distribution and estimation of area and temperature from TM data. *International Journal of Remote Sensing*, 20(10), 1935–1946.

Prakash, A., Gupta, R.P., and Saraf, A.K., 1997, A landsat TM based comparative study of surface and subsurface fires in the Jharia Coalfield, India. *International Journal of Remote Sensing*, 18(11), 2463–2469.

Prakash, A., Saraf, A.K., Gupta, R.P., Dutta, M., and Sundaram, R.M., 1995a, Surface thermal anomalies associated with underground fires in Jharia Coal Mine, India. *International Journal of Remote Sensing*, 16(12), 2105–2109.

Prakash, A., Sastry, R.G.S., Gupta, R.P., and Saraf, A.K., 1995b, Estimating the depth of buried hot feature from thermal IR remote sensing data, a conceptual approach. *International Journal of Remote Sensing*, 16(13), 2503–2510.

Prakash, A., Schaefer, K., Witte, W.K., Collins, K., Gens R., and Goyette, M., 2011, Remote sensing—GIS based investigation of a boreal forest coal fire. *International Journal of Coal Geology*, 86(1), 79–86.

Prakash, A. and Vekerdy Z., 2004, Design and implementation of a dedicated prototype GIS for coal fire investigations in North China: Challenges met and lessons learnt. *International Journal of Coal Geology*, 59, 107–119.

Quattrochi, D.A., Prakash, A., Evena, M., Wright, R., Hall, D.K., Anderson, M., Kustas, W.P., Allen, R.G., Pagano, T., and Coolbaugh, M.F., 2009, Thermal remote sensing: Theory, sensors, and applications. In Jackson, M. (ed.), *Manual of Remote Sensing 1.1: Earth Observing Platforms & Sensors*, ASPRS, 550p.

Raju, A., Gupta, R.P., and Prakash, A., 2013, Delineation of coalfield surface fires by thresholding landsat TM-7 day-time image data. *Geocarto*, 28(4), 343–363.

Ray, S.K. and Singh, R.P., 2007, Recent developments and practices to control fire in undergound coal mines. *Fire Technology*, 43, 285–300.

Rosema, A., Guan, H., van Genderen, J.L., Veld, H., Vekerdy, Z., Ten Katen, A.M., and Prakash, A.P., 1999, Manual of coal fire detection and monitoring. Report of the Project 'Development and implementation of a coal fire monitoring and fighting system in China'. Netherlands Institute of Applied Geoscience, Utrecht, the Netherlands, NITG 99-221-C, 245p.

Saraf, A.K., Prakash, A., Sengupta, S., and Gupta, R.P., 1995, Landsat TM data for estimating ground temperature and depth of subsurface coal fire in Jharia Coal Field, India. *International Journal of Remote Sensing*, 16(12), 2111–2124.

Schmugge, T., Hook, S.J., and Coll, C., 1998, Recovering surface temperature and emissivity from thermal infrared multispectral data. *Remote Sensing of Environment*, 65(2), 121–131.

Schott, J.R., Hook, S.J., Barsi, J.A., Markham, B.L., Miller, J., Paduala, F.P., and Raqueno, N.G., 2012, Thermal infrared radiometric calibration of the entire landsat 4, 5, and 7 archive (1982–2010). *Remote Sensing of Environment*, 122, 41–49.

Sobrino, J.A., Jiménez-Muñoz, J.C., Sòria, G., Romaguera, M., Guanter, L., Moreno, J., Plaza A., and Martínez, P., 2008, Land surface emissivity retrieval from different VNIR and TIR sensors. *IEEE Transactions on Geoscience and Remote Sensing*, 46, 316–327.

Sobrino, J.A., Raissouni, N., Simarro, J., Nerry, F., and François, P., 1999, Atmospheric water vapour content over land surfaces derived from the AVHRR data: Application to the Iberian Peninsula. *IEEE Transactions on Geoscience and Remote Sensing*, 37, 1425–1434.

Stracher, G.B., Prakash, A., Schroeder, P., McCormack, J., Zhang, X.M., and van Dijk, P., 2005, New mineral occurrences and mineralization processes: Wuda coal-fire gas vents of Inner Mongolia. *American Mineralogist*, 90(11–12), 1729–1739.

Stracher, G.B, Prakash, A., and Sokol, E.V. (eds.), 2010, *Coal and Peat Fires: A Global Perspective, Volume 1, Coal—Combustion and Geology*, 335 p. Elsevier, Oxford, Amsterdam.

Stracher, G.B, Prakash, A., and Sokol, E.V. (eds.), 2013, *Coal and Peat Fires: A Global Perspective, Volume 2, Photographs and Multimedia Tour*, 564p. Elsevier, Oxford, Amsterdam.

Stracher, G.B. and Taylor, T.P., 2004, Coal fires burning out of control around the world: Thermodynamic recipe for environmental catastrophe. *International Journal of Coal Geology*, 59, 7–18.

Tetzlaff, A., 2004, Coal fire quantification using ASTER, ETM, and BIRD satellite instrument data. LMU Munich PhD dissertation. http://edoc.ub.uni-muenchen.de/4398/ (last accessed, April 2014).

van Dijk, P., Zhang, J., Jun, W., Kuenzer, C., and Wolf, K.H., 2011, Assessment of the contribution of in-situ combustion of coal to greenhouse gas emission; based on a comparison of Chinese mining information to previous remote sensing estimates. *International Journal of Coal Geology*, 86(1), 108–119.

van Genderen, J.L., Prakash, A., Gens, R., Veen, B.S. van, Liding, C., Tao, T.X., and Feng, G., 2000, Coal fire interferometry. Netherlands Remote Sensing Board, Delft, the Netherlands, USP-2-99-32.

Voigt, S., Tetzlaff, A., Zhang, J., Künzer, C., Zhukov, B., Strunz, G., Oertel, D., Roth, A., van Dijk, P.M., and Mehl, H., 2004, Integrating satellite remote sensing techniques for detection and analysis of uncontrolled coal seam fires in North China. *International Journal of Coal Geology*, 59(1–2), 121–136.

Waigl, C., Prakash, A., Ferguson, A., and Stuefer, M., 2014, Chapter 24—Delineating coal fire hazards in high latitude coal basins: A case study from interior Alaska. In Stracher, G.B, Prakash, A., and Sokol E.V. (eds.), *Coal and Peat Fires: A Global Perspective, Volume 3, Case Studies*, Elsevier, Amsterdam, Oxford, Waltham. 816p.

Wessling, S., Kessels, W., Schmidt, M., and Krause, U., 2008a, Investigating dynamic underground coal fires by means of numerical simulation. *Geophysical Journal International*, 172(1), 439–454.

Wessling, S., Kuenzer, C., Kessels, W., and Wuttke, M.W., 2008b, Numerical modeling for analyzing thermal surface anomalies induced by underground coal fires. *International Journal of Coal Geology*, 74, 175–184.

Whitehouse, A.E. and Mulyana, A.A.S., 2004, Coal fires in Indonesia. *International Journal of Coal Geology*, 59, 91–97.

Yang, B., Li, J., Chen, Y., Zhang, J., and Kuenzer, C, 2008, Automated detection and extraction of surface cracks from high resolution Quickbird imagery. In *UNESCO Beijing, 2008, Spontaneous Coal Seam Fires: Mitigating a Global Disaster*, Vol. 4, 602pp. ERSEC ecological book series, Tsinghua University Press, Beijing, China, pp. 381–389.

Zhang, J. and Kuenzer, C., 2007, Thermal surface characteristics of coal fires: Results of in-situ measurements. *Journal of Applied Geophysics*, 63, 117–134.

Zhang, J., Kuenzer, C., Tetzlaff, A., Oertl, D., Zhukov, B., and Wagner, W., 2007, Thermal characteristics of coal fires 2: Results of measurements on simulated coal fires. *Journal of Applied Geophysics*, 63, 135–147.

Zhang, J., Wagner, W., Prakash, A., Mehl, H., and Voigt, S., 2004a, Detecting coal fires using remote sensing techniques, *International Journal of Remote Sensing*, 25(16), 3193–3220.

Zhang, X., Zhang, J., Kuenzer, C., Voigt, S., and Wagner, W., 2004b, Capability evaluation of 3–5 µm and 8–12,5 µm airborne thermal data for underground coalfire detection. *International Journal of Remote Sensing*, 25(12), 2245–2258.

Zhang, X.M., van Genderen, J.L., and Kroonenberg, S.B., 1997, A method to evaluate the capability of landsat-5 TM band 6 data for sub-pixel coal fire detection. *International Journal of Remote Sensing*, 18, 3279–3288.

Zheng, Y., Feng, C., Jing, G., Qian, X., Li, X., and Liu, Z., 2009, A statistical analysis of coal mine accidents caused by coal dust explosions in China. *Journal of Loss Prevention in the Process Industries*, 22(4), 528–532.

Zhukov, B., Lorenz, E., Oertel, D., Wooster, M., and Robert, G., 2006, Spaceborne detection and characterization of fires during the bi-spectral infrared detection (BIRD) experimental small satellite mission (2001–2004). *Remote Sensing of Environment*, 100(1), 29–51.

Urban Areas

23

Urban Growth Mapping of Mega Cities: Multisensor Approach

Hasi Bagan
National Institute for Environmental Studies

Yoshiki Yamagata
National Institute for the Environmental Studies

Acronyms and Definitions

AVNIR-2	Advanced Visible and Near Infrared Radiometer type 2
DN	Digital Number
ETM+	Enhanced Thematic Mapper Plus
GCPs	Ground control points
GIS	Geographic information systems
GSI	Geospatial Information Authority of Japan
LiDAR	Light detection and ranging
MODIS	Moderate-Resolution Imaging Spectroradiometer
NOAA	National Oceanic and Atmospheric Administration
PRISM	Panchromatic Remote-Sensing Instrument for Stereo Mapping
SRTM	Shuttle Radar Topography Mission
TM	Thematic Mapper
UTM	Universal Transverse Mercator
WGS 84	World Geodetic System of 1984

23.1 Introduction

Urban areas occupy a relatively small fraction of the earth's land area, but at present more than half of the global population lives in urban areas, and this proportion is expected to increase in the coming decades (United Nations, 2014). Urban areas contribute significantly to climate change as a result of the use of fossil fuels for electricity generation, transportation, and industry. Already the intensive burning of carbon fuels in the world's urban areas accounts for about 70% of global greenhouse gas emissions (Solecki et al., 2013). In addition, previous research suggests that a 10% increase in urban land cover in a country is associated with an increase of more than 11% in the country's total CO_2 emissions (Angel et al., 2011). Urban form and structure are key factors that determine urban energy use and emissions, and urbanization fundamentally changes the urban form and urban spatial structure, including the number of buildings, their geometry, pattern, distribution, and density (Frolking et al., 2013).

Remote sensing and techniques have already proven useful for mapping and modeling urban growth. Various available global datasets are used for measuring, analyzing, and, hence, understanding the complex processes of urbanization (Potere et al., 2009). Examples are global urban extent maps based on, for example, National Oceanic and Atmospheric Administration (NOAA) Air Force Defense Meteorological Satellite Program (DMSP)/Operational Linescan System (OLS) sensor nighttime lights imagery (Elvidge et al., 2007, Doll, 2008) or Moderate-Resolution Imaging Spectroradiometer (MODIS) data (Schneider et al., 2009). The distribution of cities around the world broadly corresponds to the brightness distribution of DMSP nighttime lights (Zhang and Seto, 2011, Parés-Ramos et al., 2013). However, no single DMSP brightness threshold is valid for extracting the urban extent of all cities because small settlements that are not frequently lit are likely to be excluded (Small et al., 2005). Although coarse spatial resolution (from 250 m to 2 km) monitoring provides global and national estimates of urban growth, coarse data may be less reliable for correctly estimating the urban area of cities and often results in either overestimation or underestimation of urban areas (Potere and Schneider, 2007). Townshend and Justice (2002) argued that "a substantial proportion of the variability of land cover change has been shown to occur at resolutions below 250 m," and Giri et al. (2013) reported that land parcels managed at a local scale are often smaller than the resolution of coarse spatial resolution satellite data. Thus, coarse spatial resolution products still lack appropriate temporal, spatial, and/or thematic resolution to effectively support detailed analyses on the characteristics of cities and to analyze changes over time (Taubenböck et al., 2014).

In recent years, higher resolution sensor systems are available (e.g., IKONOS, QuickBird, WorldView-1 and WorldView-2, and GeoEye-1 and GeoEye-2) for monitoring the spatial effects of urbanization that provide spatial information content that is hundreds of times better than coarse spatial resolution datasets. Small (2003) compares 14 cities at a very high geometric level using QuickBird data with a submeter geometric resolution and derives parameters such as vegetation fraction; Berger et al. (2013) extract urban land-cover information from high-spatial-resolution multispectral and light detection and ranging (LiDAR) data; and Taubenböck et al. (2012) analyze the spatial effects of urbanization over a span of almost 40 years using Landsat data. However, little attention has been paid to the quantitative analysis of relationships among land-cover category changes (Bagan and Yamagata, 2014).

Monitoring and mapping of urban growth and developing effective urban planning strategies require spatio-temporal extent and expansion trends of cities. Urban land-use and land-cover changes are linked to socioeconomic activities (Lambin et al., 2003; Small and Cohen, 2004; Doll et al., 2006; Avelar et al., 2009), and urbanization includes both the physical growth of a city and the movement of people to urban areas. Therefore, it is essential to combine remote sensing–derived parameters with socioeconomic parameters to analyze the spatio-temporal changes of urban growth (Bagan and Yamagata, 2012).

Recently, spatio-temporal analyses of land-cover changes using 1 km² grid cells have demonstrated that grid cells provide a new way to obtain spatial and temporal information about areas that are smaller than the municipal scale and uniform in size (Bagan and Yamagata, 2012; Qian et al., 2014) and to further develop the change dynamics analysis in order to better characterize the phenomena using limited available data. The relationships among changes in urban land-cover patterns can be better understood if these data are mapped onto a grid composed of square grid cells.

This chapter has two major objectives with a goal of demonstrating the value of remote sensing in urban studies. First, the past and the present patterns and trends of urban growth are studied by combining multidate remote sensing over four decades with population census data. Second, changes

in urbanization and other land cover as a result of a new economic activity, taking coal mining as an example, are provided. These objectives are illustrated taking two case studies. In the first study, we used grid cells to investigate spatial and temporal land-cover changes in the Tokyo metropolitan area, specifically combining remote sensing data with population census data to investigate the past and present patterns and trends of urban growth. With 8.5% of its land area being urbanized (GSI, 2012), Japan is one of the world's most urbanized countries. The rapid growth of the Japanese economy has concentrated its population, industry, and other economic activities in metropolitan regions to an extreme degree (Bagan and Yamagata, 2012). Thus, urban growth represents one of the most important land-cover/land-use change events in recent Japanese history. In the second study (Section 23.4.4), to assess the quality of the grid square method, we used the grid square method to investigate spatio-temporal urban land-cover changes in Holingol City, Inner Mongolia Autonomous Region, China. In recent decades, China's demand for coal for electricity generation has put the coal-rich Holingol region under strong pressure, and a surface coal-mining boom has emerged. Moreover, coal-mining activities and construction of coal power bases and coal-based industries were the major contribution to the conversion of grassland to urban land-use types (Qian et al., 2014).

23.2 Materials

The Tokyo metropolitan area includes the city and prefecture of Tokyo as well as Kanagawa, Saitama, Chiba, and parts of Ibaraki prefectures (Figure 23.1). The population of the study area reached 37.6 million in 2010, which is about 29.4% of the country's population of 128.1 million, while the area occupies less than 3% of Japan's land surface (Statistics Bureau, Japan, 2011).

Three Landsat scenes, namely, the center, west, and east, cover the study area, as shown in Figure 23.1. The center scene accounts for about 90% of the total study area; thus, as a convention, we hereafter refer to the mosaic Landsat imagery by the date of the center scene.

We acquired Landsat multispectral scanner (MSS), Thematic Mapper (TM), and Enhanced Thematic Mapper Plus (ETM+) images to interpret land-use and land-cover changes for the study area from four separate dates (nominally 1972, 1987, 2001, and 2011). All Landsat data are processed for standard terrain corrections by the U.S. Geological Survey. Table 23.1 provides information on the image data. Only images from April to November, the green vegetation season, and with low cloud cover were considered to maximize the vegetation information content for each monitoring date. All analyses were based on the optical and thermal infrared bands of the MSS, TM, and ETM+ sensors, while excluding panchromatic bands (Table 23.1).

Precise geometric registration to a common map reference and coregistration between individual images are crucial for ensuring the reliable detection of temporal changes of land cover. All Landsat imagery was geometrically rectified to a common map reference system (Universal Transverse Mercator

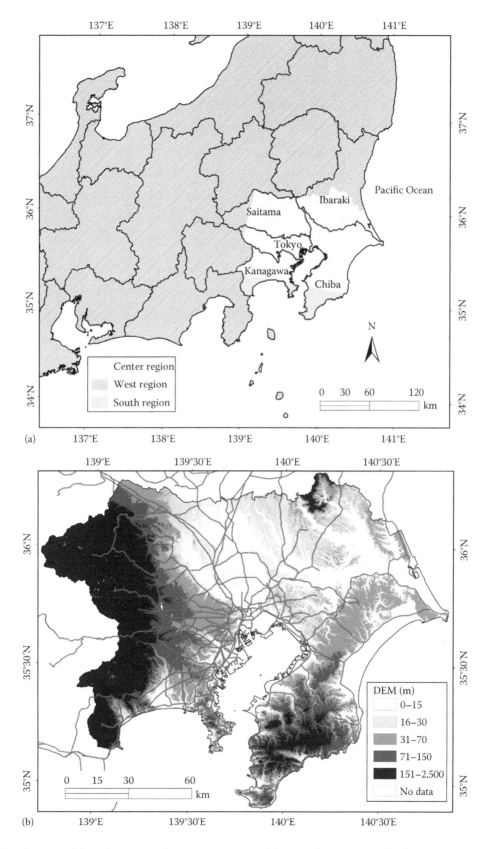

FIGURE 23.1 (a) Site location of the Tokyo metropolitan area. A mosaic of three Landsat images is also shown with the south, center, and west regions shaded in green, blue, and pink, respectively (Table 23.1). (b) The gray image is a digital elevation model (DEM) generated from SRTM data. Red lines refer to the major railway and metrolines.

TABLE 23.1 Path/Row and Acquisition Dates for the Landsat Time-Series Scenes Used in This Study

Sensor	Date Acquired	Path/Row	Location	Spatial (m)	Spectral Bands
Landsat-1 MSS	November 26, 1972	115/35	Center	60	2 bands visible
Landsat-1 MSS	November 26, 1972	115/36	South		2 bands infrared
Landsat-1 MSS	November 09, 1972	116/35	West		
Landsat-5 TM	May 21, 1987	107/35	Center	30	3 bands visible
Landsat-5 TM	May 21, 1993	107/36	South		3 bands infrared
Landsat-5 TM	December 06, 1990	108/35	West		1 bands thermal
Landsat-7 ETM+	September 24, 2001	107/35	Center	30	3 bands visible
Landsat-7 ETM+	September 24, 2001	107/36	South		3 bands infrared
Landsat-7 ETM+	November 02, 2001	108/35	West		1 bands thermal
Landsat-5 TM	April 05, 2011	107/35	Center	30	3 bands visible
Landsat-5 TM	April 05, 2011	107/36	South		3 bands infrared
Landsat-5 TM	April 25, 2010	108/35	West		1 bands thermal

[UTM] map projection Zone 54 North, World Geodetic System of 1984 [WGS 84] geodetic datum) using a different number (around 25–36) of ground control points (GCPs) for each image. The GCPs were selected using orthorectified Advanced Land Observing Satellite (ALOS) RGB color images (2.5 m spatial resolution, derived from Advanced Visible and Near Infrared Radiometer type 2 [AVNIR-2] and Panchromatic Remote-sensing Instrument for Stereo Mapping [PRISM] from 2006 to 2009; data provided by NTT Data, Japan) and the Digital Map 2500 (Spatial Data Framework) at a scale of 1:2500 (provided by the Geospatial Information Authority of Japan [GSI]) as references. The GCPs were well dispersed throughout each scene and yielded root-mean-square errors of less than 0.7 pixels.

Tokyo is a data-rich area to study the dynamics of spatial and temporal land-cover change patterns. In addition to ALOS data, we acquired and utilized numerous ancillary data for determining typical land-cover classes and selecting ground reference sites for each Landsat recording date:

1. Vegetation maps for 1973, 1983–1986, and 1994–1998 with scales ranging from 1:25,000 to 1:200,000, which were derived from field surveys and airborne images (published by the GSI)

2. Land-use geographic information system (GIS) datasets from 1976, 1987, 1997, and 2008 with a spatial resolution of 100 m (published by the GSI)
3. High-spatial-resolution remote sensing data such as IKONOS (acquired in 2003–2007), WorldView-2 (acquired in 2011), and aerial photography data (acquired in 2005–2007, provided by the GSI)
4. Shuttle Radar Topography Mission (SRTM) elevation data

Using the field investigation results, GIS datasets, and visual interpretation of the remote sensing data and considering the Landsat scene acquisition dates, we designated five to nine land-cover types in this experiment (Table 23.2).

We analyzed three different types of data to quantify the urban spatial extent: population census data (1 km spatial resolution; Basic Grid Square data), land-cover map derived from Landsat, and mean annual stable nighttime lights data from DMSP/OLS.

The Basic Grid Square population census data are available from the Statistics Bureau of Japan for every 5-year period from 1970. Each square (area about 1 km^2) is assigned an appropriate unique eight-digit ID number based on its longitude and latitude that is used as location information in the National Land Numerical Information databank (Statistics Bureau, Japan, 1973).

TABLE 23.2 Description of the Land-Cover Classification System and Training and Test Pixel Counts for Data from 1972, 1987, 2001, and 2011

Class	1972 Training	1972 Testing	1987 Training	1987 Testing	2001 Training	2001 Testing	2011 Training	2011 Testing
1. Forest	1,478	680	2,672	1,375	2,287	1,259	2,300	1,157
2. Urban/built-up	875	456	2,277	1,156	1,940	1,101	2,014	1,164
3. Cropland	872	458	2,062	1,022	1,949	1,033	1,609	903
4. Paddy	—	—	1,528	823	—	—	—	—
5. Grassland	800	427	1,140	624	1,267	542	1,080	556
6. Sparse	—	—	1,170	588	1,002	570	1,022	562
7. Water	692	331	1,582	775	1,308	606	1,138	618
8. Flooded paddy	—	—	—	—	—	—	797	433
9. Wheat	—	—	—	—	—	—	796	422
10. Bare	—	—	559	300	490	291	—	—
11. Snow	—	—	—	—	—	—	742	386
Total	4,717	2,352	12,990	6,663	10,243	5,402	11,498	6,201

23.3 Methods

23.3.1 Subspace Methods

The supervised subspace classification method was applied to each of the four dataset groups (Table 23.1). Subspace methods (Oja, 1983), widely used for pattern recognition and computer vision, have been applied to remote sensing data classification (Bagan et al., 2008; Bagan and Yamagata, 2010). In this method, high-dimensional input data are projected onto a low-dimensional feature space, and the different classes are then represented in their own low-dimensional subspace. The following is a description of the subspace method procedure (Figure 23.2).

In the preprocessing step, the training samples are normalized as follows. Let d be the input data dimension, which is equal to the number of bands. For a given pixel $\mathbf{x} = (x_1, x_2, \ldots, x_d)^T$, the normalized pixel is computed as

$$\mathbf{x} = (x_1/L, x_2/L, \ldots, x_d/L)^T, \qquad (23.1)$$

where $L = \sqrt{x_1^2 + x_2^2 + \cdots + x_d^2}$ is the pixel length. To be concise, we also use $\mathbf{x}$ to denote a normalized pixel.

Let $\varphi_{k,i}$ ($1 \leq i \leq r$, $1 \leq k \leq K$) be the basis vectors of the subspace of class C^k, which are computed from class training samples by eigenvalue and eigenvector solving algorithms; here r denotes the subspace dimension and K denotes the number of classes. The projection length of pixel $\mathbf{x}$ in a subspace of class C^k is given by

$$P_k = \sum_{i=1}^{r} (\mathbf{x}, \varphi_{k,i})^2. \qquad (23.2)$$

After computing the projection length between pixel $\mathbf{x}$ and each subspace, pixel $\mathbf{x}$ is then labeled by the class that has the largest projection length.

Misclassifications may occur when class subspaces overlap. To separate them, the subspaces are slowly rotated to reduce the overlap between them. The main steps are described as follows.

At iteration k, the conditional correlation matrix is computed by

$$P_k^{(i,j)} = \sum_{\mathbf{x}} \{ \mathbf{x}\mathbf{x}^T \mid \mathbf{x} \in C^i, \mathbf{x} \mapsto C^j \}, \qquad (23.3)$$

where the symbol $\mapsto$ denotes that the training sample $\mathbf{x}$ belonging to class C^i has been misclassified into class C^j.

Once the conditional correlation matrix is generated, the correlation matrix for class C^i is updated as follows:

$$P_k^{(i)} = P_{k-1}^{(i)} + \alpha \sum_{j=1, j \neq i}^{K} P_k^{(i,j)} - \beta \sum_{j=1, j \neq i}^{K} P_k^{(j,i)}, \qquad (23.4)$$

where α and β are learning parameters, both usually with small positive constant values. Then, the eigenvalues and eigenvectors of $P_k^{(i)}$ are calculated to generate a new subspace of class C^i. The iterations end when either all the training data are fully recognized or the maximum number of iterations has been reached (Bagan and Yamagata, 2010).

23.3.2 Grid Cell Process

The advantage of grid cell system, which is used for population census data, is that it can avoid the potential problem of changing boundaries of administrative units during the time interval of interest. Thus, grid cells with unique IDs enabled us to link census data with land-cover maps for spatial and temporal land-cover change analysis. To do this, it is necessary to represent the land-cover maps in grid square cells.

First, we merged similar categories of land cover into five main land-cover classes for analysis convenience, namely, forest, urban/built-up, cropland, grassland, and water (Bagan and Yamagata, 2012).

Second, we overlaid the reclassified images on the 1 km² grid cells to compute for each cell the percentage of the five land-cover types within it and stored the results in a new attribute table. When calculating the percentage of a land-cover type within a cell, we divided the sum of the land-cover-type area by the area of the cell.

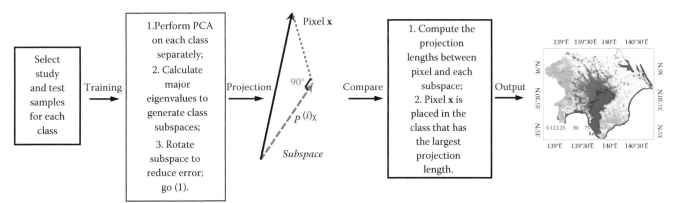

FIGURE 23.2 Subspace training and classification process.

Finally, the land-cover maps were expressed as a percentage of each of the five land-cover types for each of the grid cells. For a pixel located on a grid line (or spanning multiple grid lines), we assigned the pixel to the cell that covered the largest part of the pixel.

Thus, the grid cells enabled us to aggregate the categories for each map and to calculate their proportions in basic grid cells. Furthermore, they enabled us to evaluate the spatio-temporal changes in land-cover categories to allow a much easier statistical comparison of the land-cover changes with population density changes. In the next section, we combine the grid cell land-cover maps and grid cell statistics of population census data to explore the patterns of urban growth.

23.4 Results

23.4.1 Spatio-Temporal Changes of the Land-Cover Classes

Subspace methods proved to be well suited for classifying mosaic Landsat imagery from a heterogeneous and dynamic urban environment with accuracies ranging between 84.5% and 93.5% for the test datasets.

As described in Section 23.3.2, after aggregation, the final five categories in the land-cover maps were forest, urban/built-up, cropland, grassland, and water. Figure 23.3 shows the after aggregation land-cover classification maps for each of the 4 years. There are several major trends evident in the changes of land cover that are consistent over the period 1972–2011. The urban/built-up area increased rapidly, and there was a marked decrease in cropland area. Although the area of forest decreased between 1972 and 1987, after 1987 there was a steady increase of forest growth.

At the same time, the overall population of the study area increased by 49.5% from 25.2 million in 1970 to 37.6 million in 2010.

As explained in Section 23.3.2, we integrated the classified images with the empty basic grid cells to compute the area of each land-cover category within a cell. The generated grid cell land-cover classification maps facilitated our calculation of the percentage land-cover category changes from 1972 to 2011 at the scale of 1 km². Furthermore, since the size and location of the grid cells were exactly the same as those of the grid cells used for population census data (grid cell system census), we were able to use these to determine statistical associations between changes in land-cover classes and the variation of population densities.

Figure 23.4 shows the grid cell–based spatial changes of urban area from 1972 to 2011. The grid cell values were calculated by subtracting the urban area of 1972 from that of 2011 in each grid cell and then dividing the changed area by the cell area. As Figure 23.4 illustrates, the urban/built-up area rapidly expanded to the surrounding suburban area where it was mainly flat or along transportation lines, whereas, in contrast,

the urban/built-up area decreased in the center of the city. In fact, many high-rise (e.g., office or commercial) buildings and city-planned parks and green spaces replaced the dense low-rise buildings in the city center concurrently with the development of housing estates along railway lines in the suburbs (Okata and Murayama, 2010).

Figure 23.5 shows the grid cell–based spatial changes of cropland area from 1972 to 2011, which were calculated in the same way as in Figure 23.4. Typically, the greatest decreases in agricultural land area took place either in relatively large and flat areas or in proximity to urban regions where the demand for land for urban purposes is high (Saizen et al., 2006; Catalán et al., 2008).

To further investigate the spatial change in population density, we computed the difference in population density between 1970 and 2010 by subtracting the 1970 cell values from the 2010 values as shown in Figure 23.6. A number of trends are immediately clear: the dominant feature is a massive decentralization of population from the metropolitan core to the surrounding region. By contrast, the suburban areas surrounding these core areas of Tokyo, Kawasaki, and Yokohama have seen enormous increases in population. Figure 23.6 also indicates that the western part of the study area has experienced a much faster rate of population growth along the transportation lines, and the spatial distribution of population increased trends is similar to the urban expansion trends (Figure 23.4).

23.4.2 Relationship between Land-Cover Classes and Population Census

To investigate the relationship between the census population and the amount of land-cover category changes, we calculated the correlation coefficients of land-cover categories (i.e., forest, cropland, urban/built-up, grassland, and water) and population density changes based on the grid cells. All 15,851 grid cells in the study area were used for the statistical analysis. Table 23.3 presents a summary of the linear correlation coefficient matrix between the changes of land-cover categories from 1972 to 2011 and population density changes from 1970 to 2010 based on the 15,851 samples.

As shown in Table 23.3, the linear correlation coefficient was −0.77 between urban/built-up and cropland, 0.54 between urban/built-up and population change, and −0.44 between cropland and population change. Meanwhile, forest change was negatively correlated with cropland ($r = -0.39$) and grassland ($r = -0.45$) owing to the portion of abandoned agricultural land and grassland that transformed to forest during the past four decades. The correlations between urban/built-up, cropland, and population change were statistically significant. These results are consistent with earlier findings in the Dhaka metropolitan area of Bangladesh (Dewan and Yamaguchi, 2009) and the Barcelona metropolitan region (Catalán et al., 2008).

Figure 23.7a shows the relationship between urban/built-up changes and cropland changes from 1972 to 2011, and Figure 23.7b

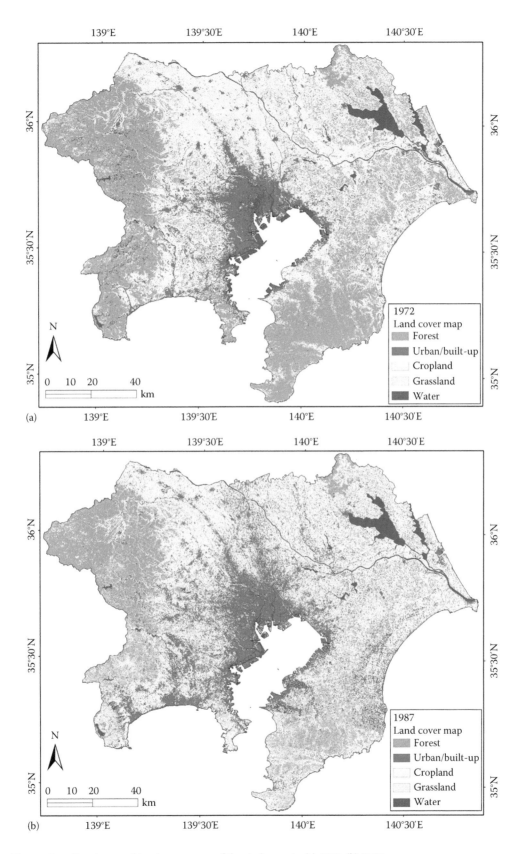

FIGURE 23.3 Time series of *land-use* and *land-cover* maps of the study area in (a) 1972, (b) 1987. *(Continued)*

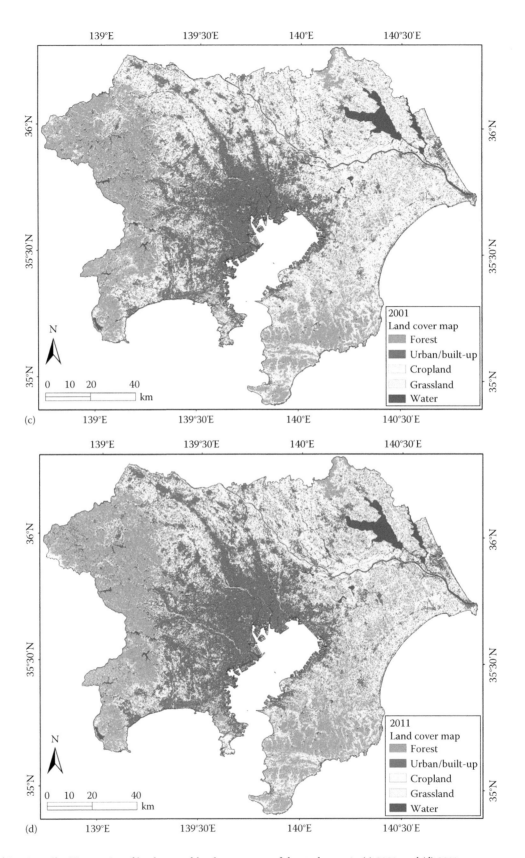

FIGURE 23.3 (*Continued*)   Time series of *land-use* and *land-cover* maps of the study area in (c) 2001, and (d) 2011.

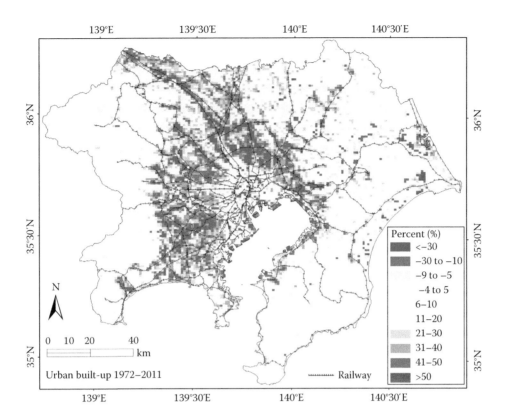

FIGURE 23.4   Percentage change of urban/built-up area in 1 km² grid cells from 1972 to 2011.

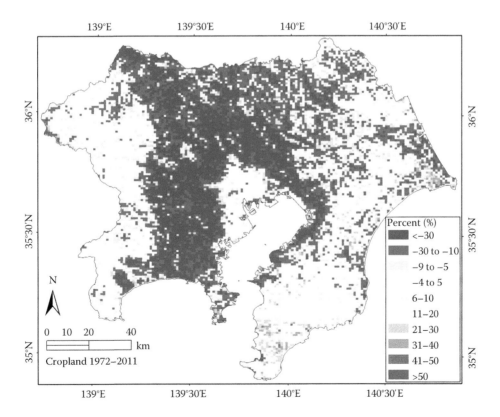

FIGURE 23.5   Percentage change of cropland area in 1 km² grid cells from 1972 to 2011.

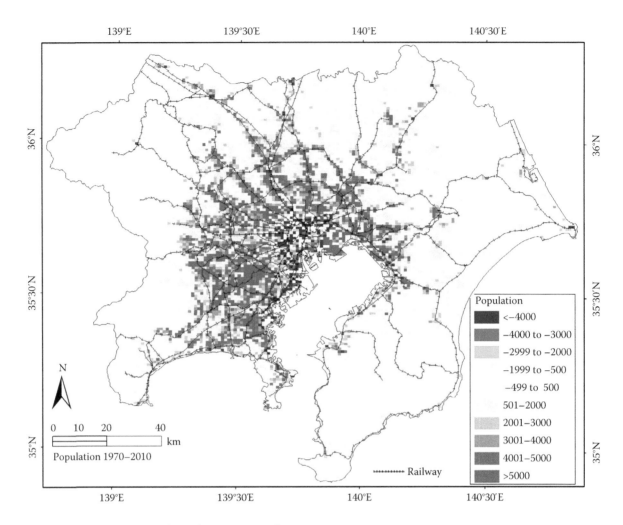

FIGURE 23.6   Change in population density between 1970 and 2010.

TABLE 23.3   Correlations (*r*) among the Changes of Land-Cover Categories in 1972–2011 and Population Density Change in 1970–2010

r	Urban/Built-Up	Cropland	Forest	Grassland	Water	Population
Urban/built-up	1					
Cropland	−0.7691	1				
Forest	0.0213	−0.3915	1			
Grassland	−0.1904	−0.0445	−0.4515	1		
Water	−0.1025	−0.1306	−0.1153	0.0431	1	
Population	0.5418	−0.4442	−0.0050	−0.0697	0.0134	1

shows the relationship between the urban/built-up changes from 1972 to 2011 and population density changes from 1970 to 2010. We found a strong, negative linear relationship between urban/built-up change and cropland change (*r* = −0.77) (Figure 23.7a), suggesting that a vast area of cropland has been converted to urban/built-up area during the last four decades.

The urban/built-up change has a significant positive correlation with the population density change (*r* = 0.54) (Figure 23.7b). However, there are areas with poor correlation caused by sparse

population density in some industrial regions (e.g., the industrial region around Tokyo Bay, where large businesses have left underutilized buildings) or dense populations in regions with overcrowded high-rise apartment buildings.

To better understand the trends of urban growth, we divided the land-cover change in 1972–2011 and the population density change in 1970–2010 into three intervals: 1972–1987 land cover vs. 1970–1985 population, 1987–2001 land cover vs. 1985–2000 population, and 2001–2011 land cover vs. 2000–2010 population.

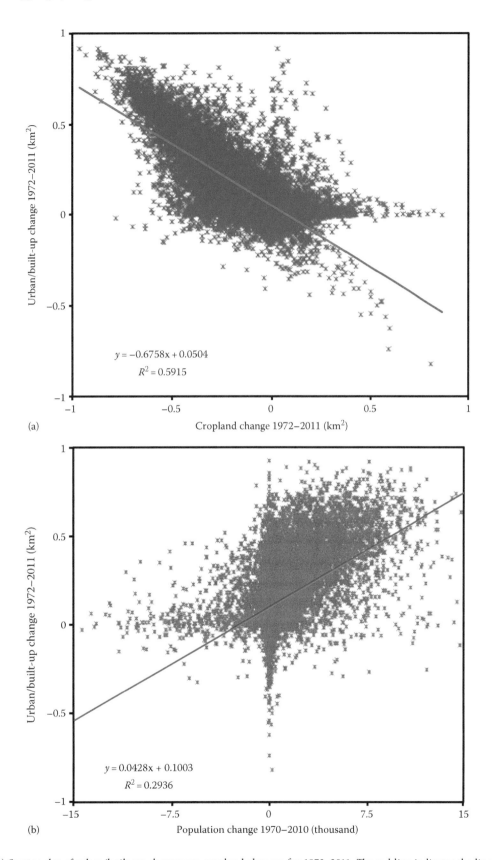

FIGURE 23.7 (a) Scatter plot of urban/built-up changes vs. cropland changes for 1972–2011. The red line indicates the linear least-squares fit and *r* is the correlation coefficient. (b) The correlation between the urban/built-up changes for 1972–2011 and population density changes for 1970–2010. The blue line indicates the linear least-squares fit.

TABLE 23.4 Correlations (*r*) between the Changes of *Land-Cover* Categories and Population Density Change: 1972–1987 vs. 1970–1985, 1987–2001 vs. 1985–2000, and 2001–2011 vs. 2000–2010, Respectively, for the Three Periods

r	Urban/Built-Up	Cropland	Forest	Grassland	Water	Population
a. Changes of land-cover categories for 1972–1987 vs. population density change for 1970–1985						
Urban/built-up	1					
Cropland	−0.4031	1				
Forest	−0.0009	−0.4966	1			
Grassland	−0.1555	0.0091	−0.6900	1		
Water	−0.1970	−0.1699	−0.1281	0.0080	1	
Population	0.3560	−0.1662	0.0622	−0.1591	−0.0034	1
b. Changes of land-cover categories for 1987–2001 vs. population density change for 1985–2000						
Urban/built-up	1					
Cropland	−0.7904	1				
Forest	−0.2001	0.0109	1			
Grassland	0.1276	−0.2116	−0.7931	1		
Water	0.1583	−0.3698	−0.0453	−0.1363	1	
Population	0.4400	−0.3204	−0.0563	0.0360	0.0313	1
c. Changes of land-cover categories for 2001–2011 vs. population density change for 2000–2010						
Urban/built-up	1					
Cropland	−0.4701	1				
Forest	−0.1450	−0.5429	1			
Grassland	−0.0827	−0.4036	−0.1612	1		
Water	−0.0271	−0.0545	−0.0426	−0.1774	1	
Population	0.0562	−0.0194	−0.0252	0.0212	−0.0338	1

We used the 1970, 1985, 2000, and 2010 population census data because they most closely matched the Landsat-derived land-cover maps.

Results from our correlation analyses between land-cover change and population density change for the three intervals are reported in Tables 23.4. The correlations between the urban/built-up and cropland are highly negative: −0.40, −0.79, and −0.47, respectively, for the three periods. The correlations between urban/built-up and population are positive for 1972–1987 (*r* = 0.36) and 1987–2001 (*r* = 0.44). It is not until 2001–2011 that almost no correlation is found (*r* = 0.06).

Forest and grassland had strong negative correlations for 1972–1987 (*r* = −0.69) and 1987–2001 (*r* = −0.79), whereas the correlation was small for 2001–2011 (*r* = −0.16) (Table 23.4). In contrast, almost no correlation appeared in the linear relationship between cropland and grassland for 1972–1987 (*r* = 0.01), but the negative correlation dramatically increased to −0.21 and −0.40 during the last two periods, respectively. There were strong negative correlations between forest and cropland for 1972–1987 (*r* = −0.50) and 2001–2011 (*r* = −0.54), but little correlation was found for 1987–2001 (*r* = 0.01).

Water change was not correlated with other land-cover changes or population changes in any of the three periods, except for 1987–2001 when the relationship between water and cropland had a negative correlation of −0.37. This slightly negative relationship may reflect the fact that the 1987 Landsat image

was acquired over flooded paddy fields on May 24 during the rice growing season; as a consequence, some paddy fields may have been misclassified into water classes.

23.4.3 Relationship among DMSP, Urban Area, and Population Census

The distribution of cities around the world broadly corresponds to the brightness distribution of DMSP nighttime lights (Small et al., 2005). Syntheses of population data, urban/built-up area, and satellite nighttime lights images can be used to identify and characterize the spatio-temporal extent and expansion trends of urban growth. Thus, the relationships among population census data, land-use data, and DMSP nighttime lights data can be better understood if these data are mapped onto a grid composed of square grid cells. Furthermore, it allows the commonly used multivariate linear regression model to be used to analyze urbanization.

The mean annual stable nighttime lights from the NOAA's DMSP/OLS nighttime lights sensor images have a 30 arc second spatial resolution. The cleaned up avg_vis contains the lights from cities, towns, and other sites with persistent lighting. Ephemeral events, such as fires, have been discarded. Then the background noise was identified and replaced with values of zero. The DMSP digital number (DN) is an integer between 0 (no light) and 63. The DMSP/OLS are available at http://www.ngdc.noaa.gov/dmsp/downloadV4composites.html.

For the link between DMSP/OLS with census data and land-cover data, we resampled DMSP maps to a 1 m spatial resolution pixel size. This process can reduce the error caused by the pixel size. Then we overlaid the resampled DMSP images on the grid cells to compute for each cell the percentage of DN values within it and stored the results in a new attribute table by using ArcGIS 10.1 software. Thus, the attribute table includes 64 new added DMSP attributes, that is, DN values from 0 to 63.

We compared the spatial distributions of DMSP DN values (2010), population density (2010), and percent density of urban/built-up area (2011) in Tokyo at the scale of 1 km² (Figure 23.8).

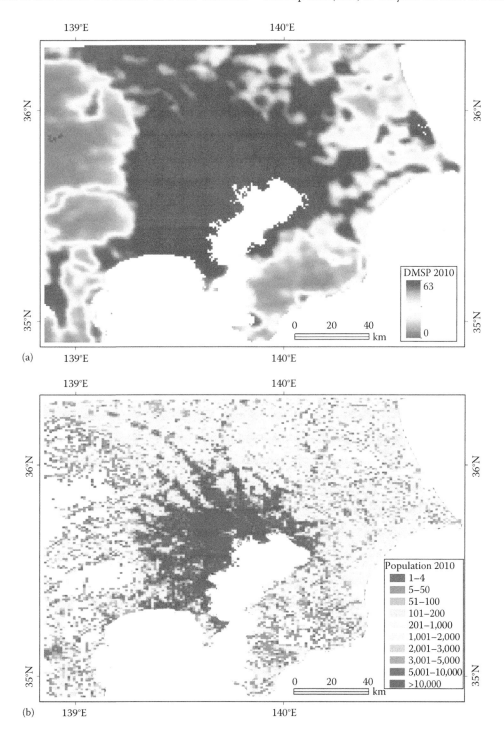

FIGURE 23.8 (a) DMSP in 2010; (b) population density in 2010 at the scale of 1 km². *(Continued)*

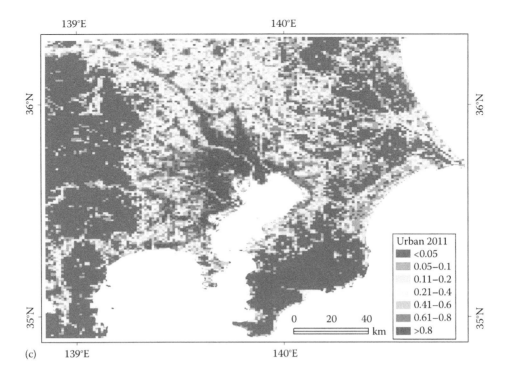

FIGURE 23.8 (*Continued*)   (c) proportion of urban/built-up area in 2011 at the scale of 1 km².

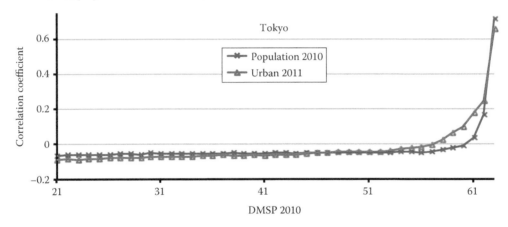

FIGURE 23.9   Correlations between DMSP DN values and urban *land-use* area and between DMSP DN values and population density in Tokyo. There are saturation issues at high DMSP values in city areas where the population density is high.

The result showed that the spatial distributions of population density, DMSP nighttime lights brightness, and urban land use are broadly consistent, suggesting a spatial correlation among them.

Furthermore, we examined the correlations of DMSP DN values with urban land-use area and population density in Tokyo (Figure 23.9).

As the DN values of DMSP nighttime lights increase, the correlation coefficient between the DMSP DN values and population density and that between DMSP DN values and urban land-use area also increase (Figure 23.9).

As population density showed a strong correlation with both land-cover and DMSP data, it is possible to use multiple linear regression to predict population density at a 1 km spatial resolution from the combination of land-cover and DMSP data.

However, as shown in Figure 23.9, the urban land-use area did not reflect the vertical component (i.e., high-rise buildings), and DMSP nighttime lights may be saturated in core city areas where the population density is very high (Frolking et al., 2013). Thus, to improve the prediction accuracy of the population density in the future, the analysis should include data of urban 3D structures.

23.4.4 Spatio-Temporal Analyses of Urban Growth in Holingol City, Inner Mongolia

In recent decades, China's demand for coal for power generation has put the resource-rich Holingol region under strong pressure, and the grazing land becomes rapidly populated urban areas.

The purpose of this section was to investigate the spatio-temporal changes of land cover due to the open-cast surface coal-mining activities and rapid urban expansion in the Holingol region from 1978 to 2011 by using 300 m by 300 m grid cells.

The study area is the Holingol region, located in the northern part of the Horqin Sandy Land of the Inner Mongolia Autonomous Region, China, and centered at 45°30′ N, 119°30′ E. We acquired Landsat MSS, TM, and ETM+ images to interpret land-use/land-cover changes for the study area from four separate dates (nominally 1978, 1988, 1999, and 2011). In addition to Landsat data, ancillary GIS datasets and other ancillary satellite data were used as reference data to assist in our field investigation in the determination of typical land-cover classes and in selecting ground reference sites for each Landsat recording date. Based on the field investigation results, GIS datasets, and visual interpretation of the remote sensing data and with consideration of the Landsat scene acquisition dates, we designated 10–12 land-cover types in this experiment. The subspace method classification was performed for each of four Landsat images with accuracies ranging between 71.4% and 84.6% for the test dataset (Qian et al., 2014).

We made 300 m × 300 m grid square cells, each given a unique ID number in the study area by using ArcGIS10.1 software. Grid squares with unique cell IDs enable us to link among land-cover maps for spatio-temporal land-cover change analysis. To do this, we merged similar categories of land cover into six main land-cover classes for analysis convenience, namely, water, forest, grassland, urban/bare, cropland, and coal. Then, the land-cover maps were expressed as a percentage of each of the six land-cover types for each of the grid cells. We calculate the linear correlation coefficient among the changes of land-cover categories from 1978 to 2011. The linear correlation coefficient was −0.30 between coal and grassland, −0.76 between urban/bare and grassland, 0.02 between coal and urban/bare, −0.19 between grassland and water, and −0.29 between grassland and cropland (Table 23.5). Figure 23.10 depicts how the land of the Holingol City has changed over time.

As can be seen in Figure 23.10, there was a remarkable increase in urban/bare and coal-mining areas and a decrease in other land-cover areas from 1978 to 2011.

As shown in Figure 23.10, the rapid opening of new mines in the last three decades was linked to the emergence of a string of

TABLE 23.5 Correlations (*r*) among the Changes of *Land Cover* in 1978–2011

r	Water	Forest	Grassland	Urban/Bare	Cropland	Coal
Water	1					
Forest	0.0231	1				
Grassland	−0.1905	−0.5240	1			
Urban/bare	−0.0327	0.1146	−0.7595	1		
Cropland	−0.0145	−0.1044	−0.2850	−0.0655	1	
Coal	0.0008	0.0667	−0.2972	0.01958	0.1298	1

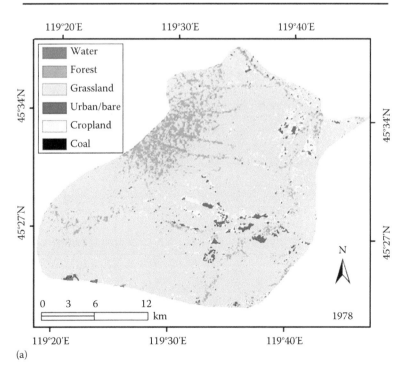

FIGURE 23.10 Coal mining propelled urban growth from 1978 to 2011 in Holingol City, Inner Mongolia. (a) 1978. *(Continued)*

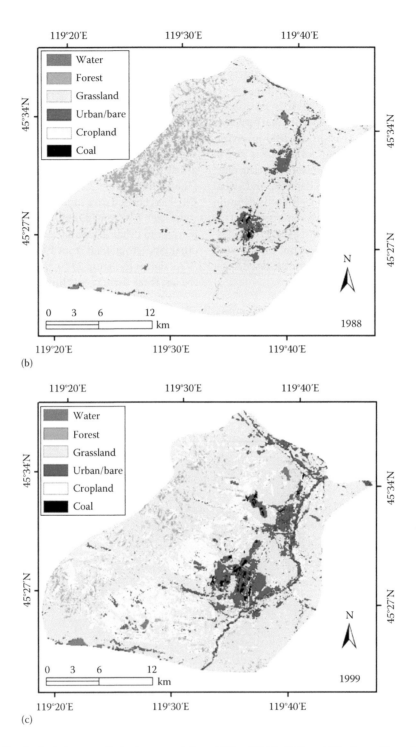

FIGURE 23.10 (*Continued*) Coal mining propelled urban growth from 1978 to 2011 in Holingol City, Inner Mongolia. (b) 1988, and (c) 1999.

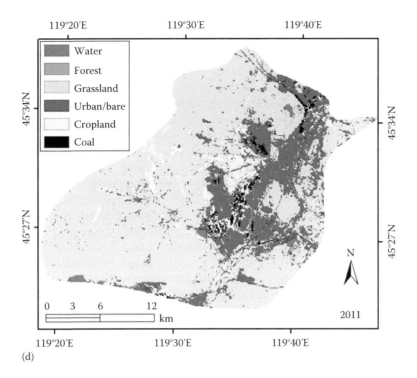

FIGURE 23.10 (*Continued*)   Coal mining propelled urban growth from 1978 to 2011 in Holingol City, Inner Mongolia. (d) 2011.

coal-based industries along the mining belt and around the edge of the city. Moreover, accompanied with coal-mining boom, the overall population of Holingol City increased from 29,897 in 1985 to 82,430 in 2010, which caused large areas of grassland transformed into coal-related industry and urban built-up area. The expansion of mining activities, emergence of coal-related industries, and increases in the population were the major contributors to the conversion of grassland to urban land-use types.

23.5   Conclusions

The first study demonstrated that rapid expansion of the area of urban land use around Tokyo metropolitan area accompanied the massive population increases that occurred in 1972–2011. The dominant changes that took place in the study area during this period were urban expansion over a wide area along transportation systems and large areas of cropland transformed into urban/built-up area, while such changes were accompanied by residents migrating to the outlying suburban areas. The second study focuses on the application of the grid square method to investigate the surface coal mining that caused urban expansion in Holingol City, Inner Mongolia, in 1978–2011. The results showed that mining area expansion was accompanied by rapid urban area expansion and caused a decline in the natural steppe grassland.

The grid cell–based method presented in this chapter is an innovative approach to the study of spatio-temporal urban growth patterns since it is capable of linking disparate data from many different agencies and organizations into a single comprehensive dataset that covers a wide range of spatial and temporal scales. The method as applied in the study has demonstrated

the capability to improve the knowledge, understanding, and analysis of urban dynamics. The results confirm that the grid square–based analysis allowed us to describe the spatio-temporal dynamics of the urban growth patterns in more detail.

References

Angel, S., J. Parent, D. L. Civco, A. Blei, and D. Potere. 2011. The dimensions of global urban expansion: Estimates and projections for all countries, 2000–2050. *Progress in Planning* 75: 53–107.

Avelar, S., R. Zah, and C. Tavares-Corrêa. 2009. Linking socioeconomic classes and land over data in Lima, Peru: Assessment through the application of remote sensing and GIS. *International Journal of Applied Earth Observation and Geoinformation* 11: 7–37.

Bagan, H. and Y. Yamagata. 2010. Improved subspace classification method for multispectral remote sensing image classification. *Photogrammetric Engineering and Remote Sensing* 76(11): 1239–1251.

Bagan, H. and Y. Yamagata. 2012. Landsat analysis of urban growth: How Tokyo became the world's largest megacity during the last 40 years. *Remote Sensing of Environment* 127: 210–222.

Bagan, H. and Y. Yamagata. 2014. Land-cover change analysis in 50 global cities by using a combination of Landsat data and analysis of grid cell. *Environmental Research Letters* 9(6): 064015.

Bagan, H., Y. Yasuoka, T. Endo, X. Wang, and Z. Feng. 2008. Classification of airborne hyperspectral data based on the average learning subspace method. *IEEE Geoscience and Remote Sensing Letters* 5(3): 368–372.

Berger, C., M. Voltersen., S. Hese, I. Walde, and C. Schmullius. 2013. Robust extraction of Urban land cover information from HSR multi-spectral and LiDAR data. *IEEE Journal of Selected Topics in Applied Earth Observations and Remote Sensing* 6: 2196–2211.

Catalán, B., D. Saurí, and P. Serra. 2008. Urban sprawl in the Mediterranean? Patterns of growth and change in the Barcelona Metropolitan Region 1993–2000. *Landscape and Urban Planning* 85: 174–184.

Dewan, A. M. and Y. Yamaguchi. 2009. Using remote sensing and GIS to detect and monitor land use and land cover change in Dhaka Metropolitan of Bangladesh during 1960–2005. *Environmental Monitoring and Assessment* 150: 237–249.

Doll, C. N. H. 2008. CIESIN thematic guide to night-time light remote sensing and its applications. Center for International Earth Science Information Network (CIESIN), Columbia University, Palisades, NY.

Doll, C. N. H., J. P. Muller, and J. G. Morley. 2006. Mapping regional economic activity from night-time light satellite imagery. *Ecological Economics* 57: 75–92.

Elvidge, C. D., B. T. Tuttle, P. C. Sutton, K. E. Baugh, A. T. Howard, C. Milesi, B. Bhaduri, and R. Nemani. 2007. Global distribution and density of constructed impervious surfaces. *Sensors* 7: 1962–1979.

Frolking, S., T. Milliman, K. C. Seto, and M. A. Friedl. 2013. A global fingerprint of macro-scale changes in urban structure from 1999 to 2009. *Environmental Research Letters* 8(2): 024004. doi: 10.1088/1748–9326/8/2/024004.

Geospatial Information Authority of Japan (GSI). 2012. Maps, aerial photographs and survey results (in Japanese). Available at: http://www.gsi.go.jp/tizu-kutyu.html (last accessed: May 15, 2013).

Giri, C., B. Pengra, J. Long, and T. R. Loveland. 2013. Next generation of global land cover characterization, mapping, and monitoring. *International Journal of Applied Earth Observation and Geoinformation* 25: 30–37.

Lambin, E. F., H. J. Geist, and E. Lepers. 2003. Dynamics of land-use and land-cover change in tropical regions. *Annual Review of Environment and Resources* 28: 205–241.

Oja, E. 1983. *Subspace Methods of Pattern Recognition.* Research Studies Press and John Wiley & Sons, Letchworth, U.K.

Okata, J. and A. Murayama. 2010. Tokyo's urban growth, urban form and sustainability. In A. Sorensen and J. Okata (Eds.), *Megacities: Urban Form, Governance, and Sustainability.* Springer, New York, pp. 15–41.

Parés-Ramos, I. K., N. L. Álvarez-Berríos, and T. M. Aide. 2013. Mapping urbanization dynamics in major cities of Colombia, Ecuador, Perú, and Bolivia using night-time satellite imagery. *Land* 2: 37–59.

Potere, D. and A. Schneider. 2007. A critical look at representations of urban areas in global maps. *GeoJournal* 69: 55–80.

Potere, D., A. Schneider, S. Angel, and D. L. Civco. 2009. Mapping urban areas on a global scale: Which of the eight maps now available is more accurate? *International Journal of Remote Sensing* 30: 6531–6558.

Qian, T., H. Bagan, T. Kinoshita, and Y. Yamagata. 2014. spatial–temporal analyses of surface coal mining dominated land degradation in Holingol, Inner Mongolia. *IEEE Journal of Selected Topics in Applied Earth Observations and Remote Sensing* 7: 1675–1687.

Saizen, I., K. Mizuno, and S. Kobayashi. 2006. Effects of land-use master plans in the metropolitan fringe of Japan. *Landscape and Urban Planning* 78(4): 411–421.

Schneider, A., M. A. Friedl, and D. Potere. 2009. A new map of global urban extent from MODIS satellite data. *Environmental Research Letters* 4: 044003.

Small, C. 2003. High spatial resolution spectral mixture analysis of urban reflectance. *Remote Sensing of Environment* 88: 170–186.

Small, C. and J. E. Cohen. 2004. Continental physiography, climate, and the global distribution of human population. *Current Anthropology* 45(2): 269–277.

Small, C., F. Pozzi, and C. D. Elvidge. 2005. Spatial analysis of global urban extents from the DMSP-OLS Night Lights. *Remote Sensing of the Environment* 96(3–4): 277–291.

Solecki, W., K. C. Seto, and P. J. Marcotullio. 2013. It's time for an urbanization science. *Environment: Science and Policy for Sustainable Development* 55(1): 12–17.

Statistics Bureau, Japan. 1973. Standard grid square and grid square code used for the statistics. http://www.stat.go.jp/english/data/mesh/02.htm. (Accessed April 28, 2015.)

Statistics Bureau, Japan. 2011. A guide to the statistics bureau, the director-general for policy planning (statistical standards) and the statistical research and training institute, pp. 85–87. http://www.stat.go.jp/english/info/guide/2011ver/pdf/2011ver.pdf. (Accessed April 28, 2015.)

Taubenböck, H., T. Esch, A. Felbier, M. Wiesner, A. Roth, and S. Dech. 2012. Monitoring urbanization in mega cities from space. *Remote Sensing of Environment* 117: 162–176.

Taubenböck, H., M. Wiesner, A. Felbier, M. Marconcini, T. Esch, and S. Dech. 2014. New dimensions of urban landscapes: The spatio-temporal evolution from a polynuclei area to a mega-region based on remote sensing data. *Applied Geography*, 47: 137–153.

Townshend, J. R. and C. O. Justice. 2002. Towards operational monitoring of terrestrial systems by moderate-resolution remote sensing. *Remote Sensing of Environment* 83: 351–359.

United Nations Department of Economic and Social Affairs, Population Division (2014) World Urbanization Prospects, the 2014 Revision: Highlights (United Nations, New York). http://www.un.org/en/development/desa/publications/2014-revision-world-urbanization-prospects.html. (Accessed April 28, 2015.)

Zhang, Q. and K. C. Seto. 2011. Mapping urbanization dynamics at regional and global scales using multi-temporal DMSP/OLS nighttime light data. *Remote Sensing of Environment* 115: 2320–2329.

24

Latest High-Resolution Remote Sensing and Visibility Analysis for Smart Environment Design

Yoshiki Yamagata
National Institute for Environmental Studies

Daisuke Murakami
National Institute for Environmental Studies

Hajime Seya
Hiroshima University

Acronyms and Definitions

2D	Two-dimensional
3D	Three-dimensional
ASTER	Advanced Spaceborne Thermal Emission and Reflection Radiometer
C1	Category 1
C2	Category 2DEM Digital elevation model
DSM	Digital surface model
EVs	Electric vehicles
GIS	Geographic information systems
JGD 2011	Japan Geodetic Datum 2011
LiDAR	Light detection and ranging
PVs	Photovoltaic panels
RD	Residential district
TM	Thematic Mapper

24.1 Introduction

24.1.1 Remote Sensing and Urban Analysis

Since the first Earth observation satellite, Landsat 1, was launched in 1972, remote sensing technology, which furnishes spatial information with high frequency of updating with low cost (see Donnay et al., 2001), has gained considerable attention. It is also true in regional/urban studies, especially, after some earlier studies clarified usefulness of remotely sensed information as a proxy of urban conditions.

For example, Forster (1983) revealed that remotely sensed imagery effectively describes urban residential qualities.

A limitation of earlier sensors was that while fine spatial resolution is vital in remote sensing–based urban analysis (Welch, 1982), they have coarse spatial resolutions (e.g., Landsat 1 multispectral scanner has a spatial resolution of 80 m). However, so-called second-generation sensors (e.g., Landsat Thematic Mapper, Advanced Spaceborne Thermal Emission and Reflection Radiometer), whose spatial resolutions are roughly between 10 and 30 m, emerge around the 1990s, and thereby, urban studies were greatly stimulated. Furthermore, after the appearance of third-generation sensors (e.g., IKONOS, QuickBird, WorldView-1, GeoEye-1, and WorldView-2), whose spatial resolution is roughly between 0.5 and 5 m, in the last one or two decades, more and more studies start using remote sensing imagery in intraurban scale analysis (see Patino and Duque, 2013). Remote sensing is now widely accepted as a technique of monitoring detailed urban environmental condition with a lower cost than their in situ observation that extremely costly in most cases (Miller and Small, 2003).

24.1.2 Remote Sensing and Smart Environment

Recently, an increasing number of urban studies focus on the concept of "smart city." While smart city lacks definitional precision yet (Hollands, 2008), according to Giffinger et al. (2007), the following subconcepts characterize it: smart economy, smart people, smart governance, smart mobility, smart environment, and smart living. Blaschke et al. (2011) discussed that, among the six characters, remote sensing plays a critical role in designing a smart environment, which is characterized by attractive natural conditions (climate, green space, etc.), pollution, resource management, and efforts toward environmental protection. Indeed, many remote sensing studies discussing urban environment appear to finish many fruitful insights toward a smart environment.

The smart environment concept emphasizes the smart use of natural resources. In other words, natural resources must be managed in a sensible way while considering their values as both market goods (goods that are traded in markets) and nonmarket goods (goods that are not traded in markets).

24.1.2.1 Remote Sensing and Natural Resources as Market Goods

Energies are principal market goods that natural resources provide. Among various natural energies, renewable energies (e.g., solar radiation and wind power) are key in sustainable development and have gained considerable attention in recent years. Regarding this topic, there is a common recognition that electric vehicles (EVs) and solar photovoltaic panels (PVs) are among the most important elements. This is because they can potentially work without emitting CO_2 provided that EVs are charged solely by using renewable-energy-based electricity, such as solar PVs (e.g., Giannouli and Yianoulis, 2012; Denholm et al., 2013); if not, use of (nonrenewable) energy to charge EVs may increase CO_2 emissions (e.g., Wu et al., 2012). Accordingly, management of solar PVs is a critical problem.

Solar PVs must be allocated considering actual amount of solar radiation. Although in situ observation of solar radiation across a region is likely to be very costly, solar radiation can be estimated accurately using a remote sensing imagery, such as light detection and ranging (LiDAR) data, which precisely describe detailed 3D urban fabrics (e.g., Chun and Guldmann, 2014; Tooke et al., 2014). Figure 24.1 is an example of the LiDAR-based estimates of rooftop solar radiations on an average day in August in the central part of Yokohama city, Japan, created using ArcGIS Spatial Analyst extension. This figure represents that high-resolution solar radiation amounts in arbitrary time can easily be estimated with the LiDAR data.

24.1.2.2 Remote Sensing and Natural Resources as Nonmarket Goods

Natural resources are important nonmarket goods that improve human amenity. Human amenity has been studied extensively by applying the hedonic approach (Rosen, 1974), which is one of the typical ways to evaluate the economic value of nonmarket goods, sometimes using a regression model. Most hedonic studies focused on the 2D structure of natural resources such as the accessibility to natural resources or scale of them (e.g., Tyrvainen and Miettinen, 2000; Irwin, 2002; Morancho, 2003; Tajima, 2003; Kong et al., 2007; Cho et al., 2008), and remote sensing imageries have often been used in these studies to acquire 2D placement data of natural resources (e.g., actual placement of green area).

However, natural resources are usually perceived through human eyes as 3D objects, and these economic values may be evaluated considering the third dimension (i.e., height). Fortunately, recent remotely sensed 3D measurement, including

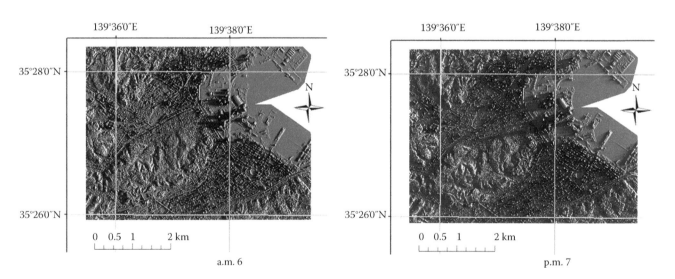

FIGURE 24.1 Example of the estimation of rooftop solar radiation for the central part of Yokohama City in August (these images are created as follows: (1) Airborne LiDAR data, which are observed in February 2006, February 2007, or April 2007, are mosaicked to one imagery and then (2) a digital surface model [DSM; spatial resolution, 0.5 m] is created using the mosaicked LiDAR data; subsequently (3) solar radiation is estimated using solar radiation analysis tools available with the ArcGIS Spatial Analyst extension): the dark color denotes poor radiation, whereas the light color denotes greater radiation.

the LiDAR measurement, accurately describes 3D objects, including buildings and natural resources (e.g., trees and mountains). As a result, an increasing number of hedonic studies start discussing economic values of 3D perception or 3D view.

24.1.2.3 Remote Sensing and Natural Resources: Summary

This section focused on energy analysis and hedonic analysis that must be discussed toward smart environment management. Remote sensing applications in energy analysis are already well summarized, for example, by Liang et al. (2010). In contrast, while remote sensing data have often been applied in hedonic analysis of 2D natural resources, application of remote sensing in hedonic analysis of 3D view is still limited, and this overview seems insufficient.

Thus, the remainder of this chapter focuses on hedonic analysis. We first review 3D view evaluation approaches, and, then, hedonic analyzes of 3D view are reviewed. Subsequently, a hedonic analysis of 3D views, including open view, green view, and ocean view, is conducted using a condominium data and LiDAR data in Yokohama, Japan.

24.2 Visibility Analysis

24.2.1 Classical Visibility Evaluation Approaches

Visibility has been discussed from earlier than the appearance of high-resolution sensors. Some earlier studies used dummy variables indicating 1 if a target object is visible and 0 otherwise. For example, McLeod (1984) used the dummy variable of river view, while Benson et al. (1998) used dummy variables of ocean view, lake view, and mountain view. Some of the other studies evaluated views based on field investigations. For instance, Luttik (2000) extracted information on environmental factors and other location factors from maps, complemented by field investigations; Tyrvainen and Miettinen (2000) and Lange and Schaeffer (2001) performed a field investigation to get window view information, and Brown and Raymond (2007) performed a mail survey.

However, both the dummy variable-based approach and the field investigation–based approach have drawbacks: The former ignores quality of view, that is, views are evaluated by 1 (visible) or 0 (invisible), and the latter is time-consuming and difficult to implement with large samples.

24.2.2 Viewshed Analysis and Isovist Analysis

There are two more sophisticated visibility evaluation approaches (Leduc et al., 2010): the isovist analysis (Benedikt, 1979), which has been discussed mainly in architectural studies and urban studies, and the viewshed analysis (Lynch, 1976), which has been discussed mainly in landscape studies.

Isovist represents the space that is visible point from an observation point (Benedikt, 1979). An isovist is usually evaluated in a 2D space and described by a 2D polygon. In other words, the typical isovist is a 2D horizontal slice of human perception (Yu et al., 2007). This approach is suitable to evaluate views in urban space, whose qualities strongly depend on the actual placement of buildings and the other objects, whose 2D structure can be described by 2D polygons. An obvious disadvantage of the isovist approach is the ignorance of the third dimension (i.e., height). The 3D isovist (e.g., Morello and Ratti, 2009) is an extension that copes with this problem.

On the other hand, the viewshed analysis quantifies 3D view by examining whether each cell in a 3D raster is visible from an observation point. This approach had rarely been adopted in urban setting because of the lack of 3D raster that precisely describes complex 3D urban geometry (Yu et al., 2007; Leduc et al., 2010). However, recent high-resolution sensors solved this problem, and now, more and more viewshed analyses start discussing the 3D urban view.

Now, both isovist-based and viewshed-based approaches are applicable to quantify 3D views in urban space if we have 3D data of describing urban geometry; high-resolution remote sensing data must have a very important role in this respect. Although the standard 2D isovist was originally a vector-based approach, which does not use any raster data, the 3D isovist approach relies on a raster data. Hence, the high-resolution remote sensing data must contribute to not only viewshed analysis, which is originally a raster-based approach, but also to 3D isovist analysis. Such a similarity between viewshed analysis and (3D) isovist analysis suggests that the difference between them is getting more and more obscure. In fact, some studies discuss them in integrated manners (e.g., Llibera, 2003; Yu et al., 2007). For more details about the isovist approach and the viewshed approach, see these papers.

24.2.3 Three-Dimensional View Indexes

Numerous indexes quantifying and characterizing 3D view have been developed under the framework of isovist or viewshed analysis.

Some indexes evaluate the volume of space that is visible from a viewpoint, just like the standard isovist. Table 24.1 summarizes such isovist-oriented indexes, including sky view factor (Ratti, 2002), sky opening index (Teller, 2003), spatial openness index (Fisher-Gewirtzman and Wagner, 2003), and viewsphere index (Yu et al., 2007). The first two measures perceived openness by evaluating the amount of view to the sky. Besides, sky view factor is now a well-accepted index in energy analysis (Yu et al., 2007); for example, solar radiation can be estimated using this index, and the result can be utilized for the solar PV allocation problem and the other problems relating the urban heat island (e.g., Chun and Guldmann, 2014). The last two measures perceived density. Spatial openness index has been adopted frequently to evaluate buildup environment (e.g., Fisher-Gewirtzman and Wagner, 2003).

TABLE 24.1   Isovist-Oriented 3D Indexes

Index	Definition	Volume of 3D View from a Site s: $V(s)$	Search Space: S (Volume: $V(S)$)
Sky view factor	$\dfrac{V(s)}{V(S)}$	Volume of the space in S that line of sights from s to the sky through	A hemisphere representing the visible area from s
Sky opening index	$\dfrac{V(s)}{V(S)}$	Volume of the area in S representing the sky	A fish-eye (2D) circle whose center is s
Spatial openness index	$V(s)$	Volume of the visible space in S	A sphere representing the visible area from s
Viewsphere index	$\dfrac{V(s)}{V(S)}$	Volume of the space in S that line of sights from s to ground objects through	A hemisphere representing the visible area from s

Note: s is a viewpoint and S is the space that is visible from the viewpoint if any obstructions are present.

Viewsphere index is available, for example, to discuss facility allocation problems. For example, Yang et al. (2007) applied it for a campus allocation problem in Singapore.

On the other hand, some other indexes examine visibility of each cell in a 3D raster, just like the viewshed analysis. This type of indexes has been discussed extensively; particularly, after the 3D analyst toolbox, a collection of 3D spatial analysis tools including viewshed analysis tools, was installed in ArcGIS (ESRI Inc.) in 1998 (Ma, 2003). Table 24.2 summarizes such viewshed-oriented indexes. The standard viewshed examines whether each cell is visible or not from an observation point. The cumulative viewshed and the total viewshed (Wheatley, 1995; Llibera, 2003) accumulate standard viewsheds from each observation point and provide a raster of recoding in each cell the number of observation points visible. The difference between the two indexes is that the former assumes a set of selected sites as observation points, while the latter assumes observation points being distributed all over the target region. On the other hand, viewshed-oriented (or raster-based) index was also proposed in the context of the isovist analysis. Iso-visi-matrix (Morello and Ratti, 2009) is such an index. It provides a 3D raster map whose cells are colored according to the number of cells in the raster that are visible from the cell. From a viewpoint of viewshed analysis, this index appears to be essentially identical to the total viewshed.

While aforementioned indexes evaluate visibility to each cell by visible or not, some other indexes explicitly evaluate quality of view to each cell. For instance, visual exposure (Llibera, 2003; Domingo-Santos et al., 2011) evaluates perceived cell size by modeling visible angle or visible span (both horizontal and vertical) of the cell, and visual magnitude (Grêt-Regamey

et al., 2007; Chamberlain and Meitner, 2013) evaluates the perceived size in a similar manner with a simpler approach. The usefulness of indexes considering perceived cell size is demonstrated, for example, by Chamberlain and Meitner (2013).

Viewshed-oriented indexes have been adopted in various purposes, including wind turbine allocation problem (Benson et al., 2004), negative visual impact assessment of marble quarry expansion (Mouflis et al., 2008), visibility assessment of landmark buildings (Bartie et al., 2010), coastal aquaculture site selection problem (Falconer et al., 2013), visibility assessment from a road (Chamberlain and Meitner, 2013), and visual exposure assessment of smoking behavior (Pearson et al., 2014). Besides, viewshed-oriented indexes have often been used in hedonic analysis (see the next section).

Some convenient indexes of that characterize view, rather than quantifying view like isovist or viewshed analysis, have been proposed too. For example, Teller (2003) proposed several indicators of characterizing sky view, including the mean distance to skyline; skyline regularity, which is defined by the standard deviation of heights of skyline; eccentricity or asymmetry of skyline; and spread, which indicates deviation of the sky line from the circle line. Llibera (2003) evaluated visual prominence of a cell by differencing visibility to the cell with visibilities to the neighborhood cells. Bartie et al. (2010) characterized view to a ground object by mapping the following indicators: the percentage of visible cells amounting for the cells describing the object; the largest horizontal angle of the target object; the visible facade area of the object; the perceived size, which is calculated based on the distance to the object; clearness, which is defined by (visible area)/(visible area if all other objects are removed from the

TABLE 24.2   Viewshed-Oriented 3D Indexes

Indexes	Description
Standard viewshed	Whether each cell is visible or not from an observation point.
Cumulative viewshed	Number of observation points that each cell can see.
Total viewshed	
Iso-visi-matrix	
Visual exposure	Perceived size of cells. They are evaluated by modeling the visual
Visual magnitude	angle or visual span (both horizontal and vertical) of the cells.

scene); and the skyline, which is defined by the percentage of the skyline of the object that is not overshadowed by the other taller more distant objects.

24.2.4 Hedonic Analysis of View

Numerous hedonic studies have examined economic values of views by analyzing their impacts on dwelling prices. Bourassa et al. (2005), and Jim and Chen (2009) summarize some important literatures.

Results of recent hedonic studies are summarized in Table 24.3. As shown in the table, many types of view, including open view (openness of visibility), ocean view (visibility of ocean), green view (visibility of open space or forest), and urban view (visibility of urban elements such as streets or buildings), have been analyzed, and many of these studies have confirmed the positive economic value of open view and water view and the negative economic value of urban view. Especially, prominent positive effect of water view is suggested in a number of studies (e.g., Benson et al., 1998; Yu et al., 2007; Jim and Chen, 2009). On the other hand, the economic value of green view is unclear, and it appears both positive and negative depending on studies. This uncertainty might partly be due to the problem of data resolution: It was difficult to get a high-resolution remote sensing data that precisely capture the actual placement and height of greens (e.g., trees). Fortunately, the third-generation sensors allow us to acquire the data of actual placement and height of trees.

The next section illustrates a hedonic analysis of 3D views, including green view, which are evaluated by the LiDAR data.

24.3 Application: 3D View Analysis in Yokohama City, Japan

24.3.1 Outline

This section evaluates economic value of 3D views, including open views, green views, and ocean views, by a hedonic analysis of apartment unit prices (source: Marketing Research Center Co. Ltd.). The target area is the seven central wards in Yokohama city (Figure 24.2), which is located about 30 km south from the central Tokyo area. The seven wards include 694 apartment buildings composed of 27,446 apartment units (see Figure 24.3). Some of the results have been reported in Yamagata et al. (2013).

Yokohama is the second largest populated city in Japan, whose population is about 3.70 million in 2014. Yokohama opened its port to foreign trade at the first time in Japan in 1959. Since then, Yokohama has rapidly developed as an entrance of Japan, while taking a lot of foreign cultures. The relics are now around Yokohama, for example, various western-style historic buildings and the largest Chinatown in Japan are in Yokohama. On the other hand, a port side area in Yokohama was redeveloped after 1980s, and the redeveloped area, which is called Minato Mirai 21, is known as a sophisticated urban area. Thus, Yokohama is a large city characterized by both foreign cultures and sophisticated urban facilities.

TABLE 24.3 Results of Hedonic Studies of Views after 2005

| | Open View | Water | | | Green | | | Urban | Data/Assessment to Estimate 3D View |
		Ocean	Lake/Wetland	River	Garden/Park	Forest	Mountain	Urban Area	
Bourassa et al. (2005)			Black						Dataset including view information
Boxall et al. (2005)								Gray	Gathered from multiple listing service
Brown and Raymond (2007)	Black								Interview
Jim and Chen (2006)						Black			Field survey (dummy)
Jim and Chen (2007)			Black			Black (Except for old town)		Insig.	Questionnaire survey
Yu et al. (2007)			Black						DSM
Hui et al. (2007)			Black						Digital map of Hong Kong (B5000)
Sander and Polasky (2009)	Black		Black		Black (Grass)	Insig.			DEM
Cavailhès et al. (2009)					Gray	Black			DEM
Jim and Chen (2009)			Black						1:5000 digital map
Hui et al. (2011)					Black				Maps
Hindsley et al. (2013)			Black						DSM
Panduro and Veie (2013) (house)			Black					Insig.	A map and aerial photo
Panduro and Veie (2013) (apartment)					Black	Insig.		Insig.	

Note: Black, positively significant at less than 10% level; gray, negatively significant at less than 10% level; Insig., insignificant.

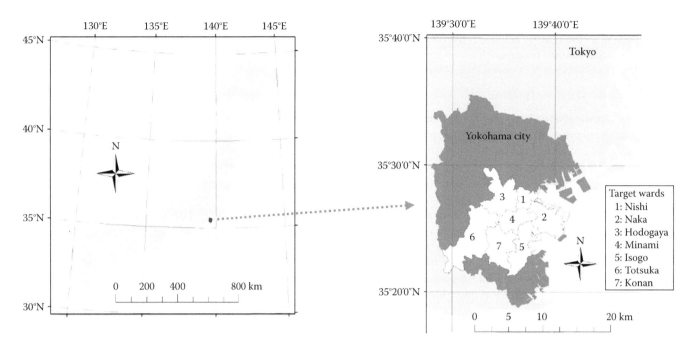

FIGURE 24.2   Target area.

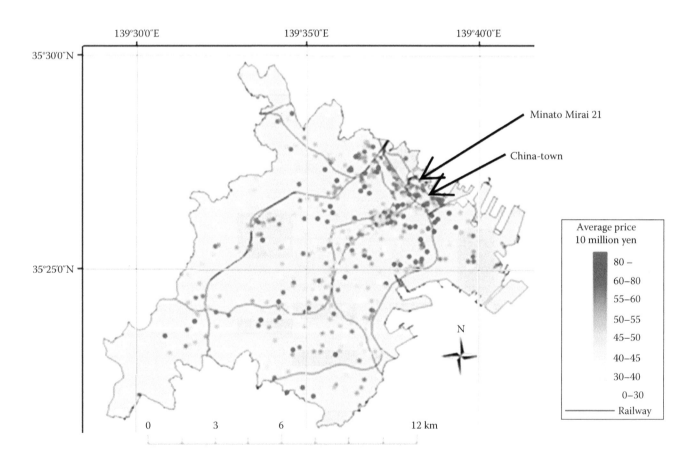

FIGURE 24.3   Average prices of the target apartments.

Note that, throughout our analysis, the Japan plane rectangular coordinate system zone 9 (Japan Geodetic Datum 2011: JGD 2011) is used for mapping, whereas, as an exception, the left side Figure 24.1 showing Japan was mapped with the geographic coordinate system, JGD 2011, which is suitable to display global map.

24.3.2 Three-Dimensional View Evaluation

We apply the cumulative viewshed to evaluate open view, green view, and ocean view. That is, numbers of cells, green cells, and ocean cells that are visible from each apartment unit are evaluated. Details of the calculation procedure are as follows:

1. The apartment data and building polygons (source: Fundamental Geospatial Data of Geographical Survey Institute of Japan) are manually associated using Google Maps and apartment web pages, and building polygons corresponding to each apartment are identified.
2. Floor heights of each unit are identified using room numbers.
3. Three-dimensional coordinates of each observation point are set for each unit.
4. The digital surface model (DSM; spatial resolution, 50 cm × 50 cm; Figure 24.4) and digital elevation model (DEM; spatial resolution, 50 cm × 50 cm; Figure 24.5) are obtained using airborne LiDAR data.
5. Visibilities from each observation point are evaluated using the ArcGIS 3D analyst, and open views, which are defined by the number of DSM cells that are visible from each observation point, are evaluated.

6. Trees are extracted and processed as follows:
   a. Actual placement of trees is identified by classifying an aerial photo, which is acquired simultaneously with the LiDAR data, using a likelihood maximization method.
   b. The extracted trees are spatially matched with the DSM, and DSM cells representing trees are identified.
   c. The heights of the trees in these cells are estimated by calculating the difference between the DSM and the DEM. Because the heights of trees are generally more than 50 cm, we included only those cells in which the tree heights are over 50 cm (Figure 24.6). This processing is necessary to remove noise.
7. Cells containing ocean views are identified from the geographical information systems data provided by Yokohama City.

Green views and ocean views are estimated by counting the cells containing trees and ocean.

In step 3, we had no data of the location and directions of windows in the units. Hence, the 3D coordinates of the observation points are defined as follows: Following Yasumoto et al. (2011), heights are defined by (height of a floor [3 m]) × ([number of stories] − 1) + (height of human eyes [1.6 m]), and longitudes/latitudes are set at four midpoints of each side of the rectangles that approximated apartment buildings. Besides, we set the maximum visible range to 500 m following Yu et al. (2007) and Yasumoto et al. (2011). Although this setting is somewhat subjective, the number of visible cells does not increase significantly even if cells more distant than 500 m are counted (Yasumoto et al., 2011).

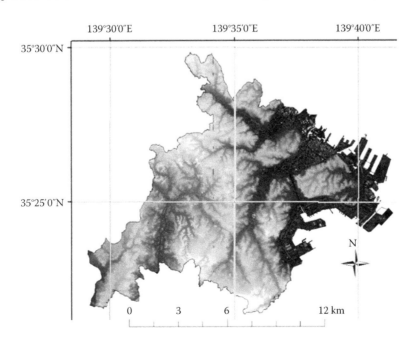

FIGURE 24.4 Digital surface model.

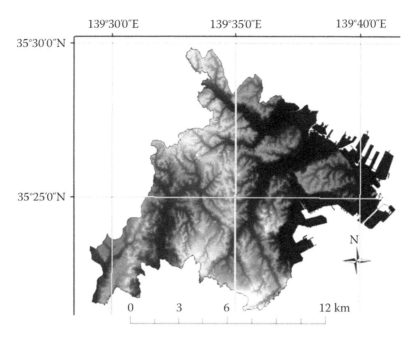

FIGURE 24.5   Digital elevation model.

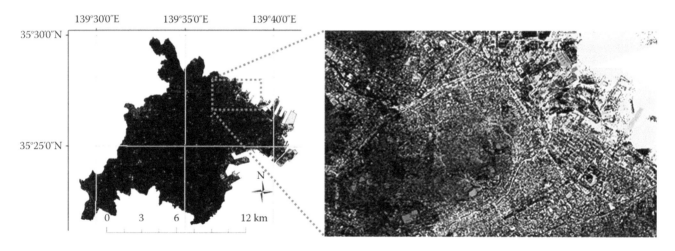

FIGURE 24.6   Extracted tree cells.

The estimated open view, green view, and ocean view are plotted in Figures 24.7 through 24.9, respectively. Open views are relatively small in the area near Yokohama station and the so-called Chinatown area, where buildings, which possibly obstruct views, are densely placed. While green views have similar tendencies, they show the greater gap between lower values in the Yokohama station and the Chinatown area and the higher values in the western woody area. Because most of the apartments were more than 500 m (i.e., the threshold distance) away from the ocean, ocean views are zero in most apartments. On the other hand, ocean views are prominent in the bayside redeveloped area, Minato Mirai 21.

24.3.3 Hedonic Analysis Results

Economic values of the view indexes are evaluated by a hedonic analysis. Here, the multilevel model (e.g., Hox, 1998), which is a regression model with unit-wise disturbance and building-wise disturbance, is applied. Explained variables are the logarithm of condominium prices, and explanatory variables are open views, green views, ocean views, and the other control variables summarized in Table 24.4. For the variable selection, the stepwise method is applied, and control variables that are not significant at the 10% level are omitted.

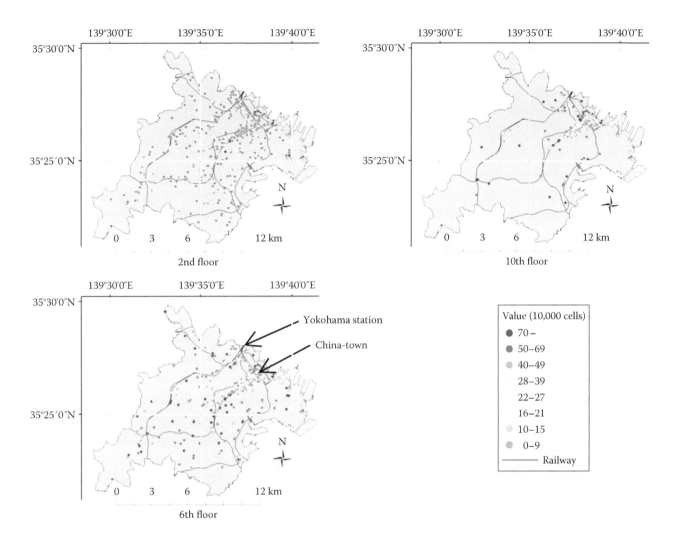

FIGURE 24.7 Open view.

Table 24.5 summarizes the estimation result. Among control variables, Area (+), Floor (+), Major dev. (+), Park (−), Ocean (−), C1 res. (−), and Semi ind. (−) are significant at more than the 5% level. These signs are intuitively consistent.

Open view and ocean view are positively significant at the 1% level. Significance of open view design and ocean view design in urban management is confirmed. On the other hand, green view is negatively significant at the 1% level, which is intuitively inconsistent. As discussed by Jim and Chen (2009), in bayside area, green view can have negative impact because it likely to be less preferred than ocean views. On the other hand, in inland area with no ocean view, green view might be preferred.

Thus, ward-wise hedonic analyzes are conducted. Estimated significance of view variables are plotted in Figure 24.10. Open view had positive impact in most wards; ocean view had positive impact only around the Minato Mirai 21 area; and, as expected, while green view is insignificant or negatively significant in bayside wards, including Nishi, Naka, and Isogo wards, it is positively significant in the western inland wards, including Totsuka and Konan wards. In other words, green

view is preferred in the inland area. Such insights regarding view to natural resources must be useful to design smart urban environment.

24.4 Concluding Remarks

This chapter first discussed roles of remote sensing in designing smart environment and indicated the importance of revealing the economic value of 3D view to natural resources. Then, after 3D view evaluation techniques are briefly reviewed, a hedonic analysis of open views, green view, and ocean view is conducted.

Throughout this section, we showed that remote sensing can provide a useful toolbox both for urban planning and design. One of the next important challenges is to use large social media–based remote sensing data like twitter. Different from traditional social economic surveys, such social media–based remote sensing data are accumulated in an almost real-time manner. Such large spatiotemporal data may contribute to the design of smart cities in the future, but at the same time, models should evolve to deal with such data efficiently and effectively.

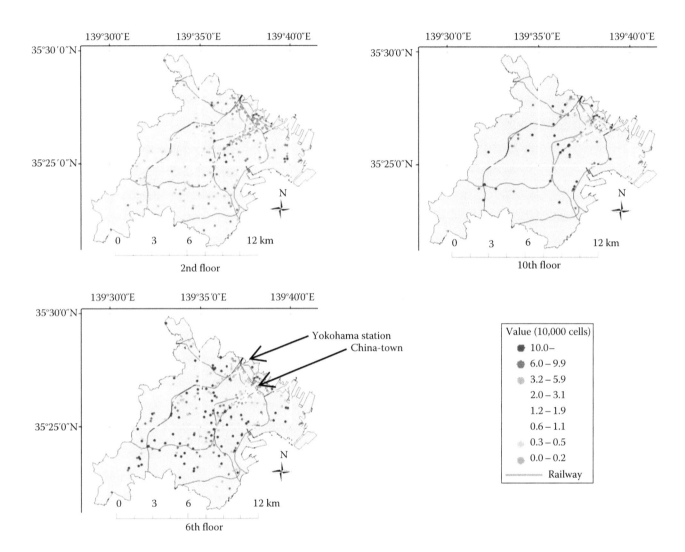

FIGURE 24.8　Green view.

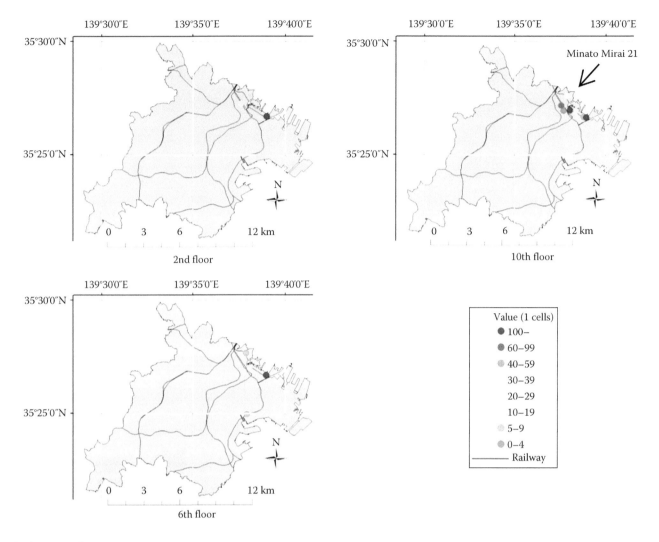

FIGURE 24.9 Ocean view.

TABLE 24.4 Control Variables

Variables	Description	Variables	Description
Const.	Constant	C1 high	Dummy of C1 medium-to-high exclusive RD
Area	Log. of unit area (m²)	C1 exclusive	Dummy of C1 exclusive RD
Floor	Log. of floor of unit	C2 res.	Dummy of category 2 (C2) RD
SRC	Dummy of the steel reinforced concrete structure	C2 high	Dummy of C2 medium-to-high exclusive RD
WRC	Dummy of the steel wall concrete structure	C2 exclusive	Dummy of C2 exclusive RD
Dev.	Numbers of related developers	Industry	Dummy of industrial districts
Major dev.	The ratio of major developers called MAJOR 8[a] accounting for dev.	Semi-ind.	Dummy of semi-industrial districts
Station	Log. of the travel time to the nearest station	Commerce	Dummy of commercial districts
Green[b]	Log. of the number of green cells within 500 m	Neigh. com.	Dummy of neighborhood commercial districts
Park	Log. of the distance to the nearest city park (km)	C1 res.	Dummy of category 1 (C1) residential districts (RD)
Ocean	Log. of the distance to the ocean (km)	C1 low	Dummy of C1 low-rise exclusive RD
Year	Year dummies (1993–2007)		

[a] MAJOR 8 includes Sumitomo Realty and Development Co., Ltd.; Tokyo Land Corporation; Mitsubishi Estate Co., Ltd.; Towa Real Estate Development Co., Ltd.; Daikyo Inc.; Nomura Real Estate Development Co., Ltd.; Mitsui Fudosan Residential Co., Ltd.; and Tokyo Tatemono Co., Ltd.

[b] Green is calculated using the tree cells shown in Figure 24.6; Park and Ocean are calculated using the GIS data provided by Yokohama city, and the other variables, except for the view variables, are from the condominium dataset of Marketing Research Center Co. Ltd.

TABLE 24.5   Estimation Result

Variables[a]	Estimate	t Value	
Const.	3.28	1.87×10^2	***
Area	1.13	4.09×10^2	***
Floor	6.19×10^{-2}	6.47×10^1	***
Major dev.	6.19×10^{-2}	4.78	***
Park	-1.16×10^{-2}	-2.39	**
Ocean	-5.22×10^{-2}	-9.65	***
C1 res.	-3.39×10^{-2}	-2.04	**
C1 exclusive	4.53×10^{-2}	1.77	*
Semi-ind.	-8.08×10^{-2}	-3.97	***
Open view	1.28×10^{-1}	2.21×10^1	***
Green view	-4.71×10^{-1}	-1.87×10^1	***
Ocean view	2.52×10^2	9.71	***
AIC	$-72,818$		

Note: *, **, and *** denote significant levels of 10%, 5%, and 1%, respectively.
[a] Estimates of year dummies are omitted.

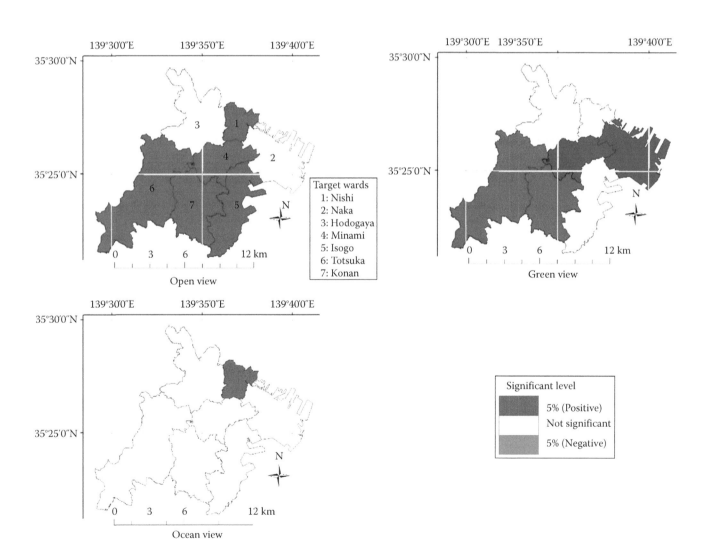

FIGURE 24.10   Significance levels of the view variables.

References

Bartie, P., Reitsma, F., Kingham, S., and Mills, S. 2010. Advancing visibility modeling algorithms for urban environments. *Computers, Environment and Urban Systems*, 34, 518–531.

Benedikt, M.L. 1979. To take hold of space: Isovists and isovist fields. *Environment and Planning B: Planning and Design*, 6, 47–65.

Benson, E.D., Hansen, J.L., Schwartz Jr., A.L., and Smersh, G.T. 1998. Pricing residential amenities: The value of a view. *Journal of Real Estate Finance and Economics*, 16, 55–73.

Benson, J.F., Scott, K.E., Anderson, C., Macfarlane, R., Dunsford, H., and Turner, K. 2004. Landscape capacity study for onshore wind energy development in the Western Isles. Scottish Natural Heritage Commissioned Report No. 42, Scottish Natural Heritage, Inverness, Scotland.

Blaschke, T., Hay, G.J., Weng, Q., and Resch, B. 2011. Collective sensing: Integrating geospatial technologies to understand urban systems—An overview. *Remote Sensing*, 3, 1743–1776.

Bourassa, S.C., Hoesli, M., and Sun, J. 2005. The price of aesthetic externalities. *Journal of Real Estate Literature*, 13, 165–188.

Boxall, P.C., Chan, W.H., and McMillan, M.L. 2005. The impact of oil and natural gas facilities on rural residential property values: A spatial hedonic analysis. *Resource and Energy Economics*, 27, 248–269.

Brown, G. and Raymond, C. 2007. The relationship between place attachment and landscape values: Toward mapping place attachment. *Applied Geography*, 27, 89–111.

Cavailhès, J., Brossard, T., Foltête, J.-C., Hilal, M., Joly, D., Tourneux, F.-P., Tritz, C., and Wavresky, P. 2009. GIS-based hedonic pricing of landscape. *Environmental and Resource Economics*, 44, 571–590.

Chamberlain, B.C. and Meitner, M.J. 2013. A route-based visibility analysis for landscape management. *Landscape and Urban Planning*, 111, 13–24.

Cho, S.-H., Poudyal, N.C., and Roberts, R.K. 2008. Spatial analysis of the amenity value of green space. *Ecological Economics*, 66, 2–3.

Chun, B. and Guldmann, J.M. 2014. Spatial statistical analysis and simulation of the urban heat island in high-density central cities. *Landscape and Urban Planning*, 125, 76–88.

Denholm, P., Kuss, M., and Margolis, R.M. 2013. Co-benefits of large scale plug-in hybrid electric vehicle and solar PV deployment. *Journal of Power Sources*, 236, 350–356.

Domingo-Santos, J.M., de Villarán, R.F., and Rapp-Arrarás, I. 2011. The visual exposure in forest and rural landscapes: An algorithm and a GIS tool. *Landscape and Urban Planning*, 101, 52–58.

Donnay, J.-P., Barnsley, M.J., and Longley, P.A. 2001. *Remote Sensing and Urban Analysis*. Taylor & Francis, London, U.K.

Falconer, L., Hunter, D.-C., Telfer, T.C., and Ross, L.G. 2013. Visual, seascape and landscape analysis to support coastal aquaculture site selection. *Land Use Policy*, 34, 1–10.

Fisher-Gewirtzman, D. and Wagner, I.A. 2003. Spatial openness as a practical metric for evaluating built-up environments. *Environment and Planning B: Planning and Design*, 30, 37–49.

Forster, B. 1983. Some urban measurements from Landsat data. *Photogrammetric Engineering and Remote Sensing*, 49, 1693–1707.

Giannouli, M. and Yianoulis, P. 2012. Study on the incorporation of photovoltaic systems as an auxiliary power source for hybrid and electric vehicles. *Solar Energy*, 86, 441–451.

Giffinger, R., Fertner, C., Kramar, H., Kalasek, R, Pichler-Milanovic, N., and Meijers, E. 2007. *Smart Cities—Ranking of European Medium-Sized Cities*. Centre of Regional Science, Vienna University of Technology, Vienna, Austria.

Grêt-Regamey, A., Bishop, I.D., and Bebi, P. 2007. Predicting the scenic beauty value of mapped landscape changes in a mountainous region through the use of GIS. *Environment and Planning B: Planning and Design*, 34, 50–67.

Hindsley, P., Hamilton, S.E., and Morgan, O.A. 2013. Gulf views: Toward a better understanding of viewshed scope in hedonic property models. *The Journal of Real Estate Finance and Economics*, 47, 489–505.

Hollands, R.G. 2008. Will the real smart city please stand up? *City*, 12, 303–320.

Hox, J.J. 1998. Multilevel modeling: When and why. In: Balderjahn, I., Mathar, R., and Schader, M. (Eds.), *Classification, Data Analysis, and Data Highways*. Springer, New York, pp.147–154.

Hui, E.C.M., Chau, C.K., Pun, C.K., and Law, M.Y. 2007. Measuring the neighboring and environmental effects on residential property value: Using spatial weighting matrix. *Building and Environment*, 42, 2333–2343.

Hui, E.C.M., Zhong, J.W., and Yu, K.H. 2011. The impact of landscape views and storey levels on property prices. *Landscape and Urban Planning*, 105, 86–93.

Irwin, E.G. 2002. The effects of open space on residential property values. *Land Economics*, 78, 465–480.

Jim, C.Y. and Chen, W.Y. 2006. Impacts of urban environmental elements on residential housing prices in Guangzhou (China). *Landscape and Urban Planning*, 78, 422–434.

Jim, C.Y. and Chen, W.Y. 2007. Consumption preferences and environmental externalities: A hedonic analysis of the housing market in Guangzhou. *Geoforum*, 38, 414–431.

Jim, C.Y. and Chen, W.Y. 2009. Value of scenic views: Hedonic assessment of private housing in Hong Kong. *Landscape and Urban Planning*, 91, 226–234.

Kong, F., Yin, H., and Nakagoshi, N. 2007. Using GIS and landscape metrics in the hedonic price modeling of the amenity value of urban green space: A case study in Jinan City, China. *Landscape and Urban Planning*, 79, 240–252.

Lange, E. and Schaeffer, P. 2001. A comment on the market value of a room with a view. *Landscape and Urban Planning*, 55, 113–120.

Leduc, T., Miguet, F., Tourre, V., and Woloszyn, P. 2010. Towards a spatial semantics to analyze the visual dynamics of the pedestrian mobility in the urban fabric. In: Painho, M., Santos, M.Y., and Pundt, H. (Eds.), *Geospatial Thinking*. Springer, Berlin, Germany, pp. 237–257.

Liang, S., Wang, K., Zhang, X., and Wild, M. 2010. Review on estimation of land surface radiation and energy budgets from ground measurement, remote sensing and model simulations. *IEEE Journal of Selected Topics in Applied Earth Observations and Remote Sensing*, 3, 225–240.

Llibera, M. 2003. Extending GIS-based visual analysis the concept of visualscapes. *International Journal of Geographical Information Science*, 17, 25–48.

Luttik, J. 2000. The value of trees, water and open space as reflected by house prices in the Netherlands. *Landscape Urban Planning*, 48, 161–167.

Lynch, K. 1976. *Managing the Sense of Regions*. MIT Press, Cambridge, MA.

Ma, J. 2003. From professional to people's software-tracing the development of 3D GIS software at ESRI. In: Ervni, S.M. and Buhmann, E. (Eds.), *Trends in Landscape Modeling: Proceedings at Anhalt University of Applied Science 2003*. Wichmann, Heidelberg, Germany, pp. 246–252.

McLeod, P.B. 1984. The demand for local amenity: An hedonic price analysis. *Environment and Planning A*, 16, 389–400.

Miller, R.B. and Small, C. 2003. Cities from space: Potential applications of remote sensing in urban environmental research and policy. *Environmental Science and Policy*, 6, 129–137.

Morancho, A.B. 2003. A hedonic valuation of urban green areas. *Landscape and Urban Planning*, 66, 35–41.

Morello, E. and Ratti, C. 2009. A digital image of the city: 3-D isovists in Lynch's urban analysis. *Environment and Planning B: Planning and Design*, 36, 837–853.

Mouflis, G.-D., Gitas, I.-Z., Iliadou, S., and Mitri, G. 2008. Assessment of the visual impact of marble quarry expansion (1984–2000) on the landscape of Thasos Island, NE Greece. *Landscape and Urban Planning*, 86, 92–102.

Panduro, T.E. and Veie, K.L. 2013. Classification and valuation of urban green spaces-A hedonic house price valuation. *Landscape and Urban Planning*, 120, 119–128.

Patino, J.E. and Duque, J.C. 2013. A review of regional science applications of satellite remote sensing in urban settings. *Computers, Environment and Urban Systems*, 37, 1–17.

Pearson, A., Nutsford, D., and Thomson, G. 2014. Measuring visual exposure to smoking behaviours: A viewshed analysis of smoking at outdoor bars and cafés across a capital city's downtown area. *BMC Public Health*, 14, 300.

Ratti, C. 2002. Urban analysis for environmental prediction. PhD thesis, University of Cambridge, Cambridge, U.K.

Rosen, S. 1974. Hedonic prices and implicit markets: Product differentiation in pure competition. *Journal of Political Economy*, 82, 34–55.

Sander, H. and Polasky, S. 2009. The value of views and open space: Estimates from a hedonic pricing model for Ramsey County, Minnesota, USA. *Land Used Policy*, 26, 837–845.

Tajima, K. 2003. New estimates of the demand for urban green space: Implications for valuing the environmental benefits of Boston's Big Dig Project. *Journal of Urban Affairs*, 25, 641–655.

Teller, J. 2003. A spherical metric for the field-oriented analysis of complex urban open spaces. *Environment and Planning B: Planning and Design*, 30, 339–356.

Tooke, T.R., Coops, N.C., and Webster, J. 2014. Predicting building ages from LiDAR data with random forests for building energy modeling. *Energy and Buildings*, 68, 603–610.

Tyrvainen, L. and Miettinen, A. 2000. Property prices and urban forest amenities. *Journal of Environmental Economics and Management*, 39, 205–223.

Welch, R. 1982. Spatial resolution requirements for urban studies. *International Journal of Remote Sensing*, 3, 139–146.

Wheatley, D. 1995. Cumulative viewshed analysis: A GIS-based method for investigating intervisibility and its archaeological application. In: Lock, G. and Stancic, Z. (Eds.), *Archaeology and Geographic Information Systems: A European Perspective*, Taylor & Francis, London, U.K pp. 171–186.

Wu, Y., Yang, Z., Lin, B., Liu, H., Wang, R., Zhou, B., and Hao, J. 2012. Energy consumption and CO_2 emission impacts of vehicle electrification in three developed regions of China. *Energy Policy*, 48, 537–550.

Yamagata, Y. Murakami, D., Seya, H., and Tsutsumi, M. 2013. A multilevel model based hedonic analysis of 3D visibility: An empirical case study of Yokohama Baycity. In: *Proceeding of the 13th Edition of the International Conference on Computers in Urban Planning and Urban Management*, Utrecht, the Netherlands.

Yang, P.P.-J., Putra, S.Y., and Li, W. 2007. Viewsphere: A GIS-based 3D visibility analysis for urban design evaluation. *Environment and Planning B, Planning and Design*, 34, 971–992.

Yasumoto, S., Jones, A., Nakaya, T., and Yano, K. 2011. The use of a virtual city model for assessing equality in access to views. *Computers, Environment and Urban Systems*, 35, 464–473.

Yu, S.-M., Han, S.-S., and Chai, C.-H. 2007. Modeling the value of view in high-rise apartments: A 3D GIS approach. *Environment and Planning B, Planning and Design*, 34, 139–153.

Summary

Remote Sensing of Water Resources, Disasters, and Urban Areas: Monitoring, Modeling, and Mapping Advances over the Last 50 Years and a Vision for the Future

Prasad S. Thenkabail
United States Geological Survey (USGS)

Acronyms and Definitions

AIRS	Atmospheric Infrared Sounder
ALEXI	Atmosphere-Land Exchange Inverse
ALOS	Advanced Land Observing Satellite "DAICHI"
AMSR-E	Advanced Microwave Scanning Radiometer EOS or Earth Observing System onboard Aqua satellite
AOD	Aerosol optical depth
ASAR	Advanced synthetic aperture radar
ASIS	Agricultural Stress Index Systems
AVHRR	Advanced very high resolution radiometer
ASTER	Advanced spaceborne thermal emission and reflection radiometer
COSMO-SkyMed	COnstellation of small Satellites for the Mediterranean basin Observation
CPW	Crop water productivity
CNES	The Centre national d'études spatiales or the National Center of Space Studies of France (CNES)
CryoSat	Europe's Satellite to Study Ice
DEM	Digital elevation model
DisAlexi	Disaggregated ALEXI
DLR	German Aerospace Center
DMSP	Defense Meteorological Satellite Program
DSS	Decision support systems
EOS	Earth observing system
ET	Evapotranspiration
ENVISAT	Environmental satellite
ERS	European remote sensing satellites
EVI	Enhanced Vegetation Index
GAC	Global area coverage
GEOSS	Global Earth Observation System of Systems
GHG	Greenhouse emissions
GHI	Global Health Index
GIS	Geographic information systems
GOES	Global Online Enrollment System
GPS	Global positioning system
GRACE	Gravity recovery and climate experiment (GRACE)
ICESat	Ice, Cloud, and land Elevation Satellite
IKONOS	A commercial earth observation satellite, typically, collecting sub-meter to 5 m data
IV	Inland valleys
JERS	Japanese Earth Resources Satellite
JRC	Joint Research Center
LIDAR	Light detection and ranging
LST	Land surface temperature
LULC	Land use land lover
MERIS	Medium resolution imaging spectrometer (MERIS)
METRIC	Mapping EvapoTranspiration at high Resolution with Internalized Calibration
METOP	Meteorological operational satellite program
MIR	Mid-infrared
MODIS	Moderate-resolution imaging spectroradiometer
MOD16	MODIS Global Evapotranspiration Dataset
MSS	Multi spectral scanner
NASA	National Atmospheric and Space Administration
NDSI	Normalized Difference Snow Index
NDVI	Normalized Difference Vegetation Index
NESDIS	National Environmental Satellite, Data, and Information Service
NOAA	National Oceanic and Atmospheric Administration
NPOESS	National Polar-Orbiting Operational Environmental Satellite System
NPP	Net primary productivity
OLI	Operational land imager
OLS	Operational linescan system
PALSAR	Phased array type L-band synthetic aperture radar
PDSI	Palmer Drought Severity Index
RADARSAT	Radar satellite
RISAT	Radar Imaging Satellite
RS	Remote sensing
SAR	Synthetic aperture radar
SCD	Snow-covered days
SCHIAMACHY	Scanning Imaging Absorption Spectrometer for Atmospheric Cartography
SEASAT	First satellite designed for remote sensing of the Earth's oceans with synthetic aperture radar (SAR)
SEB	Surface energy balance
SEBAL	Surface energy balance algorithm for land
SIR C\X	Spaceborne imaging radar-C/X
SMMR	Scanning Multichannel Microwave Radiometer
SPI	Standardized Precipitation Index
SPOT	Satellite Pour l'Observation de la Terre, French Earth Observation Satellites
SSEB	Simplified surface energy balance
SSMI/S	Special sensor microwave imager
SRTM	Shuttle Radar Topographic Mission
STARFM	Spatial and temporal adaptive reflectance fusion model
SWIR	Shortwave infrared
SWSI	Surface Water Supply Index
TCI	Temperature Condition Index
TerraSAR-X	A radar Earth observation satellite, with its phased array synthetic aperture radar
TIR	Thermal infrared
TM	Thematic mapper
TRMM	Tropical Rainfall Measuring Mission
TROPOMI	TROPOspheric monitoring instrument

NPP	Net primary productivity
NVIR	Visible and near-infrared
USDM	United States Drought Monitor
VCI	Vegetation Condition Index
VegDRI	Vegetation Drought Index
VH	Vegetation health
VHI	Vegetation Health Index
VI	Vegetation index
VIIRS	Visible Infrared Imaging Radiometer Suite
VegOut	Vegetation outlook
VUA	Vrije Universiteit Amsterdam
WF	Water footprint
WP	Water productivity
WUM	Water use mapping

This chapter provides a brief summary of all the 24 chapters appearing in this book of the Remote Sensing Handbook. This book has focus on remote sensing (RS) of water resources, disasters, and urban areas. The chapters in the Volume cover remote sensing of: (1) hydrology and water resources; (2) water use and water productivity (WP); (3) floods; (4) wetlands; (5) snow and ice; (6) nightlights; (7) geomorphology; (8) droughts and drylands; (9) disasters; (10) volcanoes; (11) fire; and (12) urban areas. Under each of these broad categories, one or more chapters provide a comprehensive coverage of the topic. For example, there are five chapters under droughts and drylands. The summaries provide a *window view* of what exists in each chapters as well as inter-linkages to various chapters. You can read this summary chapter in three ways: (1) before reading the chapters to get an overview, (2) after reading all the chapters to recap and refresh major highlights, and (3) do both (1) and (2) in order to capture the totality of all chapters. The chapter summaries also help the reader to establish interlinkages that exist between chapters, in a nutshell.

25.1 Hydrological Studies Using Multisensor Remote Sensing

Remote sensing (RS) has played a key role to better understand and characterize hydrological cycle. A wide array of hydrological parameters can be consistently characterized and mapped using a set of RS sensors gathering data in various portions of the electromagnetic spectrum such as the radar, optical, or thermal wavelengths. Some of these parameters like evapotranspiration (ET) and basin characteristics (e.g., topography, river morphology, vegetation, and land use) can be mapped with great degree of accuracies. Some other hydrological parameters like groundwater and soil moisture have greater degree of uncertainties mainly as a result of insufficient number of sensors of adequate resolution (e.g., existing sensors mostly have anywhere between 5 and 25 km spatial resolution). Chapter 1 by Dr. Sadiq Khan et al. provides us a comprehensive assessment of the key hydrological parameters (precipitation, ET, soil moisture, and groundwater) monitored and mapped by various sensors and their strengths and limitations.

The key hydrological parameters required in hydrological studies through the satellite sensors are briefly presented as follows (e.g., Figure 25.1):

Precipitation: Precipitation retrieval from satellite data comes from either visible and near-infrared (NIR) (VNIR) bands of geostationary Earth orbital (GEO) satellites gathered every 15–30 min and/or passive/active microwave images from low Earth orbital (LEO) satellites but acquired with much lower sampling frequency. GEO data are only indirectly related to surface rainfall, whereas passive microwave (PMW) sensors on LEO satellites give more direct sensing of rain clouds. Precipitation Estimation from Remotely Sensed Information using Artificial Neural Networks (Sorooshian et al., 2011) utilized both GEO and LEO data for rainfall estimation at high temporal resolution (e.g., 3 hourly [60S-60N] and 6 hourly [50S-50N]). These data are available at 0.25° (~27 km) grid. Tropical Rainfall Measuring Mission (TRMM; e.g., Figure 25.1), GEOS-8,10, geostationary meteorological satellites, Defense Meteorological Satellite Program (DMSP), and National Oceanic and Atmospheric Administration (NOAA) are some of the other satellite systems gathering precipitation data. Data are gathered in VNIR passive and active microwave.

ET: ET is one of the most widely measured hydrological parameter from remote sensing at various scales. The most widely used technique for actual ET (water use; mm/day) computation is through surface energy balance (SEB) modeling. SEB models typically required data from visible, near infrared, and thermal bands. Sensors that have thermal bands such as Landsat-5,7,8 (60–120 m), Advanced Spaceborne Thermal Emission and Reflection Radiometer (ASTER) (90 m), Terra/Aqua Moderate-resolution Imaging Spectroradiometer (MODIS) (1020 m), and NOAA Advanced Very High Resolution Radiometer (AVHRR) (1000 m) are widely used in ET modeling. The most widely used SEB models are

- Surface energy balance algorithm for land (SEBAL)
- Mapping evapotranspiration at high resolution with internalized calibration (METRIC)
- SEB system model

Soil moisture: Soil moisture (e.g., Figure 25.1) is important in (1) agricultural crop management (e.g., irrigation scheduling, droughts), (2) understanding ET flux, (3) understanding water and heat energy between the land surface and the atmosphere through evaporation and plant transpiration, and (4) amount of precipitation runoff. Soil moisture is mostly measured by passive RS in the microwave region with best frequencies of about 6 GHz (C band), but also in 1 and 3 GHz (L band) frequencies for detection of soil moisture. Soil moisture is measured by sensors such as Advanced Microwave Scanning Radiometer for the Earth Observing System

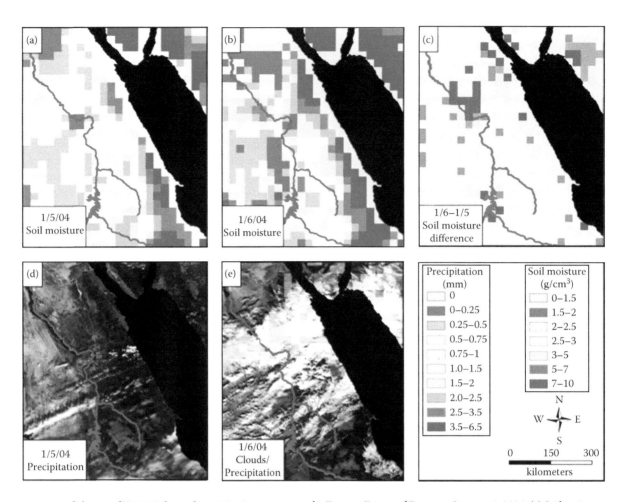

FIGURE 25.1 Validation of TRMM-derived precipitation event over the Eastern Desert of Egypt on January 6, 2004. (a) Soil moisture content extracted from Advanced Microwave Scanning Radiometer for EOS acquired on January 5, 2004. (b) Soil moisture on January 6, 2004. (c) Soil moisture difference image (January 6, 2004 image minus January 5, 2014 image). (d) Advanced Very High Resolution Radiometer (AVHRR) image showing minimal cloud coverage on January 5, 2004. (e) AVHRR image acquired on January 6, 2004 showing extensive cloud coverage (white areas). Also shown on (e) is TRMM-derived precipitation (colored areas). (From Milewski, A. et al., *J. Hydrol.*, 373(1–2), 1, 2009.)

(EOS) (AMSR-E; e.g., Figure 25.1) onboard Aqua with 12 channel, six-frequency PMW radiometer acquiring data in 6 × 4 km (89 GHz), 14 × 8 km (36.5 GHz), and 74 × 43 km (6.9 GHz); DMSP the special sensor microwave/imager (SSM/I) in 19.3–85.5 GHz in resolutions of 15 × 13 km (85.5 GHz) and 37 × 28 (37 GHz), TRMM Microwave Imager observations at 10.65 GHz with 25 km; and scanning multichannel microwave radiometer (SMMR) onboard Nimbus-7 in 6.6–37 GHz and in resolutions 27 × 18 (37 GHz) and 148 × 95 (6.6 GHz). As is clear from these sensor characteristics, the main problem with solid moisture sensors is the very coarse resolution.

Groundwater: Groundwater is very difficult to map and are often done by indirect means. Early studies used lineaments. The twin Gravity Recovery and Climate Experiment (GRACE) satellites measure Earth's gravity field, and its data are used to establish terrestrial water storage variation by detecting gravitational anomalies (Rodell, 2007).

Surface water: Surface water changes can be measured using AMSR-E. GRACE data can be used to assess surface water storage. NASA-(CNES or The Centre national d'études spatiales) plans to launch Surface Water and Ocean Topography mission in 2019 to study surface water.

Snow, ice, and glaciers: Snow, ice, and glaciers play a key role in Earth's energy and water balance. Many rivers have significant to large proportion of the flow from snowmelt runoff. Satellite sensors such as the AVHRR, MODIS, and Geostationary Operational Environmental Satellites (GOES) have been used for snow cover mapping over many years. In addition, microwave sensors such as conventional SSMI/S and AMSR-E and more recent missions and sensors such as synthetic-aperture radar (SAR), RADARSAT, ENVISAT, MERIS, ICESat, and CryoSat have played an important role in studies pertaining to snow, ice, and glaciers.

Basin characteristics: Water balance is often accounted at varying watershed and river basin scales. This requires us to obtain basin characteristics such as topography, geology, soils, vegetation, and land use and geomorphological characteristics (e.g., river length, stream orders, and stream elevation). All of these properties can be studied using a wide array of optical and radar RS.

Chapter 1 by Dr. Sadiq Khan presents the main water budget parameters discussed earlier except the basin characteristics.

25.2 Groundwater Studies Using Remote Sensing

Groundwater accounts for ~30% of the world's freshwater resources and the current withdrawal rates in the estimated range of 982 km^3/year forms the largest extraction of any raw material on planet Earth (Margat and van der Gun, 2013). Three countries, India (251 km^3/year), the United States, and China (112 km^3/year each), account for nearly 50% of this extraction. Fortunately, groundwater is replenishable and a renewable water resource. Nevertheless, many parts of the world suffer from overexploitation that is unsustainable. For example, about 60% of groundwater withdrawn worldwide is used for agriculture, and the rest is almost equally divided between the domestic and industrial sectors (Siebert et al., 2010). In many parts of the world, especially in rural areas, groundwater is the primary source of drinking water and also often the safest, given the extent of surface water pollution.

Groundwater is subterranean. Hence, there is no real direct measure of groundwater from RS, but several indirect measures exist such as through the use of several thematic maps (e.g., Figure 25.2) in a spatial modeling, to determine potential groundwater zones. These thematic maps are derived from RS and non-RS sources and provide wide array of information of an area. When this information is integrated in spatial modeling framework, it is possible to derive potential groundwater zones. RS is used as (1) a reconnaissance tool, (2) an indirect indicator requiring terrain image interpretation of various nature, and (3) an input data in geographical information system (GIS) spatial modeling framework that in turn will help establish groundwater-potential zones or groundwater stress areas or in supporting groundwater management. Long-wave radar RS data can penetrate ground under ideal conditions and can at times detect groundwater at depths of few meters. Thermal imagery can detect temperature anomalies in water bodies thus, for example, helping identify areas of discharge into lakes and rivers from groundwater. Optical and radar imageries are interpreted for various geomorphological, geological, terrain, and vegetation characteristics that help determine the potential areas of groundwater source or stress. Most of these interpretations require considerable skill of an expert groundwater scientist. In Chapter 2 on groundwater,

Dr. Santhosh Kumar Seelan shows us a number of approaches and methods used for groundwater detection, groundwater stress areas, and groundwater management issues using RS. These are summarized as follows:

Lineaments and lithology: RS plays an important and powerful role in geological structural analysis such as lineament detection and mapping. Lineaments are often the best indicators of groundwater presence, especially in hard rock landscapes (e.g., crystalline and basaltic terrains) and are mapped using image enhancement and filtering techniques using optical or radar imagery. Wells located on the lineament zones produced as much as 14 times better yield than wells located away from the lineaments.

Quantitative geomorphologic parameters: Drainage density and stream lengths have a bearing on recharge conditions and permeability of the rocks. These are studied using optical as well as radar data. One indicator, often, is the lesser the drainage density, the greater the groundwater potential. Fossil drainage in deserts and paleodrainage (old drainage), for example, can be detected using radar imagery that can penetrate up to 3 m and yield far more groundwater than elsewhere.

Vegetation, land use, and cropland (e.g., irrigated or rainfed) information: These help support water balance budgets and are studied using optical, radar, or thermal data. For example, actual ET or water use by crops and natural landscape is assessed using SEB modeling where thermal and optical RS plays a key role. Vegetation associated with fault zones helps detect near-surface groundwater. Fine-grained clayey soils generally tend to permit less recharge and more runoffs, while coarse-grained sandy soils allow more recharge to groundwater.

Groundwater discharges into lakes and rivers: These are mapped based on temperature differences of lake or river water (higher temperature) compared to groundwater discharge into lakes (cooler water). In the case of two adjacent streams, the one fed by a hot spring appears brighter.

Groundwater-fed irrigation in many cases have center pivots (e.g., in much of the southern Ogallala system) and can easily be detected and the water use pattern modeled using SEB models in which both thermal and optical RS are key inputs.

Hydrogeological studies based on faults in sedimentary rocks: These are often indicative of groundwater presence mapped using either optical or radar data.

Unconsolidated terrains such as alluvial plans, deltas, and other Pleistocene deposits: These morphological units often hold large quantities of groundwater. These landscapes are best mapped using optical or radar RS and provide a good reconnaissance of the area for further investigations.

Landforms of various natures where groundwater potential exists: These can be mapped using optical or radar or thermal RS and digital elevation model (DEM) data.

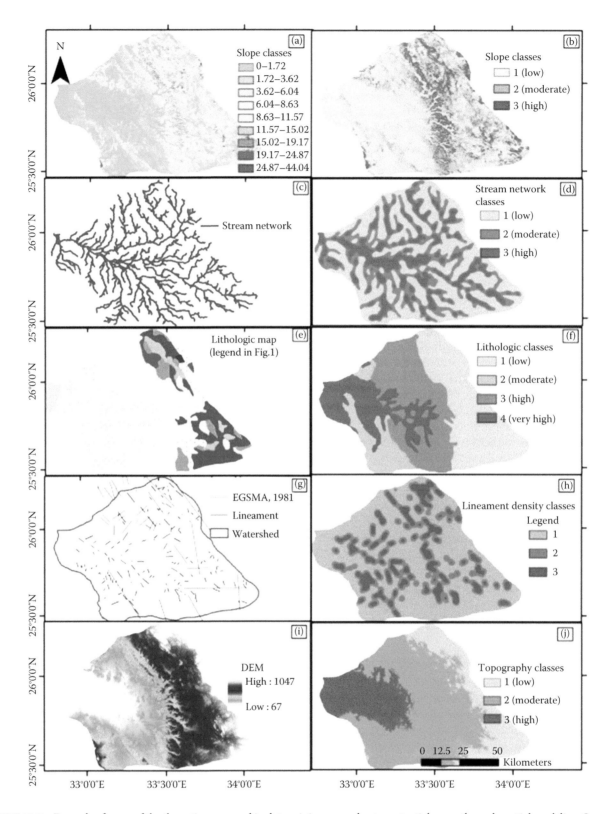

FIGURE 25.2 Example of some of the thematic maps used in determining groundwater-potential zones through spatial modeling. Some of the thematic maps include (a) slope map derived from Shuttle Radar Topography Mission 90 m, (b) slope classes map, (c) stream network map, (d) stream density map, (e) simplified geology map modified after Egyptian Geological Survey and Mining Authority (1981) (Geologic Map of Egypt, scale 1:2,000,000), (f) geologic map assigned weight factor depending on capability for holding water, (g) lineament map, (h) lineament density map, (i) DEM of the study area, and (j) DEM classified map assigned weight factor depending on water infiltration. (From Abdalla, F., *J. Afr. Earth Sci.*, 70, 8, 2012.)

These landforms include weathered rocks, landform and soil associations, karst terrain (in karst depressions using DEM data), and volcanoes (e.g., groundwater discharge in lower volcanoes).

Total water storage change quantification studies: These studies are conducted using GRACE satellite microgravity measurements (Rodell et al., 2007). Regional, national, and river basin scale total water storage changes are studied using the GRACE satellite measurements.

Groundwater modeling inputs: RS data can provide useful inputs to groundwater modeling. For example, in areas of heavy irrigated agriculture, information such as the area cropped and type of crops will be useful inputs to groundwater numerical models.

Spatial modeling for groundwater assessment and management: This involves a wide array of spatial data such as geology, geomorphology, terrain, soils, and vegetation that will help make a number of decisions on groundwater potential or stress and will help groundwater management, decide on aquifer recharge areas.

The readers with deep interest in groundwater RS are also encouraged to read the works of Ray (1960), Sabins (1987), and Meijerink et al. (2007).

25.3 Actual Evapotranspiration (Water Use) of Croplands from Remote Sensing

The actual water consumed by the crops to grow food is referred to as actual ET (ET_a expressed in mm/day). Currently, based on different estimates, anywhere between 4500 and 7500 km^3/year of water is consumed by 1.5–1.7 billion hectares of global croplands. This consumption is expected to increase by 2050 to 12,000–13,500 km^3/year (Droogers et al., 2010). Since agriculture accounts for about 80% of all human water use, determining and managing crop water use is of great importance. RS is the most powerful data source for consistent and accurate estimates of ET_a (water use) from single farm to very large areas routinely and repeatedly. For example, Figure 25.3 shows 10-day cumulative ET_a or water use in mm of irrigated crops derived using multiple-sensor time series imagery.

Chapter 3 by Dr. Trent W. Biggs et al. provides systematic steps on methods of estimating ET_a using RS data. They categorize the ET_a methods using RS into three distinct groups:

1. Vegetation-based methods include empirical crop coefficient methods that require ground-level reference data and process-based approaches that have minimal ground-level data requirements. Process-based approaches include the Priestley–Taylor Jet Propulsion Laboratory (PT-JPL) and MODIS global evapotranspiration data set (MOD16) methods. Vegetation-based methods may have difficulty predicting evaporation from wet or inundated soil, which complicates their use in some types of irrigated agriculture. However,

there is significant uncertainty that is yet to be resolved in these methods and approaches, requiring further studies.

2. Temperature- or energy-based methods include one-source models (SEBAL, METRIC), two-source models (atmosphere–land exchange inverse [ALEXI], disaggregated ALEXI [DisALEXI]), and simplified methods (simplified surface energy balance [SSEB], and operational SSEB [$SSEB_{op}$]). SEBAL and METRIC are most useful in irrigated areas in semiarid and arid landscapes where extremes of ET_a are present in the image. Two-source models are more complex but are often more accurate than one-source models, particularly over areas with partial vegetation cover. SSEB is simpler to understand and apply and often has comparative performance with other one-source models.

3. Scatterplot or triangle–trapezoidal methods use plots of surface temperature or albedo versus a vegetation index. This simple approach may provide comparable results with SEB models.

For large regions or global applications, the authors recommend the use of ensemble of models, since no one model performs the best across land cover types. However, the uncertainty in ET_a models is still very high, especially when the models are applied over large areas.

Each model has its own intricacies and complexities involving parameters derived from RS data as well as some from meteorological data as discussed in detail in various sections and subsections of the Chapter 3. In its most simplistic form, ET_a or the water used by crops (expressed m^3/ha or mm/m^2) is derived by (Platanov et al., 2009)

- Determining the ET fraction (e.g., Landsat Enhanced Thematic Mapper Plus [ETM+] thermal data)
- Calculating the reference ET (e.g., using Penman–Monteith equations)
- Computing ET_a by multiplying ET fraction with reference ET

The ET fraction (ET_f) or evaporative fraction is the ratio of ET_a over reference ET (ET_0). The SSEB model, for example, calculates ET_f based on the assumption that the latent heat flux (ET_a) varies linearly between the land surface temperature (LST) of "hot" and "cold" pixels (Platonov et al., 2008):

$$ET_f = (T_{hot} - T)/(T_{hot} - T_{cold}) \qquad (25.1)$$

where
ET_f is the fraction of ET (dimensionless)
T is the LST of any pixel
T_{hot} and T_{cold} are the LST of *hot* and *cold* pixels, respectively, the LST expressed in Kelvin or degree Celsius

The *hot* and the *cold* pixels are selected inside the irrigated fields of the investigated area for each image.

Reference ET (ET_0) is typically calculated using various methods (e.g., Priestley–Taylor, Blaney–Criddle, Hargreaves,

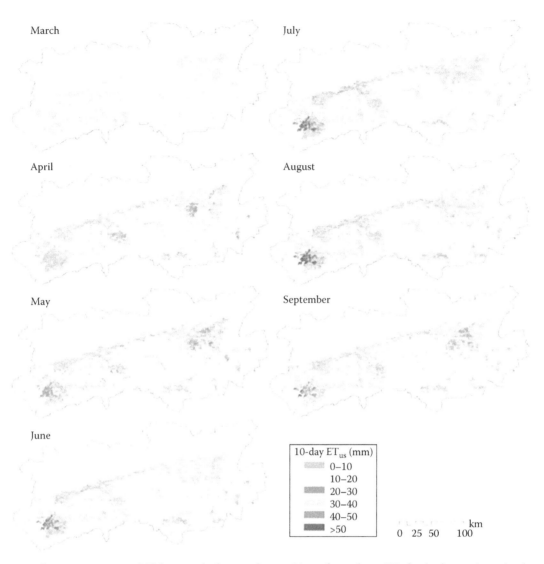

FIGURE 25.3 Actual evapotranspiration (ET) (water use) of irrigated crops. Maps of cumulative ET_{us} for the first 10 days of each 2009 campaign month over the Guadalquivir river basin irrigated area. When the ET from different months is added, we will get cumulative ET_{act} (water use) of these irrigated crops for the entire season. Note: ET_{us} (with the subscript "$_{us}$" meaning unstressed) to explicitly acknowledge that the reduction in plant transpiration due to stomata closure or water vapor pressure deficit is ignored. However, the effect of water stress on plant growth is reflected through the vegetation index; thus, this indirect effect of water stress is accounted for in ET_{us}. (From González-Dugo, M.P. et al., *Agric. Water Manage.*, 125, 92, 2013.)

Penman, Penman–Monteith) using various meteorological data. $ET_a = ET_f * ET_0$.

The SEB models (e.g., SEBAL, METRIC) compute ET_a on pixel-by-pixel basis for the instantaneous time of the satellite image, as the residual amount of energy remaining from the classical energy balance:

$$\lambda ET = Rn - G - H \qquad (25.2)$$

where

    λET is the latent heat flux (the energy used for ET)
    Rn is the net radiation at the surface
    G is the soil heat flux
    H is the sensible heat flux to the air

All fluxes are in W m^2 day^{-1} units.

ET (mm day^{-1}) is calculated from latent heat flux by dividing it by the latent heat of water vaporization (λ) (Platonov et al., 2008).

Readers should note that this equation only covers one-source energy-based methods (and not PT-JPL or MOD16). Please refer to Chapter 3 for equations pertaining to PT-JPL or MOD16.

25.4 Crop Water Productivity Studies from Earth Observation Systems

Globally, ~80% (4500–7500 km^3 $year^{-1}$) of all human water use goes toward agriculture to produce food from ~1.5 billion hectares of existing irrigated and rainfed croplands (Thenkabail et al., 2010). Also, competition for water from multiple sectors (e.g., urban, recreation, environmental flows, and industries) is rising steeply, making such large quantities of agricultural land and water use untenable. On the other hand, global population is increasing and expected to reach 9.2–10 billion by year 2050 from current 7.2 billion, further increasing demand for food and nutrition, and associated increases in land and water allocations. However, it is widely perceived that neither the increase allocation of land nor the increased allocations of water is practical. Indeed, it is highly likely that the rising demand for food and nutrition need to be met by decreasing land and water for growing food. Also, the green revolution (productivity increases per unit of land) era of the last 50 years has come to an end. Thereby, there are several attempts for increased food production through smart technologies and scientific advancement that are, hitherto, rarely explored. Improved WP of croplands around the world is considered as the most promising opportunity, of all the available measures, for increasing food productivity.

Crop WP (CWP; kg m^{-3}; productivity per unit of water or crop per drop or biomass per ET or crop yield per ET) studies are best conducted by integrating multisensor RS data with SEB modeling (ET modeling for water use by crops), agrometeorological data, water withdrawal data, and biophysical and yield data in a GIS.

Methods and protocols for CWP, typically, involve five broad steps: (1) cropland area mapping (ha), (2) crop productivity mapping (CPM, kg m^{-2}; yield per unit of land) through biophysical modeling, (3) water use mapping (WUM, m^3 ha^{-1}) through SEB modeling (ET modeling) or through water balance from hydrological models, (4) WP mapping (kg m^{-3}) through a simple ratio of CPM and WUM, and (5) developing a spatial decision support system (DSS) for government agencies and farmers to determine where, how, and by how much water can be saved through improved WP.

The steps for WP modeling are well illustrated in Chapter 4 by Dr. Antonio Heriberto de C. Teixeira by taking rainfed and irrigated crops in study areas of Brazil. They have focused their researches on physical and economic values of CWP. The physical CWP (kg m^{-3}) is the ratio of crop biomass or yield (kg m^{-2}) to the amount of water used (m^3 m^{-2}); in Chapter 4, WP is defined as biomass/ET$_{actual}$, which, after applying a harvest index, becomes the CWP. The economic WP ($ m^{-3}) relates the economic benefits per unit of water used. WP (yield per unit of water per drop; kg m^{-3}) is inverse of water footprint (WF) and is illustrated (Figure 25.4) for several main global crops by Mekonnen and Hoekstra (2014). Raising WP would mean reducing WF as

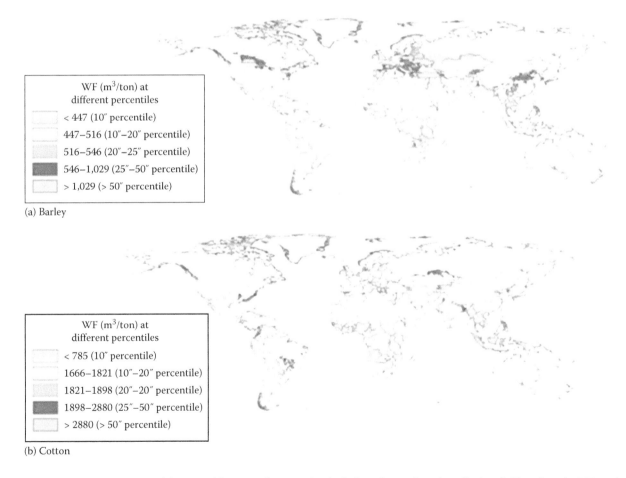

WF (m³/ton) at different percentiles
< 447 (10″ percentile)
447–516 (10″–20″ percentile)
516–546 (20″–25″ percentile)
546–1,029 (25″–50″ percentile)
> 1,029 (> 50″ percentile)

(a) Barley

WF (m³/ton) at different percentiles
< 785 (10″ percentile)
1666–1821 (10″–20″ percentile)
1821–1898 (20″–20″ percentile)
1898–2880 (25″–50″ percentile)
> 2880 (> 50″ percentile)

(b) Cotton

FIGURE 25.4　Spatial distribution of the green–blue water footprint (WF) of selected crops (in m^3 ton^{-1}), classified based on the WFs at the different production percentiles. Water productivity (WP) (productivity per unit of water or crop per drop; kg m^{-3}) is inverse of WF. Raising WP in agriculture, that is, reducing the WF per unit of production, will contribute to reducing the pressure on the limited global freshwater resources. (From Mekonnen, M.M. and Hoekstra, A.Y., *Ecol. Indic.*, 46, 214, 2014.)

(*Continued*)

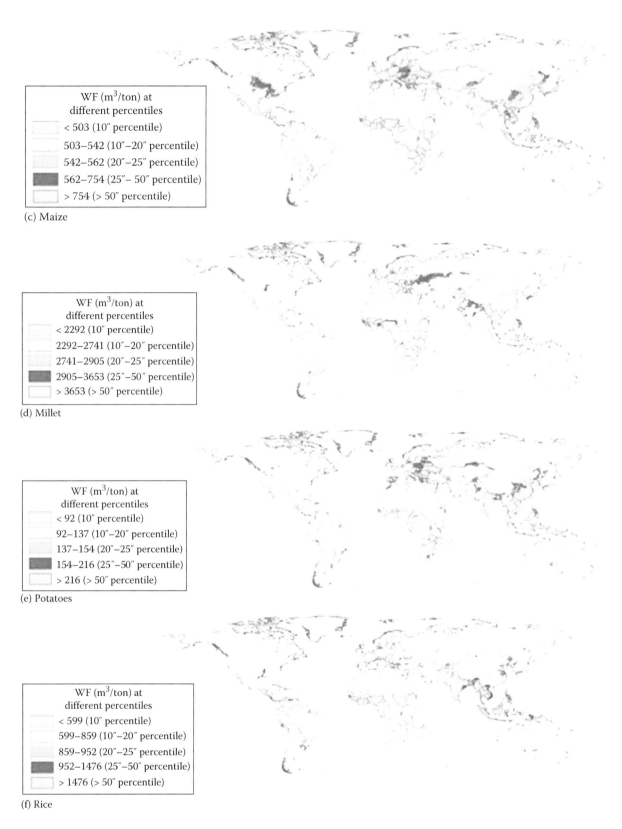

FIGURE 25.4 (*Continued*)   Spatial distribution of the green–blue water footprint (WF) of selected crops (in m³ ton⁻¹), classified based on the WFs at the different production percentiles. Water productivity (WP) (productivity per unit of water or crop per drop; kg m⁻³) is inverse of WF. Raising WP in agriculture, that is, reducing the WF per unit of production, will contribute to reducing the pressure on the limited global freshwater resources. (From Mekonnen, M.M. and Hoekstra, A.Y., *Ecol. Indic.*, 46, 214, 2014.)   (*Continued*)

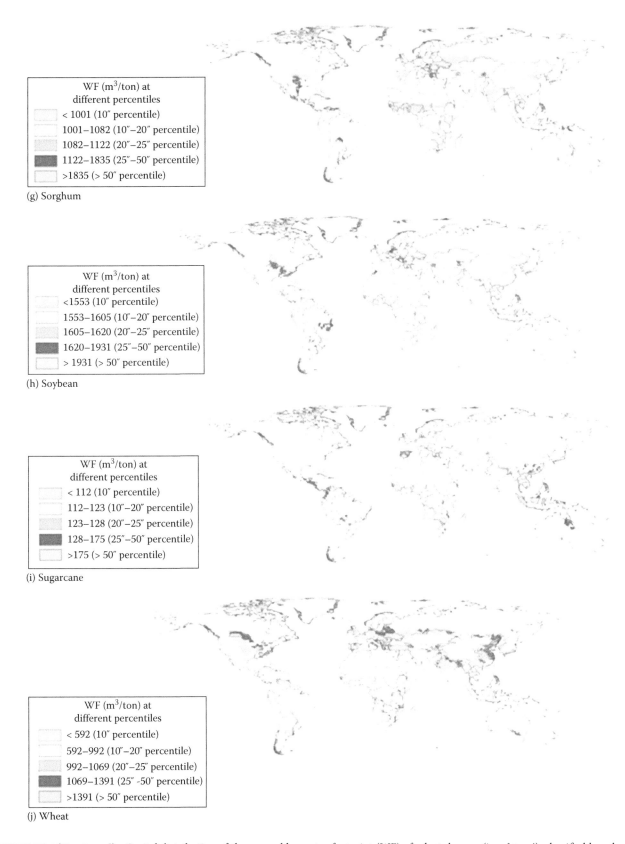

FIGURE 25.4 (*Continued*) Spatial distribution of the green–blue water footprint (WF) of selected crops (in m³ ton⁻¹), classified based on the WFs at the different production percentiles. Water productivity (WP) (productivity per unit of water or crop per drop; kg m⁻³) is inverse of WF. Raising WP in agriculture, that is, reducing the WF per unit of production, will contribute to reducing the pressure on the limited global freshwater resources. (From Mekonnen, M.M. and Hoekstra, A.Y., *Ecol. Indic.*, 46, 214, 2014.)

illustrated. RS data of various resolutions can be used to study WP at different spatial scales (e.g., global, continental, regional, local). The accuracy and precision of WP studies will depend on the resolution (e.g., spatial, spectral, radiometric, and temporal) of imagery used. Chapter 4 shows studies conducted using Landsat (30–120 m) and MODIS (250–1000 m) imagery.

25.5 Flood Studies by Integrating Remote Sensing in Hydrological Models and Other Data

Chapter 5 by Dr. Allan Arnesen shows the advantages of flood studies when SAR RS data are combined with (1) ancillary data and (2) hydrological models. First, they demonstrate the use of SAR data and algorithms to map flood extent. Second, they integrate SAR RS-derived information in hydrological models. Third, they forecast flood to minimize socioeconomic impacts of floods.

First, Chapter 5 demonstrates that in flood mapping, SAR data offer the advantage of looking through clouds, identification of water below trees and other vegetation, and help study aquatic vegetation (e.g., submerged rice fields, mangrove forests). The following critical information of SAR data for flood mapping is established by Dr. Arnesen in Chapter 5:

1. Three scattering mechanisms are dominant in floodplain areas: double bounce, volumetric (or canopy), and surface (or specular).
2. The longer the SAR wavelength, the higher will be the radiation penetration in the canopy, and high backscattering values are expected due to double-bounce occurrence.
3. Shorter wavelengths (X and C bands) have reduced penetration at the canopy, and volumetric and surface backscattering mechanisms are predominant.

4. Copolarization configuration (HH and VV) data are preferable than cross polarization (HV or VH) to identify folded forest.
5. Flood mapping includes such approaches as (a) thresholding backscatter values and (b) binary (flood vs. no flood) algorithms based on backscatter value in decibel. Both supervised and unsupervised classification methods are applied. However, pixel-based classification approaches provide low accuracies, many times as low as 30% due to speckle effects in SAR data.
6. Object-oriented classification methodologies involving segmentation procedure and fuzzy logic classifiers using TerraSAR-X and COSMO-SkyMed SARs have shown significantly improved classification accuracies of 80% or above.

Second, Chapter 5 established the advantage of integrating RS with hydrological models that allows to determine various aspects of flood that include extent, stage, frequency, and different scenarios of storms. RS data (e.g., water level data from ENVISAT altimeter, land cover from SAR) can help substantially improve the performance of hydrological models by providing a number of parameters such as land cover maps, vegetation type maps, and drainage network information needed in the hydrological models. In addition to RS, the use of parameters derived from other ancillary data (e.g., Shuttle Radar Topography Mission (SRTM)-derived DEM, TRMM precipitation, AVHRR skin temperature) will further enhance the performance of hydrological models. Flood simulations will help flood risk zone mapping, flood prevention, and preparedness.

Third, Chapter 5 shows an approach to real-time flood forecasting that uses hydrological models fed by SAR RS data, other RS data, and ancillary data. The hydrological models use several types of data (e.g., Figure 25.5) such as land use, soil type, soil moisture,

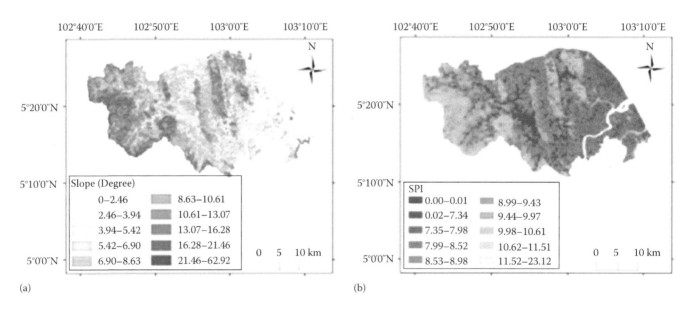

(a)

(b)

FIGURE 25.5 Some input thematic layers in flood susceptibility mapping shown for Terengganu, Malaysia: (a) slope, (b) stream power index. (From Tehrany, M.S. et al., *J. Hydrol.*, 512, 332, 2014.)   (*Continued*)

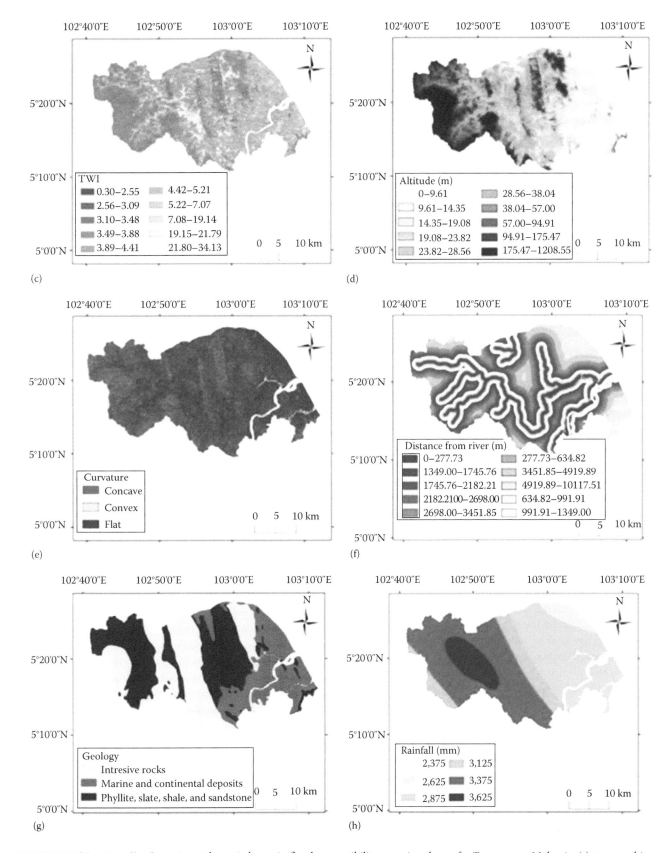

FIGURE 25.5 (*Continued*)   Some input thematic layers in flood susceptibility mapping shown for Terengganu, Malaysia: (c) topographic wetness index, (d) altitude, (e) curvature, (f) distance from river, (g) geology, (h) rainfall. (From Tehrany, M.S. et al., *J. Hydrol.*, 512, 332, 2014.)

(*Continued*)

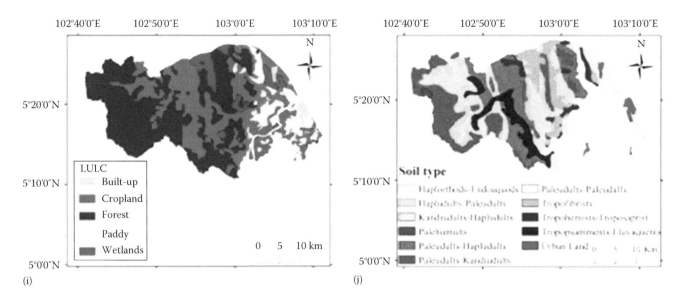

FIGURE 25.5 (*Continued*)   Some input thematic layers in flood susceptibility mapping shown for Terengganu, Malaysia: (i) land use and land cover, and (j) soil type. (From Tehrany, M.S. et al., *J. Hydrol.*, 512, 332, 2014.)

stream/river base flow, rainfall amount/intensity, snowpack characterization, DEM data, and drainage density/frequency. Most of these data (Figure 25.5) are best derived using one or the other RS data.

25.6 Flood Studies Using SAR Remote Sensing

Every year, flood affects nearly 100 million people with an average of nearly 7000 people killed and causing ~14 billion dollar in damages (EM-DAT: The OFDA/CRED International Disaster Database, University Catholique de Louvain, Brussels, Belgium, Data version: v11.08). Great floods in the past have caused immense damage to human life such as the 1991 floods of Bangladesh (139,000 deaths), 1938 floods of China (870,000 deaths), and 1530 flood of the Netherlands (400,000 deaths) (World Commission on Water for the 21st Century). Apart from looking at floods as from disaster point of view, they are also natural phenomena that are often the lifeline of several billions living along the river courses that bring water for agriculture and fertile soil from elsewhere and replenish groundwater. Floods should also be looked as *cleansing* the river systems of pollution. Flood water when harnessed can serve humanity during lean flow periods. So there are many reasons to study floods.

There are numerous aspects of floods that RS provides useful information. Some of the following are discussed in Chapter 6 by Martinis et al.:

1. Flood inundation studies that includes before-, during-, and after-flood events
2. Establishing flood-prone areas
3. Assessing areas of damage and areas benefited by floods
4. Determining storage volumes in reservoirs as a result of flooding
5. River morphological studies and changes in river meandering as a result of flooding

6. Flood prediction studies that include snowmelt runoff studies, rainfall, and soil moisture
7. Flood stage (which in turn is related to discharge)
8. Flood sedimentation studies (this last aspect not covered in this chapter, but important for readers to further explore)

Both active and passive RS are widely used in flood studies. Passive sensors are preferred when these data are available during the flood events since they often offer better spatial resolution than SAR data and the multiple spectral characteristics of optical sensors allow us to study most of the flood parameters listed earlier. Passive sensors include data from sensors such as MODIS, ASTER, Landsat, and suite of very-high-resolution imagery. Using these data is always a trade-off between spatial resolution, the large area coverage, and frequency of coverage (e.g., MODIS has coarse resolution of 250–1000 m, but has daily global coverage). Flood studies often require large area coverage such as the entire river basin to understand, model, and map the flood phenomenon rather than look at just the areas flooded.

The microwave region (1 mm to 1 m wavelength) or 300 GHz to 300 MHz (frequency) is specifically beneficial to overcome clouds. Chapter 6 by Martinis et al. focuses on flood studies from SAR data. Since most floods occur when clouds are maximum, SAR data become very useful. The microwave region of the electromagnetic spectrum can be divided into the following bands: P band (30–100 cm wavelength, 0.3–1 GHz frequency), P band (30–100 cm, 0.3–1 GHz), L band (15–30 cm, 1–2 GHz), S band (8–15 cm, 2–4 GHz), C band (4–8 cm, 4–8 GHz), X band (2.5–4 cm, 8–12 GHz), Ku** band (1.7–2.5 cm, 12–18 GHz), K band (1.1–1.7 cm, 18–27 GHz), Ka** band (0.75–1.1 cm, 27–40 GHz), V band (0.4–0.75 cm, 4–75 GHz), W band (0.27–0.4 cm, 75–100 GHz), and mm band (<0.27 cm, 110–200 GHz). Especially, L, C, X, and Ka radar bands are used in flood studies with C being most common and L being used to map inundation under vegetation. Spaceborne SAR data include data gathered from such sensors (see Chapter 6 by Martinis et al.): TerraSAR-X/TanDEM-X, COSMO-SkyMed, X- (SIR-C/X-SAR, SRTM), C band (ERS-1/2 AMI, ENVISAT ASAR, RADARSAT-1, RISAT-1, SIR-C/X-SAR), and L band domain (SEASAT-1, JERS-1, ALOS PALSAR, SIR-A/B/C/X-SAR), ERS-1, ERS-2, JERS-1, JERS-2, RADARSAT-1, RADARSAT-2, ENVISAT-1, Sentinel-1, SMMR, SSM/I, DMSP, and radar altimetry. Nevertheless, poor spatial resolutions of most of the SAR sensors limit their use. Since 2007, the successful launch of the platforms TerraSAR-X/TanDEM-X, COSMO-SkyMed, and RADARSAT-2 marks a new generation of civil SAR systems suitable for detailed flood mapping purposes up to a resolution of 0.24 m (TerraSAR-X Staring Spotlight mode).

Martinis et al. (Chapter 6) helps us understand interactions of SAR signal with water bodies including their strengths and limitation in flood mapping. For example, they establish factors that overestimate flooding as (1) shadowing effects behind vertical objects such as vegetation, topography, anthropogenic structures; (2) smooth natural surface features such as sand dunes, slat and clay pans, and bare ground; and (3) smooth anthropogenic features such as streets, airstrips, and heavy rain cells. They also enumerate the underestimating factors of flooding such as (1) volume scattering of partially submerged vegetation; (2) double-bounce scattering of partially submerged vegetation; (3) anthropogenic features on the water surface such as ships and debris; (4) roughening of the water surface by wind, heavy rain, or high flow velocity; and (5) layover effect on vertical objects such as topography and urban structures. This theoretical basis provides many insights (Chapter 6) such as the following:

1. Water monitoring using X band SAR (e.g., Figure 25.6) is considered more suitable than using longer wavelengths such as C and L bands.
2. HH polarization provides best discrimination between water and nonwater terrain.
3. Superiority of cross polarization HV over like polarization VV in monitoring roughened water surfaces.
4. Partially submerged vegetation may cause a backscatter increase over water bodies.
5. L band SAR is more effective in detecting water under forest canopies.
6. Rice crop in early to midphases increases backscatter, whereas in ripening phase backscatter decreases.
7. Likelihood to detect flooding in urban areas generally increases with decreasing incident angle.

Chapter 6 provides an exhaustive picture of flood mapping using SAR data (e.g., Figure 25.6) through visual as well as digital image analysis. They discuss in detail the three flood mapping approaches:

1. Semiautomatic object-based flood detection (RaMaFlood)
2. Automatic pixel-based water detection (WaMaPro)
3. Fully automatic pixel-based flood detection (TerraSAR-X Flood Service)

These are comprehensive and thorough clearly illustrated with case studies.

25.7 Mangrove Wetlands of the World

Mangroves occupy ~0.1% of the terrestrial Earth, but are unique and irreplaceable ecosystems with very high primary productivity, and have the most carbon-rich biomes (e.g., Figure 25.7) containing an average of 937 tC ha^{-1} (Alongi, 2014) and rich and unique flora and fauna housing many endangered species, supporting shoreline protection (e.g., during tsunamis); sustaining livelihoods through food, fodder, and several renewable resources (e.g., firewood, furniture); and encouraging coastal tourism. Even though mangroves account for approximately 1% (13.5 Gt year^{-1}) of carbon sequestration by the world's forests (due to their low percentage of area), but as coastal habitats, they account for 14% of carbon sequestration by the global ocean (Alongi, 2014). As per the United Nations (UN)'s Food and Agricultural Organization (FAO) data, they covered more than 20 million hectares (Spalding et al., 1997) of sheltered tropical and subtropical coastlines, but are disappearing at 1%–2% per year (Duke et al., 2007), but current RS estimates (Chapter 7) show the total forest cover as 13.776 million hectares. Mangrove wetlands are encroached for multiple purposes that include shrimp farming, fish farming, agriculture, deforestation for multiple reasons, and diversion of freshwater and coastal development for urbanization.

In Chapter 7, Dr. Chandra Giri discusses the rich importance and many uses of mangrove wetlands. The chapter shows us how the mangrove wetlands are mapped using RS. The chapter defines mangroves for mapping and monitoring purposes using RS and enumerates on scale/resolution issues of various RS data in mangrove mapping and monitoring. The chapter answers questions such as areal extent of mangrove forests, where they are located, their change over space and time, causes of those changes, and potential areas for regeneration. The chapter does this by using Landsat data for 1975, 1990, 2000, and 2005. The chapter presents and discusses various classification and change detection approaches and methods. The chapter also shows mangrove species discrimination methods and approaches. The challenges and opportunities in using very-high-spatial-resolution data in mangrove wetland mapping have also been presented.

25.8 Wetland Modeling and Mapping Methods Using Multisensor Remote Sensing

Wetlands are areas where water is the primary factor controlling the environment and the associated plant and animal life (Ramsar). Article 1.1 of the Ramsar Convention defines wetlands as "areas of marsh, fen, peatland or water, whether natural or artificial, permanent or temporary, with water that is static or flowing, fresh, brackish or salt, including areas of marine water the depth of which at low tide does not exceed six metres." Wetlands are found throughout the world except Antarctica and occupy anywhere between 7 and 9 million km^2 area (~4%–6% of the terrestrial area). They are major habitats for a diverse group of plants and animals.

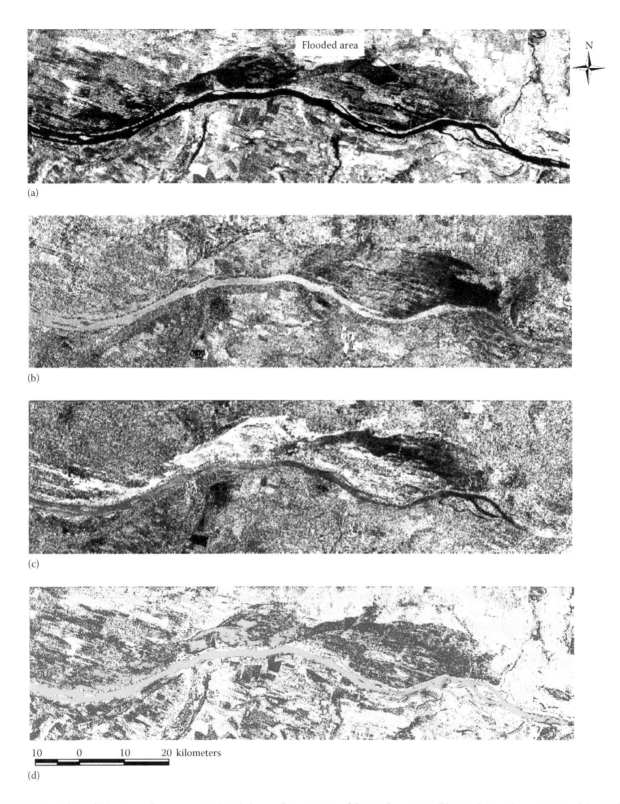

FIGURE 25.6   Plots of (a) principal component 1 (PC1) that explains 39.81% of the total variance, (b) PC2 that explains 15.47% of the total variance, and (c) PC3 that explains 12.79% of the total variance, of the principal component analysis of 7 European Remote Sensing Satellite2 synthetic-aperture radar (ERS2-SAR) images of original pixel values for the RCB reach of Danube River at Romania (using 7 ERS2 SAR images), where dark color represents permanent water (Danube River) and part of flooded area that can also be represented by gray color. Light gray and white likely represent the dryland and (d) the classification of PC1 of 7 ERS2-SAR images by the Isodata classifier (green, permanent water; blue, flooded area; yellow, dryland). (From Gan, T.Y. et al., *Int. J. Appl. Earth Observ. Geoinform.*, 18, 69, 2012.)

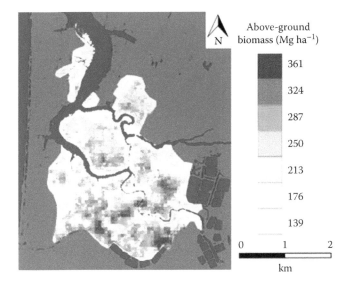

FIGURE 25.7 Map of mangrove biomass produced using high-resolution GeoEye-1 satellite imagery, medium-resolution Advanced Spaceborne Thermal Emission and Reflection Radiometer satellite elevation data, field-based biomass data, and a support vector machine regression model. The aboveground biomass (AGB) for the 151 ha Kamphuan mangrove was computed by the equation AGB = 0.16 * elevation + 0.27 * (band 1/band 2) − 0.11 * band 2 + 0.41 * band 4 − 0.03 and was 250 ± 53.4 Mg ha^{-1} with the highest carbon stocks located in the mangrove interior and the landward edge (Figure 25.7). Using the 0.38 ratio of BGB to AGB yielded an estimated BGB of 95 Mg ha^{-1}. Combined AGB and BGB at the site was 345 ± 72.5 Mg ha^{-1}. Using the 0.45 conversion factor between biomass and carbon stock (Twilley et al., 1992), the estimated above- and belowground carbon biomass was 113 and 42.8 Mg C ha^{-1}, respectively. (From Jachowski, N.R.A. et al., *Appl. Geogr.*, 45, 311, 2013.)

Since wetlands constitute various mix of water, land, and vegetation and they are of various sizes from very large areas such as the flood plains along major rivers and river deltas (e.g., Figure 25.8) to tiny streams, every type of RS data acquired in various spectral, spatial, temporal, and radiometric resolution and in various portions of the electromagnetic spectrum (e.g., optical, thermal, radar) has been used to study them. Chapter 8 by Dr. Deepak R. Mishra et al. starts by providing an overview of the evolution of wetland RS. This reveals that various widely used RS techniques such as supervised and unsupervised clustering algorithms for wetland classification, vegetation indices for modeling wetland biophysical and biochemical quantities, and advanced image processing techniques such as object-based image analysis (OBIA) and data fusion involving multiple-sensor data such as the LiDAR and hyperspectral sensors are used to quantify, model, map, and monitor wetlands. They provide three studies to illustrate the value of various aspects of wetland studies using widely varying remotely sensed data. These three studies are

1. LiDAR (1 m, 1 band, acquired at 1047 nm) and hyperspectral data (1 m, 63 bands, over 400–980 nm) fusion for accurate habitat mapping and elevation mapping in salt march environments

2. Biophysical and biochemical quantification studies of the salt marsh habitats of the entire Georgia coast using the MODIS 250/500 m time series data

3. Estimation of aboveground and belowground biomass (BGB) and foliar nitrogen (N) using satellite data and hybrid modeling in the freshwater marsh through hyperspectral and multispectral data

Some of the highlights of this comprehensive study by Dr. Deepak R. Mishra et al. in Chapter 8 are as follows:

1. Hyperspectral data were able to classify nine distinct classes of salt marshes with overall accuracy of ~90% and producer's and user's accuracies of individual species ~80%.

2. Decision trees incorporating LiDAR-derived DEM and hyperspectral narrowband normalized difference vegetation index (NDVI) significantly improved classification accuracies.

3. Vegetation classification by fusing LiDAR and hyperspectral data offered significantly higher classification accuracies, especially for the producer's and user's accuracies.

4. LiDAR tends to overestimate salt marsh elevations due to poor penetration of dense vegetation.

5. When there are multiple species within a hyperspectral pixel, species separation becomes more complicated. This indicated hyperspectral data need to be supported by hyperspatial pixel size that can capture the species spectra for better classification of the species.

6. Species separability needs to consider phenology, biomass productivity, and biochemical compositions (e.g., N) for better understanding and separation.

7. When highly accurate vegetation classification maps are available (e.g., using hyperspectral data), the DEMs of wetland obtained from LiDAR can be substantially improved.

8. Biophysical variables such as canopy chlorophyll content, green leaf area index (a ratio of green foliage area vs. ground area), green vegetation fraction (percent green canopy cover), and aboveground green biomass can be modeled using MODIS and Landsat data with reasonable consistency over large areas.

9. There were substantial difficulties in determining the aboveground and below-ground biomass (BGB) of freshwater marsh species with the best hyperspectral and multispectral models explaining up to 56% variability in data. This is mainly due to mixed pixel signatures making it difficult to acquire data of specific species. Uncertainties in data lead to uncertainty in results.

25.9 Inland Valley Wetland Characterization and Mapping

Inland valleys (IVs) are lowland ecosystems spread across landscapes. Unlike large wetlands that occur along the higher order streams and in the river deltas, IVs occur along the lower-order (e.g., first to fourth order) streams (e.g., Figure 25.9). IVs offer an

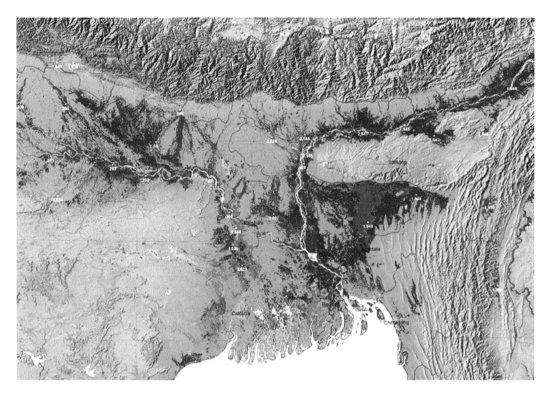

FIGURE 25.8   Wetlands along the major rivers and river deltas. Dartmouth Flood Observatory–produced spatial flood coverage (1999–2009) depicting large wetlands along the Ganges–Brahmaputra rivers (India, Bangladesh, China), and their deltas are characterized with data from three remote sensing systems (Advanced Microwave Scanning Radiometer for the EOS, SRTM, and MODIS). Colors represent different years with more recent years overlying earlier flooded areas. Often, flooded areas reoccur from year to year. (From Syvitski, J.P.M. et al., *Sediment. Geol.*, 267–268, 1, 2012.)

extensive, fairly unexploited potential for agricultural production due to rich soils, significantly higher water and moisture availability when compared with adjoining uplands. As discussed in Chapter 9 by Dr. Murali Krishna Gumma et al., the IV wetlands have high potential for growing agricultural crops due to (1) easy access to the river water, (2) significantly longer duration of adequate soil moisture to grow crops when compared with adjoining uplands, and (3) rich soils (depth and fertility) (Thenkabail et al., 2000a).

Chapter 9 focuses on characterizing and mapping IV wetlands of Africa, given highly unexploited IVs found throughout Africa and the need to understand, model, map and then prioritize the use and conservation of wetlands. The wetlands of Africa are increasingly considered "hot spots" for agricultural development and for expediting Africa's Green and Blue Revolution. Currently, these IV wetlands are unutilized or highly underutilized in Africa in spite of their rich soils and abundant water availability as a result of (1) limited road access to these wetlands and (2) prevailing diseases such as *malaria*, *trypanosomiasis* (sleeping sickness), and *onchocerciasis* (river blindness). However, the utilization of IV wetlands for agriculture is becoming unavoidable in African countries due to increasing pressure for food from a ballooning human population and difficulty finding arable land with access to water resources. Given these pressures, it is critical to recognize

important functions of wetlands, which include holding about 20% of all carbon on Earth and providing habitat for unique and increasingly rare flora and fauna. Therefore, any proposed use of wetlands for agriculture has to be carefully weighed against the social benefits provided by the inherent ecological services of conserved and intact wetlands. In order to address the issue of development versus conservation of wetlands, Chapter 9 provides a structured approach to utilizing multisensor RS data as well as various secondary data to develop methods and approaches of mapping IVs leading to informed decision making by local, regional, national, and global stakeholders. The chapter follows, earlier, pioneering work of Thenkabail and Nolte (1995), and Thenkabail et al. (2000a).

Chapter 9 first demonstrates how to identify, delineate, map, and characterize wetlands over large areas using data fusion involving satellite sensor data (e.g., Landsat, SPOT, MODIS Terra/Aqua, JERS SAR, IKONOS/Quickbird), secondary data (SRTM, FAO soils, precipitation), and in situ data. Second, the chapter shows methods and approaches to develop a DSS through spatial modeling to perform land suitability analysis in order to determine which of the IV wetland areas are best suited for (1) agricultural development or (2) preservation. Chapter 9 demonstrates the differences in detail and accuracies mapped using widely varying resolution (e.g., Figure 25.9) of RS data.

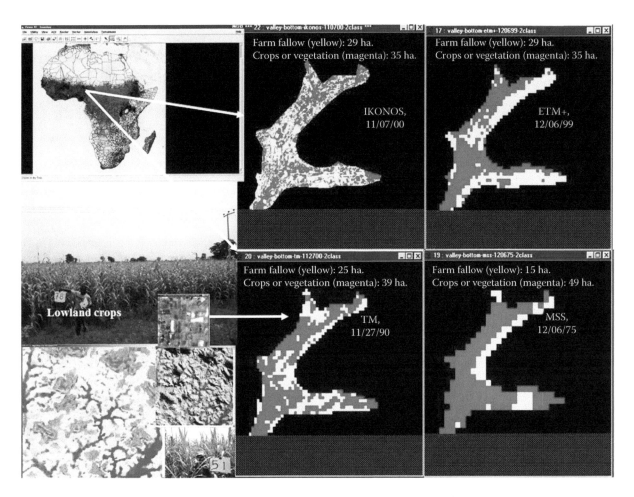

FIGURE 25.9 A typical lower-order inland valley (IV) near Kaduna in Northern Nigeria mapped using four satellite sensor data types: 4 m IKONOS, 30 m Landsat Enhanced Thematic Mapper Plus, 30 m Landsat Thematic Mapper, and 56 m Landsat multi-spectral scanner (MSS). In each case, two classes—farmland fallow and cultivated farmlands—were mapped. It is clear from these images that the spatial detail and precision seen in 4 m IKONOS are significantly higher than any other imagery. The lower-left bottom images show 4 m DEM image derived from stereo pairs of IKONOS showing distinct lowland IVs (deep blue) as opposed to the rest of the landscape. The photos (#78, #51) are two typical farms in the IVs of Northern Nigeria.

25.10 Remote Sensing of Snow Cover and Its Applications

Snowfall is an important component of hydrological cycle and has major impact on global water resources. Unlike rainfall, snowfall often has significant lag time on when it melts and flows into rivers or recharges groundwater. About 48 million km² of the Earth's surface is covered by snow of which ~98% is in the Northern Hemisphere. Snow is the largest single component of the cryosphere (ice, snow, glaciers, or permafrost) (National Snow and Ice Data Center). Snow cover is very dynamic with as much as 50% of the Northern Hemisphere land surface that can be snow covered at times during winter (Hall and Robinson, in press).

Visible wavelengths (400–700 nm) are best to study snow. This is because snow has very high albedo (80%–90% reflectivity in 400–700 nm) relative to just 10%–30% for vegetation and soils. Clouds, typically, have about 10% less albedo in 400–700 nm compared to fresh snow. However, older snow can have as low as 40% albedo. In NIR bands, the contrast between snow and no snow is poor. Cloud and snow are best discriminated in wavebands beyond 1500 nm. In this region, clouds have high reflectance (around 60%), whereas snow <10%. Active microwave or SAR is especially good in differentiating base soil, wet snow, and melting snow. Frequencies versus backscatter coefficient plots show this discrimination. Thermal data are not very useful, except in the case of detecting snow/land boundaries. This is because in order to measure snow temperature spectral emissivity and related properties of liquid, water content and grain size need to be known. Any sensor with visible and shortwave infrared (SWIR) bands can be used for snow studies. Some of the satellite sensors used in snow studies include AMSR-E, MODIS Aqua/Terra (e.g., Figure 25.10), GOES, NOAA AVHRR, SSM/I, and SMMR.

RS is used to measure snow parameters such as (e.g., Figure 25.10)

- Snow cover
- Snow depth
- Snow water equivalent

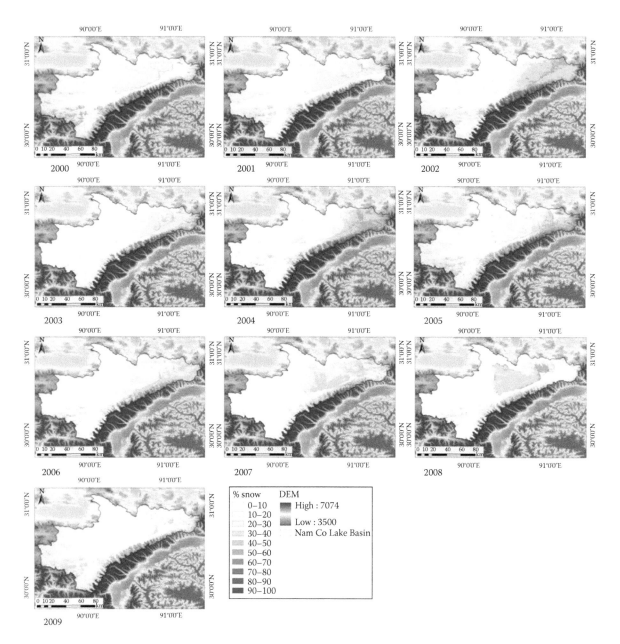

FIGURE 25.10 Monitoring changes of snow cover, lake, and vegetation phenology in Nam Co lake basin (Tibetan Plateau) using remote sensing (2000–2009). Annual snow cover from 2000 to 2009 based on MODIS/Terra Snow Cover 8-Day L3 Global 500m Grid (MOD10A2) snow cover time series from March 2000 to December 2009. The values show the percentage of time that a pixel was snow covered throughout a year during the study period (2000–2009). The decreasing trend over time of persistent snow cover and intra-annual pattern of decreasing frequency of cover from SE (higher elevations along watershed divide) to NW can be observed. *Note*: MOD10A2 data set contains data fields for maximum snow cover extent over 8 days. (From Zhang, B. et al., *J. Great Lakes Res.*, 39(2), 224, 2013.)

Normalized difference snow index (NDSI) uses data from visible and far infrared or SWIR. In Landsat Thematic Mapper (TM), for example, TM band 2 (530–610 nm) and TM band 5 (1570–1780 nm) are used to compute NDSI. If NDSI ≥ 0.4, the area is classified as snow.

Chapter 10 by Dr. Hongjie Xie et al. presents snow mapping approaches and methods using MODIS Terra/Aqua data as well as AMSR-E. The NDSI is calculated using MODIS band 4 (545–565 nm) and band 5 (1628–1652 nm) or band 7 (2105–2155 nm). In the standard MODIS snow product, four masks are also used

in classifying NDSI value as snow or no snow. These masks are dense forest stands mask, thermal mask, cloud mask, and ocean and inland water mask. MODIS Terra/Aqua also provides the fractional snow cover within pixel.

Some of the facts highlighted in Chapter 10 by Dr. Hongjie Xie et al. include the following:

1. MODIS standard snow cover products provide accuracies of >90% (clear-sky conditions) and 30–50% (all-sky conditions).

FIGURE 25.11   Map of estimated economic activity based on DMSP's OLS radiance-calibrated nighttime lights for the United States. (From Doll, C.N.H. et al., *Ecol. Econ.*, 57(1), 75, 2006.)

2. Combining terra and aqua MODIS data (acquired in 3 h gap daily) can reduce cloud cover by ~10%–20%.

3. By merging MODIS (500 m) and AMSR-E (25 km), cloud-free snow cover maps are often achieved, but by compromising spatial resolution. Accuracies of snow cover maps can be above 85% by combining these two data.

4. Flexible multiday combination of snow mapping is controlled by two thresholds: maximum cloud percentage (P) and maximum composite days (N). Based on this approach, four snow cover parameters are calculated to examine spatiotemporal variations of snow cover during hydrological year. These four parameters are snow cover index, snow-covered duration/days (SCD) map, snow cover onset dates map, and snow cover end dates map.

5. Snow products are used in snowmelt runoff models to determine discharges into lakes and reservoirs. These computations are made with high level of certainty (e.g., <90% variability explained compared to measured values), especially when multivariate analysis is performed.

6. Landsat NDSI can map snow with accuracy as high as 98% when the snow cover >60% (Burns and Nolin, 2014).

7. SCD maps and snowmelt runoff prediction assessments to predict floods and disasters have also been illustrated.

25.11 Nighttime Light Remote Sensing

Nighttime light RS provides social scientists a powerful yet convenient tool to monitor human societies from space. Specifically, applications of nighttime light RS include (1) urban extents;

(2) population studies; (3) adverse effects of light at night on human health, ecology, and astronomy; (4) economic growth center and poverty studies (e.g., Figure 25.11); (5) greenhouse gas (GHG) emissions; and (6) disaster management. Intensity of lights can be used to gather information on population concentration, and the spread of light indicates the area extent of the cities, towns, and even villages. Nightlight data from satellites are used in economic studies (e.g., location of street networks and markets and growth in general) and poverty assessment especially in data-scarce regions of the world. Cities contribute nearly 70% of global GHG emissions. Nightlight RS helps in the assessment of GHG emissions from specific locations of cities. For example, even though population dense cities are known to contribute less to GHG emissions, the benefit goes away when cities have extensive suburbs (Jones and Kammen, 2014). This is because dense settlements mean people walk to locations and drive less and use space and energy more efficiently and less consumptively per person, and denser population also means lesser land disturbance elsewhere.

Dr. Qingling Zhang et al., in Chapter 11, present how nighttime light RS is carried out by low-light imaging capabilities, for example, VNIR radiance down to 10^{-9} W/cm$^2\cdot$sr$\cdot\mu$m at night, which is more than four times fainter than the minimal detectable VNIR radiances from satellite sensors optimized for daytime observation of reflected solar radiance (Elvidge et al., 1997). This chapter shows us with clear illustrations of a wide array of applications and how nightlight RS was pioneered by a series of DMSP Operational Linescan System (OLS) data for nearly four decades since the early 1970s. Following the DMSP program,

NASA and NOAA launched in 2011 the Suomi National Polar Partnership satellite carrying the first Visible Infrared Imaging Radiometer Suite (VIIRS) instrument. The DMSP OLS nightlight data are gathered, globally, daily in the VNIR (400–1,100 nm) and thermal infrared (TIR) (10,500–12,600 nm) wavebands in 3,000 m spatial resolution and 6 bits radiometric resolution. In contrast, VIIRS data are in 740 m spatial resolution, also daily and globally in 14 bit with a VNIR band (505–890 nm). VIIRS offers a substantial number of improvements over the OLS in terms of spatial resolution, dynamic range, quantization, calibrations, and the availability of spectral bands suitable for discrimination of thermal sources of light emissions (Elvidge et al., 2013). Dr. Qingling Zhang et al., in Chapter 11, provide an overview of additional sensors offering medium and high spatial resolutions, used in nightlight RS. However, global studies are best carried out for the present (2011–present) using NOAA Suomi VIIRS and for the past (1992–2012) using DMSP OLS. Indeed, numerous applications of nighttime RS data such as from DMSP OLS have appeared in the literature over the years (e.g., Figure 25.11). This number is expected to increase in the near future.

25.12 Geomorphological Studies of Remote Sensing Studies

Geomorphology is the study of landforms and processes that shape them (Rao, 2002; Smith and Pain, 2006). RS offers a synoptic view of large areas to study landform characteristics, their classification, and their changes over space and time. Geomorphological features include such features as land topography and lithology, land use and vegetation, coastal landforms, river systems, and tidal deltas (Liew et al., 2010). Such and other information derived from RS will help enrich geomorphological maps (e.g., Figure 25.12). Chapter 12, by Dr. James B. Campbell and Dr. Lynn M. Resler, presents geomorphological studies addressing topics such as

1. Alpine and polar periglacial environments
2. Glacial geomorphology
3. Mass wasting
4. Fluvial landforms
5. Floodplain analysis
6. Channel migration
7. Stream bank retreat
8. Coastal geomorphology
9. Aeolian landforms
10. Biogeomorphology

The following are some of the important lessons we learn from Chapter 12:

Landslides: Landslides are studied through three types of information gathered from different RS approaches. These information are (1) historical inventories of landslides using, for example, Landsat or IRS imagery, (2) site-specific detection of mass wasting events using radar imagery, and (3) examination of active sites using ground-based LiDAR.

Mass wasting and change detection: Mass wasting studies often focus on before and after the event. This multisensor approach is used, depending on image availability.

Terrain deformation: This is studied using interferometric SAR (InSAR), employing phase differences derived from two or more SAR images to identify changes in local topography over time.

Characterization of steep periglacial high-mountain faces: This requires very-high-spatial-resolution imagery (<5 m).

Inaccessible or difficult to access areas: These areas, such as arctic environments and remote mountainous areas, are best studied using remotely sensed data.

Fluvial processes and floodplain analysis: Changes in river morphology, river meanders, and floodplain characteristics, including land use in these landscapes, are studied using multidate and multisensor imagery.

Biogeomorphological studies: These include fluvial geomorphology and hydrology, vegetation cover, and land use/land cover (LULC) changes, using RS.

Terrain studies with and without vegetation: LiDAR data offer the ability to observe the terrain surface free of vegetation by comparing LiDAR pulses reflected of the top of the canopy with the corresponding LiDAR data reflected from below the canopy.

Contributing factors: In many of the studies aforementioned, a number of contributing factors such as vegetation, land use, hydrology, and terrain are required. These contributing factors can be obtained from one or more of RS data from optical, radar, LiDAR, and thermal platforms.

Drs. Campbell and Resler highlight the importance of LiDAR as key to multiple geomorphological studies, with significant contributions from DEM data, and a wide array of RS data from various platforms, and integration of these data with GPS in a GIS framework.

25.13 Agricultural Droughts Using Vegetation Health Methods

Droughts are broadly classified as (1) meteorological, (2) hydrological, and (3) agricultural. Meteorological drought occurs when there is prolonged period of precipitation deficiency; hydrological droughts occur when there is below-normal water levels in lakes, reservoirs, and rivers; and agricultural drought occurs when there is insufficient soil moisture for healthy growth of croplands and rangelands. Food security is heavily dependent on healthy crop productivity, which in turn is feasible only with sufficient water availability. Extreme conditions of agricultural droughts (e.g., Figure 25.13) will result in full or partial crop failure and/or substantial decreases in crop yields, which in turn will lead to food insecurity and price rise and may even lead to famine. Conventional drought monitoring is performed using meteorological, hydrological, and soil moisture data and includes such indices as standardized precipitation index (SPI),

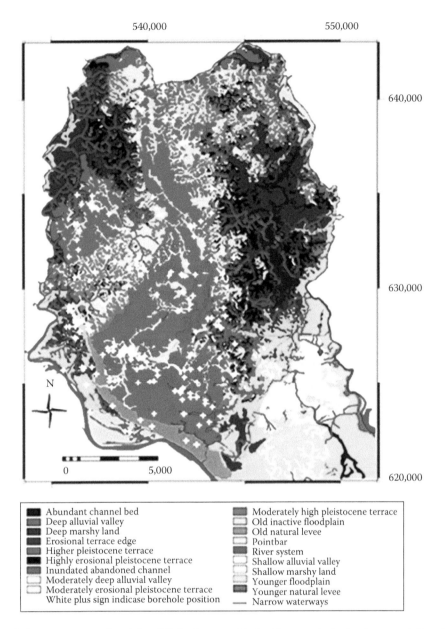

FIGURE 25.12 Geomorphology accompanied with landfill thickness in Dhaka City area derived using a spatially enhanced fused image of Indian Remote Sensing-1D PAN data and Landsat Enhanced Thematic Mapper Plus bands 5, 4, and 3, acquired on February 2000 and 2002, respectively. In addition, data were also obtained from topographic maps. (From Kamal, A.S.M.M. and Midorikawa, S., *Int. J. Appl. Earth Observ. Geoinform.*, 6(2), 111, 2004.)

palmer drought severity index (PDSI), crop moisture index, palmer hydrological drought index, surface water supply index (SWSI), reclamation drought index, deciles, and percent of normal. The U.S. Drought Monitor (USDM) integrates several of these indices along with ancillary data to provide weekly operational drought maps. In contrast, RS offers consistent global coverage and better spatial representation for drought studies. In its simplest form, drought can be monitored using RS by simple deviation of NDVI from its long-term mean.

One of the most widely used, attractive, simple, and yet powerful RS measure of drought is the vegetation health (VH) method of NOAA/NESDIS developed by Dr. Felix Kogan

and reported in Chapter 13 by Dr. Felix Kogan and Wei Guo. They identify VH-based drought products to include moisture and thermal stress, the start/end of drought, intensity, duration, magnitude, area, season, origination, and impacts. Vegetation condition index (VCI) is a proxy for moisture condition, temperature condition index (TCI) is a proxy for thermal condition, and vegetation health index (VHI) is a proxy for combination of moisture and thermal conditions. These indices are represented in scale of 0 (extreme vegetation stress) and 100 (optimal conditions of VH). VHI is somewhat limited in energy-limited environments like high elevations and high latitudes (see Chapter 15).

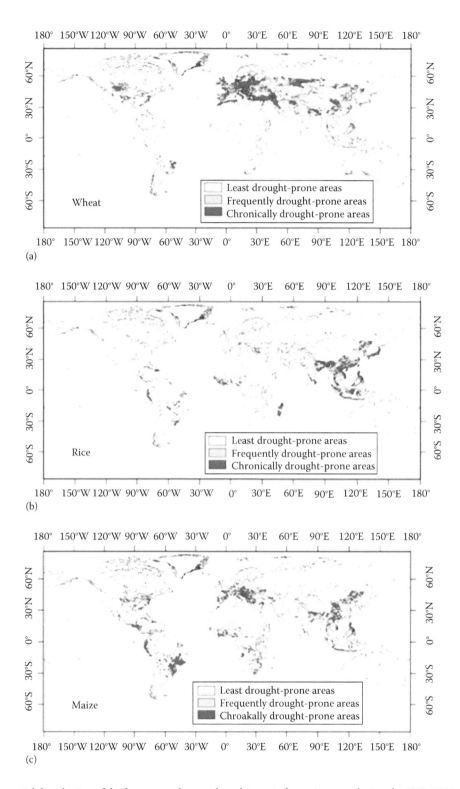

FIGURE 25.13 The spatial distribution of the frequency of severe drought events for main crops during the 1950–2008 period (a, b, c, d, e, and f show the spatial distribution of the frequency of severe drought events in wheat, rice, maize, soybean, and barley crop-planting regions, respectively) determined based on standardized precipitation evapotranspiration index. Global chronically drought-prone areas have increased significantly, from 16.19% in 1902–1949 to 41.09% in 1950–2008. (From Wang, Q. et al., *Quatern. Int.*, 349, 10, 2014.)   (*Continued*)

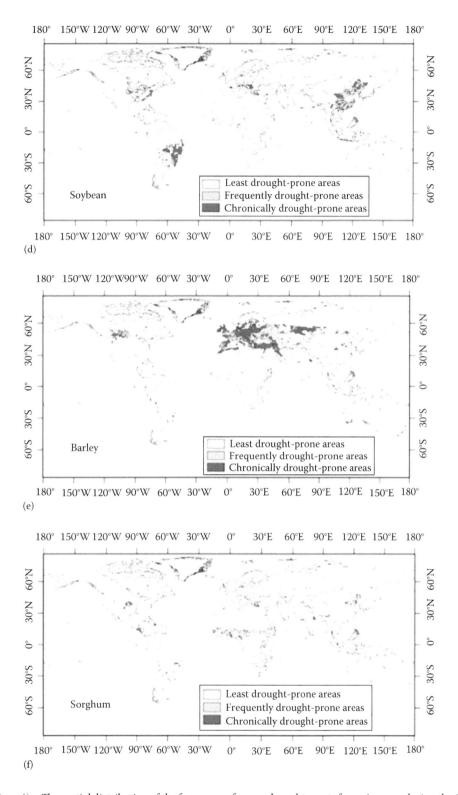

FIGURE 25.13 (*Continued*)   The spatial distribution of the frequency of severe drought events for main crops during the 1950–2008 period (a, b, c, d, e, and f show the spatial distribution of the frequency of severe drought events in wheat, rice, maize, soybean, and barley crop-planting regions, respectively) determined based on standardized precipitation evapotranspiration index. Global chronically drought-prone areas have increased significantly, from 16.19% in 1902–1949 to 41.09% in 1950–2008. (From Wang, Q. et al., *Quatern. Int.*, 349, 10, 2014.)

Emphasis in Chapter 13 is on global vegetation health (GVH) computation using 4 km resolution NOAA AVHRR GAC data, now available from 1981 to present, through three indices: VCI, TCI, and VHI. GVH maps provide sophisticated and powerful, yet simple to visualize drought conditions, severity, and progression maps and data. This is especially useful in assessing such measures as global food production and famine early warning. An exciting prospect is the continuity of this 4 km, 34-year drought record through much higher resolution of 375 m from VIIRS sensor onboard NPOESS net primary productivity (NPP). They demonstrate VH by studying droughts in the warmer world of the twenty-first century. Even though there generally accepted temperature *pause* over last 15 years (0.04°C temperature rise during 1998 and 2013 when compared with 0.18°C increase in the 1990s), climate (e.g., precipitation, temperature) variability over space and time have caused severe droughts in many parts of the world (e.g., 2012 Midwestern agricultural drought in the United States).

25.14 Agricultural Drought Monitoring Using Biophysical Parameters

Chapter 14 discusses the use of biophysical indicators derived from optical satellite RS in current agricultural drought monitoring systems. Most operational systems exploit the qualitative approach (e.g., Figure 25.14) of looking at a pixel or an area and determining the deviation of vegetation condition for the same

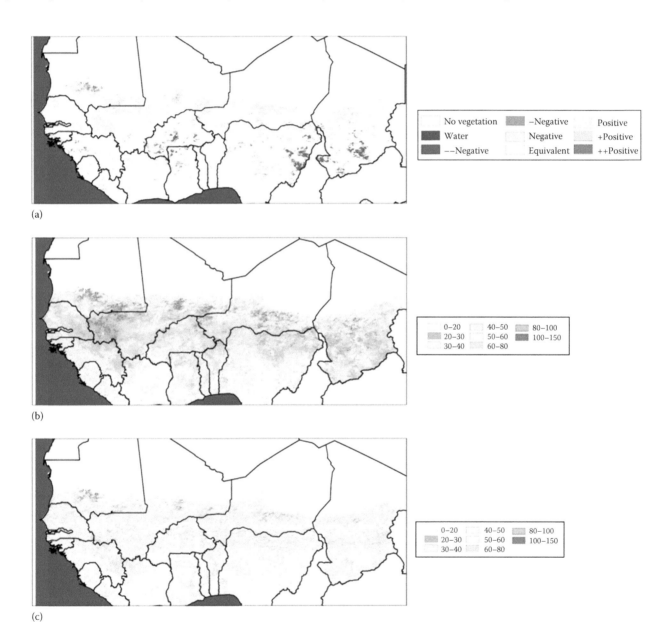

(a)

(b)

(c)

FIGURE 25.14 Indicators of vegetation status in West Africa on July 31, 2013. (a) Standardized normalized difference vegetation index. (b) Normalized growth index (indice de croissance normalize). (c) VCI. (From Traore, S.B. et al., *Weather Climate Extremes*, 3, 22, 2014.)

pixel or area with respect to its long-term mean. The deviation from long-term mean of NDVI, VCI, TCI, VHI, and other vegetation indices all constitute various forms of this approach. Typically, these studies analyze one of these vegetation indices and look for changes over time for a given pixel or area. Such a pragmatic approach is widely used and works reasonably well. Some of these methods are also discussed in Chapter 13.

However, agriculture productivity depends on a complex set of conditions that include features such as changes in

Crop type or cropping pattern over years
Agronomic inputs (e.g., high input agriculture in previously low input agriculture, irrigation on previously nonirrigated lands) leading to changes in biomass and yield
Phenology (e.g., late crop planting will result in a different phonological cycle)
Cultivar (e.g., different cultivar leading to different biomass and yield)

All these conditions lead to changes in biophysical indicators at a given time of the year or season relative to the same time during a previous year or season. As a result, the observed changes cannot be attributed univocally to drought conditions. For many of these factors, little information is available at regional and national scale, and biophysical indicators derived mainly from satellite image are the main data source.

Chapter 14 focuses on three recent approaches for monitoring agricultural drought at regional scale, implemented by different international organizations: the FAO of the UN, the nongovernmental organization Action Contre la Faim (ACF), and the Joint Research Centre (JRC) of the European Commission.

The agricultural stress index system (ASIS) of FAO aims at detecting agricultural areas with a high likelihood of drought at the global level using optical and thermal channels from METOP-AVHRR instrument. ASIS determines the likelihood of drought occurrence on the basis of three distinctive drought features: intensity, duration, and spatial extent.

The ACF approach is currently being developed to target pastures in the Sahel by considering three key elements: biomass production, water availability to animals, and livestock movements. SPOT-VEGETATION data are used to map pasture biomass production (with a light-use efficiency model) and surface water bodies.

The JRC system is based on a statistical approach aiming to provide early estimations of the probability of experiencing a critical biomass production deficit. The method is built upon the similarity between the current seasonal fraction of absorbed photosynthetic active radiation (faPAR) profile and the past ones and provides the analyst with an information that is adjusted for its reliability, as estimated from past forecasts.

Other novel and promising modeling frameworks for drought monitoring are then mentioned. These include data mining techniques, combined use of biophysical parameters and precipitation data, explicit modeling of the spatiotemporal evolution of extreme events, and developing new and advanced indices (e.g., Thenkabail et al., 2000a,b; Thenkabail et al., 2004, Thenkabail et al., 2014).

Eventually, current critical needs in agricultural drought monitoring with RS are discussed: ensuring the continuity of long-term time series from different satellite missions, developing reliable and operational unmixing techniques, achieving a better balance between timeliness and noise reduction by the use of simplified vegetation growth models and data assimilation techniques, and finally profiting from consolidated techniques such as thermal RS and emerging ones such as the analysis of fluorescence and the photochemical reflectance index.

25.15 Drought Monitoring Advances and Toolkits Using Remote Sensing

Chapter 15 by Dr. Brian Wardlow et al. provides a number of innovative advances in using RS in drought studies. Prior to data, approaches, and methods discussed in this chapter, drought studies using RS were primarily based on either looking at NDVI deviation from long-term mean and/or using VCI, TCI, and VHI discussed in Chapters 13 and 14. In contrast, Chapter 15 by Dr. Brian Wardlow et al. provides six RS-based advances in drought studies:

1. *USDM approach*: The USDM approach utilizes multiple conventional, non-RS indices such as PDSI, SPI, and SWSI along with guidance from experts such as climatologists, agricultural scientists, and water resources managers. In addition, more recently, USDM utilizes RS-based VCI, TCI, and VHI using data from such sensors as AVHRR and MODIS. USDM provides regular drought maps that depict drought intensities over large area extents such as a county or multiple counties. It is somewhat limited in depicting localized drought conditions. Nevertheless, USDM approach is also attracting considerable international attention with similar approach used in U.S. Department of Agriculture Farm Service Agency and Famine Early Warning System Network.

2. *Vegetation drought response index* (*VegDRI*): This index identifies vegetation stress by integrating RS, climatic, and biophysical data. RS data, derived either from AVHRR or MODIS, component makes use of (a) percent annual seasonal greenness, (b) start of season anomaly, and (c) out of season to represent NDVI during non–growing crop season. Climate part includes PDSI and SPI. Biophysical part takes dominant LULC type for each pixel. The VegDRI model integrates these RS, climatic, and biophysical variables. VegDRI and USDM provide comparable results. However, the higher spatial resolution of VegDRI helps in better understanding of local (e.g., subcounty level) droughts. The USDM now routinely consults VegDRI to produce more refined USDM maps.

3. Vegetation outlook (VegOut): This is akin to VegDRI in that it also integrates and uses RS and climatic and biophysical data. However, VegOut also uses seven oceanic/atmospheric indices that make it distinct from VegDRI.

4. Evaporative stress index: This index measures standardized anomalies in a ratio of actual to reference ET (ET_a/ET_{ref}) retrieved using the RS-based ALEXI two-source SEB model. ALEXI requires at least two measurements of LST measured during morning and obtained from thermal images (e.g., 10 km resolution TIR data from GOES sounder instrument). Higher resolution applications require field or sub-field scale sampling and are achieved through DisALEXI.

5. Microwave-based surface soil moisture retrievals from sensors such as SSM/I and AMSR-E: Surface soil moisture retrievals are better achieved using the low-frequency L band (1–2 GHz), but also with high-frequency (2–20 GHz) microwave observations. Since the microwave can only penetrate the top soil (<5 cm) and not the plan rooting depth (30–150 cm), microwave-based soil moisture retrievals to agricultural drought have generally been confined to their use in land data assimilation systems.

6. GRACE microgravity measurements: These are used to retrieve satellite-based soil moisture and groundwater. However, GRACE data are very coarse (150,000 km²) and are available over an area only in about 2–4 months' time frame. These spatial and temporal resolution limitations allow for only broad understanding of drought situation. However, since no other satellite offers a reasonably good understanding of the groundwater fluctuations, GRACE offers a good understanding at regional, national, and river basin scales (Ghoneim and El-Baz, 2007).

Indeed, more specific drought indices can be developed that address specific issues of crops or other vegetation (e.g., Figure 25.15). In Figure 25.15, a water deficit index is compared to vegetation water stress index (VWSI) developed using ETM+, NIR, and SWIR bands for wheat crop.

25.16 New Remote Sensing–Based Drought Indices

Can we monitor drought using RS data alone? This is a question in Chapter 16 Jinyoung Rhee et al. try to answer. The state-of-the-art drought monitoring systems like USDM use RS, non-RS, and some subjective expert opinion to come with drought maps. Chapter 16 lists the six key indicators used by the USDM: (1) PDSI, (2) percent of normal precipitation, (3) SPI (McKee et al., 1993), (4) U.S. Geological Survey (USGS) daily stream flow percentiles, (5) model-based Climate Prediction Center (CPC) soil moisture model percentiles, and (6) satellite-based VHI. A number of ancillary data are used in USDM such as the palmer crop moisture index and Keetch–Bryam drought index, as well as reservoir and lake levels, groundwater levels, and soil moisture field observations. For the western United States, snowpack telemetry is also used. Further, USDM used subjective judgment of experts. Other drought monitoring systems such as objective blend drought index of CPC and CPC seasonal drought outlook also use RS and non-RS data, methods, and approaches.

Chapter 16 by Jinyoung Rhee et al. focuses on discussing and developing new advanced drought indices that are purely dependent on RS. Their own (Rhee et al., 2010) scaled drought condition index uses only remote sensed data that include MODIS LST, NDVI, and TRMM precipitation through linear combination. This index was found useful for agricultural drought monitoring in both arid and humid regions. Another RS data–driven drought index is drought severity index that uses MODIS ET, potential ET, and NDVI products (e.g., Mu et al., 2013). Generally, RS-only-derived drought indices do as well or even better than drought indices that use either both RS and non-RS data or just the non-RS data.

Figure 25.16, for example, illustrates monitoring meteorological drought in semiarid regions using multisensor microwave RS data. With data from multiple sensors that include optical, radar, microwave, hyperspectral, LiDAR, thermal, and hyperspatial sensors, data of which are now becoming more frequently available, there are opportunities to develop newer more specific and advanced drought indices that are based purely on RS data and indices (e.g., Figure 25.16).

25.17 Remote Sensing of Drylands

Drylands cover ~40% of the global land area and support approximately one-third of the world's population (Andela et al., 2013) with food and fiber or by providing grazing resources through rangelands. (e.g., Figure 25.17a,b). Agriculture in drylands is mostly rainfed and thus highly susceptible to droughts. At present, ~25% of the drylands are affected by drought every year (e.g., Figure 25.17c), and it is expected that these areas will increase to as much as 50% (see Chapters 13 through 16) due to climate change. Besides this direct provision of ecosystem services, rangelands are also rich in biodiversity, but may be prone to wildfires depending on vegetation composition and climatic conditions. Even though many research initiatives aimed at assessing the status of drylands, the estimated rate of land degradation diverges greatly.

Global drylands cover extended areas (Figure 25.17) and are often hard to access and study; hence, RS offers an ideal platform to observe, characterize, and study drylands. Further, image acquisitions over drylands are relatively easier than over tropical areas due to fewer cloudy days in a year. Chapter 17 by Dr. Marion Stellmes et al. provides us an overview of dryland studies using RS. As the variety of studies is huge, the authors focused on vegetation-related studies based on medium- to coarse-scale satellite imagery that provides regular information of the Earth's surface since the 1980s. The overview the authors provide covers the following major themes:

1. Assessing the extent of land degradation and desertification as well as the condition of ecosystems considering climate variability
2. Monitoring conversions and modifications in dryland land cover based on biological productivity over long time periods
3. Integrating RS-derived results with the human dimension to identify drivers of land degradation

FIGURE 25.15 Advances in drought indices. Modern-day remote sensing data from different sensors can be used to derive various drought indices that address specific issues of crops and other vegetation. For example, the figure shows estimating crop water stress with Enhanced Thematic Mapper Plus, near-infrared and SWIR data water-deficit index, and VWSI. VWSI describes the canopy water stress rather than water content, and it is of interest particularly for agricultural drought monitoring. (From Ghulam, A. et al., *Agric. Forest Meteorol.*, 148(11), 1679, 2008.)

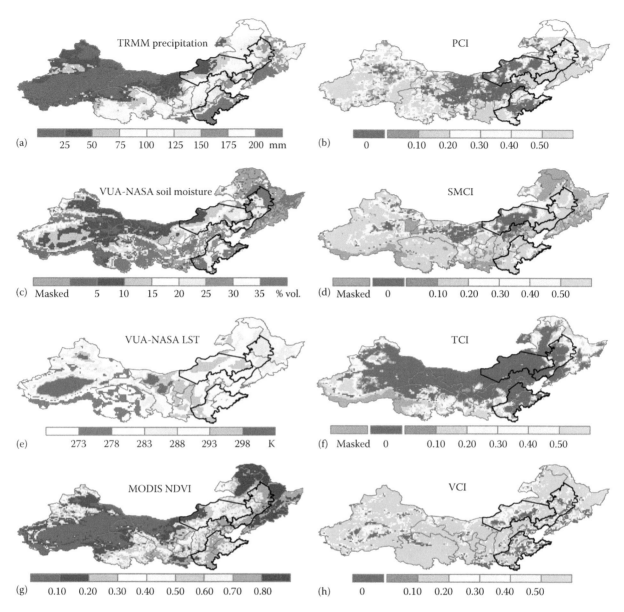

FIGURE 25.16 Remote sensing (RS) drought indices. The images illustrate detecting and monitoring drought timely at regional scale using RS data and indices: TRMM precipitation condition index (PCI), the soil moisture condition index (SMCI), and the TCI based on microwave remote sensed TRMM precipitation, Advanced Microwave Scanning Radiometer for the EOS soil moisture, and land surface temperature retrievals from 2003 to 2010. The remotely sensed variables were linearly scaled from 0 to 1 for each pixel based on absolute minimum and maximum values for each variable over time, in order to discriminate the weather-related component from the ecosystem component as done for VCI using normalized difference vegetation index (NDVI) (see Chapter 13). After normalization, the scaled value changed from 0 to 1, corresponding to the precipitation changes from extremely low to optimal. RS drought indices studied here are TRMM PCI, SMCI, TCI (land surface temperature), VCI, TRMM precipitation and soil moisture condition index (PSMCI), TRMM precipitation and temperature condition index (PTCI), soil moisture and temperature condition index (SMTCI), and microwave integrated drought index (MIDI). These drought indices are defined as follows: PCI = (TRMMi − TRMMmin)/(TRMMmax − TRMMmin); SMCI = (SMi − SMmin)/(SMmax − SMmin); TCI = (LSTmax − LSTi)/(LSTmax − LSTmin); VCI = (NDVIi − NDVImin)/(NDVImax − NDVImin); PSMCI = α * PCI + (1 − α) * SMCI; PTCI = α * PCI + (1 − α) * TCI; SMTCI = α * SMCI + (1 − α) * TCI; and MIDI = α * PCI + β * SMCI + (1 − α − β)TCI, where α and β represented the weight of single index while constituting the integrated drought indices, i is the current time period (e.g., SMCI of this week), min is the long-term minimum (e.g., SMCI minimum of this week over, say, the last 10 years), and max is the long-term maximum (e.g., SMCI maximum of this week over, say, the last 10 years). The image shows drought detected by RS indices for July of 2010 over northern China. (a), (c), (e), and (g) were, in 2003–2010, mean TRMM precipitation, VUA-NASA soil moisture, VUA-NASA land surface temperature, and MODIS NDVI for July, while (b), (d), (f), and (h) were drought detected by PCI, SMCI, TCI, and VCI for July of 2010, respectively. Soil moisture and SMCI of forest were masked out in (c) and (d), while areas of SMCI/TCI in (d) and (f) corresponding to zero soil moisture/land surface temperature in July of 2010 were masked out too. (From Zhang and Jia, 2013.)

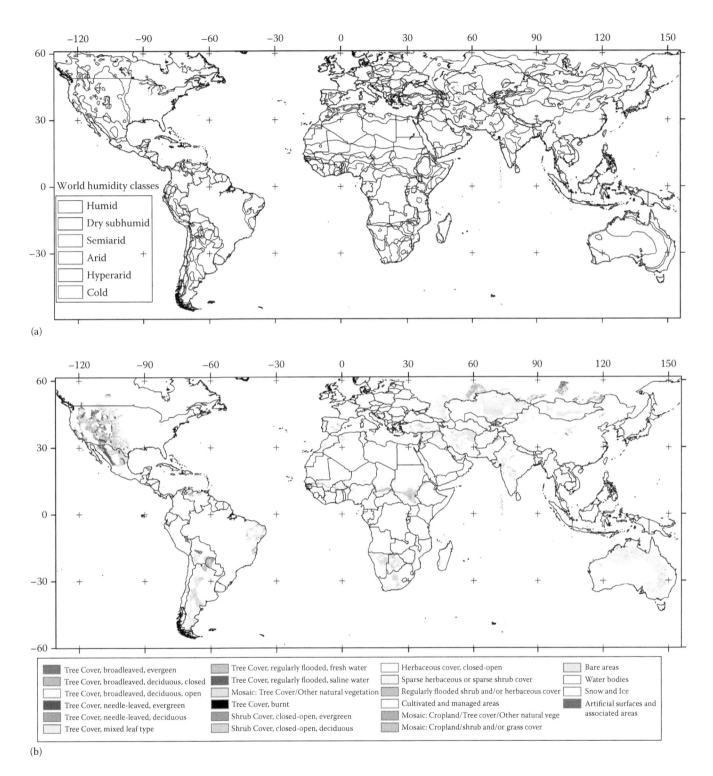

(a)

(b)

FIGURE 25.17 (a) Humidity classes as defined in the *World Atlas of Desertification* (UNEP, 1997). (b) Global land cover (GLC2000) for the semiarid areas. *Note*: For the semiarid areas overlaid by areas (white circles and lines) selected for analysis of seasonality by Fensholt et al. in their paper. (From Fensholt, R. et al., *Remote Sensing Environ.*, 121, 144, 2012.)   (*Continued*)

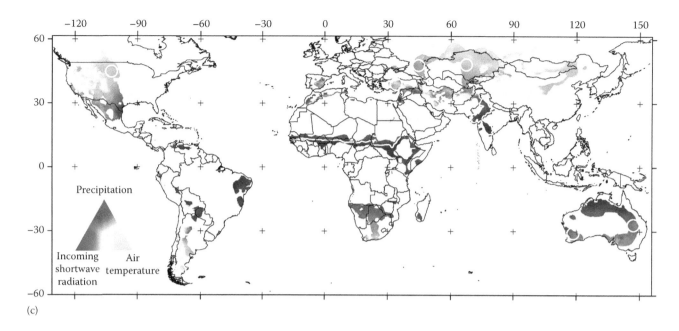

(c)

FIGURE 25.17 *(Continued)*   (c) Geographic distribution of potential climatic constraints to plant growth (Nemani et al., 2003). *Note*: For the semiarid areas overlaid by areas (white circles and lines) selected for analysis of seasonality by Fensholt et al. in their paper. (From Fensholt, R. et al., *Remote Sensing Environ.*, 121, 144, 2012.)

In reviewing these studies, Chapter 17 by Dr. Marion Stellmes et al. discusses various satellites and sensors used in dryland studies, methods and approaches adopted, and the use of various vegetation indices. Many qualitative and quantitative measures of drylands studies are based on vegetation indices related to greenness, faPAR, and NPP or on techniques like spectral mixture analysis. The chapter also gives an overview of techniques used to assess conversions and modifications in drylands and evaluates these. These comprise a large variety of change detection methods based on multitemporal classifications, annual time series approaches of vegetation cover, or phenological dynamics. Moreover, short outlooks of new techniques and sensors are given that, for example, comprise the combination of medium- and coarse-scale-resolution data such as STARFM or future satellite-based hyperspectral sensor platforms that will allow for the operational derivation of enhanced land degradation indicators such as top soil organic matter or pigment and water content. The chapter also includes a critical review of uncertainties in land degradation assessment by RS.

25.18 Disaster Risk Management through Remote Sensing

Disasters come in many forms and are both of natural origin and human induced. Natural disasters include cyclones, tsunamis, floods, droughts, volcanoes, earthquakes, fire, and landslides. Costs associated with disaster events have risen substantially since about 1995, but also vary strongly between years, currently typically ranging between 100 and 200 billion US$ per year (Figure 25.18). Casualty numbers have been even more variable, with statistics skewed toward particularly devastating individual

events that have claimed millions of lives, in particular flood and drought events in the 1920s and 1930s. Nearly 95% of all disasters also take place in economically developing world.

RS is widely used to study all phases of disasters: (1) hazard and risk assessment, (2) prevention and mitigation, (3) early warning, (4) seen-event monitoring, (5) postdisaster response and damage assessment, and (6) reconstruction and rehabilitation (e.g., Figure 25.18). In Chapter 18, by Dr. Norman Kerle, different eras of disaster risk management (DRM) with RS are identified:

1. Early disaster damage mapping starting with the pioneering effort by George R. Lawrence to study the severe 1906 San Francisco earthquake, using an airborne kite-based camera
2. The early satellite era driven by military interests, in particular the Corona, Argon, and Lanyard missions in the 1960s, imagery from which was only declassified between 1995 and 2002
3. The early civilian satellite era with the launch of TIROS in 1960, NOAA's VHRR in 1970, and Landsat in 1972
4. New advanced era of satellite Earth observation beginning in the early 1990s with the launch of numerous advanced satellites, both by governments (e.g., Landsat, SPOT, IRS, various SAR systems from such as RADARSAT to very recent Sentinels) and commercial platforms (e.g., IKONOS, Quickbird, Geoeye)

DRM concepts and approaches, and the many uses of RS data (e.g., Table 18.2), are the centerpiece of Chapter 18 by Dr. Norman Kerle. These are broadly identified as follows:

1. Hazard assessment and risk identification. This involves identifying hazard elements at risk (EaR) that are highly diverse and range from the obvious, such as buildings or

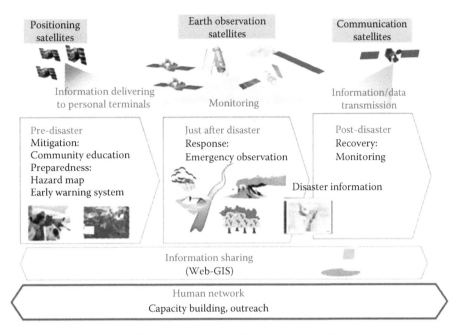

FIGURE 25.18 Concept of Sentinel Asia Step 3 with the goal to expand activities to cover all the disaster management cycle—mitigation/preparedness phase, response phase, and recovery phase—utilizing various and many satellites such as Earth observation, communication, and navigation satellite, under further collaboration for operation and human networking by Sentinel Asia Joint Project Team (Kaku and Held, 2013).

infrastructure, to features less frequently considered, such as places of cultural significance or heritage sites, national parks, and sites of biological diversity.

2. Assessment of vulnerability, which is defined as the capacity for loss and which can include aspects such as physical, social, and environmental, that is, the damage that EaR will sustain in a given hazard event.
3. Monitoring and early warning to detect potential hazardous situations once hazards are well understood through (1) and (2).

   These efforts will help in hazard identification and mitigation, adaption to present hazards, and early warning for threatening events. Yet, in many cases, hazards cannot be avoided, thus necessitating

4. Postdisaster response and damage assessment

Currently, RS provides an ideal platform to study, assess, and respond to each of the aforementioned DRM concepts and approaches. First, for most parts of the world, freely available, web-enabled baseline satellite image archives now exist. Second, numerous new satellite platforms in various spatial and spectral domains are being launched or developed. These satellites acquire data frequently, if tasked appropriately providing data of any land area on Earth on any given day, by one satellite or another, and/or have the capacity to respond at short notice to acquire images over an area of interest. Third, advances in analysis methods, computing, and processing power of modern-day RS make it feasible to respond to disasters rapidly and accurately. Finally, there have been increasing global efforts, such as the International Charter "Space and Major Disaster," or the Global Earth Observation System of Systems (GEOSS), to make coordinated use of the growing set of RS tools and instruments.

Chapter 18 does not deal with technological disasters, industrial disasters, or the complex humanitarian crisis situations associated with political or ethnic conflicts, nor with disasters such as famine that are of natural origin, but that predominantly result from political or societal failures.

25.19 Humanitarian Disasters and Remote Sensing

Humanitarian disasters can be both natural and human caused, and often, causes are interrelated. Cyclones, tornadoes, earthquakes, floods, droughts, and any number of other events (e.g., internal and external conflicts, wars, and spread of rare diseases) can lead to great humanitarian disasters. RS provides neutral and unbiased data on most of the disasters. The ability of spaceborne RS to repeatedly and consistently monitor disasters makes it most appealing tool for disaster assessment and management, comprising short-term catastrophes with a clear peak (e.g., flood events) to long-term, protracted crises with severe impact on society and the environment on the longer run. Chapter 19, written by Dr. Stefan Lang and other experts in the field, provides an excellent introduction to all these topics. The authors show us how a wide array of RS data, methods, and approaches can support humanitarian action and disaster mitigation. They provide case studies of monitoring displaced population and associated impact on the environment, detecting indications of destruction during conflicts and monitoring illegal mining and logging activities in relation to conflict situations. It is important to note that humanitarian disasters are often jointly coordinated by agencies such as the UN Office for the Coordination of Humanitarian Affairs, United Nations High Commissioner for Refugees, World Food

Programme, United Nations office for Disaster Risk Reduction, United Nations Office for Outer Space Affairs, and U.S. Agency for International Development, often through multinational, and multiagency efforts. For example, the United Nations Platform for Space-based Information for Disaster Management and Emergency Response put together a large collection on RS data gallery post the 2004 great Indian Ocean tsunami.

Disasters require quick response at every level. The chapter shows the issues and importance of using "OpenStreetMap" where online volunteers help prepare maps with real-time updates of events taking the typhoon in the Philippines in 2013 as an example. Whereas very-high-spatial-resolution (e.g., submeter to 5 m) imagery such as QuickBird or WorldView is of great importance in assessing disasters (e.g., damaged buildings, bridges, refugee settlements), all types of RS data can be useful in various types of disasters. Figure 25.19 (Dong et al., 2011) shows the extracting of damages caused by the 2008 8.0 Ms Wenchuan earthquake from the German TerraSAR-X (with X band wavelength 31 mm, frequency 9.6 GHz, spatial resolution up to 1 m) data. Other slow progression disaster events such as droughts are monitored over large areas using NDVI or EVI from coarse-resolution MODIS or moderate-resolution Landsat. Flood events often need radar imagery to look through clouds since floods occur during heavy rainfall and cloud cover, and ground penetrating radar and thermal imagery to track people trapped in collapsed buildings can be very useful.

25.20 Remote Sensing of Volcanoes

In Chapter 20, Dr. Robert Wright provides an overview of volcanoes and the role of RS in studying them. There are around 1500 active or potentially active volcanoes around the world of which roughly about 5% erupt every year.

Chapter 20 shows us how RS is used to study various volcanic processes. These include

1. Topographic deformation (e.g., Figure 25.20) associated with subsurface magma movements
2. Measurement of the gases and ash released by volcanoes
4. Geothermal and hydrothermal manifestations of active volcanism
5. Lava flows and lava domes produced during effusive volcanic eruptions

The phenomena associated with active volcanism are many, and no single wavelength region provides data at which all (e.g., high-temperature lavas, low-temperature hydrothermal activity, passive emissions of gas, explosive eruptions of ash, deformation of the volcanoes themselves) can be studied adequately. Furthermore, changes in these processes occur on both short and long time scales (e.g., daily to interannual), as well as over large spatial extents (e.g., meters to hundreds of kilometers) (see Figure 25.20). By leveraging data acquired by many sensors, at wavelengths from visible to microwave wavelengths, from both low Earth and geostationary orbits, scientists have developed techniques for quantifying much about active volcanic

processes from space, taking advantage of the synoptic, repeated coverage offered by the RS technique.

No single mission has been launched to specifically study volcanism, so volcanologists have exploited a wide range of sensors for this purpose. InSAR data, which revolutionized the study of volcano geodesy, uses SAR data sets, including the C band SARs flown on ERS-1, ERS-2, ENVISAT, **RADARSAT, the** X band SAR on board TerraSAR-X, **and the L band SAR carried on board the Advanced Land Observing Satellite** "DAICHI" (ALOS). Studies of volcanic gases have relied heavily on the Total Ozone Mapping Spectrometer series of sensors, the Ozone Mapping Instrument (OMI) on board NASA's Aura, the Scanning Imaging Absorption Spectrometer for Atmospheric Cartography (SCHIAMACHY) sensor on board ENVISAT, and the Global Ozone Monitoring Experiment 2 on board MetOp-A, in addition to the Terra ASTER and Aqua AIRS instruments, using a mixture of ultraviolet and long-wave infrared measurements, acquired at a range of spatial resolutions. Sensors such as MODIS, AVHRR, and GOES have been the workhorses of volcanic ash detection, due to their ability to discriminate silicate ash clouds from water clouds based on their contrasting long-wave infrared transmissivities, in a timely manner. The Landsat TM (and its successors), the Terra ASTER, and the EO-1 Hyperion have been widely used for measuring the thermal properties of high-temperature lavas. Low spatial (but high temporal) resolution sensors such as MODIS, AVHRR, and GOES have been utilized to detect the onset of volcanic eruptions around the globe in near real time, based on their thermal emission signatures. Although low-temperature targets utilize the long-wave infrared (8–14 μm), studies of active lavas also exploit the middle infrared (MIR) (3–5 μm) and SWIR (1.2–2.2 μm).

25.21 Biomass Burning and Greenhouse Gas Emissions Studied Using Remote Sensing

Biomass burning contributes to anywhere from 2290 to 2714 Tg C/year (Ito and Penner, 2004; Mieville et al., 2010) when compared with 8180 Tg C/year by fossil fuel combustion, cement production, and gas flaring (Barker et al., 2007). Chief sources of biomass burning are savannas (42.1%), agricultural wastes (23.1%), tropical forests (14.5%), fuelwood (16.2%), temperate and boreal forests (3.3%), and charcoal (1%) (Andreae, 1991; Andreae and Merlet, 2001). Biomass emissions release wide array of GHGs such as carbon dioxide, carbon monoxide, methane, nitric oxide, ammonia, sulfur, and hydrogen (Andreae, 1991; Andreae and Merlet, 2001). These trace gases in atmosphere are best quantified through their absorption characteristics in specific wavelengths, with wavebands in visible, NIR, and SWIR. Over the years, a number of satellite sensors have been used to study trace gases. These include GOES fire radiative power (FRP), Global Ozone Monitoring Experiment sensor on board ERS-2, MODIS (e.g., MODIS burned area product [Giglio et al., 2006]), fire counts from the TRMM (Visible and Infrared Scanner), and European Remote Sensing satellites Along-Track Scanning Radiometer sensors.

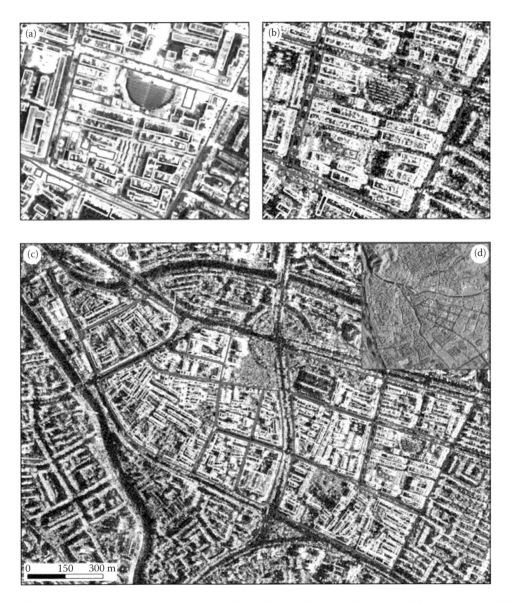

FIGURE 25.19 Building damage information extraction of sample blocks in Dujiangyan City from aerial photograph and TerraSAR-X (with X band wavelength 31 mm, frequency 9.6 GHz, spatial resolution up to 1 m) image; the GIS data of blocks and buildings are extracted from a pre-earthquake QuickBird image acquired on July 22, 2005: (a) aerial photograph of Hehuachi residential area acquired on May 18, 2008; (b) backscattering coefficient image of TerraSAR-X of Hehuachi residential area acquired on May 15, 2008; (c) building damage information of sample city blocks extracted from TerraSAR-X image; and (d) index map showing the location of study blocks in Dujiangyan City. (From Dong, Y. et al., *J. Asian Earth Sci.*, 40(4), 907, 2011.)

In Chapter 21, Dr. Krishna Prasad Vadrevu and Dr. Kristofer Lasko show us how nitrogen dioxide (NO_2) emissions in troposphere are measured using Ozone Monitoring Instrument (OMI) on board EOS Aura satellite, and SCIAMACHY and MODIS (Aerosol Optical Depth [AOD]) signal varies in relation to MODIS fire retrievals. They demonstrate how one of the trace gases, NO_2, is measured, modeled, and mapped from biomass burning of subtropical evergreen forests of Northeastern India. The concept to measure other trace gases using sensor data is similar. Specific to South Asia, atmospheric burning of the related trace gases in the atmosphere is maximum in months of March, April, and February and minimum during the month of July

(e.g., Figure 25.21). This is because of postharvest fire activity and also slash and burn of forests during summer, before the start of the rainy season in June in the tropics. Chapter 21 clearly showed that OMI performed better than SCIAMACHY by (1) detecting higher NO_2 concentrations, (2) providing stronger correlation of MODIS fire counts, and (3) establishing stronger correlation with MODIS AOD. The study highlights satellite RS of NO_2 from biomass burning sources.

In near future, the TROPOspheric Monitoring Instrument (TROPOMI) will be carried on board ESAs GMES Sentinel 5 Precursor mission, which will allow it to measure a large range of gases, including the most important trace gases, aerosols, and

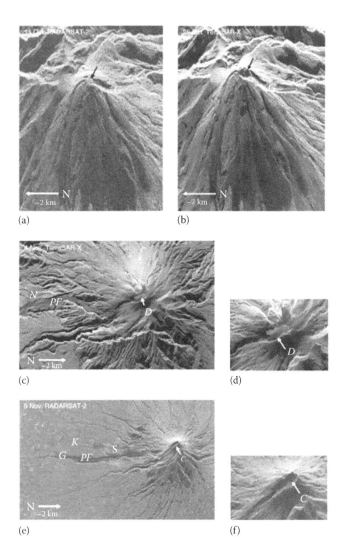

FIGURE 25.20 SAR images of the summit of Merapi volcano before and after the times of the October 26 and the November 4 explosive eruptions. For clarity, images are oriented with respect to line of sight of the radar. Arrows indicate north (N) direction and approximate scale. G (Kali Gendol), K (Kali Kuning), and Kj (Kinahrejo). (a) RADARSAT image, October 11, 2009. Arrow indicates the 2006 lava dome. (b) TerraSAR-X image, October 26, showing new summit crater (arrow) produced by the explosive eruption of October 26. (c) TerraSAR-X image, November 4, 2010, was showing large (~5 × 10⁶ m³) lava dome (d). Pyroclastic flow deposits (PF) from the October 26 eruption appear dark in the radar images. (d) Enlargement of the summit area of image (a). (e) RADARSAT image of November 5, 2010, showing PF (dark gray) and surge deposits (light gray). These deposits formed earlier during the main phase of the November 4–5 explosive eruption. An enlarged, elongate crater, produced by the November 4–5 eruption, is also evident at the summit. (f) Enlargement of the summit area of image (c). (From Surono, J.P. et al., *J. Volcanol. Geothermal Res.*, 241–242, 121, 2012.)

clouds (www.tropomi.eu/). TROPOMI is a nadir-viewing spectrometer with bands in the ultraviolet, the visible, the NIR, and the SWIR (www.tropomi.eu/). There are also airborne systems such as the Autonomous Modular Sensor-Wildfire airborne multispectral imaging system.

25.22 Coal Fires Studies Using Remote Sensing

Coal contributes to approximately one-third of today's global energy needs, meets ~40% of world's electricity needs with consumption of about ~8000 million metric tons per year (World Coal Association). Some even expect it to be the main energy source by 2025 beating petroleum. But this paradigm may shift and change if coal's reputation of being "dirty fuel" continues. Many drawbacks of coal energy include GHG emissions, acid rains, coal fires, and pollution of river waters and air. Yet the importance of coal to global energy needs can't be disputed, possibly for another 50–100 years by which time technologies may find answers for cleaner fuels to meet the global energy demand.

Coal fires are a recurring phenomenon and are caused by both natural and anthropogenic causes. One of the major contributors to GHG comes through CO_2, and overwhelmingly from burning coal. The other GHG gases released by coal include CO, CH_4, SO_x, and NO_x. Coal takes millions of years to form, and burning it will waste a nonrenewable resource forever. As a result, the study of coal fires acquires great importance. Stracher et al. (2010) provide a comprehensive global perspective of coal fires.

RS has played a key role in coal fire detection, characterization, mapping, and monitoring from the beginning of the satellite sensor era of the early 1960s. Chapter 22 by Dr. Anupma Prakash and Claudia Kuenzer presents RS data, approaches, and methods used for coal fire detection, characterization, mapping, and monitoring. The chapter begins with an overview of the key regions of the electromagnetic spectrum used in coal mining as well as coal fire applications. These include (1) visible (0.4–0.7 μm) spectrum for LULC, geomorphological faults; (2) NIR (0.7–1.3 μm) spectrum for vegetation, soil moisture, and gas emissions; (3) SWIR (1.3–3.0 μm) and MIR (3.0–8.0 μm) spectrum for soil composition, associated forest fires, and high-temperature surface coal fires; (4) TIR (8.0–14 μm) spectrum for underground coal fires, fire depths, and gas emissions; and (5) microwave (0.75–1.0 μm) spectrum for faults and subsidence. The TIR, MIR, and SWIR bands are most important in studying surface and subsurface fires. Whereas TIR is most widely used, other wavebands such as MIR (e.g., Figure 25.22) are also widely used. Often, making use of daytime and nighttime images and looking at diurnal temperature anomaly. Satellite sensors such as NOAA AVHRR, Terra/Aqua MODIS, Landsat-8 OLI, Bispectral Infrared Detection (BIRD (e.g., Figure 25.22), and ASTER carry wavebands in visible, NIR, SWIR, MIR, and TIR and can be ideal platforms for the study of coal mining as well as coal fires. However, limited spatial resolution of some of these sensors (e.g., AVHRR, MODIS) can be a problem.

Some of the key messages in the coal fire study using RS, gathered from Chapter 22 by Dr. Anupma Prakash and Claudia Kuenzer, are as follows:

1. Surface fires are best studied using SWIR and MIR images where high temperatures of fires are easily detected.
2. Subsurface coal fires are best studied using nighttime TIR images where feasible. However, given the paucity of

MODIS aqua day/night fires during July 2011

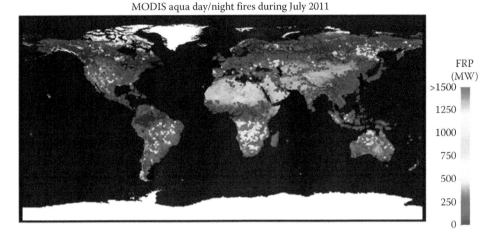

FIGURE 25.21   Fire detection from Aqua MODIS for July 2011 overlaid on a composited surface reflectance map (also from MODIS), showing the fire radiative power value ranges for the individual fire pixels. Compared to the map scale, the fire pixels are indicated with relatively large dots to enhance visualization, causing substantial fire-pixel overlap in certain regions. (From Ichoku, C. et al., *Atmospheric Res.*, 111, 1, 2012.)

nighttime TIR images, one will have to, mostly, depend on daytime TIR images. Subsurface fires are hard to map as they are detected using indirect indicators such as heated ground and smoke from cracks/vents. Also, TIR thresholds of subsurface fires can vary based on the region of study, and there is uncertainty in this aspect, especially when other surfaces (e.g., shale, coal dust) can have similar TIR-derived temperatures. But a continuous watch on coal mining areas and/or TIR images of nighttime will help detect these indirect indicators.

3. RS is a powerful tool for early detection, mapping, and continual monitoring of coal fires due to its repeated coverage of an area. Coal fires are often concentrated in an area and their movement is slow. However, they may burn for very long time (e.g., at times even decades as in Centralia, PA, USA). So RS is a great tool for their continual monitoring.

4. Key coal fire monitoring factors using RS are to (a) understand which time of the day or night (or both) and/or which season is the best to acquire images and (b) establish fire areal extent, duration, variability, and extinction.

5. Scope for further studies in global coal fires is substantial. For example, a global standardized approach to coal fire mapping and monitoring using RS does not exist, mainly as a result of too few people working in the area.

6. Further, the need for automated or semiautomated methods and approaches is required.

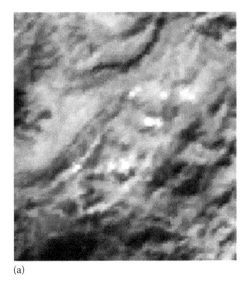

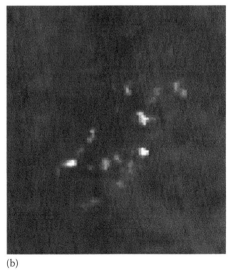

(a)   (b)

FIGURE 25.22   Coal fire detection and quantitative characterization based on the German Aerospace Center (DLRs) small satellite BIRD, an experimental fire remote sensing satellite launched in October 2001, has three nadir-looking bands at the wavelength of 0.84–0.90 μm (near infrared), 3.4–4.2 μm (MIR), and 8.5–9.3 μm (TIR) with a pixel spacing of 185 m. BIRD images obtained on September 21, 2002, at daytime (left) and on January 16, 2003, at nighttime (right). (a) Daytime MIR image, (b) nighttime MIR image, (From Voigt, S. et al., *Int. J. Coal Geol.*, 59(1–2), 121, 2004.)

(Continued)

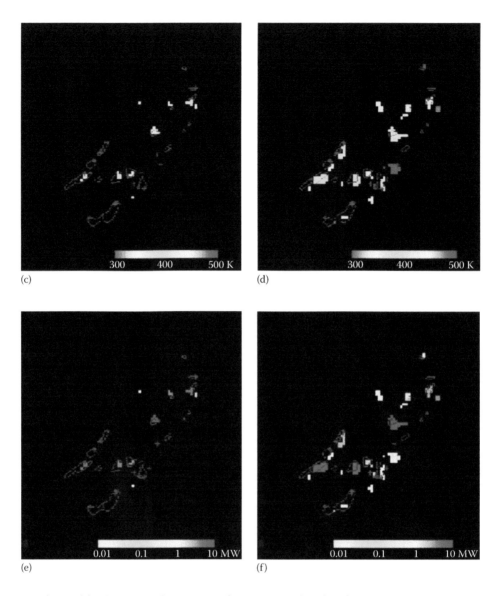

FIGURE 25.22 (*Continued*)   Coal fire detection and quantitative characterization based on the German Aerospace Center (DLRs) small satellite BIRD, an experimental fire remote sensing satellite launched in October 2001, has three nadir-looking bands at the wavelength of 0.84–0.90 μm (near infrared), 3.4–4.2 μm (MIR), and 8.5–9.3 μm (TIR) with a pixel spacing of 185 m. BIRD images obtained on September 21, 2002, at daytime (left) and on January 16, 2003, at nighttime (right). (c) effective fire temperature of detected hot spots at daytime (~11 am local time), (d) effective fire temperature of detected hot spots at nighttime (~11:00 pm local time), (e) radiative energy release of detected hot spots at daytime, and (f) radiative energy release of detected hot spots at nighttime. Blue contours and crosses show location of coal seam fire verified on ground in September 2002, while purple triangles show location of industrial chimneys. (From Voigt, S. et al., *Int. J. Coal Geol.*, 59(1–2), 121, 2004.)

25.23 Urban Remote Sensing

Global urban areal extent estimates vary widely from anywhere between 27 million hectares (vector maps), 65.8 million hectares (MODIS 500 m estimates), 72.7 million hectares (MODIS 1 km), 30.8 million hectares (global land cover 2000 [GLC2000]), 32.2 million hectares (GlobCover), and 350 million hectares or 2.7% of the terrestrial Earth area (Global Rural-Urban Mapping Project) (see Schneider et al., 2010). Such wide variations in estimates are not surprising given the various definitions, data sets, approaches, and methods used in determining the areal extent.

But what is clear is that urban areas occupy a very small portion (<3% by any estimate) of the land cover of the total terrestrial area (14.894 billion hectares) and is also very small compared to other land covers such as the tree-covered areas (27.7%), bare soils (15.2%), grasslands (13%), croplands (12.5%), snow and glaciers (9.7%), shrub-covered areas (9.5%), and sparse vegetation (7.7%) (FAO's Global Land Cover-SHARE of year 2014—Beta-Release 1.0). However, urban areas comparable to other land cover are inland water bodies (2.6%), herbaceous vegetation (1.3%), and mangroves (0.1%). The FAO map shows artificial surfaces (includes all urban) as <0.6%. However, global cities are sources as much as 70% of all GHG emissions as pointed out by Dr. Hasi Bagan and Dr. Yoshiki Yamagata in Chapter 23. Cities are also cause of much of deforestation as a result of various demands of urban dwellers (e.g., housing, furniture, paper) resulting in CO_2 release into the atmosphere. In the last decades, many great cities have grown (e.g., Figure 25.23). RS offers the best opportunity to map and asses the urban sprawl (e.g., Figure 25.23).

Wide arrays of RS data are increasingly used to study many urban issues (see Taubenböck and Esch, 2011). These include (1) areal extent, (2) structures (e.g., buildings, roads), (3) land use, (4) temperature, (5) human population, (6) peri-urban agriculture, (7) gardens and trees, (8) golf courses, (9) lawns, (10) hydrology, (11) sewage and drainage systems, (12) planning, and (13) CO_2 emissions. Based on the characteristics of RS data, its use in urban studies is determined. For example, very-high-spatial-resolution data can be used to detect and map individual buildings and road networks, thermal data in the study of

urban heat islands, and multispectral data in determining urban growth. In Chapter 23, Dr. Hasi Bagan and Dr. Yoshiki Yamagata carry out urban growth studies of Tokyo over the last 40 years using RS and link it to non-RS-derived population dynamics. They showed us the following:

1. The use of multitemporal Landsat data in urban studies. Ability to map major LULC with emphasis on urban sprawl and its change over time and space.
2. Landsat RS-derived urban/built-up changes for 1972–2011 had significant (a) positive correlation with population density of 1970–2010 and (b) significant negative correlation with croplands during 1972–2011.
3. DMSP OLS nighttime lights a sensors image (~1 km) are highly correlated with population density, but does have problems of saturation at very high population densities. DMSP data are in 0–63 digital numbers and the relationship begins to saturate over 55.

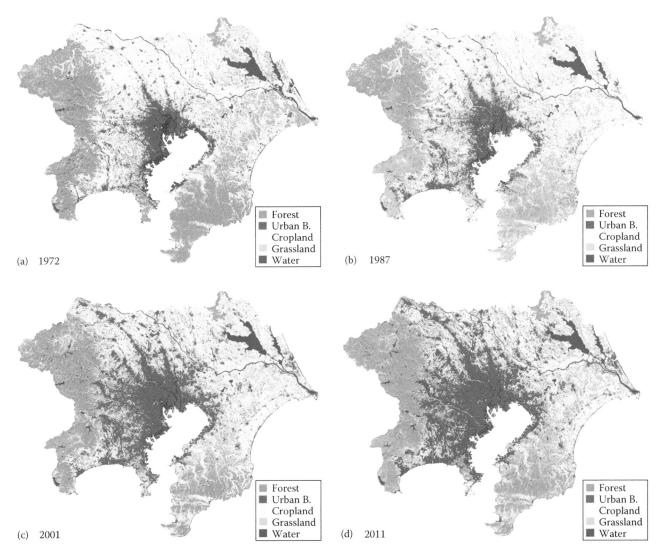

FIGURE 25.23 Time series of land use and land cover maps of Tokyo and the surroundings using Landsat images of (a) 1972, (b) 1987, (c) 2001, and (d) 2011. (From Bagan, H. and Yamagata, Y., *Remote Sensing Environ.*, 127, 210, 2012.)

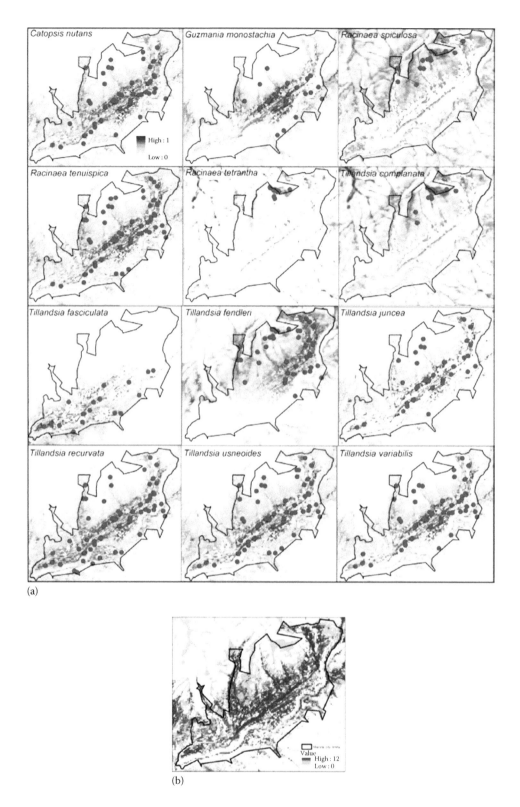

FIGURE 25.24 Smart cities need to be designed with great efficiency in terms of its planning, energy use, and a number of other factors. In this figure, increasing urbanization on biodiversity in tropical regions is illustrated for (a) occurrences (dots) and modeled distributions of Bromeliaceae species in the city of Mérida (polygon) and (b) modeled species diversity. The study combined a rapid species assessment approach with environmental niche modeling based on high-resolution Advanced Spaceborne Thermal Emission and Reflection Radiometer satellite imagery to predict species distributions of Bromeliaceae in the city of Mérida, Venezuela. (From Judith, C. et al., *Landscape Urban Plan.*, 120, 107, 2013.)

4. How human activity and growth of cities related to new economic activity can be studied routinely and accurately using multidate, multispectral imagery of ~30 m spatial resolution or better. This illustrates the growth of a new coal mining city in the last 30 years.

5. Advance methods of image classification for urban areas such as subspace methods and grid cell processing.

What is clear from their study is the presence of large archives of Landsat and other RS data over large cities like Tokyo. This in itself allows anyone to look through the archive and study various aspects of a city's growth. In many parts of the world, rapid urbanization is occurring over the last one to two decades. So changes from the past will be massive, and the availability of digital archive of RS data from sensors such as Landsat, IRS, SPOT VGT, and very-high-submeter to few-meter resolution commercial imagery make decadal urban studies feasible, consistent, and powerful. Recent advances on **OBIA** (Weng and Quattrochi, 2006; Taubenböck et al., 2010) are crucial for more refined and powerful approaches of urban studies using RS.

25.24 Remote Sensing in Design of Smart Cities

Sensors are becoming ubiquitous, and there are many expectations that the next level of connectivity will be through smart sensors. Earlier eras of connectivity are broadly classified into age of (1) e-mails and web portals and (2) networks and social media. Now, the next in connectivity is likely to be through smart sensors. Plants can have sensors when to irrigate or fertilize automatically, vehicles can have sensors for self-driving, homes can have various types of sensors to manage the house including robots driven by sensors to vacuum or clean dishes, and cities can have sensors from various platforms (ground based, airborne such as UAVs, and spaceborne satellites) gathering data for infrastructure management, CO_2 pollution, aerosols, traffic control, and city planning. Further, mega cities of today are major contributors of anthropogenic-induced aerosols and CO_2 pollution. Aerosol optical thicknesses are studies using spaceborne sensors such as MODIS. LiDAR technologies are used for 3D modeling of buildings and trees. Wide array of satellite-borne sensors are routinely used in studies such as urban sprawl and urban heat islands. In addition to RS data, GIS data such as street maps are mapped and updated using GPS through mechanisms like crowd-sourcing. Locations of sites are mapped, tracked, and updated using GPS. Even the delivery of packages may happen through UAVs. This collection of rich sets of sensor-driven management and services of cities that includes infrastructure (e.g., roads, buildings, sewage, drainage, electricity), surveillance, delivery of services, biodiversity and habitat preservation (e.g., Figure 25.24), traffic management, and other services like health care is called smart cities (see Hancke et al., 2013).

Smart cities require us to ensure preserving the richness of habitats (e.g., Figure 25.24), ensuring efficient movement of traffic, avoidance of pollution, efficient use of resources, and a host of other requirements. This is especially important in the twenty-first century when there is a swift increase in urban population around the world.

Chapter 24 by Dr. Yoshiki Yamagata et al. provides a case study of Yokohama and how the city is managed as a smart city using high-resolution RS and GIS. Wide array of RS data are used to gather information on CO_2 emissions, heat emissions from building and transport (road), green space ratio, and building density using various remotely sensed data. In Chapter 25, Dr. Yoshiki Yamagata et al. demonstrate the use of RS for urban planning through the studies of (1) land use and (2) view. Specifically, they demonstrate the following:

1. LiDAR 3D modeling will help viewshed analysis to determine which buildings have full sunlight and which have less, impacting solar radiation for solar photovoltaic (PVs) panels on rooftops. LiDAR is also used to determine the economic view value from buildings (e.g., ocean view vs. street view). Tree height in cities is another application of LiDAR RS.

2. Solar radiation estimation on the rooftop is crucial for electricity generation from PVs on the rooftop. They use high-resolution imagery to detect rooftop area and use LiDAR data to determine building heights.

3. DEM data, along with RS data, add value in many of these applications.

4. Determining the cost of land through remotely sensed-derived information (e.g., PVs are installed in low-rent suburban regions).

Acknowledgments

I thank the lead authors and coauthors of each of the chapters for providing their insights and edits of my chapter summaries.

References

Abdalla, F. 2012. Mapping of groundwater prospective zones using remote sensing and GIS techniques: A case study from the Central Eastern Desert, Egypt. *Journal of African Earth Sciences*, 70, 8–17. ISSN 1464-343X, http://dx.doi.org/10.1016/j.jafrearsci.2012.05.003.

Alongi, D.M. 2014. Carbon sequestration in mangrove forests. *Carbon Management*, 3(3), 313–322.

Andela, N., Liu, Y.Y., van Dijk, A.I.J.M., de Jeu, R.A.M., and McVicar, T.R. 2013. Global changes in dryland vegetation dynamics (1988–2008) assessed by satellite remote sensing: Comparing a new passive microwave vegetation density record with reflective greenness data. *Biogeosciences*, 10, 6657–6676. doi:10.5194/bg-10-6657-2013.

Andreae, M.O. 1991. Biomass burning: Its history, use, and distribution and its impact on environmental quality and global climate. In *Global Biomass Burning: Atmospheric Climatic and Biospheric Implications*, Levine, J.S. (ed.). The MIT Press, Cambridge, MA.

Andreae, M.O. and Merlet, P. 2001. Emissions of trace gases and aerosols from biomass burning. *Global Biogeochemical Cycles*, 15, 955–966.

Bagan, H. and Yamagata, Y. 2012. Landsat analysis of urban growth: How Tokyo became the world's largest megacity during the last 40 years. *Remote Sensing of Environment*, 127, 210–222. ISSN 0034-4257, http://dx.doi.org/10.1016/j.rse.2012.09.011.

Barker, T., Bashmakov, I., Bernstein, L., Bogner, J.E., Bosch, P.R., Dave, R., Davidson, O.R. et al. 2007. Technical summary. In: *Climate Change 2007: Mitigation. Contribution of Working Group III to the Fourth Assessment Report of the Intergovernmental Panel on Climate Change, 2007*, Metz, B., Davidson, O.R., Bosch, P.R., Dave, R., Meyer, L.A., eds. Cambridge University Press, Cambridge, U.K.

Burns, P. and Nolin, A. 2014. Using atmospherically-corrected Landsat imagery to measure glacier area change in the Cordillera Blanca, Peru from 1987 to 2010. *Remote Sensing of Environment*, 140, 165–178. ISSN 0034-4257, http://dx.doi.org/10.1016/j.rse.2013.08.026.

Doll, C.N.H., Muller, J.P., and Morley, J.G. 2006. Mapping regional economic activity from night-time light satellite imagery. *Ecological Economics*, 57(1), 75–92. ISSN 0921-8009, http://dx.doi.org/10.1016/j.ecolecon.2005.03.007.

Dong, Y., Li, Q., Dou, A., and Wang, X. 2011. Extracting damages caused by the 2008 Ms 8.0 Wenchuan earthquake from SAR remote sensing data. *Journal of Asian Earth Sciences*, 40(4), 907–914. ISSN 1367-9120, http://dx.doi.org/10.1016/j.jseaes.2010.07.009.

Droogers, P., Immerzeel, W.W., and Lorite, I.J. 2010. Estimating actual irrigation application by remotely sensed evapotranspiration observations. *Agricultural Water Management*, 97, 1351–1359.

Duke, N.C., Meynecke, J.O., Dittmann, S., Ellison, A.M., Anger, K., Berger, U., Cannicci, S. et al. 2007. A world without mangroves? *Science*, 317, 41 (edited by Etta Kavanagh).

EGSMA, 1981. Geologic map of Egypt, Scale 1:2,000,000, Egyptian Geological Survey and Mining Authority.

Elvidge, C.D., Baugh, K.E., Kihn, E.A., Kroehl, H.W., and Davi, E.R. 1997. Mapping city lights with nighttime data from the DMSP operational linescan system. *Photogrammetric Engineering and Remote Sensing*, 63(6), 727–734.

Elvidge, C.D., Zhizhin, M., Hsu, F.-C., and Baugh, K. 2013. What is so great about nighttime VIIRS data for the detection and characterization of combustion sources? *Proceedings of the Asia-Pacific Advanced Network*, 35, 33–48. http://dx.doi.org/10.7125/APAN.35.5.

Fensholt, R., Langanke, T., Rasmussen, K., Reenberg, A., Prince, S.D., Tucker, C., Scholes, R.J. et al. 2012. Greenness in semi-arid areas across the globe 1981–2007—An Earth Observing Satellite based analysis of trends and drivers.

Remote Sensing of Environment, 121, 144–158. ISSN 0034-4257, http://dx.doi.org/10.1016/j.rse.2012.01.017.

Gan, T.Y., Zunic, F., Kuo, C.C., and Strobl, T. 2012. Flood mapping of Danube River at Romania using single and multi-date ERS2-SAR images. *International Journal of Applied Earth Observation and Geoinformation*, 18, 69–81. ISSN 0303-2434, http://dx.doi.org/10.1016/j.jag.2012.01.012.

Ghoneim, E. and El-Baz, F. 2007. DEM-optical-radar data integration for palaeohydrological mapping in the northern Darfur, Sudan: Implications for groundwater exploration. *International Journal of Remote Sensing*, 28(22), 5001–5018.

Ghulam, A., Li, Z.L., Qin, Q., Yimit, H., and Wang, J. 2008. Estimating crop water stress with ETM+ NIR and SWIR data. *Agricultural and Forest Meteorology*, 148(11), 1679–1695. ISSN 0168-1923, http://dx.doi.org/10.1016/j.agrformet.2008.05.020.

Giglio, L., Werf Van Der, G.R., Randerson, J.T., Collatz, C.J., and Kasibhatla, P. 2006. Global estimation of burned area using MODIS active fire observations. *Atmos. Chem. Phys.* 6:957–974.

González-Dugo, M.P., Escuin, S., Cano, F., Cifuentes, V., Padilla, F.L.M., Tirado, J.L., Oyonarte, N., Fernández, P., and Mateos, L. 2013. Monitoring evapotranspiration of irrigated crops using crop coefficients derived from time series of satellite images. II. Application on basin scale. *Agricultural Water Management*, 125, 92–104. ISSN 0378-3774, http://dx.doi.org/10.1016/j.agwat.2013.03.024.

Hall, D.K. and Robinson, D.A. In press. Global snow cover, in *Satellite Image Atlas of Glaciers of the World*, Williams, Jr., R.S. and Ferrigno, J.G., eds. USGS Professional Paper 1386. U.S. Geological Survey.

Hancke, G.P., Silva, B.C., and Hancke, Jr., G.P. 2013. The role of advanced sensing in smart cities. *Sensors*, 13(1), 393–425.

Ichoku, C., Kahn, R., and Chin, M. 2012. Satellite contributions to the quantitative characterization of biomass burning for climate modeling. *Atmospheric Research*, 111, 1–28. ISSN 0169-8095, http://dx.doi.org/10.1016/j.atmosres.2012.03.007.

Ito, A. and Penner, J.E. 2004. Global estimates of biomass burning emissions based on satellite imagery for the year 2000. *Journal of Geophysical Research*, 109(D14), D14S05.

Jachowski, N.R.A., Quak, M.S.Y., Friess, D.A., Duangnamon, D., Webb, E.L., and Ziegler, A.D. 2013. Mangrove biomass estimation in Southwest Thailand using machine learning. *Applied Geography*, 45, 311–321. ISSN 0143-6228, http://dx.doi.org/10.1016/j.apgeog.2013.09.024.

Jones, C. and Kammen, D.K. 2014. Spatial distribution of U.S. household carbon footprints reveals suburbanization undermines greenhouse gas benefits of urban population density. *Environmental Science & Technology* (ES&T), 48(2), 895–902. doi:10.1021/es4034364.

Judith, C., Schneider, J.V., Schmidt, M., Ortega, R., Gaviria, J., and Zizka, G. 2013. Using high-resolution remote sensing data for habitat suitability models of Bromeliaceae in the city of Mérida, Venezuela. *Landscape and Urban Planning*, 120, 107–118. ISSN 0169-2046, http://dx.doi.org/10.1016/j.landurbplan.2013.08.012.

Kaku, K. and Held, A. 2013. Sentinel Asia: A space-based disaster management support system in the Asia-Pacific region. *International Journal of Disaster Risk Reduction*, 6, 1–17. ISSN 2212-4209, http://dx.doi.org/10.1016/j.ijdrr.2013.08.004.

Kamal, A.S.M.M. and Midorikawa, S. 2004. GIS-based geomorphological mapping using remote sensing data and supplementary geoinformation: A case study of the Dhaka city area, Bangladesh. *International Journal of Applied Earth Observation and Geoinformation*, 6(2), 111–125. ISSN 0303-2434, http://dx.doi.org/10.1016/j.jag.2004.08.003.

Liew, S.C., Gupta, A., Wong, P.P., and Kwoh, L.K. 2010. Recovery from a large tsunami mapped over time: The Aceh coast, Sumatra. *Geomorphology*, 114(4), 520–529. ISSN 0169-555X, http://dx.doi.org/10.1016/j.geomorph.2009.08.010; http://www.sciencedirect.com/science/article/pii/S0169555X09003390.

Margat, J. and J. van der Gun. 2013. *Groundwater around the World: A Geographic Synopsis*. CRC Press/Balkema, London, U.K.

McKee, T.B., Doesken, N.J. and Kleist, J. 1993. The relationship of drought frequency and duration of time scales. *Eighth Conference on Applied Climatology, American Meteorological Society*, Jan 17–23, 1993, Anaheim, CA, pp. 179–186.

Mekonnen, M.M. and Hoekstra, A.Y. 2014. Water footprint benchmarks for crop production: A first global assessment. *Ecological Indicators*, 46, 214–223. ISSN 1470-160X, http://dx.doi.org/10.1016/j.ecolind.2014.06.013.

Mieville, A., Granier, C., Liousse, C., Guillaume, B., Mouillot, F., Lamarque, J.F., Grégoire, J.M., and Pétron, G. 2010. Emission of gases and particles from biomass burning during the 20th century using satellite data and a historical reconstruction. *Atmospheric Environment*, 44, 1469–1477.

Milewski, A., Sultan, M., Yan, E., Becker, R., Abdeldayem, A., Soliman, F., and Gelil, K.A. 2009. A remote sensing solution for estimating runoff and recharge in arid environments. *Journal of Hydrology*, 373(1–2), 1–14. ISSN 0022-1694, http://dx.doi.org/10.1016/j.jhydrol.2009.04.002.

Meijerink, A.M.J., Bannert, D., Batelaan, O., Lubczynski, W., and Pointet, T. 2007. Remote sensing applications to groundwater. IHP-VI, Series on Groundwater No. 16. Published by the United Nations Educational, Scientific, and Cultural Organization (UNESCO), Paris, France. Composed by Marina Rubio, 93200 Saint-Denis Printed by UNESCO. IHP/2007/GW/16 © UNESCO 2007, Printed in France, SC-2007WS/43. pp. 312.

Mu, Q., Zhao, M., Kimball, J., McDowell, N., and Running, S. 2013. A remotely sensed global terrestrial drought severity index. *Bulletin of the American Meteorological Society*, 94, 83–98.

Platonov, A., Thenkabail, P.S., Biradar, C.M., Cai, X., Gumma, M., Dheeravath, V., Cohen, Y. et al. 2009. Water Productivity Mapping (WPM) using landsat ETM+ data for the irrigated croplands of the Syrdarya River Basin in Central Asia. *Sensors*, 8, 8156–8180.

Rao, D. 2002. Remote sensing application in geomorphology. *Tropical Ecology*, 43, 49–59.

Ray, R.G. 1960. Aerial photographs in geologic interpretation. USGS Professional Paper No. 373. U.S. Government Printing Office, Washington, DC.

Rhee, J., Im, J., and Carbone, G.J. 2010. Monitoring agricultural drought for arid and humid regions using multi-sensor remote sensing data. *Remote Sensing of Environment*, 114, 2875–2887.

Rodell, M., Chen, J., Kato, H., Famiglietti, J., Nigro, J., and Wilson, C. 2007. Estimating ground water storage changes in the Mississippi River basin (USA) using GRACE. *Hydrogeology Journal*, 15(1), 159–166. doi:10.1007/s10040-006-0103-7.

Sabins, F.F. 1987. *Remote Sensing: Principles and Interpretation*, 2nd edn. W. H. Freeman, San Francisco, CA.

Schneider, A., Friedl, M.A., and Potere, D. 2010. Monitoring urban areas globally using MODIS 500m data: New methods and datasets based on urban ecoregions. *Remote Sensing of Environment*, 114, 1733–1746.

Siebert, S., Burke, J., Faures, J.M., Frenken, K., Hoogeveen, J., Döll, P., and Portmann, F.T. 2010. Groundwater use for irrigation. *Hydrology and Earth Systems Science*, 14, 1863–1880. www.hydrol-earth-syst-sci.net/14/1863/2010/doi:10.5194/hess-14-1863-2010.

Smith, M. and Pain, C. 2006. Applications of remote sensing in geomorphology. *Progress in Physical Geography*, 33, 568–582.

Sorooshian, S., AghaKouchak, A., Arkin, P., Eylander, J., Foufoula-Georgiou, E., Harmon, R., and Hendrickx, J. 2011. Advanced concepts on remote sensing of precipitation at multiple scales. *Bulletin of the American Meteorological Society*, 92(10), 1353–1357. doi:10.1175/2011BAMS3158.1.

Spalding, M., Blasco, F., and Field, C. 1997. *World Mangrove Atlas*. International Society for Mangrove Ecosystems, Okinawa, Japan, 178pp.

Stracher, G.B., Prakash, A., and Sokol, E.V. (Eds.). 2010. *Coal and Peat Fires: A Global Perspective*, Vol. 1: *Coal-Combustion and Geology*. Elsevier, 335pp., ISBN 978-0444528582. New York.

Surono, J.P., Pallister, J., Boichu, M., Buongiorno, M.F., Budisantoso, A., Costa, F., Andreastuti, S. et al. 2012. The 2010 explosive eruption of Java's Merapi volcano—A '100-year' event. *Journal of Volcanology and Geothermal Research*, 241–242, 121–135. ISSN 0377-0273, http://dx.doi.org/10.1016/j.jvolgeores.2012.06.018.

Syvitski, J.P.M., Overeem, I., Brakenridge, G.R., and Hannon, M. 2012. Floods, floodplains, delta plains—A satellite imaging approach. *Sedimentary Geology*, 267–268, 1–14. ISSN 0037-0738, http://dx.doi.org/10.1016/j.sedgeo.2012.05.014.

Taubenböck, H. and Esch, T. 2011. Remote sensing—An effective data source for urban monitoring. *Earthzine IEEE Magazine*. Published on Wednesday, 20 July 2011.

Taubenböck, H., Esch, T., Wurm, M., Roth, A., and Dech, S. 2010. Object based feature extraction using high spatial resolution satellite data of urban areas. *Journal of Spatial Science*, 55 (1), 111–126.

Tehrany, M.S., Pradhan, B., and Jebur, M.N. 2014. Flood susceptibility mapping using a novel ensemble weights-of-evidence and support vector machine models in GIS. *Journal of Hydrology*, 512, 332–343. ISSN 0022-1694, http://dx.doi.org/10.1016/j.jhydrol.2014.03.008.

Traore, S.B., Ali, A., Tinni, S.H., Samake, M., Garba, S., Maigari, I., Alhassane, A. et al. 2014. AGRHYMET: A drought monitoring and capacity building center in the West Africa Region. *Weather and Climate Extremes*, 3, 22–30. ISSN 2212-0947, http://dx.doi.org/10.1016/j.wace.2014.03.008.

Thenkabail, P.S., Enclona, E.A., Ashton, M.S., Legg, C., and Jean De Dieu, M. 2004. Hyperion, IKONOS, ALI, and ETM+ sensors in the study of African rainforests. *Remote Sensing of Environment*, 90, 23–43.

Thenkabail, P.S., Gumma, M.K., Teluguntla, P., and Mohammed, I.A. 2014. Hyperspectral remote sensing of vegetation and agricultural crops. *Photogrammetric Engineering and Remote Sensing*, 80(4), 697–709.

Thenkabail, P.S., Hanjra, M.A., Dheeravath, V., and Gumma, M.A. 2010. A holistic view of global croplands and their water use for ensuring global food security in the 21st century through advanced remote sensing and non-remote sensing approaches. *Remote Sensing*, 2(1), 211–261. doi:10.3390/rs2010211, http://www.mdpi.com/2072-4292/2/1/211.

Thenkabail, P.S. and Nolte, C. 1995. Mapping and characterising inland valley agroecosystems of West and Central Africa: A methodology integrating remote sensing, global positioning system, and ground-truth data in a geographic information systems framework. RCMD Monograph No. 16. International Institute of Tropical Agriculture, Ibadan, Nigeria, 62pp.

Thenkabail, P.S., Nolte, C., and Lyon, J.G. 2000a. Remote sensing and GIS modeling for selection of benchmark research area in the inland valley agroecosystems of West and Central Africa. *Photogrammetric Engineering and Remote Sensing*, 66(6), 755–768 (Africa Applications Special Issue).

Thenkabail, P.S., Smith, R.B., and De-Pauw, E. 2000. Hyperspectral vegetation indices for determining agricultural crop characteristics. *Remote Sensing of Environment*, 71, 158–182.

Twilley, R.R., Chen, R.H., and Hargis, T. 1992. Water, Air, and Social Pollution. 64L265–258.

Voigt, S., Tetzlaff, A., Zhang, J., Künzer, C., Zhukov, B., Strunz, G., Oertel, D., Roth, A., Dijk, P.V., and Mehl, H. 2004. Integrating satellite remote sensing techniques for detection and analysis of uncontrolled coal seam fires in North China. *International Journal of Coal Geology*, 59(1–2), 121–136. ISSN 0166-5162, http://dx.doi.org/10.1016/j.coal.2003.12.013.

Wang, Q., Wu, J., Lei, T., He, B., Wu, Z., Liu, M., Mo, X. et al. 2014. Temporal-spatial characteristics of severe drought events and their impact on agriculture on a global scale. *Quaternary International*, 349, 10–21. ISSN 1040-6182, http://dx.doi.org/10.1016/j.quaint.2014.06.021.

Weng, Q. and Quattrochi, D.A. 2006. *Urban Remote Sensing*. CRC Press/Taylor & Francis, 448pp. ISBN: 0849391997. New York.

Zhang, A. and Jia, G. 2013. Monitoring meteorological drought in semiarid regions using multisensor microwave remote sensing data. *Remote Sensing of Environment*, 134, 12–23.

Zhang, B., Wu, Y., Lei, L., Li, J., Liu, L., Chen, D., and Wang, J. 2013. Monitoring changes of snow cover, lake and vegetation phenology in Nam Co Lake Basin (Tibetan Plateau) using remote SENSING (2000–2009). *Journal of Great Lakes Research*, 39(2), 224–233. ISSN 0380-1330, http://dx.doi.org/10.1016/j.jglr.2013.03.009.

Index

T - #0542 - 071024 - C708 - 279/216/31 - PB - 9780367868963 - Gloss Lamination